AGRICULTURAL
OF SOUTH ASI
THEIR MANAGEMENT

AGRICULTURAL PESTS OF SOUTH ASIA AND THEIR MANAGEMENT

A.S. ATWAL

Former Professor & Head
Department of Zoology-Entomology
Dean, College of Agriculture and Dean, Postgraduate Studies
Punjab Agricultural University
Ludhiana, India

G.S. DHALIWAL

Visiting Professor
Insect Biopesticide Research Centre, Jalandhar, India
Former Professor of Ecology
Department of Entomology, Punjab Agricultural University
Ludhiana, India

KALYANI PUBLISHERS

LUDHIANA - NEW DELHI - NOIDA (U.P.) - HYDERABAD
CHENNAI - KOLKATA - CUTTACK - GUWAHATI - KOCHI - BANGALORE

KALYANI PUBLISHERS

Head Office

B-1/1292, Rajinder Nagar, **Ludhiana**-141 008 • Ph : 0161-2760031
E-mail : kalyanibooks@yahoo.co.in

Administration Office

4779/23, Ansari Road, Daryaganj, New Delhi-110 002
Ph : 011-23271469, 23274393, 23278688 **E-mail :** kalyani_delhi@yahoo.co.in

Works

B-16, Sector-8, **NOIDA** (U.P.)

Branches

1, Mahalakshmi Street, T. Nagar, **Chennai**-600 017 • Ph : 044-24344684
110/111, Bharatia Towers, Badambadi, **Cuttack**-753 009 (Odisha) • Ph : 0671-2311391
3-5-1108, Narayanaguda, **Hyderabad**-500 029 • Ph : 040-24750368
10/2B, Ramanath Mazumdar Street, **Kolkata**-700 009 • Ph : 033-22416024
Arunalaya, 1st Floor, Saraswati Road, Pan Bazar, **Guwahati**-781 001 • Ph : 0361-2731274
Koratti Parambil House, Convent Road, **Kochi**-682 035 • Ph : 0484-2367189
No. 24 & 25, 1st Floor, Hameed Shah Complex, Cubbonpet Main Road, **Bengaluru**-560 002

KPP K 16953 3 O-26785 2

First Edition, 1976
Second Edition, 1986
Third Edition, 1997
Fourth Edition, 2002
Fifth Edition, 2005
Sixth Edition, 2008
Seventh Edition, 2013
Eighth Edition, 2015
Reprinted, 2018

Datalink Computer
(PST-SO14)

ISBN 978-93-272-5072-5

PRINTED IN INDIA
At B.B. Press, A-37, Sector-67, NOIDA-201301

ACKNOWLEDGEMENTS

We express our sincere thanks to Dr S.N. Puri, Vice-Chancellor, Central Agricultural University, Imphal; Dr K.S. Khokhar, Vice-Chancellor, CCS Haryana Agricultural University, Hisar; Dr Y.R. Sarma, FAO Consultant and Ex-Director, Indian Institute of Spices Research, Calicut; Dr Opender Koul, Director, Insect Biopesticide Research Centre, Jalandhar; Dr Ram Singh, Director, Human Resource Management, CCS Haryana Agricultural University, Hisar; Dr G.T. Gujar, Head, Division of Entomology, Indian Agricultural Research Institute, New Delhi; Dr T.B. Gour, Ex-Professor of Entomology, Acharya N.G. Ranga Agricultural University, Rajendranagar, Hyderabad; and Dr. M Jalal Asif, Professor of Entomology, University of Agriculture, Faisalabad (Pakistan), for their active support and critical suggestions at various stages during revision of the book.

We are immensely indebted to the following scientists for their generosity in providing some of the colour photographs used in the book: Dr R.K. Sharma, Principal Scientist and Dr M.K. Dhillon, Senior Scientist, Indian Agricultural Research Institute, New Delhi; Dr S.N. Sushil, Senior Scientist, Indian Institute of Sugarcane Research, Lucknow; Dr H. Basappa, Principal Scientist, Directorate of Oilseeds Research, Rajendranagar, Hyderabad; Dr Nitin Kulkarni, Scientist-G and Head, Forest Entomology Division, Tropical Forest Research Institute, Jabalpur; Dr S.P. Singh and Dr (Mrs) Saroj Jaipal, Senior Entomologists, CCS Haryana Agricultural University, Hisar; Dr D.C. Sharma, Professor and Head, Department of Entomology, CSK Himachal Pradesh Krishi Vishvavidyalaya, Palampur; Dr V. Ambethgar, Professor of Entomology, Tamil Nadu Agricultural University, Regional Research Station, Kriddhachalam; and Dr D.R. Sharma and Dr D.K. Sharma, Senior Entomologists; Dr K.S. Sangha, Dr Vijay Kumar, Dr P.S. Saraon and Dr K.S. Suri, Entomologists; Dr Sandeep Singh, Dr G.K. Taggar, Dr Sarwan Kumar, Dr P.S. Shera, Dr Jawala Jindal and Dr Beant Singh, Assistant Entomologists, Punjab Agricultural University, Ludhiana and Mr Amarjeet Batth, an art and nature lover from Ludhiana. We are thankful to the National Bureau of Agriculturally Important Insects, Bangalore and Tamil Nadu Agricultural University, Coimbatore, for some of the photographs of insect pests and their damage.

We express our heartfelt thanks to Dr Vikas Jinkal, Ms. Bharathi Mohindru and Dr Sanjeev Kumar, Assitant Entomologists, Punjab Agricultural University, Ludhiana, for their painstaking efforts and everwilling help from time to time. We express our gratitude to Col. V.M. Joneja, Datalink Computer, Jalandhar, for the commendable job of typesetting and preparation of diagrams. We express our appreciation to Mr Raj Kumar, Proprietor, Kalyani Publishers, for taking keen interest in bringing out the revised version of the book in a short time. Finally, we express our sincere thanks to various authors, publishers and organizations for some of the copyright figures and photographs used in the book.

A.S. Atwal

G.S. Dhaliwal

PREFACE (Eighth Edition)

Pest problems have been associated with agriculture ever since the domestication of plants about 10,000 years ago. In traditional agriculture, pests were kept under check by natural regulating factors and various cultural practices. The advent of synthetic organic pesticides provided a potent weapon against pests. However, the misuse and overuse of pesticides proved detrimental to natural enemies and resulted in the problems of pest resurgence and resistance. All these factors accentuated pest problems and pest outbreaks became more frequent. Further, the green revolution technology provided favourable environment for proliferation of pests. Therefore, integrated pest management (IPM) has been considered to be environment friendly and economically feasible approach to pest management. Moreover, the development and large scale adoption of transgenic crops has generated several socioeconomic and ecological problems. Futhermore, the potential climate change is likely to increase the severity of pest problems in several geographical regions of the world. All these issues have been extensively dealt with in the book and future strategies have been suggested.

The present book, in its eighth edition, has been divided into four sections, viz, Basic Principles, Pest Management Tactics, Pests of Crops and Domestic Animals, and Pest Management in Global Perspective. In section I, a new chapter on 'Pest Surveillance and Forecasting' has been added. Apart from the basic aspects, this chapter gives an account of systems analysis, modelling, database management, computer programming, simulation techniques, decision support systems and expert systems. The second section contains two new chapters entitled 'Botanical Pest Control' and 'Genetic Control'. Two new chapters, viz. 'Pests of Mushrooms'and 'Pests of Forest Trees' have also been included in the third section. The last section focuses on the global issues involved in promotion of IPM. The first chapter of this section outlines the impact of climate change on pests and pest management strategies. The progress of IPM at global and national level has been described in the penultimate chapter. The last chapter gives a critical analysis of various pest management approaches and a future roadmap leading to sustainable crop protection, has been proposed.

All the chapters of the book have been updated keeping in view the recent developments. The descriptions of several new pests have been added and a number of line diagrams of pests and their damage have been included. Recent references on different aspects of pests and pest management have been incorporated. In addition, many colour plates given at the end of the book would aid in better understanding of the activities of various categories of pests. We hope that the enlarged edition of the book would continue to find favour among students, teachers and research workers in developing countries.

March 15, 2015

A.S. Atwal
G.S. Dhaliwal

PREFACE (First Edition)

I first thought of writing such a book about 25 years ago when I started teaching Entomology to B.Sc (Agri.) classes. The inspiration behind that thought has been such books as H.M. Lefroy's (1909) *Indian Insect Life*; T.B. Fletcher's (1914) *Some South Indian Insects*; T.V.R. Ayyar's (1940) *Handbook of Economic Entomology for South India* and K.A. Rahman's (1940) *Pest Number* of the Punjab Agricultural College Magazine. Behind each of these topics and many other papers and bulletins, there are the works of numerous entomologists of India who have been interested in the systematic and the applied aspects of Entomology. As is well known, after Lefroy's monumental work, there had been a gap of knowledge and Mian Afzal Hussain was the beacon-light of a number of prominent entomologists, who either started their career under his guidance or had the privilege of working with him for a number of years. The foremost among these entomologists are Hem Singh Pruthi, Khan-A-Rahman, M.L. Roonwal, Taskhir Ahmad and S. Pradhan. No less are the contributions made independently by C.C. Ghosh, Y.R. Rao, A.P. Kapoor and M.S. Mani. After the publication of *Indian Insect Life, Some South Indian Insects, Veterinary Entomology and Acarology for India, Entomology in India*, the only other major work in applied entomology was that of H.S. Pruthi's book *Textbook on Agricultural Entomology* published by the Indian Council of Agricultural Research. All these books serve their purpose in various ways and have made landmarks in the advacement of Entomology in India. Yet, there was a need for a concise book that would give a background to the biology, life-history and control of insect pests and which would also include some of the essentials of the principles and practices of pest control. It is hoped that this book will fill the gap.

This book has been written primarily for students of Applied Zoology and Agricultural Entomology but can also be of great use to extension workers and educated farmers who have elementary knowledge of biology. Apart from discussing the principles and methods of pest control, the biology, life cycle and behaviour of individual pests have also been discussed and arranged in various chapters or sections, cropwise. It has been pointed out to the students of Entomology that they should adopt an approach of integrated pest control by the use of selective insecticides and natural enemies and through cultural practices so as to obtain sustained high yields of crops over long periods of time. While discussing the principles of ecology or the application of insecticides, a mention has also been made of the role of productive, useful and beneficial insects like parasites, predators, insect pollinators, etc.

This book is being published under my authorship simply and purely for the sake of convenience otherwise so many people have directly or indirectly helped in the preparation of this book over the last many years that it is impossible to remember the exact contributions made by each one of them. Yet, it is my pleasure and duty to mention at least those names whose contributions are uppermost in my mind and I do hope that the others will kindly forgive me for making any unintentional omissions. Such unpublished material as Annual Reports of the Entomology Department, PAU, theses of postgraduate students, and many lesser publications have been consulted wherever there was a gap in published information. All these sources have not been listed in the selected bibliogrpahy given at the end of the chapters.

I thank the following for their invaluable assistance : Dr M.S Randhawa, Vice Chancellor, PAU for his encouragement and providing facilities ; Dr S.S. Bains , Associate Professor of Ecology and Dr Balraj Singh, Insect Ecologist, for helping in critically reading the manuscript and the proofs and also in preparing the bibliography , index and spending incalculable hours checking the technical names, spellings of the animal pests ; Dr K.S. Bedi, former Professor of Plant Pathology and Joint Director of Agriculture (Research and Education), Punjab, for vetting the manuscript from the language point of view ; Mr M. Ramzan and Mr S.S. Madan for making the line drawings ; Mr Amarjit Lal for photographs of the plates; Dr G.S. Sandhu, Entomologist, for the colour photographs: Shri Surjit Singh for typing the manuscript and Shri R.G. Prasher, Assistant Librarian, for his assistance in preparing the index.

Finally, I am particularly grateful to my wife, Dr S.K. Siddoo-Atwal, M.D. for her constant encouragement and invaluable suggestions for the improvement of language and subject matter. As a gratitude for that help, I have dedicated this book to our younger daughter Chandermapatti.

February 20, 1976 **A.S. Atwal**

LIST OF FIGURES

Fig. No.	Title	Page No.
1.1	Abundance of described species in the major taxa	3
1.2	Mollusks. Snail, *Helix,* sp.; Slug, *Limax* sp.	5
1.3	Citrus nematode, *Tylenchulus semipenetrans*	6
1.4	Centipede, *Scolopendra morsitans*; Millipede, *Thyroglutus malayus*	8
1.5	Crayfish, *Cambras* sp.	9
1.6	Jumping spider	10
1.7	Red spider mite, *Tetranychus cinnabarinus*	10
1.8	Grasshopper, *Poekilocerus pictus*	13
1.9	A schematic section of typical insect integument, showing the various layers	14
1.10	Head of the grasshopper, *Poekilocerus pictus*	14
1.11	Modifications of the insect antennae	15
1.12	Insect mouthparts	16
1.13	Modifications of Insect mouthparts	17
1.14	Modifications of insect legs	18
1.15	A hypothetical primitive wing venation in insects	19
1.16	Modifications of insect wings	20
1.17	Internal organs of the grasshopper, *Poekilocerus pictus*	21
1.18	Digestive system of the grasshopper, *Poekilocerus pictus*	22
1.19	Reproductive system of the grasshopper, *Poekilocerus pictus*	23
1.20	Modifications of insect larvae	25
1.21	Modifications of insect pupae	26
1.22	Classical scheme indicating the role of hormones in moulting and metamorphosis in insects	27
1.23	Pin positions for representative insects	33
1.24	Spreading of insect specimens	34
4.1	Effect of cultural practices on biodiversity of natural enemies and abundance of insect pests	83
5.1	Common insect predators	103
5.2	Some important parasitoids	104
5.3	NPV infection of an insect host	113
5.4	Life-cycle of nucleopolyhedrovirus	114
5.5	Life-cycle of entomopathogenic nematodes	117
5.6	Stepwise procedure for establishing a classical biiological control programme	119
6.1	Behaviour and physiological effects of phytochemicals on insect pests	130
6.2	Structures of nicotine, rotenone and azadirachtin	131
6.3	Structures of major insecticidal constituents of pyrethrum	134
7.1	Chemical structure of some organochlorine insecticides	156
7.2	Chemical structure of some organophosphorus insecticides	159

Fig. No.	Title	Page No.
7.3	Chemical structure of some carbamate insecticides	164
7.4	Chemical structure of some synthetic pyrethroids	165
7.5	Two types of hand compression sprayers	174
7.6	Manually operated hydraulic knapsack sprayer	175
7.7	Pattern of pesticide consumption in India vis-a-vis world	181
7.8	Cases of resistance to pesticides in different arthropod orders	181
7.9	Pesticide contamination of foodstuffs in India	186
8.1	Classification of semiochemicals	193
8.2	Structures of naturally occurring insect juvenile hormones	203
8.3	Structures of some important chitin synthesis inhibitors	205
8.4	Structures of some moulting hormone agonists	206
8.5	Production of codling moth pheromone and area under mating disruption worldwide	214
9.1	Multitrophic interactions, showing the relationship between intrinsic and extrinsic resistance	224
9.2	Effect of carbofuran treatment on dead heart formation due to sorghum shoot fly in susceptible and moderately resistant cultivars of sorghum	241
10.1	Gamete formation with combination of translocated chromosomes	256
10.2	Diagram illustrating meiotic drive in which only one sex (female) producing sperms are formed	258
11.1	Steps involved in producing a transgenic organism	267
11.2	Steps involved in the genetic engineering of plants through *Agrobacterium*-mediated DNA transformation	268
11.3	Global area of transgenic crops under different traits, 1996-2012	284
12.1	Major components of a pest management system and their relationships	294
12.2	Relationship between economic injury level and economic threshold level	296
12.3	Three phases in the devlopment of an integrated pest management programme	304
12.4	Role of IPM in ecosystem stability	309
13.1	Brown planthopper, *Nilaparvata lugens*	314
13.2	Whitebacked planthopper, *Sogatella furcifera*	315
13.3	Green leafhopper, *Nephotettix nigropictus*	316
13.4	Zigzag leafhopper,*Recilia dorsalis*	317
13.5	Yellow stem borer, *Scirpophaga incertulas*	318
13.6	Dark-headed striped borer, *Chilo poychrysus*	320
13.7	Pink stem borer, *Sesamia inferens*	321
13.8	Rice leaf folder, *Cnaphalocrocis medinalis*	321
13.9	Rice caseworm, *Nymphula depunctalis*	322
13.10	Rice gall midge, *Orseolia oryzae*	323
13.11	Rice hispa, *Dicladispa armigera*	324
13.12	Rice grasshopper, *Hieroglyphus banian*	325
13.13	Rice bug, *Leptocorisa acuta*	326
13.14	Maize borer, *Chilo partellus*	327

Fig. No.	Title	Page No.
13.15	European corn borer, *Ostrinia nubilalis*	329
13.15	Sorghum earhead bug, *Calocoris angustatus*	331
13.16	Sorghum shoot bug, *Peregrinus maidis*	331
13.17	Sorghum shootfly, *Atherigona soccata*	332
13.18	Armyworm, *Mythimna separata*	335
14.1	Gram pod borer, *Helicoverpa armigera*	339
14.2	Lentil pod borer, *Etiella zinckenella*	341
14.3	Red gram pod fly, *Melanagromyza obtusa*	341
14.4	Blister beetle, *Mylabris pustulata*	342
14.5	Green potato bug, *Nezara viridula*	344
15.1	Mustard aphid, *Lipaphis erysimi*	348
15.2	Painted bug, *Bagrada hilaris*	349
15.3	Mustard sawfly, *Athalia lugens*	349
15.4	Groundnut aphid, *Aphis craccivora*	351
15.5	Castor semilooper, *Achaea janata*	354
15.6	Castor capsule borer, *Conogethes punctiferalis*	355
15.7	Castor slug, *Parasa lepida*	356
15.8	Til leaf and pod caterpillar, *Antigastra catalaunalis*	357
16.1	Cotton jassid, *Amrasca biguttula biguttula*	363
16.2	Cotton aphid, *Aphis gossypii*	364
16.3	Cotton whitefly, *Bemisia tabaci*	365
16.4	Dusky cotton bug, *Oxycarenus laetus*	366
16.5	Red cotton bug, *Dysdercus koemigii*	366
16.6	Pink bollworm, *Pectinophora gassypiella*	368
16.7	Spotted bollworms, *Earias vittella* and *E. insulana*	370
16.8	Cotton leafroller, *Sylepta derogata*	371
16.9	Cotton semilooper, *Tarache notabilis*	372
16.10	Green semilooper, *Anomis flava*	372
16.11	Cotton grey weevil, *Myllocerus undecimpustulatus*	374
16.12	Jute semilooper, *Anomis sublifera*	376
16.13	Beet armyworm, *Spodoptera exigua*	376
17.1	Sugarcane pyrilla, *Pyrilla perpusilla*	381
17.2	Sugarcane whitefly, *Aleurolobus barodensis*	383
17.3	Sugarcane woolly aphid, *Ceratovacuna lanigera*	385
17.4	Sugarcane top borer, *Scirpophaga excerptalis*	386
17.5	Sugarcane early shoot borer, *Chilo infuscatellus*	388
17.6	Sugarcane root borer, *Emmalocera depressella*	390
17.7	Gurdaspur borer, *Acigona steniellus*	392
18.1	Citrus psylla, *Diaphorina citri*	396
18.2	Citrus whitefly, *Dialeurodes citri*	397
18.3	Citrus caterpillar, *Papilio demoleus*	401

Fig. No.	Title	Page No.
18.4	Citrus leafminer, *Phyllocnistis citrella*	402
18.5	Citrus mite, *Oligonychus citri*	406
18.6	Grapevine girdler, *Sthenias grisator*	410
18.7	Mango hopper, *Amritodus atkinsoni*	412
18.8	Mango mealybug, *Drosicha mangiferae*	413
18.9	Mango stem borer, *Batocera rufomaculata*	414
18.10	Mango stone weevil, *Sternochetus mangiferae*	415
18.11	Mango fruit fly, *Bactrocera dorsalis*	415
18.12	Mango gall insect, *Apsylla cistella*	417
18.13	Ber fruitfly, *Carpomyia vesuviana*	420
18.14	Pomegranate butterfly, *Deudorix isocrates*	422
18.15	Banana scale moth, *Nacoleia octasema*	423
18.16	Banana stem borer, *Odoiporus longicollis*	424
18.17	Banana weevil, *Cosmopolites sordidus*	425
19.1	San Jose scale, *Quadraspidiotus perniciosus*	438
19.2	Woolly apple aphid, *Eriosoma lanigerum*	*439*
19.3	Peach leafcurl aphid, *Brachycaudus helichrysi*	440
19.4	Codling moth, *Cydia pomonella*	441
19.5	Peach fruit fly, *Bactrocera zonata*	443
20.1	Cabbage caterpillar, *Pieris brassicae*	452
20.2	Diamondback moth, *Plutella xylostella*	*453*
20.3	Cabbage semilooper, *Thysanoplusia orichalcea*	454
20.4	Tobacco caterpillar, *Spodoptera litura*	454
20.5	Crucifer leaf webber, *Crocidolomia binotalis*	455
20.6	Cabbage borer, *Hellula undalis*	456
20.7	Cabbage flea beetle, *Phyllotreta cruciferae*	456
20.8	Potato tuber moth, *Phthorimaea operculella*	458
20.9	Onion thrips, *Thrips tabaci*	460
20.10	Brinjal lacewing bug, *Urentius sentis*	463
20.11	Brinjal fruit borer, *Leucinodes orbonalis*	463
20.12	Brinjal hadda beetle, *Henosepilachna vigintioctopunctata*	465
20.13	Melon fruit fly, *Bactrocera cucurbitae*	466
20.14	Red pumpkin-beetle, *Aulacophora foveicollis*	468
20.15	Singhara beetle, *Galerucella birmanica*	469
21.1	Sciarid fly, *Bradysia tritici*	473
21.2	Phorid fly, *Megaselia agarici*	474
21.3	Cecid fly, *Heteropeza pygmaea*	474
21.4	Mushroom springtail, *Seira iricolor*	475
21.5	Mushroom mite, *Tyrophagus* spp.	476
21.6	Mushroom nematode, *Aphelenchoides sacchari*	476
22.1	Rose aphid, *Macrosiphum rosaeiformis*	479

Fig. No.	Title	Page No.
22.2	Hollyhock tingid bug, *Urentius euonymus*	481
22.3	Ak butterfly, *Danais chrysippus*	481
22.4	Jasmine leaf webworm, *Nausinoe geometralis*	483
22.5	Lily moth, *Polytela gloriosae*	485
23.1	Arecanut mirid bug, *Carvalhoia arecae*	489
23.2	Black-headed caterpillar, *Opisina arenosella*	491
23.3	Rhinoceros beetle *Oryctes rhinoceros*	492
23.4	Red palm weevil, *Rhynchophorus ferrugineus*	493
23.5	Coconut weevil, *Diocalandra frumenti*	493
23.6	Coconut white grub, *Leucopholis coneophora*	494
23.7	Striped mealybug, *Ferrisia virgata*	495
23.8	Coffee stem borer, *Xylotrechus quadripes*	496
23.9	Coffee berry borer, *Hypothenemus hampei*	497
23.10	Tea mosquito bug, *Helopeltis theivora*	498
23.11	Red borer, *Zeuzera coffeae*	500
23.12	Red crevice tea mite, *Brevipalpus phoenicis*	500
23.13	Yellow tea mite, *Polyphagotarsonemus latus*	501
24.1	Banana aphid, *Pentalonia nigronervosa*	504
24.2	Chillies thrips, *Scirtothrips dorsalis*	506
24.3	Pollu beetle, *Longitarsus nigripennis*	507
24.4	Skipper butterfly, *Udaspes folus*	508
24.5	Cinnamon butterfly, *Chilasia clytia*	510
25.1	Teak defoliator, *Hyblaea puera*	512
25.2	Pink gypsy moth, *Lymantria mathura*	513
25.3	Deodar defoliator, *Ectropis deodarae*	514
25.4	Shisham defoliator, *Plecoptera reflexa*	515
25.5	Kadam defoliator, *Arthroschista hilaris*	516
25.6	Gamhar defoliator, *Craspedonta leayana*	517
25.7	Sal borer, *Hoplocerambyx spinicornis*	517
25.8	Babul borer, *Celosterna scabrator*	518
25.9	Semul borer, *Xystrocera globosa*	519
25.10	Shisham borere, *Aristobia horridula*	520
25.11	Bamboo shoot weevil, *Cyrtotrachelus longipes*	521
25.12	Deodar beetle, *Scolytis major*	521
25.13	Gallery system of deodar beetle, *Scolylus major*, on the inner surface of bark of a log of *Cedrus deodara* showning egg gallery and larval galleries	522
25.14	Subabul psyllid, *Heteropsylla cubana*	523
25.15	Gamhar lace bug, *Tingis beesoni*	523
26.1	Desert locust, *Schistocerca gregaria*	528
26.2	Surface grasshopper, *Chrotogonus trachypterus*	532
26.3	Different castes and the termitarium of *Odontotermes obesus*	533

Fig. No.	Title	Page No.
26.4	Red hairy caterpillar, *Amsacta moorei*	537
26.5	Bihar hairy caterpillar, *Spilarctia obliqua*	538
26.6	Gram cutworm, *Ochropleura flammatra*	539
26.7	Greasy cutworm, *Agrotis ipsilon*	540
26.8	Turnip moth, *Agrotis segetum*	541
27.1	Angoumois grain moth, *Sitotroga cerealella*	543
27.2	Rice moth, *Corcyra cephalonica*	544
27.3	Indian meal moth, *Plodia interpunctella*	545
27.4	Khapra beetle, *Trogoderma granarium*	546
27.5	Rice weevil, *Sitophilus oryzae*	547
27.6	Red flour beetle, *Tribolium castaneum*	548
27.7	Lesser grain borer, *Rhyzopertha dominica*	549
27.8	Gram dhora, *Callosobruchus chinensis*	550
27.9	Commonly used grain storage receptacles	553
28.1	House fly, *Musca nebulo*	556
28.2	Mosquitoes, *Anopheles* sp. and *Culex* sp.	*560*
28.3	Cockroach, *Periplaneta americana*	*563*
28.4	Cockroach, *Blatella germanica*	563
28.5	Cockroach, *Blatella orientalis*	564
28.6	House cricket, *Grylloides sigillatus*	565
28.7	Bed-bug, *Cimex lectularius*	565
28.8	Human louse, *Pediculus humanus*	566
28.9	Rat flea, *Xenopsylla cheopis*	567
28.10.	Wasp, *Vespa orientalis*	568
28.11	Silverfish, *Lepisma saccharina*	570
28.12	Greater wax moth, *Galleria mellonella*	570
29.1	Poultry shaft louse, *Menopon gallinae*	573
29.2	Sand fly, *Phlebotomus minutus*	574
29.3	Gad fly, *Tabanus striatus*	575
29.4	Bot fly, *Oestrus ovis*	576
29.5	Stable fly, *Stomoxys calcitrans*	578
29.6	Blow fly, *Calliphora vicina*	579
29.7	Blow fly, *Lucilia sericata*	580
29.8	Horny fly, *Hippobosca maculata*	580
29.9	Poultry stickfast flea, *Echidnophaga gallinacea*	581
29.10	Cattle tick, *Hyalomma* sp.	582
29.11	Fowl tick, *Argas persicus*	582
29.12	Mange mite, *Sarcoptes scabei*	583
29.13	Dog mite, *Demodexis canis*	584
33.1	Impact of pesticide subsidy on rice production in Indonesia	629
33.2	Intensity of insecticide use and cotton yields in Colombia, 1967-1987	630

LIST OF TABLES

Table No.	Title	Page No.
1.1	Number of herbivorous species in different insect orders	12
1.2	Classification of insects into various subclasses and orders	28
1.3	Distinguishing characteristics of moths and butterflies	31
2.1	Global losses due to various categories of pests in major crops	57
2.2	Losses caused by insect pests to major agricultural crops/ commodities in India	57
3.1	Different decision support systems developed in various parts of the world	70
3.2	Some important expert systems developed in various parts of the world	71
4.1	Selected examples of intercropping systems that help to prevent insect pest outbreaks	90
4.2	Examples of trap cropping practices applied in agroecosystems	92
5.1	Important examples of successful biological control of insect pests of agricultural crops by use of introduced natural enemies in Asian countries	101
5.2	*Bacillus thuringiensis* (Bt)-based commercial pesticides	107
5.3	Bt-based microbial pesticides marketed in India	108
5.4	Main target pests of commercially available mycoinsecticides	111
5.5	Baculovirus preparations registered for pest control in various parts of the world	115
5.6	Relative strengths and weaknesses of parasitoids and predators compared to pathogens	125
6.1	Important plant families that have a number of species evaluated for anti-insect properties	129
6.2	Some of the commercially produced neem-based pesticides in India	132
7.1	Salient events in evolution of chemical control of insects	145
7.2	Droplet spectrum of various types of spray discharges	172
7.3	Consumption of pesticides in different states of India	180
7.4	Ten most resistant arthropod species to insecticides or acaricides	182
7.5	Classification of pesticides on the basis of their toxicity to honey bees	185
7.6	List of pesticides not approved, withdrawn or banned in India	187
7.7	List of pesticides restricted for use in India	188
8.1	Practical uses of semiochemicals in pest management	201
8.2	Naturally occurring insect antifeedants in host plants	210
8.3	Examples of large scale use of semiochemicals in pest management	212
8.4	Global use of pheromones for mating disruption	213
9.1	Dual role of allelochemicals in insect-plant interactions	228
9.2	Genetics of resistance to major insect pests of crop plants	229
9.3	Biotypes of insect pests of agricultural crops	230
9.4	Insect-resistant varieties of different crops released in India	232
9.5	Insect and disease reactions of IR varieties of rice	233

Table No.	Title	Page No.
9.6	Selected examples of use of varieties with resistance to insect pests as the principal method of pest management	239
10.1	Estimated doses of γ-rays and x-rays for male sterilization	251
10.2	SIRM model of Knipling	252
10.3	Number of generations needed to eradicate a population of 1000 insects by insecticides envisaging a control of 90 per cent and an annual population increase of five-fold	252
10.4	Number of generations needed to eradicate a population of 1000 insects by SIRM envisaging a control of 90 per cent and annual increase of five-fold	253
10.5	Major field trials for control of insect pests using sterile insect release technique	259
11.1	Biotechnological methods employed in crop improvement	265
11.2	Transgenic crops carrying Bt genes for insect resistance	272
11.3	Transgenic crops expressing insecticidal plant genes	275
11.4	Global area under transgenic crops in 2013	283
12.1	Economic threshold levels (ETLs) of major insect pests of agricultural crops in India	298
19.1	Common species of defoliating beetles in the western Himalayas	449
26.1	Locust plague cycles and upsurges in India	526
28.1	Differences between the various life stages of *Anopheles* and *Culex*	561
31.1	List of important insect vectors of plant viruses	601
31.2	List of important mite vectors of virus diseases of crop plants	602
31.3	List of important vectors of phytoplasma-transmitted diseases	604
31.4	List of major vectors of bacterial diseases of crop plants	606
31.5	List of principal vectors of fungal diseases of crop plants	606
32.1	Effect of environmental factors on the efficacy of microbial pesticides in pest management	617
33.1	Landmarks in the history of pest management	621
33.2	Impact of IRM/IPM strategy on cotton in Punjab, 2002-2010	631
34.1	Approaches to pest management : Retrospect and prospect	642

CONTENTS

S. No.	Chapter	Pages

PART-I : BASIC PRINCIPLES

1. ANIMAL WORLD: DIVERSITY PATTERNS **3–38**

Introduction 3; Animal Diveristy 4, Phylum Chordata 4, Phylum Mollusca 5, Phylum Nemathelmenthes 6, Phylum Arthropoda 8; Class Insecta 11, Insect Structure 13, Insect Growth and Development 24, Hormonal Control of Metamorphosis 26, Insect Classification 27, Nomenclature 32, Mounting, Preserving and Lebelling Insects 33, Economic Importance of Insects 35, Entomology in India 37.

2. PEST POPULATIONS AND CROP LOSSES **39–58**

Introduction 34; Attainment of Pest Status 40; Invisive Pests in Agriculture 41; Factors Affecting Pest Populations 42, Abiotic Factors 42, Biotic Factors 46; Measurement of Pest Populations 47, Absolute Estimates 48, Relative Estimates 49, Population Indices 51; Changing Status of Pests 52; Yield Loss Assessment 54, Types of Losses 54, Estimation of Losses 55, Crop Losses : Global and Indian Scenario 56.

3. PEST SURVEILLENCE AND FORECASTING **59–72**

Introduction 34; Pest Surveillance 59, Offectives of Surveillance 60, Components of Surveillance 60, Identification of Pest 60, Determination of Pest Population 61, Estimation of Abundance of Natural Enemies 61, Estimation of Yield Loss 62; Factors Affecting Survey 62, Nature of Sample 62, Size of Sample 63, Number of Samples 63, Sampling Habitat 63, Sampling Pattern 64, Type of Sampling 64; Pest Forecasting 65, Types of Pest Forecasting 66, Methods of Forecasting 66; Systems Analysis and Modelling 66; Database Management and Computer Programming 69; Simulation Techniques 72; Conclusions 72.

PART-II : PEST MANAGEMENT TACTICS

4. LEGISLATIVE, CULTURAL AND MECHANICAL CONTROL **75–98**

Introduction 75; Legislative Measures 75, Insecticides Act, 1968, 77, Central Bodies and Laboratories 78, Registration of Insecticides 79, Licences for Manufacutre and Sale 79, Central Insecticides Laboratory 79, Packing and Labelling 80, Enforcement Machinery 81, Prevention of Food Adulteration Act, 1954, 81, Pesticides Management Bill, 2008, 82; Cultural Control 82, Tillage 84, Clean Seed 85, Seed Rate 85, Irrigation 85, Fertilizers 86, Clean Culture 86, Crop Spacing 87, Crop Rotation 88, Intercropping 89, Trap Cropping 91, Presence of Weeds 93, Pruning and Thinning 94, Time of Sowing and Harvesting 94, Destruction of Crop Residues 94; Mechanical and Physical Control 95, Manual Labour 95, Manipulation of Physical Factors of Environment 96; Potential and Constraints 97.

S. No.	Chapter	Pages
5.	**BIOLOGICAL CONTROL**	99–127

Introduction 99; Historicalk Developments 100; Biological Control Agents 102; Vertebrates 102; Arthropods 102, Spiders 102, Mites 103, Insects 103; Microorganisms 104, Bacteria 105, Fungi 110, Protozoa 112, Rickettsiae 112, Viruses 112, Nematodes 115; Techniques in Biological Control 118, Classical Biological Control 118, Conservation 121, Augmentaion 122, Aumentative Biocontrol in India 123; Synergism among Biocontrol Agents 125; Potential and Contraints 126.

6. BOTANICAL PEST CONTROL 128–143

Introduction 128; Promising Pesticidal Plants 128, Neem 129, Chinaberry 133, Chrysanthemum 133, Tobacco 134, Rotenone Plants 135, Pongram 135, Custard Apple 136, Sabadilla 136, Ryania 136, Quassia 137, Essential Oil Bearing Plants 137; Effects on Non-target Organisms 138, Natural Enemies 138, Other Organisms 139, Man 139; Environmental Impact 140; Pest Resistance to Phytochemicals 141; Integration with Other Tactics 141; Potential and Constraints 142.

7. CHEMICAL CONTROL 144–191

Introduction 144; Historical Aspects 144; Classification of Insecticides 147; Stomach Poisons 147, Systemic Poisons 148, Inert Dusts 148, Contact Poisons 148, Fumigants 149, Miscellaneous Chemicals 149; Factors Influencing Effectiveness of Insecticides 150, Physical Properties of Formulations 150, Penetration of Insecticide through Cuticle 152, Specific Susceptibility to Insecticides 153, Weather Conditions 153, Conditions in the Field 153, Compatibility 154, Application of Pesticides 154; Major Groups of Pesticides 155; Insecticides 155, Chlorinated Hydrocarbons 155, Organophosphates 158, Carbamates 164, Synthetic Pyrethroids 165, Miscellaneous Insecticides 165, Fumigants 166; Acaricides 167; Nematicides 168; Rodenticides 170; Molluscides 171; Pesticide Application Equipment 171, Dusters 171, Sprayers 172, Agricultural Aircrafts 177, Miscellaneous Machinery 178; Pattern of Pesticide Consumption 179; Environmental Impact of Pesticides 181, Insecticide Resistance 181, Pest Resurgence 183, Effect on Non-target Organisms 184, Pesticide Residues 185, Pesticide Poisoning 188; Potential and Constraints 190.

8. BIORATIONAL APPROACHES 192–219

Introduction 192; Semiochemicals 192; Pheromones 193, Types of Pheromones 194, Strategies for Exploitation of Pheromones 195, Monitoring 196, Mass Trapping 197, Mating Disruption 199; Allelochemicals 199, Allomones 199, Kairomones 200, Synomones 200; Push-Pull Strategy 201; Development Inhibitors 202, Brain Hormones 203, Juvenile Hormones 203, Chitin Synthesis Inhibitors 204, Moulting Hormones 206, Sclerotization Disruptors 207; Miscellaneous Approaches 207, Propesticides 207, Avermectins 208, Spinosyns 208, Polynactins 208, Pyrrole Insecticides 208, Phenylpyrazoles 208, Pyridine Insecticides 209, Oxadiazines 209, Antifeedants 209, Repellents 209; Role in Pest Management 209, Allelochemicals 209, Pheromones 211, Parapheromones 216, Juvenile Hormone Analogues 216, Moulting Hormone Analogues 217, Chitin Synthesis Inhibitors 218; Resistance to Biorationals 218; Potential and Constraints 219.

S. No.	Chapter	Pages

9. HOST PLANT RESISTANCE 220–245

Introduction 220; Concept of Plant Resistance 220; Types of Resistance 221, Intensity of Resistance 221, Ecological Resistance 222, Evolutionary Concept 222, Genetic Resistance 223, Multitrophic Interactions 224; Mechanisms of Resistance 224, Nonpreference/Antixenosis 224, Antibiosis 224, Tolerance 225; Bases of Resistance 225, Biophysical Bases 225, Biochemical Bases 226; Genetics of Resistance 228; Concept of Biotypes 230; Breeding for Insect Resistance 231; Durable Plant Resistance 234, Major Gene Resistance 234; Polygenic Resistance 235; Induced Resistance 235; Economic Impact 235; Factors Affecting Host Plant Resistance 237, Temperature 237, Soil Moisture 237, Photoperiod 237, Nutrients 237, Soil pH 238, Air Pollution 238, Plant Factors 238, Insect Factors 238; Host Plant Resistance in IPM 239, Principal Method 239, Integration with Other Tactics 240, Chemical Control 240, Biological Control 242, Cultural Control 243; Potential and Constraints 244.

10. GENETIC CONTROL 246–263

Introduction 246; Requirements for Genetic Control 247, Colonization and Mass Rearing 247, Population Dynamics 247, Post-Production Processes 247, Field Monitoring 247, Economic Analysis 248; Sterile Insect Technique 248; Chemosterilants 248, Alkylating Agents 249, Antimetabolites 249, Miscellaneous Compounds 249, Field Trials 249; Ionising Radiations 250, Concept of Knipling's Technique 250, Eradication of Screwworm Fly 253; Hybrid Sterility 254; Cytoplasmic Incompatibility 255; Chromosome Translocation 256; Lethal Factors 257; Sex-Ratio Distortion 257; Species Replacement 258; Successes in Genetic Control 258, Screwworm Fly 258, Fruit Flies 260, Pink Bollworm 260, Codling Moth 260, Tsetse Flies 261, Onion Fly 261; Integration with Other Tactics 261; Potential and Constraints 262.

11. BIOTECHNOLOGICAL APPROACHES 264–291

Introduction 264; Biotechnolgical Methods 264; Tissue Culture 265, Micropropagation 266, Somaclonal Variation 266, Protoplast Culture and Somatic Hybridization 266, *In Vitro* Production of Haploids 266; Recombinant DNA Technology 267, Vector-mediated Gene Transfer 267, *Agrobacterium*-mediated Gene Transfer 268, DNA Viruses as Vectors 269, Direct Gene Transfer Methods 269, Direct Uptake of DNA 269, Electroporation 269, Microprojectile Bombardment 269, Microinjection 270, RNA Interference 270; Transgenic Crop Protection 270, Bt Endotoxins 271, Cotton 271, Maize 273, Rice 273, Potato 274, Vegetables 274, Plant-derived Genes 274, Protease Inhibitors 274, α-amylase Inhibitors 277, Lectins 277, Enzymes 278, Alarm Pheromones 278, Other Novel Genes 278; Resistance in Pests to Transgenics 279, High Expression Level 279, Refuge Strategy 279, Tissue and Temporal Expression of Toxins 280, Rotations 281, Gene Pyramiding 281, Mosaics 282, Trap Plant Strategy 282, Integration of Tactics 282; Commercialization of Transgenic Crops 282; Transgenic Crops : Global Scenario 283, Bt Cotton in India 284, Economic and Ecological Impact 285, Future Outlook 285; Impact on Non-Target Organisms 286, Non-target Insects 286, Predators 286, Parasitoids 287, Pollinators 287, Secondary Insect Pests 287, Wild Relatives of Crops 288; Soil Biota 288; Food Safety and Human Health 289, Marker Genes 289, Bt Toxins 289; Potential and Constraints 290.

S. No.	Chapter	Pages
12.	**INTEGRATED PEST MANAGEMENT**	**292–310**
	Introduction 292; Origin of IPM 292; Concept of Injury Levels 295, Economic Injury Level 295, Economic Threshold Level 296, Environmental Economic Injury Levels 297, General Equilibrium Position 300, Cost : Benefit Ratios 300; Integration of Tactics 301; Pre-Requisites for Decision Making 302; Decision Making 303; Development of an IPM Programme 304, Problem Definition Phase 304, Research Phase 305, Implementation Phase 305; Constraints in IPM Implementation 307, Institutional Constraints 307, Informational Constraints 308, Sociological Constraints 308, Economic Constraints 308, Political Constraints 308; Framework of IPM Programme 308; Potential of IPM 309.	

PART-III : PESTS OF CROPS AND DOMESTIC ANIMALS

S. No.	Chapter	Pages
13.	**PESTS OF CEREALS AND MILLETS**	**313–337**
	Introduction 313; Rice 314; Maize 327; Sorghum 331; Wheat 334.	
14.	**PESTS OF PULSE CROPS**	**338–346**
	Introduction 338; Insects Attacking Reproductive Structures 338, Pod Borers 338, Pod Sucking Bugs; Insects Feeding on Vegetative Parts 344.	
15.	**PESTS OF OILSEED CROPS**	**347–361**
	Introduction 347; Brassica Crops 347; Groundnut 351; Castor 353; Sesame 356; Linseed 359; Safflower 359; Sunflower 361.	
16.	**PESTS OF FIBRE CROPS**	**362–379**
	Introduction 362; Cotton 363; Jute 375; Sunnhemp 378.	
17.	**PESTS OF SUGARCANE**	**380–394**
	Introduction 380; Sucking Pests 380; Sugarcane Borers 386; Miscellaneous Pests 393.	
18.	**PESTS OF TROPICAL AND SUB-TROPICAL FRUITS**	**395–436**
	Introduction 395; Citrus 395; Grapevine 407; Mango 411; Guava 418; Ber 419; Pomegranate 422; Banana 423; Fig 425; Jackfruit 427; Loquat 429; Pineapple 429; Date Palm 429; Jamun 430; Tamarind 431; Litchi 433; Papaya 434; Sapota 435; Phalsa 436.	
19.	**PESTS OF TEMPERATE FRUITS**	**437–450**
	Introduction 437; Sap Suckers 438; Frugivorous Pests 441; Xylophagous Pests 444.	
20.	**PESTS OF VEGETABLE CROPS**	**451–471**
	Introduction 451; Winter Vegetables 452, Cole Crops 452, Potato and Tomato 457, Onion 460, Peas 461; Summer Vegetables 463, Brinjal 463, Cucurbits 466, Okra 468, Water Nuts 469, Sweet Potato 470; Miscellaneous Pests 470.	

S. No.	Chapter	Pages
21.	**PESTS OF MUSHROOMS**	472–477
	Introduction 472; Insect Pests 472; Non-Insect Pests 475.	
22.	**PESTS OF ORNAMENTAL PLANTS**	478–487
	Introduction 478; Insect Pests 478; Non-Insect Pests 486.	
23.	**PESTS OF PLANTATION CROPS**	488–502
	Introduction 488; Arecanut 489; Cashewnut 489; Coconut 490; Coffee 494; Tea 498; Rubber 502.	
24.	**PESTS OF SPICES**	503–510
	Introduction 503; Cardamom 503; Large Cardamom 506; Chillies 506; Black Pepper 507; Turmeric 508; Ginger 509; Coriander 509; Cinnamon 509.	
25.	**PESTS OF FOREST TREES**	511–524
	Introduction 511; Defoliators 511; Borers 517; Sap Suckers 522.	
26.	**POLYPHAGOUS PESTS**	525–541
	Introduction 525; Locusts and Grasshoppers 525, Locusts 525, Anti-Locuat Organization 529; Termites 532, Productive Castes 532, Sterile Castes 533, Management of Termites 536; Hairy Caterpillars 537; Cutworms 539.	
27.	**STORED GRAIN PESTS**	542–554
	Introduction 542; Environment and Storage Pests 542; Lepidopteran Pests 543; Beetles and Weevils 546; Bruchids 549; Management of Storage Pests 551.	
28.	**HOUSEHOLD PESTS**	555–571
	Introduction 555; Pests Associated with Man 555; Pests of Household Materials 563.	
29.	**PESTS OF FARM ANIMALS**	572–584
	Introduction 572; Lice 572; Flies 574; Fleas 581; Ticks 581; Mites 583.	
30.	**NON-INSECT PESTS**	585–599
	Introduction 585; Birds 585; Rats and Mice 589; Damage by Rodents 591; Management of Rodents 592; Fruit Bats 593; Other Mammals 594; Snails and Slugs 598.	
31.	**INSECT VETORS OF PLANT DISEASES**	600–607
	Introduction 600; Viruses 600, Types of Viruses 602, Mechanism of Transmission 603, Virus-Vector Relationship 603; Phytoplasmas 603, Mechanism of Transmission 604, Virus-Phytoplasma Relationship 605; Bacteria 605; Fungi 605; Management of Vectors and Diseases 606.	

S. No.	Chapter	Pages

PART-IV : PEST MANAGEMENT IN GLOBAL PERSPECTIVE

32. CLIMATE CHANGE AND PEST MANAGEMENT 611–619

Introduction 611; Geographical Distribution of Insects 612; Effect of Temperature on Insects 613; Effect of CO_2 on Insects 614; Pest Management Strategies 614, Host Plant Resistance 615, Transgenic Crops 616, Natural Enemies 616, Microbial Pesticides 617, Botanical Pesticides 618; Future Outlook 618.

33. PEST MANAGEMENT: GLOBAL SCENARIO 620–640

Introduction 620; Salient Events in Pest Management 620, Era of Traditional Approaches (Ancient-1938) 620, Era of Pesticides (1939-1975) 624, Era of IPM (1976 onwards) 624; IPM in Developed Countries 625, IPM in Apple Orchards 625, Huffaker Project 625, Consortium for IPM 625, National IPM Initiative 626, Integrated Fruit Production System 626, IPM in Greenhouses 626; IPM in Developing Countries 627, Rice 628, Cotton 630, Vegetable Crops 632, Plantaion Crops 633; Pest Management Organizations 634, International Organizations 634, National Organizations 637; Future Outlook 640.

34. PEST MANAGEMENT: FUTURE PERSPECTIVES 641–654

Introduction 641; IPM : Springboard to Sustainable Agriculture 642, Economic Mandate 642, Environmental Mandate 643, Social Mandate 643; Pespectives in IPM 643, Parasitoids and Predators 643, Microbial Pesticides 644, Botanical Pesticides 645, Semiochemicals 646, Chemical Control 647, Host Plant Resistance 648, Transgenic Crops 649, Integrated Pest Management 651; Future Outlook 653.

SELECTED BIBLIOGRAPHY 655–666

SUBJECT INDEX 667–678

PHOTO GALLERY (i)-(xviii)

PART-I

BASIC PRINCIPLES

- Animal World: Diversity Patterns
- Pest Populations and Crop Losses
- Pest Surveillance and Forecasting

ANIMAL WORLD: DIVERSITY PATTERNS

INTRODUCTION

The diversity of life is one of the most striking aspects of our planet. Hence knowing how many species inhabit Earth is among the most fundamental yet elusive questions in science. However, the answer to this question remains obscure as efforts to sample the world's biodiversity to date have been limited. Moreover, obtaining an accurate number is constrained by the fact that most species remain to be described and because indirect estimates rely on assumptions that have proven highly controversial. The higher taxonomical classification of species (*i.e.* the assignment of species to phylum, class, order, family and genus) follows a consistent and predictable pattern, from which the total number of species in any taxonomic group can be estimated.

The total number of species on Earth has been estimated to vary between 3 and 100 million. Recent estimates put the total number of species on Earth at 8.7 million, with 6.5 million on land and 2.2 million (about 25% of total) dwelling in the ocean depths. About 86 per cent of the species on land and 91 per cent of those in the seas have yet to be discovered, described and catalogued. Among the estimated 8.7 million species, 7.77 million are animals, 298,000 are plants, 611,000 are fungi, 36,400 are protozoa and 27,500 are chromists (including *e.g.* brown algae, diatoms, water moulds, etc.). Of all the 1.72 million species described, approximately 63861 are vertebrates, 3,07,674 are plants and 1,000,000 (57.86%) are insects **(Fig. 1.1)**.

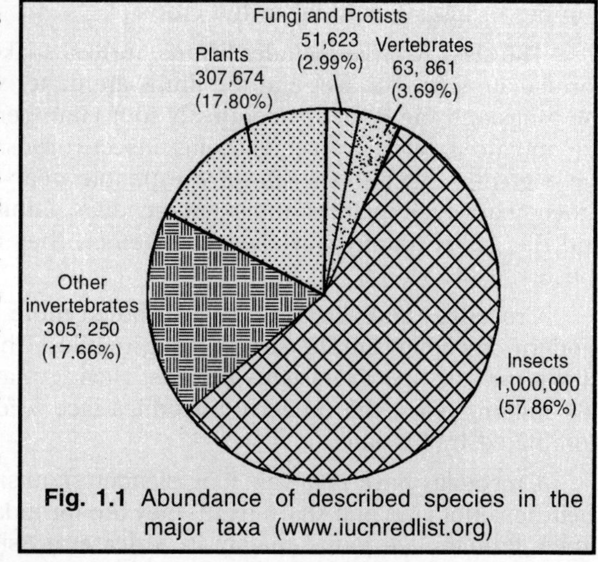

Fig. 1.1 Abundance of described species in the major taxa (www.iucnredlist.org)

The identified animal species have been placed in ten phyla. Many of these species are pests of crops, domestic animals, and household goods and materials. In the economy of nature, these species have close association with a number of other animals that act as predators and exercise a check on their population increase. Animal pests belonging to these phyla are further placed under different classes, orders and families. Their classification is necesary not only for their identification, but also for the understanding of their feeding and breeding habits, their modes of multiplication, their seasonal activity and hibernation, the duration of their life-cycles and the number of generations in a year. All this information can be further utilized in determining the severity of pest outbreaks and their periodicity.

ANIMAL DIVERSITY

Practically all the pests and most of their enemies are included in the four phyla, *viz.* Chordata, Mollusca, Nemathelminthes and Arthropoda.

Phylum Chordata

The chordates are bilaterally symmetrical animals, having a single dorsal tubular nerve-cord, a notochord and gill slits in the pharynx. However, in the adult stage, any one of the characters may be altered or may disappear. The lower chordates are worm-like creatures and are mostly aquatic. The higher chordates comprise the greatest number and are generally known as vertebrates. Typically, their body has four regions, *viz.* head, neck, trunk and tail. The brain is enlarged and is enclosed in the skull or cranium. The cranium also forms a part of the internal axial skeleton which makes the framework and gives support to the body. With only a few exceptions, the sexes are separate. The chordates are divided into various classes and orders.

The fishes as chordates were formerly included in a single class, *Pisces,* but are now placed in three separate classes, *Cyclostomata,* the lampreys and hagfishes; *Chondrichthyes*, the dogfish and sharks; *Osteichthyes,* the bony fishes that are mostly found in fresh, brackish or salt water. The animals belonging to all the three classes have elongated bodies and they breathe through gills. Their heart is two chambered and they possess scales on their body.

The class *Amphibia* includes frogs, toads, salamanders, etc. and they possess two pairs of limbs for walking or swimming. Their heart is three chambered and at some stage of developement they possess gills. Depending upon the mode of life of the adults, respiration may be by the gills, lungs, skin, or the mouth-lining. The frogs and toads are very useful to the farmers as they eat a large number of insects harmful to the crops.

The class *Reptilia* includes lizards, turtles, snakes, crocodiles, etc. The body is covered with dry corniculate skin. The feet and the limbs are either very small or are absent. Respiration is through the lungs and the heart is imperfectly four-chambered. The reptiles are mostly terrestrial but some are aquatic. Lizards are useful, being insectivorous. Crocodiles, even though prized for their hide, are a great danger to the life of the people near the rivers. The crocodiles belong to the order Crocodilia, which includes the true crocodiles (family Crocodylidae), alligators (family Alligatoridae) and the gavial (family Gavialidae). Together, they are called crocodilians and there are about 20 species in the world.

Crocodiles and alligators are distributed more widely in tropical and semitropical regions. The modern crocodiles are the largest living reptiles. They are what remains of a once abundant group in the Jurassic and Cretaceous periods. Having managed to survive virtually unchanged for about 160 million years, the modern crocodiles face a forbidding and perhaps short future in a world dominated by humans.

Crocodiles have relatively long slender snouts; alligators have short and broader snouts. With their powerful jaws and sharp teeth, they are formidable antagonists. The members of the group that attack humans are found mainly in Africa and Asia. The estuarine crocodile, *Crocodylus porosus* Schneider, found in South Asia grows to a great size and is very much feared. It is swift and aggressive and, will eat any bird and mammal it can drag from the shore to water, where the prey is violently torn to pieces. The crocodiles are known to attack animals such as cattle, deer and people.

The class *Aves* includes birds, such as the crow, sparrow, pigeon, parrot and the domestic fowl. There are approximately 8000 to 25,000 species of birds which are placed in 170 families and 28 orders. In fact, there are 33 orders of birds, but 5 of these are extinct. **Tinamiformes** is the most primitive order and the **Passeriformes** is the most advanced. Their body is covered with feathers, the forelimbs are modified to form wings used in flying, and they have well developed powers of hearing, seeing and producing sounds. They are warm-blooded animals, as their temperature remains constant. The heart is four-chambered and respiration is through lungs. The mouth projects into a beak or bill having an external horny sheath. Their legs also have a horny skin. They lay eggs on which they sit to hatch chicks and rear them till they can fend for themselves. Most birds are either useful or harmless, but a number of species are pests of crops or they are carriers of diseases in man and domestic animals.

The class *Mammalia* includes warm-blooded animals having mammary glands in which milk is secreted for the nourishment of the young ones. Parental care is highly developed in this group. Mammals have four limbs which are used for locomotion and the heart is four-chambered and respiration is by the lungs. Their senses of smell, vision and hearing are highly developed and they produce characteristic sounds. In feeding habits, they may be herbivorous, carnivorous, insectivorous or omnivorous. There are approximately 4000 species of mammals in the world which are placed in 130 families and 19 orders.

The important orders of this class are :

Primates. It includes monkey, lemur, chimpanzee, *gorilla,* orangutan, man, etc.

Carnivora. It includes cat, tiger, lion and cheetah all of which belong to the family *Felidae.* In the family *Canidae* are included the wolf, jackal, fox and dog. Hyaena, civet and mongose are other close relatives and are placed in separate families. The bear belongs to the family *Ursidae.*

Insectivora. This order includes the insect-eating mammals, *i.e.* the shrew, hedgehog, mole, etc.

Chiroptera. This order includes flying foxes and bats, which are the only mammals that can fly.

Rodentia. It includes the squirrel, rat, mouse, porcupine, etc.

Lagomorpha. It includes the hare, rabbit, etc.

Probascides. It includes the animals having a long trunk, *i.e.* the elephant.

Ungulata. There are two suborders, *viz. Artiodactyla* and *Perissodactyla.* The former suborder includes even-toed animals, e.g. deer, antelope, sheep, goat, cattle, water buffalo, bison, yak, pig, etc. The latter suborder includes odd-toed animals such as the horse, ass, rhinoceros, etc.

Cetacea. It includes the marine mammals, namely the whale, dolphin, porpoise, sea cow, etc.

Phylum Mollusca

The mollusks are quite a large group of animals and are widely distributed. They have an external slimy shell into which they retract at the time of danger or under adverse climatic conditions (**Fig. 1.2**). A typical mollusk has a distinct head, a segmented soft body and a ventral muscular foot with which it crawls, leaving behind a trail of slime, which it uses as a protection against the rough surface. Some mollusks are

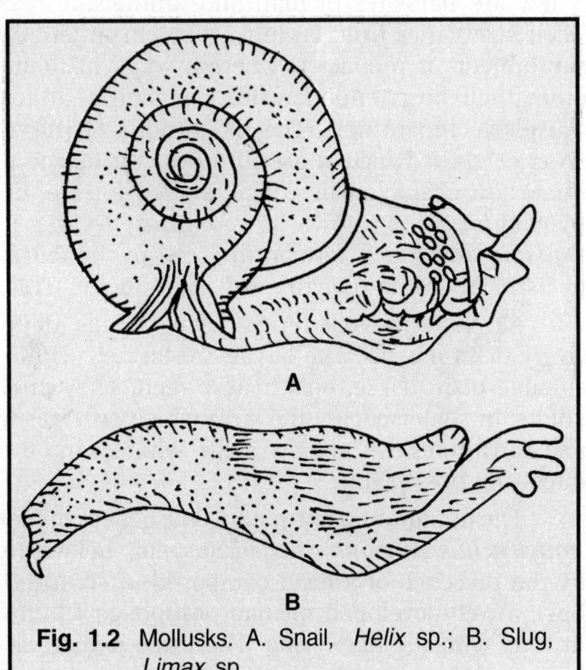

Fig. 1.2 Mollusks. A. Snail, *Helix* sp.; B. Slug, *Limax* sp.

terrestrial whereas others are aquatic. There are a number of classes in this phylum, but two are most important. The class *Pelecypoda* or the bivalves comprises the clams, mussels and oysters, including the pearl oyster which can be cultured in sea water in protected areas. The class *Gastropoda* includes slugs and snails which sometimes damage or disfigure economic and flowering plants. The snails usually have spiral shells into which the viscera is retracted under adverse environ- mental conditions, or when the animals are at rest. In slugs, the shell is either internal or absent.

Phylum Nemathelminthes

The nematodes belong to the class *Nematoda* of the phylum *Nemathelminthes.* They are minute worm-like animals without a true body cavity. They are unsegmented and are bilaterally symmetrical **(Fig. 1.3)**. The body consists of two tubes, one outer and one inner. The outer is composed of cuticle, hypodermis and nerve muscle cells, acting as the skeleton. The inner is the digestive tract, divided into a muscular oesophagus (pharynx), mid-intestine and a short rectum. The gonads are composed of a pair of tubes, one of which is often not developed fully. The female genital aperture or vulva opens typically near the centre of the body. The male genital aperture opens along with the anus into a common cloaca and is armed with copulatory spicules.

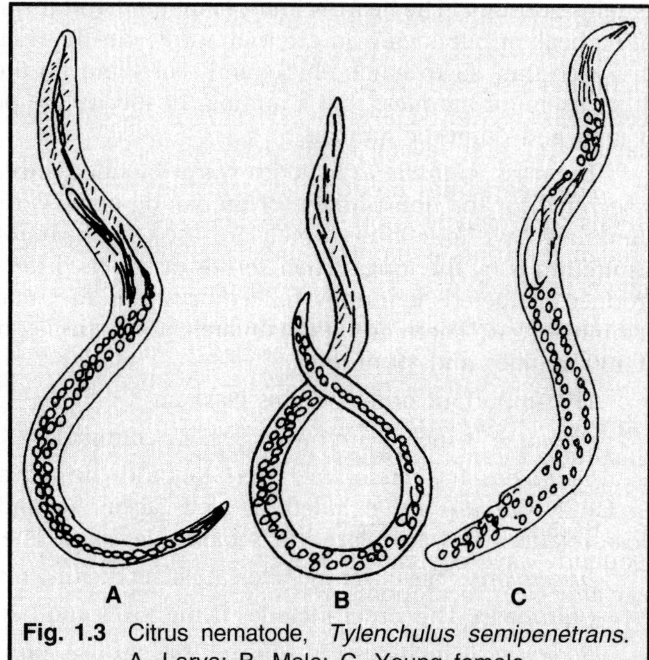

Fig. 1.3 Citrus nematode, *Tylenchulus semipenetrans.* A. Larva; B. Male; C. Young female

The nematodes are world-wide and are found in salt or fresh water and in soil. A few are parasites of man and animals. Their size varies from 20 mm to 30 cm in length. Most of the species found in the soil are either saprophytic or predatory, whereas some of them are parasites of plants. The plant parasites may enter their host (endoparasites) or simply attach themselves to plant tissues and feed from the outside (ectoparasites). The mouthparts of these nematodes are armed with a hollow protrusible stylet or *spear* which is used for puncturing the plant cells and extracting the sap. The spear is an important characteristic which is used for identification of various groups of nematodes. In order *Tylenchida,* it is known as *stomatostyle* and it develops in the pharynx. In the superfamily *Dorylaimordea,* it is formed in a cell in the left submedial wall of oesophagus and moves forward to its place in the phayrnx with each moult. This is known as *odontostyle.*

All plant parasitic forms begin as eggs, deposited by the female in plant tissue or in the soil, singly or in masses. The larvae are formed within the eggs and, on hatching , they go in search of suitable plant tissue, on which to feed. They undergo several moults, increase in size and become adults. In some species, the larva which emerges from the egg is the only stage that can move about and constitutes the infective stage, whereas in others, all stages (except the egg) can move about and infect the host plants.

The important plant-pathogenic genera of nematodes, viz. *Meloidogyne, Heterodera, Tylenchulus, Pratylenchus, Anguina, Ditylenchus,* etc., belong to the order Tylenchida. This order is characterized by the presence of a head composed of six fused lips, inconspicuous amphids (sensory organs on lips), a well-developed median oesophageal bulb, annulated cuticle and an excretory pore, almost in level with the nerve ring. The three nematode genera, *Xiphinema, Longidorus* and *Trichodorus* belong to the order Dorylaimida. They are characterized by a smooth cuticle, a papillated head and a cylindrical oesophagus with 3, 5 or 7 oesophageal glands. These nematodes are known to be vectors of viruses. They are free-living, migratory and ectoparasitic.

The main characterisitics of the important genera are briefly discussed below :

***Meloidogyne* (Root-knot nematodes).** They show marked sexual dimorphism. Vermiform males have a well-developed oesophagus and no bursa (cuticular extension over the tail), whereas the females are swollen and flask-shaped, with two ovaries and an anus. The eggs are deposited in a mass within the root galls. The first stage larva which is found inside the egg shell, and the secondstage larva which is free, are the infective stages. On entering a young root, the larva establishes itself and begins to feed and grow, hardly showing any movement, except that of the head while feeding. During its developement, the larva becomes swollen. It first becomes flask-shaped and after several moults, develops into an adult.

The nematodes produce galls on roots which may range from a string of beads on smaller laterals to a heavily knotted root mass. With the formation of galls, the flow of water and food through the roots is blocked and the plants show symptoms of starvation. These nematodes are known to attack over 1,700 species of plants and are most damaging in warm climates.

***Heterodera* (Cyst-forming nematodes).** These nematodes resemble *Meloidogyne* in morphology. In addition, the cuticle of the female *Heterodera* forms a tough, resistant, brown sphere as in *H. rostochiensis* Wollenweber, or a pear shaped cyst containing embryonated eggs, as in *H. avenae* (Filipjev). These cysts may remain dormant in the soil for several years. The second stage larva constitutes the motile infective stage and the subsequent development takes place within the roots of the host plant.

Because of their attack, the roots proliferate into a shallow bushy system. The infested fields frequently have circular or oval patches of plants exhibiting poor growth. These patches increase year after year, if cropping continues.

Tylenchulus. The citrus nematode, *Tylenchulus semipenetrans* (Cobb) (**Fig. 1.3**), is an example of this genus. The female nematode is *saccate* with a well-developed stylet having basal knobs, and with the excretory pore located just anterior to the vulva. It is a partly exposed sedentary endoparasite. The young larva feeds on the outer cortical layer of the young roots. As the larva destined to be a female develops through succesive stages, its head penetrates deep into the rootlet, while the body outside the root enlarges and becomes saccate. At maturity, the female deposits eggs in a jelly-like mass. The males remain small and worm-like and they neither feed nor is their presence necessary for reproduction.

The signs of the affected trees are death of the terminal buds, chlorosis and dying of foliage, early wilting and die-back of twigs. Soil is seen adhering to infested roots at places where the sticky eggs are located.

***Pratylenchus* (Root-lesion nematodes).** These nematodes are small, vermiform and have an annulated cuticle and a sclerotized lip region. Their styles are well-developed and bear basal knobs and, the median oesophageal bulb is ovate to spheroid. These are the most important nematodes associated with the destruction of plant roots. They are endoparasites of fibrous roots on which they cause lesions, browning or necrosis. They move from root to root, feeding and laying eggs. The eggs may also be laid in the soil. All the active larval and adult stages are capable of entering the roots.

***Anguina* (Seed-gall nematodes).** These are obligate plant parasites, giving rise to galls in leaves, stem and flowers of a number of grasses. A. *tritici* (Steinback), which causes ear-cockles in wheat, is the best known species in this group. The hard round galls containing nematode larvae replace the seed in the ears. The nematode is characterized by a narrow, flattish head with irregularly shaped oesophageal glands. The ovary consists of cells arranged about a rachis and ending in a cap cell. The adults are slightly swollen and ventrally curved.

The second stage larvae, liberated from the galls in moist soil, infest wheat seedlings at an early stage. For some time, they live ectoparasitically between the young leaves and as the flower primordia develop, the nematodes enter the tissue. They moult successively to reach the adult stage. After pairing, the females lay hundreds of eggs in the galls. The eggs develop into second stage larvae and in this state, the nematode has been observed to resist desiccation for as long as 28 years.

The laboratory examination of the plant roots and the soil is necessary to make sure of the presence of injurious species of nematodes. The Baermann funnel technique with its various modifications is commonly used for nematode extraction, which essentially consists in extracting nematodes in water, from plant or soil samples. The filtrate is passed through folded muslin, making it settle to the bottom of the funnel stem. After some hours or overnight, a small quantity of water containing the nematode is run off for examination and studied under a microscope.

Phylum Arthropoda

This phylum includes the largest number of pests of crops and domestic animals. Nemathelminthes is the only other phylum which comes anywhere near it. The arthropods are also the largest group of the animal kingdom. The class *Insecta*, alone, constitues a larger number of species than all the other classes and phyla of the animal kingdom. As the very name implies, the arthropods possess jointed legs, each terminating into a claw, a characteristic not to be found in any other group of invertebrates. Moreover, these animals have a chitinous exoskelton and their body is divided into segments which are discernible externally, but may be fused internally. Many of the internal organs, such as nerve ganglia, breathing-tubes and sets of muscles are also repeated segmentally. There are five principal classes of this phylum.

The class *Chilopoda* includes centipedes or the hundred legged worms. They are flattened worm like creatures, with a distinct head bearing two antennae. They have one pair of legs on every segment of the body. The first pair, just behind the head, has poison glands whose secretion has the property of paralysing the prey on which it feeds. Some of the larger tropical species which may be 45 cm long may give a painful bite to man. The common species in northern India is *Scolopendra morsitans* Linnaeus (**Fig 1.4A**), which is approximately 10-15 cm long and has 21 segments.

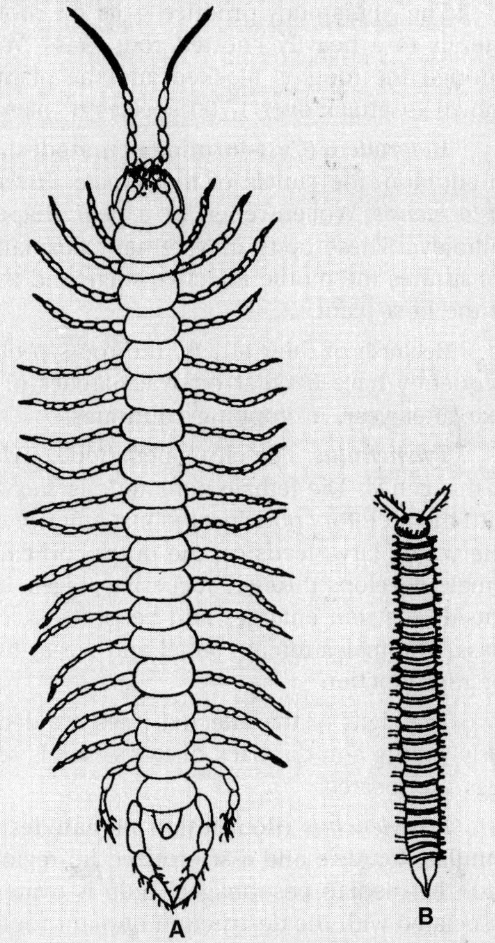

The class *Diplopoda* includes the millepedes or thousand-legged worms. These are also elongated worm-like creatures, but unlike the centipedes, their body is rounded and the segments are internally fused in twos. The large number of legs gives an erroneous idea that there are perhaps a thousand legs, but in fact, neither they nor the centipedes have the implied number of legs.

A B

Fig. 1.4 A. Centipede, *Scolopendra morisitans*; B. Millepede, *Thyroglutus malayus.*

In millipedes, the antennae are short and there are no poison claws. The reproductive organs open on the anterior end of the body, close to the head. They generally feed on decaying vegetable matter, but a few of them may attack live plants in moist soil. The common species in northern India is *Thyroglutus malayus* Attems (**Fig. 1.4B**), and a large number of their colonies are seen above the ground in the monsoon season.

The class *Crustacea* is represented by a bulk of small zooplankton, such as cyclops, which serve as food for fish. Some of the larger forms include woodlice and other frill bugs and sow bugs which live in moist places and sometimes become pests of greenhouse plants and flowers. The still larger forms include crayfish (**Fig. 1.5**), shrimps, crabs and lobsters, all of which live in the sea or in fresh water and are a source of food for man. The barnacles are sessile marine forms which get fixed to the bottom of the ships and impede their speed.

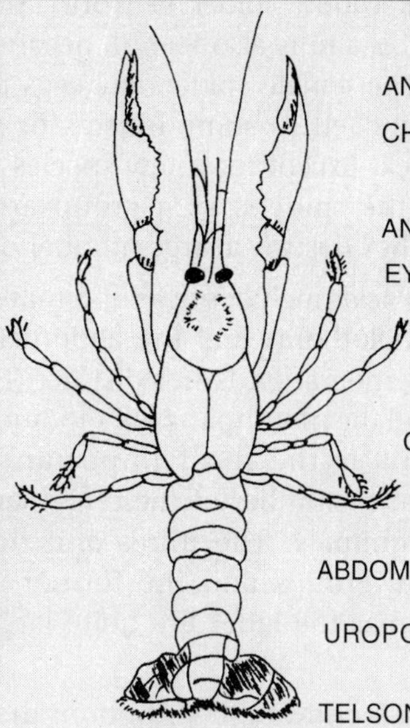

The crustaceans have two distinct regions of the body, viz. the *cephalothorax* (head and thorax fused) and the *abdomen*. They have two pairs of *antennae*, and at least five pairs of legs some of which are modified for swimming. They breathe by means of gills, except the small forms which breathe through the body wall. The sexes are separate and the genital openings are generally in pairs.

Fig. 1.5 Crayfish, *Cambras* sp.

The class *Arachnida* includes scorpions, spiders, ticks and mites. Their body also has two distinct regions, cephalothorax and abdomen. They do not have antennae and compound eyes, but go about with the help of simple eyes. They have well-developed palps on the head and possess four pairs of legs on the thorax and use them for walking. They are terrestrial animals and breathe by means of air-tubes or book-lungs. The sexes are separate and the reproductive organs are near the anterior portion of the abdomen and open through a single orifice. The arachnids are generally small, but some of them are quite large; the smallest is the mite measuring 0.09 mm (eriophyid mite), the largest is the scorpion measuring 160 mm. Some of the the species have poison glands and a sting, with which they inject poison and cause severe pain. This class is divided into a number of orders.

The order *Scorpiondia* includes the true scorpions in which the abdomen is well segmented and there is a tail like prolongation at the end, terminating in a sting or the *telson*. The scorpions are widely distributed in temperate and tropical climates, and are found in rotten logs, under stones or bricks in deserted buildings. They feed on insects, spiders and other small animals, which are first paralysed with a sting. The female produces 30-40 young ones, rears them for some time and then leaves them. She may even eat some of them. The yellow scorpion, *Palamnaeus* sp. is the commonest species in the plains of northern India and other adjoining dry areas. Its sting is very painful because of the neurotoxin it injects, but it is never fatal. The large black or dark brown *Buthus* sp. is more common in temperate climates. Its poisonous sting may even cause the death of a child.

The order *Araneida* includes the spiders which are universally predators and abound in the temperate and tropical climates. They are capable of secreting four types of silk through tubes located at the ventral surface of the abdomen. With the help of silken threads or webs, the spiders have the remarkable ability of floating in the air. With wind, they are carried over long distances and hence, many of the genera of spiders are cosmopolitan. They make various types of webs, each one of which is characteristic of the spider group.

The spiders have distinct cephalothorax and abdomen, which are joined by a short stalk (**Fig. 1.6**). They breathe with the help of book-lungs. The sexes are separate, males being smaller than the females. They are carnivorous and predatory in habit. Courtship is generally exhibited, and, in many cases, the male is eaten up by the female after mating. The black widow spider of North America is well known for this habit and is also feared, because its bite can be fatal to man. The female spiders lay eggs in silken sacs, which may be attached to some objects or may even be carried on the back. Except for a few species whose bite is painful to man, the spiders, as a group are very useful animals because they destroy a large number of insect pests.

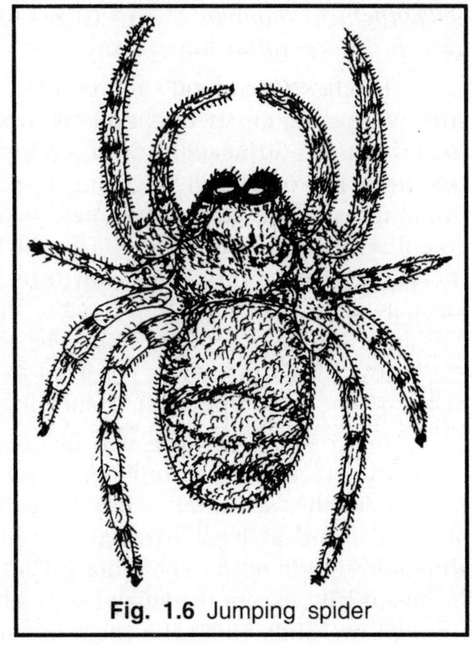

Fig. 1.6 Jumping spider

The subclass *Acari* includes ticks and mites, the animals in which the cephalothorax and the abdomen are broadly joined and the segmentation is not distinct (Fig. 1.10). The body is sac-like and the mouthparts are modified for piercing and sucking. This is the most important subclass of Arachnida, as a large number of them are serious pests of crops and farm animals. The mites and ticks are quite distinct biological groups and the former are generally microscopic, whereas the latter are quite large and have a leathery skin.

Mites have a worldwide distribution; they are rival to insects in their host range and varied habitats. They live in salt and fresh water, in organic debris of all kinds, and on plants and animals. In forests they greatly outnumber all other arthropods. They have been known to cause serious damage to livestock, agricultural crops, ornamental plants and stored products. Although many mite species are injurious, some are beneficial as they act as predators of phytophagous mites. Many mites that live in soil and water assist in the breakdown and decay of organic matter. Some of them spin silk with which they make webs, whereas others, such as the itch mite, live in subcutaneous tissue in the skin of man and animals. Most of the species deposit eggs, although some retain them inside till they hatch. A newly emerged larva is six-legged and after feeding and moulting transforms itself into nymph having eight legs. It feeds and moults one or more times and changes into an adult, when it becomes somewhat larger and acquires a genital pore.

Plant feeding mites belong to the families Tetranychidae, Eriophyidae, Tenuipalpidae, Tarsonemidae and Tuckerellidae. Important genera of plant feeding mites in different families are: *Bryobia, Eotetranychus, Eutetra-nychus, Oligonychus, Petrobia* and *Schizotetranychus* (Tetranychyidae); *Acalitus, Aceria, Cecidophyopsis, Colomerus* and *Eriophyes* (Eriophyidae); *Brevipalpus, Cenopalpus* and *Tenuipalpus* (Tenuipalpidae); *Polyphagotarsonemus* (Tarsonemidae); and *Tuckerella* (Tuckerellidae).

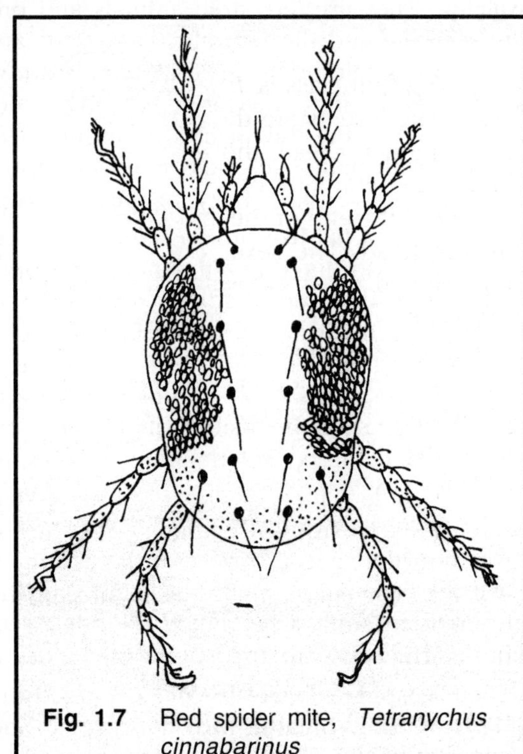

Fig. 1.7 Red spider mite, *Tetranychus cinnabarinus*

The red spider mite, *Tetranychus cinnabarinus* (Boisduval) of vegetables **(Fig. 1.7)** is the typical example of a phytophagous parasite and the mange mite,

Sarcoptes scabei (Linnaeus) is a typical example of an animal parasite. Another parasitic mite is *Acarapis woodi* (Rennie) that lives in the thoracic tracheae of honey bees and causes acarine disease.

The ticks are blood suckers and when in need of a meal, they reach their specific hosts among the mammals, birds and reptiles. They are also important carriers of diseases. Texas fever is transmitted by the cattle tick; rocky mountain fever of man is caused by the wood tick; fowl spirochetosis is carried by the fowl tick. There are two main types of ticks. The cattle tick, dog tick, and others like them, belonging to the family Ixodidae possess a dorsal shield or *scutum* and exhibit marked sexual dimorphism, and are commonly known as the **hard ticks**. The other type, belonging to the family Argasidae, is devoid of scutum and, therefore, it is known as the **soft tick.** In this group, sexual dimorphism is not marked, as for example, in the fowl tick, *Argas persicus* (Oken).

The life-cycle of various species of ticks is similar, although some complete it on one host and others on two, three or even many hosts. Consequently, they are known as one-host, two-host, three-host and many-host ticks. The cattle tick belongs to the one-host category, and the fowl tick to the many-host category.

The sheep tick, *Ixodes ricinus* (Linnaeus) is an example of the three-host species. The female usually lays eggs on the ground in sheltered places. On hatching, the six-legged larva rises on top of a grass blade, a crop plant or a bush and waits for its host to pass by and gets attached to it. Only a few, however, reach their hosts and the rest perish. After feeding, the larva drops to the ground, moults and transforms into a nymph which, in turn, climbs the nearby plant waiting for the host. The nymph (four pairs of legs) after taking a meal of blood, again drops to the ground to moult a second time to become adult. The female also needs blood before the eggs can mature; therefore, it needs the host for the third time. After feeding and mating, the female drops to the ground where it lays eggs and thereafter dies. The tick is very prolific, laying about one thousand eggs; this biological adaption compensates the scant possibility of its reaching a host.

CLASS INSECTA

The fifth class is *Insecta,* to which the insects belong. *Hexapoda* is the original name of this class. The insects are those tracheate arthropods in which the body is typically divided into three parts, namely, head, thorax and abdomen. They possess one pair of compound eyes, one pair of antennae, two pairs of wings and three pairs of legs, and they have their reproductive apertures placed at the end of the abdomen. A variable numher of simple eyes or the ocelli (1-3) are present on the head. The sexes are separate, but in some species males may be rare or absent and reproduction may take place parthenogenetically. The females generally lay eggs and the young that hatch are different from the adults in body form. They grow and change form by a series of moulting, a process known as *metamorphosis*. This is the largest group of the animal kingdom. Of all the species described, about 57 per cent are insects. Of the estimated number of species in the world, about 64 per cent are thought to be insects Nearly half of all existing insect species feed on living plants. Thus, more than 400,000 phytophagous insect species live on approximately 300,000 vascular plant species.

According to recent estimates, the total number of insect species is considerably larger than was previously thought and may range from 4 to 10 million. Herbivory does not occur to the same extent in all insect groups. The members of some orders of insects are almost exclusively herbivores whereas in other orders herbivory occurs less frequently or is even absent. The prominent among the herbivores are the Lepidoptera, Hemiptera, Orthoptera and some small orders such as Thysanoptera and Phasmida. A large proportion of the herbivorous insects belongs to Coleoptera, Hymenoptera and Diptera, all three of which also include numerous species with predatory and parasitic habits **(Table 1.1)**

Table 1.1 Number of herbivorous species in different insect orders

S. No.	Insect order	Total No. of species	Herbivorous species	
			Number	%
1.	Coleoptera	349,000	122,000	35.0
2.	Lepidoptera	119,000	119,000	100.0
3.	Diptera	110,000	35,700	32.5
4.	Hymenoptera	95,000	10,500	11.1
5.	Hemiptera	59,000	53,000	89.8
6.	Orthoptera	20,000	19,900	99.5
7.	Thysanoptera	5,000	4,500	90.0
8.	Phasmida	2,000	2,000	100.0

Source: Schoonhoven *et al.* (2006)

Certain biological attributes are associated with this predominance.

(*i*) **Small size.** The great majority of insects are quite small, require little food and can easily seek shelter from adverse weather and enemies by entering crevices, hiding under the bark of trees and fallen leaves, and in the undergrowth. Minuteness, however, is not always advantageous and some of the Hymenoptera that are merely 0.2 mm long, if covered with water, cannot free themselves from the film of water owing to the surface tension. Some of the species may be quite large and measure 120 mm. As their size becomes larger than 20 mm in diameter, the tracheal system of respiration limits the supply of oxygen required, causing them to become sluggish, which is to their disadvantage.

(*ii*) **Strong exoskeleton.** Combined with their small size, the chitinous exoskeleton gives the insects much strength without becoming too heavy. It gives them great physical strength and allows numerous modifications in the configuration of the body. In contrast, the body endoskelton of the vertebrates is too heavy and weak.

(*iii*) **High mobility through flight.** Insects are very mobile creatures and can easily seek food and mates, escape from their enemies, and disperse for fresh colonization. Apart from flying they can also run, jump or take long leaps with great agility and strength. A flea with 0.125 cm long legs can make a high jump of 20 cm and if a man with his onemetrelong legs were to show the same strength, he would have to jump 150 metres.

(*iv*) **Efficient water conservation.** Insects resist desiccation by various modifications of the exoskeleton and also by efficient water conservation. They can retain metabolic water in the body and avoid liquid excretion, and void crystalline uric acid instead. Excessive transpiration from the body of an insect takes place only when the waxy coating on the outside melts at high temperature.

(*v*) **Rapid reproduction.** Most of the insects feed on plants, weeds, animals, etc., which have a short seasonal growth. They are adapted for completing the life-cycle within that period and produce progeny prolifically. For example, it has been estimated that if all the progeny of a single house fly female were to survive, within half a year she would produce enough flies to form a layer 15 metres thick all over the surface of the earth. Likewise, the female of an aphid in one year would produce a biomass equal to the entire population of India. However, in nature there are many adversities that check insects from multiplying to such an extent.

(*vi*) **Adaptability.** No other group of animals is adapted to live in so diverse a habitat as the insects. They live in all corners of the earth from the poles to the equator, in soil, in fresh water, in hot springs, on all sorts of plants and animals (dead or alive), in pools of crude petroleum, in argol, opium, pepper or strychnine. Insects have no mechanism for regulating their body temperature and whenever the environmental temperature becomes too hot or too cold, they hibernate and become temporarily inactive. Many of the species that are subjected to extremes of temperature or drought have the physiological adaptation of diapause in which their growth is arrested and the metabolic activity is reduced to the bare minimum and tolerance to the extremes of weather is increased manifold. The most remarkable feature of diapause is that its initiation in many cases is triggered off by the day-length which naturally is precise to the day of the calendar. When the physiological changes required for the termination of diapause and the secretion of the 'growth hormone' have taken place, growth is resumed. Quite often, the photoperiod provides the required stimulus for the termination of diapause. Thus, the life-cycle of the species synchronizes with the seasons and the availability of food.

Insect Structure

The internal anatomy and external morphology vary a great deal in the various orders of insects. There are, however, many common features of the integument, body sclerites, arrangement of appendages and of various internal organs. In general, the form of body, mouthparts, antennae, genitalia, ocelli, sense organs and wing venation are so different that these characters are used for grouping insects into orders, families and genera. Some of these characters can be conveniently described through a diagram of a typical insect, whereas others can be explained adequately through the diagram of an insect which might be closest to a generalized form, *i.e.* a grasshopper **(Fig 1.8)**. Still other characters can be better described from certain species of insects. In this section, unless otherwise mentioned, the various anatomical structures described are those of the grasshopper, *Poekilocerus pictus* (Fabricius) (Acrididae : Orthoptera.)

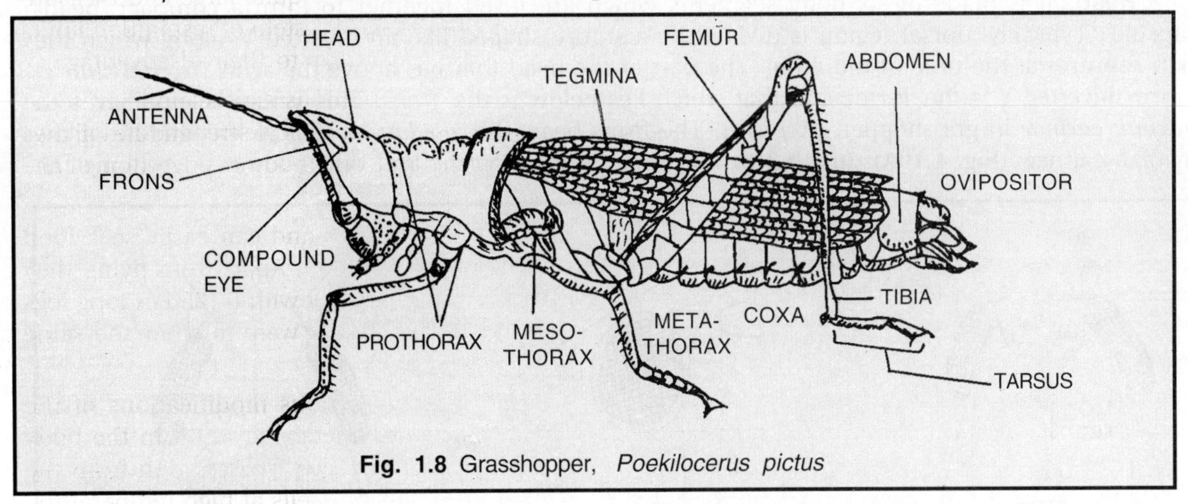

Fig. 1.8 Grasshopper, *Poekilocerus pictus*

The integument or the body wall consists of a layer of cells and is known as the hypodermis. Outside, there is a non-cellular chitinous layer secreted by the epidermis-the *cuticle* which forms the outer layer of the exoskeleton. Typically, the cuticle is composed of three layers **(Fig. 1.9)**:

(*i*) *Epicuticle* is the outermost layer, less than 4 microns thick and contains waxes which reduce water loss from the body.

(*ii*) *Exocuticle* is the middle thick rigid layer consisting mostly of chitins and protein, and is particularly prominent in hard-bodied insects.

(*iii*) *Endocuticle* is the innermost and thickest layer but is soft and flexible.

The cuticle is resistant to alkalies and dilute mineral acids. The surface of the cuticle bears two types of processes: one rigid, which includes spines and minute hair-like structures, and the other, articulate or movable such as *spurs*, general clothing hair, *sensory setae* and glandular setae.

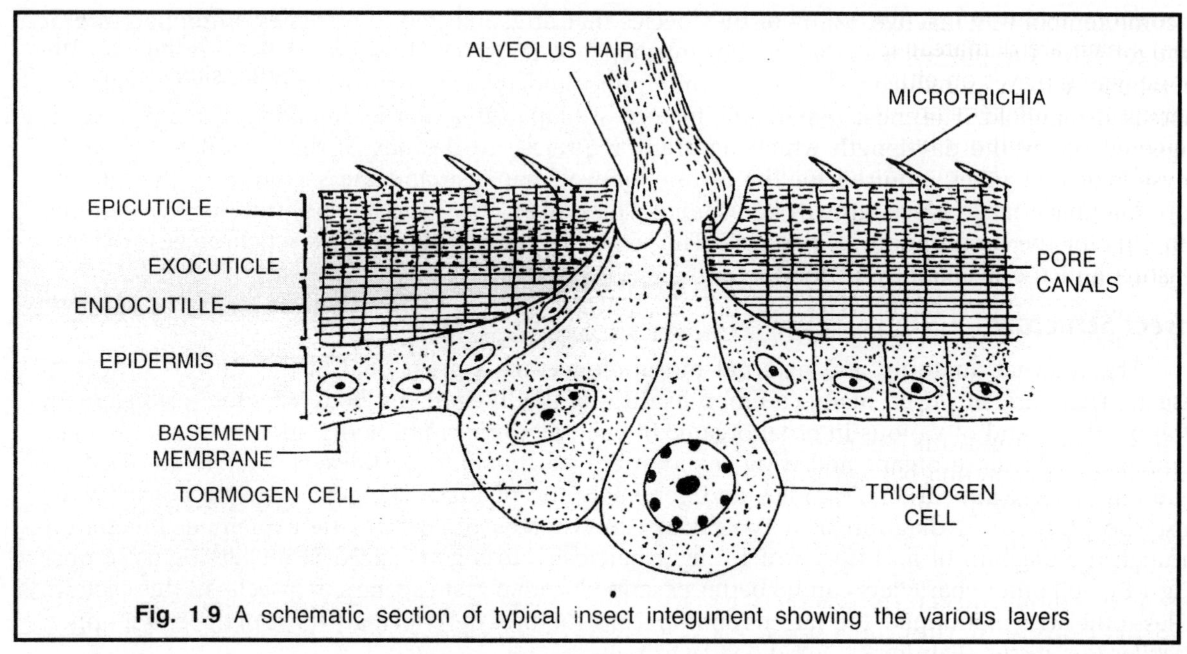

Fig. 1.9 A schematic section of typical insect integument showing the various layers

Head. It is made of six body segments which are fused together to form a compact 'head-capsule'. Typically, dorsal region is divided by a suture shaped like an inverted Y, along which the skin ruptures at the time of moulting. The part of the head that lies above the arms (*frontal suture*) of this inverted Y is the *vertex* and that which lies below is the *frons*. This suture is modified to a *median carinae* in grasshopper, *P. pictus*. The frons bears the *median ocellus* in the middle of the median carinae (**Fig. 1.10A**). Just below the frons, is the *clypeus* and on its lower edge along the

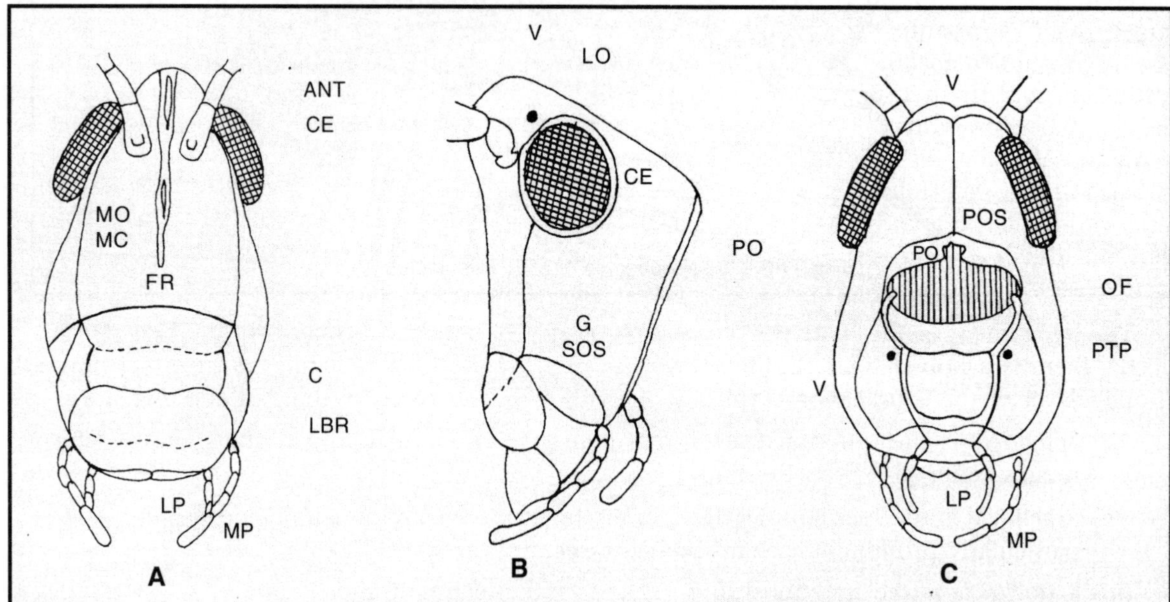

Fig. 1.10 Head of the grasshopper, *Poekilocerus pictus*. A. Anterior view; B. Lateral view; C. Posterior view; ANT, Antenna; CL, Clypeaus; CE, Compound eye; FR, Frons; G, Gena; LBR, Labrum; LO, Lateral ocellus; LP, Labial palp; MC, Median carinae; MO, Median ocellus; MP, Maxillary palp; OF, Occipital foramen; PO, Post occiput; POS, Post occipital suture; V, Vertex.

clypeolabral suture is the *labrum*. The sides of the head or the cheeks are known as *genae* (singular gena) which are demarcated by the *subocular suture*. Above each gena is located a *compound eye*. In line with the eyes on either end of the upper frons is located an *antenna*. On either side of *clypeus*, where it joins the gena are deep invaginations known as the *anterior tentorial pits*. On the pit's facet is attached the anterior basal articulation of the *mandible* (tooth). The posterior *basal condyle* of the mandible is fixed in a cavity located in the back and at the base of the gena **(Fig. 1.10B)**.

The hind wall of the head-capsule has a perforation the *occipital foramen* **(Fig. 1.10C)** through which the oesophagus or the feeding-tube and nerve cord enter the thorax. Behind the mandibles, attached on each side, are the *maxillae*, on the back of which is a single *labium* or lower lip, which has a paired origin. Arising from the floor of the mouth cavity is a tongue-like lobe, the hyphopharynx whose body or *lingua* is supported by suspensory sclerites.

Antennae among insect groups have various modifications, although their sensory functions remain the same. A typical antenna has the *scape* as the first segment, *pedicel* as the second segment and the remaining divisions, together form the *flagellum*. Antennae among insects have many modifications and they also form important distinguishing characteristics of the various groups **(Fig. 1.11)**.

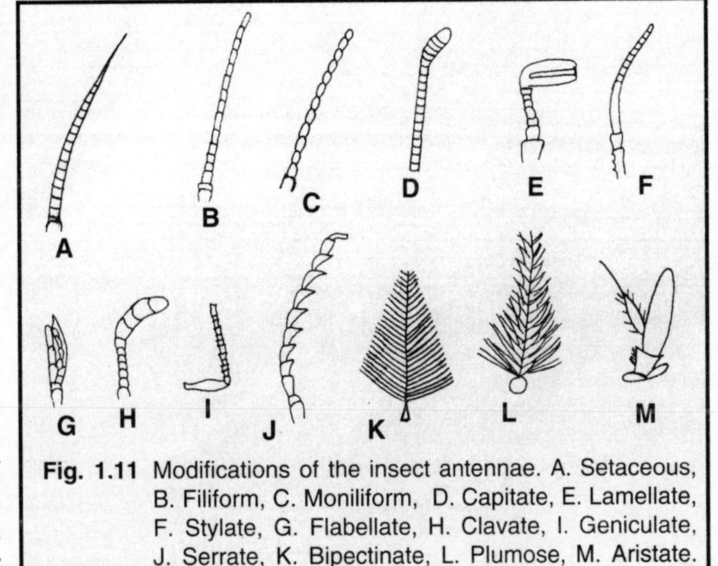

Fig. 1.11 Modifications of the insect antennae. A. Setaceous, B. Filiform, C. Moniliform, D. Capitate, E. Lamellate, F. Stylate, G. Flabellate, H. Clavate, I. Geniculate, J. Serrate, K. Bipectinate, L. Plumose, M. Aristate.

Setaceous type of antenna is found in the cockroach, *Periplaneta americana* (Linnaeus); the antennae in this type are straight fine structures, successively reduced in width, till the apex becomes pointed. The *filiform* antenna is found in the *Ak* grasshoper, *Poekilocerus pictus* (Fabricius); *annuli* in this type are almost uniform in width and the antenna gives a straight rod-like appearance. In *moniliform* antenna the divisions of flagellum are globose or ovoid ; the typical example is that of the white ant, *Microtermes obesi* Holmgren. The *capitate* type is found in *khapra* beetle, *Trogoderma granarium* Everts, and the distal annuli in this type are unusually enlarged. In the *ber* beetle, *Adoretus pallens* Arrow, the distal annuli are drawn out into leaf-like compact plates and this type of antenna is called *lamellate*. In the *stylate* form of antenna found in the robberfly, *Philonicus albiceps* (Meigen), the pedicel is enlarged and the one segment flagellum is modified into a bristle like process at the apex. The *flabellate* type of antenna is a characteristic feature of the male *Strepsiptera* (insect order) in which the basal annulus is modified into a fan-like process. In the cabbage butterfly, *Pieris brassicae* (Linnaeus), the distal annuli are enlarged gradually forming a club like structure and this type is called *clavate*. The *geniculate* antenna is similar to the filiform but in this case the pedicel is elbow jointed; it is found in *Apis* spp. *Serrate* type of antenna is typical of the *dhora* beetle, *Callosobruchus chinensis* (Linnaeus) and the annuli possess a row of lateral projections. *Bipectinate* type of antenna has annuli with two rows of long processes and it is typical of the silkworm, *Bombyx mori* (Linnaeus). *Plumose* antenna has a whorl of hairs at the base of each annulus and this type is found in mosquitoes; the hair are longer in males than in females. *Aristate* antenna has a single enlarged segment, bearing a bristle like outgrowth, and it is typically found in the housefly, *Musca nebulo* Wiedemann.

The mouthparts typically constitute two *mandibles*, one *labrum*, two *maxillae*, one *labium* and a *hypopharynx*. The mouthparts are variously modified according to the feeding habits. In grasshoppers, they are typically of the *chewing type* whereas in wasp, they are of biting and chewing type (**Fig. 1.12**). The maxilla consists of the hinge or the *cardo* with which it is joined to the head, and the distal portion *stipes*, which bears two lobe-like structures, the *galea* and *lacinia*, and the lateral 5- segmented *maxillary palpi*. The labium is, in fact, a pair of second maxillae fused together and it has at its base, the *postmentum*, which is sometimes subdivided into *submentum* and *mentum*. There are two distal lobes of labium, the *prementum*, and laterally, on each side, the three segmented *palpi* arising from special lobes the *palpigers*. The lobes of prementum are also known as *ligula* which, in some primitive insects, consist of the outer lobes known as *paraglossae* and the inner lobes, *glossae*.

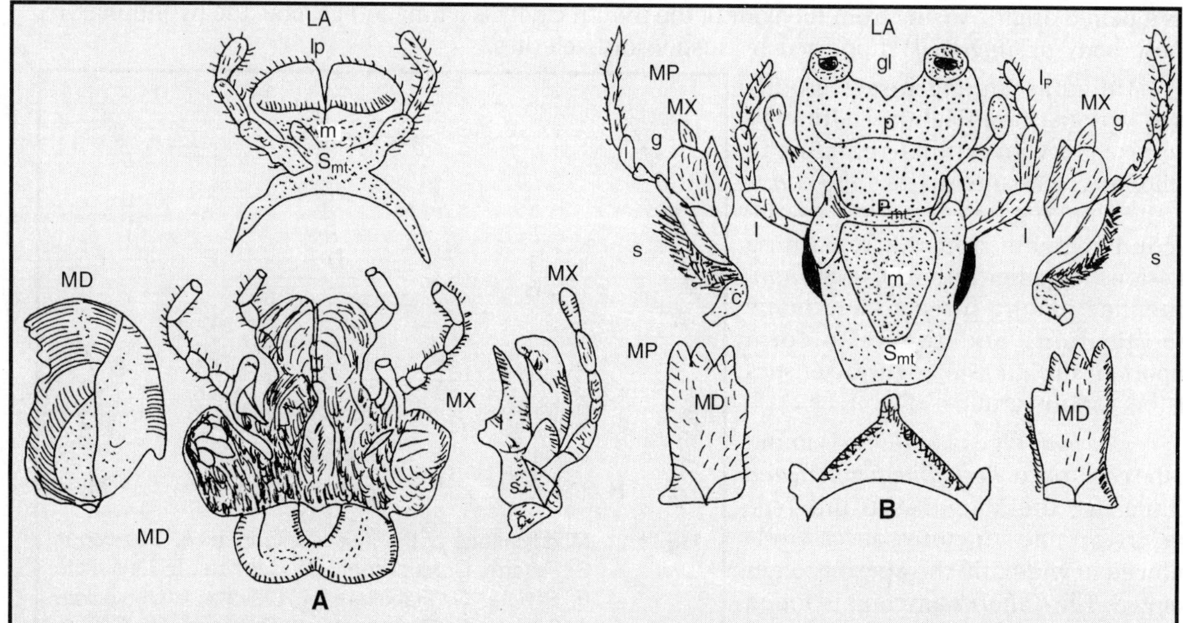

Fig. 1.12 Insect mouthparts. A. Typical chewing type; B. Biting and chewing type; c, Cardo; e, Epipharynx; g, Galea; gl, Glossa; l, Lacinia; LA, Labium (la); Li, Ligula; lp, Labial palpi; lr, Labrum; m, Mentum; Smt, Submentum; Pmt, Premantum; MD, Mandible; MP, Maxillary palpi; MX, Maxilla; p, paraglossa; s, Stipes.

The *rasping-sucking* mouthparts are intermediate in structure between the *piercing-sucking* and *chewing* types, and are found in thrips. With these mouthparts, the insects lacerate the plant epidermis and suck the sap as it exudes. In this type, the left mandible, maxillae and hypopharyx are elongated and move in and out through an opening in the apex of the cone-shaped head. No regular food channel is formed by joining of the stylets.

The *chewing-lapping* type of mouthparts are found in honeybees (**Fig. 1.13A**). The labrum and the mandibles are of the chewing type, but the maxillae and the labium are united and elongated to form the tongue, with which they collect nectar from flowers.

Typically, the *piercing-sucking* type of mouthparts are found in bugs and mosquitoes (**Fig. 1.13B**). They consist of an outer tube formed by the labium which is merely a protective covering for the stylets inside. Depending upon the insect species, the mandibles, maxillae, labrum, epipharynx and hypopharynx are modified to form the stylets which are actually used for piercing and sucking. The paired stylets are grooved on the inside and, as they join together, the food channel is formed through which the plant sap or blood is sucked.

In the *siphoning* type of mouthparts, the labrum is reduced, the maxillary palpi are rudimentary, the mandibles are absent, and the labium may be large, heavy, or scaly. The *galeae* of the maxillae form the major part of the mouthparts **(Fig. 1.13C)**. They are very much elongated and join to form a hollow slender tube which is coiled up under the head like a watch-spring. At the time of feeding, it is unwound, elongated and, nectar or juice is sucked through it. Typically, butterflies feed in this manner.

The *sponging* type of mouthparts are common in flies, the typical example being that of the housefly **(Fig. 1.13D)**. On the lower side of the head, this insect has a fleshy, elbowed proboscis which

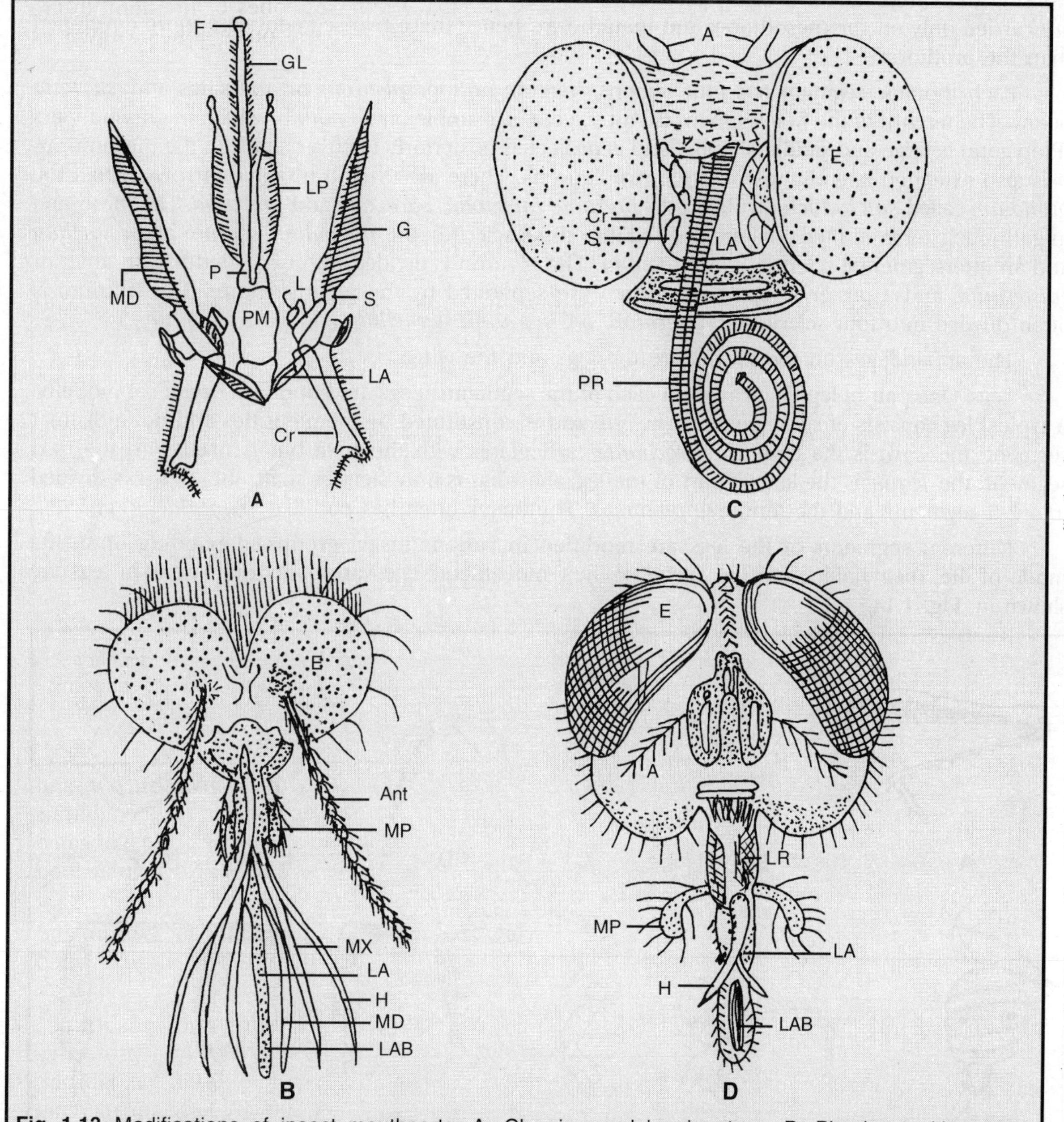

Fig. 1.13 Modifications of insect mouthparts. A. Chewing and lapping type; B. Piercing-sucking type; C. Siphoning type; D. Sponging type; Ant. Antennae; Cr. Cardo; E, Compound eye; F, Flagellum; G, Galea; GL, Glossa; H, Hypopharynx; LAB, Labellum; LP, Labial palp; MP, Maxillary palp; MX, Maxilla; L, Lacinia; LA, Labium; LR, Labrum; P, Paraglossa; PM, Prementum; PR, Proboscis; S, Stipes.

can be folded and withdrawn. It is the modified labium and has at its base the rostrum which is modified clypeus, and the distal *haustellum*, at the end of which is the sponge like structure *labellum*.

Thorax. This is the most muscular portion of the insect body and consists of three segments, *prothorax, mesothorax* and *metathorax*, which are more or less immovably attached with one another, except in certain groups. The muscles located in the thorax are used for moving the legs and for working the wings. There is one pair of legs on each of the segments, but the two pairs of wings are carried only on the mesothorax and metathorax; hence these two segments are more developed than the prothorax.

Each thoracic segment has four regions, *tergum* on top, *pleurons* on the sides and *sternum* below. The tergum of the first segment in Apterygota is a simple plate, *notum*, which in grasshoppers (Pterygota) is large and saddle-shaped, and is projected posteriorly to cover a part of the mesothorax. It is also extended laterally to cover the propleurons. There are three transverse furrows across the *pronotum* called *sulci* which divide it into *prezona, interzona, parazona* and *postzona*. The meso-and metathoracic terga in Pterygota are divided into three sclerites, the *prescutum, scutum* and *scutellum* and an intersegmental scleritie, *postscutellum*. The pleuron is divided into two scleritiеs, an anterior *episternum*, and a posterior *epimeron* which are separated by the pleural suture. The sternum is often divided into four sclerites, *presternum, basisternum, sternellum* and *poststernellum*.

The appendages on the thorax are the legs and the wings.

Legs. One pair of legs is located on each of the segments. Legs are important organs of mobility. A typical leg consists of six segments. The *subcoxa* is constituted by the pleurities and not a distinct segment; the *coxa* is the base; the *trochanter* articulates with the coxa but is fixed with the next segment; the *femur* is the largest part of the leg; the *tibia* is only slender shaft; the *tarsus* is divided into 2-5 segments and the terminal *pretarsus*. The tarsus often has pad-like *plantulae* or *pulvilli*.

Different segments of the legs are modified in various insect groups, depending upon the mode of life, their habits and the speed of their movement. The various modifications of legs are shown in **Fig. 1.14**.

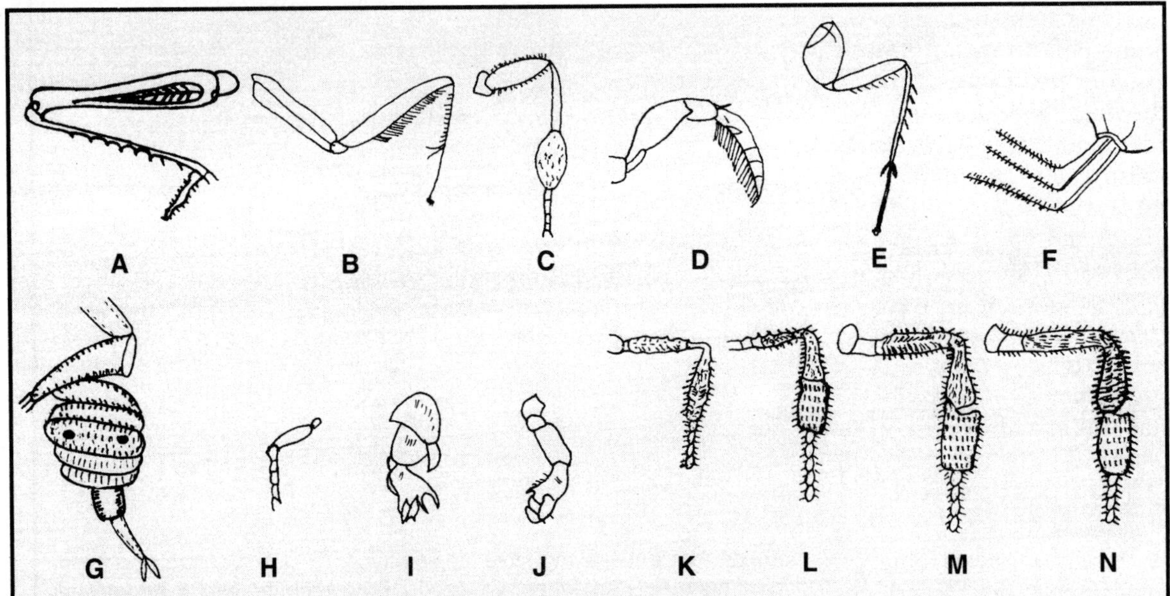

Fig. 1.14 Modifications of insect legs. A. Saltatorial; B. Rap-torial; C. Web-spinning; D. Notatorial; E. Ambula-torial; F. Scooping; G. Grasping; H. Bladder-footed; I. Fossorial; J. Clinging; K. Antenna-cleaning; L. Pollen-brushing; M, N. Pollen collecting.

The *saltatorial* type of legs are found in the grasshopper, *Poekilocerus pictus* (Fabricius). In this case, the femur is greatly enlarged and it contains strong muscles. Tibia is elongated and is spined dorsally; when it strikes against the ground surface the insect takes a leap in the air. The *raptorial* type of legs are the modified fore-legs in the superfamily Mantoidea. These legs are adapted for catching and holding the prey in between the greatly enlarged and ventrally grooved femur and the blade like curved tibia. The fore legs in the order Embioptera are adapted for *web-spinning*. The metatarsus of foreleg is swollen to accommodate the silk gland, out of which comes the silk thread. The *notatorial* legs are typical of the giant waterbug, *Belostoma indica* Linnaeus. These legs are flattened and are marginally fringed with dense hair, forming oar-like structures. *Ambulatorial* or running type of legs are typical of the cockroach, *Periplaneta americana* (Linnaeus). The legs are cyclindrical, with well developed coxae and well defined tarsomeres. The *scooping* type of legs are found among dragonflies (sub-order Anisoptera); the legs are long having rows of stiff bristles along the inner margins. In the *grasping* type of legs, found in male giant diving beetle, *Dytiscus marginalis* (Linnaeus), the basal tarsomeres are enlarged forming circular disc-bearing suction cups on the inner surface. The *bladder-footed* legs are the characteristic feature of the order Thysanoptera, the common example being the onion thrips, *Thrips tabaci* Lindeman; in these legs the distal tarsomeres bear vesicle to provide a firm hold on to the surface on which the insects feed. *Fossorial* legs are adapted for digging into the ground and are typically represented as the fore egs of the mole cricket, *Gryllotalpa* spp. In this case, the tibia and tarsi are flattened and are shaped like shovels for digging the soil. The *clinging* type of legs are found in the human louse, *Pediculus humanus* Linnaeus; here a padded projection is present at the distal end of tibia and the pretarsus is modified into an enlarged claw-shaped structure which fits on the pad. The fore egs of honey bees, *Apis* spp., are adapted for *antennae-cleaning* by a tibial spur which overlaps the bristled notch. The middle pair of legs in honeybees is modified for *pollen- brushing*; there are transverse rows of bristles on the inner side of metatarsus, which are used for brushing pollen into a heap. The *pollen-collecting* legs are also found in honeybees; these are the hind legs in which tibia are modified to form cavities (pollen baskets), fringed with spines on the outer surface, and a concave pollen cavity fringed with spines at the proximal end of metatarsus.

Wings. Among insects, wings are the principal organs of flight, one pair being on mesothorax and one pair on metathorax. Each wing is attached to the thorax with an *anterior notal process* and a *posterior notal process.* In the grasshopper, there are four axillary or wing sclerites at the base of each wing; these are the first, second, third and fourth axillary sclerites which are also called *pteralia.* In addition, there is a median plate lying between the *veins* and *pteralia.* Each wing of an insect has the anterior *costal margin,* the outer *apical margin* and the posterior *anal margin.* The angle at the base of the costal margin is called *humeral angle* and that between the costal and apical margins is called the *anal angle.*

The wings are strengthened by the *veins,* in which the blood flows. Wing venation is extremely variable among insects and the principal veins are shown in a hypothetical drawing giving their names **(Fig. 1.15)**. *Reticulation* of veins is seen very much

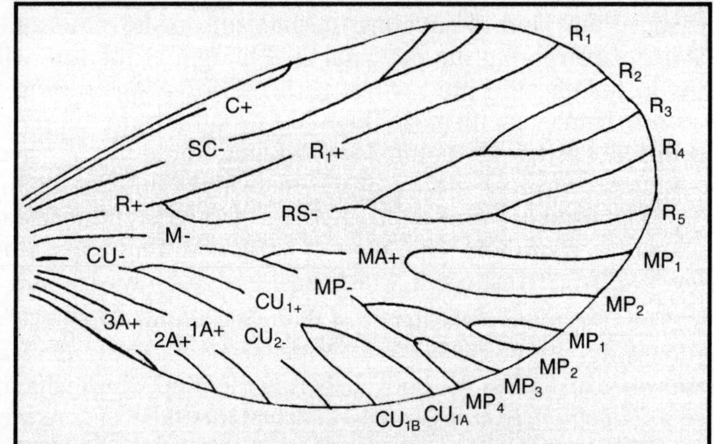

Fig. 1.15 A hypothetical primitive wing venation in insects. C, Costa; SC, Subcosta; R, Radius; Rs, Radial sector; M, Median; MA, Anterior median; MP, Posterior median; CU, Cubitus; IA to 3A, Anal veins; +, Convex veins; -, Concave veins.

developed in the order Odonata, for example, in the dragonflies, *Aeshna* spp. This modification provides a strength to the wings of these most versatile and powerful fliers. In certain insects, veins along the costal margin are bifurcated, which are known as *marginal accessory veins.* They provide strength to the otherwise delicate wings found in the green lacewing, *Chrysoperla* spp. There are various types of wings and secondary peculiarities like the scales and the wing-coupling mechanisms as shown in **Fig. 1.16.**

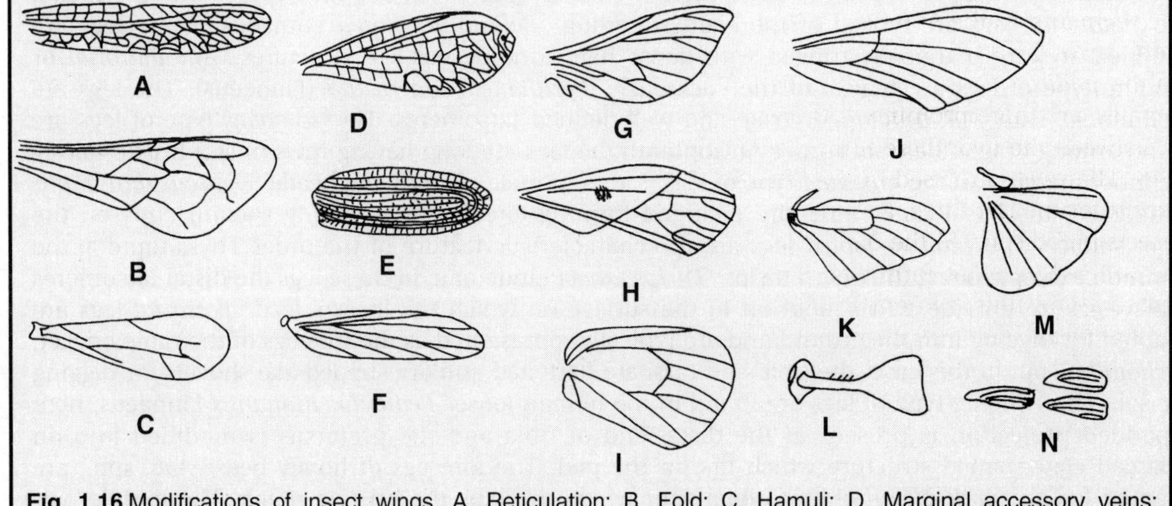

Fig. 1.16 Modifications of insect wings. A. Reticulation; B. Fold; C. Hamuli; D. Marginal accessory veins; E. Elytra; F. Tegula; G. Jugum; H. Hemelytron; I. Humeral lobe; J. Hook or retinaculum; K. Frenulum; L. Haltere; M. Pseudohaltere; N. Scales on the wings.

Wings among the Lepidoptera have *scales* on them which are, in fact, modified setae; these scales are attached loosely and can be easily rubbed off. Most of the insects when in flight have the fore-and hind-wings held together by various devices in the form of coupling apparatus. Thus, the insects can beat their wings in unison. Among hymenopteran insects, the posterior edge of the anal margin on the fore wing is curved into a *fold* and this fold engages with the *hamuli,* which are in the form of a row of small hooks present on the costal margin of the hind wings. Honey bees, *Apis* spp., have this kind of coupling mechanism. In the suborder Monotrysia (swift moths), there is a *lobe-like* expansion at the proximal anal margin of the fore wings. It is known as the *jugum* and it helps in holding the hind wings with the forewings during flight. Among the butterflies, *Pieris brassicae* (Linnaeus), there is a lobe-like expansion on the proximal costal margin of the hind wings and this lobe is known as the *humeral lobe.* Some moths have a catching flap on the underside of the anterior proximal surface of the forewings. This flap is known as *hook* or *retinaculum,* which receives a bristle like projection present at the base of the costal margin of the hind wing. The bristle is known as *frenulum.* This type of wing coupling apparatus is found in the male hawk moth, *Acherontia styx* (Westwood) which has a large body size and is a powerful flier.

The shape and consistency of the wings among insects is quite varied. In the order Coleoptera, the forewings are horny and are arched and are known as *elytra*; when the insects are not in flight these wings lie flat on the body, covering the hind wings which are folded underneath. The red flour beetle, *Tribolium castaneum* (Herbst) has this type of fore wings. In the sugarcane pyrilla, *Pyrilla perpusilla* (Walker), the forewings are strengthened by scale like sclerites present at the base of costa of the forewing which are known as *tegula.* In the order Heteroptera, *i.e.* red cotton bug, *Dysdercus koenigii* (Fabricius), the forewing is thickened proximally but is membranous distally and this type of wing is known as *hemelytron.* The hind wings in the insect order Diptera, i.e. *Musca nebulo* Wiedemann, are extremely small and are very slender proximally but knobbed distally. These wings are known as *halteres* and are used for balancing the body while in flight. Similarly, the fore wings in the order Strepsiptera are also very small and are club-shaped; these types of wings are known as *pseudohalteres.*

Abdomen. This is the third and terminal part of the insect body, the posterior end of which has the anus and an aperture of the reproductive organs. There are only a few appendages on the abdominal segments, typically 11, which are movably joined together by the intersegmental membranes. In many of the immature stages of aquatic forms, there is one pair of gills on each of the segments. In the Apterygota, each segment carries a pair of plate-like limb bases or *coxites* which carry *styles.* In a grasshopper **(Fig. 1.17)**, the abdomen has 11 segments, the last one carries *cerci* on the *podical plates.* The 9th and 10th segments carry the paired genital appendages which in the females, are more prominent and are known as the *upper valves* and the *lower valves,* respectively. In between the two, there is a *middle valve* carrying a finger-like forked organ used for placing the eggs. The valves are used by the female for digging a hole in the soil, in which she deposits the eggs. Except for the last three, there is one pair of *spiracles* on each of the segments, through which these insects breathe air. In the male, the 9th tergum is convex and hood-shaped. The 9th sternum carries the *sub-genital plate,* forming a deep cavity for the genitalia. On top, it has a membranous roof, the *pallium.* The 10th segment and the telson are like those in the female. In both sexes, on the lateral

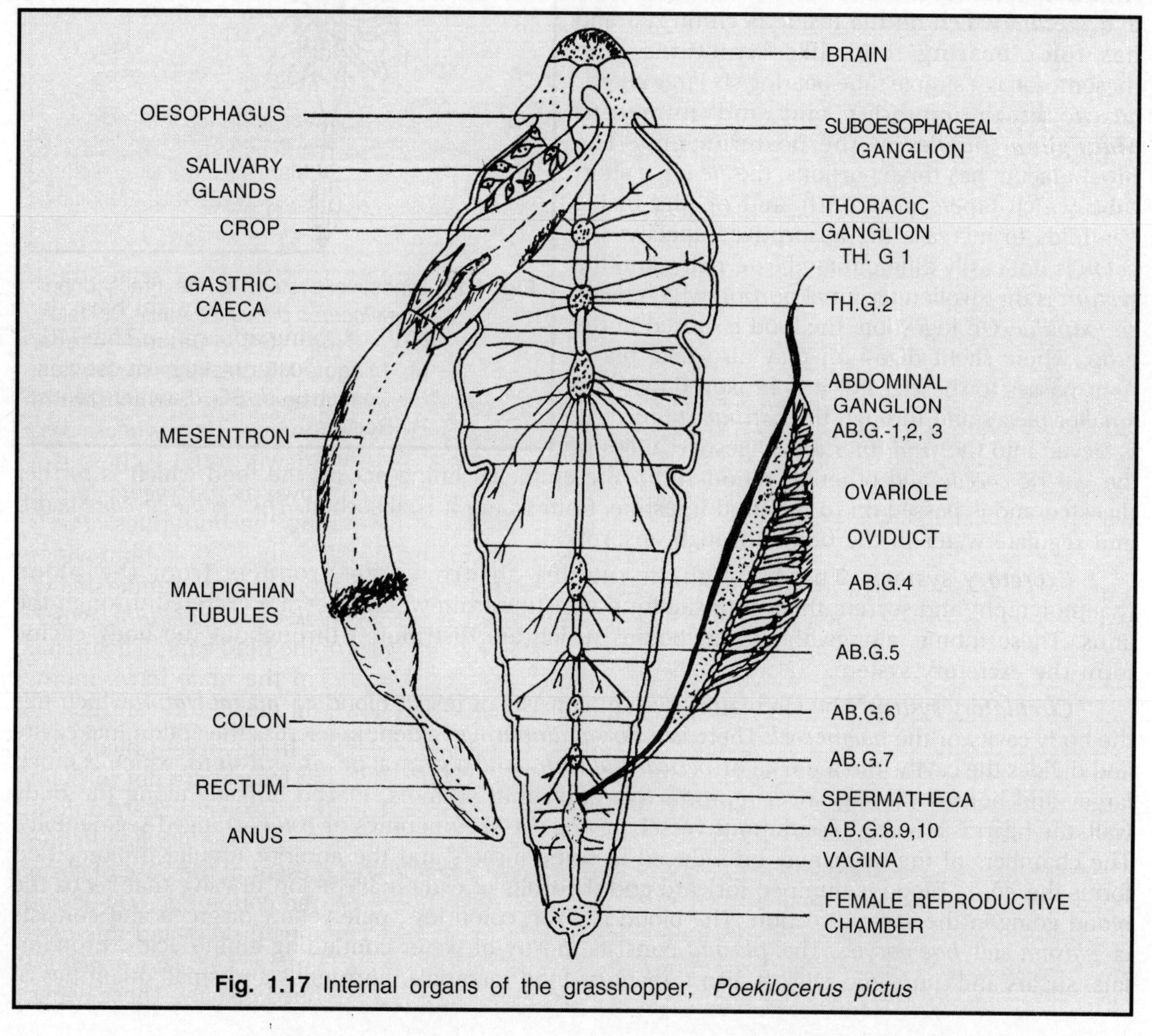

Fig. 1.17 Internal organs of the grasshopper, *Poekilocerus pictus*

aspects of the first tergum are the *tympanal organs* covered over with a tense membrane, the *tympanum* or hearing organ and a spiracle in front of each tympanum.

The abdomen is the major seat of the internal organs in which most of the physiological functions take place. However, some of the internal organs or parts thereof commence in the head, pass through the thorax and extend as far as the abdomen **(Fig. 1.17)**. The main systems are the digestive, excretory, circulatory, nervous, respiratory and reproductive systems. These systems and their functions are discussed briefly.

Digestive system. The digestive system consists of the *stomodaeum* or foregut, the *mesenteron* or midgut, and the *proctodaeum* or hindgut **(Fig. 1.18)**. The stomodaeum starts at the mouth or buccal cavity into which open the salivary ducts, bringing in secretion from the salivary glands which resemble a bunch of grapes. Then follows the *pharynx* and *oesophagus* which is dilated at the posterior end to form the *crop*, that is provided with transverse ridges internally. The foregut ends in a *gizzard* which on the inside is chitinized and has folds bearing tooth-like structures. The mesenteron is a simple tube bearing six large *enteric caecae* at the anterior end and numerous *Malpighian tubules* at the posterior end. The proctodaeum has three portions; the *ileum*, a short tube which tapers posteriorly and on the inside has folds to increase the absorptive capacity; the *colon* is not easily differentiated externally; and the *rectum* is the swollen terminal portion which bears six *papillae*. On ingestion, the food is stored in the crop, where slight digestion may also take place. As it passes to the gizzard, it is broken down into smaller pieces and through the *cardiac sphincter* it is sieved into the *mid intestine*. Digestive juices of the *gastric caecae* and other secretions from the epithelial lining act on the food which is further digested and is passed on to the hind intestine, from where it is absorbed. The *rectal caecae* absorb and regulate water in the body, through absorption.

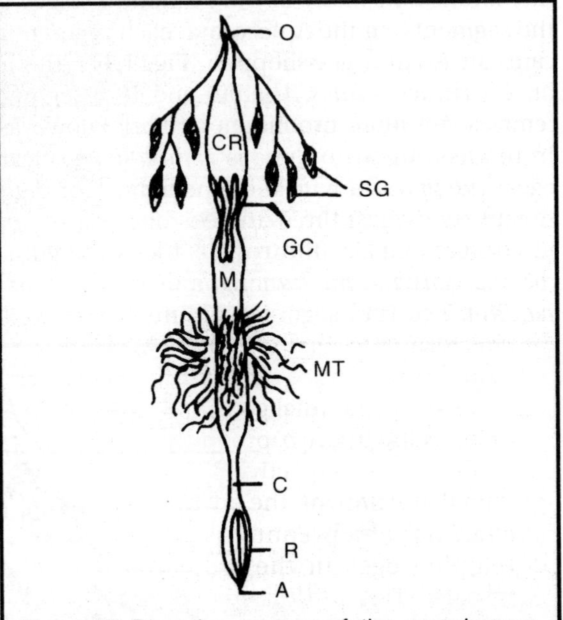

Fig. 1.18 Digestive system of the grasshopper, *Poekilocerus pictus*. A, Anus; C, Colon; CR, Crop; GC, Gastric caeca; M, Midgut; MT, Malpighian tubules; O, Oesophagus; SG, Salivary glands; R, Rectum.

Excretory system. The Malpighian tubules absorb waste products from the blood (haemolymph) and secrete them into the hind intestine from where they are excreted through the anus. These tubule, alongwith the fat bodies which are distributed throughout the body cavity, form the excretory system.

Circulatory system. The circulatory system consists of insect blood or *haemolymph* which fills the body cavity or the *haemocoel*. There is a *dorsal diaphragm* extending across the abdominal cavity and divides the cavity into a dorsal or *pericardial sinus*, and a ventral or *visceral sinus*, which is much larger and houses the digestive, reproductive and other systems. Placed dorsally along the body wall, the heart is a narrow continuous vessel, having slit-like openings or *ostia* arranged segmentally. The chambers of the heart may be reduced in some insects and the anterior tubular prolongation forms the *aorta*. Blood is pumped forward and the walls of ostia make a sort of valve that keeps the blood going in the same direction. The blood is clear, colourless, pale yellow or green and consists of *plasma* and *haemocytes*. The *plasma* consists mostly of water containing amino acids, proteins, fats, sugars and inorganic salts. It also transports food materials, hormones and small quantities of oxygen and carbon dioxide. The haemocytes have amoeboid mobility. They surround foreign bodies and congregate around the wounds to heal them.

Reproductive system. The reproductive system in a female grasshopper (**Fig 1.19A**) consists of a vagina into which open the *accessory glands*, and the *spermatheca* which receives and holds sperms after copulation. There are two ovaries, each having a number of *ovarioles* or egg tubes. The ovarioles open into the *egg calyx* that leads to the oviduct. The oviducts join together to form the common oviduct which meets the vagina. The distal end of an ovariole is the *terminal filament* which joins other similar filaments to form the *suspensory ligament*. Two ligaments of the respective ovaries join to form a *median ligament* which keeps the ovaries in position. The filament at the terminal portion of an ovariole extends from the germarium, which contains the primordial germ cells and the nutritive cells, the two producing *oocytes* and *trophocytes*, respectively. The third or proximal portion of the ovariole is the *vitellarium* which contains a series of developing eggs in the *follicle* or egg chambers. The *follicular epithelium* surrounding each egg secretes the chorion or egg-shell, which is completed after fertilization in the ovariole or the oviduct, depending upon the the species.

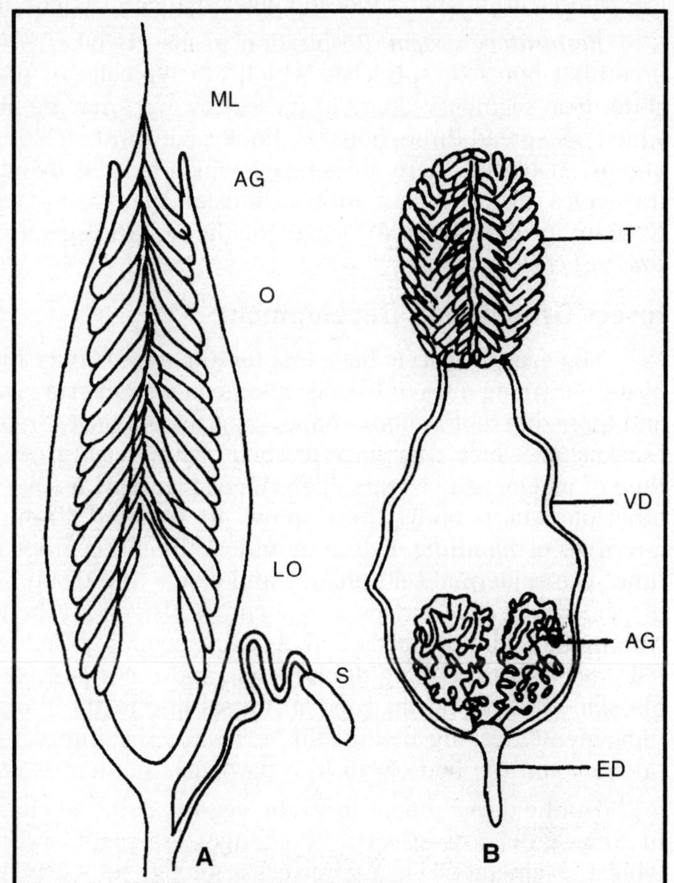

Fig. 1.19 Reproductive system of the grasshopper, *Poekilocerus pictus*. A. Female; B. Male; AG, Accessory glands; ED, Ejaculatory duct; LO, Lateral oviduct; ML, Median ligament; O, Ovariole; S, Spermatheca; T, Testes; V, Vagina; VD, Vas deferens.

The male reproductive system (**Fig. 1.19B**) consists of two gonads or *testes*, both of which have lateral ducts or *vasa deferentia* and the two join to form a median duct, the *ductus ejaculatorius* and open to the outside through the *aedeagus*. There are two pairs of accesory glands.

Nervous system (**Fig. 1.17**). The nervous system consists of a brain formed by the fusion of *ganglia* of the head segments, and a pair of *ventral ganglia* in each body segment. The ganglia are joined by longitudinal double tracts or the *connectives* which run throughout the length of the body on the inside of the ventral body wall. The ganglia, in pairs, are also connected across by transverse nerve fibres or *commissures*. In the grasshopper, *P. pictus*, there are three thoracic and eight abdominal ganglia, the first three abdominal ganglia being fused with the metathoracic ganglia.

The *supra-oesophageal ganglion* or the brain is a combination of the presegmental and two anterior head segments. The three portions of the brain corresponding to these three ganglia are known as the *protocerebrum*, the *deutocerebrum* and the *tritocerebrum*. In addition, there is a *suboesophageal ganglion* which is connected with the brain by *paraoesophageal connectives*. It is formed by the fusion of ganglia of the mandibular, maxillary and labial segments.

The protocerebrum consists of *protocerebral lobes* and *optic lobes*. The protocerebral lobes are interconnected by a *median commissural system*. The optic lobes reach the compound eyes. The deutocerebrum represents the ganglia of antennary segment and innervates the antennae. It is with the *tritocerebrum* that suboesophageal ganglion is connected.

Respiratory system. Respiration in insects takes place through tracheae which originate from breathing pores or spiracles, which are typically in pairs and one on each of the thoracic and abdominal segments. Some of these, however, may be absent. The tracheae branch into tracheoles which are spread throughout the body cavity and are sometimes seen as silvery lines on the internal organs. At places, there are air-sacs which help in flying. The diffusion of gases takes place in the tracheoles. In the aquatic forms, diffusion may take place through the spiracles or the integument. In many of the immature stages of the aquatic species, respiration may also take place through *tracheal gills.*

Insect Growth and Development

The eggs in insects have less food materials than those in birds and reptiles. Therefore, unlike birds, the young ones of insects emerge at an early stage of development and are much smaller than, and quite different in body shape, from the adults. Moreover, the young have a covering of inflexible exoskeleton which they have to shed a number of times in order to grow. The young larva at the time of its emergence tears open the egg-shell with spines or *egg-bursters* located on its head or on other parts of its body. These spines are retained till the first skin is cast off by the process known as *ecdysis* or *moulting*. Before moulting, the insect stops feeding and becomes quiescent for a short time. The epidermal cells enlarge and divide mitotically. The old cuticle is detached and a new layer of epidermal cells secretes a new cuticle. The space between the old and the new cuticles is filled with moulting fluid which contains two enzymes, *protease* and *chitinase*. These enzymes dissolve the old cuticle, not touching the new one and the dissolved fluids are absorbed by the epidermis. The old skin splits along the median dorsal line in the thorax, and the insect frees itself by muscular movements, leaving behind the *exuviae*. The interval between two moultings is known as the *stadium* and the body form in a particular stadium is called an *instar.*

During development from the egg to adult, the changes in form are known as *metamorphosis*. In some groups of insects, the changes are slight and only the size of the body increases, being typical of ametabolism, e.g. silverfish of the sub-class Apterygota. In others the changes are very drastic and the immature stages look quite different from the adult. The direct development or incomplete metamorphosis is known as *hemimetabolism* and the indirect development or complete metamorphosis is known as *holometabolism*. The former is most prevalent in the Exopterygota and the young ones are commonly called *nymphs*. The latter is characteristic of Endopterygota and the young ones are commonly known as *larvae.*

Among *hemimetabolous insects,* most of the external structures and internal organs remain essentially the same and gradually transform into those of the adult; the young ones are called *nymphs,* and in some cases, *naids* (Odonata). In the *holometabolous insects,* there is the pupal stage in between the larva and adult. In this stage, there is extensive destruction of the larval organs through *histolysis* and the tissues are broken up and dissolved through the activity of phagocytic blood-cells. The break down products are absorbed into the blood and provide material for the reconstruction of adult organs which are initiated from certain latent cells whose activity remains suppressed throughout the larval period.

Among the *holometabolous insects,* there are four distinct developmental stages; the egg stage, the active larval stage with variable number of legs, the resting stage or the pupa, and the winged adult. The larvae of this group of insects display an enormous range of structural variations and are adapted for life in a wide variety of environments. The *endopterygote* larvae may be divided into a number of forms **(Fig. 1.20)**. The highly specialised form is known as the *protopod* larva. It is

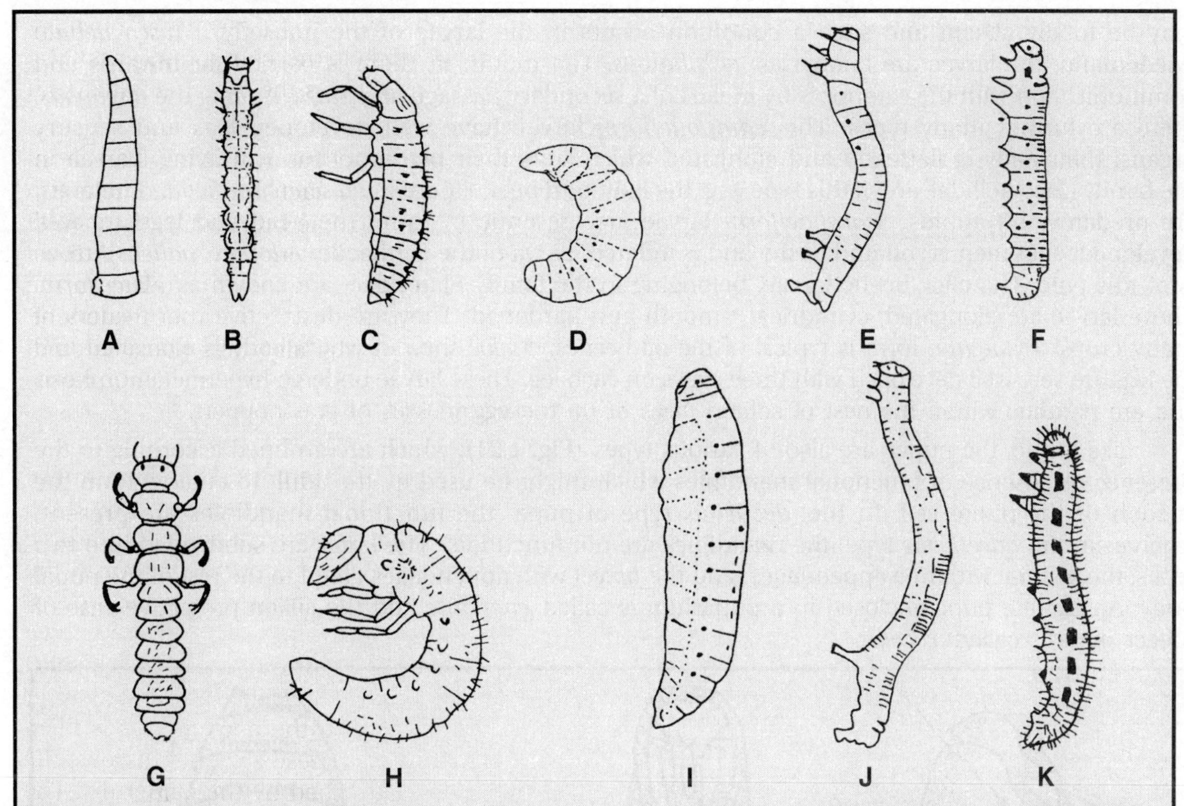

Fig. 1.20 Modifications of insect larvae. A. Acephalus; B. Elateriform; C. Campo-deiform; D. Hemicephalous; E. Semi-looper; F. Pseudo-caterpillar; G. Triun-gulin; H. Scarabaeiform; I. Eucephalous; J. Looper; K. Caterpillar

common among the endoparasitic Hymenoptera and Diptera. This form, in a way resembles the embryonic stage because the body segmentation is ill-defined and the appendages are either rudimentary or absent. The *polypod* type of larva is structurally of the more advanced form and is provided with thoracic legs and abdominal *prolegs* (unjointed). The *apodous* form lacks all the legs and its head may be well developed, reduced or absent altogether. The *oligopod* type is well-developed with three pairs of legs on thorax.

Larvae of insects in the orders Lepidoptera and Hymenoptera are polypods. Larvae of the sawfly, *Athalia lugens proxima* (Klug), have 6-8 pairs of prolegs including those on the last abdominal segment. This type of larva is a defoliator and is known as a *pseudocaterpillar.* Among lepidopteran larvae, the *caterpillars* bear five pairs of prolegs on third to sixth, and on the tenth abdominal segments. Many caterpillars are pests of crops; the larva of cabbage butterfly, *Pieris brassicae* (Linnaeus), is of this type. The *semiloopers* bear three pairs of prolegs on the fifth, sixth and tenth abdominal segments. These larvae are known as semiloopers because as they crawl they form a semiloop in their body and then they stretch straight. These larvae are normally defoliators and are pests of crops, for example, larva of the cabbage semilooper, *Thysanoplusia orichalcea* (Fabricius). Looper bears two pairs of prolegs on the sixth and the tenth abdominal segment, and it moves on a leaf surface by forming a full loop. Larvae of the inchworm (Family Geometridae) are the typical loopers. They feed on the roots of grasses.

Apodous larvae with the reduced head portion are known as *hemicephalous*. Larvae of the honey bees, *Apis* spp., are of that type. Larvae of the red wasp, *Vespa orientalis* Linnaeus, on the other hand, have fully developed head and are known as *eucephalous.* In some cases, however, the head may be totally absent and such a condition occurs in the larvae of the housefly, *Musca nebulo* Wiedemann; the larvae are known as *acephalous*. The mouth in them is carried far inwards and communication with the exterior is by means of a secondary passage or *atrium*. Among the *oligopods*, there are further many types. The *campodeiform* larvae have well developed legs and sensory organs; their body is flattened and elongated which suits their predatory form of living. Larvae in the family Coccinellidae are of this type, e.g. the lady-bird beetle, *Coccinella septempunctata* Linnaeus, are predatory on aphids. *Scarabaeiform* larvae are cresentic in form; the head and legs are well developed; abdomen is soft and fleshy and is inflated. Larva of the *ber* beetle, *Adoretus pallens* Arrow, is of this type. The click-beetle larvae, belonging to the family Elateridae, are known as *elateriform*. These larvae are elongated, cylindrical, smooth and hardened. They are destructive root feeders of many crops. *Triungulin* form is typical of the oil beetles, *Meloe* spp., in which body is elongated and the legs are very well developed with three claws on each leg. These larvae undergo hypermetamorphosis and are parasitic within the nest of solitary bees or on the egg masses of grasshoppers.

Like larvae, the pupae are also of various types (**Fig. 1.21**), which are grouped according to the presence or absence of functional mandibles which might be used by the adult to emerge from the cocoon or the pupal cell. In the *decticous* type of pupa, the functional mandibles are present, whereas in the *adecticous* type, the mandibles are not functional. The latter are subdivided into two types, the *exarate* with free appendages, and the *obtect* with appendages glued to the rest of the pupal body. An exarate pupa enclosed in a puparium is called *coarctate* and the silken protective case of obtect pupa is called *cocoon*.

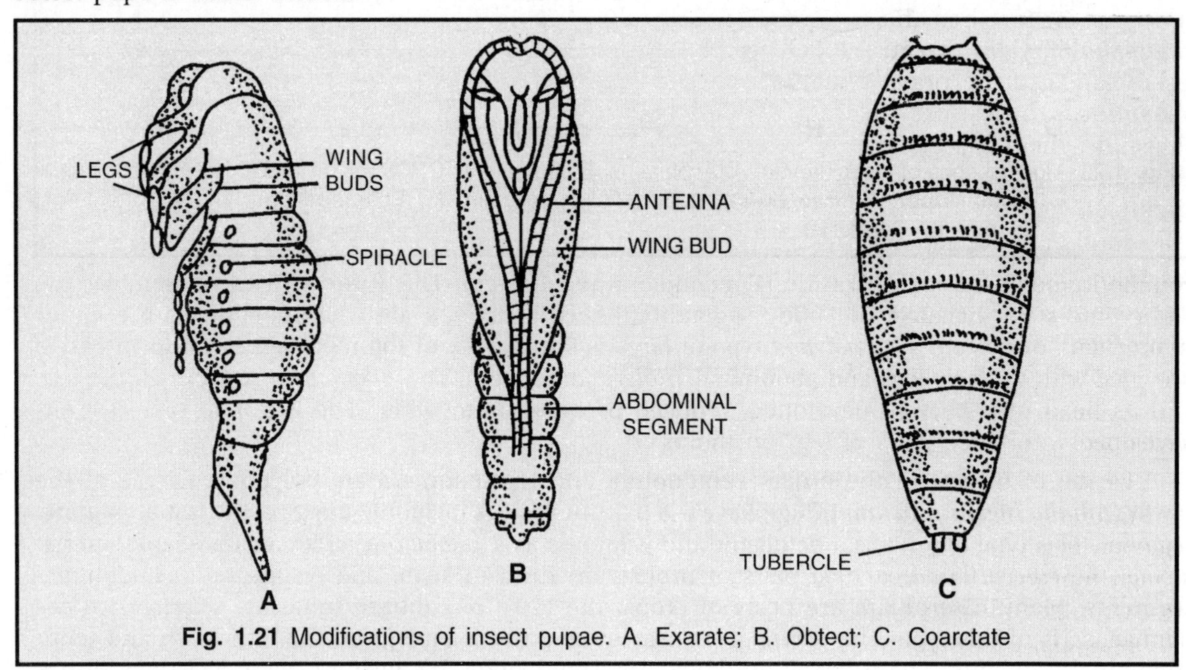

Fig. 1.21 Modifications of insect pupae. A. Exarate; B. Obtect; C. Coarctate

Hormonal Control of Metamorphosis

The growth and moulting in insects are controlled by hormones. The neurosecretory cells in the brain, under certain stimuli, secrete a hormone prothoracicotropic hormone (PTTH), which activates the prothoracic glands. As a result of this activation, the growth and differentiation hormone (GDH) or *ecdysone* is secreted. This hormone induces the insects to moult. The corpora alata produce the *juvenile hormone* (JH) or *neotenin*. High concentration of JH results in larval-pupal moult, lower concentration results in larval-pupal moult and absence of JH results in pupal-adult moult (**Fig. 1.22**). The two other hormones which also play a role in development are the eclosion and tanning hormones. The eclosion hormone released by the brain controls the process of eclosion in insects while the tanning hormone or burscion regulates the process of tanning and sclerotization of the new cuticle.

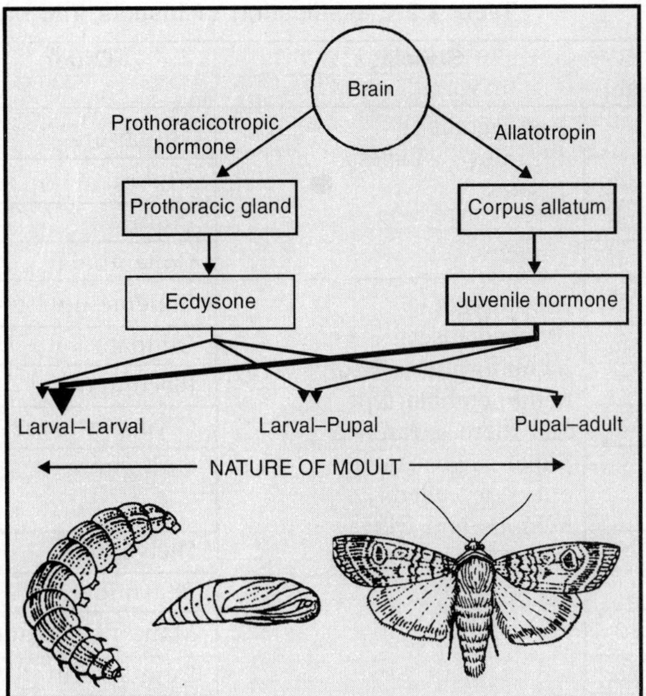

Insect Classification

The classification of insects into various orders is based on the presence or absence of wings and their venation, type of mouth parts, type of metamorphosis and characteristics of antennae and tarsi. The class Insecta has two subclasses, *Apterygota*, the primitive wingless insects, and *Pterygota*, insects with wings (although some of them may be secondarily apterous). **(Table 1.2)**

Initially, there were 29 orders of insects. Recently, a new insect order (with only 3 members) has been added in 2002 and the order has been named as *Mantophasmatodea*, because of superficial resemblance to the praying mantids and phasmids, the stick insects. Unlike phasmids, these insects lack an elongate mesothorax and unlike mantids, these lack raptorial forelegs. Of these 30 orders, only the more important ones are discussed below :

Fig. 1.22 Classical scheme indicating the role of hormones in moulting and metamorphosis in insects

Thysanura (*thusanos*, a fringe; *oura*, a tail). This order includes the spring-tails. The median tail has bristle-like processes and on each side there are long cerci. The body is covered with scales; antennae are long and mouthparts are meant for biting. The two common species are *Lepisma saccharina* Linnaeus, often seen behind calendars on the walls, and *Thermobia domestica* (Packard), the fire-brat, which is found in warmer places.

Ephemeroptera (*ephemeros*, living a day; *pteron*, a wing). Mayflies are short-lived, have soft bodies and large eyes. The wings are membranous. The nymphs are aquatic, with tracheal gills, and form an important food for fish, particularly trout.

Odonata (*odontos*, a tooth). This order includes dragonflies and damselflies, These are large insects with elongated and brilliantly coloured bodies. Their eyes are large and mouth parts are strongly toothed. The wings are membranous with numerous cross veins. The nymphs are aquatic. The suborder Anisoptera includes dragonflies and Zygoptera, the damselflies.

Plecoptera (*plekein*, to fold; *pteron*, a wing). The stoneflies are soft-bodied insects and have long thread-like antennae. Their wings are membranous and mouthparts are of the biting type. The nymphs are aquatic and breathe through gills.

Orthoptera (*orthos*, straight; *pteron*, a wing). It includes grasshoppers (**Figs. 1.8, 13.12**), locusts, crickets, etc. The insects of this order have biting and chewing mouthparts and the fore wings are modified into tegmina that lie straight. The hind legs are modified for jumping. The females have well-developed ovipositors which are often used for digging holes in the soil where eggs are deposited. Stridulatory and auditory apparatuses are well developed in many of the species. There are three main groups of this order and they are abundant in the tropical areas. Those with 4-segmented tarsi, long antennae and laterally compressed ovipositors are the long-horned

Table 1.2 Classification of insects into various subclasses and orders.

S. No.	Subclass	Order	Example(s)	Remarks
1.	Apterygota (Primitive wingless insects)	Thysanura	Bristle-tails	
2.		Diplura	Japygids	
3.		Protura	Proturans	
4.		Collembola	Spring-tails	
5.	Exopterygota (Winged insects, metamorphosis is of hemimetabola type and there is rarely a pupal stage. Young ones are called nymphs and wings develop externally.)	Ephemeroptera	Mayflies	Paleopteran orders
6.		Odonata	Dragonflies	
7.		Plecoptera	Stoneflies	Orthopteriod orders
8.		Grylloblattodea	Grylloblatta	
9.		Orthoptera	Crickets, grasshoppers	
10.		Dictyoptera	Cockroaches	
11.		Phasmida	Stick insects	
12.		Mantophasmatodea	Gladiators	
13.		Dermaptera	Earwigs	
14.		Embioptera	Web spinners	
15.		Isoptera	Termites	
16.		Zoraptera	Zorotypus	
17.		Psocoptera	Psocids, booklice	Hemipteriod orders
18.		Mallophaga	Biting lice	
19.		Phthiraptera	Human lice	
20.		Hemiptera	Bugs	
21.		Thysanoptera	Thrips	
22.	Endopterygota (Winged insects, metamorphosis is of holometabola type and there is a pupal stage, wings develop internally)	Neuroptera	Antlions	Panorpoid orders
23.		Mecoptera	Scorpion flies	
24		Lepidoptera	Butterflies, moths	
25.		Trichoptera	Caddisflies	
26.		Diptera	True flies	
27.		Siphonaptera	Fleas	
28.		Hymenoptera	Bees, wasps	
29.		Coleoptera	Beetles, weevils	
30.		Strepsiptera	Stylops	

grasshoppers, *e.g.* katydid (Tettigoniidae). Those with 3-segmented tarsi, long antennae and long straight ovipositors are the crickets (Gryllidae). The mole-cricket (Gryllotalpidae) also belongs to this group, but in this case the ovipositors are vestigial and the fore legs are fossorial, modified for digging. They generally live a subterranean life and may do serious damage to the roots of crops. In both these groups, sound is produced by rubbing the wings and the auditory organs are in the fore tibia.

The third group is that of the short-horned grasshoppers and locusts (Acrididae), with 3-segmented tarsi, short antennae and short stout ovipositors. They produce sound by rubbing the hind femur against the fore wings. The auditory organs are located on the first abdominal segment. The desert locust, *Schistocerca gregaria* (Forskal), is the best-known example and it is the most notorious pest and scourage throughout the ages. It has two phases. In the solitary phase, it lives like other grasshoppers in its natural habitat in Arabia and northern Africa. In the gregarious phase, it moves in swarms, many kilometres long from one country to another, leaving not a single blade of green vegetation behind. In the past, it has been the cause of famines and the change of kingdoms. There are about 20,000 species of insects belonging to this order.

Phasmida (*phasma,* an apparition). The stick-insects and leaf-insects are well known for their imitation in body form by resembling twigs or green leaves. They are usually large, elongated, cylindrical and often apterous. Their mouthparts are of the biting type, the legs are similiar and ovipositors are concealed. They live mostly in the tropics.

Dermaptera (*derma,* skin; *pteron,* a wing). The earwigs have leathery fore wings. The hind wings are large, membranous and , when at rest, lie folded underneath the fore wings. Their mouthparts are for biting and there is one pair of forceps at the end of the body. They feed on humus and live in decaying wood. The females show maternal instinct. There are about 900 species in the temperate and tropical parts of the world.

Embioptera (*Embia,* genus; *pteron,* a wing). These insects are commonly known as the web-spinners and they live in silken galleries made on the bark of trees. They are elongated and soft-bodied, the males having two pairs of smoky wings which are more or less alike. The mouthparts are for biting. The tarsi are 3-segmented and the first segment of the first pair is swollen and carries silk glands. The females are wingless. They generally live in warmer parts of the world and comprise about 140 species.

Dictyoptera (*dictyon,* a network; *pteron* , a wing). It includes the cockroaches **(Fig. 28.3)** and mantids. Their fore wings are thickened to form tegmina, and the tarsi are 5-segmented. The mouthparts are for biting; the ovipositors are concealed. Their eggs are contained in ootheca. The cockroaches (Blattidae) comprise about 3,500 species, mostly in the tropics. They are nocturnal and feed on stuff left in the kitchen and pollute human food with their excrements and impart a foul smell which is not lost even on cooking. The praying mantids (Mantidae) are carnivorous and have raptorial fore legs with which they catch and hold their prey. They abound in warmer parts of the world.

Isoptera (*isos,* equal ; *pteron,* a wing). The termites or whiteants are common in the tropics. They are social insects and live in large colonies, each colony having a single queen. Their body is soft which easily gets desiccated and, hence, they live in earthern galleries or mounds. Their mouthparts are of the biting type and they have two pairs of similar elongated wings placed in sockets which are easily shed after the first flight. Caste system among them is highly developed. The most common forms are workers and soldiers which are both sterile females and males. The sexual forms may be winged (*macropterous*) or short-winged (*brachypterous*) or may be wingless (*apterous*) **(Fig. 26.3)**. The whiteants (Termitidae) may be subterranean (*Coptotermes*), may live in dry wood (*Mastotermes*), may form mounds (*Odontotermes*), or they may make huge termitaria (*Nasutitermes*). The queen is the life of the colony and if she dies the colony may perish. In certain species, however, the death of the queen is not necessarily the end of the colony, as brachypterous forms may start laying eggs to maintain the colony. Practically, all termites feed on the cellulose obtained from dead plants, wood, wooden structures, books, etc. They also attack crops, particularly sugarcane and wheat sown under unirrigated conditions.

Psocoptera (*Psocus,* genus; *pteron,* a wing). The book lice and their allies are small soft-bodied winged or apterous insects. Their mouthparts are of the biting type and they have no cerci. About 1,000 species are known.

Mallophaga (*mallos*, hair; *phagein*, to eat). The bird-lice or the biting lice have the biting type of mouthparts. They are ectoparasites of birds and a few of them are also found on mammals. The thoracic segments are more or less free and the tarsi may be 1- or 2-segmented. They attach eggs to the feathers of their hosts and complete life-cycle without dropping off. Separated from the host, the lice cannot live for long. The chicken louse (shaft louse), *Menopon gallinae* (Linnaeus) (Menoponidae) or the chick 'mite', as it is called sometimes, is an important pest of poultry.

Siphunculata (*siphunculus*, a little tube). The sucking lice are important ectoparasites of man and domestic animals. They are wingless insects having piercing- and-sucking type of mouthparts, with which they suck the blood of their host. Their thoracic segments are fused and they have 1-segmented tarsi, and a single claw with the help of which they cling to the hair of their host. The body louse of man, *Pediculus humanus* Linnaeus (Pediculidae) **(Fig. 28.8)** and the cattle louse, *Haematopinus* (Haematopinidae) are the common examples of this order.

Hemiptera (*hemi*, half; *pteron*, a wing). The plant-bugs and their allies have piercing and sucking type of mouthparts and their fore wings are generally corneous. Many of the forms are apterous and in some, there may be an incipient pupal instar. There are two suborders, Heteroptera and Homoptera. In the former, the basal half of the fore wing is of a much harder consistency than the rest of it. The plant bugs are medium insects and the important crop pests include the rice bug, *Leptocorisa acuta* (Thunberg) (Coreidae) **(Fig. 13.13)**; the green potato bug, *Nezara viridula* (Linnaeus) (Pentatomidae); the red cotton bug. *Dysdercus koenigii* (Fabricius) (Pyrrhocoridae); the lace bug, *Urentius sentis* Distant (Tingidae) **(Fig. 20.11)**; the dusky cotton bug, *Oxycarenus laetus* Kirby (Lygaeidae) **(Fig. 16.4)**. The common bed bug, *Cimex lectularius* Linnaeus (Cimicidae) **(Fig. 28.7)** although wingless, also belongs to this suborder. In the second suborder, Homoptera, the fore wings, when present, are of a uniform consistency and it includes such crop pests as leafhoppers (Cicadellidae) **(Figs. 13.3, 16.1)**, aphids (Aphididae) **(Figs. 15.1, 15.4)**, pyrilla (Fulgoridae), citrus psylla (Psyllidae) **(Fig. 18.1)**, the whiteflies (Aleyrodidae) **(Fig. 18.2)** and the scales (Diaspididae) **(Fig.19.1)**. Such commonly known insects as cicadas (Cicadidae) and the lac insect (Lacciferidae) also belong to this suborder which is of great importance in agriculture. This order is known to contain 59.000 species.

Thysanoptera (*thusanos*, a fringe; *pteron*, a wing). The thripses have very characteristic narrow wings, with long fringes. They are minute insects with 6-9 segmented antennae. The mouthparts are stylet-like and are used for rasping and sucking plant sap. The tarsi are short-each ending in a vesicle. In the life-history of these insects, there is an incipient pupal instar. The thripses are commonly seen in flowers, but they may also feed on leaves. The well-known pest species are the grapevine thrips, *Rhipiphorothrips cruentatus* Hood (Heliothripidae) and onion thrips, *Thrips tabaci* Lindeman (Thripidae) **(Fig. 20.9)**

Neuroptera (*neuron*, a nerve ; *pteron*, a wing). This order includes the heterogeneous group of alder flies, lace wings, ant lion, etc., which have two pairs of membranous wings, with many accessory branches and cross veins. They are small to large insects with soft bodies, and their larvae are campodeiform with biting or suctorial mouthparts, and are generally predacious. The larvae of lace wings destroy a large number of aphids, and those of the ant lions dig typical funnel-shaped pits in the sand and wait for stray ants to fall in and serve as their prey.

Lepidoptera (*lepidos*, a scale; *pteron*, a wing). The moths and butterflies whose bodies are covered over with scales belong to this order. They are small to very large insects in which the galeae are modified to form a spirally coiled suctorial proboscis **(Fig. 1.16)** used for sucking nectar from flowers and in some cases the fruit juice. Their larvae are active, mostly phytophagous, and some also feed on wax and cloth. Their pupae are obtect or partially free and are usually enclosed in cocoons. It is very large order of 1,19,000 species and many of them are brilliantly coloured. The lepidopteran larva has a well-developed head, 3 thoracic and 10 abdominal segments. Abdominal feet or prolegs are present on segments 3-6 and 10. In some, the number of pairs are reduced and consequently, as they walk they form a loop of the body and are known as semiloopers or loopers.

The distinguishing features of moths and butterflies are listed in **Table 1.3**

Table 1.3 Distinguishing characteristics of moths and butterflies

Characteristic	Moths	Butterflies
Eggs	Generally flat and round	Cigar shaped and cylindrical
Larvae	Generally covered with hairs	Smooth and naked
Ocelli	Two ocelli present	Absent
Type of antennae	Thread-like, comblike or feathery	Club-shaped
Pupation	Pupae are mostly enclosed in silken cocoons	Pupae are naked
Shape of body	Relatively large body	Slender body
Pattern of resting position of the wings	The wings lie in a slanting roof-like disposition on either side of the body-length	The wings are held over the back in the vertical position
Activity time	Usually night-fliers	Day-fliers

As pests of crops, this is the most important order and in a tropical country, such as India, several hundred spcies are known pests of economic plants. Out of these, over 50 are major pests. Some of the commonest pests are the cutworms, *Agrotis* spp. (Noctuidae); the sugarcane top borer, *Scirpophaga (=Tryporyza) nivella* (Fabricius) (Pyralidae); the maize borer, *Chilo partellus* (Swinhoe) (Pyralidae); the clothes moth, *Tineola bisselliella* (Hummel) (Tineidae); the red hairy caterpillar, *Amsacta moorei* Butler (Arctiidae); the cabbage butterfly, *Pieris brassicae* (Linnaeus) (Pieridae); the citrus caterpillar, *Papilio demoleus* Linnaeus (Papilionidae); the hawk moths (Sphingidae) and numerous others. The silk moth, *Bombyx mori* (Linnaeus) (Bombycidae), has a great economic importance as a producer of commercial silk.

Diptera (*dis*, two; *pteron*, a wing). The flies and its allies have two wings; the hind pair is modified to form halteres used in balancing the body. Their mouthparts are modified into a proboscis used for piercing and sucking or sponging liquid foods. The typical larva is worm-like, legless, and has an indistinct head. It has a terrestrial, aquatic or parasitic mode of life, The pupa is weakly obtect or exarate, enclosed in a puparium made of the last larval skin. There are three suborders : *Nematocera,* in which the larva has a well-developed head; *Brachycera,* in which the head is incomplete and retractile; and *Cyclorrhapha,* in which the head is vestigial. The mosquito (Culicidae) belongs to the first suborder; horse flies (Tabanidae) **(Fig. 29.2)** and robber flies (Asilidae) belong to the second; the hover flies (Syrphidae), leafminers (Agromyzidae), blow flies (Calliphoridae) and the house fly (Muscidae) **(Fig. 28.1)** belong to the third suborder. This is a very large and important order comprising about 1,19,000 species.

Siphonaptera (*siphon*, a tube; *apterous,* wingless). The fleas which are included in this order have piercing-and-sucking mouthparts, with which they suck blood from the body of warm-blooded animals. The adults are ectopara-sites with the body laterally compressed. The rat flea, *Xenopsylla cheopis* (Rothschild) (Pulicidae) frequently migrates to man and transmits the bacillus of bubonic plague.

Hymenoptera (*hymen*, a membrane; *pteron*, a wing). The ants, bees, wasps, sawflies, etc. have membranous wings, the hind pair being smaller than the fore wings. Their mouthparts are of the biting type and sometimes of the lapping or sucking type. The first segment of the abdomen is fused with thorax, often like a narrow petiole. The pupae are generally enclosed in cocoons. Hymenoptera is divided into two suborders. *Symphyta* includes the wood-boring wasps. The female has ovipositors adapted for sawing and boring, as in the metallic-blue *Sirex noctilio* Fabricius (Siricidae), an important pest of timber in many parts of world. *Apocrita* includes the majority of Hymenoptera and the members are recognized from their constricted or petiolated abdomen. The honey bee, *Apis cerana* Fabricius (Apidae), the red wasp, *Vespa orientalis* Linnaeus(Vespidae), the ants (Formicidae) and the parasitic wasps (Ichneumonidae and Chalcididae) are the common examples. The honey bees produce honey, which is a source of income of many apiculturists. With the introduction of *Apis mellifera* Linnaeus, all four species of *Apis* are now found in India. The parasitic insects of this order are of great importance to agricultural entomologists interested in the biological control of insect pests. About 1,95,000 species of this order have been described so far.

Coleoptera (*koleos*, a sheath; *pteron*, a wing). This is the largest order of the animal kingdom and over 2,20,000 species have so far been described. The beetles are minute to large and have their fore wings modified to form elytra (**Fig. 1.16**), which are hard and lie flat on the body, the membranous hind wings are folded underneath. Their prothorax is large and mouthparts are of the biting type. Metamorphosis is complete and the larvae are compodeiform or eruciform (**Fig. 1.20**).

Most of the beetles are ground dwellers, some are aquatic and others are important pests of stored grains and other products. There are two main suborders, *Adephaga* and *Polyphaga*; the predacious tiger beetles (Cicindelidae) belong to the former. Most of the commonly known beetles belong to Polyphaga, which includes the *ber* beetle, *Adoretus pallens* Arrow (Scarabaeidae); the *khapra* beetle, *Trogoderma granarium* Everts (Dermestidae); the ladybird beetle, *Coccinella septempunctata* Linnaeus (Coccinellidae) ; wheat flour beetle, *Tribolium castaneum* (Herbst) (Tenebrionidae); the gram *dhora*, *Callosobruchus chinensis* (Linnaeus) (Bruchidae) ; the red pumpkin beetle, *Raphidopalpa foveicollis* (Lucas) (Chrysomelidae); the granary weevil, *Sitophilus granarius* (Linnaeus), and the rice weevil, *Sitophilus oryzae* (Linnaeus) (Curculionidae). Many of the pests of stored grains are now distributed throughout the world. This is the most diverse order containing 3,49,000 species.

Strepsiptera (*Stylopids*). *Stylops* sp. is the most important parasite of the nymphs of the sugarcane pyrilla. Its larvae are endoparasitic and the male adults are free living. In the adults, fore wings are modified into small clubs and are known as pseudo-halteres. The hind wings are large and fan-shaped. The females live in the host inside puparia. It is a small order of about 300 species.

Nomenclature

According to the modern concept, where two kinds of insects or animals differ from each other in definite but relatively minor structural characters, they are said to be of distinct *species*. These differences are either known or presumed to be the indicators of a barrier to inter-breeding. This barrier may be operating owing to incompatibility in mating, in seeking the host or because of chromosome differences. A given species has two parts to its name; the first indicating the genus and the second the species to which it belongs. The name of the *genus a*nd the species are Latinized and are so written that phonetically they are harmoneous. In publications, these are printed in italics or they are underlined separately when typed or written in long hand. The name of the genus starts with a capital letter and that of the species with a small letter. At the end of each name, the name of the author who described the species first, is given. The name may be given either in full form or in an abbreviated form with a full stop at the end. Sometimes, the name of a species has three parts. In that case, the third part indicates the subspecies. The various species belonging to a genus have certain common characters of close relationship. The allied *genera*, displaying the same important characters, are grouped in a *family*. The allied individuals that show major features linking them together to form a natural assemblage are placed in an *order*. The orders are grouped in a *class* and the classes are placed in a *phylum*.

This method of naming and classification is known as the *binomial system of nomenclature*. Aristotle was the first scientist to give names to animals and plants according to their habits, habitats and body characters. However, Linnaeus, the great Swedish naturalist, is given the credit of forming a sound basis for the modern system of classfication. The detailed rules and the priority in accepting the name of a species that might have been described more than once were framed in various meetings of the International Commission on Zoological Nomenclature, which was set up in the 5th

International Congress of Zoology, held in Berlin in 1901. The final draft of the International Code of Zoological Nomenclature was completed in 1958 and published in 1961.

According to the current Code of Nomenclature, the various *synonyms,* are replaced by the *senior synonym,* that is, that name published first, provided that *junior synonym* has not been in use for more than 50 years. Where more than one kind of animal has been given the same name by different authors because of the lack of communiation, the senior *hymonym,* according to the date of publication, is to be accepted. This is called the *Law of Priority.*

Mounting, Preserving and Labelling Insects

Insects are collected by nets and are killed with potassium cyanide gas slowly released in a capped wide mouth jar. Insects can be mounted and preserved in various ways. Most specimens are pinned, and, once dried, will keep indefinitely. Specimens too small to pin can be mounted on "points," on tiny "minuten" pins or on microscope slides. Large and showy insects, such as butterflies, moths, grasshoppers, dragonflies or damselflies, may be mounted in various types of glass-topped display cases.

Pinning. Pinning is the best way to preserve hard-bodied insects; pinned specimens keep well, retain their normal appearance and are easily handled and studied. Insects should be pinned with a special type of steel pin known as an insect pin. Insect pin sizes range from 00 to 7, size 2 and 3 being the best for common use.

Insects are usually pinned vertically through the body (**Fig. 1.23**). Bees, wasps, flies, butterflies and moths are pinned through the thorax between the bases of the front wings; with the flies and wasps it is desirable to insert the pin a little to the right of the midline. Bugs are pinned through the scutellum, a little to the right to the midline. Grasshoppers are pinned through the posterior part

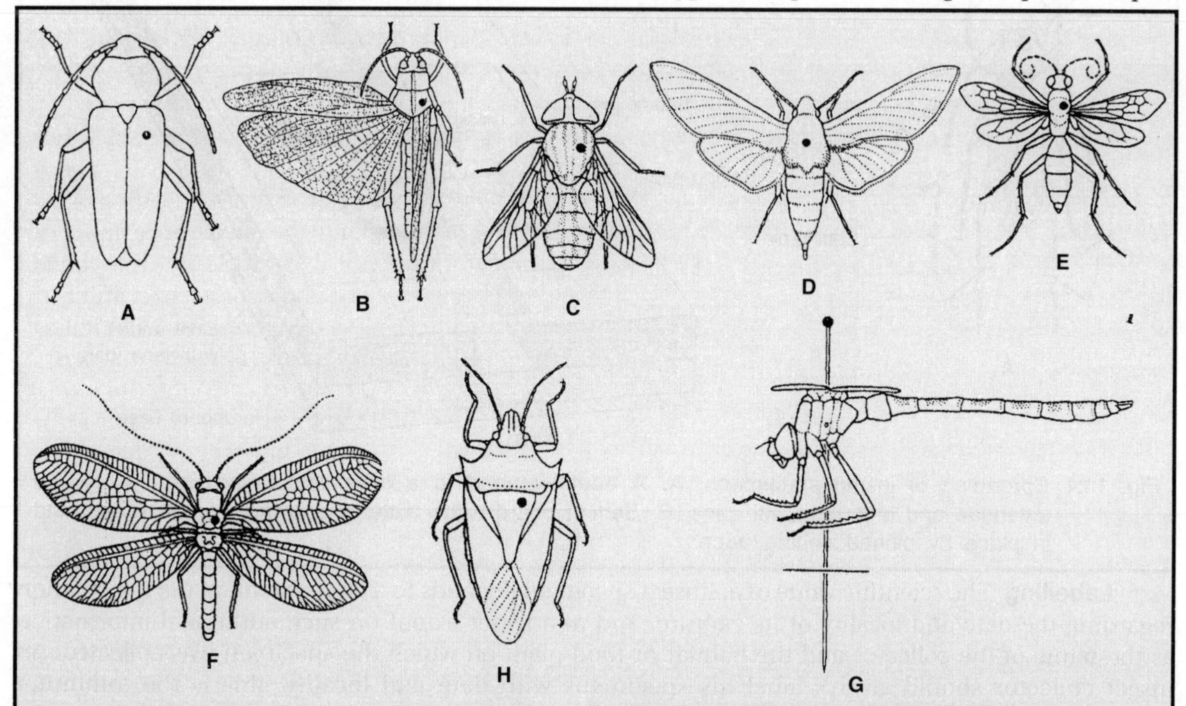

Fig. 1.23 Pin positions for representative insects : A. Beetles (Coleoptera); B. Grasshoppers, katydids, and crickets (Orthoptera); C. Files (Diptera); D. Moths and butterflies (Lepidoptera); E. Wasps and sawfiles (Hymenoptera); F. Lacewings (Neuroptera); G. Dragonfiles and damselfies (Odonata), lateral view; H. Bugs, cicadas, and leaf-and planthoppers (Hemiptera)

of the pronotum, just to the right of the midline. Beetles should be pinned through the right elytron, about halfway between the two ends of the body, the pin should go through the metathorax and emerge through the metasternum so as not to damage the base of the legs. The easiest way to pin an insect is to hold it between the thumb and fore finger of one hand and insert the pin with the other. All specimens should be mounted at a uniform height on the pin, leaving about 2.5 cm free above the insect surface.

Mounting small insects. Insects too small to pin may be mounted on a card point or on a "minuten" pin. The points are elongated triangular pieces of light cardboard or celluloid, about 8 or 10 mm long and 3 or 4 mm wide at the base; the point is pinned through the base and the insect is glued to the tip of the point.

Putting an insect on a point is a very simple process. The point is put on the pin and the upper side of the tip of the point is touched to the glue and then touched to the insect.

Spreading insects. An insect that is to be a part of a collection is pinned and then spread on a spreading board with the dorsal side up **(Fig. 1.24)**. There are certain standard positions for the wings of a spread insect. In the case of butterflies and moths, and mayflies, the rear margins of the front wings should be straight across, at the right angles to the body; and the hind wings should be far enough forward that there is no large gap at the side between the front and hind wings. With grasshoppers, dragonflies, damselflies and most other insects, the front margins of the hind wings should be straight across, with the front wings far enough forward that they just clear the hind wings. The front and hind wings of a butterfly or moth are always overlapped, with the front edge of the hind wing under the rear edge of the front wing; in other insects the wings are usually not overlapped.

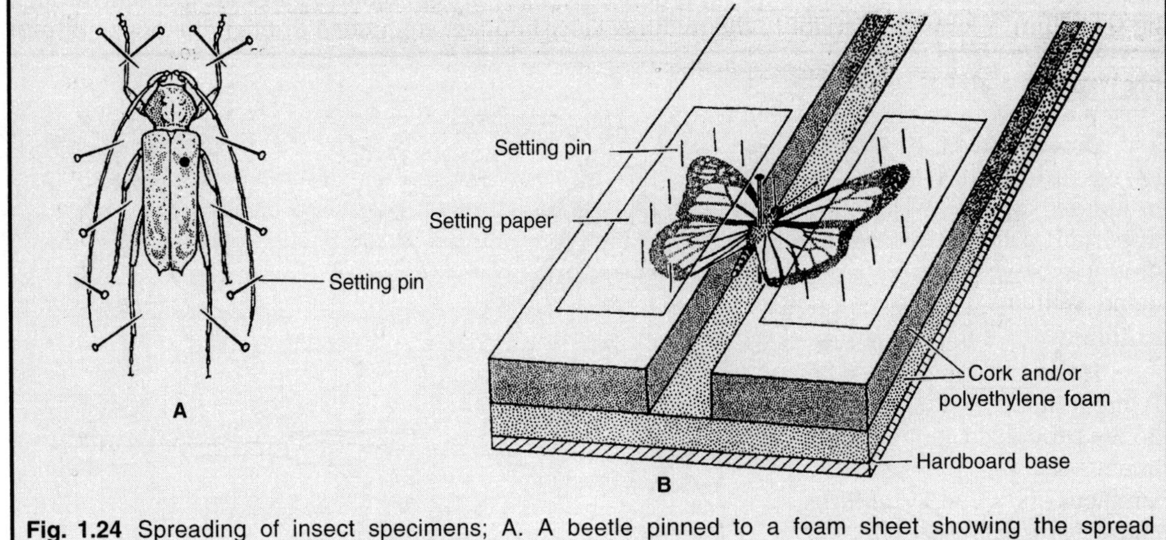

Fig. 1.24 Spreading of insect specimens; A. A beetle pinned to a foam sheet showing the spread antennae and legs held with pins; B. Setting board with butterfly showing spread wings held in place by pinned setting paper

Labelling. The scientific value of an insect specimen depends to a large extent on the information regarding the date and locality of its capture, and to a lesser extent on such additional information as the name of the collector and the habitat or food plant on which the specimen was collected. An insect collector should always label his specimens with date and locality; this is the minimum amount of data for a specimen; additional data are desirable, but optional.

Pinned insects should be kept in boxes having a soft bottom that will permit easy piercing. Large collections in institutions are frequently kept in cabinets containing drawers similar in construction to Schmitt boxes.

All insect collections are subject to attack by dermestid beetles, ants and other museum pests, and if the collection is to last any length of time, certain precautions must be taken to protect it from these pests. Various materials may be used for this purpose, but one material commonly used is naphthalene (in flake or ball form). Naphthalene flakes can be put into a small cardboard pillbox that is firmly attached to the bottom of the insect box, usually in one corner, and has a few pin holes in it. Paradichlorobenzene can also be used, but it volatilizes more rapidly than naphthalene and must be renewed at more frequent intervals.

Preservation of insects in fluids. Some insects cannot be pinned and must be preserved in fluids. All soft-bodied adult insects, such as mayflies, stoneflies, caddisflies, aphids, most insect nymphs, all larvae, myriapods, crustaceans and arachnids should be preserved in fluid. Many minute insects, such as springtails, lice, fleas and minute flies are usually preserved in fluid until they are mounted on microscope slides. Several chemical solutions may be used but the most important one is X.A. mixture: Xylene, one part and ethyl alcohol (95%) one part. Larvae killed with this mixture should be transferred to 75 per cent alcohol after 24 hours. This killing agent is likely to remove the bright colours of larvae, especially greens, yellows and reds.

Identification. Identification services are provided by the Zoological Survey of India, Calcutta; Division of Entomology, Indian Agricultural Research Institute, New Delhi; Indian Forest Research Institute, Dehradun; the British Museum, Queens Gate, London, U.K., and many other national and semi-official organizations in various countries throughout the world. Before sending the specimens by post they are pinned and packed in strong and light boxes specially made for the purpose.

Economic Importance of Insects

Of an estimated 5-10 million species of insects, probably not more than a fraction of 1 per cent interact, directly or indirectly, with humans. Insects have been associated with man's interests in many ways. Their ability to both benefit and harm mankind makes this relationship a unique one. They play a silent role in maintaining the balance of nature; some feed on other plants and animals, and some are eaten by others. Many of them are scavengers and convert the dead plant and animal tissues into humus and enrich the soil. Many species of insects live both as parasitoids and predators on insect pests of crops and help in suppressing their numbers. They also play an important role in the pollination of various crops and thus increasing their yield. Some of them produce materials useful to man, *i.e.* silk, honey, wax, lac, etc. They also serve a useful purpose in being aesthetically beautiful and attractive. However, insects are also powerful competitors of man as they cause injury to crops and animals, and also act as vectors of many diseases.

Honey bees. There are about 20,000 kinds of bees, all belonging to the insect order Hymenoptera. Of them, three families of social bees, *i.e.* Bombidae, Meliponidae and Apidae, are honey producing. Among these, Apidae is the main honey producing family. All the four species of honey bees, viz. *Apis carana* Fabricius, *Apis dorsata* Fabricius, *Apis florea* Fabricius and *Apis mellifera* Linnaeus are found in India. Of these, the first three are indigenous, and the last has been introduced and acclimatized in India for the last about 50 years (Atwal, 2013).

A. cerana makes parallel combs on trees, termitaries, hollows of rocks and all kinds of cavities. Its honey yield is 3.6-4.5 kg in hills and 1.3-2.2 kg in plains. *A. dorsata* makes large hives (15 cm long and 7.5 cm deep) that hang from high rocks and tall trees. Their colonies shift from place to place to avoid extremes of climate or in search of honey. They make single vertical comb, are very industrious and produce about 36 kg honey per colony per year. They are most ferocious and do not spare their victim even inside water. *A. florea* is the smallest of all these bees. It makes a single vertical comb of the size of palm suspended from branches of bushes, hedges, buildings, caves, chimneys, empty cases, etc. Its honey yield is poor, about 0.5 kg and the consistency of honey is thin. *A. mellifera* makes its nest in enclosed spaces in parallel combs and is endowed with all the good qualities of a hive bee. It has adapted itself well to modern methods of movable frame hives and, therefore, is the favourite of beekeeping industry the world over. Its honey yield averages 50-100 kg per colony per year.

The total global production of honey is about 1.54 million tonnes a year, gathered by about 60 million honey bee colonies. There are about 1 million *A. mellifera* and *A. cerana* colonies in India, maintained by about 2.76 lakh beekeepers. The present production of honey in India is estimated at about 90,000 tonnes, out of which about 25,000, tonnes is exported. In Punjab, 8500 tonnes of honey is produced by about 2,50,000 *A. mellifera* colonies maintained by about 25,000 beekeepers.

Silkworm. The silkworms belong to the insect order Lepidoptera. While all Lepidopterans secrete silk, only some species weave their silk into cocoons to protect their vulnerable pupae from weather and predators. There are four species of silk worms in India, which can be exploited for commercial silk production. These are mulberry silkworm, *Bombyx mori* (Linnaeus); tasar silkworm, *Antheraea paphia* (Linnaeus); muga silkworm, *Antheraea assama* Westwood and eri silk worm, *Philosamia* (= *Attacus*) *ricini* (Donovan). Among these, *B. mori* and *P. ricini* are domesticated, *A. assama* semi-domesticated and *A. paphia* wild.

Silk is the product of a pair of silk glands (salivary glands) of the larva. The secretion is produced in liquid form but on coming in contact with air, it solidifies into a fibre called silk. When mature, the 5th instar larva makes quick round movements of its head at the rate of 65 movements per minute while spinning its cocoon. The weight in grams of 900 m long silk filament is called a 'denier' and the size of a normal cocoon is 1.8-3.0 deniers. A single cocoon weighs 1.8-2.0 g and its shell (without the enclosed pupa) about 0.45 g. About 2500 cocoons yield 0.45 kg of silk. The total production of raw silk from all the four species of silk worms in India is about 24,000 tonnes, of which about 90 per cent is mulberry silk.

Lac insect. The lac insect, *Kerria* (= *Laccifer*) *lacca* (Kerr)) belongs to the order Hemiptera. There are six genera of lac insects, of which five secrete lac and only one, *i.e. Kerria* secretes recoverable or commercial lac. The lac insects live in cavities or cells made in the resin or lac secreted by them on their host plant. Two distinct strains of lac insects are known in India, *i.e. kusumi* and *rangeeni.* The *kusumi* strain is grown on kusum (*Schleichera oleon*) or on other hosts using *kusumi* brood while the *rangeeni* strain is grown on hosts other than kusum. Each of these strains gives two crops in a year. The lac of *rangeeni* crop is collected while it is not fully mature. About 90 per cent of the production is contributed by *rangeeni* crop and only 10 per cent from the *kusumi* crop. *Kusumi* is considered superior because of the lighter colour of the resin. The annual production of lac in India is about 16495 tonnes.

Pollinators. Many species of insects particularly the bees, butterflies, moths and thrips are the major pollinators of grain crops, cotton, fruits, vegetables and flowering plants. Some plants like many fruits particularly figs, peas, beans, tomatoes, many vegetables, sunflower, seasonal flowers, chrysanthemum and many ornamental plants would produce no fruits and seeds unless pollinated by insects. It has been estimated that 50-70 per cent of the grain crops are pollinated by insects, a major portion of it by bees alone. For every Rs 100 worth of honey collected, the bees make Rs 2000 worth of seeds and fruits by pollinating the flowers.

Entomophagous insects. Entomophagous insects are those insects which feed on other insects destroying our crops and stored grains. Such entomophagous insects are either parasitoids (a parasitoid is usually much smaller than its host and a single individual usually does not kill the host) or predators (a predator is usually larger than its prey, kills its prey and requires more than one prey to complete its development). The parasitic insects mainly belong to Hymenoptera and a few to Diptera. The insect predators are spread over a wide range of orders including Coleoptera, Neuroptera, Diptera, Hemiptera, Hymenoptera, etc. It is due to the activities of entomophagous insects that many harmful insect pests have been kept under check under natural conditions.

Pests of crops and animals. The greatest importance of insects lies in their being pests of crops and animals and as carriers of diseases. A large number of insects cause injury to economic plants by feeding on them externally and chewing their leaves or other parts; by sucking the cell sap with the help of piercing mouth parts and reducing the vitality of the plant; by disseminating various plant diseases and by living a sub-terranean life and damaging roots. The recent estimates put the total annual losses caused by insect pests to major field crops and the foodgrains in storage in India, to the tune of Rs 900 billion (Dhaliwal *et al.,* 2013).

At times, insects make it impossible for human beings to live in a locality by causing them untold misery. They annoy man by biting, cause him pain by injecting venoms, live on his body and on farm animals as ecto-and endo-parasites causing ill-health. They spread many pathogens by acting as their carriers or as intermediary hosts. To mention only a few, the house fly spreads cholera and typhoid; mosquitoes transmit malaria, filaria and encephalitis; tsetse fly transmits the parasite causing sleeping sickness; the rat flea transmits the bacillus of bubonic plague, etc. Thus, insects as a group of the animal kingdom clash directly with the interests of man for survival.

Entomology in India

The heavy losses caused by insect pests have attracted the attention of scientists and administrators to initiate work on plant protection. The earliest available record of Indian insects is that of Linnaeus, who included 12 Indian insects in the 10th edition of his famous book *Systema Nature* published in 1750 A.D. However, the work on Indian insects was initiated by S.G. Koenig, a pupil of Linnaeus, in 1779 in South India. With the growing realization of the importance of insect pests, an Asiatic Society was established in Bengal in 1785. The scientific studies received greater attention and impetus with the establishment of Indian Museum at Calcutta in 1875 and the *Bombay National History Society* in 1883. Since 1888, with the publication of the *Journal of Bombay Natural History Society,* several contributions on Indian insects were published by various workers. The most noteworthy event of this period was the starting of the publication of the *Fauna of British India* in 1892, published by the Government of India in London. Further, the Indian Museum, Calcutta, published five volumes of *Indian Museum Notes* between 1889 and 1903. These are monumental contributions in economic entomology and may even be said to contribute the birth of applied entomology in India.

Lionel de Niceville was appointed the first Entomologist to the Government of India in 1901 with headquarters at the Indian Museum, Calcutta. He was succeeded by H. Maxwell Lefroy in 1903 and posted at Surat to study the cotton pests. In 1905, when the Imperial (now Indian) Agricultural Research Institute was established at Pusa in Bihar, Lefroy was transferred there as the first Imperial Entomologist. He published two very useful books, *Indian Insect Pests* (1906) and *Indian Insect Life* (1909). At the same time, several State Governments also initiated entomological work, viz. Madras (1902-05), Punjab (1905) and U.P. (1906). To coordinate work in these states, separate entomologists were appointed, *i.e.* Coleman in Mysore (Karnataka) in 1908, T.B. Fletcher in Madras (comprising of present Tamil Nadu, Kerala and Andhra Pradesh) in 1912, M. Afzal Husain in Punjab in 1919 and P.B. Richards in U.P. in 1921.

Based on his extensive studies, T.B. Fletcher published his book *Some South Indian Insects* in 1914. Subsequently, several textbooks were published, among which some prominent ones are *Handbook of Economic Entomology for South India* by T.V. Rama Krishna Ayyar (1940, revised 1963), *Textbook on Agricultural Entomology* by H.S. Pruthi (1969), *Insects and Mites of Crops in India* by M.R.G.K. Nair (1975, revised 1986), *General and Applied Entomology* by K.K. Nayar, T.N. Ananthakrishnan and B.V. David (1976, latest edition 2000), *Agricultural Pests of India and South-East Asia* by A.S. Atwal (1976,1986), revised as *Agricultural Pests of South Asia and their Management* by A.S. Atwal and G.S. Dhaliwal (1997, latest edition 2015), *Integrated Pest Management: Concepts* and *Approaches* by G.S. Dhaliwal and Ramesh Arora (2001, revised 2006, latest edition 2015), *Essentials of Agricultural Entomology* by G.S. Dhaliwal, Ram Singh and B.S. Chhillar (2006, revised 2015), *Biopesticides and Pest Management: Conventional and Biotechnological Approaches* by G.S. Dhaliwal and Opender Koul (2007) *Quest for Pest Management. From Green Revolution to Gene Revolution* by G.S. Dhaliwal and Opender Koul (2010), and *A Textbook of Integrated Pest Management* by G.S. Dhaliwal, Ram Singh and Vikas Jindal (2013).

Plant protection in India got boost with the establishment of the Central Plant Protection Organization (now called the Directorate of Plant Protection, Quarantine & Storage) in 1946, under the Ministry of Agriculture, Government of India, New Delhi. H.S. Pruthi was appointed the first Plant Protection Adviser to the Government of India. The Directorate was shifted to Faridabad in 1968. The Directorate is functioning with the support of 31 Integrated Pest Management Centres (CIPMCs), 35 Plant Quarantine Stations (PQSs), 10 circle offices of the Locust Warning Organization, One Central Insecticides Laboratory, 2 Regional Pesticides Testing Laboratories (RPTLs) and 68 Pesticides Testing Laboratories in States/Union Territories.

At present, several agencies including the Indian Council of Agricultural Research Institutes. Directorates of Plant Protection and Extension, Government of India, State Agricultural Universities, State Departments of Agriculture, Pesticide Industries and many Non-Government Organizations (NGOs) are involved in development and implementation of IPM in various agricultural crops. Outstanding success has been achieved in devising IPM strategies in principal agricultural crops, viz. cotton, rice, vegetables, fruits, pulses, oilseeds and plantation crops. However, much more needs to be done so that farmers are able to adopt IPM on a large scale to avoid pest-related losses and increase agricultural production.

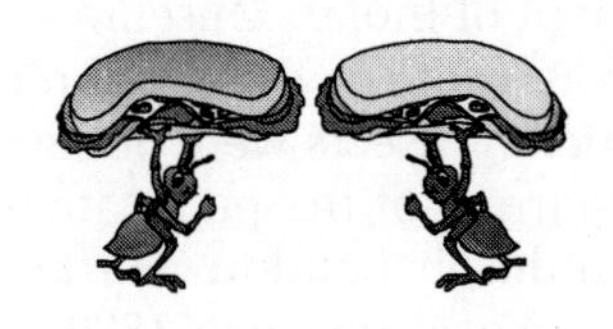

PEST POPULATIONS AND CROP LOSSES

INTRODUCTION

Man has inhabited the earth for about 1 million years, but was a hunter-gatherer for nearly 9,90,000 years, depending exclusively on the natural ecosystem. About 10,000 years ago, humans began to settle and cultivate crops. This required that the diverse natural vegetation be removed and usually replaced by few species of crops. The loss of diversity and stability of the natural ecosystems also destroyed a major part of the natural fauna, while enabling a few species of insects, pathogens and weeds to survive and multiply in the new agroecosystems. These species began to compete with man for food, feed, forage, fibre, etc., and were, thus, described as pests.

A pest is any organism whose population increases to the extent that it starts causing annoynance, inconvenience or injury to man, his animals, plants and material possessions. A pest may be an insect, mite, nematode, bird, rodent, fungus, bacterium, virus, weed or any other organism. The pest status of a species is determined by a number of criteria, viz. increse in the number of individuals, change in the type of damage, more pronounced damage due to conducive climatic conditions, changes in the methods of cultivation or harvesting, and fluctuations in the market value of the crop. Pest infestations are sometimes referrered to as epidemic or *endemic*. When the infestation occurs in a severe form in a region or locality at a particular season or time it is known as ***epidemic***. If the infestation is a regular feature and confined mostly to a particular area or locality, it is referred to as ***endemic***. The outbreak of an animal pest or a disease or the spread of noxious weeds quiet often becomes a major concern of the forester, the farmer or public health custodian. Two major classes of pest outbreaks have been identified, viz. eruptive and gradient. The *eruptive outbreak* spreads out from local epicentres to cover large areas. The eruptive outbreak is self-perpetuating, *i.e.* once initiated positive feedback processes operating at high population densities maintain the outbreak. On the other hand, ***gradient outbreak*** does not spread from local epicentres to cover large areas. It is not self driven and is entirely dependent on external environmental or internal genetic conditions. This type of outbreak arises and subsides as its driving forces change in time and space.

There are three categories of pests :

(*i*) *Regular pests.* These are generally found in abundance during a crop season, *e.g.* aphids, jassids and thrips.

(*ii*) *Sporadic pests.*These assume pest status occasionally in certain years and include locusts, grasshoppers, hairy caterpillars, crickets and cutworms.

(*iii*) *Potential pests.* These pests normally cause negligible damage but may become highly destructive resulting from some disturbance in the environment and the consequent increase in their number, *e.g.* armyworm on wheat.

ATTAINMENT OF PEST STATUS

A species may attain the status of a pest by the following ways :

(*i*) ***Entry into a new habitat.*** A species may become pest by entering into a previously uncolonized habitat, e.g. introduction of San Jose scale and cottony cushion scale of citrus into India; gypsy moth and European corn borer into North America; and insect pests of *Eucalyptus* into New Zealand and South Africa.

(*ii*) ***Changes in species characteristics.*** A species may asssume the status of a pest due to certain changes in its characteristics that did not previously compete directly with man, e.g. the sudden appearance of previously unknown and only recently involved species of cynipid wasp on chestnut plants in Japan.

(*iii*) ***Change in habitat.*** The change in the habitat of the species or in the activities of man himself may increase his sensitivity to the species, i.e. increased urbanization in India has provided conditions conducive to breeding of *Culex quinquefasciatus* Say and filariasis has thereby spread to new areas.

(*iv*) ***Increase in abundance.*** A species whose interactions with man were previously negligible because of low numbers, may increase in abundance and assume the status of a pest. Such an increase may result from the greater supply of a limited resource, decreased severity or frequency of repressive environmental factors or by the simultaneous occurrence of both these changes. The introduction of a susceptible and nutritious host plant into the environment of a species may lead to its multiplication and abundance, *e.g.* the Colorado potato beetle attained pest status with an increase in the crop area in North America. Similarly, cotton jassid was not so serious a problem on the indigenous cotton, *Gossypium arboreum* in India, but with the introduction of American cotton, *G. hirsutum*, in the beginning of the century, this insect became the dominant pest of cotton.

It implies that the general evolutionary processes of nature (including man's influence) provide situations for an organsim to assume the status of a pest. It may not be possible for man to direct the evolutionary processes but the ecological events can be modified to his advantage. About 10,000 years ago, before man started living in communities and began to organize agriculture, the vegetation on earth and the animals it supported, presumably, lived more or less in a natural balance. The supposition may be supported by the fact that the perennial forests today, untouched by man's influence, exhibit a balanced ecosystem. It can be said that the plants and the animals in a stable ecosystem live and die fitting in the energy cycle like cogs of a wheel and no one species increases in numbers out of proportion to cause a threat to the others.

Due to geological upheavals or due to interference by man, however, conditions can change and in the process of adjustment to the new climatic and other conditions, many species might be unable to adjust themselves and, hence, become extinct; numerous examples are available in literature.

Another agency causing upheavals is man himself. To provide more and more food to the increasing population, man had to cut down forests and clear the savannas to bring more land under cultivation. In this process, he might have disturbed the balance of nature with the result that original fauna receded to whatever natural vegetation was left. On cultivated lands, a faunal vacuum was created as new niches were formed in the newly established agroecosystems. Perhaps, the herbivora that survived were those which could do well on the crops grown or on the original vegetation that persisted.

In general, the destruction of vegetation and the cultivation of soil provided harsh conditions to those animal species, because they were exposed to weather in various developmental stages.

However, some of the indigenous species that lived on weeds might have found the succulent and high yielding crops very attractive and consequently they multiplied at a fast rate. Gradually, their population probably increased to the level of a regular pest. In time, they established their own less stable balance with the new environment of agroecosystems at a higher level of population which often rose above the economic threshold. This level might have been reached every year or once every few years and the economic threshold varied according to the nature of the injury. For instance, a few dozen aphids might go unnoticed on one type of fruit or vegetable, whereas a single caterpillar boring into an apple may make it unfit for marketing.

While growing crops, man either practised monoculture–growing the same crop season after season–or mixed cropping, i.e. growing a number of crops in a locality in rotation with one another. These two practices presented entirely different situations. Under monoculture, the same crop was available in a season over large stretches of land. The favourable food was well suited to the fast multiplication of a phytophagous animal feeding on it and, year after year, its population was built up. Naturally, such a species either synchronized its life-cycle with the host crop and did not require food throughout the year or was able to sustain itself in an off-season in lower numbers, on weeds or on other vegetation of the original ecosystem that was still available in uncultivated fields, along the hedges, in forest reserves, etc. Under multiculture conditions, the sequence of cropping provided was such that for a given pest species, food was always available in greater or smaller quantity depending upon climatic and other conditions, being favourable or unfavourable to the development and multiplication of the species.

INVASIVE PESTS IN AGRICULTURE

Global trade has brought prosperity and benefits to societies around the world, much of which would be impossible without the exchange of goods. However, the movement of commodities through such trade has provided pathways for many insect species to spread and colonize new areas. Alien species are non-native or exotic organisms which occur outside their natural adapted habitat. In contrast, invasive species is any organism that is outside of its native geographical range that may or has become injurious to animal or human health, the economy and/or natural environment. In other words, alien species with potential to threaten ecosystem habitats or species, which establish a viable population in an area where it does not occur normally, is known as invasive species. The invasive insects, pathogens and weeds cause annual crop losses to the tune of US$ 58 billion.

It is well-known that in agriculture, crops have been introduced from far-off places, even from different continents. Many times pests were introduced from the homeland along with plant materials or with the household goods. In the early stages, the introductions were due to neglect and, in more recent times since the imposition of quarantine, due to accidents. These species were either known pests in the site of origin, or they were of no consequence there and became serious pests in the new country where there were few natural enemies. San Jose scale, *Quadraspidiotus perniciosus* (Comstock), is an example of the former category, being introduced into southern Asia from the United States of America. The Japanese beetle, *Popillia japonica* Newman is of the latter type. It was quite a harmless insect in Japan but when it was introduced into the United States, it became a serious pest of many economic crops and plants.

Another example of invasive pest which became threat to wine industry is grape phylloxera, *Daktulosphaira vitifoliae* Fitch, which was introduced from USA in France in 1859. Introduction of boll weevil, *Anthonomus grandis* Boheman from Mexico at the beginning of the 20th century resulted in billions of dollars damage and almost complete eradication of the cotton crop in US. The import of potatoes and associated packing material lead to invasion of Colorado potato beetle, *Leptinotarsa decemlineata* (Say) in Germany from USA in 1874, due to which the importation of potatoes was banned. The fire ants, *Solenopsis invicta* Buren were introduced in US in 1930-40*s*, as stowaways in cargo shipped from South America range. The estimated costs of control, medical treatment and damage to property due to this insect in US alone is greater than $6 billion annually. The East Asian sawfly, *Aproceros leucopoda* Takeuchi, a pest of elm trees, was earlier detected in Poland and Hungry. It invaded into Italy in 2009 through the road traffic. The completely defoliated elms were observed in motorway parking areas where cars and trucks often stop in transit from central Europe to Italy.

In India, woolly apple aphid, *Eriosoma lanigerum* (Hausmann) was introduced during 18th century with imported rootstock from China. San Jose scale, *Quadraspidiotus perniciosus* (Comstock), a native of Chna reached India in 1911. Cottony cushion scale, *Icerya purchasi* Maskell, was accidently introduced from Australia into India in 1921. Potato tuber moth, *Phthorimaea operculella* (Zeller), was introduced with imported potatoes from Italy to India in 1937. During transit of coffee from Sri Lanka, coffee berry borer, *Hypothenemus hampei* Ferrari, was introduced in India in 1990. The polyphagous serpentine leaf miner, *Liriomyza trifolii* (Burgess), accidently invaded India in 1990-91 and caused infestation of several host plants like cucurbits, tomatoes, castor, ornamental plants, etc.

In an era of globalization, we cannot survive without the global trade, but the risk of invasive species can be limited by the implementation of various international/national agreements governing trade. One of the agreements is Sanitary and Phytosanitary Agreement, which deals directly with the trade related invasive species risk. Similarly, many countries have developed various organizations and quarantine programmes to check the introduction of invasive insects. The preventive measures like quarantine, trade regulations, inspection, etc. have been identified as quite effective means to reduce introduction of invasive pests. However, more global coordination and cooperation is still required for effective implementaion of various international agreements and national policies to cope up with the problem of invasive pest species.

FACTORS INFLUENCING PEST POPULATIONS

The pest populations have a tendency to fluctuate as a result of their inherent characteristics as influenced by the environmental factors. The degree of influence of various environmental factors determines the magnitude of increase or decrease in numbers of a pest population. The rate of change in a pest population is determined by the fecundity, speed of development and survival among its numbers. The environmental factors that favour fecundity or speed of development and are not inimical to survival, promote increase, and those having a reverse influence cause a decline in numbers. The same factor may be favourable in case of one population but may become unfavourable for another. It is thus necessary to consider the influence of various factors with respect to a particular pest population.

The environmental factors may be grouped into two main categories, *i.e.* abiotic and biotic. Among the abiotic factors, it is primarily the physical factors such as temperature, moisture and light that have a direct influence on the populations of insect pests. These factors also influence the pest populations indirectly by modifying the biotic factors. The biotic factors include food and other populations, primarily the natural enemies of pest populations (Dhaliwal and Arora, 2012).

A. ABIOTIC FACTORS

The climatic factors exercise a dominating influence on the development, longevity, reproduction and fecundity of insect pests. It is well known that densities of pest populations fluctuate with the prevailing weather conditions such as temperature, moisture, light and wind. Extremes of temperature, humidity or rainfall cause mortality among the pest and its natural enemies. The chances of an insect population to survive and reproduce first increase and then decrease as the population is exposed to unfavourable low range through the optimum into the unfavourable high range.

Temperature

The insects are all poikilothermic, *i.e.* they have no precise mechanism for regulating the temperature of their bodies. Their body temperature, therefore, follows more or less closely that of the surrounding medium. There is a fairly well defined favourable range of temperature for every insect species within which it is able to survive. This temperature range is determined by the prevailing temperature at which the normal physiological activities of the insect take place. This narrow band of temperature has been called the ***preferred temperature*** or ***temperature preferendum***. Exposure to temperatures beyond the favourable range, whether low or high, may retard growth and development oı the insect or may even cause its death. The upper lethal limits are usually between 40° and 50°C, but some insects, such as stored product and desert insects, can withstand temperatures in the neighbourhood of 60°C. The lower lethal limits vary widely and may lie below the freezing point of water. In the laboratory, some insects have been reported to tolerate temperatures as low as –70°C.

The departure from the optimal range on both sides is tolerated to some extent, depending upon the physiological adaptations of the conceived populations. The reaction to changed temperature depends upon the suddenness of the change. In case of a gradual change, the insects become *conditioned* or *acclimatized.* The rate of acclimatization is dependent on the duration for which they are conditioned. The total heat required for the completion of physiological processes in the life-history of a species is considered as constant, a *thermal constant.* Within the favourable range, the thermal constant is not affected by the level of temperature.

Exposure to lethal low or high temperatures may result in instant killing and even the survivors may fail to grow and reproduce normally. The duration of exposure to such a condition is important and the harmful effects of exposure to sub-lethal temperatures may be manifested at some later critical stage, *i.e.* ecdysis and pupation. Some of the insect species when exposed to extremes of temperatures beyond the favourable range may become dormant, *i.e.* undergo hibernation or aestivation, which are reversible processes as the individuals may resume activity on being exposed to favourable temperature.

The insect populations can be grouped into three categories according to their responses to low temperature:

- Those which cannot survive for any considerable time if the temperature falls below the lower limit of the favourable range. They cannot become dormant and hence must either develop or die. Such species have originated from the tropical or subtropical climates, *e.g.* locust.
- Species which have a stage in the life-cycle adapted to survive exposure to low temperature and the other stages resemble those in the above category in lacking the capability to become dormant at low temperature. Such species have originated from the temperate climates, *e.g. Helicoverpa* sp. and *Agrotis* sp.
- This group has a diapause stage and they have also originated from the temperate climates, *e.g.* many lepidopteran borers.

The species adapted to live in temperate climates, hibernate in a particular developmental stage and frequently this very stage and none other is capable of undergoing diapause. Some other species depend for their survival during winter on the insulating protection of the hibernacula such as debris, soil, plant remnants, snow and ice, and also on their ability to withstand exposure to low temperature through under-cooling. The combination of these two protective mechanisms prevents mortality, but in the case of incomplete protection exposure to severe cold period may lead to eradication of the insect population. This may be more relevant in the case of introduced natural enemies.

Moisture

A constant supply of moisture is essential for metabolic reactions as well as for the dissolution and transport of salts. The water content in insects varies from less than 50 per cent to more than

90 per cent of the total body weight. Variation occurs between different species and even between different stages in the life-cycle of the same species. Soft bodied insects such as caterpillars tend to have comparatively large amount of water in their tissues, whereas many insects with hard bodies tend to have somewhat lesser amounts. Active stages commonly have higher water content than dormant stages.

The range of moisture required is not so broad as in the case of temperature. Most of the insect species are capable of maintaining their body water at fairly constant levels while living under varying conditions. For most of the species, food in the shape of plant or its products is the source of water, and they have the adaptation to cope with conditions of excessive moisture and shortage of water. The other sources of gaining moisture are direct drinking of water or absorption through the integument. The loss of water from the body is prevented by insect cuticle having a waxy layer. A number of adaptations-morphological, biological and physiological-in nature help insect populations in overcoming unfavourable conditions of excessive moisture or acidity.

As in temperature, the phenomenon of ***humidity preferendum*** also operates in insects, and it helps insects to congregate in suitable places. The humidity preference is influenced by the prevailing temperature. Most adverse effects of moisture are due to its scarcity or absence. Exceptionally dry air may prove lethal because some insect species may not survive the loss of even a small percentage of body water for a long time. Those in aestivation or diapause lose a large proportion of body water without adverse effects. One insect, the larva of an African chironomid midge, *Polypedilum vanderplanki* Hint, can tolerate dehydration and suspension of metabolism for several years.

The exposure to excessive moisture can prove harmful to insect populations in the following ways :

- By adversely affecting the normal development and feeding activity of insects.
- By encouraging disease-causing microorganisms such as fungi, bacteria and mycoplasma, and thereby causing mortality among insects.
- Excessive moisture in insect body during winter reduces its capability to withstand exposure to low temperature and thus leads to an adverse effect on its cold hardiness.

Moisture also influences the speed of development and fecundity of most insect species. In some species, these activities are accelerated by excessive moisture, while in others they get retarded. Studies conducted in Punjab, India showed that sugarcane black bug *Cavelerius excavatus* (Distant), multiplied more rapidly at high humidity (90% R.H.), whereas relative humidity above 70 per cent was harmful for multiplication of cotton jassid, *Amrasca biguttula biguttula* (Ishida). For eggs and pupae of spotted bollworm, *Earias insulana* (Boisduval), both very high (around 100%) and low (<40%) relative humidities were not conducive for development of these stages.

Light

Light is an essential ecological factor for many biological processes such as orientation or rhythmic behaviour of insects, bioluminescence, periodicities of occurrence and periods of inactivity. Light acts as a token stimulus by enabling insects to regulate and synchronize their life-cycles with the change in seasons. Unlike temperature, light is a non-lethal factor and it has specific direction in its flow. A characteristic feature of light is its quantitative shift from a minimum to a maximum and vice-versa in a short time. The properties of light that influence insect life are intensity or illumination, quality or wavelength and duration of light hours or photoperiod.

The insects orientate to the source of light and thus reach the right place at the right time. Such phototactic behaviour of insects is altered and modified by a number of factors such as temperature, humidity or moisture and food. Several species of moths, leafhoppers and beetles are attracted to light at dusk or during the night, and this behaviour has been used extensively through light traps for observing brood-emergence or fluctuations in their populations. Direct sunshine may injure or kill an exposed insect largely because of heat and desiccation.

Photoperiodism influences the motor activity rhythms of insects such as locomotion, feeding, adult emergence, mating and oviposition, and also moulting and growth in some species. The reproductive cycle in most of the temperate-zone insects is so tuned that they reproduce only during favourable periods and the remaining period is passed in diapause state. The induction of diapause, a genetically determined state of suppressed development and the manifestation of which is induced by environmental factors, is influenced by photoperiod in most of the lepidopteran insects. Photoperiodic responses in insects also influence polymorphism, ecological adaptations and phenological synchronization with the sources of food.

Oxygen and Carbon Dioxide

Insects can tolerate a wide range of oxygen and carbon dioxide. Some insects can survive several days in the absence of oxygen by reducing their metabolic rates and utilizing the oxygen in their tissues. Excessive carbon dioxide causes varied reactions in different insects. Some insects can live in an atmosphere with high carbon dioxide for several days. However, an excess of this gas in the atmosphere causes growth retardation in many insects. If the environment is high in carbon dioxide, the spiracles of insects tend to remain open, which may lead to excessive water loss.

Air Currents

Air currents are of great value to insect displacement and, therefore, affect population changes by influencing the numbers into or out of an area. Most insects will not undertake flight when the speed of wind exceeds the normal flight speed. However, insects are rarely blown about at random but have evolved such patterns of behaviour which enable them to exploit the wind to achieve their migratory needs.

Strong flying insects tend to fly with the winds during migrations and are displaced long distances as in case of spruce budworm moths. Many insects which are weak fliers also have specific behaviour patterns enhancing their opportunity to migrate to specific areas with the help of wind. The air currents carry aphids, leafhoppers and scale insects to far-off places and thus are instrumental in their dispersal.

Air currents may be directly responsible for the death of insects. Firstly, severe wind coupled with heavy rains may cause mortality. Secondly, movement of air above a surface where evaporation is occurring (*e.g.* insect cuticle) increases the gradient of water vapour concentration and hence tends to increase the rate of evaporation.

Water Currents

Water currents often determine which species of insects would inhabit a particular area. The various genera of mayflies may be classified into still- and rapid-water forms. The legs and bodies of these insects are appropriately adapted. Black fly larvae fasten themselves to stones or other stationary material in the water. Caddisflies attach their cases to submerged objects. The mosquito larvae are unable to survive in moving water.

Another important feature of currents in the aquatic environment involves the circulation of dissolved gases, salts and nutrients. For example, caddisfly and mayfly larvae may be found under conditions of relatively high oxygen concentration, and midge and black fly larvae at somewhat lower concentration, while certain mosquito and other fly larvae at very low concentrations.

Edaphic Factors

Edaphic factors include the structure, texture and composition of soil along with its physical and chemical characteristics. Each soil has a distinctive flora as well as fauna of fungi, bacteria, algae, protozoa, rotifers, nematodes, molluscs, arthropods, etc. Some of these organisms help in maintenance of soil fertility through nitrogen fixation while others are responsible for return of the essential elements back to the soil by decomposition of dead organic matter. In humus formation, earthworms, millipedes, dipteran larvae, slugs and snails play important role in breaking up and division of litter.

Several properties of the soil like texture, moisture, drainage, chemical composition and physiography (topography) affect the distribution and abundance of insects. Soil texture varies from hard-packed clays to loose sands. Few insects dwell in hard- packed types, because they are unable to push or dig their way through them. The loams allow digging and burrowing operation and are usually favourable in their characteristics like moisture, drainage and organic matter. The cutworm, *Agrotis ipsilon* (Hufnagel), larvae live in soil of fairly light texture in which they move around freely in response to daily or seasonal temperature and moisture changes. Drainage and texture together exert considerable influence on the distribution of insects which pass part of their life in soil. The wireworms in wet arid land become important pests of potato, onion, lettuce and many other crops grown in irrigated fields.

Chemicals naturally present in the soil affect both the abundance and distribution of phytophagous insects. Deficiencies of mineral elements, resulting in similar plant deficiencies, inhibit the growth of some insects. Nitrogen deficiency lowers the productivity of some species of insects but results in outbreak of others.

Major topographic factors which affect biogeography of insects include height of mountain chains, steepness of the slope, directions of mountains and valleys, and exposure of the slope. These features also affect the climate of an area, thus influencing the distribution of certain insects. The mountain range and large bodies of water such as seas act as physical barriers to the spread of insects.

B. BIOTIC FACTORS

Under natural conditions, organisms live together influencing each other's life directly. Such vital processes as growth, nutrition and reproduction depend upon the interaction between the individuals of the same species (intraspecific) or between those of different species (interspecific). The interdependencies between insects themselves as well as between insects, other animals and plants exist throughout the whole life, or are casual and temporary. However, interdependency may exist between species which are taxonomically widely different such as between insects and bacteria, screw-worms and cattle, etc. The relationship between species may be beneficial to both, harmful to both or beneficial or harmful to one and neutral for the other.

The most important biotic components of the insect life-system are natural enemies and food.

Natural Enemies

The natural enemies of insect pest populations include predators, parasites/parasitoids and disease causing microorganisms such as fungi, bacteria, viruses and rickettsiae. The natural enemies are also influenced by various environmental factors such as weather and hyper-parasitism. The degree of influence of various natural enemies on the pest population would thus vary. The abundance of predators influences the abundance of their prey in field conditions. Predators respond to an increase in prey population density through *numerical response* (increase in the density of predators in a given area) and *functional response* (increase in consumption by individual predators).

Weather is known to determine the effectiveness of a natural enemy. For example, in the control of woolly apple aphid, with *Aphelinus mali* (Haldeman), the prevalent temperature in spring would determine the effectiveness of the parasitoid. Both the host and the parasitoid remain dormant during winter but the activity of host is resumed at a slightly lower temperature, and thus it is able to breed in the absence of the parasitoid in the early spring. In such a situation, the parasitoid may fail to press heavily on the host population as compared with the one when relatively higher temperature prevailed during early spring. Similarly, the larval parasitoid, *Apanteles flavipes* (Cameron), of maize borer, after successful overwintering in the host larvae, resumes activity in the spring earlier than its host but is unable to parasitize this particular host during dry and hot summer. This is because of lethal effects of high temperature (above 35°C) and low humidity (below 60% R.H.) on the parasitoid.

Food

Insects being heterotrophic depend directly or indirectly on plants for food. The quantity and quality of food play an important role in insects' survival, longevity, distribution, reproduction, speed of development, etc.

The quantitative aspects of food may be expressed as *absolute food shortage* (when food is completely destroyed within a localized region) and *effective food shortage* (when food shortages may occur in patches throughout the distribution of a given insect population). The causes of absolute and effective food shortages are numerous :

- There may be large number of the same individuals per unit quantity of food (intraspecific competition).
- There may be more than one species consuming the same food materials (interspecific competition).
- There might be other species that influence the food of a particular species without actually consuming it. For example, an epizootic of the fungus, *Entomophthora,* so reduced the number of the caterpillars of *Plutella xylostella* (Linnaeus), that the predator, *Angitis* sp. suffered severely from a shortage of food when it had been abundant before the outbreak of the disease.
- Any environmental factor causing reduction in populations of particular plants or insects will probably cause reduction of not only the insect populations that utilize these plants or animals for food but also of parasitoids or predators of these animals. Monophagous insects are more likely to be affected by such reductions in food supply than polyphagous species.

The quality of available food greatly influences egg production, larval development, longevity and size of insects. Many insect species store sufficient nutrients during the larval stage to accomplish adult activities. The adult longevity of these species is of comparatively short duration and they commonly do not feed at all. Mayflies which live long enough to copulate and egg laying, depend entirely on these reserves. In other species, larvae may store nutrients sufficient for egg production, but the adults must ingest water and carbohydrates to survive.

There are differences with regard to suitability of a host at the species or even at the varietal level within the same crop. The differences in varieties/species as regards their suitability for a particular insect population are governed by their acceptability as host, nutritional adequacy and absence of metabolic inhibitors or agents toxic to pest species.

MEASUREMENT OF PEST POPULATIONS

Pest population studies are helpful in pinpointing the factors that bring about numerical changes in the natural population and also in understanding the functioning of the life system of the pest species

Pest population studies are of two types :

(*i*) **Extensive studies.** These studies are spread over a large area and are needed to understand the distribution pattern of a population, to predict the damage it is likely to cause, to initiate control measures and to relate changes in the population to certain climatic or edaphic factors. A particular area is observed once or at the most a few times during the season and counts are made of a particular developmental stage of the pest.

(*ii*) **Intensive studies.** These studies involve repeated observations in a given area when it is desired to determine the contribution of various age intervals to the overall rate of change in the population or the dispersal of species. In this case the number of successive developmental stages are counted, and life-tables and budgets are prepared for determing the key factor(s).

For a correct and scientific understanding of a pest population, it is of fundamental importance to develop sound methods of population estimation. This involves two considerations. First, the life stage (egg, larva, pupa or adult) at which counting can be made most advantageously; and secondly, the actual process of counting. The ideal approach to population estimation would be to count all the individuals. However, it is not possible to count most of the pest species over an area large enough to be of use in a practical study and hence some method of sampling becomes necessary.

The amount of time and effort required to obtain absolute counts even on a limited area is so large that it is often uneconomical and unproductive. Thus, although we wish to have information on the true population, we are forced to take smaller collections (samples) and use these to make inferences about the total population. Based on the goal of sampling population, sampling plans in IPM can be grouped into three categories. First, detection sampling is used when pest is not present in the field. These are planned to avoid the chance that the organism is erroneously missed. Second, estimation sampling when the actual population of pest is estimated with desired levels of precision. It is mainly used in research and also to evaluate the effectiveness of any IPM module or pest control strategy developed to manage the insect pests in the farmers' field. Third, decision sampling based on which the decisions are made when to intervene with management tactics. In this sampling, objective is not to quantify the actual abundance of pest, but to decide the correct timings when control measures should be adopted or not. Various sampling techniques used for this purpose have been developed.

Measurements taken to estimate pest population density fall into three categories, viz. absolute estimates, relative estimates and population indices.

A. ABSOLUTE ESTIMATES

The total number of insects per unit area (1 ha, 1 m row length, 1 m^2 quadrat, etc.) is the absolute estimation. The numbers per unit of the habitat (per plant, shoot or leaf) indicate the density of population. The estimates of absolute population and population density are used for preparing life tables, study of population dynamics of field populations and to calculate oviposition and mortality rates.

The following methods are commonly employed for estimating absolute population :

(*i*) **Quadrat method.** Small areas or quadrats are chosen at random from a large area which contains the population. A large number of quadrats are required for over-dispersed population than that in the randomly distributed population. Stratified sampling is followed in the case of aggregated distribution, i.e. the population is divided into different strata and varying number of samples is taken from each stratum. From a quadrat, the insects may be counted or collected directly as in the case of fairly immobile but relatively large insects such as cutworms, caterpillars and grasshoppers. In case of tissue borers such as sugarcane borers, maize borer, etc., the estimation is done by first removing the infested plants from the quadrats and then counting them after splitting open the plant parts.

(*ii*) **Capture, marking, release and recapture technique.** This technique is generally used for estimating the population of flying insects. The losses or gains in a population over a period can be determined with the help of this method. The estimate of density fluctuations and the death rate can also be made for making comparisons between different forms of pests under varying environmental conditions. The different types of markers used are paints and dyes, labels, mutilation, radioisotopes, etc. This method has been used for estimating the absolute population of butterflies, grasshoppers and beetles. The total population is estimated by using the following formula :

$$P = \frac{N \times M}{R}$$

Where P = Population of insects

N = Total number of insects caught

M = Number of marked individuals released

R = Number of marked individuals recaught

For effective application of capture-recapture technique in population estimation, the following assumptions have to be met :

- The marking of individuals does not lead to changes in their behaviour or longevity and the marks do not get lost easily.
- The marked individuals, after being released, become completely mixed up with the unmarked individuals of the population.
- The population is sampled randomly with respect to its mark status.
- The method of marking should be such that it should distinguish between different dates of capture.
- The population under study must be reasonably stable and not subject to rapid fluctuations in numbers.

B. RELATIVE ESTIMATES

In relative population estimates, the samples usually represent an unknown constant proportion of the population. A given amount of labour and equipment is utilised to yield much more data than is possible for absolute estimates. Such estimates are useful in making comparisons in space or time. These are useful for studying the activity patterns of a species or for determining the constitution of a polymorphic population. The relative estimates are influenced by a number of factors like variation in (*a*) behaviour of an insect with change in age, (*b*) level of activity of the pest as influenced by its diurnal cycle, (*c*) responsiveness of sexes to trap stimuli, and (*d*) efficiency of the trap or the searching method, besides the pest population.

The methods employed for relative estimates include the catch per unit time or effort and the use of various types of traps.

(*i*) **Catch per unit time or effort.** Various types of collection nets are available for use in different habitats and the sweep net is the most widely used for sampling insects from vegetation. Only those individuals on the top of the vegetation and those that do not fall off or fly away on the approach of the collector can be caught with the sweep net. The efficiency of catch with a sweep net will be influenced by changes in the habitat, species and vertical distribution of the pest species, variation in the weather conditions and effect of diet cycle on verticle movements. The number of sweeps necessary to obtain a mean that is within 25 per cent of the true value will depend upon the pest species, its spatial distribution and the diurnal periodicity.

(*ii*) **Line-transect method.** If one walks in a straight line at a constant speed through a habitat, the number of individuals can be counted. This technique is used for quantitative comparisons both between different species, and between different occupiers of habitats. This technique was originally employed by botanists for estimating the population of plants, but could not be easily used in animals owing to their mobility from place to place. However, the data based on number of encounters have been used for estimating the absolute population of locusts and grasshoppers.

The number of organisms per unit area or their density can be calculated by the formula :

$$D = \frac{Z}{2R\,(\overline{V} + \overline{W})^{1/2}}$$

Where D = Density

Z = Number of encounters between the observer and the organism in a unit time

R = Radial distance within which the organism must come in contact with the observer to affect an encounter

$\overline{V}$ = Average speed of the observer

$\overline{W}$ = Average speed of the organism

The following pre-requisites are necessary for application of this formula in the study of natural populations :

- The number of organisms to be counted should remain reasonably constant during the period of observation.
- The average speed of the observer ($\overline{V}$) needs to be determined before the actual observations. The time spent in locating the organisms and in making records has to be taken into account while working out the average speed.
- The average speed of the organism ($\overline{W}$) should be determined by taking a large number of small samples.
- The effective radius (R) of the organism is determined on the basis of recognition distance which can be visual or auditory.

Further, the following conditions must be satisfied for getting valid estimates and comparisons:

- The observations should be made in the same phase of the life-cycle.
- Weather conditions, time of the day and season of the year, should be reasonably comparable.
- The features of the environment influencing recognition must remain uniform.

(*iii*) **Shaking and beating.** Some insects can be collected on ground by shaking or beating the plants. A piece of cloth or polythene may be laid out under the plants and the plants are vigorously shaken. The insects fall on the cloth or polythene and need to be counted immediately before they disperse. The gram pod borer, *Helicoverpa armigera* (Hubner) larvae may be sampled by vigorously shaking the chickpea plants.

(*iv*) **Knockdown sampling.** The insecticides such as pyrethrum or other pesticides with safe chemistry may be sprayed on plants enclosed in a polythene envelope. The insects will be knocked down due to the insecticides. The plants are shaken and the insects which fall on the ground or cover sheet may be counted to estimate the population. Other approach is heating the sample which forces the insects to come out of their dwellings. The plant sample may be placed in a special device, the Berlesse funnel, that heats the sample. The insects come out, fall through the funnel and can be counted in a receptacle.

(*v*) **Use of traps.** In this case, it is the insects rather than the observer that would make the action leading to their enumeration. Traps are of two types, viz. *interception traps* (that catch the insects randomly) and *attraction traps* (which attract the insects in some way). The interception traps provide indices of absolute population more easily than those of attraction traps, as there is no variation due to attraction. Various types of interception traps are flight traps, aquatic traps, pitfall, light and other visual traps. The flight traps that combine interception and attraction include sticky traps and water traps. The traps that attract the insects by some natural stimulus or substitute include the shelter traps, trap crops, bait traps, chemical attractants and pheromones.

(*vi*) **Remote sensing.** Remote sensing technology has long been used for monitoring insect infestation in field crops. It is based on the principle that the absorbance and reflectance of plants in response to pest attack changes and these changes are recorded by a device from far away. Remote sensing platforms can be aircraft, satellites or ground based. The remote sensing techniques include full-colour photography, infrared (IR) wavelength and multiband spectrometers. In full colour photography, the coloured photographs are examined for chlorosis or other symptoms. In IR wavelength, the changes in leaf temperature in relation to differential moisture content induced by pest activity are measured using remote thermometer. In multiband spectrometers, the reflectance at specific wavelength is measured to record different types of vegetation.

The insect infestations on beet and cabbage plants were monitored by aircraft based multispectral imaging. The negative correlation between beet armyworm, *Spodoptera exigua* (Hubner), hits and transformed normalized differences in vegetation index values was reported. Based on these, future sampling plans and site specific management techniques can be developed. The hyperspectral remotely sensed data with an appropriate pixel size have the potential to portray greenbug, *Schizaphis graminum* (Rondani), density on winter wheat and discriminate its damage to wheat with repeated accuracy and precision. The early identification of damage due to two-spotted spider mite, *Tetranychus urticae* Koch, on greenhouse pepper (*Capsicum annuum*) can be obtained by multispectral means as it can be spectrally detected in the reflectance of the visible and near-infrared regions. The hyperspectral reflectance data of greenhouse pepper leaves were transformed into vegetation indices allowing early two-spotted spider mite damage detection by separation between leaf damage levels.With advancement in spatial information technologies such as Global Positioning Systems (GPS) and Geographical Information Systems (GIS), remote sensing is finding more practical application for monitoring and management of insect pests. The three-dimensional real time observations on insect population can be achieved using remote sensing in conjunction with 'silicone oil-carbon black powder suspension squeeze' (3S) technique.

C. POPULATION INDICES

Population indices do not count insects at all, but rather they are measures of insect products or effects. Under field conditions, it is not possible to estimate the absolute population in most of the cases. It, therefore, becomes necessary to establish a relationship between absolute estimates and population indices or the relative estimates so that the latter two types of estimates could be converted to absolute terms by using certain correction factors.

(*i*) **Insect products.** In some cases, a species that is difficult to sample creates products directly that are easily sampled by absolute methods. The insect product most often sampled is frass or excrement of lepidopterous defoliators. The rate at which frass is produced can be estimated from the amount falling into a box or funnel placed under the trees. The size and shape of the frass pellets is rather constant for a given species and instar; this allows one to identify the species and age composition of defoliators.

(*ii*) **Plant damage.** The amount of damage caused by insects to crop plants is a function of the pest density, the characteristic feeding or oviposition behaviour of the species and the biological characteristics of the plants. Different methods have to be adopted for measuring damage by direct and indirect pests.

(*a*) *Direct pests.* These pests attack the produce directly, destroying a significant part of its value. Bollworms on cotton and fruit borers in fruit and vegetable crops are examples of such pests. The damage by direct pests is sampled on the basis of absolute or relative numbers of damaged unit, e.g. number of damaged bolls per plant, apples per tree, pods per meter row length, etc.

(*b*) *Indirect pests.* Damage by indirect pests may be measured by estimating the extent of defoliation in case of defoliating pests like lepidopterous caterpillars, leaf beetles, grasshoppers, etc.

Many devices like planimeters, photoplanimeters, leaf area meters, etc. are now available for quantifying the extent of defoliation. In case of sap-sucking insects like aphids, jassids, white flies, etc. damage may be estimated by change in coloration of the leaves, plant vigour and size. For root feeders, damage to root system may be evaluated by percentage of nodes below the soil surface having injury, force needed to pull a plant from the soil, etc.

For a correct and scientific understanding of a pest population, it is of fundamental importance to develop sound methods of population estimation. This involves two considerations. First, the life stage (egg, larva, pupa or adult) at which counting can be made most advantageously, and secondly, the actual process of counting. The ideal approach to population estimation would be to count all the individuals. However, with most of the pest species it is not possible to count them over an area large enough to be of use in a practical study and hence some method of sampling becomes necessary. For devising a satisfactory and successful sampling programme, consideration has to be given to the nature of the sample, mode of sampling, the size and number of samples. The precision of a pest population estimate based on a given sampling technique would depend upon both the properties of the population in terms of its density and degree of aggregation and upon the charactersitics of the sampling plan, i.e. the number and size of samples.

CHANGING STATUS OF PESTS

Agriculture in India started around 2000 BC during the period of Indus valley civilization in Sind and West Punjab, and spread to East Punjab, West U.P., Rajasthan and Madhya Pradesh. At around the same time, agriculture was also started in South and East Asia. The implements used during this period included plough, seed-drill, unwheeled cart, etc. The only references regarding pest problems during this period pertain to wild birds and animals, which were scared away by use of terracota sling-balls. The advent of Aryans brought the iron technology but there was still no indication of any pest problems. Only around 900 AD, locusts are listed alongwith birds and mammalian pests as causing damage to crops. In ancient Indian literature, there are indications of a sound study of useful insects but the knowledge about harmful ones in the same period was expectedly poor. Although pest problems were quite serious, the knowledge about their nature and control was very inadequate. This situation continued almost till the end of the nineteenth century.

Traditionally, the crops were grown only during the monsoon period, and winter served as a closed period for crops as well as for pests in India. With no assured irrigation and little use of chemical fertilizers, the yields obtained were also low. However, even at that time a number of insect pests were known to cause serious damage to different crops. Balfour in 1887 in his book *Agricultural Pests of India and of Eastern and Southern Asia,* reported many important groups of pests like cutworms, aphids, bugs, caterpillars and beetles damaging different field crops, viz. *Leptocorisa* and greenfly on rice; sugarcane borer, *Diatraea saccharalis* (Fabricius) and sugarcane bug on sugarcane, and pink bollworm and *Helicoverpa* on cotton. Harold Maxwell Lefroy, who was appointed Entomologist to the Government of India in 1903, published a book on *Indian Insect Pests* in 1906, wherein he described pests injurious to cotton, rice, wheat and other crops. T.B. Fletcher, who succeeded Maxwell Lefroy as Imperial Entomologist (1913-1932), produced major treatises on cotton bollworms, sugarcane borers and stored grain pests. Fletcher in 1920, at the Third Entomological Society Meeting presented a comprehensive list of several hundred species of insects associated with different crops in India. Of these, 45 species were reported as major pests on important crops. These included 8, 7, 7, 6, 6 and 1 species causing serious damage to paddy, cotton, oilseeds, pulses, vegetables and sugarcane, respectively.

Following the introduction of high yielding varieties (HYVs) and associated technology, there has been a tremendous increase in the number of insect pests damaging various crops, viz. pigeonpea (250), rice (225), sugarcane (215), cotton (170), etc. Many minor insects have started assuming serious proportions, whereas several pests have spread to new areas.

The incidence of brown planthopper, yellow stem borer and leaf folder has increased manifold with the cultivation of modern HYVs of rice. Gall midge, originally being confined to monsoon season, has, of late, become a pest in summer season also. The spread of rice cultivation into the non-traditional north-western states like Punjab and Haryana, has brought up the problem of whitebacked planthopper. The continuous cropping of maize in northern parts of India has led to the emergence of shoot fly, *Atherigona* spp., as important pest in spring sown maize. *Pyrilla perpusilla* (Walker), a serious pest of sugarcane has developed a perference for hybrids of maize and sorghum. Sorghum midge, earhead bug and aphids are gaining importance in many sorghum growing regions.

Severe outbreaks of whitefly and American bollworm have been observed on cotton during the last decade. American bollworm appeared in serious form in parts of Andhra Pradesh and Tamil Nadu during 1987-88, where it caused about 66 per cent loss in yield of seed cotton. The pest caused severe losses to cotton crop in Haryana, Punjab and Rajasthan during 1990s. Incidence of this pest on cotton has considerably declined following the introduction of Bt cotton. However, mealybug, *Phenacoccus solenopsis* Tinsley has started appearing in serious proportions.

A number of polyphagous pests have also attained serious status. Most dreaded of these is *Helicoverpa armigera* (Hubner), which has been recorded on more than 180 cultivated and uncultivated plant species, of which chickpea, pigeonpea, cotton, tomato and tobacco are considered to be the key source crops. Similarly, *Spodoptera litura* (Fabricius) has been recorded on more than 115 species of cultivated plants in India.

Insect pest problems in agriculture are probably as old as agriculture itself. However, under subsistence agriculture, the pest numbers were generally low as the productivity was poor. The insect pests were kept under check by cultural and mechanical practices developed by farmers largely through trial and error. Rapidly increasing population during the last century has necessitated intensification of agriculture, which has also resulted in severe outbreaks of insect pests in agricultural crops.

High yielding varieties. When altering the genetic make up of the crop plant to increase yield with little or no attention to pest attack, natural resistace may be lost or greatly reduced. Efforts to breed plants more palatable to human taste by elimination of such factors as bitterness and hairiness may also result in development of more susceptible varieties. Moreover, the fields planted to modern varieties, which are short and heavy tillering, develop a distinctly different microclimate conducive for proliferation of insect pests.

Monoculture. In agriculture, the natural plant community is removed, destroyed and usually replaced by a single crop species. Larger the area that is planted to a single crop, the greater the potential for pest problems. Generally, same crops are maintained in a given area year after year and this helps in survival and multiplication of the pest throughout the year. Large scale monoculturing of cotton, paddy, sugarcane and vegetable crops in India in contiguous blocks has resulted in rapid increase in number and intensity of pest attacks.

Nutrients. Altering the nutrient level in the soil affects the concentration of nutrients in the host plant, which, in return, influence the pests that are feeding on the plant. The HYVs are nitrogen responsive and require extensive use of nitrogenous fertilizers to provide maximum yield. High levels of nitrogenous fertilizers significantly increase the incidence of most insect pests. Increase in soil moisture level due to frequent irrigations also helps in build up of insect pests.

Pesticides. The advent of synthetic organic pesticides during 1940s enabled us to gain an upper hand in the struggle against insect pests. However, the large scale indiscriminate use and abuse of these insecticides has resulted in increased attacks of insect pests on many crops. This may be due to development of insecticide resistance, resurgence of insect pests or outbreaks of secondary pests. The insecticides may kill natural enemies of target pest or potential pest. They may also alter crop physiology to make the plant more susceptible to attack by insects or they may even directly stimulate reproduction of surviving insects.

Cultural practices. New agronomic recommendations like closer spacings, sowing times, rotations, harvesting procedures, etc. are essentially meant for exploiting the full genetic potential of HYVs, but are simultaneously favourable to most of the insect pests. In cultivated fields, crop plant densities are carefully controlled to obtain the maximum population possible for optimal growth resulting in maximum economic yield. The new plant spacings often result in an ecological situation that encourages pest outbreaks.

Climate Change. The incidence of insect pests is likely to change as a result of global warming. Current estimates of changes in climate indicate an increase in global mean annual temperatures of 1°C by 2025 and 3°C by the end of next century. The doubling of CO_2 is estimated to be attained between 2025 and 2070 depending on the level of emission of greenhouse gases. Such changes are likely to result in expansion of geographical range of insect pests, increased overwintering and rapid population growth, changes in insect-plant and insect-natural enemy interactions, increased risk of invasion by migrant pests, changes in synchrony between insect pests and their crop hosts, and reduced effectiveness of crop protection technologies.

YIELD LOSS ASSESSMENT

The yields of crops are governed by a variety of factors including pests. It is not always possible to manipulate the factors that determine crop yield. The assessment of crop losses due to insect pests is of significance from the planning point of view, as it is helpful in prediction of crop production and planning for higher production in the future. Thus, information on crop losses would serve as a guide for research programmes in crop improvement.

The loss suffered by a crop is a function of the pest population, behaviour of the pest and the crop plants. Damage to the plant occurs because of the effect of injury by the insect. A simple damage to the plants by an insect may or may not lead to crop loss. The reduction in quantity/ quality of the produce is the crop loss. The loss in quality may affect the appearance of the crop produce, its nutritive value or it may result in the produce being rendered unfit for use.

Insect pests damage crop plants either by feeding or during the process of oviposition. Some of the insect pest species are host specific and they feed on plants of a single species and are termed *monophagous*. Others attack plant species belonging to the same family and are known as *oligophagous*. Others are capable of infesting plant species belonging to several diverse families and are called *polyphagous*. Some of the pests are strictly specific as regards their site of feeding and oviposition, as for example leafhoppers, leafminers, fruit borers or root borers. They cause damage to only one part of plant. There are others, like the locust and some species of beetles, that can attack several parts of the same plant simultaneously.

Types of Losses

The losses due to insect pests can be categorised in many ways, depending upon the significance of pests and their management.

Direct losses. These relate to decrease in productivity (quantitative) or intrinsic value/ acceptability of the produce (qualitative). Direct quantitative losses include killing of flowers, buds, twigs of whole plant because of infestation by a pest having either chewing or piercing-sucking mouth parts, e.g. locust and grasshoppers, bollworms, fruit borers, root borers, etc. Examples of direct qualitative losses include light infestation of fruits by the scales, puncturing of normal fruits immediately before harvest owing to feeding or ovipositional activity. Damage by the pests to the fruit trees from the blooming to harvesting periods results in quantitative losses in the earlier phase and qualitative ones in the later phase.

Indirect losses. These are primarily of economic interest as for example decreased purchasing power of the agriculturists and those depending upon agriculture owing to reduced production. This would lead to decrease in related activities, reduced productivity of agro-based industries, expenses incurred for importation of agricultural produce and also forced acceptance of less desirable substitute products.

Actual losses. These include the total value of losses, both direct and indirect, the cost of control measures alongwith the amount spent on researches for developing pest control knowledge among the agriculturists.

Recognized and hidden losses. These are subjective terms showing as to whether the factors determining the losses are known or not.

Avoidable and unavoidable losses. These are also subjective terms related to the belief as to whether a certain pest can be controlled or not.

Estimation of Losses

The amount of damage caused by insect pests of crop plants is a function of the pest population– its characteristics of feeding or oviposition behaviour and the biological characteristics of the host plants. Each of these factors is affected differently by the environmental factors, both biotic and abiotic. It is, sometimes, rather difficult to establish correlations between the levels of pest population and plant damage. The estimation of damage is, however, critical to pest management. The evaluation of damage is helpful in pest management in the following ways :

- Defining the economic status of a pest species
- Estimating the effectiveness of control measures
- Helping in assigning priorities on the basis of relative importance of different pests.
- Evaluating crop varieties for their resistance to pests
- Helping in deciding the allocations for research and extension in plant protection.

The techniques adopted for the assessment of crop losses caused by insect pests fall into the following categories :

Mechanical protection. The crop is grown under enclosures of wire gauge or cotton cloth. The enclosures keep the pest away from the crop. The yield under such enclosures is compared with that obtained from the infested crop under similar conditions. The technique has been used with various modifications for estimating the losses caused by jassid and whitefly to cotton. In the case of non-flying insects, sometimes, the barriers are substituted for the cages. The limitation in the case of enclosures is that the plants generally become pale and weak due to changes in microenvironment. This technique cannot be adopted on an extensive scale because it is very time consuming and impracticable on field scale.

Chemical protection. The crop is protected from pest damage through the application of pesticides. The yield of treated crop is compared with the one which has been subjected to normal infestation. This technique has been very widely used and it can be adopted on a large scale in cultivators' fields. It should, however, be ensured that the treated and untreated fields/plots are as similar as possible in soil type, manuring, variety and cultural practices. The flaw in this approach is that the crop treated with chemicals can be physiologically affected and it may give either more or less yield.

Pest incidence in different fields. The yield is determined per unit area in different fields carrying different degrees of pest infestation. The correlation between the crop yield and degree of infestation is worked out to estimate the loss in yield. This technique can be used for estimating crop loss due to different pests and diseases over a larger area. The drawback in this technique is that the yield in different fields can be influenced by soil heterogeneity.

Pest incidence on individual plants. In this case, individual plants from the same field are examined for the pest incidence and their yield is determined individually. The loss in yield is estimated by comparing the average yield of healthy plants with that of plants showing different degrees of infestation. The same data can also be used for working out a correlation equation between yield and infestation on the basis of individual plants.

The advantage in this technique over the preceding one is that soil heterogeneity factor is considerably reduced in the same field. However, different plants showing varying degrees of infestation in itself is a proof that plants differ from one another in some unknown factors due to which they carry different degrees of infestation. This factor may be genetic or physiological or it may be mere soil heterogeneity in the same field. Moreover. this method is very time consuming and involves lot of labour.

Damage by individual insect. Preliminary information on the damage caused by individual insect is obtained from studies on biology of pest species. The details regarding the amount of damage caused by different stages or ages of the pest, and the exact nature and amount of loss caused are then worked out. This technique is quite easy in the case of leaf-feeding insects. However, it is very difficult to use this technique over large areas since it is very time consuming.

Simulated damage. This technique involves simulation of pest injury by removing or injuring leaves or other parts of the plant. The simulated damage may, however, not always be equivalent to the damage caused by an insect. Insects may persist over a period of time or inject long acting toxins rather than producing their injury instantly. Feeding on margins of a leaf may not be equivalent to tissue removal from the centre of the leaves. Insect feeding is usually extended over a period of time and it is rather difficult to incorporate the concept of the rate of injury in simulation studies. Furthermore, the kind of leaf removed may be important, as for example, quality and position of leaf on the plant. In addition, the time of simulating damage with respect to the stage of plant growth is also critical.

Manipulation of natural enemies. The manipulation of natural enemies of a pest species offers a means of evaluating plant damage. The pest is controlled by introducing predators/parasitoids into the field and the yield of such a crop is compared with that on which no such pest control measures have been undertaken. This technique has, however, not been widely used. This method is feasible only in small plots and is not practicable on a field scale.

Any of the above methods can be suitably modified and used for estimating loss in yield of a given crop. The degree of pest infestation and the damage caused by it may differ from field to field in the same season, and from season to season in the same field. It is, therefore, imperative to work out the average values. In case the crop losses have to be worked out on the regional/state basis, the number of places from where estimations have to be made are more important than the degree of precision of the technique employed.

CROP LOSSES : GLOBAL AND INDIAN SCENARIO

Despite the annual investment of US$ 40 billion for the application of 3 million metric tonnes of pesticides, plus the use of various biological and other non-chemical controls worldwide, global crop losses remain a matter of concern. Although crop protection aims to avoid or prevent crop losses or to reduce them to economically acceptable losses, the availability of quantitative data on the effect of different categories of pests is very limited. The generation of experimental data is time-consuming and laborious, and losses vary from season to season due to variation in pest incidence and weather conditions, and estimates of loss data for various crops are fraught with problems. The assessment of crop losses is required for demonstrating where future action is to be needed and for decision making by farmers as well as government agencies.

The first attempt to estimate crop losses due to various pests on global scale was made by Cramer (1967), who put the global losses due to insect pests at 10.8 per cent. Subsequently, Oerke *et al.* (1994) made extensive study to estimate losses in principal food and cash crops and total losses due to all categories of pests animals, weeds and pathogens) were estimated at 42.1 per cent. In spite of the wide spread use of synthetic pesticides and other control measures, the losses due to insect and mite pests increased in post-green revolution era than in pre-green revolution era. Worldwide total pre-harvest losses for post-green revolution era (1988 through 1990) period value at US$ 90 billion for eight principal food and cash crops (barley, coffee, cotton, maize, potato, rice, soybean and wheat).

Since mid nineties, the production systems have undergone a tremendous change, particularly in crops like maize, soybean and cotton in which the advent of transgenic technology has modified the strategies for pest management in some major production regions. Therefore, Oerke (2006) updated the loss data for major food and cash crops for the period 2001-03 according to which the total losses due to all categories of pests were calculated as 32.1 per cent. The actual losses due to various pests have been estimated as 26-29 per cent for soybean, wheat and cotton, and 31, 37 and 40 per cent for maize, rice and potatoes, respectively (**Table 2.1**). Thus, there is a decline in global crop losses due to various pests from 42.1 per cent during 1988-90 to 32.1 per cent during 2001-2003. The corresponding decline in loss due to animal pests, weeds and pathogens was from 15.6 to 10.8, 13.2 to 8.8 and 13.3 to 12.5 per cent, respectively.

Table 2.1 Global losses due to various categories of pests in major crops

Crop	Losses (%)				
	Animal pests	Weeds	Pathogens	Viruses	Total
Cotton	12.3	8.6	7.2	0.7	28.8
Maize	9.6	10.5	8.5	2.7	31.3
Potatoes	10.9	8.3	14.5	6.6	40.3
Rice	15.1	10.2	10.8	1.4	37.5
Soybean	8.8	7.5	8.9	1.2	26.4
Wheat	7.9	7.7	10.2	2.4	28.2
Average	**10.8**	**8.8**	**10.0**	**2.5**	**32.1**

Source : Oerke (2006)

Losses due to insect pests in Indian agriculture have been estimated from time to time. Extensive surveys carried out during early 1960s revealed that fruits, cotton, rice and sugarcane suffered 25, 18, 10, and 10 per cent yield losses, respectively. However, during the early 2000s, the losses increased considerably due to intensification of agriculture. The highest losses were reported in cotton (50%), followed by sorghum and millets (30%), and rice, maize and oilseeds (each 25%). In general, the losses in post-green revolution era have shown an increasing trend than in the pre-green revolution era. Overall, the losses increased from 7.2 per cent in early 1960s to 23.3 per cent

Table 2.2 Losses caused by insect pests to major agricultural crops/commodities in India

Crop/Commodity	Production (million tonnes)	Estimated loss in yield		Hypothetical production (million tonnes) in the absence of losses	Monetary value of estimated loss (₹ billion)
		Per cent	million tonnes		
Cotton	44.0	30	18.9	62.9	339.7
Rice	96.7	25	32.2	128.9	240.1
Maize	19.0	20	4.8	23.8	29.4
Sugarcane	348.2	20	87.1	435.3	70.7
Rapeseed-mustard	5.8	20	1.5	7.3	26.1
Groundnut	9.2	15	1.6	10.8	25.2
Other oilseeds	14.7	15	2.6	17.3	35.8
Pulses	14.8	15	2.6	17.4	43.5
Coarse cereals	17.9	10	2.0	19.9	11.9
Wheat	78.6	5	4.1	82.7	41.4
Other crops (fruits, vegetables, spices, plantation crops, etc.)	–	12	–	–	19.2
Stored grains (losses due to insects, rodents, fungi, etc.)	–	10	–	–	17.1
Total		**16.4**			**900.1**

Source : Dhaliwal *et al.* (2010, 2013)

in early 2000s. The maximum increase in loss occurred in cotton (18.0 to 50.0%), followed by other crops like sorghum and millets (3.5 to 30.0), maize (5.0 to 25.0) and oilseeds (other than groundnut) (5.0 to 25.0).

There has been a paradigm shift in the crop management scenario of Indian agriculture since the beginning of this century. Bt cotton was released in the country in 2002 and the area under Bt cotton increased from 50,000 ha in 2002 to 11 million ha in 2013. Currently, Bt cotton occupies about 90 per cent of the total area under this crop. Secondly, concerted efforts were made to implement integrated pest management programmes in principal food and cash crops. As a result of these developments, losses due to insect pests in several agricultural crops have shown a declining trend (**Table 2.2**). However, in terms of monetary value, the decline in losses does not appear to be prominent. This is due to both the increase in production levels as well as the increae in prices of different commodities.

Thus, the Indian agriculture suffers annual losses to the tune of ₹ 900 billion due to the ravages of arthropod pests in field and during storage. This is a colossal loss and all out efforts should be made to bring the losses to the minimum so that the access to food is increased for the expanding population. This has to be done in such a way that environmental quality is maintained and long term sustainability of the agroecosystem does not suffer. Integrated pest management (IPM) appears to be the only viable option to achieve these objectives.

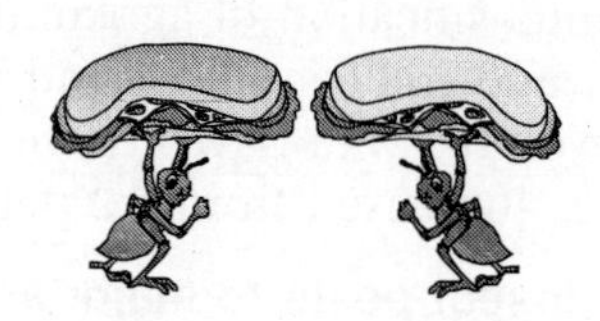

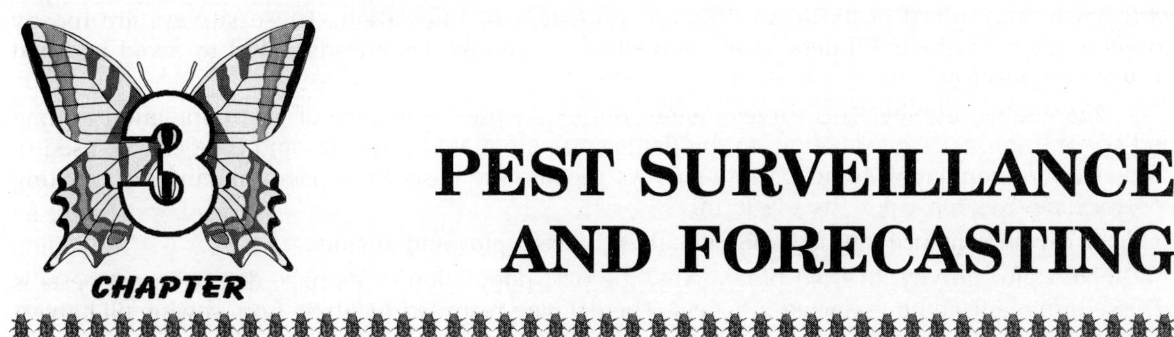

PEST SURVEILLANCE AND FORECASTING

INTRODUCTION

The recent developments in pest management tactics, like new molecules, biocontrol agents, microbials, genetically engineered crops, etc. opened new vistas for integrated pest management (IPM). The fundamental principles of IPM must be followed for successful and sustainable use of these tactics for management of insect pests. The basic information required for success of any IPM programme is about the population fluctuations of pests and beneficial organisms in a particular habitat. The use of control tactics, without any prior information about whether the potential pests are abundant enough to cause economic damage to the crop, will lead to various problems. The ill effects of non-judicious use of insecticides lead to several problems like pesticide resistance, pest resurgence, secondary pest outbreaks, destruction of natural enemies, contamination of food, soil, environment, etc. Therefore, interventions to manage the pests must be undertaken when the pest population is large enough, which could not be brought down by natural enemies and other climatic factors, and causes economic damage. Hence, the regular monitoring/survey of pest situation in field is basic step for implementation of IPM. With advancement in knowledge about biology of pests, factors affecting pest abundance, and environmental factors for growth of pest, pest forecasting came into being. Pest forecasting provides an opportunity for the farmers to take timely action to apply biopesticides or chemicals for management of insect pests. The forecasting models have been developed for many pests and validated, which provide useful information and lead to increase in productivity (Dhaliwal *et al.,* 2013).

PEST SURVEILLANCE

An insect pest survey is a detailed collection of insect population information at a particular time in a given area. The regular surveys of same place or locality at consistent intervals to assess changes in pest species over a time is called 'surveillance'. The word 'surveillance' and monitoring have been used in literature synonymously. It has also been mentioned sometimes that monitoring involves determining number and life stages of pest present in a location only, however, surveillance in addition includes the loss assessment and economic benefits by adopting control measures. The basic components of pest surveillance include identification of pest, determining the stage and population density of pest and natural enemies, estimating loss the pest species may cause, the economic and other benefits that pest control will provide. The planned procedure to determine these characteristics over a defined period of time is known as survey. In general, it is a method of collecting quantitative information about insect population in a particular location at a particular time. The survey may be done in a small field, a pond, or it may be as extensive as a state or region of the country. These are carried out at certain critical periods in the insect life cycle.

Pest surveys can be grouped into two main categories, i.e., qualitative survey and quantitative survey.

Qualitative survey. It is generally aimed at pest detection and provides list of pest species present along with reference to density like common, abundant, and rare. These are usually employed with newly introduced pests to understand the extent of infestation. These surveys are mostly adopted at international borders where agricultural commodities are inspected to avoid invasion of any new species.

Quantitative survey. This survey defines numerally the abundance of pest population in time and space. It provides information on the damaging potential of a species and data can be used to predict future population trends. These surveys provide the basis to decision making for adopting control measures for a pest by the farmers.

Survey/surveillance can also be classified as fixed plot and roving.

Fixed plot survey. In fixed plot survey, the pest population or damage due to insect pests is assessed from a fixed plot selected in a field. The data are recorded regularly from sowing till harvest of the crop from the same fixed plot in a particular field. The data collected in these surveys are used to develop forecasting models. The direct counting of population on plant, light traps, sticky trap, pheromone traps, etc. are techniques which can be used to monitor population in this survey. Counting total tillers and number of tillers affected by stem borer from 10 randomly selected plants from fixed five spots of $1m^2$ in one ha is an example of fixed plot survey.

Rapid/Roving survey. This survey includes assessment of pest population or damage from randomly selected spots in a short period of time over a large area. It provides information on pest level which helps in determining the timing of adopting appropriate control measures. The surveys are made to monitor the initial development of pests in endemic areas in the beginning of crop season. Based on these surveys, the farmers are instructed to monitor the pest incidence in their respective fields by the agriculture extension specialists and take interventions to manage the pests accordingly. The counting of whitefly adults from lower surface of leaf from randomly selected cotton plants on a predetermined route after a definite period of intervals is an example of roving survey.

Objectives of Surveillance

- To monitor, when pest population/ damage at different growth stages of crop reaches the economic threshold levels.
- To estimate the crop losses caused by pests.
- To study changing pest status from minor to major in a given crop ecosystem which enables to determine the research priorities.
- To monitor the development of biotypes, resistance to insecticides, resurgence, etc.
- To monitor the invasion of new pest species in a local ecosystem and determining the rate of spread of exotic pest that has already been established.
- To study the influence of weather parameters on pest by recording the changes in density of pest population throughout the year.
- To assess natural enemies population and their influence in a particular cropping system and in different seasons.

Components of Pest Surveillance

1. Identification of Pest

The first most important step for surveillance/survey is correct identification of pest. If the identification of a pest is incorrect, the decision for taking intervention for the management of pest will not be reliable. The incorrect identification may occur when one known species is confused with other or when a previously unknown species is grouped into known species. Therefore, the person deputed for survey must have expertise on identification of all the stages of pest and their common visible morphological characters. Otherwise, the samples may be collected and reared in laboratory for all stages and specialist/expert/taxonomist may be consulted. With the advent of molecular techniques, the data on conserved gene sequences is being generated for all the biodiversity of the world through International Barcode of Life (iBOL) project. The data will be publicly available which can be useful for identification of pest specimen. The simple procedure includes sequencing of conserved genes amplified from total DNA of insect species and alignment of the sequence with the gene sequence databank. It would help to establish the identity of new unknown species and other species, which are generally confused with other species.

2. Determination of Pest Population

The second basic component of surveillance is estimation of pest population. Most of the economic threshold levels for different pests depend on number of pest population present in the field. The study on pest population is helpful in pinpointing the factors that bring about numerical changes in the natural population, and also in understanding the functioning of life-system of the pest species.

For a correct and scientific understanding of a pest population, it is of fundamental importance to develop sound methods of population estimation. This involves two considerations. First, the life stage (egg, larva, pupa or adult) at which counting can be made most advantageously; and secondly, the actual process of counting. The ideal approach to population estimation would be to count all the individuals. However, it is not possible to count most of the pest species over an area large enough to be of use in a practical study and hence some method of sampling becomes necessary.

The amount of time and effort required to obtain absolute counts even on a limited area is so large that it is often uneconomical and unproductive. Thus, although we wish to have information on the true population, we are forced to take smaller collections (samples) and use these to make inferences about the total population. Based on the goal of sampling population, sampling plans in IPM can be grouped into three categories (Moon and Wilson, 2009). First, detection sampling is used when pest is not present in the field. These are planned to avoid the chance that the organism is erroneously missed. Second, estimation sampling when the actual population of pest is estimated with desired levels of precision. It is mainly used in research and also to evaluate the effectiveness of any IPM module or pest control strategy developed to manage the insect pests in the farmers' field. Third, decision sampling based on which the decisions are made when to intervene with management tactics. In this sampling, objective is not to quantify the actual abundance of pest, but to decide the correct timings when control measures should be adopted or not.

3. Estimation of Abundance of Natural Enemies

The importance of natural enemies in regulating the populations of herbivorous insects was recognized much before the concept of IPM was developed. Unfortunately, few IPM programmes at the commercial level attempt to estimate the abundance and impact of these agents on insect pest populations.

The procedure generally employed to study the role of natural enemies, requires gathering different stages of the target organisms from the field, and subsequent emergence of parasitoids is recorded in the laboratory. Unless repeated frequently, this procedure underestimates total mortality since organisms are protected from any mortality factors while in the laboratory. Ovipositional probing and host adult parasitoid feeding is often responsible for more mortality than is caused by developing parasitoids, and is not well estimated except by detailed field observations.

For studying the impact of natural enemies on the rate of increase of pest population, it is essential to undertake field level studies so that potential for and degree of control exerted by the whole complex of parasitoids and predators can be quantified by comparing growth rates under a range of natural enemy levels. There is an urgent need to develop efficient and cost reliable estimation procedures and forecasting models which incorporate the role of natural enemies in the decision making process; experimentation aimed at estimating functional relationships and not just significant differences. The ratios of pests to natural enemies estimated in the field could be used to predict trends in pest populations.

4. Estimation of Yield Loss

One of the objectives of the surveys is to estimate the yield loss due to insect pest species in different areas and type of farming systems. The crop loss estimation holds significance to justify the control measures, which should be taken to manage the insect pest species. In general, surveys to assess the crop loss due to insect pests can be done directly by recording the yield or by recording infestation of pests.

Direct loss surveys. When no information on relationship between different infestation levels of pest and yield is available and the pest distribution and farming systems are quite variable, these types of surveys are adopted to asses the crop loss. The actual crop cutting yields in field with different infestation levels and field with no pest attack are recorded. The correlation between crop yield and degree of damage of infestation is worked out to estimate the loss in the yield. These survey techniques can be used to estimate the crop loss due to different pests over a large area. The experienced people can also evaluate the damage or loss through visual observations while surveying the crops.

Surveys of infestation/pest damage. Damage by pest tends to vary, both geographically and seasonally, as well as through the life of an individual plant. We depend on surveys to assess the severity of damage due to major insect pests. Different parameters are recorded to assess the damage based on the plant part the insect attacks, e.g. dead hearts due to rice stem borer damage, boll damage in case of cotton bollworm complex, leaf foliage damage due to defoliators, etc. The per cent damage is calculated from these surveys, which is used to assess the crop yield loss. It is possible to assess the loss at different stages of the crop even when the crop is in the field. The uniform pest infestation and farming system, and reliable models of yield infestation relationship are the pre-requisites to estimate the loss with more accuracy. The basic information on effect of different levels of insect pests on crop yield is determined through research trials. Based on the data generated, models are developed to estimate the yield loss based on the different damage levels in the field. Through infestation surveys and using the particular models, the yield loss can be evaluated in larger areas. It is possible to further refine the sample survey techniques and to translate the observed infestation level to actual crop loss by further experimentation at research stations. The different techniques adopted for assessing the avoidable crop losses are discussed in chapter 1.

Factors Affecting Survey

The actual number of insect pests present in a particular field is not possible to count. Therefore, survey is conducted to record the approximate pest population in the crop. The expertise of the surveyor is the primary factor on which the efficiency of the survey depends. The surveyor must have thorough knowledge of morphological characters of insect pests, based on which these are identified and biology of the pest. For example, one must be able to differentiate the larvae of spotted, pink, and American bollworms in the field. Some other factors which affect survey are discussed below.

1. Nature of Sample

The mechanics of sampling a given species depends upon its life-history and habits, which determine among other things the best time to sample. When only a single stage is being sampled in an extensive study, it is most important to coincide this operation with the peak of population. The timing of sampling is still more critical in cases where the rate of development is faster. Regular sampling is required throughout the season in case of intensive studies designed for the construction of life-tables.

The time of the day may also affect sample pattern considerably. The insects are known to move from one part of the habitat to another because of their diurnal rhythms, as in the case of Moroccan locust, *Dociostaurus maroccanus* (Thunberg). Many grassland insects move up and down the vegetaion not only because of changes in weather, but also at certain times of the day and night. A majority of the insects may be airborne during the day, as in case of adults of *Pieris brassicae* (Linnaeus) and other butterflies, but there are still others that are active at night, *e.g.* the defoliating beetles. It is thus necessary to resort to sampling at the time when the pest is active.

The developmental stage to be sampled depends on the objectives of the study. If the objective desires only one estimate in a generation, a quiescent and easily approachable stage is often selected. When the objective demands frequent sampling during each generation, it is often necessary to sample active stages that are repeatedly appearing in the population rapidly. If the purpose of sampling is to determine the necessity for taking up control measures, the timing should be such as to give advanced information of an outbreak. The estimation of population levels at any one point in the life-history of an insect would give only the total effect of various factors on all the earlier stages in the generation. The total effect may often be confusing and it is desirable to estimate the population level at as many suitable points in the life-cycle as desirable or practicable.

2. Size of Sample

The sampling unit for its size in population estimation should meet the following criteria.

- The size of sample be such that all units of the universe have an equal chance of selection.
- The sampling unit must be easily delineated in the field.
- The sampling unit should be of such size as to provide a reasonable balance between the variance and cost.

The optinum size should be based on variance components and cost function formulae, the latter being generally expressed in terms of man hours and would include the time required to select and take a sample, and do the counting together with the time spent in moving from one sampling site to another. The determination of the size of the sample unit is not always easy because the most efficient size would also depend on the way in which the sampled organisms are spatially distributed.

3. Number of Samples

The total number of samples to be taken for estimating pest population depends on the degree of precision required. For all practical purposes, an error of 10 per cent of the mean is usually considered as the standard. Accordingly at the chosen probability level, the estimated mean will have the chance of being within 10 per cent of the true mean. Within an ecologically homogeneous habitat, the number of samples required can be worked out by the formula :

$$N = \left(\frac{ts}{D\bar{x}}\right)^2$$

Where N = Number of samples

t = It is a constant, depending upon the number of samples and is obtained from the statistical tables for 'student t' (for more than 10 samples, t approximates to 2 at 5% level)

s = Standard deviation

D = Desired level of accuracy expressed as a decimal (normally 0.1)

$\bar{x}$ = Mean number of individuals per sample.

Generally, a fixed number of samples have to be taken every time the population density needs to be determined. For assessing pest density in relation to need based control measures, we can resort to sequential sampling in which case the total number of samples taken is variable. Very few samples would be required in the case of extremely high or extremely low density populations and this would result in considerable saving of time and effort. For devising such a plan, however, extensive preliminary information is required about the type of distribution pattern and the level of densities that can be tolerated and those that are associated with the economic damage to the crop.

4. Sampling Habitat

The different insect pests survive and feed in various parts of the plants, e.g, leaves, stem, inside the stem, inside fruiting bodies, inside flowers, in soil, etc. Thus the specific habitat must be defined for different pests before starting survey. For example, the survey of white grubs should be done by digging out the soil sample and for whiteflies one should take the samples from leaves of plants. Depending upon the potential damage, the ETLs of various insect pests vary in different habitats, viz. root (rice)-root weevil, *Hydronomidus molitor* Faust (2 grubs/hill); stem (rice)-yellow stem borer, *Scirpophaga incertulas* (Walker) (5% dead hearts or white ears); leaf sheath (rice)-brown planthopper, *Nilaparvata lugens* (Stal) (10 hoppers/hill); leaf (cotton, sucking pest)-whitefly, *Bemisia tabaci* (Gennadius) (6-8 adults/leaf); leaf (groundnut, chewing pest)-tobacco caterpillar, *Spodoptera litura* (Fabricius) (20-25% defoliation); floral form (cotton)-American bollworm, *Helicoverpa armigera* (Hubner) (5-10% infestation); shoot (sorghum)- shoot fly, *Atherigona soccata* Rondani (15% dead hearts); pod (pigeonpea)-pod borer, *H. armigera* (5 eggs or 3 small larvae/plant); fruit (citrus)-fruit fly, *Carpomyia vesuviana* (Costa) (1-2% incidence)

5. Sampling Pattern

To collect samples, the surveyor follows a particular path which is called as sampling pattern. Various sampling patterns are followed to fulfill different objectives, however, the important thing is that the pattern should be unbiased. Several patterns are followed for insect pest monitoring.

Random sampling. This is the simplest and most widely used design and provides an unbiased estimate of the population. The visual selection of sample is difficult and may result in non-randomness. For example, it is a natural tendency of human being to select damaged plant or plant part. Therefore, some pre-determined pattern such as U, X, W, etc. should often be used to walk in a field.

Stratified sampling. When there is great variability in sampled area, the habitat is divided into different strata based on variation with respect to a particular character, from which random samples are taken. This design is powerful alternative to simple random sampling. This technique assures coverage, use sampler's knowledge and gives precision.

Systematic sampling. This pattern of sampling provides the best combination of reliability and cost. The first sample is taken at one reference point and subsequent sampling is done at fixed intervals. A particular route is usually followed through the growing area to collect the samples.

6. Type of Sampling

To estimate the insect population in a particular area, sampling is done from a limited area. The sampling techniques have already been discussed in previous section. However, different types of sampling are described herein.

Sequential sampling. An increasingly important type of sampling programme in IPM is the sequential sampling or sequenctial decision programme. Such programmes are based on insect dispersion pattern and economic decision levels. In this type of sampling, the total number of samples are variable and depend on the population density and distribution pattern. A sequential sampling plan (SSP) requires only a few samples to be taken at very high or low populations and hence expenditure in terms of time and efforts is minimized. More samples are required when pest population density is near the dividing line between endemic and outbreak levels so that a management decision can be taken confidently. Besides reducing the cost of sampling, sequential sampling has also helped to reduce insecticide use by 25-35 per cent for the control of pink bollworm on cotton. However, the utility of the method diminishes as the number of species or age classes being monitored increases. The greater this number the more likely are one or more species to be close to the action threshold requiring that the scouts or farmers sample upto an upper sample number limit.

Variable intensity sampling. Variable intensity sampling (VIS) provides either a classification of density or an estimate of density if the density is within a fixed distance from threshold value. It is designed to ensure a representative sample throughout the sample area and sampling is done more intensely when mean pest population is close to economic threshold level. Because it is designed to meet both objectives, it is very appealing and when possible should be more widely used. The biggest drawback of VIS is that it is more difficult to implement than other sampling procedures because the decision to continue sampling is not simply yes or no; instead the number of samples to be taken at a particular sampling location in a field is a function of the cumulative number of samples and the current estimate of mean density. This generally requires the use of programmable calculator or computer.

Double sampling. Double sampling is occasionally more convenient and useful than sequential sampling. The basic principle is to take an initial sample and use the information so obtained to decide about the size of a second sample, if required.

Binomial sampling. Sampling can sometimes be made easier and less time-consuming by substituting binomial counts for complete counts. Binomial sampling is found on defining a relationship between the density of organisms (m) per sample unit and the proportion of sample unit with more than T organisms ($1 - P_T$), where T = 0, 1, 2. Historically, T = 0 has been used most often, but there are compelling reasons to use some other tally threshold. Binomial sampling is the most economical and feasible field sampling method for many organisms. The most critical aspect of binomial sampling is the knowledge of the formulae relating to the binomial proportion (P_T) obtained from field sampling to the mean density (m).

PEST FORECASTING

Pest forecasting is the perception of future activity of biotic agents, which would adversely affect crop production. In other words, it is the prediction of severity of pest population which can cause economic damage to the crop. The systematically recorded data on pest population or damage over a long period of time along with other variable factors, which affect the development of pest, may be helpful in forecasting the pest incidence. The prediction of a particular pest depends upon characteristics/ biology of a pest and the meteorological factors. These meteorological factors may affect the pest either directly affecting their survival, development, reproduction, emergence and behaviour, or indirectly by their action on host plants or on natural enemies. These factors also determine the geographical limits of distribution and the time of appearance and abundance of pests. The forecasting of pests guides the farmers about the timing and biology of insect incidence, and to eliminate blanket applications, reduce pesticide amounts, and achieve quality results. The farmers can take to timely action of applying various pest control measures to harvest maximum returns. Several studies are required to generate the basic information, which is required to develop forecasting models (Mahal *et al.,* 2012).

(*i*) **Quantitative seasonal studies.** Using appropriate sampling techniques, the pest abundance must be studied over several years along with seasonal range, variability in number and distribution. The seasonal counts in relation to climate and topography need to be provided.

(*ii*) **Life-history studies.** The detailed bioecology of pest under a range of temperature, humidity, etc. should be known. The duration of different instars, number of generations, survival rate, amount of food eaten, overwintering, host range, number of eggs laid, etc and other parameters can be studied in laboratory.

(*iii*) **Ecological studies.** Life-table studies of pest are important for better understanding of pest population build-up, natural mortality factors, intrinsic growth rate, etc. Life-table of a pest can be helpful in finding mating and emergence period which are quite useful for predicting population dynamics of the pest. The migration and immigration of pests can also be used for forecasting of pests.

(*iv*) **Field studies.** Climatic factors not only affect the pest abundance but also affect the natural enemy population which is an important natural factor in controlling pest population. In field situations, the natural enemy abundance under a range of temperature and humidity should be studied. The other cultural practices like fertilizer application, irrigation, plant spacing, etc affect the crop phenology which directly influence the population build-up of a pest.

Types of Pest Forecasting

Pest forecasting may be divided into two categories, viz., short-term forecasting and long-term forecasting.

A. Short-term Forecasting

The short-term forecasts are often based on current or recent past conditions that form a basis for, or an enhancement to, the forecast. These may cover a particular season or one or two successive seasons only. The pest population is sampled from a particular area within a crop using appropriate sampling technique and the relationship is established between weather data and progress in pest infestation. The laboratory studies on the effect of temperature on emergence and egg laying can be used to forecast the pest situation in the field. The short term forecasting can be completely empirical, such as use of environmental cues reported from Japan, where the date of first blooming of cherry blossom and the mean March temperature were used to predict the peak emergence of rice stem borer. Based on multiple regressions, short term forecasting of wheat grain aphid, *Sitobion avenae* (Fabricius), has been done. The peak population density on each field was positively correlated with the population densities at the end of ear emergence, mid-anthesis and the end of anthesis. Based on two counts on the crop, the accuracy increased from ear emergence to the end of anthesis, however, the forecast at mid-anthesis of peak density was much more accurate.

B. Long-term Forecasting

These forecasts are based on possible effect of weather on the pest population and cover a large area. The data are recorded over a number of years on wide seasonal range and from different areas. Long-term forecasting is based on knowledge of the major aspects of the pest insect's life-cycle, and of how it is regulated. The data recorded are analyzed and models are developed based on the available information. The models help in forecasting pest population in various geographical areas based on common weather parameters. Long-term population forecast based on Markov chain theory was developed for effective management strategies for *Nilaparvata lugens* (Stal) and *Sogatella furcifera* (Horvath). A transition probability matrix of 5-yr steps of Markov chain theory was constructed based on 31-yr light-trapping data of the two pests from 1977 to 2007 in Jiangkou County, Guizhou, China. The models accurately forecasted field occurrence in 2008 in Jinangkou County for both species. This model is an effective method for long-term population forecasting of *N. lugens* and *S. furcifera,* and thus provides plant protection agencies and organizations with valuable information in implementing appropriate management strategies. Long-term forecasting of brown and whitebacked planthoppers in Japan was based on the assumption that both the hoppers overwinter as diapausing eggs on winter grasses. After it was discovered that the brown planthopper migrates in Japan from outside, the short-term forecasting was adopted. In China, long-term forecasting of the peak day of immigration, adult numbers caught on the peak day and accumulated adult numbers in the immigrant generation of rice leaf roller, *Cnaphalocrocis medinalis* Guenee, was done, one year ahead of practical occurrence. In Kenya, data of 23 years of rainfall was correlated with the number of outbreaks of armyworm, *Spodoptera exempta* (Walker), in the following season. Based on these data, local or countrywide forecasting of the outbreaks of this pest was done.

Methods of Forecasting

Pest forecasting has generally based on environmental factors, climatic areas and empirical observations.

Environmental factors. The population development of a particular pest mainly depends on the favourable environmental conditions available in a particular geographical region. The pest attack occurs in epidemic form only when the favourable environmental conditions for multiplication of pest prevail for longer duration. Therefore, the factors responsible for environmental conditions are the major criteria on the basis of which the forecasting can be done

The sugarcane pyrilla, *Pyrilla perpusilla* (Walker), outbreak is predicted based on high temperature during monsoon. The population per 30 plants (Y) is predicted based on the mean maximum temperature (X) of week preceding the data of observation of field population. It is given as

$$Y = (3.47536 \times 10^{-28})\ (33.4965)^{X}\ .\ (0.9552)^{X}$$

Every insect requires a consistent amount of heat accumulation to reach certain life stages, such as egg hatch or adult flight, which can be interpreted in terms of degree days. One degree day is an accumulation of heat units above some threshold temperature for a 24 hour period. Degree days (often referred to as "growing degree days") are accurate because insects have a predictable development pattern based on heat accumulation. Insects are exothermic (cold-blooded) and their body temperature and growth are affected by their surrounding temperature. Biological development of insects over time in correlation to accumulated degree days has been studied, discovering information on key physiological events, such as egg hatch, adult flight, etc. There is a threshold temperature for each insect; for example, 48ºF for the alfalfa weevil, *Hypera postica* (Gyllenhal). No development occurs when temperatures are below that level. Insects have an optimum temperature range in which they will grow rapidly. Then, there is maximum temperature (termed upper cutoff) above which development stops. These values can be used in predicting insect activity and appearance of symptoms during the growing season. Therefore, the degree days would be useful in pest management programme to time the scouting of insect pests. This predictive information is known as an insect model. Models have been developed for a number of insect pests.

$$\text{Degree days} = \frac{\text{Maximum temperature} + \text{Minimum temperature}}{2} - \text{Development threshold}$$

As an example, codling moth, *Cydia pomonella* (Linnaeus), pheromone monitoring traps are placed in the apple orchard at 100 degree days after March 1 in northern Utah to determine initiation of adult moth flight.

A temperature range of 50° to 85°F is most comfortable for European corn borer, *Ostrinia nubilalis* (Hubner). Below 50°F, it will not develop, and above 85°F, development will slow dramatically. A degree day for European corn borer is one of degrees above 50°F over a 24-hour period. For example, if the average temperature for a 24-hour period was 70°F, then 20 degree-days would have accumulated (70 - 50 = 20) on that day. These accumulations can be used to predict when corn borers will pupate, emerge as adults, lay eggs, and hatch as larvae.

A forecasting model of spring emergence of *Carposina sasakii* Matsumura in apple orchards in Korea was constructed based on degree-days. The two peaks for adult spring emergence were recorded, first major peak in late June and the second smaller peak in late July. A bimodal distribution model was developed to describe this emergence pattern. The bimodal model predicted more accurately *C. sasakii* spring emergence times than the Weibull model.

The generation time of *C. pomonella* populations was predicted using a degree day model. The model was developed after studying 176 generations in walnuts, apples, pears, and other hosts at several locations throughout California. Out of five models, a degree-day model using 10°C as the lower threshold, 31.1°C as the upper threshold (horizontal cutoff), and a generation time of 619 DD provided an adequate fit.

In Japan, three generations of rice leaf roller, recently emerged as serious pest, were recorded in each season. It was estimated that the second generation requires 210 degree-days for development, whereas third generation requires 300 degree-days. The number of degree-days required for development of many insect pests is known in several different countries like Canada, USA, Japan and Europe.

Forecasting of pest emergence, epidemics, etc has also been reported based on other environmental factors like humidity, rainfall and sunlight. In Tanzania, outbreaks of red locust, *Nomadacris septemfasciata* (Serville), have been forecast from an index of the previous year's rainfall. Severity of *Spodoptera exempta* (Walker) outbreak seasons was highly dependent upon the number of rainy days during November in central Tanzania. In China, the econometric analyses showed that rise and fall of mirids are largely related to local temperature and rainfall.

Observations of climatic areas. The distribution of insects throughout the world is based on evolutionary history which includes main important factor, i.e. climate of the geographical region. There are three distinct zones of abundance of each insect species.

Zone of natural abundance (endemic). In this zone, the pest species is often in large number, regularly breeds and is a regular pest of some importance. The climate conditions are most favourbale for its development and pest is seen all the time.

Zone of occasional abundance. The insect species emerge in epidemic occasionally in this zone because the climatic conditions are either less suitable or the suitable conditions exist only for a short period of time followed by unsuitable conditions. Sometimes, the climatic is severe to destroy the entire population, which is then re-established by dispersal from zone of natural abundance.

Zone of possible abundance. The pest species in this zone can be seen only after migration from zone of natural and occasional abundance outbreaks. The climatic conditions are drastic for their breeding and development. The population is destroyed by the severe climatic conditions within a short period of time. Three different regions Orlando, Naples and Ankara corresponding to zone of natural abundance, occasional abundance and possible abundance, respectively are known for Mediterranean fruit fly, *Ceratitis capitata* (Wiedemann).

The observation on the climatic areas where critical infestations are likely to occur can be predicted for some insects. Combination of climatic factors like temperature, rainfall, humidity, etc. existing in a geographical region gives an indication of possibility of establishment of pest in that region. The other factors like biotic and topography may also be used for prediction of insect pests.

Empirical observations. This type of pest forecasting is based on estimating the number of insects available during a particular time. In other words, it is nothing but the sampling of insect or monitoring of pest population. It involves forecasting the population in the next season by counting the pest in the previous seasons. In many cases, the number of pests in the early part of cropping season will give an indication as to the extent of its likely multiplication in the season. From the counting of immature stages of insects, approximate estimations of later stages can be made. For example, in UK taking soil cores for insect eggs of carrot fly, *Psila rosae* (Fabricius) and cabbage root fly, *Delia radicum* (Linnaeus), is successful for estimating the later population of root maggots. The adult catch in the traps especially pheromone traps can be used to estimate the approximate abundance of pest population later in the season. The sampling of insect pest on alternate host/weeds during non-availability of main crop can be quite useful to forecast the pest population development in the coming season, e.g. counting overwintering eggs of blackbean aphid, *Aphis fabae* Scopoli, on spindle trees helps in estimating the aphid population on peach-potato crop. In many lepidopteran species, pest forecasting is based on estimating the number of eggs and young larvae on the crop, e.g. cotton bollworms, stem borers, pulse moths, etc.

SYSTEMS ANALYSIS AND MODELLING

In general, systems analysis is defined as a systematic approach to help a decision maker to choose a course of action by investigating the full problem, searching out objectives and alternatives, and comprising them in the light of their consequences, using an appropriate framework-in so far as possible analytic- to bring expert judgement and intuition to bear on the problem. It is a holistic approach to the evaluation of natural, dynamic systems (such as host-pest system) under "normal" or imposed conditions.

For insect pest management, the systems analysis focuses on decision making for intervening the pest control tactic taking into account the pest ecosystem. It involves the detailed studies and data collection on various biotic and abiotic factors. Therefore, experimentation is an integral part of systems analysis through which data are generated. A series of controlled experiments over a period of time may produce complete description of a natural system or pest ecosystem. Both qualitative and quantitative techniques are applied for understanding the crop-pest system and their relationship with management practices.

The systems analysis consists of mathematical, statistical, and mechanical techniques for analyzing the pest ecosystem, based on which various pest models have been developed. Models are conceptual or mathematical devices that aim to describe or stimulate natural processes. They can be used to predict the outcome of hypothetical eventualities and as management tools to predict or establish the optimal tactics required to achieve a particular result within the constraints of a model. The population models are useful for developing appropriate pest management strategies such as optimal timing of insecticide applications. The pest modelling has become a key component of most of the systems analyses. Models have been categorized by various workers based on their intended function, e.g. an insect dispersion model, a yield-loss model, management model, etc; tactical, strategic, or policy models; ecological or economic models. The most commonly used classification of models is statistical, analytical, simulation and optimization models (Dhaliwal *et al.,* 2013).

DATABASE MANAGEMENT AND COMPUTER PROGRAMMING

An organized collection of data for one or more purposes usually in digital form is known as database. The use, creation and maintenance of the database using software packages with computer programmes is called as database management system (DBMS). During 1960s with evolution of computers, this concept of 'database' has emerged to manage the large, complex and diverse data with ease to designing, building and maintaining complex information, which is otherwise quite difficult. It has evolved together with database management systems which enable the effective handling of databases. Though the terms database and DBMS define different entities, they are inseparable, as database's properties are determined by its supporting DBMS and vice-versa.

Computer programming is the process of designing, writing, testing, debugging, and maintaining the source code of computer programmes. It is considered as one phase of software development. It involves writing source code in programming languages with an aim to create a set of instructions that computers use to perform specific operations or to exhibit desired behaviours. In precise way to understand the three terms, comprehensive collection of data in organized way in computer (database), is coded by professional in different languages (computer programming) to develop a software through which the data can be manipulated, maintained and used for decision making (database management system). The acronym "DBMS" is universally understood with information technology (IT).

The DBMS has become an important part of today's modern life as all the information can be assessed in any discipline. In the field of biology especially insect pest management, DBMS is also being used extensively to address the problems from deciphering genetic codes to decision making by farmers. An elegant example of application of DBMS is collection of data on genetic codes for all the organisms at one platform, i.e. National Centre for Biotechnology Information. The data generated by different workers from different geographical locations in the world is submitted to a common platform, processed though computer programming and published online (http://www.ncbi.nlm.nih.gov/) which is accessible to all the research personnel. Another outstanding database being developed by collaborative efforts of worldover scientists is DNA barcode of all living species. The nucleotide sequence of the mitochondrial cytochrome oxidase 1 (COX1) region in animal DNA, developed for all animal species is submitted in Barcode of Life Data Systems (BOLD), which is an online workbench that aids collection, management, analysis, and use of DNA barcodes. The database generated in the system will be useful in future for precise identification of any insect/animal species based on cytochrome oxidase 1 gene sequences.

Database management has been widely used in pest identification, managing pest monitoring data, population modelling, exploring control strategies, biology of pests and decision support system to take decision for adopting intervention for pest management in field which are important in IPM. The database has been developed in the form of CD-ROM, internet based decision support systems, expert systems, etc. For example, the data on biology of various pests of one crop like cotton along with photographs of all stages of insects and damaging symptoms are programmed and developed into software and distributed among the farmers on CD ROM for their direct use, which will enhance their knowledge about the pests in cotton crop. Due to rapid adoption of internet, most of DBMS permits applications through web-based development. "Database of IPM Resources" (DIR), a key component of IPMnet has been developed to reduce the gap between accumulated information and its accessibility to users. DIR is a continuously updated and expanding information management exchange, retrieval, and referral system for global IPM information resources accessible through the internet.

The National Centre for Integrated Pest Management (NCIPM), New Delhi has developed database management system (www.ncipm.org.in/agroweb) where the information is available on pest, disease management, nutritional deficiency and physiological disorder in different crops being grown in India. Based on electronic processing of the data, Decision Support System (DSS) has been designed. A DSS is an interactive system that helps decision-makers utilize data and models to solve unstructured or semi-structured problems. In agriculture, these systems have been designed to address complex tasks. In IPM, DSSs are widely used for identification, recommendation of insecticides, emergence of pest outbreaks, etc. In 2010, a web-based decision support system was developed for Washington tree fruit growers that integrates environmental data, predictions for ten insects, four diseases and a pesticide database that provides information on non-target impacts on

Table 3.1 Different decision support systems developed in various parts of the world

Decision Support System	Crop/ Commodity	Target pest	Country	Function
EntomoLOGIC	Cotton	*Helicoverpa* spp., two-spotted spider mite, *Tetranychus urticae* Koch	Australia	Predicting future pest numbers
Stored Grain Advisor Pro (SGA Pro)	Wheat grains	Various coleoptern pests	USA	Predicting future risks
proPlant	Oilseed rape	Different pests	Europe	Predicts the start of pest infestation and provides selection, date and rate of application of suitable chemicals.
Southern Pine Beetle DSS (SPBDSS)	Pine forest	Southern pine beetle, *Dendroctonus frontalis* Zimmermann	USA	Decision making about IPM
SIMLEP DSS	Potato	Colorado potato beetle, *Leptinotarsa decemlineata* (Say)	Europe	Forecasted the first occurrence of young and old larvae
Spruce Budworm Decision Support System (SBWDSS)	Forests	Spruce budworm, *Choristoneura fumiferana* (Clemens)	Canada	Assess the effects of SBW outbreak

Source: Dhaliwal *et al.* (2013)

other pests and natural enemies. A spatial decision support system (sDSS) MedCila was developed in Israel for controlling Mediterranean fruitfly, *C. capitata*, in citrus in 2004. The number of flies and presence of a 'blue eye' in the nearest trap, the host-species susceptibility, the relative development of the Medfly based on accumulative day-degree model, the history of trapping, and the Medfly population in the nearby traps are factors on which the sDSS is based. The recommendations of MedCila are generally accepted and reduce the unnecessary sprays. A number of decision support systems have been developed in different countries on different crops **(Table 3.1)**. A general DSS website (www.dssresources.com) provides information on basic concepts, development, deployment, and evaluation of DSSs.

Another computer programme based on database management is Expert Systems (ES). It is a problem-solving computer programme that achieves good performance in a specialized problem domain that is considered difficult and requires specialized knowledge and skill. Though both DSS and ES seek to improve the quality of the decision, these are distinguished based on objectives and intents, operational differences, users, and development methodology. The different Expert Systems (ESs) have been designed which help in identification of insect pests, estimating risk from pests, control measure recommendations, etc **(Table 3.2)**. The comprehensive database once generated requires little efforts to maintain it. If the database is maintained, it can be useful for various studies over a number of years.

Table 3.2 Some important expert systems developed in various parts of the world

Expert System	Crop/ Commodity	Target pest	Country	Function
GyMEs (Gypsy Moth Expert System)	Different plant species in forests	Gypsy moth, *Lymantria dispar* (Linnaeus)	North America	Estimate risk to forest from pest
PIES (Potato Insect Expert System)	Potato	Colorado potato beetle, *Leptinotarsa decemlineata* (Say)	Virginia	Determine when the insecticide is to be applied
TEAPEST	Tea	Pests of tea	India	Identify and suggest appropriate control measures
SOYPEST (Soybean Pest Expert System)	Soybean	120 pests	India	Identification and decision in IPM
SMARTSOY	Soybean	Different insect pests	USA	Recommend insect pest management
Qpais	Stored grains	150 species of quarantine stored insect pests	China	Identification of quarantine stored insect pests
Expert System for BPH	Rice	Brown planhopper, *Nilaparvata lugens* (Stal)	China	Identify infestation and recommend insecticides
Expert System for Bulb Fly	Wheat	Bulb fly, *Delia coarctata* (Fallen)	UK	Determine egg number, control effectiveness
TBIS	Fruits and vegetables	Tephritid fruit flies	China	Identification based on DNA barcode
BOLD	Various crops	All insect pests	All countries	Identification based on DNA barcode

Source : Dhaliwal *et al.* (2013)

SIMULATION TECHNIQUES

Simulation is the imitative representation of the functioning of one system or process by means of functioning of another. In other words, it is the act of imitating the behaviour of some situation or some process by means of something suitably analogous. In different domains like computing, health, law, administration, aerospace, environment, military, physics, religion, statistics, etc., special definitions of simulation have been quoted. The simulation techniques are used to design a model for a real system, which provides users with the approximated behaviour of that real system.

A Monte Carlo simulation technique, based on rates of insect development, has been produced for forecasting the timing of attack by pest insects. The technique was successful because it uses a fixed number of individuals from one generation to the next and simulates the timing of events rather than the population dynamics of the insects. The effectiveness of this technique has been demonstrated for cabbage root fly, *Delia radicum* (Linnaeus); the carrot fly, *Psila rosae* (Fabricius); the bronzed blossom beetle, *Meligethes aeneus* (Fabricius) and the large narcissus fly, *Merodon equestris* Fabricius. The simulated insect defoliation technique can be used to understand the insect injury-yield loss relationships. Understanding the mechanisms underlying yield loss is essential for better explanations of insect-plant interactions and for developing economic injury levels (EILs) for practical use in pest management. In soybean, manual sequential defoliation based on insect consumption model was done to compare the EIL level in three different cultivars, viz. Dunbar, Corsica, and Clark. A significant yield reduction (15-70%) in all cultivars was reported due to defoliation. The significant EIL difference was also observed among three cultivars which suggested the revision of existing EIL for defoliation pests.

CONCLUSIONS

Insect populations in nature are regulated by various biotic and abiotic factors. The estimation of pest populations and their economic impact is essential for developing suitable pest management strategies. Although a number of techniques have been developed for pest surveillance, there is need to further refine and define some of the techniques so that these are easily adopted by farmers. The knowledge about bioecology and identification based on morphological characters are pre-requisites for making reliable survey, based on which decisions can be made. The precise and reliable surveillance holds significance for prediction of pest population, which is essential to advise the farmers to adopt suitable control measures. A number of pest forecasting methods based on environmental factors, climatic zones and empirical observations have been devised. Recent advancements in systems analysis consisting of mathematical, statistical and mechanical techniques for analyzing the pest ecosystem have enabled us to develop several pest forecasting models. The various decision support systems (DSS) and expert system (ES) are providing quite useful information for enhancing awareness among farmers about insect pests of crops, requirements of crops, taking decisions to initiate control measures, etc. A close collaboration between computer programmers and plant protection scientists is essential to develop farmer friendly systems. Further advances in information technology and computer programming would lead to precision in pest forecasting and develop sustainable pest management programmes in the twenty-first century.

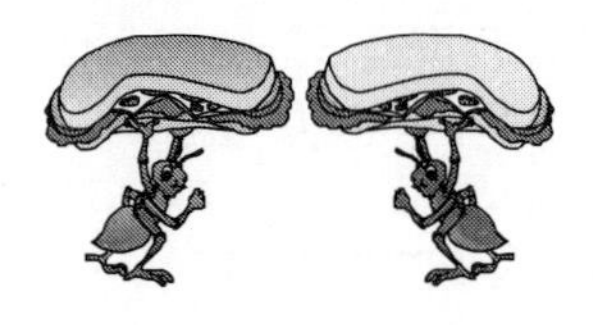

PART-II

PEST MANAGEMENT TACTICS

- Legislative, Cultural and Mechanical Control
- Biological Control
- Botanical Pest Control
- Chemical Control
- Biorational Approaches
- Host Plant Resistance
- Genetic Control
- Biotechnological Approaches
- Integrated Pest Management

LEGISLATIVE, CULTURAL AND MECHANICAL CONTROL

INTRODUCTION

For thousands of years, man could do nothing about the pests except to appeal to the power of magic and a variety of Gods. For the most part, early humans had to live with and tolerate the ravages of insects and pathogens, but gradually learned to improve their condition through trial and error experiences. These improvements led to the beginning of pest control. Prior to the emergence of crop protection sciences and even before the broad outlines of the biology of pests were understood, humans evolved many cultural and physical control practices for protection of their crops. Some of these control practices are still valid and useful to-day.

Historically, cultural practices were the farmer's most important method of preventing crop losses. The first reference to use of cultural practices for insect control in India is found in the book *The Agricultural Pests of India and of Eastern and Southern Asia*, by Balfour (1887). He advocated the use of some cultural practices like crop rotation involving cereals and pulses, and clean cultivation for minimising damage by insect pests. Maxwell Lefroy (1906) suggested some other practices like mixed cropping, use of trap crops, hoeing, etc., in addition to clean cultivation. As early as 1911, removal of cotton sticks by 1st of August every year was made compulsory by law and this measure helped in reducing the pink bollworm incidence in the erstwhile Madras State.

LEGISLATIVE MEASURES

Legislative control involves enactment of laws to regulate the entry, establishment and spread of pests. Man has often realized that when insects and other animals appear as pests in epidemic form, they become formidable enemies and in most cases, cannot be controlled with individual effort. When a pest like the desert locust appears in a country, it becomes a national calamity. The entire population and the government of the country have to make a united effort to meet the threat. The massive operations required have to have the sanction of the society through legislation and resources have to be mobilized for financial backing.

Legislation is also imperative to stop the accidental entry, from outside the country, of certain pests, which may not be present in that country. Discipline must be enforced among citizens not to bring in certain prohibited material which they might attempt because of ignorance of the danger involved or because of sheer temptation. The legislation is of four kinds :

(*i*) Legislation for foreign quarantine to prevent the introduction of new pests from abroad.

(*ii*) Legislation for domestic quarantine to prevent the spread of established pests within the country or within a particular state.

(*iii*) Legislation for notified campaigns of control against pests.

(*iv*) Legislation to prevent the adulteration and mishandling of insecticides or other devices used for the control of pests.

The history of pest-control legislation in India is very interesting. The first Act in this country was passed in 1906 under the Sea Customs Act of 1878 to stop the entry of the Mexican cotton boll weevil. It was followed by the present Destructive Insects and Pests Act No.II passed on 3rd February 1914 at the instigation of Bombay Chamber of Commerce. Thus, provisions were made for preventing the entry of foreign insect pests, like the American cotton boll weevil, *Anthonomus grandis* Boheman, and others that might be harboured in agricultural products. The entry of such plant material as sugarcane and unginned cotton was banned totally. Plants and seeds, in general, could be imported under certificates of health through notified ports where they could be inspected by qualified staff, fumigated properly and kept in quarantine before being released. There, they could be examined for the presence of eggs of pests or for the presence of bacteria, fungi and other pathogens, which are normally not killed by fumigation. The import of useful insects such as parasites was permitted only to scientific institutions.

According to various amendments of the Government of India Act, 1914, provisions were made for adopting control measures against local and exotic pests in centrally administered territories and states after first obtaining sanction from the President of India or from the Governor. Before control measures could be adopted by the state, it was considered necessary to (*i*) declare the organism to be injurious, (*ii*) place the infested area under quarantine, and (*iii*) ensure that preventive and remedial measures were prescribed. In this Act, provisions were also made for the state governments to pass their own legislation for adopting remedial measures. Thus, the East Punjab Agricultural Pests, Diseases and Noxious Weeds Act was passed in 1949. Other states have passed similar legislation.

According to notification No. 1581, dated 1st October, 1931, under the Destructive Insects and Pests Act, 1914, provisions were made to restrict the imports of cotton from America and the West Indies, and to bring air freight within the scope of the Act. The import of cotton was allowed, provided it was fumigated and disinfected at the port of entry. The import of plant material by air was prohibited except under strict quarantine for scientific purposes, with prior permission of the Government of India. Similarly, fruits and vegetables from Afghanistan were allowed to be imported, provided they were accompanied with certificates of health issued by a competent authority. The import of seeds by air was exempted, but the entry of some of the seeds, namely cotton, Egyptian clover, flax, rubber, coffee and sunflower was banned. The import of potato seed from areas infested with golden nematode and wart disease was prohibited.

Under this Act, notifications were also issued from time to time for the inspection, disinfection or destruction of plant material to be transported by road or by rail within the country. This checked the further spread of pests and diseases that might be present in certain plant parts. A number of such plants and plant material harbouring various pests were listed and the station masters of railway stations and other inspection authorities were given legal powers. The owners causing hindrance to the inspection of material at the destination were liable to be prosecuted and punished with fine up to Rs 1000. These provisions were made to restrict the spread of San Jose scale, *Quadraspidiotus perniciosus* (Comstock); cottony cushion scale, *Icerya purchasi* Maskell, the potato-wart disease, water hyacinth, lantana, spike disease of the sandal tree and the bunchy-top disease of banana.

According to the provision of the East Punjab Agricultural Pests, Diseases and Noxious Weeds Act 1949 (and other similar State Acts), the state government could enforce, when necessary, control measures for the eradication of pests, diseases or weeds, such as locust and grasshoppers, hairy caterpillars, rats, pyrilla and the Gurdaspur borer of sugarcane, *Acigona steniellus* (Hampson), ergot of pearl-millet, *pohli* (*Carthamus oxycantha*), water hyacinth (*Eichhornia crassipes*) and other weeds.

The Government would issue a notification, declaring the pest or weed to be noxious, specifying the area and the period for which it would remain in force, giving proper directions for carrying out control measures, and for adopting other practices against the introduction, reappearance or spread of the pest, disease or the weed concerned. The sowing of a particular crop for a given period could also be prohibited. The farmers of the notified area would be bound to carry out the preventive or remedial measures. If, on inspection, it was found that a given farmer did not carry out these instructions and the pest was still present, the Inspector would issue a notice that the operations must be carried out within a specified period, failing which the Inspector with his staff would perform this job at the expense of the farmer. The amount involved would be recoverable as arrears of land revenue. It would appear that the procedures laid down were lengthy and it would be rarely possible to enforce, in case of default, the measures recommended during a crop season. For that resaon, the implementation of the Act has not been very effective and *pohli*, for example, remained in the fields after the harvest of wheat fields throughout North India. It was only because of double cropping and intensive agriculture since 1965 that this weed gradually disappeared in Punjab and other states.

In other cases where free pesticides were supplied for the control of pests, the success was greater. Anti-rat campaigns have been carried out by the Department of Agriculture and the Community Development Organization, and the cutting of sugarcane tops for the control of the Gurdaspur borer has been adopted in the factory zones (10 km of sugar factory) of Punjab and Haryana with considerable success.

It has been noticed that when the entire government machinery and the public were mobilized and the effective instruments, such as the dusting machines and insecticides, were made available to combat locust swarms, the operations were very successful. It is to the credit of the Governments of Punjab and Haryana, and to the rural farmers that, since 1963-64, the locust has not been allowed to breed and not a single swarm originated from these areas, although conditions have been favourable for its breeding.

It is only natural to believe that in spite of best quarantine measures to restrict the import of new pests from abroad or to restrict the spread of an organism from one part of the country to another, new problems will continue to appear and, therefore, notifications or amendments to the 1914 Act will be needed from time to time. The East Punjab Act 1949 and other similar Acts of various states also need amendments to make the Governments effective in their implementation.

The unadulterated sale of pesticides under proper labels, giving instructions for their safe usage, also needs the attention of the Govenment. In the absence of any legislation, certain pesticides were brought under the purview of the Poisons Act 1919, and the Drugs Act 1940. However, in view of the fast expanding pesticide industry and the widespread use of these highly poisonous chemicals, more specific provisions were needed to safeguard public interests and even human life. Consequently, a special legislation to regulate the manufacture, transport, handling, sale and use of pesticides was envisaged and a new Bill, namely the Insecticide Bill was passed by the Indian Parliament in September 1968. Under this Act, provisions have been made for the compulsory registration of pesticides and for the establishment of pesticide laboratories to carry out various functions under the legislation. Earlier, the Prevention of Food Adulteration (PFA) Act, 1954 was enacted on September 1954, for prevention of adulteration of food. Also, Pesticides Management Bill, 2008 is currently under active consideration of the Government of India.

I. Insecticides Act, 1968

The Insecticides Act was passed by Parliament on September 2, 1968 to regulate the import, manufacture, sale, transport, distribution and use of insecticides with a view to prevent risks to human beings or animals, and for matters connected therewith. The Act was enforced throughout the country on August 1, 1971 and the rules were framed thereunder on October 30, 1971.

The Insecticides Act, 1968 was amended in 1972 and 1977 to overcome teething troubles in implementation of the Act. However, several practical difficulties were experienced during administration and implementation of the Act during the last three decades. Consequently, a bill was introduced in the Parliament and several amendments to the Insecticides Act, 1968 were notified in the Gazette of India on August 7, 2000.

The term 'insecticide' has been defined under the Act to include any substance in the Schedule or such other substances or preparations intended for the purposes of preventing, destroying, repelling or mitigating any insects, rodents, fungi, weeds and other forms of plant and animal life not useful to human beings. The Central Government on the recommendation and approval of the Central Insecticides Board can include substances in the Schedule from time to time. Applications for the registration of insecticides can be made for those chemicals, which are included in the Insecticides Schedule.

The Insecticides Act has been enforced in the country to achieve some of the under mentioned objectives :

- To register only safe and efficacious pesticides.
- To ensure that the farmers/users get quality product for controlling the pests.
- To prescribe usages of pesticides both from ground and air, and also important precautions for their handling and use.
- To minimise health hazards from the pesticide residues through contaminated food, water and air.
- To ensure that the pesticide industry manufacture, transport, distribute, store and sell the pesticides as per the prescribed regulations, failing which legal action is taken.
- To ensure that the pesticides are properly packed and labelled to avoid any leakage of the hazardous pesticides in transit and to provide enough instructions for their safe handling and use.

In order to achieve these objectives, the Government of India and state governments have created suitable machinery both at the Centre and the State-level for implementing this Act. Under this Act, both the Central and the State Governments have powers to make rules in consultation with the Central Insecticides Board in the manner prescribed. Some of the most important functionaries prescribed and the regulatory procedures laid down under this Act are as follows:

Central Bodies and Laboratories

(*i*) **Central Insecticides Board.** This Board advises the Central and State Governments in all technical matters such as manufacture, formulation, storage, transport, distribution, sale, and safe use of insecticides. The Board also fixes tolerance limits for insecticide residues, safety period and shelf life, based on the data provided by the manufacturers and also based on research findings of the scientists in the country. The Board has 29 members representing various Ministries and Departments of the Government of India, and some specialists. The Director General of Health Services is the Chairman of the Board.

(*ii*) **Registration Committee.** This Committee consists of a Chairman, and not more than five persons from amongst the members of the Central Insecticides Board, including the Drugs Controller and the Plant Protection Advisor. The person appointed as Secretary by the Central Government functions as Secretary of the Board and of this Committee. The Secretary is assisted by seven technical officers namely, an Entomologist, a Plant Pathologist, an Agronomist, a Medical Toxicologist, a Senior Chemist, a Packaging Engineer, and a Law Officer and other ministerial staff.

Before these can be sold in India, all insecticides have to be approved and registered by this Committee. Registration is done after scrutinizing and verifying the formulae and their efficacy, and safety to human beings and animals. In addition, the Committee may also specify the precautions to be taken against poisoning by the use and handling of the particular insecticide.

(*iii*) **Other Committees.** The Board has constituted the following Committees and Expert Panels:

(*a*) Pesticides Environmental Pollution Advisory Committee, to advise the Board on all hazards emanating from the use of pesticides for agriculture, public health, grain storage or for any other purpose. The Committee has 20 members including the Chairman.

(*b*) Six different Expert Panels have been constituted to finalise the approved usages of insecticides (organophosphates, carbamates and organochlo-rinated hydrocarbons), fungicides, herbicides and fumigants, and rodenticides. The manufacturers of pesticides are obliged to give these usages and bio-effective qualities on leaflets.

Registration of Insecticides

A prospective manufacturer or an importer of a given insecticide, first applies to the Secretary, Registration Committee and Central Insecticides Board for consideration and acceptance of the product by the Registration Committee. The Committee, after making enquiries on the claims of the manufacturer or the importer as to the effectiveness and safety of the insecticide to human beings and animals, registers the insecticide on payment of prescribed fee and allots its registration number and issues a Certificate of Registration. This certificate is given within a period of 12 months from the date of receipt of the complete application. This period can be further extended to a maximum of six months when it is to be decided to register or reject application.

The general policy of the Government of India is that the insecticides approved should, as far as possible, be manufactured in the country. The new insecticides which are introduced in the country for the first time are generally registered provisionally for import and formulations for a period of 2 years. This period can be further extended in exceptional cases. During this time, the complete data are required for regular registration of the insecticide. Simultaneously, the laboratory and the field tests are also made in the Central Insecticides Laboratory or the Regional Laboratories and other scientific institutions. After an insecticide has been tried and found to be effective and safe, a regular registration is granted for its import or indigenous manufacture.

Licences for Manufacture and Sale

Any person desiring to manufacture or to sell, stock or exhibit for sale or distribute any insecticide or to undertake commercial pest control with the use of any insecticide, must first get a licence from the Licensing Officer of the State Government. Such a licence can be granted for a specific period on receipt of an application on prescribed forms and on the payment of prescribed fee. The licence can be renewed from time to time, and may even be cancelled if it is found that there was some misrepresen-tation made as to some essential facts or in the event of failure to comply with the conditions on which the licence was granted.

Central Insecticides Laboratory

Under the provision of the Act, the Government of India has set up a Central Insecticides Laboratory under the control of a Director. The Director of this Laboratory is responsible for the following duties and functions :

- To analyse the samples of insecticides sent by any officer or authority of the Central or State Government.
- To analyse samples of materals for insecticide residues.
- To carry out such investigations as may be necessary for ensuring the conditions of the registration of insecticides.
- To determine the efficacy and toxicity of insecticides
- To carry out such other functions as may be entrusted to him by the Central Government or by State Governments with the permission of the Insecticides Board.

State Insecticides Testing Laboratory

Every State and Union Territory of India is required to establish one or more insecticides testing laboratories to test the quality of pesticide samples sent by Insecticide Inspectors. The required staff and suitable laboratory equipment to carry out the analysis of the pesticide samples have been provided for each such laboratory. The Insecticide Analyst acts as the Incharge of the Laboratory and is required to deliver a test or analysis report, in duplicate, duly signed within a period of 60 days, for insecticide samples submitted to him by the Insecticide Inspector.

Packing and Labelling

After the enforcement of the Act, the insecticides are not allowed to be sold, unless they are packed and labelled in accordance with the provisions of the rules. The shape, size and appearance of every container of insecticide must be approved by the Registration Committee.

The labels on these packages must also be pasted in a prescribed manner. The wording on the package is either printed or written in indelible ink on the label of both the inner-most container as well as the outer-most covering in which the container is packed. Depending upon the poisonous qualities of the insecticides, the word 'Poison' printed in red (with a symbol of skull across and cross bones) is written for extremely toxic insecticides, the word 'Poison'printed in red for highly toxic, the word 'Danger' for moderately toxic and the word 'Caution' for slightly toxic chemicals. It is always written on the packages 'Keep out of the reach of children' and also 'If swallowed or symptoms of poison occur, call a physician immediately'. Some of the other important instructions to be printed are given below :

(*i*) The label should be so fixed that it cannot be ordinarily removed.

(*ii*) The label should be pasted on a prominent place on the packing. One sixteenth of the area is demarcated as a square which is divided into two equal triangles, the upper portion having the symbol and the words as approved, and the lower triangle shall have the specified colour (red, yellow, blue, green) according to the degree of toxicity of the insecticide.

(*iii*) The upper triangle of the square shall carry the symbol of a skull and cross bones, and the word 'Poison' printed in red, etc. depending upon the category of the insecticide.

(*iv*) In case an insecticide is likely to catch fire, it is to be indicated that the liquid is inflammable and container should be kept away from heat or open flame.

(*v*) The insecticide containers should not carry any such labels as 'safe', 'non-poisonous', 'non-injurious' or 'harmless'.

(*vi*) The labels and the leaflets to be enclosed alongwith the containers should be printed in Hindi, English and one or two regional languages.

(*vii*) The main label on the insecticide containers should have the following information printed : (*a*) name of the manufacturer, and the packer and the distributor, if the last two are different; (*b*) name of the insecticide, giving the brand name or the trade mark under which the insecticide is sold; (*c*) registration number of the insecticide; (*d*) kind and name or the active and other ingredients accepted by the International Standards Organization or the Bureau of Indian Standards; (*e*) net weight or volume of the quantity; (*f*) batch number and expiry date, and; (*g*) antidote statement.

(*viii*) The leaflet which must be given alongwith the containers of the insecticides should have the following instructions printed: (*a*) The plant diseases, insects and noxious animals or weeds for which the insecticide is to be applied and the adequate directions concerning the manner in which the insecticide is to be used at the time of application; (*b*) particulars regarding chemicals harmful to human beings, animals and wild life, warning and cautionary statements including the symptoms of poisoning, suitable and adequate safety measures and emergency or first-aid treatment where necessary; (*c*) cautions regarding storage and application of insecticides with suitable warnings relating to inflammable, explosive, or other substances harmful to the skin; (*d*) instructions concerning the decontamination or safe disposal of used containers; (*e*) a statement naming the antidote for the poison shall be included in the leaflet and the label; (*f*) if the insecticide is irritating to the skin, nose, throat or eyes, a statement shall also be included to that effect.

Enforcement Machinery

The Insecticide Inspectors appointed by the Central or the State Governments have the following duties :

- To ensure that all conditions of the licence are being complied with.
- To conduct an enquiry into any complaint given to him in writing.
- To inspect not less than three times a year, all the establishments selling insecticides within the area of his jurisdiction.
- To maintain a record of all inspections made and actions taken by him, including the taking of samples, the seizure of stocks, etc. and to send copies of these reports to the Licencing Officer.
- To make inspections and to conduct enquiries which may be considered necessary to detect the sale of any insecticide contrary to the provision of the Act .
- To institute prosecution in the local courts for breaches of rules and regulations made under the Act.

For the execution of the above mentioned duties, the Insecticide Inspectors have been given the following powers :

(*i*) To enter and search at reasonable times, with or without assistance, any premises in which any offence is being or is likely to be committed against conditions to the certificate of registration or licence for manufacture.

(*ii*) To inspect, examine and to make copies of records and registers and other documents kept by the manufacturer, distributor, carrier, dealer or any other person. In case the records and registers are taken into custody, he is to inform the magistrate and obtain orders for keeping those papers in custody.

(*iii*) To make examinations and to conduct enquiries in order to ascertain that the rules and regulations and other provisions of the Act are being complied with; any vehicle may be stopped for that purpose.

(*iv*) To stop the distribution, sale or use of a particular insecticide, for a period of thirty days, which, in his opinion, is being distributed in contravention of the provisions of the Act. The Inspector may also seize the stocks of insecticides and inform the magistrate who would issue orders for that purpose.

(*v*) To lift samples of insecticides and send those samples to the Insecticide Analyst for prescribed tests and analysis. The Inspector must give fair price of the samples taken and intimate, in writing on the prescribed form, the purpose for which the samples were taken.

II. Prevention of Food Adulteration Act, 1954

The Prevention of Food Adulteration Act (PFA), 1954, was enacted on September 29, 1954. The Central Committee for Food Standards (CCFS) under the Ministry of Health & Family Welfare recommends the quality of food commodities under the PFA Act. Among its various sub-committees. Pesticide Residue Sub-Committee advises CCFS on the following :

(*i*) Tolerance limits for pesticides in different articles of food based on use.

(*ii*) Restriction on sale of insecticides by the persons manufacturing and pattern, dietary habits and nutritional status of our population, storing/selling food

Before tolerance limits for any pesticide are considered, the same needs to be approved and registered by the Registration Committee, constituted under the Insecticides Act, 1968, administered by the Ministry of Agriculture. Based on the recommendations of the Pesticide Residue Sub-Committee and after the approval of CCFS, the Ministry of Health and Family Welfare has so far notified maximum residue limits (MRLs) for 71 pesticides under Rule 65 (2) of PFA Rules, 1955. The limits have been prescribed on different raw agricultural products, which move in commerce like foodgrains, milled foodgrains, milk and milk products, fruits and vegetables, meat, eggs, fish, okra, leafy and other vegetables, maize cob (kernels), rice, chillies, cotton seed and cotton seed oil, pulses, groundnuts, dry fruits, tea (dry manufactured), coffee (raw beans), safflower, etc.

III. Pesticides Management Bill, 2008

A Bill seeking to encourage production of good quality and safe pesticides at affordable rates while curbing sale of spurious products was introduced in the Rajya Sabha on 20th October, 2008. The new Bill, which would replace the Insecticides Act, 1968, was needed to provide for better management of pesticides to respond to the need for faster agricultural growth. The proposed law aims to regulate the import, manufacture, export, sale, transport, distribution, quality and use of pesticides in the country. The new law is being enacted with a view to minimize the contamination of agricultural commodities by pesticide residues. The government will allow the use of a pesticide only after assessing its efficacy and safety. The Bill enables the government to constitute the Central Pesticides Board to advise on scientific and technical matters arising out of the administration of the proposed law.

For the first time, the government has defined household pesticides to enable delicensing of their retail sales for easy availability to the consumers. The Bill will also prohibit field application of household pesticides.

Under the Bill, there is provision to accredit private laboratories to carry out all functions of the Central Pesticides Laboratory. it has also provided an elaborate procedure for drawing pesticide samples and their inspection. Under the proposed law, the government has made punishment more stringent to check production and sale of mis-branded, sub-standard and spurious pesticides.

The salient features of the proposed Bill are as follows:

- To improve the quality of pesticides available to the Indian farmers and introduce new, safe and efficacious pesticides.
- More effective regulation of import, manufacture, export, sale, transport, distribution and use of pesticides, to prevent risk to human beings, animals and the environment.
- Detailed categorization of offences and punishment for greater deterrence to violators.
- De-licensing of retail sale of household pesticides.
- Timely disposal of time-barred pesticides in an environmentally safe manner.

CULTURAL CONTROL

The cultural methods of insect control comprise regular farm operations, which are so performed as to destroy the insects or to prevent them from causing injury. A large number of insects are normally killed by farmers unconsciously when they expose them to adverse climatic or biological conditions through agricultural operations like ploughing, hoeing, weeding, etc. A still more effective kill can be obtained by following improved agricultural practices or so synchronizing the existing practices with the life cycles of the pests, that the weakest links in the life cycle of pests are subjected to adverse environmental conditions. Moreover, many cultural practices enhance the functional biodiversity of the natural enemies whereas some others negatively affect it (Fig. 4.1).

By adopting various cultural practices such as preventive measures, the pests may either be killed directly or indirectly. Since these methods must be employed far in advance of the actual appearance of the pest and the consequent damage, the control achieved is not so spectacular and thus has not a strong appeal to the farmers. These methods cost hardly anything because all that is required is to adjust the time of ploughing, sowing, irrigating, harvesting, etc. Crop rotation and improved management of the farm are also important.

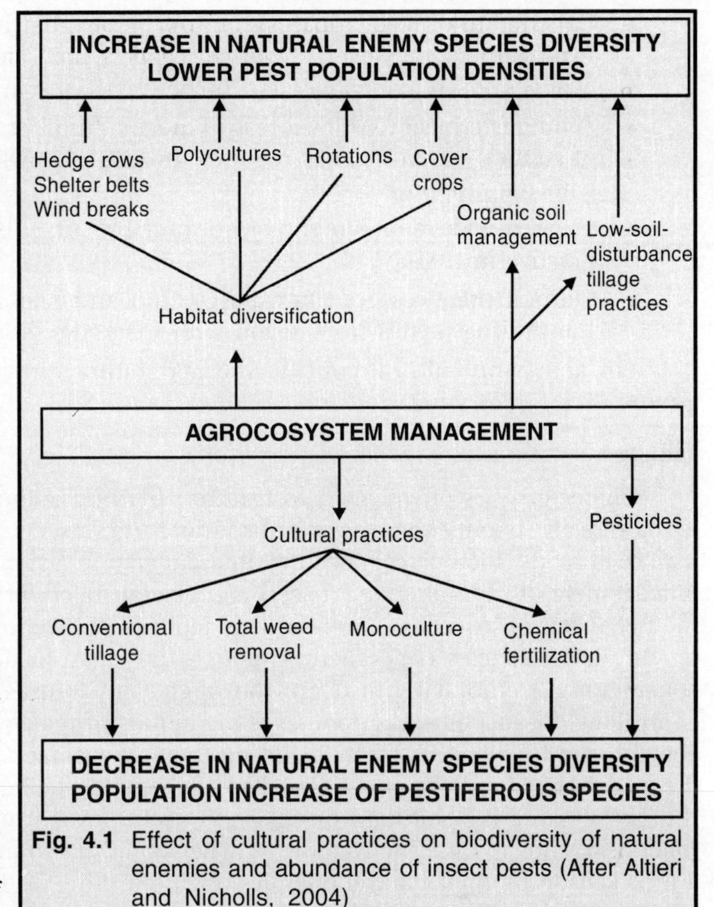

Fig. 4.1 Effect of cultural practices on biodiversity of natural enemies and abundance of insect pests (After Altieri and Nicholls, 2004)

The effective cultural practices can be further improved or new ones devised if the life-history, behaviour, habitat and ecology of the pest concerned are fully understood. A method, effective against one species, might not be effective against another owing to variation in its biology. Proper timing of the practices is the keynote of success. It is not useful to try to destroy the shelters of hibernating insects after they have emerged from them and reached the host plants. Cultural methods of control are particularly important for the destruction of tissue-borers which are not hit by insecticides or easily attacked by natural predators and parasites.

The following are the main characteristics of cultural practices which should be viewed in the context of IPM :

- Cultural practices are simple modifications or adaptations of regular farm operations. The extra cost of their incorporation into pest management system is minimal in most cases. These are often the only control options economically feasible for low-value crops.
- Cultural practices generally produce no or negligible undesirable ecological consequences.
- Cultural control is primarily aimed at prevention and reduction of pest outbreaks.
- The results of cultural practices are often difficult to quantify primarily because ecological relationships within crop systems are poorly understood.
- Cultural control tactics are an effective means of pest control. These do not result in total elimination of the pest thus allowing for conservation of beneficial insects.
- Most cultural practices indirectly affect the pests and these are relatively slow acting and thus cannot resolve a pest outbreak. These are, however, important in minimizing pest damage by preventing pest build-up, rather than relieving an already existing pest problem.
- Cultural practices make cropping systems less friendly to the establishment and proliferation of pest populations.

- Cultural practices are designed to have positive effects on farm ecology and pest management, however, negative impacts may also result due to variations in weather, changes in crop management or perturbations in agroecosystems.
- Timing is critical to the success of most cultural practices, accordingly, the implementation of cultural control tactics requires thorough knowledge of pest ecology and its interaction with the cropping system.
- An area-wide deployment of cultural practices is essential for the effectiveness of the practice in IPM.
- Cultural practices are often pest-, crop and region-specific. Care should be exercised in transferring tactics to a region with markedly different agroecological conditions.

For achieving cultural control, the agricultural practices can be categoised under various groups :

Tillage

Characterisitics of soil such as texture, chemical composition, moisture, temperature and soil fauna directly inflence the survival of soil-infesting insects. These characteristics also influence the quality of food which determines the abundance of a given pest. Thus, with the proper stirring and management of the soil, many insects can be controlled, provided the details of the life history and the behaviour of insects are known. The depth of cultivation, its frequency and timing will depend on the insect species concerned. Although the value of deep tillage in insect control cannot be considered all embracing, a deep, thorough, and frequent cultivation of the fields is useful in controlling, for instance, the root-infesting aphids of maize and apple. The soil-infesting insects, in general, can be killed easily by exposing them to weather. The insects are particularly helpless in the pupal stage, and can be easily killed, an example being the Oriental fruit moth, *Cydia molesta* (Busck). Packing the soil hard or caking is useful in controlling cutworms. Packing the soil with a roller also tends to raise the water level. The subterranean insects rise to the soil surface, where they are picked up by their natural enemies. Light earthing at the early stage of sugarcane crop during May-June is helpful in checking the shoot borer.

It is a common sight in India to see a number of birds, such as the king-crow, the myna, the starling, etc., following the plough and picking up a number of insects, when exposed. If ploughing is done early in the spring, a number of hibernating insects which would have emerged later are exposed. Some of the common hibernating species found in the soil are the cotton semilooper, *Tarache notabilis* (Walker) ; the lucerne caterpillar, *Spodoptera exigua* (Hubner); Bihar hairy caterpillar, *Spilarctia obliqua* (Walker); cutworms, *Agrotis* spp., etc.

Most of the surface grasshoppers and field crickets lay eggs in the summer or during the autumn, in the upper 7-10 cm of soil. When a field is cultivated, the eggs in the soil are brought to the surface where they are either desiccated, or picked up by birds and other predators. A majority of the remaining eggs do not hatch in the spring because their emergence holes are blocked. Similarly, the larvae of the pink bollworm, *Pectinophora gossypiella* (Saunders) and the spotted bollworms, *Earias* spp., of cotton found in the soil in damaged bolls or cotton seed, if buried deep by ploughing, cannot emerge as moths in the spring.

The type of implement used for ploughing or cultivating the soil is also important. The mouldboard ploughs are, in general, more useful in burying the hibernating insects and at the same time bringing eggs and grubs of other species to the surface. A disc plough treatment caused upto 73 per cent mortality of overwintering larvae of *Dectes texanus* Le Conte, a pest of sunflower as compared to 40 per cent mortality recorded using a sweep plough. With frequent hoeing and interculture, the soil-dwelling insects, such as the grubs of the *ber* beetle, *Adoretus* spp.; the cotton grey weevil, *Myllocercus undecimpustulatus* Faust, are disturbed and killed.

Clean Seed

A number of insect pests are carried over from one crop to the next through seeds, cuttings or other infested plant parts. Certified seeds, free from pests and diseases, should be used for raising a new crop. Wheat galls should be separated by dipping the wheat seed in 5-10 per cent salt solution.

The hibernating larvae of the pink bollworm are found in the double seeds of cotton and can be killed easily by fumigating the seed in ginneries. Methyl bromide, phosphine and carbon bisulphide are the effective fumigants. If unfumigated seed is sown, the pest is carried over to the new crop.

In order to protect new orchards from infestation by the San Jose scale, *Quadraspidiotus perniciosus* (Comstock), it is advisable to fumigate nursery plants before despatching them. Citrus plants should be similarly treated, so that the red citrus scale, *Aonidiella aurantii* (Maskell), is not allowed to spread.

Seed Rate

Adoption of appropriate seed rate ensures proper stand, spacing and crop canopy that helps in adoption of proper spray technology and checks the unwanted growth of crop. The traditional practice of using a high seed rate helps to maintain optimum plant stand and reduce insect damage in cereals. Use of high seed rate is recommended in those crops where removal of infested plants is helpful in minimising the incidence of insect pests, viz. maize borer in maize and shoot fly in sorghum and other crops. The farmers in north India use 25-50 per cent less seed rate in cotton to avoid thinning. Low seed rate along with untimely rain and high temperature during sowing, drought conditions, termite attack and incidence of root rot, further reduce the plant stand. The plant stand estimated at the time of harvest is 12-30 per cent less than the recommended plant population. The use of appropriate seed rate in cotton leaf curl virus affected areas helps to maintain proper crop stand even after regueing of the diseased plants at an early stage. Besides this, too low or high seed rate adversely affects spraying operation.

Irrigation

When water is applied to the fields by flooding, a large number of insects present in the soil are drowned. Some of the others are driven out and exposed to their natural enemies. After lucerne is harvested, many lucerne caterpillars remain in the fields. A newly harvested field should be irrigated and the pest is killed by drowning. The cotton bollworms can be shaken off by dragging a rope over the plants and throwing them into standing water.

Sugarcane and wheat crops can be saved from the attack of white-ants by irrigating them. The potato tuber moth, *Phthorimaea operculella* (Zeller), is effectively suppressed by frequent overhead irrigation of potatoes. The stalk borer of sugarcane, *Chilo auricilius* Dudgeon, is more serious in flooded areas and, therefore, water should be controlled.

Flooding of fields has been recommended for reducing the attack of cutworms, armyworms, termites, whitegrubs, etc. On the other hand, draining the rice fields for 3-4 days during infestation controls brown planthopper and whorl maggot. Early termination of irrigation (by end of September) to *hirsutum* cotton in Punjab helps in reducing diapausing larval population of pink bollworm.

Sap sucking insects like aphids, jassids and whitefly are especially sensitive to changing water levels in their host plants. Maximum fecundity of mustard aphid reared on *sarson (Brassica compestris* Linnaeus cv. yellow *sarson*) and *raya* [*B. juncea* (Linnaeus) Czern.] host plants was recorded when the water level was maintained continuously above the field capacity. The concentration of acidic amino acids and amides increased significantly under lower soil water regimes. Nutritional imbalance created by this higher concentration coupled with increasing burden on the excretory mechanisms of the aphid were mainly responsible for reduction in the fecundity of the aphid under such conditions. Due to this reason, economic threshold of the pest in the field was reached only at the highest soil water regime.

Indian mustard irrigated once harboured significatly higher aphid population than the unirrigated one. However, increase in number of irrigations from one to two did not result in a further increase in aphid incidence. In case of *B. campestris* cv. Chinese cabbage, the incidence of mustard aphid increased considerably with the increase in number of irrigations from 3 to 6. The build up of aphid in relation to irrigation levels may, thus, vary in different *Brassica* spp. Waterlogging enhances the multiplication of a number of borers including stalk borer, internode borer, plassey borer and whitefly on sugarcane crop. Increase in soil moisture level due to frequent irrigation also helped in build up of *Pyrilla* population on this crop, while black bug damage is maximum in unirrigated fields.

Fertilizers

Healthy and vigorous plants are able to resist the attack of a given pest better and for a longer period than the sickly, undernourished plants. Plant growth can be stimulated with proper manuring. According to the requirements of a particular soil, NPK should be applied in the right porportions. It has been observed that by the application of nitrogenous manures and increased irrigation, the incidence of whitefly of cotton is reduced significantly. However, in another locality, where the whitefly attack is not so serious, the excessive application of nitrogen might increase the attack of cotton jassid.

It is desirable to stimulate the growth of sugarcane during April-May with the application of ammonium sulphate, particularly after the shoots attacked by the top-borer have been cut mechanically. The application of manure at this stage also induces tillering. Beneficial effect on the growth of potato crop by the application of Bordeaux mixture is well known. The crop so stimulated, becomes less susceptible to the attack of jassid, blight and other fungal diseases. Wheat grown in well-fertilized fields has a good stand and is comparatively free from the attack of *molya* disease, caused by a nematode.

High levels of nitrogenous fertilizers increase incidence of yellow stem borer, rice leaf folder and gall midge on rice; leaf folder, whitefly and bollworms on cotton; and internode borer, stalk borer and pyrilla on sugarcane. On the other hand, application of potash and sometimes phosphorous, either singly or in combination results in lower incidence of many insect pests, viz. *Empoasca kerri* Pruthi and *Spdoptera litura* (Fabricius) on cowpea; brown planthopper and whitebacked planthopper on rice; aphid and thrips on chillies, and *Hypera variabilis* Herbst and *Aphis craccivora* Koch on lucerne.

Clean Culture

Clean cultivation as a method for minimising the incidence of insect pests was advocated more than a century ago. Destroying or removing crop residues from fields is one of the basic ways to eliminate pest overwintering sites and reduce the spread of infestation. This may be achieved by ploughing directly or shredding and chopping, burning residues or raking and scooping them into piles for burning. Removal of cotton sticks by 1st August every year was enforced under the Cotton Pests Act, 1911 which helped in reducing the incidence of pink bollworm in erstwhile Madras state. The incidence of boll weevil, *Anthonomus grandis* Boheman on cotton; European corn borer, *Ostrinia nubilalis* (Hubner) and South-western corn borer, *Diatraea grandiosella* Dyar in corn and Hessian fly, *Mayetiola destructor* (Say) in wheat in USA was reduced by sanitation measures.

Collecting and burning of stubble and chaffy panicles reduces the carryover of spotted stem borer, *Chilo partellus* (Swinhoe) and midge, *Stenodiplosis sorghicola* (Coquillett) in sorghum. Stalks from the previous season should be fed to cattle or burnt before the onset of monsoon rains to reduce the carryover of *C. partellus.* Piling and burning of trash at dusk in the field attracts the adults of white grubs, *Holotrichia consanguinea* (Blanchard), and the red hairy caterpillar, *Amsacta moorei* (Butler), and kills them. This helps to reduce the oviposition and damage by these insects. Sanitation measures have helped to reduce the incidence of planthoppers, leafhoppers, leaf folder, rice hispa, gall midge and stem borer on rice crop. Removal and destruction of rice stubble has been found highly effective in minimising overwintering populations of many species of stem borers. It has been estimated that stubble left in a field on one hectare with 6.8 per cent population of immature stages is a potential source of emergence of 1,20,000 moths. Burning of rice stubble has been reported to reduce the stem borer infestation on the subsequent crop by 67 per cent. Removal of damaged plant parts and uprooting of infested plants including those showing deadhearts at thinning time have been recommended to lower maize borer, shoot fly, armyworm, pyrilla, aphids, cutworms and termite incidence in maize crop.

Detrashing of dry leaves from August onwards reduces the attack of pyrilla and scale insect on sugarcane crop. Detrashing at fifth, seventh and ninth months checks internode borer damage and detrashing in October and November results in the reduction of stalk borer damage. Destruction of water shoots protects the crop from the ravages of stalk borer and internode borer. Burning of trash after harvest of the crop up to specific period before the pest migrates to nearby crop kills the carryover population of pyrilla, black bug and scale insect to some extent. Avoiding ratooning of pigeonpea during off-season helps in reducing the carryover of pod fly and eriophyiid mite, *Aceria cajani* Channabasavanna, an important vector of pigeonpea sterility mosaic virus. Selective pruning of pigeonpea can be useful in reducing the incidence of leaf folder and spotted caterpillar.

Cotton sticks followed by cotton seeds are the major sources for the carryover of pink bollworm in the Punjab. Following measures have been recommended to reduce the carryover population of pest :

- After the last picking, sheep and goats should be allowed to graze on unopened left over green bolls and shed material in the field.
- The left over bolls, wherever feasible, should be removed manually before cutting the sticks. These bolls may be kept on roofs of houses, in thin layer for opening to extract good quality seed cotton and the remaining unopened and small bolls should be burnt.
- During summer months, on an average, 50.7 per cent (31.6–71.4%) of the larvae in cotton sticks died due to high atmospheric temperature which may go as high as 49°C in some years. The mortality of larvae in cotton stacks put in the open was higher (53.5%) than those kept under tree shade (32%). Similarly, the mortality of larvae in stacks stored in bundles in upright position was higher (53.6%) than in sticks stored untied horizontally (37.7%). Therefore, cotton sticks with left-over bolls should be stacked in the open away from tree shade in the villages, preferably after tying them in bundles and storing in upright position.
- The ginning of seed cotton and extraction of oil from cotton seeds should be accomplished by end of April and May, respectively, so that diapausing larvae in the seed cotton and cotton seeds are killed during these processes well before the onset of fruiting bodies in the ensueing cotton crop.

Undesirable plants in gardens give protection and provide food to the newly emerged nymphs of the mango mealybug, *Drosicha mangiferae* (Green), in January and February. For about one month, they feed on a number of weeds and then ascend the mango trees when fresh leaves appear. The eradication of *baru* grass, *Sorghum halepense,* is very helpful in controlling sugarcane mites. The moths of the red hairy caterpillar, *Amsacta moorei* (Butler), lay eggs on the weed *lahni, Heliotropium eichwaldi,* and on emergence, the caterpillars also feed on it.

Melon, guava, peach, tomato fruits, etc. which harbour maggots of fruit flies and tissue-borers should not be allowed to remain on the farm after marketable fruits have been harvested. If allowed to remain, the adults emerge and reinfest the standing crop.

Crop Spacing

The major objective in spacing crop plants is to obtain maximum high quality yield per unit area per unit time. But spacing may also influence the population and damage of many insect pests by modifying the micro-environment of the crop or affecting health, vigour and strength of the crop plants or pattern and duration of crop growth and development. Closer spacing has been reported to increase the incidence of planthoppers, viz. brown planthopper and whitebacked planthopper, gall midge and leaf folder in rice crop. The percentage of plants showing hopper burn symptoms ranged from 100 in closer spacing (10 × 10 cm) to 7-67 in wider spacing (23 × 10 cm). Detailed multilocation trials under All India Coordinated Rice Improvement Project showed that stem borer and gall midge were more in closer spacings (10 × 10 cm and 10 × 15 cm) than in wider spacings (20 × 15 cm, 20 × 20 cm and 30 × 30 cm). On the other hand, closer spacing resulted in lower incidence of green leafhopper, rice hispa and whorl maggot.

The closer spacing in cotton results in bushy growth of the crop that affects penetration of light, results in vertical growth of the plant, hinders spraying operation and also results in higher relative humidity that favours higher incidence of sucking pests and bollworms. In closer plant spacing (45 × 30 cm), jassid population has been observed to be high as compared to wider spacing (60 × 45 cm). Incidence of bollworms is also higher in closer plant spacing (15 cm) in *hirsutum* cotton varieties LH 886 and F 505 than wider plant spacing (30 cm). On the other hand, closer spacing resulted in lower incidence of green leafhopper, rice hispa and whorl maggot.

Closer spacing in groundnut lowered the incidence of thrips, jassids and leafminer, and increased parasitism in the latter. On the other hand, in cotton it increased tha damage by jassid, whitefly and bollworms. In *hirsutum* cotton, the incidence of pink bollworm was 5.2 and 17.1 per cent at 75 × 20 cm and 50 × 20 cm, respectively. Larval population of pod borer, *H. armigera* on chickpea was 4 times as large at closest (33 plants/m^2) as at widest (3 plants/m^2) spacing. There was little advantage of more than 8 plants/m^2 in chemically protected crop. In soybean, damage by a number of insect pests including *Spilarctia obliqua* (Walker), *Melangromyza sojae* (Zehntner) and *B. tabaci* was greater in closely planted crop. In sugarcane, closer spacing resulted in higher incidence of shoot, internode and stalk borers. In contrast to all these studies, population of *Aphis craccivora* Koch on chickpea was more in widely spaced crop (60 × 20 cm) than in closer spacing (30 × 10 cm).

Crop Rotation

There are two distinct systems of cropping, viz. specialized farming and mixed farming. Crops provide food for insect pests and, if food is abundant all the year round, they flourish and soon increase in number. Their abundance depends on the fecundity, hibernation, number of generations completed in a year, dispersion ability, etc. Therefore, pest problems in an area where there is specialized farming will be different from that in the areas of mixed farming. Where there is a single-cropping pattern, the specific pests having a limited power of migration, breed slowly, particularly those spending a long time in the feeding stage and having only a few generations per year. They take number of years to become abundant and their multiplication can be checked by introducing crop rotation.

Some insects, however, which have greater mobility and which are omnivorous, can appear as pests in the very first year of planting a new crop. Crop rotation in that case does not materially affect multiplication. If the potato tuber moth, *Phthorimaea operculella* (Zeller), and nematodes become serious in a given locality, it is recommended to discontinue the cultivation of potatoes for three or four years. In the case of the golden nematode of potato, *Heterodera rostochiensis* Woll., this is the only practical method of control known so far. The underlying principle in crop rotation is to starve the pest. It is, therefore, essential that there should be a discontinuity in the supply of food. In other words, the cropping scheme should be such that alternative hosts are eliminated in one season or another.

From the entomological point of view, the following rotations are unsound :

(*i*) *Guara*-wheat-*toria* (*Brassica campestris*). Any delay in the burial of *guara* (*Cyamopsis tetragonoloba*) will attract termites and may jeopardize the chances of growing a successful wheat crop.

(*ii*) Sugarcane-maize-cotton. Termites are attracted to the dried roots of cotton and they attack sugarcane setts later on.

(*iii*) Maize should not be sown on grassland (both belong to the family Gramineae), because soil insects, white grubs, cutworms, wireworms, etc., are common to both.

The following rotations are considered useful :

(*i*) Maize-*senji* (*Melilotus parviflora*)-sugarcane is a good rotation from the entomological point of view.

(*ii*) For protection against cattle ticks, the pastures should be kept free from animals for a year, so that the ticks may be starved to death in 8-9 months.

(*iii*) The incidence of attack of the sugarcane black bug, *Cavelerius excavatus* (Distant) and tissue borers is greater on the ratoon crop than on the planted crop.

(*iv*) Rotation of groundnut with non-leguminous crops in recommended for minimising the damage by leafminer.

(*v*) Cotton should be rotated with non-preferred hosts like *ragi*, maize, rice, groundnut, cowpeas or soybean to minimise the incidence of insect pests.

Intercropping

A carefully selected cropping system (intercropping or mixed cropping) can be used to reduce pest incidence, and minimize risks of monocultures. Some of the examples of intercropping systems that help in preventing pest outbreaks are listed in **Table 4.1.**

In central and southern India, intercropping of cotton with black gram, green gram, onion, cowpeas, etc. is reported to divert the population of sucking pests and American bollworm from cotton. Monoculture of cotton was also found to harbour more insect pests than cotton intercropped with groundnut, cowpea and soybean. The intercrop of cowpeas in cotton helped in the colonization of coccinellids and also enhanced the parasitism of spotted bollworm. On the other hand, okra, *mung* and pigeonpea as intercrops with cotton increased the population build up of jassid, whitefly, spotted bollworms and American bollworm.

Intercropping of groundnut with pearlmillet reduces the incidence of thrips, jassid and leaf-miner. Intercropping of *taramira* in *raya* reduces the incidence of mustard aphid in the latter crop due to allelopathic influence of the former. Intercropping of chickpea with barley, wheat, linseed, mustard and sunflower is also known to reduce infestation of *H. armigera* in chickpea.

The damage by sorghum shoot fly, *Atherigona soccata* Rondani and midge, *Stenodiplosis sorghicola* (Coquillett) is reduced when sorghum is intercropped with leguminous crops. Intercropping sorghum with cowpea or lablab reduced the damage by spotted stem borer, *Chilo partellus* (Swinhoe) by 50 per cent and increased the grain yield by 10-12 per cent over a single crop of sorghum. Intercropping sorghum with pigeonpea reduces the damage by *H. armigera* in pigeonpea. Intercropping red clover with maize also reduces the damage by the European corn borer, *Ostrinia nubilalis* (Hubner). Intercropping of groundnut with pearlmillet reduced the incidence of thrips, jassid and leaf miner whereas intercropping with sunflower and castor increased the incidence of thrips and jassid, respectively. When pearlmillet was grown as an intercrop in groundnut, the parasitic activity of *Goniozus* sp. was considerably enhanced. The pollen grains of the millet were preferably used as food by the adult parasitoids.

Tomato intercropped with cabbage has been reported to inhibit or reduce egg laying by diamondback moth. A planting pattern of one row of cabbage and one row of tomato (cabbage planted 30 days later than tomato), caused maximum reduction of diamondback moth and leafwebber larvae on cabbage. Reduction in insect incidence was attributed to possible release of volatile substances from late crop growth stages of tomato which inhibited oviposition by incoming moths.

Table 4.1. Selected examples of intercropping systems that help to prevent insect pest outbreaks

Intercropping system	Pest(s) regulated	Factor(s) involved
Beans grown in relay inter-cropping with winter wheat	*Empoasca fabae* (Harris) and *Aphis fabae* Scopoli	Impairment of visual searching behaviour of dispersing aphids
Brassica crops and beans	*Brevicoryne brassicae* (Linnaeus) and *Delia radicum* (Linnaeus)	Higher predation and disruption of oviposition behaviour
Cabbage intercroped with white and red clover	*D. radicum*, cabbage aphids, and imported cabbage butterfly, *Pieris rapae* (Linnaeus)	Interference with colonization and increase of ground beetles
Intercropping of pigeonpea with red, black and green gram	Pod borers, jassids, and mem-bracids	Delayed colonization of herbivores
Cassava intercropped with cowpeas	Whiteflies, *Aleurotrachelus socialis* Bondar and *Trialeurodes variabilis* (Quaintance)	Changes in plant vigour and increased abundance of natural enemies
Maize intercropped with fava beans and squash	Aphids; mite, *Tetranychus urticae* (Koch) and *Macrodactylus* sp.	Enhanced abundance of predators
Maize intercropped with soybean	European corn borer, *O. nubilalis*	Differences in corn varietal resis-tance
Maize intercropped with sweet potatoes	Leaf beetles, *Diabrotica* spp.; leafhoppers, *Agallia lingula* Van Duzee	Increase in parasitic wasps
Intercropping maize and beans	*Dalbulus maidis* (Delong & Wolcott)	Interference with leafhopper movement
Cotton intercropped with forage cowpea	Boll weevil. *Anthonomus grandis* Boheman	Population increase of parasitic wasps (*Eurytoma* sp.)
Intercropping cotton with sorghum or maize	Corn earworm, *Helicoverpa zea* (Boddie)	Increased abundance of predators
Cucumbers intercropped with maize and broccoli	*Acalymma vittatum* (Fabricius)	Interference with movement and tenure time on host plants
Groundnut intercropped with field beans	*Aphis craccivora* Koch	Aphids trapped on epidermal hair of beans
Groundnut intercropped with maize	Corn borer, *Ostrinia furnacalis* (Guenee)	Abundance of spiders (*Lycosa* sp.)
Maize-bean intercropping	*S. frugiperda* and *Diatraea lineolata* (Walker)	Lower oviposition rates, trap cropping
Strip cropping of muskmelons with wheat	*Myzus persicae* (Sulzer)	Interference with aphid dispersal
Oats intercropped with field beans	*Rhopalosiphum* sp.	Interference with aphid dispersal
Sesame intercropped with corn or sorghum	Webworms (*Antigastra* sp.)	Shading by the taller companion crop
Tomato and tobacco intercro-pped with cabbage	Flea beetles, *Phyllotreta cruciferae* (Goeze)	Feeding inhibition by odours from non-host plants
Tomato intercropped with cabbage	Diamondback moth, *Plutella xylostella* (Linnaeus)	Chemical repellency or masking

Source : Altieri and Nicholls (2004)

Even the female moths which entered the mixed crop laid fewer eggs probably because of tomato foliage spread over every row of cabbage.

The cowpea intercropped with sorghum significantly decreased the incidence of aphid, *Aphis craccivora* Koch and thrips, *Megalarothrips sjostedi* Trybom. Two varieties of cowpea, Kanannado white and Kanannado brown were sown solo and intercropped with sorghum. This reduction in population may be due to the micro-environmental effect of the associated crop which may attact predators and or disruption of insect visual search for preferred host.

Carrot intercropped with lucerne has been shown to suffer less damage by the rust fly, *Psila rosae* (Fabricius). Intercropping bean with collards decreases flea beetle, *Phyllotreta cruciferae* (Goeze) densities on collards and minimizes the leaf damage. Intercropping of *taramira* in *raya* reduced the incidence of mustard aphid in the latter crop due to the allelopathic influence of the former. Similarly, trials conducted under the All India Coordinated Pulses Improvement Project at several locations demonstrated that the sole crop of chickpea attracted more *H. armigera* compared to intercrops with wheat, barley, linseed, mustard and safflower. On the other hand, lentil and field peas as inter-crops enhanced infestation in chickpea. Crop mixtures were more effective than row plantings.

In Kenya, intercropping maize with the non-host molasses grass, *Melinus minutiflora* Beauv decreased levels of infestation by stem borers, *C. partellus* and *Busseola fusca* Fuller, and also increased larval parasitization of stem borers by *Apanteles sesamiae* Cameron. In field trials, *M. minutiflora* planted in alternate rows with maize significantly reduced stem borer infestation of the main crop (damaged maize plants : single crop 39.2%, intercropped with *M. minutiflora* 4.6%). There was also a significant increase in parasitization by the larval parasitoid, *A. sesamiae* (parasitized larvae in maize : single crop 5.4%, maize with *M. minutiflora* intercrop 20.7%). Volatile agents produced by *M. minutiflora* repelled female stem borers and attracted foraging female *A. sesamiae*.

Trap Cropping

Trap cropping is the planting of a trap crop to protect the main cash crop from a certain pest or several pests. The trap crop can be from the same or different family group, than that of the main crop, as long as it is more attractive to the pest. Even early or late plantings of the same crop in the main crop may also serve as traps. Thus trap crops are plant stands that are grown to attract insects or other organisms so that the target crop escapes pest attack. Protection is achieved either by preventing the pests from reaching the crop or by concentrating them in certain part of the field where they can easily be destroyed. The attractiveness of trap crops may be enhanced by use of insect pheromones, plant kairomones or insect food supplements. Depending on the seasonal cycle, insects may be left to develop in the trap or killed with an insecticide.

There are two types of planting the trap crops; perimeter trap cropping and row intercropping. Perimeter trap cropping (border trap cropping) is the planting of trap crop completely surrounding the main cash crop. It prevents a pest attack that comes from all sides of the field. It works best on pests that are found near the borderline to the farm. Row intercropping is the planting of the trap crop in alternating rows within the main crop.

The major benefit of trap cropping is that insecticides are seldom required to be used on the main crop and this enhances the natural control of pests. Moreover, trap crops may also attract natural enemies thus enhancing natural control. The overall use of pesticides is clearly less than in conventional farming, making the strategy environmentally attractive. However, it needs to be emphasized here that the trap crop may also serve as a pest nursery for the target pest or some other pest(s). Some highly mobile natural enemies may be attracted to and aggregate on the trap crop just as on their host or prey. They are likely to be destroyed by insecticide sprays. A thorough understanding of the agroecocystem is, therefore, essential for recommending trap cropping as a means for minimising pest damage.

Globally, trap cropping has proved successful in experimental studies in a number of crops and has been recommended for large scale utilization (**Table 4.2**). Trap cropping has been successfully employed on a large scale in four crop ecosystems, viz. cotton and soybean in USA, potatoes in CIS (erstwhile USSR) and Bulgaria, and cauliflower in Finland. In cotton/sesame intercrop trials in USA, row strips of sesame, constituting 5 per cent of the total acreage, were used as a trap crop to attract

Table 4.2 Examples of trap cropping practices applied in agroecosystems

Trap crop	Main crop	Method of planting	Pest(s) controlled
Alfalfa	Cotton	Strip intercrop	Lygus bug
Castor	Cotton	Border crop	*Heliothis/Helicoverpa* sp.
Chinese cabbage, mustard, and radish	Cabbage	Planted in every 15 rows of cabbage	Cabbage webworm, flea hopper, mustard aphid
Beans and other legumes	Maize	Row intercrop	Leafhopper, leaf beetles, stalk borer, fall armyworm
Chickpea	Cotton	Block trap crop at 20 plants/ m^2	*Heliothis/Helicoverpa* sp.
Collards	Cabbage	Border crop	Diamondback moth
Cowpea	Cotton	Row intercrop in every 5 rows of cotton	*Heliothis/Helicoverpa* sp.
Green beans	Soybean	Row intercrop	Mexican bean beetle
Horse radish	Potato	Intercrop	Colorado potato beetle
Indian mustard	Cabbage	Strip intercrop in between cabbage plots	Cabbage head caterpillar
Marigold (French and African marigold)	Solanaceous crucifers, legumes, cucurbits	Row/strip intercrop	Nematodes
Maize	Cotton	Row intercrop, planted in every 20 rows of cotton or every 10-15 m	*Heliothis/Helicoverpa* sp.
Napier grass	Maize	Intercrop, border crop	Stem borer
Nasturtium	Cabbage	Row intercrop	Aphids, flea bettle, cucumber beetle, squash vine borer
Okra	Cotton	Border crop	Flower cotton weevil
Onion and garlic	Carrot	Border crops or barrier crops in between plots	Carrot root fly, thrips
Soybean	Maize	Row intercrop	*Heliothis/Helicoverpa* sp.
Sudan grass	Maize	Intercrop, border crop	Stem borer
Sunflower	Cotton	Row intercrop in every 5 rows of cotton	*Heliothis/Helicoverpa* sp.
Tobacco	Cotton	Row intercrop, planted in every 20 rows of cotton	*Heliothis/Helicoverpa* sp.
Tomato	Cabbage	Intercrop (tomato is planted 2 weeks ahead at the plots' borders)	Diamondback moth

Source : www.oisat.org

Heliothis spp. from the main crop of cotton. Sesame, which is highly attractive to *Heliothis* species from the seedling stage to senescence, attracted large numbers of insects away from the cotton. It also attracted the parasitoid, *Campoletis sonorensis* (Cameron) which ultimately parasitized large numbers of *Heliothis* insects. Sesame, sunflower, marigold and carrot can be used as trap crops for *H. armigera.* In rice, trap cropping of rice for green leafhopper control resulted in 12 per cent higher economic return than chemical control and 29 per cent higher than the untreated control. The possibility of use of Indian mustard as preferred crop of *Plutella xylostella* (Linnaeus) was investigated. Indian mustard was found to be suitable trap crop for *P. xylostella.*

Planting Sudan grass (a commercial fodder grass), *Sorghum vulgare sudanense* (Piper) Hitchc. around maize fields reduced infestation on maize by *C. partellus* and *B. fusca* in Keyna, as considerable number of stem borers were trapped on Sudan grass. There was increase in maize yield (4.45 t/ha) due to reduction in stem borer damage as compared to control (3.44 t/ha). Planting grass around maize fields also increased parasitization of borers by *Apanteles flavipes* (Cameron) and *A. sesamiae* Cameron. As compared to maize mono field where only 4.8 per cent *C. partellus* and 0.5 per cent *B. fusca* larvae were parasitized, 18.9 per cent *C. partellus* and 6.17 per cent *B. fusca* larvae were parasitized in maize surrounded by grass.

Okra can be used as a trap crop around cotton for cotton jassid, American bollworm and spotted bollworms. The infested fruits that have large population of bollworms should be removed periodically and destroyed, and jassid can be easily controlled by spraying with any systemic insecticides. Marigold and *Nicotiana rustica* Linnaeus grown as trap crops are also preferred hosts of *H. armigera.* Planting of castor as trap crop diverts the population of *Spodoptera litura* (Fabricius) from cotton.

Some early crops are sown in narrow strips around a major crop to serve as a trap for the pests that might be common to both. The trap crop can either be harvested early or otherwise cut and used as fodder. The preferred host plants can also be grown around valuable crops, and when the pest has appeared, they can be cut and destroyed. Sesame can be sown around the cotton field to attract the red hairy caterpillar, *Amsacta moorei* (Butler) and the Bihar hairy caterpillar, *Spilarctia obliqua* (Walker). The cotton grey weevil, *Myllocerus undecimpustulatus* Faust, has a marked preference for *arhar* (*Cajanus cajan*), which can be sown as a trap crop. Bold seeded mustard (*Brassica juncea*), when used as a trap crop in cabbage, attracts 80-90 per cent diamond back moth population. The use of African marigold (*Tagitus erecta*) as a trap crop is quite effective for the control of tomato fruit borer.

The major benefit of trap cropping is that insecticides are seldom required to be used on the main crop and this enhances the natural control of pests. Moreover, trap crops may also attract natural enemies thus enhancing natural control. The overall use of pesticides is clearly less than in conventional farming, making the strategy environmentally acceptable. However, it needs to be emphasized that the trap crop may also serve a pest nursery for the target pest or some other pests. Some highly mobile natural enemies may be attracted to and aggregate on the trap crop just as on their host or prey and they are likely to be destroyed by insecticide sprays. A thorough understanding of the agroecosystem is, therefore, utmost essential for recommending trap cropping as a means for minimising pest damage.

Presence of Weeds

The presence of diverse vegetation within or near the field may add essential resources for predators or parasitoids and so enable them to find all their requirements near the pest population. Such resources include food, cover or alternate prey. Conversely, weeds may also adversely affect the orientation of predators and parasitoids to their prey. Many common weeds also act as hosts for oviposition, and provide a better ecological niche for the insects to hide, thus shielding them from natural enemies and insecticide sprays The common vetch or garden vetch, *Vicia sativa* Linnaeus is a common leguminous weed associated with chickpea in northern India. Removal of the weed at a time when maximum eggs are laid substantially reduces the incidence of pod borer,

H. armigera. Weeds may also aid in chemical repellence, interference in movement of insects or provide alternate host for natural enemies.

Pruning and Thinning

Some pests are normally carried from the old crop to the new one. It is particularly so in the case of perennial plants, such as the fruit-trees. Proper pruning of the undesirable portions of citrus plants is useful for keeping under check the citrus leaf-miner, *Phyllocnistis citrella* Stainton; citrus psylla, *Diaphorina citri* Kuwayana; the citrus red scale, *Aonidiella aurantii* (Maskell); San Jose scale, *Quadraspidiotus perniciosus* (Comstock); woolly apple aphid, *Eriosoma lanigerum* (Hausmann) and peach leafcurl aphid, *Brachycaudus helichrysi* (Kaltenbach) of stone-fruits.

Time of Sowing and Harvesting

By adjusting the time of planting, infestation by some pests can be prevented ; the egg-laying period of a particular pest can be avoided; young plants can be established before the attack starts; short duration crops can allow the minimum possible time for pests to multiply or they can mature before the pest appears. Similarly, by adjusting the time of harvesting, a crop can be saved from attack of the pest which might become abundant rather late in the season or the pests can be killed before they have completed their life-cycle. Where it is desirable to have a crop mature early, the purpose can be achieved by fertilizer application, by adjusting irrigation or by making a varietal selection for early maturity.

Early sowing can be used to minimise the damage to chickpea from *Helicoverpa armigera* (Hubner) in northern India. Two peaks of *H. armigera* occur during December and March in the *rabi* season. During the second peak, the pest inflicts severe damage to chickpea crop. Early (October) sown crop escapes with least damage. Late-sowing (December and January) matures during late March to April and suffers heavy damage. November sown crop also suffers moderate damage.

If maize meant for seed is sown after the 15th of August in the Punjab, it escapes a heavy attack of the maize-borer. Moreover, the borers in August and September are heavily attacked by their parasites and, as a result, the pest remains under control. The early sowing of rice in the Punjab between the 3rd week of May and mid-June is helpful in protecting it from the attack of rice-borer, *Scirpophaga incertulas* (Walker). The incidence of whitefly, *Bemisia tabaci* (Gennadius), is also severer on the early sown cotton than on the late-sown crop.

Another example to illustrate the importance of the time of sowing is that of wheat in America to save it from attack of the Hessian fly, *Mayetiola destructor* (Say). Hessian fly eggs are laid on the upper surface of young winter wheat leaves. By delayed planting, suitable hosts are not available until most of the overwintering adult flies emerge and die.

It is recommended that sugarcane should be harvested before mid-February, when moths of the top-borer appear on wings. If it is not possible to harvest the crop, the terminal portions of cane should be cut and fed to cattle. The maize borer, *Chilo partellus* (Swinhoe), which hibernates in the stalks of the crop should be killed before its emergence in spring, by chopping up stalks which are kept for use as fodder.

The attack of mustard aphid on the seed pods of *sarson, toria* (*Brassica campestris*) and other cruciferous crops is severe in January-February. In certain years, the attack is so heavy that seeds are not formed at all. Early maturing varieties or the early sowing of these crops is the best solution for protection against this fast-reproducing pest. Damage of pod-borer, *H. armigera*, in chickpea and pigeonpea can be reduced by growing early maturing varieties.

Destruction of Crop Residues

A great majority of insect pests, particularly those which hibernate, are found in various portions of the plants remaining in the fields or orchards. After spending their inactive period in such crop residues, they emerge as adults, more or less at the same time as the new crop is grown. Therefore, it is most logical to destroy such remnants of the crops thoroughly and systematically.

After the crops are over, the stubble of rice and sugarcane should be ploughed up, collected and burnt, so that the hibernating borers are destroyed. The cotton stem borer, *Sphenoptera gossypii* Cotes, hibernates in the lower portion of the stalks. If cotton sticks are cut below the ground level and are used as firewood, the pest can be destroyed. Deep harvesting of sugarcane fields which are to be ratooned, provides protection to the crops from root borer and scale insect damage.

The old and dry cucurbit creepers and other debris are sometimes allowed to remain in the fields. A number of insects, like the brinjal *hadda* beetle, *Epilachna* spp.; red pumpkin beetle, *Raphidopalpa foveicollis* (Lucas); the dusky cotton bug, *Oxycarenus laetus* Kirby, and a host of other pests hibernate in crop remnants. Their destruction, immediately after the crop is over, will go a long way in reducing insect population.

Infested fruits of guava, melon, peach, plum, apricot, etc. which contain maggots of fruit flies and the fruits of walnut which contain grubs of the walnut weevil should be collected and burnt or buried deep in the soil.

MECHANICAL AND PHYSICAL CONTROL

Mechanical and physical control measures involve the use of force or physical factors of the environment with or without the aid of special equipment. The physical control measures give immediate tangible results and are generally popular and convincing to the farmers, even though they are time-consuming, laborious and are often applied when much damage has already been done. These control measures are generally ineffective on a large scale and cannot be applied commercially.

Manual Labour

It means working with hands, sometimes with the aid of some simple equipment, like bags, nets, etc.

(*i*) **Hand-picking.** This is the most ancient method employed by man and is still being used for picking out lice from human hair. In the field, insects can be hand-picked if they are: (*a*) easily accessible to the picker, (*b*) large and conspicuous, and (*c*) present in large numbers. This method is recommended for dealing with adults and egg-clusters of the lemon butterfly, grubs of the mustard sawfly, *Athalia lugens* (Klug) and all the developmental stages of *Epilachna* spp.

(*ii*) **Handnets and bagnets.** The collection of adults with hand-nets is recommended for pyrilla, *Pyrilla* spp., when these insects are migrating in April-May from maize to sugarcane. The field-bag is a strong cloth bag, 2 metres long with its mouth measuring 1 × 1.5 metres and supported with bamboo sticks and two strings on the upper side. It is scraped on the surface of the ground by two men and is recommended against surface grasshoppers, rice grasshopper, crickets, etc. Six men can cover one hectare in six hours. A one-man field bag can also be devised by reducing the size of its mouth.

(*iii*) **Beating and hooking.** Killing houseflies with fly-flappers and locusts with brooms or thorny bushes, is effective. On coconut palms, the rhinoceros beetle can be picked out of the holes with the help of crooked hooks made of iron.

(*iv*) **Shaking or jarring.** Shaking small trees or shrubs, particularly early in the morning in the cold season when the insects are benumbed, and collecting them in open tubs containing kerosenized water or simply burying them in pits is effective against locust and the defoliating beetles, *Adoretus* spp.

(*v*) **Sieving and winnowing.** These are commonly employed against insect pests of stored grains. A good number are removed with these operations, particularly the grubs of *Tribolium castaneum* (Herbst) and *Trogoderma granarium* Everts, which infest wheat.

(*vi*) **Clipping.** Clipping off the top of rice seedlings containing egg masses of yellow stem borer and grubs of hispa would reduce carry over of infestation from seed bed to field.

(*vii*) **Mechanical exclusion.** Mechanical exclusion consists of the use of devices by which insects are physically prevented from reaching crops and agricultural produce. The various methods include :

- The application of a fluffy cotton band 15 cm wide or a band of a sticky material like 'Ostico' or a band of a folded slippery sheets like alkathene around the trunk of a mango tree to prevent the upward movement of the mango mealybug, *Drosicha mangiferae* (Green).
- Screening windows, doors and venltilators of houses to keep away house flies, mosquitoes, bugs, etc. is quite helpful. In the morning and at dusk when mosquitoes gather on the screen they can be squashed with a piece of cloth.
- Wrapping individual fruits of pomegranate and citrus with butter paper envelopes to save them from attack of the pomegranate butterfly, *Deudarix isocrates* (Fabricius) and fruit-sucking moths, *Ophideres* spp., respectively. Maize cobs can be protected from the attack of crows if the nearest leaf is wrapped around the exposed portion of the cob.
- Trenching fields or erecting barriers, 30 cm high, in order to save crops from the invading bands of locust hoppers or the red hairy caterpillars, is quite helpful.
- Placing four legs of a meat-safe in vessels containing water to prevent ants from ascending, is also useful.
- Using red light in the monsoons to keep away most of insects, and to keep the fields well lit with white light at night to protect against certain insects has been quite effective. Light reflection by aluminium foil is effective against aphids. Similarly, light reflected by plastic ribbon bands or plastic flags hung in the ripening rice fields will protect the crop from bird attack.
- Scaring birds by creating noise with explosives is quite effective; an automatic device is available in which an explosive gas catches fire intermittently and a loud noise is produced.

(*viii*) **Mechanical traps.** Various types of traps have been devised for collecting and killing different types of insects.

(*a*) *Cricket trap.* This is a deep cylindrical vessel containing beer as a bait and having wooden splinters to aid crickets to reach the bottom.

(*b*) *House-fly trap.* The house-fly trap is a box containing a piece of stale cake, with a side opening for the insects to get in only to be trapped in a wire-gauze cage on the top.

(*c*) *Light traps.* Light traps against the red hairy caterpillar, *Amsacta moorei* (Butler) and the *ber* beetle, *Adoretus* spp. have given good results. An electric bulb or a petromax lamp is placed in the centre of a wide flat vessel containing kerosenized water in which the moths or the beetles get drowned. Trapping adults through light traps has proved useful in controlling plassey borer, root borer and white grub damage in sugarcane.

(*d*) *Air suction traps.* These may be fixed in godowns against stored grain pests.

(*e*) *Electric traps.* These are live metal screens on which birds or insects are electrocuted.

(*ix*) **Burning.** The burning of locust adults or hoppers with the help of flame torches and flame-throwers, although costly, has a good psychological effect in mobilizing the public for locust control operations.

Manipulation of Physical Factors of Environment

(*i*) **Application of heat**

(*a*) Superheating of empty godowns to a temperature above 50°C for 10-12 hours will kill the hibernating stored-grain pests.

(*b*) Exposing infested grain to the sun on a *pucca* floor in June also kills stored grain insects in the adult stage.

(*c*) If cotton seed is exposed to 52°C for 5 minutes, the hibernating larvae of the pink bollworm, *Pectinophora gossypiella* (Saunders), are killed.

(*d*) By steaming woollen clothes, the woolly bear, *Antherenus vorax* (Waterhouse), is killed

(*ii*) **Application of cold**

(*a*) Refrigeration at 5-10°C of all eatables, including dry fruits and woollen clothes, will kill insects.

(*b*) When stored grains are exposed to subzero temperatures by opening doors and windows of godowns, the insects are killed.

(*iii*) **Manipulation of moisture**

By raising or lowering the moisture content of food and other materials, unfavourable conditions are created for insect pests :

(*a*) Draining stagnant water kills the breeding mosquitoes.

(*b*) Reducing moisture content of grains below 8 per cent will make these unfit for the consumption of stored grain pests.

(*c*) Soaking logs in water over extended periods (15 days) drowns the boring weevils and larvae of the wood wasps.

POTENTIAL AND CONSTRAINTS

A number of physical, mechanical and cultural control measures are in use ever since the evolution of agriculture. Some of these methods were modified and refined over time to increase their efficacy. These practices are generally prophylactic in nature and are frequently the first line of defence against crop pests. The major advantages of the use of these practices are as below :

- Most of these practices result in little or no added production cost, because they are merely variations in the timing or manner of performing operations that are normally necessary to the production of the crop. Some of the techniques are, however, labour intensive, and, therefore, become highly expensive if the labour is costly.
- These tactics are environmently safe as they do not have significant distruptive effects on the non-target organisms and environmental quality.
- These are usually compatible with other components of IPM like resistant varieties, bicontrol agents and even insecticides.
- Usually, the chances of development of resistance to these practices are considered remote.

In spite of these advantages, these practices are not widely used in modern agriculture due to a number of limitations.

- A thorough knowledge of the ecology of the host plant, insect pest, natural enemies and their interactions is required for developing suitable cultural control practices. The generation of this information requires long term exhaustive studies.
- The measures have to be taken long in advance of the pest attack and in many cases even before sowing the crop.
- In most cases, these practices alone are not sufficient to bring insect pest populations below economic injury level. So, these have to be supplemented by other methods.
- These practices have to be developed locally as their effect on the pest may vary with various climatic and biotic factors.
- Many of the practices are obstacles to farmers' aspirations to intensify and mechanise their holdings.
- A cultural control practice directed against one pest may well increase the chances of survival and multiplication of another pest. This is akin to secondary pest outbreaks in case of pesticide application.

- Many of the cultural practices are effective only when undertaken on an area-wide basis. Ideally, all farmers may adopt a useful cultural practice voluntarily but unfortunately some farmers generally hold out. Such farmers called 'spoiler holdouts' may impair the success of a programme by failing to adopt a necessary practice. This can only be corrected by enacting and enforcing suitable legislative measures. This is often difficult in a democratic set up.

Many of the traditional cultural practices may not be useful under modern, intensive agriculture. Concerted research efforts are needed to develop improved cultural control measures which have great potential for use in sustainable agroecosystems. A combination of several cultural control practices can help to lower the general equilibrium position of many insect pests. If properly managed, some of these practices may obviate the need for insecticidal applications. Research efforts are needed to substantiate the usefulness of these methods as an economical way of controlling pests at different stages of crop growth. There is a need to identify and apply these measures in specific situations, where other measures like the application of pesticides is not possible or uneconomical. Some of the legislative meausures have played a pivotal role in preventing the introduction of exotic pests or spread of indigenous pests to new areas. There is a need to apply these laws more stringently to avoid potential pest problems.

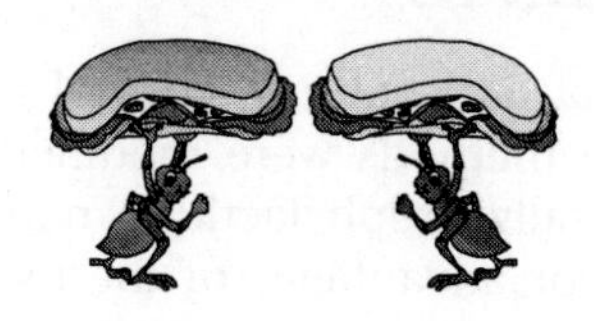

CHAPTER 5 BIOLOGICAL CONTROL

INTRODUCTION

A large number of predators, parasitoids, bacteria, fungi and viruses regulate the populations of insect pests under natural conditions. This regulation has been termed as 'biological control', and has been defined as the action of parasitoids, predators or pathogens in maintaining another organism's population density at a lower average than would occur in their absence. In the applied sense, it may be defined as the utilization of natural enemies to reduce the damage caused by noxious organisms to tolerable levels. Success in biological control is measured as complete, substantial or partial according to the extent of population regulation achieved.

Complete success is achieved when the populations of the target pest are maintained below economic threshold levels in an extensive area so that pesticide treatments become rare, if ever necessary. Substantial success is achieved if only occasional pesticide treatments are required to maintain pest populations below economically damaging levels. Partial success refers to cases in which pesticide application remains necessary but less frequent or where complete success is achieved in only a major portion of the pest-infested area.

A *parasite* is an organism which at one time or other lives on the body of the host which may or may not be killed after it has completed development. A *parasitoid* is an organism which completes its life on one host only and kills it. A *predator*, on the other hand, is a free living animal and kills the host immediately. *Superparasitism* is a type of parasitism where more individuals of the same species are present in a single host than can complete development in a normal way. *Multiple parasitism* is a type of parasitism where the host is attacked by two or more species of parasitoids. *Hyperparasitism* is a type of parasitism in which a parasitoid attacks another parasitoid.

In 1949, E.A. Steinhaus coined another term 'microbial control' in which pathogens are employed to control insect pests. Since then some scientists feel that the scope of biological control should be further widened by including host plant resistance, control by agricultural practices, genetic control by the use of lethal genes, control by causing sterility in the pest either by radiation or by chemicals. These methods, including the destruction and control of weeds by insects, are also termed by some as para-biological control methods or bio-ecological methods. The control of pests by chemicals is quite distinct from others.

HISTORICAL DEVELOPMENTS

Biological control came into prominence in recent times owing to some spectacular successes achieved in various parts of the world. Many practices of biological control have, no doubt, been passed down from times immemorial. In olden time, the Chinese used Pharaohs' ant, *Monomorium pharaonis* (Linnaeus) for the control of pests of stored grain and other products. Another predaceous ant, *Oecophylla smaragdina* (Fabricius), was used by the Chinese in citrus groves to control the foliage feeders. This very species was also used by the datepalm growers of Yemen to control insect pests that appeared at various stages of development of the fruit.

In more recent times, the Indian Mynah, *Acridotheres tristis* (Linnaeus), was introduced in Mauritius in 1762, to control the red-locust. In 1873, Riley sent the predaceous mite, *Tyroglyphus phylloxerae* Riley, from the United States to Europe for the control of the grape phylloxera, *Daktulosphaira vitifloliae* (Fitch), which was causing havoc to the vines.

The first significant success in controlling a pest was achieved on the suggestion of C.V. Riley of California in 1888. The Vedalia beetle, *Rodolia cardinalis* (Mulsant), was introduced from Adelaide (Australia) into California for the control of the cottony cushion scale, *Icerya purchasi* Maskell, on citrus which had been introduced earlier from Australia accidentally. This very scale has been controlled in southern India by using the same method and, in addition, the woolly apple aphid, *Eriosoma lanigerum* (Hausmann), has been controlled in the western Himalayas by the introduction of its parasitoid, *Aphelinus mali* (Haldeman), from USA and subsequently from Canada and South Africa also. Likewise, the American strain of *Encarsia perniciosi* (Tower) was imported in 1958 and was successfully released in the western Himalayas for the control of the San Jose scale, *Quadraspidiotus perniciosus* (Comstock). Subsequently, the Russian strain was also introduced and it gave even better results.

Some of the successful examples of biological control of insect pests by introduction of natural enemies in Asia are listed in **Table 5.1.**

It has been reported that the eggs of *Scirpophaga nivella* (Fabricius) are parasitized by *Telenomus beneficiens* (Zehntner) up to 90 per cent in Tamil Nadu. *Isotima javensis* Rohwer. parasitizes the larva of S. *nivella* up to 70 per cent in the tropical areas. Species of the hymenopterus genus *Trichogramma* occur world-wide on a wide range of crops and hosts. Research on the use of this tiny wasp in biological control started at the turn of the 20th century and a method of mass-producing *Trichogramma* on eggs of the Angoumois grain moth, *Sitotroga cerealella* Olivier, was developed in 1930 in the USA. Later, similar mass-production methods using the eggs of this species or other lepidopteran stored product pests gave rise to a world-wide use of this parasitoid as a biological control agent. About 18 different species of this egg parasitoid are being used to control pests on maize, sugarcane, rice, soybean, cotton, sugarbeet, vegetables and pine.

Releases of *Epiricania melanoleuca* (Fletcher) egg masses @ 4-5 lakh and 4000-5000 viable cocoons per hectare were found effective for the control of sugarane pyrilla in parts of Andhra Pradesh, Kerala, Karnataka and Madhya Pradesh. *Trichogramma chilonis* Ishii has been successfully established in Tamil Nadu after it was released at the rate of 1,25,000 adults per ha in 2-3 split doses during the months of March and April, and it reduced the incidence of *Chilo infuscatellus* Snellen by 10-15 per cent in the Punjab. When released in Tamil Nadu, it also successfully parasitized *Chilo sacchariphagu indicus* (Kapur).

Releases of T. *chilonis* from March to June have been used for the control of sugarcane borers in an area of 30,000 ha in Pakistan where the damage has been reduced from >16 per cent to <5 per cent. Similarly, releases of *Trichogramma japonicum* Ashmead at $50{,}000^{-1}$ ha during egg laying period of rice stem borer have reduced borer damage. Releases of T. *japonicum* @10 adults per m^2, 4-8 times during the season reduced rice leaf folder damage by 12-60 per cent. In 1962, the predaceous snail, *Euglandina roses* Ferussac, was introduced from Bermuda and was released in Orissa to control a new pest, the giant African snail, *Achatina fulica* Ferussac.

Table 5.1 Important examples of successful biological control of insect pests of agricultural crops by use of introduced natural enemies in Asian countries

Year	Crop	Insect pest	Natural enemy	Region or country	Country from where imported
1925-26	Citrus	Spiny black fly, *Aleurocanthus spiniferus* (Quaintance)	*Encarsia smithi* (Silvestri)	Kgushu Island (Japan)	Japan Mainland
1925-29	Coconut	Coconut moth, *Levuana iridiscens* Bethume-Baker	*Bessa remota* Aldrich	Fiji	Malaysia
1928-29	Coconut	Coconut scale, *Aspidiotus destructor* Signoret	*Cryptognatha nodiceps* Marshall*	Fiji	Trinidad
1920s	Apple	Woolly apple aphid, *Eriosoma lanigerum* (Hausmann)	*Aphelinus mali* (Haldeman)	Assam (India)	England
1929-31	Wattle of Commerce, *Acacia decurrens*	Cottony cushion scale, *Icerya purchasi* Maskell	*Rodolia cardinalis* Mulsant*	India	USA
1930s	Citrus	Cottony cushion scale, *I. purchasi*	*R. cardinalis**	Cheju Island (Korea)	Australia
1933-34	Coconut	Coconut leaf-mining beetle, *Promecotheca coeruleipennis* Blanchard	*Pediobius parvulus* (Ferriere)	Fiji	Java
1934-35	Apple	Woolly apple aphid, *E. lanigerum*	*A. mali*	Korea	Japan
1940s	Citrus	Redwax scale, *Ceroplastes rubens* Maskell	*Anicetus beneficus* Ishii & Yasumatsu	Various islands in Japan	Uncertain observed in Japan
1958-60	Apple	San Jose scale, *Quadraspidiotus perniciosus* (Comstock)	*Prospaltella perniciosi* (Tower)	India	China
1960	Apple	*Q. perniciosus*	*Aphytis diaspidis* (Howard)	India	USA
1964	Castor	*Achaea janata* (Linnaeus)	*Telenomus* sp.	India	New Guinea
1965	Coconut	*Oryctes rhinoceros* (Linnaeus)	*Platymeris laevicollis* (Distant)*	India	Zanzibar
1975	Citrus	Wax scale, *C. rubens*	*Anicetus beneficus* Ishii	Korea	Japan
1980-86	Citrus	Arrowhead scale, *Cinaspis vanonensis* (Kuwana)	*Physcus fulvus* Compere & Annecke	Japan	Taiwan
			Aphytis yanonensis DeBach & Rosen		Taiwan

Natural enemies with asterisk (*) are predators, others are parasitoids.

Source : DeBach and Rosen (1991)

At global level, there have been 5634 recorded releases of 2119 species of entomophagous arthropods to control 597 pest species. However, complete control (no other measures needed) was obtained in 212 cases and substantial control (other measures rarely required) was recorded in 419 cases. The primary target pests belonged to Homoptera (2420 releases) followed by Lepidoptera (1772), Coleoptera(596), Diptera (533), Hymenoptera (137) and Heteroptera (96). The primary agents in classical biological control programmes have belonged to Hymenoptera (3887), Coleoptera (1168) and Diptera (457); the success rates recorded were 9.2, 8.1 and 12.3 per cent for these groups, respectively. In India, 163 exotic bioagents have been studied for utilization against crop pests and weeds. Of these, 69 species were recorded after releases, 4 provided partial control, 4 substantial and 6 recurring benefits worth millions of rupees (Dhaliwal and Koul, 2007)

BIOLOGICAL CONTROL AGENTS

Biological control agents can be grouped under vertebrates, arthropods, protozoa and microorganisms.

I. VERTEBRATES

Fishes, tadpols, frogs and toads destroy a large number of mosquito larvae and other harmful aquatic insects. On land also, frogs, toads, salamanders and lizards are active all day eating insects, many of which are crop pests. Toads and the wall lizards live almost exclusively on insects and they are particularly active near the street lights to which such insects as temites, crickets grasshoppers, leafhoppers, bugs and defoliating beetles are attracted at night.

It is very interesting to note that the wall lizard, *Gecko* sp., which normally hangs around the sources of light to catch photopositive insects, becomes a ground lizard when not many insects come to light or when the houses are properly screened to keep insects away. Under the latter situations, naturally they subsist on ground insects like crickets, fleas, ants, cockroaches, etc.

The majority of the snakes are harmless and feed on rats, mice and frogs. It is not well known to the public at large that of the total number of snakes observed by humans, 75 per cent are non-poisonous and only 25 per cent are poisonous, but all of them receive the same treatment from man, i.e. to kill the snake at first sight !

There are many useful birds which destroy pests of crops and are often seen following the plough and they pick up grubs and other insects that get exposed. The useful birds include the Indian roller, *Coracius benghalensis* L.; starling, *Sturnus vulgaris* L.; quail, *Coturnix coturnix* L.; patridges, *Francolinus* spp; king crow, *Dicrurus adsimilis* Bech; and even myana, *Acridotheres tristis* (L.) and house crow, *Corvus splendens* Vieillot. Among the mammals, th most active predators are shrew, mole, insectivorous bats, squirrel, hedghog, mongoose, etc.

II. ARTHROPODS

From the point of view of biological control, three groups of arthropods are important, i.e. spiders, mites and insects.

Spiders

The spiders universally live a predatory life and are constantly on the look out for insects as their food. They catch them directly or with the help of various types of snares made out of webs. *Ascyltus pterygodes* (Koch) is known to have controlled the outbreak of coconut moths in Fiji. The important species found actively feeding on sugarcane pyrilla in India are *Clubiona atwali* Singh and *Clubiona drassodes* Cambridge, belonging to the family Clubionidae. The populations of these spiders showed a positive correlation with the population of their prey.

The important genera of spiders are *Atypus, Scytodes, Pholeus, Lycosa, Argiope* and *Marpissa*. Since the spiders spin threads and are carried away by air over long distances, they are more or less universal in distribution and have acquired specialized habits. The climate inside the houses is controlled for human comfort and the same spider genera are found internationally and they are adapted to feed on household pests. For example, the spiders, *Scytodes* and *Pholeus,* which make irregular cobwebs inside houses can feed on a large variety of insects attracted to light or which feed near the ceiling. Thus, they trap houseflies, blow flies, moths, mosquitoes, etc. and feed on them.

Among the tree dwelling spiders the most spectacular is the triangular web forming, *Hyptiotes*, which generally lives on insects. It's snare is most complicated but uniform. The two ends of snare are stuck with two twigs of tree and third end is held by itself. When an insect strikes the snare, it loosens the end held by itself thereby engulfing and trapping the insect into the snare.

Mites

Vegetable mites are notorious pests of many crops. Some of them also infest man and domestic animals and cause scabies. *Acarapis woodi* (Rennie) lives in the thoracic tracheae of the honey bee and causes the acarine disease. Many other species have also acquired a parasitic life on insect pests. *Allothrombium* sp. lives as an ectoparasite on many small insects and *Entrombidium* sp. on the eggs of locust and grasshoppers. *Bdellodes lapidaria* (Kramer) was introduced from Europe into Australia and it exercised a good check on *Sminthurus viridis* (Linnaeus) (Collembola), a pest of clovers.

Insects

Insects form the single largest and the most important group of predators and parasitoids **(Figs. 5.1, 5.2)**.It is believed that 25-33 per cent of the insects are useful to farmers, as they control or suppress populations of known or potential pests. They belong to 15 orders and more than 240 families. The total number of species recorded in India alone would run into thousands. Praying mantis (Mantidae) **(Plate 5.1A)**, even though few in numbers, devour a large number of insects of all sorts. Among Neuroptera, the green lacewing, *Chrysoperla* spp. **(Plate 5.1B)**, and the brown lacewing flies feed voraciously on aphids and other soft-bodied insects at the rate of 160 individuals per day. Their larvae are very active and have large grooved mandibles. The antlions, *Myrmeleon* spp., also belong to this order and their larvae catch ants and other insects in cone-shaped pitfalls made in sand. Among the predaceous bugs of the family Reduviidae, *Reduvius cincticrus* Reuter and *Acanthaspis rama* Distant, may be mentioned. They suck body juices from other bugs. Pirate bugs (Anthocoridae) **(Plate 5.1C)** prefer thrips larvae, but also feed on eggs and adult thrips, spider mites, insect eggs and small caterpillars. Among beetles, the tiger beetle, *Cicindela sexpunctata* Fabricius is very common in the North and Western India.

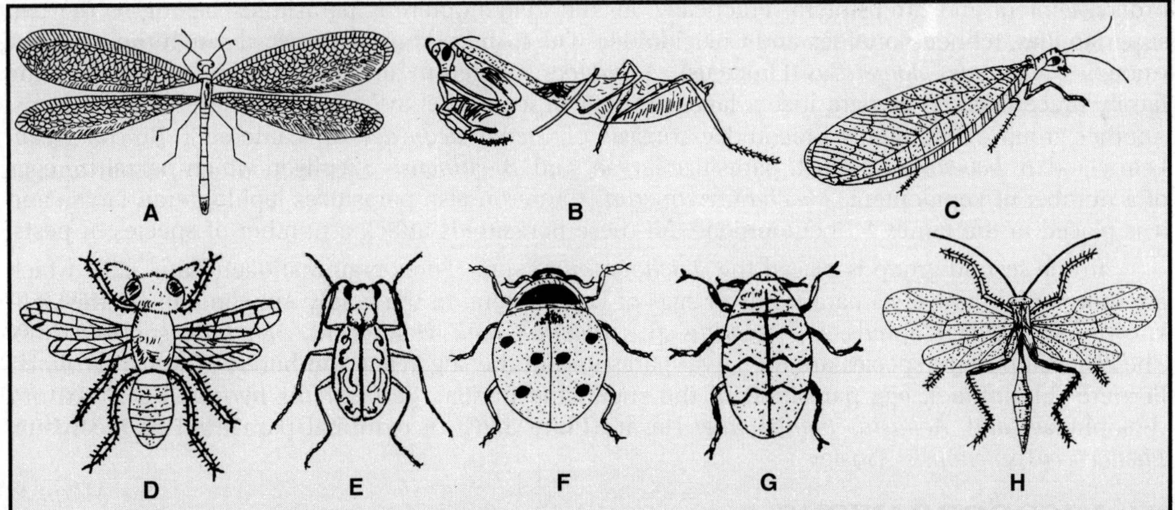

Fig. 5.1 Common insect predators. A. Antlion, *Dendroleon* sp. (Order Neuroptera); B. Praying mantid, *Stagmomantis* sp. (Family Mantidae); C. Common lacewing, *Chrysoperla* sp. (Family Chrysopidae); D. Bee killer wasp, *Philanthus ventilabris* Fabricius (Family Vespidae); E. Tiger beetle, *Cicindela* sp. (Family Cicindelidae); F. Ladybird beetle, *Coccinella septempunctata* Linnaeus (Family Coccinellidae); G. Coccinellid beetle, *Menochilus sexmaculatus* (Fabricius) (Family Coccinellidae); H. Ichneumonid wasp, *Phobocampe disparis* (Viereck) (Family Ichneumonidae)

The ladybird beetles form a very important group of predators of aphids in the larval stage as well as in the adult stage. The more well-known species are *Menochilus sexmaculatus* (Fabricius) **(Plate 5.1D)**, *Brumoides suturalis* (Fabricius) and *Coccinella septempunctata* Linnaeus **(Plate 5.1E)**, the last mentioned being very common on cabbage aphid in the spring. It diapauses in summer and also has an irrepresible tendency to migrate to the higher altitudes and congregate near the snowline. On the southern slopes of the Dhaula Dhar range (about 4260 metres above sea level), as many as 200,000 beetles were counted per square metre. The beetles can be collected from there and released in the plains, otherwise they get buried under snow and perish. The larvae of the syrphid fly, *Episyrphus balteatus* (DeGeer) and *Episyrphus viridaureus* Wiedemann **(Plate 5.1F)**, are also very important predators of aphids.

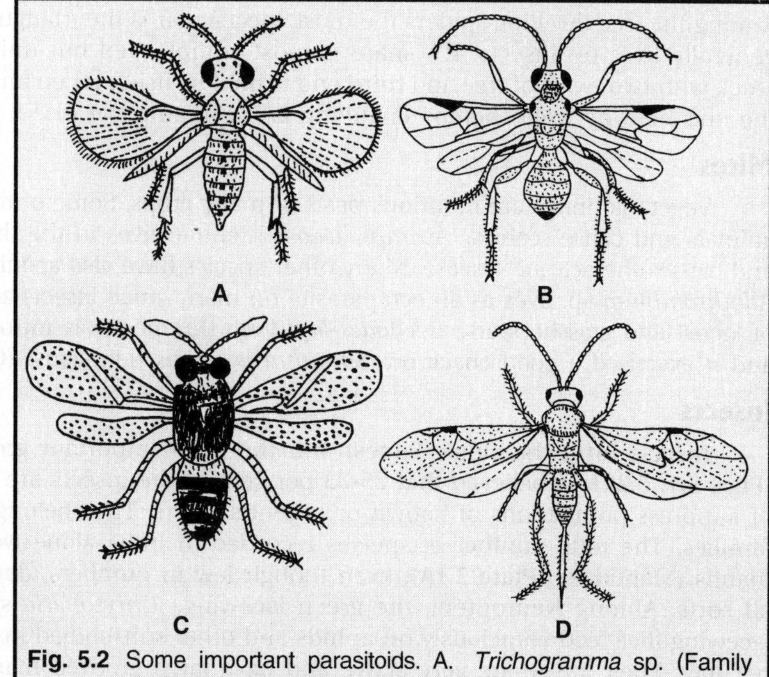

Fig. 5.2 Some important parasitoids. A. *Trichogramma* sp. (Family Trichogrammatidae); B. Braconid male; C. *Aphelinus jucundus* Gahan (Family Eulophidae); D. *Apanteles* sp. (Family Braconidae)

The vast majority of the parasitoids belong to Hymenoptera and the two other important orders, Diptera and Strepsiptera. Practically, all the hymenopterous parasitoids belong to the two superfamilies, Ichneumonoidea and Chalcidoidea. The former group is known through the common examples, *Apanteles glomeratus* (Linnaeus), *A. flavipes* (Cameron) and many others belonging to the family Braconidae. They parasitize a large number of lepidopteran larvae which are pests of crops. Another group of the family Braconidae consists of *Stenobracon deesae* (Cameron.), *Bracon greeni* Lefroy and *B. hebetor* Say, which parasitize larvae, and *B. chinensis* Szepligeti which parasitize eggs of a number of Lepidoptera. *Melcha ornatipennis* Cameron also parasitizes lepidopteran larvae and it is placed in the family Ichneumonidae. All these parasitoids attack a number of species of pests.

In the second group is placed the *Trichogramma* spp. (Trichogrammatidae)**(Plate 5.2B)** which are more or less universal parasitoids of eggs of Lepidoptera. In the family Aphelinidae are the well-known parasitoids of aphids, e.g. *Aphidius* sp., *Aphelinus mali* (Haldeman), *Aphelinus gossypii* Timb. The other important species are the larval parasitoid of the sugarcane top borer, *Elasmus zehntneri* Ferriere (Elasmidae), egg parasitoid of the sugarcane pyrilla, *Tetrastichus pyrillae* J.C. Crawford (Eulophidae) and *Aenasius bombowalei* Hayat **(Plate 5.2C)**, a nymphal parasitoid of mealybug, *Phenacoccus solenopsis* Tinsley.

III. MICROORGANISMS

More than 3000 microorganisms have been reported to cause diseases in insects. The honey bees and silkworms were the first arthropods which were observed and recorded to suffer from various diseases, since their products were very important in the man's economy. Pasteur's detailed studies on the 'Flacherie', as well as 'Pebrine' diseases of the silkworms and Agostino Bassi's excellent contribution towards the establishment of a relationship between a microorganism, the fungus, *Beauveria bassiana* (Balsamo) Vuillemin and the infectious white muscardine of silkworm, led to the development of insect pathology as a specific section or discipline in the science of Entomology.

Metchnikoff is generally credited with actually initiating the experimental work to demonstrate that a disease may be intentionally caused in insect pests of economic importance. In 1879 he had published his work on the role of green muscardine, *Metarhizium anisopliae* (Metchnikoff) Sorokin in the control of the wheat cockchafer, *Anisoplia austriaca* Herbst. Krassilstchik in 1888, mass produced spores of the fungus and applied them in field tests against insect pests. Subsequently, contributions on insect diseases by Steinhaus (1963) and his co-workers at the University of California, Berkeley, USA, gave the real impetus to researches on the use of insect pathogens as pesticidal agents.

Steinhaus (1949) coined the term 'microbial control' to express the pest population management through disease causing microorganisms. Microbial control includes all aspects of utilization of microorganisms or their by-products in the control of pest species. For the control of insect pest populations, many of the principles governing the role of insect parasitoids and predators also apply to entomogenous microorganisms.

An ideal microbial insecticide should have the following characteristics :

- It should ensure consistent suppression of pest populations to acceptable low densities.
- It should be economical for mass production.
- It should be available in formulations which should have long shelf life, remain stable in the target insect and disseminate quickly and uniformly.
- It should be safe to non-target organisms and man.
- It should be acceptable in the market to make development and production attractive to potential producers.

In nature, viruses, bacteria, protozoa, fungi and rickettsiae perform important role in the dynamics and natural regulation of insect and mite population. Their effect is most evident in epizootics of diseases which occur at intervals and under some circumstances decimate the host populations. Less obvious is the effect of insect pathogens in the enzootic stage or those which cause chronic or low grade infections. Other than causing the outright death, pathogens may interfere with insect development, alter reproduction, lower insect resistance to attack by parasitoids, predators and other pathogens, and influence the susceptibility of insects to control by chemical insecticides or other artificial methods.

Insect pathogens may be divided into two groups according to the means by which they enter and infect their hosts. One group which includes bacteria, protozoa and viruses, must be ingested in order to cause infection and mortality, and can be considered similar to chemical insecticides which act as stomach poisons. Some of these microorganisms, such as the viruses, are quite specific in their sites of development and they multiply only in certain tissues within the body of the host. Others, including bacteria, may cause a spreading septicaemia by growing diffusely in the tissues and body fluids. Some bacteria may kill their hosts purely by the activity of toxins which they produce during growth.

The second group, which includes the pathogenic fungi enter their hosts through the outer integument of the insect's body. These invasive organisms can be equated to contact chemical insecticides since they need not be ingested to cause infection. They are more subject to regulation by physical factors in the environment, since their penetrative stages generally are not very resistant to adverse effects of external conditions.

The successful use of diseases for insect control depends upon the biology and characteristics of both the host insects and the pathogenic microorganisms as well as the environment. Host insects must occupy the habitats suitable for introduction of a pathogen and they must have habits that enhance the possibilities of infection. Since disease is generally considered as a density-dependent factor of mortality, the insects that live in aggregations or which form large populations are more susceptible to epizootics as compared with the species which generally maintain low population densities.

The major emphasis in the application of microbes has been to field-collect or artificially mass culture a specific insect pathogen and disseminate it when the host is most susceptible to its effect. One approach is to introduce and colonize pathogens as permanent mortality factor in the host populations. This approach is called the *microbial introduction.* The success of this method has been demonstrated for (*i*) the milky disease bacteria, *Bacillus popilliae* Dutky and *B. lentimorbus* Dutky, to control the Japanese beetle, *Popillia japonica* Newman, in USA, (*ii*) the accidental introduction and establishment of a polyhedrosis virus which lead to the control of devastating European spruce sawfly, *Gilpinia hercyniae* (Hartig), in Canada, and (*iii*) the importation and successful dissemination of a nuclear polyhedrosis virus for the control of the European pine sawfly, *Neodiprion sertifer* (Geoffroy) in Canada.

Another microbial technique is to make the repeated applications of a pathogen as microbial insecticide for temporary suppression of insect pests. The best illustration of this is the development of a bacterium, *Bacillus thuringiensis* Berliner. This bacterium is produced by fermentation and is formulated as a dust, wettable powder or emulsion. The material is handled and applied in the same manner as a chemical insecticide. Its effect is short-lived and many applications are needed. In general, microbial agents meant for introduction must spread from relatively small inocula and should persist in the environment, allowing the nature to take its course. Those unable to spread and persist must be used as microbial insecticides, to be applied repeatedly.

The microbial control has numerous advantages. The relatively high degree of specificity of most pathogens tends to protect the beneficial insects. They are harmless and non-toxic to other forms of life because they do not possess any toxic residues. Many pathogens are compatible with insecticides to the degree that they may be used concurrently and in some cases, synergistically; the infection may cause the insect to be more susceptible to chemical poisons. Microbial pathogens are highly versatile in so far as the method of their application is concerned. Some of the pathogens may be introduced and then colonized, so that the control brought about may be permanent; while other pathogens may be used as sprays or dusts, just like insecticides. It is also easier and comparatively inexpensive to produce some of the pathogens. The low dosage required in some instances to get the desired control and the slow development of resistance to a microbial pathogen, further make the insect pathogens as ideal pesticidal agents for their use in pest management programmes.

Besides these advantages, microbial control methods also have certain disadvantages or limitations. These limitations include the correct timing of application with respect to the incubation period of the disease, the specificity of the pathogens, narrowing the spectrum of effectiveness where several pests are involved, the necessity to maintain the pathogens in a viable condition, the difficulty of culturing in large quantities and the required favourable climatic conditions for obtaining best efficiency. In addition to this, a thorough understanding of the interactions within the ecosystem is necessary for attempting the microbial control efficiently, effectively and economically. The major problems include the short residual effectiveness of pathogens under field conditions and the inadequate methods for obtaining their dissemination.

Bacteria

Approximately one hundred species of pathogenic bacteria have been recorded from various species of insects which could be either obligate or crystalliferous or facultative. Spores of obligate pathogens like *Bacillus popilliae* Dutky and *B. lentimorbus* Dutky, which are the causal agents of 'milky' diseases- Types A and B, respectively, in the population of Japanese beetle, *P. japonica,* are commercially available under different trade names, viz. 'Doom', 'Japidemic,', etc. Some of the selected soil sites, as for example a patch in a house lawn, can be spot-treated with the spores of these pathogens.

The soil offers an excellent habitat for the pathogen which is mainly carried from one site to another either by the irrigation water or by the movement of diseased grubs. Hence soil once treated ensures the persistence of pathogens and the control of the beetle on a long term basis. These bacteria being obligate pathogens, have to be mass-produced on living insect hosts. White grubs, belonging to the genus, *Holotrichia*, have been found to be susceptible to an infection caused by *B.popilliae.* Field trials with the milky disease spore formulations have been attempted in Gujarat State with considerable success and as many as 20-25 per cent of the grubs were found infected with the disease in subsequent years, in areas where the pathogen had been applied earlier. There is a considerable potential for the use of this bacterium against the white grubs of groundnut in India which are not easily controlled by insecticides.

Bacillus thuringiensis Berliner(Bt) is a crystalliferous spore former and unlike milky disease organisms, this bacterium can only be used for short term pest control. This is a unique microbial control agent in the sense that it is free from pollution, residual toxicity, biomagnification in non-target organisms, besides having several other advantages. *B. thuringiensis* has a wide spectrum of insecticidal activity in the class Insecta. More than 525 insect species belonging to 13 orders have been found to be infected by Bt around the world. It can be easily produced using ordinary fermentation technology. Standardized formulations have already been produced by several firms which sell them under their own trade names **(Table 5.2).**

Table 5.2 *Bacillus thuringiensis* (Bt)-based commercial pesticides

Bt strain	Trade name	Manufacturer(s)	Uses
Bt var. *aizurai*	Florback	Novo Nordisk	Diamondback
	Centari	Abbott Labs	moth
Bt var. *galleriae*	Certan	Sandoz Inc.	Wax moth larvae in honey combs
Bt var. *israelensis*	Bactimos	Philips Duphar	Larvae of mosquitoes and black flies
	Bactis	CRC	
	Skeetal	Novo Biokontrol	
	Teknar	Sandoz Inc.	
	Thurimos	Novo Biokontrol	
	Vectobac	Abbott Labs	
Bt var. *kurstaki*	Bactucide	CRC	Lepidopteran
	Bactospene	Philips Duphar	larvae on many agricultural crops
	Biobit	Novo Biokontrol	
	Bt	Korea Explosives	
	Condor & Cutlass	Ecogen Biotechnology	
	Delfin	Sando Inc.	
	Dipel	Abbott Labs	
	Javelin	Sandoz Inc.	
	Larve Bt	Knoll Labs	
	Sok	Nor-American	
	Thuricide	Sandoz Inc.	
Bt var. *sandiego*	M- One Plus	Mycogen Biotechnology	Beetles and weevils
Bt var. *sandiego tenebrionis*	Diterra	Abbott Labs	Beetles and weevils
	Trident	Sandoz Inc.	
	Gnatrol	Abbott Labs	
	Novodor	Novo Nordisk	
	Foil	Ecogen	
	M-Track	Mycogen	
Bt var. *thuringienis*	Muscabac	Formos	Flies

The bacterium, *B.thuringiensis* also produces toxins which are poisonous to insects. As many as 25 varieties/serotypes are known with different capacities to produce various toxins and poisons. This is the reason for variable toxicity of this bacterium towards different species of insects. The pH of midgut of an insect is an important factor in determining its susceptibility to the crystalliferous bacteria. In general, susceptible insect species have a high midgut pH value. Under such circumstances the endotoxin releases the toxic principle which further damages the tissues, thus resulting in an increase in the potassium ions in the haemolymph, the condition responsible for the paralysis of the host insect. These toxins alone can be formulated without the live bacteria and such preparations can be even more useful as control agents in areas wherever silkworm rearing is carried out; the chemicals will be free from the infectivity found in *B. thuringiensis.*

This bacterium is also reported to be safe to the honey bees wherever it has been used for treating the agricultural crops against various pests. It is non-toxic to man and vertebrates. The reason for this lack of toxicity is that, in mammals, the primary digestion of proteins is at low pH. The stomach enzyme pepsin, which has optimum pH value of 2, degrades the endotoxin/crystal toxin into an atoxic compound. But the exotoxin produced by some of the strains of this bacterium has been found to be toxic to mice when given through injection. It has also been observed that the thermostable exotoxin kills several dipterous species of insects in animal droppings. The exotoxin is a nucleotide and thus it is quite possible that it may produce mutagenic effects. Fortunately, not all the strains of B. *thuringiensis* are capable of producing the exotoxin. The presence of endotoxin alone would be sufficient for the consideration of this bacterium as the most powerful microbial control agent yet developed for the control of insect pests.

In India, different commercial preparations of *B. thuringiensis* have been tested against more than 50 species of crop pests. For example, 'Thuricide' has been used in an attempt to control sugarcane Gurdaspur borer in Punjab, reducing its infestation from 11.34 to 4.8-5.8 per cent. The effective control of *Pieris brassicae*(Linnaeus) and *Plutella xylostella* (Linnaeus) was achieved, when cauliflower crop was treated with the bacterium and the control thus achieved was only next to carbaryl. 'Dipel'- a commercial preparation when applied on Bengal gram gave effective control of the pod borer, *Helicoverpa armigera* (Hubner). Several Bt-based products are being marketed in India by various agrochemical industrial concerns **(Table 5.3).**

Table 5.3 Bt-based microbial pesticides marketed in India

Trade name	Target pest(s)	Crop(s)
Halt	*Plutella xylostella* (Linnaeus)	Cabbage
Biolep	*Helicoverpa armigera* (Hubner), *Pectinophora gossypiella* (Saunders), *Earias* spp.	Cotton
Bioasp	*H. armigera, P. gossypiella, Earias* spp.	Cotton
Delfin WG	*H. armigera, P. gossypiella, Earias* spp., *Spodoptera litura* (Fabricius)	Cotton
	S. litura	Castor
	P.xylostella	Cabbage, cauliflower
Biopit	Lepidopteran caterpillars	Various crops
Spicturin	*H. armigera*	Cotton
	Cnaphalocrocis medinalis (Guenee)	Rice
	P. xylostella	Cabbage, cauliflower
	S.litura	Chillies

While spraying *B. thuringiensis* it is important to maintain the pH of the spray fluid at neutrality. The performance of this bacterium has been enhanced by spray additives, Plyac, Triton-X, corn oil, mollasses, etc. In addition to these agricultural pests, *B.thuringiensis* has also been used against pests of forests, shade trees and ornamental plants. Effective control of the Gypsy moth is reported in the field tests with Thuricide 90 TS. Use of the bacterium for the control of stored grain pests is only in an experimental stage, because it is too expensive and pehaps it would not kill the whole range of insects found in the godowns or bins.

Recent reports indicate that insects have the capacity to develop resistance to *B. thuringensist.* Within the last few years, at least 16 insect species have been selected for resistance to Bt *delta*-endotoxins. These include *Aedes aegypti* (Walker), *Choristoneura fumiferana* (Clemens), *Chrysomela scripta* (Fabricius), *Culex quinquefasciatus* Say, *Ephestia cautella* (Walker), *Ephestia kuehniella* Zeller, *Heliothis virescens* (Fabricius), *Homoeosoma electellum* Hulst, *Leptinotarsa decemlineata* (Say), *Ostrinia nubilalis* (Hubner), *Pectinophora gossypiella* (Saunders), *Plodia interpunctella* (Hubner), *Plutella xylostella* (Linnaeus), *Spodoptera exigua* (Hubner), *Spodoptera littoralis* (Boisduval) and *Trichoplusia ni* (Hubner). Among these reported cases of resistance, only three involve resistance among wild populations. The Indian meal moth, *P. interpunctella,* evolved low levels of resistance in grain bins because of treating the grain with Bt. The diamondback moth, *P. xylostella* is possibly the most notable because it evolved high levels of resistance in the field as a result of repeated use of Bt in intense control programmes. Recently, the cabbage looper, *T. ni* has been reported to develop resistance to Bt in vegetable greenhouses in British Columbia, Canada (Dhaliwal and Koul, 2010).

The reports of development of resistance in the field populations of *P. xylostella* are essentially from the countries where *B. thuringiensis* is extensively used, viz. China 1000 tons, Philippines 300 tons, Malaysia 250 tons, and North America about 1000 tons per annum. In Japan also, high level of resistance to *B. thuringienis* was observed in *P. xylostella* collected from watercress plant. This occurred because watercress was grown throughout the year with a frequent use of *B. thuringiensis* formulations (15-20 times a year). In India, the populations collected from Tamil Nadu showed significantly high resistance to *B. thuringiensis,* whereas populations obtained from Delhi, Uttar Pradesh, Punjab and Maharashtra showed a high level of susceptibility despite their high level of resistance to many synthetic insecticides. Increased susceptibility to *B. thuringiensis* has been reported to be associated with synthetic insecticide resistance possibly involving monooxygenases in increasing susceptibility of resistant insects to *B. thuringiensis.*

The first report of development of field resistance to Bt in *T. ni* populations came from the commercial vegetable greenhouses in British Colombia, Canada. In a 3-year survey initiated in 2000, several greenhouse populations were surveyed multiple times within a growing season to monitor the rate at which resistance developed within a year in response to the grower Bt sprays. Bt resistance levels were directly correlated to the amount of Bt applied, and Bt resistance was observed to evolve repeatedly within one year as a consequence of grower spray programmes. The application of high doses of Bt greatly intensified the rates of resistance evolution. It is likely that the greenhouse environment plays a significant role in contributing to the development of resistance to Bt. The favourable environmental conditions and longer growing season in greenhouses increase the exposure period of *T. ni* to Bt, thus increasing the selection intensity for Bt resistance. Greenhouses probably also enhance Bt persistence by protecting Bt from sunlight degradation and rain. Furthermore, once resistance is present, the problem may be exacerbated by the application of high doses.

Resistance management strategies have been proposed as a means to decrease the rate at which resistance evolves and hence, to keep the frequency of resistance genes sufficiently low for insect control.

- An effective monitoring programme to detect as early as possible, shifts in pest susceptibility that could be abated in the initial stages.
- Use of mixtures of toxins with different mechanisms.
- Use of synergists to increase toxicity.
- Mosaic application to resort to time alterations rather than space alterations.

- Rotation of toxins to reduce the frequency of resistant individuals.
- Ultra-high doses of toxins that kill resistant heterozygotes and homozygotes.
- Refuges to facilitate survival of susceptible individuals.
- Generally, resistance to Bt is the consequence of a mutation(s) that alters an insect midgut receptor protein(s), so that it no longer binds to the Cry protein. However, if a toxin gene was engineered, so that toxin is bound to other midgut cell surface proteins, then resistance might be less likely to arise.

Fungi

There are over 72,000 described species of fungi and the total number of species in the world may be as high as 1.5 million. Fungi belonging to four groups, viz. Phycomycetes, Ascomyetes, Basidiomycetes and Deuteromycetes attack different species of insects. In contrast to other pathogens like bacteria and viruses that pass through the gut wall from contaminated food, fungi mainly infect their hosts through the integument. If ingested by the insect, the fungal spores do not germinate in the gut and are voided in the faeces. Infection, therefore, results form contact between a virulent infectious inoculum and a susceptible insect cuticle, its germination, the penetration of the germ tube through the integument and finally spread of the fungus through the host tissues. Usually high humidity is required for the successful germination of the spores and sunlight has an adverse effect on them. Therefore, the atmospheric humidity or more precisely, the microclimate surrounding the fungal spores is a very important factor for the successful use of fungal pathogens. Some of these fungi are known to produce toxic substances like 'aflatoxin' from *Aspergillus flavus* Link, 'beauvericin' from *Beauveria bassiana* (Balasamo) Vuillemin, and 'destruxin A and B' from *Metarhizium anisopliae* (Metchnikoff) Sorokin, which are also responsible for causing death of the host insects, but they are equally harmful to the mammals. There are, however, no fungal toxins under development at present for pest control.

With most entomopathogenic fungi, disease development involves nine steps :

- Attachment of infective units like conidia or zoospores to the insect epicuticle.
- Germination of the infection unit on the cuticle.
- Penetration of the cuticle, either directly by germ tubes or by infection pegs from appresoria.
- Multiplications of the yeast phase-hyphael bodies in the haemocoel.
- Death of the host.
- Growth in the mycelial phase with invasion of virtually all host organs.
- Penetrations of hyphae from the interior through the cuticle to the exterior of the insect.
- Production of infective units on the exterior of the insect.

More than 500 species of insects have been observed to be infected with fungi. Fungal epizootics have often been observed in the field among pest populations **(Plate 5.2D)**. In fact, the first record of insect disease is that of *Metarhizium anisopliae* (Metschnikoff) Sorokin in the wheat cockchafer, *Anisoplia austriaca* Herbst, in USSR in 1879. The common fungal diseases are caused by the green (*Metarhizium*) and the white (*Beauveria*) muscardine fungi and they have a wide host range. Species of *Entomophthora* and *Coelomomyces* have been tried against aphids and mosquitoes in different parts of the world. During the past few years, concerted efforts have been made to develop fungi as microbial insecticides, viz. *Verticillium lecanii* (Zimmerman) Viegas and *Hirsutella thompsonii* Fisher for the control of aphids and scales, and also the citrus rust mite, *Phyllocoptruta oleivora* (Ashmead). The latter is a specific fungal pathogen of Acarina, particularly of the eriophyid and tetranychid mites infesting citrus plants. Abbott Laboratories, USA have developed a safe commercial formulation of the conidia of *H. thompsonii*. Another fungus, *Nomuraea rileyi* (Farlow) Samson, is also being developed as microbial insecticide for many species of caterpillar pests. Some of the commercially available mycoinsecticides alongwith their uses are lised in **Table 5.4.**

Table 5.4 Main target pests of commercially available mycoinsecticides

Fungus	Product	Target pests
Benuveria bassiana	BotaniGard, Boverol. Naturalis-L, Proecol, Mycotrol, Beauverin, Bio-Power	Lepidoptera (diamondback moth, beet armyworm, cabbage looper, cutworm, etc.), Coleoptera (scarab beetle grubs, weevils, coffee berry borer, cutworms, etc.), Heteroptera (psyllids, stinkbug, plant bugs, leafhoppers, mealybugs, aphids, whitefly, etc.), Thysanoptera (western flower thrips)
Beauveria brongniartii	Betel, Schweizer, Beauveria	Lepidoptera (diamondback moth, beet armyworm, cabbage looper, cutworm, etc.)
Lecanicillium lecanii	Mycotal, Bio-Catch, Vertalec	Heteroptera (stinkbug, aphids, whitefly, plant and leafhoppers, mealybugs), Thysanoptera (western flower thrips, onion thrips)
Mctarhizium anisopliae	Bio-Magic, Bio-Catch-M	Coleoptera (scarab beetle grubs, weevils), Blattodea (termites), Heteroptera (leafhoppers), Orthoptera (grasshoppers), Lepidoptera (cutworms)
Mctarhizium flavoviridc var. flavoi'iride	BioGreen, BioCane	Coleoptera (scarab beetles, weevils), Orthoptera (grasshoppers and locusts), Blattodea (termites)
Isaria futnosorosea	Preferal, Priority, FuturEco, Nofly	Heteroptera (whiteflies, aphids, etc.), Lepidoptera (tomato moth), Acari (rust mites, spider mite, etc.)
Paecilomyces lilacinus	Bio-Nematon	Nematodes (Root knot, cyst, lesion, burrowing)

The fungal spores can be applied as dusts, sprays or granules. The methods of culturing fungi have been studied in detail. The spores of *B. bassiana* (Balsamo) Vuillemin and *M. anisopliae* can easily be mass-produced on simple media containing wheat corn and/or potato products. There have been some difficulties in the mass production of the fungi. *Entomophthora* require some special media like coagulated egg yolk. Colonization is usually attempted with relatively small inocula in the form of diseased insects or cultured material. For the use of microbial insecticides large amounts of material are needed, which can be applied satisfactorily with equipment designed for application of chemical insecticides. Timing of application must, of course, coincide with the presence of susceptible stages of the hosts, but timing should also be related to the environment. Application immediately after rain or irrigation would be better than in the dry weather. Besides, treatment in the evening will save the fungi from sunlight.

There seems to be a wide scope for the utilization of insect pathogen fungi in India. Locally available materials can be used for mass production of fungal pathogens and then for extensive field trials, particularly on pests of rice, cole crops and other irrigated crops. It has been determined that the conidia of *M. anisopliae* can initiate mycosis resulting in the death of host insects among the field population of *Pyrilla perpusilla* (Walker) when sprayed immediately after irrigating the sugarcane crop. The use of *Entomophthora* could cause 2.5-28.0 per cent infection among the mustard aphid, *Lipaphis erysimi* (Kaltenbach) under field conditions. *Cephalosporium lecanii* Zimmermann was found highly infective for controlling the coffee green bug, *Coccus viridis* (Green) in the field.

Before any wide-scale operation involving the use of fungal pathogens in any country is planned, these organisms must be thoroughly evaluated for their safety to vertebrates. Many of these fungi can be a potential source of allergenic and toxic reactions.

Protozoa

These are unicellular microscopic organisms which form cysts or spores and these adaptations play an important role in the epizootiology of infections among insect populations. Several species of protozoa are associated with insects with different degrees of infection. *Nosema apis* Zander causes nosema disease in honey bee and *Nosema bombycis* Naegeli causes the pebrine disease in silkworm.

More than 1000 species of protozoa pathogenic to insects have been described. Protozoan diseases are generally chronic in nature and take quite sometime to kill their hosts, hence they are often debilitative on insect populations and also the spread of the disease is slow. Since the efficiency of host's reproduction, and other physiological functions are generally reduced, protozoan disease ultimately leads to reduction in pest populations. *Glugea pyraustae* (Paillot), pathogenic to *Ostrinia nubilalis* (Hubner), is the most important factor maintaining corn borer populations at a level which facilitates economic control by other means. In USA, storage pests are excellent hosts for protozoan infections and this fact mostly goes unnoticed. Moreover, such pathogens cannot be utilized in a country like India where deliberate mixing of any insecticidal agent with foodgrains is not permissible and yet people do mix them.

Protozoa are generally non-specific in nature. Further, most protozoa are difficult to culture on artificial media. In spite of such limitations, considerable efforts have been made in the past few years to develop a microsporidian, *Nosema locustae* Canning, for the long time control of grasshoppers. This can be produced and applied effectively and efficiently, and can be regarded as safe microbial control agent for use. Similarly, another microsporidian, *Vairimorpha necatrix* (Kramer), has the high virulence, wide host range (covering as many as 36 lepidopteran pests), can also be mass produced at a reasonable cost and can be stored for short periods. Thus, *V.necatrix* is primarily, if not exclusively, a pathogen of phytophagous Lepidoptera. Though there are quite a few protozoan pathogens of insect pests reported from India yet none of them has been tried under field conditions for pest control.

Rickettsiae

Rickettsiae are the microorganisms which like viruses, are known to be obligate pathogens, though they also have many features characteristic of bacteria. They possess an active metabolism, which is so heterotrophic that their cultivation on simple artificial media is not usually possible. *Rickettsiella melolonthae* Wille & Martignoni, and *Rickettsiella popilliae* Dutky & Gooden are known to cause disease in the populations of *Melolontha melolontha* (Linnaeus) and *Popillia japonica* Newman, respectively. Because of their low host specificity and pathogencity for vertebrates, the use of these organisms in microbial control is not considered possible at present.

Viruses

A virus is an entity whose genome is an element of nucleic acid, either DNA (deoxyribonucleic acid) or RNA (ribonucleic acid) which is reproduced inside the living cells. It uses its host's synthetic machinery to direct the synthesis of the specialized particle called the virion (nucleic acid & coat) which contains viral genome.

There are six main groups of viruses recognized as causing diseases in insects and mites. These are the baculoviruses (Baculoviridae), cytoplasmic polyhedrosis viruses (Reoviridae), entomopoxviruses (Poxviridae), irridoviruses (Iridoviridae), densoviruses (Parvoviridae) and small RNA viruses (unclassified). These viruses occur naturally and produce diseases in Lepidoptera, Hymenoptera, Coleoptera, Diptera and several other smaller groups. Of these known viruses, many are closely related to those which are pathogenic to man, domestic animals, a wide range of invertebrates and plants. Only viruses in the Baculovirus group have no such dangerous relationships. Approximately 60 per cent of the 1200 known insect viruses belong to the family Baculoviridae and it is estimated that such viruses could be used against nearly 30 per cent of all the major pests of food and fibre crops. Baculovirus infections have been described in over 700 species of invertebrates including Lepidoptera (455), Hymenoptera (31), Diptera (27), Coleoptera (5), Neuroptera (2), Trichoptera (1), Thysanura (1), Siphonaptera (1), besides Crustacea.

Most of the insect viruses, unlike the plant viruses, have an inclusion body around the virions. Insect viruses, being obligate pathogens, need to be cultivated on live insect hosts or in live cells through tissue culture.

Most baculoviruses infect only the larval stages of susceptible insects, exceptions are NPVs that infect Hymenoptera and the nonoccluded virus of *Oryctes*. These viruses infect the adult as well as the larval stage. Baculoviruses must be ingested to infect the larvae, so they are used mainly to control open-feeding species. The occlusion bodies dissolve in the highly alkaline host mid-gut and the virus infects the host epithelial cells. The basic virus infection process is shown in **Fig. 5.3**. The virus replicates within the nuclei of susceptible tissue cells. Tissue susceptibility varies greatly between viruses with some NPVs being capable of infecting almost all tissue types and most GVs being tissue-specific replications (e.g. fat body cell only). The budded virus initiates infection to other tissues in the hemolymph, i.e. fat bodies, nerve cells, haemocytes, etc. The cells infected in the second round of virus replication in the insect larva also produce budded virus, but in addition occlude virus particles within polyhedra in the nucleus. The accumulation of polyhedra within the insect proceeds until the host consists almost entirely of a bag of virus.

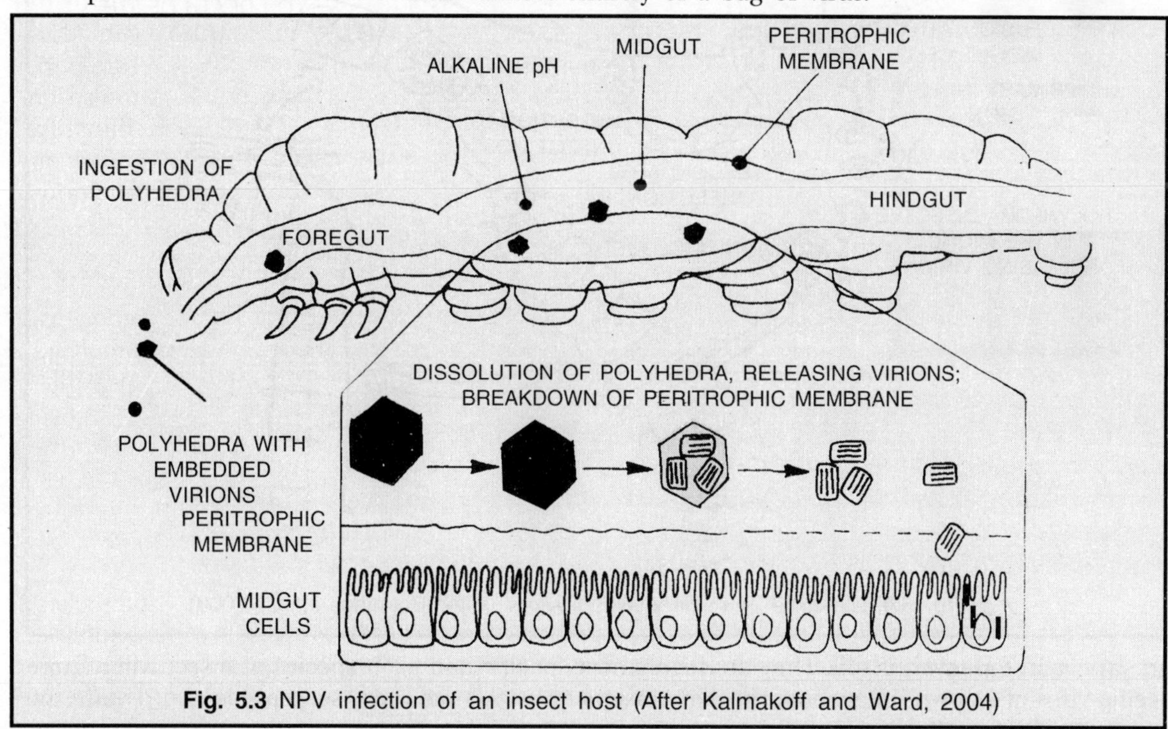

Fig. 5.3 NPV infection of an insect host (After Kalmakoff and Ward, 2004)

In the terminal stages of infection, the insect liquefies and thus releases polyhedra, which can infect other insects upon ingestion. A single caterpillar at its death may contain over 10^9 occlusion bodies from an initial dose of 1000. The infected larvae exhibit negative geotropism before succumbing to the virus infection, thereby facilitating widespread dissemination **(Fig. 5.4)**. The speed with which death occurs is determined in part by the environmental conditions. Under optimal conditions, target pests may be killed in 3-7 days, but death may be caused in 3-4 weeks, when conditions are not ideal.

The viruses, in general, lack quick knockdown effect and they take about one week for causing mortality. They are easily inactivated by sunlight and ultraviolet rays but they are known to accumulate

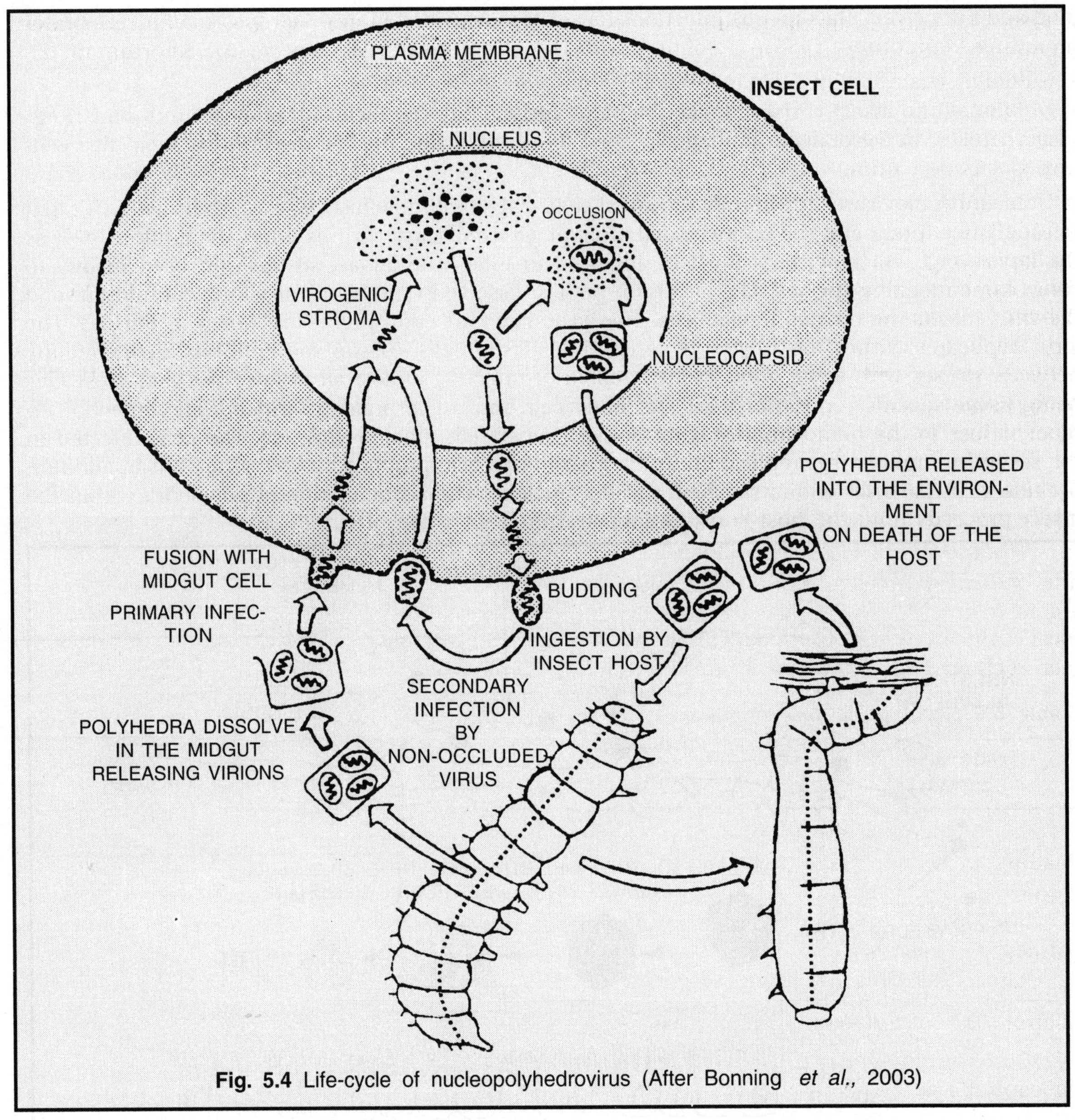

Fig. 5.4 Life-cycle of nucleopolyhedrovirus (After Bonning *et al.*, 2003)

and survive in top layer of soil. They are highly host specific and each species of insect will require specific virus of its own. In general, the early instars of insects are more susceptible and require the correct timing of application. From the point of view of microbial control, the baculoviruses (nuclear polyhedrosis viruses and granulosis viruses) seem to have more value. To some extent cytoplasmic polyhedrosis viruses too have been investigated but these are treated with caution. During the last decade, considerable work has been carried out on the practical use of these viruses, including application technology, safety testing, field testing and their use in pest management.

The use of nuclear polyhedrosis virus (NPV) for the control of soybean caterpiller, *Anticarsia gemmatalis* (Hubner), in Brazil, represents the world's largest programme for the use of an entomopathogen to control a pest on a single crop. The use of the viral biopesticide has increased from 500,000 ha in 1986-87 to 1 million ha in 1989-90, 1.5 million ha in 1995 and 2.0 million ha at present. The current annual savings at the soybean grower level are over US$ 14 million. Most importantly, this programme has avoided spraying of 23 litres of chemical pesticides resulting in considerable ecological benefits to the society. With recent achievements in commercial production of virus under controlled conditions, it is expected that the use of the biopesticide may very soon reach 4 million ha per year.

Spectacular results have been achieved through the use of baculoviruses in sawfly populations in USA and Canada. The other important viruses which have undergone extensive field testing include nuclear polyhedrosis viruses of cotton bollworm, *Helicoverpa* sp.; cabbage loopper, *Trichoplusia ni* (Hubner); cotton leaf worm, *Spodoptera littoralis* (Boisduval); alfalfa caterpillar, *Colias eurytheme* Boisduval; the spruce budworm, *Choristoneura fumiferana* (Clemens); gypsy moth, *Lymantria dispar* (Linnaeus); codling moth, *Cydia pomonella* (Linnaeus); tent caterpillars, *Malacosoma* spp.; the European spruce sawfly, *Gilpinia hercyniae* (Hartig); the European pine sawfly, *Neodiprion sertifer* (Geoffroy), and a few other species of *Neodiprion*; granulosis viruses of the cabbage butterfly, *Pieris rapae* (Linnaeus); codling moth, *C.pomonella,* and the spruce budworm, *C. fumiferana*. Several baculoviruses have been incroporated into integrated control programms for certain crops throughout the People's Republic of China. These are the nuclear polyhedrosis viruses of *Spodoptera litura* (Fabricius) **(Plate 5.2E)** and *Helicoverpa armigera* (Hubner) (cotton), *Sphrageidus similis* (Fuessly) (mulberry) and *Euproctis pseudoconspersa* Strand (tea)

The first baculovirus comprehensively tested for safety was that of the nuclear polyhedrosis virus of *Helicoverpa* spp., during 1965-71, which had paved the way for granting exemption from a residue tolerance in 1973 by the Environment Protection Agency of USA. Subsequently, the product, 'Elcar' containing this virus was registered in 1975. At present more than 10 viral insecticides have been registered in various parts of the world **(Table 5.5).**

Table 5.5 Baculovirus preparations registered for pest control in various parts of the world

Trade name of product	Year of registration	Insect species	Country
		Granulosis Viruses	
Madex	1987	*Cydia pomonella* (Linnaeus)	Czechoslovakia
Capex	1989	*Adoxophyes orana* (Fischer von Roesierstamm)	Czechoslovakia
Agrovir	1990	*Agrotis segetum* (Denis & Schiffermuller)	Germany
		Nuclear Polyhedrosis Viruses	
Biotrol VHZ & Viron H	1973	*Helicoverpa* spp.	USA
Elcar	1975	*Helicoverpa* spp.	USA
Biocontrol-1 & Virtuss	1976	*Orgyia pseudotsugata* (McDunnough)	USA
Gypcheck	1978	*Lymantria dispar* (Linnaeus)	USA
Lecont-virus	1982	*Neodiprion lecontei* (Fitch)	USA
	1983	*N. lecontei*	Canada
Monisarmiovirus (Kemira Sertivirus)	1983	*N.sertifer* (Geoffroy)	Finland
Virox	1984	*N.sertifer*	UK
Mamestrin	1988	*Mamestra brassicae* (Linnaeus)	Finland

Due to abundant and inexpensive labour, baculoviruses have been introduced into developing countries and are thus ideally suited for production. There are four strategies for using viral insecticides.

(*i*) The virus spreads from limited applications and permanently regulates the insect population through a classical biological control.

(*ii*) An epizootic is established through vertical and horizontal transmission, but reapplication may be necessary because control is not permanent.

(*iii*) A vertical inoculum in the environment is conserved and reactivated through environmental manipulation.

(*iv*) Repeated applications are used to control an insect population because there is no horizontal transmission of the virus–a strategy, which is widely used because of its effectiveness.

The prospects of microbial control of agricultural pests through the use of baculoviruses seem to be very good in India. This is evidenced by the fact that more than 30 species of insect pests have been reported to suffer from either the nuclear polyhedrosis virus or the granulosis virus infections. Extensive fieldscale testings have been carried out with nuclear polyhedrosis viruses of *S.litura, Mythimna separata* (Walker), *Spilarctia obliqua* (Walker) and *H. armigera.*

To develop these viruses for their further use as viral insecticides the three requisites that have to be fulfilled are (*a*) characterization of the virus, (*b*) field efficacy, and (*c*) safety evaluation. Once these requirements are fulfilled then the candidate virus can be commercially exploited. Research work has already been done on the development of nuclear polyhedrosis viruses for the control of *S. litura* and *S. obliqua* which are serious polyphagous pests in India. Various aspects like symptomatology, extent of natural incidence, production of the virus inclusion bodies, transovarial transmission, gross infectivity, effect of storage, surface disinfections, resistance to alkalies and temperature, and virulence of these viruses, besides their characterization, have already been investigated in detail. The nuclear polyhedrosis virus of *S. obliqua* is the only baculovirus which has so far been exhaustively safety-tested keeping in view the recommendations on safety evaluation of viral agents laid down by the World Health Organization.

Results of preliminary studies on the safety testing of nuclear polyhedrosis viruses of *M. separata, Amsacta albistriga* (Walker) and *H. armigera* also indicated the safe nature of these viruses. It is thus likely that the viruses might be developed in future as viral insecticides in India. The three *summum bonum* of the application and practice of viral control are (*a*) it is safe, (*b*) it can be easily and economically produced with locally available material and skill, and (*c*) it gives desired level of control.

Nematodes

The entomopathogenic nematodes have received increased attention as biological control agents in recent years. Beneficial nematodes are microscopic (0.6 mm) roundworms found associated with most of the insect orders. The two major groups of entomopathogenic nematodes that attack insects are *Steinernema* and *Heterorhabditis* **(Plate 5.2F)**. There are about 55 species of *Steinernema* the world over and a dozen of them occur in India. Among *Heterorhabditis*, about 12 are on record and the most common in India is *H. indica*. The nematode-bacterium complex has attained the status of a potential biopesticide because of their impressive attributes. They are unique due to their symbiotic relationship with bacteria in the genera *Xenorhabdus* or *Photorhabdus*. All species of *Steinernema* are associated with bacteria of the genus *Xenorhabdus* and all *Heterorhabditis* with *Photorhabdus* species.

Steinernema and *Heterohabditis* are obligate pathogens in nature. The major difference between *Steinernema* and *Heterorhabditis* is that *Heterorhabditis* adults are hermophrodites in the first generation but amphimictic in the following generations, whereas *Steinernema* adults are always amphimictic **(Fig. 5.5)**. The only stage that survives outside of host is the non-feeding third stage infective juvenile (IJ) or dauer juvenile. The IJs carry cells of their bacterial symbiont in their intestinal tract. After locating a suitable host, the IJs invade it through natural openings (mouth, spiracles, anus) or thin areas of the host cuticle (common only in *Heterorhabditis*) and penetrate into

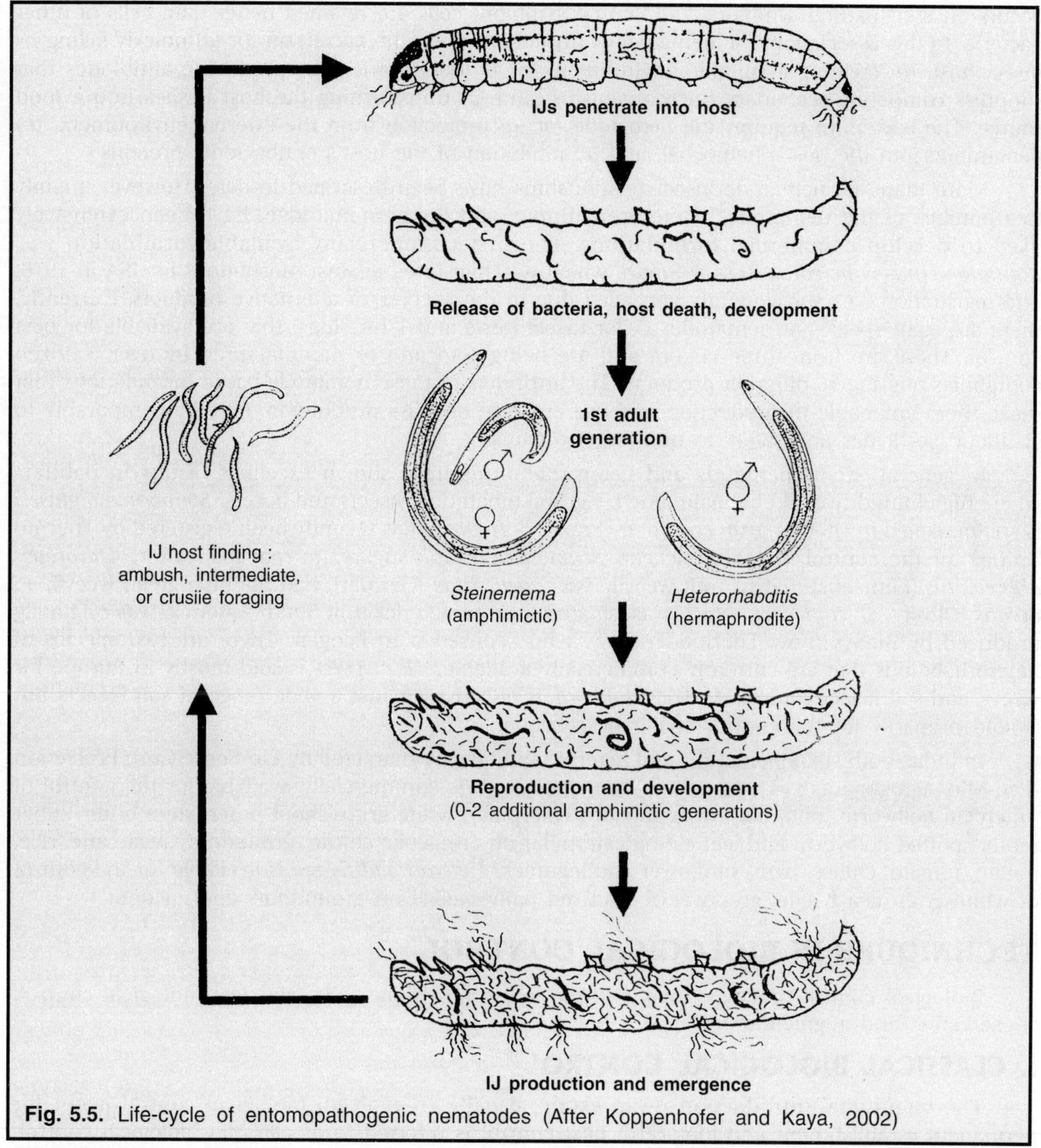

Fig. 5.5. Life-cycle of entomopathogenic nematodes (After Koppenhofer and Kaya, 2002)

the host hemocoel. The IJs recover from their developmental arrestment, release the symbionts, and bacteria and nematodes cooperate to overcome the host's immune response. The bacteria propagate and produce substances that rapidly kill the host and protect the cadaver from colonization by other microorganisms. The nematodes start developing, feed on the bacteria and host tissues metabolized by the bacteria, and go through 1-3 generations. Depleting food resources in the host cadaver leads to the development of a new generation of IJs that emerges from the cadaver in search of a new host.

Each nematode species is specifically associated with one symbiont species, although a symbiont species may be associated with more than one nematode species. This specificity has been demonstrated to operate at two levels. (*i*) Although the nematode can develop on other bacteria, but reproduction occurs on their natural symbiont. (*ii*) Natural symbiont cells are retained better than cells of other bacteria. In this association, the nematode is dependant upon the bacterium for (*a*) quickly killing its insect host, (*b*) creating a suitable environment for its development by producing antibiotics that suppress competing secondary microorganisms, and (*c*) transforming the host tissues into a food source. The bacterium requires the nematode for (*a*) protection from the external environment, (*b*) penetration into the host's hemocoel, and (*c*) inhibition of the host's antibacterial proteins.

More than 300 nematode-insect relationships have been described to-date. However, mainly two families of the nematodes, Heterorhabditidae and Steinernematidae, have been extensively used to develop commercial formulations. The first commercially available formulation was *Romanomermis culicivorax* Ross & Smith, which was registered against mosquitoes in USA in 1976. The registration was subsequently cancelled due to the success of alternative products. Currently, there are eight species of nematodes (7 for insect pests and 1 for slugs) that are available for pest control. These are from three genera and are being sold/and or manufactured by over a dozen companies offering 36 different products. The attributes of these nematode based formulations that make them amenable to production and use are ease of mass production, efficacy comparable to synthetic pesticides and safety to non-target organisms.

In general, steinernematids and heterorhabditids have shown excellent results in habitats where high humidity could be maintained, e.g. soil inhabiting insects and borers. *Steinernema glaseri* is still marketed in USA for grub control by Praxis. *S. riobrave* has recently been registered by Thermo Trilogy for the control of mole crickets, *Scapteriscus* spp.; sugarcane rootstalk borer, *Diaprepes abbreviatus* (Linnaeus) ; citrus root weevil, *Pachnaeus litus* (Germar), and the blue-green weevil, *P. opalus* (Oliver). *S. scapterisci,* which was isolated from mole cricket in South America, was originally produced by BioSys (now Thermo Trilogy), who licensed it to Ecogen. There are two species of heterorhabditids that are currently commercially available. *H. megidis* is sold mostly in Europe for weevil and soil insect control. *H. bacteriophora* is effective against a wide range of soil insects but is sold primarily for the control of Japanese beetle.

In India, both steinenematidis and heterorhabditids are marketed by Bio-Sense Crop Protection (Eco-Max Agrosystems Ltd.), Mumbai. *Steinernema* sp. is commercially available for the control of Amercian bollworm, pink bollworm, tobacco caterpillar, white grub, shoot borer, stem borer, tuber moth, spotted bollworm and leaf eating caterpillar on crops like cotton, groundnut, sugarcane, rice, potato, tomato, chilies, okra, sunflower and legumes. *Heterorhabditis* sp. is available for the control of white grub, flea beetle, grey weevil, and red palm weevil on groundnuts and coconut.

TECHNIQUES IN BIOLOGICAL CONTROL

Biological control practices involve three major techniques, viz. classical biological control, conservation and augmentation.

I. CLASSICAL BIOLOGICAL CONTROL

The intentional introduction of an exotic, usually co-evolved, biological control agent for permanent establishment and long-term pest control is referred to as classical biological control. The basic approach in this technique is to identify natural enemies that control a pest in its home location and introduce these enemies in the pest's new location. It is usually used to control species that have attained pest status following invasion of a new geographical location. Importation of biological control agents has been widely practiced since the initial spectacular success of the Vedalia beetle in 1889-90. The approach has yielded some of the best known examples of biological control. Procedures followed in natural enemy introduction programmes are fairly standard and generally include exploration for agents in areas of pest origin, pre-introduction studies (such as species identification, biological and ecological characterization, and rearing procedures), quarantine of agents for introduction, release and evaluation **(Fig. 5.6)**.

It is advisable to introduce an exotic species of a natural enemy either when there is an unoccupied niche in the life system of the pest which needs to be filled or when an inefficient natural enemy occupies a niche and is required to be displaced by a more efficient exotic species that fulfills the conditions of an ecological homologue. The former is a common situation in newly introduced pests in a country. Foreign explorations for parasitoids and natural enemies have been made primarily to introduce parasitoids from the place of origin of the pest and, sometimes also, to introduce exotic natural enemies of the indigenous pest species. In the latter case, attempts have not been very successful, because the indigenous parasitoids or predators which are already adapted to the host concerned are not easily replaced by the introduced species.

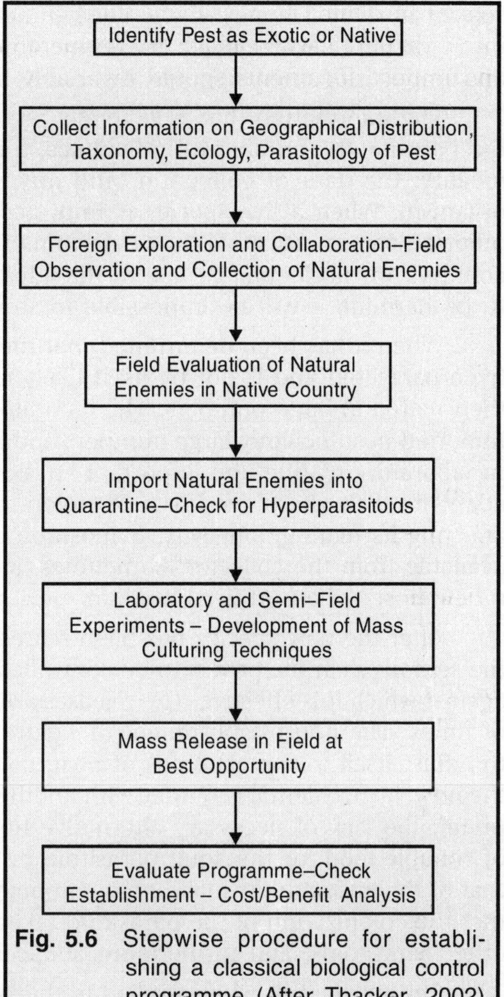

Fig. 5.6 Stepwise procedure for establishing a classical biological control programme (After Thacker, 2002)

When an introduced pest is to be controlled naturally, the place of its origin should be determined first, because it would be best to find a suitable natural enemy only in its original homeland. Before on entomologist goes abroad, he should know the basic habitat, behaviour and biology of the host. His job abroad is to find the pest, to collect its natural enemies, if any, and to tranship them to the receiving country, in sufficient numbers, in viable condition. The collector should also remember that the pest species concerned might be quite scarce in the country where he might be going. The duration for which search ought to be made should be extended to at least one full season of the pest's activity, to search natural enemies in various stages of development or in different generations found in various seasons of the year.

The actual shipment of a parasitoid or a predator follows its collection and rearing. Since it is a very valuable and perishable commodity, all possible care should be taken so that it reaches its destination within the minimum possible time during the period of the year when temperature is favourable. In the modern jet age of travel, it should not be necessary to send the consignment by steamers. Before a packing is prepared and handed to the airline concerned, definite instructions should be given that it is to be placed in a pressurized compartment at the desired range of temperature. Personal contacts with the crew are important at both ends, from the origin of the consignment to its destination. As soon as the parcel has been handed to the airline staff, a fax message intimating the date and time of its arrival should be conveyed to the recipient. At the receiving end, the parcel should be located in the mail bag, if sent by air parcel post. If the consignment is sent as an air-freight under special arrangements, it might be necessary to meet and contact the crew as soon as the aeroplane arrives.

Needless to say that the parcel should be made of light but strong material which can withstand handling and sufficient air holes should be provided, so that the insects do not die. Sometimes, it is necessary to make provision for maintaining humidity or to make emergence holes in case the adults are expected to come out of puparia in transit, so that they can move from the chamber containing the pupae to that of the adults, where food is kept. The food consists of a cotton swab soaked in diluted honey. If the consignment is not very large, it is advisable to pack the material into a vacuum flask in which the temperature remains fairly constant even during transit. The export and import documents should invariably be attached to the waybill or stitched to the package.

The information that is generally provided along with the consignment includes the name of the parasite, the number of individuals, the host and plant from which collected, the name of the locality, the date of collection, and any other instruction that could be useful for rearing the organism. When a live species is imported from abroad, it is kept in quarantine. Its biology and other attributes are studied to determine whether or not that species could be useful in the new country. Any accidental escape of the adults must not be allowed, because if the species turns out to be harmful, it will be impossible to stop its spread.

When it has been determined that the imported species is of the desirable type, free from any hyperparasitoid and is not by itself likely to become a pest of any economic plant or animal, it is then reared in large numbers. The laboratory rearing can sometimes be bypassed if the insects are imported in sufficiently large numbers and if, on their release, they easily get established in the field. In laboratory rearing, the important studies to be made are: the ability of the imported species to adapt itself to alternatve hosts, its tolerance to prevailing temperature and humidity, the factors affecting its mating, longevity, oviposition and diapause. Some of this information may already be available from the collector. Sometimes, it is also desirable to try to rear the imported species on a new host, on an artificial medium or a storable plant product.

After the new species has been reared in sufficient numbers, it is released in the field during the season when the pest is to be controlled, and during the availability of the stage of development against which it is effective. The releases are made only in a few localities and are repeated a number of times. The non-establishment or failure of the imported species will depend upon its inability to adjust itself to the extremes of heat, cold or aridity; the non-acceptance and incompatibility of the host; an accidental ecounter with another undetected host which the introduced parasitoid might prefer; the lack of necessary alternative hosts when the pest is not available; the non-availablility of suitable food for the adult parasitoid or predator; the non-synchronization of the life cycle with that of the host; and its inability to compete with the existing natural enemies. It has been observed that the colonization of the parasitoid takes place more readily inside the cages where there are no other parasitoids, and furthermore, its accidental dispersal is minimized.

Certain species, when studied in the laboratory, show promise but, on release in the field, they do not give the same performance and do not get established. The next step, if established, is to determine its success in reducing the pest population. The parasitized hosts are collected from the field and kept in the laboratory for the parasitoids to emerge. Parasitization would indicate that the host has been accepted under field conditions, but it does not necessarily mean that the parasitoid has been established in that locality. Sometimes, it completes a number of generations, but as soon as it faces the extremes of weather, it perishes. This has been observed repeatedly in the case of parasitoids introduced into northern India for the control of tissue-borers of sugarcane, maize and sorghum. Invariably, the parasitoid species have been brought from the tropical regions of the world and after getting an initial establishment, they perish during the winter, when most of these borers are either in hibernation or in diapause. It is, therefore, essential that the species introduced must also have a synchronization of the life-cycles with that of the host. For northern India, the parasitoids collected from subtropical or near-temperate regions of the world are likely to be more successful whereas in the south these should be introduced from tropical areas.

Recently, many international organizations have been established to facilitate the movement of beneficial species from one country to another, and the largest of the organizations is the International Institute of Biological Control (previously called Commonwealth Institute of Biological Control), established in 1927, which also has laboratories in Switzerland, Trinidad, Malaysia and Pakistan. About 40 per cent of the introduced natural enemies have established in the introduced countries and provided partial to complete control of important insect pests at the global level. In India, since the launching of All India Co-ordinated Research Project on Biological Control of Crop Pests and Weeds (AICRPBC) in 1977, 79 species of natural enemies have been imported, out of which 53 have successfully multiplied and 21 have been established in the field.

The most recent outstanding example of biological control by introduction is the control of cassava mealybug, *Phenacoccus manihoti* Matile-Ferrero by a tiny wasp, *Epidinocarsis lopezi* (De Santis) in Africa. The mealybug was devastating the cassava plant in late 1970s, destroying as much as 80 per cent of the crop in some areas and widespread famine was a real possibility in Sub-Saharan Africa. *E. lopezi* was imported from Paraguay and released in Nigeria in 1981. By mid 1980s, *E. lopezi* was multiplied in millions and released @ 100 wasps per second across Africa from the aeroplane, in addition to many releases from the ground. Over the next 7-8 years, mealybug problem was effectively eliminated in 30 countries. Dr Hans R. Herren was awarded the 1995 World Food Prize for developing and implementing the world's largest biological control project.

II. CONSERVATION

Conservation of biological control agents refers to modification of the environment or existing practices to protect and enhance specific enemies or other organisms to reduce the effects of pests. In contrast to classical biological control and augmentative releases, conservation does not involve collection and rearing of natural enemies for release. Rather, this approach involves creating conditions under which natural enemies are able to enhance their efficacy as biocontrol agents. Conservation is often the least emphasized of the three major approaches to biological control. However, it can constitute the most important approach because of two reasons; (*i*) The practices conducive to natural enemy conservation influence the success of native, imported and periodically released natural enemies. (*ii*) Conservation of natural enemies enables farmers to utilize beneficial species that exist in the agroecosystem. Thus, it serves to strengthen farmers' appreciation of biological control and their understanding of the natural enemies' ecology in relation to the crop production systems.

Agroecosystems are characterized by many agricultural practices and designs that have the potential to enhance the functional biodiversity of natural enemies and many others that negatively affect it. In general, there seem to be reluctance on the part of farmers to modify their production practices to accommodate or promote natural enemies. This may be because many a time, such practices conflict with conventional crop production or protection methods. The farmers are also not fully convinced about the potential of natural enemies for pest suppression. Most importantly, there is often very little knowledge about the environmental factors that limit or encourage the effectiveness of natural enemies. Conservation requires detailed knowledge of the natural enemy's phenology and resource requirements. Moreover, knowledge of interactions of various management practices, both agronomic as well as pest management techniques, with natural enemy populations is essential. At present, such information is available only for a few species and systems. The most important strategies to conserve natural enemies within an agroecosystem are cultural control by habitat manipulation and rational use of pesticides.

Conservation means the avoidance of measures that destroy natural enemies and the use of measures that increase their longevity and reproduction or the attractiveness of an area to natural enemies. The conservation and enhancement of natural enemies should be the first consideration. If they are properly conserved the need for other control measures is greatly reduced.

(*a*) **Preservation of inactive stages.** This is most critical when there is a small reservoir of natural enemies outside the cropped area. Pupae of *Epipyrops* are found in large numbers on the trashes of sugarcane leaves at the time of harvesting. If these are not burnt but left around the harvested fields, the adults emerge to augment the supply of natural enemies in the pre-monsoon season against *Pyrilla perpusilla* (Walker) on the young crop of sugarcane.

(*b*) **Avoidance of harmful cultural practices.** Cultural practices like ploughing, mowing or burning can be harmful to natural enemies. For example, burning of sugarcane trash destroys the resting stages of *Epipyrops*. Such practices can be modified to avoid harmful effects.

(*c*) **Maintenance of diversity.** The concept "more the diversity more is the stability", holds true here also, since diverse system may provide alternate hosts as source of food, overwintering sites, refuges and so on. Moreover, natural enemies have evolved in natural diverse communities than agroecosystem so these are expected to perform in diverse systems, e.g. mixed cropping, intercropping, etc.

(*d*) **Natural food, artificial food supplements and shelters.** Many parasitoids and predators require foods frequently not available in monocultures. The availability of predatory mite, *Euseius hibisci* (Chant) in California, was related to the availability of pollen, even when prey mites were absent. In irrigated desert areas of California, artificial honeydew and pollen in the form of food sprays induced early oviposition of the antlion, *Chrysoperla* spp. and Coccinellids, in the alfalfa fields. In North California, marked reduction of tobacco hornworms was achieved by predacious wasps, following erections of nesting shelters near the fields.

(*e*) **Control of honeydew feeding ants.** These ants make biological control of honeydew producing insects like scales, mealy bugs and aphids difficult because of interference with natural enemies.

(*f*) **Protection from pesticides.** Almost all kinds of pesticides and their formulations have adverse effects on natural enemies. The most striking case of interference is the long standing biological control by the Vedalia beetle, *Rodolia cardinalis* (Mulsant) of cottony cushion scale, *Icerya purchasi* Maskell, on citrus in California by the indiscriminate use of DDT in 1946 and 1947. Vedalia beetles were killed, while the cottony cushion scales were not killed, resulting is serious outbreaks of the pest. Later, adjustment of spray treatments allowed Vedalia beetles to regain control in subsequent years. Similarly, several predaceous mites are killed by the pesticides meant for the control of phytophagous mites. The solution lies in the use of relatively resistant strains of predators, selective use of pesticides and manipulation of prey and predator population through certain management practices.

III. AUGMENTATION

Augmentative biocontrol is focused on enhancing the numbers and/or activity of natural enemies in agroecosystems. This strategy involves mass multiplication and periodic release of natural enemies so that they may multiply during the growing season. The processes of augmentation or inoculation with natural enemies occur where natural enemies are absent or where they exist at levels that are ineffective for pest control. The aim is to establish a natural enemy population that is able to suppress pest numbers below economically damaging levels until harvest. The rationale for this approach to biological control is that many effective natural enemies cannot survive on a year-round basis on a crop. However, they are able to survive and affect control at certain times of the year, hence the need for annual releases. The critical difference between this approach and that of classical biological control is, therefore, that it is not long term. Ecologically, the general equilibrium of the pest population is not altered as in case of classical biological control. The releases have to be repeated periodically and only temporary suppression of the pest is achieved (Koul and Dhaliwal, 2003).

Several issues must be considered before releasing natural enemies for augmentation programmes.

- The cost and the anticipated benefits of using biocontrol relative to alternative tactics (e.g. pesticides), must be weighed carefully.
- The overall crop production system should be considered; releases may not be of any use if crop production practices are inimical to the use of natural enemies.

- The information on the most appropriate species or strain, most effective means of application (e.g. method of release, number of releases per unit area or per pest, evaluation of efficacy, etc) must be available.

More than 125 species of natural enemies are commercially available at global level for augmentative biological control. This form of control is applied in the open field in crops that are attacked by only a few pest species, and it is particularly popular in greenhouse crops, where the whole spectrum of pests can be managed by a suite of natural enemies. When compared with chemical control, augmentative control does not cause any phytotoxic effects on young plants, premature abortion of flowers and fruits does not occur, release of natural enemies takes less time, several key pests can be controlled only by natural enemies, and there is no safety or re-entry period after release of natural enemies, this allows continuous harvesting without danger to the health of greenhouse personnel.

Augmentation includes all activities designed to increase numbers or effect of existing natural enemies. These objectives may be achieved by releasing additional numbers of a natural enemy into a system or modifying the environment in such a way as to promote greater numbers or effectiveness. These releases differ from introduction of imported natural enemies in that these have to be repeated periodically. Further, they result only in temporary suppression of the pest rather than permanent lowering of the general equilibrium position as in introductions. The periodic releases may be either inoculative or inundative.

(*a*) *Inoculative releases* may be made as infrequently as once a year to re-establish a species of natural enemy which is periodically killed out in an area by unfavourable conditions during part of the year, but operates very effectively the rest of the year. Here control is expected from the progeny and subsequent generations, not from the release itself.

(*b*) *Inundative releases* involve mass culture and release of natural enemies to suppress the pest population directly as in the case of conventional insecticides. These are most economical against pests that have only one or at the most a few discrete generations every year. Massive releases have been attempted in several programmes involving natural enemies like *Trichogramma* spp., a tiny wasp that parasitizes insect eggs, and general predators like green lace wings, *Chrysoperla carnea* Stephens and ladybird beetles, *Hippodamia convergens* (Guerin-Meneville).

Augmentation may also result from environmental manipulations to increase the effectiveness of natural enemies. This may be achieved by providing alternate nutrients, nesting habitats, overwintering sites, etc. The possibilities for increasing the population of predaceous arthropods through habitat management are many. Small changes in agricultural practices can cause great increases in key predators by affecting the availability of alternate non-economic prey or other foods. Some of the weeds can also be excellent habitats for predators.

Augmentative Biocontrol in India

The status of augmentative biocontrol in selected crops is briefly presented herein (Dhaliwal and Koul, 2007, 2010).

Rice. Releases of mirid bug @ 100 bugs or 50-75 eggs/m^2 at 10-day intervals have been found effective for the control of brown planthopper. Inundative releases of exotic *Trichogramma japonicum* Ashmead at 50,000 ha^{-1} during egg laying period of rice stem borer reduced borer damage and increased crop yield. It has been found that 7-9 releases of *T. chilonis* Ishii and *T. japonicum* @ 1,00,000/ha, starting at 30 days after transplanting proved as effective as the standard insecticide treatment for the control of stem borer and leaf folder. A total of 11 releases of *T. japonicum* and *T. chilonis* @ 50,000 per ha per week on a long duration variety of rice (Swarna) reduced the tiller damage caused by yellow stem borer and folded leaves by rice leaf folder, to the tune of 50.1-61.3 and 63.8-75.5 per cent, respectively. Similarly, reduction in tiller damage and folded leaves varied from 78.1 to 81.6 and 72.6 to 81.8 per cent, respectively, when the egg parasitoids were released @ 1,00,000 per ha per week during the season.

Sugarcane. The relatively stable sugarcane agroeco-system provides ideal conditions for the colonization of natural enemies. During recent years, releases of *T. chilonis* have been found effective for the management of various shoot and stalk borers in different parts of the country. The parasitoid releases @ 1,25,000 ha^{-1} are recommended against shoot borer in Andhra Pradesh. Weekly releases at 1,25,00 parasitic wasps per ha from 4th to 11th week stage of the crop provided effective control of internode borer, *Chilo sacchariphagus indicus* Kapur in Tamil Nadu. The release of 12 ml of parasitoid adults (3,00,000) per ha in six split doses effectively checked internode borer infestation in an area of about 500 ha. In Punjab, the sequential releases of *T. chilonis* (12 releases @ 50,000/ha at 10-day intervals) and *Cotesia flavipes* (Cameron) (6 releases @ 10,000/ha at 10-day intervals) during July to October proved very effective for the control of stalk borer. For the control of early shoot borer, *Chilo infuscatellus* Snellen, 6-9 releases of *T. chilonis* during April to June at 10-day intervals @ 50,000/ha proved effective. Similarly, stalk borer and Gurdaspur borer have been controlled in parts of sub-tropical India by large-scale releases of *T. chilonis.*

In addition to egg parasitoids, field releases of the larval parasitoid, *Isotima javensis* (Rohwer) exercised effective control of top borer in Tamil Nadu and Karnataka. Releases of *Sturmiopsis inferens* Townsend at 312 gravid females per ha provided effective control of *C. infuscatellus* in coastal areas of Tamil Nadu. Sugarcane pyrilla has been effectively controlled by redistribution and periodic release of 8000-10,000 cocoons or 8,00,000-10,00,000 eggs per ha of parasitoid *Epiricania melanoleuca* (Fletcher). In 1999, there was an outbreak of pyrilla in 6 lakh ha in north India, perhaps due to indiscriminate burning of trash. However, the parasitoid overpowered the pest within 20 days and the spraying cost of about Rs 480 million crores was saved.

Cotton. A number of promising natural enemies have been found attacking cotton pests in the country. However, large scale use of insecticides has reduced the population of most of these natural enemies to insignificant levels. In the fields where biocontrol was practiced, 25 natural enemies were recorded as compared to only 2 in the fields sprayed with insecticides. Releases of *Trichogramma* spp. at 1,50,000 parasitized eggs ha^{-1} at weekly intervals have proved promising for bollworm control. The green lacewing, *C. carnea,* can be released @ 2 larvae per plant during the peak egg hatching of cotton bollworms. Sucking pests may be checked by releasing chrysopids at 1 lakh per ha at fortnightly intervals. Recently, a new parasitoid, *Aenasius bambawalei* Hayat, has been reported to cause 50-80 per cent parasitization of the nymphs of mealybug, *Phenacoccus solenopsis* Tinsley. The parasitoid could kill the mealybugs within a week, which turned into reddish dark brown mummies. The parasitoid population could be augmented in cotton through collection of mealybug mummies from various host plants and releasing them onto the cotton plants.

Horticultural crops. Augmentative and inoculative release of two exotic parasitoids, *Encarsia perniciosi* (Tower) and *Aphytis* spp. *proclia* (Walker) group @ 2000 per infested tree have given promising results for the suppression of San Jose scale on apples. The Russian strain of *E. perniciosi* proved effective in Himachal Pradesh while Chinese and American strains proved better in Uttar Pradesh. *Aphelinus mali* (Haldeman) was found effective against woolly apple aphid in Kullu Valley but subsequent releases in Shimla Hills, Chaubattia, Conoor and Shillong were not successful. Similarly, releases of exotic parasitoid, *Leptomastix dactylopii* Howard have proved effective for the control of citrus mealy bug, *Planococcus citri* (Risso) in Karnataka. Mass rearing and release of the Australian coccinellid predator, *Cryptolaemus montrouzieri* Mulsant have been found effective against citrus mealybug, grape mealybug, guava scale and other scales in a number of plantations.

In case of vegetable pests, inundative releases of egg parasitoid, *Trichogramma brasiliensis* (Ashmead) @ 4000 adults ha^{-1} for six weeks suppressed the attack of fruit borer on tomato. Phytophagous mites on okra and brinjal can be effectively checked by augmentative releases of phytoseiid mites, *Phytoseiulus persimilis* Athias-Henriot and *Amblyseius tetranychivorus* (Gupta) @ 10 adults $plant^{-1}$.

Plantation crops. Inundative release of native parasitoids, *Goniozus nephantidis* (Muesebeck) and *Bracon brevicornis* Wesmael at 3000 and 4500 per ha, respectively, has given encouraging results for the management of *Opisina arenosella* Walker on coconut. The inoculative release of coccinellid, *Curinus coerulens* Mulsant, against subabul psyllid, *Heteropsylla cubana* D.L. Crawford has proved very effective in the management of this pest.

SYNERGISM AMONG BIOCONTROL AGENTS

Two or more biological control agents can be said to act synergistically in the regulation of a pest population when the control effects they exert in combination exceed the sum of effects that would be expected when they act completely independently. Biological control involving parasitoids and pathogens is most likely to be a sound strategy when (*i*) the pathogen produces external stages

Table 5.6 Relative strengths and weaknesses of parasitoids and predators compared to pathogens

Control agent	Strengths	Weaknesses
Parasitoids	• Efficient searchers • In some cases strong intrinsic competitors (produce antimicrobial factors) • Generally effective at low population density • May be able to detect and reject living hosts infected with pathogens • Not important intraguild predators • May synergise activity of pathogens	• Slow mode of action • Low reproductive potential • In some cases, weak intrinsic competitors (susceptible to premature death of diseased host) • Generally not effective against pest outbreaks • Generally costly to augment
Predators	• Efficient searchers • Fast mode of action • May be effective at low host density • May be effective at high host density through augmentation • Some species may be applied strategically • May synergise action of pathogens	• Low productive potential • Generally costly to augment • May be important intraguild predators
Pathogens	• Potential for moderately rapid mode of action • Potential for extremely high rate of reproduction • Generally strong intrinsic competitors • Many species easily and economically augmented • Generally easy to apply strategically • May be effective at low host density • May be effective against pest outbreaks • May synergise activity of parasitoids and predators	• Generally unable to search for hosts • Infectivity, speed of action, reproductive potential, effective field life and efficacy are generally dependent upon environmental conditions • May be important intraguild predators

that may span the intergenerational gap, (*ii*) parasitoid and pathogen attacks are moderately clumped, (*iii*) both natural enemies have high rates of search or transmission, and (*iv*) there is some degree of overlap in the timing of attacks of the enemies, and/or competition with the host is not invariably one-sided.

The relative strengths and weaknesses of pathogens, parasitoids and predators in terms of various traits such as reproductive potential, speed of control action and potential to function as synergists are listed in **Table 5.6.** These characteristics suggest some additional considerations with respect to the combination of pathogens and entomophagous insects for biological control.

It may be beneficial to select natural enemy combinations in which the pathogen functions as the synergist. Most parasitoids have long developmental cycles (and thus slow modes of action). Pathogens generally exhibit more rapid lethal action ; however, most pathogens, including the fastest-acting (toxin producers) are most effective against the early instars of their hosts. Therefore, it is generally recommended, in case of both pathogens and parasitoids, that augmentative releases be made during the early stages of a pest cycle. Thus, optimal integrated use of these agents will likely depend on discovery of pathogens and parasitoids capable of coexistence and joint action.

POTENTIAL AND CONSTRAINTS

The potential of biological control has largely remained untapped because it has been under used, underexploited, underestimated and often untried and, therefore, unproven. In fact, the use of parasitoids and predators should be a primary consideration in any pest management programme. It is generally said that if it works, biological control is the best method of control based on ecological and environmental considerations.

- Biological control is highly economical. It rarely costs more than a few lakh rupees as against billions of rupees required to develop an insecticide.
- Biological control method is usually selective with no side effects.
- Biological control is self-propagating and self-perpetuating.
- There are no harmful effects on humans, livestock and other organisms.
- Biological control is virtually permanent. The successes achieved more than half a century ago continue to work to this day. Biological control agents possess the ability to search for their prey (pest).
- The use of parasitoids and predators is usually compatible with most other tactics of pest management except the use of broad-spectrum synthetic organic insecticides.

Although biological control has been successful in a number of cases, these still represent a small proportion of all pest control situations. It has been estimated that only about 4 per cent of insect pests worldwide have been controlled by biological methods. The reasons for the relatively low level of success with biocontrol agents are numerous as this method suffers from several drawbacks.

- Usually, the natural enemies do not control the pest completely and a low population of the pest continues to exist.
- Biocontrol is a specific form of control and attracts little interest from companies who want to market products that can be used to control a wide range of pest species.
- The success of biological control limits the subsequent use of pesticides for the same or other pests in the agroecosystem.

- Biocontrol agents take time to work and they do not have the immediate and dramatic effects that many pesticides do.
- Biological control attempts may require more labour and more specialized knowledge than is required with chemical pesticide applications.

Biological control can be elegant, sustainable, non-polluting and inexpensive, but the success rate has not been particularly high, especially for biological control of arthropods. A lot of work needs to be done to optimise the utilisation of predators and parasitoids in IPM. Better understanding of the ecological aspects of natural enemies is vital for improving the performance of predators and parasitoids, especially in tropical regions. Little is known regarding the biodiversity, ecology, community structure and host relationships in most tropical ecosystems and assessment of the impact of biocontrol agents is difficult under these conditions. Environmental implications of the release of these organisms must be studied, especially in case of introductions and genetically engineered organisms. There is an urgent need to establish a network of large scale multiplication units, so that bioagents are made available to the farmers. With increasing concerns about the impact of conventional insecticides, biocontrol agents are likely to play a prominent role in future IPM progammes.

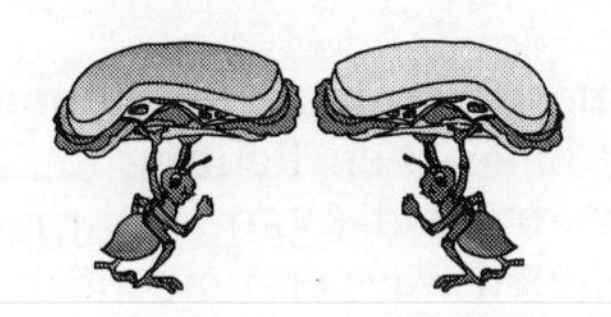

BOTANICAL PEST CONTROL

INTRODUCTION

Insects and plants originated almost simultaneously about 500 million years ago, and ever since both these groups of organisms have been fighting for their survival. This fight is still on between the attacker (insect) and the attacked (plant). During this long evolutionary history of attacker and attacked, plants have developed ways and means to combat attacker. Plants cannot run away nor can they fight physically, hence nature during this evolutionary period provided plants with highly sophisticated in-built defense mechanisms to resist/desist attack by insects and other pests. The greatest testimony that they are not helpless is that even the plant species considered most susceptible survive well in the nature. Of the various defense mechanisms developed, chemicals elaborated by the plants are the most important for defense. Plants are nature's 'chemical factories', providing the nature's richest source of chemicals on earth (Dhaliwal and Koul, 2011).

Botanical pesticides are either naturally occurring plant materials or products derived rather simply from such plant materials. The chemicals that plants produce to protect themselves against insect attack belong to a group that includes compounds known as secondary plant substances. These chemicals are a subset of what are known as phytochemicals (plant-based chemicals). Within the context of pest control, they are referred to as botanical pesticides. Secondary plant substances are produced as byproducts of major biochemical pathways and chemically, they include alkaloids, terpenoids and phenolics as well as a number of other compounds. These chemicals repel approaching insects, deter feeding and oviposition on the plants, disrupt behaviour and physiology of insects in various ways and even prove toxic to different developmental stages of many insects (Koul and Dhaliwal, 2001; Koul, 2012).

PROMISING PESTICIDAL PLANTS

Plants are biochemists *par excellence.* During their long evolution, plants have synthesized a diverse array of chemicals to prevent their colonization by insects and other herbivores. It is estimated that there are about 2,50,000 to 5,00,000 different plant species in the world today. Only 10 per cent of these have been examined chemically indicating that there is enormous scope for further work. Over the years, more than 6000 species of plants have been screened and more than 2500 plant species belonging to 235 families were found to possess biological activity against various categories of pests (**Table 6.1**). The highest number of pesticidal plants belong to Meliaceae

Table 6.1 Important plant families that have a number of species evaluated for anti-insect properties

Plant family	No. of plant species	Plant family	No. of plant species
Annonaceae	12	Meliaceae	>500
Apoaceae	23	Moraceae	26
Apocyanaceae	39	Myrtaceae	72
Asteraceae	147	Pinaceae	52
Bignoniaceae	13	Piperaceae	14
Cryptogams	58	Poaceae	27
Cupressaceae	22	Ranunculaceae	55
Euphorbiaceae	63	Rosaceae	34
Fabaceae	157	Rubiaceae	38
Labiatae	52	Rutaceae	42
Lamiaceae	24	Solanaceae	52
Leguminosae	60	Verbenaceae	60

Source: Koul (2012)

(>500), followed by Fabaceae (157), Asteraceae (147), Myrtaceae (72), Euphorbiaceae (63), Leguminosae (60), Verbenaceae (60), Cryptogams (58), Ranunculaceae (55), Labiatae (52) and Solanaceae (52). This number seems to be far less than the actual number of naturally occurring pesticidal plants as this is just 0.77 per cent of the total 3,08,000 species of plants or 0.87 per cent of 2,75,000 species of flowering plants. It is thus likely that novel and potent molecules that can be used for pest suppression remain to be discovered from many plant species.

Plants are known to produce a diverse range of secondary metabolites such as terpenoids, alkaloids, polyacetylenes, flavonoids, unusual amino acids, sugars, etc. Various isolated chemically from plants include 350 compounds that are insecticidal and more than 900 isolates that are feeding deterrent alone. The structures of more than 600 alkaloids, 3000 terpenoids, several thousands of phenylpropanoids, 1000 flavonoids, 500 quinones, 650 polyacetylenes, and 4000 amino acids have already been elucidated. Many of these chemicals protect the plants from pests and pathogens. But in addition to high insecticidal activity, plant species must possess some other characteristics for development into an ideal botanical insecticide, viz. safety to plant and animal life, biodegradability with sufficient residual action, ready availability of the plant or capability for cultivation with a reasonably short gestation period, economical isolation procedures for the active component(s) or capacity for formulation of crude extracts obtained from plant parts, and yield products of consistent quality.

Neem

Neem, *Azadirachta indica* A. Juss. (Fam. Meliaceae) is indigenous to India from where it has spread to many Asian and African countries. For centuries, the tree has been held in esteem by Indian folk because of medicinal and insecticidal value. A breakthrough in the insecticidal application of neem was made by Pradhan *et al.* (1962) who successfully protected the standing crops at Indian Agricultural Research Institute, New Delhi, by spraying them with 0.001 per cent neem seed kernel suspension during a locust invasion. Due to its legendry insect-repellent and medicinal properties, it has been identified as the most promising of all plants by the National Research Council, Washington, USA. Neem has assumed the status of an international tree which is evident from the fact that it has been a subject of discussion at several global conferences, viz. Rottach-Egern, Germany (1980), Rauischolzhausen, Germany (1983), Nairobi, Kenya (1986), Bangalore, India (1993), Queensland, Australia (1996), Vancouver, Canada (1999), Mumbai, India (2002), Kunming, China (2006) and Coimbatore, India (2007).

All parts of the neem tree possess insecticidal activity but seed kernel is the most active. Neem bark, leaf, fruit and oil as well as extracts with various solvents especially ethanol have been found to exhibit activity against insect pests. Neem products exhibit almost every conceivable type of activity against insects (**Fig. 6.1**). In addition, neem possesses fungicidal, nematicidal, bactericidal, molluscidial, diueretic, and antiarthritic properties. It also exhibits immunomodulatory, antiinflammatory, antihyperglycaemic, antiulcer, antimalarial, antiviral, antioxidant, antimutagenic, and anticarcinogenic effects. Azadirachtin has systemic effects in certain crop plants, greatly enhancing its efficacy and field persistence (Khater, 2012).

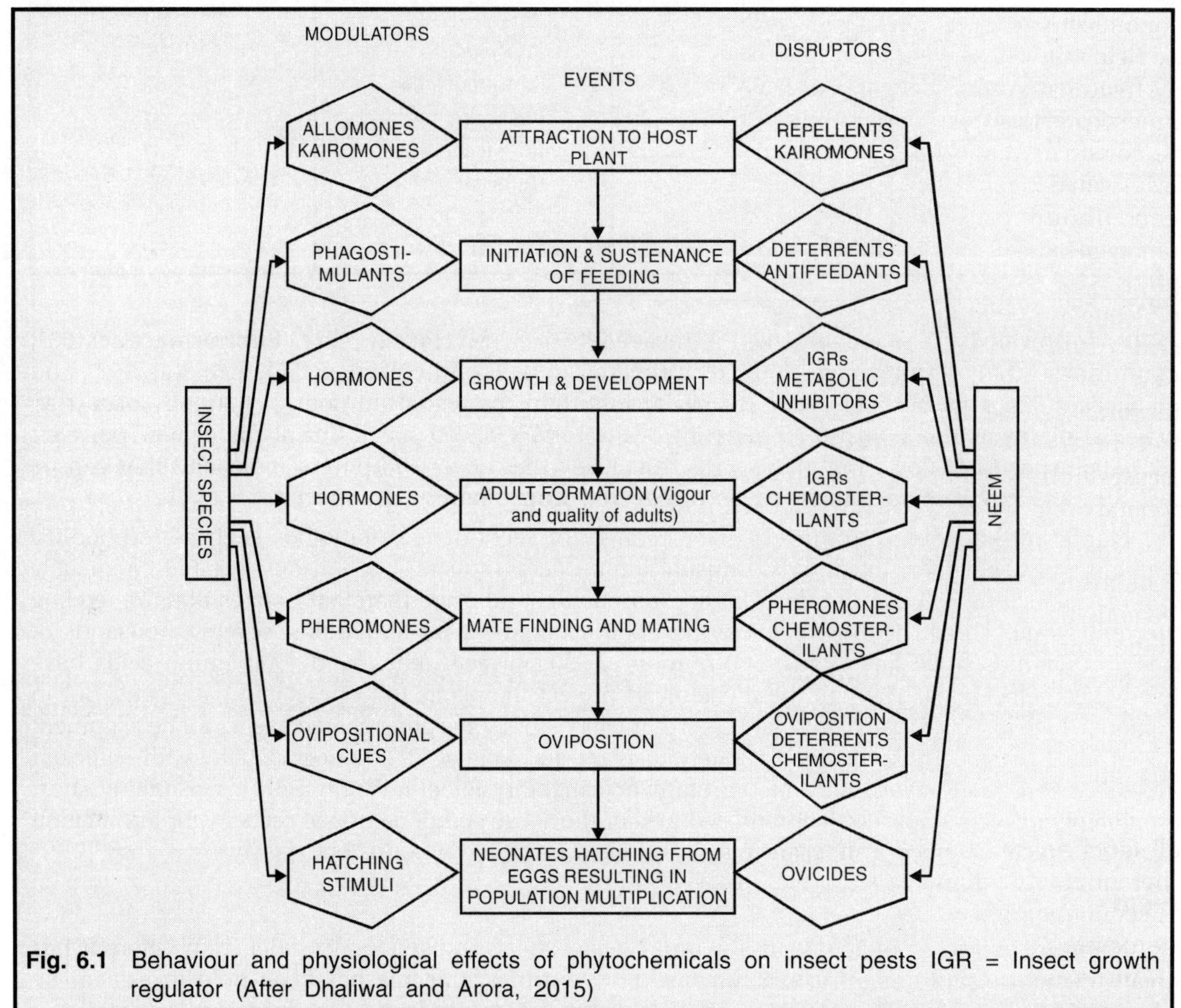

Fig. 6.1 Behaviour and physiological effects of phytochemicals on insect pests IGR = Insect growth regulator (After Dhaliwal and Arora, 2015)

Neem products also affect insect vigour, longevity and fecundity. Females of *E. varivestis* and *Leptinotarsa decemlineata* (Say) were sterilized by neem compounds, while reproductive maturation was inhibited in *N. lugens* males. At higher concentrations, most females did not emit normal male eliciting signals. During the last decade, neem products have been found to act as ovipositional deterrents for *Bactrocera cucurbitae* (Coquillett), *H. armigera, Spodoptera litura* (Fabricius), *Callosobruchus* spp., etc. Ovicidal activity of neem products has been reported in *Corcyra cephalonica* (Stainton), *Earias vittella* (Fabricius) and *S. litura.* Strong contact effects of neem oil leading to transformation of gregarious nymphs to intermediate and solitary forms have been recorded in locusts. Direct contact toxicity of neem products has been demonstrated against termites, *Macrotermes* sp. and aphids, *Lipaphis erysimi* (Kaltenbach), *Rhopalosiphum nymphae* (Linnaeus), etc.

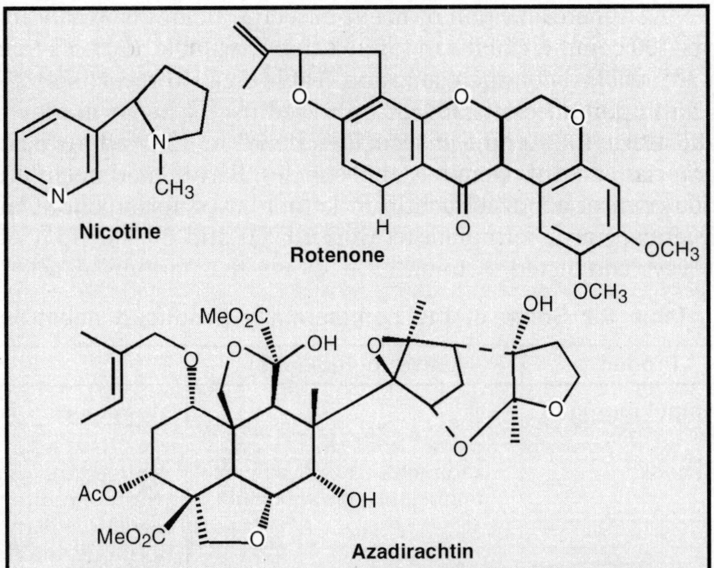

Neem biopesticides have been extensively used in IPM modules for the management of insect pests of various crops. Studies have indicated that in rice IPM, neem treatments did not allow the occurrence of brown planthopper, green leafhopper, yellow stem borer and gall midge to exceed economic injury levels. On an average, rice yield was 20.5 per cent more under IPM treatments using neem than under farmers' practice treatment. In one of the cotton IPM modules, neem oil mixed with detergent (2.5 litres/ha) has been found effective for the management of jassids, whitefly and aphids. Use of neem oil 0.5 per cent and teepol 0.1 per cent and neem seed kernel extract (NSKE) 5 per cent have been effective against cotton bollworms and sucking pests. An IPM module for groundnut resulted in 24-46 per cent control of major insect pests (white grub, leaf miner, tobacco caterpillar, gram pod borer, jassids, thrips and aphids) and diseases from 24-48 per cent. It also resulted in average increase in yield by 19 per cent and higher monetary returns by 40 per cent as against farmers' practice.

Nicotine

Rotenone

Azadirachtin

Fig. 6.2 Structures of nicotine, rotenone and azadirachtin

Nearly 100 protolimonoids, limonoids or tetranor-triterpenoids, pentanortriterpenoids, hexanortriterpenoids and some non-terpenoids have been isolated from various parts of the neem tree and still more are being isolated. Azadirachtin (**Fig. 6.2**), the most important biologically active component of neem shows phagorepellent and toxic effects at 0.1 to 1000 ppm when incorporated into diets of different insect species. Azadirachtin is safe to mammals-the rat oral acute LD_{50} is > 5000 mg kg^{-1}. A 90-day oral feeding of rats with 10,000 ppm of azadirachtin did not show chronic toxicity. Azadirachtin was synthesised in the laboratory by scientists of the University of Cambridge (UK) in 2007, after 22 years of research when structure of azadriachtin was determined in 1985.

Inspite of these impressive strides, large scale practical utilization of neem in pest management still faces several hurdles. There is an urgent need to characterize best ecotypes of neem tree for different environmental conditions. Intensive breeding and selection work needs to be undertaken for economic production of various high quality raw materials required for insecticide production. The development of new techniques for harvesting, depulping, drying and storage of neem seeds is a necessary pre-requisite for large scale production of neem-based pesticides.

Ironically, the neem tree itself is attacked by 60 species of insects besides mites, nematodes, mammals and 16 phytopathogens. Some of these like *Pulvinaria maxima* Green, *Aonidiella orientalis* (Newstead), *Helopeltis antonii* Signoret and *Rhizoctonia solani* Kuhn are already causing serious damage in some areas. As commercially high quality genotypes are introduced on a large scale, the pest and pathogen problems may be aggravated. It is, therefore, imperative that the neem ecosystem is thoroughly studied to manipulate various biological components and cultural practices to keep the damaging organisms in check. Another problem facing the utilization of neem is the phytotoxic effects recorded in several crops, viz. cabbage, onion, muskmelon, potato, tobacco, tomato, etc. Dose-response relationships need to be worked out for all the crops where neem products appear to be useful for use in IPM (Dhaliwal and Arora, 2012).

Commercialization of neem-based pesticides is expanding at a rapid rate. Worldwide, there are over 100 commercial neem formulations available. In India many products are either being marketed or are awaiting commercialization (**Table 6.2**). However, lack of standardized and reliable formulations is an important constraint in increased use of neem products for pest control. To overcome these difficulties, the Central Insecticides Board in 1991 approved the guidelines and data requirements for registration of neem based pesticides. Formulated neem products are required to contain at least 1500 ppm of a.i. azadirachtin in kernel based formulations and 300 ppm in neem oil based ones. Recently, neem formulations with 10,000 and 50,000 ppm of azadirachtin have been introduced. Surveys conducted in India have shown that farmers usually buy commercial products, which are

Table 6.2 Some of the commercially produced neem-based pesticidal products in India

Product	Active ingredient(s)	Spectrum of activity
Amitul mosquito oil	Oil	Mosquitoes
Achook	Azadirachtin, azadiradione, nimbocinol, epinimbocinol, etc.	Antifeedant, repellent/deterrent male sterility, molt/ chitin inhibitor, change in sexual behaviour, growth regulator/disruptant, ovipositional deterrent, juvenile hormone, interference in biosynthesis, toxicant, ovicidal, oviposition deterrent, altered egg hatchability
Field Marshal	Azadirachtin	Antifeedant/repellent
Jaeevan Crop Protector	Extracts containing azadirachtin	Antifeedant/repellent, disturbs growth and reproduction
Margoside –CK (20 EC)	Azardirachtin	Antideedant, growth inhbitory, ovicidal, oviposition deterrent, nematicidal, chemosterilant
Margoside –OK (80 EC)	Azadirachtin (major ingredient)	Insecticidal
Moskit	Oil	Mosquito repellent
NeemAzal	Azadirachtin	Antifeedent, repellent, toxic
Neemark	Azadirachtin	Antifeedant/repellent, nematicidal, synergistic
Neemgold	Azadirachtin, kernel extract	Antifeedant
Neemguard	Extract concentrate	Repellent, metamorphosis disruptor
Neemrich	Extracts	Neemrich I-warehouse pests Neemrich II- antifeedant
Neemta 2100	Extract concentrate	Repellent, metamorphosis disruptor
Nethrin	Oil	Pesticidal .
Nimba	Kernel based powder	Pesticidal i
Nimbecidine	Azadirachtin	Antifeedant, repellent, metamorphosis disruptor, synergist
Rakshak Gold	Azadirachtin	Antifeedant, repellant, toxic
Nimlin	Extract concentrate	Repellent, metamorphosis disruptor
RD-9 Repelin	Neem, *Pongamia, Annona* and castor products (azadirachtin 3000 ppm)	Antifeedant, repellent
Sukrina	Azadirachtin, meliantriol and other actives of *A.indica & Galedupa indica*	Antifeedant, repellent
TRIC	Oil	Household pests
Wellgro	Neem kernel powder	Repellent, fungus inhibitory, antiviral, plant nutrition, N-loss prevention

Source : Modified from Dhaliwal *et al.* (2013)

also difficult to obtain in the local market. Improved local production units, which incorporate inexpensive equipment and chemicals would probably make neem pest control attractive to many more farmers.

Many times neem products have shown inconsistent results in the field. The products available in the market and/or prepared locally vary widely in quality and quantity of different active components. A survey of neem oil samples in Canada revealed that 2 samples completely lacked detectable amounts of azadirachtin while remaining samples contained 200 to 4000 ppm (0.02-0.4%) azadirachtin. In another study, one sample containing 6800 ppm was also detected. There is an urgent need to monitor market samples of neem products so that farmers are able to get consistent and reliable results while using neem based pesticides.

Chinaberry

Melia azedarach Linnaeus, commonly known as *Dharek*, Chinaberry tree, China tree, Persian lilac or 'Pride of India' is a close relative of neem tree. Very strong antifeedant effects of the tree against locusts, *Schistocerca gregaria* (Forskal) were discovered during the locust invasion of Palestine in 1915. It was observed by an adolescent, Rachel Shpan-Gabrielith, that while all other vegetation was nearly completely devoured, the Persian lilac, was almost undamaged. Subsequently, laboratory studies by her established that the foliage remained untouched by locusts even after one week of starvation. The hot water extracts of the plant applied to wheat bran also prevented feeding by the locusts. She subsequently successfully used this technique to save several crops during later invasions (1945, 1951) by locusts.

The work on active principles established that several tetranortriterpenoids (limonoids) related to azadirachtin were present in *M. azedarach*. The compounds isolated included meliacarpinins, azedarachins, ammorastatins, amoorastatone, trichlins and nimbolidins. Among the meliacarpinins, 1-cinnamoyl-3, acetyl-11-methoxymelia-carpinin with antifeedant index (AI_{50}) of 50 ppm was the most active. Among the four azedarachins, azedarachin–A was twice as active (200 ppm) as other azedarachins against *Spodoptera eridania* (Cramer) larvae.

A wide range of behavioural, physiological and toxic effects were observed against several insect species including *Epilachna varivestis* Mulsant, *Nilaparvata lugens* (Stal), *Mythimna separata* (Walker), *Plutella xylostella* (Linnaeus), etc. Extracts in different solvents as well as pure compounds were found to exhibit phagodeterrent, oviposition deterrent, fecundity and longevity reducing, development disrupting and toxic effects. Limited work on natural enemies also revealed that Chinaberry products are comparatively safe to coccinellids and predatory mites, *Amblyseius sewsami* Evans.

At least 36 limonoids including toosendanin have been isolated from stem and root back of *Melia toosenden* Siebold & Zucc. as antifeedant constituents. Based on toosendanin, a product has been developed and registered in China for use against fruit and vegetable pests. A number of toxic tetranortriterpenoids namely meliantriol, 12-hydroxy amoorastatone, 12-hydroxy amoorastatin and 12-acetoxy amoorastatin have been isolated from the dried stem bark. Since these meliatoxins exhibit mammalian toxicity, it needs to be ensured that any potential product based on *M. toosendan* must be devoid of such harmful toxicants.

Chrysanthemum

Pyrethrum, derived from the dried flowers of *Tanacetum* (*Chrysanthemum*) *cinerariaefolium* (Treviranus) (Fam. Asteraceae), has been used as an insecticide since ancient times. The original home of the plant is Middle and Near East. Its commercial use originated in Persia from where it was introduced to Europe, America and Japan in the nineteenth century. After the first world war, its cultivation was taken up in Africa and majority (>75%) of the world supply of pyrethum was produced in Kenya and Tanzania. However, its production began in Tasmania (Australia) in 1996 and it produces almost one-half of the world supply. A related plant, Persian insect flower, *T. roseum*

(Adams), is the basis of the pyrethrum industry in Papua New Guinea. Worldwide annual production of pyrethrum now averages 30,000 tonnes.

Pure pyrethrins are moderately toxic to mammals (rat oral acute LD_{50} values range from 350 to 500 mg kg^{-1}) but technical grade pyrethrum is considerably less toxic (almost 1500 mg kg^{-1}). Pyrethrum is a highly effective insecticide against common household insects like house flies, mosquitoes, fleas and lice. It is safe to mammals and is easily broken down to non-toxic metabolites. The insecticidal principals in pyrethrum comprise pyrethrins I and II, cinerins I and II, and jasmolins I and II (**Fig. 6.3**). Pyrethrin I is the most effective. Pyrethrins act quickly on the insect central nervous system, causing a knockdown effect. Addition of sesame oil or synergists like piperonyl butoxide enhances the insecticidal activity of pyrethrins, which causes most flying insects to drop almost immediately upon exposure. Because of its low toxicity to vertebrates, pyrethrum finds wide acceptance worldwide. Like other natural pesticides, pyrethrins have limited stability and shelf-life, and under field conditions are rapidly degraded by sunlight and heat. The use of pyrethrum in agriculture is mainly restricted to some vegetable and fruit crops due to its high cost and photo-instability. Efforts were, therefore, made from 1940s onwards to develop photostable analogues resulting in the development of synthetic pyrethroids.

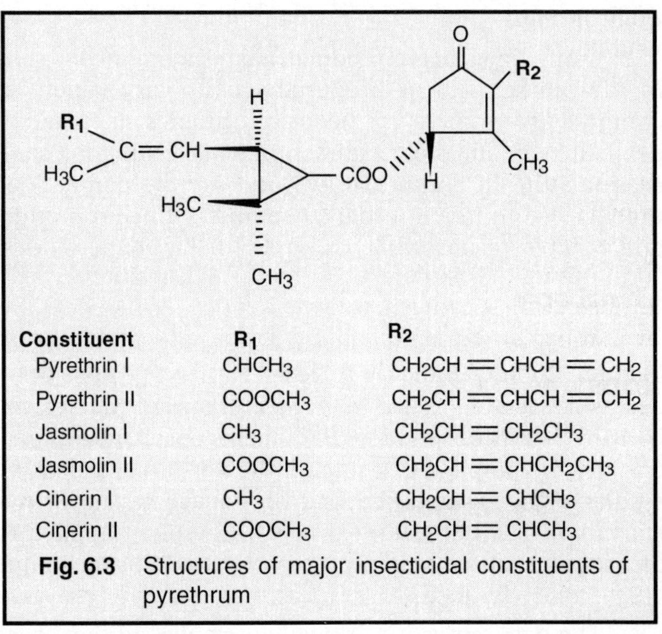

Constituent	R_1	R_2
Pyrethrin I	$CHCH_3$	$CH_2CH=CHCH=CH_2$
Pyrethrin II	$COOCH_3$	$CH_2CH=CHCH=CH_2$
Jasmolin I	CH_3	$CH_2CH=CH_2CH_3$
Jasmolin II	$COOCH_3$	$CH_2CH=CHCH_2CH_3$
Cinerin I	CH_3	$CH_2CH=CHCH_3$
Cinerin II	$COOCH_3$	$CH_2CH=CHCH_3$

Fig. 6.3 Structures of major insecticidal constituents of pyrethrum

Tobacco

Tobacco, *Nicotiana tabacum* Linnaeus, has been cultivated by the American Indians for at least 1000 years and it remained a part of their religious ceremonies. Long before people knew of nicotine alkaloid (**Fig. 6.2**) in tobacco, the latter was being used as a dust or water extract to control phytophagous insects, some three hundred years ago. Nicotine has by now been isolated from at least 18 species of plants. *N. rustica* Linnaeus containing 18 per cent nicotine is a better source than the more familiar *N. tabacum* containing 6 per cent nicotine. Many of these plants also contain the related alkaloids, nornicotine and anabasine. *N. glauca* Graham grown in Argentina and Uruguay contains a higher amount of anabasine. Systematic use of nicotine sulphate started with the introduction of a standardized pesticide formulation containing 40 per cent actual nicotine around 1910. Before the second world war, nicotine sulphate was a very popular insecticide around the globe. With the advent of synthetic insecticides, it lost its ground due to less persistence and high cost. Now that interest in botanicals is being revived, nicotine sulphate has also resurged as a preferred pesticide.

On a world-wide basis, about 600,000 kg of nicotine sulphate and 75,000 kg of pure nicotine were being produced annually till recently, mostly in UK, India, Germany and the Netherlands. India manufactures around 800 tonnes of nicotine sulphate annually and exports the entire quantity to Japan and Europe. The cost may be a prohibitive factor in its use in India. But the pesticide could be obtained from 35-40 million kg of tobacco waste produced in the country annually. Recently, cost effective methods have been developed for extraction of nicotine sulphate from tobacco waste.

Nicotine sulphate is effective against a wide range of pests. Its efficacy against soft bodied insects like aphids is well known, but it has also been found effective against whitefly, thrips and bollworms in cotton; brown planthopper and green leafhopper in rice; grubs in brinjal, potato and cauliflower. Recently, nicotine sulphate (0.2 and 0.4% a.i.) has been found highly toxic to eggs and neonate larvae of *H. armigera* and *S. litura*. It was also found highly effective against *Bemisia tabaci* (Gennadius) under field conditions.

Nicotine sulphate is safe to coccinellids but toxic to chrysopids. It does not leave any residue on the crop. However, its low persistence necessitates a repeated number of sprays for effective control of insects. It may prove very useful for use on fruit, vegetable and edible oilseed crops, where residues of pesticides are not acceptable. However, because of its high mammalian toxicity (oral human lethal dose is 60 mg) and detrimental effect on human health, the use of nicotine as an insecticide has decreased tremendously. As a result, it is seldom used today in North America or Europe, although it continues to be used in China and crude tobacco extracts are used in Africa.

Rotenone Plants

Rotenone **(Fig. 6.2)** and related alkaloids occur in the roots of tropical legumes, *Derris* and *Lonchocarpus* plants and in leaves and seeds of *Tephrosia* plants, all belonging to family Fabaceae. *Derris* is native to East Asia, *Lonchocarpus* to American tropics and *Tephrosia* to Eastern and Southern Africa. Traditionally, all these plants were used as fish poisons by the people for centuries. The commercially important species *Derris elliptica* (Wallich) Benth. contains 4-5 per cent rotenone while *Lonchocarpus nicou* (Aubl.) DC contains 8-10 per cent rotenone in dried roots. The vogel tephrosia, *Tephrosia vogelii* Hook f. is regarded as a more promising source of rotenone than *Derris* and *Lonchocarpus*. Rotenone is commonly sold as dust containing 1 to 5 per cent active ingredients for home and garden use, but liquid formulations used in organic agriculture may contain as much as 8 per cent rotenone and 15 per cent total rotenoids.

In addition to rotenone, these plants contain a number of other flavonoids like deguelin, tephrosin, elliptone, sumatrol, toxicarol, malaccol, etc., which show toxic as well as various behavioural and physiological effects on the insects. Rotenone is mainly active as a contact poison with some action as a stomach poison. It mainly acts as a site I respiratory inhibitor. Its rapid degradation in sunlight has limited its utility in crop protection. Synthetic rotenoids have so far proved very costly because of the complex chemical structure. Its use has also been discontinued due to high fish and/or mammalian toxicity. Pure rotenone is quite toxic to mammals, the oral acute LD_{50} in rats is 132 mg kg^{-1}.

Pongram

Pongamia pinnata Linnaeus (Syn. *P. glabra* Vent.), variously known as *karanja*, puna oil tree, Indian beech or Pongram is also indigenous to India. Karanj seed oil is rich in karanjin, a furanoflavonoid and a host of other polyphenolics like pongamol, pongapin, glabrin, karanj ketone, karanjone and pongaglabrone. Karanjin has been found effective against mites, scales, chewing and sucking insect pests. It is a potent deterrent to many different genera of insects and mites, and is effective against whiteflies, thrips, leafminers, caterpillars, aphids, jassids, beetles, mealybugs, etc. on a wide range of crops. Karanjin has a dramatic antifeedant and repellent effect with many insects avoiding treated crops. Insect antifeedant, growth reduction and miticidal activity of karanj oil have been attributed to the presence of high concentration of karanjin and pongamol in the oil.

Karanja oil applied as a surface protectant effectively checked the infestation of pulse beetles, *Callosobruchus maculatus* (Fabricius) and *C. chinensis* (Linnaeus) and other storage pests like *Rhyzopertha dominica* (Fabricius) and *Sitotroga cerealella* (Olivier). A concentration of 1 per cent afforded complete protection even after 150 days and did not alter taste and smell of the grains. Pongamia cake was found effective in controlling the attack of ground beetles on tobacco. It also did not leave any of harmful residues in the soil. Pongamia cake water extract was found effective for protecting tobacco seedlings from *S. litura* damage.

The oil of karanja repelled brown planthopper in rice and significantly reduced its ingestion and assimilation of food. Both brown planthopper and whitebacked planthopper suffered heavy mortality but green leafhopper was less susceptible. Other pests which have been reported susceptible to powders or extracts of Pongamia are *Henosepilachna vigintioctopunctata* (Fabricius), *Amsacta moorei* (Butler), *Chilo partellus* (Swinhoe), *Papilio demoleus* Linnaeus, *Leucopholis lepidophora* Blanchard, etc. Various types of bioactivity observed were antifeedant, fecundity curtailing and toxicity.

Custard Apple

Custard apple, *Annona squamosa* Linnaeus and other *Annona* species are well known for their pesticidal activity. Annonins (acetogenins) and related compounds namely squamocin, asimicin and annonacins occur widely in twigs and branches, unripe fruits and seeds of several *Annona* species. Most of the acetogenins performed better than the conventional insecticides. Powdered seeds applied to wheat and rice grains act as a protectant against *Sitophilus oryzae* (Linnaeus) and *C. chinensis.* The plant extracts act as a feeding deterrent against *A. moorei, Oncopeltus fasciatus* (Dallas), *N. lugens, Dicladispa armigera* (Olivier), *Nephotettix virescens* (Distant), *S. litura* and *H. vigintioctopunctata.* As with other botanical insecticides, disruption of growth, reduced oviposition, reduced adult emergence and moderate toxicity has also been observed in different species. Annonine, an alkaloid found in the stems and leaves of custard apple, has been found effective in checking the infestation by termites, root grubs, etc.

Sabadilla

Sabadilla formulations were originally obtained from *Sabadilla officinarum* Brandt, a lily that grows wild in Central and South America. The commercial source of sabadilla was, howevers a related plant cevadilla or caustic barley, *Schoenocaulon officinale* (Schitdl. & Chem.) (Liliaceae), which subsequently became widely cultivated in Venezuela. *S. officinale* and several other plant species including the false hellebore, *Veratrum album* (Verat) (Melanthaceae) produce insecticidal veratrine alkaloids. The veratrine is commonly sold under the trade names 'Red Devel' or 'Natural Guard'. This compound gained popularity during Second World War, when other botanicals like pyrethrum and rotenone were in short supply. The veratrine alkaloids comprise about 0.3 per cent of the weight of aged sabadilla seeds; of these alkoloids, cevadine and veratridine are most active insecticides. Other alkaloids present in the seed and in insecticidal extracts include sabadinine, sabadiline and sabadine.

Sabadilla was used historically for the control of insect pests on crops, animals and humans. Since the advent of synthetic insecticides, sabadilla's use has declined and organic gardeners currently provide the major market for sabadilla products. Sabadilla acts as a contact and stomach poison for the control of a variety of pest species. Grasshoppers, house flies, jassids, lice, thrips and various caterpillars have been reported to be controlled with crude extracts. The pests for which sabadilla, in particular, is considered effective include hemipteran bugs such as the squash bug, *Anasa tristis* (DeGeer); chinch bug, *Blissus leucopterous* (Say); and the stink bugs. Although not persistent, sabadilla is known to be toxic to some soil bacteria and honey bees. The active ingredient veratrine with an oral LD_{50} of 4000-5000 mg/kg is considered among the least toxic of botanical pesticides.

Ryania

The compound ryanodine has been derived from the woodly stem tissue of the shrub, *Ryania speciosa* Vahl. (Flacourtiaceae), a plant native to South America and is sometimes referred to as ryania. Ryania represents one of the first examples of a commercially successful natural insecticide discovered during 1940*s*, by randomly screening plant extracts for activity. It began to be marketed for pest control from 1945 onwards. Among the 11 compounds identified with insecticidal activity, the most active constituents were ryanodine and 9, 21-dehydroryanodine.

Ryanodine is a contact and stomach poison that is more stable than many other botanical pesticides. The residual toxicity of ryanodine has been reported to be for over one week after application. It is used mostly for the control of caterpillar pests of fruits and foliage. The codling moth, *Cydia pomonella* (Linnaeus) in apples and pears; citrus thrips, *Scirtothrips citri* (Moulton) in citrus ; and the European corn borer, *Ostrinia nubilalis* (Hubner) in corn are among the most common targets of ryania used by organic farmers. However, high costs associated with large scale production and processing of extracts are now becoming prohibiting except for home gardeners and organic producers.

Quassia

Quassia was originally extracted from *Quassia amara* Linnaeus (Simaroubaceae), a central American tree with a characteristically bitter bark and wood. However, in the eighteenth century, the commercial extracts of quassia were obtained from the related shrub, *Aeschrion excelsa* (Sw.) Kuntze. The yellowish white wood is the source of quassia chips from which a bitter extract with insect killing activity is prepared. The active component within extracts is quassin, a water soluble molecule that acts as a contact and stomach poison. Quassin has been shown to possess systemic activity. it is a potent aphicide and toxic to a number of lepidopteran pests.

In India, farmers have been seen to use extracts from *Picrasma excelsa* (Sw.) Planchon (shrub closely related to *A. excelsa*) for pest control. At least 31 quassinoids from *P. ailanthoides* Planchon have been reported to be potent antifeedant and insecticidal compounds against *Plutella xylostella* (Linnaeus). The quassinoids reported from *Simaba multiflora* A. Juss. and *Soulamea soulameoides* (Gray) Nooteboom are feeding deterrents against *Heliothis virescens* (Fabricius) and *Spodoptera frugiperda* (J.E. Smith), and induce toxic and bioregulatory effects in *H. virescens.*

Essential Oil Bearing Plants

Essential oils are volatile oils that have strong aromatic components which give distinctive odour, flavour or scent to a plant. These are the by-products of plant secondary metabolites. Essential oils are found in glandular hairs or secretory cavities of plant cell wall and are present as droplets of fluid in the leaves, stems, bark, flowers, roots and/or fruits in different plants. Plant essential oils are produced commercially from several botanical sources, many of which are members of the mint (Lamiaceae), carrot (Apiaceae), myrtle(Myrtaceae) and citrus (Rutaceae) families. Among higher plants, there are 17,500 aromatic plant species and approximately 3,000 essential oils are known out of which 300 are commercially important for cosmetics, perfume, and pharmaceutical industries, apart from pesticidal potential. The oils are generally composed of complex mixtures of monoterpenes, biogenetically related phenols, and sesquiterpenes (Khater, 2012).

There are several examples of essential oils, which are known for their pest control properties. Eugenol from cloves, *Eugenia cryophyllus* (Sprengel) Bullak & Harr. (Myrtaceae); 1, 8-cineole from eucalyptus, *Eucalyptus globulus* Labill. (Myrtaceae); citronellal from lemon grass, *Cymbopogon nardus* (Linnaeus) Rendle; pulegone from pennyroyal, *Mentha pulegium* Linnaeus (Labiatae); thymol and carvacrol from thyme, *Thymus vulgaris* (Linnaeus) (Labiatae) are among the most active constituents against insects. In addition to their toxic effects, most of these compounds act as larval growth inhibitors, antifeedants and repellents to a wide range of insects, mites and even nematodes.

Some terpenoids like thymol and carvacrol were more effective for deterring oviposition by *Aedes aegypti* (Linnaeus) mosquitoes than N, N-diethyl-m-toluamide (DEET). The essential oil from the rhizomes of sweet flag, *Acorus calamus* Linnaeus (Araceae) is known for its insecticidal and antigonadal actions associated with its most abundant constituent β-asarone Vulgarone B, isolated from *Artemisia douglasiana* Besser; apiol, isolated from *Ligusticum hultenii* Fernald; and cnicin **(Fig. 6.5J)** isolated from *Centaurea maculosa* Lamarck, exhibited high termiticidal activity against the Formosan subterranean mite, *Coptotermes formosanus* Shriaki.

Ginger oleoresin extracted from fresh rhizomes of *Zingiber officinale* Roscoe (Zingiberaceae) is a complex mixture of several closely related phenolic alkalones such as gingerols, shogaols, gingerones, paradols, gingerdiols and diarylheptanoids. Ginger based products have been found to exhibit insect growth regulatory and antifeedant activity as well as antifungal activity. Turmerone and (ar) turmerone (dehydroturmerone), the major constituents of turmeric, *Curcuma longa* Linnaeus (Zingiberaceae) rhizome powder oil, are strong repellents to stored grain pests. The turmeric leaves, the unutilized part of turmeric plant, on hydrodistillation, yielded oil rich in 2-phellandrene that inhibited the growth of *Spilarctia obliqua* (Walker), *Plutella xylostella* (Linnaeus) and several stored product beetles. The active compounds in garlic, *Allium sativum* Linnaeus (Amaryllidaceae) have antibacterial, antifungal, nematicidal, amoebicidal, insecticidal and insect repellent properties. It is the storehouse of a large number of bioactive molecules which include allin, allicin, garlicin, opine and allyl sulfides. The garlic extract has been found to be toxic to house fly, mosquitoes and storage pests, and highly repellent to adults of cockroach, *Blatella germanica* (Linnaeus). It has also proved to be oviposition deterrent, and toxic to eggs and larvae of *P. xylostella.*

Today, essential oils represent a market estimated at US$ 700 million and a total world production of 45,000 tons. Almost 90 per cent of this production is focused on mint and citrus plants. Several private companies produce essential oil-based insecticides for controlling greenhouse pests and diseases, and for controlling domestic and veterinary pests. In general, essential oils and their major constituents are relatively non-toxic to mammals, with acute oral LD_{50} values in rodents ranging from 800 to 3000 mg kg^{-1} for pure compounds and > 5,000 mg $kg^{=1}$ for formulated products (Khater, 2012).

EFFECTS ON NON-TARGET ORGANISMS

Natural Enemies

The application of NSKE for the control of *S. litura* did not affect the emergence of the egg parasitoid, *Telenomus remus* Nixon, and predator, *Brinckochrysa scelestes* (Banks). However, longevity of the parasitoid was reduced in case of oviposition on pretreated egg masses. Topical application of seed oil of neem, Chinaberry and custard apple showed no adverse effect on predatory spider, *Lycosa pseudoannulata* (Bosenberg & Strand) and was only slightly toxic to the mirid bug predator, *Cyrtorhinus lividipennis* Reuter at and above 10 μg/female.

Neem formulations, Repelin and Neemguard, were relatively safe at lower concentrations to the egg, larval and pupal parasitoids, viz. *Trichogramma australicum* (Girault), *Bracon hebetor* Say and *Tetrastichus israeli* (Mani & Kurian) of *Opisinia arenosella* Walker. Repelin, Neemark and nicotine sulphate were safe to the predatory coccinellid, *Menochilus sexmaculatus* (Fabricius) but highly toxic to its hyperparasitoid, *Tetrastichus coccinellae* Kurdyumov. Neemark and Repelin worked as good acaricides against *Tetranychus macfarlanei* Baker & Pritchard on okra, but were safer to predatory mites.

In laboratory trials with NSKE on bean leaf discs, predatory phytoseiid mite, *Phytoseiulus persimilis* Athias-Henriot was less affected than the casmine spider mite, *Tetranychus cinnabarinus* (Boisduval). In the same trials, predatory clubionid spider, *Chiracanthium mildei* Koch was unaffected. Neem formulations, viz. Neemark, Repelin, Wellgro, neem seed kernel suspension (NSKS) and Neemrich did not show toxic and ovicidal effects against the green lacewing, *B. scelestes* but acted as oviposition repellents against the females. On the other hand, nicotine sulphate was toxic to adults and showed ovicidal activity against the eggs of the lacewing. Parasitization of whitefly, *Bemisia tabaci* (Gennadius) nymphs by *Eretmocerus* spp. was not affected by application of various botanical insecticides, viz. mineral oil, neem oil, nicotine sulphate, saradine oil, etc.

There have been reports indicating moderate to high adverse effects of neem products on natural enemies. Coccinellid predator, *Delphastus pusillus* (LeConte) preferred untreated eggs of

B. tabaci. The emergence of aphelinid parasitoid, *Eretmocerus californicus* Howard from treated *B. tabaci* was reduced by more than 50 per cent. Untreated whiteflies were attacked at a rate three times more than that for treated ones. Emergence of eulophid parasitoid, *T. howardi* from pupae of *M. patanalis* treated with 1000 ppm neem seed bitters decreased. A dose of 50 μg/female of neem seed bitters proved lethal to the parasitoid, *Goniozus triangulifer* Kieffer. The emergence of parasitoids from treated leaf folder, *M. patanalis* hosts as well as the fecundity of such parasitoids were reduced. Treatment of cocoons at 2.5 per cent or higher concentrations of neem oil reduced emergence of braconid parasitoid, *Apanteles plutellae* Kurdyumov in the laboratory.

Other Organisms

Recent studies have shown that neem leaf extract completely stopped aflatoxin production by *Aspergillus flavus* Link and *A. parasiticus* Speare. These fungi which grow widely on various foods are one of the most deadly organisms on earth and the aflatoxins produced by them are highly carcinogenic. The use of neem for inhibiting aflatoxin production may open the door to a simple, inexpensive method for protecting stored foods using locally produced materials, even in the remotest villages.

Neem oil extracts at 0.005 per cent used as a mosquito larvicide were non-toxic to insectivorous fish, *Gambusia* sp. Concentrations upto 0.01 per cent were essentially non-toxic to the fish. However, at 0.04 per cent concentration, 80 per cent tadpole mortality was obtained within 24 hours and 100 per cent in 48 hours. The LC_{50} values of Margosan-0 to rainbow trout, *Salmo gairdneri* Richardson and blue sunfish, *Lepomis macrochirus* Rafinesque, 96 hours after treatment were 8.8 and 37 mg/l of water, respectively. The 96 hours no effect concentrations for the two species were 5 and 20 mg/litre, respectively. Young guppies, *Lebistes reticulatus* (Peters) tolerated 100 ppm AZI-VR-K/litre of water. It has been reported that toxicity of Margosan-0 to fish and other aquatic organisms is caused by its petroleum oil content (15%) or probably another compound used for its formulations. Margosan-0 is also toxic to the water flea, *Daphnia magna* Straus and other invertebrates that inhabit stagnant water.

The feeding of water extract of neem berries to poultry birds resulted in toxicity symptoms like sluggish movement, dropping head, etc., and in many cases even death. The liver underwent degenerative changes with focal congestions, retention of bile in gall bladder and congestion of kidney with localized haemorrhages. The hepato- and nephro-toxic effects may prove lethal. Trial feeding of neem seed to starter chicken caused severe heptatitis with necrotic patches, mild to severe nephritis with congestion and slight inflammation in intestine. However, compared to synthetic insecticides, the toxic effects were produced at higher dosages. The acute LD_{50} of Margosan-0 to mallard duck, *Anas platyrhynchos* Linnaeus is 16.0 mg/kg. Acute oral LC_{50} of this compound to bobwhite quail, *Colinus virginianus* (Linnaeus) and mallard duck is in excess of 7000 ppm in a 5-day test period.

Traditionally, neem-based preparations have been used for the treatment of a wide range of disorders in domestic animals and livestock. Neem leaves contain appreciable amounts of protein, minerals and carotene, and adequate amounts of trace minerals except zinc. Incorporation of 20 per cent neem cake in sheep diet increased the growth rate. But neem leaves have also been reported to cause toxic effects on sheep, goats and guinea pigs. The acute oral LC_{50} of Margosan-O against rat, *Rattus* sp. is 5.0 ml kg^{-1} and acute inhalation $LC_{50} < 43.9$ mgl/hours. *Lantana camara* Linnaeus is toxic to sheep when given orally at 60 mg/kg body weight. Affected animals show depression and anorexia, a few days after ingesting. Death occurs a few weeks later and may be due to renal failure.

Man

Neem preparations have been used since ancient times in unani medicine for the treatment of a wide range of human ailments due to their anti-inflammatory, concoctive, blood purifying, anti-leprosy, anti-arthritic, anti-pyretic, anti-microbial and antihelminthic actions. Injections of sodium nimbidinate at 1 g and its oral administration up to 7 g to human beings did not produce any harmful effects. Cases of neem seed oil intoxication have, however, been reported. The intake of oil resulted in vomitting, drowsiness, metabolic acidosis, etc. It has been reported that the oil uncouples mitochondrial oxidative phosphorylation, thus inhibiting respiratory chain. Normal human cells in culture were not affected by neem extracts at 5 mg/ml while tumor-originated cells degenerated.

In USA, Margosan-O underwent comprehensive toxicological tests prior to registration by US Environmental Protection Agency (EPA). Tests for skin irritation, inhalation, mutagenicity, and immune response were low enough to allow EPA registration. There have been a number of reports of death of children and adult human beings after consuming neem oil in South India. These poisonings apparently resulted because the neem seeds from which the oil was extracted had been contaminated with aflatoxin producing fungus, *Aspergillus flavus* Link. When extracted from clean and fungal-free seed kernels, neem oil did not cause any oral toxicity in laboratory rats at 500 mg/kg body weight.

Neem preparations have been found to have no teratogenic and carcinogenic effects. In the standard Ames mutagenicity test, azadirachtin showed no mutagenic activity on strains of *Salmonella thyphimurium* (Loeffler) castellanichalmers. There are extensive records of human exposure and response to pyrethrum used as an insecticide and these establish the negligible hazards to the users of pyrethrum products. On the other hand, several other plants are known to be highly toxic to humans and other mammals. A single leaf of yellow oleander, *N. oleander,* is potentially lethal to humans. Affected persons become dizzy and drowsy, heart beat becomes progressively weaker and irregular leading to dyspnoea and coma.

ENVIRONMENTAL IMPACT

The use of botanical pesticides for plant protection has assumed greater importance in recent years all over the world due to environmental deterioration and health hazards associated with the use of synthetic pesticides. It is hoped that extensive use of plant-based pesticides in integrated pest management will help in conserving environmental quality.

The neem tree has answers to several environmental problems such as the rehabilitation of degraded ecosystem and wastelands, reduction in the use of agro-chemicals such as fertilizers and synthetic pesticides, and generation of income for all small farmers with limited resources. The neem-based pesticides are relatively safe and do not leave any residues on agricultural produce. As the pesticidal preparations using neem seeds, leaves or cake are very simple, farmers may make direct use of these products locally. The use of neem products for plant protection will help in minimizing atmospheric pollution and prevent food poisoning. It will also reduce the demand for costly chemical pesticides.

Neem oil has found a wide range of industrial applications in India. Due to its antiseptic properties, it has been a major ingredient in soaps for atleast 50 years. In addition, pharmaceutical preparations like emulsions, ointments, poultices and liniments as well as cosmetics such as creams, lotions, shampoos, hair tonics and gargles have been prepared. Neem toothpaste and neem sticks are also widely used for cleaning of teeth and as a mouth freshner. Neem oil is non-drying and is, therefore, used to grease cast wheels.

Neem is a valuable forestry species in India. Being a hardy species, it is ideal for reforestation programmes and for rehabilitating degraded, semiarid and arid lands and coastal areas. During a severe drought in Tamil Nadu during June-July, 1987, it was witnessed that neem trees remained luxuriant while all other vegetation dried up. Neem is useful as windbreaks and in areas of low rainfall and high windspeed, it can protect crops from desiccation. Neem is also a preferred tree along avenues, in markets and near homelands because of the shade it provides. It can grow in and even neutralize acid soils that plague much of the tropics. It may also act as an important source of fuel. Neem oil is burned in lamps. The wood has long been used in firewood. The husk from seeds produced as a waste during pesticide manufacture can also be used as fuel. The neem timber is durable and resistant to attack of termites and other pests. It can be used for making fence posts and poles for house constructions.

It is, thus, clear that all parts/products of the neem tree find a variety of uses in agriculture, and as raw materials for the manufacture of household articles, pharmaceutical preparations and other products. Even the industrial by-products have found application in agriculture and as fuel. Due to its successful cultivation even under adverse environmental conditions and utilization of all its products so that no waste material is left, neem growing can be an important component in environmental protection programmes.

PEST RESISTANCE TO PHYTOCHEMICALS

The naturally occurring phytochemicals exert a wide range of behavioral and physiological effects on insects and there is, thus less likelihood of development of resistance to these pesticides. Ironically, the neem tree itself is attacked by 60 species of insects besides mites, nematodes and 16 phytopathogens. Some of these like *Pulvinaria maxima* Green, *Aonidiella orientalis* (Newstead), *Helopeltis antonii* Signoret and *Rhizoctonia solani* (Kuhn) are already causing serious damage in some areas. This indicates that some insects might adopt to limonoids in the future but in laboratory tests two genetically different strains of the diamondback moth, *Plutella xylostella* (Linnaeus), treated with neem seed extract showed no signs of resistance in feeding and fecundity tests up to 35 generations. In contrast, deltamethrin-treated lines developed resistance factors of 20 in one line and 35 in the other. There was no cross resistance between deltatamethrin and neem seed extract in the deltamethrin resistant lines. However, two lines of *Myzus persicae* (Sulzer) of the same origin, when treated with pure azadirachtin, developed a 9-fold resistance to azadirachtin compared to non-selected control line after 40 generations.

Some resistance to pyrethrins has been reported among a few agricultural pests, particularly those with resistance to organochlorines, organophosphates and carbamates. However, the diversity of neem compounds and their combined effects on insects seem to confer a built-in resistance prevention or delay mechanism in neem. Even then, the farmers should refrain from exclusive and extended application of single bioactive materials such as azadirachtin. Also, for durability sake, even novel neem-rich insecticides should be applied within the framework of integrated pest management programmes.

INTEGRATION WITH OTHER TACTICS

The use of pesticides of plant origin for the control of agricultural pests has a long history but has assumed greater importance in recent years due to environmental deterioration and health hazards associated with the use of synthetic pesticides. Botanical pesticides exert a range of behavioural and physiological effects on the colonization, development, growth, survival and multiplication of insects. In view of their environmental safety, these pesticides offer an attractive alternative to synthetic pesticides for use in IPM. Botanical pesticides can be integrated with the use of parasitoids and predators, microbial pesticides and even synthetic chemical pesticides to achieve greater efficiency in pest control.

The application of neem seed kernel extract (NSKE) for the control of *S. litura* did not affect the emergence of the egg parasitoid, *Telenomus remus* Nixon, and predator, *Brinckochrysa scelestes* (Banks). Topical application of seed oil of neem, Chinaberry and custard apple showed no adverse effect on predatory spider, *Lycosa pseudoannulata* (Bosenberg & Strand) and was only slightly toxic to the mirid bug predator, *Cyrtorhinus lividipennis* Reuter at and above 10 μg/female in rice crop. An active neem seed kernel fraction evaluated against sorghum pests was found to be safe to midge, *Contarinia sorghicola* (Coquillett) parasitoid, *Tetrasictus* sp. and predator, *Orius* sp. It was, however, toxic to *Apanteles ruficrus* (Haliday), a larval parasitoid of *Mythimna separata* (Walker). Field trials with neem oil for the control of sorghum aphid, *Melanaphis sacchari* (Zehntner) did not show any adverse effect on syrphids and coccinellids.

Botanical pesticides have been found to be compatible with synthetic chemical insecticides. The use of neem formulations in combination with conventional insecticides resulted in better control of pink bollworm and spotted bollworm on cotton. The use of neem pesticides and alphamethrin on cotton proved as good as the sprays of alphamethrin alone in checking the menace of bollworms and increasing the seed cotton yield. Similarly, alternate application of high potency neem-based insecticides and synthetic insecticides led to the effective management of cotton whitefly. Neem and other plant products have also been found to increase the efficacy of microbial control agents like baculoviruses, *B. thuringiensis* and fungi.

POTENTIAL AND CONSTRAINTS

At present, many plant species are used for the production of pest control formulations. Some estimates put the number of plant species in use as source of botanical pesticides at over 2500. Despite this, it has been argued that resource poor farmers in many developing countries could make far more use of these free, naturally occurring products. The use of botanical pesticides offers several advantages over the synthetic pesticides.

- As plants have developed these chemicals in response to the combined selection pressure of phytopathogens, insects and other herbivores, many of these pesticides are effective against diseases, nematodes and other organisms in addition to phytophagous insects.
- The naturally occurring phytochemicals exert a wide range of behavioural and physiological effects on the insects, and, therefore, there is less possibility for insects to develop resistance to these pesticides.
- The studies conducted so far indicate that neem and other botanical pesticides are comparatively safe to natural enemies and higher organisms.
- The available evidence indicates that botanical pesticides are biodegradable in contrast to persistent synthetic insecticides.
- Many botanical pesticides can be developed from indigenous plants, and, therefore, it will save valuable foreign exchange. Moreover, village cooperatives can take up the formulation of locally available plants and, thus, farmers will be saved from spending large sums of money for the purchase of costly synthetic agrochemicals.
- There is great demand in international market for residue-free cotton garments, fruits, vegetables and beverages. The large scale utilization of botanical pesticides will certainly help us in meeting international standards of quality and safety in these products.

However, the large scale utilization of botanical pesticides in IPM is limited by several factors.

- Numerous species show some antifeedant/repellent activity. The identification of promising plant species suitable for utilization as botanical pesticides is, therefore, a difficult task.
- Standardized bioassay procedures need to be developed for efficient screening of plants for different types of toxic, morphogenetic, behavioural and physiological effects.
- The results obtained with botanical pesticides are usually inconsistent. In many cases, this problem is due to a lack of quality control and could be overcome by developing suitable guidelines for registration of botanical pesticides.
- The mortality obtained is usually moderate and the required degree of control is not achieved in many cases.

- The trees like neem and *dharek* have to be planted many years in advance. There is a long gestation period and only after that can these trees be used for obtaining pesticidal products.

A lot of work needs to be done before large scale utilization of botanical pesticides in IPM becomes a reality. Plants with pesticidal potential should be identified and grown with an industrial approach in order to obtain the raw material with greater ease and at a lower cost. For example, scores of leading enterprises are engaged in producing neem products developed in Yunnan province of China. There are over 400,000 plantations of neem in the Yunnan province, developed with the support of government institutions and enterprises as well as local villagers. This makes Yunnan province the bigest artificial area of neem planting globablly and the raw material centre of neem products in China. This type of approach will need to be adopted globally for any plant that may have the potential to be developed as a biopesticide. Intensive breeding and selection work will have to be undertaken for economic production of various high quality raw materials.

The biodiversity of a given plant like neem can be enhanced by developing neem clones with required characteristics like faster maturation, high yield of seed with high oil content and yield of azadirachtin. This will provide the growers with a 'menu', from which they can choose the desired characteristics according to their needs. Quality control in botanical pesticides is a major problem. There is wide variation in the quality and quantity of extractives obtained from a plant due to variation in ecotypes, environmental factors, etc. Such variations affect the performance and shelf-life of formulated products. There is an urgent need to develop and prescribe suitable standards for registration of these products. Recent advances will certainly enable us to identify selective, diverse, renewable, cheap and environmentally acceptable plant products from the repository of 'mother nature' for use in IPM programmes.

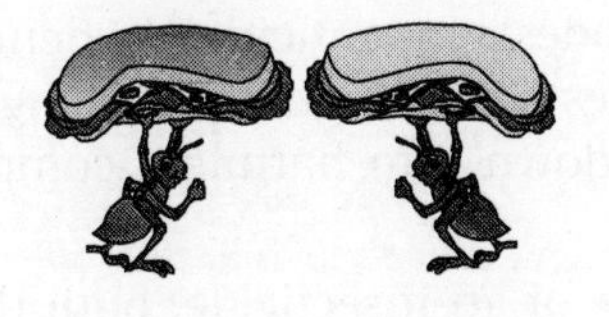

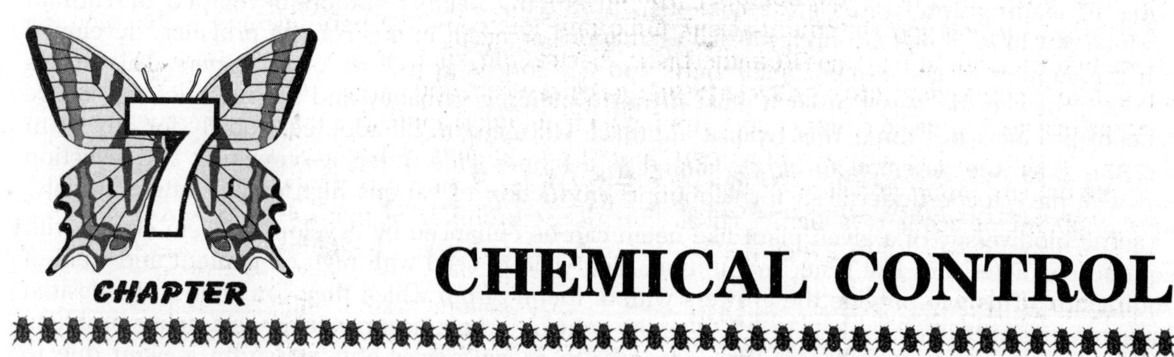

CHEMICAL CONTROL

INTRODUCTION

The cultural and other agro-technological approaches are quite often not sufficient to keep the pests at such a level that economic losses could be avoided. Therefore, chemical agents are resorted to both as preventive and curative measures to minimize the pest incidence and damage. The chemicals to combat different kinds of pests are called pesticides. based on the target organisms, these pesticides are grouped into various categories, viz., insecticides (insects), herbicides (weeds), fungicides (fungi), nematicides (nematodes), rodenticides (rodents), etc. A good pesticide should be cheap to produce and potent against pests. It should not endanger the health of man and domestic animals, and should ultimately break down into harmless compounds, so that it does not persist in the environment.

In order to determine the potency of an insecticide, both the relative toxicity and the specific toxicity are to be worked out. The value of median lethal dose (LD_{50}) of various poisons against a given insect species indicates the relative toxicity. LD_{50} is expressed in terms of dosage in mg/kg of body weight necessary to kill 50 per cent of a population of male white rats/rabbits/guinea pigs. The specific toxicity may vary a great deal from species to species. A poison may be highly effective against one and quite ineffective against another species. There are very few compounds which have all the qualities considered desirable and, therefore, we have to accept those which are comparatively cheap, more effective and less dangerous.

HISTORICAL ASPECTS

Since the early nineteenth century, certain inorganic compounds, like lead arsenate, sodium fluosilicate, Paris green and zinc phosphide have been used as stomach poisons for the control of insects or other pests. These poisons were administered along with food and they were equally potent against mammals, including man and domestic animals. Some of them, when not properly applied were even phytotoxic. With the discovery of insecticidal properties of certain plant extracts, the insecticides like nicotine sulphate, pyrethrum and rotenone came into prominence. These insecticides were comparatively safe for mammals. This important fact was kept in mind by the organic chemists while looking for other safer organic compounds which could be used as insecticides. DDT (Dichlorodiphenyl trichloroethane) synthesized in 1874, but discovered as an insecticide in 1939, was one such compound. In fact, this was the first wonder insecticide that appeared

Table 7.1 Salient events in evolution of chemical control of pests

Year	Event(s)
Pre-Scientific Era	
1000 BC	Homer referred to sulphur, used in fumigation and other forms of pest control.
100 BC	The Romans applied hellebore for the control of rats, mice and insects.
900 AD	Chinese used arsenic to control garden insects.
1649	Rotenone used to paralyse fish in South America.
1690	Tobacco extracts used as contact insecticide.
1773	Nicotine fumigation by beating tobacco and blowing smoke on infested plants.
1787	• Soap mentioned as insecticide. • Turpentine emulsion recommended to repel and kill insects.
1800	• Persian louse powder (pyrethrum) known to the Caucasus. • Sprays of lime and sulphur recommended for insect control.
1825	Quassia used as insecticide in fly baits.
1848	Derris (rotenone) reported being used in insect control in Asia.
1854	Carbon disulphide tested experimentally as grain fumigant.
1867	Paris green used as an insecticide.
1868	Kerosene emulsions employed as dormant sprays for deciduous fruit trees.
1877	Hydrogen cyanide (HCN) first used as fumigant in museum cases.
1880	Lime sulphur used against San Jose scale in California.
1882	Naphthalene cakes used to protect insect collections.
1892	• Lead arsenate first prepared and used to control gypsy moth in Massachusetts, USA. • First use of a dinitrophenol compound, the potassium salt of 4-6-dinitro-o-cresol, as insecticide.
1897	Oil of citronella used as a mosquito repellent.
Pre-Green Revolution Era	
1902	The value of lime sulphur for apple scab control discovered in New York, USA.
1907	Calcium arsenate in experimental use as an insecticide.
1909	First tests with 40% nicotine sulphate made in Colorado, USA.
1912	Zinc arsenite first recommended as an insecticide.
1922	Calcium cyanide begins commercial use.
1931	Thiram, first organic sulphur fungicide, discovered.
1932	Methyl bromide first used as a fumigant in France.
1936	Pentachlorophenol introduced as wood preservative against fungi and termites.
1938	TEPP, first organophosphate insecticide, discovered by Gerhardt Schrader.
1939	DDT discovered as an insecticide by Paul Muller in Switzerland.
1941	Hexachlorocyclohexane (HCH) (benzene hexachloride, BHC) discovered to be insecticidal in France.
1942	Introduction of 2,4-D, the first of the hormone (or phenoxy) herbicides.
1943	First dithiocarbamate fungicide, zineb, introduced commercially.
1944	Introduction of 2,4,5-T for brush and tree control and warfarin for rodent control.
1945	Chlordane, the first of the persistent, chlorinated cyclodiene insecticides, introduced.
1946	First resistance in house flies to DDT observed in Sweden.

Table 7.1 contd...

Year	Event(s)
1947	• Toxaphene insecticide introduced, to become the most heavily used insecticide in U.S. agricultural history. • Passage of Federal Insecticide, Fungicide and Rodenticide Act (FIFRA).
1948	Aldrin and dieldin, first produced; the best of the persistent soil insecticides.
1949	• Captan, first of the dicarboximide fungicides, appears. • Synthesis of first synthetic pyrethroid, allethrin.
1950	• Malathion introduced, probably the safest organophosphate insecticide. • Maneb fungicide introduced.
1951	First carbamate insecticides, viz. isolan, dimetan, pyramat and pyrolan, introduced.
1952	Fungicidal properties of captan first described.
1953	Insecticidal properties of diazinon described in Germany.
1956	Introduction of carbaryl, the first successful carbamate insecticide.
1958	Atrazine, first of the triazine herbicides; and paraquat, first of the bipyridylium herbicides, introduced.
1962	Publication of *Silent Spring* by Dr Rachel Carson.
1964	Fungicidal properties of thiabendazole described.
1965	Development of Temik[â], first soil-applied insecticide-nematicide.
Green Revolution Era	
1966	• Carboxin, the first systemic fungicide, developed. • Methomyl insecticide and chlordimeform acaricide-ovicide introduced.
1967	Introduction of second group of systemic fungicides with benomyl.
1968	Discovery of tetramethrin, resmethrin and bioresmethrin, synthetic pyrethroids with greater activity than natural pyrethroids.
1971	Glyphosphate herbicide first introduced.
1973	• Development of first photo-stable synthetic pyrethroid, permethrin. • Cancellation of virtually all uses of DDT by the EPA.
1975	Cancellation of all uses of aldrin and dieldrin, except as termiticides.
1983	EPA cancels most uses of ethylene dibromide.
1984	• EPA cancels most registrations for endrin. • Discovery of first neonicotinoid, imidacloprid.
1985	First registration of azadirachtin as an insecticide for non-food use.
Post-Green Revolution Era	
1986	EPA cancels all agricultural uses of toxaphene.
1988	Chlordane and heptachlor use as termiticides cancelled.
1991	The first neonicotinoid, imidacloprid, marketed for the first time.
1993	Neemix® granted EPA exemption on food.
1996	The neonicotinoid-acetamiprid, commercially developed.
2001	The neonicotinoid, thiamethoxam, commercially developed.
2005	Flonicamid, a novel insecticide with a rapid inhibitory effect on aphid feeding was launched in the world market
2007	Azadirachtin synthesised in the laboratory by Dr Steven V. Ley and colleagues, University of Cambridge (UK).

Source : Modified from Dhaliwal and Koul (2010)

commercially and it was soon followed by BHC (Benzene hexachloride), which is a misnomer for HCH (Hexachlorocyclohexane) (Table 7.1).

By now, thousands of compounds have been synthesized and tested for their insecticidal properties. Quite a few are appearing in the market. Those which are finally accepted have to be subjected to certain rigorous tests. A good insecticide should be cheap to produce and potent against insect pests. It should not endanger the health of man and domestic animals, and should ultimately break down into harmless compounds so that it does not persist in the human environment. It is not easy to find new compounds having all these desirable qualities; hence fewer chemicals are being added to the list of good insecticides than one would expect.

Modern synthetic pesticides gave mankind unprecedented manipulative power to suppress the pest populations. For the first time in world history, little more than 50 years ago, man had curative tools to fight agricultural pests and vectors of human diseases. In fact, green revolution was driven by a wide array of different chemical technologies including the synthetic organic pesticides. Availability of pesticides made cultivation of crops profitable in areas where this was previously not possible because of pest problems. Increased agricultural productivity associated with high levels of pest control, improved crop and animal genetics, availability of fertilizers and mechanised production systems allowed food supplies to increase and standard of living to improve in many regions of the world.

CLASSIFICATION OF INSECTICIDES

Insecticides can be grouped in various ways, viz. according to the mode of their entry (stomach poisons, contact poisons and fumigants), their mode of action (physiological poisons, protoplasmic poisons, respiratory poisons and nerve poisons) or according to their chemical composition. As an insecticide may enter and act in an insect body in more than one way, it may be desirable to group them according to their chemical nature.

Chemically, the types of insecticides are : (*i*) elements, such as sulphur, phosphorus, thallium, mercury, etc.; (*ii*) inorganic compounds such as lead arsenate, sodium fluosilicate, zinc phosphide, Paris green, etc. (Chapter 6); (*iii*) organic compounds; (*a*) compounds of plant origin such as pyrethrum, nicotine, rotenone, etc. (chapter 6); (*b*) animal and mineral oils such as fish oil, diesel oil, etc., (*c*) synthetic organic compounds such as DDT, malathion, carbaryl, etc.; (*iv*) poisonous gases such as hydrogen cyanide, ethylene dichloride, carbon tetrachloride, methyl bromide, phosphine, etc., used as fumigants.

For those interested in insect control, however, it is more meaningful to list these compounds according to their mode of entry in the insect body and lethal action on the pests. The above mentioned compounds could, therefore, be rearranged under the following categories :

Stomach Poisons

They are generally used against insects with chewing type of mouthparts and under certain conditions against those with sponging, siphoning, lapping or sucking mouthparts. Some of these poisons are also mixed with food for killing higher animal pests, such as rodents, jackals and birds.

The stomach poisons are applied in many ways. The foliage or other natural food is thoroughly covered with poison or the poison is mixed with food along with an attractant which makes the mixture very acceptable. Sometimes, these poisons are sprinkled in the runways of insects. As the insects move about, they pick up the poison on their feet or antennae and while cleaning these parts with mouth, they ingest the poison.

Certain organophosphates act as systemic insecticides, which are taken up by the plant or animal system, and are translocated into tissues. When insect pests feed on these tissues, they die because of the poisoned cell sap.

For the stomach poisons to be acceptable in common use, they should have certain pre-requisites:

- The concentration which is applied to the plant for insect control should not be injurious to the foliage.
- The non-systemic compound should generally be insoluble in water, otherwise it might be absorbed through leaves and roots and, thus, poison the plants.
- It should be potent so as to kill the pest quickly and, thus, save the crop.
- It should be inexpensive and be easily available in large quantities.
- The compound should not be distasteful to the insects, so that they may not be repelled.
- Chemically, the compound should be stable, so that it does not lose its toxicity while in shipment, in storage or after it has been applied to the plants.
- It should also not break down too easily when mixed with other chemicals.
- The insecticide should have the physical properties of adhering to the plant surface on application and yet after a certain period it should disintegrate or be washed away by rain, etc., so that it does not persist on the plants for long periods and, thus, endanger the health of farm animals or of human beings.

The stomach poisons are applied as sprays, dusts, dips or baits. These poisons include inorganic and organic compounds. Among the inorganic compounds are the lead arsenate, calcium cyanide, Paris green (copper acetoarsenite), sodium flouride, sodium fluosilicates and sodium aluminofluoride (cryolite) and borax.

Systemic Poisons

An insect with piercing mouthparts sucks cell-sap through its proboscis which it embeds into the plant or animal tissue. The only way a poison can reach its stomach is through the plant or animal system. It is only in recent years that some highly effective systemic insecticides have been discovered. A systemic insecticide when applied to seeds, roots, stems or leaves of plants, is absorbed and translocated to various parts of the plant in amounts lethal to insects which feed on them. Some of them are quite specific to certain groups of insects. The systemic insecticides have a great value in an integrated approach to pest control, because through their application the parasites, predators and other natural enemies of the pests escape damage. Most of these insecticides act primarily as stomach poisons, and they include oxydemeton methyl, dimethoate, phosphamidon, aldicarb, phorate, carbofuran and others.

Inert Dusts

Fine dusts of various compounds have been used for a long time to control insect pests of crops and stored grains. These dusts are mixed with the grain or dusted on a surface from where the insects pick up the fine particles which cause irritation and abrasions in the cuticle, particularly at the body articulations. It results in excessive loss of moisture from the body of an insect which ultimately dies of desiccation. These dusts are based on fine sand, powdered cowdung ash (which also acts as a hygroscopic compound), calcium carbonate, silica, aerogel, etc.

Contact Poisons

The other most effective and most prevalent method of insect control lies in the use of contact poisons, a large number of which are now available. The contact poisons are applied as sprays or dusts, either directly onto the body of insects or to the places frequented by them. These poisons kill the insects either by clogging spiracles and respiratory system or by entering through the cuticle into the blood and acting as nerve or general tissue poisons. It has been shown that the insect cuticle possesses very high absorptive properties, so that the lethal dose applied externally is almost the same as the quantity required to kill by injecting it into the body. The contact insecticides, even though applied in the form of emulsions, suspensions or dusts are highly lipophilic and are readily absorbed by the lipids present in the epicuticle of insect exoskeleton. The residual film of a contact insecticide may kill an insect by its action on the sensory organs present in the tarsi of its legs.

Like the stomach poisons, the contact poisons also must have certain qualities before they can be accepted for use. Some of them are also phytotoxic. Others, like mineral oils, are so lethal to the foliage that they can only be applied in winter as dormant sprays when there are no leaves on the plants. Even among the milder insecticides which are used as summer sprays or dusts some have phytotoxic effect. For instance, high doses of DDT and HCH are known to cause phytotoxicity in plants. Most of the insecticides damage citrus inflorescence and, therefore, they cannot be applied when plants are in blossom. Some of these compounds are also toxic to mammals and are readily absorbed through the skin, for instance, parathion and endrin, when sprayed on crops in the field during the monsoon can easily cause death of the operators, if due precautions are not taken. Insecticides like DDT and HCH, which do not act as acute poisons, may cause very serious chronic ailments among mammals. Because of their slow breakdown over time, they are preferably deposited in the adipose tissue and are secreted through milk. Thus, the babies are exposed to the risk through contaminated food.

There is a wide range of contact poisons available in the market today. Among the inorganic compounds, sulphur, sodium fluoride and arsenicals may be mentioned but these are very limited in use. The mineral oils, including diesel oil, kerosene oil, crude oil and animal fats, including fish oil or the soaps prepared from them are also used as winter and summer sprays. A number of complex organic compounds which are extracted from various plants have the remarkable property of being very safe for man and domestic animals. These compounds include nicotine in sulphate form, anabasine, rotenone, pyrethrum, *neem* extract, *dharek* extract, etc.

A wide spectrum of new synthetic organic compounds have revolutionized the modern practices of pest control. The major groups include chlorinated hydrocarbons (DDT, HCH, aldrin, etc.), carbamates (carbaryl, aldicarb, carbofuran, etc.) and the organophosphates (malathion, oxydemeton-methyl, monocrotophos, etc.).

Fumigants

Poisonous gases, derived from either solids or liquids, are used as fumigants to kill insect pests of stored grains and other products in warehouses, museums, godowns, etc. They are also used to kill insect pests found in animal sheds or human dwellings; soil-infesting grubs and nematodes, the scales infesting nursery stock, the borers found in trees or wooden structures; to control all kinds of greenhouse pests and even worms found inside the intestines of animals. Since practically all the fumigants are deadly poisons, great care is needed in their use. Some of the common fumigants, both old and new include nicotine, hydrogen cyanide, carbon bisulphide, sulphur dioxide, paradichlorobenzene, naphthalene, ethylene dichloride and carbon tetrachloride mixture, methyl bromide, phosphine, DBCP, D-D (dichloropropane-dichloropropene) mixture, etc.

The kill obtained is determined to a great extent by the temperature and atmospheric pressure at which the fumigation is done. In a partial vacuum, the penetration is very much enhanced and at lower temperatures a longer exposure is required to kill the pests than at room temperatures (21-37°C). The duration of effective exposure depends upon the gas used. This is determined by its toxicity, dose, rate of penetration, its absorption by the food stuffs, the tainting or discoloration of products, the chemical composition and the moisture content of the materials to be fumigated.

Miscellaneous Chemicals

Under this group are a number of chemicals, both new and old, which cause changes in the physiology and behaviour of insects and are, thus, used either to repel them or to kill them. Chemosterilants such as apholate, tepa and metapa are fed to the insects along with food and through their physiological action they cause sterility and ultimately the pest population declines.

Antifeedants, such as neem seed extract or 4′-(dimethyl triazene acetanilide), when sprayed on plants or mixed with food make them distasteful to the insects who stop eating and ultimately die of starvation.

Repellents are another group of chemicals which are only mildly poisonous but prevent damage by making the food unattractive or offensive. These substances generally have a specific effect and are effective against certain groups of insects only. Trichlorobenzene is effective against crawling insects such as termites and ants, and is applied in the basement to provide protection to buildings. Adhesive tapes, carrying bichloride of mercury, when fixed around the legs of a table, also provide protection against ants. Heavy oils applied at the base of poultry roosters act as barriers against poultry lice and mites. Bordeaux mixture and lime-sulphur wash, repel leafhoppers and certain chewing insects. Oil citronella or dimethyl phthalate (Odomos) when applied as liquid or mixed with a vanishing cream, keeps mosquitoes away from human beings. Smoke keeps biting flies away from cattle. Oil of cedar, naphthalene or dry leaves of *neem* (*Azadirachta indica* A. Juss.) protect warm clothes and carpets against the attack of moths and beetles. There are still other chemicals which repel insects from sitting on wounds and laying eggs there. For example, pine-tar oil and diphenylamine, when applied to the skin near the wounds of animals, keep the screwworm flies away so that they do not oviposit. The wheat crop can be protected against damage by birds like sparrows, by spraying it with Thiuram (tetramethyl thiuram disulphide).

When insects are to be poisoned through baits, it is advantageous to mix an attactant with the food which gives olfactory stimulation. Many such attractants occur naturally either in the food, or as pheromones, which are secretions released by other members of the same species that promote aggregation, mating and other types of behaviour. Apart from these natural attractants, a number of other chemicals are also known to be effective. In United States, a mixture of 9 parts of geraniol and 1 part of eugenol serves as a food lure for the Japanese beetle, *Popillia japonica* Newman. Fermented sugar or syrups act as attractants for moths and butterflies. Sometimes, the addition of a chemical makes this food even more attractive. For example, the essential oil anethol is particularly luring to the codling moth, *Cydia pomonella* (Linnaeus) and isoamyl salicylate to the tomato and tobacco hornworm moths, *Protoparce* spp. Ammonia strongly attracts many insects and its slow liberation from a mixture of glycine and sodium hydroxide has been used in United States for luring the walnut husk fly, *Rhagoletus completa* Cresson. Sometimes, closely related species of insects exhibit a marked preference for one chemical over the other. The Oriental fruitfly, *Bactrocera dorsalis* (Hendel), is attracted to methyl eugenol, whereas the melon fruitfly, *Bactrocera cucurbitae* (Coquillett), is attracted to anisyl acetone. Metaldehyde which is lethal by itself also acts as an attractant in poison baits for snails and slugs.

There is still another group of chemicals which, when fed to rodents and other mammals, produces physiological changes in the blood and inhibits coagulation. When chemicals, like warfarin, fumarin, coumafuryl (Rodafarin) are fed to rats, they die of internal haemorrhage.

FACTORS INFLUENCING EFFECTIVENESS OF INSECTICIDES

Insecticides are applied in various forms, depending upon the type of pest, the situation under which they are to be applied and the hazards involved. Apart from the physical properties of these formulations, a number of factors, both inherent in the pest and those of the environment, also influence the effectiveness of chemicals.

Physical Properties of Formulations

The insecticides are available as dusts, wettable powders, emulsifiable concentrates, water soluble concentrates, granules, solutions, soluble powders, low volume concentrates (LVC), aerosols, mists, fumigants, etc. For common use, the insecticides are rarely, if ever, marketed in pure form. They are always marketed in a reduced concentration which can be further diluted at the time of application.

(*i*) **Dusts.** Dusts are fine powders mixed with various types of carriers, i.e. walnut-shell flour or pulverized minerals such as talc, pyrophyllite, bentonite, diatomite, etc. The finished product is generally available in the concentration varying from 0.1 to 25 per cent of the active material. In general, the toxicity of insecticides increases as the particle size decreases.

When used as dusts, a number of properties affect the efficiency of the insecticides. In the powdered insecticides, the particle shape varies. It may be spherical, polyhedral, tubular, irregular, and if ground further, the shape changes and becomes more spherical and, therefore, desirable because when various shapes of particles are present in a particular preparation, they may segregate during application and act in different ways.

Particle size is also important in stomach poisons. It is believed that 10 micro-metre (μm) is the right size, as it gives a sufficiently large surface area and provides good adherence to the foliage or other parts dusted. If the particle size is too small, the dust settles down very slowly and has a tendency to drift at the time of application. The particle density determines the carrying power, segregation of the dust components and settling of the suspension of particles. If the particles are too dense, they do not spread easily.

The particle size, shape, density and chemical composition of the mixture determine the flowability of the dust. A good dust should have a higher feed-rate through the dusting equipment. Dusts with fibrous or needle-shaped particles have a slow feed-rate, whereas those with spherical particles have a faster rate. These properties also determine the sticking quality of the dust, even though dusting is done when the plants are wet with dew. If the particles have a high sorptive capacity, they might have a tendency to form aggregations which will not be desirable. If the particle size is too small, again there will be a tendency for the dust to granulate.

When a dust is applied with a machine, an electrical charge is produced by friction among the particles, and between the particles and the dusting equipment. A high charge will make the dust to stick to the machine and not have a good flow. The magnitude of the charge depends chiefly on the composition of the dust, material of which the equipment is made, amount of agitation of the particles, temperature and humidity at the time of application.

(*ii*) **Sprays.** Spray fluids are prepared by dissolving insecticides in water or oil or by the dilution of wettable powders (WP), emulsifiable concentrates (EC) and water soluble concentrates (WSC) in water. In wettable powders, it is not desirable that the solid particles should settle down too rapidly. Therefore, some dispersing agent or deflocculator is added to the powder to keep its particles in suspension. The common materials used are gelatin, glue, gum, etc.

Insecticides are also available in the form of wettable powders which have to be mixed with water at the time of application and diluted to the desired concentration. The particle size in such cases must be less than 10 μm in diameter and in some preparations, a particle size up to 1 μm is also available.

Emulsions are made by the dispersion of oil droplets in water, there being a protective film of emulsifier around the oil droplets. The size of droplet is important in many ways and it is governed by the kind and amount of the emulsifier used. If the size is large, the emulsion is unstable and the oil tends to rise and float on the surface, when allowed to stand for sometime. If the droplet size is very small, the emulsion becomes very stable and the oil does not rise even on standing for hours. However, in insecticides it is desirable to have an emulsion that is sufficiently stable in the spray tank but on leaving the nozzle or upon contact with the surface being treated, the film of the emulsifier should break and the oil droplet should stick to the surface or to the insect pests. In case the film does not break, the oil droplet drains off. The effectiveness of a spray is influenced by a number of other factors as well like the leaf surface, the prevailing temperature, the insect species involved, the rate of break down of the chemical, etc.

On spraying, a surface becomes wet when the liquid comes into direct and persistent contact with the surface of a solid, there being no layer of air in between. Immediately after wetting, the liquid forms a persistent liquid-solid intersurface, solely by surface tension and, thus, the spread of the liquid is obtained. The surface of the foliage, in some cases, is wetted more readily than that in others, depending upon the waxy layer on the cuticle. A high-quality spreader is used in the preparation of an insecticide when maximum coverage is desired in the field. Where it is desirable to have the residual effect, the spreader is replaced by an adhesive such as protein or a similar colloid, which is absorbed on the particle of the insecticide and provides adhesion to a solid surface. Other adhesives used are hydrophilic inorganic compounds, molasses, dextrin, sugar, starch, wheat flour, gum, linseed oil, fish oil, etc.

Where solid particles are suspended in a spray, the size of the particle determines the efficieny of an insecticide. The principle involved is the same as discussed under dusts. When the fluid is actually sprayed, the droplet size varies according to the fineness of the spray. It is 400 μm or above in a coarse spray, 100-400 μm in a fine spray and 50-100 μm in a mist spray. An emulsifiable concentrate of an insecticide available in the market has to be diluted to the desired concentration by adding water. The percentage of active material is written on the packing and the dilution is done as follows:

The concentration of the active material in the product used is divided by the concentration required in the spray fluid. The figure obtained indicates how many parts of water are to be added to one part of the product, either by weight or by volume. Malathion which is available as 50 per cent emulsifiable concentrate is generally applied to a crop at the concentration of 0.05 per cent, i.e. one part of the insecticide is added to 1000 parts of water.

(*iii*) **Granules.** The granular formulations are discrete aggregates of dust particles which disperse gradually and, thus, last longer and are very effective when placed at strategic sites, i.e. in the soil near the root-zone or in the whorl of leaves in maize or sorghum plants. The granule size is expressed as 30/60 or similar figures for other sizes. This figure indicates that all the granules pass through a 30-mesh sieve (30 meshes per linear inch or 2.5 cm) and only a negligible number of particles will pass through a 60-mesh sieve.

(*iv*) **Aerosols.** The aerosol 'bombs' (non-explosive) are produced for use in houses or stores for killing flying insects. An aerosol consists of minute particles of insecticides suspended in the air as smoke, fog or mist. The particle size varies from 0.1 to 50 μm. The dispersion of the insecticide is achieved by burning a dry insecticide material or an insecticide in a volatile oil-base, by atomizing mechanically a liquid insecticide, or by pushing with force an insecticide through an orifice 0.03-0.06 cm in diameter, with the help of a liquid converible into gas. In the latter case, the liquid evaporates and the insecticide particles are dispersed in the air as a fine mist. These 'bombs' are very useful in the kitchen, pantry, living- rooms, bedrooms, cellars, etc. and before use the doors and windows are closed. If an aerosol is opened for 6 seconds in a room of 30 m^3, mosquitoes, house-flies, sand flies, moths, etc. are killed and, if released for 15 seconds, fleas, wasps, hornets, etc. are also killed. The particle size is important for obtaining the maximum effectiveness. If it is too small, the particles are deflected from the flying insects, and if too large their dispersion is poor and they settle down too rapidly. A particle size of 10-15 μm is the optimum.

Penetration of Insecticide through Cuticle

Apart from its entry through spiracles and the alimentary canal, the insecticide is also absorbed directly through the cuticle. The penetration and effectiveness of an insecticide depend upon the quality of the cuticle and chemical composition of the insecticide. This is so, because most of the insecticides are lipophilic and are absorbed by wax in the epicuticle. However, some lipid-insoluble materials, such as arsenicals and flourides may also find their way through the cuticle. If the thin waxy layer of the epicuticle is dissolved in some organic solvent like ether, chloroform, xylene, etc., penetration becomes even faster. The most vulnerable regions for action of contact insecticides are the head, neck, thorax or other intersegmental parts where the cuticle is thin and there are vital nerve-centres. Other such parts are antennae, bases of wings, trochanters, tarsi and intersegmental membranes. The body characters that provide protection against contact insecticides are the hard elytra of beetles, hair-coating of hairy caterpillars, thick waxy secretions of mealy bugs, some aphids and coccids. All these mechanisms are normally meant to keep water away.

After penetrating the cuticle, a contact insecticide like pyrethrum, may directly affect the hypodermal cells, reach the haemolymph and destroy haemocytes. It may be deposited in various tissues as in the case of DDT or may be transported along the lipid sheaths of the nerves, starting from the nerve-endings, as in pyrethrum.

Whether the insecticide is lipophilic or not, the undissociated molecule has a much faster penetration than the dissociated ionic component. The type of carrier used also determines to a great extent the rate of penetration. An oil solution or an oil-in-water emulsion has better penetration than an aqueous suspension. Water, because of its high surface tension, is unable to enter the tracheae, whereas mineral oils or aqueous solutions of the wetting agent having reduced surface tension can enter the breathing tubes.

Specific Susceptibility to Insecticides

Different species of insects have a wide range of susceptibility to insecticides. An insecticide can act only if it hits the vital part of an insect. Lepidopteran tissue-borers or weevils that feed while hiding inside plant tissues, naturally escape the topical application. The habitat and the behaviour of the species play an important role. A very mobile insect like the cockroach or another having a strong sense of smell will stay away from a treated area.

Different developmental stages of a pest species are also affected differently by an insecticide. In the pupal stage, an insect neither respires very much nor feeds and is often protected by a cocoon or by an earthen cell and hence, it escapes being hit directly. In contrast, a larva, which feeds exposed, is very highly vulnerable to stomach as well as to contact insecticides.

Weather Conditions

The various components of weather influence the effectiveness of insecticides. If the wind is blowing at a speed of 8-10 km per hour, it will interfere with spraying. Dusting from an aircraft is best done early in the morning or in the evening when there is only a slight wind. On the other hand, when the sun is up, the rising currents tend to carry the dust away from the crop. If aerosols are to be applied, wind velocity of 1.5-6.0 km per hour is useful and again cooler temperatures near the ground in the morning or after sunset are suitable. A few meters higher up, the atmosphere is warm and the aerosol cloud has a tendency to rise and drift. In general, extremes of temperature are not favourable. Certain chemicals, when applied in hot weather, may injure the plants. The rate of metabolism is fast at high temperatures so the insects have an inherent capacity to resist the poison. If the weather is too cold, the oil sprays may cause injury to foliage.

If insecticides do not contain a good adhesive and the spray does not stick to the plants effectively, it may be washed off by rain. Dusts are more easily washed-off than sprays but if dusting is done when there is dew on the plants, the insecticide sticks quite well.

Conditions in the Field

The condition of the plant, the location of the pest, the time of application and the amount of insecticide applied, all determine the effectiveness of an operation. Some plants grow faster than others, still others grow faster in one season than in another, and the fruits grow much faster in the early stages than in the later stages. In these cases if the entire surface is to be kept covered with an insecticide, naturally the frequency of application has to be changed accordingly. Even where it is not necessary to maintain a deposit, it is desirable to time the application properly. Spraying or dusting is to be done when the pest is in a vulnerable stage; if done too early, the maximum kill may not be obtained; if too late, much damage to the crop may have already been caused. In general, crops should not be sprayed while in blossom, because the flowers in which seed is to set may be damaged or the bees and other insects so important for the pollination of crops, may be killed, thus causing reduction in yield instead of an expected increase. The structure of the leaf is also important; hairy leaves are more easily covered while those having smooth and waxy surface have to be sprayed with more sticky insecticides.

Since the pest is to be directly hit, the method or site of application of the poison would depend on its location. Insects feeding on the lower surface of the leaves have to be reached by the spray, those feeding on the roots have to be killed by mixing insecticides with the soil, and those rising up the trunks of the trees have to be intercepted with sticky or poison bands.

Various insecticides are effective for different durations. As a rule of thumb, the duration of effectiveness of dusts is for about one week and that of the sprays for about two weeks. The excessive use of insecticides is expensive and wasteful, and can be injurious to plants or non-target animals in the soil. Even pond life can be affected by the run-off from treated fields. The selection of an insecticide and the frequency of its application should be such that the near maximum kill is obtained with the least quantity, so that the natural enemies of the pests can be spared as far as possible.

Compatibility

Since the farmer is interested in protecting the crop both from pests and diseases, the insecticides and fungicides or even some weedicides might have to be mixed together before spraying. Some chemicals are compatible whereas others are not; the former should not be mixed because they might become ineffective or damage the foliage. The mixing should be done with the full knowledge of their compatibility or in consultation with an expert. Based on the compatibility of various compounds, a miscibility chart which should always be consulted at the time of mixing the chemicals.

Application of Pesticides

Apart from precautions for spraying and dusting, there are many other points that should be considered for obtaining the best results. A given preparation of insecticide has instructions printed on the packing and these should be followed strictly. For example, an oil-base insecticide meant for use on walls, furniture, wooden structures and floors should not be sprayed on plants and again the one meant for use on dormant trees cannot be used on those having foliage. Arsenicals (and chlorinated hydrocarbons) persist in the soil for many years and may accumulate in quantities sufficient to injure the plants under temperate conditions. The organophosphates are generally very toxic to man and domestic animals, but they have the advantage that they do not persist for long and are detoxified in the soil after 2 or 3 months. They are also generally destroyed when the food is cooked.

The old saying that "If little is good, more is better" does not apply in the case of insecticides. The right dose (quantity per hectare) or the right potency (concentration of an insecticide) should be applied at right intervals. To get the best results a preparation has to be applied in optimum quantity. This is possible if the machine is correctly calibrated. It is particularly important with the power-operated sprayers and dusters. With the manually operated high-volume sprayers, generally 600-1000 litres of the fluid is required per hectare, whereas with the low-volume sprayers (carried on the back or on a trolley) only 125-200 litres is sufficient, hence the concentration is increased. Since the rate of flow is important in power sprayers and power dusters, it should be measured by noticing the amount of the insecticide used on a known crop area. The rate of flow can then be increased or decreased by making adjustments in the apertures.

Certain safeguards are needed to protect the consumers of food (vegetable or animal origin) and other animals in the environment such as wildlife, domestic animals, honeybees, fish, etc,. When applying insecticides inside buildings, we have to be sure that no food lies exposed or that the toys of children do not get excessive deposits. The unused insecticide should be kept out of reach of children or illiterate people and also away from food products. Empty containers should not be used for other purposes.

Strict precautions are to be taken at the time of application. All insecticides are poisons and, therefore, not to be swallowed. Many of them can also be absorbed through the skin in lethal quantities, or inhaled while applying them. Therefore, the operators should wear gloves, goggles, overalls, gum boots and gas masks, and should never dip their hands into the preparation. While spraying or dusting they should not work against the wind.

The left-over spraying fluid should not be dumped into a pond or a stream, because fish and other aquatic animals will die through contamination; nor should the equipment be washed there after use. The left-over fluid and the empty containers should be buried in soil, about half a metre deep at some isolated place and away from any water source. It is important that the equipment after use is washed, otherwise there can be a mistake, particularly if it is used for the application of weedicides and then unknowingly, for applying insecticide or a fungicide. Even small residues of a weedicide like 2,4-D can cause very serious damage to valuable crops like cotton and potato.

While fumigating, the operators using a particular gas should be thoroughly trained in its use. They should use gas-masks and protect their hands with gloves. The warehouse or the store to be fumigated should be made airtight by closing all cracks and crevices with mud, leaving open the main entrance which is closed after opening the outlet of container of the gas. If the room to be fumigated is close to human dwellings, the persons should be asked to vacate them temporarily. After allowing the required exposure for fumigation, the room is opened, so that the gas may escape and only then should any person be allowed to enter it or to live nearby.

The dose to be used is generally expressed as litres/kilograms per 100 cubic metres space. This dose should be very carefully applied to plants in a greenhouse, because plants are sensitive to excessive doses. Moreover, the fumigation should be done at night owing to the possible injury to green plants in the presence of light. The duration is rarely more than one hour and the optimum temperature is 5-25°C.

Fruit trees like citrus, apple, walnut, etc. can also be fumigated in the field by enclosing them in canvas or nylon tents lined with polythene. The same procedure can be followed for fumigating open heaps of grain, a pile of furniture or a stack of timber.

The soil fumigants are very costly and, therefore, their application should be made with proper equipment which is generally tractor-driven or power-driven. If the fumigant is not applied at the right depth, at the right soil moisture, and is not properly packed afterwards, the gas can escape rendering the effort and money futile. For killing soil nematodes, D-D mixture, DBCP (Nemagon) and fensulfothion (Dasanit) are quite effective.

MAJOR GROUPS OF PESTICIDES

A farmer is generally interested in protecting his crops against ravages of insects and other animal pests, diseases and weeds. Many times insecticides or other pesticides are mixed with fungicides before spraying them on the crops. If a pesticide is sprayed when the deposit of a weedicide is present on the plants, damage may be caused to the crop itself. Familiarity with the chemical nature, properties and the effectiveness of some of the important compounds is, therefore, essential for plant protection.

I. INSECTICIDES

1. Chlorinated Hydrocarbons

(*i*) **DDT (Dichlorodiphenyltrichloroethane).** DDT ($C_{14}H_9Cl_5$) **(Fig. 7.1)** was introduced in 1942 by J.R. Geigy under the trade names Gesarol, Guesarol and Neocid. Its insecticidal properties were, however, discovered by Paul Muller in 1939.It is a potent non-systemic stomach and contact insecticide, effective against a wide range of insects, including pests of crops and household goods. It is also of importance in public health. It has no action on phytophagous mites and, when used

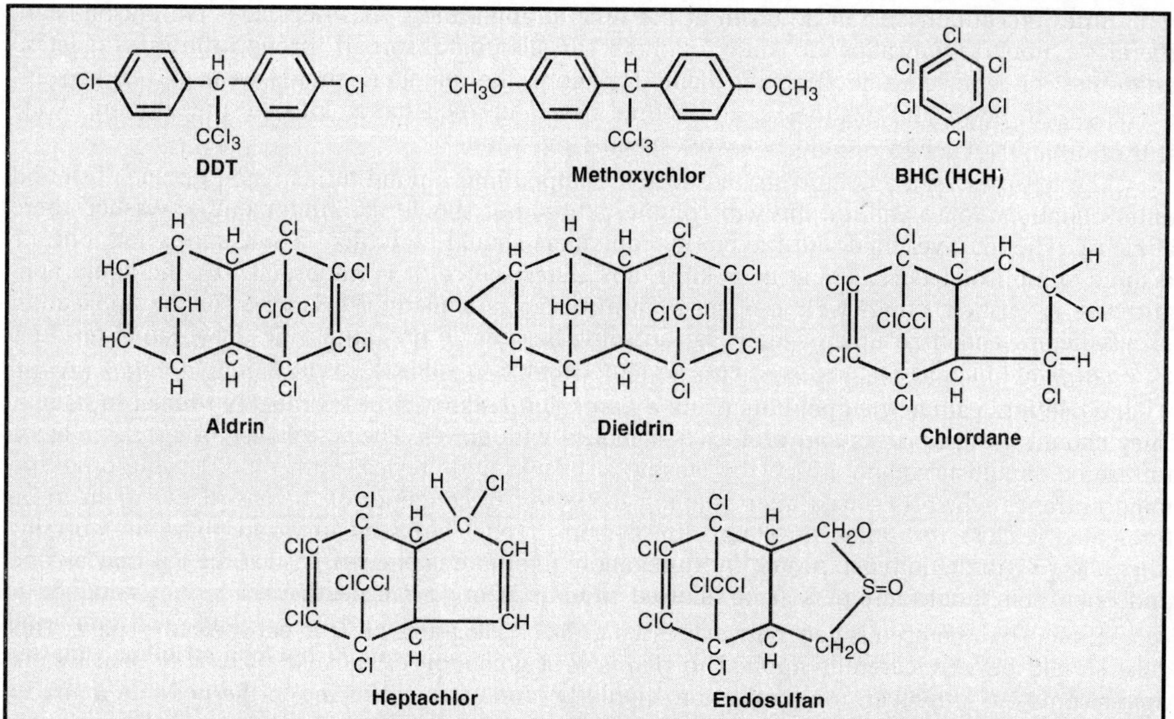

DDT Methoxychlor BHC (HCH)

Aldrin Dieldrin Chlordane

Heptachlor Endosulfan

Fig. 7.1 Chemical structure of some organochlorine insecticides

on crops against insects, the mite population increases invariably. DDT is generally non-phytotoxic, but it damages cucurbits.

At low concentrations, it stimulates the plant growth in crops like potato, tomato, cabbage and brinjal. Because of its wide spectrum effectiveness it is not safe to most natural enemies of insect pests. Its acute oral LD_{50} for male rat is 113 mg/kg, for female rats 118 mg/kg; the acute dermal LD_{50} for female rats is 2,510 mg/kg. In India it is being manufactured by M/s Hindustan Insecticides Limited.

DDT is sold under the trade names of Chlorophe-nothene, Guesarol, Teffidex, etc. and is available in various formulations including EC, WP and dust. It is applied at the rate of 0.50-2.50 kg a.i. per ha in concentrations varying from 0.1-0.25 per cent on practically all crops except the cucurbits. It is effective against jassids, white flies, bollworms, pyrilla, leafminers, capsule and pod borers, beetles, but should not be used on farm animals or inside the houses.

It persists for quite a long time on the plants, in the soil, and it also accumulates in the body fat of birds, fishes and mammals. It is then secreted in the milk. Owing to its repeated use, even though some of the insects have developed resistance, it still continues to be a leading protective insecticide for outdoor use. The use of DDT in agriculture has recently been banned in India.

(*ii*) **Methoxychlor [1, 1, 1-trichloro-2, 2, di-(3 metho-xyphenyl ethane)].** The insecticidal properties of methoxychlor (**Fig. 7.1**) were first discovered in 1944 and it was introduced in 1945 under the trade name 'Marlate'. It is also known as dimethoxy-DT, DMDT, dianisyl trichloroethane. It is a non-systemic, contact and stomach insecticide but has little aphicidal and acaricidal activity. Its range of activity is as wide as that of DDT but unlike DDT, it shows little tendency to be stored in body fat. It is, therefore, used for fly control in dairy barns. It is available as a WP of different strengths and also as EC.

(*iii*) **Toxaphene.** Toxaphene (approximately $C_{10}H_{10}Cl_8$) was first described by W. Le Roy Parker and J.R. Beacher in 1947 but was introduced in 1948 by Hercules Incorporated under the code number 'Hercules 3956'. It is a non-systemic and persistent contact and stomach chlorinated terpene with some acaricidial action. It is non-phytotoxic except to cucurbits. However, it is harmless to bees. The acute oral LD_{50} for male rats is 90 mg/kg, for female rats 80 mg/kg ; the acute dermal LD_{50} for male rats is 1,075 mg/kg, for female rats 780 mg/kg.

Toxaphene has had by far the greatest use of any single insecticide in agriculture. It is a mixture of more than 177 polychlorinated derivatives, which are 10-carbon compounds including Cl_6, Cl_7, Cl_8, Cl_9 and Cl_{10} constituents. No single component makes up more than a small percentage of the technical mixture. Toxaphene is manufactured by the chlorination of camphene, a pine tree derivative. These materials are persistent in the soil, though not long lasting as other cyclodienes. It disappears in three to four weeks from the surface of most plant tissues. The modes of action of toxaphene and strobane are similar to other cyclodiene insecticides by acting on the neurons and by causing an imbalance in sodium and potassium ions.

Toxaphene is also sold under the trade name of Anatox, and is available in formulations of EC and dust. It is applied at the rate of 2.0 kg a.i. per ha or 0.1 to 0.2 per cent concentration against hairy caterpillars and armyworms, feeding on weeds and field crops. It is also effective against termites and cutworms.

(*iv*) **BHC (Benzene hexachloride).** The insecticidal properties of 'gamma' isomer of BHC (misnomer for HCH) (**Fig. 7.1**) were discovered in the early 1940s and the compound was introduced by the Imperial Chemical Industries under the trade name 'Gammexane.' HCH is a persistent stomach poison and a contact insecticide with some fumigant action also. It is a mixture of several isomers of which *gamma* is the most toxic to insects and is the main (13 %) insecticidal component of HCH. It is non- phytotoxic, except to cucurbits, at the normal insecticidal concentrations, but at higher concentrations it may cause root deformation and polyploidy. It taints certain foods, such as potatoes. Pure *gamma* HCH commonly known as lindane, was introduced to overcome the tainting effects. It is quite persistent on the plants and accumulates in the soil and in the animal system. Both HCH and lindane are formulated as dust, WP, granules, EC and smoke. The use of HCH on vegetables, fruits, oilseed crops and preservation of foodgrains in India had been banned earlier. Now, there is a total ban on manufacture and use of HCH in India with effect from Ist April, 1997.

(*v*) **Chlordane.** Octachloro-4, 7 methano-tetra-hydroindane ($C_{10}H_6Cl_8$) (**Fig. 7.1**) was introduced as an insecticide in 1945 by Velsicol Chemical Co. under the name of Chlordane. It is a contact and stomach poison and also shows fumigant properties with long residual effects. It is a non-systemic and non- phytotoxic insecticide with LD_{50} value of 335 mg/kg.

Chlordane is sold under the trade names of Octachlor, Intox, Chlordan, etc. as EC, WP and dust. It is applied at the rate of 1.0-4.0 kg a.i. ha. It gives termite control upto five years. It has also been used against grasshoppers, Japanese beetles, field crickets, spiders and mosquitoes.

(*vi*) **Heptachlor.** Heptaclorotetrahydro-4,7-metha-noindane ($C_{10}H_7Cl_7$) (**Fig. 7.1**) was introduced as Heptachlor by Velsicol Chemical Co. in 1945. It is a contact and stomach poison with long residual effects. It is a non-systemic and non-phytotoxic insecticide with LD_{50} value of 100 mg/kg

It is also sold under the trade names of Heptagran and Drinox as EC, WP, dust and granules. It is applied at the rates of 1.0-4.0 kg a.i. per ha. It has been recommended for use against termites, Japanese beetle, grasshoppers, mosquitoes, etc. It has been used to protect the turfs from the damage of white grubs.

(*vii*) **Aldrin and dieldrin.** Aldrin ($C_{12}H_8Cl_6$) (**Fig. 7.1**) was introduced in 1948 by J. Hyman & Co. as 'Compound 118' under the trade name of Octalene and its insecticidal action was first described by Kearns and co-workers in 1949. It is a non-systemic and persistent contact cyclodiene insecticide. It is usually non-phytotoxic. The acute oral and dermal LD_{50} values for rat are 40-50 and 100 mg/kg, respectively. It is manufactured in India by M/s National Organic Chemicals Industries Ltd.

Aldrin is sold under the trade names of HHDN, Octalene and Aldrex, and is available as EC and dust. It is basically a soil insecticide and is applied at the rate of 1.75 kg a.i. per ha in concentrations varying from 0.02 to 0.2 per cent. Its soil application is recommended in fields of wheat, sugarcane and cotton. It is effective against termites, black ants, hairy caterpillars, root weevils and other soil dwelling pests. The use of aldrin has recently been banned in India.

Dieldrin (**Fig. 7.1**) was introduced as an insecticide in 1949. It is a non-systemic and persistent insecticide with high contact and stomach action. It is also non-phytotoxic. This insecticide is highly stable and compares well with most other pesticides. Like aldrin it also has a high dermal toxicity and should be used with care. It is available as an emulsifiable concentrate.

(*viii*) **Endrin.** Hexachloroepoxyoctahydro-endo, endo-dimethanonaphthalene ($C_{12}H_8Cl_6O$) was introduced as an insecticide by Shell Chemical Co. in 1950 as Endrin. It is an isomer of dieldrin. It is a contact and stomach poison. Usually it is non-phytotoxic except on corn and cucumber. It is a non-systemic insecticide with LD_{50} value as low as 10 mg/kg for mammals. It is highly toxic to fish and is also absorbed through skin.

It is marketed as EC, WP, dust and granules. It is effective against a wide range of sucking and biting insects and has very long residual effect. Because of its high mammalian toxicity and long residual effects, its use has now been banned in many countries of the world, including India.

(*ix*) **Endosulfan.** The insecticidal properties of endosulfan ($C_9H_6Cl_6O_3S$) (**Fig. 7.1**) were first described by W. Finkenbrink in 1956 and it was introduced in 1956, by Farbwerke, Hoechst A.G. under the trade name Thiodan. It is a non-systemic contact and stomach insecticide. It is highly toxic to fish but, in practical use, is comparatively safer to wildlife and bees. Its acute oral and dermal LD_{50} values for rat are 43 and 130 mg/kg, respectively. Both M/s Hoechst Pharmaceuticals Ltd. and M/s Excel Industries Limited manufacture endosulfan in India.

Endosulfan is sold under the trade names of Thiodan, Hexasulfan and Thionex, and is available in various formulations including EC, dust and granules. It is applied at the rate of 225-875 g a.i. per ha in concentrations ranging from 0.02 to 0.1 per cent on cotton, redgram, wheat, rapeseed and mustard, brinjal, cabbage, tomato, okra, peas, sugarcane and mango.

It is effective against tissue borers, stemfly, bark-eating caterpillars, semiloopers, armyworms, cutworms, leaf-rollers, hairy caterpillars, bugs, chaffer beetles, leafhoppers, pyrilla, aphids, thrips, white flies and gall midges. On May 13, 2011, the Supreme Court of India had banned the manufacture, sale and use of endosulfan in the country.

2. Organophosphates

(*i*) **Parathion and parathion-methyl.** Ethyl parathion, commonly known as parathion, was the first organophosphorus insecticide introduced in 1948 under the trade name of Folidol. It is a potent contact and stomach insecticide with some fumigant action but is non-systemic. It is highly effective against scale insects, aphids and mites, and its residues do not persist for long. It is non-phytotoxic, except to certain ornamentals and fruit plants. As its dermal toxicity is very high, it is highly toxic to mammals, although it is non-cumulative. It should not be used in houses or on livestock. Its use should also be avoided on crops in blossom, firstly because the flowers are damaged and secondly because the bees and other pollinators are killed. It is available as a WP, EC, dust, smoke and aerosol.

Parathion-methyl ($C_8H_{10}NO_5PS$) (**Fig. 7.2**) was first described by G. Schrader and was introduced in 1949 by Farbenfabriken Bayer A.G. under the trade names of Dalf, Folidol-M, Metacid, Nitrox 80, the latter two being registered by Chemagro Corporation. Parathion-methyl is a non-systemic contact and stomach phosphorothionate insecticide with some fumigant action. It is non-phytotoxic. It is hazardous to wildlife but is of short persistence. The acute oral LD_{50} for male rats is 14 mg/kg, for female rats 24 mg/kg; the acute dermal LD_{50} for rats is 67 mg/kg. M/s Bayer (India) Ltd. manufacture these insecticides in India.

Parathion-methyl (O,O-dimethyl O-p-nitrophenyl phosphorothioate) is an organophosphoro-thioate phenyl derivative. It became available in 1949 and proved to be more useful than ethyl parathion because of its lower toxicity to humans and domestic animals, and with broader range of insects that it controlled. Its shorter residual life also makes it more desirable in certain situations.

Parathion-methyl is sold under the trade names of Metron, Nitrox, Metacid and is available in the form of EC. It is applied at the rate of 500 g a.i per ha in 0.05 per cent concentration on a number of field crops except those eaten as green vegetables. It is effective against tissue borers, scale insects, armyworms, hispa, coccids and mealy bugs.

(*ii*) **Malathion.** Malathion ($C_{10}H_{19}O_6PS_2$) (**Fig. 7.2**) was introduced in 1950, by the American Cyanamid Company under the code number 'Experimental Insecticide 4049'. It is a non-systemic contact phosphorothiolothionate insecticide and acaricide of low mammalian toxicity with short to moderate persistence. It is generally non-phytotoxic but may damage cucumber, spring bean and squash under glasshouse conditions. It is highly toxic to bees. The acute oral LD_{50} for rat is 1,375 mg/kg, the acute dermal LD_{50} for rabbit is 4,400 mg/kg. M/s Excel Industries Limited manufacture malathion in India.

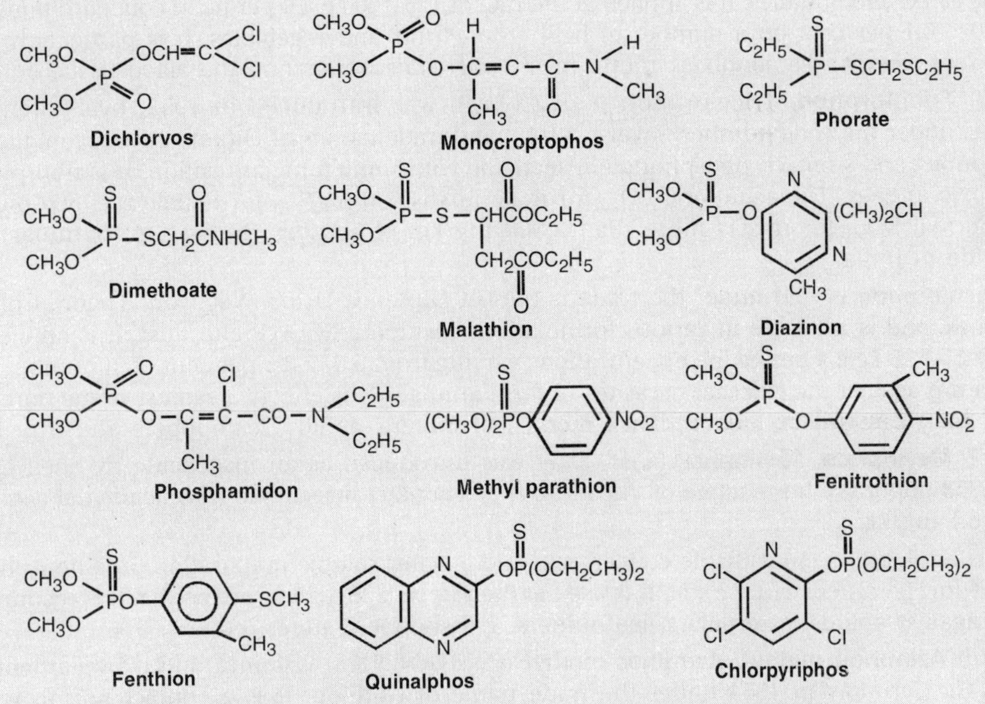

Dichlorvos, Monocroptophos, Phorate, Dimethoate, Malathion, Diazinon, Phosphamidon, Methyl parathion, Fenitrothion, Fenthion, Quinalphos, Chlorpyriphos

Fig. 7.2 Chemical structure of some organophosphorus insecticides

Malathion is sold under the trade names of Cythion, Malamar, Malathion and is available in various formulations including EC, LVC, dust, etc. It is applied at the rate of 0.5-2.5 kg a.i. per ha in concentrations varying from 0.05-0.15 per cent on a number of crops particularly those which are edible in green or ripe stage of growth. It is effective against tissue borers, caterpillars, bugs, leafhoppers, aphids, thrips, whiteflies, mites, fruit-flies, grasshoppers, midges, pests of household and farm animals.

(*iii*) **Schradan (OMPA).** It is one of the first systemic compounds whose insecticidal properties were discovered in 1941. It is effective against sap-sucking insects and mites, but has little effect as a contact poison. It is non-phytotoxic at insecticidal concentrations, but its mammalian toxicity is rather high. Its use on edible crops is also restricted. It is a selective insecticide and is less destructive to parasites, predators and insect pollinators. Schradan is formulated as an aqueous solution.

(*iv*) **TEPP (Tetra-ethyl pyrophosphate).** The aphicidal properties of this compound were discovered in 1938. In 1943, a derivative of hexaethyl tetraphosphate (HETP) was introduced and was shown to contain TEPP as the main active compound. TEPP was introduced as a substitute for nicotine sulphate in Germany during the Second World War. Due to its high mammalian toxicity it should be handled carefully. It is also hazardous to insect pollinators. In the commercial products, the mixture of polyphosphate contains 40 per cent of pyrophosphate and is marketed either in a non-water solvent solution or for use as an aerosol.

(*v*) **Diazinon.** Diazinon ($C_{12}H_{21}N_2O_3PS$) (**Fig. 7.2**) was introduced in 1952 by J.R. Geigy, S.A. under the code number G-24480 and trade name of Basudin. It is non-systemic phosphorothionate insecticide with some acaricidal action. It is a contact and stomach poison, and has a fumigant effect. It is non-phytotoxic but highly toxic to bees. The acute oral LD_{50} for male rats is 108 mg/kg, for female rats 76 mg/kg, for mice 82 mg/kg; the acute dermal LI_{50} for male rats is 900 mg/kg, for female rats 455 mg/kg. It is manufactured in India by M/s Rallis (India) Ltd.

Diazinon is sold under the trade names of Basudin, Spectracide, Tik- 20, Ditaf, Diazol and is available as EC and granules. It is applied at the rate of 125-1000 g a.i. per ha in concentrations ranging from 0.025-0.1 per cent on a number of field crops, fruits and vegetables. It is particularly effective against root weevils, diamondback moth, mealy bugs, rice borers and the allied tissue borers.

(*vi*) **Trichlorphon.** Tricolorphon ($C_4H_8Cl_3O_4P$) was introduced in 1952, by Farbenfabriken Bayer AG under the code number Bayer L 13/59, and trade names of Dipterex, Neguvon and Tugon. It is a contact and stomach phosphonate insecticide with some fumigant action. It is non-phytotoxic but is toxic to bees. The acute oral LD_{50} for male rats is 630 mg/kg, for female rats 560 mg/kg; the acute dermal LD_{50} for rats is more than 2,000 mg/kg. M/s Bayer (India) Ltd. manufacture this insecticide in India.

Trichlorphon is sold under the trade names of Dipterex, Dylox, Neguvon, Tugon, Chlorophos and Danex, and is available in various formulations including EC, WP and dust. It is applied at the rate of 0.25-1.25 kg a.i. per ha in concentrations varying from 0.075-0.1 per cent on a number of crops and as a dip against the external parasites of farm animals. It is effective against tissue borers, army worms, hairy caterpillars, diamondback moth, leaf miners, aphids, leafhoppers and fruit flies.

(*vii*) **Mevinphos.** Mevinphos ($C_7H_{13}O_6P$) was introduced as an insecticide by Shell Chemical Co. in 1953 under the trade name of Phosdrin. It is a contact insecticide with acaricidal activity. The LD_{50} is 6.1 mg/kg.

It is available as emulsifiable concentrate and a water soluble preparation. It is non-phytotoxic with very low persistence in the soil. It is used at the rate of 0.25-1.25 kg a.i. per ha. It is recommended for use against aphids, caterpillars, leaf-miners, grasshoppers, mites, etc.

(*viii*) **Azinphos methyl.** Azinphos methyl ($C_{10}H_{14}O_3 N_3PS$) was introduced by Farbenfabriken Bayer A.G., Germany in 1954 under the trade name of Guthion. It is a contact poison with long residual activity. LD_{50} is 13 mg/kg. It has acaricidal properties also. It is a non-phytotoxic insecticide.

It is also sold under the trade names of Gusathion and DBB, as EC, WP, dust and granules. It is applied at the rate of 1.0 - 2.5 kg a.i. per ha. It is used against aphids, mites, caterpillars, grasshoppers, etc. It is prohibited from use on greenhouse insect pests.

(*ix*) **Acephate.** Acephate ($C_4H_{10}NO_3P_5$) is a systemic insecticide with moderate persistence (10-15 days) and is effective against lepidopterous larvae, sucking pests, etc. It is formulated as 75% SP. The acute oral LD_{50} for rat is 886-945 mg/kg and dermal LD_{50} for rabit is 2000.

(*x*) **Thiometon.** Thiometon ($C_6H_{15}O_2PS_8$) was introduced by Sandoz Limited in 1953 under the trade name of Ekatin. It is a contact as well as systemic insecticide-acaricide and is usually non-phytotoxic. The LD_{50} is 100 mg/kg and it is absorbed through skin.

It is marketed as EC and dust. It is applied at the rate of 0.1 per cent as spray. It is reported to be effective against aphids, mites, sawflies, psyllids, thrips, etc., and is applied on ornamental plants also. It persists for 2-3 weeks on plants.

(*xi*) **Phorate.** Phorate ($C_7H_{17}O_2PS_3$) (**Fig. 7.2**) was introduced in 1954, by the American Cyanamid Company under the code number 'Experimental Insecticide 3911' and the trade name of Thimet.

It is a systemic phosphorothiolo-thionate insecticide with little contact and fumigant action. It has also nematicidal and acaricidal action. The acute oral LD_{50} for male rats is 2.3 mg/kg, for female rats 1.6 mg/kg; the acute dermal LD_{50} for male rats is 6.2 mg/kg, for female rats 2.5 mg/kg. M/s Bayer (India) Ltd. manufacture it in India.

Phorate is sold under the trade name of Thimet, as granules which have a concentration of 10 per cent of insecticide. In terms of actual ingredient it is applied at the rate of 0.75-3.0 kg per ha. It is mixed in the soil near the plant root zone and is also applied topically at strategic spots in a number of crops. It is effective against root weevils, tissue borers, maggots, leafhoppers, aphids, white flies and it acts both as a contact insecticide and as a stomach poison.

(*xii*) **Oxydemeton-methyl.** Oxydemeton-methyl ($C_6H_{15}O_4PS_4$) was introduced in 1960, after tests since 1956, by Farbenfabriken Bayer A.G. under the code numbers Bayer 21097 and R 2170. It is a systemic and contact phosphorothiolate insecticide with acaricidal properties also. It is non-phytotoxic and safe to parasitoids and predators but toxic to bees. The acute oral LD_{50} for male rats is 65 mg/kg, LD_{50} for female rats is 75 mg/kg, the acute intraperitoneal LD_{50} for male rats is 250 mg/kg. M/s Bayer (India) Ltd. manufacture it in India.

Oxydemeton-methyl is sold under the trade names of Metasytox, Methyl-O-demeton, and is available in the form of EC. It is applied at the rate of 125-250 g a.i. per ha in concentrations ranging from 0.025-0.05 per cent in various field crops except those eaten as green vegetables. It is very effective against leafhoppers, aphids, white flies, thrips, leaf-miners and tingid bugs.

(*xiii*) **Carbofenothion.** Carbofenothion ($C_{11}H_{16}O_2PS_3$) was introduced in 1955 by Stanffer Chemical Co. under the trade names of Trithion and Garrathion. It is non-systemic acaricide and insecticide having a long residual action. It is used on deciduous fruits, in combination with petroleum oil as dormant spray for the control of scale insects, mites and aphids. It is less toxic to honeybees than DDT and to natural enemies (parasites) than parathion. It is formulated as dust, WP and EC.

(*xiv*) **Dichlorvos (DDVP).** Insecticidal properties of dichlorvos ($C_4H_7Cl_2O_2P$) **(Fig. 7.2)** were first described in 1951, by Ciba AG in BP 775,085 but it was given an incorrect structure in the patent. It was later intoduced by Ciba AG under the trade names of Nogos and Nuvan. It is a contact and stomach insecticide with fumigant and penetrant action because of its high vapour pressure. It is non-phytotoxic, moderately toxic to fish but highly toxic to bees. The acute oral LD_{50} for albino male rats is 80 mg/kg, for female rats 56 mg/kg; the acute dermal LD_{50} for albino male rats is 107 mg/kg, for female rats 75 mg/kg. M/s Bayer (India) Ltd. manufacture it in India.

Dichlorvos is sold under the trade names of DDVP, Vapona, Nuvan, Nogos, Dentavepon, Marvex and Divipan, and is available in the form of EC. It is applied at the rate of 200-500 g a.i. per ha in concentrations varying from 0.03-0.1 per cent on a variety of crops ranging from evergreens to deciduous trees. It is effective against hairy caterpillars, armyworms, tissue borers, tingid bugs, white flies, woolly apple aphid, etc.

(*xv*) **Dimethoate.** Dimethoate ($C_5H_{12}NO_3PS$) **(Fig. 7.2)** was first described by Hoegberg and Cassaday in 1951 and was introduced in 1956 by the American Cyanamid Company under the code number 'Experimental Insecticide 12,880'. It is a systemic and contact phosphorothiolo-thionate insecticide and acaricide. It is phytotoxic to certain varieties of sorghum, hops, olives, figs, chrysanthemum, begonia, jacobinas and sunnhemp. The acute oral LD_{50} of the pure compound for rats is 250-265 mg/kg, for mice 285 mg/kg; the acute oral LD_{50} of the technical product for mice is 155 mg/kg. It has a low mammalian dermal toxicity of 610 mg/kg. M/s Rallis (India) Ltd. manufacture it in India.

Dimethoate is sold under the trade names of Rogor, Dimethoate, Perfekthion, Roxion and Cygon as EC. It is applied at the rate of 375-750 g a.i. per ha in concentrations ranging from 0.03-0.1 per cent in practically all the field crops including fruit trees. It is effective against tissue borers, leaf feeding caterpillars, leafhoppers, aphids, thrips, psylla, mites, white flies, mealy bugs and tingid bugs.

(*xvi*) **Phosphamidon.** Phosphamidon ($C_{10}H_{19}ClNO_5P$) (**Fig. 7.2**) was introduced in 1956 by Ciba AG under the code number Ciba 570. It is a systemic and contact phosphate. Phosphamidon is less toxic to fishes but toxic to bees. It is non-phytotoxic except to certain cherry varieties. The insecticidal activity of the *beta*-isomer is several times greater than that of the *alpha*-isomer. The acute oral LD_{50} for rats is 23.5 mg/kg; the acute dermal LD_{50} for rats is 143 mg/kg. M/s Ciba Geigy (India) Ltd. manufacture this insecticide in India.

Phosphamidon is sold under the trade name of Dimecron and is available in the form of WSC. It is applied at the rate of 165 g a.i. per ha in concentrations ranging from 0.02-0.5 per cent on most of the field crops. It is effective against tissue borers, caterpillars, bugs, leafhoppers, aphids, white flies and mites.

(*xvii*) **Menazon.** Menazon ($C_5H_{12}N_5O_2PS_2$) was introduced by Plant Protection Ltd., England, under the trade name of Sayfos. It is a selective, systemic insecticide-cum-acaricide. It is reported to be phytotoxic to cruciferous vegetables. It has low mammalian toxicity and LD_{50} for rats is reported to be 1200 mg/kg.

It is sold under the trade names of Sayfos, Saphicol, Saphizon and Saphos. It is available as EC and WP. It is applied at the rate of 0.50-1.25 kg a.i. per ha. Tuber and seed treatments are made at 0.5 per cent concentrations by weight. In the field, it is recommended for use on apples, potato, sugarbeet, tobacco, peas, etc. against aphids, mites, leafhoppers, etc. It is most effective as an aphicide.

(*xviii*) **Formothion.** Formothion ($C_6H_{12}O_4PS_2$) was introduced by Sandoz Ltd. in 1962 as Anthio. It is a contact as well as systemic insecticide-cum-acaricide. It is usually non-phytotoxic except on chrysanthemum and hops. The dermal toxicity to mammals at LD_{50} is 310 mg/kg.

It is also sold under the trade name of Aflix and is colourless and odourless. It is available as EC and is applied at the rate of 175 g a.i. per ha or at a concentration of 0.05 per cent on green vegetables, tea, coffee and ornamental plants. It is recommended for use against mites, aphids, psyllids, leafhoppers, thrips, etc. It is toxic to bees and fish.

(*xix*) **Phenthoate.** Phenthoate ($C_{12}H_{18}O_4PS_2$) is a wide spectrum contact organo-phosphatic insecticide and acaricide. It is phyto toxic on some vines, peach and apple varieties. Its acute oral and dermal LD_{50} to rats are 200-300 and 700-1400 mg/kg, respectively. In India, M/s Bharat Pulverising Mills (P) Ltd., Mumbai and M/s Motilal Pesticides (India) Pvt. Ltd., New Delhi manufacture and formulate this product.

Phenthoate is sold under the trade names of Elsan, Cidial and Phendal, and is available in the form of EC. It is applied at the rate of 500 g a.i. per ha, generally at the concentration of 0.05 per cent, on a number of crops. It is effective against the bollworm complex, tissue borers and leafhoppers.

(*xx*) **Fenitrothion.** Fenitrothion ($C_9H_{12}P_5NPS$) (**Fig. 7.2**) is a low toxic organophosphorus contact insecticide discovered in 1959 by the research group of Sumitomo Chemical Co. Ltd., Osaka, Japan. It is toxic to bees. It is also selective acaricide but of low ovicidal activity. Phytotoxicity may be caused on Brassica crops, certain susceptible apple varieties and cotton when applied at high dosages. Its acute oral and dermal LD_{50} for rats are 250-673 and 1500-3000 mg/kg, respectively. M/s Cyanamid India Ltd., Mumbai; M/s Bayer (India) Limited, New Delhi and M/s Rallis India Limited, Bangalore, are the manufacturers of fenitrothion in India.

Fenitrothion is sold under the trade names of Sumithion, Folithion, Accothion, Bayer 41-831 and Hexafin, and is available in the form of EC, WP and dust. It is applied at the rate of 0.5-1.0 kg a.i. per ha in varying concentrations at 0.025-0.05 per cent on a number of crops except those which are tall or those which are used as green food or fodder. It is highly effective against the white grub adults, armyworms, thrips, aphids, tissue borers, bollworms, tingid bugs, weevils, white flies, webworms, leafhoppers and leaf-feeding caterpillars.

(*xxi*) **Ethion.** Ethion ($C_9H_{22}O_4P_2S_4$) is a non-systemic insecticide effective against aphids, scales and thrips, and an acaricide. It can be used on tea and in combination with petroleum oils on dormant fruit trees for the control of scales and as an ovicide. It is available in the form of 50% EC. The oral LD50 for rat is 65 mg/kg, whereas the dermal LD_{50} for rabbit is 245 mg/kg.

(*xxii*) **Monocrotophos.** Monocrotophos ($C_7H_{14}NO_5P$) (**Fig. 7.2**) is a contact and systemic organophosphorus insecticide. It was introduced in 1965 by Ciba A.G. under the code number C 1414 and trade name of Nuvacron and by the Shell Development Co., under the trade name of Azodrin. On certain varieties of cherry and some varieties of sorghum it causes scorching of the edges of the leaves. The formulations corrode iron, tin, plate and aluminium. The acute oral LD_{50} for rats is 21 mg/kg, the acute dermal LD_{50} for rabbits is 354 mg/kg. It is toxic to bees. M/s Ciba-Geigy India, Mumbai and M/s NOCIL India, New Delhi, manufacture monocrotophos in India.

Monocrotophos is sold under the trade names of Azodrin, Nuvacron, Monocron and Monocil, and is available in the form of EC. It is applied at the rate of 0.50-1.25 kg a.i. per ha in concentrations varying from 0.03 to 0.1 per cent on a large number of crops. It is effective against the bollworm complex of cotton, tissue borers, tingid bugs, aphids, thrips, whiteflies, leaf-miners, planthoppers, hairy caterpillars, mealy bugs, whorl maggots and leafhoppers.

(*xxiii*) **Quinalphos.** Quinalphos ($C_{12}H_{15}O_3H_2PS$) (**Fig. 7.2**) is a contact and stomach organophosphorus insecticide and acaricide. It was introduced by Bayer in 1969 under the code No. Bayer 77049. Its acute oral LD_{50} and acute dermal LD_{50} for rats is 62-137 and 1250-1400 mg/kg, respectively. In India, M/s Sandoz (India) Ltd., Mumbai, manufacture quinalphos.

Quinalphos is sold under the trade names of Ekalux, Bayrusil, Sandoz 6538 and Bayer 77049 and is available in various formulations including EC, granules and dust. It is applied at the rate of 125-500 g a.i. per ha in concentrations varying from 0.025 to 0.2 per cent on a large number of crops. It is effective against the bollworm complex, armyworms, diamondback moth, hairy caterpillars, tissue borers, thrips, leaf miners, web-worms, mealy bugs, *hadda* beetles, etc.

(*xxiv*) **Phosalone.** Phosalone ($C_{12}H_{15}$ $ClNO_4PS_2$) was discovered in 1960. It is a contact organophosphorus insecticide and is moderately toxic to bees. It is considered safe for parasitoids and predators. Its dermal LD_{50} is 1390 mg/kg. M/s Voltas Ltd., Mumbai, are manufacturing it in India.

Phosalone is sold under the trade names of Zolone, Rubitox, RP 11 and RP 974 in the form of EC, WP and dust. It is applied at the rate of 0.5-1.0 kg a.i. per ha in concentrations varying from 0.025-0.1 per cent on a number of crops. It is effective against the bollworm complex, hairy caterpillars, semiloopers and leaf miners.

(*xxv*) **Fenthion.** Fenthion ($C_{10}H_{15}O_3PS$) (**Fig. 7.2**) was introduced in 1957, by Farbenfabriken Bayer A.G. under the code numbers Bayer 29493 and 51752. It is a systemic and contact organophosphorus insecticide. By virtue of its low volatility and stability to hydrolysis it has high persistence. It may be phytotoxic on sensitive cotton, and apple varieties such as 'golden' or 'delicious'. The acute oral LD_{50} for male rats is 215 mg/kg and the acute dermal LD_{50} per male rats is 350 mg/kg. M/s Bayer (India) Ltd., New Delhi, manufacture it in India.

Fenthion is sold under the trade names of Lebaycid, Baytex, Entex and Tiguvon, and is available in the form of EC. It is applied at the rate of 625-1250 g a. i. per ha in concentrations of 0.05 to 0.1 per cent on many field crops. It is effective against the tingid bug, leaf-folders, web-worms, thrips and leafhoppers.

(*xxvi*) **Triazophos.** Triazophos ($C_{12}H_{16}N_3O_3P_5$) is a broad spectrum insecticide/acaricide with contact and stomach action. It is effective against aphids, thrips, lepidopterous larvae, etc. on crops like cotton, vegetables, etc. The registered formulations are 20% EC and 40% EC. The acute oral and dermal LD_{50} for rat is 57-68 and 2000 mg/kg, respectively. It is toxic to honey bees.

(*xxvii*) **Chlorpyriphos.** Chlorpyriphos ($C_9H_{11}Cl_3NO_3PS$) (**Fig. 7.2**) was introduced in 1965 by the Dow Chemical Co. under the code number Dowco 179. It is a contact organophosphorus insecticide and acaricide especially for the control of soil pests. It persists in soil for 2-4 months. The acute oral LD_{50} for male rats is 155 mg/kg and the acute dermal LD_{50} for rabbits is about 2100 mg/kg. It is rapidly detoxified in the animal body. Motilal Pesticides (India) Pvt. Ltd., New Delhi, formulate this insecticide at Mathura.

Chlorpyriphos is sold under the trade name of Dursban and is available in the form of EC and granules. It is applied at the rate of 200-250 g a.i. per ha. As a spray it is used in the concentration of 0.06 per cent on field crops. As a soil insecticide it is used in higher doses against termites and white grubs. It is effective against black bug, tobacco caterpillar, termites, white grubs and ground weevils.

(*xxviii*) **Disulfoton.** Disulfoton ($C_8H_{19}O_2PS_3$) was introduced in 1956 by Farbenfabrikien Bayer A.G. under the code numbers Basu 19639 and 5276. It is a systemic organophosphorus insecticide and acaricide possessing residual action for 6-12 weeks. The acute oral LD_{50} for male rats in 7.0 mg/kg and the acute dermal LD_{50} for male rats is 15 mg/kg. M/s Bayer (India) Ltd., New Delhi, manufacture it.

Disulfoton is sold under the trade names of Disyston, Solvirex, Dithiodemeton, Dithiosystox, Thiodemeton and Disystox, and it is available in the form of granules. It is applied in the soil as a systemic insecticide at the rate of 1.5-2.0 kg a.i. per ha on a number of field crops. It is effective through the plant system against aphids, leafhoppers, leaf-miners, thrips, stem weevils and shoot flies.

3. Carbamates

(*i*) **Carbaryl.** Carbaryl ($C_{12}H_{11}NO_2$) (**Fig. 7.3**) was introduced in 1956 by the Union Carbide Corporation under the code number Experimental Insecticide 7744. It is a contact insecticide with slight systemic properties. There is no evidence of phytotoxicity at the recommended dosage. Two distinct qualities have made it the most popular material : very low mammalian oral and dermal toxicity, and a rather broad spectrum of insect control. The acute oral LD_{50} for male rats is 850 mg/kg, the acute dermal LD_{50} for rats is more than 4,000 mg/kg. M/s Union Carbide India Ltd., New Delhi, manufactured carbaryl at Bhopal in India. However, the factory has been closed since December 1984, when the leakage of methyl isocyanate gas resulted into the death of more than 3000 persons.

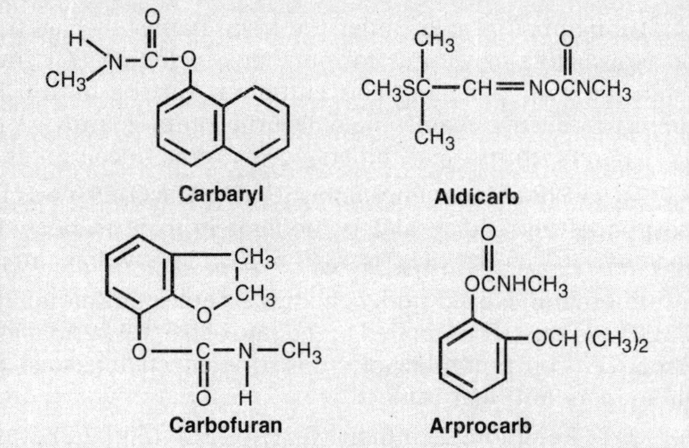

Carbaryl Aldicarb

Carbofuran Arprocarb

Fig. 7.3 Chemical structure of some carbamate insecticides

Carbaryl is sold under the trade names of Sevin, Hexavin, Carbavin and Ravagon, and is available in various formulations including WP, LVC, dust and granule. It is applied at the rate of 2.50-6.25 kg a.i per ha in concentrations of 0.1-0.3 per cent on a variety of crops including maize, cotton, gram, paddy, sugarcane, peas, jute, wheat, onions, grapes, coconut, cardamum, etc. It is effective against tissue borers, bugs, leafhoppers, midges, thrips, termites, white flies, maggots, hairy caterpillars, pests of household and external parasites of farm animals.

(*ii*) **Carbofuran.** Carbofuran ($C_{12}H_{15}NO_3$) (**Fig. 7.3**) is systemic phenylcarbamate insecticide and nematicide. Its acute oral and dermal LD_{50} to rat are 8.2-14.1 mg/kg and 10-20 mg/kg, respectively.

Carbofuran is sold under the trade name of Furadan and is available in the form of EC and granules. It is used at the rate of 0.75- 1.00 kg a.i. per ha and is generally used in the concentration of 0.025 per cent on various fruit crops. It is effective against tissue borers, fruit flies, aphids, white flies, thrips, leaf-miners, leaf-rollers, white grubs and ground weevils.

4. Synthetic Pyrethroids

Natural pyrethrins were quite effective against a variety of stored grain pests but were unstable in light and very expensive. The first synthetic pyrethroid was described in 1973 and many new photostable pyrethroids were synthesized between 1973-77. Synthetic pyrethroids exhibit high activity against insects, low mammalian toxicity, greatly increased stability, effectiveness at very low dosages, rapid action and degradation to innocuous residues. The activities of pyrethroids to insects, mammals and other groups such as fish, depend on the optical and geometrical configurations of their acidic and alcoholic components. The mobility in soil is very small. The mammalian toxicity of pyrethroids is generally lower than that of other classes of insecticides. These are more effective as contact insecticides and to a lesser extent as stomach poisons.

These can be used for the control of a wide range of pests even when used at a very low dose, and being biodegradable leave no residue to accumulate in the biological systems. These can be safely applied on agricultural crops, pests of foodgrains, household pests, parasites of domestic animals and livestock. Their field trials indicate that the tissue borers of many crops which are not controlled by other powerful insecticides easily succumb to the application of pyrethroids.

Cypermethrin (**Fig. 7.4**) is sold under the trade names of Ripcord, Cymbush, Cyperkill, etc. and is available as emulsifiable concentrate. It is applied at the rate of 50 g.a.i. per ha in low concentrations on cotton, arhar, sugarcane, gram, lentil, coffee, etc. It is effective against the bollworm complex of cotton, pod borers of arhar, gram, lentil, etc.

Permethrin (**Fig. 7.4**) is sold under the trade names of Permasect, Ambush, etc. as emulsifiable concentrate. It is applied at the rate of 50 ga.i. per ha in low concentrations on cotton, arhar, vegetables and pulses. It is effective against the bollworm complex of cotton, tissue borers and pod borers.

Deltamethrin (**Fig. 7.4**) is sold under the trade name of Decis, as an emulsifiable concentrate. It is applied at the rate of 10 g a.i. per ha against the bollworm complex of cotton, pod borers and tissue borers of fruits, vegetables and pulses.

Cypermethrin

Fenvalerate

Deltamethrin

Permethrin

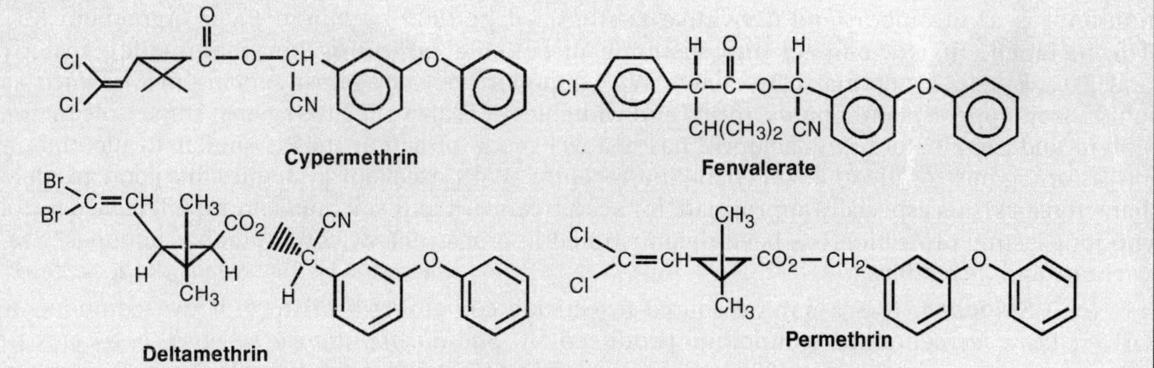

Fig. 7.4 Chemical structure of some synthetic pyrethroids

Fenvalerate (**Fig. 7.4**) is sold under the trade names of Sumicidin, Fenval, etc. as an emulsifiable concentrate. It is applied at the rate of 50 g a.i. per ha in low concen-trations against the bollworm complex of cotton, pod borers and tissue borers of fruits, vegetables and pulses.

5. Miscellaneous Insecticides

(*i*) **Pirate.** The pyrroles were discovered by the American Cyanamid Co., and the lead compound of this group is Pirate. This broad-spectrum insecticide/acaricide is highly active by ingestion, exhibits contact activity, and provides moderate residual activity. Because of different mode of action (uncoupler of oxidative phosphorylation), Pirate is effective against insects such as diamondback moth and tobacco budworm which have developed resistance to carbamates, cyclodienes, organophosphates and pyrethroids. In field studies, it has provided effective control of several agricultural pest species from the orders Acari, Coleoptera, Diptera, Heteroptera, Homoptera and Lepidoptera. It is classified as moderately toxic to mammals based on acute toxicity studies.

(*ii*) **Diflubenzuron.** It is an insect growth regulator and interferes with moulting process in the larvae leading to abnormal developments. It is effective against most leaf feeding larvae at 25-75 g a.i/ha and at 50-150 g a.i/ha against cotton boll worms. It is non-toxic to predators and honey bees. It is registered as 25% WP in India. Its oral LD_{50} for rat is 4640 mg/kg and dermal for rabbit 2000 mg/kg.

(*iii*) **Novaluron.** Novaluron [1-[3-clhlaro-4-(1, 1, 2-trifluoro-2-2-trifluoromethoxyethoxy) phenyl]-3-(2, 6-difluorobenzoyl) urea] is an insect growth regulator that inhibits the chitin formation in larvae of Lepidoptera, Coleoptera, Homoptera and Diptera. It has potent insecticidal activity against several foliage feeding insect pests. It has very low toxicity to mammals, birds and earthworms.

(*iv*) **Cartap.** It is commonly known as cartap hydrochloride ($C_7H_{16}ClN_3O_3S_2$). Its insecticidal properties were reported in 1967. It is systemic insecticide with stomach and contact action. It acts on the central nervous system by ganglionic blocking action resulting in paralysis, cessation of feeding and death due to starvation. It is effective against rice stem borer and leaf- folder, sugarcane shoot borer, cabbage diamondback moth, etc. It is registered in India as 50% SP and 4G. Its acute oral and dermal LD_{50} for rat is 325-345 and 1000 mg/kg, respectively.

(*v*) **Fipronil.** Fipronil is a new phenylpyrazole insecticide that has excellent activity against many soil and foliar insect pests. Its insecticidal properties were discovered by Rhone-Poulenc Ag. Co. in 1987. This broad-spectrum insecticide is highly active via ingestion, contact and systemic routes. It has shown a great promise both as a soil and foliar insecticide against several important insect pests of cotton. Its unique action as a potent blocker of the GABA-gated chloride channel makes it effective against insects resistant to carbamate, organophosphate and pyrethroid insecticides. Acute oral LD_{50} for rats is 100 mg/kg and decimal 72000 mg/kg.

(*vi*) **Imidacloprid.** Imidacloprid (1-[(6-chloro-3-pyridinyl)-methyl]-N-nitro-2- imidazolidinimine) is a nitromethylene derivative synthesized in 1985 by Nihon Bayer Agrochem K.K. (Tokyo, Japan). It is a contact and systemic insecticide exhibiting low mammalian toxicity (oral LD_{50} = 424–475, dermal LD_{50}=5000). With superior activity against sucking insects such as aphids, leafhoppers, planthoppers, thrips and whiteflies, it is also effective against some Coleoptera, Diptera and Lepidoptera. Imidacloprid has a novel mode of action that is similar to nicotine; it acts as an agonist of nicotine acetylcholine receptor. With excellent systemic and good residual characteristics, it is especially appropriate for seed treatment and soil application. Effective control with long lasting protection has been demonstrated in crops such as cotton, maize, potatoes, rice, sorghum and vegetables.

(*viii*) **Spinosad.** It is a natural source insecticide containing a mixture of two components derived from fermentation technology produced by *Saccharopolyspora spinosa,* a species of actinomycete. It is formulated as 48SC and is active against *Helicoverpa armigera* (Hubner), leafhopper, aphid and whitefly on cotton at 75-100 g a.i./ha.

6. Fumigants

(*i*) **Hydrogen cyanide.** Hydrocyanic acid (HCN) was introduced as an insecticide by American Cyanamid Co. in 1783. It is an effective fumigant for stored grains and other commodities kept in enclosed structures. It is a highly toxic substance and at a dose of 0.04 g it is lethal to human beings. The threshold limit is 10 ppm and the LD_{50} is 4 mg/kg.

It is also available under the trade names of Cyclon, Cyclone B, Zacon-Discoids, Prussic acid, etc. It is available as dust, granules, and the conventional packing is in the form of pressurized cylinders. It is applied at the rate of 3.5-5.3 kg/100/m^3. It is also used for tent-fumigation of citrus and other plants which should be exposed for 45 minutes at night. It is recommended for use in silos, warehouses, ships and other bulk storage structures against pests of almonds, beans, peas, peanuts and other dry fruits. It can kill all insects and rats when fumigation is allowed for 12 hours.

(*ii*) **Methyl bromide.** Monobromo methane (CH_3Br) was introduced as an insecticide in 1932 by Dow and other chemical companies of the USA. It is a gaseous fumigant used in the soil as well as on commodities. It is highly phytotoxic. An exposure to 2000 ppm for one hour may be lethal and it also causes burns. Acute vapour toxicity is 200 ppm.

It is available as 26, 69, 98 and 100 per cent liquids and 98 per cent compressed gas. It is sold under the trade names of MB, Bromothane, Weedfume, EDCO, Tribrome and Profume. It is used for controlling several annual and perennial weeds like Johnson grass, several fungi, nematodes, parasitic plants, insects and mites. It is used in quarantines against *khapra* beetle and pink bollworm larvae. It is usually applied at the rate of 1.6-6.4 kg per 100 m^3. It is recommended not to carry out fumigation with methyl bromide under high moisture conditions. It does not impair germination of dry seeds when fumigated.

(*iii*) **Ethylene dichloride and carbon tetrachloride.** Carbon tetrachloride (CCl_4) was introduced in 1908 by Diamond Shamrock and other chemical companies for nursery fumigation. It is mixed, usually, with other chemicals which suppress its flammability. It is generally mixed with 1, 2-dichloro-ethane ($C_2H_4Cl_2$) which is primarily a fumigant and was introduced by Union Carbide Chemical Co., USA in 1927; the LD_{50} of carbon tetrachloride is 5730 mg/kg. Repeated exposures of the mixture to human beings are dangerous specially when alcohol is consumed. Dichloroethane is irritating to skin. Its threshold value is 100 ppm and LD_{50} is 890 mg/kg.

Carbon tetrachloride and dichloroethane are used in the ratio of 1 : 3 to reduce fire hazards. This mixture is effective against insect pests of stored products. It should not be used on growing plants, although it has been used successfully against peach twig borer and other wood boring insects.

(*iv*) **Ethylene dibromide.** 1,2-dibromoethane ($C_2H_4Br_2$) was introduced by Dow Chemical Co. and others in USA in 1925 as EDB. It is soil as well as stored product fumigant. The threshold value of toxicity is 25 ppm and LD_{50} is reported to be 146 mg/kg. It can cause burns on the skin and is phytotoxic to plants at higher doses.

It is also marketed under the trade names of Fumo-gas, E-D-Bee, Bromofume and Dowfum W-85 as 100, 83, 75, 40 and 20 per cent solutions and 75 per cent water miscible liquid. It is recommended for use as soil fumigant on beans, crucifers, cucurbits, tobacco, tomato, etc. against nematodes which do not form cysts and against soil insects, including termites at the rate of 30-190 litres per ha. It is also used for spot fumigation of stored products as it does not penetrate deep into the grains. It is not recommended to grow crops for 7-15 days after soil fumigation. The stores are fumigated for 24 hours and then aerated.

(*v*) **Aluminium phosphide.** Aluminium phosphide (AlP) was introduced as a pesticide around 1955 by Degesch A.G. under the trade name of Phostoxin. It is highly insecticidal and is also a potent mammalian poison and is mainly used for the fumigation of grain.

Aluminium phosphide is sold under the trade names of Phostoxin, Celphos and Delicia, and is available in the form of solid tablets which on contact with moist air gradually release the deadly phosphine gas. It is used at the rate of 25 tablets per 100 m^3 for the fumigation of infested grains, furniture, carpets or other industrial goods. If is effective against stored grain insect pests, wood borers, carpet beetles, woolly bear and other insects infesting leather, silk, cotton, jute, etc.

II. ACARICIDES

(*i*) **Chlorobenzilate.** Ethyl 4,4′-dichlorobenzilate ($C_{16}H_{14}C_3Cl_2$) was introduced as an acaricide by Geigy Chemical Co. in 1952. It is a chlorinated hydrocarbon, closely related to DDT in structure. It is usually not phytotoxic but sometimes it may cause injury to peach, plum, rose and prunus. Unlike DDT, it is not accumulated in body fat. The mammalian toxicity as LD_{50} is reported to be 1000 mg/kg.

It is also marketed under the trade names of Folbex, Akar and Acarben as 25 per cent wettable powder or emulsifiable concentrate. It is recommended for use on fruits, cotton, cucurbits, turfs and ornamental plants against all species of phytophagous mites. It is applied at the rate of 0.75-2.50 kg per ha. Being a selective acaricide, it is used against the tracheal mites causing acarine disease in honeybees. The hives are fumigated by burning Folbex paper strips.

(*ii*) **Dicofol.** 1, 1-bis (p-chlorophenyl)-2, 2, 2-trichlor-ethanol ($C_{14}H_{13}OCl_5$) was introduced as an acaricide in 1955 by Robin and Haas Chemical Co., under the trade name of Velthane. It is also related to DDT in structure. It differs from DDT only with respect to bridge 11 which is replaced by 'OH'. It is non-systemic with long residual activity. It is non-phytotoxic. The mammalian toxicity, LD_{50}, is 684 mg/kg.

It is marketed as EC, WP and dust. It is applied at the rate of 1.0-2.5 kg per ha on all types of plants against mites. It gives good initial kill and long residual action. Its residue disappears in soil rapidly but traces may persist for a year or more. It is compatible with most other pesticides.

(*iii*) A number of insecticides are equally effective against mites. These insecticides are thiometon, phorate, menazon, formothion, chlorpyriphos, etc.

III. NEMATICIDES

These chemicals are used for controlling plant parasitic nematodes. Some of the common nematicides are mentioned here.

(*i*) **Dibromochloropropane (DBCP).** 1, 2-dibromo-3-chloropropane ($C_3H_6Br_2Cl$) as soil fumigant was introduced in 1955 under the trade names of Nemagon and Fumazon by Shell Chemical Co. and Dow Chemical Co. It is also marketed as Memapaz and Namafume. As nematicide, it is particularly effective against root-knot nematodes when the soil temperature is 21-27°C. It is non-phytotoxic to most of the plants at normal concentrations. Some plants such as potato and tobacco are, however, sensitive to its action. It is irritating to eyes and its LD_{50} is reported to be 173 mg/kg.

It is available as EC and granules. The dose of its application varies from 12-30 litres per ha. It can be applied around the basin area of woody perennials. Also, it can be injected into soil or applied with irrigation water to the standing crops which are tolerant to its action. This nematicide is very effective for the control of root-knot nematodes, cyst nematodes, stylet nematodes, root lesion nematodes and other ectoparasitic nematodes present in the rhizosphere of roots.

(*ii*) **Dichloropropane-dichloropropene (D-D-mixture).** DD mixture is a by product of plastic manufacture which was first used as a soil fumigant in 1943 by Dow Chemical Co. and Shell Chemical Co. Since then, it has been in use as an effective nematicide in cooler climates (4-27°C). It is a mixture of 1,3-dichloropropane and 1,2-dichloropropene and is phytotoxic. It is a dark liquid at ordinary room temperature. It does not accumulate in soil and is safe for the soil micro-organisms. It has quite a high mammalian toxicity, LD_{50} being 140 mg/kg. Its contact with skin or the inhaling of its fumes should be avoided. Its strong odour warns the operator and it should be stored away from food. Being inflammable, it should be used with great care.

The commercial names of this nematicide are DD, Vidden D and Telone. It is a liquid fumigant and can be applied to the properly tilled soil by injecting it and later plugging the holes. A crop should be sown or transplanted after a period of three weeks so as to avoid phytotoxic effects. The dosage varies from 200 to 400 litres per ha. It is very effective for the control of cyst nematodes, *Heterodera* and *Globodera* spp., the root-knot nematodes, *Meloidogyne* spp, the burrowing nematodes, *Rapopholus similis* and the root-lesion nematodes, *Pratylenchus* spp.

(*iii*) **Ethylene dibromide.** Chemical and biological properties have been described in the section on 'Fumigants'. Its trade name is Dowfume W-85. This fumigant is in liquid form and should be applied by injection method upto a depth of 15 cm in rows 25-40 cm apart. The dose per hectare is 250 kg. After soil fumigation, a crop should not be sown or transplanted before a waiting period of 3 weeks in order to avoid phytotoxic effects. It can be employed for the control of the root-knot, the root- lesion and cyst nematodes.

(*iv*) **Ethoprop.** It is marketed under the commercial name of Mocap by Mobile Chemical Co. and is available in granular or EC form. It is applied at the rate of 2.4 kg/ha.

(*v*) **Phenamiphos.** It is commercially available under the name of Nemacur, and its formulations are both granular and EC. It is very effective for the control of root-knot and cyst nematodes. Being a systemic nematicide, it is capable of killing both ecto and endo-parasitic nematodes. The application rate varies from 5 to 10 kg a.i. per ha and it can be applied before or at the time of sowing or transplanting a crop.

(*vi*) **Metham sodium.** Sodium N-methyl dithiocar-bamate dihydrate ($C_2H_4NS_2Na.2H_2O$) was introduced in 1954 by Stauffer Chemical Company and E.I. Dupont de Nemours, under the trade name of Vapam. It is a soil fumigant for the control of fungi, bacteria, nematodes, weeds and soil insects. It is phytotoxic and in gaseous form, is irritating to eys, skin and mucous membranes. The LD_{50} is reported to be 1260 mg/kg.

It is also sold under the trade names of Vitafume, Unifume, Sistan, Trimaton and Mapsol. It is available in water miscible form and the breakdown product is methyl isothiocyanate. The soil application is by injection method and it should be applied 2-3 weeks before transplanting the crop. The dose varies upto 1000 litres per ha. It controls the root knot and lesion nematodes. It is not advisable to use this chemical in green houses. It is corrosive to brass and copper.

(*vii*) **Methyl bromide (CH_3Br).** Chemical and biological properties of this compound have been described under the section on 'Fumigants'. It is a gas fumigant and is available under the trade names of Embafume-C and Dowfume MC-2. Dowfume MC-2 is a formulation of bromomethane and 2 per cent chloropicrin. It should be applied in the soil before transplanting or sowing of a crop at the rate of 100-440 kg per ha under a gas-tight cover. It is a broad spectrum nematicide and kills all types of plant parasitic nematodes.

(*viii*) **Qamyl.** The commercial name of the chemical is Vydate. It is available in the form of granules and it is an emulsifiable concentrate. The granules can be applied by broadcasting in the field before sowing or at the time of sowing. Emulsifiable concentrate can also be sprayed on plants for the control of stem and bulb nematodes and also root-knot nematodes infesting vegetable crops.

(*ix*) **Aldicarb.** Aldicarb ($C_7H_{14}N_2O_2S$) **(Fig. 7.3)** was introduced as an insecticide-acaricide - nematicide in 1962 by Union Carbide Corporation, USA. It is systemic in nature and at the recommended dose it does not damage seeds. It is highly toxic to mammals and the LD_{50} is only 0.93 mg/ kg. It is marketed under the name Temik and is formulated as granules having 5 and 10 per cent active ingredient. It is very effective at the dose of 2-4 kg a.i. per ha and can be broadcast or drilled at the planting sites. The root-knot, the cyst, the root-lesion and citrus nematodes can be controlled effectively. It has no phytotoxic effects and hence can be applied on vegetable crops quite safely.

(*x*) **Carbofuran.** Chemical and biological properties of carbofuran **(Fig. 7.3)** have been described under the section on Carbamates. It is available in granular and flowable formulations under the commercial name Furadan. The recommended dose is 2 to 4 kg a.i. per ha for the control of root-knot and lesion nematodes infesting vegetable crops. Flowable powder can be applied as seed treatment to provide protection to the newly emerging seedlings.

(*xi*) **Dazomet.** Dazomet ($C_5H_{10}N_2S_2$) was introduced by Union Carbide Corporation, USA, in 1952, as a pre-plant soil fumigant under the trade name of Mylone. It is reported to be phytotoxic to growing plants. It is the break down product, methyl isothiocyanate, which is toxic to nematodes. Dazomet is irritating to skin and the mammalian toxicity as LD_{50} is reported to be 500 mg/kg.

It is also marketed under the trade names of Basamid, DMTT and Soil-Kare. It is available in the form of 50 and 85 per cent WP, 50 per cent granules, and 2 and 5 per cent dusts. It is recommended for use on tobacco and ornamental plants at the rates of 200-500 kg per ha. It controls all nematodes except the cyst forming nematodes. Since it is phytotoxic, it is applied 4 weeks before planting. It also kills some weeds.

(*xii*) **Fensulfothion.** Fensulfothion ($C_{11}H_{18}O_4PS_2$) was introduced as a soil insecticide and nematicide by Farbenfabriken Bayer, A.G. of Germany in 1964 under the trade name Dasanit. It is systemic and long-persistent. It is non-phytotoxic at the doses recommended for use. It is a highly toxic compound and the mammalian LD_{50} is reported to be 2.2 mg/kg.

It is also sold under the trade names of DMSP and Terracur-P. It is available as emulsifiable concentrate and granules. It is recommended for use on cruciferous vegetables, tobacco, tomato, potato and ornamental plants. It is recommended for the control of root-knot and golden nematodes, when applied at the rate of 1-16 kg a.i. per ha. It protects the plants against soil insects and nematodes for 4 months. It should not be applied on the foliage or in the green houses. It is also toxic to fish in water.

(*xiii*) **Phorate.** The chemical and the biological properties have been described under the section on 'Organophosphorus Insecticides'. It is also a granular nematicide and is available as Thimet. For the control of root-knot and other ectoparasitic nematodes, it is recommended at the rate of 10 kg a.i. per ha. It is applied by broadcast method and then mixed into the soil.

IV. RODENTICIDES

(*i*) **Strychnine.** Strychnine ($C_{20}H_{20}N_2O$) was used as a rodenticide in Europe around 1930. Chemically, it was synthesized in 1954 in USA and was distributed by a number of companies. It is of botanical origin and was extracted from the seeds of *Strychnos* spp. It is a strong mammalian poison; the LD_{100} for rats is 1-30 mg/kg and for man is 30-60 mg/kg. The threshold limit is 0.15 ppm.

It is sold under the trade names of Sanaseed, Mousetox, Kwik-kill, Rodex, etc. It kills mice, squirrels, dogs, porcupines, rabbits, jackals, birds and other animals. It is marketed as a concentrated powder which is used for making baits. It is mixed with raw meat as a bait to kill jackals. It can also be used by mixing it with grains, vegetables or fruits at the rate of 0.5 per cent by weight. Since it is a deadly poison, its handling needs utmost care by the users. One application is usually sufficient to kill the animals.

(*ii*) **Zinc phosphide.** Zinc phosphide (Zn_3P_2) was introduced by Hooker Chemical Co. in 1943. This rodenticide belongs to inorganic phosphide group and has long been in use. It is an acute poison for mammals and birds. It is stable when dry, but decomposes slowly to phosphine in moist air. It is a general poison and the mammalian toxicity at LD_{50} is reported to be 45.7 mg/kg. It is sold under the trade names of Rumetan, Mous-con and Kilrat as 80-95 per cent concentrated powder.

It is usually used in the concentration of 2.5 per cent as a bait for killing field rats, mice and other rodents. The bait is prepared by mixing one part of zinc phosphide with 40 parts of wheat-flour, adding sufficient water to make pellets. Alternatively, the poison can be coated on grains of wheat, maize, sorghum or *bajra* (the pearl millet, *Pennisetum typhoides*) after they are rinsed with a small quantity of some edible oil. It is an acute poison and only one dose is needed to kill the animal. The disadvantage of this compound is that rats come to know of its presence and they develop bait shyness.

(*iii*) **Warfarin.** Warfarin ($C_{19}H_{16}O_4$) is the first anti-coagulant coumarine derivative which was synthesized by Wisconsin Alumni Research Foundation in 1950. It was sold by S.B. Penick and Co. and Prentis Chemical Co. The mammalian toxicity as LD_{50} is reported to be 186 mg/kg as a single dose. On a multiple dose basis, the LD_{100} for rats is reported to be 0.2 mg/kg/day, for 5 days.

It is sold under the trade names of Rat-B-Gon, Warf, Warfarat, Dethmor and Duocide in numerous formulations such as concentrated powder, dust, bait, water miscible liquid, etc. It is used to kill rats and mice. The rats die of internal haemorrhage after eating it for about 5 days. It is applied in cereal baits at the rate of 0.025-0.05 per cent by weight. The 0.5 per cent powder is mixed with dry bait material at a 19 : 1 ratio to get the required concentration. It should be kept out of the reach of children, pets and other domestic animals. Vitamin K is an antidote. The other commonly used anticoagulant is Fumarin sold as Rodafarin and Ratafin.

V. MOLLUSCICIDES

Molluscicides may be classified into two broad categories, viz. aquatic and terrestrial. *Aquatic molluscicides* may be botanical, e.g. endod, or chemical, e.g. copper sulphate, niclosamide, sodium pentachlorophenate, trifenmorph,etc.*Terrestrial molluscicides* include carbamates, i.e. aminocarb, carbaryl, isolan, methiocarb, mercaptodimethur, mexacarbate, etc. and others, i.e. metaldehyde.

Metaldeyde (plymer aldehyde) is also known as metacetaldehyde and by the trade name Meta. The slug killing properties of metaldehyde were reported in 1936, slugs are immobilized before they die. It is formulated as slug bait in a protein rich milling offal such as bran. Ordinary lime is also used as a molluscicide by sprinkling it on the ground, around the cropped area.

PESTICIDE APPLICATION EQUIPMENT

Pesticides are applied on crops to keep the pests under minimum biological activity. Insecticides are used against insect pests, fungicides are used against fungal crop diseases and herbicides are used against weeds in order to protect crops and avoid losses. It is necessary to select the most efficient equipment for securing a uniform deposit on the target in the least time with minimum labour and without appreciable wastage of material. The concentration and quantity of the spray fluid or dust material to be applied will depend upon the type of machinery used.

The efficacy of a machine depends upon its ability to give the maximum pest control per unit area and time with the minimum dose of actual ingredient of the poison. The wastage and pollution of the environment should also be the minimum. Pesticides may be applied as dusts, sprays, mists, aerosols, smokes, etc. Various types of equipment used include dusters, sprayers, vapour and smoke generators, agricultural aircrafts, granule applicators, soil injectors, etc.

I. DUSTERS

Appliances that are used to distribute dust formulations are known as dusters. These dusters may be manually operated or power operated.

1. Manually Operated Dusters

There are several types of manually operated dusters that are available in the market. Some of the important ones are described here.

(*i*) **Plunger duster.** It is made up of air pump, dust chamber and a discharge assembly. The duster is held by one hand and is pumped with the other hand. The air pumped in creates a dust cloud which passes through the delivery vent. The amount of dust to be applied can be controlled by the speed at which plunger is forced and by adusting orifice at the vent. These are easy to operate and are quite inexpensive and are useful for dusting in households and kitchen gardens. These are also used for spot-treatments against ants, poultry pests and external parasites of farm animals.

(*ii*) **Bellows duster.** These can be carried in hand as well as on the back and the air blast is created by operating the bellows. It has a small container for dust, which is fed to the air stream, thus producing a dense cloud.

(*iii*) **Rotary or fan duster.** It has an enclosed fan geared to a hand crank and a hopper holding the dust. The high speed gear transmits the velocity to a blower and the dust cloud enters the atmosphere through spreader nozzle. The dust is carried by wind and it falls on the target. These machines are useful for dusting on crops like paddy growing in standing water.

(*iv*) **Wet dusting equipment.** In this type, while the dust is passing, a fine discharge of water mixes with it before reaching the target. This dust adheres better on the crop surface than the ordinary dust and drift is also reduced. This machine is more suitable for semi-arid zones.

2. Power Operated Dusters

(*i*) **Tractor mounted duster.** It is also known as power take off duster. This duster is mounted on a tractor and the power to operate the duster is taken from tractor with the help of a V-belt. The dust flows through eight nozzles mounted on a boom. The speed of fan producing the blast varies with that of the tractor; so it is difficult to maintain a uniform discharge. The machine can carry 20-45 kg dust.

(*ii*) **Engine-operated duster.** These dusters are operated with an internal combustion four stroke engine of 1-3 hp. Larger dusters are also available with power up to 25 hp. These can be held in hand, can be carried on back (knapsack type) or can be carried on a stretcher or trolley, depending upon the size and the use. In these dusters the fan normally has a speed of 2,200- 3,400 rpm and delivers 14.56-28.00 m^3/min of air at velocities of 80-260 kmph. These dusters are suitable for dusting trees and large fields. Some of the tractor trailor dusters have booms up to 9 metres with as many as 18 delivery nozzles.

II. SPRAYERS

A sprayer is an appliance which atomizes the spray fluid, which may be a suspension, an emulsion or a solution. The fluid is ejected with some force for proper distribution. The various types of sprayers can be grouped into four categories, depending upon the volume of spray fluid discharged to cover a unit area.

1. Categories of Sprayers

(*i*) **High volume (HV).** These require 300-500 litres per ha of spray fluid normally and some times even up to 800-1000 litres per ha, depending upon the size of the crop. These are the most versatile sprayers. The size of spray droplets ranges between 300-500 μm and the density is 10-20 droplets per cm^2. There are many such sprayers, for example, foot sprayer, stirrup sprayer, knapsack sprayer, hand compression sprayer, rocker sprayer, etc.

(*ii*) **Low volume (LV).** These sprayers require 50-100 litres of spray fluid per ha. They are normally motorized knapsack sprayers. The size of droplets ranges between 100-300μm and the density is 30-50 droplets per cm^2.

(*iii*) **Ultra low volume (ULV).** These sprayers require 1-5 litres per ha of spray fluid. Normally special formulations are available which are used without any dilution. The size of spray droplet ranges between 60-100 μm and their density is 50-70 droplets per cm^2. The fogair, knapsack mist blowers, with restrictors, are inlcluded in this category.

(*iv*) **Aerosols.** Less than 1.0 litre per ha of spray fluid is required to create fog. Aerosols are applied normally in enclosed places. The size of droplets ranges between 1-50 μm and density is 70-350 droplets per cm^2. The examples are pressurised containers, swing fog machine, etc.

Various types of droplet spectra are adopted for different situations. Most of the flying insects require very fine spray having particle size 10-30 μm. For locusts and forest insects, a spectrum of 20-60 μm is most suitable. Most herbicidal sprays have a spectrum of 400-1000 μm which gives minimum drift. The droplet spectra are generally classified as given in **Table 7.2.**

Table 7.2 Droplet spectrum of various types of spray discharges

Size of droplets (μm)	Type of spray	Equipment
400-1000	Coarse spray	Hydraulic sprayer
100-400	Fine spray	Mist blower
50-100	Mist spray	Micron sprayer, ULVsprayers
1-50	Fog spray	Aerosols, Fogging machine
0.001-1	Smoke	Smoke generator
<0.001	Vapour	Vapour generator

2. Types of Nozzles

There are various types of nozzles to disperse the spray fluids. Nozzles are usually classified and named after the energy used to form droplets.

(*i*) **Hydraulic nozzles.** These nozzles are mostly used in sprayers in which large quantities of water are used in spray fluid. The liquid under pressure is fed through a nozzle. The pressure determines the rate of discharge, throw of the fluid and its atomization. These nozzles usually produce coarse sprays and are of different types.

(*a*) *Impact or floodjet nozzles.* These operate at low pressures of 0.5-1 kg per cm^2 giving a coarse spray without any drift. These are used for spraying herbicides and liquid fertilizers.

(*b*) *Flat fan nozzles.* These are supplied with variable discharge rates varying from 500 to 3,000 ml per min at different angles (60-110°). Those most commonly used have a discharge of 1,500 ml per min at a pressure of 3 kg per $cm^2$and an angle of 80°. These also give fine to coarse sprays without much drift and are more suitable for insecticidal and fungicidal spray for a good coverage.

(*c*) *Cone nozzles.* Generally, hollow cone nozzles rather than solid cone nozzles are used in agriculture. These are suitable for the application of insecticides and are most widely used because of their low price.

(*ii*) **Gaseous/Pneumatic nozzles.** These are usually used with mist blowers. The atomization occurs by the impact of air blast on the drop of a pesticide falling with gravity. The ratio of fluid discharge to air is 1 to 1000 times for appropriate atomization. These nozzles provide fine sprays suitable for low volume spraying.

(*iii*) **Centrifugal nozzles.** The spray fluid is fed to a rotating disc and the centrifugal force disintegrates it into droplets of very fine size. The nozzle is used in rotatory disc ULV sprayers. The droplet size is inversely proportional to the speed and the size of the disc.

(*iv*) **Thermal or hot tube nozzles.** A fog composed of very fine droplets can be produced by condensing a pesticide which has been injected into a stream of hot gas to shear the liquid into droplets which are immediately vaporized. These vapours later on condense into very fine droplets which are suspended in the air for sometime till they finally settle on a surface. This method is very suitable for controlling flying insects in enclosed spaces.

3. Types of Sprayers

A. Manually Operated Sprayers

These are of two types, i.e. pneumatic and hydraulic.

(*i*) **Pneumatic Sprayers**

(*a*) *Hand compression sprayer.* The spray fluid is filled in a tank above which there is an air pressure, created with the help of a built-in air pump (**Fig. 7.5**). After all the solution is discharged through the nozzle, the air is released before the tank can be refilled. Since pressure drops with the lowering of fluid level in the tank, occasional pumping is required. These are small spray pumps and are suitable for kitchen gardens and houses. These pumps carry 0.5-1.0 litre fluid. The body of the pump can be made of polyvinyl carbonate or of a metal with an anti-corrosion coating.

(*b*) *Pressure retaining pumps.* There is an air chamber which is filled with air under pressure and then liquid is pumped with the help of a charge pump. Because of continuous and constant pressure at the nozzle, a uniform discharge and droplet size is maintained throughout the operation. The lance is also provided with a regulation to have uniform discharge. In a knapsack battery sprayer, for example, a pressure of 4-5 kg per cm^2 is developed with a pump mounted on the sprayer itself. The tank capacity is 10-20 litres and the liquid is discharged at a constant rate. These are low volume sprayers requiring 250-300 litres spray fluid for one hectare. One person can spray one ha per day with 15-20 refillings.

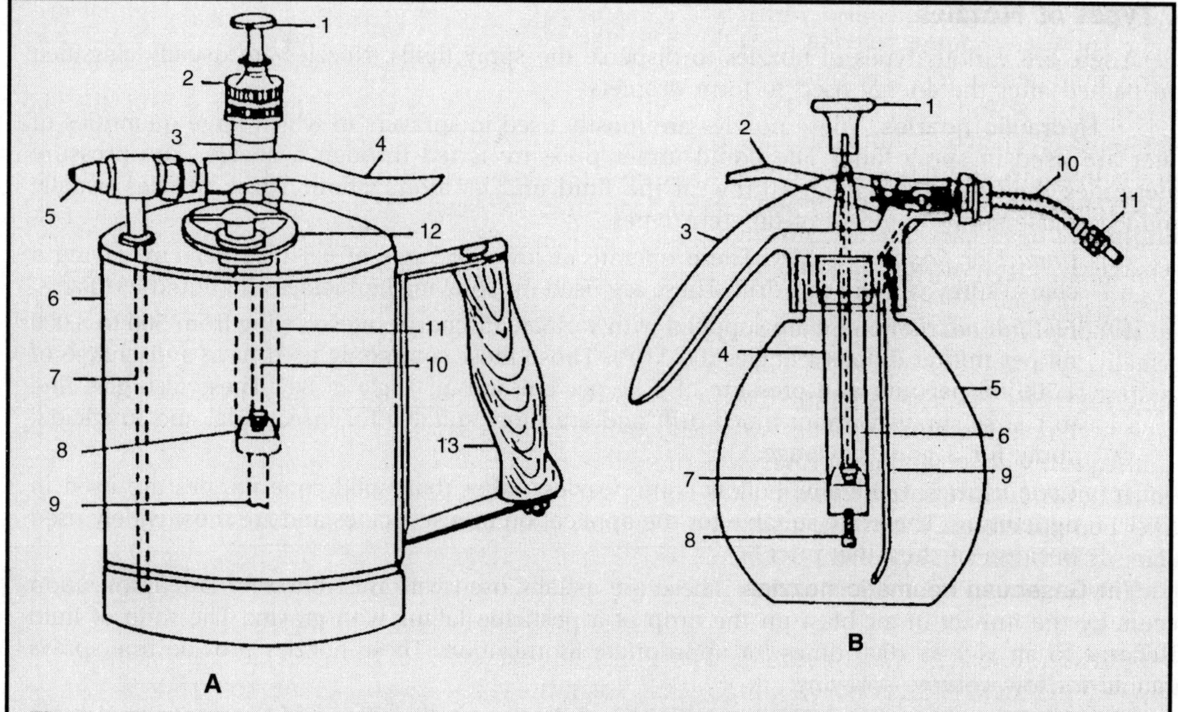

Fig. 7.5 Two types of hand compression sprayers. **A.** 1-Knob, 2-Cap, 3-Air pump, 4-Cut-off lever, 5-Nozzle, 6-Tank, 7-Delivery tube, 8-Bucket assembly, 9-Air-check valve assembly, 10-Pulnger rod, 11-Pump barrel, 12-Filter hole, 13-Handle. **B.** 1-Knob, 2-Cut-off lever, 3-Handle, 4-Pump barrel, 5-Tank, 6-Plunder rod, 7-Bucket assembly, 8-Air-check valve assembly, 9-Delivery tube, 10-Spray lance, 11-Nozzle.

(*c*) *Small pneumatic atomizers.* These pumps have a small container of 250 ml and there is a pump mounted on it externally. On working the pump handle there is a direct fine discharge. These machines are most suitable for controlling household pests like mosquitoes, houseflies, etc. by keeping the doors and windows shut for a while. These are made of polyvinyl carbonate as well as of tin metal. The fluid should be removed from container after use.

(*ii*) **Hydraulic Sprayers**

(*a*) *Knapsack sprayer.* The tank in this sprayer is not pressurized but is provided with a shut off regulator. It has a plunger or a diaphragm pump which is mounted either outside or inside the tank, and is immersed in the spray fluid (**Fig. 7.6**). The pump is operated by a hand lever under the arm. Newer models are now equiped with solid piston type pumps instead of the plunger type. Continuous pumping is necessary to have a uniform discharge. The capacity of the tank varies from 15 to 20 litres. One person can work this sprayer independently and can cover one ha of normal crop in one day, with 15-20 refillings. It is a medium volume sprayer, requiring 350-450 litres of spray fluid per ha.

(*b*) *Foot sprayer.* It is a high volume sprayer and is also known as the pedal pump. It is worked by up and down movements of the foot of an operator. It has one suction outlet but two delivery tubes, lances and nozzles which are handled by two men. Constant pedalling is required to develop a continuous pressure up to 17-21 kg/ cm^2. This sprayer is also suitable for tall crops. It requires 500-1000 litres water per ha depending upon the growth of a crop and four persons are needed to operate it. One person pedals, two persons hold the lances for spray and fourth person prepares the spray fluid and also acts as a reliever to the pedaller. With this pump, 2 hectares can be sprayed in a day.

(*c*) *Rocker sprayer.* It is also a high volume sprayer requiring 500-750 litres of spray fluid per ha. It is very similar to the foot sprayer but is operated by a long hand lever. The pump may be either of single action or of double action type but in both cases continuous pumping is necessary for a uniform discharge. It can develop a pressure up to 14-18 kg per cm^2 and can also be used for spraying trees up to 5 m height. Normally, two persons are needed for this sprayer and they can cover about 1-2 ha in one day.

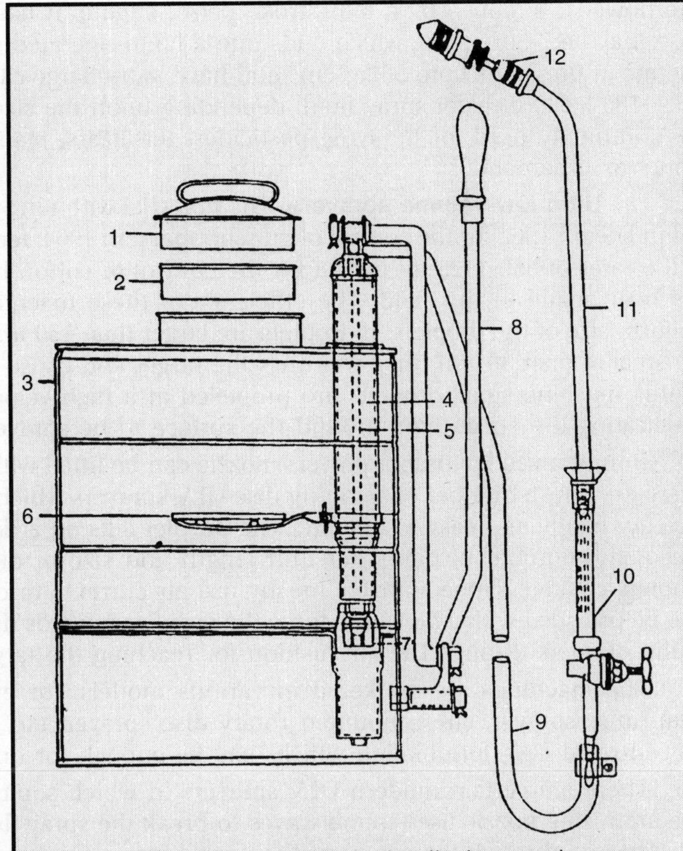

Fig. 7.6 Manually operated hydraulic knapsack sprayer. 1-Filter-hole cap, 2-Strainer, 3-Tank, 4-Pressure chamber, 5-Delivery tube, 6-Agitator, 7-Delivery-valve assembly, 8-Pump lever, 9-Delivery hose, 10-Wheel cut-off valve with strainer, 11-Spray lance, 12-Nozzle.

(*d*) *Stirrup pump.* It is a high volume sprayer and is commonly known as bucket pump, because it is worked after placing it in a bucket full of the spray fluid. The pump may have a single barrel or a double barrel. The pump is worked with hand by up and down movements. It can develop a pressure up to 14 kg per cm^2 and can be provided with two delivery tubes, two lances and two nozzles. If there are two lances, then four persons are required to carry out the spray operations and they can cover 2 ha in a day. It requires 500-875 litres water per ha depending upon the crop size.

B. Power Operated Sprayers

(*i*) **Mist blower or motorized knapsack sprayer.** It is a low volume sprayer, but with a restriction in the nozzle. It can be converted into an ultra low volume sprayer. A two-stroke petrol engine drives the blower; a part of the air also pressurizes the chemical tank. An impeller blows the air through the sprayer nozzle, while the solution separately flows from the liquid tank. The fluid gets atomized at the point of contact with the air blow. The tank capacity is 7-12 litres and the air blast is delivered at the rate of 2.7-9.1 cm^3/ min, at velocities ranging from 175-320 kmph. The discharge rate is variable from 0.5 to 5 ml per min. When used as a low volume sprayer one man can cover 2 ha in one day, using 125-200 litres spray fluid per ha. The swath width is 4-5 m horizontally and 3-4 m vertically. Trolley mounted mist blowers are also available with a much larger tank, having a capacity of 50 litres.

(*ii*) **Tractor mounted sprayer.** The pump is driven by the power take-off shaft of a tractor. The sprayer may be mounted on the power lift or on a trailor. It has a big liquid tank of 500-600 litre capacity. The swath width is about 18 m and may carry as many as 18 nozzles on a single boom lance. Two persons are required to conduct the spraying and they can cover upto 4-5 ha in a day. The crop should be in rows and it should be just as tall as the clearance of body of tractor allows. These pumps are comparatively less expensive but are very versatile in use.

(*iii*) **Power sprayers.** These are high volume sprayers and can be mounted on a stretcher, a wheel barrow, a tractor or a truck. These sprayers usually have hydraulic pumps of the piston type.

The power is supplied by a four-stroke petrol engine. It has one or two delivery hoses. A delivery hose may be 7-30 m long which ends into a hand-operated spray gun with a regulator. The pumps operate at pressures upto 56 kg/cm^2 and have a discharge capacity of 6.8 to over 273 litres/min and carry 180-3,600 litres of spray fluid, depending upon the size of tank and the engine. The sprayers are commonly used for applying pesticides, fertilizers, plant hormones, etc. on shade trees, fruit plants or ornamentals.

(*iv*) **Ultra low-volume sprayers.** The first trial with ultra low volume (ULV) sprayers was carried out in USA in 1962 for the control of grasshoppers. In 1964, technical malathion was used successfully at the rates of 850-1125 ml per ha for the control of cotton boll weevil. Since then a lot of progress has been made in this field. The efficiency of these machines depends upon the regulation of a uniform size of the droplets. If droplets are larger than 150 μm, the coverage will be poor. If droplets are smaller than 30 μm, they will miss the target and cause environmental pollution through drift. In this machine small droplets are propelled at a high velocity so that they reach the target after penetrating the air cushion around the surface to be sprayed.

In motorized knapsack sprayers, nozzle can be fitted with a suitable restrictor for ULV spraying. There are now a number of spinning disc ULV sprayers which run on 12-volt battery. These sprayers are very handy and easy to operate. The droplet falls on a dentate disc spinning at 3,000 rpm. The speed, the number of dents per unit length and size of disc determine the ultimate size of the droplets created. These sprayers mostly use air currents to disseminate the droplets but these can also be provided with a blower to force the droplets towards the target. This increases the penetration of the droplets through the air cushion for reaching the target surface.

The machines are marketed in various models, for example, overhead sprayer, hand held rotary disc sprayer, chest-mounted rotary disc sprayer, etc. The container can carry up to 1 litre concentrated ULV formulation which may be enough for one hectare.

There are certain modern ULV sprayers in which sonic energy is used to break the droplets. The atomizing nozzle uses sound waves to break the spray fluid into fine droplets. The compressed air is passed through the centre of the nozzle into a resonator cavity; such nozzles are called soni-core nozzles. The sonic-energy field explodes the passing fluid into a mist of 50-100 μm. It is a very efficient sprayer, and is being marketed by Buffalo Turbine Agricultural Equipment Co., New York, USA. This ULV sprayer is recommended for use in enclosed spaces like a warehouse or a living quarter, for the control of flies, mosquitoes, grain pests and other indoor pests.

The drift and droplet spectrum can be studied by placing stretched parafilms or manganese oxide-coated plates or cellophane tapes at varying distances from the sprayer. The droplets so collected can be used to determine the density and the size of droplets under a microscope, using occular and stage micrometers.

(*v*) **Aerosol dispensers.** These dispensers have metallic containers to withstand the pressure of liquified propellent. The droplet size being small (1-50 μm), the particles remain suspended in the air for a long time, which is desirable for controlling flying insects like flies and mosquitoes. These can be used successfully in greenhouses also. A dispenser is fitted with a delivery tube which remains dipped in the pesticide. When a button having a small orifice is pressed at the top of this tube, the propellent forces its way out alongwith a fine spray of the pesticide. The propellent most frequently used is an inert gas, dichlorodifluoromethane (Freon-12) or methyl chloride. Aerosol dispensers for pyrethrum, allethrin, DDT, lindane, DDVP, etc. are available. The popular capacity of an aerosal dispenser is 300-400 g of formulation but 2-5 kg sizes are also available. Aerosols are used at the rate of 7-14 g/100 m^3 of space, at 5-10 per cent concentration in the dispenser. Other ways of generating aerosols are through centrifugal energy, compressed air or high velocity hot air pulse jet.

(*vi*) **Smoke generator.** Solid particles in a smoke are in the range of 0.001 to 0.1 μm. The smoke rises to the top and no enclosure remains untreated. The smoke particles settle down later and the crawling insects usually pick up the lethal doses. The smoke produced by burning insecticides loses killing capacity to some extent. In relatively stable insecticides like DDT and lindane, the loss due to heat may be as much as 30 per cent. Smoke formulations are prepared by mixing the technical material with some slow burning material. For example, DDT or lindane formulated for slow burning contains 58-60 per cent technical material, a burning mixture of sucrose and potassium chlorate (30-40 %) and a retardent such as clay or diatomaceous earth (2- 10 %). A container having 450 g of this mix will burn for 3 minutes, and treat 450 m^3 of enclosed space. These are also commonly called as 'smoke bombs'. Smokes are used to kill pests of greenhouses, warehouses, industrial establishments and cargo ships.

(*vii*) **Vapour generators.** These are popularly known as pesticide vaporizers or thermal vapour generators or electric-vaporising devices. These are used in enclosed spaces. Mostly hydrocarbon pesticides like lindane and DDT are held in an electrically heated container at a flash temperature which permits slow but continuous vaporization. It can convert 28 g of pure lindane crystals into vapours in 5 minutes which would be sufficient to kill insects in 280 m^3 of space. The vapours act as fumigants.

III. AGRICULTURAL AIRCRAFTS

An aircraft was first used for pest control in 1920. Since then these have been used for applying pesticides in 40 countries. It is estimated that there are 20,100 aircrafts, which cover 171.4 million hectares annually, throughout the world. The use of an aircraft enables the coverage of vast areas rapidly, timely and economically. Both the fixed wing aircrafts and the rotary wing aircrafts (helicopters) have been employed for applying pesticides on field crops, orchards, forests, pastures and wastelands. It is also employed most successfully for locust control in desert areas.

There are three types of fixed wing aircrafts.

(*i*) **Light aircraft.** It is usually monoplanes with ground speed of 100-150 kmph. It has single engine of 90-125 hp.

(*ii*) **Medium aircraft.** It is usually biplanes with ground speed of 130-300 kmph. It has single engine of 100-450 hp.

(*iii*) **Heavy aircraft.** It is a biplane with ground speed of 150-250 kmph and has usually two engines of 125-500 hp each. It has the advantage of maintaining altitude even when one of the engines fails. It is a desirable factor for low-level operations.

The rotary wing-aircraft may be a light machine with a single rotor or a heavy machine with large twin rotors. The advantages of helicopters over aeroplanes are ability to land close to the fields to be treated, capacity to hover-over the 'hot' spots of pests, accuracy to confine pesticide application to a given field, and greater manoeuvrability.

Operational conditions greatly influence the choice of aircraft. A fixed plane requires 270-450 m of clear landing strip whereas a helicopter can land on a clear piece of land about 30 m^2 Further, the spray pattern of a helicopter is better than that of a fixed wing plane in which the wing tips and propeller create air columns and turbulences in the micro-environment that is being sprayed. In addition to this, the spray from helicopter drops at a speed of 48-50 kmph which results in a rebound from the soil surface and thus foliage from top as well as from bottom gets treated. In other words, coverage of foliage is more thorough in case of helicopter spray and also there is lesser drift and more accurate spraying of the target as compared with that of the fixed-wing aeroplanes. There is also a greater crash safety in case of helicopter for pilots. The major drawbacks of a helicopter are the higher initial cost and high maintenance investment. The cost of a helicopter is 3-4 times that of a fixed wing.

In India, aircraft was used first in 1945 to spray insecticides for the control of mosquitoes in Delhi. Against locusts, the aircraft was used for applying aldrin on ground in 1951 and on flying swarms in 1961 in Punjab. Till 1955, *Piper-super cruisers* used to be requisitioned for spraying purposes from the flying clubs. Two *Beaver* aeroplanes were then procured from Canada under the Colombo Plan in 1956-57 for general plant protection and locust control. Later, one *Auster Autocar* was purchased for this purpose. *Basant,* indigenously developed at Hindustan Aeronautics Ltd., Bangalore is an agricultural aircraft which is in extensive use currently. It has a 400 hp engine and can carry a load of 825 kg. It has proven its versatility in aerial spraying. Currently, there are more than 65 aircrafts engaged in aerial spraying in India.

Aircrafts can be used for conventional as well as for ULV spraying. For conventional spraying, the aircraft is flown at 2-4 m above the crop. It gives a swath width of 20-30 m and the plane uses 20 litres per ha of spray fluid. It can spray 300-400 ha in a day. In case of ULV application, the flying height is increased to 7.5-10 m. It gives a swath width of 30-45 m. It can cover about 1200 ha in a day. The spray fluid is undiluted or special LVC formulations are used. The tank capacity varies from 160 to 900 litres depending upon the type of aircraft. The spray pump is normally centrifugal type generating 3.5 kg per cm^2 pressure; it may be motor driven, propeller driven or it may derive power from the aircraft. The size of a boom varies between 10-15 m and there may be 30-35 nozzles to cover a swath width of 30-45 metres. The volume median diameter (VMD) of droplets varies between 100-150 μm. Hydraulic, pneumatic or centrifugal rotary nozzles are used in aeroplanes.

Besides spraying, aircrafts can be used for applying dusts also. An aircraft can carry 90-900 kg of dust. Wind powered agitator is provided for uniform flow of the dust. There is an adjustable air-vent in the hopper for discharge regulation. Dusting is carried out at greater height than spraying, for obtaining a uniform distribution. Normally, aerial dusting is not as effective as spraying.

IV. MISCELLANEOUS MACHINERY

(*i*) **Granule applicator.** Placement of insecticidal granules in the whorls of crops like maize, jowar, etc. is very effective for the control of stem borers. To place granules at the right spot, a hand held granule applicator is required. The capacity of container is 500 g to 1 kg. The lid of container is perforated and is adjustable to deliver granules. Sometimes, the container is provided with a long lance which ends into an orifice which is provided with a springhook for keeping the hole in closed position. A clutch through a wire is used to open the hole and a lance guides the granules to the desired place. It is also good for applying granules in burrows. One person can cover 0.25 ha in a day. There is one knapsack granule applicator which has a similar device except that the hopper is large and is mounted on the back. The capacity of hopper is 10 kg.

There is a special applicator designed for palm trees in which case the hopper is at the top of the lance in inverted position. There is a sickel blade alongwith hopper to cut leaf axil so that granules can be dispensed at the correct place in palm trees.

(*ii*) **Soil injectors.** Most soil fumigants are liquids and are applied with soil injectors. Hand-held soil injectors with a capacity of 2-3 litres are commonly used to inject fumigants up to a depth of 15-22 cm, for controlling soil insects and nematodes. One person can treat about 0.25-0.5 ha in a day by selective spot treatment.

Tractor-drawn soil injectors are in regular use in advanced countries. The injectors are placed alongwith the harrow to direct the liquid into the furrow. These injectors get the supply from a tank mounted on the tractor. Similar injectors can be used for applying liquid ammonia in the soil as a fertilizer.

There are certain injectors which are used to control rats. These injectors have a container which is connected with plunger pump. There is a rubber tube about 1 m in length attached with the container. When plunger is worked, the air escapes through the tube connected with container. This carries a load of dry fumigant. The rubber tube is inserted in a live burrow and pitted with soil. After injection, the rubber tube is pulled out and the hole is plugged with soil. For rat control, calcium cyanide dust is used which on coming in contact with soil moisture releases hydrogen cyanide gas, which is a deadly poison.

PATTERN OF PESTICIDE CONSUMPTION

The use of synthetic pesticides in agriculture is the most widespread method of pest control. Farmers spend approximately $ 4.1 billion on pesticides annually on global basis. This high cost is justified by a direct return of $ 3-5 for every dollar spent on pesticides. The world's pesticide consumption is estimated to be about 3.0 million tonnes, which includes 1500 active ingredients and 50,000 or more commercial products. About 46 per cent of the total pesticide use are herbicides, 26 per cent insecticides, 23 per cent fungicides and 5 per cent other pesticide groups. About 34 per cent of the total pesticide is consumed in USA, 45 per cent in Europe and 20 per cent in developing countries. The Asia-Pacific region accounts for 16 per cent of the total pesticide consumption, out of which 75 per cent is used on rice, cotton and vegetable crops. About 85 per cent of all pesticide use in the world is for agriculture. However, the relative amount of each type of pesticide varies from country to country. For example, 75 per cent of the pesticides used in Malaysia are herbicides, and insecticides account for only 13 per cent. In the Philippines, insecticides account for 55 per cent of the total pesticide use, and fungicides account for 20 per cent. India with about 4 per cent of the world cropped area has a share of around 2 per cent of the global pesticide consumption. About 60 per cent of the pesticides used in India are insecticides.

India was one of the first countries in the third world to start large scale use of pesticides for the control of insect pests of public health as well as of agricultural importance. At present, 238 pesticides have been registered under Insecticides Act 1968 in India. Some of the most toxic and persistent pesticides have been banned or restricted for use. The consumption of pesticides for pest control in agriculture picked up after the introduction of high yielding varieties in 1966-67. The total amount of pesticides used in the country increased from 154 metric tonnes in 1953-54 to 61,357 metric tonnes in 1994-95. Earlier projections had put the pesticide demand at nearly 100,000 metric tonnes by the year 2000. But in view of the ban on DDT, HCH, aldrin, etc., high potency (and consequently lower required dosages) of new insecticides especially synthetic pyrethroids and high priority being accorded to IPM, the pesticide consumption has shown a decreasing trend during the last few years. The pesticide consumption in India had come down to 39,773 metric tonnes during 2005-06. Thereafter, there is a slight increase in consumption reaching 64380 and 41822 metric tonnes during 2008-09 and 2009-10, respectively **(Table 7.3)**.

India is the third largest consumer of pesticides in the world and highest among the South Asian countries. Up to 1995-96, the major group of chemicals used in agriculture was insecticides (80%), followed by fungicides (10%), herbicides (7%) and others (3%). Thereafter, the consumption of insecticides declined with simultaneous increase in the consumption of herbicides and fungicides. The consumption of insecticides in 1999-2000 was 60 per cent, fungicides 21 per cent, herbicides 14 per cent and others 5 per cent. During this period, types of insecticides used also changed, the percentage of organochlorines decreased from 40 to 14.5 per cent, carbamates from 15 to 4.5 per cent and synthetic pyrethroids from 10 to 5 per cent, but there was a sharp increase in percentage of organophosphates from 30 to 74 per cent. A modest consumption (2%) of natural pesticides (neem and Bt formulations) was also registered during this period. According to latest estimates, insecticides account for 61.39 per cent, followed by fungicides (19.06%), herbicides (16.75%) and others (2.80%) **(Fig. 7.7)**.

Along with the change in the amount of pesticides used, the potency of some of the new chemicals is also much higher. DDT was applied at dosages of 1-2 kg a.i/ha for the control of different pests ; the OPs, monocrotphos and quinalphos were effective at 250-500 g a.i./ha and the SPs, fenvalerate and cypermethrin at only about 50 g a.i./ha. In case of deltamethrin, the dosage has been further reduced to 10 g a.i. per ha. Thus, there has been more than 100- fold increase in the potency of new insecticides.

TABLE 7.3 Consumption (metric tonnes, technical grade) of pesticides in various states of India

S.No.	Name of State/U.T.	2000-01	2004-05	2005-06	2006-07	2007-08	2008-09	2009-10
1.	Andhra Pradesh	4000	2133	1997	1394	1541	1381	1015
2.	Arunachal Pradesh	13	17	2	17	16	10	10
3.	Assam	245	170	165	165	158	150	19
4.	Bihar	853	850	875	890	870	915	828
5.	Chhatisgarh	NA	486	450	550	570	270	205
6.	Goa	6	5	5	9	2.3	8.9	10.3
7.	Gujarat	2822	2900	2700	2670	2660	2650	2750
8.	Haryana	5025	4520	4560	4600	4391	4288	4070
9.	Himachal Pradesh	302	310	300	292	296	322	328
10.	Jammu & Kashmir	1	12	1433	829	1248	2679.3	1640
11.	Jharkhand	150	69	70	82	81	85	88.5
12.	Karnataka	2020	2200	1638	1362	1588	1675	1647
13.	Kerala	754	360	571	545	880	272.1	631
14.	Madhya Pradesh	871	749	787	957	696	663	645
15.	Maharashtra	3239	3030	3198	3193	3050	2400	4639
16.	Manipur	20	26	28	26	26	30.36	30.36
17.	Meghalaya	6	8	6	9	6	–	6.1
18.	Mizoram	8	25	25	40	44	44.25	39.05
19.	Nagaland	8	5	5	5	5	17.83	13.58
20.	Orissa	1006	692	963	778	NA	1155.75	1588
21.	Punjab	7005	6900	5610	5975	6080	5760	5810
22.	Rajasthan	3040	1628	1008	3567	3804	3333	3527
23.	Sikkim	4	–	–	2	6	2.68	4.22
24.	Tamil Nadu	1668	2466	2211	2048	3940	2317	2335
25.	Tripura	11	17	14	19	27	38	55
26.	Uttar Pradesh	7023	6855	6671	7414	7332	8968	9563
27.	Uttrakhand	99	132	141	207	270	221.10	222
28.	West Bengal	3250	4000	4250	3830	3945	4100	N.A.
29.	Andaman & Nicobar Islands	3	3	3	NA	NA	6.24	14
30.	Chandigarh	2	0.78	0.78	NA	NA	–	NA
31.	Delhi	55	53	39	NA	57	57	49
32.	Dadra & Nagar Haveli	6	5	4	NA	NA	–	NA
33.	Daman and Diu	2	1	1	NA	NA	–	NA
34.	Lakshadweep	2	1	1	NA	NA	–	NA
35.	Pondicherry	65	42	41	40	41	39	39.29
	All-India	**43584**	**40672**	**39773**	**41515**	**43630**	**43860**	**41822**

Source : www.indiastat.com

India's consumption of pesticides is quite low when compared to many other countries. The highest consumption is reported by the Republic of Korea (16.56 kg a.i. ha^{-1}), followed by Italy (13.35 kg) Hungary (12.57 kg) and Japan (10.80 kg), whereas India consumes only 380 g per ha. Japan spends $633 per ha on pesticides, South Korea $255 per ha and Philippines $24 per ha as compared to $3 per ha in India. Based on 2009-2010 data, four states, viz. Haryana, Punjab, Maharashtra and Uttar Pradesh consumed more than 4000 metric tonnes (technical grade) pesticides annually. Seven states, viz. Andhra Pradesh, Gujarat, Jammu & Kashmir, Karnataka, Rajasthan, Tamil Nadu and Orissa consumed pesticides between 1000 and 4000 metric tonnes. Six states, viz. Bihar, Chhatisgarh, Himachal Pradesh Kerala, Madhya Pradesh, and Uttrakhand consumed pesticides between 100 and 1000 metric tonnes. Nine states (Arunachal Pradesh, Assam, Goa, Assam, Jharkhand, Manipur, Mizoram, Nagaland, Tripura and Delhi) and union territories of Pondicherry, and Andaman &

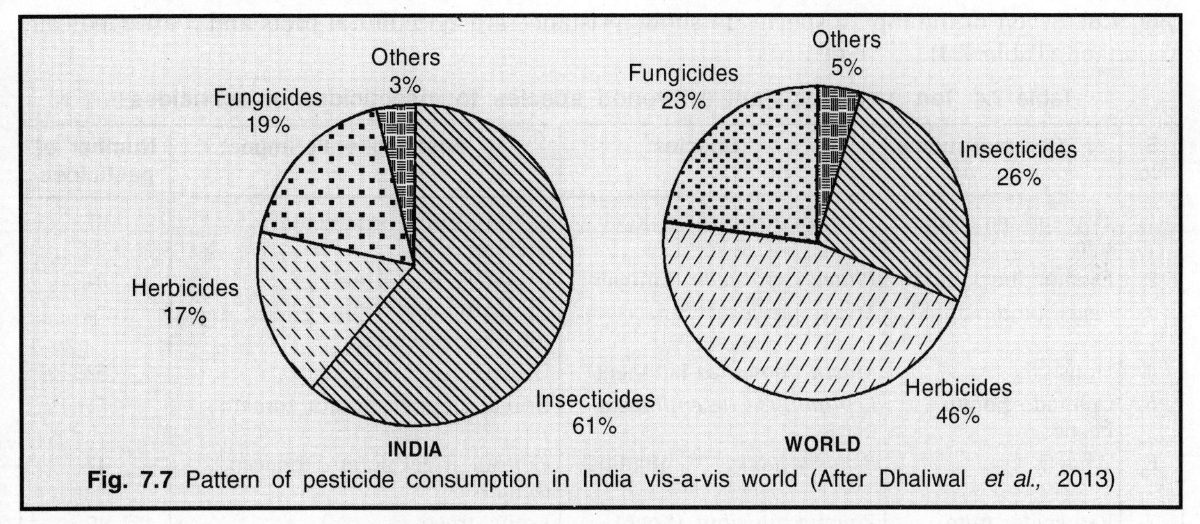

Fig. 7.7 Pattern of pesticide consumption in India vis-a-vis world (After Dhaliwal *et al.,* 2013)

Nicobar Islands consumed pesticides between 10 and 100 metric tonnes annually. Two states (Meghalaya and Sikkim) and four union territories of Chandigarh, Dadra & Nagar Haveli, Daman & Diu, and Lakshadeep consumed less than 10 metric tonnes pesticides annually (www.indiastat.com).

The consumption pattern of pesticides on different crops is also highly uneven. Both at the national as well as at the global level, cotton crop receives a disproportionately large share of pesticides. In India, cotton crop receives about 45 per cent of the total consumption of pesticides, followed by rice (23%), jowar (9%), vegetables and fruits (7%), wheat (6%) and pulses (4%). On the other hand, consumption of pesticides on many important crops like barley, gram, jute, rapeseed-mustard, soybean, sunflower and tobacco is even less than 1 per cent. At the global level, horticultural crops receive the highest proportion of pesticides.

ENVIRONMENTAL IMPACT OF PESTICIDES

The large scale use of pesticides has caused many environmental problems.

Insecticide Resistance

Resistance is the development of an ability to tolerate a dose of an insecticide, which would prove lethal to the majority of individuals in the normal population of the same species. Large scale use of pesticides to control pests has resulted in the development of resistance which is most serious bottleneck in the successful use of pesticides. Since the first report of development of resistance in San Jose scale to lime sulphur in 1908, more than 577 species of insects and mites have developed resistance to insecticides by 2012. The resistant species represent 15 different arthropod orders and among these the top seven orders (Diptera, Homoptera, Lepidoptera, Thysanoptera, Coleoptera, Hemiptera and Acari) comprise approximately 93 per cent of the resistant species

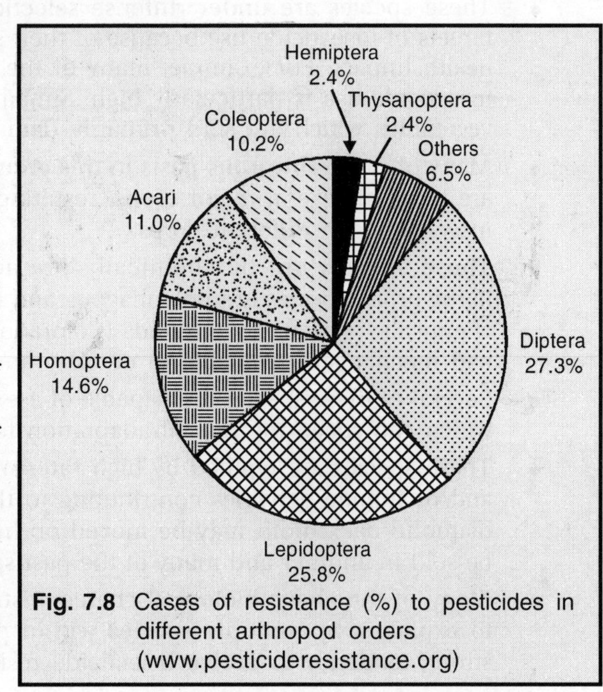

Fig. 7.8 Cases of resistance (%) to pesticides in different arthropod orders (www.pesticideresistance.org)

(Fig. 7.8). Seven of the top 10 species to show resistance are agricultural pests and 3 are medically important **(Table 7.4)**.

Table 7.4 Ten most resistant arthropod species to insecticides or acaricides

S. No.	Common name	Species	Host/economic impact	Number of pesticides
1.	Two-spotted spider mite	*Tetranychus urticae* Koch	Cotton, fruits, ornamentals, walnuts	91
2.	Diamondback moth	*Plutella xylostella* (Linnaeus)	Crucifers, nasturtium	84
3.	Peach-potato aphid	*Myzus persicae* (Sulzer)	Crops, flowers, fruits, grains, trees, tobacco, vegetables	74
4.	House fly	*Musca domestica* Linnaeus	Urban	52
5.	Colorado potato beetle	*Leptinotarsa decemlineata* (Say)	Brinjal, pepper, potato, tomato	51
6.	Whitefly	*Bemisia tabaci* (Gennadius)	Cotton, sweet potato, tobacco, tomato	47
7.	Red spider mite	*Panonychus ulmi* (Koch)	Fruits, trees	45
8.	Southern cattle tick	*Boophilus microplus* (Canestrini)	Cattle	43
9.	German cockroach	*Blatella germanica* (Linnaeus)	Urban	43
10.	American bollworm	*Helicoverpa armigera* (Hubner)	Berseem, chickpea, cotton, cowpea pigeonpea, soybean, sunflower, tomato	43

Source : www.pesticideresistance.org

Most of these pest species belong to a relatively small number of families of arthropods, viz. Tetranychidae (mites), Culicidae (mosquitoes), Noctuidae (moths) and Aphididae (aphids). A number of biological factors help to explain why these species have repeatedly evolved insecticide resistance.

- These species are under rintense selection for resistance. All of these pests are major targets of insecticide use because of their significant economic and in some cases, human health impact. For example, many of the harbivorous species are pests of cotton, where insecticide use is particularly high. Similarly, diamondback moth is a pest of cruciferous vegetables which are sold primarily damage-free, requiring significant insecticide use.
- Many of the herbivorous pests in this group, such as the heliothines and the sucking pests are pests of several major crops, resulting in their being exposed to the same or similar insecticides on different crops.
- These species are biochemically pre-adapted to evolve insecticide resistance. The herbivorous species are polyphagous and evolved to deal with a variety of plant defensive chemicals, particularly alkaloids (Colorado potato beetle and cotton pests) and, therefore, had mechanisms available to detoxify and excrete novel toxins.
- Several of these species are capable of asexual reproduction (aphids, whiteflies and mites), which can speed the rate of adaptation to insecticides.
- These species are typified by high rates of dispersal, with the adults being highly mobile and/or human activities contributing to their long distance movement, *e.g.* whitefly and diamond-back moth may be moved on host plants that have been grown in one area to be sold in another and many of the pest species may be found inside ships or aeroplanes.
- Many of the above biological characteristics are common to related species which helps to explain why multiple species within particular arthropod families appear to have a strong tendency to evolve insecticide resistance.

Resistance to organochlorine pesticides is most common and comprises approximately 66 per cent of the top 10 insecticides/acaricides or formulations that have a arthropod species resistant to them. Organophosphorous products make up an additional 29 per cent and the carbamates are 5 per cent of this total.

Develoment of resistance has become more prevalent during the last 50 years, but the number of new resistant species recorded recently is less than in previous decades. This is due to many recent reports adding to the resistance spectrum in species already reported as resistant to other compounds. This phenomenon has also appeared in 100 species of plant pathogens and 55 species of weeds, as well as in some nematodes and rodents. In India, 14 insect pests of public health and household importance, 7 insect pests of agricultural crops and 6 stored grain insects have developed resistance to insecticides.

In Asian countries, the first report of development of pesticide resistance was reported from India in 1963, when Singhara beetle, *Galerucella birmanica* (Jacoby) was found resistant to DDT and HCH. Since then, 14 other pests have been demonstrated to become resistant to different insecticides in one or more countries. Widespread occurrence of resistance in this pest has also been reported from Indonesia and Thailand. Similarly, another polyphagous pest, *Spodoptera litura* (Fabricius) has become resistant to nearly all the available groups of pesticides in India. In *Plutella xylostella* (Linnaeus) high levels of resistance to quinalphos (>600 fold), fenvalerate (2700 fold), cypermethin (2880 fold) and other insecticides have been reported. Brown planthopper of rice has been reported to have developed resistance to one or more pesticides in Fiji, Malaysia, the Philippines, Sri Lanka and Vietnam. Moderate to high levels of resistance to most of the insecticides have been reported in *Bemisia tabaci* (Gennadius) in Pakistan and India. In case of other pests, the problem is not yet widespread and can be tackled by undertaking timely remedial measures.

The problem of pesticide resistance can be contained by following one or the combination of the following methods: (*i*) Pesticides should be used only if their use is essential and is based on monitoring the pest population in the field. (*ii*) Increase the dose applied so that even potentially resistant genotypes are killed. (*iii*) Use of synergist which will enhance the toxicity of a given pesticide by inhibiting the detoxification mechanism. (*iv*) Alteration of pesticides with unrelated mode of action. (*v*) Incorporation of non-insecticidal strategies in an integrated pest management approach.

Pest Resurgence

Resurgence refers to an abnormal increase in pest population or damage following insecticide application often far exceeding the economic injury level. Pest resurgence may broadly be classified into two categories, *i.e.* primary pest resurgence and secondary pest resurgence (replacement).

Primary pest resurgence. Primary pest resurgence occurs when the target pest population responds to a pesticide treatment by increasing to a level atleast as high or higher than in an untreated control or higher than the population level observed before the treatment. The resurgence may occur after the first application or after several applications of the pesticide. Pest population outbreaks can be caused by many factors, but pest resurgence occurs after a treatment of the crop with a chemical, targeted at the pest population that is intended and expected to control the targeted pest.

Secondary pest resurgence. Secondary pest resurgence refers to the replacement of a primary pest with a secondary pest or a secondary pest outbreak occurs when a non-target, but injurious pest population increases in a crop, after it is treated with a pesticide to control a primary pest population. The increase is an unintended and unexpected consequence of the pesticide treatment. For example, pesticide sprays to control the codling moth, apple maggot and plum curculio on apple lead to resurgence of populations of white apple leafhopper, spotted tentiform leafminer, and European red mite. Season long sulphur sprays for control of powdery mildew on grapes often lead to resurgence of the red spider mite, *Tetranychus pacificus* McGregor.

Resurgence of insect pests following application of insecticides has been known for a long time. As early as 1956, more than 50 species of insect pests and mites whose populations showed resurgence after insecticidal treatments with diverse chemicals were known. Maximum cases of resurgence belong to Homoptera (44%) followed by Lepidoptera (24%) and phytophagous mites (26%). It is interesting to note that homopteran insects are protected from contact insecticides by a waxy covering and many of the Lepidoptera exhibiting resurgence are borers and leafminers which also escape direct contact with the insecticides. Interestingly, Homoptera (66%) and Lepidoptera (18%) head the list for classical biological control successes associated with each insect order. It thus appears that resurgence is associated with pests which have effective natural enemies and which are also less affected by contact pesticides than their natural enemies.

Maximum reports of resurgence pertain to planthoppers especially brown planthopper (BPH) of rice. BPH was a minor pest of rice crop during 1950s in India but extensive use of insecticides has elevated it to the status of a major pest not only in India but whole of Asia. Reports of insecticide-induced resurgence in BPH have been received from Bangladesh, India, Indonesia, and Solomon Islands. As many as 27 insecticides have been reported to cause resurgence in BPH. The factors implicated in BPH resurgence include reduction in duration of nymphal stage, longer oviposition period, shortened life cycle, enhanced reproductive rate, higher feeding rate and destruction of natural enemies especially *Cyrtorhinus lividipennis* Reuter.

The use of synthetic pyrethroids on cotton during the last two decades has resulted in increasing incidence of serveral sucking pests including whitefly, aphid, red mite and mealybugs. Over-reliance on SPs for the control of bollworms has been the major factor responsible for whitefly outbreaks in Andhra Pradesh, Gujarat, Tamil Nadu and other parts of the country. SPs have also been reported to cause resurgence of mustard aphid infesting Indian mustard. It was found that the application of SPs to the host plants increased the concentration of glucose and some amino acids especially arginine, lysine, isoleucine, leucine and cystine. These changes might be responsible for stimulation of aphid reproduction and increased weight of such aphids. On the other hand, endosulfan did not affect the quality of leaf sap of the host plants but still caused enhanced reproduction in the aphid. It was found that endosulfan reduced the excretion of some amino acids in the honeydew of the aphid indicating that better utilization of these amino acids was responsible for the increased fecundity of such aphids. It is, thus, clear that different insecticides may cause resurgence of the same insect by different mechanisms and even the same insecticides may produce variable effects at different sub-lethal concentrations.

The use of right type of insecticide, dosage, time and method of application can play a significant role in reducing the risk of insecticide resurgence. The agronomic practices like date of sowing, judicious use of fertilizer and irrigation water can help in reducing the insecticide-induced resurgence. The cultivation of healthy crop, conservation of natural enemies with the use of ecofriendly insecticides, proper surveillance and making farmers IPM experts can help in avoiding insecticide-induced resurgence.

Effect on Non-target Organisms

Repeated use of pesticides on cotton, fruits, tobacco and other crops has disruptive effects on beneficial insects like pollinators, biocontrol agents, soil, wild and aquatic life. Many invertebrates take up pesticides from soil into their bodies and may concentrate pesticides several times greater in their tissues than those in the surrounding soil. The animals that feed upon these invertebrates may, in turn, concentrate these residues to levels that may kill them or affect their normal activities. Soil microrganisms which cause breakdown of cellulose, nitrification, turn-over of organic matter and other biological materials may also be adversly influenced by pesticides.

Indiscriminate use of insecticides on the field crops has resulted in widespread mortality of honey bees and wild bees which are essential for pollination. Studies conducted in Punjab have revealed that application of carbaryl, endosulfan, fluvalinate and monocrotophos for bollworm control caused 94, 74, 44 and 100 per cent mortality, respectively of the bees present in cotton fields. In *raya* crop, quinalphos, anthio and endosulfan caused 100, 98 and 25 per cent bee mortality, respectively by direct spray. A list of pesticides according to their toxicity to honey bees is given in **Table 7.5.**

Table 7.5 Classification of pesticides on the basis of their toxicity to honey bees

Highly toxic (LD_{50} 0.001 – 1.99 μg/bee)	
Aldrin	Dimethoate
Clacium/lead arsenate	Fenvalerate
Carbaryl	Monocrotophos
Carbofuran	Oxydemeton-methyl
Carbophenothion	Parathion
Chlorpyriphos	Permethrin
Cypermethrin	Phorate
Deltamethrin	Phosphamidon
Dichlorvos	Quinalphos
Dicrotophos	Thiometon
Moderately toxic LD_{50} 2.0 – 10.0 μg/bee)	
Insecticides	
DDT	Heptachlor
Diazinon	Hinosan
Dieldrin	Lindane
Endrin	Malathion
Ethyl parathion	Metasystox
Fenitrothion	Methyl demeton
Fenthion	Methyl parathion
Formothion	Mevinphos
HCH	Trichlorphon
Fungicides	
Bavistin	Difolitan
Carbendazin	Foltaf
Dithane	Hexcacap

Relatively non-toxic (LD_{50} > 11.0 μg/bee)	
Insecticides	
Bacillus thuringiensis	Methoxychlor
Chlorobenzilate	Morestan
Dicofol	Nicotine
Dimite	Nuclear polyhedrosis virus
Endosulfan	Phosalone
Ethion	Pyrethrum
Menazon	Sabadilla
Fungicides	
Anilazine	Dinocap
Benomyl	Dodine
Bordeaux mixture	Folcid
Catafol	Polyram
Captan	Sulphur
Cuprous oxide	Thiram
	Ziram
Herbicides, defoliants and desiccants	
Alachlor	Cynazine
Amitrole	Diuron
Ammate	Methazole
Atrazine	Nitrofen
Bromocil	Oil sprays

Source : Modified from Abrol (1997)

Pesticide Residues

Pesticide residue is any substance or a mixture of substances in or on any substrate resulting from the use of pesticides. It has been demonstrated that less than one per cent of the pesticide applied to a crop reaches the target pests and the remaining quantity gets into different components of the environment. Since most of the chlorinated pesticides are non-biodegradable, they leave excessive residues in various food commodities. The presence of residues of these pesticides in food commodities and other components of the environment is a matter of serious concern.

Monitoring surveys in different parts of India have revealed widespread pesticidal contamination of all types of food materials including cereals, pulses, vegetables, fruits, animal products, vegetable oils, spices, honey, milk and milk products (**Fig. 7.9**). Twenty per cent of the market samples of non-fatty food commodities have been found to contain residues above the legal maximum residue limits (MRLs). A broader picture of the magnitude of contamination of milk has been provided by the analysis of 458 samples collected from different parts of the country under the All India

Coordinated Research Project (AICRP) on Pesticide Residues. About 87 per cent samples were contaminated with DDT, with 43 per cent above the legal limit of 0.05 mg/kg. Similarly, 90 per cent of samples were contaminated with HCH, 78 per cent of which had total HCH above the legal limit of 0.1 mg/kg. In contrast, only 1-2 per cent of the samples of food commodities have been found to be contaminated with pesticide residues above MRL at the global level. Even human milk samples have been found to be contaminated with high levels of DDT and HCH.

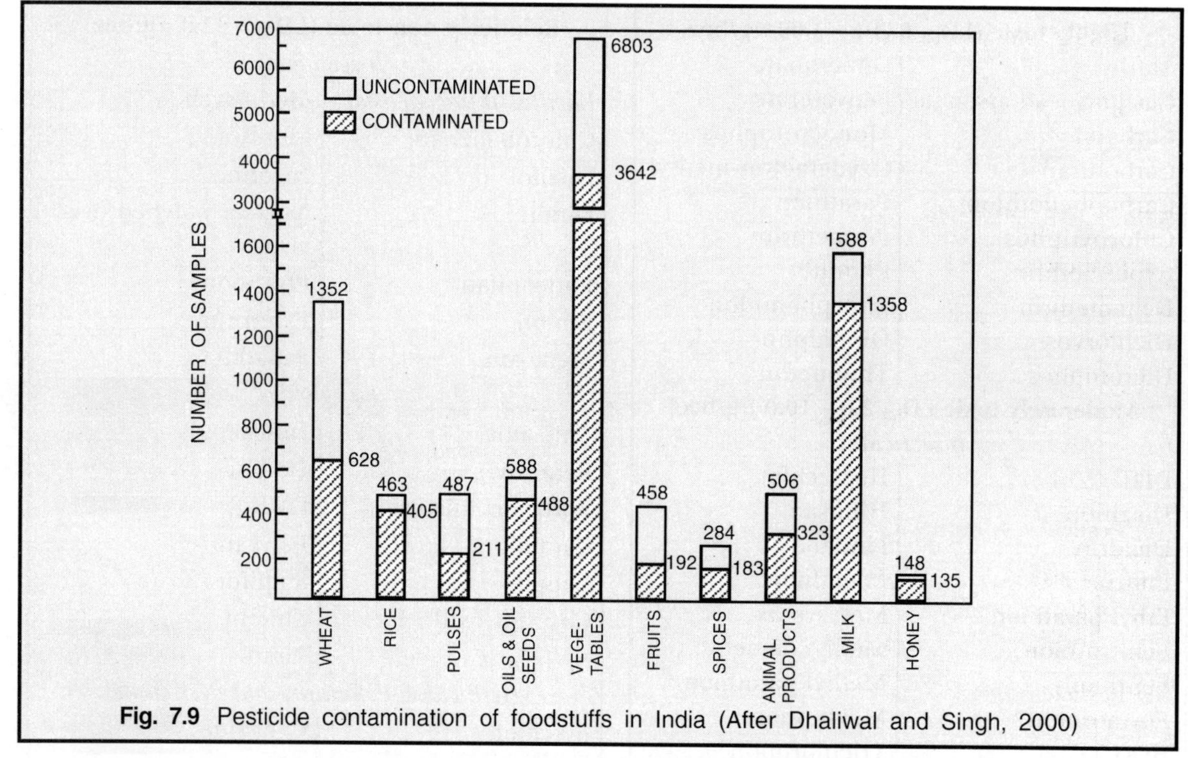

Fig. 7.9 Pesticide contamination of foodstuffs in India (After Dhaliwal and Singh, 2000)

Recent surveys conducted in Punjab reveal significant reduction in the residues of these insecticides in food commodities. However, lindane residues were found to be present in 53 per cent of milk above the legal limit. The presence of commonly used organophosphorus insecticides in 71 per cent samples of vegetables and 85 per cent of fruits is a matter of concern.

The population of developing countries is known to carry heavy burden of pesticides in their bodies. The principal source of these residues is believed to be the diet which contains significant quantities of the persistent pesticides. It has been estimated that the estimated dietary intake of DDT and HCH exceeds the acceptable daily intake (ADI) in India. In addition, pesticide residues in agricultural commodities have a significant influence in the area of international trade. Commodities bearing pesticide residues above the legally permitted levels may be rejected by the importing country, thereby causing appreciable economic loss to the exporting country.

The Government of India has withdrawn the use of DDT in agriculture in 1989 **(Table 7.6).** In very special circumstances warranting the use of DDT in plant protection work, the State or Central Government may purchase it directly from M/s Hindustan Insecticides Ltd, to be used under expert Government supervision. The use of DDT for the public health programme has been restricted to 10,000 metric tonnes, except in case of any major outbreak of epidemic. The manufacture and use of HCH stands totally banned with effect from Ist April, 1997. With this, the pesticide residue scenario is likely to change if this recommendation is strictly adhered to in the country.

The levels of pesticide residues can be minimized if the following precautions are followed :

- Pesticides should be used only when it is absolutely essential. The non-chemical methods of pest control should be encouraged.
- Only recommended pesticides should be applied at the right time and at the prescribed dosages.
- Preference should be given to the use of less persistent pesticides. Never use the banned (**Table 7.6**)
- Restricted pesticides should be used only under specific conditions (**Table 7.7**).

Table 7.6 List of pesticides not approved, withdrawn or banned in India

S.No.	Pesticide	S.No.	Pesticide	S.No.	Pesticide
(*a*) Pesticides refused registration					
1.	Ammonium sulphamate	2.	Azinphos ethyl	3.	Azinphos methyl
4.	Binapacryl	5.	Calcium arsonate	6.	Carbophenothion
7.	Chinomethionate	8.	Dicrotophos	9.	EPM
10.	Fentin acetate	11.	Fentin hydroxide	12.	Lead arsenate
13.	Leptophos	14.	Mephosfolan	15.	Mevinphos
16.	2, 4, 5-T	17.	Thiodemeton/Disulfoton	18.	Vamidothion
(*b*) Pesticides withdrawn					
1.	Dalapon	2.	Ferbam	3.	Formothion
4.	Nickel chloride	5.	Pentachlorobenzene (PCB)	6.	Simazine
7.	Warfarin				
(*c*) Pesticides banned for manufacture, import and use					
1.	Aldicarb	2.	Aldrin	3.	Calcium cyanide
4.	Chlordane	5.	Chlorobenzilate	6.	Chlorofenvinphos
7.	Copper acetoarsenite	8.	Dibromochloro-propane (DBCP)	9.	Dieldrin
10.	Endrin	11.	Ethyl mercury chloride	12.	Ethyl parathion
13.	Ethylene dibromide	14.	Heptachlor	15.	Hexachlorocyclohexane (HCH)
16.	Lindane (Gamma-HCH)	17.	Maleic hydrazide	18.	Menazon
19.	Metoxuron	20.	Nitrofen	21.	Paraquat dimethyl sulphate
22.	Pentachloronitro-benzene (PCNB)	23.	Pentachlorophenol (PCP)	24.	Phenyl mercury acetate (PMA)
25.	Sodium methane arsonate (MSMA)	26.	Tetradifon	27.	Toxaphene
28.	Trichloro acetic acid (TCA)				
(*d*) Festicide formulations banned for import, manufacture and use					
1.	Carbofuran 50% SP	2.	Methomyl 12.5% L	3.	Methomyl 24% formulation
4.	Phosphamidon 85% SL.				

Source : www.cibrc.nic.in

Table 7.7 List of pesticides restricted for use in India

S.No.	Pesticide	Restrictions
1.	Aluminium phosphide	Pest control operations to be carried out by Govt./Govt organizations/undertakings/pest control operators under strict supervision of experts.
2.	Captafol	Use as foliar spray banned, to be used only as seed dresser.
3.	Cypermethrin	Cypermethrin 3% smoke generator to be used only through pest control operators, not allowed to be used by general public.
4.	Dazomet	Use not permitted on tea.
5.	DDT	Use for domestic public health programme is restricted upto 10,000 metric tonnes per annum, except in case of any major outbreak of epidemic.
6.	Diazinon	Banned for use in agriculture except for household use.
7.	Fenitrothion	Use banned in agriculture except for locust control in scheduled desert area and public health.
8.	Fenthion	Use banned in agriculture except for locust control, household and public health.
9.	Methoxy ethyl mercuric chloride (MEMC)	Use banned completely except for seed treatment of potato and sugarcane.
10.	Methyl bromide	May be used only by Govt./Govt. organizations/undertakings/ pest control operators under the strict supervision of experts.
11.	Methyl parathion	Use permitted only on those crops approved by Registration Committee where honey bees are not acting as pollinators.
12.	Monocrotophos	Banned for use on vegetables.
13.	Sodium cyanide	Use restricted for fumigation of cotton bales under expert supervision.

Source : www.cibre.in.

- Ripe fruits and vegetables should be plucked before pesticide application. After pesticide use, the crop should be harvested only after the recommended waiting period.
- Pesticide residues on the produce can also be reduced by washing alongwith rubbing and peeling of vegetables.

Pesticide Poisoning

Human pesticide poisoning is a major concern in producing and using pesticides in agricultural and public health programmes. The World Health Organization has estimated that more than 3 million occupational or accidental poisoning cases occur in the world annually with about 2,20,000 deaths. More than half of these poisoning cases and three-fourth of the documented deaths take place in the third world, even though these consume only 15 per cent of the total world pesticide output. India accounts for one third of the total pesticide poisoning cases in the world.

In the first major accident involving pesticides in India, more than 100 people died in Kerala in 1958 due to consumption of wheat flour and sugar contaminated with parathion leakage during shipment from Bombay to Cochin. More than 3000 people died due to inhaling of vapours of methyl isocyanate, leaked from a carbaryl manufacturing plant in Bhopal in 1984. More than 30,000 people were disabled to varying degrees. The surviving population is still expressing teratogenic, mutagenic, carcinogenic and other effects involving vital body organs. Cases of blindness, cancer, diseases of liver and nervous system from pesticide poisoning have been identified in cotton growing areas of Maharashtra and Andhra Pradesh.

In case the poison is swallowed accidently, the victim should be taken immediately to the nearest physician or preferably to a hospital. It is most important that the insecticide container from which the poision was swallowed should be shown to the physician. The medical treatment for each pesticide is given below :

DDT and HCH. (*i*) Give universal antidote (activated charcoal 2 parts, magnesium oxide 1 part, tannic acid 1 part ; 15 g of the mixture in half a glass of warm water) to absorb or neutralize poisons, to be followed by gastric lavage to evacuate stomach contents. If patient is vomiting do not give an emetic but give large amounts of warm water. If an emetic is needed give 15 g sodium chloride in warm water and repeat until vomit fluid is clear. (*ii*) Give magnesium sulphate 30 g in water as a cathartic and force fluid. (*iii*) Give caffeine sodium benzoate 0.5 g subcutaneously or intravenously or give hot tea or coffee drink. (*iv*) Give calcium gluconate 10 ml of 10% solution intravenously for incoordination and tremors or pentobarbital sodium 0.1 g intravenously if necessary. (*v*) To prevent any liver damage use diet having high carbohydrate and calcium.

Cyclodienes. (*i*) If ingested, give universal antidote. (*ii*) Give a saline purgative but not oil laxative. (*iii*) Give phenobarbital up to 0.7 g or pentobarbital 0.25-0.50 g per day to induce sedation and control convulsions. (*iv*) Give calcium gluconate 10% intrave-nously. (*v*) Oxygen therapy and artificial respiration may be needed during depression.

Organophosphates. (*i*) Atropinize the patient immediately and repeat doses of 2-4 mg at 5-10 minutes intervals till the continuance of symptoms. As much as 50 mg of atropine may be given per day without any side effects. (*ii*) Administer 1g of 2-pyridine–2 aldoxine–N-methyliodide (2 P.A.M.) intravenously. This should be done very slowly in 5-10 minutes. (*iii*) Give gastric lavage with 5 per cent sodium bicarbonate. (*iv*) Wash contaminated skin and irrigate eyes with normal saline. (*v*) Maintain electrolyte balance but avoid large amounts of intravenous fluids. (*vi*) Apply artificial respiration if necessary to remove upper respiratory obstruction. (*vii*) Do not give morphine, theophyline or aminophy-line.

Zinc phosphide. (*i*) Give gastric larvage with 1 : 5000 potassium permanganate solution. (*ii*) Give 0.02% cupric sulphate. This may act as an emetic and produce insoluble cupric phosphide which should be removed by lavage. (*iii*) Use morphine for sedation and salivation. (*iv*) Give large doses of vitamin K.

Mercurial fungicides. (*i*) Give universal antidote as for DDT followed by gastric lavage. (*ii*) Give high colonic irrigation with sodium formaldehyde sulphoxylate solution allowing small amounts to remain inside. (*iii*) Inject 100–200 ml of freshly prepared 5-10% sodium formaldehyde sulfoxylate solution intravenously. For later treatment give sodium citrate 1-4 g every 4 hours by mouth. (*iv*) Give calcium gulconate 10 ml of 10% solution intramuscularly or intravenously for muscle spasm.

Strychnine hydrochloride. (*i*) Give universal antidote followed by gastric lavage with 240 ml of potassium permanganate 1 : 1000 solution leaving one fourth in the stomach. (*ii*) Give pentobarbital sodium 0.1 g intravenously. (*iii*) Do not use morphine or its derivatives.

Metaldehyde. (*i*) Keep the patient lying and warm. (*ii*) Give sodium chloride 15 g in a glass of warm water, repeat until vomit fluid is clear and give 30 g of magnesium sulphate. (*iii*) Give strong tea or coffee or one tablespoonful of aromatic spirit of ammonia in water.

Methyl bromide. (*i*) Correct the metabolic acidosia. (*ii*) Give artificial respiration and guard against respiratory infection. (*iii*) Inject caffeine sodium benzoate 0.5 g and if necessary epinephrine 0.5 ml 1 : 1000. (*iv*) Keep the mouth rinsed with 10% solution of sodium thiosulphate. (*v*) When revived give hot tea or coffee.

Aluminium phosphide. (*i*) Take the patient immediately into air and let him lie down in a comfortable position. (*ii*) Apply artificial respiration, if necessary. (*iii*) Inject glucose solution.

Ethylene dibromide. (*i*) Give oxygen therapy, respiratory stimulants such as nikethamide or caffeine. (*ii*) After acute nervous symptoms subside, give adrenergic blocking agents like dibenzyl betachloroethyl amine hydrochloride to prevent damage to liver and kidneys. (*iii*) If the poison has been swallowed, give gastric lavage and saline purgatives.

POTENTIAL AND CONSTRAINTS

Pesticides have played a major role in raising world's food production. Availability of pesticides made cultivation of crops profitable in areas, where this was previously not possible because of pest problems. Pesticide use almost universally brought about increased crop production by minimising pest damage and thereby, encouraging farmers to adopt better agronomic practices and obtain high yields. The development of high yielding varieties of crops which ushered in the green revolution was possible only under the protection provided by the pesticide umbrella. In spite of their glorious past and great potential in future, pesticides face mounting public pressure due to increasing environmental problems caused by their use. Despite the various environmental problems caused by their use, pesticides continue to be the single most widely used method of insect pest control. Pesticides will continue to play an important part in most IPM programmes in the foreseeable future in view of a number of advantages over alternative means of pest control.

- In view of their persistent action, most of the pesticides provide a long lasting pest control.
- Pesticides provide control of some pests where no other effective tactics were previously available.
- When an increasing insect population approaching ETL is observed in the field, pesticides are often the only means of preventing economic damage.
- Pesticides are easily available across the counter in convenient and ready to use packings.
- Pesticides often provide rapid remedial action against the targeted pest and allow rapid control of an existing pest problem.
- Pesticides are easy to apply and large areas can be covered in a relatively short time.
- Even if there is a complex of several insect pests causing economic losses, a single insecticide or a combination of two insecticides in a single application may control the whole pest complex.
- When properly used under appropriate conditions, pesticides provide a relatively predictable level of control. There is often a greater level of uncertainty associated with the use of other tactics.
- Use of insecticides protects the crops from the ravages of insect pests and thus provides stability in yield as well as in farmers' income.
- The use of insecticides is compatible with other components of IPM under intensive agriculture and modern farming conditions.
- Pesticide production, quality, handling, storage, transportation, safe use, consumer safety, etc. are governed by legal enactments in each country as well as globally.

Many problems arise with pesticide use and misue. The magnitude and importance of these problems vary considerably among pesticide categories. As with the advantages, some problems are real, but others are only perceived.

- The repeated usage of a single pesticide may lead to the selection of pests that are resistant to the pesticide. This is an enormous problem for use of all types of pesticides, but has been particularly important with insecticides, herbicides, fungicides, and bactericides (antibiotics).

- When the pesticide (usually an insecticide) kills the targeted pest, but also kills beneficial insects, the pest population often will increase to a level, higher than the level preceding the application. The phenomenon is called pest resurgence. When the survivors of the targeted population begin to reproduce, their number grows exponentially because the beneficials that once limited population growth are no longer present.
- When the pesticide kills the key pest, but not a minor (secondary) pest, the minor pest population may increase and become important. This is a problem for both insecticide and herbicide use.
- Pesticides may harm useful organisms, such as honey bees, or beneficial insects that are critical in the biologial control of pest populations. Some pesticides are toxic to wildlife.
- Pesticides may move from the place where they were applied, resulting in contamination of surface water or groundwater, and pesticide accumulation in the food chain.
- Pesticide residues remain in the soil, and on or in harvested produce after the application of a pesticide. Residues may be of particular concern if pesticides are applied incorrectly.
- Due to their toxicity, pesticides have the potential to cause illness in farm workers, especially for those working in hand-harvested fresh market crops.
- Although low cost may be a reason for the use of pesticides, in other situations pesticides are an expensive choice; this is especially true for management of certain insects where biological control is a feasible alternative.
- Misuse of pesticides may lead to more frequent applications and require higher rates of the product needed to control the same pest. This phenomenon has been called the pesticide treadmill and it is the result of severe ecosystem disruption.

The development of modern synthetic pesticides has had an enormous impact on the lives of people all around the globe. Pesticides are likely to remain an integral and necessary component of pest management systems in the foreseeable future, also particularly in the annual cropping systems with multiple pests. Furthermore, the use of pesticides has made possible the profitable production of many commodities in the presence of potentially devastating pests. However, it is critical that the use of pesticides be viewed as a single component of an ecologically and biologically based system in order to conserve and prolong the utility of the declining pool of pesticides. New pesticides with more specific modes of action and reduced spectrums of activity may be more suited for large scale agricultural systems utilising a vast arsenal of chemical control options. The efficacy of existing pesticides can also be enhanced by developing more efficient formulations.

There is a need to regulate the use of pesticides in developing countries. A number of developing countries promote pesticide reduction through official policies or regulation. China introduced the Green Certificate programme and banned highly toxic pesticides from vegetable crops. Biological control is a national priority for Cuba; the new policy is intended to make integrated pest management (IPM) biointensive, with 80 per cent of pests managed through biological control. A National IPM Committee in Malaysia, is reducing pesticide use and increasing farmer knowledge of other pest management techniques. Iran constituted the High Council of Policy and Planning for Reduction of Agricultural Pesticides. India, Indonesia, Nepal, Philippines and Sri Lanka have adopted national IPM policies. The Food and Agriculture Organization and the World Health Organization are collaborating with the international comunity and developing nations to formulate pesticide policy. All these efforts would go a long way in regulating the use of pesticides in sustainable IPM programmes.

BIORATIONAL APPROACHES

INTRODUCTION

In view of the adverse effects of synthetic organic pesticides on non-target organisms and the environment, efforts have been made during the last three decades to develop safer and more selective pesticides. Accordingly, many conventional pesticides have been replaced by low-risk or 'biorational pesticides' which have been defined as those chemicals that affect the growth, development, biology and ecology of pest species, differently from that of beneficial species. In other words, 'biorational or 'reduced risk' pesticides are synthetic or natural compounds, that effectively control insect pests, but have low toxicity to non-target organisms (such as humans, animals and natural enemies and the environment. Most of the biorational pesticides are preferable to the conventional pesticides because of their specificity to the targeted pests, their effectiveness at lower rates and their non-persistent characteristics in the environment.

Insects and vertebrates differ with respect to structure of the integument, endocrine and chemical communication systems. Insect behaviour, in contrast to a vertebrate, is much more dependent on chemical cues than any other sensory attribute (visual, hearing, touch, etc.), and chemicals more often than physical cues (visual, mechanical) have been identified with successful host plant finding and acceptance. Many efforts have been made to manipulate these differences for development of selective pest control approaches. These approaches work through blockage, disruption or inhibition of any of the events from biosynthesis, storage, release, transport and reception to degradation which result in profound behavioural or physiological disturbances which may ultimately prove lethal to insects. Since a wide spectrum of chemicals and mechanisms are involved in each of these systems, the possibilities of disruption of processes involved are very wide.

SEMIOCHEMICALS

Semiochemicals are chemicals that are able to modify the behaviour of a perceiving organism at sub-micro/nanogram levels. The term is derived from the Greek word *simeone*, meaning a mark or a signal. Semiochemicals are divided into intraspecific and interspecific communication chemicals (**Fig. 8.1**). Such behaviour-modifying chemicals (BMCs) include pheromones, kairomones, allomones, attractants, deterrents and other para-pheromones. Sometimes, a single chemical secreted may have both intra- and inter-species activities. Semiochemicals are becoming increasingly valuable in insect pest management and offer enormous potential for controlling many insect pests through a range

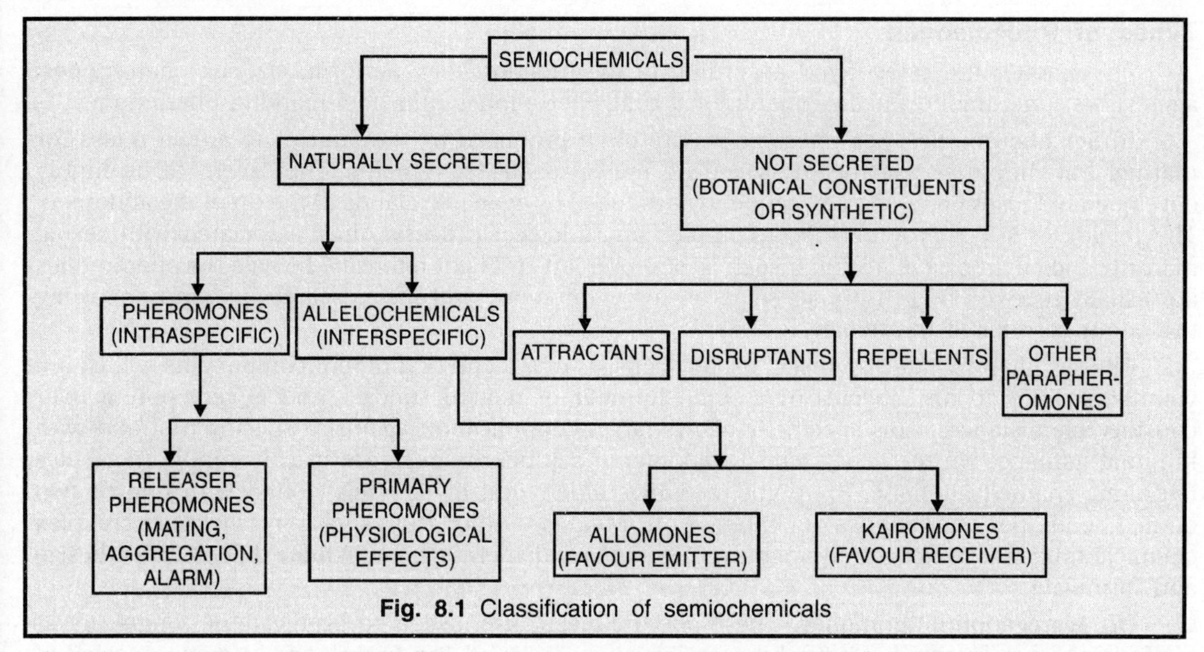

Fig. 8.1 Classification of semiochemicals

of environmentally and ecologically sound pest management technologies. The chemicals used for intraspecific communication are called *pheromones*. The term is derived from the Greek *pherein*, meaning to transfer and *hormone*, meaning to excite. Thus, a pheromone can be defined as a chemical or a mixture of chemicals that is released to the exterior by an organism and that causes one or more specific reactions in a receiving organism of the same species. When a chemical not found in an insect has a pheromone-like action, it is often referred to as a *parapheromone*.

Chemicals involved in interspecific communication are known as *allelochemicals* or *allelochemics*. These are non-nutrient substances originating from an organism which affect the behaviour, physiological condition or ecological welfare of organisms of a different species. These chemicals are further divided into different categories :

Allomone. A compound released by one organism which evokes a reaction in an individual of a different species that is favourable to the emitter but not to the receiver.

Kairomone. A compound released by one organism which evokes a response beneficial to a member of another species but not to the emitter.

Synomone. A substance released by organism which benefits both the sender and the receiver.

Antimone. A substance produced or acquired by an organism that when it contacts an individual of another species in the natural context, evokes in the receiver a behavioural or physiological reaction that is maladaptive to both the emitter and the receiver.

Apneumone A chemical released by non-living substances that is beneficial to the receiver but detrimental to other organism in the substance.

I. PHEROMONES

A dazzling variety of behaviour, from sexual attraction and dispersal to caste determination, is controlled by pheromones. It is thus natural that pheromones have attracted the attention of applied entomologists for exploitation in the control of insect pests. Pheromones may be divided into two categories, viz. *releasers,* which induce an immediate behavioural change, and *primers,* which initiate changes in development, such as sexual maturation, and so do not result in immediate behavioural changes, but predispose to them.

Types of Pheromones

Pheromones are categorized according to the function they perform, viz. sex pheromones, aggregation pheromones, alarm pheromons, trail pheromones and host-marking pheromones.

(i) **Sex pheromones.** Sex pheromones are often produced by the females to attract males for mating, but they may also be produced by males to attract females. They seem to be highly developed in Lepidoptera and are frequently produced by eversible glands at the tip of the abdomen. The release of sex pheromones is a complex physiological process, often associated with sexual maturity and environmental stimuli such as photoperiod and light intensity. Female sex pheromones are usually received by sensory sensillae on male antennae, and males search up wind, following the odour corridor of the female.

The sex pheromone of an insect usually consists of a blend of different components, which are volatile, specific to one species or a small number of related species, and is very potent over considerable distances. This specificity allows targeted application against a specific pest and with minimal influence on the ecosystem. Lepidopteran sex pheromones are usually simple molecules (e.g. long-chained aliphatic, lipophilic, acetates, aldehydes, or alcohols), often with one or two double bonds. Sex pheromones of Diptera, Coleoptera and other groups have usually more complex chemical structures, which are comparatively unstable and, therefore, much more diffcult to synthesize and formulate.

(ii) **Aggregation pheromones.** These pheromones cause insects to aggregate or congregate at food sites, reproductive habitats, hibernation sites, etc. and are prominent in some species of beetles. They are particularly well understood in bark beetles, *Ips* spp. and *Dendroctonus* spp., which are involved in tree attacks. They are attractive to both sexes and tend to operate over a long range and have the potential of attracting thousands of individuals. Generally, aggregation pheromones have more complex chemical structures, not very stable or amenable to synthesis and deployment, and elicit a much more complex behaviour that is less open to manipulation.

(iii) **Alarm pheromones.** Alarm pheromones are common in social insects such as ants and bees, and aphids, which are usually found to occur in aggregation. The function of this type of pheromone is to raise alert in conspecifics, to raise a defense response, and/or to initiate avoidance. For instance, the remains of the sting apparatus of a honey bee left in the victim's body releases an alarm pheromone that attracts other bees and stimulates them to sting. A dose response of attractancy and repellency has been demonstrated for several pure volatiles from the venom of common wasp, *Vespa vulgaris* Linnaeus and the German wasp, *V. germanica* Fabricius. The alarm pheromones are usually highly volatile (low molecular weight) compounds such as hexanal, 1-hexanol, sesquiterpenes [*e.g.* (E)-β-farnesene for aphids], spiroacetals or ketones.

(E)-β-farnesene is naturally occurring chemical found in about 400 plant species. It is synthesized from its precursor, farnesyl pyrophosphate by the action of the enzyme, (E)-β-farnesene synthase. (E)-β-farnesene repels the aphids, but attracts their natural enemies like the ladybird beetles and the parasitic wasps. Recently, the scientists at the Rothamsted Research in Harpenden, Hertfordshire (UK), have developed genetically modified (GM) wheat by inserting the peppermint (*Mentha piperata* Linnaeus) into the DNA of a spring wheat strain Cadenza. The GM wheat would go in a long way to protect the crop against the grain aphid, *Sitobion avenae* (Fabricius) ; bird cherry-oat aphid, *Rhopalosiphum maidis* (Fitch) ; and rose grain aphid, *Metopolophium dirhodium* (Walker), which are three main aphid pests of wheat in U.K.

The applications of (E)-β-farnesene alongwith host plant volatiles caused the aphids to move away from feeding sites and may be utilized to prevent colonization of the host plants by the aphids. It also increased the mobility of aphids, which is useful for increasing the pick up of contact insecticides or microbial pathogens. (E)-β-farnesene is now commercially available for use against a number of aphid species including *Aphis gossypii* (Glover) and *Lipaphis erysimi* (Kaltenbach). Farnesol and nerolidol are the alarm pheromones from the two-spotted mite, *Tetranychus urticae*

Koch, and are released under natural conditions when the population is threatened or is being attacked by a mite predator. The result is an increase in activity of mites with consequent greater exposure to a co-applied miticide.

(*iv*) **Trail pheromones.** Trail pheromones are produced by foraging ants, termites and larvae of some lepidopteran insects. They are essentially used to indicate sources of requisites to other members of the colony. While trail pheromones are frequently associated with walking insects such as ants, they are also known from other insects. Bees use trail pheromones during foraging for making attractive foraging sites as well as for scent marking of unproductive food sources. Trail pheromones are characteristically less volatile than alarm pheromones. The trails are continuously replenished through traffic, otherwise they dissipate. Identification and synthesis of trail pheromone of bumble bees could lead to increased efficiency in their use for pollination. It is also possible to manipulate trail following and recruitment of tent caterpillars, *Malacosoma americanum* (Fabricius).

(*v*) **Host-marking pheromones.** Host marking, spacing or epidietic pheromones elicit dispersal away from potentially crowded food sources, thereby reducing numbers. These pheromones reduce intraspecific competition by disrupting landing, feeding or oviposition of pests on their host plants. They are thus one of the few pheromones that serve to repel rather than attract the insects. They are known from a number of insect orders, viz. Coleoptera, Lepidoptera, Diptera, Homoptera, Orthoptera and Hymenoptera. The females of apple maggot, *Rhagoletis pomonella* (Walsh) and cherry fruit fly, *R. cerasi* (Linnaeus), ovipositing in fruit, mark the surface to deter other females. Mating deterrent pheromones are known from a number of insects including tsetse flies, house flies and other Diptera. There is also exploitation of prey host marking and sex pheromones by parasitoids, which use signal persistence of these intraspecific cues to find their hosts. The host-marking pheromones of the bark beetles, *Dendroctonus* spp., such as verbenone for *D. frontalis* Zimmermann and 3,2-methylcyclohexane (MCH) for *D. pseudotsugae* Hopkins, are well known. The beetles that have successfully colonized a host release these pheromones. These compounds, in combination with aggregation pheromones, tend to keep beetle densities within an optimal range (6-8 per 1000 cm^2 of bark surface).

(*vi*) **Recruitment pheromones.** Recruitment pheromones are common in social insects, which are used to maintain order, and recruit members and coordinate the activities of the group. One example is a pheromone called the Nasonov pheromone in honey bees which is released by worker bees to orient returning forager bees back to the colony, as well as to recruit other workers outside the hive. To broadcast this scent, bees raise their abdomens, which contain the Nasonov glands, and fan their wings vigorously. These pheromones are also produced by bark beetles when they find a suitable tree for colonization.

(*vii*) **Recognition pheromones.** The insects have sex and species recognition pheromones that act only at close range. In case of social insects like ants, bees and termites, these are used to distinguish colony members from non-colony members. These pheromones tend to be simple straight-or branched-chain hydrocarbons and are a blend of compounds. The termite egg recognition pheromone (TERP) has been one of the most important pheromones to be identified, which strongly evokes the egg-carrying and grooming behaviour of workers. In five species of males of carpenter bees, the secretion has been from the mandibular glands.

(*viii*) **Retinue pheromones.** These chemicals were reported to be released by queen honey bee and, therefore, termed as queen retinue pheromones. These invoke the retinue attraction which encourages workers to feed and groom the queen and acquire and distribute her pheromone messages to other workers throughout the colony. Nine different compounds have been identified in queen retinue pheromone.

Strategies for Exploitation of Pheromones

The discovery, isolation and chemical identification of sex pheromone (bombycol) of the silk worm, *Bombyx mori* (Linnaeus) in 1959, provided impetus for the exploitation of pheromones in pest management. Upto 1970s, pheromones and pheromone mimics for more than 200 insects were known and by the end of 1980s the number rose to over 2000. Some of the important sex pheromones with a potential in pest management are disparlure (gypsy moth), gossyplure (pink bollworm),grandlure (cotton grey weevil) and frontalin brevicomin (pine beetles).Besides pheromones, chemical attractants have been utilized for management of insect pests, viz. methyl eugenol (Oriental fruit fly), cue-lure (melon fly) and amlure (European chafer). These chemicals have been employed for disrupting the activities of insect pests in three ways :

A. Monitoring

Pheromone baited traps for monitoring pest populations provide a highly sensitive means of detecting the insect pests with many advantages over conventional methods such as light traps and scouting programmes. Pheromone traps can be used to detect both the presence as well as the density of pest species. Insect population can thus, be estimated and new areas of infestation detected at a very early stage. The trap catches may be used to forewarn regarding outbreaks of important pests. Monitoring of quarantine pests such as gypsy moth, *Lymantria dispar* (Linnaeus) and Mediterranean fruit fly, *Ceratitis capitata* (Wiedemann), is being successfully accomplished by use of pheromone traps.

Detection of pest. Pheromone-baited traps provide a relatively simple and reliable means of detecting the presence of insects. They are one of the primary tools employed in quarantine surveys, where the aim is to determine the presence of a species and prevent its establishment and spread. Traps are routinely deployed around airports and harbours to detect potential introductions of exotic pests at these high-risk sites. Similarly, large number of traps is used in regional survey programmes to determine the distribution of specific pests and provide the required information for preventing spread to new areas. The use of attractant-baited traps to demonstrate pest-free production and storage zones is assuming significance. The process often requires implementing a defined monitoring programme to allow for the export of various agricultural commodities to specific countries.

Measurement of pest density. Traps baited with sex pheromones are commonly used to monitor the population density of pests. Pheromone traps have also been used in area-wide studies of pest disruption and dispersal. Stored product pests have been successfully monitored from food processing plants and warehouses, using pheromone or food traps. Sex pheromones are used for moths, and beetles usually use aggregation pheromones and/or food baits. Quantitative relationships between adult captures and counts of larval stages or signs of larval feeding, such as feces or damage, have been found for pests of tree fruits, annual crops and forests.

Assessment of density of natural enemies. Sex pheromones have been employed to trap the insect biological control agents. Pheromone traps have also been used for monitoring the establishement of a biological control agent, *Cydia succedana* Denis & Schiffermuller (Tortricidae), introduced for the control of a weed, gorse, *Ulex europeaus* Linnaeus, in New Zealand. This new tool would permit monitoring of presence, phenology and relative abundance of the biocontrol agents and could give an indication to the growers whether the population in future might be high enough for successful control of the particular pest.

Assessment of pest phenology. The combination of pheromone traps and predictive phenology models can provide a reliable method for predicting the timing of flight activity or life stages. The first moth catch (referred to as 'biofix') has been used as a predictor of the beginning of adult emergence. The precision of this method has been well documented for codling moth, a key pest of pome fruits throughout the world. The phenology model for this pest is based on accumulating degree-days (base 10°C) beginning on the day the first moth is captured in a pheromone trap, provided moths are captured on two successive trapping dates. In 6 of the 10 years in the state of Washington, USA, this model predicted the start of egg hatch on the same day that it was observed in the field and there was never a discrepancy of more than two days between the predicted and observed event. Traps have generally been less effective in predicting peak emergence and the emergence of later generations probably because of trap saturation, pesticide use, female moth-trap competition, and the variability associated with the degree of trap and lure maintenance. This approach is likely to be particularly useful for biorationals such as IGRs, which require precise timing as they are primarily active against specific instars or life stages.

Assessment of effectiveness of mating disruption. An increasingly important use of attractant-baited traps is to measure the efficacy of mating disruption formulations. Capture of zero (complete shutdown) or very few moths in a pheromone-baited trap has been used to indicate successful disruption of the target pest. However, it is not uncommon to record low moth catches in traps and still have less than adequate pest control in pheromone-treated plots. In some cases, it is also possible to greatly inhibit catches in pheromone traps but still detect substantial numbers of females mating. Two approaches have been suggested to improve the utility of pheromone-baited traps as monitoring tools in pheromone-treated plots. Firstly, a lure with an emission rate closer to the natural rate would seem to be the most suitable for measuring the efficacy of a disruption treatment. The second approach is to use lures with very high release rates as a means following changes in adult population densities in spite of air permeation with pheromone (Gut *et al.*, 2004).

Monitoring of insecticide resistance. Pheromone-baited traps are one of the more widely adopted methods for monitoring insecticide resistance in several lepidopteran pests. The technique involves collecting large number of males in traps and testing for expression of resistance by topical application of insecticides or through incorporation of the insecticide through glue. The major advantage of pheromone-trap bioassays is the rapid collection and determination of resistance for large samples of the pest without incurring the costs in time and money associated with rearing of large numbers of insects. However, the technique may overestimate the impact of resistance in the field because only males are captured and assayed. Sex-related differences in tolerance to insecticides in insects are known, with females more susceptible than males. Moreover, this approach cannot be used to monitor for resistance to materials whose primary mode of action requires ingestion.

Decision support. Pheromone-based monitoring systems can be used to assess population trends and determine the need to treat. Thresholds of catch have been developed for a large number of insects and used as the basis for conventional pest management interventions. Basing management decisions on adult catches rather than taking a preventive or calender-based approach is a key step in many efforts to reduce the number of insecticide applications. The approach works on the principle that an intervention with a spray is required only if certain defined sampling threshold is exceeded. The decision may be based on a single weekly catch, consecutive catches or cumulative catches over an extended period, such as a generation. A threshold of apple leafroller, *Epiphyas postvittana* (Walker), pheromone trap catch has been determined from a correlation of catch with fruit damage at harvest. Thus catches greater than the threshold led to recommen-dation for application of a selective insecticide. Treatment thresholds for codling moth, *Cydia pomonella* (Linnaeus), based on moth captures in pheromone traps, have been developed for most pome fruit producing regions of the world. A threshold of 1-5 moths per trap per week has been established as the point at which pesticides need to be applied depending upon location. This need-based application has led to 50-75 per cent reduction in pesticide use.

B. Mass Trapping

Mass trapping aims at catching substantial proportion of a pest population before mating, oviposition or feeding and thus preventing damage to the crop. Success with this technique requires the combination of a very attractive lure and a highly efficient trap. The lure should be very attractive, eventually out-competing the naturally occurring attractant. For Lepidoptera, it is essential that males are trapped before mating, and it is most likely to succeed with insects that mate only once. In case of Coleoptera, trapping based on aggregation pheromones aims to reduce the number of both sexes before eggs are laid or damage is done by feeding adults. It is most important that there is minimal influx of the pest from outside the protected areas. In addition, mass trapping is rather cost-and labour-intensive because of trap maintenance. As with other traps, there may also be problems with the blend, change of release rate, or trap efficiency over time. The ability to attract and retain very high numbers will be affected by trap design, placement and maintenance.

Mass trapping is considered most effective for pests, which are geographically isolated and/ or at low densities. Four year of mass trapping with a sex and floral lure reduced a small pocket of Japanese beetles in a city park by 97 per cent. Male removal using sex pheromone traps was shown to be an effective means of controlling Chinese tortrix, *Cydia trasias* (Meyrick), on street-planted Chinese scholar trees. Success of these efforts in urban or park settings was, in part, due to the isolation of the sites and relatively low population densities. Similarly, food warehouses and other enclosed situations provide a high level of isolation, which should enhance the prospects for mass trapping. Perhaps the most successful use of mass trapping has been for the control of several species of beetles on forest trees. One of the most effective uses has been for the control of ambrosia beetles in timber-processing facilities in British Columbia. In this case, the programme probably benefited from the trapping being somewhat isolated from beetle populations in the forest. Control of some forest beetles may be enhanced by use of deterrents to push the target beetles away from a host, combined with attractant baited traps or trap trees to 'pull' them away. Recently, the potential of using lures containing the aggregation pheromone components in combination with ethyl acetate, cut sugarcane and insecticide (permethrin), was demonstrated for mass trapping of New Guinea sugarcane weevil, *Rhabdoscleus obscurus* (Boisduval), in Guam.

Lure and kill. The lure and kill approach is a modification of mass trapping, where instead of being trapped, the responding insects come in contact with a toxicant and get killed. In many ways, this method also suffers the same constraints as for mass trapping, e.g. population density, attractiveness of the lure and efficiency of the method of killing. However, the problem of trap-saturation may be eliminated and this may improve the effectiveness of control in high-density situations. The problems of trap maintenance and high cost of the control programme may also be mitigated to some extent, especially if the system relies on attracting the insects to a plant surface treated with an insecticide rather than to some kind of target device.

The lure and kill formulations have been developed for the control of various beetles, moths and flies. Some of the earliest applications of attractants in combination with insecticides have been for the control of tephritid fruit flies. The most successful example is the control of olive fly, *Bactrocera oleae* (Gmelin) in Greece. Protein/insecticide-bait sprays have been used to control this pest in most Mediterranean olive-growing areas for a number of years. However, due to toxicity to natural enemies, a system was developed based on the use of target traps baited with either a food-attractant or a sex pheromone dispenser. This target-device method of controlling *B. oleae* was effective in reducing fruit infestation, especially when applied on area-wide basis. The most recent development for fruit fly control has been microencapsulated sprayable formulation, comprising of the sex pheromone of this species, 1,7-diozaspiro, and an insecticide (either dimethoate or malathion). Recently, biodegradable or wooden spheres laced with a low dose of imidacloprid have shown promise for the control of *Rhagoletis pomonella* (Walsh) in apple and *R. mendax* Curran in blueberry.

Lure and infect. This innovative and promising approach combines an attractive lure with an entomo-pathogen. This technique is also called 'autodissemination'. In this case, the insects that arrive at the source are not killed, but are inoculated with the pathogen with the idea to magnify the treatment by spreading the disease to other individuals. Different pathogens could be used with slightly different pathways including viruses, bacteria, fungi and nematodes. This approach can generate disease outbreaks that can multiply in the area and affect the pest populations. Fewer insects may need to be directly attracted to the pathogen stations, which could reduce the cost and labour required. However, the critical requirements for success with pathogens may be difficult to achieve and include biological factors as well as operational factors such as formulation and delivery systems.

The lure and infect approach has been explored with nucleopolyhedrosis virus against tobacco budworm; a granulosis virus against codling moth; a protozoan against stored-product insect, *Trogoderma glabrum* (Herbst); and fungi against diamondback moth, Japanese beetle and termites. Fungi seem to be the best candidates as they are transferred between adults and larvae, and do not require consumption or copulation to be pathogenic. Once an appropriate pathogen is selected, a formulation must be developed that protects the organism from environmental degradation. A major constraint with these systems, as with mass trapping, is likely to be the ability to make them cost-effective as many bait stations may need to be developed for the approach to be effective.

C. Mating Disruption

Control of insect pests by mating disruption technique is achieved by widespread application of synthetic pheromone over the treated crop. Various slow-release pheromone formulations have been developed which either permeate the air with relatively high levels of pheromone so as to achieve sensory adaption or provide numerous discrete point sources so as to mask trail following or to create false trail.

The development of hand applied, slow release dispensers for season long control has undoubtedly contributed to make the technique effective, reliable and economical. Pink bollworm,*Pectinophora gossypiella* (Saunders), has been successfully managed in a number of countries including USA, Egypt and Pakistan by using controlled release formulations of its female sex pheromone gossyplure. In Egypt, 150,000 ha of cotton were treated with pheromones against *P. gossypiella.* In Pakistan, a single twist-tie formulation containing pheromones of both pink and spotted bollworms has been successfully used to obtain season-long control of bollworm complex. Other successful examples include control of Oriental fruit moth, *Cydia molesta* (Busck), over 2000 ha stone fruit area in South Africa and *Chilo suppressalis* (Walker) in 2500 ha rice area in Spain.

Another area which has received considerable attention during the last decade is the utilization of a variety of semiochemicals for aphid management. (E)-β-farnesene (EBF) has been identified as alarm pheromone of aphids. The application of alarm pheromone alongwith host plant volatiles caused the aphids to move away from feeding sites and may be utilized to prevent colonization of the host plants by the aphids. It also increased the mobility of the aphids and this is useful for increasing the pick up of contact insecticides or pathogens by the pest. EBF is now commercially available for use against a number of aphid species including *Aphis gossypii* Glover and *Lipaphis erysimi* (Kaltenbach).

Proteinaceous artificial honeydews sprayed onto foliage have been found to attract aphid natural enemies like chrysopid and syrphid adults and arrest the movement of a number of coccinellids. Among the volatiles in artificial honeydew attractive to *Chrysoperla carnea* (Stephens) adults was a tryptophan degradation product, indolė acetaldehyde. Parasitism by several aphid parasitoids including *Aphidius* spp., *Dacus rapae* and *Praon volcure* (Haliday) is increased by spraying nepetalactone which is a component of the male sex pheromone of several aphid species.

II. ALLELOCHEMICALS

Among the most effective weapons developed by plants against phytophagous insects are the various noxious phytochemicals which adversely affect the growth, survival, development and behaviour of these pests. Plants may also provide chemical cues for the parasitoids and predators of insects.

Allomones

The primary defense of plants against insect pests is the possession of toxic or repugnant allomones. These substances repel, deter or harm many potential phytophagous insects. Plant allomones might be applied to crops to act as long or short distance insect repellents or contact feeding deterrents. Therefore, allomones which reduce pest injury by rendering plants unattractive or unpalatable offer a novel approach in vector and disease management.

There are a number of examples where allomones have been found to affect the growth and development of insects. The level of hydroxamic acid in a number of graminaceous plants including maize and wheat, affects the survival and development of *Ostrinia nubilalis* (Hubner), *Schizaphis graminum* (Rondani) and *Sitobion avenae* (Fabricius). The toxic effects of gossypol present in cotton plants have been demonstrated against *Spodoptera exigua* (Hubner), *Helicoverpa zea* (Boddie), *Heliothis virescens* (Fabricius), *Trichoplusia ni* (Hubner), *Spodoptera littoralis* (Boisduval) and *Pectinophora gossypiella* (Saunders).

Pentadecanal in rice exhibits allomonal properties against *Nilaparvata lugens* (Stal), *Sogatella furcifera* (Horvath) and *Chilo suppressalis* (Walker). Another chemical, 2-tridecanone, which is found in the leaves of tomato plants is toxic to the larvae of *Manduca sexta* (Johannsen) and *H.zea*, and the adults of *Aphis gossypii* Glover, whereas a derivative, tridecanyl acetate, is active against stored product beetles.

Kairomones

Plant-produced substances that serve as attractants, arrestants, and oviposition and feeding stimulants for herbivores are undoubtedly the best known kairomones. These plant-derived substances also influence organisms of the third trophic level when they are sequestered by herbivores and used by natural enemies to locate the herbivore. The aphid parasitoid, *Diaeretiella rapae* (M'Intosh) is attracted by the allylisothiocyanate released by cruciferous plants harbouring various aphid species including *Brevicoryne brassicae* (Linnaeus), *Lipaphis erysimi* (Kaltenbach) and *Myzus persicae* (Sulzer). *Chrysoperla carnea* (Stephens) and *Collops vittatus* (Say) are attracted to caryophyllene, a terpenoid released by damaged cotton leaves. The spined soldier bug, *Podisus maculiventris* (Say), orients to soybean plants damaged by *Trichoplusia ni* (Hubner).

Synomones

Some plants respond to feeding or tissue damage by emitting synomones attractive to insect's natural enemies. Corn seedlings attacked by *Spodoptera exigua* (Hubner) release large amounts of terpenoid volatiles which serve as cues for females of the parasitoid wasp, *Apanteles marginiventris* (Cresson). The females of parasitoid, *Eucelatoria bryani* Sabrosky respond positively to the extracts of 19 plants which serve as food source for *Helicoverpa* spp. Lima bean leaves, upon infestation by two spotted spider mites, *Tetranychus urticae* Koch, produce the terpenoids which attract predators of the herbivores. Corn plants contain the chemical tricosane and the corn earworm, *Helicoverpa zea* (Boddie) incorporates tricosane unchanged into its eggs. This chemical attracts egg parasitoid, *Trichogramma evanescens* Westwood, to find its host.

Interestingly, a number of chemicals released by phytophagous insects serve as cues for their natural enemies. Aphid odours are both an arrestant and an oviposition stimulant for the syrphid fly, *Syrphus corollae* Fabricius and a short range attractant for *Aphidoletes aphidimyza* (Rondani). Larvae of *Chrysoperla carnea* (Stephens) respond to chemicals emanating from the scales left behind by the ovipositing females. The odour of frass or webbing from herbivorous mites is highly stimulatory to predatory mites.

Some arthropod predators have evolved the ability to attract their prey chemically. The bolas spider, *Mastophora* sp. attracts males of two noctuid moth species by producing an allomone similar to the sex pheromone of the female moth. The assasin bug, *Apiomerus pictipes* Herrich-Schaeffer releases a substance attractive to the stingless bee, *Trigona fulviventris* Guerin-Meneville.

Although the majority of successful uses of semiochemicals are for monitoring pest activity, there is an increasing number of examples of direct control with pheromones and other behaviour-modifying compounds. The various approaches in which semiochemicals are used in pest management are listed in **Table 8.1.**

Table 8.1 Practical uses of semiochemicals in pest management

Monitoring

- Detect the presence of a species.
- Measure seasonal activity and provide decision support.
- Evaluate the effectiveness of mating disruption.
- Assess levels of insecticide resistance.

Direct Control

- Mass deployment of attractant-baited traps.
- Application of attract-and-kill formulations or devices.
- Pheromone-mediated mating disruption.
- Manipulation of natural enemies using allelochemicals.
- Pheromone-based interference with host location or acceptance.
- Plant allomone based deterrence of feeding or oviposition.
- Application of pheromones to enhance pollination.

PUSH-PULL STRATEGY

The push-pull strategy involves behavioural manipulation of insect pests and their natural enemies by integration of stimuli that act to make the protected resource unattractive or unsuitable to the pests (push), while luring them towards an attractive source (pull), from where the pests are subsequently removed (Cook *et al.,* 2007). The pests are repelled or deterred away from the resource (a crop or a farm animal) by using stimuli that mask host apparency or are repellent or deterrent. The pests are simultaneously attracted, using highly apparent and attractive stimuli, to other areas such as traps or trap crops, where they are concentrated, facilitating their elimination.The term 'push-pull' was first conceived by Pyke *et al.* (1987) to explore the use of repellent and attractant stimuli, deployed in tandem, to manipulate the distribution of *Helicoverpa* spp. in cotton in Australia. Subsequently, the concept was formalized and refined by Miller and Cowles (1990), who termed the strategy as stimulo-deterrent diversion for management of onion maggot, *Delia antiqua* (Meigen). However, the original terminology has been favoured by the scientific community and has been well accepted.

The stimuli for push components include visual cues (host colour, shape or size), synthetic repellents, non-host volatiles, host-derived semiochemicals, antiaggregation pheromones, alarm pheromones, antifeedants, oviposition deterrents and oviposition deterring pheromones. The stimuli for pull components include visual stimulants (traps or trap crops), host volatiles, sex and aggregation pheromones, and gustatory and oviposition stimulants. The principles of the push-pull strategy are to maximize control efficacy, efficiency, sustainability and output, while minimising negative environmental effects. This strategy maximises the efficiency of behaviour-manipulating stimuli through the additive and synergistic effects of integrating their use. The efficacy and efficiency of population reducing methods can also be increased by concentrating the pests in a predetermined site. Population reduction by biocontrol methods or highly selective botanical pesticides is preferred to broad-spectrum synthetic pesticides. The deployment of renewable sources, particularly plants, for the production of semiochemicals is encouraged. In agricultural systems, the goal is to maximise output from the whole system while minimising cost, and using harvestable trap crops or intercrops, rather than sacrificial crops, wherever feasible. The development of reliable and sustainable push-pull strategy requires a clear understanding of the pest's biology and the behavioural/chemical ecology of the interactions with its hosts, conspecifics and natural enemies.

The push-pull strategies are under development or used in practice in the major areas of pest control. However, the most successful example currently used in practice, was developed in Africa for the control of lepidoptern stem borers, viz., *Chilo partellus* (Swinhoe), *Eldana saccharina* Walker, *Busseola fusca* (Fuller), and *Sesamia calamistis* Hampson in maize and sorghum. The strategy involves the combined use of intercrops and trap crops, using plants that are appropriate to the farmers and that also exploit natural enemies. The stem borers are repelled from the crops by repellent non-host intercrops, particularly molasses grass, *Melinis minutiflora* Beauv; silverleaf desmodium, *Desmodium uncinatum* (Jacq.) DC or greenleaf desmodium, *D. intortum* (Mill.) Urb. (push). These are concentrated on attractive trap plants, primarily Napier grass, *Pennisetum purpureum* (Linnaeus) or Sudan grass, *Sorghum vulgare sudanense* (Piper) Hitch (pull). Molasses grass, when intercropped with maize, not only reduced stem borer infestation, but also increased parasitism by *Cotesia sesamiae* Cameron. A trap crop of Sudan grass also increased the efficiency of stem borer natural enemies. The push-pull strategy has contributed to increased crop yields and livestock production, resulting in a significant impact on food security in the region.

Molasses grass is also known to contain attractive compounds similar to those found from maize. In addition, five other compounds including (*E*)-b-ocemene and (*E*)-4, 8-diethyl-1, 3, 7-nonatriene, which were repellent to stem borers. Desmodium intercrops also produce these compounds, together with large amounts of sesquiterpenes. When intercropped with maize or sorghum, desmodium suppresses the parasitic African witchweed, *Striga hermonthica* (Del.) Benth. Six host volatiles, viz. octanal, nonanal, naphthalene, 4-allylanisole, eugenol, and (*R*, *S*)-linalool, have been found to be attractive to gravid stem borers. Recent studies have indicated that the differential preference of moths between maize and sorghum, and Napier grass trap crops is related to a large burst of four electrophysiologically active green leaf volatiles released from the trap crop plants within the first hour of the scotophase, the time at which most oviposition occurs (Cook *et al.*, 2007).

The deployment of push-pull strategies is advantageous over the use of individual components in isolation. Individual elements may not be able to provide effective control on their own. For example, trapping strategies using attractive baits may cause substantial effect on species with low reproductive rates, but not for species with high reproductive rates. By incorporating other element with negative effects on host selection, the preference differential is increased and the additive effects may reduce pests below economic threshold levels. Moreover, the efficiency of push and pull components is often not only additive but also synergestic. The use of antifeedants and oviposition deterrents is often limited because of habituation, or host deprivation, in the absence of more suitable hosts. The pull stimuli provide a choice situation for alternative feeding or ovipositional outlets. As the pest populations are concentrated in pre-determined areas (either traps or trap crops), less chemical or biological material is required to treat the pest population, thereby reducing costs. The semiochemicals in push-pull strategies are used in combination and do not select strongly for resistance. Moreover, the reduction in the use of conventional insecticides reduces the chances of pests to develop insecticide resistance.

DEVELOPMENT INHIBITORS

The growth and development of insects is regulated through the secretion of internal ductless glands, i.e. hormones. Basically, three types of hormones are involved in insect development (Fig. 1.25). The *brain hormone* or *prothoracicotropic hormone*, secreted by the neurosecretory cells of the brain, activates the prothoracic glands to secrete another hormone called *moulting hormone* (MH) or *ecdysone*. Moulting hormone is necessary for each of the immature moults which must take place for insects to grow. The third hormone, *juvenile hormone* (JH), is secreted by the gland corpora allata and regulates the juvenile characters in insects. In the presence of high titer of JH, a larva moults into a larva, while at a low titer of JH, it moults into a pupa. In the absence of JH, pupa moults to become adult. Two other hormones which also play a role in development are the eclosion and tanning hormones. The *eclosion hormone*, released by the brain, controls the process of eclosion in insects, while the *tanning hormone* or *bursicon* produced by neurosecretory cells of the brain or abdominal ganglia regulates the process of tanning and sclerotization of the new cuticle.

The potential of exploitation of hormonal regulation of growth of insects for pest management is discussed here briefly.

Brain Hormones

Chemical messengers produced by the central nervous system to regulate various events in the body are known as brain hormones or neurohormones. The first neuropeptide, proctolin, was isolated from *Periplaneta americana* (Linnaeus) in 1975 and since then 50 such structures have been isolated from different insects. There are several promising approaches in which the information on insect neuropeptides may be utilized to the detriment of insects, viz. (*i*) the design of peptide mimics that can penetrate the insect cuticle or gut and block or overstimulate the peptide-mediated response at the target cell, (*ii*) development of control agents that interfere with the secretion of a neuropeptide, and (*iii*) incorporation of neuropeptide producing genes into microorganisms will make large scale production economical and also help in overcoming stability and penetration problems.

Juvenile Hormones

The role of juvenile hormone (JH) in the growth of insects was recognized at least 60 years ago when Prof. C.M. Williams reported the possibility of its use in upsetting insect development. Till now, six closely related naturally occurring JHs have been isolated from insects. However, the natural JHs *per se* have limited scope in pest management because these are unstable to UV light and are rapidly metabolized by insects. Therefore, the juvenile hormone analogue (JHA) synthesis was directed towards the stabilization of the molecules. Some of the earlier compounds showing promise as pest control agents were methoprene and hydroprene, and later on fenoxycarb and pyriproxyfen were commercialised **(Fig. 8.2)**.

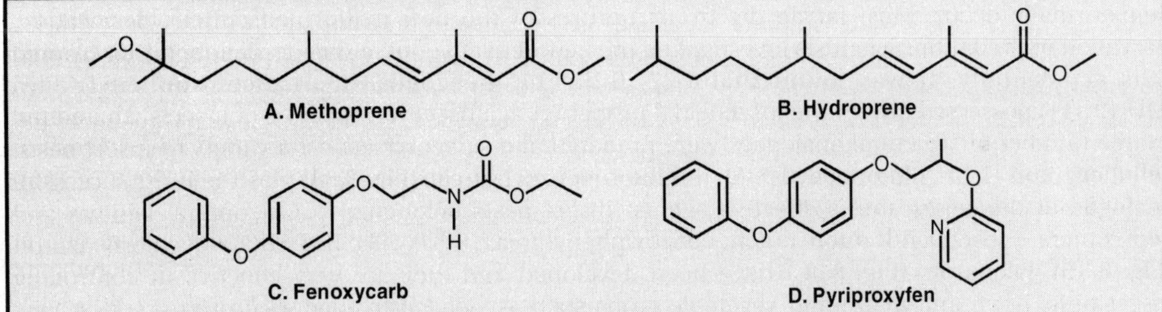

A. Methoprene

B. Hydroprene

C. Fenoxycarb

D. Pyriproxyfen

Fig. 8.2 Structures of some important juvenile hormone analogues

The JHAs have been tested in large scale trials for the control of various pests of agriculture, forestry and public health importance. Five different formulations of methoprene are available for the control of mosquitoes, manure breeding flies, cigarette beetle, pharaoh's ant, fungus gnat and fleas. Another JHA, fenoxycarb is registered in Switzerland for the control of several fruit pests. Foliar application of fenoxycarb on cotton has provided satisfactory control of *Helicoverpa zea* (Boddie) and *Heliothis virescens* (Fabricius) in USA. Pyriproxyfen has proved effective against a number of sucking pests including whitefly, *Dialeurodes citri* (Ashmead); *Thrips palmi* Karny, California red scale, *Aonidiella aurantii* (Maskell) and cottony cushion scale, *Icerya purchasi* Maskell.

A newer analogue, diofenolen, has been found highly effective for the control of scale insects and lepidopterous pests attacking a number of fruit crops. NC-196, a derivative of benzyl pyridazinone has shown good systemic and contact activity against brown planthopper of rice.

The exogenous applcation of JHAs is effective only when the endogenous JH titer in insects is low. This leaves the larval/nymphal stages of most of the insects unaffected. This limitation could be overcome by blocking the biosynthesis of natural JHs. The chemicals exhibiting this property are called *antijuvenile hormone agents* (AJHAs) or *JH antagonists*. Interestingly, AJHAs were also initially isolated from a plant, *Ageratum houstonianum* Mill. These compounds designated as precocene I and II induced precocious metamorphosis in the milkweed bug, *Oncopeltus fasciatus* (Dallas). Another related compound precocene III was isolated from *Ageratum conyzoides* L.

Apart from precocenes, several other compounds like ETB and EMD act as JH antagonists. More recently, many synthetic furanyl compounds have been reported to possess AJH activity against *O. fasciatus*. The exogenous application of AJHAs to insects not only shortens the life cycle of immature stages but also results in diminished feeding. Further, precocious males are unable to mate and inseminate normal females, while precocious females are sterile. AJHAs are thus effective against different developmental stages of insects which generally exist simultaneously in the field.

Chitin Synthesis Inhibitors

A new class of insecticides was developed accidently when the insecticidal activity of benzoylphenyl urea (BPU) analogues was discovered around 1970 by the Philips-Duphar Company. One of the first analogues was code-named DU 19.111 which resulted from the combination of the herbicide dichlobenil with the urea herbicide diuron and possessed interesting insecticidal properties against several insect species. Structural optimization resulted in the production of diflubenzuron [1-(4-chlorophenyl)-3-2-(2, 6-difluorobenzoyl) urea] which was commercialized under the name of Dimilin. Later, other bioactive molecules like BAY SIR 8514 and IKI 7899 (chlorfluazuron) were successfully commercialised.

Chitin synthesis inhibitors interfere with production of chitin, the structural polysaccharide found in insect cuticle, and thus affect the integrity of insect exoskeleton. Exposure causes improper attachment of the new cuticle during moulting and produces a cuticle that lacks some of the layers that normally occur. Most larvae die from ruptures of the new malformed cuticle, desiccation, starvation or predation. Scientists investigating the new derivatives of herbicide, diclobenil, discovered CSIs accidentally. It was found that 1-(2, 6-dichlorobenzoyl)-3-(3, 4-dichlorophenyl) urea (DU 19.111) possessed interesting insecticidal properties against several species of insects. Thereafter, a large number of structural analogues were prepared and screened against a variety of insect pests. Diflubenzuron [1-(4-chlorophenyl)-3-(2, 6-difluorobenzoyl) urea] **(Fig. 8.3A)** was the most successful analogue of the series and is effective against insect pests belonging to Coleoptera, Diptera and Lepidoptera. Subsequently, more potent benzoylphenylureas (BPUs), like BAY SIR 8514, chlorfluazuron **(Fig. 8.3B)**, lufenuron **(Fig. 8.3C)** have been developed and they are very effective in controlling insect pests of cotton, maize and vegetable crops such as *Spodoptera* and *Helicoverpa* spp. A new BPU insecticide, novaluron **(Fig. 8.3D)**, has somewhat more contact and translaminar activity as compared with other BPUs, and cyromazine **(Fig. 8.3E)** thereby affecting whiteflies and leafminers, in addition to lepidopteran and coleopteran larvae.

The benzoyl phenyl urea (BPU) analogues are compounds with selective properties, affecting the larval stage. They mainly act as stomach toxicants and kill insects at the time of moulting. However, in some species they suppress fecundity and exihibit ovicidal and contact toxicity. The BPU analogues block the terminal polymerization step catalyzed by the enzyme chitin synthase during the process of biosynthesis of chitin. However, the exact site of action is not yet clearly understood and a number of interesting hypotheses have been put forward to explain the effects produced by the application of BPU analogues. The most plausible explanation so far is that BPU analogues block the availability of substrates at the active sites of the membrane bound enzyme, chitin synthase, probably by altering membrane permeability. These compounds also inhibit a number of other enzymes and DNA biosynthesis in larval epidermal cells. BPUs generally affect the larval stages of insects, which synthesize chitin during their moulting processes. Hence, the adults of beneficial species, predators and parasitoids, are seldom affected. For this reason, BPUs are considered important components in IPM programmes.

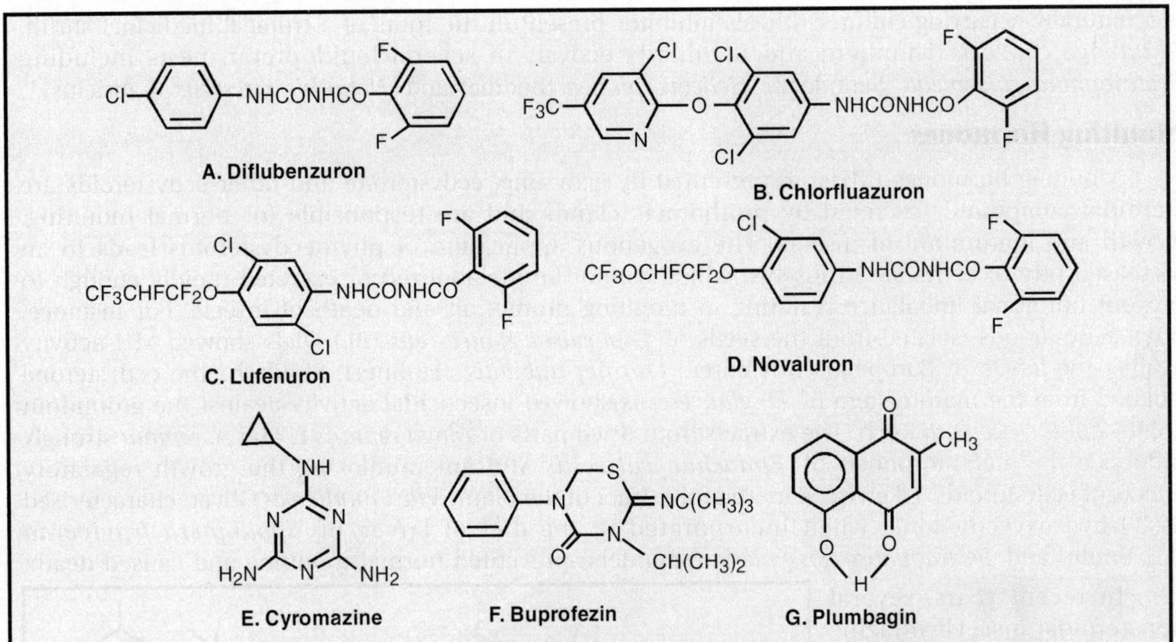

A. Diflubenzuron

B. Chlorfluazuron

C. Lufenuron

D. Novaluron

E. Cyromazine

F. Buprofezin

G. Plumbagin

Fig. 8.3 Structures of some important chitin synthesis inhibitors

Another compound, buprofezin (**Fig. 8.3F**), is a thiadizine like compound with long residual activity that also acts as chitin synthesis inhibitor. It has both contact and vapour activity, and acts on nymphal stages of sucking insects such as leafhoppers, planthoppers and whiteflies. Its mode of action resembles that of BPUs, although its structure is not analogous. The compound inhibits incorporation of 3H-glucose and N-acetyl-D-3H-glucosamone into chitin. Because of chitin deficiency, the procuticle of the whitefly nymph loses its elasticity and the insect is unable to molt. As in case of BPUs, buprofezin is effective against immature stages and not adults. The compound has a mild effect on natural enemies and is an important component of IPM programmes for managing whiteflies in cotton, vegetables and ornamental plants.

In addition to BPU analogues, several other groups of compounds have also been reported to act as CSIs. Plumbagin (**Fig. 8.3G**) is a naturally occurring CSI present in the roots of a tropical medicinal shrub, *Plumbago capensis* Thunb. It has been reported to possess novel mode of action like chitin and ecdysteriod inhibition thereby leading to growth inhibition in insects. Recent investigations have revealed that plumbagin possesses high ovicidal activity against eggs of *Dysdercus koenigii* (Fabricius), *Spodoptera litura* (Fabricius), *Corcyra cephalonica* (Stainton) and *Plutella xylostella* (Linnaeus). Besides its activity as a CSI, plumbagin is toxic to *S. litura, P. xylostella* and *Helicoverpa armigera* (Hubner) larvae, and adults of *Myzus persicae* (Sulzer). Toxicity of plumbagin was at least 3-fold higher for winged than apterous adults in case of topical assays, and at least 1645-fold higher for pterous than that for apterous aphids in residual film assays.

Chitin synthesis inhibitors belonging to the acyl urea group have been used commercially for the control of a number of foliage feeders and tissue borers. Satisfactory control of cabbage looper, *Trichopulsia ni* (Hubner); cotton leaf perforator, *Bucculatrix thurberiella* Busck and Egyptian cotton leafworm, *Spodoptera littoralis* (Boisduval), damaging cotton crop has been obtained with diflubenzuron. Foliar application of diflubenzuron has shown excellent residual activity against eggs of *Helicoverpa armigera* (Hubner) and *Spodoptera litura* (Fabricius). Some new chitin synthesis inhibitors like XRD- 473, IKI-7899 and IGR-1055 have shown excellent larvicidal activity against *H. armigera.* Teflubenzuron has been found to posses high ovicidal action against *Cydia pomonella* (Linnaeus) on apple. Buprofezin has been found promising against several sucking pests. Plumbagin is a naturally occurring chitin synthesis inhibitor present in the roots of a tropical medicinal shrub, *Plumbago capensis* Thunberg and it inhibits ecdysis in several lepidopteran pests including *Pectinophora gossypiella* (Saunders), *Helicoverpa zea* (Boddie) and *Heliothis virescens* (Fabricius).

Moulting Hormones

Moulting hormones (MHs), represented by ecdysone, ecdysterone and other ecdysteroids are steroidal compounds secreted by prothoracic glands and are responsible for normal moulting, growth and maturation of insects. The exogenous application of phytoecdysteroids leads to an increased titre of ecdysone in insects, which cannot be metabolized or excreted rapidly enough to prevent hormonal imbalance resulting in moulting promotion and death of insects. For instance, phytoecdysteriods isolated from the seeds of *Diploclisia glaucescens* (Bl.) Diels showed MH activity against the larvae of European corn borer, *Ostrinia nubilalis* (Hubner). Similarly, the ecdysterone isolated from the mature stem of *D. glaucescens* showed insecticidal activity against the groundnut aphid, *Aphis craccivora* Koch. The extracts from dried parts of *Ajuga reptans* L. and *A. remota* strongly influence the metamorphosis of *Epilachna varivestis* Mulsant mimicking the growth regulatory effects of ecdysteroids. Likewise, a methanol extract of the plant, *Vitex madiensis* Oliver, characterised as 20-hydroxyecdy-sone, when incorporated in the diet of larvae of *Spodoptera frugiperda* (J.E. Smith) and *Pectinophora gossypiella* (Saunders) prevented normal moulting and caused death.

In recent years, several nonsteroidal bisacylhydrazine ecdysone agonists have been synthesized. The first ecdysteroid agonist was discovered in 1983 and subsequent chemical modification of this early lead produced a simple and slightly more potent analogue, RH-5849, followed by other more potent and cost-effective analogues like tebufenozide (**Fig. 8.4A**), halofenozide (**Fig. 8.4B**), methoxyfenozide (**Fig. 8.4C**) and chromafenozide (**Fig. 8.4D**). These chemicals are much more potent than 20-hydroxyecdysone in inducing molting. They are also known to reduce feeding and weight gain. In lepidopteran insects, a lethal moult is induced following administration of the ecdysone agonist and the insect dies trapped within the exuvial cuticle. Feeding stops 4-6 hours after ingestion of toxic doses of the agonist, and molting is initiated in the absence of an ecdysteroid increase. Usually, the insect dies in the slipped head capsule stage following onset of apolysis. However, supernumerary larval molts may also occur, when the JH titer is high.

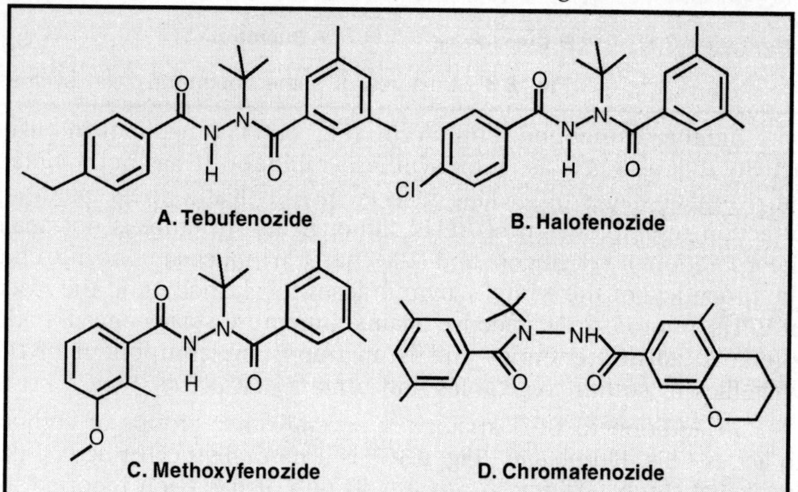

A. Tebufenozide B. Halofenozide C. Methoxyfenozide D. Chromafenozide

Fig. 8.4 Structures of some moulting hormone agonists

Ecysteroid agonists have been demonstrated to be active against many lepidopterans including *Manduca sexta* (Johannsen), *Pieris brassicae* (Linnaeus), *Plodia interpunctella* (Hubner), *Spodoptera exempta* (Walker), *S. littoralis* (Boisduval) and *S. litura* (Fabricius). RH-5992 (tebufenozide) is more toxic to lepidopteran larvae than RH-5849. RH-0345 (halofenozide) has an overall insect control spectrum similar to that of RH-5849 but with accentuated soil systemic efficacy against scarabacid beetle larvae, cutworms and webworms. RH-2485 (methoxy-fenozide) is more potent than tebufenozide against lepidopteran pests of cotton, maize and other major agricultural crops.

Methoxyfenozide induces an immediate and fatal moult in *S. littoralis* when added to the diet of 2nd and 4th instar larvae at 1 ppm and to that of 6th instar larvae at 0.001 ppm concentration. Ten times lower doses fed to the larvae continuously allow an apparently normal larval development that is terminated by a supernumeracy larval moult. The other effects of methoxyfenozide include death during metamorphosis and impaired fertility of emerged adults. The number of progeny is reduced even with low doses, *e.g.* insects fed 0.0001 ppm since 2nd, 4th and 6th instars produce 72, 62, and 22 per cent, respectively, less progeny than the controls.

The ecdysone agonists have been tested in larvae and adults of more than 16 different insect orders, but these compounds have produced lethal effects mostly in lepidopteran, dipteran and coleopteran larvae. The high binding affinity of tebufenozide and methoxyfenozide to proteins in nuclear extracts of lepidopteran cells is correlated with their selective action on lepidopteran insects. By contrast, the ecdysteroid receptors of coleopteran insects bind tebufenozide with low affinity. This difference thereby explains the specificity of this compound for lepidopteran insects. Ecdysone agonists affect all larval stages but the effect induced depends on when the insect ingests or is treated with the compounds during the time of application within a stadium. If treatment occurs early in an instar, an immediate lethal molt is induced, but if the insect is treated towards the end of an instar, first a normal moult will occur, which is then followed by the lethal moult. In adult stage insects, egg production and spermatogenesis may be adversely affected by exposure to the ecdysone agonists.

Sclerotization Disruptors

Sclerotization is a complex process used by insects to confer stability and mechanical versatility to their cuticular exoskeletons and certain other proteinaceous structures. Inhibitors of sclerotization may disrupt the metabolism and/or deposition of phenolic compounds, proteins or other components that participate in cuticular stabilization mechanism. The only commercially available sclerotization inhibitor is MON 0585 developed by Monsanto. It is a di-tertiary butyl alcohol compound which is highly toxic to mosquitoes and other dipterous insects. It causes mortality at the time of pupation and dead pupae are white instead of tan colour because of failure of the organism to harden the pupal skin.

Another group of compounds with capacity to inhibit sclerotization are inhibitors of the enzyme 3, 4-dihydroxy-phenyl-alanine decarboxylase (DDC). DDC inhibitors like α-methyl DOPA cause mortality in dipteran larvae at the time of moulting. Accelerators of sclerotization process also exert a toxic effect on insects. Cyromazine, a substituted diaminotriazine produces necrotic lesions in the cuticle of blowfly and stiffens housefly larval cuticle by inserting an extra layer between the endo-and exocuticle, causing rod like puparia. The cuticle of *Manduca sexta* (Johannsen) is unable to expand following treatment with cyromazine.

MISCELLANEOUS APPROACHES

A number of other approaches have been evolved which hold a great potential for managing pest populations under certain situations.

Propesticides

A propesticide is a compound which is inactive in its original form, but is transformed into a pesticidally active state by a plant, animal or microorganism. Classical pesticides can be modified to yield propesticides retaining their insecticidal activity but lowering mammalian toxicity and acquiring plant systemic properties. The technique has led to the development of new derivatives of toxic methyl carbamate insecticides with improved toxicological properties. Carbosulfan is a derivative of carbofuran with a similar activity spectrum but is substantially less toxic to mammals.

This technique has also been applied for developing new organophos-phorus compounds. Acephate, the acetylated product of methamidophos is 45-fold less toxic to rat than the parent compound but retains approximately the same insecticidal activity. In insects, acephate is converted to methamidophos which is believed to be responsible for high toxicity. Similarly, cartap is a proinsecticide which is rapidly converted to nereisotoxin in the insect body. Nereisotoxin is a dithiolane compound found naturally in the marine annelid, *Lumbriconereis heteropoda* Marenz.

Avermectins

Avermectins are macrocyclic lactones which were originally isolated in 1976 by scientists at Merck & Co., Inc. (Rahway, New Jersey), from a culture of *Streptomyces avermitilis* from Japan. The avermectins seem to exert their toxicity by disrupting the action of both ligand-gated (*i.e.* GABA) and voltage-gated chloride channels. The end result is functional disruption of GABA-gated chloride channels. These are among the most potent anthelminthic, acaricidal and insecticidal compounds. Among the eight analogues identified, avermectin B (commercialized as Abamectin) is insecticidally most active. Emamectin benzoate is a novel macrocyclic lactone insecticide derived from the avermectin family. Another group of related compounds called milemycins have been obtained from *S. hygroscopicus aureolacrimosces.*

Spinosyns

A new class of insect control molecules, the spinosyns, was discovered in 1994 by Dow Elanco (Indianapolis, 1N). Naturally derived from a new species of actinomycetes, *Saccharopolyspora spinosa,* they are very active against many pests of crops, ornamentals, forestry, greenhouse, garden and households. Spinosad, a mixture of spinosyn A and spinosyn D, is the lead compound and it has shown contact and stomach activity against Coleoptera, Diptera, Hymenoptera, Isoptera, Lepidoptera, Siphonoptera and Thysanoptera. Spinosad causes persistent activation of nicotine acetylcholine receptors in the insect nervous system, a unique mode of action with no known cross resistance with other insecticides. Spinosad has been approved for registration in 1997 in India under the Insecticides Act, 1968.

Polynactins

Polynactins are secondary metabolites from the actinomycete *Streptomyces aureus* strain S-3466. They are quite effective in controlling spider mites under wet conditions. It is believed that the mode of action is through a leakage of basic cations (such as potassium ions) through the lipid layer of the mitochondrian membrane. Water is considered essential to this toxic effect by either assisting penetration or accelerating ion leakage. They are particularly recommended for the control of spider mites on fruit trees.

Pyrrole Insecticides

Pyrrole Insecticides have been derived from a natural product, dioxapyrrolomycin, isolated from a strain of *Streptomyces.* Chlorfenapyr is a promising pyrrole, which has been commercially developed because of its broad spectrum of activity against many species of Coleoptera, Lepidoptera, Thysanoptera and Acarina. Chlorfenapyr acts at the mitochondrial level by uncoupling oxidative phosphorylation. It is mainly a stomach toxicant, but has some contact actively. Field trials have demonstrated that foliar applications of chlorfenapyr are effective in controlling more than 70 insect pests and mites on cotton, cereals, vegetables, orchard trees and ornamentals plants.

Phenylpyrazoles

The phenylpyrazoles comprise a new class of biorational pesticides, which exhibit insecticidal and herbicidal activities. The first highly successful member of this class is fipronil, which is active at the neuroinhibitory GABA-gated chloride channels Fipronil exhibits broad activity against various insect pests including soil insects, foliar feeding pests such as *Plutella xylostella* (Linnaeus), *Helicoverpa armigera* (Hubner) and *Spodoptera* spp; sucking pests such as thrips (but not aphids or whiteflies) and household pests.

Pyridine Insecticides

The most important chemical of this group is pymetrozine, which affects the nerves controlling the salivary pump and causes irreversible cessation of feeding due to an obstruction of stylet penetration, followed by starvation and insect death. It is highly specific against sucking insects such as aphids, whiteflies and planthoppers.

Oxadiazines

Indoxocarb is the first commercialized insecticide of the oxadiazine group. It acts by inhibiting sodium ion entry into nerve cells, resulting in paralysis and death of target pest species. This insecticide is active against lepidopteran as well as certain homopteran and coleopteran pests on vegetables, cotton and other field and orchard crops. Efficacy of this product has been demonstrated against important pests such as *Heliothis* sp., *Helicoverpa* sp., *Spodoptera* sp., *Plutella* sp. and *Trichoplusia* sp. (Lepidoptera), *Lygus* sp. and *Empoasca* sp. (Hemiptera) and also the Colorado potato beetle, *Leptinotarsa decemlineata* (Say) (Coleoptera).

Antifeedants

The antifeedants are chemicals which inhibit or deter the feeding of insects due to their presence on the natural food of the species concerned. In their absence, the species would otherwise feed normally. An antifeedant acts by suppressing the gustatory receptors. It is a type of feeding deterrent which deprives an insect from continued feeding on the host. Death in the end is due to starvation. These compounds belong to five major categories: (*i*) Triazines– Compound 24,055, (*ii*) Organotins– Stenous chloride, (*iii*) Carbamates–Baygon, (*iv*) Botanical extracts like pyrethrum, Neem oil and (*v*) Miscellaneous–Copper stearate, Phosphon, Cycocel, etc. The compounds like 4′–(dimethyl triazene) acetanilide, when used against surface feeders on cabbage and cotton, and Phosphon or Cycocel against leaf eaters on pepper and cotton, have been found to be very effective. A fungicide, Brestan is found to be a potent antifeedant against potato tuber moth larvae and the larvae of cutworms and cotton leaf worms. Naturally occurring antifeedants exist for many insects and play a major role in host selection and specificity **(Table 8.2)**.

Repellents

The repellents elicit avoidance and thus, lead the insect pests to move away from the source. The plants are rendered unattractive, unpalatable or offensive. Except for the naturally occurring feeding repellents, the artificial use of such chemicals has not been successful. When used as foliage repellents these chemicals need a thorough coverage of the crop to be effective. Since the growing points remain uncovered the chemicals do not show their full effectiveness. However, the repellents have been used scccessfully against mosquitoes (dimethyl phthalate), flies, fleas, mosquitoes (2-ethyl-1,3-hexanediol and N, N-diethyl m-toluamide), mites (benzyl benzoate, benzil and dibutyl phthalate) and flies on catttle (dibutyl succinate).

ROLE IN PEST MANAGEMENT

Biorational methods have currently emerged as viable alternatives to monitor and suppress pest populations. The estimated annual production of semiochemicals for monitoring and mass trapping is to the tune of tens of millions, covering at least 10 million ha. Insect pest populations are controlled by air permeation and attract-and-kill techniques on at least one million ha. Besides the problems that have occurred in the practical application of biorationals, there are a number of examples of their large scale use in pest management.

Allelochemicals

Although a number of kairomones have been isolated and identified from plants and animals, only a few of them have been exploited for pest suppression.

Table 8.2 Naturally occurring insect antifeedants in host plants

Plant	Feeding deterrent(s)	Insect(s)
Cruciferae	Sinigrin	*Pieris brassicae* (Linnaeus)
Cucumis sativus Linnaeus	Cucurbitacin	*Tetranychus urticae* Koch, *Phyllotreta nemorum* (Linnaeus)
Gossypium spp.	Gossypol	*Heliothis virescens* (Fabricius), *Spodoptera littoralis* (Boisduval), *Earias insulana* (Boisduval)
	Isoquercitrin, quercitrin, quercetin	*Helicoverpa zea* (Boddie), *Pectinophora gossypiella* (Saunders)
Hordeum spp.	Gramine	*Schizaphis graminum* (Rondani)
Lycopersicon esculentum Mill.	Rutin, chlorogenic acid, α- Tomatine	*H. zea*
Lycopersicon hirsutum Humb. & Bonpl. f. *glabratum*	2–Tridecanone	*H. zea*
Medicago spp.	Coumarin	*Hypera postica* (Gyllenhal), *Sitona cylindricollis* Fahraeus
	Dicoumaril	*Acyrthosiphon pisum* (Harris)
Melilotus spp.	Coumarin	*Listroderes costirostris* Schoenherr
Momordica charantia Linnaeus	Momordicine II	*Raphidopalpa foveicollis* (Lucas)
Solanaceae (potato, tomato)	Demissine, solacauline, tomatine, leptine I and II	*Leptinotarsa decemlineata* (Say), *Manduca sexta* (Johannsen)
Solanum spp.	Tomatine, solanidine, α-chaconine	*Choristoneura fumiferana* (Clemens)
Solanum tuberosum Linnaeus	Tomatine	*Empoasca fabae* (Harris)
Sorghum bicolor (Linnaeus) Moench	*p*-Hydroxybenzaldehyde, dhurrin, procyanidin	*S. graminun*
Triticum aestivum Linnaeus	Hydroxamic acid	*Metopolophium dirhodum* (Walker)
Zea mays Linnaeus	DIMBOA	*Ostrinia nubilalis* (Hubner)

Source : Panda and Khush (1995)

Corn rootworms. Three species of corn rootworms, viz. the western corn rootworm, *Diabrotica virgifera* Le Conte; the southern corn rootworm, *D. undecimpunctata howardi* Barber, and the northern corn rootworm, *D. barberi* Smith & Lawrence, are known to attack corn. Two features of diabroticite chemical ecology are helpful in manipulation of their behaviour. Firstly, the association between these beetles and cucurbitacin, a potent arrestant and feeding stimulant, may be so strong that other behaviours such as sex attraction are masked. Secondly, several plant-derived volatile attractants appear to mediate host selection by *Diabrotica* beetles. Powdered gourd roots and squash fruits have been evaluated for improving the capture of beetles in monitoring programmes. In addition, palatable baits containing cucurbitacins and an insecticide have been employed for short-term suppression of rootworm populations. The addition of a volatile plant-derived attractant also enhanced the suppression of *D. undecimpunctata howardi.* The first insecticide-laced bait was registered for use against corn rootworms in 1993 and it was applied to 12,000 ha in 1994. This approach reduced the insecticide use in corn by 90-95 per cent. However, three factors may limit further adoption of this approach; (*i*) adult management requires more intensive scouting and better timing than conventional approaches; (*ii*) adult management is feasible in only corn-corn rotations; and (*iii*) in view of difficulties in extracting cucurbitacin from the buffalo gourd, *Cucurbita foetidissima* Kunth, this approach is costly.

Tsetse flies. Traditionally, tsetse flies, *Glossina* spp., have been managed by clearing suitable habitat and by area-wide insecticide applications. Early experimentation with semiochemicals made use of livestock treated with toxicants to attract and kill flies. Specific attractants, such as 1-octen-3-ol, isolated from cattle, when combined with acetone and carbon dioxide, were similar to ox odour in attracting the flies. The addition of cattle urine further enhanced the attraction. The commonly used bait to attract the flies is a combination of acetone, 1-octen-3-ol, 4-methylphenol and 3-propylphenol. Baited traps have been used to monitor the presence of tsetse fly. Baited targets have also been used in combination with insecticides and a setrilant for pest suppression. This approach has resulted in eradication of tsetse fly from a number of areas, but continual reintroductions of flies demand that a yearly trapping programme be maintained. Recently, the introduction of artificial cow that attracted tsetse flies saved thousands of cattle from nagana infection (caused by tsetse flies) in Zimbabwe. The kairomones mimic the smell of real cattle and attract the tsetse flies to fake cow, which is loaded with pesticide.

Blow flies. The Australian sheep blow fly, *Lucilia cuprina* (Wiedeman) and related species are potentially controlled by a rather selective synthetic kairomone attractant, Traditionally, blow fly traps have been baited with liver and sodium sulfide. However, a synthetic kairomone, consisting of 2-mercaptoethanol, indole, butanoic acid, and a sodium sufide solution is far more effective and selective for *L. cuprina,* than the standard liver attractant. More importantly, the synthetic mix can be packaged in controlled-release dispensers to generate constant, prolonged release of the attractant. Field studies have confirmed that kariomone traps are a useful component of a blow fly control programme (Witzgall *et al.,* 2010).

Pheromones

Considerable progress has been made in the use of pheromones for pest control in agriculture, forestry, stored products and households. It has been estimated that about 20 million pheromone lures are produced for monitoring or mass trapping every year **(Table 8.3)**. The worldwide area treated with pheromones for mating disruption has surpassed 770,000 ha **(Table 8.4)**.

Pink bollworm. Pink bollworm, *Pectinophora gossypiella* (Saunders), a key pest of cotton, has been successfully managed in a number of countries including USA, Egypt, India and Pakistan by using controlled release formulations of its female sex pheromone, a 1:1 mixture of (Z, E) and (Z, Z)-7, 11-hexadecadienyl acetate known as gossyplure. In Egypt, 1,50,000 ha of cotton (about 36% of the country cotton acreage) were treated with pheromones against *P. gossypiella* in 1994. Mating disruption has been successfully employed in San Joaquin Valley, Imperial Valley of California and Parker Valley of Arizona of USA. Continual immigration of pink bollworms into the Imperial Valley and severe outbreaks of whiteflies, have reduced cotton production and adoption of mating disruption in this region. The growers in Parker Valley formed an informal pest control district in 1990 and adopted the use of mating disruption over 1,20,000 ha of cotton. The programme has been very successful, but problem of whiteflies has increased and a more expensive pheromone-insecticide strategy is required.

Sex pheromones of cotton bollworms have been extremely useful for monitoring, survey and surveillance, and assessing damage to cotton in India. Gossyplure is commonly used for monitoring the pink bollworm throughout the cotton belt. In Punjab, pheromone traps used for monitoring the activity of cotton bollworms on *hirsutum* cotton indicate that the maximum activity of *P. gossypiella, Helicoverpa armigera* (Hubner) and *Earias* spp. was in July, September and mid October to the end of season, respectively. Monitoring the population of bollworms can help in their early detection and build up, and a need-based control measure can be adopted depending upon the severity of

Table 8.3 Examples of large scale use of semiochemicals in pest management

Species	Purpose	Region	Lures/ Year
A. Agriculture			
Cotton boll weevil, *Anthonomus grandis* Boheman	MT, AK	North and South America	2600000
Pink bollworm, *Pectinophora gossypiella* (Saunders)	M, AK	North and South America, South Asia	–
American bollworm *Helicoverpa armigera* (Hubner)	M, MT		830000
Cotton leafworm, *Spodoptera litura* (Fabricius)	M, MT		480000
African armyworm, *Spodoptera exempta* (Walker)	D	East Africa	–
Spotted bollworm, *Earias vittella* (Fabricius)	M, MT		280000
Yellow rice stem borer, *Scirpophaga incertulas* (Walker)	M, MT	India	100000
Southwestern corn borer, *Diatraea grandiosella* Dyar	D	USA	–
Potato tuber moth, *Phthorimaea operculella* (Zeller)	AK	South Africa	–
B. Horticulture			
Red palm weevil, *Rynchophorus ferrugineus* (Olivier)	MT	Asia	1175000
American palm weevil, *Rhynchophorus palmarus* (Linnaeus)	MT	Central and South America	25000
Palm fruit stalk borer, *Oryctes elegans* Prell	MT	Asia	125000
Banana weevil, *Cosmopolites sordidus* (Germar)	MT	Worldwide	120000
Coffee white stem borer, *Xylotrechus quadripes* Chevrolet	MT	India	40000
Olive fruit fly, *Bactrocera oleae* (Gmelin)	MT, AK	EU	–
Grapevine moth, *Lobesia botrana* (Denis & Schiffermuller)	M	EU, Mediterranean countries, USA	–
Codling moth, *Cydia pomonella* (Linnaeus)	M, AK	Worldwide	
Oriental fruit moth, *Cydia molesta* (Busck)	M, AK	Worldwide	
Tomato leafminer, *Tuta absoluta* Meyrick	M, MT	South America, EU, North Aftica	2000000
Brinjal fruit and shoot borer, *Leucinodes orbonalis* Guenee	MT	India, Bangladesh	400000
Fall armyworm, *Spodoptera frugiperda* (J.E. Smith)	MT	Central America	50000
C. Forestry			
Spruce bark beetle, *Ips typographus* (Linnaeus)	MT	Europe, China	800000
Mountain pine beetle, *Dendroctonus ponderosae* Hopkins	MT	North America	–
Douglas-fir beetle, *D. pseudotsugae* Hopkins	MT	North America	–
Gypsy moth, *Lymantria dispar* (Linnaeus)	D	USA, EU	250000
Spruce budworm, *Choristoneura fumiferana* (Clemens)	D	Canada, USA	–
D. Stored products			
Cigarette beetle, *Lasioderma serricorne* (Fabricius)	M, MT	Worldwide	2500000
Indian meal moth, *Plodia interpunctella* (Hubner)	M, MT	Worldwide	2000000
E. Households			
House fly, *Musca domestica* Linnaeus	MT	Worldwide	2000000
German cockroach, *Blattella germanica* (Linnaeus), American cockroach, *Periplaneta americana* (Linnaeus)	MT	Worldwide	1000000

D: Detection ; M: Population monitoring ; MT: Mass trapping ; AK: Attract and kill.
Source : Witzgall *et al.* (2010)

Table 8.4 Global use of pheromones for mating disruption

Insect pest	Main crop(s)	Region	Area (ha)
Gypsy moth, *Lymantria dispar* (Linnaeus)	Forest	USA	230.000
Codling moth, *Cydia pomonella* (Linnaeus)	Apple, pear	Worldwide	210.000
Grapevine moth, *Lobesia botrana* (Denis & Schiffermuller)	Grape	EU, Chile	100.000
Oriental fruit moth, *Cydia molesta* (Busck)	Peach, apple	Worldwide	50.000
Pink bollworm, *Pectinophora gossypiella* (Saunders)	Cotton	USA, Israel, South America, EU	50.000
Grapeberry moth, *Eupoecilia ambiguella* (Hubner)	Grape	EU	45.000
Leafroller moths, Tortricidae	Apple, pear, peach, tea	USA, EU, Japan, Australia	25.000
Striped stem borer *Chilo suppressalis* (Walker)	Rice	Spain	20.000
Other species	Fruits, vegetables		40.000
Total			770.000

*Usage dropped significantly upon widespread adoption of transgenic cotton varieties
Source : Witzgall *et al.* (2010)

a particular species of bollworms. The pheromone trap catches of *H. armigera* moths have been successfully used to predict the egg and larval populations of this insect and damage to cotton plant reproductive bodies.

In Pakistan, a single twist-tie formulation containing pheromones of both pink and spotted bollworms has been successfully used to obtain season long control of bollworm complex. A single application of the pheromone either alone or in combination with a conventional insecticide provided effective control, while the conventional spray programme required 4-5 sprays per season to achieve a similar level of control (Dhaliwal and Arora, 2015).

Gypsy moth. The identification of a female produced sex pheromone, disparlure, has led to spectacular advances in the management of gypsy moth, *Lymantria dispar* (Linnaeus). Mating disruption programmes were established on about 7000 ha from 1979 to 1987 in the eastern United States. In 1993, a federal programme 'Slow the Spread' was launched to slow the rate of gypsy moth expansion by following IPM strategies over an area of 3,000,000 ha spread in four states. About 13,000 ha are treated each year and mating disruption is used on 15 per cent of this area. The greatest use of disparlure has been in surveys to determine the need for management tactics. In the western United States, incipient infestations of gypsy moth were detected in mid 1970s. In Washington state, gypsy moth adults were trapped only in a few sites, but by early ninties, 13-35 new outbreak sites were located each year. The standard protocol for monitoring gypsy moth in Washington state each year is to place about 10,000 traps at a density of one trap per square mile in all high-risk areas. When a moth is captured, a grid of traps is placed within a 3-mile radius around each site. If additional moths are caught and other life stages are also found, three sprays of *Bacillus thuringiensis* Berliner are applied. This strategy has prevented the establishment of gypsy moth in western Unites States and British Columbia (Dhaliwal and Koul, 2007).

Codling moth. Codlemone, the major sex pheromone component of codling moth, *Cydia pomonella* (Linnaeus), has been used extensively for monitoring and timing of pesticide applications. It was also used for mass trapping in some of the earliest evaluations. Mating disruption with codlemone has been tested in several apple-growing regions, including Australia, USA and parts of

Europe. The effectiveness of mating disruption in these trials was determined by moth density, moth immigration, amount of pheromone released, and the number and positioning of dispensers used per area. A three-component blend housed in a polyethylene dispenser was registered in USA in 1991 and was applied to treat about 10,000 ha in 1994.

A fully integrated attract-and-kill product, containing 0.16 per cent pheromone and 6 per cent permethrin, has provided control of codling moth at economic levels of less than 1 per cent harvest infestation in apple orchards in Switzerland, Based on reduction in trap catch and mating frequency of tethered moths, efficiency of the attract and kill droplets lasted 5-7 weeks, requiring two seasonal applications. Subsequent experiments replaced permethrin with an insect growth regulator, fenoxycarb, which has a sterilizing effect. Field tests showed that autosterilization, i.e. transfer of insect growth regulator from a contaminated male to the female moth at mating, contributes to the control effect. World wide annual production of codlemore is about 25,000 kg for codling moth control on about 21,000 ha **(Fig. 8.5)** and the price of codlemone is now well below US$ 1000 per kg. (Witzgall *et al.,* (2010).

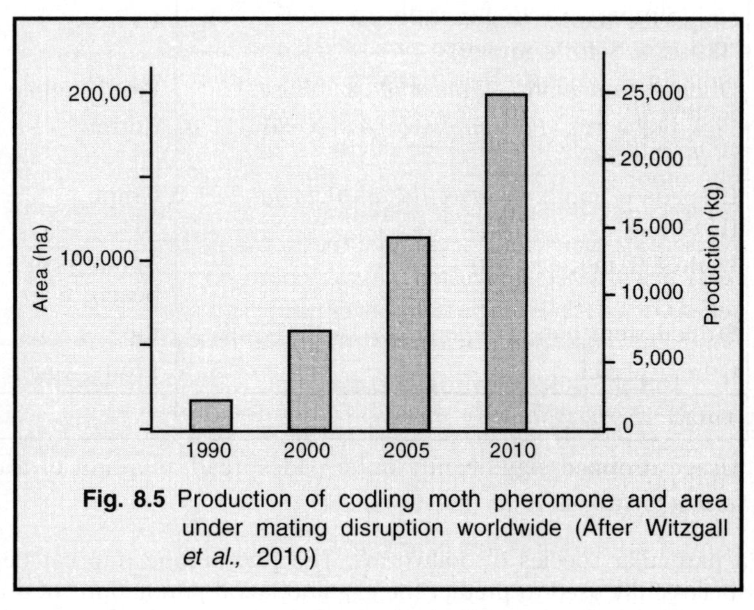

Fig. 8.5 Production of codling moth pheromone and area under mating disruption worldwide (After Witzgall *et al.,* 2010)

Brinjal fruit and shoot borer. Eggplant is an important vegetable in South Asia and brinjal fruit and shoot borer, *Leucinodes orbonalis* Guenee, is a key pest of the crop which causes enormous loss. A pheromone-based mass trapping strategy has been developed, from optimization of the pheromone blend and dose, trap design, and placement to field implementation. Mass trapping, without the use of insecticides, has led to a 50 per cent and higher increase in marketable fruit, which has been attributed to the combined effects of mass trapping and enhanced impact of natural enemies. Additionally, secondary pests, such as mites and whitefly, were reduced in the pheromone plots. The yield increase translate to earnings of $1,000 US$ per ha and year for resource-poor families.

Tomato pinworm. The tomato pinworm, *Keiferia lycopersicella* (Walsingham) has been amenable to control with mating disruption due to several features of the pest: (*i*) the host range of the tomato pinworm is restricted to plants in the family Solanaceae; (*ii*) as the pest feeds inside the calyx of the fruit, the insecticides have not proved to be highly effective in its control; and (*iii*) the pest has developed high levels of resistance due to the overuse of pesticides. The sex pheromone of the tomato pinworm was identified in 1979 and the first large scale trials (500 ha) were conducted in Culican, Mexico, in 1981. The results of this trial demonstrated that the programme reduced fruit injury with less use of insecticide and was economical as compared to conventional insecticide programmes. These results gave impetus to adoption of an IPM programme for tomatoes using pheromones, biological control, *B. thuringiensis* and cultural control.

Rice stem borers. Two species of rice stem borers, viz. *Chilo suppressalis* (Walker) and *Scirpophaga incertulas* (Walker) have been effectively controlled with sex pheromones in large scale field trials. In Spain, *C. suppressalis* has been successfully controlled for upto 100 days in 1500 ha of rice in both the Ebro Delta and Valencia rice-growing areas. Deployed at the rate of 2500 polymer dispensers ha^{-1} (8 mg active pheromone each) translating into 20 g ha^{-1}, they performed as well as conventional insecticide (fenitrothion) by controlling stem borers down to a damage level of 0.2 per cent. Confusion efficacy rates were approaching 100 per cent. Similarly, high rates and control levels were achieved by reducing the point sources to 500 or 100 ha^{-1}, depending on infestation history and using, respectively 1.6 or 8.0 g pieces of polymer to give a total dosage of 40 g ha^{-1}, a system which is now recommended commercially in Spain.

Similarly, a single application of a polymer formulation of the pheromone of yellow stem borer, *S. incertulas* of rice provided effective, season-long control of the pest in Andhra Pradesh, India in large scale field trials. Under normal pest pressure, the control was at least as good as achieved with several applications of conventional insecticides, but under high pest pressure both approaches failed to provide adequate control. Field trials on 20 ha of rice in West Bengal showed pheromone control to be equal to that achieved with insecticide regimes against yellow stem borer, *S. incertulas.* In Uttarakhand, mass trapping resulted in 5.78 per cent white ears due to *S. incertulas*, 70 days after installation of the pheromone traps, as compared to 17.90 per cent white ears in plots treated with insecticides, suggesting that sex pheromone mediated male annihilation technique is self-sufficient to keep pest under control. The use of pheromones for mating disruption of rice stem borers has also been made in Indonesia, China, Korea and Thailand.

Cotton boll weevil. The cotton boll weevil, *Anthonomus grandis* Boheman, is a major pest of cotton in the Americas. Males produce an aggregation pheromone, grandlure, that has been successfully incorporated into a pheromone-baited killing station known as "Boll Weevil Attract and Control Tubes." These tubes are produced in large numbers every year. A density of 14 traps per ha achieves a strong reduction in weevil population, at minimal crop damage. After successful control and eradication programmes in the USA, boll weevil trapping is now also used in South America on at least 250,000 ha.

Red palm weevil. The red palm weevil, *Rhynchophorus ferrugineus* (Olivier) is widely distributed in Asia, Africa and Oceania. It infests a range of tropical palms, including date, oil, and coconut palms. Mating in red palm weevil is mediated by an aggregation pheromone produced by the male weevil, composed of the main compound (4S, 5S)-4-methyl-5-nonanol (ferrugineol) and 4-methyl-5-nonanone. Traps loaded with ferrugineol, supplemented with ethyl acetate and plant volatiles, and a fermenting mixture of dates and sugarcane molasses, are placed at densities of up to 10 traps per ha for monitoring and mass trapping. Pheromone traps have played a significant role in the suppression of red palm weevil populations, *e.g.* in date palm plantations in Isreal.

Fruit flies. Several species of fruit flies, viz. guava fruit fly, *Bactrocera dorsalis* (Hendel) ; peach fruit fly, *B. zonata* (Saunders) and ber fruit fly, *Carpomyia vesuviana* Costa, are known to damage fruit crops in Punjab, Fruit flies cause almost 100 per cent damage to rainy season crop of guava, 78 per cent to peach, 84 per cent to pear and 85 per cent to kinnow. The population of fruit flies starts building up in April and continues up to November on different fruit crops. The Punjab Agricultural University (PAU) has developed a fruit fly trap using methyl eugenol based one litre mineral water bottle (Singh and Sharma, 2014). The trap consists of a plywood dispenser, suspended vertically inside the bottle, aligning with four vents that allow entry of fruit flies inside the bottle **(Plate 18.2D)**. The plywood blocks (5×5×1 cm) are immersed in a solution of ethyl alcohol, methyl eugenol (98%) and malathion in ratios of 6:4:1 (*v*/*v*) for 72 hours. The bented bottles are hanged with the trees at equidistant @ 40 traps per ha in the first week of May in peach, first week of June in pear, first week of July in guava and second week of August in kinnow. A single trap has the capacity to trap about 6000 male fruit flies. In case of rainy season guava, the yield in 40 traps per h was 20 t per ha as against 0.55 t per ha in untreated control. The corresponding yield in peach was 12.75 and 1.05 t per ha. The traps have provided effective control of fruit flies and have been recommended to farmers for use in guava, peach, pear and kinnow orchards.

Parapheromones

Parapheromones have a high potential as alternative material in integrated pest management programmes, particularly when the natural pheromones are expensive to produce, present longevity problems, or are quickly degraded under field conditions. Three important criteria need to be taken into consideration to include them in synthetic formulations, viz. the efficiency of the blend must be optimized, the specificity must be observed, and the stability must be improved. A number of field experiments have demonstrated that parapheromones meet these prerequisites for their use in insect monitoring or control.

The most extensive application of parapheromones has been the use of lures to attract tephritid fruit flies. Parapheromones and plant volatiles as attractants (methyl eugenol, trimedlure, cuelure, angelica seed oil, enriched ginger oil, raspberry ketone) and hydrolysed protein have been widely used for monitoring and annihilation of several fruit flies, including Oriental fruit fly, *Bactrocera dorsalis* (Hendel); melon fly, *B. cucurbitae* (Coquillett); and Mediterraneon fruit fly, *Ceratis capitata* (Wiedemann), for almost 50 years. Early detection of pest introductions using traps baited with male lures allows quick eradication of incipient populations. New infestations of the Oriental fruit fly, in California have been eradicated 12 times since 1966 using methyl eugenol in an insecticide-laced bait. Another parapheromone, trimedlure, has been developed for monitoring the Mediterranean fruit fly, and has been used in traps to pinpoint the location and size of infestation sites. The flies can then be eradicated through the use of a sprayable formulation of protein hydrolysate and an insecticide.

The standard monitoring programme for *C. capitata* in California is five traps baited with trimedlure per square mile in urban areas. After flies are detected, additional 1700 traps are baited in a 4.5 mile radius for three generations. In Los Angeles, the release of sterile flies in combination with helicopter spraying of malathion mixed with protein hydrolysate was successful in eradicating infestations of *C. capitata.* However, this programme was not favoured due to perceived environmetal and health risks. The baited traps were first developed in 1970 and are still in use in many countries today. The lure has proven particularly useful for monitoring pest presence within import/export crops at airports and seaports.

Structural modification of the pheromone generally reduces field activity, but a significant increase in catches has sometimes been reported. More males of *Cydia pomonella* (Linnaeus) were caught in traps baited with (*E,E*)–10, 11–difluro-8, 10-dodecadienol or (*E,E*)-11-chloro-8, 10-undecadienol than in traps containing codlemone. In large scale field trials, pheromone communication was disrupted in the navel orangeworm, *Amyelois transitella* (Walker), by air permeation with (Z, Z)-9,11-tetradececadien-1-ol formate, the formate analogue of the main pheromone component (*Z, Z*) –11-13-hexadecadienal. A novel olefin analogue was also found to disrupt mating communication in this insect. Both compounds are more stable than the natural aldehyde pheromone component.

Juvenile Hormone Analogues

The JHAs have been tested in large scale trials for the control of various pests of agriculture, forestry and public health importance. Five different formulations of methoprene are available commercially for the control of mosquitoes, manure breeding flies, cigarette beetle, pharaoh's ant, fungus gnat and flies. In house flies, adult emergence may be prevented by earlier treatment with methoprene. When administrated as a feed additive to cattle, methoprene controls horn flies and other veterinary pests that breed in dung. Other uses of methoprene include coleopteran and lepidopteran pests in stored tobacco, sciarid flies in mushrooms, flea larvae indoors, Homoptera on houseplants, pharaoh ants indoors and leaf-mining flies in vegetable and flower crops. Hydroprene is used indoors against cockroaches and in this case, exposure of nymphs to the material causes them to become sterile adults. Kinoprene is a strong highly selective JHA effective against aphids, whiteflies, scales and mealy bugs. In view of its environmental instability, it has been most successfully used on ornamental plants and vegetable seed crops in greenhouses and shade houses (Pedigo, 2002).

Different formulations of fenoxycarb have been developed for different groups of pests and used against number of coleopteran and lepidopteran pests of stored wheat and rice. It has also been found effective against several fruit pests such as grape moths, *Lobesia botrana* (Denis & Schiffermuller) and *Eupoecilia ambiguella* (Hubner); plum moth, *Cydia funebrana* (Treitschke); codling moth, *Cydia pomonella* (Linnaeus); summerfruit tortix, *Adoxophyes orana* (Fisher von Roeslerstamm); pear sucker, *Cacopsylla pyricola* (Forster); fruit tree tortix, *Archips podanus* (Scopoli); tufted apple bud moth, *Platynota idaeusalis* (Walker); and a number of leaf rollers, pear psyllids and diaspidid scales. Foliar application of fenoxycarb on cotton provided satisfactory control of *Helicoverpa zea* (Boddie) and *Heliothis virescens* (Fabricius). It is also effective for the control of fire ants, fleas, cockroaches and mosquito larvae (Dhadialla *et al.* 1998; Dhaliwal and Arora, 2012).

Pyriproxyfen has proved effective against a number of sucking pests including whitefly, *Dialeurodes citri* (Ashmead); *Thrips palmi* (Karny); California red scale, *Aonidiella aurantii* (Maskell); chaff scale, *Parlatoria oleae* (Colvee); cottony cushion scale, *Icerya purchasi* Maskell; sweet potato whitefly, *Bemisia tabaci* (Gennadius); greenhouse whitefly, *Trialeurodes vaporariorum* (Westwood); green peach aphid, *Myzus persicae* (Sulzer); and pear psylla, *Cacopsylla pyricola* (Forster). It is also active against a number of mosquito species and as a feed-through compound in poultry, cattle and swine for the control of the house fly, *Musca domestica* (Linnaeus); face fly, *M. autumnalis* DeGeer and horn fly, *Haematobia irritans* (Linnaeus). Pyriproxyfen also disrupts the development of horn fly following direct application to the fly. Pyroproxyfen has also been found to impair larval activity comprising hatching and leaf mining of tomato leaf miner, *Tuta absoluta* (Meyrick), leading to nearly 50 per cent mortality. This JHA also has an adverse effect on the development and metamorphosis of the citrus swallowtail, *Papilio demoleus* Linnaeus.

A newer analogue, diofenolan is highly effective for the control of scale insects and lepidopteran pests attacking citrus, pome fruit, stone fruit, grape, mango, olive, nut, tea and ornamental plants (Pedigo, 2002). It shows excellent activity at 300 g a.i./ha against *A. aurantii, Quadraspidiotus perniciosus* (Comstock), *Saissetia oleae* (Olivier), *Planococcus citri* (Risso), *I. purchasi, Aulacapsis tubercularis* Newstead, *C. pomonella* and *C. molesta* (Busck). It is also effective against tea leaf roller, *Caloptilia theivora* (Walsingham). NC-196, a derivative of benzyl pyridazinone shows good systemic and contact activity against brown planthopper of rice. In Japan, application of NC-196 @ 300-600 g a.i./ ha effectively suppressed brown planthopper populations for nearly one month.

Moulting Hormone Analogues

Both tebufenozide and methoxyfenozide are highly active in the field against a variety of vegetable, fruit and ornamental pests. Tebufenozide has been found to be highly effective against *Cydia pomonella* (Linnaeus) and miscellaneous leafrollers in apples, various leafrollers in treefruit, *Spodoptera exigua* (Hubner) in cotton, *Choristoneura fumiferana* (Clemens) and *Lymantria dispar* (Linnaeus) in forestry, *Lobesia botrana* (Denis & Schiffermuller) in grapes, *Cnaphalocrocis medinalis* (Guenee) and stem borers in rice, *Diatraea saccharalis* (Fabricius) in sugarcane, *S. exigua, Helicoverpa* spp. and various other lepidopteran pests in vegetables and ornamentals. The use rates vary from 100 to 300 g a.i./ha, depending on the target pest and crop.

Halofenozide is highly effective against the soil-dwelling larval stages of scarabaeid beetles such as *Popillia japonica* Newman, *Phyllophaga* spp., *Cyclocephala* spp. and *Hyperodes* spp. as well as various soil and sod-dwelling caterpillars such as cutworms and web worms. Halofenozide was effective against all stages of *P. japonica,* but its effects were stage dependent in case of European chafer, *Rhizotrogus majalis* (Razoumowsky) and the Oriental beetle, *Anomala orientalis* (Waterhouse). Generally, the effects were more pronounced when the applications were made in the egg or first larval stages. Methoxyfenozide has exhibited high field efficacy against lepidopteran pests of apple, cotton, grape, maize, rice, fruits and vegetables.

Chitin Synthesis Inhibitors

Chitin synthesis inhibitors belonging to the acyl urea group are also being used commercially for the control of a number of foliage feeders and tissue borers. Satisfactory control of cabbage looper, *Trichoplusia ni* (Hubner) ; cotton leaf perforator, *Bucculatrix thurbeviella* Busck; Egyptian cotton leafworm, *Spodoptera littoralis* (Boisduval) and boll weevil, *Anthonomus grandis* Boheman damaging cotton crop has been obtained with diflubenzuron. Foliar application of diflubenzuron has shown excellent residual activity against eggs of *H. armigera* and *S. litura.* In addition to its well known stomach toxicity, it has significant contact activity against some insect species including *H. armigera* and *S. littoralis.* Diflubenzuron is promising against a variety of other pests on several crops including soybean, citrus, vegetables and forests, and against nuisance and medical pests like flies, gnats, midges and mosquitoes.

Some new CSIs like XRD-473, IKI-7899 and IGR 1055 have shown excellent larvicidal activity against *H. armigera.* Teflubenzuron possesses high ovicidal action against the tortricid, *C. pomonella* on apple. Buprofezin has been found promising against homopteran pests of rice, cotton, citrus, potato and vegetables. Hexaflumuron is used primarily for termite control around houses and other buildings. Termites feed on hexaflumuron bait and die within 6-8 weeks during moulting. It is also effective against certain Lepidoptera, Coleptera, Homoptera and Diptera on cotton, potatoes and fruit trees. Lufenuron is prescribed by veterinarians for flea control on dogs. When a female flea bites a treated dog, lufenuron enters the flea's system, and prevents some eggs from hatching and prevents any enclosed larvae from becoming adults.

In India, several CSIs have been evaluated for their efficacy against *Helicoverpa armigera* (Hubner) infesting chickpea under field conditions. Lufenuron (0.006%) was found to be the most effective which recorded the least pod damage (11.9%) with highest efficacy (97.4%) and grain yield (2064.3 kg/ha). This was followed by flufenoxuron (0.008%), novaluron (0.01%), diflubenzuron (0.05%) and standard insecticide endosulfan (0.07%) with average efficiency of 94.8, 92,8, 89.9 and 65.9 per cent, respectively.

RESISTANCE TO BIORATIONALS

Insects were believed to be less prone to the development of resistance to biorationals because of their novel mode of action and high level of safety to predators and parasitoids. However, several instances of resistance have been documented. Pyriproxyfen resistance has been generated in house flies selected for 17 generations for resistance. Although the house fly which possessed 880-fold resistance to pyriproxyfen had no cross resistance to diflubenzuron, it showed medium cross-resistance to two other juvenile hormone analogues, fenoxycarb and methoprene. Resistance to pyriproxyfen has been observed in the whitefly, *Bemisia tabaci* (Gennadius), on cotton and rose. The ineffectiveness of diflubenzuron in controlling the tufted apple bud moth, *Platynota ideausalis* (Walker), was attributed to the increased levels of enzymatic detoxifi-cation, which were also observed in organophosphate resistant insects.

A loss in efficacy of tebufenozide was observed in a laboratory strain of codling moth, *Cydia pomonella* (Linnaeus), originally collected in southeast France and amplified for resistance to chitin synthesis inhibitor, diflubenzuron. However, no such loss in tebufenozide susceptibility was observed in a different strain of the codling moth resistant to benzophenylurea. Similarly, no loss in tebufenozide susceptibility was observed in a pyrethroid resistant strain of cotton leafworm, *Spodoptera littoralis* (Boisduval), and organophosphate resistant strain of the tufted apple bud moth, *P. idaeusalis* (Dhaliwal and Koul, 2011).

POTENTIAL AND CONSTRAINTS

The use of selective, biorational approaches in place of broad spectrum conventional insecticides offers several advantages in IPM programmes.

- Most of these chemicals are more or less non-toxic to man and domestic animals. Methoprene, for example, has an acute oral LD_{50} (rat) of > 34600 mg/kg.
- These compounds do not persist or accumulate in the environment and are degraded to simple molecules that are unlikely to cause problems of environmental contamination.
- Many of the semiochemicals are species specific and have no adverse effect on parasitoids and predators. Among the IGRs, studies conducted so far indicate that these are also comparatively safe to natural enemies.
- Most of these compounds are active at very low concentrations. In case of pheromones, it has been shown that even a single molecule landing on the receptor of a perceiving individual is capable of generating a response.

Biorational pesticides also suffer from several disadvantages which limit their large scale utilization.

- Each of these compounds is effective against a single pest or a closely related group of pests. Therefore, the market potential of these chemicals is very limited.
- These chemicals disrupt the development and bahaviour of insects and do not provide immediate control of pests. In case of IGRs, the treated larvae/grubs continue to cause damage for a number of days.
- Many of these compounds are photodegradable and, therefore, they rapidly loose their effect following their application to the crop. This problem has been solved in a number of cases by developing suitable protected, controlled release formulations which retain their effectiveness for a considerable length of time.
- As with other techniques, insects are capable of developing resistance to these chemicals. In many cases, insects already possess the ability to metabolize these compounds in the course of their normal development. Any selection pressure would, therefore, result in rapid development of resistance to such compound.
- Semiochemicals have to be used on an area-wide basis in order to achieve desired results.
- In many cases products available in the market are not of uniform quality and give inconsistent results. This problem could be overcome by establishing suitable standards for quality and performance.

These problems are not insumountable and could be solved by undertaking further research. There is a need for fundamental research on controlled release systems which are cheap, non-toxic and biodegradable. Suitable application technology also needs to be developed for these formulations. On a long term basis, the process of regulation of insect development needs to be understood in a sequential manner. The relative importance of different communication systems in the life of insect needs investigation. The behavioural responses of insects and effects of various meteorological and physicochemical factors on these responses need to be elucidated. Registration procedures for these products need to be simplified. Last decade has witnessed significant advances in most of these areas. The biorational products are becoming more reliable and look set for a promising future in IPM programmes.

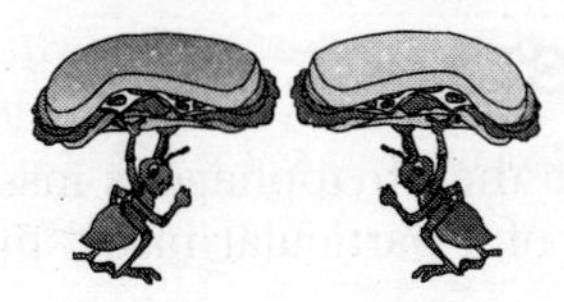

HOST PLANT RESISTANCE

INTRODUCTION

Although man showed keen interest in the relationship of arthropods with their host plants ever since silkworm was domesticated in China about 6000 years ago but the first documented record of an insect resistant variety was reported in 1782, when wheat variety Underhill was found to resist the attack of the Hessian fly, *Mayetiola destructor* (Say) in New York. However, the scientific investigations into the mechanism of host plant resistance date back to approximately 100 years, and the first significant economic contribution of host plant resistance in agriculture was made in 1890 when European grape vines were successfully grafted on to resistant rootstock to save the French wine industry from grape phylloxera, *Daktulosphaira vitifoliae* (Fitch). Another early instance of host plant resistance was that of the apple variety Winter Majetin found to be resistant to the woolly apple aphid, *Eriosoma lanigerum* (Hausmann) in England.

It is only during the last three decades that insect-plant interactions have been extensively investigated from the behavioural, ecological and physiological points of view. More than 400 public breeding programmes have released about 8000 modern varieties of 15 major crops in more than 100 countries. Today more than 500 insect-resistant cultivars of rice, maize, cotton, sorghum, alfalfa and wheat are available, which are providing substantial control of more than 50 key insect pests world-wide. These varieties are grown on millions of hectares annually and help the farmers to save billions of dollars in insecticide costs.

The development and use of insect resistant crop cultivars has led to significant crop improvements in the major food producing areas of the world over the pest 50 years. These improvements include increased food production, alleviation of hunger and better human nutrition. One of the most spectacular successes of insect resistant crops occurred during the green revolution era in tropical Asia during 1960s when high yielding pest resistant cultivars of rice were introduced into production agriculture. The cultivation of such cultivars has made significant improvement in the economies of several South and Southeast Asian countries with the result chat many countries that were previously food importers are now food exporters.

CONCEPT OF PLANT RESISTANCE

A plant is neither susceptible to all the phytophagous insects nor any insect species is the pest of all the species of plants. Host range of a particular insect may be wide or narrow whereas some insects like locust are feeders on all types of plants. However, such insects are usually not considered in host plant-insect interactions. Plant species which are fed by an insect are called host plants while those which are not fed at all are non-host plants. The inability of the insects to attack a non-host plant is termed immunity and such a plant is not considered a host of that insect. The terms host plant and *immunity* exclude each other. Plants which are not fed at all would not generally be considered for resistance and, therefore, would be classified as immune. A host plant can be resistant, more or less, but not totally immune. Any degree of host reaction short of immunity is, thus, resistance. Considering all the flora and fauna in nature and host plant-insect interactions, it may be said that immunity is the rule and susceptibility is an exception.

In every plant species there exists a great deal of diversity with respect to the extent of damage done by an insect. Individual plants which show lesser damage are called resistant and those showing more damage are called susceptible, thus, these terms are relative. Host plant resistance is the result of interactions between two biological entities, the plant and the insect under the influence of various environmental factors.

Since the host plant resistance is the result of interactions between the plant and the insect, it is, therefore, assumed that optimum conditions under which a plant species is grown are also favourable enough for the growth and development of the insect so that the plant species is accepted by the insect. The concept of host plant resistance should, therefore, be developed by comparing the performance of a variety under optimum conditions for the growth and development of the plant in the absence and presence of insect populations capable of causing maximum loss to the host plant.

Painter (1951) described plant resistance as the "relative amount of heritable qualities that influence the ultimate degree of damage done by the insect. In practical agriculture, resistance represents the ability of a certain variety to produce a larger crop of good quality than do ordinary varieties at the same level of insect population".

Maxwell *et al.* (1972) extended the definition of Painter (1951) by considering level of insect infestation and environmental conditions. According to them, resistance is "those heritable characteristics possessed by the plant which influence the ultimate degree of damage done by the insect. From a practical point of view, resistance is the ability of certain variety to produce larger yield of good quality than other varieties at the same initial level of infestation and under similar environmental conditions." According to Kogan (1982) resistance to insects is the "inheritable property that enables a plant to inhibit the growth of insect populations or to recover from injury caused by populations that were not inhibited to grow. Inhibition of population growth generally derives from the biochemical and morphological characteristics of a plant which affect the behaviour or the metabolism of insects so as to reduce the relative degree of damage these insects can potentially cause."

In other words, host plant resistance refers to the heritable qualities of a cultivar to counteract the activities of insects so as to cause minimum per cent reduction in yield as compared to other cultivars of the same species under similar conditions (Dhaliwal and Dilawari, 1993). The emphasis in this definition is not on the absolute yield obtained from the so called resistant variety as compared to the susceptible one, but on the per cent decrease in yield vis-a-vis the yield obtained without the attack of the insect. It means that a cultivar may yield poor but carries the genes for resistance and on the contrary a cultivar may yield good without having any genes for resistance.

TYPES OF RESISTANCE

Host plant resistance to insects may be divided into different categories based on several parameters :

Intensity of Resistance

Interactions between host plants and insects are spread over a wide spectrum of intensity. In terms of the host plant, lesser the population of the insect and/or lesser the damage they cause to the plant, more resistant the plant is likely to be. On the other hand, from the point of view of the insect, interaction varies from totally unsuitable host to completely suitable for growth and development of the insect. Therefore, intensity of resistance is a relative term and should be discussed in relation to a susceptible cultivar of the same species. Painter (1951) used the following scale to classify degrees of resistance based on intensity :

(*i*) *Immunity.* An immune variety is one which a specific insect will never consume or injure under any known conditions. There are thus few, if any, cultivars immune to the attack of specific insects which are, otherwise, known to attack cultivars of the same species.

(*ii*) *High resistance.* A variety with high resistance is one which possesses qualities resulting in small damage by a specific insect under a given set of conditions.

(*iii*) *Low resistance.* A low level of resistance indicates the possession of qualities which cause a variety to show lesser damage or infestation by an insect than the average for the crop under consideration.

(*iv*) *Susceptibility.* A susceptible variety is one which shows average or more than average damage caused by an insect.

(*v*) *High susceptibility.* A variety shows high susceptibility when much more than average damage is done by the insect under consideration.

These terms are relevant to express resistance vis-a-vis screening of varieties under field conditions and have nothing to do with the mechanism of resistance. An intermediate level of resistance is sometimes referred to as *moderate resistance.*

Ecological Resistance

Sometimes a plant or a variety may be classified as resistant due to unfavourable environmental conditions for the insect and no heritable trait is involved. In this case, there may be differential impact of the environment on the host and on the insect which affects the expression of resistance. Painter (1951) called this type of resistance as *pseudoresistance,* which refers to apparent resistance resulting from transitory characters in potentially susceptible host plants. Pseudoresistance is generally classified into three broad categories :

(*i*) *Host evasion.* Under some circumstances, a host may pass through the most susceptible stage quickly or at a time when number of insects is less. Some varieties evade injury by early maturing. Late planting of an early maturing variety or other special experiments will indicate whether true resistance is present or not.

(*ii*) *Induced resistance.* This term may be used for increase in resistance temporarily as a result of some changed conditions of plants or environment, such as change in the amount of water or nutrient status of the soil (Kogan, 1982). Such induced resistance may be of great significance especially in the field of horticulture, but should not be confused with inherent differences in resistance which exist between varieties or individual plants.

(*iii*) *Escape.* Escape refers to the absence of infestation or injury to the host plant because of transitory circumstances such as incomplete infestation. Thus, an uninfested plant located in a susceptible population does not necessarily mean that it is resistant. Even under very heavy infestation, susceptible plants will occasionally escape and only studies of their progenies will establish their true expression of resistance or susceptibility.

The terms host evasion and escape seem to be synonyms but critical analysis reveals that host evasion pertains to whole population of the host and insect is absent or insignificant while escape pertains to one or a few individuals in the presence of insects causing damage to other plants.

Evolutionary Concept

Resistance to an insect is evolved either due to long host plant and insect association at the gene centers or due to pleiotropic effects of genes which are present as a result of selective forces unrelated to the insect. Based on these factors host plant resistance to insects can be divided into sympatric and allopatric resistance.

(i) *Sympatric resistance.* Sympatric resistance may be defined as those heritable qualities possessed by an organism which influence the ultimate degree of damage done by a parasitic species having a prior continuous, coevolutionary history with that species of organism. This type of resistance evolves at original home of plants and insects. Association at the gene centres results in natural selection for resistance in plants. The resistance is evolved as a result of gene-for-gene nature of coevolution of plants and herbivores.

(ii) *Allopatric resistance.* Allopatric resistance may be defined as those heritable qualities possessed by an organism which influence the ultimate degree of damage done by a parasitic species having no prior continuous coevolutionary history with that species of organism. The resistance to insects in plants is evolved in the absence of insects to which the host is resistant. Allopatric resistance is not the result of coevolution, but rather due to fortuitous, pleiotropic effects of genes which are present as a result of selective forces unrelated to the pest insect.

Though not essential, in general, sympatric resistance is governed by major genes and allopatric resistance is polygenic in nature.

Genetic Resistance

Genetic resistance may be grouped under various categories :

A. Number of Genes

(i) *Monogenic resistance.* When resistance is controlled by single gene, it is called monogenic resistance.

(ii) *Oligogenic resistance.* When resistance is governed by few genes, it is called oligogenic resistance.

(iii) *Polygenic resistance.* When resistance is governed by many genes, it is referred to as polygenic resistance. The term horizontal resistance is also used to denote the resistance governed by polygenes.

B. Major or Minor Genes

(i) *Major gene resistance.* The resistance controlled by one (monogenic) or a few (oligogenic) major genes is called major gene resistance. This is also called *vertical resistance.* Major genes have a strong effect and these can be identified easily.

(ii) *Minor gene resistance.* When resistance is controlled by a number of minor genes, each contributing a small effect, it is called minor gene resistance. This is also referred to as *horizontal resistance.* In certain crops, the cumulative effect of minor genes is expressed when the plants grow older and this phenomenon is termed as *adult resistance, mature resistance* or *field resistance.*

C. Biotype Reaction

(i) *Vertical resistance.* This type of resistance is effective against certain specific biotypes of the insect but not against others. It is also called *specific resistance.* Vertical resistance is qualitative as the frequency distribution of resistance and susceptible plants is discontinuous.

(ii) *Horizontal resistance.* This type of resistance is effective against all the known biotypes of the insect. It is also called *nonspecific resistance.* Horizontal resistance is quantitative as the degree of resistance depends on the number of minor genes each contributing a small effect.

Multitrophic Interactions

Interaction among host plants, insect pests and their natural enemies (tritrophic interaction) leads to effective defense and attack at each level (**Fig. 9.1**). On this basis, two types of plant resistance have been recognised.

(*a*) *Intrinsic resistance.* Here the plant alone produces defense through physical means (trichomes or toughness) or through production of chemicals (toxins or digestibility reducers) or both (glandular trichomes or resins).

(*b*) *Extrinsic resistance.* Here the natural enemies (third trophic level) of insect pests (second trophic level) benefit the host plants (first trophic level) by reducing the pest abundance.

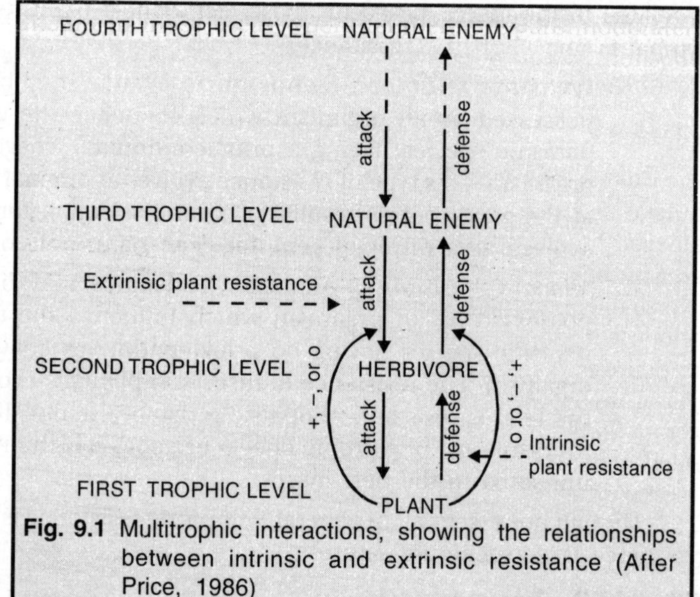

Fig. 9.1 Multitrophic interactions, showing the relationships between intrinsic and extrinsic resistance (After Price, 1986)

MECHANISMS OF RESISTANCE

Painter (1951) grouped mechanisms of resistance into three main categories, viz. nonpreference, antibiosis and tolerance.

Nonpreference/Antixenosis

Nonpreference refers to the response of the insect to the characteristics of the host plant which make it unattractive to the insect for feeding, oviposition or shelter. As the term 'nonpreference' pertains to the insect and not to the host plant, Kogan and Ortman (1978) proposed the term *antixenosis* to describe the host plant properties responsible for nonpreference. Antixenosis signifies that the plant is considered an undesirable or a bad host. Antixenosis may result from certain morphological characteristics or the presence of allelochemicals in the host plant (Kogan, 1982).

Under certain circumstances the nonpreference response of the insect can be quite important, especially when light infestations cause severe damage, e.g. infestation by insect vectors of plant diseases or insects which sever growing parts or peduncles of the plants such as stem borer infestations resulting in white heads. In field plantings, nonpreferred varieties frequently escape infestation and even when insects are caged on nonpreferred hosts, they lay fewer eggs and thereby develop smaller populations than those caged on susceptible varieties.

Antibiosis

Antibiosis refers to the adverse effect of the host plant on the biology (survival, development or reproduction) of the insects and their progeny infesting it. All these adverse physiological effects of permanent or temporary nature following ingestion of a plant by an insect are attributed to antibiosis. The insects feeding on resistant plants may manifest antibiotic symptoms varying from acute or lethal to subchronic or very mild. The most commonly observed symptoms in insects include larval death in first few instars, abnormal growth rates, disruption in conversion of ingested food, failure to pupate, failure of adults to emerge from pupae, abnormal adults, inability to concentrate food reserves followed by failure to hibernate, decreased fecundity, reduction in fertility, restlessness and abnormal behaviour. These symptoms may appear due to various physiological processes, viz. presence of toxic substances, absence or insufficient amount of essential nutrients, nutrient imbalances, presence of antimetabolites and enzymes adversely affecting food digestion and utilization of nutrients.

Tolerance

Tolerance refers to the ability of the host plant to withstand an insect population sufficient to damage severely the susceptible plants. It is generally attributable to plant vigour, regrowth of damaged tissues, resistance to lodging, ability to produce additional branches, utilization of non-vital parts by insects and compensation by growth of neighbouring plants. However, tolerance has no adverse effect on the insect pest. The ability of tolerant varieties to support insect infestation for longer periods without loss in yield or quality than the susceptible varieties enables them to frequently escape insect damage through compensation by the plants. As tolerance is not likely to provide a high level of resistance, it could be useful in combination with other mechanisms of resistance. Moreover, tolerant varieties do not depress insect populations nor do they provide any selection pressure on the insects. Thus, these can prove very useful to prevent the development of insect biotypes.

Although the above widely recognised classification of mechanisms appears to provide a generally acceptable breakdown of the phenomenon of host plant resistance, however, some overlap may occur between antixenosis and antibiosis, and a problem may arise in the separation of these two mechanisms. Antixenosis refers to undesirability, i.e. avoidance by insect whereas antibiosis refers to unsuitability, i.e. adverse effects on the insect after feeding on the host plant. However, sometimes it becomes difficult to separate the two mechanisms unless the insect-plant relationship is fully examined. For example, *Eruca sativa* Lam. (*taramira*) is not a preferred host of mustard aphid, *Lipaphis erysimi* (Kaltenbach). The growth and development of this insect was observed to be slower on *E.sativa* as compared to that on *Brassica* species in confinement. The mechanism appeared to be antibiosis, but it had been found that the poor development was due to reduced feed uptake because of the presence of certain allelochemicals in *E. sativa*, indicating antixenosis (Dhaliwal *et al.*, 1993).

BASES OF RESISTANCE

A number of plant characteristics are known to render the cultivars less suitable or unsuitable for feeding, oviposition and development of insect pests. Broadly, these characteristics can be classified into two categories, i.e. biophysical and biochemical.

Biophysical Bases

The plant resistance is controlled by several morphological factors like remote factors, e.g. colour, shape, size, etc. and close range or contact factors, *e.g.* thickening of cell walls and rapid proliferation of plant tissues, solidness and other stem characteristics, trichomes, incrustation of minerals in cuticle, surface waxes and anatomical adaptations of organs. The resistance mechanisms related to morphological or structural plant features that impair normal feeding or oviposition by insects or contribute to the action of other mortality factors are together called *phenetic resistance.* The morphological characteristics of the host plant may also influence the nutrition of the insect by limiting the amount of feeding due to shape, colour or texture which may limit the ingestion of the nutritive material and influence the digestibility and utilization of food by the insect.

A general association between resistance to stem borers and several morphological and anatomical characteristics of the rice plant, viz. tightly wrapped leaf sheaths, closely packed vascular bundles, thick sclerenchyma and high silica content has been recorded. The solid stem of Rescue and other wheat varieties is considered to be the major cause of resistance to the wheat stem sawfly, *Cephus cinctus* Norton. The sawfly eggs tend to be mechanically damaged and desiccated in resistant varieties and the hatching larvae are restricted in their movements. The minimum stem solidness required to obtain field control of sawfly has been established. A highly significant association between resistance to brown planthopper, *Nilaparvata lugens* (Stal), and red pericarp in rice has been found. There is a highly significant and positive correlation between leaf colour and percentage of dead hearts caused by shoot fly, *Atherigona soccata* Rondani on sorghum, and negative and highly significant correlation with leaf length/breadth ratio, plant height, glossiness of leaves, trichome density and length. The stem tips of cotton varieties tolerant to aphid, *Aphis gossypii* Glover were nearly twice as stiff as those of susceptible cultivars and, therefore, indicated that hardness for piercing the proboscis into the stems of tolerant strains, as one of the main causes for nonpreference by aphids. The resistance in sorghum to the sorghum midge, *Stenodiplosis sorghicola* (Coquillet), was positively correlated with the size of floral parts, viz. glume, lemma, palea, lodicule, anther, style and stigma.

Many leafhoppers are unable to establish on plants whose epidermis is covered with a thick layer of long cellulose hairs. Pubescent varieties of soybean are known to be highly resistant to potato leafhopper, *Empoasca fabae* (Harris). Trichomes affected the behaviour, oviposition, and growth and development of a number of insect pests. Okra varieties resistant to *Amrasca biguttula* (Ishida) had more and longer hairs on mid rib and lamina of leaves than susceptible varieties. Insect resistance was influenced by the number and length of hairs rather than their density. The resistance in clones of beach strawberry to black vine weevil, *Otiorhynchus sulcatus* (Farbicius) was found to be due to dense covering of simple hairs on the abaxial surface of the leaves. Trichomes on the pods of *Vigna vexillata* partly accounted for resistance to pod-sucking bug, *Clavigralla tomentosicollis* Stal. Pubescence in sugarcane adversely affected oviposition and larval movement by the pyralid, *Diatraea saccharalis* (Fabricius). However, tobacco budworm, *Heliothis virescens* (Fabricius), moths oviposited twice as many eggs on the hairy lines as on glabrous cottons. Leaf hairs also do not alter attack by the saddle gall midge, *Haplodiplosis marginata* (Roser) on wheat.

Biochemical Bases

A wide array of chemical substances including inorganic chemicals, primary and intermediary metabolites and secondary substances are known to impart resistance to a wide variety of insect pests. Broadly, the chemicals imparting resistance to insects can be classified into two main categories.

Nutrients. The host plant may be deficient in certain nutritional elements required by the insect and hence prove resistant. The nutritionally deficient plant may cause antibiotic and antixenotic effects on the insect. The antibiosis may result from the absence of certain nutritional substances in the host plant, deficiency of some nutritional materials and/or imbalance of available nutrients. Pea varieties resistant to pea aphid, *Acyrthosiphon pisum* (Harris), were generally deficient in amino acids and hence were less nutritious than the susceptible varieties. Resistance in bean varieties to Mexican bean beetle, *Epilachna varivestis* Mulsant, has been attributed to lower amounts of carbohydrates and reducing sugars.

The occurrence of asparagine in minute quantities in rice variety Mudgo was considered to be the primary cause of resistance to brown planthopper. Young females of brown planthopper caged on Mudgo had underdeveloped ovaries containing few eggs while those caged on susceptible varieties had normal ovaries full of eggs. It has been suggested that selection for low total free amino acids and high surface wax may lead to increase in resistance to cereal aphids in barley. Most of the resistant lines of oats and barley to *Rhopalosiphum padi* (Linnaeus) contained less asparagine, but higher amount of glutamic acid. The basis of resistance in maize to larval leaf-feeding of fall armyworm, *Spodoptera frugiperda* (J.E. Smith) has been explained in terms of amino acids. Although ratios of essential amino acids in susceptible and resistant lines were similar, there were differences in non-essential amino acids particularly aspartic acid, which was higher in resistant lines.

Allelochemicals. Allelochemicals are non-nutritional chemicals produced by an organism of one species and affect the growth, health, behaviour or population biology of individuals of another species. The allelochemicals have been broadly classified into two categories, viz. *allomones*-tending to confer an adaptive advantage to the producing organism, *i.e.* the host plant, and *kairomones*–

tending to give an adaptive advantage to the receiving organism, *i.e.* the phytophagous insect. Allomones are considered to be a major factor of insect resistance in plants and these have been exploited to increase levels of resistance in several agricultural crops.

One of the most classical examples of exploitation of allelochemicals in an economic crop is that of resistance in maize to first generation of the European corn borer, *Ostrinia nubilalis* (Hubner). It was shown that 2,4-dihydroxy-7-methoxy-1,4-benzoxazin-3-one (DIMBOA) is a resistance factor in maize to first brood borers. A significant correlation existed between the concentration of DIMBOA in the leaf whorl tissue and resistance to first brood borers. The chemical analysis of plant tissue for DIMBOA was employed as a method of screening for resistance to the first brood corn borer larvae. A correlation between the DIMBOA content of maize varieties and resistance to *O.nubilalis* larvae has been established and it has been suggested that it could form the basis of a rapid method for assessing resistance.

The major allelochemical imparting resistance to several insect pests in cotton has been found to be gossypol (8,8′-dicarboxaldehyde-1,1′,6,6′,7,7′-hexahydroxy-5,5′ diisopropyl-3,3′-dimethyl-2,2′-binaphthalene), which occurs in much higher quantities in glanded than in glandless varieties. The survival and development of major insect pests of cotton on high gossypol containing varieties is much less as compared to those containing lower amounts of gossypol. The efforts to transfer high gossypol into good agronomic varieties have met with success and some good agronomic types with high gossypol content have been developed.

The resistance of an isogenic strain of barley to greenbug, *Schizaphis graminum* (Rondani), was reported to be governed by benzyl alcohol which was also found in the resistant parent strain Omugi but absent in the susceptible parent Rogers. Gramine (N,N-dimethyl-3-aminomethyl- indole) has been suggested to be one of the factors responsible for resistance of barley seedlings to *R. padi.*

A wide array of chemicals appear to play a dominant role in host plant resistance, e.g. terpenoids including sesquiterpene lactones and heliocides; phenolic compounds including flavonoids and aromatic acids, proteinaceous compounds including protease inhibitors, glycosidase inhibitors and phytohemagglutinins, lectins; nitrogeneous compounds including amino acids and amides; toxic seed lipids including fatty acids, acetylenic and allenic lipids, fluolipids and cyanolipids, saponins, lignins and tannins.

The same plant allelochemical may play the dual role of repellent as well as attractant to different insects. For example, glucosinolates and their hydrolysis products are highly toxic to the unadapted lepidopteran, the swallowtail butterfly, *Papilio polyxenes* Fabricius, but are a feeding stimulant and provide host plant recognition clues for adapted insect, *Pieris brassicae* (Linnaeus). Such a phenomenon suggests that a chemical messenger can, therefore, be a 'double agent' (Table 9.1). Hence, classification of a chemical as a repellent, deterrent, feeding suppressant, toxin or digestibility reducer, may be situation and dose-dependent.

The defensive role of glandular trichomes of certain members of the Solanaceae plants against herbivorous insects has been studied. In certain wild potato species, *i.e. Solanum polyadenium* Greenm, *S. berthaultii* Hawkes and *S. tarijense* Hawkes, an exudate is discharged from the four-lobed head of the glandular hairs when aphids, *Myzus persicae* (Sulzer) or *Macrosiphum euphorbiae* (Thomas) mechanically rupture the cell wall. On contact with atmospheric oxygen, the clear, water-soluble exudate is changed into an insoluble black substance that hardens around the aphid's tarsi and seriously impedes its movement. Further accumulation of glandular material sticks the aphid firmly to the plant and starvation leads to death.

The nature and effect of exudates of trichomes against a number of insect pests have been elucidated. The trichomes exude a sticky substance, then sesquiterpenoids are released which disturb the insect and cause agitated movements. Subsequently, polyphenoloxidase and phenolic substrate react to form quinones. These events lead to insect immobilisation, cessation of feeding and ultimately death of the insect.

Table 9.1 Dual role of allelochemicals in insect-plant interactions

Allelochemical	Insect reaction	
	Stimulant	**Deterrent**
Cucurbitacin	*Diabrotica undecimpunctata* Mannerheim	*Epilachna tredecimnotata* (Latreille)
Cucurbitacin E and I	Diabroticites	*Phyllotreta nemorum* (Linnaeus)
Cyanogenic glycoside	*Epilachna varivestris* Mulsant	Many phytophagous insects (Fabricius)
Furanocoumarins	*Papilio polyxenes* Fabricius	*Spodoptera exempta* (Walker), *S. litura* (Fabricius), *S. eridania* (Cramer)
Glucosinolate	*Pieris rapae* (Linnaeus)	*P. polyxenes, Phyllotreta cruciferae* (Goeze)
Gossypol	*Anthonomus grandis* Boheman	*Helicoverpa zea* (Boddie)
Iridoid	*Euphydryas editha* (Edwards)	*Locusta migratoria* (Linnaeus)
Lignin	*Bootettix argentalus* Bruner	*Ligurotettix coquilletti* McNeill
Lupanin	*Macrosiphon albifons* Essig	*Acyrthosiphon pisum* (Harris)
Tannin	*Anacridium melanorhodon* Walker	*H. zea*
Tomatine	*Pieris brassicae* (Linnaeus)	*Leptinotarsa decemlineata* (Say), *Empoasca fabae* (Harris)

Source : Panda and Khush (1995)

Usually, a complex of allelochemicals is involved in imparting resistance to insect pests in agricultural crops. A number of chemical substances including phenols, alkaloids and methyl ketones have been demonstrated to be involved in host plant resistance in *Lycopersicon* to several insect pests. The growth of tobacoo budworm larvae is known to be retarded by several compounds, viz. gossypol and related compounds, several flavonoids, catechin condensed tannins, cyanidin, delphinidin and their glucosides. A number of components including volatiles, amino acids and *trans*-aconitic acid are involved in host plant resistance in rice to several insect pests. One of these chemicals, pentadecanal, has been isolated from TKM6, a *Chilo suppressalis* (Walker)-resistant rice variety. This chemical has exhibited semiochemical properties against several agricultural insect pests, but bioassays with crude extracts of rice varieties have demonstrated that some more allelochemicals may be involved. Thus, for breeding varieties with strong 'inhibitory biochemical profiles', a detailed chemical analysis of all the constituents of resistant varieties should be carried out.

GENETICS OF RESISTANCE

Information on the number of genes involved in resistance of plants to a particular insect pest, has great practical significance in identifying diverse sources of resistance and using these for breeding broad-based resistant plants. The inheritance of Hessian fly resistance in wheat has been most thoroughly investigated and there are 32 genes identified that are dominant or partially dominant for resistance. The resistance to first brood European corn borer in corn inbreds has been found to result from one gene pair; two or more gene action controlling resistance has been variously reported as dominant or partially dominant, primarily additive or having a significant epistatic component. The inheritance of resistance in corn to the corn leaf aphid, *Rhopalosiphum maidis* (Fitch), is determined by many genes with varying degrees of dominance and additivity.

The resistance to brown planthopper and green leafhopper in rice is simply inherited. Nine dominant [*Bph-1, Bph-3, Bph-6, Bph-9, Bph-10, Bph-14, Bph-16, Bph-17* and *Bph-18* (*t*)] and 12 recessive (*bph-2, bph-4, bph-5, bph-7, bph-8, bph-11, bph-12, bph-13, bph-15, bph-19(t), bph-20* and *bph-21*) genes for brown planthopper resistance have been identified from rice varieties. Ten dominant (*Glh-1, Glh-2, Glh-3, Glh-5, Glh-6, Glh 7, Glh-9(t), Glh-11(t), Glh-12(t)* and *Glh-13(t)*) and three recessive (*glh-4, glh-8* and *glh-10*) genes for green leafhopper resistance have also been identified. The resistance to whitebacked planthopper, *Sogatella furcifera* (Horvath), has been found to be governed by seven dominant genes designated as *Wbph-1, Wbph-2, Wbph-3,Wbph-5, Wbph-6, Wbph-7(t)* and *Wbph-8(t)* and one recessive gene, *wbph-4*. The resistance of barley to the greenbug is attributed to two dominant genes. A single incomplete dominant gene or dominant genes at more than one locus, account for resistance to greenbug in sorghum (**Table 9.2**).

Table 9.2 Genetics of resistance to major insect pests of crop plants

Crop	Insect pest	Gene(s) for resistance
Barley	Corn leaf aphid	*s*-1, *s*–2 (complementary)
	Greenbug	*Grb (Rsg-1a), Rsg-2b*
	Hessian fly	*H-f, Hf-1, Hf-2*
	Russian wheat aphid	*Dnb-1, Dnb-2*
Cowpea	Cowpea aphid	*Rac-1, Rac-2*
	Cowpea seed beetle	*rcm-1, rcm-2*
Raspberry	Raspberry aphid	A-1, A-2, A-3, A-4, A-5, A-6, A-7, A-8, A-9, A-10, A-k4a, A-cor1, A-cor2
Rice	Brown planthopper	*Bph-1, bph-2, Bph-3, bph-4, bph-5, Bph-6, bph-7, bph-8, Bph-9, Bph-10, bph-11, bph-12, bph-13, Bbh-14, bph-15, Bph-16, Bph-17, Bph-18(t), bph-19(t), bph-20, bph-21*
	Gall midge	*Gm-1, Gm-2, gm-3, Gm-4, Gm-5, Gm-6, Gm-7(t), Gm-8(t)*
	Green leafhopper	*Glh-1, Glh-2, Glh-3, glh-4, Glh-5, Glh-6, Glh-7, glh-8, Glh-9(t), glh-10, Glh-11(t), Glh-12(t), Glh-13(t)*
	Whitebacked planthopper	*Wbph-1, Wbph-2, Wbph-3, wbph-4, Wbph-5, Wbph-6, Wbph-7(t), Wbph-8(t)*
	Zigzag leafhopper	*Zlh-1, Zlh-2, Zlh-3*
Sorghum	Chinch bug	Monogenic, 1-2 dominant genes
	Greenbug	Monogenic, 2-3 genes
	Shoot fly	Polygenic, *tr*
	Sorghum midge	Polygenic, 2-3 genes
	Stem borer	Polygenic
Soybean	Soybean aphid	*Rag*-1, *Rag*-2, *Rag*-3, *rag*-1, *rag*-2
Wheat	Greenbug	*gb-1, Gb-2, Gb-3, Gb-4, Gb-5, Gb-6*
	Hessian fly	*H-1, H-2, H-3, H-4, H-5, H-6, H-7, H-8, H-9, H-10, H-11, H-12, H-13, H-14, H-15, H-16, H-17, H-18, H-19, H-20, H-21, H-22, H-23, H-24, H-25, H-26, H-27, H-28, H-29, H-30, H-31, H-32,*
	Russian wheat aphid	*Dn-1, Dn-2, dn-3, Dn-4, Dn-5, Dn-6, Dn-7, Dn-8, Dn-9*

***Source*:** Modified from Smith and Clement (2012)

The significance of genetic analysis of resistance is evident from the success of host plant resistance programme in rice for the brown planthopper (BPH). The first BPH-resistant variety with *Bph-1* gene, IR26, was released in 1973. It was widely accepted in Indonesia, Philippines and Vietnam, but became susceptible in 1976-77 due to the development of biotype 2 of BPH. IR36 with *bph-2* gene replaced IR26 and is still widely grown. Meanwhile, when a biotype capable of damaging IR36 appeared in small pockets in the Philippines and Indonesia, IR56 and IR60 with *Bph-3* gene for resistance were released. IR66 with *bph-4* gene for resistance was released in 1987, while IR68,

IR70, IR72 and IR74, all with *Bph-3* gene were released in 1988. These varieties are now extensively grown in tropical and subtropical rice-growing countries.

CONCEPT OF BIOTYPES

The continuous growing of insect-resistant varieties may lead to certain physiological and behavioural changes in insect pests so that they are capable of feeding and developing on the resistant varieties. The term biotype is generally used to describe a population capable of damaging and surviving on plants previously known to be resistant to other populations of the same species. More specifically, biotype refers to the populations within a species which can survive on and destroy varieties that have genes for resistance. Broadly speaking, the term biotype is an intraspecific category referring to insect population of similar genetic composition for a biological attribute.

Although the occurrence of biotypes among insects is comparatively less frequent than in plant pathogens, however, biotypes have been recorded in a number of insect species (**Table 9.3**). Biotype selection is, in fact, one of the major constraints encountered in breeding programmes of varietal resistance. Most biotypes do not arise *de novo* due to cultivation of resistant varieties but are present at a very low level in natural populations and increase in frequency as a result of continuous cultivation of resistant varieties. The concept of biotypes involves gene for gene relationship between the genotype for resistance in the host plant and the genotype for virulence in the insect. This phenomenon is well illustrated in case of brown planthopper in which the occurrence of five biotypes has been established. Biotype 1 destroys varieties that do not possess any gene for resistance (TN1). Biotype 2 damages varieties with *Bph*1 resistance gene (Mudgo). Biotype 3 thrives on varieties with *bph*2 gene for resistance (ASD 7). Biotype 4 damages varieties with *Bph*3 gene (Rathu Heenati), whereas biotype 5 destroys varieties with *bph4* gene for resistance.

Table 9.3 Biotypes of insect pests of agricultural crops

Insect species	Common name	Number of biotypes	Crop(s)
Acyrthosiphon pisum (Harris)	Pea aphid	9	Alfalfa
Amphorophora rubi (Kaltenbach)	Raspberry aphid	6	Raspberry
Bemisia tabaci (Gennadius)	Whitefly	9	Cotton, okra, cassaova
Eriosoma lanigerum (Hausmann)	Woolly apple aphid	3	Apple
Mayetiola destructor (Say)	Hessian fly	16	Wheat
Nephotettix virescens (Distant)	Green leafhopper	3	Rice
Nilaparvata lugens (Stal)	Brown planthopper	5	Rice
Orseolia oryzae (Wood-Mason)	Gall midge	6	Rice
Rhopalosiphum maidis (Fitch)	Corn leaf aphid	5	Maize
Schizaphis graminum (Rondani)	Greenbug	11	Wheat
Therioaphis trifolii f. *maculata* (Buckton)	Spotted alfalfa aphid	6	Alfalfa

***Source*:** Srivastava and Dhaliwal (2010)

The development of insect biotypes has posed a serious threat to the success of plant resistance for management of insect pests. Biotypes are known to occur in more than 36 arthropod species belonging to 17 families of six orders. Aphids constitute about 50 per cent of these species with known biotypes. Since most of the aphid species are parthenogenic, even one mutant capable of feeding on resistant variety can result into a new biotype. Biotypes are known to develop on varieties where antibiosis is the mechanism of resistance and they rarely develop on varieties where nonpreference or tolerance is the mechanism of resistance.

The future breeding programmes should be reoriented to cope with the problem of development of biotypes. These include the sequential release of varieties with major genes, pyramiding of major genes, development of horizontal resistance, combining major and minor genes, rotating major genes and breeding tolerant varieties that exert no selection pressure on the insect. A systematic surveillance programme for monitoring the shift to new virulent biotypes should be developed. The techniques to develop biotypes in the laboratory should be established so as to predict the stability of resistance in the field.

BREEDING FOR INSECT RESISTANCE

The first requirement of a breeding programme is the need and potential for plant resistance to insects within the pest management and cropping system. This will determine the type and level of resistance required, which in turn, will depend upon the pest biology and the production and commercial requirements for the crop. For example, if the insect feeds directly on a cosmetically important part of the plant, either antixenosis or a high level of antibiosis to prevent noticeable damage to the commercial product will be required. If, however, the insect population takes several generations on the crop to build-up to the economic threshold levels, moderate levels of any of the three types of resistance would be sufficient.

Development and standardization of screening techniques is pre-requisite to any effective resistance breeding programme. Information about the periods of greatest insect activity and hot spots is the first step to initiate work on resistance screening. Other effective means of augmenting insect populations, viz. delayed plantings and use of infester rows of a susceptible cultivar may also be employed. These techniques have been effectively used for gall midge, planthoppers and leafhoppers in rice; shootfly, stem borers, midge and head bugs in sorghum; leafminer and jassids in groundnut; pod borer and pod fly in pigeonpea; pod borer in chickpea; shootfly in pearlmillet, etc.

Screening for insect resistance under natural conditions is a long term process. Because of variations in insect populations in space and time, it is difficult to identify reliable and stable sources of resistance under natural multi-choice field conditions. In order to overcome these problems, it is essential to develop and standardize multi- or no-choice screening techniques where test cultivars can be subjected to uniform insect pressure at the most susceptible stage of the crop. This is done by placing relatively immobile stages, such as eggs, young larvae or apterous adults directly on the plants. These insects may be collected from the field, but more often, they are reared in the laboratory/screen house on susceptible host plants or artificial media. The results of the test are then rated in a standard manner, evaluating either the damage done by the insect to the plant or the effects of the plant on the attraction, growth, survival or reproduction of the insect. Such techniques have been developed in India for leafhoppers, gall midge and borers in rice; shoot fly, stem borer, midge and head bugs in sorghum; armyworm in pearl millet and sorghum; leaf miner, aphids, jassids and *Spodoptera* in groundnut; pod borer in chickpea and pigeonpea; stem borer in maize; etc.

The traditional breeding approaches have generally aimed to developing durable major gene resistance to single dominant pest based on some morphological/phenological/biochemical characteristics of the host plant. This type of resistance is known as vertical resistance in contrast to horizontal resistance which confers resistance against a broad range of genotypes but has low heritability. However, even partial horizontal resistance offers another advantage in the form of reduced selection pressure on a pest so that chances of breaking down of resistance are minimum. This form of resistance is, therefore, most desirable from the point of view of its stability.

Considerable progress has been made in India in identification and utilization of resistance for crop pests (**Table 9.4**). Resistance breeding programmes are underway only for a few crop pests. Insect resistance should be one of the major components in the development and release of new

Table 9.4 Insect-resistant varieties of different crops released in India

Crop	Insect	Varieties
Rice	Gall midge	Phalguna, Shakti, Surekha, Kakatiya,
	Stem borer	Ratna, Sasyasree, Saket
	Brown planthopper	Co42, Co46 Jyoti
	Green leafhopper	IR20, Vani
Maize	Maize borer	Ganga 9, Ganga Safed 2
	Pink stem borer	Deccan 101, 103
Sorghum	Shoot fly	M35-1, SPV 491, Swati
	Midge	ICSV 197, ICSV 745
Mung	Galerucid beetle	Jawahar 45, Gujarat 1, PIMS 4
Chickpea	Pod borer	C235, Anupam, ICCV 10, Pusa 261
Brinjal	Fruit and shoot borer	Doli 5, Pusa Purple Long, Pusa Purple Round
Okra	Fruit and shoot borer	AE 57, PM 58, Parkins Long Green, Nasnaul Special
Tomato	Fruit borer	T 32, T 27, Punjab Kesri, Pant Bahar
Pigeonpea	Pod borer	ICPL 332
Groundnut	Tobacco caterpillar & leaf miner	ICGV 86031, FDRS-10
	Jassids & thrips	M13, ICG5043
Rapeseed and mustard	Aphid	Regent, Laha 101, Pusa Kalyani
Cotton	Bollworms	Sanguineum
	Jassids	MCUS, Krishna
	Whitefly	Kanchan, Supriya
Sugarcane	Top borer	Co 67, Co 1158, Co 7224
	Stalk borer	Co 7302, CoS 767
	Scale insect	CoS 671, Co 8014, Co 611
Tobacco	Stem borer	SBR 1, SBR 2
	Tobacco caterpillar	GT4, DWFC
Pea	Pod borer	Boveville, T 6113
	Leaf miner	PS40, PS41-6, KMPR9

Source **:** Dhaliwal and Singh (2005)

crop varieties. Insect resistant varieties have been developed for rice (gall midge, stem borers, brown planthopper and green leafhopper), maize (maize stem borer and pink stem borer), sorghum (shoot fly and midge), pigeonpea (*Helicoverpa*), groundnut (*Spodoptera,* leafminer, jassids and thrips), rapeseed and mustard (aphid), cotton (bollworms and jassids), sugarcane (top borer, scale insects, mealy bugs and whitefly), tobacco (stem borer and *Spodoptera*) and pea (pod borer and leafminer) and these are being currently grown by farmers in India (Dhaliwal and Singh, 2005).

Varieties possessing multiple resistance to a number of insect pests and diseases are ideal in IPM programmes. The progress in breeding for resistance to multiple pest species varies among different crops and depends on a number of factors including the importance of the crop, importance of pests as constraints to production and the availability of resistant donors to use as parents in the breeding programme. Cultivars with multiple resistance to insects, nematodes, pathogens and tolerance to abiotic stresses (drought, soil mineral toxicity, etc.) have been developed in various crops, viz. wheat, alfalfa, cowpeas, maize, pearlmillet, sorghum, soybean and rice.

A spectacular success in development of improved varieties that possess resistance to as many as four insect pests and five diseases has been achieved at the International Rice Research Institute, Manila, Philippines. For example, IR36 is resistant to brown planthopper, green leafhopper, stem borers, gall midge, blast, bacterial blight and tungro **(Table 9.5).** The yield of IR8 fluctuates widely

Table 9.5 Insect and disease reactions of IR varieties of rice

Variety	Insect or disease reaction*									
	Brown planthopper biotype			Green leaf-hopper	Stem borer	Gall midge	Blast	Bacterial blight	Grassy stunt	Tungro
	1	2	3							
IR5	S	S	S	R	MS	S	MR	S	S	S
IR8	S	S	S	R	S	S	S	S	S	S
IR20	S	S	S	R	MR	S	MR	R	S	MR
IR22	S	S	S	S	S	S	S	R	S	S
IR24	S	S	S	R	S	S	S	S	S	S
IR26	R	S	R	R	MR	S	MR	R	MR	MR
IR28	R	S	R	R	MR	S	R	R	R	R
IR29	R	S	R	R	MR	S	R	R	R	R
IR30	R	S	R	R	MR	S	MS	R	R	MR
IR32	R	R	S	R	MR	R	MR	R	R	MR
IR34	R	S	R	R	MR	S	R	R	R	R
IR36	R	R	S	R	MR	R	R	R	R	R
IR38	R	R	S	R	MR	R	R	R	R	R
IR40	R	R	S	R	MR	R	R	R	R	R
IR42	R	R	S	R	MR	R	R	R	R	R
IR43	S	S	S	R	MR	S	R	R	S	S
IR44	R	R	S	R	MR	S	R	R	S	R
IR45	S	S	S	R	MR	S	R	R	S	S
IR46	R	S	R	MR	MS	S	R	R	S	MR
IR48	R	R	S	R	MR	NK	R	R	R	R
IR50	R	R	S	R	MR	NK	MS	R	R	R
IR52	R	R	S	R	MR	NK	MR	R	R	R
IR54	R	R	S	R	MR	NK	MR	R	R	R
IR56	R	R	R	R	MR	NK	R	R	R	R
IR58	R	R	S	R	MR	NK	R	R	R	R
IR60	R	R	R	R	MR	NK	R	R	R	R
IR62	R	R	R	R	MS	NK	MR	R	R	R
IR64	R	MR	R	R	MR	NK	MR	R	R	R
IR65	R	R	R	R	MS	NK	R	R	R	R
IR66	R	R	R	R	MR	NK	MR	R	R	R
IR68	R	R	R	R	MR	NK	MR	R	R	R
IR70	R	R	R	R	MS	NK	R	R	R	R
IR72	R	R	R	R	MR	NK	MR	R	R	R
IR74	R	R	R	R	MR	NK	R	R	R	R

*S: Susceptible; MS, Moderately susceptible ; MR, Moderately resistant ; R, Resistant ; NK, Not known. Reactions are based on tests conducted in the Philipinnes for all insects and diseases, except those for gall midge, which were conducted in India.

Source : Khush and Virk (2005)

due to the pressure of insect pests and diseases, whereas the yield of IR36 and IR42, having multiple resistance, shows little variation from year to year. Thus, insect-resistant cultivars have greater yield stability and ensure food security. IR36 is alone planted on about 11 million ha of area in the world and yields an additional income of one billion dollars annually to rice growers and processors. Rice varieties resistant to brown planthopper and green leafhopper are grown over 20 million ha of riceland in Asia. These resistant varieties can be grown with the minimum use of insecticides and are an important component of integrated pest management programmes. Dr Henry M. Beachell and Dr Gurdev S. Khush were awarded the 1996 World Food Prize for developing many high yielding varieties of rice, including the multiple pest resistant variety, 1R36.

Insect-resistant cultivars with desirable agronomic backgrounds have been developed in several crops, and cultivars with multiple resistance to insect pests and diseases will be in great demand in future. This requires concerted efforts from scientists involved in crop improvement programmes worldwide. There is a need to look into the following aspects to make plant resistance sustainable (Sharma, 2009).

- Emphasis laid on plant resistance in crop improvement programmes.
- Availablability of cost-effective and reliable screening techniques.
- Identification and utilization of sources of resistance to insect pests.
- Multilocational testing to understand genotype-environment interactions.
- Emphasis given to insect resistance in identifying and releasing new crop cultivars.
- Efforts to spread and popularize insect-resistant varieties.

DURABLE PLANT RESISTANCE

Various strategies can be adopted to prolong the useful life of the resistant varieties or to develop varieties with different genes so that the farmers may have access to new varieties when the resistance of the current varieties breaks down.

Major Gene Resistance

Several strategies can be employed to maintain varietal resistance and to prolong the useful life of major genes.

Sequential release of varieties. The sequential release of varieties with major genes involves the incorporation of a single major gene into commercial varieties. This strategy has been successfully followed for controlling the brown planthopper (BPH) of rice in Asia. Widespread outbreaks of BPH occurred in 1973-74 in several rice-growing countries. The rice varieties with *Bph-1* gene for resistance such as IR26, IR28 and IR30 were released in several countries such as Indonesia, Vietnam and the Philippines. By 1977-78, a new biotype capable of attacking these varieties appeared and varieties with *Bph-2* gene such as IR36, IR38 and IR42 were released. These were widely grown for about a decade when a new biotype appeared. Varieties with *Bph-3* gene, IR68, IR70, IR72 and IR74 were then released and are now widely grown. This strategy has also been followed for resistance to the Hessian fly in wheat.

Gene pyramiding. Pyramiding of major genes aims to combine two or more major genes into the same variety. Varieties with two or more major genes are likely to have a longer useful life as the development of new biotypes will be slower. It has been postulated that a simultaneous release of varieties with two genes for resistance to the Hessian fly of wheat would result in better durability than a sequential release. Pyramiding of two *Bt* genes through genetic engineering has also been proposed to prolong the useful life of transgenic resistance. Similarly, combining genes encoding a toxin and a repellent may offer longer lasting resistance than either approach alone.

Rotation of varieties. The process of adaptation of an insect can be interrupted by growing a resistant variety in one season and another resistant variety with a different gene during the next season. This strategy was followed to protect the rice crop from tungro virus epidemics in South Sulawesi, Indonesia, in 1970s. Varieties with one gene for green leafhopper (vector for virus) resistance were planted in one season and another variety with a different gene for resistance was planted in the next season. This strategy has been very effective in prolonging the useful life of the vector-resistant varieties.

Multiline varieties. The multiline approach envisages the incorporation of several major genes into an isogenic background and the mixing of these lines to form a multiline variety. This strategy was successfully employed for breeding oats with crown rust resistance. However, the effectiveness of this strategy in developing insect-resistant varieties is little known.

Varietal mixtures. This strategy employs the use of varietal mixtures consisting of 80-90 per cent resistant plants and 10-20 per cent susceptible plants of similar varietal background. Such varietal mixtures exert lower selection pressure on the insect as they are able to survive and reproduce on the susceptible plants.

Polygenic Resistance

Polygenic resistance is a quantitative trait that is governed by a large number of genes, each with a small contribution to resistance. The level of resistance is not generally high and it does not exert strong selection pressure on the insect, hence a virulent biotype rarely if ever develops and the resistance is more durable. However, parents with polygenic resistance are generally landraces with poor agronomic traits. In the process of selecting plants with better agronomic traits in crosses involving such parents, not all the polygenes are transferred and the level of resistance is diluted. Alfalfa germplasm resistant to the spotted alfalfa aphid has been developed through mass selection for polygenic variation.

INDUCED RESISTANCE

Induced resistance is the qualitative or quantitative enhancement of a plant's defense mechanisms against pests in response to external physical or chemical stimuli. Induced resistance results in change in a plant that produce a negative effect on herbivores. This is a non-heritable resitance where host plants are induced to impart resistance to tide over pest infestation. Induced resistance offers considerable promise to increase the levels of resitance to insect pests.

Plant allelochemical production may be induced by any injury to the plant, such as herbivore feeding or even through the autolysis of plant cells. Mechanical disruption of plant tissues whether by shaking or rubbing can also affect the plants physicochemically and thus, the development of insects associated with them. These mechanical disruptions evoke phytochemical responses which have been classified as (*i*) cellular chemical changes, (*ii*) changes in cells adjacent to the damaged tissues, and (*iii*) generalized changes apparent in a plant part or the entire plant. Many studies of induced responses have indicated changes in the levels of tannins and phenols, which are products of shikimic acid pathway. The relative activity of the enzyme phenylalanine ammonia lyase (PAL) can determine the production of phenolics, including lignin. Hence, PAL activity is considered an important indicator of induced resistance. Herbivore damage also affects the concentration of available nitrogen as well as other important nutrients in foliage.

Wounding plant tissues may induce changes in protein, lipid and phenol metabolism. Phytochemicals produced by damaged plants may be determinantal to insects. Mechanically wounded tomato leaves have been found to stimulate the release of a proteinase inhibitor inducing factor (PIIF) into vascular transport system of damaged plants. The feeding of larvae of *Spodoptera littoralis* (Boisdual) on damaged potato leaves decreased by nine fold within 8h after damage, and within 24h, leaves adjacent to initially damaged leaves promoted similar adverse effects on larval feeding. Wounding induced the oxidation of plant phenols to produce toxic quinones and synthesis of mono-and diphenols. Phenol levels also increased following damage by lygus bugs, *Lygus disponsi* Linnavuori, to Chinese cabbage, sugarbeet and cotton. One of the most exciting reports is about the damage to popular and sugar maple tree foliage that increased the total phenol content of foliage of adjacent, non-connected trees, suggesting that plants are capable of communicating through wounding. Therefore, mild wounding via defoliation, abrasion or infection appears to elicit a general plant response that is beneficial to the plant but determental to the insect.

The potential of the use of plant growth regulators (PGRs) to induce resistance in plants has recently attracted considerable attention. The PGR (2-chloroethyl trimethyl-ammonium chloride (CCC), limits fecundity or survival of the cabbage aphid, *Brevicoryne brassicae* (Linnaeus) and green peach aphid, *Myzus persicae* (Sulzer) on treated susceptible cultivars of Brussels sprout. PGRs affect the chemical bases of insect-plant interactions in two ways : (*i*) PGRs can alter the nutritional quality of the host plant by reducing the amount of available protein or amino acids; (*ii*) PGRs can trigger the biosynthesis of allelochemicals in the plant which will be both toxic and feeding deterrent to herbivorous insects.

ECONOMIC IMPACT

There are several examples which reveal distinct advantages to farmers by growing insect-resistant varieties. The Hessian fly, *Mayetiola destructor* (Say), used to be a serious pest of wheat in USA, but its incidence was reduced from nearly 100 per cent to below 1 per cent in certain areas by the cultivation of resistant varieties. By 1974, nearly 6.5 million ha were planted to Hessian fly-resistant wheat cultivars and by 1980, more than 28 Hessian fly-resistant varieties had been released to farmers in USA. The cultivation of resistant varieties of wheat has saved the growers at least $10 million annually in production losses from Hessian fly and wheat stem saw fly, *Cephus cinctus* Norton. The significance of resistance in this case is even more prominent because alternate control measures against these pests are ineffective or impractical, owing to large area involved and relatively small per unit area value of the crop. The yield losses from wheat stem saw fly alone could exceed 75 per cent when resistant varieties are not used.

Similarly, the cultivation of European corn borer-resistant maize has made it possible a reduction in the use of insecticides by about 22,000 tonnes per year against this insect and has increased maize yields considerably. The resistant maize inbreds grown in mid western U.S. reduced losses by the borer from $350 million in 1949 to $10 million in the 1960s. The use of resistant varieties to the spotted alfalfa aphid, *Therioaphis trifolii* f. *maculata* (Buckton) has also led to an annual saving of about 300 tonnes of insecticides in USA. The aphid-resistant varieties saved growers at least $35 million annually in the southwestern U.S. during the 1960s. It has been estimated that about 319,000 tons of insecticides (approximately 37% of the total insecticides applied during 1960s) were saved annually through planting of insect resistant cultivates of alfalfa, barley, maize and sorghum in USA.

The success in developing insect-resistant rice cultivars has been outstanding and cultivars resistant to brown planthopper, green leafhopper, yellow stem borer, striped stem borer and gall midge have been developed and are extensively grown. As an example, a rice variety 1R36, which is resistant to all the above five insects, is planted annually in over 10 million ha of riceland of the world. Its cultivation alone has yielded an additional income of one billion dollars annually to rice growers and processors. Rice varieties resistant to the brown planthopper and green leafhopper are planted over 200 million ha of riceland in Asia. Dr Gurdev S. Khush was awarded the 1996 World Food Prize for developing IR36 and other rice varieties.

Despite the cost and time of development, the ultimate returns from insect-resistant varieties are quite impressive. It has been estimated that the costs for development of cultivars resistant to Hessian fly, wheat stem sawfly, spotted alfalfa aphid and European corn borer were about $9.3 million. However, the total savings to cultivators using these varieties were about $308 million anually. Thus, after 10 years of use, these varieties provided a net saving of about $3 billion, *i.e.* a

300 : 1 return on each research dollar invested. The total estimated global value of insect resistant cultivars is approximately $ 2.0 billion. This is in addition to the other ecological advantages such as pest suppression in adjacent susceptible crops, absence of secondary pest outbreaks, reduced mortality of beneficial arthropod populations. and minimum disturbance of the agroecosystem.

FACTORS AFFECTING HOST PLANT RESISTANCE

The most desirable form of insect resistance is one which is stable across location and seasons. However, the level and nature of insect resistance are influenced by several climatic and edaphic factors. Several environmental factors are known to influence the inherited characters, especially those involving physiological characteristics of crop plants.

Temperature

Temperature is one of the most important physical factors of the environment affecting the behaviour and physiological interactions of insects and plants. Temperature-induced stress can cause changes in plant physiology and affect the expression of genetic resistance, resulting in changes in morphological defenses and/or changes in the levels of biochemicals or nutritional quality of the host plant. Temperature also has pronounced effect on plant growth and indirectly influences the extent of damage. Temperature also affects the biology and behaviour of insects and thus influences insect growth and population build-up. In general, low temperatures cause a negative effect on resistance. Differences between resistant and susceptible genotypes of sorghum to green bug increase with an increase in temperature. The level of resistance to pea aphid and alfalfa aphid in alfalfa is greater at higher temperatures. On the other hand, progressive loss of resistance to Hessian fly in wheat, has been reported at temperatures higher than 18°C.

Soil Moisture

The level of plant resistance is influenced by water stress. High levels of water stress reduce the damage by sorghum shoot fly, *Atherigona soccata* Rondani. The populations of *Aphis fabae* Scopoli have lower rates of reproduction on water stressed plants. However, water stressed plants suffer greater damage due to aphid feeding. Thus, moisture stress can alter the plant reaction to insect damage leading either to increased or lower susceptibility to insect damage. Atmospheric humidity also influences the insect-plant interaction. High humidity increases the ease of detecting of odours which may influence host finding and antixenosis mechanism of resistance to insects.

Photoperiod

Photoperiod affects the development of both the crop plants and their insect pests. It may also bring about changes in the physico-chemical characteristics of crop plants and thus influence the interaction between insects and plants. Intensity and quality of light have been reported to influence biosynthesis of anthocyanins and phenylpropanoids. Continuous high intensity light is reported to induce susceptibility in resistant PI 227687 soybean plants to cabbage looper, *Trichoplusia ni* (Hubner). Reduction in light intensity leads to loss in resistance of some wheat cultivars to wheat stem saw fly by decreasing stem solidness and density. Shade-induced loss of resistance has also been reported in sugarbect genotypes resistant to the green peach aphid. Similarly, shading has been known to reduce resistance in potato genotypes to Colorado potato beetle, which is due to reduced levels of steroidal glycosides, some of which are known to limit growth and development of beetle.

Nutrients

Plant nutrition exerts pronounced effect on resistance of plants to insects. In several cases, high levels of nutrients increase the susceptibility, whereas in others they increase the level of resistance. Application of nitrogenous fertilizers decreases the damage by shootfly and stem borer, *Chilo partellus* (Swinhoe) in sorghum. Application of phosphatic fertilizers is reported to decrease the damage of shoot fly. Similarly, application of potash decreases the incidence of sugarcane top borer, *Scirpophaga excerptalis* (Walker), and several rice insect pests like yellow stem borer, brown planthopper, green leafhopper, leaf folder, whorl maggot and thrips. However, high levels of nitrogen are known to increase the damage by cotton jassid, *Amrasca biguttula biguttula* (Ishida); rice stem borer, *Scirpophaga incertulas* (Walker) and many other insect pests of agricultural crops. In alfalfa, excess levels of nitrogen and magnesium, and low levels of calcium and potassium, have been associated with loss of resistance to the spotted alfalfa aphid.

Soil pH

Plants can grow in soils over a pH range of 3.0 to 9.0, but extreme conditions stress the plants. Foliar damage to whorl-stage sorghum by the fall armyworm, *Spodoptera frugiperda* (J.E. Smith) was greater in acidic soils (pH 5.4) than in plants grown in soils with a pH higher than 6.0. Salinity induces plant responses that can alter the suitability of a plant as a host for insects. Salinity stress at 1.25 m^{-1} increased nitrogen, decreased potassium and reduced the production of allelochemicals in rice-plant leaf sheaths. The salinity-stressed rice plants were conducive to the feeding and survival of the whitebacked planthopper, *Sogatella furcifera* (Horvath). The salinity stress led to increase in insects' growth and development, longevity, fecundity and population build-up.

Air Pollution

Air pollution refers to the presence in the outdoor atmosphere of one or more contaminants such as dust, fumes, gas, mist, odour, smoke or vapour in such quantities and duration as to be injurious to human, plant, or animal life or to property. Air pollution has been demonstrated to influence insect-plant interactions. When Mexican bean beetles were fed on soybean fumigated intermittently with SO_2, the mean number of beetle progeny was 1.5 times greater than that of beetles feeding on the control plants. Mexican bean beetles under laboratory conditions preferred to eat soybean leaves pre-fumigated with ozone-enriched ambient air. Substantial modifications in the form and content of plant nitrogen and sugars have been reported following plant exposure to moderate levels of ozone. Such nutritionally enriched plants are better hosts for insect herbivores; especially aphids.

Plant Factors

Insects have often exhibited feeding preferences for particular plant organs or for foliage of certain specific age. Moreover, variations in physical growth conditions such as plant canopy and plant height may influence host selection and subsequent population development. For example, *Myzus persicae* (Sulzer) was relatively more successful in feeding on older leaves of Brussels sprouts than *Brevicoryne brassicae* (Linnaeus) and vice-versa. The early-maturing genotypes of soybean show greater defoliation by soybean caterpillar, *Anticarsia gemmatalis* (Hubner); soybean looper, *Chryspdeixis includens* (Walker) and beet armyworm, *Spodoptera exigua* (Hubner). The damage by the sorghum midge, *Stenodiplosis sorghicola* (Coquillett) on sorghum hybrid CSH5 was higher in plots with low plant densities. In contrast, infestation by wheat stem sawfly, *Cephus pygmeus* (Linnaeus) was higher in low plant densities. Wheat plants sown at low densities (40 × 40 cm, 2.5 kg seed ha^{-1}) had a longer interval between plant development stages and had higher stem solidness than those sown at higher densities (10 × 3 cm, 133 kg seed ha^{-1}). Stem solidness was negatively correlated with the percentage of sawfly-infested stems.

Insect Factors

The age, sex and levels of infestation of insects affect the preference for host plants with varying levels of resistance. Young (3-day old) female Mexican bean beetles showed no preference for leaves of susceptible Deane soybean over resistant soybean leaves. In contrast, 14-day old Mexican bean beetle females showed a marked preference for leaves of susceptible plants. In feeding bioassays, the female beetles demonstrated a distinct preference for leaves of susceptible soybean over the resistant soybean leaves, but males did not. The size of sorghum shoot fly, *Atherigona varia* (Meigen) population had a pronounced effect on both absolute and relative rates of oviposition in various sorghum genotypes. The resistance was partially dominant when evaluated under low shootfly populations, however, when evaluated undcr high populations, susceptibility appeared dominant.

HOST PLANT RESISTANCE IN IPM

Host plant resistance as a method of insect pest control in the context of IPM has a great potential than any other method of pest suppression. The use of insect resistant cultivars improves the efficiency of other pest management practices, including the synthetic insecticides. In some cases, it is the only practical and effective method of pest management. Insect-resistant varieties need to be carefully fitted into pest management programmes in different agroecosystems. The nature of deployment, alone or in combination with other methods of insect control, depends on the level and mechanism of resistance, and the cropping system.

A. Principal Method

Host plant resistance to insects has been used as a primary method of pest control long before the advent of synthetic organic insecticides. A few insect pests have been controlled for many years by the use of resistant crop varieties alone(**Table 9.6**). The first deliberate use of plant resistance to control a major insect pest was the importation in 1873 of American phylloxera-resistant rootstock into France for the control of grape phylloxera, *Daktulosphaira vitfoliae* (Fitch). This insect had destroyed over 1 million ha of vineyards and caused tremendous loss to the French wine industry, but the use of resistant rootstock allowed a quick recovery, resulting in effective control of phylloxera for more than 100 years. This is still the principal method of phylloxera control worldwide.

Table 9.6 Selected examples of use of varieties with resistance to insect pests as the principal method of pest management

Crop	Insect pest		Region
	Common name	Scientific name	
Alfalfa	Spotted alfalfa aphid	*Therioaphis maculata* (Buckton)	USA
	Pea aphid	*Acyrthosiphon pisum* (Harris)	
Chickpea	Pod borer	*Helicoverpa armigera* (Hubner)	India
Corn	European corn borer	*Ostrinia nubilalis* (Hubner)	USA
	Corn earworm	*Helicoverpa zea* (Boddie)	
Cotton	Cotton jassid	*Amrasca biguttula biguttula* (Ishida)	India
	Cotton jassid	*Jacobiella facialis* (Jacobi)	Africa
Grapes	Grape phylloxera	*Daktulosphaira vitifoliae* (Fitch)	Worldwide
Pigeonpea	Pod borer	*H. armigera*	India
Rice	Brown planthopper	*Nilaparvata lugens* (Stal)	Worldwide
	Gall midge	*Orseolia oryzae* (Wood-Mason)	India
	Green leafhopper	*Nephotettix virescens* (Distant)	Worldwide
Sorghum	Greenbug	*Schizaphis graminum* (Rondani)	USA
	Shoot fly	*Atherigona soccata* Rondani	India
	Sorghum midge	*Stenodiplosis sorghicola* (Coquillett)	India
Wheat	Hessian fly	*Mayetiola destructor* (Say)	Worldwide
	Wheat stem saw fly	*Cephus cinctus* Norton	

Source : Dhaliwal *et al.* (2013)

The resistant varieties of wheat in USA have provided the principal method of control of the Hessian fly, *M. destructor*. Pawnee, Ponca, Poso 42 and Big Club 43 developed in Kansas and California, were the first Hessian fly resistant varieties released. These were followed by the release of the resistant varieties Dual and Benhur in Indiana. By 1974, 8 million ha of wheat in USA were planted with 42 varieties resistant to Hessian fly and 0.6 million ha included five varieties resistant to wheat stem sawfly, *Cephus cinctus* Norton. As a consequence of the use of resistant varieties, losses inflicted by Hessian fly were reduced to less than 1 per cent and this insect was reduced to the status of a minor pest.

The control of European corn borer, *Ostrinia nubilalis* (Hubner), a major pest of corn in USA also depends on the use of resistant varieties. The control of borer by insecticides and cultural methods generally has not been satisfactory and the insect caused $350 million worth of damage in 1949. A major research effort was directed towards the development of borer resistant varieties. Since 1940s, more than 100 insect-resistant lines of corn have been developed. Widespread planting of hybrids with partial, polygenic resistance reduced corn borer survival and damage to the plant over a wide area and reduced annual losses to an average of $100 million annually in the late 1960s. All the field corn hybrids grown on 33 million ha in USA are routinely screened to maintain at least a moderate level of resistance.

The brown planthopper, *Nilaparvata lugens* (Stal), is the single most important pest of rice in Asia. Extensive research to develop varieties resistant to this pest have been carried out at the International Rice Research Institute (IRRI) in the Philippines since early 1960s. The first brown planthopper-resistant variety was released in 1963 and since then several varieties have been released to counter the development of new biotypes by this insect. These varieties are currently planted over millions of hectares in Asia and have proved effective to reduce the damage by this insect. At present, IRRI is operating a sequential release strategy of varieties with new genes for resistance.

In India, insect resistant varieties have been developed for the control of a number of insect pests. Several insect pests have been kept under check through the use of insect resistant cultivars, e.g. *A. biguttula biguttula* - Krishna, Mahalaxmi, Khandwa 2 and MCU5; rice gall midge, *O. oryzae* - IR36, Kakatiya, Surekha and Rajendradhan ; Sorghum shoot fly, *A. soccata*-Maldandi, Swati and ICSV705; sorghum midge, *S. sorghicola*–ICSV745, DJ6514 and AF28 ; *H. armigera*- ICPL332 and ICPL88039 (pigeonpea), and ICC506 and ICCV10 (chickpea).

B. Integration with Other Tactics

Host plant resistance can be effectively used in combination with other control tactics of IPM. High levels of resistance are not necessary for a crop variety to be of practical value in IPM. Varieties with low or moderate levels of resistance can be used to good advantage for pest suppression. Deployment of insect-resistant cultivars should be aimed at conservation of natural enemies and minimizing the number of insecticide applications.

Chemical Control

The most common form of integrated control involves the use of insect-resistant cultivars and insecticides. The pest numbers are reduced in each generation, and this process slows the rate of population growth of target insects. Even a moderately resistant cultivar in combination with insecticides can bring about a substantial reduction in pest numbers, and minimize the losses in grain yield. Plant resistance enhances the effectiveness of insecticides through :

- Better insecticide coverage of the plant parts through modified plant canopy, for example, loose panicles in sorghum and frego-bract in cotton.
- Imbalanced nutrition or toxic substances having an adverse effect on insect growth and development, which may increase insect susceptibility to insecticides.
- Easy access to parasitoids and predators through changes in plant canopy.

There are many examples of host plant resistance enhancing the efficacy of insecticides . Evaluations of rice insecticides indicate that they cause higher mortality of planthoppers and leafhoppers feeding on resistant than on susceptible rice varieties. Mortality of brown planthopper when reared on either a moderately resistant ASD7 or a highly resistant cultivar Sinna Sivappu was higher than when feeding on a susceptible TN1 cultivar. The LD_{50} of whitebacked planthopper was 9.4 on the susceptible variety TN1 treated with ethylan, but only 2.8 on moderately resistant N22. The combination of moderate varietal resistance and low dose of pesticide resulted in effective hopper control. The integration of host plant resistance and insecticides has cumulative effect on *Nephotettix virescens* (Distant), the vector of rice tungro virus. There was no tungro virus infection on the resistant cultivar IR28 even without the application of the insecticides. On the moderately resistant cultivar, IR36, tungro virus infected plants decreased from 42 per cent in the 0 kg a.i./ha treatment to 10 per cent in the 0.5 kg a.i./ha rate. On the susceptible cultivar, IR22, virus infestation was higher at all insecticide rates decreasing from 92 per cent in the control to 74 per cent in the 1 kg a.i./ha rate.

Cotton cultivars exhibiting the frego-bract and okra (thin) leaf traits allow more than 30 per cent penetration of insecticides into the cotton foliage canopy, increasing the efficiency and decreasing the amount of insecticide required for control. In frego-bract cottons, the square has a rolled, twisted, and open bract (unlike in normal cotton, where the bract is flat and encloses the square). Insecticide aplication is not required for boll weevil control on frego-bract cotton varieties, where up to 94 per cent of boll weevil population was suppressed. This also reduces the over-wintering population of the weevil. A high level of oviposition suppression can be very useful in eradication programmes. As boll weevils feed and oviposit on cotton buds, the exposed buds in the frego-bract cotton can ideally be covered with insecticides. When sprayed with methyl parathion, frego-bract buds have seven times more deposits of insecticide residue than those with normal bracts.

Generally, a lower concentration of insecticide is needed to control insects feeding on a resistant variety than those feeding on a susceptible variety. In this regard, nymphs of the wheat grain aphid, *Sitobion avenae* (Fabricius), reared on resistant wheat variety Altar possessing the antibiosis compound DIMBOA were significantly more susceptible to the insecticide deltamethrin than nymphs on the susceptible wheat variety Dollarbird. The LD_{50} was reduced by 91 per cent for nymphs reared on the cultivar with a high DIMBOA content. Although the population of the aphid, *M. persicae* on the partially resistant variety of Brussels sprouts was about 85 per cent of that on the susceptible variety, the LD_{50} of malathion was only about 55 per cent, *i.e.* the insecticide requirement was much less on the partially resistant variety. A half dose of chlorfenvinphos gave equal or better control of turnip root fly, *Delia floralis* (Fallen) on resistant cultivar of Swede (Cruciferae) S7790 than the full dose on the susceptible cultivar, Ruta.

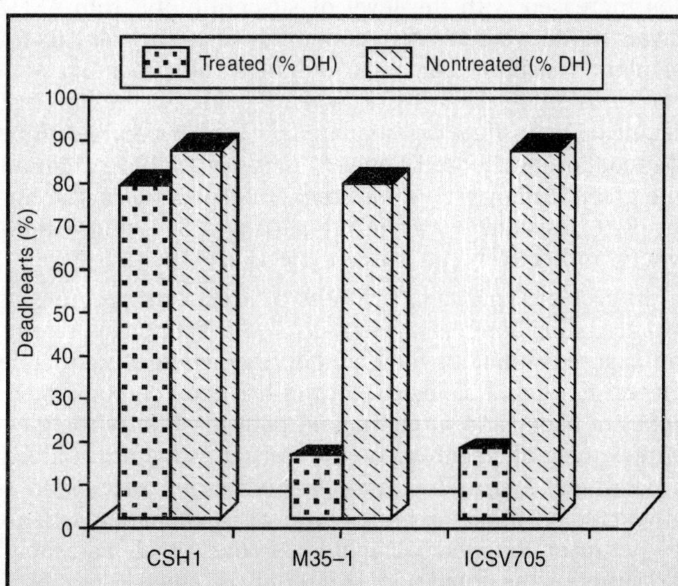

Fig. 9.2 Effect of carbofuran treatment on dead heart formation due to sorghum shoot fly, *Atherigona soccata* Rondani in susceptible (CSH1) and moderately resistant (M 35-1 and ICSV705) cultivars of sorghum (After Sharma, 2009)

A combination of moderate levels of plant resistance and insecticide application can be used for effective control of insect pests. Application of carbofuran granules results in a significant reduction in deadheart formation due to shoot fly, *Atherigona soccata* Rondani, in sorghum cultivars with moderate levels of resistance (M 35–I and ICSV705) (**Fig. 9.2**), but there is no effect of carbofuran application on shoot fly damage on the susceptible cultivar, CSHI Maximum grain yield in sorghum has been realized with four sprays of demeton-S-methyl against sorghum head bug, *Calocoris angustatus* Lethiery on IS 9692 and CSH11, whereas only 1-2 sprays are sufficient to realise the maximum yield potential of the head bug-resistant genotypes, IS17610 and IS21443 (Sharma, 2009).

In spite of many positive interactions between plant resistance and insecticides, negative interactions between the two tactics do occur. This is mainly due to the enhanced detoxification of insecticides when pest insects are fed foliage containing high levels of allelochemicals that mediate insect resistance in several agricultural crops.

Biological Control

Plant resistance and biological control are the key components of integrated pest management. Varieties with moderate levels of resistance that allow the insect densities to remain below the economic threshold levels are best suited for use in pest management in combination with natural enemies. Insect-resistant varieties also increase the effectiveness of natural enemies because of a favourable ratio between the densities of the target pest and its natural enemies. Such a combination is more effective in crops with a tolerance mechanism of resistance. This combination also reduces the insect population's genetic response to selection pressure from either plant resistance or the natural enemies. In general, the rate of insect adaptation to a resistant cultivar is lower when the suppression is achieved by the combined action of plant resistance and natural enemies than by high levels of plant resistance alone.

By reducing pest numbers, resistant varieties help to shift the pest : predator ratios in favour of biological control. In field studies at International Rice Research Institute (IRRI), Philippines, the brown planthopper, *Nilaparvata lugens* (Stal) : spider, *Lycosa pseudoannulata* (Rosenberg & Strand) ratios increased with the level of susceptibility from ASD7 and IR36, both highly resistant rice cultivars to IR42 and Triveni, moderately resistant cultivars to IR8 and TN1, susceptible cultivars. The host plant resistance may also enhance the predatory activity. Predation rate of the mirid bug, *Cyrtorhinus lividipennis* Reuter when feeding on the first instar *N. lugens* nymphs increased on the resistant cultivar, IR36 as compared to the susceptible IR8. Combinations of host plant resistance and predation by the mirid bug, *C. lividipennis* have a cumulative effect on the population increase of the green leafhopper, *N. virescens*. In cage studies, the number of green leafhopper reached only 6 on IR29 (resistant) with the predator and 31 without the predator, while there were 91 and 220 hoppers, respectively, on susceptible IR22.

In case of sugarcane, tightness of leafsheath and increased hardness of stalk tissues conferring resistance to the borer, *Diatraea saccharalis* (Fabricius) prolong the time during which the pest remains in an exposed situation, vulnerable to the action of natural control agents. The compatible nature of plant resistance and biological control has been demonstrated in case of interaction between resistant varieties of barley and sorghum, and parasitization of the greenbug, *Schizaphis graminum* (Rondani). The movement of greenbugs on the resistant sorghum cultivars exposes them to greater parasitization. The parasitoid, *Lysiphlebus testaceipes* (Cresson) was able to keep the biotype C greenbug population nearly static on both susceptible and resistant barley, when the initial population of the aphid was three per plant. But when 12 aphids and one female parasite were introduced per plant, the parasitoid could suppress the aphid population only on the resistant barley. Thus, damage to barley was reduced by the combined effect of varietal resistance and parasitoids.

There are many ways natural enemies could be adversely affected by plant characteristics, such as pubescence or sticky glandular trichomes that interfere with searching; high levels of allelochemicals that accumulate in the body of the pest and affect natural enemies feeding on the pest, or changes in plant characteristics, such as colour structure or volatile chemicals used for habitat location by the natural enemy. High trichome density in insect-resistant cotton, potato and tomato has been shown to be detrimental to predators and parasitoids. The effects of the parasitoid, *Trichogramma pretiosum* (Riley) and the predator, *Chrysoperla rufilabris* (Burmeister) on the larvae of the bollworm, *Heliothis virescens* (Fabricius) are reduced with increasing degrees of cotton leaf pubescence. Genotypes of tobacco with glandular trichomes severely limit the parasitization of the eggs of tobacco hornworm, *M. sexta* by *Telenomus* sp. and *Trichogramma minutum* Riley. In case of pigeonpea and chickpea, trichomes and trichome exudates hamper the parasitization of eggs of *H. armigera* by the egg parasitoid, *Trichogramma chilonis* Ishii. The density of trichomes on the stems of tomato plants affects the ability of the predatory mite, *Phytoseiulus persimilis* Athias-Henriot to control the phytophagous, two-spotted spider mite, *Tetranychus urticae* Koch. Similarly, densely pubescent genotypes of cucumber impede the movement of *Encarsia formosa* Gahan to locate the greenhouse whitefly, *Trialeurodes vaporariorum* (Westwood).

Suppression of the effects of beneficial arthropod populations may also result from the effects of allelochemicals in resistant plant cultivars being transferred to the predators or parasitoids. Some genotypes of insect-resistant potato, tomato and soybean contain levels of toxic allelochemicals that have negative effects on beneficial insects, entomopathogenic fungi and insect viruses. The first example of negative interaction between host plant resistance and natural enemies was the toxicity of a-tomatine, an alkaloid from resistant tomato cultivars, to *Hyposoter exiguae* (Viereck), an endoparasitoid of *Helicoverpa zea* (Boddie). Tomatine also affects the egg predators, *Coleomegilla maculata* (DeGeer) and *Geocoris punctipes* (Say), when *H. zea* is fed on the folliage of wild tomato line, PI 134417. The reduction in parasitism of *H. zea* and *M. sexta* eggs by *Telenomus sphingis* (Ashmead) and *T. pretiosum* on accession PI 134417 is due to the effects of exposures to methyl ketones (2-tridecanone and 2-undecanone), as well as reduced parasitoid mobility after entrapment in trichome adhesive exudates. Similarly, high levels of isoflavone-based resistance in soybean are detrimental to a number of natural enemies. The pathogenicity of the fungus, *Nomuraea rileyi* (Farlow) Samson is reduced if *H. zea* larvae ingest tomatine from tomato plants. Thus, a better understanding of the evolution of crop plants, pests and pest biological control agents is required to understand how plant resistance and biological control can be integrated for more durable insect pest management.

Cultural Control

Cultural practices cause specific physiological changes that reduce the suitability of host plants for phytophagous insects. Most of these practices have long been associated with subsistence farming and are compatible with other pest control tactics, including host plant resistance. Insect-resistant cultivars, including those that can escape pest damage, are highly useful in pest management in combination with cultural practices. Cultural control by itself may not reduce the pest populations below economic threshold levels, but aids in reducing the losses through interaction with plant resistance. Plant resistance in concert with cultural control can also drastically reduce the need for insecticide application.

Insect-resistant varieties in combination with early planting, early maturity, defoliation, destruction of stalks and deep ploughing can be used effectively to control boll weevil, *Anthonomous grandis* Boheman and bollworms, *Heliothis virescens* (Fabricius) and *Pectinophora gossypiella* (Saunders) in cotton. The nectarless cotton varieties reduce pink bollworm infestation by 50 per cent, and this in combination with cultural practices can reduce the pink bollworm infestation by 16-fold.

It has been demonstrated that *Nilaparvata lugens* (Stal) populations and *N. lugens*: predator ratios are significantly lower on very early-and early-maturing rice cultivars than those on mid-season maturing cultivars. Thus, the incorporation of *N. lugens* resistance into early maturing cultivars can enhance the crop protection from this pest. Infestation of the greenbug, *Schizaphis graminum* (Rondani) , has been shown to be effectively reduced by combining the planting of a *S. graminum* resistant sorghum cultivar at a later than a normal planting date, in combination with no tillage cultivation of the preceding crop stubble.

Trap crops that attract insect pest populations (so that they may be destroyed) are synergistic when used in combination with insect resistant cultivars of several agricultural crops. The growing of antixenotic cotton cultivars resistant to boll weevil, *A. grandis* in combination with early maturing cotton cultivars, is effective in suppressing boll weevil populations. Treatment of boll weevil on the trap crop, causes a 20 per cent reduction in overall insecticide application and increases yield by 14-33 per cent. Rice trap crop planted 20 days earlier than the main crop (a brown planthopper resistant cultivar), attracts more hopper population, preserves more natural enemies and yields significantly more than the fields without trap crop.

POTENTIAL AND CONSTRAINTS

Utilization of plant resistance as a control strategy has enormous practical relevance and additional appeal. It is in this context that host plant resistance assumes a central role in our efforts to increase the production and productivity of crops. Host plant resistance holds a great promise for exploitation in IPM programmes because of the following unique features :

(*i*) **Specificity.** Host plant resistance is usually specific to a particular insect pest or pest complex and does not possess any adverse effect on non-target or beneficial organisms.

(*ii*) **Cumulative effect.** The effects of host plant resistance on insect population density are cumulative over successive generations of the target pest because of reduced survival, delayed development and reduced fecundity.

(*iii*) **Ecofriendly.** Host plant resistance does not cause pollution in any component of the environment nor does it have any deleterious effect on man or wildlife.

(*iv*) **Easily adoptable.** An insect resistant variety with high yield potential can easily be adopted by farmers, as it does not require much higher cost and seed is to be procured otherwise also. Also, the farmers do not need to have knowledge of any application techniques.

(*v*) **Effectiveness.** Resistant varieties increase the susceptibility of insect pests to insecticides and many natural enemies of insect pests are more effective on resistant varieties.

(*vi*) **Compatibility.** When host plant resistance alone cannot maintain the insect population below the economic threshold, it can easily be combined with other methods of pest control, and also improves the efficiency of other pest management tactics.

(*vii*) **Decreased pesticide application.** Resistant varieties usually need less frequent treatment with a pesticide or may require low rates of application or pesticide application may not be necessary at all.

(*viii*) **Persistence.** Most of the insect-resistant varieties express moderate to high levels of resistance to the target insect pests throughout the crop-growing season. Some varieties with durable resistance are likely to maintain their resistance for long periods.

(*ix*) **Unique situations.** Host plant resistance may serve as a means of control in unique niches where other controls are not feasible or difficult to use, *e.g.* (*a*) when there is a critical timing in which the insect is exposed for only a brief period of its life cycle, (*b*) crop is of low economic value, and (*c*) pest is continuously present and is a single limiting factor over a large area.

In spite of the above advantages of host plant resistance, the following limitations have usually been recognized.

(*i*) **Time consuming.** Host plant resistance is not favourable for solving sudden or localized pest problems as the time taken for identification of sources of resistance and breeding of resistant varieties is usually longer, and varies from 5 to 10 years. However, this time can be shortened by employing modern biotechnology tools.

(*ii*) **Biotype development.** The use of varieties with vertical resistance may lead to development of insect biotypes. However, this problem may be tackled by employing polygenic resistance or breeding varieties resistant to specific biotypes.

(*iii*) **Complexity of pests.** Most of the crops are attacked by a complex of pests and there is a potential problem of resistance to one pest being linked to susceptibility to another. For example, pubescence in plants may be attractive to some pests and provide resistance to others. Similarly, high levels of allelochemicals may provide protection from one species, but may be attractive to other species.

(*iv*) **Effects on non-targeted species.** Plants that are resistant to arthropod pests, may have adverse impact on non-targeted species in two ways. Firstly, elevated toxin levels may be unpalatable, allergenic or even dangerous for consumers. For example, potatoes with higher glycoaldehyde levels that were resistant to the Colorado potato beetle were withdrawn because of customer complaints. Secondly, some secondary plant substances may be toxic to beneficial species such as parasitoids.

(*v*) **Genetic limitations.** The use of host plant resistance may be limited by the absence of preadaptive resistance genes among available germplasm. Moreover, resistant cultivars may lack the potential to produce high yields. Farmers will not accept resistant varieties unless their yield is atleast as good as the susceptible varieties that need chemical protection from pests.

(*vi*) **Interdisciplinary approach.** Dedicated inter-disciplinary efforts are required to combine insect resistance with all the desirable traits needed in an elite, commercially competitive variety. Sometimes, the failure of entomologists and plant breeders to proceed systematically after locating or developing an arthropod resistant germplasm becomes a major bottleneck.

The co-operative efforts of plant breeders, entomologists, molecular biologists and biochemists to identify, quantify and develop insect resistant crops cultivars during past several decades have led to the most significant accomplishment of modern agricultural research in the development of insect resistant varieties of crop plants. The use of resistant varieties provides crop protection that is biologically, ecologically, economically and socially acceptable. The resistant cultivars provide a better control of insect pests, but do not give complete protection to the crops from the pests. Therefore, resistant cultivars should be used in combination with other methods of pest control to achieve sustainable pest management.

Host plant resistance to arthropods has been successfully integrated with allied pest management tactics in many of the world's major food and fibre crops. Host plant resistance represents the key for opening up major advances in insect pest management. It would have a major impact on the need or rather the lack of need for other control tactics. Plant resistance to insects should form the backbone of pest management programmes for sustainable crop production.

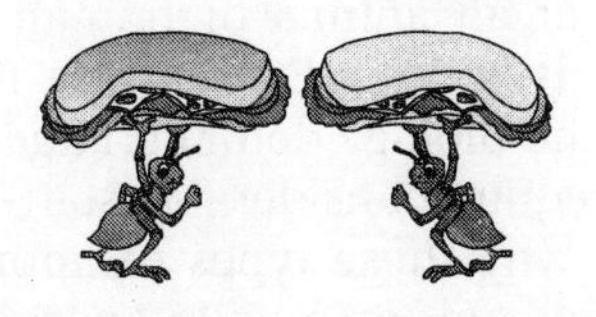

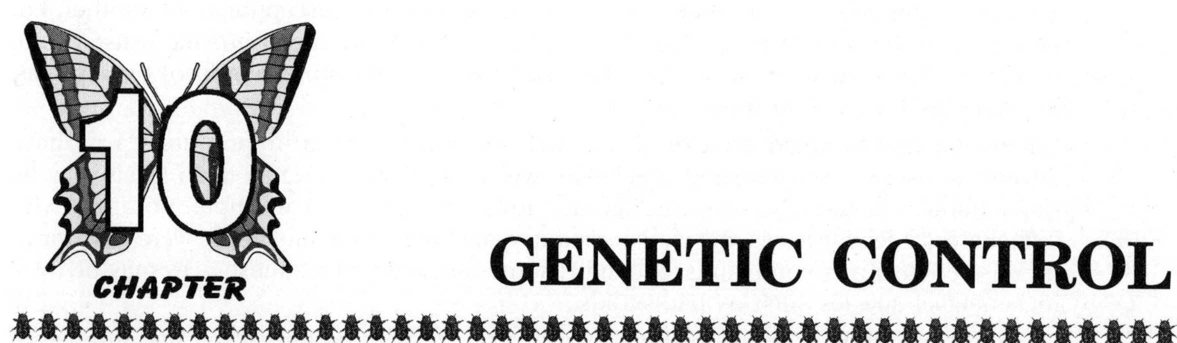

GENETIC CONTROL

INTRODUCTION

Genetic control strategies involve alterations of a targeted pest's ability to reproduce or the insertion of some deleterious character into a pest population. Genetic control is a form of biological control of pest species which exploits the insect's mate-seeking expertise to introduce genetic abnormalities (typically, but not necessarily, dominant lethal mutations) into the eggs of the wild population. Since the genetic control methods utilise insect pests for their own destruction, these are also known as autocidal methods. Genetic approaches for managing arthropod pests have been considered by scientists for more than half a century. Yet, because of recent and current emerging technologies, genetic approaches are often regarded as novel and unproven. Certainly, current advances and theory in this field of study remain to be implemented, but some applications of these methods, such as the sterile insect technique, are well known in their effectiveness and have great potential for managing arthropod pests. Research and development for genetic autocidal programme demand a long-term commitment before potential benefits can be realised. Therefore, genetic approaches are usually developed only for the most economically important pest species. However, future costs of developing, implementing, and operating genetic control may be reduced substantially as a result of the technological gains from current programmes using genetic approaches and from recent developments in molecular genetic techniques.

The majority of the genetic control techniques have the unique property of becoming more effective as the target population is reduced in numbers. However, they tend to be less effective at high population densities. This contrasts sharply with the use of insecticides where net effectiveness decreases when populations become small. Genetic control in most cases has to be viewed as an area-wide approach in which a crop or an animal or human population is protected from insect attack over a large geographical area. It is not suitable for a field by field or even a farm by farm approach as both the economics and the biology demand large-scale application. This implies that effective genetic control programmes require considerable start-up funds and the large scale financial resources required is a major reason why these types of approaches do not find favour with the funding agencies. However, in the long run, area-wide approaches have a much better return on investment than do conventional farmer by farmer approaches. A key element in area-wide economics is the mobilisation and organization of the beneficiaries. In the long term, genetic control techniques will only be successful, if they become commercially viable and are able to compete economically with other control methods (Dhaliwal *et al.,* 2013).

REQUIREMENTS FOR GENETIC CONTROL

The detailed knowledge of the population dynamics, ecology and behaviour of the target pest is essential for the success of any genetic control technique. The level of knowledge required is much greater than for most other insect control strategies.

Colonization and Mass Rearing

All types of genetic control require the colonization and to some extent the mass rearing of the target species with individual species differing in the ease with which they accept these two processes. All developmental stages of the insect have to be provided with an environment that not only enables them to reproduce in a predictable and efficient manner, but also produces individuals with a certain level of quality at an acceptable economic price. The effects of laboratory colonization on many aspects of insect behaviour are incremental, heterogeneous and to a certain degree unpredictable. Many quality parameters can be effectively monitored in the laboratory, *e.g.*, size, survival, etc., but the assessment of behavioural parameters would seem to be of little value under these conditions. As all genetic control techniques require the mating of the released insects with the wild population, any change in mating behavioural patterns will have an immediate deterimental effect on the efficiency of the technique. This aspect of quality has to be monitored in a representative and meaningful way, *i.e.* probably in the open field or in field cages.

Population Dynamics

An understanding of the population dynamics, ecology and behaviour of a candidate species is a pre-requisite to the application of a genetic control method. Information crucial to the economic assessment and development of a genetic control method includes the following parameters :

- Time of the year when the pest population is at its lowest level, and methods of reducing the native pest population to minimal number.
- An estimate of the number of insects in the native population and how this number changes over time.
- Intrinsic rate of increase and long-range movement capabilities of a pest during various seasons and in different cropping systems.
- All the host plants capable of supporting a pest species, and the relative abundance and economic significance of each host plant.
- Whether secondary pests would be benefitted or deterred by the reduction or absence of the target pest.

Post-Production Processes

For any area-wide genetic control programme, a large number of insects have to be prepared for release. This involves marking the insects so that they can be recognized in the field, sterilizing them if necessary, transport to the field area, and then their dispersal over the treatment area. All these processes have to be carried out within a defined and generally short time frame, and have to be simple, economical and cause little damage to the insect. In general, adult insects are released as they are mobile and less prone to the attack of predators. Being mobile, they can also aid in the dispersal process and they are usually released from aircraft for large scale programmes. Aerial release is often much cheaper than ground release and ensures a much better distribution of insects at a relatively low cost.

Field Monitoring

A continuous evaluation of a field programme is essential both in terms of monitoring effectiveness and in making programme adjustments. Released insects must be clearly distinguishable from field insects in a rapid and secure way, and methods must be available to monitor the released and wild population. The current methods that rely on fluorescent dust are not optimal. The wrong classification of a single fly as wild as opposed to released can have a major impact on a programme where eradication is the goal. The use of genetic transfer technology to introduce benign genetic markers will provide a high degree of security for the determination of the origin of the trapped insect. If any form of sterility technique is being used for control, a measure of the population fertility before and during the programme is highly informative. It is the only direct evidence that the released insects have interacted with the wild population.

Economic Analysis

Before any genetic control programme is implemented, an in-depth economic analysis should be conducted. Such an analysis should take into consideration damage estimates by the pests, current cost of their control, cost estimate of a genetic control programme, and potential benefits to the environment. The advantages of various methods and combinations of methods of control should be analysed and compared for multiple years. Knipling (1979) observed that reducing pest species to low populations may be difficult and costly by the use of insecticides or by a combination of several methods. However, once this is accomplished, continuous pest population management by genetic approaches may be most economical, effective and ecologically acceptable method. The cost/benefit ratio is not a static figure, but it should continue to decrease with each passing year. Cost/benefit ratios, generated from increased production and the absence of conventional pest control, estimate that the screwworm programme saved $ 4 billion through 1987, and that melon fly programme saved more than $ 100 million per year.

STERILE INSECT TECHNIQUE

The idea of controlling insects by sterilising technique, was first conceied by E.F. Knipling as far back as 1937 in a laboratory in Texas, USA. While studying the biology of the screwworm fly, *Cochliomyia hominivorax* (Coquerel), he observed that the female fly mated only once (monogamous). From this, he argued that if the males of this fly could somehow be sterilised, the females would fail to produce viable eggs and the insect's population could thus, be controlled. While entomologists were looking for sterilising agents to test Knipling's idea, geneticists were accidently sterilising insects in their own search for chemical mutagens immediately after World War II.

E.F. Knipling conceived a new approach for insect control in which the pest's natural reproductive processes would be disrupted by physical or chemical means. This concept, as it was applied to the screwworm fly starting in the early 1950s, eventually came to be known as the sterile insect technique (SIT), or the sterile insect release method (SIRM). Interestingly, about the time that Knipling was formulating his concept of insect control, a Russian geneticist, A. Serebrousky, suggested the use of chromosomal translocations to reduce reproduction in harmful species.

Studies of insect reproduction, from the early 1900s to the present time, have demonstrated that the insects treated with certain mutagenic chemicals (chemosterilants) or x-ray or gamma radiation, are unable to produce a normal number of living progeny. Treated insects (which may be completely or only partially sterile) are released in large number into a field environment and are expected to mate with the normal insects, thus interfering with reproduction. If the matings between treated insects and normal insects are successful, reproduction of the field populations will be disrupted, and the population will decline.

A. CHEMOSTERILANTS

Any chemical that can inhibit the growth of gonads or interfere with the reproductive capacity of an insect in any other way (such as prevent copulation and production of viable eggs, induce dominant lethal mutations, inhibit development of progeny at any stage (egg, larval or pupal) is called a ***chemosterilant.*** Chemicals that are currently employed as chemosterilants can be grouped as alkylating agents, antimetabolites and miscellaneous compounds.

1. Alkylating Agents

These compounds were discovered by LaBrecque in 1961 in the USA. They contain several groups of aziridinyls (aziridines or ethylenimines) and can be classified into three categories : apholate, aphomide and aphoxide whose examples are described below :

Apholate. Chemically, this compound is 2, 2, 4, 4, 6, 6-hexakis (1-aziridinyl) 2, 2, 4, 4, 6, 6-hexahydro-1, 3, 5, 2, 4, 6-triazatriphosphorine. It is a white odourless and crystalline substance, partially soluble in water and organic solvents (alcohol, chloroform, acetone), stable in cool dry condition and affected by humidity and high temperature.

Aphomide. Chemically, this compound is N, N´-ethylenebis, [p, p-bis (1-aziridinyl) N´´-methylphosphinic amide].

Aphoxide. Chemically, this compound is tris (1-aziridinyl) phosphine oxide, also called TEPA. It is a crystalline, colourless and odourless substance, extremely soluble in water, and highly soluble in alcohol, ether and acetone and hygroscopic and unstable. Its other derivatives are METEPA, phosphine oxide, tris (2-methyl 1-aziridinyl) and thio-TEPA, (phosphine sulphide, tris (1-aziridinyl). The former is a straw coloured liquid completely soluble in water and solvents, and the latter, a white crystalline solid relatively less soluble.

2. Antimetabolites

Antimetabolites are chemicals, which are structurally related to biologically active substances. Due to their structural similarity, the biological system fails to distinguish them from their own natural substances and commits the mistake of utilising them in place of the latter. For example, if bacteria are grown in a medium containing 5-fluorouracil, the latter will replace a large percentage of the normal metabolite, uracil, in the RNA of the former. Three groups of antimetabolites are more important : purine compounds, pyrimidine compounds and folic acid compounds.

3. Miscellaneous Compounds

Under this are the chemosterilants which are structurally unrelated. They are of the following categories:

Non-alkylating analogues of aziridinyl compounds. These chemicals were discovered in an attempt to find safer chemicals. They are HEMPA (hexamethylphos-phoramide), a dimethylamine of TEPA and HEMEL (hexamethylemelamine) **(Fig. 10.1)**, a dimethylamine analogue of tretamine, both being male housefly sterilants.

Organotin compounds. Numerous triphenyltin derivatives such as triphenyltin hydroxide, triphenyltin chloride (chlorotriphenyltin), triphenyltin acetate (acetoxy triphenyltin), **(Fig. 10.1)** bis (triphenyltin) sulfide, alkyltriphenyltin, etc., have found to be reproduction inhibitors in the house flies. They sterilise both males and females, but the females are more sensitive because they are affected at a much lower concentration than the males.

Field Trials

Agricultural pests. Not much effort has been made to control agricultural pests by chemosterilants. Most of the work that has been carried out, relates to cage experiments performed in various laboratories. Such experiments have been tried with lepidopteran insects like cabbage looper, *Trichoplusia ni* (Hubner); pink cotton bollworm, *Pectinophora gossypiella* (Saunders), etc. However, some actual field trials have been attempted in California on Mexican fruit flies (with TEPA) and in Alabama on the cotton boll weevil (with apholate). In spite of the success in the case of the former, chemosterilants were abandoned in favour of irradiation for the sake of safety. The latter could not meet with complete success due to insufficient isolation of the treated areas.

Public health pests. The house flies were the first to be tried with TEPA mixed in food bait and used in a half area refuse dump in Florida in 1962. Nine weekly treatments brought the insect population down. Measurements were done by counting the number of flies landing per minute on a grid 45 × 18 cm kept at ten scattered localities. However, the population soon increased because the area was not isolated. Attempts to control house flies have also been made in Italy (1964), Sicily (1967) and Japan (1968). While some success was achieved, the problem of reinvasion from the nearby untreated areas was always there to annul the effects of the treatment. Evidences that the house flies are getting resistant to chemosterilants are now coming in.

Testse flies were chemosterilised in Zimbabwe and 98 per cent control was achieved but complete eradication could not be possible due to reinvasion from the adjoining areas. Mosquitoes (*Aedes, Culex*) were also put to chemosterilant treatments. The first field release of chemosterilised mosquitoes was made in Seahorse Key islands off the coast of Florida in 1968, another, in Kenya (1971), and then in India at villages Bamnauli (1971), and Dhulsiras (1972), near Delhi. Controls were only partial again due to the lack of isolation of these places (Srivastava, 1996).

B. IONISING RADIATIONS

Sterilisation of insects by irradiation can be brought about by exposing them to γ-radiation (gamma-radiation or alpha and beta particles), X-rays and neutrons. Of these, γ-radiation by ^{60}CO with its half-life of 6 years, is the most common method of sterilisation. In India, sterile insect technique for red palm weevil, *Rhynchophorus ferrugineus* (Olivier) has been developed at the Bhaba Atomic Research Centre (BARC), Trombay. Use of ionizing radiation has been the main method of inducing sterility in mass-reared insects for area-wide IPM programs for the past five decades, because of the mutagenic or teratogenic effects of chemosterilants which leads to human health and environmental issues especially the integrity of ecological food chains, and chances of development of resistance in insects. Other advantages of the irradiation are : insects can be irradiated inside packaging materials, the sterile insects may be released immediately after the irradiation, radiation does not leave residues that could be harmful to humans or the environment, temperature rise in the irradiation process is usually insignificant. Millions of sterile insects per week for national Area-Wide Integrated Pest Control programs or making research on SIT against screwworms, fruit flies, moths and tsetse flies are being produced at 41 different facilities worldwide (www.dirsit.iaea.org).

Radiation doses differ for different species and even for different developmental stages of the same species. A high dose will provide complete sterilisation (*i.e.* all generations produced by the sterilised insects will be sterile), while lower (sub-sterilisation) dose will provide a partial sterility which may not continue beyond the F_1 generation. However, higher doses adversely affect longevity and competitiveness and, therefore, partial rather than complete sterility is preferred (atleast in the cases of pests, affecting costly crops or those of more serious diseases), so that some control (atleast up to F_1 generation) could be achieved. Competitiveness being more crucial for males, allowing residual fertility (partial sterility) in males and complete sterility in females, will maximise the impact in a control (not eradication) programme (Srivastava and Dhaliwal, 2010).

In every insect species, a radiation dose that does not adversely affect longevity and competitiveness has to be determined separately by experimentation because such a dose differs for different species. Sterilising doses of some of the more common insects are given in **Table 10.1.** The more detailed information on doses of radiation applied for insect disinfection and sterlization of mites and insects are available with International Database on Insect Disinfection and Sterlization (IDIDAS) (www.ididasiaea.org).

Concept of Knipling's Technique

The concept of E.F. Knipling's technique involves two systems, viz. sterile male technique and sterile insect technique. Although both the procedures envisage the common principle of sterility, they greatly differ in the manner in which populations are affected.

Table 10.1 Estimated doses of γ-rays and x-rays for male sterilization

Scientific name	Common name	Stage	Sterilizing dose (rads)
Musca domestica Linnaeus	House fly	2-3 days pupae	3000
Cochliomyia hominivorax	Screwworm fly	5-days pupae	2500
(Coquerel)		1-day adult	5000-7000
Drosophila melanogaster (Meigen)	Fruit fly	Adult males	11000-12000
Culex quinquefasciatus Say	House mosquito	Pupae	7700
Apis mellifera Linnaeus	Honey bee	Adults	7000-11000
Sitophilus oryzae (Linnaeus)	Rice weevil	7-days adults	7500-10000
Tribolium confusum Duval	Confused flour beetle	Old pupae	4000

Source : Srivastava and Dhaliwal (2010)

Sterile male technique. According to this technique, if fully competitive sterile males are released in the natural (*i.e.* wild) populations, it will reduce the reproductive potential of the natural populations in the same ratio as sterile to fertile. If, for instance, the sterile to fertile (S:F) ratio is 1:1, then according to the law of chances, both sterile and fertile males will have an equal (fifty-fifty) chance of mating with the wild females. The matings with the sterile males will be infructuous (eggs will be non-viable or sterile), which will mean that the reproductive capacity of the natural population is reduced by 50 per cent. In other words, a control of 50 per cent has been achieved. Thus, if the S:F ratio is increased to 9:1, then the control (atleast theoretically) will be 90 per cent. This technique is now synonymous with the sterile insect release method (SIRM), wherein both males and females, mass reared and sterilised in the laboratory, are released at every generation until the pest is eradicated or controlled.

Sterile insect technique. This technique envisages the sterilisation of a portion of the wild population itself (both male and females) by chemosterilants. The effect then will be two-fold : (*i*) the percentage sterilised, cannot reproduce and, in effect, will amount to killing the same proportion of insects (same as in the sterile male technique) and additionally, (*ii*) the sterilised males and females of the wild population will, in turn, nullify the reproductive capacity of a proportionate number of the fertile individuals in the population by competing with them. This method, in a way, combines chemical and biological control methods : the effect of the chemosterilant will be chemical, and that of the sterilised insects, biological.

Thus, if 90 per cent of the given population is sterilised, this percentage cannot reproduce, which will amount to 90 per cent killed (first effect). Of the remaining fertile 10 per cent, only 1 per cent be expected to have fertile mating because of the existence of 90 per cent sterilised insects in the population (second or *bonus effect*, as termed by Knipling). Thus, a total of 90 + 9 = 99 per cent of the insect population, will fail to have fertile mating, amounting to 99 per cent control. An added effect produced by released insects on the natural population, could be to exert an additional pressure on their struggle for existence by competing with them for food and shelter. Besides, this method does not cause hazards to man and his environment.

Knipling's SIRM Model. The SIRM modified model of Knipling has been illustrated in Table 10.2. The assumed number of insects in the wild population is 1000, and that of the sterile inects released in each generation, is 2000.

If for an assumed natural population of 1000 insects (1000 males and 1000 females), 2000 fully competitive males are released in each generation, the ratio of sterile to fertile (S:F)in the population will be 2:1. This means that only one-third of the natural wild females can be expected to mate with

Table 10.2 SIRM model of Knipling

Generation	Natural population	Sterile insects released	S:F matings	Infertile (%) progeny	No. fertile
1	1000	2000	2 : 1	66.7	333
2	333	2000	6 : 1	85.7	47
3	47	2000	42 : 1	97.7	1
4	1	2000	2000 : 1	99.9	0

the wild fertile males and two-thirds with the released sterile males resulting in 1000 × 2/3 = 66.7 per cent infertile mating. Thus, only 333 progeny of each sex will be produced instead of 1000. In the next generation, the release of 2000 sterile males will result in a 6:1 sterile to fertile ratio and only one-seventh of the 333 wild fertile females will be expected to mate with the fertile males, producing 47 progeny and six-seventh with the released sterile males resulting in 100 × 6/7 = 85.7 per cent infertile matings. The release of 2000 sterile males in the third generation will raise the S:F ratio to 42:1, resulting in only 1 progeny which, in the fourth generation, will be eliminated because of its having 2000 times greater chance of mating with the sterile than with the fertile males.

The aforestated model, assumes a constant insect population in each generation, which is not the case in nature. In each generation, the insect population tends to increase. And though the potential rate of increase of most insect populations is very high, casualties due to various factors (such as parasitism, predation, natural hazards), bring the net increase down to a lower rate. According to Knipling's own estimate, an insect population tends to increase five-fold in each generation. Therefore, to offset (neutralise) this increase, a constant number of sterile insects has to be released every generation. Since the control brought about by the sterile insect is more than the annual increase, a time comes when the sterile to fertile ratio becomes so disproportionate that, there may not be any fertile matings, resulting in complete control or eradication.

The significant feature of the SIRM is that, each release of sterile insects achieves an increasingly higher sterile to fertile ratio and thus becomes progressively more efficient, which is an advantage over the insecticides, where the kill in each generation remains constant. The results shown in **Tables 10.3** and **10.4** will bear this out.

Table 10.3 Number of generations needed to eradicate a population of 1000 insects by insecticides envisaging a control of 90 per cent and an annual population increase of five-fold

Generation	Natural population treated with insecticides	Survivors	Five-fold increase
1	1000	100.0	500
2	500	50.0	225
3	225	22.8	114
4	114	11.4	56
5	56	5.6	28
6	28	2.8	14
7	14	1.4	6
8	6	0.0	0

When both the tables are compared, it becomes obvious that while it takes eight generations to eradicate the insects by the insecticide control system, it takes only four by the SIRM, the latter getting progressively more efficient.

Table 10.4 Number of generations needed to eradicate a population of 1000 insects by SIRM envisaging a control of 90 per cent and an annual population increase of five-fold

Generation	Natural population	Sterile insects released	S:F matings	Progeny	Five-fold increase
1	1000	9000	9 : 1	100.0	500.0
2	500	9000	18 : 1	26.0	130.0
3	130	9000	67 : 1	1.9	9.5
4	10	9000	900 : 1	0.0	0.0

Eradication of Screwworm Fly

The screwworm fly, *Cochliomyia hominivorax* (Coquerel) was the first insect pest that was put to this method of control. This is a muscid fly parasitising on warm-blooded animals from the southern parts of the USA to South America and certain West Indian islands. The females of this insect lay eggs in skin lesions where the larvae hatch and feed for 5-6 days before falling off to the ground for pupation in the soil. Due to the severity of infestation, the animals die. Before 1958, the estimated loss due to this pest used to be to the tune of $ 12 million a year. The monogamous habit of this insect led Knipling in 1937 to develop his sterile male and sterile insect techniques. The first successful attempt to control this insect by the release of irradiated insects was made in 1952 on a small island named Sanibel, 3.2 km off the west coast of Florida (USA). Irradiated flies at the rate of 200 (100 of each sex)/2.59 km^2 (each 2.59 km^2 is estimated to have 10 insects) were released once a week. After two weeks, 80 per cent of the egg masses sampled proved to be sterile and after three months, the fly population was almost zero. However, the fly could not be eradicated completely because of the island's proximity to the uncontrolled mainland (Srivastava, 1996)

Subsequently, spectacular success was achieved in the eradication of screwworm on the 440.3 km^2 Island of Curacao during 1954-55. The males sterilised at a dose of 2500r were released at the rate of 400 per 2.59 km^2 per week. By concerted efforts, screwworms were eradicated from Florida (eastern coast) by 1960, from Texas and New Mexico (mid-mainland) by 1964, and from Arizona and California (west mainland) by 1965. To neutralise the effect of invasion from untreated neighbouring areas, barrier releases (*i.e.* additional releases) between the treated and untreated zones were carried out.

In 1972, the population of the screw worm fly increased beyond the effective management by the sterile fly releases. Some of the reasons could be (Knipling, 1979):

- Unusually favourable conditions for winter survival of screwworms in northern Mexico, south of the sterile-fly release area and in the sterile fly barrier.
- More favourable weather conditions in the spring and summer in the south western United States.
- Relaxation in animal husbandry practices which result in the breeding of animals virtually the year round (earlier, livestock owners followed breeding practices to assure the birth of most calves, lambs and other livestock when screwworms were absent or scarce).
- Failure of livestock owners to follow other animal management practices such as minimal surgical operations and careful surveillance of animals during the fly season to look for and treat infested animals before larvae can mature.
- A general increase in deer populations throughout the normal range of the screwworm.
- An upsurge in the Gulf Coast tick, *Amblyomma maculatum* Koch, which is one of the principal natural predisposing causes of screwworm cases in livestock.
- Changes in the behaviour and competitiveness of the released sterile flies because of genetic deterioration.

- Changes in the competitiveness of flies due to modifications in the rearing procedures.
- Changes in the behaviour of native population through genetic selection pressure making the adults prone to avoid matings with the released strain.

The situation, however, improved in 1973 to 1975 because the sterile fly production rate was doubled to about 200 million per week. New strains of screwworms were established to release flies that were not maintained as laboratory cultures for long. Other modifications of the programme were also made including better distribution of sterile flies.

Dr Edward F. Knipling and Dr Raymond C. Bushland were awarded the 1992 World Food Prize for developing eco-friendly SIT to control or eradicate insect pests that threaten vast sources of food, especially livestock and wild life populations.

The following requirements must be met before developing and applying SIT for pest suppression (Knipling, 1979) :

- Practical procedures must be developed for rearing enough insects to overflood the natural population.
- Methods of inducing sterility or strains possessing appropriate genetic defects must be available.
- Reasonably accurate estimates of absolute numbers of the insects in the natural population are essential to determine how many insects must be reared for release, considering the relative competitiveness of the released and native insects.
- The released insects must be distributed so that they will be in reasonable spatial competition with the natural population for mating.
- Information on the normal rate of increase of the natural population is desirable as a guide to the rate of overflooding required to achieve the necessary results.
- The degree of infiltration of the target pest and its impact on the effectiveness of the technique must be considered.
- The numbers of released insects required for control or elimination must not be unduly hazardous to crops, animals or man.
- A critical analysis of the candidate insect pests, including costs, effectiveness and ecological effects of available alternative control methods, is essential in appraising the value of SIT as a replacement or supplement for other methods of control.

HYBRID STERILITY

Hybrid sterility refers to sterility that occurs when certain strains, races, or closely related species are crossed and either one or both sexes of F_1 progeny are viable but cannot produce viable progeny. In hybrid sterility, mating is induced between closely related species under laboratory conditions. If such mating leads to fertilization, the hybrid individuals may die in the embryonic stage, in one of the pre-adult stages or early in the adult stage. Alternatively, apparently normal adults produced may show partial or complete sterility in one or both the sexes. The various defects in the mating between such insects is either an imperfect copulation due to structural differences in their genitalia or if the copulation is perfect, it may lack insemination and if inseminated, the fertilization may not take place and even if fertilized, the individuals may die before hatching, any time between larval and adult stages or they may even survive, but only as hybrid adults. The hybrid adults may look quite normal externally or may even show heterosis (*i.e.* hybrid vigour with increased longevity, increased sexual aggressiveness, increased competitiveness, etc.), but one or both sexes will be partially or completely sterile or will have defective gametogenesis resulting in premature death of the progeny at any stage of their development. This phenomenon has been investigated and used in the management of several pest species, (*e.g.* between species of the tsetse flies, *Glossina morsitans* Westwood and *G. swynnertoni* Austen, between races of the gypsy moth, *Lymantria dispar* (Linnaeus), and between the moth species *Heliothis virescens* (Fabricius) and *H. subflexa* (Guenee).

The hybridisation of *H. virescens* and *H. subflexa*, stimulated interest in reducing field populations of *H. virescens*, using released backcross insects. In an experiment on St. Croix, US. Virgin Islands, using both male and female backcross insects, sterility was infused into a field population with release ratios of 20 sterile backcross : 1 feral insect. The frequency of sterile male progeny increased for one generation after release, and the distribution of backcross frequencies became homogeneous througout the population. During a 6-week period, 94 per cent of trapped males were sterile progeny from released or field-reared backcross females and the native males. Isolated populations of *H. virescens*, probably can be eradicated, using this method, given a sufficient number of released hybrid individuals.

Many attempts have been made to produce sterile hybrids from other pest species. For example, *Helicoverpa zea* (Boddie) has been mated with *Helicoverpa armigera* (Hubner) from Australia, Russia, and China and *H. assulta* (Guenee) from Pakistan and Thailand. These hydridisation attempts failed to produce a sterile hybrid. Similarly, no measurable hybrid sterility has been found in crosses between the pink bollworm from areas within the United States, Mexico, Puerto Rico, or St. Croix. There was no incompatibility between a strain of the pink bollworm from southern India and two strains (one long-term laboratory strain, the other a newly colonised strain) from Arizona.

CYTOPLASMIC INCOMPATIBILITY

In this case, a cross between two apparently conspecific populations results in only partial embryonation in some ova. Spermatozoa enter the egg cytoplasm, but no fusion occurs between the nuclei of spermatozoa and ova with the result, that zygote is not formed and the egg remains sterile.

This sterility appears to be due to a cytoplasmic factor and not to chromosomal incompatibilities. It has been found that crosses between individuals from certain populations of the same species of mosquito, *Culex pipiens* Linnaeus are sterile. Crosses between other populations are sometimes completely fertile or partially sterile. The phenomenon has also been discovered in a number of other mosquito species and has been used as a control measure in experimental releases. The phenomenon is best known in the crosses between the subspecies of *C. pipiens*, collectively referred to as the *C. pipiens*–complex, viz. *C. pipiens* (temperate region), *C. quenquefasciatus* Say (tropical and sub-tropical) and *C.p. australicus* Dobrotworsky & Drummand (restricted to Australia and nearby regions). The same subspecies existing in a different part of the wild forms a strain reproductively incompatible with a strain of another country. At least 20 reproductive types amongst these three subspecies are known which show three types of incompatibilities (Srivastava and Dhaliwal, 2010):

(*i*) *Partial incompatibility*, where crossing of two strains produces low hatches of eggs.

(*ii*) *Unidirectional incompatibility*, where only one of the sexes of a particular strain is capable of producing sterile eggs.

(*iii*) *Bidirectional incompatibility*, where the reciprocal crosses are incompatible.

Field trials of this method were carried out in a village near Rangoon in Myanmar in 1971 and at Delhi in 1972 under a WHO/ICMR programme. The method involved release of incompatible strains of the mosquito, *C. quinquefasciatus* from other (temperate) regions. Before actual release, suitable backcrosses in tropical and temperate strains were carried out to produce a variety that would survive tropical conditions without losing its incompatibility. This is possible in backcrosses between unidirectional incompatibilities. The trials at both the places demonstrated the efficacy of the method, but the invasion by mosquitoes from untreated regions was the obstacle preventing a complete eradication of the species.

Cytoplasmic incompatibility also, has been observed in the plum curculio, *Conotrachelus nenuphar* (Herbst) ; cherry fruit fly, *Rhagoletis cerasi* (Linnaeus) ; and in *Ephestia cautella* (Walker) leading to the hope that it might occur in many economically important insects. The use of sterility induced by cytoplasmic incompatibility has one serious drawback. Released strains of insects must be separated by sex before release, so that they might not interbreed and become established in the field.

CHROMOSOME TRANSLOCATION

Translocation is the breakage of two non-homologous (dissimilar) chromosomes (as in a heterozygote) and the re-attachment of the broken parts to the wrong partners as shown in **Fig. 10.1.** During meiosis in gamete formation, the translocated chromosome segregates out into six types of gametes as indicated in the figure. Of the six types of gametes, only two are balanced (*orthoploid*), having a full gene complement and, therefore, viable and four are unbalanced (*aneuploid*), having more genes of one chromosome and less of the other and, therefore, non-viable or lethal. Eggs produced by gametes having lethal genes will be sterile. This phenomenon can be utilised in the control of insects.

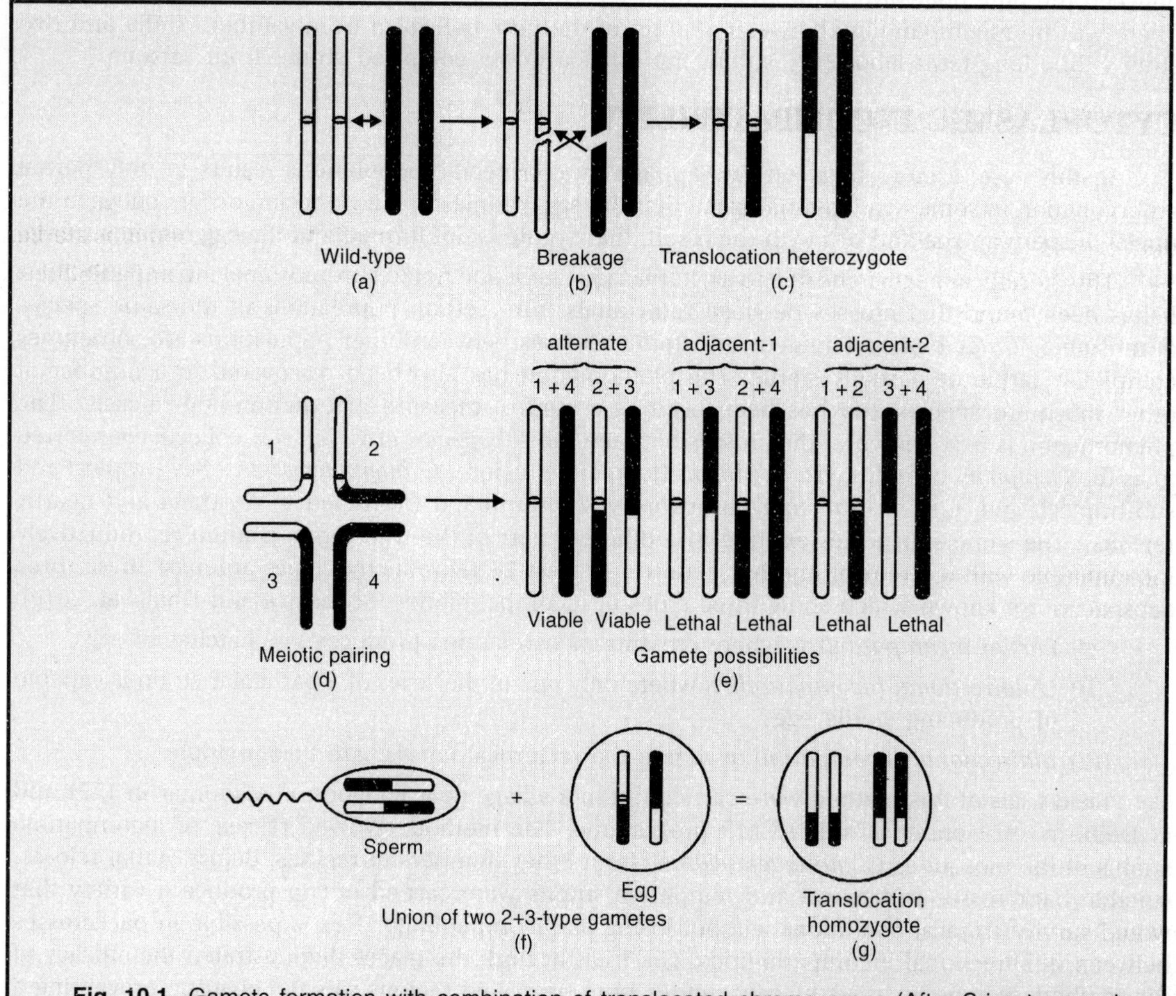

Fig. 10.1 Gamete formation with combination of translocated chromosomes (After Srivastava and Dhaliwal, 2010)

Translocations in insects can be produced by exposure to irradiation and to radiomimetic chemical (such as EMS, MMS, bleomycin, mitomycin-c, nitrogen mustard, etc.). Such translocated insects could be released into wild population. Mating between wild and translocated heterozygote individuals will result in the reduction of population of the wild type by a fall in the viable egg production. By persisting in the release of more and more translocated insects, the wild population (theoretically atleast) can be eliminated.

This method has been tried only in laboratory in cage experiments on mosquitoes, viz. *Culex*, *Aedes* and *Anopheles*; house fly, *Musca domestica* Linnaeus ; lepido-pteran pests codling moth, cabbage looper, pink bollworm, etc. At first, translocated and wild population is kept in small cages under laboratory conditions. If there is a decline in the wild population, the same experiment is repeated with large population in large-sized cages kept under field conditions.

LETHAL FACTORS

Natural population carries a considerable load of deleterious genes which are detrimental to their survival. Two such types of genes can exist, *i.e.,* straight forward lethal genes and conditional lethal genes. The *straight forward lethal genes* are recessive, being kept suppressed by suppressor genes, but can be isolated (made dominant) by inbreeding in the laboratory. When released in homozygous or heterozygous conditions, such insects could produce progeny that would soon be eliminated by a normal process of natural selection. The *conditional lethal genes* are dominant, but have deleterious effects under some conditions, but not under others.

There are temperature sensitive (hot or cold) lethal genes. Insects with such genes could be mass reared and released in wild population in conditions favourable to them, so that they could, by mating with wild population, produce a progeny that would not survive a later change in climatic conditions (temperature). For example, where diapause is necessary for the survivial of the pink bollworm during host-free or environmentally unsuitable periods, its inability to enter diapause would be a conditional lethal trait. A non-diapausing strain could be reared readily in the laboratory, but progeny produced by this strain in the field would not diapause, and it could not be able to reproduce during the host-free period. Similarly, a cocoon-producing insect could be made to produce a progeny incapable of producing cocoons. In either case, the progeny will die to the vicissitudes of the climatic changes (Srivastava and Dhaliwal, 2010).

SEX-RATIO DISTORTION

Some insects produce progeny of one sex more than the other-these can be male producing or female producing. Such insects when mass reared in the laboratory and released in nature, compete with wild population to mate and produce offsprings that are predominantly of one sex. In course of time, imbalance in the sex-ratio leads to a reduction in the population and ultimately to the extinction of the species itself. The mechanism by which the above distortion is brought about is called meiotic drive where (*i*) a factor called segregation distorter, present on a chromosome, drives only one of the two chromosomes of the XY (sex chromosome) pair to the poles to form gametes during meiosis, (*ii*) the driven chromosome undergoes supplementary replication with a concomittant loss of its homologue, or (*iii*) only one of the XY pair of chromosomes survives to end up in functional gametes, the other one, degenerating and being reabsorbed (**Fig. 10.2**).

The sex-ratio distortion, it is believed, can also be caused by a differential sperm behaviour. Depending upon the genetic constitution, one of the two kinds (X or Y) of sperms may compete better than the other to fertilize the ovum or even by a selective acceptance by the ovum of the two kinds of sperms. In nature, meiotic drive sex-ratio distortion has been found in the mosquito, *Aedes aegypti* (Linnaeus). Some strains of this mosquito are predominantly male producing. Such males, when mass reared and released into wild populations, will produce mostly male progeny which, in course of time, will not only lead to their extinction but will also have an additional effect in reducing disease transmission, since it is only the female of this species that transmits the disease. If the genetic constitution of an insect is known, it is also possible to produce sex distortion artificially by a selective breeding. A strain of *Musca domestica* Linnaeus has now been produced in the laboratory that is predominantly male producing. Thus, the method of sex-ratio distortion offers a great promise in insect pest control.

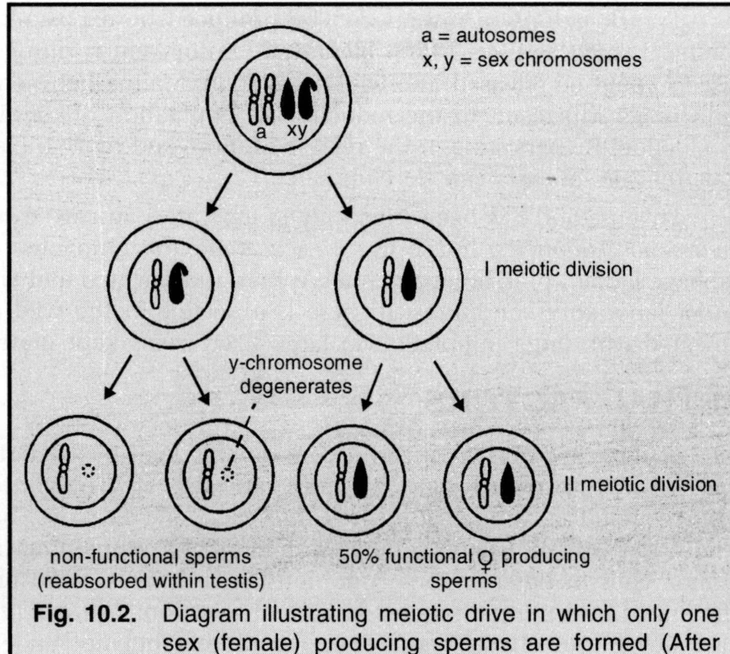

Fig. 10.2. Diagram illustrating meiotic drive in which only one sex (female) producing sperms are formed (After Srivastava and Dhaliwal, 2010)

SPECIES REPLACEMENT

It has been observed that of the two species or strains of a single species living in the same ecological niche, one is more competitive than the other. The males of the better competitor can be mass reared and released to mate with the female of the other. Such mating results in the production of sterile eggs and ultimately leads to the replacement of the species. An attempt has been made to replace *Aedes polynensiensis* Marks, the natural vector of periodic *Wuchereia bancrofti* (Cobbold) (filaria producing nematode) in the South Pacific with the non-vector allopatric species *A. albopictus* (Skuse) from Pune, Maharashtra. The males of the latter, readily mate with the female of the former even in presence of their own males. The result of such a mating is sterile eggs.

SUCCESSES IN GENETIC CONTROL

The SIT and other genetic methods are species specific and have no negative impact on the environment. But their complexity puts them beyond the reach of individual farmers and the programmes have to be planned, organized and implemented by the government agencies. Perhaps due to this reason, hardly any SIT programmes have been undertaken in Asian countries where farms are very small and large-scale, centralized pest control strategies are difficult to implement. The only exceptions to this are fruit fly, *Bactrocera cucurbitae* (Coquillett) eradication programme undertaken extensively in Japan from 1972 onwards and the *Bactrocera dorsalis* (Hendel) suppression programme in Taiwan from 1975 onwards. On the other hand, Mediterranean fruitfly, *Ceratitis capitata* (Wiedemann) eradication programme was a huge success in Mexico and other Central American Countries. For a number of other pests including onion fly, *Delia antiqua* (Meigen); European cockchafer, *Melolontha melolontha* (Linnaeus); European cherry fruit fly, *Rhagoletis cerasi* (Linnaeus) ; coding moth, *Cydia pomonella* (Linnaeus) and grape moth, *Eupoecilia ambiguella* (Hubner), large scale programmes are in operation in a number of European countries. A list of the more notable trials that have been carried out to date, is given in **Table 10.5** and some selected examples are discussed herein.

Screwworm Fly

The first major use of SIT was for the eradication of the primary screwworm from the island of Curacao in 1954. Following this success, the pest was then eradicated from the southeastern

Table 10.5 Major field trials for control of insect pests using sterile insect release technique

Insect		Trial locations
Common name	**Scientific name**	
Diptera		
Screwworm fly	*Cochliomyia hominivorax* (Coquerel)	North America, Central America, Libya
Mediterranean fruit fly	*Ceratitis capitata* (Wiedemann)	Central America, Hawaii, Canary islands
Mexican fruit fly	*Anastrepha ludens* (Loew)	Mexico
Queensland fruit fly	*Bactrocera tryoni* (Froggatt)	Australia
Cherry fruit fly	*Rhagoletis cerasi* (Linnaeus)	Switzerland
Oriental fruit fly	*Bactrocera dorsalis* (Hendel)	Guam
Melon fly	*Bactrocera cucurbitae* (Coquillett)	Various Japanese islands
Onion fly	*Delia antiqua* (Meigen)	Netherlands
Olive fly	*Bactrocera oleae* (Gmelin)	Various European countries
House mosquito	*Culex quinquefasciatus* Say	Florida
Malarial mosquito	*Anopheles ludens* Wiedemann	El Salvador
Stable fly	*Stomoxys calcitrans* (Linnaeus)	St. Croix (Virgin Islands)
Tsetse fly	*Glossina palpalis* (Robineau-Desvoidy)	Nigera, Upper Volta
Lepidoptera		
Grape moth	*Eupoecilia ambiguella* (Hubner)	Various European countries
Codling moth	*Cydia pomonella* (Linnaeus)	USA, Canada, Switzerland
Coleoptera		
European cockchafer	*Melolontha melolontha* (Linnaeus)	Switzerland
Pink bollworm	*Pectinophora gossypiella* (Saunders)	California, USA
Boll weevil	*Anthonomus grandis* Boheman	Louisiana, USA

***Source*:** Thacker (2002)

United States in 1959 and from Puerto Rico in 1975. Eradication of the screwworm was initiated in the south-western United States in 1962 and completed in 1982. Because of the extensive damage that the screwworm causes in Mexico, and because it was impossible to prevent reinfestation into the United States, a Mexican-American Commission was created in 1972 to eradicate the pest from northern and western Mexico and to establish a "sterile fly barrier" at the Isthmus of Tehuantepec. In 1986, the commission extended its eradication activities to the Yucatan Peninsula and countries of Central America. As a result of this programme, Mexico, Belize, and most of Guatemala are now free of the screwworm. Operations are in progress to eradicate the screwworm from Honduras and El Salvador. The present goals of the programme are to eradicate the pest from Central America and Panama and to establish a sterile fly barrier at the Darien Gap to prevent its reinfestation. Also, efforts will be made to eradiate the screwworm from Caribbean islands which are still infested.

The screwworm invaded North Africa in the 1980s and became established in the Libya. After detecting the screworm in 1988 and confirming its presence in 1989, a joint decision was made by the Food and Agriculture Organization/International Atomic Energy Agency (FAO/IAEA) and the government of the Libya to initiate an eradication campaign using the sterile insect technique. Sterile pupae were transported from the Mexico-US Commission production plant at Tuxtla Gutierrez in Mexico, emerged in Libya, and released over a treatment area of 40,000 km^2. Between 1990 and 1992,

1300 million sterile insects were shipped from Mexico to Libya for release. The campaign was implemented late in 1990 and completed in October of 1991. North Africa was declared free of the screwworm in June 1992.

Fruit Flies

Genetic control programmes using the SIT have been successful against several species of fruit flies (Diptera : Tephritidae). Limited field application of the SIT has resulted in population suppression or eradication of the Caribbean fruit fly, *Anastropha suspensa* (Loew) in Florida; cherry fruit fly, *Rhagoletis cerasi* (Linnaeus), in Switzerland; Oriental fruit fly, *Bactrocera dorsalis* (Hendel), in the Mariana Islands; Queensland fruit fly, *Bactrocera tryoni* (Froggatt) ; and Chinese citrus fly, *B. minax* (Enderlein), in China. However, areawide use of the SIT has been successful in the eradication or control of several species of fruit flies. The most spectacular successes have been with eradication of the Mediterranean fruit fly, *Ceratitis capitata* (Wiedemann), from southern Mexico, and of the melon fly, *B. cucurbitae* (Coquillett) from Japan, and the prevention of Mexican fruit fly from invading California and Texas.

After the first detection of the Mediterranean fruit fly in Mexico, the Ministries of Agriculture of Mexico and Guatemala and the United States Department of Agriculture, formed the Moscamed programme to combat this pest. The objectives of the programme were to stop the northern advance of the pest, to eradicate it from southern Mexico and Gautemala, and in the long-term, to eradicate it from Central America and Panama. By combining the discrete use of malathion bait spray with the release of sterile flies, the medfly was eradicated from Mexico in 1982. The programme has also been successful in Guatemala.

The Japanese government initiated a project to eradicate the melon fly in 1972. Using the SIT, following one or more treatments of a lure/toxicant, the eradication effort began on Kume Island and was expanded until the melon fly had been eradicated from all the infested islands of the Kagoshima and Okinawa Prefectures. Japan was declared free of the melon fly in 1992. Although the eradication programme required an investment of about $100 million, the benefits from eradication of the melon fly should be over $100 million per year.

Pink Bollworm

The pink bollworm, *Pectinophora gossypiella* (Saunders), is a serious pest of cotton in many parts of the world. Although the pink bollworm is occasionally recovered from states east of Texas, this pest is most destructive in Arizona, southern California, and the adjacent northwest Mexican desert. After its introduction and establishment in central Arizona in the mid 1950s and in the Colorado River Basin of western Arizona, southern California, and northwestern Mexico in 1965, a programme was initiated in 1968 to protect the 500,000 ha of cotton in the San Joaquin Valley. This programme has prevented the establishment of the pink bollworm through the use of sterile insect-release technology, minor use of pheromones as a mating disruptant, and adequate cultural control. The California cotton industry considers the 27-year old San Joaquin Valley Exclusion Project to be a major success.

Codling Moth

Codling moth, *Cydia pomonella* (Linnaeus) is the key pest of most apple and pear growing areas in the world. A sterile insect release programme to eradicate codling moth was initiated in 1992 in the Okanagan region of British Columbia, Canada, by the year 2000. The programme was jointly developed and implemented by the British Columbia Fruit Growers' Association, the provincial government of British Columbia and the federal government of Canada. The whole programme was divided into three distinct phases, viz. (*i*) A pre-release sanitation phase was designed to reduce wild populations to the maximum possible by using cultural control methods and insecticide applications for two years. (*ii*) The second phase involved mass-rearing and release of sterile moths for three years. (*iii*) After eradication, the third phase was intended to protect against reinfestation by monitoring for the presence of wild moths, releasing sterile moths at border sites to prevent the invasion of wild moths, and controlling the transfer of infested fruit containers. A mass-rearing facility was built which produces 15 million moths per week. After several years of operation, and inspite of some initial failures, most growers in the programme area report no damage and no longer have to spray against codling moth.

Tsetse Flies

Tsetse flies of the Genus *Glossina* represent a major threat to agricultural development in Sub-Saharan agriculture as they transmit protozoan parasites of the Genus *Trypanosoma,* which cause a debilitating sickness in livestock and sleeping sickness in humans. The first genetic control attempts of any insect were conducted with this species in the 1940s, where hybrid sterility between the three subspecies of *G. morsitans* Westwood group was used. Over 100,000 field collected *G.m. centralis* Machado pupae were released into an isolated population of *G. swynnertoni* Austen over a 7 month period. The sterility generated led to the replacement of the latter species by the former. Due to the arid conditions, *G. m. centralis* population rapidly disappeared and the area became tsetse free. Large scale field trials using the release of sterilized males have been successfully carried out in Nigeria and Burkino Faso. However, in both cases reinvasion occurred when the programmes were terminated and the tsetse free areas were recolonized. Recently, the same technique has been used with complete success to eradicate *G. austeni* Newstead from Unguja Island, Zubland, Republic of Tanzania. The island has now been declared tse tse free and is free to develop its agriculture without the threat of trypanosomosis.

Onion Fly

Onion fly, *Delia antiqua* (Meigen), is the single insect pest of onions in the temperate regions of the world. A small commercial SIT programme started in 1981 is currently operating in the Netherlands for the control of this pest. This is probably the only truly commercially run programme of its kind in the world. About 400 million flies are produced annually and are used for the control of the pest on about 2600 ha of onion, representing about one sixth of the Dutch onion crop. The programme is technically very successful but suffers from the poor farmer uptake and in some way the selfish behaviour of a minority of farmers who shy to benefit from sterile flies released on their neighbour fields. This situation illustrates the need that all potential beneficiaries participate in such area-wide programmes. Given the right political and social support, the programme could be expanded to cover the whole of the Dutch onion crop.

INTEGRATION WITH OTHER TACTICS

The advantage of combining genetic control methods with other pest control methods has been recognized from the inception of genetic control. All the successful genetic control programmes mentioned above have non-genetic components that have served vital roles in the control or eradication of pests, as well as preventing re-infestation of pests. Most often, genetic control methods have been integrated with insecticide applications, cultural controls, and quarantines. However, population models constructed to predict the potential advantages of combining genetic control methods with other methods, such as inundative releases of parasitoids, host plant resistance, pheromones for mating disruption, and insect pathogens, have suggested that these combinations would yield synergistic effects. Because the ratio between irradiated and non-irradiated insects is the most critical factor in regulating the efficacy of SIT or F_1 sterility release programmes, any mortality agent (i.e. resistant host plants, insecticides) applied during a continuous release of genetically altered insects, would benefit a release programme. Although the mortality agent would reduce the number of both released and wild insects, it would not change the ratio. Therefore, subsequent to the application of the mortality agents, wild population would be lower and the ratio of released to wild insects would be increased by continual releases of genetically altered insects.

Integration of genetic techniques and inundative releases of parasitoids may be more complementary than most other pest control combinations because their optimal actions are at opposite ends of the host density spectrum and do not interfere with each other. Although the use of parasitoids and sterile insect techniques have different modes of action, the effectiveness of the sterile insect technique increases the ratio of adult parasitoids to adult hosts, and the effectiveness of the parasitoids increases the ratio of sterile to fertile insects. Greater pest suppression could be expected if parasitoid releases were combined with the F_1 sterility technique. Not only is F_1 sterility in lepidopterans more effective than full sterility in reducing population increases, the F_1 sterility technique produces eggs and sterile F_1 larvae that would provide an increased number of hosts for the parasitoids.

Several laboratory and field studies have been conducted to determine the compatibility and effectiveness of combining different types of genetic control techniques with other control techniques. The effects of F_1 sterility and host plant resistance on *Helicoverpa zea* (Boddie) and *Spodoptera frugiperda* (J.E. Smith) development were investigated. It was found that larvae resulting from irradiated male by non-irradiated female crosses were equally competitive with normal larvae for all measured parameters. Ovipositional acceptance tests and parasitism studies with *Glabromicroplitis croceipes* (Cresson) in the laboratory and field using *Heliothis virescens* (Fabricius) and *H. virescens–H. subflexa* (Guenee) backcross larvae as hosts were conducted. It was concluded that augmentative releases of *G. croceipes* during or following a sterile backcross release should not adversely affect the backcross release ratio, and that the two control techniques possibly could be used effectively together in an area-wide management programme for controlling *H. virescens*. The effect of concurrent parasitoid and sterile fly releases on wild *Ceratitis capitata* (Wiedemann) populations in the Kula area of Maui, Hawaii was investigated. It was concluded that concurrent SIT and parasitoid augmentation programme may interact synergistically, producing a greater suppression in targeted insect population than either method used alone.

POTENTIAL AND CONSTRAINTS

The potential of genetic control of insect pests has yet to be fully exploited despite the fact that the theoretical considerations have been with us for more than 40 years.

- Genetic control is most successful in situations where a crop or an animal or human population is to be protected from insect attack over a large geographical area.
- In most of the cases, genetic approaches result in the complete kill or even eradication of the target pest. In the long run, area-wide genetic approaches provide a much better return on investment than do conventional farmer by farmer approaches.
- Genetic control methods are highly specific to the pest species and hence do not cause any adverse effect on non-target organisms.
- The application of genetic techniques does not pollute the environment and is safe to pest control workers.
- Genetic methods are compatible with most other pest management strategies such as chemical control, cultural control, natural enemies, host plant resistance, etc.

Some obstacles act as bottlenecks which will slow the rate of progress in the use of genetic methods for controlling pests.

- Research and development requirements for genetic approaches often demand a long-term commitment before benefits can be realized.

- Also, technological constraints, such as the difficulty and high cost of insect rearing, may impede progress.
- The major factor which has limited the success of genetic methods is the lack of information about the behaviour and quality of sterile insects.
- The background ecological information, so important to the success of these programmes, is also wanting in many cases.
- Genetic control in most cases is successful as an area-wide approach, which requires large scale mobilisation and organization of the beneficiary farmers.

Before genetic control methods could be more fully implemented for pest management, the concepts of preventative and area-wide pest management must gain wider acceptance. Future pest management strategies for major pests and pest complexes should be developed with definable, long-term goals. These exercises in strategic planning should consider the long-range economic and ecological advantages that genetic approaches may provide as components of integrated pest management. With the generation of additional information, genetic methods are likely to fit into a broader concept of IPM to achieve a range of control from suppression or eradication on the one hand to preventive quarantine on the other.

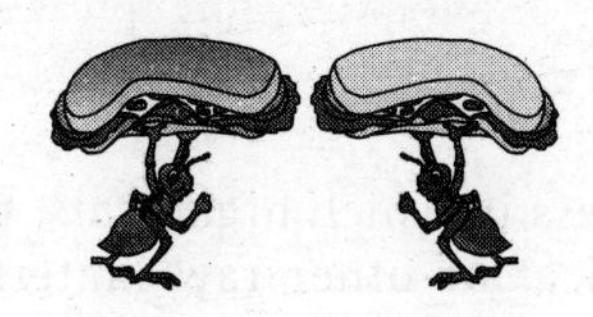

BIOTECHNOLOGICAL APPROACHES

INTRODUCTION

Biotechnology includes any process in which organisms, tissues, cells, organelles or isolated enzymes are used to convert biological or other raw materials to products of greater value. Biotechnology involves the production, isolation, modification and use of substances derived by means of biosynthesis. Biotechnology relies on the integrated use of molecular genetics, biochemistry, microbiology and process technology to supply goods and services, employing microorganisms, part of microorganisms, or cells and tissues of higher organisms. The Convention on Biological Diversity has defined biotechnology as any technological application that uses biological systems, living organisms, or derivatives thereof, to make or modify products or processes for specific use. This definition includes medical and industrial applicatons as well as many of the tools and techniques that are used in agriculture and food production. In a narrow sense, this definition covers a range of different technologies such as gene manipulation and gene transfer, DNA typing, and cloning of plants and animals. In pest management, it is the use of genetically modified organisms in the production of crops or animals including the production of insect suppressive agents.

Biotechnology offers direct access to a vast pool of useful genes not previously available for crop improvement. Current biotechnological techniques allow the simultaneous use of several desirable genes in a single event, thus allowing coordinated approaches for introduction of novel genes or traits into the elite background. Biotechnology offers the possibility of introducing a desirable character from closely related plants without associated deleterious genes or from related species that do not readily cross with the crop of interest, or from completely unrelated species, even in other taxonomic phyla. As a result of advances in biotechnology during the last two decades, there has been rapid progress in using biotechnology in various facets of crop improvement like yield, quality, draught, chilling, salinity and herbicide tolerance, and resistance to lodging, diseases, aflatoxins and insect pests.

BIOTECHNOLOGICAL METHODS

Biotechnology, as an aid to crop improvement, became scientifically established following developments in our understanding of the genetics that took place in the early 1900s. These developments led to the production of high yielding crop varieties that substantially boosted global agricultural output throughout the twentieth century. However, developments in cell and molecular biology that have taken place since the early 1970s have permitted scientists to shift their attention from yield to pest control. The main biotechnological methods that can be used to produce genetically modified organisms that are of relevance to pest control can be split into two (not mutually exclusive) categories, i.e. (*i*) those that involve tissue culture techniques, and (*ii*) those that involve the use of recombinant DNA technology (**Table 11.1**). The former techniques have been extensively developed in relation to the production of crop plants while the latter have been employed with plants, pests and natural enemies. In many cases, both of these techniques have been used to produce genetically modified organisms. For example, tissue culture techniques are often used to regenerate plants that have been produced using recombinant technology.

Table 11.1 Biotechnological methods employed in crop improvement

Technique	Application	Example(s)
Tissue Culture Techniques		
Protoplast fusion	Production of somatic hybrids-novel plants from species that would not cross in the wild	(i) Herbicide-resistant potato plants. (ii) The pomato-cross between potato and tomato plants
Clonal propagation	Culturing protoplasts to produce uniform plants	Disease-free potatoes and strawberries
Somaclonal variation	Culturing protoplasts with useful traits that are normally hidden	Research with a range of crops currently under way
Mutant selection	Culturing protoplasts and stressing them. Surviving cells are selected	Herbicide-resistant maize
Recombinant Techniques		
Agrobacterium-based plant transformation	Ti-plasmid (tumour-inducing plasmid) used to carry novel DNA into plants	Bt-insect-resistant crop plants (tobacco, corn, cotton)
Particle acceleration	DNA-coated gold particles fired into growing tissue	Used to produce transgenic soybean
Electroporation	Electric current used to alter protoplast membranes permitting DNA uptake	Used to produce transgenic rice
Microinjection	DNA injected into the nucleus or cytoplasm of a protoplast	Used to produce transgenic tomato
RNA interference	Blockage of gene function by inserting short sequences of ribonucleic acid.	Potential for protecting cotton, rice and maize against insect pests

Source : Modified from Thacker (2002)

A. TISSUE CULTURE

Tissue culture is a technique of growing plant tissues on synthetic medium under controlled and aseptic conditions. A German plant physiologist, G. Haberlandt, is considered to be the father of plant tissue culture who in 1902 conceived the idea of *totipotency*, which refers to the capability of a cell to give rise to a complete plant under suitable cultural conditions. Such a property of the cell has far reaching implications to manipulate plant cells for rapid multiplication of plants, to cross plants at the level of somatic cells by overcoming the limit of crossability, and also to regenerate adult plants after modifying the DNA molecule at the cellular level. One of the principle advantages of using tissue culture is that the desirable traits can be selected in the laboratory and the screening process can, therefore, be completed relatively quickly in contrast to consider whole plants in a field-based situation. Plant tissue culture includes several specialized areas, like micropropagation, somaclonal variation, prototoplast and anther culture.

Micropropagation

Micropropagation involves the production of plants from very small (0.2-1.0 mm) plant parts through tissue culture techniques. Micropropagation of selected plant species is one of the best and most successful examples of commercial application of tissue culture technology for mass multiplication of plants that has four district advantages, viz. (*i*) It is independent of seasonal constraints and hence ensures year-round rapid propagation. (*ii*) Micropropagated plants are generally true to type. (*iii*) Micropropagated plants are disease free. (*iv*) Micro-propagated field grown plants usually exhibit vigorous growth, better quality and higher yields.

Somaclonal Variation

Somaclonal variation refers to variation among tissues or plants derived from the *in vitro* somatic cell cultures, i.e. callus and suspension cultures. Historically, it was accepted that all plants arising from tissue culture should be exact copies of the parental plants. However, phenotypic variability was observed among regenerated plants, which was usually ignored. Plants regenerated from stem callus have been referred to as *calliclones* and those from protoplasts as *protoclones*. A general term *somaclone* has been proposed for plants derived from any form of cell culture and the variation among such plants termed as somaclonal variation. The somaclonal variation may be genetic or it may result from culture induced epigenetic changes. The epigenetic changes are expressed at cell culture stage but usually disappear when plants are regenerated or they reproduce sexually. Variation arising out of anther/pollen culture is more precisely known as *gametoclonal variation* and that through protoplast culture is called *protoclonal variation.*

The major steps involved in the isolation of somaclones for insect resistance are (*i*) growing calli or cell suspension cultures for several cycles from a high yielding and well-adapted susceptible variety, (*ii*) regeneration of plants from such long-term cell lines, and (*iii*) evaluation of large population of regenerated plants for insect resistance. Using this procedure, somaclonal variants of sugarcane and sorghum have been obtained with good levels of resistance to sugarcane borer, *Diatraea saccharalis* (Fabricius) and fall armyworm, *Spodoptera frugiperda* (J.E. Smith), respectively.

Protoplast Culture and Somatic Hybridization

Somatic hybridization is an effective approach to hybridize the sexually incompatible species. Complete fusion of the nuclei and cytoplasms of somatic cells from both species leads to the formation of somatic hybrid cell and plant. Likewise, the fusion of cytoplasm from two species and nuclear genes from any one leads to the development of cybrid. The plant cells are surrounded by a thick cell wall which does not allow the two cells to fuse to get somatic hybrid cell/plant. However, the protoplast can be easily fused and employed in several experiments aiming at the genetic modification of the plants. Protoplast is a naked cell without cell wall surrounded by plasma membrane and potentially capable of cell wall regeneration, growth and division, The techniques of isolation, culture and regeneration of protoplasts have been established in more than 100 plant species including major field, vegetable and fruit crops.

In vitro Production of Haploids

In self-pollinated crops, an inordinately long period is required to recombine desirable gene combinations from different sources in homozygous form. Generally, it takes 8-10 years to develop stable, homozygous and ready to use material from a fresh cross of two or more lines. Likewise, due to inbreeding depression in cross-pollinated crops, it is difficult to obtain vigorous inbreds for hybrid seed production programmes. In this regard, haploids possessing gametic chromosome number are very useful for producing instant homozygous true breeding lines. Besides, haploids constitute an important material for induction and selection of mutants particularly for recessive genes. The period required to develop homozygous true breeding line is less through haploid breeding as compared to conventional breeding.

B. RECOMBINANT DNA TECHNOLOGY

One of the most significant breakthroughs in modern science is the development of techniques to transfer genes from unrelated sources into crop plants. Until recently, scientists could manipulate only the primary and secondary gene pools of the cultivated species for crop improvement. However, recent advances in molecular biology have made it possible to introduce genes from diverse sources such as unrelated plants, bacteria, viruses, fungi, insects, higher animals and even from chemical synthesis in the laboratory. The recombinant DNA technology involves the use of genetic engineering so that the modified plants carry the functional foreign genes. These novel genes either reinforce the existing functions or add new traits to the transformed plants. These developments have provided the opportunity to develop crops with novel genes for insect resistance.

The whole process of introduction, integration and expression of the foreign gene(s) in the host is called genetic transformation or transgenesis. In fact, transgenesis has emerged as a novel tool to carry out single gene breeding or transgenic breeding of crop plants. Unlike conventional plant breeding, in this case only the cloned gene(s) of agronomic importance are introduced into the plants without the co-transfer of other undesirable genes from the donor. The recipient genotype is least disturbed thereby setting aside the need for any backcross. This approach has the potential to serve as an effective means to remove certain defects of an otherwise well adapted cultivar, which are not easily manageable through conventional breeding approaches. The steps or processes involved in producing a transgenic organism are summarized in Fig. 11.1.

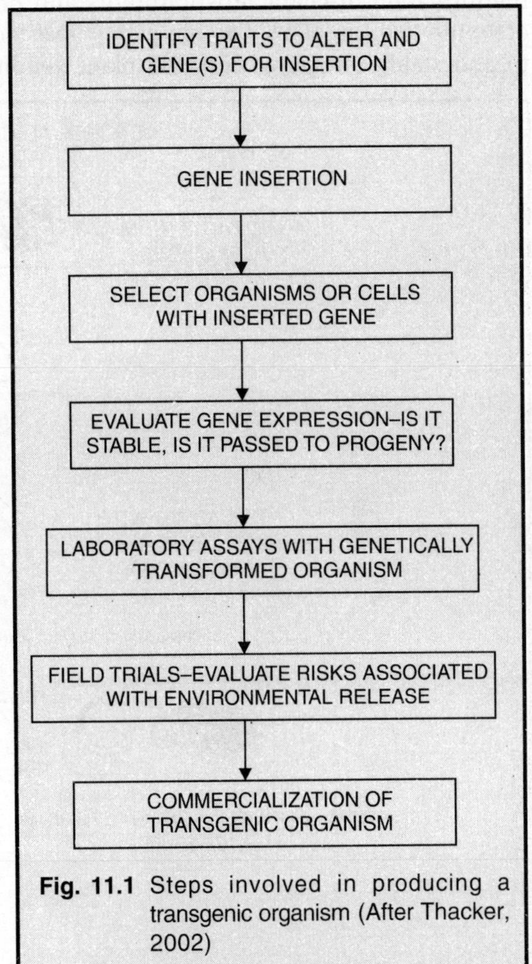

Fig. 11.1 Steps involved in producing a transgenic organism (After Thacker, 2002)

Foreign genes can be transferred through vector-mediated or direct DNA transfer methods using protoplasts or other tissues. In vector-mediated approach, the transgene is combined with a vector, which takes it to the target cells for integration. In direct gene transfer, on the other hand, the gene is physically delivered to the target tissue.

Vector-mediated Gene Transfer

A vector is a vehicle that transports the foreign genes into the recipient cells, protoplasts or intact plant. It is a DNA molecule, capable of replication in a host organism, into which a gene is inserted to construct a recombinant DNA molecule. This method is also called indirect method of gene transfer. A vector could be either a DNA virus such as caulimovirus or geminivirus or could be plasmids such as tumor inducing (Ti) and root inducing (Ri) plasmids of *Agrobacterium tumefaciens* and *A. rhizogenes*, respectively. A plasmid is usually a circular piece of DNA, primarily independent of host chromosome often found in bacterial cells. The cells to be used for transformation must be replicating DNA, which is available in wounded or dedifferentiated cells or protoplasts.

(*i*) ***Agrobacterium*-mediated gene transfer.** *Agrobacterium tumefaciens* is a gram negative soil bacterium which infects a wide range of dicot plant species causing the crown gall disease. It has been demonstrated that a virulent bacterium, in addition to its chromosomal DNA, carries Ti plasmid. The Ti plasmid has two major regions of interest in transformation, *i.e.* T-DNA and the *vir* region. During infection, Ti plasmid transfers a protein of its transfer DNA (T-DNA) into the plant cell, which becomes integrated into the chromosomal DNA of plant (**Fig. 11.2**). T-DNA segment of Ti plasmid carries a number of genes encoding enzymes for the synthesis of phytohormones such as cytokinins and auxins, which stimulate the growth and division of the plant tissue resulting in the formation of characteristic tumors and production of specific metabolites called opines such as octopine and nopaline. Foreign genes inserted within the T-region of the Ti plasmid are transferred to and stably integrated into the plant genome. *Agrobacterium*-mediated transformation has been

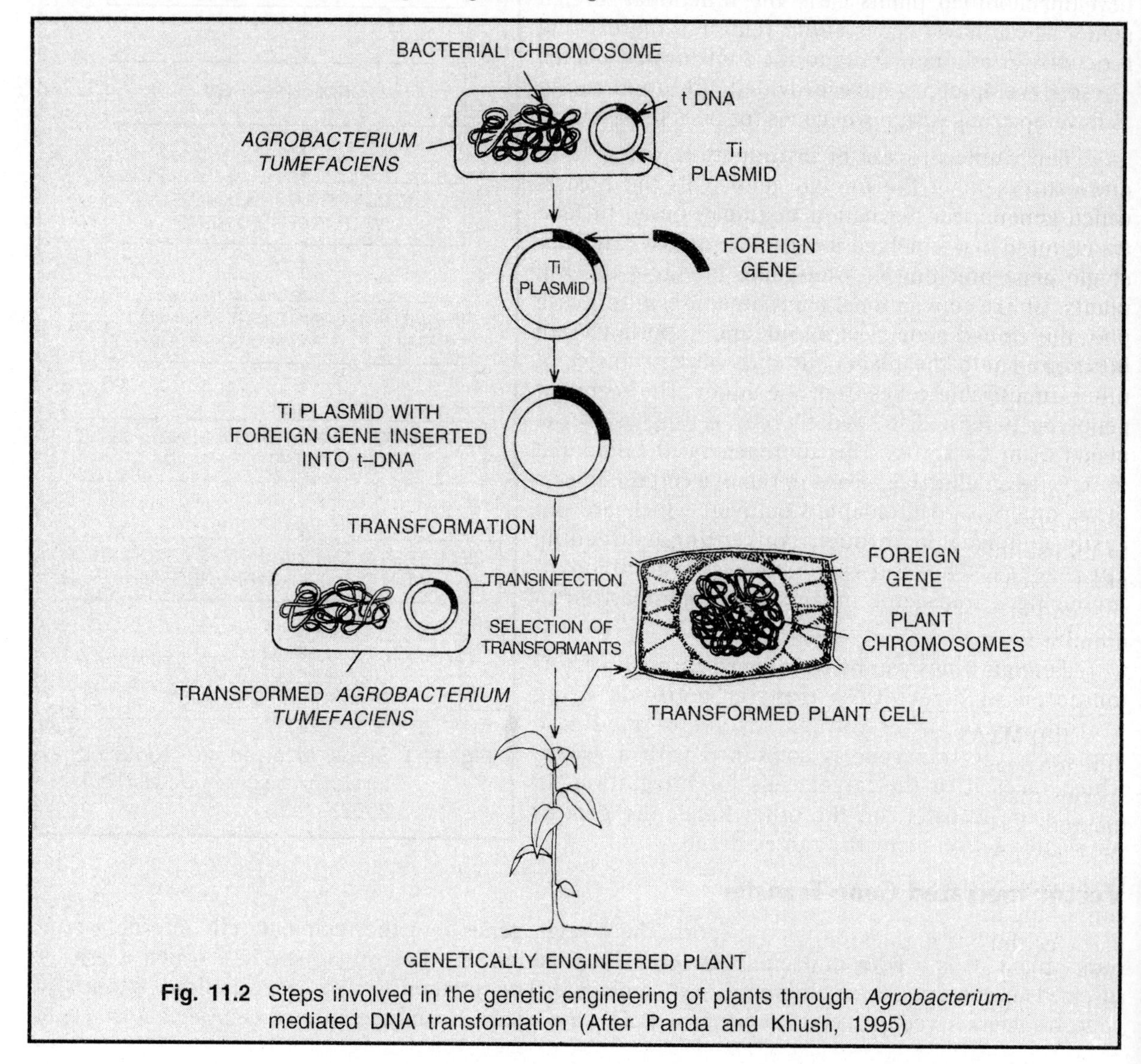

Fig. 11.2 Steps involved in the genetic engineering of plants through *Agrobacterium*-mediated DNA transformation (After Panda and Khush, 1995)

a method of choice in dicotyledonous plant species, where plant regeneration systems are well established. However, there are several reports of successful transformation of cereals using *Agrobacterium.*

(*ii*) **DNA viruses as vectors.** Due to their ability to cause systemic infections, the viruses have been investigated as vectors for gene transfer in plants. Genetic engineering of the genomes of DNA and RNA viruses has been accomplished with the introduction of foreign DNA sequences. The foreign genes replace a part of the viral genome and create a defective viral particle that can infect the target plant only in the presence of a helper virus. The genomes of caulimoviruses such as cauliflower mosaic virus (CaMV) and geminivirus such as tomato golden mosaic virus (TGMV) are double stranded DNA which make these viruses as potential transformation vector. However, viral vectors have not been developed to a stage where these can be routinely used for plant transformation.

Direct Gene Transfer Methods

Since the host range of *Agrobacterium* has been largely limited to dicotyledonous plant species, some other methods of transfer of genes to monocotyledonous plants such as cereals have been developed. Many grain legume species and woody plants are also not amenable to *Agrobacterium*-mediated transfor-mation. For genetic transformation of such recalcitrant crops, techniques of direct gene delivery have been attempted. Direct gene transfer methods make use of chemical, physical or electrical methods to introduce DNA into the cells.

(*i*) **Direct uptake of DNA.** This method is based on the ability of the protoplasts to uptake the foreign DNA from the surrounding solution. An isolated plasmid DNA (vector) is mixed with the protoplasts in the presence of polyethylene glycol (PEG), polyvinyl alcohol and calcium phosphate, which enhance the uptake of DNA by protoplasts. After 15-20 minutes of incubation, the protoplasts are cultured in the presence of appropriate selective agent. Upon treatment with PEG, the permeability of the plasma-membrane is increased, leading to transient formation of pores, thus facilitating the uptake of DNA by protoplasts. Foreign gene of interest can be as naked DNA such Ti or *E. coli* plasmids or can be encapsulated into liposomes. Although the frequency of integration of DNA into plant genome is relatively low, regeneration of transferred protoplasts selected for the presence of the particular selectable marker, yields transgenic plants. This method, however, depends on the plant regeneration ability of the protoplasts and has been successfully used to produce transgenic plants in brassica, strawberry, lettuce, rice, wheat and maize.

(*ii*) **Electroporation.** In this technique, a high voltage current is applied in a pulsed manner, which creates tiny holes in the plant cell membrane. These holes are large enough for DNA molecule to diffuse into the cell. The cells recovered from the electric shock can be regenerated into whole plants. Generally, protoplasts are used since they have exposed plasma membrane. Protoplasts are suspended in buffered saline solution containing plasmid DNA in a cuvette and an electrical impulse is applied across two platinum electrodes in the cuvette. The protoplasts are then cultured to regenerate the plants. Electroporation has been successfully used for obtaining transgenics in tobacco, maize, rice, wheat and sugarcane.

(*iii*) **Microprojectile bombardment.** This method is also called *biolistic approach* and has been a major development in direct gene transfer of DNA. It has enabled transformation of many plant species that were not amenable to *Agrobacterium* or protoplast based gene transfer techniques. The method is considered independent of genotype and the tissue. This method consists of delivering DNA into cells of intact plant organs, cultured tissues or cell suspensions via projectile bombardment. High-density particles, usually gold or tungsten (1-2 μm in diameter) coated with plasmid DNA containing the foreign gene are accelerated to high velocity by a particle gun apparatus. These dense particles thus acquire sufficient kinetic energy to penetrate the membranes, and thus to deliver the DNA into the cells. The transgenic plants have been recovered in several important crops including banana, barley, bean, canola, cassava, cotton, maize, papaya, peanut, poplar, rice, soybean, squash, sugarbeet, sugarcane, sunflower and wheat, by using more regenerable tissues/organs such as embryogenic calli, immature embryos and shoot meristems.

(*iv*) **Microinjection.** This method involves the injection of DNA directly into the plant cell nucleus or intact plant organ. The foreign DNA is delivered into protoplasts using a glass micropippette having an orifice diameter of less than one μm. It is a useful technique for the precise delivery of DNA into target cell. In this method, the location of DNA delivery can be controlled with micromanipulator and the volume of injection can be controlled by the microinjection apparatus. It is a direct and precise method of delivery of DNA into specific cell compartments. Microinjection of DNA into the nucleus can give high efficiency of transformation. This method raises the possibility of microinjecting a variety of materials such as chromosomes, and even chloroplasts and mitochondria can also be transferred by microinjection. This technique has been successfully used in alfalfa, tobacco and brassica crops.

C. RNA INTERFERENCE

RNA interference (RNAi) is a method of blocking gene function by inserting short sequences of double-stranded ribonucleic acid (dsRNA) that match part of the target mRNA sequence, thus no proteins are produced. It is a cellular process in which exogenous dsRNA with a strand complementary to a fragment of such mRNA causes degradation of the complementary mRNA, i.e. knock down the expression of genes. This technology is proving a prized functional genomics tool in insects to ascertain the function of the many newly identified genes accumulating from genome sequencing projects. RNAi can unveil the functions of new genes, lead to the discovery of new functions for old genes, and find the genes for old functions.

The dsRNA, introduced into the cell, is cleavage into short 21–25 nucleotide RNAs, by an enzyme called dicer that has RNase III domains, which are known small interfering RNAs or short interfering RNAs (siRNAs). These siRNAs are unwound and the antisense siRNA strand couples to RNA-induced silencing complex (RISC). This RISC complex finds the complementary sequence of a target messenger RNA molecule and induces cleavage by Argonaute, the catalytic component of the RISC. The cleaved target mRNA fragments are released from the RISC complex and are subsequently degraded by cytoplasmic nucleases.

RNAi-related phenomenon was initially elucidated in plants (*Petunia*) and was named post-transcriptional gene silencing (PTGS). In animals, the first clue of RNAi is demonstrated in *Caenorhabditis elegans* Maupus in 1995. *Drosophila melanogaster* Meigen was the first among the insects in which the RNAi was used *in vivo* to study the functions of genes *frizzled* and *frizzled 2*. Later, RNAi has been used to study the function of various genes in number of insect pests like *Tribolium castaneum* (Herbst), *Oncopeltus fasciatus* (Dallas), *Periplaneta americana* (Linnaues), *Blattella germanica* (Linnaues), *Spodoptera litura* (Fabricius), *Helicoverpa armigera* (Hubner), *Plutella xylostella* (Linnaues), *Bemisia tabaci* (Gennadius), etc. Three techniques, viz. direct injection, feeding and egg soaking have been adopted to deliver the dsRNA into different test organisms. RNAi technology holds significant importance in functional genomics. The first successes of development of transgenic plants expressing dsRNA against insect pests have opened new vistas for pest management. Thus, RNAi technology is emerging as an alternate biotechnological tool in the ongoing task of developing pest-resistant crops.

TRANSGENIC CROP PROTECTION

A transgenic plant is defined as one that has had foreign genetic material purposefully introduced and stably incorporated into the plant genome through means other than those that naturally occur in the environment. In other words, transgenic plants are those which carry additional stably integrated and expressed foreign gene(s), usually transferred from unrelated organisms. The major components for the development of transgenic plants are (*i*) development of reliable tissue culture and regeneration systems, (*ii*) proportion of gene constructs and transfor-mation with suitable vectors, (*iii*) efficient techniques of transformation for introduction of genes into the crop plants, (*iv*) recovery and multiplication of transgenic plants, (*v*) molecular and genetic characterization of transgenic plants for stable and efficient gene expression, (*vi*) transfer of genes into elite cultivars by conventional breeding methods, and (*vii*) evaluation of transgenic plants for their effectiveness in alleviating the biotic and abiotic stresses without being an environmental hazard.

Before taking up any attempt to produce transgenic plants to counter insect attack, the following requirements and priorities need to be identified. (*i*) The factors for resistance should be controlled by single genes. (*ii*) Standardization of methods for transfer of such genes can easily be accomplished. (*iii*) Expression of transferred gene should occur in the desired tissues at the appropriate time. (*iv*) The transgenic plant should be safe for consumption. (*v*) Inheritance of the gene in the successive generations should be very stable. (*vi*) There should be no penalty for yield in terms of other quantitative characters.

Remarkable achievements have been made in the production, characterization and field evaluation of transgenic plants. Both *Agrobacterium*-mediated gene transfer and direct DNA transfer methods have been used to produce transgenic plants with new genetic properties. Genes conferring resistance to insects have been inserted into crop plants such as maize, rice, wheat, sorghum, sugarcane, cotton, potato, tabacco, broccoli, cabbage, chickpea, pigeonpea, cowpea, groundnut, tomato, brinjal and soybean, Development of insect-resistant transgenic crop cultivars has focused on two district approaches : (*i*) integration of bacterial genes encoding for production of toxic proteins, especially from *Bacillus thuringiensis* (Bt), and (*ii*) integration of plant genes encoding for production of enzyme inhibitors and sugar-binding lectins. Both the approaches were pioneered in the mid-1980s and thus have developed in parallel. However, the first approach, based in particular on integration of δ-endotoxin genes derived from various subspecies of Bt, has undoubtedly received more attention and thus enjoyed greater progress. To date, all commercially available insect-resistant transgenic cultivars express semi-active Bt toxins, whereas cultivars expressing insecticidal plant proteins are not currently available outside research institutions.

A. Bt ENDOTOXINS

Bacillus thuringiensis Berliner (Bt) is a gram positive entomocidal spore-forming bacterium. Bt synthesizes an insecticidal crystal protein (Cry) which when ingested by insect larvae is solubilized in the alkaline conditions of the midgut and processed by midgut proteases to produce a protease resistance polypeptide toxic to the insect. Bt endotoxins are attractive candidates for insect-resistant crop development using transgenic technology because (*i*) they have an established safety record, (*ii*) they act rapidly and are completely biodegradable and proven safe to humans and non-target organisms and the environment, and (*iii*) the endotoxins are the products of single genes and are highly effective against the larvae of Lepidoptera, a major group of destructive insect pests. Therefore, the genes encoding the δ-endotoxins were cloned in the early 1980s and expression of modified toxin genes in transgenic tabacco and tomato provided the first examples of genetically engineered insect resistance in plants. Since then, Bt genes have been introduced into and expressed in a wide range of crop species (**Table 11.2**).

Cotton

Considerable progress has been made in developing cotton cultivars with Bt genes for resistance. The first generation Bt cotton varieties were developed by Monsanto and their seed partners to express the CryIAc protein. Worldwide, several different Cry1Ac events are grown commercially. In the USA and some other countries, Bollgard® event MON531 was used by seed company breeding programmes to develop the first commercial Bt cotton varieties. In Australia and other countries, Ingard® event MON757 was used. The tobacco budworm, *Heliothis virescens* (Fabricius) and pink

Table 11.2 Transgenic crops carrying Bt genes for insect resistance

Transgenic crop	Transgene(s)	Target insect pest(s)
Alfalfa	*Cry1C*	*Spodoptera littoralis* (Boisduval)
Broccoli	*Cry1C*	*Plutella xylostella* (Linnaeus), *Trichoplusia ni* (Hubner), *Pieris rapae* (Linnaeus)
Canola	*Cry1A(c)*	*Helicoverpa zea* (Boddie), *Spodoptera exigua* (Hubner)
Chickpea	*Cry1A(c)*	*Helicoverpa armigera* (Hubner)
Cotton	*Cry1A(b), Cry1A(c)*	*H. armigera, H. zea, Heliothis virescens* (Fabricius), *Pectinophora gossypiella* (Saunders,), *S. exigua, T. ni*
Groundnut	*Cry1A(c)*	*Elasmopalpus lignosellus* (Zeller)
Maize	*Cry1A(b), Cry1A(c), Cry9C*	*Ostrinia nubilalis* (Hubner), *Chilo partellus* (Swinhoe), *Busseola fusca* Fuller, *H. zea*
Potato	*CryIA(b), Cry1A(b)6*	*Phthorimaea operculella* (Zeller)
	CryIIIA, CryIIIB	*Leptinotarsa decemlineata* (Say)
	Cry1Ac9, CryV-Bt	*P. operculella*
Rice	*Cry1A(b), Cry1A(c), CryII(a)*	*Chilo suppressalis* (Walker), *Cnaphalocrocis medinalis* (Guenee), *Scirpophaga incertulas* (Walker)
Sorghum	*Cry1A(c)*	*C. partellus*
Soybean	*Cry1A(c)*	*H. virescens, H.zea*
Sugarcane	*Cry1A(b)*	*Diatraea saccharalis* (Fabricius)
Tobacco	*Cry1A(b), Cry1A(c)*	*H. virescens, M. sexta*
	CryIIa5	*H. armigera*
	Cry1Aa2	*H. zea*
Tomato	*Cry1A(c)*	*M. sexta*
	Bt(k)	*H.zea, M. sexta, Keifera lycopersicella* (Walsingham)

Source : Dhaliwal and Koul (2007)

bollworm, *Pectinophora gossypiella* (Saunders) are very well controlled by Bollgard® cotton whereas the cotton bollworm, *Helicoverpa zea* (Boddie) is controlled satisfactorily except during the bloom stage when feeding on reproductive parts of flowers which tend to have a lower concentration of Cry protein than other plant parts. The introduction of Bt cotton in 1996 and 1997 was very timely since *P. gossypiella* and *H. virescens* in the USA and *H. armigera* in China and Australia had become resistant to many of the conventional insecticides and Bt cotton offered a reasonable solution to the problem.

Cotton cultivar Coker 312, transformed with the *cry1A(c)* gene (having 0.1% toxin protein), has shown high levels of resistance to cabbage looper, *Trichoplusia ni* (Hubner), tobacco caterpillar, *Spodoptera exigua* (Hubner) and cotton bollowworms, *H. zea* and *H. virescens.* In transgenic cotton, bollworm damage was reduced to 2.3 per cent in flowers and 1.1 per cent in bolls compared to 23 per cent damage in flowers and 12 per cent damage in the commercial cultivar, Coker 312.

The second-generation Bt crops, developed with stacked *cry* genes, have been shown to improve the level of control and broaden the spectrum of caterpiller pests controlled. China has developed the new events, GK (expressing fused Cry1Ab and Cry1Ac proteins) and SGK (expressing the fused Cry1Ac-Cry1Ab and stacked CpTi proteins). Monsanto has created Bollgard® II by inserting a synthetic gene from *B. thuringiensis, cry2Ab,* expressing Cry2Ab protein, into the Bollgard® cotton variety, DP50B, already expressing Cry1Ac through event MON531, creating the new stacked gene event 15985. Bollgard® II is much more effective than Bollgard® at controlling all the caterpiller pests of cotton worldwide, especially *H. armigera, S. exigua, S. frugiperda* and *H. zea.* Laboratory and field studies have shown that Bollgard® II increased mortality of *H. zea* from 84.2 to 92.2 per cent, *S. frugiperda* from 16.1 to 100 per cent, *S. exigua* from 50.1 to 94.9 per cent and *T. ni* from 1.2 to 97.4 per cent. This high level of pest control is due, in part, to the very high concentration of Cry2Ab protein in Bollgard® II.

Maize

The first generation Bt proteins engineered into maize were Cry1Ab (from *B. thuringiensis*) and Cry9c (from *B. thuringiensis* subsp. *tolworthi*) and these were highly effective against European corn borer, *Ostrinia nubilalis* (Hubner). The dominant Cry1Ab events are Monsanto's YieldGard® maize, event MON810; Syngenta's BT11 sold under the NK-brand as YieldGard®, and Syngenta's event 176, known as Knockout®. The level of expression of Bt proteins tends to be lower in fresh leaves of 176 than MON810 and BT11. Recently, Cry1F, has been introduced in maize (Dow's Herculex I®) and maize with a *cry3Bb* gene or a binary toxin genetic system for control of the maize rootworm (*Diabrotica*) complex is being developed.

Maize plants transformed with Bt genes have also been found to be effective against the spotted stem borer, *Chilo partellus* (Swinhoe) and the maize stalk borer, *Busseola fusca* Fuller in Southern Africa. Maize plants with *cry1Ab* gene are also resistant to the sugarcane borers, *Diatraea grandiosella* Dyar and *D. saccharalis* (Fabricius). Transgenic tropical maize inbred lines with *cryIAb* or *cryIAc* genes with resistance to corn earworm, *H. zea;* fall armyworm, *S. frugiperda;* Southwestern corn borer, *Diabrotica undecimpunctata howardi* Barber and Western maize rootworm, *D. virgifera virgifera* LeConte have also been developed.

More recently, SmartStax maize has been developed by combining eight genes, six for insect resistance and two for herbicide tolerance. SmartStax has been created by crossing four transgene varieties (MON89034 × 1507 × MON88017 × 59122), rather than using genetic transformation of a single maize strain. SmartStax is the first transgenic crop with as many as eight traits; the current transgenic crops in the market have only up to three traits each. The eight genes in SmartStax maize include three for lepidopteran (moth) resistance (*cry1A.105, cry2Ab, cry1Fa 2),* three for coleopteran (corn rootworm) (*cry3Bb1, cry35Ab1, cry34Ab1*) and two for herbicide tolerance, *i.e. Pat* (glufosinate tolerance) and *CP4 epsps* (glyphosate tolerance). SmartStax has proved effective against eight primary pests, viz. six lepidopterans, *i.e. H. zea, O. nubilalis, S. frugiperda, D. grandiosella, A. ipsilon,* and *Loxagrotis albicosta* (Smith), and two coleopterans, *i.e. D. virgifera virgifera* and *D. barberi* Smith & Lawrence (www.i-sisorg.uk).

Rice

Various Chinese universities and research institutes, in cooperation with the International Rice Research Institute (IRRI) and universities in many countries have transformed rice by inserting the *cry1Ab* gene alone and a fused *cry1Ab/cry1Ac* into conventional rice varieties and hybrids. Initial field and laboratory tests of these Bt lines against lepidopteran pests have shown good season long control. Rice plants having 0.05 per cent toxin of the total soluble leaf protein have shown high levels of resistance to the striped stem borer, *Chilo suppressalis* (Walker) and rice leaf folder, *Cnaphalocrocis medinalis* (Guenee). Field testing of transgenic rice lines showed high protection against *Scirpophaga incertulas* (Walker) and *C. medinalis.* The percentage of plants with whiteheads was significantly lower on the Bt Shanyou 63 (11%) as compared to control Shanyou 63 (44%) plants. Similarly, transgenic plants showed no damage by *C. medinalis* as compared to 47.9 per cent damaged plants in non-transgenic rice. Scented varieties of rice (Basmati 370 and M7) have been transformed with *cryII(a)* and are resistant to *S. incertulas* and *C. medinalis.* Selected Bt-lines of IR64 and Pusa Basmati 1, having Bt-titres of 0.1 per cent (of total soluble protein) showed 100 per cent mortality of *S. incertulas* larvae within 4 days of infection in cut-stems as well as at the vegetative stage in whole plant assays.

Potato

A modified *cry3A* gene has been expressed in potato plants with resistance to Colorado potato beetle, *Leptinotarsa decemlineata* (Say). Transgenic potato plants containing the *cry1A(b)* gene (*Bt 884)* and a truncated gene *cry1A(b)6* resulted in less damage to the leaves by the potato tuber moth, *Phthorimaea operculella* (Zeller). However, the size of the leaf tunnels increased over time in plants containing only the *Bt 884* gene, while there was no increase in tunnel length in those containing *cry1A(b)6*. The latter also resulted in 100 per cent mortality of the insects in tubers stored up to six months. Several other Bt potato lines showing complete mortality of larvae of *P. operculella* have also been developed. However, the marketing of Bt potatoes was stopped in 2001 due to opposition from several food producers not to use Bt potatoes in their products.

Vegetables

Transgenic tomato was one of the first examples of genetically modified plants with resistance to insects. Tomato plants expressing *cry1A(b)* and *cry1A(c)* genes are highly effective against *Helicoverpa armigera* (Hubner). Transgenic broccoli containing *cry1A(c)* is resistant to *Trichoplusia ni* (Hubner) and *Pieris rapae* (Linnaeus). Synthetic *cry1A(c)* gene introduced into broccoli provides protection not only from susceptible *Plutella xylostella* (Linnaeus) larvae, but also from those selected for moderate levels of resistance to Cry1A(c). Synthetic *cry1A(b)* gene inserted into broccoli cultivar Pusa Broccoli KTS-1 and *cry1A(b)* in cabbage have shown resistance to *P. xylostella.* Transgenic cauliflower plants transformed with synthetic *cry9A(a)* have also shown high levels of activity against *P. xylostella.* Transgenic brinjal plants expressing *cry1A(c)* gene have also shown insecticidal activity against the fruit borer, *Leucinodes orbonalis* Guenee. The Genetic Engineering Approval Committee (GEAC) of the Government of India, had approved four BT brinjals, viz., MHB-4Bt, MHB-9Bt, MHB-80Bt and MHB-99Bt, for environmental release in 2009. However, the Government of India withheld the commercial release of Bt brinjal in 2010 till more scientific data on biosafety aspects are generated (Dhaliwal and Koul, 2010).

B. PLANT-DERIVED GENES

Largely, plant transformation involving plant genes has focused on (*i*) protease inhibitors, which disrupt amino acid metabolism; (*ii*) α-amylase inhibitors, which target carbohydrate metabolism; (*iii*) lectins, which cause agglutination and cell aggregation; and (*iv*) enzymes such as chitinase, which target insect exoskeleton. The varied modes of action and levels of specificity of these gene products increase the potential target-pest range of transgenic cultivars and allow the possibility of combining (pyramiding) genes that are active at various target sites within a pest insect or against various pests. However, levels of protection provided by genes of plant origin are typically lower than those provided by genes expressing Bt toxins. Frequently, effects on target insects are sublethal, including reductions in feeding, weight gain, developmental rates and fecundity. Even then, transgenic cultivars with sublethal or chronic effects on target pests may be more attractive components of IPM strategies than cultivars with acute toxic effects because they are more likely to be compatible or act synergistically with other biopesticide strategies.

Thus far, a number of economically important crop plants, such as oilseeds, potato, rice, sugarcane, tobacco, among others, have been genetically transformed to express various genes of plant origin (**Table 11.3**). In many of these cases, significant effects of the plant genes were evident on target-pest mortality rates and/or developmental and reproductive parameters.

Protease Inhibitors

The presence of antimetabolic proteins, which interfere with the processes of digestion in insects, is a strategy for defence that plants have used extensively. Proteins can occur constitutively in tissues that are particularly vulnerable to attack, such as seeds, or mechanical wounding in tissues

Table 11.3 Transgenic crops expressing insecticidal plant genes

Transgenic crop	Transgene(s)	Origin of transgene	Target insect pest(s)
Apple	*CpTi*	Cowpea	*Cydia pomonella* (Linnaeus)
Adzuki bean	a - *AI*	Bean α-amylase I	*Callosobruchus chinensis* (Linnaeus), *C.maculatus* (Fabricius)
Lettuce	*Pot PI-II*	Potato	*Teleogryllus commodus* (Walker)
Maize	*WGA*	Wheat	*Ostrinia nubilalis* (Hubner), *Diabrotica* spp.
Mustard	*WGA*	Wheat	*Lipaphis erysimi* (Kaltenbach)
	OC-1	Rice	Coleoptera
Pea	a-*AI*	Bean α-amylase I	*Bruchus pisorum* (Linnaeus) *Zabrotes subfasciatus* (Boheman)
Potato	*CpTi*	Cowpea	*Lacanobia oleracea* (Linnaeus)
	GNA	Snowdrop	*L. oleracea, Myzus persicae* (Sulzer), *Aulacorthum solani* (Kaltenbach)
	BCH	Bean chitinase	*M. persicae, A. solani*
	OC-1	Rice	*Leptinotarsa decemlineata* (Say)
Rice	*CpTi*	Cowpea	*Chilo suppressalis* (Walker), *Sesamia inferens* (Walker)
	Pot PI-II	Potato	*C. suppressalis, S. inferens*
	Pin 2	Potato	*C. suppressalis*
	GNA	Snowdrop	*Nilaparvata lugens* (Stal), *Nephotettix virescens* (Distant)
	SBTI	Soybean	*N. lugens*
	CC	Corn	*Sitophilus zeamais* Motschulsky
Tobacco	*CpTi*	Cowpea	*Heliothis virescens* (Fabricius), *Helicoverpa armigera* (Hubner), *H. zea* (Boddie), *Manduca sexta* (Johannrsen)
	Pot PI-II	Potato	*M. sexta, Chrysodeixis eriosoma* (Doubleday)
	α-*AI*	Bean α-amylase	*Agrotis ipsilon* (Hufnagel), *Tenebrio molitor* Linnaeus
	GNA	Snowdrop	*H. virescens, M. persicae*
	p-Lec	Pea	*H. virescens*
	SBTI	Soybean	*H. armigera, H. zea, H. virescens, S. littoralis, M. sexta*
	SpTi	Sweet potato	*S. litura*
Tomato	*CpTi*	Cowpea	*L. oleracea*
	GNA	Snowdrop	*L. oleracea*
	Pot PI-I	Potato	*H. armigera, T. commodus*
	Pot PI-II	Potato	*H. armigera, T. commodus*
Wheat	*CMe*	Barley	*Sitotroga cerealella* (Olivier)
	GNA	Snowdrop	*Sitobion avenae* (Fabricius)

a-*AI* = a-Amylase inhibitor, *BCH* = Bean chitinase, *CC* = Corn cystatin, *CpTi* = Cowpea trypsin inhibitor, *GNA* = Snowdrop lectin, *OC-I* = Oryzacystatin I, *p-lec* = Pea lectin, *Pot PI-I* = Potato proteinase inhibitor I, *Pot PI-II* = Potato proteinase inhibitor II, *SBTI* = Soybean Kunitz-type trypsin inhibitor, *SpTi* = Sweetpotato trypsin, *WGA* = Wheat germ agglutinin

Source : Dhaliwal and Koul (2007)

attacked by chewing insects can induce them. Analysis of the effects of dietary protease inhibitors has shown that these are detrimental to the growth and development of insects from a variety of genera including *Helicoverpa, Spodoptera, Diabrotica* and *Tribolium.* Plants can now be transformed with protease inhibitor genes with strong promoters to express the inhibitor proteins in relatively high levels at specific times.

Several classes of protease inhibitors corresponding to different types of insect gut proteases have been characterized. Many insects, particularly members of Lepidoptera, depend on serine proteases (trypsin, chymotrypsin and elastase like endoproteases) as their primary protein digestive enzymes and genes encoding members of various serine protease inhibitor families have been cloned and introduced into transgenic plants. Insects also produce their own serine protease inhibitors for the regulation of their digestive proteases. It has been suggested that these could be turned against the insects by expressing them in transgenic plants. Other pests rely on cystein proteases (thiol proteases) as their primary digestive proteases, which act on papain and cathepsin. These have been targeted with cysteine protease inhibitors, which have been shown to exert chronic effects on important pests such as corn root worm, *Diabrotica* spp., against which there are no effective Bts.

Serine protease inhibitors. The first gene of plant origin to be used in transgenic crop protection was that isolated from cowpea encoding a double-headed trypsin inhibitor (CpTi) and transferred in tobacco. A simple construct was prepared in which a full length coding sequence derived from a cDNA clone was placed under the control of the constitutively expressed cauliflower mosaic virus (CaMV) 35S promoter. Transformants were screened for CpTi expression, which showed that many of the resulting plants expressed CpTi at levels greater than 0.1 per cent of total soluble protein. Subsequent experience has shown that this is generally the case for expression of genes of plant origin encoding defensive proteins in transgenic plants, in contrast to the very low level of expression observed for unmodified toxin genes of bacterial origin. CpTi is a small polypeptide of about 80 amino acids, homologous sequences are encoded by a moderately repetitive gene family in the cowpea genome. This protein is considered to be particularly suitable candidate for transfer to other species through genetic engineering because (*i*) it is an effective antimetabolite against a wide range of field and storage pests belonging to Lepidoptera, Coleoptera and Orthoptera, (*ii*) has no deleterious effects on mammals, and (*iii*) also has a small polypeptide of about 80 amino acids.

Biossays against *H. virescens* caterpillars showed that transgenics expressing CpTi at the highest levels (about 1% of total soluble protein) caused increased mortality, reduced growth and reduced plant damage. The antimetabolic effects of CpTi expressed in transgenic tobacco have also been observed with other lepidopteran pests including *H. zea, S. littoralis* and *M. sexta.* Subsequent trials carried out in California showed that expression of CpTi in tobacco afforded significant protection against *H. zea* in the field. Following on from the study using tobacco as a model system, the gene encoding CpTi has been expressed in a range of different crops (**Table 11.4**). For example, constitutive expression of CpTi in rice conferred significantly enhanced levels of resistance towards two species of rice stem borer, viz. *Chilo suppressalis* (Walker) and *Sesamia inferens* (Walker) in the field. Furthermore, the trials with CpTi transgenic strawberry plants suggested that these plants were highly resistant to the vine weevil, *Otiorhynchus sulcatus* (Fabricius).

Cysteine protease inhibitors. Cysteine proteases are used by plants for protein mobilisation and by animals for intracellular lysozomal protein digestion, and protein inhibitors of cysteine proteases (cystatins) are widely distributed throughout all living organisms to regulate these endogenous proteases, even if they are usually present in small amount. The genes encoding cysteine protease inhibitors have been suggested for use in transgenic plants for the control of coleopteran pests. Although there have been several studies carried out demonstrating *in vitro* inhibition of insect digestive proteases by cysteine protease inhibitors, with a few examples of their deleterious effects against insects when incorporated into artificial diets, as yet there are few published reports describing their insecticidal effects *in planta.*

A gene encoding a rice cystein proteinase inhibitor oryzacystatin, has been expressed constitutively in transgenic poplar trees, conferring resistance to the coleopteran pest, *Chrysomela tremulae.* Corn cystatin (CC) has been introduced into protoplasts of rice and cystatin activity of the transgenic rice plants was assayed against a crude midgut proteinase fraction from *Sitophilus zeamais* Motschulsky. The results showed that 50 per cent of the midgut protease activity in *S. zeamais* was inhibited by 2 μg and completely inhibited by 5 μg of transgenic seed protein fraction, whereas untransformed rice seeds had no significant effect.

α-amylase Inhibitors

Plants produce inhibitors of insect gut alpha-amylases, which are required for the digestion of starch, a major energy source, particularly for the weevils. The alpha-amylase inhibitors produced by plants have different types of structure and different mode of action and target specificity, and hence can be used for insect control in transgenic plants. Transgenic tobacco plants expressing amylase inhibitors from wheat (wheat alpha-amylase inhibitor, WAAI) increase the mortality of lepidopteran larvae by 30-40 per cent. Similarly, transgenic pea seeds expressing alpha-amylase inhibitor derived from common beans (BAAI) were found to exhibit increased resistance against bruchid beetles, *Callosobruchus* spp. and pea weevil, *Bruchus pisorum* (Linnaeus).

Enhanced levels of resistance to the bruchids have been observed in transgenic adzuki beans expressing the alpha-amylase inhibitor of common bean. While even low levels of the amylase inhibitor were sufficient to provide resistance to the adzuki bean weevil, higher levels of the protein make the seeds resistant to the cowpea weevil, *C. maculatus* (Fabricus) and pea weevil, *C. chinensis* (Linnaeus) as well. Two alpha-amylase inhibitors, α-AI-1 and α-AI-2, from the common bean inserted into pea, are effective in protecting peas against the pea weevil, *B pisorum* under field conditions. α-AI-1 provided complete protection from *B. pisorum,* by inhibiting alpha-amylase by 80 per cent, while α-AI-2 inhibits the enzyme by 40 per cent. α-AI-1 results in larval mortality, whereas α-AI-2 delays the maturation of the larvae.

Lectins

Lectins are plant-derived proteins that bind to oligo-and polysaccharides, and cause agglutination and cell aggregation. Lectins have been isolated and characterized from a wide variety of plants such as pea, rice, wheat, snowdrop, castor, soybean, mungbean, garlic, sweet potato, tobacco, chickpea and groundnut. A number of plant lectins exhibiting insecticidal characteristics are being evaluated as alternatives to Bt δ-endotoxins. The precise mode of action of insecticidal lectins is unknown. However, binding to specific carbohydrates and agglutination (fusion by adhesive substances) in the insect midgut has been clearly demonstrated. Specifically, lectins may interfere with development and structural integrity of the midgut peritrophic membrane; bind to glycosated targets in the insect midgut, thereby inhibiting nutrient absorption or cell disruption in the midgut; or bind or block the peritrophic membrane protecting the insect midgut surface.

Lectins from wheat (wheat germ agglutinin, WGA) and the snowdrop plant (*Galanthus nivalis* agglutinin, GNA) are inhibitory to the sap-sucking homopteran pests such as aphids, leafhoppers and planthoppers, which feed on the phloem exudates and against which there are no known Cry proteins. In addition, lectins have also exhibited inhibitory activity against several lepidopteran and coleopteran pests. A gene encoding the pea lectin (*p-lec*) has been expressed in transgenic tobacco and the plants expressing pea lectin upto 1 per cent of total protein reduced the larval biomass of *Heliothis virescens* (Fabricius) and leaf damage. Transgenic tobacco plants containing both CpTi and and P-Lec were obtained through hybridization of two primary transformed lines. The plants expressing two insecticidal proteins reduced the insect biomass by 90 per cent as compared to 50 per cent reduction in plants expressing either CpTi or P-Lec. This study demonstrated that not only the products of lectin genes could enhance resistance to insect attack in transgenic plants, but also showed that additive protective effects could be obtained from different plant-derived insect-resistant genes.

Survival of the brown planthopper, *Nilaparvata lugens* (Stal) decreased to 60 per cent in transgenic rice expressing GNA protein (2% of total protein), under the control of phloem tissue-specific sucrose synthase promoter which directs the expression of the gene in the phloem of leaves, stems, petioles and roots for protection from the phloem feeding pests. Wheat germ agglutinin (WGA) is antimetabolic, antifeedant and insecticidal to the mustard aphid, *Lipaphis erysimi* (Kaltenbach). Bioassays using leaf discs showed that feeding on transgenics induced high mortality and significantly reduced fecundity of aphids. Thus, plant lectins have shown biological activity against a wide range of insects. However, consideration should be given with regard to their deployment in transgenic plants because of their known toxicity to mammals and humans.

Enzymes

Transgenic expression of various enzymes has been proposed as a crop protection strategy. The most obvious candidate is chitinase, which is an important structural component of insects. Expression of an insect chitinase in transgenic tobacco enhances resistance to some lepidopterans. Similar marginal protective effects have been observed from expression of bean chitinase (BCH) in transgenic tobacco. Transgenic potato plants expressing a gene encoding BCH were found to reduce fecundity of the glasshouse potato aphid, *Aulacorthum solani* (Kaltenbach), though this reduction was not statistically significant. However, nymphs produced on these BCH expressing plants were significantly smaller compared to those on control, non-transformed plants.

Alarm Pheromones

Alarm pheromones are volatile substances released by certain species of insects, which alert them about the potential danger from their predators. Many aphid species produce the sesquiterpene, (E)-β-farnesene (EBF) as the principal component of the alarm pheromone. EBF is released when aphids are attacked by enemies and it leads aphids to undertake predator avoidance behaviours and to produce more winged offspring that can leave the plant. Many plants also release EBF as a volatile substance and this chemical could act to defend plants against aphid infestation by deterring aphids from settling, reducing aphid performance due to frequent interruption of feeding, and inducing the production of more winged offspring.

Scientists at Rothemsted Research (UK) have developed genetically modified (GM) wheat, by transferring EBF synthase gene from peppermint (*Mentha piperita* Linnaeus) to the genome of a spring wheat strain, Cadenza. Laboratory trials have shown that EBF emitting wheat not only repels aphids, but also attracts their natural enemies. This GM wheat is currently being evaluated in field trials at Rothamsted. This is the world's first GM crop which repels insects instead of killing them, reducing the chances of the pests developing resistance to it. Scientists have created GM wheat to combat aphid attacks that can cause loss of more than £120 million each year to the UK's most important cereal crop, which has an annual value of more than £1.2 billion.

Other Novel Genes

The genes derived from other sources such as chicken, scorpion and spider are also being screened for their insecticidal potential. Avidin, a glycoprotein found in chicken egg white, sequesters the vitamin biotin. Transgenic maize containing *avidin* gene has been produced. The avidin at ≥ 100 ppm is toxic to and prevents development of insects that damage grains during storage. Insect-specific neurotoxin AaIT from the venom of the scorpion, *Androctonus australis* (Linnaeus), in tobacco has shown insecticidal activity against *H. armigera* larvae (up to 100% mortality after 6 days). Transgenic plants of tobacco have been obtained containing an insecticidal spider peptide gene, and some of these plants have exhibited resistance to *H. armigera*. The role of neurotoxins from insects and spiders need to be studied in greater detail before they are deployed in other organisms and plants because of their possible toxicity to mammals.

RESISTANCE IN PESTS TO TRANSGENICS

The insects are exposed to the toxin proteins in transgenic crops throughout the feeding cycle and season, and as a result, the insect populations are under continuous selection pressure. As the toxins are expressed in all parts of the plant, there is a serious threat of rapid development of resistance in insect pests. Toxin production may also decrease over the crop-growing season. Low doses of the toxins eliminate the most sensitive individuals of a population, leaving a population in which resistance can develop much faster.

By far the biggest problem that is likely to occur following the use of transgenic crop plants for pest control is the development of resistance in pest species. This phenomenon, of course, occurs with conventional insecticides. However, the critical difference is that a pesticide application represents a selective force that is time-limited, i.e. to the time the application is made. The current transgenic crops express toxins throughout their tissues continuously and selective pressure on the target population to adapt will thus be substantial. The strategies for managing insect resistance to toxins expressed in transgenic plants are comparable to those used for managing resistance to conventional insecticides, with only a few exceptions.

High Expression Level

High expression level of toxins in transgenic crops is comparable to the high dose strategy in case of resistance to conventional pesticides. The premise in this case is that toxin concentration is so high that all individuals of the target species are killed. This approach seems to be very effective in theory and practice provided the toxin expression is consistently high and stable over a period of time or among different plant issues. The high dose strategy assumes that resistance to transgenic plants is recessive and is conferred by a single locus with two alleles, resulting in three insect genotypes, viz. susceptible homozygotes (SS), heterozygotes (RS) and resistant homozygotes (RR). It also assumes that there will be random mating between resistant and susceptible adults. Under ideal circumstances, only rare RR individuals will survive a high dose produced by the transgenic crop and both SS and RS individuals will be susceptible to the toxin.

The production of toxin is generally expected to decrease over time due to plant senescence. If the level of production of Bt toxin decreases toward the reproductive phase, heterozygous individuals, which may often be slightly more resistant than susceptible homozygotes, might be able to survive and transmit resistance alleles to the offspring. As Bt transgenic cotton plants matured Bt toxin concentrations decreased, increasing the chances that pests such as *Heliothis virescens* (Fabricius) will encounter sublethal or low doses of Bt toxins, leading to development of resistance more rapidly. Moreover, different levels of toxins were found in various plant parts, thus increasing the chances that target pests will receive a lower dose.

Similarly, if wild host plants of herbivorous pests or non-transgenic host plants of the same or different species are located in close proximity to the transgenic plants, then pests can move to these non-toxic plants before they receive a lethal dose. Also, an insect displaying a high frequency of movement may feed alternatively on toxic and non-toxic plants thus diluting the dose of the toxin. In fact, such cases would actually increase the rate of resistance by allowing the survival of partially-resistant genotypes. For protease inhibitors, amylase inhibitors and lectins, the doses that can be engineered into plants are generally insufficient to control a target pest and for these compounds, high dose strategy cannot be implemented. Although some of these substances may be engineered at high doses in the future, many of these are harmful to man, other mammals and birds, particularly at high doses.

Refuge Strategy

Refuges or refugia are areas of crops or host plants free of insecticidal toxins that allow part of the pest population to survive and to act as a reservoir of wild-type susceptible alleles. By maintaining a refuge area close to the transgenic field, surviving individuals (RR) that have been exposed to insecticidal toxins will mate with unselected individuals (SS) coming from the refuge and produce RS heterozygotes that will be killed by the transgenic crop. This strategy would dilute resistant (R) alleles, reduce the intensity of selection and delay the evolution of resistance. Spatial organization of refuges has been considered in various ways. One type of spatial refuge includes mixtures of transgenic and non-transgenic plants; the mixture may be obtained by mixing seeds before planting or by planting a smaller proportion of the field with the non-transgenic strain. Another type of spatial refuge results from the expression of toxins only in critical areas of the plant. A third type of spatial refuge, plantings of Bt-transgenic crops with non-Bt transgenic crops, such as Bt-transgenic corn with non-transgenic cotton, could be applied for resistance management with polyphagous pests such as *Helicoverpa zea* (Boddie). Temporal refuges involve alternating transgenic plants with non-transgenic plants over growing seasons.

Another aspect of resistance management with refuges is that there must be random mating between individuals from the refuge and the transgenic plants, which means that the refuge must be located close to the transgenic planting. At the same time, for individual pests that tend to feed on several plants rather than just one, the refuge and toxic plant plots must be located far enough apart so that susceptible individuals do not move from non-toxic to toxic plants. Similarly, if refuge fields are less attractive than transgenic plantings to female insects for oviposition, the effective size would be decreased.

According to the recent refuge requirements approved by Environmental Protection Agency (EPA) of USA for Bt corn grown outside cotton-growing areas, growers have to have a minimum of 20 per cent non-Bt corn refuge (treatable with other pest control products). In cotton-growing regions, a minimum of 50 per cent non-Bt corn refuge has to be planted. Refuges planted as external blocks should either be adjacent or in proximity to the Bt corn field. The current requirement for corn is that the refuges should be within approximately 800 meters of the Bt field, although within approximately 400 meters is the preferred distance.

Tissue and Temporal Expression of Toxins

The tissue and temporal expression of toxins is based on the principle that specific gene promoters could be used to express genes only in (*i*) the most important tissue (tissue or structure specific expression) or (*ii*) critical growth periods (temporal specific expression) or (*iii*) be environmentally benign chemical. Thus the production of toxins is limited to the most economically sensitive or most vulnerable parts of plant or to specific time. This strategy does not require external refugia as the plant itself acts as such. However, efficient tissue or time-specific promoters are not yet available. This strategy is dependent on the feeding behaviour of the pests and can be influenced by the feeding deterrent effect of Bt transgenic plants. Also, an *in planta* refugia may be relevant for protection against a given member of a group of pests but may fail to provide protection against a secondary pest attacking the plant at the same time but feeding on a different part. Moreover, several potential economical and sociological problems are related to the use of chemical-inducible promoters. The chemicals to be used for induction might be costly and negatively perceived with respect to environmental protection due to potential pollution hazards.

Tissue-or structure-specific expression could perhaps leave refuges within each plant. This has been suggested for a maize genotype where the Cry toxin is under the control of a phosphoenolpyruvate carboxylase (PEPC) promoter, which provided high levels of expression in green tissue and much lower expression in kernels. However, the level of expression in the kernel is still 4-6 times the LC_{50} for *Ostrinia nubilalis* (Hubner). Damage to corn kernels is tolerable for corn grown for feed, but this level of expression suggests that the ear will not prove to be a very hospitable site for susceptible corn borers. A slight variation of this idea is to deter insects from feeding on the most economically important plant structures without being killed, thereby lowering selection pressure while protecting the crop. For example, in case of cotton, considerable damage can be tolerated to leaves but not flower buds and bolls. However, to be fully effective, expression in the transgenic structures may have to be higher than in current constitutive varieties, and will have to provide sufficiently high expression in all important tissues without affecting the tissues that can be sacrificed. In cotton, for example, both bolls and terminal meristems must be protected.

Rotations

In case of transgenic crops, rotations refer to the alteration over time of two or more varieties containing different toxins. Rotation is based on the assumption that the frequency of resistance alleles will decrease when the selection pressure is reduced. However, many reports indicate that resistance could remain stable or decrease slowly after removal of selection pressure, making the use of rotation inefficient. The resistance of individuals to one toxin must decline during the use of second toxin for rotations to be effective. However, assuming no occurrence of cross-resistance between toxins and that resistance does not increase in the absence of toxins, then alternating two toxins over a period of time would actually at least double (if the increase were linear) the time to resistance. Similarly, alternating three toxins would at least triple the time to resistance, and so on. However, if a fitness cost to resistance or negative cross-resistance exists, rotations can be particularly effective. For insect-resistant transgenic plants, there are limited possibilities for the use of rotations. Bt toxins have similar modes of action and hence there is a problem of cross resistance, thus rotating between different toxins from year to year is unlikely to slow down resistance. Other insect toxins such as enzyme inhibitors and lectins have not so far given adequate protection when used alone and hence would not be useful in rotations.

Gene Pyramiding

Gene pyramiding or gene stacking refer, to a single crop variety expressing several different toxins, known as pyramiding or gene stacking. This strategy assumes that resistance to each toxin is monogenic, there is no cross-resistance among toxins used, resistant pest individuals are rare so that no one individual is resistant to both toxins and the toxins have equal persistence. If this is achieved, then the other toxin will kill pests that survive one of the toxins. In contrast to chemical insecticides, at least two features of transgenic plants suggest that mixtures of two or more factors pyramided into the same variety will greatly delay resistance. Transgenic plants offer a way to get consistently high control of SS homozygotes, and Cry toxins provide at least one good candidate gene with consistently fairly recessive inheritance of resistance. It has been observed that Cry1Aa has overcome a 500-fold resistance to Cry3Aa already established in cotton wood leaf beetle, *Chrysomela scripta* Fabricius.

Pyramids also have several advantages over single gene strategies :

- Fitness costs should have a greater impact on delaying resistance to mixtures than single toxins.
- Pyramids are less sensitive to initial resistance allele frequency; even resistance frequencies of 10^{-3} could allow significant benefits.
- Pyramids can be greatly improved by manipulating the mortality of susceptible homozygotes, which can be measured.
- Pyramids are more robust to survival of heterozygotes, which cannot be measured with certainty until resistance has evolved.
- Pyramids can be very effective with a smaller refuge.

Ideally, the two toxins to be used in a pyramided variety should be very different to avoid cross-resistance. It is believed that mixtures of plant protease inhibitors may be the best way to prevent insect resistance to these compounds. Plants naturally improve the efficacy of protease inhibitors by exhibiting multidomain or multimeric variants. Feeding studies have confirmed that inhibitor combinations act synergistically to reduce insect herbivore growth. Since genes for different proteases have been found in nature and engineered into crop plants, the success rates for use of protease and amylase inhibitors in plant resistance to insects can be greatly improved, especially through pyramiding genes that produce a variety of these compounds. The activity of Bt genes in transgenic plants is enhanced by serine protease inhibitors and tannic acid. Similarly, protease inhibitors engineered into cotton with high gossypol and/or tannic content may achieve greater protection against *H. armigera.*

Mosaics

Mosaics are a spatial patchwork of two or more varieties of the same crop species expressing different toxins at the same time. For mosaics to delay resistance, it is assumed that there is no cross-resistance among the different toxins, so that the individuals resistant to one toxin are susceptible to other toxins. Thus, susceptible individuals are maintained and are able to move among different patches. The mosaic theory is not considered appropriate to deploy two toxin genes. It is assumed that each of the varieties can serve as a refuge for the other, however, they cannot since neither is producing many susceptible insects. Even with an untreated refuge, the mosaic system essentially selects for resistance to each of the toxins simultaneously.

Trap Plant Strategy

This approach is closely related to the refuge system. In this approach, the transgenic crop is considered as a trap crop and not as the main source of protection. The Bt-transgenic plants are planted earlier than the non-transgenic plants and owing to their advanced maturation attract the pests that are killed after feeding, thus leaving the non-transgenic varieties relatively unharmed. There is, however, no evidence that the basic key assumptions on movement and absence of ovipositing preference will be verified in the field for the target species or for another one. However, a computer model designed on slightly different assumptions about feeding preference and movements demonstrated that this strategy was not successful. Indeed, avoidance of transgenic plants and ovipositing preferences were demonstrated. Even if an insect species may be suitable for the trap plant strategy owing to its feeding preferences, the presence of Bt toxins may adversely alter its feeding behaviour.

Integration of Tactics

The most important approach to manage resistance to transgenic plants is to diversify pest management techniques as well as resistance management techniques. This is particularly so in case of multiple crops and multiple pests prevalent in tropical and sub-tropical countries like India. A toxin may be highly active against a given pest but be moderately active against another pest also present in the same crop or in the same area, and for which the dose delivered by the transgenic plant does not correspond to a high dose. In that case, the second pest will be exposed to a moderate dose and may develop resistance. As there is no single answer or strategy to delay resistance, only a logical combination of various tactics will provide sustainability.

The multiple tactics for Bt resistance management in transgenic crops are as follows : (*i*) provide refuge for susceptibles, (*ii*) do not use pesticides that kill susceptibles, (*iii*) do not use pesticides that kill natural enemies, (*iv*) enhance natural control through crop rotation or tillage, and rotate between susceptible and resistant crop genotypes over large areas. The high-dose refuge strategy, and especially the association of this approach with combinations of toxins is advocated by scientists, seed companies and administrators as the best approach of controlling evolution of resistance. This composite strategy allows flexibility in terms of spatial organization of refuge areas and choice of toxins that will help adaptation and implementation in various agronomic systems.

COMMERCIALIZATION OF TRANSGENIC CROPS

The People's Republic of China was the first country to commercialize transgenic crops in the early 1990s, with the introduction of a virus resistant tobacco, followed by virus resistant tomato.

In 1994, the USA followed when the Calgene Company got the first approval to commercialize a genetically modified tomato 'Flavr Saver TM', the delayed ripening tomato. From then onwards, the development and use of transgenic crops gained momentum.

Transgenic Crops : Global Scenario

The global area under transgenic crops has increased more than 100-fold from 1.7 million ha in 1996 to over 175 million ha in 2013. This makes the transgenic crops the fastest adopted crop technology in recent history. In the 18 year period (1996-2013), millions of farmers in about 30 countries worldwide, planted an accumulated area of more than 1.6 billion ha. This is an area equivalent to more than 150 per cent the size of the total land mass of USA or China.

In 2013, transgenic crops were planted in 27 countries (19 developing and 8 developed countries) by 18 million farmers (**Table 11.4**). More than 90 per cent of these were risk-averse small,

Table 11.4. Global area under transgenic crops in 2013

Rank	Country	Area (million hectares)	Transgenic crop(s)
1.	USA	70.1	Maize, soybean, cotton, canola, sugarbeet, alfalfa, papaya, squash
2.	Brazil	40.3	Soybean, maize, cotton
3.	Agrentina	24.4	Soybean, maize, cotton
4.	India	11.0	Cotton
5.	Canada	10.8	Canola, maize, soybean, sugarbeet
6.	China	4.2	Cotton, papaya, poplar, tomato, sweet pepper
7.	Paraguay	3.6	Soybean, maize, cotton
8.	South Africa	2.9	Maize, soybean, cotton
9.	Pakistan	2.8	Cotton
10.	Uruguay	1.5	Sobyean, maize
11.	Bolivia	1.0	Soybean
12.	Philippines	0.8	Maize
13.	Austrailia	0.6	Cotton, canola
14.	Burkina Faso	0.5	Cotton
15.	Myanmar	0.3	Cotton
16.	Spain	0.1	Maize
17.	Mexico	0.1	Cotton, soybean
18.	Colombia	0.1	Cotton, soybean
18.	Colombia	0.1	Cotton, maize
19.	Sudan	0.1	Cotton
20.	Chile	< 0.1	Maize, soybean, canola
21.	Honduras	< 0.1	Maize
22.	Portugal	< 0.1	Maize
23.	Cuba	< 0.1	Maize
24.	Czech Republic	< 0.1	Maize
25.	Costa Rica	< 0.1	Cotton, soybean
26.	Romania	< 0.1	Maize
27.	Slovakia	< 0.1	Maize
	Total	175.2	

Source : James (2013)

poor farmers in developing countries. In India, 7.3 million farmers benefitted from transgenic crops and in China there were 7.5 million beneficiary farmers. The latest economic data (1996-2012) indicated that farmers in India gained US$ 14.6 billion and in China US$ 15.3 billion. During 1996-2012, cumulative economic benefits in developing countries were US$ 57.9 billion compared to US$ 59 billion generated by developing countries. The five lead developing countries in transgenic crops in the three continents of the South are China and India in Asia, Brazil and Argentina in Latin America, and South Africa on the continent of Africa. These five countries collectively grew 47 per cent of the global transgenic crops and have about 41 per cent of the world population.

Herbicide tolerance deployed in soybean, maize, canola, cotton, sugarbeet and alfalfa, occupied 100.5 million ha or 59 per cent of the total transgenic crop area (**Fig. 11.3**). In 2012, the stacked double and triple traits occupied a larger area (43.7 million ha or 26% of global transgenic area), than insect resistant varieties (26.1 million ha, 15%). The stacked trait products were the fastest growing trait group between 2010 and 2011 at 31 per cent growth, compared with 5 per cent for herbicide tolerance. The area under transgenics with insect tolerance trait decreased by 9.0 per cent from 2011 to 2012. A total of 43.7 million ha of stacked transgenic crops were planted in 2012 compared with 42.2 million ha in 2011. Double stacked (pest resistance and herbicide tolerance) and triple stacked (two insect pests and herbicide tolerance) maize were the fastest growing components in 2010 in USA and Philippines, respectively. Transgenic maize, SmartStax^{TM}, has been released in USA and Canada in 2010, with eight different genes coding for several pest resistant and herbicide tolerant traits.

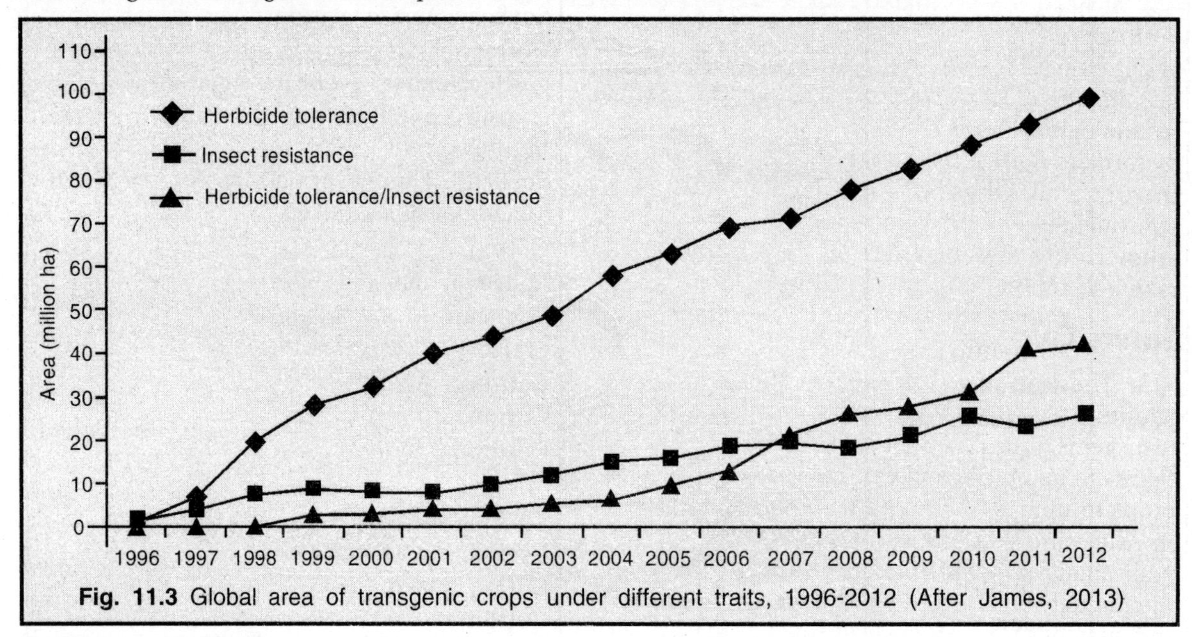

Fig. 11.3 Global area of transgenic crops under different traits, 1996-2012 (After James, 2013)

Bt Cotton in India

Bt cotton was first commercialized in USA in 1996 and subsequently in other countries like Australia (1996), Argentina (1997), China (1997), Mexico (1998), South Africa (1998) and Colombia (2002). The Government of India's Genetic Engineering Approval Committee (GEAC) approved the commercial cultivation of Bt transgenic cotton in the country in 2002. Two Bt cotton varieties, viz. MECH-162 and MECH-184 were recommended for the Central Zone and MECH-12 for the South Zone. Bt cotton was planted in an area of 50,000 ha in 2002, which went upto 1.3 million ha in 2005. In 2008, Bt Bikaneri Narma, the first public sector transgenic crop was released for commercial cultivation in India. This variety has been developed by the Central Institute of Cotton Research, Nagpur; the University of Agricultural Sciences, Dharwad and the Indian Agricultural Research Institute, New Delhi. In 2012, 10.8 million ha area came under Bt cotton from which more than

7.2 million farmers were benefited. Bt cotton occupies 93 per cent of the total area under cotton (11.6 million ha) in India. A total of 1097 Bt cotton introductions (1095 hybrids with the discontinuation of a hybrid and one variety) were approved for planting in 2012. The major states growing Bt cotton in 2012 were Maharashtra (3.99 million ha) representing 36 per cent of all Bt cotton, followed by Gujarat (2.01 million ha, 19%), Andhra Pradesh (1.93 million ha, 18%), Northern zone (1.39 million ha, 13.6%), Madhya Pradesh (0.65 million ha, 5.6%) and the balance in Karnataka, Tamil Nadu and other states. In 2013, the area under Bt cotton increased to 11 million ha.

India enhanced farm income from Bt cotton by US$ 12.6 billion in the period 2002-11 and $ 3.2 billion in 2011 alone. Typically, yield gains in India are approximately 31 per cent, a significant 39 per cent reduction in the number of insecticide sprays, leading to an 88 per cent increase in profitability, equivalent to a substantial increase of about $250 per ha. Thus, Bt cotton has revolutionized cotton production in India by increasing yield, decreasing insecticide applications and through welfare benefits contributed to the alleviation of poverty for over 7 million small resource poor farmers. With the boom in cotton production, India has become transformed from an importer to a major exporter of cotton. India celebrated a decade of Bt cotton growing in 2011, which has been a boon to cotton, Indian agriculture and the country. Bt cotton is expected to achieve new heights in the future due to the availability of a wide variety of hybrids with stacked traits.

Economic and Ecological Impact

It has been estimated that transgenic crops have caused substantial economic benefits at the farm level amounting to a cumulative total of US$ 116.9 billion during 1996-2012. Of these gains, about 58 per cent were due to reduced production costs (less ploughing, fewer pesticide sprays and less labour) and 42 per cent due to subtantial yield gains of 377 million tons. The transgenic technology has also resulted in 497 million kg of less pesticide use (a saving of 8.7% in pesticides) by farmers during 1996-2012 and 18.5 per cent reduction in the environmental impact of pesticides. In 2012 alone, there was a reduction of 36 million kg of pesticides (a saving of 8% in pesticides), equivalent to 23.6 per cent reduction in the environmental impact of pesticides. In addition, transgenic crops contributed to reducing greenhouse gas emissions by 24.6 billion kg in 2012, equivalent to taking ~ 10.9 million cars off the road (James, 2013).

Future Outlook

The future of transgenic crops appears to be encouraging with the number of countries adopting these crops expected to grow, and their global area and the number of farmers planting transgenic crops expected to increase. The outlook for the second decade of commercialization points to an increase in area up to 200 million ha, with at least 20 million farmers growing transgenic crops in upto 40 countries or more. These new countries are likely to include three more countries in Asia, upto 7 countries in Sub-Saharan Africa (subject to regulatory approval), and possibly some additional countries in Latin/Central America and Western/Eastern Europe. Thus, the second decade of commercialization is likely to feature significantly more growth in Asia and Africa, compared with the first decade, which was the decade of the America. By far, the most important of the new transgenic crops that are now nearing commercial approval and adoption is rice. The pro-vitamin A rich Golden Rice is expected to be available shortly in the Philippines and some other countries. Also, Bt rice could be available in China within a few years. Several other medium hectarage crops such as potato, sugarcane and banana are expected to be approved in the near future. Some orphan crops such as Bt brinjal may become available in India, Bangladesh and the Philippines. Vegetable crops such as tomato, broccoli, cabbage and okra, which require heavy application of pesticides are also under development. Pro-poor crops such as cassava, sweet potato, pulses and groundnut are also under development. The transgenic crops have enormous potential for contributing to the humanitarian Millenium Development Goals (MDG) of ensuring a secure supply of affordable food, and reduction of poverty and hunger by 50 per cent by 2015.

EFFECT ON NON-TARGET ORGANISMS

One of the major public concerns about transgenic crops is their effect on non-target organisms. The results of a number of studies have demonstrated that the effects of transgenic crops on non-target organisms including natural enemies and other arthropods are likely to be much less severe than those of the broad-spectrum insecticides (Dhaliwal and Koul, 2010; Dhaliwal *et al.,* 2013).

Non-target Insects

Within any agricultural system, several non-target species are expected which are related to the target species and which may be susceptible to the Bt protein expressed in the Bt crops. The first report on the adverse effects of transgenic crops on non-target insects appeared in 1999. It was reported that the caterpillars of monarch butterfly, *Danais plexippus* (Linnaeus), fed on the milkweed plants sprinkled with Bt maize pollen, grew slower and had higher death rates than caterpillars fed on leaves sprinkled with pollen from non-Bt maize. On the basis of the laboratory data, the authors developed a scenario in which they hypothesized that there could be potentially profound implications for the conservation of monarch butterflies with the widespread use of Bt maize. This report was criticized for its inappropriate design, methodology and interpretation; the major criticism was the fact that pollen densities on milkweed leaves were not quantified.

In another study, mortality of larvae of monarch butterfly from Bt maize pollen was recorded but there were large discrepancies between the toxin levels in pollen that they measured and those from replicated measurements accepted by EPA. This discrepancy was due to the use of pollen samples containing 43 per cent plant debris. This debris was known to cause significant mortality and reduced weight gain by more than 80 per cent. This debris (mostly other parts) was an artifact of the collection method and unrelated to the Bt maize pollen that may fall on milkweed plants. However, these reports have tremendous implications about how science is conducted and communicated. In other studies, there was no relationship between pollen deposition from transgenic maize and mortality of the black swallowtail butterfly, *Papilio polyxenes* Fabricius, and the milkweed tiger moth, *Euchatias egle* Drury. There was no adverse effect on overall development of non-target herbivore, *Mamestra brassicae* (Linnaeus), when reared on *Plutella xylostella* (Linnaeus)-resistant Bt Chinese cabbage.

Predators

The effects of transgenic plants on the activity and abundance of predators vary across crops, insect species and the transgenes in question. Cry1Ac was detected in *Chrysoperla carnea* (Stephens) larvae fed on resistant *Plutella xylostella* (Linnaeus) larvae reared on Bt oilseed rape. However, no Cry1Ac could be detected in *C. carnea* larvae when the lacewings were transferred to *P. xylostella* larvae reared on non-transgenic plants, indicating that *C. carnea* is able to metabolize plant-produced Cry1Ac. There was no effect on preimaginal development or mortality of *C. carnea* when reared on *Rhopalosiphum padi* (Linnaeus) fed on Bt-maize. Similarly, survival, aphid consumption, development and reproduction of *Hippodamia convergens* (Guerin-Meneville) are not influenced when fed on *Myzus persicae* (Sulzer) reared on potatoes expressing δ-endotoxin. Feeding *C. carnea* on *Tetranychus urticae* Koch (which ingested Bt toxin from the transgenic plants) or *R. padi* (which did not ingest the Bt toxin) did not affect survival or development of the predator. Field surveys have shown little impact of Bt maize on predator species numbers or densities. In tritrophic studies with the hemipteran predator, *Orius insidiosus* (Say), there was no effect when feeding on Bt-intoxicated European corn borers. In this case, the results were confirmed with direct feeding studies on Bt corn silks and observations of populations in Bt and non-Bt maize fields (Dhaliwal and Koul, 2011).

A significant increase in the mortality and delay in development of *C. carnea* was observed when fed on *Spodoptera littoralis* (Boisduval) and *Ostrinia nubilalis* (Hubner) which had ingested Bt toxins from transgenic corn. The mean total immature mortality for *C. carnea* raised on Bt-fed prey was 62 per cent, as compared with 37 per cent on Bt-free prey. However, experimental design in this study did not make distinction between a direct effect due to the Bt protein on the predator and indirect effect of consuming a sub-optimal diet consisting of sick or dying prey that had succumbed to the Bt toxin. Thus, the effects observed appear to reflect the poor nutritional quality of Bt-susceptible prey rather than any toxic effect of the Bt protein on lacewings.

Parasitoids

In general, transgenic plants seem to have little or no effect on parasitoids of insect pests. In fact, increased levels of parasitism by *Campoletis sonorensis* (Cameron) on *H. virescens* have been observed on transgenic tobacco as compared to non-transgenic plants. Physiological mechanism was put forth to support this phenomenon, i.e. toxic plants generally caused larvae to grow more slowly which may increase the duration of attack by natural enemies. In another study, activity of *Cordiochiles nigriceps* Viereck on *H. virescens* was not influenced by transgenic plants, which may be due to behavioural mechanism, i.e. toxic plant increased movement of larvae, which may alter their chances of encountering by parasitoids. Similarly, transgenic corn was observed to have no adverse effects on the parasitization of *O. nubilalis* by *Eriborus terebrans* (Gravenhorst) and *Macrocentrus grandii* Goidanich. However, the larval development and mortality of the parasitoid, *Parallorhogas pyralophagus* (Marsh), was adversely affected, when reared on Bt-susceptible insects that had fed on Bt maize, but the fitness of the emerging adults was not impacted.

No adverse effect was found on diamondback moth parasitoid, *Cotesia plutellae* Kurdjumov by feeding on Cry1Ac-resistant larvae. No Cry1Ac protein was detected in newly emerged larvae of the parasitic wasp, *Cotesia vestalis* Haliday, fed on diamondback moth larvae, which had fed on Bt oilseed rape. Similarly, no significant changes were observed in the parasitization rate, larval period, pupal period, cocoon weight or adult emergence rate when the parasitoid, *Microplitis mediator* (Haliday), was reared on the *M. brassicae* larvae fed with Bt transgenic Chinese cabbage. Studies on the impact of Bt broccoli plants on *Pteromalus puparum* (Linnaeus), an endoparasitoid of *Pieris rapae* (Linnaeus), indicated that there was adverse effect on parasitism rate, developmental time, total number and longevity of *P. puparum*. However, no Cry1C toxin was detected in newly emerged *P. puparum* adults developing in Bt-fed hosts. Moreover, no negative effect was found on the progeny of *P. puparum* developing from the Bt plant-fed host when subsequently supplied with a healthy host (*P. rapae* pupae). The negative impact of Bt broccoli on *P. puparum* resulted from poor quality of the host rather than direct effects of the Bt toxin.

Pollinators

Pollination is another factor that must be considered in terms of possible effects of transgene products on beneficial insects. Some reports indicate that transgenic plants seem to have low or no harmful effects on the lifespan and behaviour of honey bees. Trypsine inhibitor, wheat germ agglutinin, serine protease inhibitor from soybean, cysteine protease inhibitor from rice, chicken egg white cystatin, and Bowman-Birk type SBTI do not produce harmful effects on honey bees at the concentrations expressed in transgenic plants. The chitinase transgene in genetically modified oilseed rape did not affect learning performance of honey bees; beta-1, 3-glucanase affected the level of conditioned responses (the extinction process occurring more rapidly as the concentration increased), and CpTi induced marked effects in both conditioning and test phases, especially at high concentrations. Thus, it can be assumed that transgenic crops do not pose a major threat to the activity and abundance of pollinators (Sharma, 2009).

Secondary Insect Pests

The large-scale cultivation of transgenic crops with resistance to certain insect pests may result in secondary insect pest problems becoming a serious constraint in crop production. It may, therefore, become necessary to resort to spraying in order to control the secondary insect pests, which would adversely affect the natural enemies. The Bt toxins may be ineffective against certain insect pests, e.g. leafhoppers, mirid bugs, root feeders, etc. and this may offset some of the advantages of the insect-resistant transgenic crops. The transgenic Bt cotton, which is resistant to bollworms, is susceptible to sucking pests like the jassid, *Amrasca biguttula biguttula* (Ishida); whitefly, *Bemisia tabaci* (Gennadius); mealybug, *Phenacoccus solenopsis* Tinsley, and less effective against *Spodoptera litura* (Fabricius). There are also no differences in the susceptibility of transgenic and non-transgenic cotton varieties to boll weevil and aphids. Effective and timely control measures should be adopted for the control of secondary pests on transgenic crops. There is a need to deploy protease inhibitor and lectin genes that are effective against sucking pests, along with the Bt genes, to make genetically modified plants to be more effective against insect pests for sustainable crop protection.

Wild Relatives of Crops

One of the concerns of deploying transgenic crops is the possibility of vertical and horizontal gene flow. Gene flow can be defined as movement of a gene, via pollen or seed, followed by gene establishment in a new population. Gene movement via pollen occurs in space and time, and can result in gene flow to wild relatives, to other crops (including other genetically modified crops, i.e. gene stacking) and to feral populations. Conversely, gene survival via seeds occurs in time (i.e. seed persistence) leading to genetically modified volunteers emerging in later years. There is evidence of hybridization to wild relatives in majority of the world's principal crops, including banana, cassava, cotton, maize, millet, oats, potato, oilseed rape, rice, soybean and wheat. Unintended lateral transfer of a transgene between related and unrelated species is a potentially worrisome aspect of transgenic technology. It is feared that the escape of a transgene to its related species or weeds growing near the transgenic crop may occur by pollen dispersal, thereby creating 'super weeds', endowed with, for instance insect resistance, which may eventually invade new habitats. While introgression of transgenes resulting in enhanced weediness is unlikely to happen in many cases, especially the gene flow between different species, it is theoretically possible (Dhaliwal and Koul 2011).

The report of a possible gene flow between maize and teosinte in Mexico is of great concern. The report recommends that quantitative studies be carried out on the potential gene flow to the genus *Zea* before liberating transgenic maize varieties, and that experimentation with transgenic crops take place under the strictest security measures to prevent gene flow. The introgression events are relatively common in maize, and the transgenic DNA constructs are maintained in the population from one generation to the next. Therefore, there is a need to study the impact of gene flow from commercial hybrids to the traditional land races in the centers of origin in order to know the period for which the integrity of transgene construct is retained and the increase and/or decrease in the abundance of the transgene construct over time. Once large-scale cultivation of transgenics is undertaken, the possibility of genetic exchange between land races and transgenic material cannot be summarily ignored and it is essential to secure germplasm in the gene banks globally (Koul and Dhaliwal, 2004).

Soil Biota

The levels of Bt proteins found in the soil of Bt crop fields will be low even after several consecutive years of growing Bt crops. Even if substantial amounts of Bt protein were to persist and accumulate in soil, no activity is expected against the invertebrate species that are important to soil processes. It has been demonstrated that two species of Collembola are not susceptible to a variety of Cry1, Cry2 and Cry3 Bt proteins. Similarly, it has been found that representative species of earthworms, nematodes, protozoa, bacteria and fungi were not impacted by the Cry1Ab protein found in most commonly used Bt corn products. Several of the non-target field studies on Bt rice, Bt corn, Bt cotton and Bt potatoes have included sampling of either ground dwelling insects and/ or pit fall trapping, and no adverse impacts on ground-dwelling or soil-dwelling taxa have been detected. Transgenic cotton leaves had no significant acute toxicity on the earthworm, *Eisenia fetida*

(Savigny), from oral exposure to the transgenic line, GK19. No significant differences were observed between Bt and non-Bt rice variety in either decomposition dynamics or in the soil microbial communities associated with residue decay (Sharma, 2009).

FOOD SAFETY AND HUMAN HEALTH

The main concern over transgenic plants with regard to food safety and human health is whether the transgenic plant is likely to pose a greater risk than the non-transgenic variety it is derived from. Food safety risks associated with transgenic plants include the spread of antibiotic resistance, changes in nutrient composition of the plant, and the production of toxic proteins and allergens.

Marker Genes

Marker genes are used by genetic engineers to select plants that have been transformed. One of the most widely used marker genes has been the *kan-r* gene, which encodes an enzyme providing resistance to the antibiotic kanamycin. The use of these marker genes has led to the suggestion that they may be transferred to gut epithelial cells, to gut bacteria and to organisms in the environment. There is a concern that their use may ultimately enhance the development of bacterial resistance to antibiotics. Although there is no evidence that this occurs, this phenomenon needs to be carefully assessed to rule out any eventual negative health impact.

The scientists have responded and begun to use alternative marker genes. A marker gene such as GFP (green fluorescent protein) can be introduced at no fitness cost, into the host plant alongwith the agronomically-important gene for the infield monitoring of the expression of the transgene. For the development of golden rice engineered to synthesize provitamin A, mannose has been used as a selective agent. Positive selection strategies use cytokinins, xylose isomerase gene and phosphomannose isomerase gene for selection of transformed plants. Recently, gene switches that regulate the expression of transgenes through ecdysone against insecticides, have been reported to be useful in reducing the risks associated with transgenic crops. Ligands that are suitable for regulation of biopesticide genes in transgenic crops are the commercially available non-steroidal ecdysone agonists, tebufenozide, methoxyfenozide, halofenozide and chromafenozide.

Bt Toxins

Most Bt toxins are specific to insects, as they are activated in the alkaline medium of the insect gut. There are no specific receptors for Bt protein in the gastrointestinal tract of mammals, including humans. The Bt proteins are rapidly degraded by stomach juices in vertebrates. The concentration of Cry proteins in transgenic plants is usually well below 0.1 per cent of the plant's total protein, and none of the Cry proteins have been demonstrated to be toxic to humans nor have they been implicated to be allergens. The transgenic Bt tomatoes are considered to pose no additional risk to human and animal health as compared to conventional tomatoes. Similarly, the seed from the Bt-transformed cotton lines is compositionally equivalent to and as nutritious as the seed from the parental lines and other commercial cotton varieties. Both protein and DNA are destroyed during the processing of highly refined foodstuffs such as oils and sugars. This is particularly true for cottonseed oil, which must be heavily refined to remove toxic secondary plant compounds. Cry1Ab and Cry1Ac become inactive in processed corn and cottonseed meal, but Cry9C is stable when exposed to stimulated gastric digestion and to temperatures at 90°C. The Bt corn containing Cry9C (StarLink®) was, therefore, not permitted for human consumption, although it was allowed for animal consumption.

In a study using Bt maize silage on the performance of dairy cows, it was found that there were no significant differences between Bt and non-Bt maize hybrids in lactational performance or ruminal fermentation. The Cry1Ab protein as a component of post-harvest transgenic maize plants dissipates readily and has not been detected in silage prepared from transgenic plants. There were no differences in the survival and body weight of broilers reared on meshed or pelletted diets prepared with Bt transgenic and non-transgenic maize. On the basis of studies on Bt crops fed to chicken-broilers, chicken-layers, catfish, swine, sheep, lactating dairy cattle, and beef cattle, it was concluded that there are no detrimental effects on growth, performance, observed health, composition of meat, milk and eggs, etc. On the basis of extensive studies on the safety issues associated with DNA in animal feed derived from genetically engineered crops, it has been concluded that consumption of milk, meat and eggs produced from animals fed genetically modified crops should be considered as safe as traditional practices (Sharma, 2009).

POTENTIAL AND CONSTRAINTS

Biotechnology has provided new avenues for management of insect pests and it holds great potential to be included in IPM system. If this technology leads to improved control of pests and vectors of diseases, then the benefits will be enormous.

- The low toxicity of protenase inhibitors and Bt-δ endotoxin as compared to conventional insecticides would reduce the selection pressure and may slow down the development of resistance.
- Since all plant parts including growing points would remain covered with toxins, dependence on weather for efficacy of the sprays would be eliminated.
- Since toxins will always be there, so there will be no need of continuous monitoring of pests.
- Transgenic plants would also provide protection to those plant parts, which are difficult to be treated with pesticides. Thus, transgenics may prove useful for controlling bollworms and borers, which are difficult to control by means of insecticides.
- The cost of application in the form of equipment and labour will be nil or negative.
- The development cost is only a fraction of the cost of development of a conventional pesticide.
- There would be no problem of contamination in the form of drift and ground water contamination or risk to the field workers. By using transgenic technology, the toxin is delivered directly to the target organism. The fact that the plant delivers the toxin and not a person also serves as a means to reduce operator exposure to pesticides.
- Insecticidal activity would be restricted to those insects which actually attack the plants. Transgenic plants would be safe to non-target species and human beings.
- Transgenic plants eliminate the problems of shelf life and field stability faced by pesticide formulations as they provide on site biosynthesis of the toxins. Such crops are easy to adopt because no new practices need to be learned for the basic use of technology. The whole technology is 'all in the seed' and the only challenge is to get the seed into the hands of the farmers.
- Transgenic plants will have inbuilt resistance to various insects replacing some of the current pesticide usage with protection which is intrinsically biodegradable, thus reducing the use of chemical insecticides and minimizing the problem of environmental pollution.

In spite of all these advantages, biotechnology does face uncertainties.

- Gene manipulation is a gray area and needs to be addressed systematically with pragmatic approach, especially in developing countries, where most of the farmers are either poor or marginal.
- The resistance against δ-endotoxin is not ruled out, though it may be slow as compared to insecticides because of continuous exposure of the insect to the toxin. Transgenic plants that are resistant to pests have a selective advantage that may lead them to become weeds.

- Some weeds may introgress an insect resistance gene from related transgenic plants and make them less susceptible to their usual herbivores, thus exhibiting greater reproductive success and may create a weed problem.
- If lepidopteran herbivores were removed from plant species, other insects might experience competitive release and become more common.
- The use of transgenic plants possessing resistance factor in one locality may affect the insect population dynamics in other areas.
- It has been claimed that there are a number of risks to humans associated with eating transgenic food crops. These risks include novel proteins acting as allergens or toxins, altered host metabolism producing new or unknown allergens or toxins, and reduced nutritional quality leading to dietary deficiencies or health problems, e.g. by reduction of antioxidants in plants.

The transgenic crops are likely to play a dominant role in pest control in the twenty-first century. However, it is necessary to adopt a rational approach with respect to transgenics and follow certain criteria.

- The existence of new transgenic crops and the possible consequences of their use should be converted into a documented information for regional policy makers, regulators, law makers, private industry, extension services, NGOs and consumers. Special attention should be given to food safety, resistance management and gene flow. Local systems, their constraints and socio-economic implications should be strictly considered before adoption of any transgenic material.
- Regional policies should be developed towards transgenic crop use. Specific attention is required to the agroecology of the cropping systems in the region. Policies, laws and regulations should be harmonized.
- The risks of gene flow and its effects need to be studied carefully.
- Given the difficulties of enforcement of regulations, the best option might be to regulate seed distribution. This will help in resistance management as well.
- Cost/benefit analysis is necessary for the introduction of transgenic crops.
- It is necessary to undertake a proactive approach to stimulate the use of the right gene for the right season and in the right way. Regional research institutions can play an important role in developing, testing and recommending management practices appropriate to the transgenic crop technology, local production systems and pests.

Biotechnology has the potential to move farming closer to ecologically sustainable practices, both in developed and developing countries and thus could make a considerable impact on agricultural systems in the future. The strategy to maximize the utility of this approach should involve the use of gene combinations whose products are targeted to different biochemical and physiological processes within the insect. In this way, it is expected to provide a multimechanistic form of resistance, which can be tailored to different crops and prevailing insect pests. However, the sustainability of transgenic crops from the point of view of pest control remains to be seen. There is a potential danger that, just as farmers became trapped in 'pesticide treadmills' during the 1950s and 1960s, so farmers in the twenty-first century may be trapped in 'gene treadmills'.

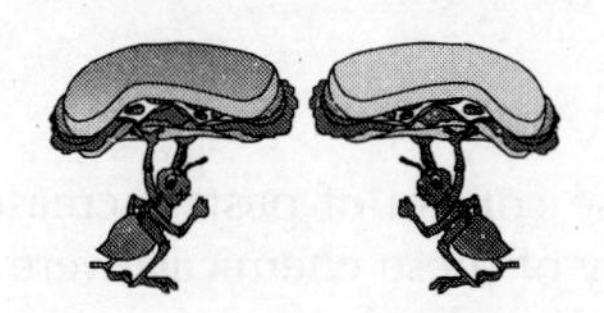

INTEGRATED PEST MANAGEMENT

INTRODUCTION

Integrated pest management (IPM)is based on ecological principles and involves the integration and synthesis of different components/control tactics into a pest management system. Many components of IPM were developed in late 19th and 20th century. By early 1920s, a highly complex and sophisticated system involving the use of multiple component suppression techniques, viz. resistant varieties, sanitation practices and chemical treatments with calcium arsenate at fixed population levels, was clearly developed for the control of boll weevil on cotton in USA. However, during the period from 1920s to 1940s, the emphasis in crop protection shifted from cultural and biological control techniques to inorganic chemical pesticides.

The discovery of insecticidal properties of DDT rapidly followed by the manufacture of other broad spectrum synthetic organic pesticides during 1940s and 1950s virtually eclipsed all other techniques. These insecticides were easy to apply and produced an almost immediate kill. Therefore, they became our first and only line of defence or attack against all insects. However, even at that time, many scientists had warned regarding the consequences of exclusive reliance on chemical insecticides ignoring ecological principles. Unfortunately, their sane voices were drowned in the euphoria generated by the initial success of synthetic insecticides.

The use of toxic chemicals for the control of pests increased tremendously during the green revolution era. At that time, their use was considered a necessity for increasing agricultural production at a reasonable cost. It was later realized that many of these chemicals were not biologically degradable and they not only persisted in the environment but also became concentrated through food chains. This realization came only when the recurrence of pests with even a greater severity was evidenced as a result of the death of natural enemies along with the pests. With the consciousness of using the chemicals judiciusly to minimize the pollution hazards, the scientists recommended that pests should be controlled by integrating the use of biological agents with the use of pesticides.

ORIGIN OF IPM

The use of toxic chemicals for the control of pests increased tremendously during the last few decades. It was realized later that many of these chemicals were not biologically degradable and they not only persisted in the environment but also became concentrated through the food chains. With the consciousness of using the chemicals judiciously to minimize the pollution hazards, the scientists recommended that pests should be controlled by integrating the use of biological agents with the use of insecticides. Based on this concept, Bartlett (1956) coined the term 'Integrated Pest Control' which was defined as the blending of biological control agents with chemical control measures.

Later on in 1961, Geier and Clark advocated the integrated use of all available techniques for the control of insects and not confining only to the biological and chemical methods of control. They suggested that the methods which are considered promising should first be evaluated and, if found effective,be consolidated into a unified programme to manage pest populations. Subsequently, the term 'pest management' was advocated by Geier (1970). Thus, pest management may be considered, an intelligent selection and use of pest control actions, that will ensure optimal economic, ecological and sociological benefits. Pest management includes all approaches ranging from single component control method to the most sophisticated and complex control method. A number of definitions have been proposed for the twin terms of integrated pest control (IPC) and integrated pest management (IPM). According to the expert panel of the Food and Agriculture Organization, integrated pest control may be defined as a system that in the context of the associated environment and the population dynamics of pest species, utilizes all suitable techniques and methods in as compatible a manner as possible and maintains the pest populations at level below those causing economic injury (FAO, 1967).

According to the National Academy of Sciences, IPM refers to an ecological approach in pest management in which all available necessary techniques are consolidated in a unified programme, so that population can be managed in such a manner that economic damage is avoided and adverse side effects are minimized (NAS, 1969).

Smith (1975) defined IPM as a multidisciplinary ecological approach to the management of pest populations, which utilizes a variety of control tactics compatibly in a single co-ordinated pest management system. Dr Ray F. Smith and Dr Perry Adkisson have been awarded the 1997 World Food Prize for their pioneering work on development and implementation of IPM concept.

Pedigo (1991) expanded the FAO definition to lay stress on the importance of socio-economics, and defined IPM as a pest management strategy that, in the socio-economic context of farming systems, the associated environment and the population dynamics of the pest species, utilizes all suitable techniques and methods in as compatible a manner as possible, and maintains the pest population levels below those causing economic injury.

Dhaliwal and Arora (2015) defined IPM as a dynamic and constantly evolving approach to crop protection in which all the suitable management tactics and available surveillance and forecasting information are utilized to develop a holistic management programme as part of a sustainable crop production technology. It is a systems approach to pest management based on an understanding of pest ecology and begins with steps to accurately diagnose the nature and source of pest problems, and then relies on a range of preventive and curative measures **(Fig. 12.1)**.

Based on an analysis of 64 definitions spanning the past 35 years, Kogan (1998) defined IPM as a decision support system for the selection and use of pest control tactics, singly or harmoniously co-ordinated into a management strategy, based on cost/benefit analyses that take into account the interests of and impacts on producers, society and the environment.

A special committee of the National Research Council's Board of Agriculture (NRC, 1996) proposed 'ecologically based pest management' (EBPM), also called 'ecologically based integrated pest management' (EBIPM), emphasizing on some key issues :

- In EBIPM, programmes should emphasize on an understanding of the ecological relationships between the host plant and the management practices like cultural control, biological control and host plant resistance.

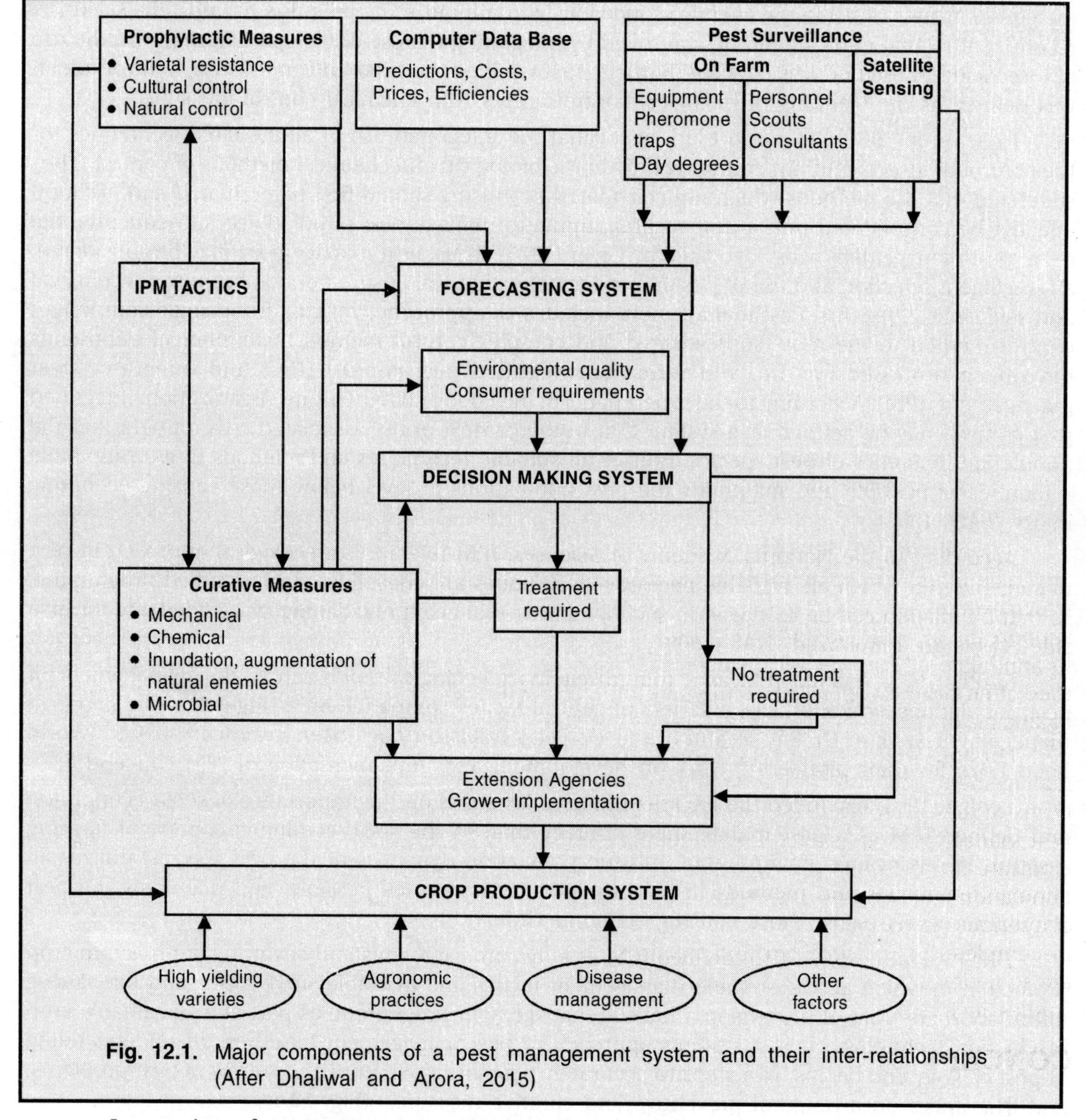

Fig. 12.1. Major components of a pest management system and their inter-relationships (After Dhaliwal and Arora, 2015)

- Integration of management practices involves biological (e.g. parasitoids, predators and microbials), chemical (e.g. selective pesticides and pheromones) and cultural (e.g. crop rotation, planting date and aeration)
- Sustainability implies durability over time.
- EBIPM programmes should minimize economic, environmental and health risks.

The idea behind EBIPM is to shift the IPM paradigm from focussing on pest management strategies relying on pesticide management to a systems approach relying primarily on biological knowledge of pests and their interaction with the crops. Hence EBIPM programmes should represent a sustainable approach to manage pests combining biological, chemical, physical and cultural tools to ensure favourable economic, ecological and sociological consequences.

Huffaker and Croft (1976) have described a series of phases in the evolution of an IPM programme:

(*i*) *Single tactic phase.* Emphasis is generally placed on a single pest utilizing a single tactic. This phase does not represent IPM, but the limitations in this approach may lead to its development.

(*ii*) *Multiple tactic phase.* This phase embraces a variety of tactics (cultural, mechanical, physical, chemical, biological, host resistance, regulatory, etc.) in manipulating pest populations.

(*iii*) *Biological monitoring phase.* This phase introduces monitoring of pest, natural enemies and host plant (phenology) populations as the basis for timing the application of various control tactics.

(*iv*) *Modelling phase.* This involves the conceptuali-zation of the processes involved in pest management systems through mental, pictorial, flowchart and methematical models. As the volume and complexity of data increase, more sophisticated modelling techniques become necessary.

(*v*) *Management or optimization phase.* This process involves the construction of a functional IPM system utilizing compatible subsystems in optimizing the integration of this IPM system with the overall crop production system.

(*vi*) *Systems implementation phase.* This is the ultimate phase through which the optimal systems are unified for delivery to and utilization by the farmer.

The ultimate aim of scientific pest management is to maintain a low level of pest population which would not only maintain the damage lower than the economic injury level but will also support the growth and survival of its natural enemies. The concept is to suppress the pest but not to annihilate it. For that very reason, the broad spectrum insecticides should not be used because they often have the effect of eliminating the pest as well as its natural enemies, thus upsetting the balancing of natural system of insect-parasitoid relationship.

For application in the field it is essential in the first instance to understand the concept of pest management and then to disseminate the knowledge among the practising farmers, translated in terms of their own local conditions and specific farm operations. In other words, the philosophy of pest management is to maintain the population of a potential pest at a sub-threshold level than to eradicate it. This philosphy is based on the observation that every plant can withstand a level of population without showing loss in yield or vigour. However, sometimes an insect may be a vector of a serious plant disease and in that case even extremely low levels of population can be instrumental in complete loss of yield. To understand these concepts more clearly, quantitative measurements are sometimes undertaken which define clearly the degree of damage and allowable damage. These studies include:

CONCEPT OF INJURY LEVELS

The critical factor which determines the damaging capacity or otherwise of an insect is its population level. The concept of injury level was propounded to enable us to identify the population level at which an insect would cause damage to a crop.

Economic Injury Level (EIL)

The critical factor that determines the damaging capacity or otherwise of an insect is its population level. The concept of injury level was propounded to enable us to identify the population at which an insect could cause damage to a crop.

According to Stern *et al.* (1959), it is the lowest pest population density that will cause economic damage. It is the level at which damage can no longer be tolerated and, therefore, at that point or before reaching that level, it is desirable to initiate deliberate control operations.

Although expressed as numbers of insects per unit area, the EIL, in reality, is a level of injury. Because injury is difficult to measure in a field situation, however, number of insects are used as an index of that injury. It may, therefore, be more useful to express EIL in standard units of injury. The standard units of injury are the injury equivalent, *i.e.* the amount of injury that could be produced by one pest through its complete life cycle, and equivalency, *i.e.* total injury equivalents (for a population) at a point of time. If management action (insect supression) can be taken quickly and loss averted completely, EIL may be expressed as follows :

$$EIL = \frac{C}{VID} \quad ...(1)$$

where EIL = No. of injury equivalents per production unit (insects/ha)
C = Cost of management activity per unit of production (Rs/ha)
V = Market value per unit of product (Rs/kg)
I = Crop injury per pest density
D = Damage per unit injury (kg reduction/ha)

These primary variables are affected by a number of complex variables.

In instances, where some loss from the insect is unavoidable, the relationship becomes

$$EIL = \frac{C}{V \times I \times D \times K} \quad ...(2)$$

where K represents proportionate reduction in injury (e.g. 0.6 for 60%)

Economic Threshold Level (ETL)

It is the pest density at which control measures should be applied to prevent an increasing pest population from reaching the ecnomic injury level. Control measures are taken at this stage so that the pest does not exceed the ecnomic injury level.

ETL is the best known and most widely used index in making pest management decisions. Although expressed in insect numbers, ETL is, in fact, a time parameter, with pest numbers being used as index for when to implement management strategies. Just as with EILs, ETLs can also be expressed in insect equivalents.

ETL is a complex value based on EIL, population dynamics of the pest, weather forecasting, and pest's potential for injury. The relationship between ETL and EIL is shown in **Fig. 12.2.** When no action is taken at ETL, population exceeds EIL, while when management steps for pest suppression are taken as the population crosses ETL, the population is forced down before it could reach EIL. ETL is a direct function of EIL and as such is subject to changes in EIL variables. In addition, ETL

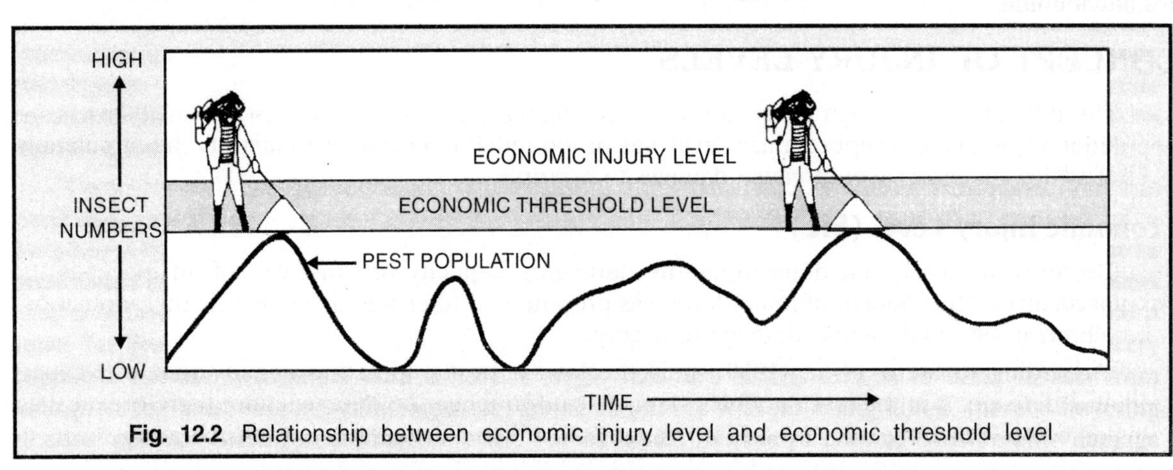

Fig. 12.2 Relationship between economic injury level and economic threshold level

varies with logistical considerations associated with time delays that may vary from one situation to another.

The concept of EIL and ETL gained wide acceptability from the time it was presented. However, implementation of the concept in practice has been very slow. This is due to a number of serious limitations in the concept. Some of these limitations are given below :

(*i*) The terms EIL and ETL are themselves misleading because both are defined in terms of population densities, while former represents an injury level and the latter the time for taking control measures. This limitation may be overcome by defining these levels in terms of injury equivalents. Moreover, it would then be possible to describe the same type of injury for many pest species.

(*ii*) There is a lack of rigorous definition of economic damage, i.e. the amount of injury that will justify the cost of control.

(*iii*) The EIL concept overlooks the influence of other production factors that can affect the crop/pest systems. The externalities left out include interseasonal dynamics, biological relationships with other pests and natural enemies, environmental contamination by pesticides, resistance to pesticides, effect of control in neighbouring fields and health problems relating to pesticides.

(*iv*) Decision levels for management of some types of pests cannot be determined with EILs. Besides medical and veterinary pests, it includes most vectors. It is very difficult to place a monetary value on the reduction in aesthetic value associated with a given type of injury. A similar problem exists with respect to forest pests. Almost all components of EILs are difficult to estimate for forest pests; accurate market values are a problem, management costs may vary greatly and frequently include mere environmental and social costs and the injury/crop response relationships may be difficult to determine because the growth of the crop spans many years.

(*v*) The concept is unsuitable in case of attack of multiple pests on a single crop at the same stage.

However, inspite of these limitations, EIL concept continues to offer a practical approach to pest related decision making in a broad sense. The ETL values for major insect pests of agricultural crops in India are presented in **Table 12.1.**

Environmental Economic Injury Levels (EEIL)

The challenge of attempting to decrease pesticide inputs further can be met by developing environmentally based EILs and their concomitant ETLs. An environmental EIL is an EIL that evaluates a management tactic based on not only its direct costs and benefits to the user but also its effects on the environment. The EIL equation (equation 2) integrates many management elements, each of which may have a role in making pest management environmentally most sustainable.

(*i*) *Assigning realistic management costs* (*C*). Component C of the EIL equation represents costs associated with taking management action against a pest population, and increased costs cause EIL to increase proportionally. Generally, C does not take into account the environmental costs associated with environmental risks; it is possible to include these costs in variable C of EIL.

One approach for estimating environmental costs of pesticides through economic techniques of contigent valuation was presented by Higley and Wintersteen (1992). They estimated the level of risk posed by 32 field crop insecticides to different environmental elements (surface water, ground water, aquatic organisms, birds, mammals, beneficial insects, etc.) and to human health (acute and chronic toxicity). They also estimated from survey data the relative importance of avoiding risk to each of these elements. Additionally, survey respondents (producers) indicated how much they would be willing to pay, in either higher pesticide costs (for safer pesticides) or yield losses, to avoid

Table 12.1 Economic threshold levels (ETLs) of major insect pests of agricultural crops in India

Crop	Insect pest		ETLs
	Common name	Scientific name	
Cotton	American bollworm	*Helicoverpa armigera* (Hubner)	5-10% infestation in floral forms
	Pink bollworm	*Pectinophora gossypiella* (Saunders)	5-10% infestation in floral forms
	Spotted bollworms	*Earias* spp.	5-10% infestation in floral forms
	Whitefly	*Bemisia tabaci* (Gennadius)	6-8 adults/leaf or appearance of honeydew on 50% plants
	Jassid	*Amrasca biguttula bguttula* (Ishida)	Appearance of yellowing and curling on the leaf margins in the upper plant canopy
	Aphid	*Aphis gossypii* (Glover)	Appearance of honeydew on 50% plants
	Tobacco caterpillar	*Spodoptera litura* (Fabricius)	4 egg masses/50m row length
Sugarcane	Early shoot borer	*Chilo infuscatellus* (Snellen)	18-22% shoot damage at tillering phase
	Scale insect	*Aulacaspis tegalensis* (Zehntner)	20-30% cane with scale incidence
	Top borer	*Scirpophaga excerptalis* (Fabricius)	5% damage
	Pyrilla	*Pyrilla purpusilla* Walker	3-5 insects/leaf
	Black bug	*Cavelerius excavatus* (Distant)	25 insects/plant
Maize	Stem borer	*Chilo partellus* (Swinhoe)	5-10% infestation
	Shoot fly	*Atherigona* spp.	5-10% dead hearts
	Earworm	*H. armigera*	25-30% damage to cobs
Rice	Yellow stem borer	*Scirpophaga incertulas* (Walker)	5% dead hearts/white ears/one egg mass per m^2.
	Brown planthopper	*Nilaparvata lugens* (Stal)	10 hoppers/hill
	Gall midge	*Orseolia oryzae* (Wood-Mason)	5-10% silver shoots
	Leaf folder	*Cnaphalocrocis medinalis* (Guenee)	10-15% infested plants
	Root weevil	*Hydronomidius molitor* Faust	2 grubs/ hill
	Hispa	*Dicladispa armigera* (Olivier)	1 adult or 1-2 damaged leaves/hill
	Gundhi bug	*Leptocorisa oratorius* Fabricius	1-2 bugs/hill
Wheat	Aphid	*Schizaphis graminum* (Rondani)	5-10% infested plants
Sorghum	Shoot fly	*Atherigona soccata* Rondani	1 egg/plant or presence of eggs on 5% plants or 15% dead hearts
	Stem borer	*C. partellus*	10% plants showing dead hearts/ shot holes/ unfilled ear-heads
Groundnut	Aphids	*Aphis craccivora* (Koch)	5-10 aphids/terminal at seedling stage
	Tobacco caterpillar	*S. litura*	20-25% defoliation at 40 days
	Leaf miner	*Aproaerema modicella* (Deventer)	5 mines/plant at 30 days of crop age
	Thrips	*Scirtothrips dorsalis* Hood	5 thrips/terminal at seedling stage

Table 12.1 contd.....

Crop	Insect pest		ETLs
	Common name	Scientific name	
Rapeseed	Aphid	*Lipaphis erysimi* (Kaltenbach)	• 50-60 aphids/10 cm terminal portion of central shoot or • 0.5-1.0 cm terminal portion of central shoot covered by aphids • 40-50% infested plants
Sunflower	Gram pod borer	*H. armigera*	One larva/head
Chickpea	Pod borer	*H. armigera*	3 eggs or 2 small larvae/plant
	Cut worm	*Agrotis ipsilon* (Hufnagel)	5% plant mortality
Pigeonpea	Pod borer	*H. armigera*	5 eggs or 3 small larvae/plant
	Pod fly	*Melanagromyza obtusa* (Malloch)	Presence of 5% oviposition on pods
	Pod sucking bug	*Clavigralla gibbosa* Spinola	One egg mass/plant
Soybean	Stem fly	*Ophiomyia phaseoli* (Tryon)	5% plant infestation
	Girdle beetle	*Obereopsis brevis* (Swed)	5% incidence
	Hairy caterpillar	*Spilarctia obliqua* (Walker)	5 larvae per meter row
Brinjal	Fruit and shoot borer	*Leucinodes orbonalis* Guenee	1-5% shoot or fruit infestation
Cabbage and cauliflower	Diamondback moth	*Plutella xylostella* (Linnaeus)	1-5% incidence
	Tobacco caterpillar	*S. litura*	1-5% incidence
Tomato	Fruit borer	*H. armigera*	1-5% fruit damage or one larva/head or 3 eggs or 2 small larvae/plant
	Cutworm	*A. ipsilon*	5% plant mortality
Apple	San Jose Scale	*Quadraspidiotus perniciosus* (Comstock)	Appearance of pest on 5% treses
	Codling moth	*Cydia pomonella* (Linnaeus)	1-2% incidence
Grapes	Thrips	*Retithrips syriacus* Mayet	20% foliar damage
	Flea beetle	*Scelodonta stricollis* (Motschulsky)	20% foliage damage
	Mealybug	*Maconellicoccus hirsutus* (Green)	1% bunch infestation
Mango	Hopper	*Amritodus atkinsoni* (Lethiery)	20% hopper damage in inflorescence
Citrus	Citrus caterpillar	*Papilio demoleus* Linnaeus	20-30% foliar damage
	Mealybug	*Planococcus citri* (Risso)	5-10% infested fruits
	Whitefly	*Dialeurodes citri* (Ashmead)	5-10 nymphs/leaf
	Leaf miner	*Phyllocnistis citrella* Stainton	1 larva/leaf

Source : Dhaliwal *et al.* (2013)

different levels of risk (high, moderate, low) from a single application of a pesticide. With these data, the individual environmental costs for each insecticide were calculated as below :

$$\text{Environmental EIL} = \frac{PC + EC}{VDIK}$$

(*ii*) *Manipulating crop market value* (*V*). This could be achieved by putting a higher market value for a pesticide–free produce. The extent of increase would depend on the consumer's willingness to pay for a safer product.

(*iii*) *Reducing damage per pest* (*D*). Reducing D implies that less loss of yield occurs for a given amount of injury. This is possible if plant is able to tolerate and compensate for injury. Plants that can tolerate or compensate for injury do not place selection pressures on pest populations. Therefore, the benefits of tolerance and compensation in plant are sustainable and permanent. Even partial tolerance will increase EILs (by decreasing D). The need for pesticides and the risks to environment will be reduced correspondingly.

(*iv*) *Developing environmentally responsible K value*. Modified K is the proportion of total pest injury averted by timely application of a management tactic. Increasing the EIL to improve environmental quality implies that we are willing to tolerate more pests. But this is not always the case. By reducing D or K, EIL can be increased even without causing increased losses or costs.

General Equilibrium Position (GEP)

It is the average population density of a pest over a long period of time unaffected by the temporary interventions of pest control. The population fluctuates around a mean level as an outcome of the influence of density dependent factors, such as parasitoids, predators, diseases, etc.

It may be understood that EIL may be at any level from well below to well above the GEP. In certain insects GEP is well below the EIL or even ETL and never reaches the latter two parameters. Such insects are rarely noticed physically but the damage caused by them through the introduction of a virus or any other disease can be most significant.

GEP touches EIL and ETL, approximately 2 to 5 years for many insect pest species. Such insects are called 'occasional pests'. The increase in population may be due to the injurious effect of pesticides or due to favourable weather conditions. For example, in Punjab, outbreaks of the armyworm, *Mythimna separata* (Walker), on wheat, are recorded every 2 to 5 years.

Sometimes, the control measures are required frequently to bring down the GEP well below the EIL and ETL. Such a situation has been observed in the tobacco caterpillar, *Spodoptera litura* (Fabricius), which is a 'regular pest' of cruciferous vegetables.

The known severe pests form another category altogether and, in their case, EIL and ETL are below the GEP level. The maize borer, *Chilo partellus* (Swinhoe), is a severe pest of maize and its severity increases in some Himalayan valleys where the climate is mild. Moreover, it is much more severe on the hybrid corn varieties than on the local, comparatively low yielding maize.

Cost:Benefit Ratios

Cost:benefit ratio is an indicator, used in cost benefit analysis, that attempts to summarize the overall value for money of a project or proposal. It is the ratio of the cost of a project or proposal, expressed in monetary terms, relative to the benefits, also expressed in monetary terms. Cost:benefit analysis can be made with the help of a partial budget. A partial budget is an estimate of the changes in income and expenses that would result from carrying out a proposed change. It quickly establishes an estimated cost of the application of a technology and compares it to the estimated change in income that results from the use or application of the technology. If the difference is a positive change in net income, it is usually recommended that the technology be adopted. A simple partial budget form has four categorical parts, viz. increased returns, decreased costs, decreased returns and increased costs.

As we improve the capability for predicting pest appearance, we can determine precisely the ETLs, and know exactly when to apply control measures. There is a need to emphasize costs and benefits. The preparation of crop life tables provides a solid foundation for analysis of pest damage and, cost/benefit ratio in pest management. If a crop is grown more than once in a year in the same field we should work out the crop-season life tables. In most pest control activities, the benefits are usually not known, because those cannot be measured, hence the cost of prevention becomes the cost of production. In other words, the use of pesticides can rarely contribute to increase in yield and, at best, it can prevent the loss of yield, making the benefit both indirect and incalculable.

Benefit/risk analysis provides the means for assessing relevant economics versus risks in pest control. The judicious use of insecticides should be the philosophy of pest management. It is estimated that, generally, 1 per cent of the insecticide applied,reaches the target pest and the rest merely contaminates the environment as residue or causes mortality of the non-target or even useful species. Some non-degradable insecticides, such as the chlorinated hydrocarbons, attain biological magnification in the environment through food chains. Thus, a grower while ensuring safety in handling and applying a higher toxic pesticide, should also consider its injurious effects in the environment.

INTEGRATION OF TACTICS

The pest management tactics are either preventive or therapeutic. Preventive practice utilizes tactics to lower environmental carrying capacity (reduce the general equilibrium position) or increase tolerance of the host to pest injury. Prevention relies on an intimate understanding of the pest life cycle, behaviour and ecology. The preventive tactics involve natural enemies, host resistance and cultural practices.In addition, quarantines are also an important component of preventive tactics. Therapeutic tactics are applied as a correction to the system when necessary. The objective of therapy is to dampen pest population below EIL. The only widely used therapeutic tactic is the use of conventional insecticides but other approaches like microbial agents, augmentation of natural enemies, use of insect growth regulators, etc. also play a vital role.

Actual integration involves proper choice of compatible tactics and blending them so that each component potentiates or complements the other. Probably, the earliest example of integration of techniques was the use of a combination of resistant varieties and sanitation practices as prophylactic measures combined with application of calcium arsenate at high population level in case of boll weevil on cotton in USA in early 1920s. Similar programmes were being developed for other pests also but the advent of synthetic organic insecticides intervened and these techniques were relegated to the background. The misuses and abuses of insecticides have again focused our attention on integrated control measures.

Rhodes grass scale, *Antonina graminis* (Maskell) is a cosmopolitan insect feeding on over 100 hosts including 38 species of range grasses in Texas. A parasite introduced from India, *Neodusmetia sangwani* (Subba Rao), was successfully established against the pest in Texas. However, during the June population peak of the pest, natural enemy population was lower due to general host unsuitability during May. The resistant Rhodes grass variety 'Bell' was found to tolerate pest attack without appreciable damage. This variety, thus, provided relief until the parasite effectively reduced scale populations. The predation rate was highest on resistant cultivars and this was attributed to the greater movement of hoppers in search of suitable feeding sites.

It has been found that the rate of parasitism by *Bracon mellitor* Say, the most important native parasite of boll weevil, *Anthonomus grandis* Boheman was higher on frego bract (resistant) than on normal bract (susceptible cotton). The mortality of *Nephotettix virescens* (Distant) has been found to be highest and population lowest when resistant varieties IR 26 and IR 56 were combined with predators, myrid bug, *Cyrtorhinus lividipennis* Reuter and spider, *Lycosa pseudoannulata* (Bosenberg & Strand). The feeding of *Helicoverpa zea* (Boddie) on ED73-371, a soybean genotype resistant to the Mexican bean beetle, *Epilachna varivestis* Mulsant increased its susceptibility to *Bacillus thuringiensis* Berliner. Thus, *Bt* which is normally ineffective against *H.zea* might be effective on resistant soybean.

Natural control by parasitoids and predators can be greatly strengthened by use of a large number of cultural practices like intercropping, trap cropping, strip harvesting, etc. Modification of the crop environment by manipulation of irrigation, fertilizer, row spacing, seed rate and tillage operations, etc. may also lead to substantial improvement in benefits of biological control. A combination of moderate resistance to carrot fly, *Psila rosae* (Fabricius), with specific sowing and harvesting dates, has enabled satisfactory yield of marketable carrots in heavily infested fields. The effect of plant resistance,planting dates and tillage practices was complementary in reducing greenbug, *Schizaphis graminum* (Rondani) population on sorghum; the resistant hybrids coupled with late planting dates and no tillage have been consistently effective for controlling the pest.

Combining plant resistance with well timed lower dosages of insecticides can sometimes achieve adequate pest suppression while reducing otherwise high insecticide inputs. Sweet corn hybrids resistant to corn earworm require less insecticide than susceptible hybrids to obtain an equivalent reduction in pest incidence. The insecticide rates on an insect-resistant groundnut cultivar (NC 6) can be reduced by 75-80 per cent against *Diabrotica undecimpunctata howardi* Barber and 60 per cent against *Frankliniella fusca* (Hinds).

Insect pest management in cotton in Texas is a good example of integration of different tactics. The foundation of the programme begins with preventive tactics aimed at boll weevils. The basic tactic in prevention is a return to early planting of short season cotton cultivars, moderate fertilizer use and well timed irrigation. Plant thinning is delayed or not implemented, which suppresses vegetative growth and stimulates early fruiting. These practices shorten the production season and period of vulnerability to insects. Early harvesting, stalk destruction and use of defoliants late in the season prevent further weevil production and weaken or starve weevils going into hibernation. Pest surveillance and therapeutic treatments with organophosphates against boll weevil and flea hopper, *Psallus seriatus* (Reuter) are used at ETLs. Pyrethroids are applied in case of *Heliothis* outbreaks.

The Indian Institute of Horticultural Research has developed IPM in cabbage and tomato by employing trap cropping and biopesticides. Growing of Indian mustard in paired rows at the beginning and after every 25 rows of cabbage attracted more than 80 per cent of diamondback moth infestation besides almost entire population of leafwebber, stem borer, bugs and aphids. To control the remaining attack of diamondback moth, a 4 per cent neem seed kernel extract (NSKE) is applied at primordial or head initiation stage of the crop. Similarly, trap cropping of marigold after every 8 rows of tomato attracts most of the ovipositing moths of *Helicoverpa armigera* (Hubner) to the former crop. The use of conventional insecticides on the trap crop reduces their attractiveness to the pest. Therefore, the pest on the trap crop has to be removed mechanically. The residual pest population on both the crops is controlled by sprays of *H. armigera* NPV @ 500 larval equivalent per ha.

PRE-REQUISITES FOR DECISION MAKING

There are certain essential pre-requisites which must be followed before deciding to employ the pest management options.

1. Correct identification of pest. The correct identification of the pest is the first most important step on which the next course of action depends. Mis-identification of a pest can lead to complication of the problem rather than solving it. Efforts should be made to identify a pest up to species level as closely related species differ with respect to several biological parameters.

2. Life-cycle. The knowledge of the sequence of developmental stages (egg to adult), their duration, number of generations and method of overwintering is essential to know the 'weakest link' in the life cycle. This would help to aim control measures effectively at the most vulnerable stage of the pest.

3. Habits. The important features of habits of pests which have a bearing on their control include the developmental stages responsible for plant injury, mode of feeding, parts of the plants attacked and whether the pest feeds externally or bores into the plant. Moreover, it is essential to know whether the pest transmits any disease in addition to causing direct injury to the plant.

4. Host range. It is essential to know whether the pest is monophagous (feeds only on one host plant), oligophagous (feeds on several plant species) or polyphagous (feeds on many plant species). Moreover, the non-cultivated plants which may act as reservoir of infestation should also be known. Plant host range can help to decide the proper crop rotation to be followed to keep the pest under check.

5. Natural regulating factors. Some pests can be kept under check by the activity of natural controlling factors like parasitoids, predators and disease organisms. Their precise role in specific circumstances should be understood to make the best use of natural mortality factors and their integration with other control measures.

6. Reinfestation. It is important to know the capacity of a pest to re-infest an area following its elimination. Some pests are highly mobile and quick fliers, and capable of rapidly re-invading the area. In contrast, some pests have very limited capacity to move and are very slow to reinvest.

7. Crop value. Expenditure on control measures can only be justified when increase in marketable yield of the crop produced is worth more than the cost of the control. Control measures should, therefore, only be initiated when they are economically justifiable. Usually, costly controls can be applied more logically to florist and fruit crops than to field crops.

8. Consumer pressure. Consumer demands have an important bearing on pest control. Certain fruits and vegetables may be rejected by the processor by mere presence of one or more insects in the samples of the produce. Similarly, even slight blemishes may not be tolerated on certain top-grade fruits.

9. Survey and detection. It is essential to carry out regular monitoring surveys to detect low-level pest infestations before they become damaging. It would help in devising strategies to prevent the pests to reach damaging levels. If the pest population tends to grow beyond damage threshold, appropriate control measures can be undertaken.

10. Selection of management options. The choice of appropriate option is very critical for the successful management of the pest. Before resorting to chemical control, the possibility of employing natural control agents, cultural practices and resistant varieties should be explored. If insecticides are to be used, information on their toxicity, cost, effectiveness, formulation, method and time of application, and their impact on beneficial arthropods in agroecosystem must be known.

DECISION MAKING

Decision making in pest management, like all other economic problems, involves allocating scarce resources to meet human needs. Initially, there is the choice of whether, when and how to attempt to manage insects and other pests with scarce capital or labour. Other resources may be scarce, however, such as pesticide-susceptible insect strains, an uncontaminated environment or information on the extent of pest infestation may also affect the desirability of particular choices.

Decision makers in IPM exist at many levels, from farmers, managers and contractors, each of whom are concerned with single fields, to industry and government personnel who are concerned with regional or national policies. Decision making at any of these levels would require generation of information on the following aspects:

(*i*) *Fundamental information*, concerned with basic technical, biological and ecological processes that affect the damage caused by pests and the effectiveness of control measures.

(*ii*) *Historical information*, in the form of records of previous pest incidence and damage, used to indicate trends in pest development and to assess the probability of future attacks.

(*iii*) *Real-time information*, collected by on-farm monitoring schemes or by regional surveillance and concerns on current pest status. Real time information on pest attack and damage can be obtained by direct assessment or indirectly through meteorological measurements.

(*iv*) *Forecast information*, involving estimates of future levels of attack and damage. This can be obtained by combining the previous categories of information often by means of a regression or more complex model.

The information so generated is analysed for the purpose of decision-making. This requires the use of decision making techniques like linear programming, goal programming, simulation modelling, decision trees and dynamic programming. As the information based on various aspects of pest management increasd dramatically, computers were used first to process data and information followed by knowledge processing and decision making. Computer based systems could be used to provide access to information by serving as a cache, integrate and synchronize the information for use, interpret the information and serve as implementation coach.

DEVELOPMENT OF AN IPM PROGRAMME

The development of an integrated pest management programme involves three distinct phases, viz. the problem definition phase, the research phase and the implemen-tation phase (**Fig. 12.3**).

Problem Definition Phase

The successful integration and use of control options in an integrated pest management programme depends to a very large extent on the time and effort apportioned to the definition phase of the programme. This is a critical phase that is rarely given the amount of consideration needed for the subsequent development of an appropriate management programme. It is essential that the true dimensions of a pest problem be understood, not just in terms of perceived damage or yield losses caused by a pest, but in terms of actual yield losses and the socio-economic context of the farming system, in which these losses are known to occur. An understanding of the impact of a pest attack in this wider context, will point the way to an appropriate selection of control measures. To take this approach, however, would involve a significant effort, since yield loss evaluations and a socio-economic exploratory survey would be required, with the former possibly taking a number of years to carry out properly.

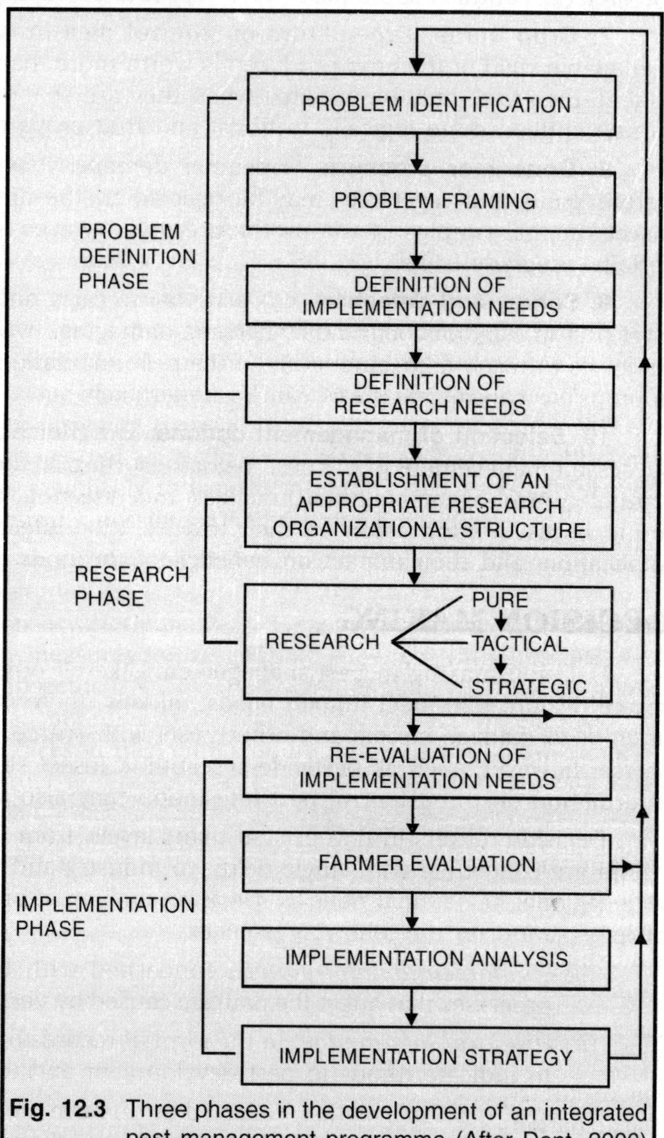

Fig. 12.3 Three phases in the development of an integrated pest management programme (After Dent, 2000)

Such long-term approaches to simply defining a problem are clearly ideal, but rarely practical, especially where pest outbreaks are having a major impact on crop production and immediate control action is required. It will rarely be possible for all of the information required to be obtained, but it is becoming increasingly obvious that the least that is required in these initial stages is a socio-economic evaluation of the situation. This should take the form of an exploratory survey. While other information may have to be obtained from the literature, this information and that obtained from an exploratory survey, will need to be carefully analysed in order that the best possible representation of the pest problem can be produced. This problem framing part of the procedure should aim to define the limits of the problem, *i.e.* it needs to identify the key processes and variables, provide a synthesis of what is already known, and determine exactly what needs to be known if a pest management programme is to be developed. It is at this stage of the definition phase that modelling can play an important role, with the construction of a conceptual model of the problem. The model may or may not be developed as a mathematical model as the programme progresses.

Research Phase

The emphasis on specialisation in research has meant that research projects are considered for funding on a piecemeal basis and on individual merit. There is little or no co-ordination between projects and in some cases, different institutes working on similar projects do not know of the other's work. While this approach to funding research may be highly appropriate for promoting innovative, specialist projects within a single discipline, it is much less suitable for promoting the objectives of a co-ordinated multidisciplinary research effort such as that required for a research programme in IPM. Even if the separate components needed for an integrated approach were funded, there would still be little hope for integration because no organizational structure exists to co-ordinate the efforts and direct them towards a common goal. This lack of an appropriate organizational structure for management of research projects has important implications during both the definition and the research phases of a programme. During the definition phase, it is important that a common paradigm of the problem be developed to integrate the individual goals of each research group and to provide a common objective. In the absence of this, each group will carry out their own research in isolation from every other group. This will result in a number of separate solutions to a number of unrelated problems, because each group will perceive their objectives differently according to their own perceptions and expertise. Thus, organizational structures have an important role to play in pest management research to provide an appropriate institutional framework within which, integration can take place.

Implementation Phase

The development of an implementation strategy should not be left to the final phases of any IPM programme. It should initially be considered during problem formulation and then continually readdressed throughout the research phase of the programme. Various factors essential for the implementation of most IPM programmes are discussed below.

Farmers' participation. To be effective and economically feasible, implementation of IPM at the farm level requires a degree of training, experience and attention to individual field conditions beyond the capabilities of most farmers, who must devote most of their time and energy to other aspects of crop production. On the other hand, top-down extension perpetuates an ideology of elitism, paternalism/passivity and social control. There is, thus, a need for a relationship of equality and dialogue between extension agent and farmer such that the partial knowledge of each is combined to solve problems and bring about positive change more effectively.

Since it is relatively complex, location specific and management-intensive, IPM is an educational challenge. The farmer must learn the principles and acquire the knowledge and skills necessary to make autonomous decisions based on specific farm conditions. There is a need for farmers' participation at every step of the R & D process in order to draw on farmers' intimate understanding of local conditions and constraints, their innovativeness and their skill at making the best possible living using limited resources.

Agricultural specialists should not dominate but instead should act as consultants, facilitators and collaborators, simulating and empowering farmers to analyse their own situation, to experiment and to make constructive choices. Extension agents should teach them analysis and decision making process. Placing the farmer at the centre of the technology development process is wholly consistent with the IPM goal of making the farmer a confident manager and decision maker, free from dependence on a constant stream of pest control instructions from outside.

An innovative approach recently applied for the implementation of IPM programmes is the 'Farmer First' approach which is being used for IPM on rice in Indonesia. In this method, farmers are divided into small groups to monitor the crop and then each group analyses the field situation by identifying the key factors. Group members then decide whether any action is required. At a combined meeting, each group presents and defends its summary to other trainees. The trainer facilitates by asking leading questions or adding technical information if necessary. This process allows farmers to integrate and practise their skills and knowledge, and gives trainers an oppurtunity to evaluate the trainees' ability. Thousands of farmers have been trained utilizing this approach and it is being tried on a pilot scale. A survey among these farmers during the first post-training season revealed that they really decreased their frequency of pesticide sprays to a level consistently lower than that of non-IPM farmers. The percentage of farmers not applying pesticides was also significantly higher among the trained ones. Inspite of lower pest control expenditures, these farmers obtained higher yield than the non-IPM farmers. Other viable and easily implementable approaches need to be developed locally taking into account small holding size, low income and general literacy level of the average farmer in the area.

Legislative measures. IPM is an information system and its adoption reduces pest control costs. The alternative to IPM is the indiscriminate use of broad spectrum synthetic organic pesticides. Unfortunately, while pesticide manufacturers and users (farmers) derive the full benefits from the use of these chemicals, they pass on the environmental and ecological costs of their use to the society as a whole. If they are made to bear the full cost of the use of these toxicants, they may find IPM a more economical and attractive alternative. This could be achieved by enforcing suitable legislative measures.

Secondly, in order for an IPM programme to be successful, it must be followed by most, if not all, farmers in a geographical area. Ideally, all farmers may adopt an IPM programme voluntarily but some farmers may hold out. Such farmers called 'spoiler holdouts' may impair the success of a programme by failing to adopt a necessary practice. Legislative measures are required to impose the programmes upon an unwilling minority.

Thirdly, the importance and benefits of pesticides are being overemphasized by a multibillion dollar industry utilizing the services of not only their salesmen but also agricultural scientists, administrators and planners. There is not yet a strong market in IPM information. Much important information which might induce a farmer to adopt IPM is not immediately observable and is, therefore, not sought by him. A manufacturer has no incentive to recommend a programme that uses less pesticides or even selective pesticides that kill a limited range of pests. This distortion could only be corrected by legislative action.

Government support. Both the national programmes of developing countries and the donor agencies must have policy commitment to IPM in the context of national economic planning and agricultural development. The costs to developing countries of not bringing their policies in line with the objectives of IPM are relatively greater than the costs of developed countries. National policies to promote IPM require close regulation at all stages related to the importation and/or manufacture, distribution, use and disposal of pesticides. In the case of pesticides which do not meet prescribed standards for safety, persistence, etc., import and manufacturing bans should be enacted. At a minimum, the conditions laid out by the *FAO Code of Conduct on the Regulation, Distribution and Use of Pesticides* should be adopted. Pesticide subsidies need to be eliminated in order to make IPM an attractive alternative. The funds so saved may be utilized for the implementation of IPM. Funds may also be diverted from some of the current research programmes to IPM-oriented plant protection programmes. Additional monetary resources may be generated through cooperation with bilateral/multilateral agencies willing to support such programmes.

Improved institutional infrastructure. IPM cannot be implemented unless there is a basic infrastructure for plant protection in a country. There is a need to develop and support national programme capabilities for on-farm testing and technology extrapolation. At the international level, establishment of *IPM Working Group* to coordinate and monitor funding of IPM projects is bound to provide impetus to the implementation of IPM. IPM is predominantly a knowledge technology, the use of which requires training of the many groups involved. There is currently little training material for most of these groups including farmers, extension personnel and researchers. If IPM is to become the major approach for pest management in the developing world, this deficiency must be remedied urgently. Another aspect requiring greater attention is co-ordination of effort within and between countries, between national research, training and implementation institutes/ programmes, and amongst international development agencies.

Lack of a reliable database has also hampered progress of IPM programmes. A reliable source of accurate information on the status of crops and pests in farmer's fields is necessary for many IPM activities. Most of the successful IPM programmes both in developed and developing countries have a reasonably accurate system of monitoring and evaluating various biological and environmental parameters in the agroecosystem. A reliable data base on crop yield and pest losses is required for planning and resource tool for IPM in developed-country cropping system and may be used in the developing countries as well.

Improved awareness. Increased education and awareness regarding the objectives, techniques and impact of IPM programmes are required at all levels including policy makers, planners, farmers, consumers and general public. The improtance and benefits of pesticides are being over-emphasized by a multibillion dollar industry utilizing the services of not only their salesmen but also agricultural scientists, administrators and planners. There is not yet a strong market in IPM information. Policymakers and planners need to be convinced that without IPM current agricultural production systems are not sustainable. Similarly, much important information which might induce a farmer to adopt IPM is not immediately observable and is, therefore, not sought by him. A manufacturer has no incentive to recommend a programme that uses less pesticides, or even selective pesticides that kill a limited range of pests.

Consumer groups and the general public may also be able to support the implementation of IPM programmes by demanding residue-free commodities. There is now a distinct market for organically produced food and other products. Non-government organizations and consumer groups need to be strengthened in developing countries, so that there is a public-oriented movement for implementation of IPM.

CONSTRAINTS IN IPM IMPLEMENTATION

The Consultant Group of the IPM Task Force has conducted an indepth study of the constraints on the implementation of IPM in developing countries, which can be categorised into the following five main groups:

Institutional Constraints

IPM requires an interdisciplinary, multi-functional approach to solving pest problems. Fragmentation between disciplines, between research, extension and implementation, and between institutes, all lead to a lack of institutional integration. Secondly, both the national programmes of developing countries and the donor agencies have lacked a policy commitment to IPM in the context of national economic planning and agricultural development. This has resulted in a low priority for IPM from national programmes and donors alike. Thirdly, the traditional top-down researth in many cases does not address the real needs of farmers, who eventually are the end-users, and who select to adopt or reject the technology based on its appropriatene. Institutional barriers to research scientists in national programmes conducting on-farm research in developing countries are real, and need to be addressed.

Informational Constraints

The lack of IPM information which could be used by the farmer and by extension workers is a major constraint in implementaion. In a study regarding implementation of IPM in Haryana, India, it was found that more than three-fourth of the farmers were not even aware of the concept of IPM. Even those aware of the concept reported that they lacked the skills necessary to practise IPM. While the individual control techniques are well known, little knowledge is available on using these in an integrated fashion under farm conditions. The lack of training materials, curricula and experienced teachers on the principles and practice of IPM is another major constraint. In many cases, the field level extension workers are not sufficiently trained in IPM to instil confidence in the farmers.

Sociological Constraints

The conditioning of most farmers and farm level extension workers by the pesticide industry has created a situation where chemicals are presented as highly effective and simple to apply. This acts as a major constraint in IPM implementation. There appears to be a direct conflict between industry's objective of more sales, and the IPM message of rational pesticide use, in the eyes of farmers. There is a need for private industry and public sector extension agencies to work in a more complementary manner. A majority of the farmers in a study in Haryana, India, expressed their lack of faith in IPM. They considered IPM practices to be risky as compared to the use of chemical pesticides.

Economic Constraints

A major constraint, even if IPM is adopted in principle, is the funding for research, extension and farmer training needed for an accelerated programme. IPM must be viewed as an investment, and as with other forms of investment, requires an outlay. In the long run, IPM programmes may become self-generating due to saving on resource inputs for production. A majority of the farmers purchase pesticides on credit and depend on shopkeepers and pesticide dealers for information about the pest control methods.

Political Constraints

The relatively low status of plant protection workers in the administrative hierarchy is a constraint to general improvement in plant protection. Associated with the above are the morale and financial standing of these workers. The continuance of pesticide subsidy by the government for political reasons and its tie up with the government-provided credit for crop production, acts as a major constraint to farmers' acceptance of IPM. Various vested interests associated with the pesticide trade also act as a political constraint on the implementation of IPM.

FRAMEWORK OF IPM PROGRAMME

There has been a widespread tendency to work within specialized subject areas of IPM and this movement towards specialization among researchers has now reached its peak of ascendancy. The need now is to place this specialist knowledge, abilities and skills within a broader scientific framework. There ought to be better coordination and exploitation of this valuable resource. This means that work of specialist groups dealing with different aspects of a common problem of insect pest control needs to be coordinated and placed in the context of the framework of an integrated pest management programme. This integration needs to be started at the top, in funding policy and reach up to the lowest level, in implementation.

- The funding authorities need to develop a coherent policy and ensure that all individual research groups working on similar cropping systems coordinate their approaches. There is a need to allocate funds for development of complete pest management programmes.
- Secondly, there is an urgent need to develop conceptual and theoretical framework for IPM. The present situation is that IPM is made up of a great many isolated parts, each of which can be developed internally but which is not clearly connected to anything else, *e.g.*

farming systems, host plant resistance, natural enemies and decision-making behaviour. These pieces ought to be combined in an integrated programme. This requires a theoretical framework that will provide guidelines for pest managers. The framework will incorporate pest outbreak theory, a classification of pest types, their hosts and farming systems, and to identify options most appropriate for management strategies.

- Thirdly, integrative level of research in the form of field trials to test combination of control options for their compatibility and effectiveness is essential. This integrative research will require the combined input of all relevant disciplines to design, carry out and analyze the data from suitable factorial or multifactorial experiments. These experimental designs will be necessary to assess the interaction between the various treatments under test and arrive at a combination of options that produce higher yields.
- Fourthly, integration is required at the level of organizational behaviour. It is important that appropriate organizational structures are developed because they are fundamental to good management. Without them, integration of research will be less likely to occur and at a different level it may be found to affect motivation, innovation, morale and decision making, and exacerbate conflict and poor coordination.
- Ideally, IPM should involve integration of control options for the management of all types of pests and not just insects. Insect pest management will then just be a subsystem of integrated pest management (IPM). This will require multidisciplinary research encompassing insects, pathogens, weeds and other pests. The importance of integrating the control of insects and pathogens is illustrated by an early story involving control of grape phylloxera on grapes in Europe. This example is often cited to illustrate the importance of plant resistance. The other part of the story is not so well known. Ironically, plants introduced from North America carried the pathogen of *Plasmopara viticola*, the causal agent of downy mildew to which European grapes were highly susceptible. The American root stock saved French vineyards from grape phylloxera but exposed them to an even more dangerous risk. The devastating epidemic of downy mildew that followed threatened wine production throughout Europe. Ultimately, the development of Bordeaux mixture,an early fungicide against the pathogen, saved European vineyards. This example underlines the need to consider the whole pest complex and the implications of any management strategy.
- Lastly, the development of an implementation strategy should not be left to the final phases of a research programme. It should initially be considered during problem formulation and then continually readdressed througout the research phase of the programme because it is often at the point of implementation that many pest management programmes fail.

POTENTIAL OF IPM

Initially, IPM programmes evolved as a result of the pest problems caused by repeated and intensive use of pesticides and increasing cases of past resistance to these chemicals. It is only during the past few years that economic and social aspects of IPM have also received increasing attention (**Fig. 12.4**). Some of the

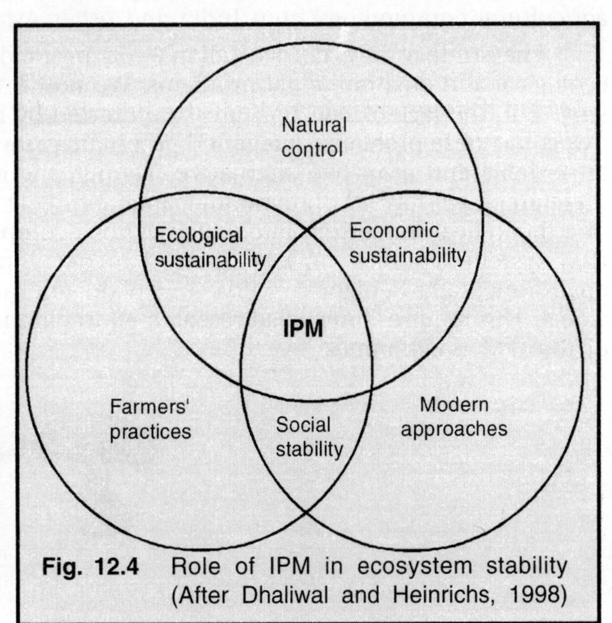

Fig. 12.4 Role of IPM in ecosystem stability (After Dhaliwal and Heinrichs, 1998)

important advantages offered by IPM over the pesticide-based plant protection programmes are listed here.

Sustainability. It is now being increasingly recognized that modern agriculture cannot sustain programmes that present productivity levels with the exclusive use of pesticides. Increasing pest problems and disruptions in agroecosystems can only be corrected by use of holistic pest management programmes.

Economics. If the environmnetal and social costs of pesticide use are taken into account, IPM appears to be a more attractive alternative with lower economic costs.

Health. Production, storage, transport, distribution, and application of pesticides involves greater health hazards than the safer inputs used in IPM. In developing countries, it is almost impossible to implement residue limits or waiting periods for pesticides on food products and other commodities. This endangers the safety of the entire population of these countries.

Environmental quality. The IPM programmes do not endanger non-target organisms, nor do they pollute the soil, water and air. The clean air, water and soil are now being recognized as non-renewable resources which once polluted are almost impossible to purify.

Social and political stability. The pesticides used by the farmers are obtained from the corporate houses and even from other countries. The inputs used in IPM are usually based on local resources and outside dependence is minimized. This helps in mainting social and political stability.

Local knowledge. IPM builds upon indigenous farming knowledge, training traditional cultivation practices as components of location specific IPM practices. This is especially important for the farmers in developing countries where traditional agricultural systems are based on indigeneous farming practices. The incorporation of IPM into these practices helps the farmers to modernise while maintaining their cultural roots.

Export of agricultural commodities. The presence of pesticide residues is affecting our exports of agricultural and horticultural commodities. There is a growing demand for organically cultivated, fresh and processed fruits and vegetables. The current consumption of organically produced fruits and vegetables at the global level is valued at US$ 27 billion. The pesticides in beverages like tea and coffee have affected our exports of these commodities during the last few years. There is also a considerable export market for cotton fabrics and garments devoid of pesticide residues in Japan and Western countries. Residue-free *basmati* rice is also highly prized in the interntional market. Thus, implementation of IPM in these crops will give boost to export of fresh and processed agricultural commodities from India and other Asian countries.

The strategy of exclusive reliance on pesticides for all pest problems created a number of ecological and environmental problems. We now know that insects possess a remarkable ability to survive in the face of selection pressure exerted by insecticides and other forms of pest control. To overcome these problems, integrated pest management based on ecological principles was developed as a viable and attractive alternative. Though the concept of IPM has been universally accepted, there are as yet few IPM programmes functioning at the farmers' level. A major limitation is the lack of a theoretical framework into which various components of IPM can be fitted to develop a viable IPM system. There is an urgent need to develop IPM systems for different crops, which are environmentally benign, conserve our plant and animal genetic resources and are economically viable. This requires intensified research efforts in formulation, research and implementation phases of the IPM programmes.

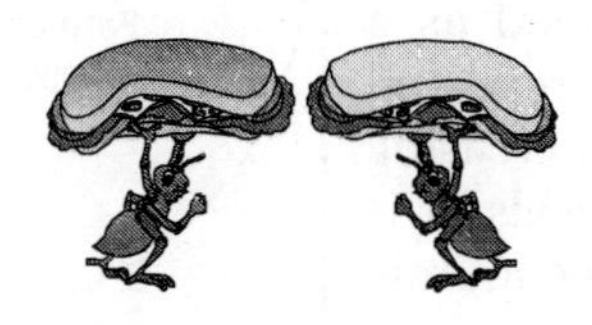

PART-III

PESTS OF CROPS AND DOMESTIC ANIMALS

- Pests of Cereals and Millets
- Pests of Pulse Crops
- Pests of Oilseed Crops
- Pests of Fibre Crops
- Pests of Sugarcane
- Pests of Tropical and Sub-Tropical Fruits
- Pests of Temperate Fruits
- Pests of Vegetable Crops
- Pests of Mushrooms

- Pests of Ornamental Plants
- Pests of Plantation Crops
- Pests of Spices
- Pests of Forest Trees
- Polyphagous Pests
- Stored Grain Pests
- Household Pests
- Pests of Farm Animals
- Non-Insect Pests
- Insect Vectors of Plant Diseases

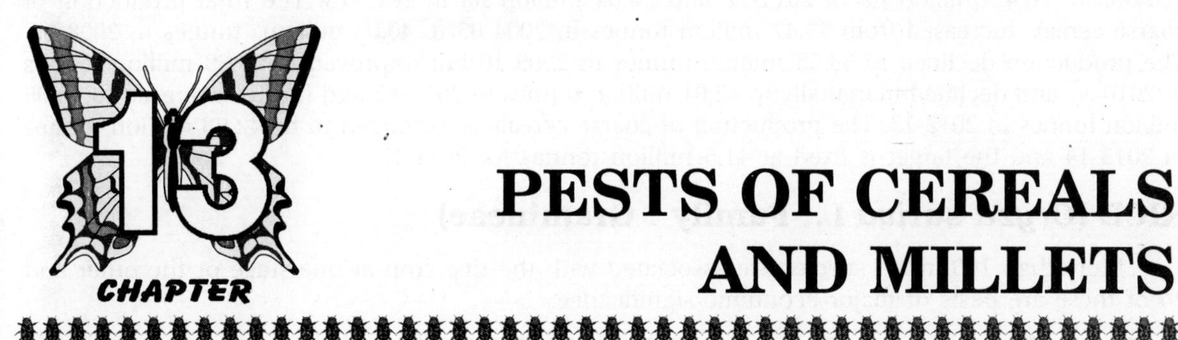

PESTS OF CEREALS AND MILLETS

INTRODUCTION

Cereals constitute the staple food of the people of Asia. Whereas rice,millets and sorghum are essentially crops of the sub-tropical areas of South-East Asia, maize and wheat are grown in the intermediate climates. India produced 234.47 million tonnes of foodgrains in 2008-09 and production declined to about 218.11 million tonnes in 2009-10. The foodgrain production further increased to 244.78 million tonnes in 2010-11 and 259.29 million tonnes in 2011-12, and fell marginally to 255.36 million tonnes during 2012-13. India is estimated to have achieved an all-time high production of foodgrains at 264.38 million tonnes during 2013-14. The total foodgrain target has been lowered to 261 million tonnes in 2014-15.

Rice is comparatively a more adaptable crop and is grown throughout India as far north as Kashmir and as far east as Assam, Bengal, Tripura and Manipur. In the Peninsular India and other South-East Asian countries, more than one crop of rice is grown in a year. In India, the area under rice decreased from 45.54 million ha in 2008-09 to 44.01 million ha in 2011-12, and further decreased to 42.41 million ha in 2012-13. The production of rice has shown an upward trend during the period 2005-06 to 2008-09 and it reached a level of 99.1 million tonnes in 2008-09. However, the production declined to 89.09 million tonnes in 2009-10 and increased to 105.30 million tonnes in 2011-12, but decreased to 104.40 million tonnes in 2012-13. India is estimated to have harvested 106.29 million tonnes of rice in 2013-14 and the target has been fixed at 106 million tonnes in 2014-15. After China, India is the largest producer of rice in the world.

Wheat is grown in winter throughout western Asia and the Indo-gangetic plains, except in the extreme eastern parts of Bengal and Assam, where this crop is grown scantily. In the southern parts of Indian Sub-continent not much wheat is grown except in certain parts of Andhra Pradesh, Gujarat and Karnataka, where it is a new introduction. In India, production of wheat has increased significantly from 68.84 million tonnes in 2004-05 to 94.88 million tonnes from an area of 29.9 million ha in 2012-13, but decreased to 92.5 million tonnes from an area of 29.7 million ha during 2012-13. The production of wheat is estimated to be 95.85 million tonnes in 2013-14 and the target for 2014-15 has been set at 94 million tonnes. Comparatively, maize is a minor crop in India. The area under this crop in 2010-11 was 8.55 million ha with production of 21.73 million tonnes. The area under maize increased to 8.78 million ha in 2011-12, with production of 21.76 million tonnes. The production increased to 22.23 million tonnes on an area of 8.71 million ha in 2012-13. The target has been fixed at 23 million tonnes for 2014-15. In the Indo-gangetic plains, and western and central India, maize is grown under irrigated conditions whereas in the Himalayas it is primarily grown under rainfed conditions.

The coarse cereals, viz. jowar, bajra, ragi, small millets and barley, are grown throughout India, particularly in the drier parts. Some of the minor millets are grown in the Himalayas up to a height of 2400 metres. There has been a decline in the area under coarse cereals from 29.03 million ha in 2004-05 to 26.42 million ha in 2011-12 and 24.64 million ha in 2012-13. The total production of coarse cereals increased from 33.47 million tonnes in 2004-05 to 40.03 million tonnes in 2008-09. The production declined to 33.55 million tonnes in 2009-10 but improved to 43.68 million tonnes in 2010-11 and declined marginally to 42.01 million tonnes in 2011-12 and further decrased to 40.06 million tonnes in 2012-13. The production of coarse cereals is estimated to be 42.68 million tonnes in 2013-14 and the target is fixed at 41.5 million tonnes for 2014-15.

RICE (*Oryza sativa* L.; Family : Gramineae)

More than 100 insect species are associated with the rice crop at one stage or the other and 20 of these are pests of major economic significance.

1. Brown Planthopper, *Nilaparvata lugens* (Stal) (Hemiptera : Delphacidae) (Fig. 13.1, Plate 13.1A)

The brown planthopper is the most destructive pest of rice in South and South-east Asia, China, Japan and Korea. In India, it has become very serious on the high-yielding varieties of paddy in many States including Uttar Pradesh, Madhya Pradesh, West Bengal, Andhra Pradesh and Tamil Nadu.

Both adults and nymphs feed on paddy, sugarcane and grasses by sucking the cell sap. The brownish adults (**Fig. 13.1**) with brown eyes are 3.5-4.5 mm in length. Their legs are light brown and the tarsal claws are black. The wings are hyaline with brown markings and dark veins. The nymphs are brownish-black in colour and have greyish-blue eyes.

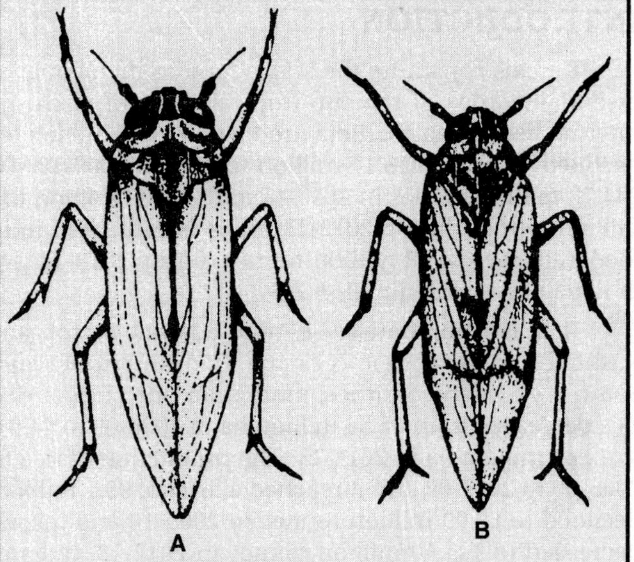

A B

Fig. 13.1 Brown planthopper, *Nilaparvata lugens*. A. Female adult; B. Male adult

Life-cycle. In tropical areas, this pest breeds on paddy crop throughout the year and its population may reach the maximum any time between October and February, depending upon the climatic conditions. In North India, however, the population of this planthopper becomes high in September-October.

The adults remain most active from 10 to 32°C. The females start laying eggs within 3-10 days of their emergence and they deposit eggs in masses, by lacerating the parenchymal tissue. The number of eggs per mass varies from 2 to 11 and a female lays, on an average, 124 egg-masses.

The eggs are somewhat dark and cylindrical, having two distinct spots. The incubation period ranges between 4 to 8 days. The nymphs, on emergence, start feeding on young leaves and after moulting 5 times, they become adults in 2-3 weeks. In South India, the life-cycle is completed in 18-24 days during June-October, 38-44 days during November-January and 18-35 days during February-April.

The coccinellid beetles, spiders, mirid bugs, black ants (*Camponotus* sp.) and red ants feed on the nymphs of this pest and exercise a limited natural control. Among the predators, mirid bug, *Cyrtorhinus lividipennis* Reuter (Miridae) and spider, *Lycosa pseudoannulata* (Bosenberg & Strand) (Lycosidae : Araneae) are quite important. Releases of mirid bug @ 100 bugs or 50-75 eggs/m^2 at 10-day intervals have been found effective for the control of brown planthopper. The presence of 3 predatory spiders/hill has been found to check the population of brown planthopper. A number of egg parasitoids namely, *Anagrus optabilis* Perk. (Mymaridae), *Oligaosita* sp. (Trichogrammatidae), *Tetrastichus* sp. (Eulophidae) and *Haplogonatopus* sp. (Dryinidae) have also been recorded. *Haplogonatopus* sp. parasitizes both the nymphs and the adults. The white muscardine fungus, *Beauveria bassiana* (Balsamo) Vuillemin, is pathogenic on *N.lugens*.

Damage. Both the nymphs and adults cause damage by sucking cell sap from the leaves which turn yellow. If the insect attacks during the early stages of growth, the entire plant may dry up. It has been noticed that even when the infestation is rather low, the tillering is adversely affected and there is diminished vigour and a decrease in plant height. Under the favourable conditions of high humidity, optimum temperature, high nitrogen application and no wind, the population increases very rapidly and the hopper-burn is observed in various localities, giving a brownish hue to the entire countryside **(Plate 13.1B)**. The loss in yield may range from 10 to 70 per cent. This insect is known to transmit the grassy stunt virus disease of rice.

Management. (*i*) Closer spacing of 15 × 10 cm creates favourable microclimate in field for rapid development of hopper population. Hence, a spacing of 20 × 15 cm should be followed. (*ii*) Alternate drying and wetting the field during peak infestation, and draining out the standing water from the field 2-3 times checks the population of the hopper to a large extent. (*iii*) Alleys 30 cm wide after every 3 metres of rice planting provide proper aeration to the crop which ultimately restricts the multiplication of the pest. Making of alleys also helps in insecticidal spraying as applicator can move freely in the field. (*iv*) Grow resistant varieties. (*v*) Spray at economic threshold level of 5 insects per hill, 100 ml of imidachloprid 17.8SL or 625 g of carbaryl 50WP or 2.0 litres of quinalphos 25EC or 2.5 litres of chlorpyriphos 20EC or monocrotophos 36SL in 250 litres of water per ha. Repeat application if hopper population persists beyond a week after application. While spraying nozzle should be directed at the basal portion of the plants. If the damage is noticed at hopper burn stage, treat the affected spots along with their 3-4 metre periphery immediately as these spots harbour high population of the insect.

2. Whitebacked Planthopper, *Sogatella furcifera* (Horvath) (Hemiptera : Delphacidae) (Fig. 13.2, Plate 13.1C)

In recent years, a number of leaf-and planthoppers have assumed pest proportions on the paddy crop in various parts of India. In all, 21 species have been identified and the white-backed planthopper is the most widespread of them all. In 1966, this insect was reported for the first time as a serious pest in the Punjab and other adjoining States of India.

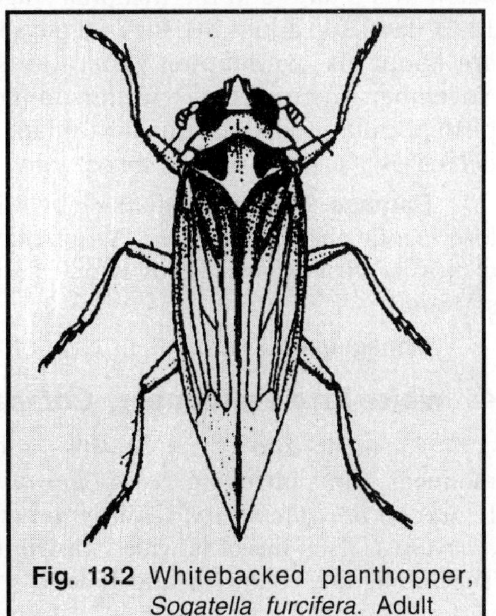

Fig. 13.2 Whitebacked planthopper, *Sogatella furcifera*. Adult

The nymphs and the adults of *S. furcifera* are very active and they can easily jump from one leaf to another on a slight disturbance. The adult is a straw-coloured, wedge-shaped insect, with white back **(Fig. 13.2)**. The nymph is greyish-white and turns dark grey when it nears maturity.

Life-cycle. The planthoppers breed continually in southern parts of India where paddy crop is available throughout the year. In northern parts, the planthopper becomes active in May in the paddy nursery, from where it shifts to the transplanted crop. By August and September, the population becomes quite high. The adults lay eggs generally on the leaf sheath. The eggs hatch in 3.4-4.6 days. The nymphs feed on leaves and are transformed into adults within 8.9-13.1 days. The life-cycle is completed in 12.3-17.7 days. The adult females live for about a week. There are several generations in a year. In northern parts of India, the planthoppers remain inactive in the winter and their population becomes noticeable only from May onwards on the paddy nursery.

Damage. The nymphs and the adults suck cell-sap from the leaf surface and tend to congregate on the leaf-sheath at the base of the plant. The leaves of attacked plants turn yellow and later on rust red. These symptoms start from the leaf tips and spread to the rest of the plant. Numerous brownish spots also appear on the feeding sites. The attacked plants ultimately dry up without producing ears. The insect also excretes honeydew on which a sooty mould appears,imparting a smoky hue to the paddy fields.

Management. Same as in case of brown planthopper. The economic threshold level is 10 insects per hill.

3. Green Leafhoppers, *Nephotettix nigropictus* (Stal) (Fig. 13.3) and *N.virescens* (Distant) (Plate 13.1D) (Hemiptera : Cicadellidae)

The green leafhoppers are found in all the rice growing regions of India, although they assume pest proportions only during certain years in Madhya Pradesh, Andhra Pradesh, Orissa and West bengal. These species are also known pests of rice in Japan, the Philippines, Formosa and Sri Lanka. Both the species of leafhoppers are greenish and are smaller and slanderer than *S. furcifera* (Fig. 13.3). The nymphs and the adults suck sap from the leaves of plants, turning them yellow and ultimately brown.

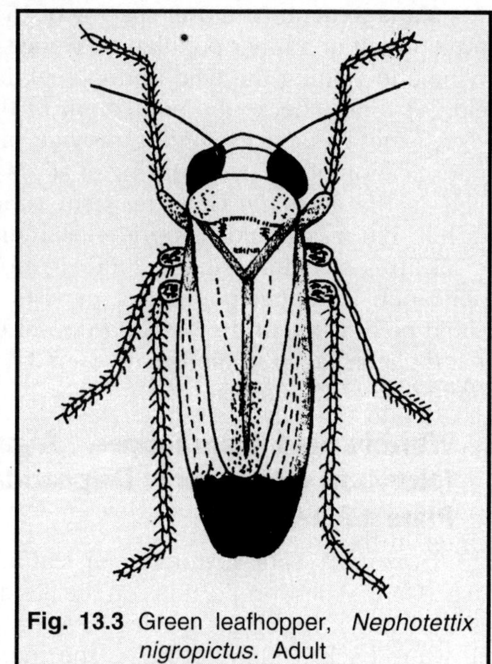

Fig. 13.3 Green leafhopper, *Nephotettix nigropictus*. Adult

Life-cycle. The females, after undergoing a pre-oviposition period of 6-9 days, lay eggs on the inner surface of the leaf-sheath in groups of 3-18. The eggs hatch in 3-5 days and the nymphal stage is completed in 12-21 days. The adults live for 7-22 days in summer. There are about six overlapping generations from March to November. The insect over-winters in the adult stage. The pest population is the maximum in July-August and decreases markedly after a heavy rain.

Damage. As a result of attack by this pest, the plants lose vigour and turn yellow. *N. virescens* is also known vector of virus diseases, of which tungro is the most serious.

Management. Same as in case of brown planthopper.

4. White Rice Leafhopper, *Cofana spectra* (Distant) (Hemiptera : Cicadellidae)

The adults and nymphs of this insect have been reported as pests of rice and a number of other monocot plants including *Leersia hexandra* Sw., *Triticum aestivum* L., *Zea mays* L., *Cyprus rotundus* L., *Saccharum officinarum* L., *Sorghum vulgare* L., *Sorghum halepense* L., *Cyperus tria* L., and *Cynodon dactylon* L. This insect is widely distributed in the sub-tropical and tropical regions in South Asia. The adults are yellowish and have a rounded, rather swollen head and four black spots on the vertex. The adult measures 5.7-7.4 mm in length. The nymphs are paler in colour.

Life-cycle. The insect is active throughout the year. The adults, on emergence, take approximately 1.8-6.8 days before mating. The females lacerate the leaf sheath with the help of their ovipositor and lay greenish eggs in masses. There may be 5-17 eggs in a mass and the female deposits a total of 17 masses in her life time of 30-35 days.

The eggs hatch in 6-7 days and the nymphs, on emergence, first feed inside the leaf sheath and, later, they move on to the leaf blade. There are 5 nymphal instars which are completed in 18 days. In South India, this pest is particularly active from July to March and completes 12 generations in a year.

Damage. Both the adults and nymphs suck cell sap and cause yellow discoloration of the leaves. When infestation is heavy, the leaves turn brown and the plants fail to produce ears. Even in moderate infestation,there is a significant reduction in the tillering of the plants.

Management. Same as in case of brown planthopper.

5. Zigzag Leafhopper, *Recilia dorsalis* (Motschulsky) (Hemiptera : Cicadellidae) (Fig 13.4)

The adults and nymphs of this insect are important pests of paddy in many countries including Japan, Korea, the Philippines, Vietnam, Malaysia, Bangladesh, India and Sri Lanka. The adult is a whitish-grey hopper (**Fig. 13.4**) which has V-shaped and zig-zag brown lines on its forewings and measures 3.1-3.8 mm in length.

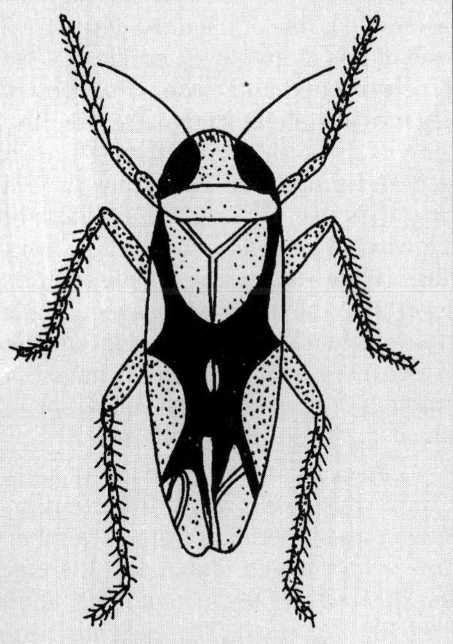

Life-cycle. In tropical areas of the world, this insect breeds throughout the year. The females lacerate the leaf surface with the help of their ovipositors and deposit eggs in the exposed leaf tissue. Sometimes, however, the eggs are also seen simply scattered on the leaf sheaths. The female has a life span of approximately 155 days, and during that period she lays upto 90 eggs.

The eggs hatch in six days and the nymphs, on emergence, feed on the leaf blades. There are 5 nymphal instars and this developmental stage is completed in 120 days. The entire life-cycle from egg to the adult stage is completed in 180 days. In temperate regions such as Southern Japan and Korea, this pest is known to pass winter in the egg stage.

Fig. 13.4 Zigzag leafhopper, *Recilia dorsalis.* Adult

Damage. As a result of the insect's feeding on cell sap, the mature leaves acquire an orange discoloration at the margins and become dry at the tips. As younger leaves grow they also show the same signs. The female hoppers are also known to be vectors of the virus disease 'orange leaf' and of the mycoplasma 'yellow dwarf'.

Management. Same as in case of brown planthopper.

6. Rice Blue Leafhopper, *Typhlocyba maculifrons* (Motschulsky) (Hemiptera : Cicadellidae)

This leafhopper is primarily a pest of rice nurseries throughout India and South-east Asia. Apart from paddy, it also feeds on sorghum, ragi,maize, sugarcane and wild grasses. It is a small bluish insect having a yellowish vertex with a black patch and a black spot in the middle of the pronotum. The adult measures 1.9-2.5 mm in length.

Life-cycle. This insect is found practically throughout the year and is particularly noticed in the nurseries before the planting season. The females lacerate the surface of the leaf sheath or the upper midrib of the leaf blade and deposit eggs in batches of 2, 3 or 4. A female lives for 45-50 days and, during that period, she lays an average of 48 eggs. The eggs hatch in 6-12 days and the young nymphs, on emergence, start feeding inside the leaf sheath. The nymphs moult five times and become mature in 11-20 days and the entire life cycle is completed in 19-32 days.

Damage. Both nymphs and adults suck cell sap from the leaves which, in the early stages of attack, exhibit whitish, waxy lines. As the damage progresses, the leaves show symptoms of withering. If the pest is not controlled, the plants die and the entire nursery may be lost.

Management. Same as in case of brown planthopper

7. Yellow Stem Borer, *Scirpophaga incertulas* (Walker) (Lepidoptera : Pyralidae) (Fig. 13.5)

The yellow stem borer is a specific pest of rice and is common throughout the Orient as one of the most destructive pests of this crop. The caterpillars alone are destructive and, when full-grown, they measure about 20 mm and are dirty white or greenish yellow, having brown head and pronotum. The adults have a wing expanse of 25-45 mm and are yellowish white with orange yellow front wings **(Plate 13.1E)**. The female moth is bigger than the male and has a centrally situated black spot on each of the forewings. The females have a prominent tuft of brownish yellow silken hair at the tip of their abdomen.

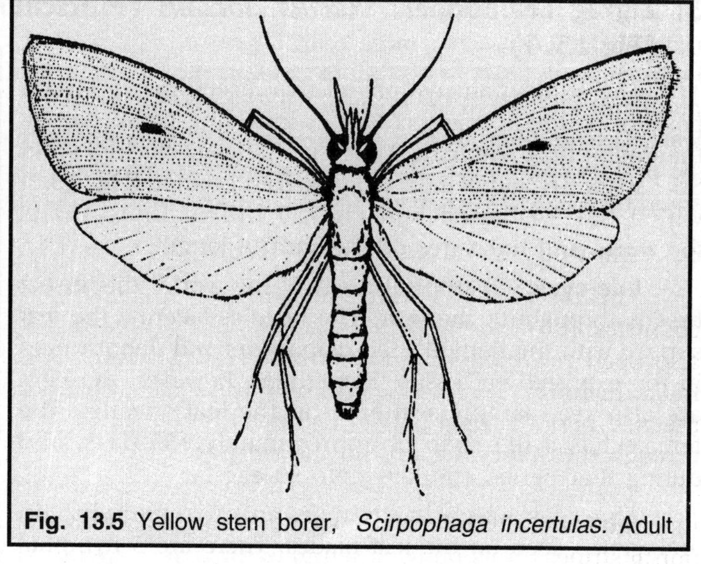

Fig. 13.5 Yellow stem borer, *Scirpophaga incertulas.* Adult

Life-cycle. In the northern regions of India, this pest is active from April to October and hibernates from November to March as a full-grown larva in rice stubble. The pupation starts sometimes in March and the emergence of moths begins in April. The moths become active after dusk when they mate and lay about 120-150 eggs on the underside of the leaves in 2-5 clusters of 60-100 eggs each. The eggs are covered with yellowish brown hair of the female tuft. The eggs are oval, flattened, pearly white at the time these are laid, but turn black before hatching. They hatch in 6-7 days and the tiny black-headed caterpillars soon bore into the stem from the growing points downwards. When a tiller is killed, the caterpillar inside migrates to another tiller of the same or of a different plant. The larva grows in 6 stages and is full-fed in 16-27 days. It then constructs an emergence hole which is always located above the water level and pupates inside the attacked plant. Within 9-12 days, it emerges as a moth. The life cycle is completed in 31-46 days. There are 3 broods in Bengal, 2 in Orissa and 5 in Andhra Pradesh and Tamil Nadu where it is active throughout the year. In the Punjab, there are 4-5 generations noticed from April to October and the caterpillars of the last brood make silken hibernacula in the stubble inside which they hibernate for the winter.

Several egg parasitoids have been found to be quite effective, viz. *Telenomus dignoides* Nixon (Scelionidae) was observed to parasitize as high as 70.8 per cent eggs at Ludhiana, *Telenomus rowani* (Gahan) 67 per cent in October and *Tetrastichus schoenobii* Ferriere (Eulophidae) 75 per cent in May at Coimbatore. Inundative releases of exotic *Trichogramma japonicum* Ashmead at 50,000 ha^{-1} during egg laying period of rice stem borer reduced borer damage and increased crop yield.

An allied species, *Scirpophaga innotata* (Walker), the white stem borer **(Plate 13.1F)**, is also present in most of the paddy growing areas in India. The life-cycle and damage of this species is almost similar to *S. incertulas.*

Damage. Damage due to stem borer is much higher in southern States where the pest multiplies throughout the year and shifts from one crop to the next. In Northern India the pest has recently started appearing in serious proportions. *Basmati* varieties suffer heavy damage than coarse varieties. The plants attacked in early stages produce ears devoid of grain **(Plate 13.1G)** and are known as the 'white ears'. An average loss of 100-500 kg of paddy per hectare has been reported from Andhra Pradesh. In Orissa, the cumulative loss has been estimated at 44G per cent; with an increase of one per cent incidence of the pest at the earing stage, the yield is reduced by 0.604 per cent.

Management. (*i*) The removal and destruction of stubble at the time of the first ploughing after harvesting the crop decreases the carry-over to the next crop. (*ii*) Ploughing and flooding the field is also effective in killing the larvae. (*iii*) Since the eggs of stem borer are laid near the tip of leaf, clipping of tips of seedlings before transplanting can reduce the carry over of eggs to the field. (*iv*) The fields showing more than 5 per cent deadhearts should be sprayed with 425 g of cartap hydrochloride 75SG or 150 ml of chlorantriniliprole 20SC or 875 ml triazophos 40EC or 1.4 litres of monocrotophos 36SL or 2.5 litres of chlorpyriphos 20EC or 37.5 g of fipronil 80WG in 250 litres of water per ha. Alternatively, apply 10 kg of chlorantriniliprole 0.4GR or 25 kg of cartap hydrochloride 4G or 15 kg of fipronil 0.3G or 10 kg of chlorpyriphos 10G or 7.5 kg of phorate 10G per ha in the standing water in the field. Same chemical should not be used repeatedly.

8. Pale-headed Striped Borer, *Chilo suppressalis* (Walker) (Lepidoptera : Pyralidae)

This insect is also known as the rice stem borer in Japan and Asiatic rice borer in some other countries. It is a serious pest in Japan, the Philippines, Southern Indonesia, Spain and China. It has also been recorded from Formosa, Korea, Malaysia, the Hawaii Islands, Northern Australia and India. In India, it is not an economc pest. Except for a few wild grasses, this borer is confined to the rice plant throughout its habitat. Injury to the plant is caused by caterpillars which tunnel through the stem and feed on the soft tissues.

The moth is 12 mm long, with pale-yellow wings of 26 mm expanse. The male moth is smaller than the female. The newly emerged caterpillar is about 1.2 mm long and has five grey-brown linear dorsal stripes on its body. When full-grown, it is 26 mm long and 2.5 mm wide, and has a yellowish brown head. The middle dorsal stripe is lighter than the two along each side.

Life-cycle. The insect passes through 1-4 generations, depending upon the climate of its habitat. In temperate regions of Japan, China and Korea, it has two broods and the full-grown larvae of the second brood remain dormant in the stubble during the winter.

A female lays about 300 eggs in masses of variable sizes on the leaf blades near the tip or on the leafsheath. The eggs hatch in 4-10 days, depending on the weather. The newly emerged larvae feed on the leafsheaths in groups. The 2nd and 3rd instar larvae disperse to the adjacent plants. They bore into the culm, through the nodes by the tenth day. As the larvae grow and plants wither, they disperse gradually to the neighbouring plants. The larvae of the second generation bore into the stem through the stalk of the panicle and travel downwards until they reach the lowermost part of the stem. The larva of the first generation becomes full-grown in 33-50 days after emerging from the egg. The mature larva pupates within the rice stalk either near the middle or in the basal internode. Pupation is completed in 5-10 days. The hibernating larvae remain in the stems or stubble when the crop is harvested. They resume development in the spring; their pupation being followed by the emergence of moths.

Damage. The attacked leafsheaths first show transparent patches, and later turn yellow-brown and eventually dry up. The caterpillars, as a result of their feeding inside the stem around the nodes, weaken the stems which easily break. Seedlings which have been attacked at the base show dead-hearts, *i.e.* the drying up of the central shoot. Attacked plants bear white heads, indicating empty panicles or those with a few filled grains.

Two other allied species, namely, *Chilo indicus* (Kapur) and *Chilo infuscatellus* Snellen, have also been recorded infesting paddy in certain areas in India.

Management. Same as in case of yellow stem borer

9. Dark-headed Striped Borer, *Chilo polychrysus* (Meyrick) (Lepidoptera : Pyralidae) (Fig 13.6)

The dark-headed striped borer is distributed widely in India, Pakistan, Bangladesh, Myanmar, Malaysia, Indonesia, Thailand, Vietnam, Laos, Sabah and the Philippines. In India, it is also known as the Malayan borer and has been reported from Assam, WestBengal, Orissa, Tamil Nadu and Kerala. In addition to paddy, the larvae of this pest also attack maize, sugarcane and several species of grasses.

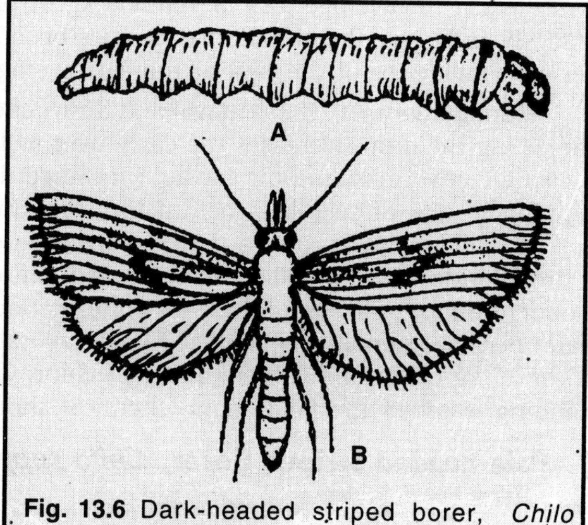

Fig. 13.6 Dark-headed striped borer, *Chilo polychrysus.* A. Larva; B. Adult

The yellowish white caterpillar has a black head capsule, a black thoracic plate and five longitudinal stripes on its body. It is 18-24 mm in length. The forewings of the male moth are brown ochreous with a cluster of dark spots covered with golden scales in the middle. The hind wings are white. In comparison, the wings of the females are paler with smaller metallic spots. The wing span of male is 16-25 mm and that of the female is 22-30 mm.

Life-cycle. The female moths lay flat scale-like eggs in rows, arranged in groups of 30-200, on the underside of leaves. Up to 488 eggs may be laid by a female moth during an oviposition period of 3 days. The eggs hatch in 4-7 days. The young caterpillar on emergence has the habits of a tissue borer and it tunnels within the leaf sheath, mid-rib or the stem. The larva moults 5 times within a period of 23-36 days. It makes a tunnel in the stem, forming an exit hole at the upper end and then pupates inside. The yellowish brown pupa has distinct abdominal stripes and when fully formed, is 10 mm long and 2 mm wide. The adults emerge in 4-6 days and have a life span of 2-5 days. The total life-cycle is completed in 26-61 days and there are six generations in a year.

The parasitoid, *Bracon albolineatus* Cameron (Braconidae), attacks the larvae and the parasite, *Tetrastichus* sp. (Eulophidae), feeds on pupae.

Damage. The caterpillars bore into the central shoot for feeding. As many as 7 larvae have been noticed in a single shoot. Since the larva bores into the outer leaves and the leaf sheaths first, they are the first to die, followed by the inner whorl, and finally the entire plant from its core. When there is severe damage, the loss may go up to 60 per cent of the crop.

Management. Same as in case of yellow stem borer.

10. Pink Stem Borer, *Sesamia inferens* (Walker) (Lepidoptera : Noctuidae) (Fig. 13.7)

The pink stem borer is a polyphagous insect and is distributed throughout India and Pakistan. In some parts of the country, it is a minor pest of sugarcane whereas in others, it is common on *ragi* (*Elusine coracana*), and in still others on wheat. In northern India, it is recorded on rice, sugarcane, maize, sorghum and wheat, but its damage is significant on rice and maize only. The damage is caused by the caterpillars which are pinkish brown and have a smooth cylindrical body, measuring about 25 mm. The moths are straw-coloured and have a stout body **(Fig. 13.7)**.

Life-cycle. The pest breeds actively from March-April to November on rice and then migrates to the wheat crop. The moths are nocturnal and lay eggs on leaves or on the ground. The eggs hatch in 6-8 days and the young caterpillars bore into the epidermal layers of the leaf sheath. Later on, they bore into the stem as a result of which the growing shoot dries up producing dead-hearts. When the attacked plants die, the larvae move on to adjoining plants. They are full-fed in 3-4 weeks and pupate inside the stem or in between the stem and leaves. The pupal stage lasts about a week and the life-cycle is completed in 6-7 weeks. There are 4-5 generations of the pest in a year.

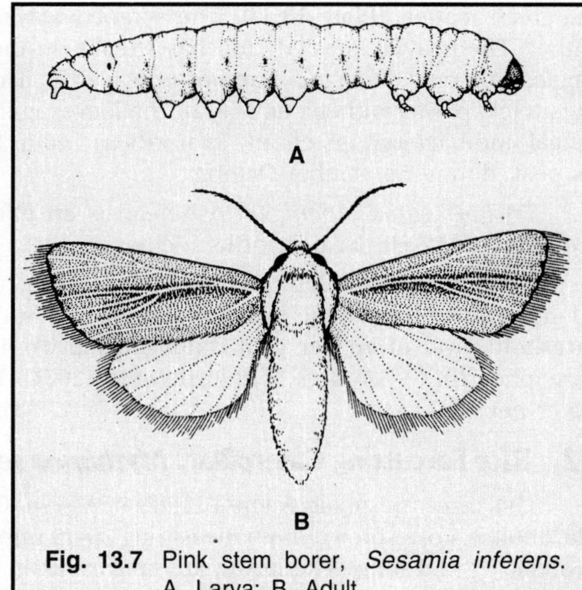

Fig. 13.7 Pink stem borer, *Sesamia inferens*. A. Larva; B. Adult

Damage. This pest is common during the dry pre-monsoon period. The attacked young plants show dead hearts and are killed altogether. The older plants are not killed, but they produce a few grains only.

Management. Same as in case of yellow stem borer.

11. Rice Leaf Folder, *Cnaphalocrocis medinalis* (Guenee) (Lepidoptera : Pyralidae) (Fig. 13.8)

The rice leaf folder is a sporadic pest. Its greenish caterpillars are very agile and they feed inside the fold made by fastening together the edges of a leaf. The moths are golden or yellowish brown and measure 8-10 mm in length and 16-20 mm in wing expanse. The wings have 2-3 wavy lines characterized by dark bands (**Plate 13.1H**). In case of heavy infestation, the plants appear whitish and scorched. The pest is distributed in Indonesia, Korea, Malaysia, the Philippines, Pakistan and India.

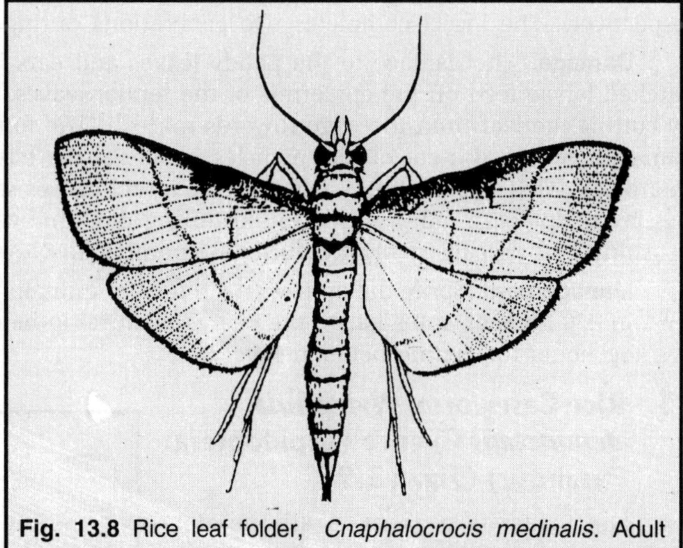

Fig. 13.8 Rice leaf folder, *Cnaphalocrocis medinalis*. Adult

Life-cycle. The moths which are nocturnal rest on the undersurface of the leaves during the day. They lay oval, creamy white eggs singly or in pairs on the leaves and leafsheaths. The eggs hatch in 3-4 days. A newly emerged larva is dull white or light yellow with a brown head. Soon it starts feeding and turns green. The full-grown larva is slender and measures 20-25 mm in length. It is very active and moves quickly in the leaf fold, when disturbed. The larval stage is completed in 15-25 days. Pupation takes place in loose silken webs in between the leaves or in the leafsheaths. The pupal stage lasts 6–8 days during the active season. The life-cycle is completed in 25-35 days.

Damage. The young larvae feed on tender leaves without folding them. The older larvae fasten the longitudinal margins of leaf together with a sticky substance and feed inside the fold by scraping the green matter (**Plate 13.1I**). The scraped leaves become membranous, turn white and finally wither. The heavily infested crop has streaks on the leaves and appears whitish from a distance. A single larva may damage a number of leaves as it migrates from one leaf to another. As a result of the attack, photosynthetic activity of the leaves is interfered with and the plants are predisposed to fungal and bacterial infections. In northern India, the pest is active form July to October and is at its peak during September-October.

Trichogramma japonicum Ashmead is an effective egg parasite of rice leaf-folder, causing a reduction of 12-60 per cent of its incidence.

Management. (*i*) Remove grass weeds from bunds around paddy fields. (*ii*) Light-trapping of adults helps to reduce the pest population. (*iii*) Spray any of the following insecticides at economic threshold level of 10 per cent damaged leaves: 425 g of cartap hydrochloride 75SG, 875 ml of triazophos 40EC, 2.5 litres of chlorpyriphos 20EC, 1.4 litres of monocrotophos 36SL in 250 litres of water per ha.

12. Rice Earcutting Caterpillar, *Mythimna separata* (Walker) (Lepidoptera : Noctuidae)

This insect occurs in Andhra Pradesh, Assam, Orissa, Punjab, West Bengal and peninsular India. The adult is a pale-brown moth measuring 35-40 mm across stretched wings. The full-grown caterpillar is about 39 mm long. The larvae are gregarious in habit and are commonly known as 'armyworm'.

Life-cycle. The eggs are laid in overlapping rows. In Uttar Pradesh, egg, larval and pupal periods have average durations of 2.4, 21.4 and 9.8 days, respectively under field conditions during July-December when the pest is present in the fields. The full-grown larva pupates in clumps of paddy, in cracks and crevices in ground or in loose soil. The total life-cycle is completed in 48.2 days by female and 45.4 days by male. The longevity of the moth is 12.6 and 9.8 days for female and male, respectively. The insect completes two generations during July to December.

Damage. The damage to the paddy leaves and ears is caused by the caterpillars. The newly hatched larvae feed on the epidermis of the tender leaves. The second and third instar larvae feed by cutting the leaf from the edge towards mid-rib. The fourth, fifth and sixth instar larvae besides damaging leaves also cut off the panicles mostly at the base and hence the name 'rice ear-cutting caterpillar'. This stage (4th-6th instar) of the insect causes serious loss to the paddy crop. The larvae are shy of sunlight, hide in the ground during day-time and generally feed at night. The damage to paddy crop is caused mostly during September to November.

Management. Spray the crop with 1.0 litre of quinalphos 25EC or 1.4 litres of monocrotophos 36SL in 250 litres of water/ha. As the pest is noctural in behaviour, the spray should be done in the evening hours for getting better results.

13. Rice Caseworm, *Nymphula depunctalis* Guenee (Lepidoptera: Pyralidae) (Fig. 13.9)

The insect is present in all the rice-growing tracts and may assume serious proportions in certain seasons in fields under swampy conditions. The adult is a small white moth with a wing expanse of 16 mm with pale brown wavy markings. The larva is light green with a light brownish-orange head.

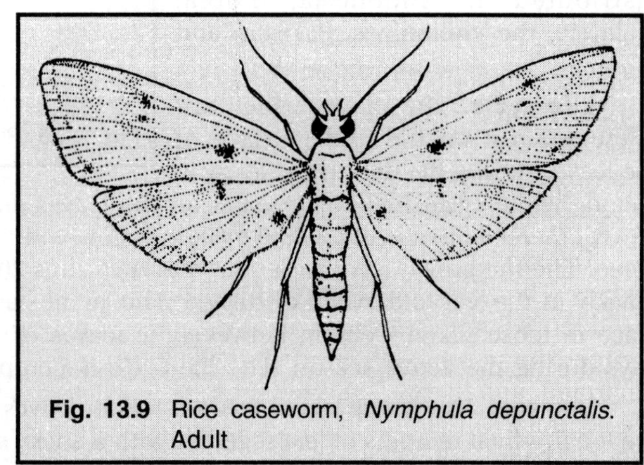

Fig. 13.9 Rice caseworm, *Nymphula depunctalis*. Adult

Life-cycle. The female lays about 50 eggs singly or in clusters of 4 on the undersurface of the leaves or grasses. The eggs hatch in

4-6 days. The young larvae feed by scraping the leaf surface. The larva makes a cylindrical tubular case out of a portion of the leaf cut out from it and remains inside it, moving with the case on the leaves. It feeds on the leaves and becomes full-grown in 18-22 days. It pupates inside the case which is attached to the base of the tiller and the adult moth emerges in 5-6 days. The life-cycle is completed in about 5 weeks.

Damage. The caterpillars cause the damage by feeding upon the leaves. They lead a semi-aquatic life floating in their tubular cases on the surface of water or attached to the stems at or above the water level. The early stages of the crop are damaged by the pest. The leaf blades are eaten away completely leaving only the midribs. In case of severe attack, the tillers become stunted and loose their vigour and often the plants are killed.

Management. (*i*) Drain-off the water from the field to kill the floating larvae. (*ii*) Put some kerosene in the field water and dislodge the leaf-cases by shaking the plants by passing rope or by branches of thorny plant. (*iii*) Chemical control measures are same as in case of rice ear-cutting caterpillar.

14. Rice Gall Midge, *Orseolia oryzae* (Wood-Mason) (Diptera: Cecidomyiidae) (Fig. 13.10)

The rice gall midge or gall midge is found in most of the paddy growing areas in the southern and eastern parts of India. The pest is also present in Pakistan, South-east Asia and Africa. It is, however, extending the area of its distribution and it appeared in Uttar Pradesh for the first time during 1971. The maggots of this fly (**Fig. 13.10**), as a result of their feeding, produce galls and kill the growing shoots.

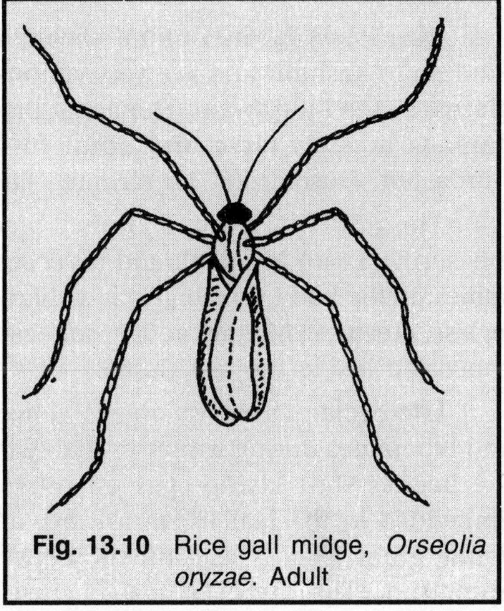

Fig. 13.10 Rice gall midge, *Orseolia oryzae*. Adult

Life-cycle. The insect breeds on a number of wild grasses and migrates to the paddy crop when it is in the tillering stage. It attacks the crop from May to September in Bihar and from mid-August to October in West Bengal. In the southern States, its maximum attack is between the third week of August and mid-September.

The tiny pink fly lays reddish elongated eggs about 0.5 mm in length, singly or in clusters, on the leaves or on stems. During August to November, they hatch in 1-3 days. The tiny maggots crawl down to the base of the shoot and enter a young bud where they feed for 10-13 days and then pupate. The pupa wriggles its way up with the help of abdominal spines and cuts a hole at the tip of the gall from which the fly emerges after 4-7 days. The life-cycle is completed in 15-23 days and the adults live for 3-4 days and lay up to 250 eggs. There are 3-5 overlapping generations on the same crop and 5-8 in a year. The insect probably over-winters in the larval stage in stubble in the northern and eastern parts of India. The period of its activity varies in different regions depending upon the weather during August-November.

Platygaster oryzae Cameron (Platygasteridae) is the most important parasitoid of gall midge maggots, causing 50-90 per cent parasitization depending upon humidity and rainfall.

Damage. In India, this pest attacks the paddy crop with varying intensity in most of the paddy regions. The maggot of this fly enters the stem and reaches the apical point of the central shoot or the tiller, where it develops. Owing to its feeding, some physiological changes take place which hamper the normal growth of the plant. The central leaf of an attacked tiller becomes hollow and deformed, and there is a swelling or gall formation on the basal portion. Its growth is stopped and the central leaf ultimately turns into a hollow outgrowth, giving a shining silvery colour called 'silver shoot'. The infested tillers do not bear ears. Infestation in the early growth phase is known to induce subsidiary tillering after the death of primary tillers. These tillers are also infested. In the long-duration varieties, infestation during the late vegetative phase results in branching of the tillers which bear ears and, thus compensate, to some extent, for the loss in yield. In case of severe infestation, losses upto 50 per cent have been reported.

Management. (*i*) Careful timing of planting can avoid damage; once past the tillering stage, the plant is not suitable as a host. (*ii*) Considerable build up of midge population on grasses near the rice crop can be avoided by removing the grasses. (*iii*) For gall midge endemic areas, seedling root dip in 0.02 per cent emulsion of chlorpyriphos for 12 hours before transplanting protects the crop for 25-30 days. (*iv*) Apply insecticides at economic threshold level of 1 gall per m^6 in endemic areas or 5 per cent affected tillers in non-endemic areas. Spray 20-45 days after transplanting 625 ml of chlorpyriphos 20EC or 500 ml of quinalphos 25EC in 750-1000 litres of water per ha. Alternatively, apply granules as mentioned for the control of yellow stem borer.

15. Rice Hispa, *Dicladispa armigera* (Olivier) (Coleoptera: Chrysomelidae) (Fig. 13.11)

Rice hispa is distributed throughout India from Andhra Pradesh to Kashmir and is a very serious pest of paddy at certain places. In the Punjab and Himachal Pradesh, this species causes damage both as larva and adult in the districts of Kangra, Gurdaspur, Hoshiarpur, Kapurthala, Ludhiana and Shimla.

The adult is a small bluish black beetle (**Fig. 13.11**), measuring 5 mm in length and is recognized by numerous short spines on the body, which give it a characteristic appearance. The legless, creamy-white larvae are not easily seen, because they are concealed inside the leaf tissue.

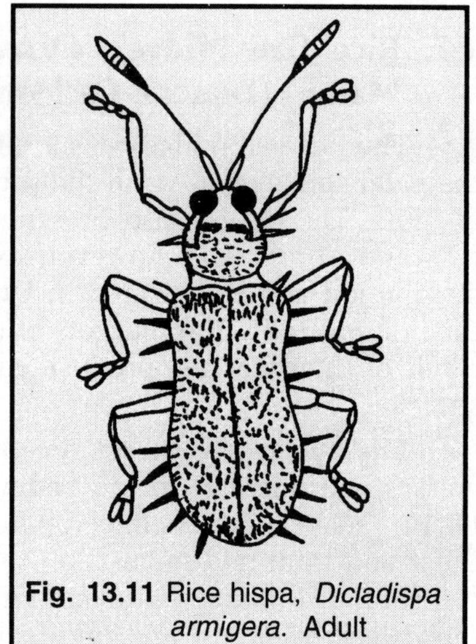

Fig. 13.11 Rice hispa, *Dicladispa armigera.* Adult

Life-cycle. This pest breeds actively from May to October and hibernates during winter probably in the adult stage. In May, the beetles start laying eggs on nursery plants. The eggs are embedded in the leaf tissue towards the tip. On hatching, the young grubs feed as leaf-miners, between the upper and lower epidermis. The attacked leaves turn membranous, showing characteristic blisters or blotches. Later on, the attacked leaves wither and die. When the larvae are full-grown, they pupate inside and finally emerge as black beetles. In Bihar, the pest is known to complete 6 generations in a year, but in the Punjab it probably completes 2-3 generations during the paddy season.

Damage. Apart from the damage caused by larvae as leaf-miners, the adults also feed on green matter and produce parallel whitish streaks on the leaves. The damage starts in nurseries and spreads to the rice fields. The infestation varies from 6 to 65 per cent.

Management. (*i*) The pest is suppressed if the infested leaf tips are clipped off and destroyed, while transplanting. (*ii*) If the nursery beds are flooded, the beetles float and can be swept together with brooms and then destroyed. (*iii*) Spray at economic threshold level (1 adult or 1-2 damaged leaves per hill) with 300 ml of methyl parathion 50EC or 1.4 litres of monocrotophos 36SL or 2.0 litres of quinalphos 25EC or 2.5 litres of chlorpyriphos 20EC in 250 litres of water per ha. If the attack continues, repeat spray after two weeks.

16. Rice Root Weevil, *Echinocnemus oryzae* (Marshall) (Coleoptera : Curculionidae)

The rice root weevil is a serious pest of rice in southern India. It was first recorded in the northern parts in 1953 at Sirsa (Haryana). It has also been observed to cause damage to the rice crop in Patiala District (Punjab). Besides paddy, it feeds on the roots of certain grasses.

Damage is caused by the grubs which feed on rootlets of paddy plants. They are translucent white and measure about 6 mm in length. There are six pairs of prominent tubercles on the dorsal side of the abdomen.

Life-cycle. This pest is active only from July to September and passes the rest of the period as pupa in the soil at depths of 8-20 cm. The weevils emerge in July with the first shower of rain and are seen sitting in large numbers on rice plants at this time. The eggs which are laid on the plant hatch in a few days. The grubs lead an aquatic life and feed on the root-hairs. Tubercles on the abdomen help them in respiration and they obtain oxygen from the air-spaces inside the roots of the host plants. Grubs are full-grown by the middle of September when they bury themselves deep into the soil for pupation. The pupae emerge next year in July and, thus, the pest completes only one generation in a year.

Damage. The grubs feed on root hairs of the transplanted crop, thereby affecting plant growth. The infested crop remains stunted and a large number of plants are killed. The crop transplanted in July is more heavily attacked than the one transplanted in August.

Management. Apply 7.5 kg of phorate 10G per ha in standing water.

17. Rice Grasshoppers, *Hieroglyphus banian* (Fabricius) (Fig. 13.12) and *H. nigrorepletus* Bolivar (Orthoptera : Acrididae)

Various species of grasshoppers are widely distributed in India. They are polyphagous and feed on leaves of rice, maize, millets, sugarcane, grasses, sunnhemp, *arhar*, etc. *H. banian* (Fig. 13.12) is sporadic pest of rice and other *kharif* cereals in the Punjab. Another species, *Oxya nitidula* Walker, which is smaller, also appears in pest proportions on the paddy crop in certain years. The other important species in northern India are *Aeolopus tamulus* Fabricius and *Acrida exaltata* Walker.

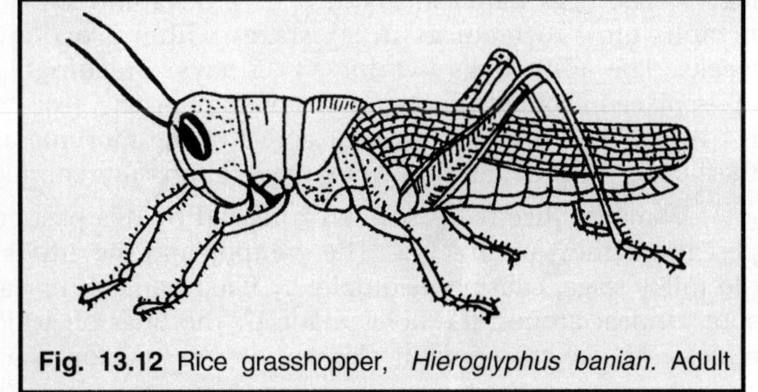

Fig. 13.12 Rice grasshopper, *Hieroglyphus banian.* Adult

Damage is caused by both adults and nymphs. *H. banian* and *H.nigrorepletus* are somewhat like locusts but are smaller. The adults are 40-50 mm long and are shining greenish yellow, having three black lines running across the pronotum. Nymphs are yellowish, with many reddish brown spots in the early stages, but become greenish as they grow older.

Life-cycle. *Hieroglyphus* spp. have one generation in a year and pass the winter and dry part of summer in the egg stage. The eggs are found in the soil and they hatch in June or in early July, a few days after the first shower of the monsoon. On emergence, the nymphs start feeding actively and complete their development in seven stages, within 3 weeks. The adults are seen feeding voraciously during August and September. When they are two months old, they mate. The female starts laying eggs by inserting her abdomen in the soil. The eggs are laid 5-8 cm deep, in pods, each containing 30-40 eggs. The egg-laying continues from September to November and the adults die soon after, sometime in the winter.

Damage. The greatest amount of damage is caused during August-September when both adults and nymphs feed on paddy and other crops, causing defoliation. In certain years, they cause extensive damage, moving from field to field over large areas.

Management. Dust carbaryl 5 per cent or malathion 5 per cent @ 25 kg per ha.

18. Rice Bug, *Leptocorisa acuta* (Thunberg) (Hemiptera: Coreidae) (Fig. 13.13)

The rice bug, commonly known as *gundhy* bug, is widely distributed in India, the Orient and Australia. It is a serious pest of rice in Uttar Pradesh and some other parts of India. Apart from rice, it also feeds on maize, millets, sugarcane and some grasses. Both the adults and the nymphs cause damage by sucking sap from the leaves and ears of rice.

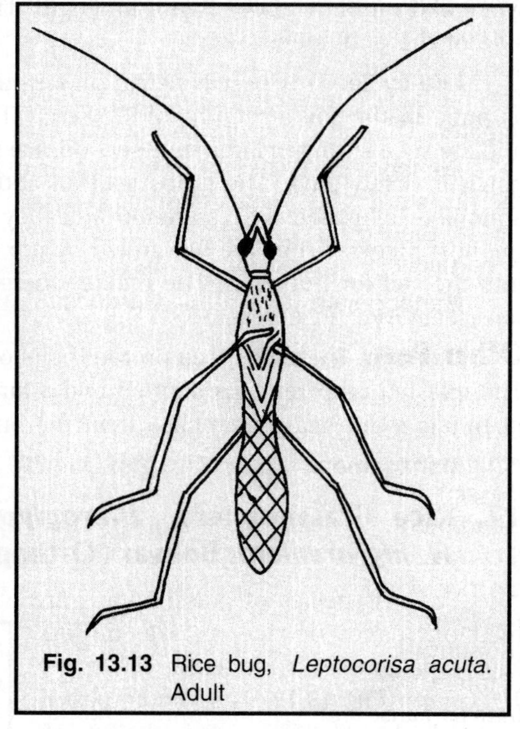

Fig. 13.13 Rice bug, *Leptocorisa acuta*. Adult

The adults (**Fig. 13.13**) are slender, about 20 mm long and greenish brown. They have long legs and antennae with four joints. The newly hatched nymph is about 2 mm long and is pale green. However, as it grows, the green colour deepens. The grown up nymphs are very similar to the adults in colour and size, but they are wingless.

Life-cycle. The rice bug breeds all the year round on grasses and various other green plants and appears in paddy fields generally in August and is most active from the middle of that month to November. The females lay 24-30 round yellow eggs in rows on the leaves. The eggs hatch in about 6 or 7 days and the nymphs grow to maturity in six stages within 2 or 3 weeks. The adult bugs live for 33-35 days. Breeding takes place in winter also, but at a slower rate. The pest is essentially diurnal and is most active in the morning and in the evening, seeking shelter during the hotter parts of the day. Many generations are completed in a year.

Damage. Rice fields severely attacked by this pest emit a repugnant smell which gives to this pest the name 'gundhy' bug. The nymphs and the adults suck juice from the developing grains in the milky stage, causing incompletely filled panicles or panicles with empty grains. Black or brown spots appear around the holes made by the bugs on which a sooty mould may develop. This pest attacks the rice crop periodically and the reasons for its abundance in certain years are not known.

Management. The population can be suppressed by killing the bugs by using light traps, collecting the adults with nets and destroying the weeds to remove alternative hosts. Chemical control measures are same as in case of rice grasshoppers.

19. Rice Mealybug, *Ripersia oryzae* Green (Hemiptera : Coccidae)

The rice mealy bug occasionally causes heavy damage to the rice crop in many countries including India, Bangladesh and Thailand. In India, the bug is found in Andhra Pradesh, Mysore, Orissa, Madhya Pradesh and West Bengal. Apart from rice, this insect can also survive on various graminaceous weeds growing in standing water or in swampy areas. The nymphs and the adult females, being wingless, look alike. They are plump, oblong, reddish white, soft bodied and have a distinct powdery coating on their body.

Life-cycle. In tropical and sub-tropical regions, this insect is found throughout the year in one developmental stage or the other. They reproduce parthenogenetically and the females either lay eggs or simply deposit nymphs in the outer leaf sheath. In India, the females have been reported to lay 126-319 eggs or nymphs, but in Thailand, the number varies from 60 to 280. The mature females have a life span of about 5 days.

The eggs and the young nymphs are whitish in colour and measure 0.3 and 0.4 mm in length, respectively. The duration of the egg stage may range from a few minutes to 24 hours. The nymphs instinctively stay within the egg sac or under the body of the mother for about 2 days. Then, they move away and establish themselves between the leaf sheath and the stem. There, they feed and complete the nymphal period in 15-34 days. After the last moulting, the female nymphs become sessile and start depositing eggs or nymphs at the same spot. The nymphs which are destined to become males, transform into winged adults which fly off without feeding and some also mate with females. Thus, the asexual and sexual cycles are completed side by side.

Damage. Damage is caused both by the nymphs and adult females by sucking plant sap from the rice stem. This results in stunted plant growth and yellowish curled leaves. When the attack is severe, the ear heads become smothered and are unable to grow out of their sheaths. The damage is particularly intense during drought conditions.

Management. Chemical control measures are same as in case of rice grasshoppers.

Other Pests of Rice

Other pests recorded in various parts of India and adjoining countries are the grasshoppers, *Oxya* spp. (Orthoptera : Acrididae); *Thaia subrufa* (Motschulsky) (Hemiptera : Cicadellidae); *Peregrinus maidis* Ashmead (Hemiptera : Delphacidae); the rice root aphid, *Rhopalosiphum rufiabdominalis* (Sasaki) and *Schizaphis graminum* (Rondani) (Hemiptera : Aphididae); *Fabrictilis australis* (Fabricius) (Hemiptera : Coreidae); *Nezara viridula* (Linnaeus) (Hemiptera : Pentatomidae); the rice thrips, *Stenchaetothrips biformis* (Bagnall) and *Haplothrips ganglbauri* Schmutz (Thysanoptera : Thripidae); the rice skipper, *Pelopidas mathias* (Farbricius) (Lepidoptera : Hesperiidae); the paddy caseworm, *Parapoynx stagnalis* Zeller (Lepidoptera : Pyralidae); paddy swarming caterpillars, *Spodoptera mauritia* (Boisduval), *S. exempta* (Walker) and *S. litura* (Fabricius) (Lepidoptera : Noctuidae); *Atherigona* spp. (Diptera : Muscidae); *Leucopholis* spp. (Coleoptera : Scarabaeidae); the root weevils, *Tanymecus indicus* Faust and *Hydronomidius molitar* Faust (Coleoptera: Curculionidae).

MAIZE (*Zea mays* L.; Family : Gramineae)

More than 130 insects have been recorded causing damage to maize in India. Among these, about half a dozen pests are of economic importance.

1. Maize Borer, *Chilo partellus* (Swinhoe) (Lepidoptera : Pyralidae) (Fig 13.14)

It is the most destructive pest of maize and sorghum in Sri Lanka, India, Pakistan, Afghanistan, Uganda, Central and East Afria. It is found throughout India. This insect has also been recorded on *bajra* (*Pennisetum typhoides*), sugarcane, Sudan grass, *baru* (*Sorghum halepense*), *sarkanda* (*Saccharum munja*) and some other grasses. Its caterpillars damage maize and sorghum by boring into the stems, cobs or ears. The grown up caterpillars are about 20-25 mm long and dirty greyish white, with black head and four brownish longitudinal stripes on the back. The adults are yellowish-grey moths, about 25 mm across the wings when spread (**Fig. 13.14**).

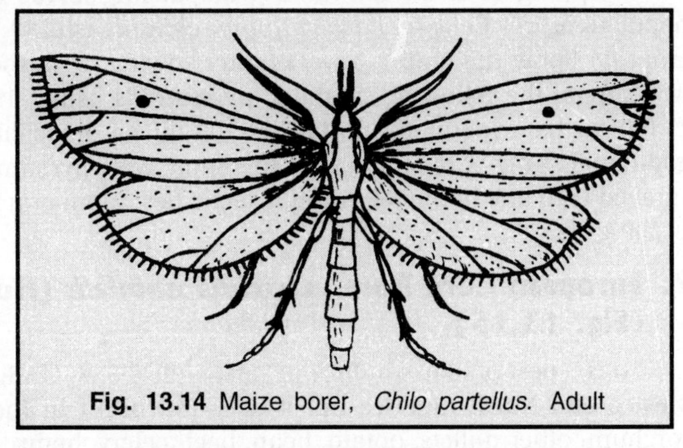

Fig. 13.14 Maize borer, *Chilo partellus.* Adult

Life-cycle. The insect breeds actively from March- April to October and for the rest of the year it remains in hibernation as a full-grown larva in maize and sorghum stubble, stalks or unshelled cobs. The larvae pupate sometime in March and emerge as moths in the end of that month or in early April. They are active at night, when they mate and lay eggs on the underside of the leaves of various host plants, particularly the early crops of maize and sorghum grown as fodder. The eggs are flat, oval, yellowish and are laid in overlapping clusters each containing up to 20 eggs. A female lays over 300 eggs during its life-span of 2-12 days and the eggs hatch in 4-5 days in summer. The young larvae first feed on the leaves, making a few shot holes and then bore their way downwards through the central whorl as it opens. More shot holes become visible, indicating an earlier attack and the plant also shows dead-hearts. In sorghum, the midribs are attacked by the young larvae, more than one plant being attacked by a larva. The larva becomes full-fed in 14-28 days, passing through six stages and after making a hole in the stem pupates inside it. The life-cycle is completed in about 3 weeks and there are probably 5 generations in a year. The full-grown caterpillars of the last generation hibernate in stubble, stalks, etc., and remain there till the next spring.

Damage. The freshly hatched larvae begin to feed on the tender leaves, making pin holes and leaf windowing (**Plate 13.2A**). After feeding for few hours on leaves, they enter through the mid whorl into the shoot. The caterpillar attacks all parts of the plant except the roots. In case of younger plants, the growing point and the base of the central whorl are infested resulting in the drying up of the central whorl and forming the dead heart (**Plate 13.2B**). In case of older plants, the stem is riddled with. The attacked plants remain stunted in growth and produce no grain. More than one caterpillars are found in a plant. Maximum damage is caused in the month of August to maize while during September and October to sorghum. The pest remains active in the field from March to November.

In Rajasthan, Delhi and southern districts of Haryana, it is a very serious pest of sorghum, particularly of the grain crop. Damage to sorghum fodder is perhaps equally great, but it is not generally noticed. In the central and northern districts of the Punjab, it is the most serious pest of maize. From 25-40 per cent of the young plants are distroyed and around Solan (Himachal Pradesh) up to 90 per cent of the plants have been found infested. Sorghum and *bajra* are also attacked at the time of grain formation.

Management. (*i*) The potential for carry over may be reduced by destroying the stubble, weeds and other alternate hosts of the stem borer by ploughing the field after harvest. (*ii*) Removal and destruction of dead-hearts and destruction of infested plants showing early pin-hole damage has been found to be successful practice in reducing the pest incidence. (*iii*) Destruction of crop residues and chopping of stems harbouring diapausing larvae could be very effective in reducing borer population. (*iv*) Release *Trichogramma chilonis* Ishii @ one lakh per ha on 10-15 days old maize crop. (*v*) Spray the crop 2-3 weeks after sowing or as soon as borer injury to the leaves is noticed with any of the following synthetic pyrethroids using 150 litres of water per ha: fenvalerate 20EC @ 100 ml/ha, cypermethrin 10EC @ 100 ml/ha or deltamethrin 2.8 EC @ 200 ml/ha. Usually, no additional spray is required after the spray with pyrethroids. (*b*) Alternatively, the crop should be sprayed with any of the following insecticides: monocrotophos 36SL @ 275 ml/ha or carbaryl 50WP @ 250 g/ha.

2. European Corn Borer, *Ostrinia nubilalis* (Hubner) (Lepidoptera : Pyralidae) (Fig. 13.15)

It is a pest of maize in Europe, erstwhile USSR, UAR, Japan and other countries of the Far-East, Mexico and North America but it does not occur in Inda. Besides maize, it feeds extensively on sorghum, other millets, potato, bean, beet, celery, hemp, cowpea, soybean, aster, dahlias and some weeds. Its host range includes over 200 species of plants.

The eggs of this insect are flattened, scale-like and about 1 mm in diameter. They are white, when freshly laid, but change to pale yellow later on. The newly emerged larva is 1.5 mm long, has black head and pale yellow body, with rows of black or brown spots on the dorsal surface. The full-grown larva is 25.4 mm long and flesh or cream-coloured, with a faintly spotted dorsum. The pupa

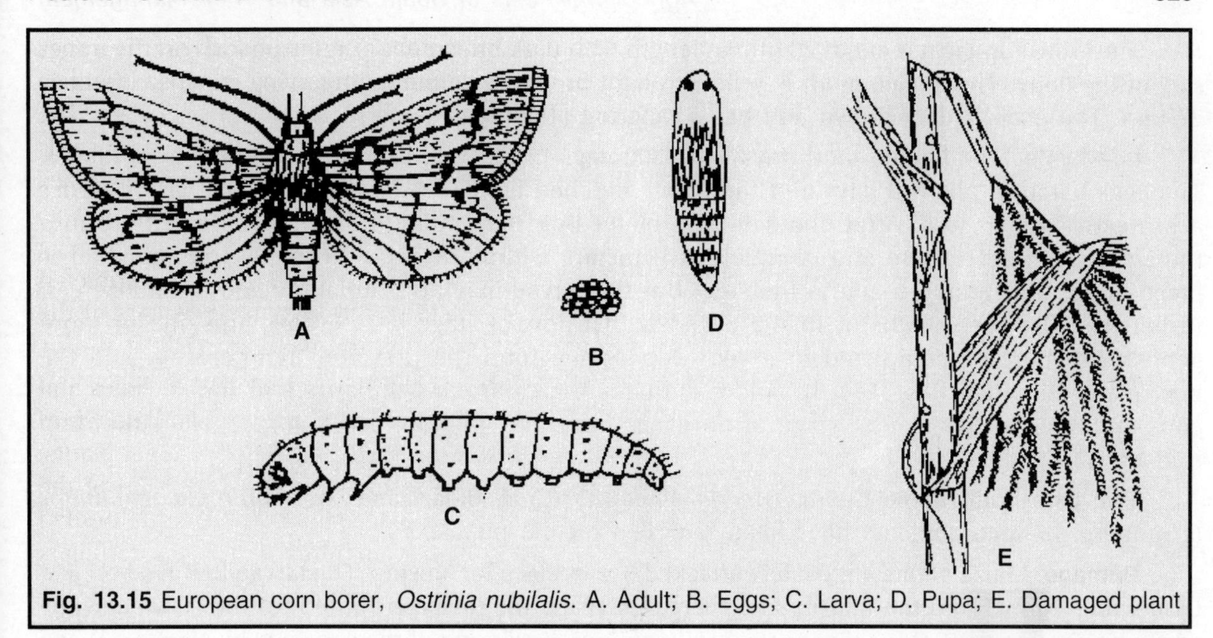

Fig. 13.15 European corn borer, *Ostrinia nubilalis*. A. Adult; B. Eggs; C. Larva; D. Pupa; E. Damaged plant

is 12.7-19.05 mm long and light to dark reddish brown. The female has pale yellow to light brown wings with an expanse of 31.75 mm. The male is slender, and is slightly smaller and has darker wings than those of the female.

Life-cycle. This insect passes through 1-5 generations depending upon the climatic conditions. The full-grown larvae pass the winter in diapause inside tunnels made in the stalks, stubble and cobs. Before changing into pupae, the larvae make circular exits for the moths to escape in the late spring or early in summer.

The female lays about 400-600 eggs in clusters of 3-50 on the underside of the leaves. The eggs hatch in 3-12 days, depending upon the temperature. The newly emerged larvae of the first generation wander and feed sparingly on the surface of the leaves in a whorl. In the third and fourth instars, they feed on the sheath and the mid-rib. Later in the fifth and sixth instars, they bore into the stalk. All the larvae do not survive owing to various causes. More than one larva may often be found in a single plant. As many as 167 eggs and 42 larvae have been recorded in a single stalk and 42 larvae in one cob. The larvae pass through 5 or 6 instars and become full- grown in 25-44 days. Pupation takes place within the tunnels made in the stalk, stubble or cobs. The pupal stage lasts 8-11 days. The female moth lives for 6-24 days, depending upon the weather.

Damage. The attack by the first-generation larvae results in a severer damage than that by larvae of the second or the subsequent generations. As a result of the attack, the plant weakens and the cobs get stunted. Tunnelling by larvae of the second generation weakens the grown up plants. This results in stalk breaking (**Fig. 13.15**), ear-dropping and shrivelled ears.

Management. The use of resistant or tolerant varieties of maize will minimize losses caused by this borer. The ploughing-under of the crop residues and sanitary measures result in the reduction of pest population, as the overwintering larvae are destroyed to a great extent.

3. Asian Maize Borer, *Ostrinia furnacalis* (Guenee) (Lepidoptera : Pyralidae)

Only the larvae of this pest are destructive and they make tunnels into the stalks and the ears. The pest is widely distributed in Sri Lanka, China, Taiwan, parts of India, Bangladesh, Indonesia, Japan, Korea, Vietnam, Laos, Cambodia, Malaysia, Myanmar, Thailand and the Philippines. It feeds on maize, sorghum, millets, Indian hemp, hops, *Artemisia* and many wild grasses.

The grown up larva is about 20 mm in length with dark brown head and is pinkish on the upper side of the body. The female moth is yellow to light brown in colour, with a wing expanse of about 27 mm. The male is dark brown and has a tapering abdomen.

Life-cycle. The female moth lays 500-1500 eggs in flat masses on the underside of leaves. The eggs hatch in about 3 days and the newly hatched larvae feed within the leaf whorl, under a protective silken web. After about 9 days of feeding they migrate downwards and bore into the stem. They feed there and complete five instars within 18 days. Normally, pupation takes place inside the tunnels made in the stem but the larvae may also pupate inside the maize ear, in-between the ear sheaths or in the ear stalk. The pupal stage lasts 6 days and one or more generations may be completed in a year depending upon the prevalent temperature and the host plant on which they feed. In colder climates the caterpillars hibernate in the stalks or the cobs and over-winter in that stage of development. They pupate next summer when the crop is ready in the field.

Two parasitoids namely, *Brachymeria obscurata* (Chalcididae) and *Xanthopimpla stemmator* (Thunberg) (Ichneumonidae) have been recorded on the pupae.

Damage. Maize plants are usually attacked 3-4 weeks after sowing. The larvae first feed on leaf tissue, then tunnel into the midrib, and thereafter reach the tassels or bore into the stalks and the ears. The development of the plants, the production of tassels and of the ears may be affected. When the pest attack is towards the base of the stalk the plant collapses, dries up and dies. At some places there are heavy crop losses.

Management. Same as in case of maize borer.

4. Sorghum Shoot Fly, *Atherigona soccata* Rondani (Diptera : Muscidae)

The shoot fly is a serious pest of maize, particularly on the crop grown during summer season. It attacks the very young (3-4 days old) seedlings, producing deformed, twisted and dead hearted plants. The pest has been described in detail under 'Sorghum'.

5. Hairy caterpillars, *Amsacta moorei* (Butler) and *Spilarctia obliqua* (Walker) (Lepidoptera : Arctiidae) (Chapter 26)

Other Pests of Maize

The maize crop is also attacked by Deccan wingless grasshopper, *Colemania sphenarioides* Bol. (Orthoptera : Acrididae); the aphids, *Rhopalosiphum maidis* (Fitch) **(Plate 13.2C)** and rusty plum aphid, *Hysteroneura setariae* (Thomas) (Hemiptera : Aphididae); maize jassid, *Zygnidia manaliensis* Singh (Hemiptera : Cicadellidae); sugarcane leafhopper, *Pyrilla perpusilla* (Walker) (Hemiptera : Lophopidae); thrips, *Anaphothrips sudanensis* Trybon, *Caliothrips graminicola* (B&C) and *Haplothrips gowdeyi* (Franklin) (Thysanoptera : Phlaeothripidae); termites, *Odontotermes obesus* (Rambur) and *Microtermes anandi* Holmgren (Isoptera : Termitidae); the pink borer, *Sesamia inferens* (Walker); *Helicoverpa armigera* (Hubner), *Spodoptera exempta* (Walker) and *Mythimna separata* (Walker) (Lepidoptera : Noctuidae); cutworms, *Agrotis* spp. and *Euxoa segetum* Denis & Schiffermuller; hairy caterpillars, *Amsacta albistriga* (Walker) and *A. lactinea* Cramer (Lepidoptera : Arctiidae); *Euproctis subnotata* (Walker) (Lepidoptera : Lymantriidae); cob caterpillar, *Cryptoblabes angustipennelta* (Hamp.) (Lepidoptera : Pyraustidae); cob borer, *Anatrachyntis simplex* (Walsingham) (Lepidoptera: Cosmopterygidae); leaf weevil, *Myllocerus discolor* Bohemann (Coleoptera : Curculionidae); blister beetles, *Mylabris macilenta* (Marshall), *M. phalerta* Pall., *M. pustulata* Thunberg and *M. liflensis* Billb. (Coleoptera : Meloidae); white grubs, *Holotrichia consanguinea* Blanchard, *H. serrata* (Fabricius) and *H. insularis* Brenske (Coleoptera : Melolonthidae).

SORGHUM [*Sorghum bicolor* (L.) Moench.; Family : Gramineae]

More than 150 species of insects have been reported to damage sorghum. However, over a dozen species are very serious and constitute a major constraint in sorghum production.

1. Sorghum Earhead Bug, *Calocoris angustatus* Lethiery (Hemiptera : Miridae) (Fig. 13.16)

It is one of the most destructive pests of sorghum in southern India. Both the nymphs and the adults feed in the green earheads. The adult is a small, slender, greenish yellow bug, measuring 5-8 mm in length and over 1 mm in width (**Fig. 13.16**). This bug has been recorded feeding on a number of cereals, millets and grasses, but its breeding is mainly restricted to sorghum, on which it assumes the status of a pest.

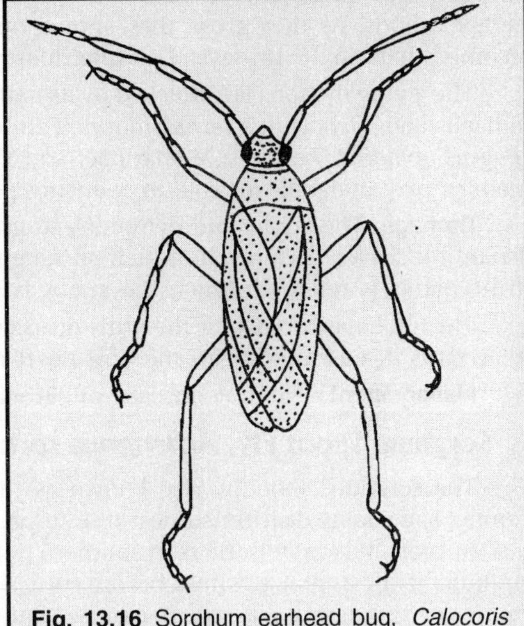

Fig. 13.16 Sorghum earhead bug, *Calocoris angustatus.* Adult

Life-cycle. The adults appear on sorghum crop as soon as the ears emerge from the leaf sheaths. The bug lays eggs under the glumes or in between anthers of florets, by inserting its ovipositor. The female lays 150-200 eggs which are cigar- shaped and measure about 1.5 mm.

The eggs hatch in 5-7 days and the nymphs start feeding on developing grains in the milk stage. The nymphs pass through 5 instars and develop into adults in about 3 weeks. The adults of the second generation are again ready to oviposit in the ears having developing grains which might be available on the same crop. As soon as the grains are ripe, the bugs stop multiplying on that crop. The insect completes its life-cycle in about one month and produces a number of generations in a year.

Damage. As a result of feeding by the bugs, the grains remain chaffy or shrivelled. When a large army of tiny nymphs feeds, the whole ear may become blackened at first and may eventually dry up, producing no grains.

Management. Spray 625 ml of malathion 50 EC or 3 kg of carbaryl 50 WP or 200 ml of phosphamidon 85 WSC in 500 litres of water per ha.

2. Sorghum Shoot Bug, *Peregrinus maidis* (Ashmead) Hemiptera : Delphacidae) (Fig. 13.17)

The nymphs and adults of this insect are serious pests of millets, sorghum and maize in south India. The insect has also been reported feeding on sugarcane, oats and various species of grasses. It is pan-tropical in distribution and has been recorded in Barmuda, West Indies, Hawaii, Cuba, East Africa and the Philippines.

The adult hopper is yellowish brown, with translucent wings and measures 3.2-3.8 mm in length (**Fig. 13.17**). The full-grown nymphs are light brown with prominent eyes and wing pads.

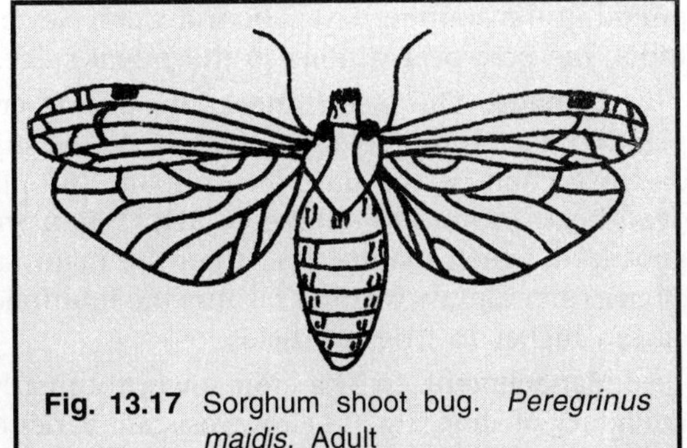

Fig. 13.17 Sorghum shoot bug. *Peregrinus maidis.* Adult

Life-cycle. This pest is active from September to January and is rather scarce from March to June. The adults, on emergence, mate and the females take 1-3 days before they start depositing eggs. The female bugs make slits in the midrib of the leaves with the help of their ovipositors and lay white, elongated and cylindrical eggs in groups of 1-4. Within the oviposition period of about 7 days, she can lay, on an average, 97 eggs.

The eggs hatch in 7-10 days and the young nymphs feed at first within the leaf sheaths and the leaf whorls. As they grow, they spread out to wider areas. After undergoing 5 moultings, they become adults in 16-18 days. The total life-cycle is completed in 3-4 weeks.

The young nymphs are attacked by its natural enemies, the lady birds, *Coccinella septempunctata* Linnaeus and *Menochilus sexmaculatus* (Fabricius) (Coccinellidae). The eggs are also parasitized by *Anagrus optabilis* (Perkins) (Mymaridae) and *Ootetrastichus indicus* Gir. (Eulophidae). These natural enemies play a significant role in reducing population of the insect.

Damage. The adults and nymphs feed gregariously within the leaf whorls, the leaf sheaths and also on the leaves. As a result of their sucking the cell sap, the leaves become yellow and the growth of the plants is retarded. When the attack is severe, the ears fail to emerge.

. The adults also transmit the virus diseases, 'corn moasic' and 'freckled yellow' of sorghum and other virus diseases of sugarcane and pearl millet.

Management. Same as in case of earhead bug.

3. Sorghum Shoot Fly, *Atherigona soccata* Rondani (Diptera : Muscidae) (Fig. 13.18)

The sorghum shootfly, also known as the sorghum stemfly, is a widely distributed pest in Europe, Africa and Asia. In India, it is more serious in southern parts. Besides sorghum, it infests maize, wheat, broom corn, small millets (*Panicum* spp.) and grasses. It causes damage to the seedlings as well as to the early stages of the crop. The maggots bore into the stem and cut the main shoot. The high-yielding hybrids are more susceptible to the attack of this fly **(Fig. 13.18)**.

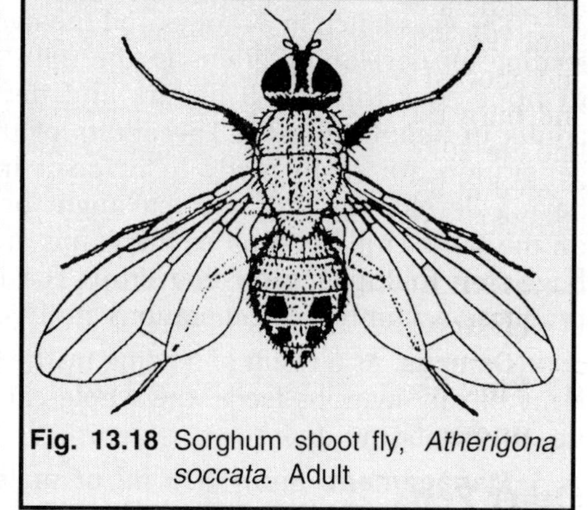

Fig. 13.18 Sorghum shoot fly, *Atherigona soccata.* Adult

Life-cycle. The female fly lays approximately 40 eggs singly on the underside of the leaves during its life span of about one month. The eggs are elongate, flattened and somewhat boat- shaped and are provided with two wing like lateral projections. The eggs hatch in 1-2 days and the tiny maggots creep out and reach in between the sheath and the axis, and bore into the stem. They feed inside the main shoot for 6-10 days and, when full grown, they may pupate either inside the stem or come out and pupate in the soil. The pupal period in the summer lasts about a week. Several generations are completed in a year. In northern India, the pest over-winters in the pupal stage.

Damage. The insect attacks the young crop when it is in the six leaf stage. Six weeks after planting, the crop is seldom attacked. As the maggots feed on the main shoot, the growing point is destroyed and by the time they pupate, the plant is almost dead. The young plants show typical dead-heart symptoms **(Plate 13.2D)**. When the attacked plants are somewhat older, tillers are produced, which mature later than the main crop. The total loss in yield is sometimes as high as 60 per cent. Cloudy weather favours the multiplication of this insect and it is believed that infestation is also higher in irrigated fields.

Management. (*i*) The crop sown from early June to the second week of July normally escapes the attack of shoot fly. (*ii*) Grow resistant varieties like CHS-7, CHS-8, Indian sorghum types IS-5566, 5285 and 5613. (*iii*) Seed coating with imidacloprid 600FS @15 ml/kg seed or isofenphos

5G @ 300 g/kg seed provides protection against shootfly up to 2 weeks. (*iv*) In case seed treatment has not been done, apply 12.5 kg of carbofuran 3G or 10 kg of phorate 10G per ha in furrows before sowing. (*v*) Alternatively, spray 1.25 litres of malathion 50EC or 2.0 kg of carbaryl 50WP in 500 litres of water/ha.

4. Sorghum Midge, *Stenodiplosis sorghicola* (Coquillett) (Diptera : Cecidomyidae)

This insect is distributed in all the sorghum growing tracts of the country and causes considerable losses. The adult fly is a very small (2 mm) fragile mosquito-like insect with a bright orange abdomen and a pair of transparent wings.

Life-cycle. The female inserts the eggs singly into developing florets at the time pollen is being shed and lays about 30-100 eggs. The eggs hatch in 2-3 days and the maggot feeds inside the developing grain and pupates there itself. The larva has four instars with a total duration of 10-11 days. It emerges between the tip of the glumes leaving the white pupal case attached to the tip of the floret which is very characteristic. The pupal stage lasts about a week and the total life-cycle is completed in 14-22 days.

Damage. The damage is caused by the maggots which feed on the ovaries and thus preventing the formation of grains. The infestation is dependent on humidity (more severe in low lying humid areas), initial infestation and duration of flowering of sorghum. The loss in yield varies from 20 to 50 per cent.

Management. (*i*) Both early and late maturing varieties of sorghum should not be grown in the same area as it would provide the pest a continuous supply of flowers. The varieties having the same flowering and maturity time would reduce midge damage considerably. (*ii*) Since the damaged and aborted seeds are the main source of carry over of the pest from one season to another, collect and burn the panicle and post-harvest trash. (*iii*) Spray 1 litre of malathion 50EC or 1.25 litres of lindane 20EC or 2.0 kg of carbaryl 50WP in 500 litres of water/ha. Only the earheads should be treated at 90 per cent panicle emergence followed by second spray after 4-5 days. (*iv*) Dusting the earheads with endosulfan 4% or carbaryl 5% or lindane 2% @ 12 kg/ha has also proved effective.

5. Stem Borer, *Chilo partellus* (Swinhoe) (Lepidoptera : Pyralidae) (Plate 13.2E, F) (Discussed under 'Maize')

6. Pink Stem Borer, *Sesamia inferens* (Walker) (Lepidoptera : Noctuidae) (Discussed under 'Rice')

Other Pests of Sorghum

The jowar and other millets are also attacked by a number of other insects which include Deccan wingless grasshopper, *Colemania sphenarioides* Bol. and *Hieroglyphus nigrorepletus* Bolivar (Orthoptera : Acrididae); *Amsacta* spp. (Lepidoptera : Arctiidae); the aphids, *Rhopalosiphum maidis* (Fitch), *Melanaphis sacchari* (Zehntner) and *Hysteroneura setariae* (Thomas) (Hemiptera: Aphididae); mealy bug, *Heterococcus rehi* (Lindinger) (Hemiptera : Pseudococcidae); whitefly, *Neomaskelliia bergii* (Signoret) (Hemiptera : Aleyrodidae); rice *gundhy* bug, *Leptocorisa acuta* (Thunberg) (Hemiptera: Coreidae); leafhopper, *Pyrilla perpusilla* (Walker) (Hemiptera : Lophopidae); plant bugs, *Nezara viridula* (Linnaeus), *Dolycoris indicus* Stal and *Menida histrio* (Fabricius) (Hemiptera : Pentatomidae); the thrips, *Anaphothrips sudanensis* (Trybom), *Sorghothrips jonnaphilus* (Ramakrishna) and *Florthrips traegardhi* Trybom (Thysanoptera : Thripidae); *Sesamia inferens* (Walker); *Spodoptera mauritia* (Boisduval) and *Helicoverpa armigera* (Hubner) (Lepidoptera: Noctuidae); the white borer, *Saluria inficta* W. (Lepidoptera: Phycitidae); the leaf-roller, *Marasmia trapezalis* (Guenee) (Lepidoptera : Pyraustidae) hairy caterpillars, *Amsacta albistriga* (Walker) and *A. moorei* (Butler) (Lepidoptera : Arctiidae); the blister beetles, *Zonabris phalarata* Pallas and *Lytta tenuicollis* (Pallas) (Coleoptera : Meloidae) and the white grubs, *Holotricha consanguinea* Blanchard, *H. insularis* Brenske and *Anomala bengalensis* Blanchard (Coleoptera : Scarabaeidae), and termites, *Odontotermes obesus* (Rambur) and *Microtermes obesi* Holmgren (Isoptera : Termitidae) **(Plate 13.2G).**

WHEAT (*Triticum* spp.; Family : Gramineae)

Wheat is comparatively less susceptible to insect pests in the field. However, in recent years, about half a dozen pests have become quite serious.

1. Wheat Aphid, *Sitobion miscanthi* (Takahashi) (Hemiptera : Aphididae) (Plate 13.2H)

Wheat aphid attacks wheat, barley, oats, etc., and is widely distributed in India. Like other aphids, the nymphs and adults suck the sap from plants, particularly from their ears. The insects are green, inert, louse like and appear on young leaves or ears in large numbers during the cold and cloudy weather. The nymphs and the females look alike, except that the latter are larger. The winged forms appear only in early summer.

Life-cycle. Like other winter aphids, the wheat aphid breeds at fast rate during the cold weather and reaches the height of its population in February-March when the ears are ripening. The females give birth to young ones and are capable of reproducing without mating. During the active breeding season, there are no males and the rate of reproduction is very high. When the wheat crop is ripe and the summer is approaching, the winged forms of both males and females are produced and they migrate to other plants like *doob* grass (*Cynodon dactylon*). It is not known how the pest passes the summer and the monsoon season. In October-November, the aphids again appear on wheat. If available, barley is preferred to wheat. The losses due to aphids have been reported upto 36 per cent.

Damage. These plant lice suck sap from the ears and tender leaves, and decrease yield of the crop. The damage is particularly severe in years of cold and cloudy weather. A heavily manured, well-irrigated and succulent crop will harbour the pest for a longer period and suffer greater damage.

Management. Spray 100 ml of imidacloprid 200SL or 50*g* of thiamethoxam 25 WG or 30 *g* of clothianidin 50WDG or 375 ml of dimethoate 30EC or oxydemeton methyl 25EC in 250 litres of water per ha. Control aphids at ear head stage at economic threshold level of 5 aphids/ear head. Since the aphids appear first on the borders of the crop, spray only the infected strip to check their further spread.

2. Armyworm, *Mythimna separata* (Walker) (Lepidoptera : Noctuidae) (Fig. 13.19, Plate 13.2I)

The armyworm is a pest of graminaceous crops all over the world. In India, it is a sporadic pest of wheat, sugarcane, maize, *jowar, bajra* and other graminaceous crops. It has gained prominence as a pest of wheat only recently, particularly after the introduction of Mexican varieties. The larvae feed voraciously and migrate from one field to another.

Life-cycle. The adult moths of armyworm are pale brown **(Fig. 13.19)**. They live for 1-9 days and lay eggs singly in rows or in clusters on dry or fresh plants or on the soil. The eggs are round, light green, when freshly laid, and turn pale yellow and finally black. In the Punjab, they hatch in 4-11 days from March to May, and in 19 days in December-January. Freshly emerged larvae are very active, dull white and later turn green. In the spring, the larval stage is completed in 13-14 days, but in the winter it is prolonged to 88-100 days.

In the pre-pupal stage, the insect spins a cocoon. The pre-pupal stage lasts 1-11 days during January to May. Pupation usually takes place in the soil at a depth of 0.5-5 cm, but it may also occur under dry leaves among the stubble or fresh tillers. Generally, the larvae before pupation seem to select sites near the water-channels.

The pupal stage is completed in 9-13 days in May and 36-48 days in the winter months. The survival of the pupae depends on the soil moisture. In one study made in the Punjab, the maximum

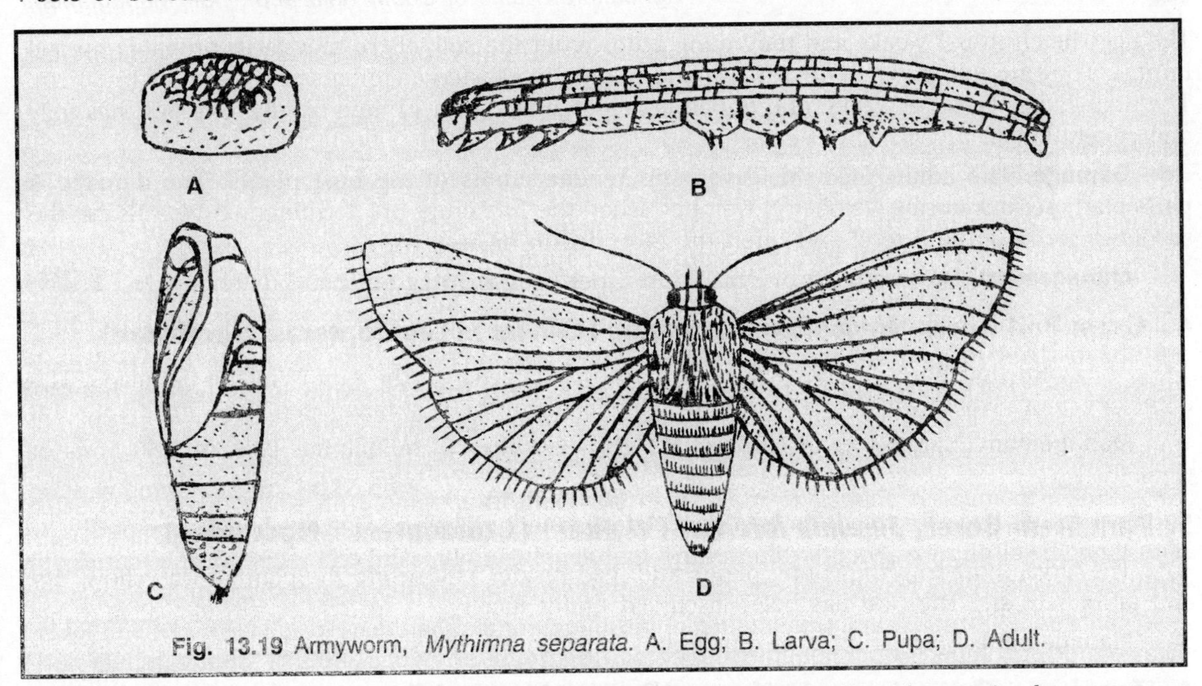

Fig. 13.19 Armyworm, *Mythimna separata*. A. Egg; B. Larva; C. Pupa; D. Adult.

population of caterpillars on wheat was found in March. From May to February in the next year, the population remains rather low and the insects feed on maize, sugarcane, *baru* grass (*Sorghum halepense*) and other crops. The population build-up starts in the beginning of March and increases with the rise of temperature in the spring.

Damage. The freshly emerged larvae spin threads from which they suspend themselves in the air and then with the help of air currents reach from one plant to another. In the early stages, they feed on tender leaves in the central whorl of the plant. As they grow, they are able to feed on older leaves also and skeletonize them totally. The grown-up caterpillars throw out faecal pellets, which are quite prominent.

In the case of a severe attack by the armyworms, whole leaves, including the mid-rib, are consumed and the field looks as if grazed by cattle. The pest may also eat away ears, including the awns and immature grains (Plate 13.2F). The yield losses upto 42 per cent have been reported from Ludhiana in Punjab.

Management. (*i*) The pest can be suppressed by collecting and destroying the caterpillars. (*ii*) Spray 500 ml of dichlorvos 85SL or 3 kg of carbaryl 50WP or one litre of quinalphos 25EC in 250 litres of water per ha.

3. Ghujhia Weevil, *Tanymecus indicus* Faust (Coleoptera : Curculionidae)

Ghujhia weevil is widely distributed in the Indian Sub-continent and is a sporadic pest of considerable importance, feeding on germinating *rabi* crops, particularly wheat, barley, gram and mustard.

The damage is caused by the adult weevils only and they cut the germinating seedlings at the ground levels. Weevils are earthen grey and measure about 6.8 mm in length and 2.4 mm in width. Their fore wings are oblong and hind wings are more or less triangular, but they cannot fly. Their mouthparts are light brown and are practically concealed beneath the head.

Life-cycle. The pest is active from June to December and passes the rest of the year as a grub or pupa in the soil. Weevils emerging in June mature sexually sometime in October. They mate frequently and lay 6-76 eggs in 5-11 instalments in the soil under clods or in crevices in the ground.

The eggs hatch in 6-7 weeks and the young grubs enter the soil where they feed, probably on soil humus. They are full grown in 10-18 days and pupate in earthen chambers at a depth of 15-60 cm. The pupal stage lasts 7-9 weeks, and the adults emerge next year in June or July. The pest has only one generation in a year.

Damage. The adults feed on leaves and tender shoots of the host plants. The damage is particularly serious during October-November, when the *rabi* crops are germinating. Weevils cut the seedlings at the ground level and often the crop has to be resown.

Management. Dust carbaryl or malathion 5 per cent @ 25 kg per ha.

4. Gram Pod Borer, *Helicoverpa armigera* (Olivier) (Lepidoptera : Noctuidae)

The gram pod borer attacks wheat at maturity. It feeds on the grains in the earheads. The damage is more where wheat follows cotton. The details of the pest have been given in Chapter 14.

Management. Spray 3 kg of carbaryl 50WP or 2.0 litres of quinalphos 25EC in 250 litres of water/ha.

5. Pink Stem Borer, *Sesamia inferens* (Walker) (Lepidoptera : Noctuidae)

The larvae bore into the young plants and feed inside. As a result, the plants turn yellow, dry and ultimately die. The pest has been discussed under Rice.

Management. Spray 2 litres of quinalphos 25EC in 250 litres of water/ha.

6. Termites, *Odontotermes obesus* (Rambur) and *Microtermes obesi* Holmgren (Isoptera: Termitidae) (Chapter 26)

Termites damage the wheat crop soon after sowing and near maturity. The damaged plants dry up completely and are easily pulled out. The plants damaged at later stages give rise to white ears.

Management. Treat the seed @ 4 ml of chlorpyriphos 20EC or 6 ml of fipronil 5SC or 3 ml of imidacloprid 600FS per kg of seed. Spread 100 kg of seed in thin layers on cemented floor, tarpaulin or plastic sheet and spray 400 ml of chlorpyriphos 20EC or 600 ml of fipronil 5SC or 300 ml of imidacloprid 600FS in 2.5 litres of water.

7. Molya Nematode, *Heterodera avenae* Wollen Weber (Tylenchida : Heteroderidae)

This nematode is widely distributed in Europe and Australia and has recently been recorded in Rajasthan, Haryana and the Punjab. It infests wheat, barley, oats and rye, and the attacked plants remain stunted and give a shrivelled unhealthy appearance.

Life-cycle. This nematode passes unfavourable season in the form of cysts, mostly in the soil. A cyst consists of the dead body of a female, containing a large number of eggs. When the conditions are favourable, eggs hatch within the cysts and the larvae are set free into the soil in the second stage of growth. The larvae may invade any underground part of a susceptible plant but most of them enter it at or near the root tips. After moving a short distance through the cortex, they assume a position, more or less parallel to the main axis of the root, with the head away from the tip.

The male increases in girth, until the width is equal to about 1/5th of its length and during this period it undergoes the second and third moultings. The body begins to elongate and becomes folded or coiled within the cuticle during the third stage. After assuming the final cylindrical shape, it moults for the fourth time and becomes an adult. The female does not undergo such metamorphosis, but after the second and third moultings it continues to increase in girth until it becomes ovate. It then undergoes the fourth or final moulting and emerges as a full grown adult. After mating, the eggs mature inside the body of the female and it dies, the body being converted into a cyst.

Damage. Although the nematodes cause very little mechanical injury to the roots, yet their presence stimulates the formation of branched rootlets. The main root remains short or bunchy, bearing small galls. In case of severe infestation, the seedlings may fail to come out of the soil and even if they grow a little, the infested plants remain stunted. The plants that escape the early damage, produce short stalks and ears, yielding a poor harvest.

8. Wheat Gall Nematode, *Anguina tritici* (Steinback) (Tylenchida : Tylenchidae)

The ear-cockle or *mamni* disease is caused by the nematode throughout the wheat-growing parts of the world. If the black rounded *mamni* galls are soaked in water overnight, the coat softens and a large number of larvae are set free. They can be observed under a microscope as wriggling thread-like creatures. Besides wheat, the nematode produces galls in rye, spelt and emer, but oats and barley are immune to its attack. The nematode is also the carrier of the bacterial yellow slime ear-rot (*tundu* disease) caused by *Corynebacterium tritici.*

Life-cycle. Under natural conditions, the dry galls either fall to the ground from the ripe ears or they are harvested and find their way to the stores along with the healthy produce. The galls may remain dry for long periods and yet the larvae within remain viable. A gall may contain from 800 to 30,000 larvae which revive and become active when the gall is moistened.

When wheat is sown, the galls become soft on imbibing moisture and the larvae are set free into the soil. From there, they reach the host plants, if available within a distance of one-third of a metre. They rise up the plant and find a site for feeding as free parasites on the young leaves and the growing-points. Later on, as the plants approach the earing stage, they penetrate into the primordia of the flower-buds and form the galls instead of normal seed.

In the developing galls, the larvae mature into males and females, as the case may be. A single gall at this stage may contain 40 females and an equal number of males. They mate within the gall and the gravid females lay a large number of eggs. The young larvae on emerging from the eggs develop up to the second stage and then become dormant. They remain in that state in the dry galls till the next sowing season. There is only one generation in a year.

Damage. Affected plants are more or less stunted and their leaves are wrinkled, rolled or twisted. A variable number of grains in an infested ear may produce galls. The diseased ears are shorter and thicker than the healthy ones and the glumes are spread farther apart.

Management. (*i*) The wheat gall nematode can be controlled by separating the galls from the wheat seed by floating them on water in a tub. The galls, being lighter, float on the surface and may be skimmed off. The seed should then be dried before sowing.

(*ii*) The pest can also be suppressed by sowing clean seed in uninfested soil. Only one year's fallowing is sufficient to eradicate this nematode from the fields.

Other Pests of Wheat

The other pests of wheat are the aphids, *Sitobion avenae* (Fabricius), *Schizaphis graminum* (Rondani), *Rhopalosiphum maidis* (Fitch), *R. padi* (Linnaeus) and *R. rufiabdominalis* Sasaki (Hemiptera: Aphididae); *Laodelphax striatella* (Fallen) (Hemiptera : Delphacidae); *Pyrilla perpusilla* (Walker) (Hemiptera : Lophopidae); the jassids, *Amrasca* spp. (Hemiptera : Cicadellidae); the wheat bug, *Eurygaster maura* (Linnaeus) (Hemiptera : Pentatomidae); the wheat thrips, *Anaphothrips favicinctus* Karny (Thysanoptera : Thripidae); the cut worms, *Agrotis* spp. (Lepidoptera: Noctuidae); *Marasmia trapezalis* (Guenee) (Lepidoptera : Pyraustidae); the shootfly, *Atherigona naqvii* Steyskal and *A. orzae* Malloch (Diptera : Muscidae); *Hydrellia griseola* (Fallen) (Diptera : Ephydridae) and flea beetle, *Chaetocnema basalis* Baby (Coleoptera : Chrysomelidae).

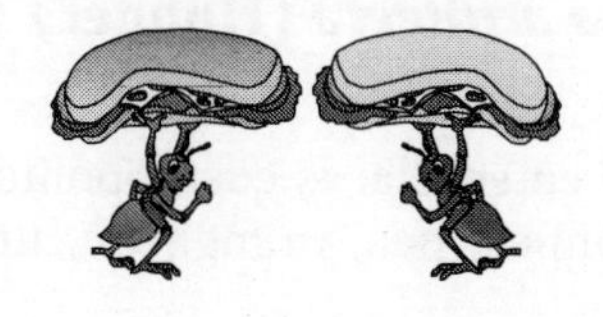

PESTS OF PULSE CROPS

INTRODUCTION

There are a large number of pulse crops, most of which belong to family Leguminoseae. Pulses are rich in proteins, particularly in the essential amino acid, lysine, which is rather deficient in the cereals. Therefore, in the vegetarian diet, a mixture of cereals and pulses along with some milk and eggs gives a balanced diet. In India, the area under pulses has increased from 22.76 million ha in 2004-05 to 24.46 million ha in 2011-12 but decreased to 23.47 million ha in 2012-13. The total production of pulses was 18.24 million tonnes in 2010-11 and was marginally lower at 17.09 million tonnes in 2011-12, but increased to 18.45 million tonnes in 2012-13. The country is estimated to have produced 19.57 million tonnes of pulses in 2013-14 and the target has been slightly lowered to 19.50 million tonnes for 2014-15.

India is the largest producer and consumer of pulses in the world accounting for 33 per cent of the global area and 22 per cent of world production of pulses.Pulses as a whole have a wide range of climatic requirements and different pulses are grown under various climatic conditions throughout the country. Yield levels of these crops are not very encouraging and average yield of pulses in the country is low at around 694 kg/ha as against 1856 kg in USA, 1814 kg in Canada, 1583 kg in China and a world average of 900 kg/ha. One of the major constraints for low yield of pulse crops is the extensive damage caused by insect pests.

More than 250 insects have been recorded feeding on pulse crops. Of these, about one dozen insects including pod borers, stem borers, leaf-miners, foliage caterpillars, cutworms, jassids, aphids and whiteflies are the most important. Some polyphagous insects also feed on these crops and cause considerable damage.

INSECTS ATTACKING REPRODUCTIVE STRUCTURES

Pod Borers

1. Gram Pod Borer, *Helicoverpa armigera* (Hubner) (Lepidoptera : Noctuidae) (Fig 14.1, Plate 14.1A)

The gram pod borer or the gram caterpillar is cosmopolitan and is widely distributed in India. It is a serious pest of chickpea, pigeonpea, pea, mungbean, urdbean, lentil, soybean and cowpea.

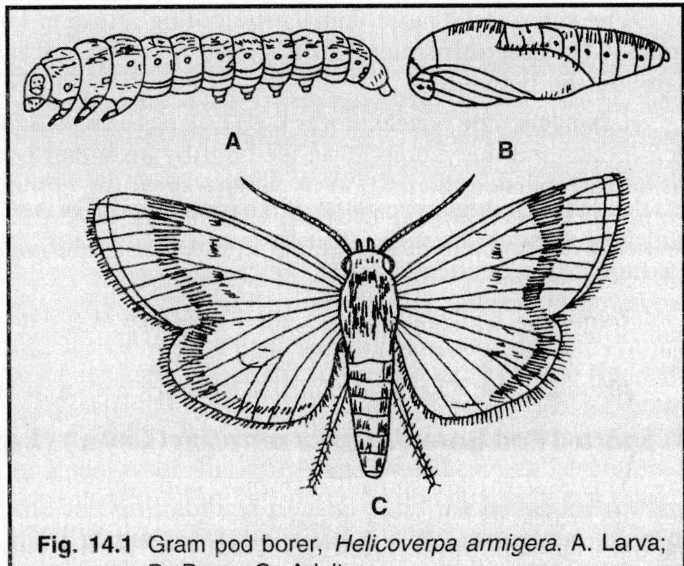

Fig. 14.1 Gram pod borer, *Helicoverpa armigera*. A. Larva; B. Pupa; C. Adult

The insect has also been found damaging cotton, sorghum, okra, maize, tomato, berseem and sunflower. In the United States of America, it is a well-known pest of corn (corn earworm) and cotton (cotton bollworm).

The moth is stoutly built and is yellowish brown. There is a dark speck and a dark area near the outer margin of each fore wing. The fore wings are marked with greyish wavy lines and black spots of varying size on the uppper side and a black kidney shaped mark and a round spot on the underside. The hind wings are whitish and lighter in colour with a broad blackish band along the outer margin. The caterpillars cause damage and, when full-grown, are 3.5 cm in length, being greenish with dark broken grey lines along the sides of the body.

Life-cycle. The females lay eggs singly on tender parts of the plants. A single female may lay as many as 741 eggs in 4 days. The eggs are shining greenish yellow and are round. They hatch in 2-4 days in April to October and 6 days in February and the young larvae feed on the foliage for some time and later bore into the pods and feed on the developing grains, with their bodies hanging outside. They move from pod to pod and are full-fed in 13-19 days and measure 35 mm in the last instar. The full-grown larvae come out of the pod and pupate in the soil. The pupa is dark brown and has a sharp spine at the posterior end. In the active season, the pupal period lasts 8-15 days, but in winter the duration is prolonged, particularly in northern India. There may be as many as 8 generations in a year. The caterpillars feed on their fellows if suitable vegetation is not available.

Campoletis chlorideae Uchida (Ichneumonidae) is a larval parasitoid of major importance. In vegetative phase, during the peak of its activity, it may parasitize 50-60 per cent of *H. armigera* larvae, whereas during podding phase, 30-40 per cent parasitization has been recorded.

Damage. Although they prefer food plants like gram and red gram, the larvae are polyphagous. They feed on the foliage, when young, and on the seed in later stages, and thus reduce yield. A single larva may destroy 30-40 pods before it reaches maturity.

Management. (*i*) Timely sowing, *i.e.*, upto mid October or growing early maturing cultivars which complete podding by first week of March in northern region helps in escaping peak activity period of *H. armigera*. (*ii*) Use of *Helicoverpa*-tolerant varieties like JG 315 and JG 74 for central zone and ICCV 7 for southern zone is recommended. (*iii*) Mixed intercropping with non-preferred host plants like barley, wheat, mustard and linseed should be preferred over sole crop. (*iv*) Apply nuclear polyhedrosis virus (NPV) @ 250-500 larval equivalents/ha. Spraying should be carried out in the evening hours. (*v*) Spray the crop at the appearance of the larvae at pod initiation with 500 ml of indoxacarb 14.5 SC or 250 ml of fenvalerate 20EC or 400 ml of deltamethrin 2.8EC or 200 ml of cypermethrin 25EC or 2 kg of acephate 75SP or 150 ml of spinosad 4.5 SC in 200-250 litres of water per ha and or dust 25 kg of malathion 5% per ha at the time of pod formation. Repeat treatment after two weeks if necessary.

2. Plume Moth, *Exelastis atomosa* (Walsingham) (Lepidoptera : Pterophoridae)

This insect is a specific pest of pigeonpea in many parts of India, particularly in Andhra Pradesh, Assam, Madhya Pradesh, Punjab, Tamil Nadu, Maharashtra and Karnataka. Outside India, it has also been recorded in Nepal and New Guinea.

The caterpillars cause damage by boring into and cutting through the pods while eating the grains. The full-grown caterpillar is greenish brown and measures 1.25 cm in length and has short hairs on the body.

Life-cycle. The female moths lay 17-19 eggs singly on tender parts of the plants. The eggs hatch in 2-5 days and the young larvae feed on the pods and become full-grown in 10-25 days. Pupation takes place outside the pod on its surface or in the entrance hole itself. The pupal period extends from 3 to 12 days. The life-cycle is completed in 17-42 days.The pest remains active throughout the year, provided suitable host plants are available, although its incidence remains greater in the monsoon season.

Damage. The larvae first scrape the surface of the pods and finally make holes into them and feed on the seeds, reducing crop yield.

Management. Chemical control measures are same as in case of gram pod borer.

3. Spotted Pod Borer, *Maruca testulalis* (Geyer) (Lepidoptera : Pyralidae) (Plate 14.1B)

This pest is wide spread in tropical and sub-tropical regions of the world. It is an important pest of cowpea, lablab, green gram, black gram, red gram, soybean, etc. The moth has a white cross band on the dark brown forewings and a dark border on the white hind wings. The larva is green with a brown head, short dark hairs and black warts on the body.

Life-cycle. The eggs are laid singly in the flowers or buds or on the pods of the host plants. After hatching, the young caterpillar enters the bud, flower or the pod. It feeds on the seeds within pods. The entrance hole is plugged with excreta. It pupates within debris or near surface of the ground. The moths shelter among lower leaves.

Damage. The larvae web together the flowers and feed on them and also bore into pods and feed on the seeds resulting in appreciable loss in yield of seeds.

Management. Chemical control measures are same as in case of gram pod borer.

4. Field Bean Pod Borer, *Adisura atkinsoni* Moore (Lepidoptera : Noctuidae)

This is a cold weather pest of lablab and red gram having a wide distribution in India. The adult is a pale yellowish-brown moth, with V-shaped specks on fore wings and pale brown markings on hind wings. The full-grown larva is brownish-green and about 2.5 cm in length.

Life-cycle. The eggs are laid in the setting flowers and very young pods. A female may lay as many as 290-950 eggs. The eggs hatch in 2-3 days and the larvae bore into the pods and feed on the seeds. The larval stage lasts 14-15 days and pupate in the soil. The pupal period is 8-16 days. During February to November, it hibernates in the pupal stage.

Damage. The larvae feed on flower buds and bore into the pods feeding on the developing seeds and cause considerable loss.

Management. Chemical control measures are same as in case of gram pod borer.

5. Lentil Pod Borer, *Etiella zinckenella* Treitschke (Lepidoptera : Phycitidae) (Fig. 14.2)

It is a serious pest of lentils (*Lens esculentus* Moench.) and green peas (*Pisum sativum* L.) in northern India and is also found on a variety of other pulses in various parts of the country, Myanmar and Sri Lanka. The tiny greenish caterpillars enter the pods and eat away the young grains. The full-grown larvae are rosy, with a purplish tinge. The moths are grey with a wing expanse of 25 mm (**Fig. 14.2**). The fore wings have dark marginal lines and are interspersed with ochreous scales.

Life-cycle. The moths emerge in February and March, and are nocturnal. The eggs are laid both singly and in clusters on various parts of the plant, including the pods. Under laboratory conditions, the eggs hatch in 5 days at 25°C, although in nature they may take up to 33 days, depending upon the climatic conditions. The newly emerged larvae feed on floral parts and subsequently, they bore into the pods to feed on the seeds. The larval stage is completed in 10-27 days. Pupation takes place in the soil at a depth of 2-4 cm and the pupal development is completed in 10-15 days. The pest breeds throughout the year and passes through 5 generations.

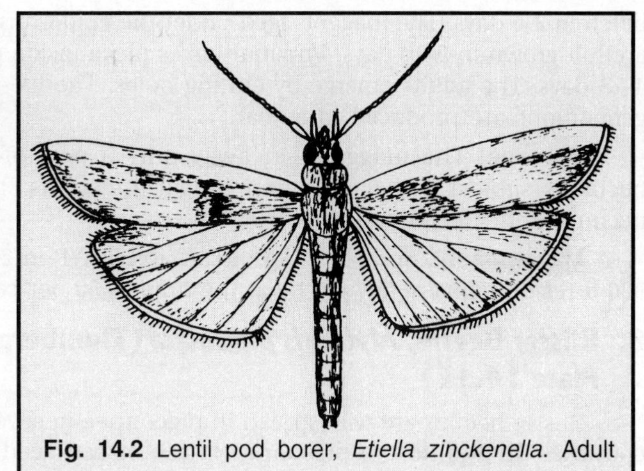

Fig. 14.2 Lentil pod borer, *Etiella zinckenella.* Adult

Damage. The larvae consume floral parts, newly formed pods and seeds inside the developing pods. The reduction in yield may be up to 5 per cent.

Management. At flower initiation, spray the crop with 2.25 kg of carbaryl 50WP in 200-250 litres of water per ha and repeat the treatment after three weeks, if necessary. Carbaryl should be used if the husk is to be fed to the cattle.

6. Pea Blue Butterfly, *Lampides boeticus* (Linnaeus) (Plate 14.1C) and *Catochrysops strabo* (Fabricius) (Lepidoptera : Lycaenidae)

Both the species of pea blue butterfly are found in Europe, Africa, Australia, and South and Southeast Asia. The adult of *L. boeticus* is a strong flyer with a jurky and rapid flight, and measures 24-34 mm in wing expanse. The male is dull purple with two black tornal spots on each hindroing. The female is brown with wing bases pale shining blue. Faint marginal spots and two tormal spots can be found on each hindwing of the female. The larva is pale green with a roughtened skin and measures 14-15 mm when full-grown.

Life-cycle. The eggs are laid singly on young shoots, flower buds or leaves of the host plant. The egg is disc like (0.5 mm dia) with a depressed micropyle. The eggs hatch in 4-7 days. The newly hatched larva makes its way to a flower bud and spends much of its time protected by the petals. It stays within the flower bud and feeds on the flower parts. The larva passes through four instars and the total larval duration is 9-27 days. Pupation takes place after one day of pre-pupal stage. The mature larva pupates on the leaves, twigs or on pods and the pupal period is completed in 7-19 days. The pupae can be found amongst the dry flowers and developing seed pods. The adults last for 2-6 days and the the entire life-cycle is completed in 20-55 days.

Damage. The larvae feed on flowers, buds, seeds and pods of red gram, cowpea, pea and dew bean and do considerable damage. The holes are seen in the damaged pods.

Management. Same as in case of lentil pod borer.

7. Red Gram Pod Fly, *Melanagromyza obtusa* (Malloch) (Diptera : Agromyzidae) (Fig.14.3, Plate 14.1D)

This pest occurs wherever red gram (*arhar*) is grown in India but is most common in northern India. It is a small metallic-black fly (**Fig. 14.3**), whose tiny maggots bore into the pods and feed on seeds.

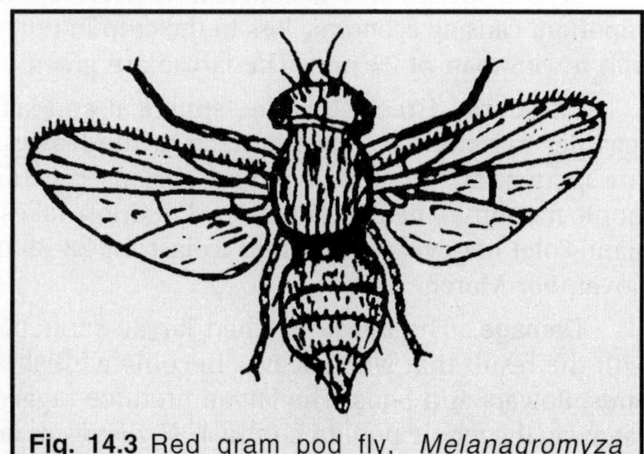

Fig. 14.3 Red gram pod fly, *Melanagromyza obtusa.* Adult

Life-cycle. The adult female fly thrusts its minute eggs into the shell of a tender pod. They hatch in 2-4 days. The maggots feed under the epidermis for some time and then enter the seed. They are full-grown in 5-10 days. Pupation takes place inside the damaged pods and the pupal period lasts 4-13 days. The adults emerge by cutting holes. The life-cycle is completed in 11-27 days and several generations are produced in a year.

Damage. The maggots eat away only a part of the seed and the partially damaged seed becomes subject to bacterial and fungal infections. The damaged grains do not germinate and become unfit for human consumption.

Management. Spray 1.75 kg of carbaryl 50WP in 200-250 litres of water on dwarf varieties and 300 litres on tall varieties per ha at the time of 50 per cent flowering. Repeat the spray after 15 days.

8. Blister Beetle, *Mylabris pustulata* (Thunberg) (Coleoptera : Meloidae) (Fig. 14.4, Plate 14.1E)

Blister beetles are widespread in pigeonpea in several Asian countries. In addition to pigeonpea, these insects feed on floral parts of several other plants. *M. pustulata* adults measure about 25 mm in length and have red and black alternate bands on the elytra. Other species may vary in size but all are brightly coloured.

Life-cylce. Eggs are usually laid in the soil and the diet of the larvae consists of other soil insects, including major pests. Thus while the adults may cause considerable damage, the larvae are beneficial. The pest has been found to be active during August to October, particularly in southern India.

Damage. Adult beetles feed voraciously on flowers and tender pods greatly affecting pod setting. This may have a significant impact on yield, particularly of short duration varieties In locations where pigeonpea is a primary crop, the after effect of blister beetles is inconsequential because their numbers are diluted over a large area. Pigeonpea genotypes that flower early or crops cultivated on small holdings may suffer substantial injury inflicted by these insects.

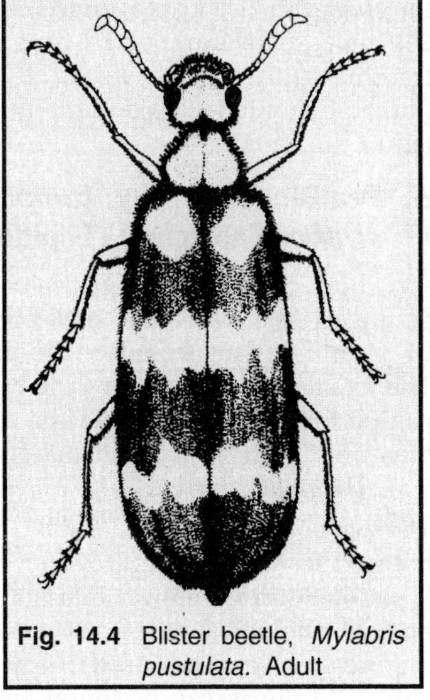

Fig. 14.4 Blister beetle, *Mylabris pustulata.* Adult

Management. (*i*) Manual picking and destruction of adult beetles is quite useful. (*ii*) Spray 500 ml of indoxacarb 14.5SC or 2 litres of acephate 75SP or 500 ml of deltamethrin 2.8EC in 200-250 litres of water/ha.

9. Semilooper, *Autographa nigrisigna* (Walker) (Lepidoptera : Noctuidale)

Several species of semiloopers have been reported to feed on chickpea, but *A. nigrisigna* is the most important causing economic loss to the crop in northern India. Adults have typically patterned forewings with a wingspan of 25 mm. The larvae are green and reach a length of 25 mm when full-grown.

Life-cycle. Greenish white, spherical and sculptured eggs are laid singly or in clusters of 2-15 eggs per mass on the lower surface of the leaves or on the tender parts of stem of the host plant. The incubation, larval and purpal periods last for 3-4, 9-27 and 6-8 days, respectively. The larvae moult five times before pupation. Pupation takes place in the silken cocoon hanging on the host plant Total life-cycle is reported to last for 28-39 days. Four generations have been recorded during November-March.

Damage. The newly hatched larvae scratch the chlorophyll from the leaves of the host plant with the result that whole leaves become whitish and skeletonized. Grown up larvae feed on leaves, buds, flowers and pods. The larvae produce ragged, irregular damage to the pod walls and normally eat away the whole pod. In contrast, *H. armigera* larva makes a neat, round hole in the pod and feeds on seeds inside

Management. Spray 500 ml of dichlorvos 100 in 200-250 litres of water/ha.

Pod Sucking Bugs

10. Tur Pod Bug, *Clavigralla gibbosa* Spinola (Hemiptera: Coreidae)

This insect is widely distributed in the Indian Sub-continent and adjoining countries including Myanmar. It feeds on pigeonpea, lab-lab, cowpea, chickpea, kidney bean, cluster bean, etc. In India, it has been recorded from Delhi, Orissa, Tamil Nadu, Uttar Pradesh, Maharashtra, Madhya Pradesh, Andhra Pradesh, Karnataka and Bihar. Both the nymphs and adults suck cell sap and are harmful to the crops.

The adult bugs are greenish brown in colour, having a spined pronotum and the femur, swollen at the apical end. The bugs are about 20 mm long. The young nymphs are reddish and show prominent lateral spines on the prothoracic and abdominal segments.

Life-cycle. The bugs appear on crops in October. They mate several times in end to end position and the copulation peroid varies from an hour to nearly 24 hours. After a lapse of about 11 days, which is the pre-oviposition period, the females lay eggs. The eggs are usually laid on pods and, less frequently, on leaves or the floral buds, in clusters of 5-25 each. A female, on an average, lays 60 eggs during an oviposition period of 15 days.

The eggs hatch in about 8 days and the newly-hatched nymphs move away from the egg shells within 10-15 minutes and gather together at a suitable feeding spot. They are gregarious in nature and are seen feeding in groups. The nymphs take about 17 days to complete development, after passing through five nymphal stages. The pest is active from the middle of October to the end of May and completes six overlapping generations during this period.

A minute insect, *Hadronotus antestiae* Dodd (Scelioni-dae), parasitizes the eggs of this pest. It has been observed in nature that upto 55 per cent of the eggs might be attacked and destroyed.

Damage. Both the adults and the nymphs suck cell sap from the stem, leaves, flower-buds and pods. As a result of this damage, the pods show pale yellow patches and later on shrivel up. The grain inside remains small in size and the yield may be reduced significantly.

Management. Spray 1.25 litres of monocrotophos 40EC or one litre of trichlorphon 50EC in 625 litres of water per ha. Two sprays during flowering and three during pod formation are quite effective.

11. Pod Bug, *Riptortus pedestris* (Fabricius) (Hemiptera: Coreidae)

This is a pest of red gram, cowpea, lablab, soybean and other pulses, and is widely distributed. The adult is a dark-brown bug measuring 15 mm long.

Life-cycle. The adults mate 2-3 days after emergence and repeat it throughout life. Egg laying starts 12-14 days after emergence. A female lays, on an average, 115 eggs during on ovipositional period of 30 days. Eggs are laid singly on pods at their base and hatch in 3-4 days. There are 5 nymphal instars in about 16 days. The adult lives for 45-47 days.

Damage. The damage is caused by nymphs and adults by sucking juice from seeds piercing the pods in the process. The seeds within the pods become rough and rugged. Tender pods when attacked do not develop.

Management. Same as in case of tur pod bug.

12. Green Potato Bug, *Nezara viridula* (Linnaeus) Hemiptera : Pentatomidae) (Fig. 14.5)

The green slink bug is cosmopolitian and is believed to have originated in Ethiopia. It is now widely distributed in tropical and subtropical regions of Europe, Asia, Africa and the Americas. The adult is shield-shaped with an overall dull green colour (**Fig. 14.4**). The adult has three white spots on the pronotum and measures about 15 × 8 mm. Newly hatched nymphs are deep orange and later instars turn into shades of black, green with multicoloured spots. It is highly polyphagous, feeding on plant species in more than 30 families. It causes considerable damage to various crops including cucurbits, crucifers, avocado, cocoa, pecan, sunflower, groundnut, okra, tomato, potato, rice, black gram, mungbean, ricebean, pigeonpea, cowpea and soybean. Stink bugs are well known for the obnoxious odour produced by the scent glands located on the metasternum.

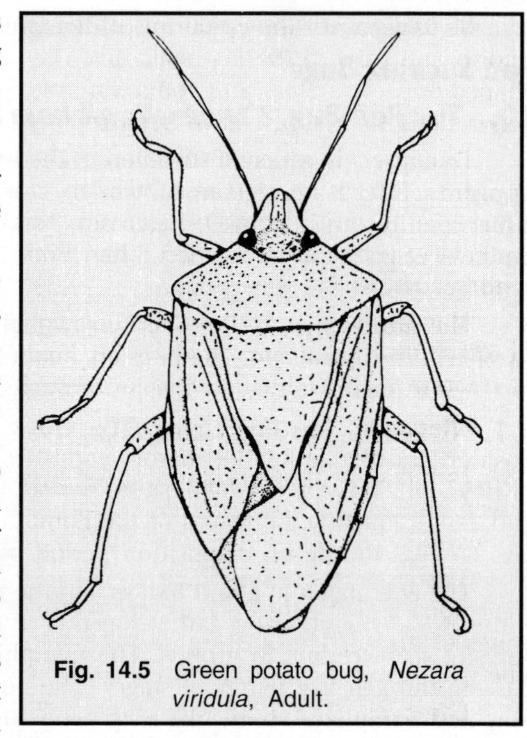

Fig. 14.5 Green potato bug, *Nezara viridula*, Adult.

Life-cycle. Eggs are laid in tightly packed, single layered rafts in the upper canopy of the herbaceous host plants, mostly under leaves or fruiting structures. The pre-oviposition, oviposition and postoviposition periods average 4.5, 13.0 and 7.6 days, respectively. Each female lays 165-290 eggs, with an average of 260. The eggs hatch in 6-7 days. The first instar nymphs do not feed and form light clusters at their natal site. Second and third instar nymphs also cluster, perhaps for protection, but they disperse if disturbed. Fourth and fifth instar nymphs do not aggregate. In all, there are 5 nymphal instars, which average 12.3, 9.0, 4.3, 10.9 and 10.6 days, respectively. The total nymphal period is 45.1 days and the total life-cycle from egg to adult emergence is completed in 53 days. The adult lifespan is 8.5 days for males and 25.1 days for females. The insect overwinters as an adult and hides in the bark of trees, leaf liter or other locations to protect from the weather. Adults leave their hibernation sites in spring, start feeding, mainly nocturnally, and soon mate and oviposit.

Damage. Both adults and nymphs cause damage and may attack all parts of the plant including the stems and leaf veins, but the bugs feed mostly on fruiting structures and growing shoots. The piercing and sucking mouthparts puncture the plant tissues and form minute, hard, brownish or blackish spots. Feeding punctures also provide access to fungal and bacterial infections. The bugs reduce seed yield, increase the proportion of seedless pods and decrease germinability of the seeds. Pods punctured during early endosperm formation are largely drained of their contents. Developing seeds from which the bugs have fed usually do not grow to full size, and are shrivelled and deformed.

Management. (*i*) Same as in case of the pod bug. (*ii*) Fourth and fifth instar nymphs and adults bask outside the plant canopy until about mid day, so application of insecticides is most effective at that time.

INSECTS FEEDING ON VEGETATIVE PARTS

13. Bean Fly, *Ophiomyia phaseoli* (Tryon) (Diptera: Agromyzidae)

The bean fly has been found to infest the streak bean (*Phaseolus vulgaris*), cowpea (*Vigna sinesis*), lima bean (*Phaseolus lunatus*), soybean (*Soja hispida*), *Cajanus, Canavalia* and *Dolichos* in Sri Lanka, India, China, Indonesia, Malaysia, the Philippines, Singapore and many African countries. It is a very minute insect, body length of the male being about 1.9 mm and that of the female 2.2 mm. The general colour is shiny black except for eggs, wing veins and antennae which are light brown. The maggot is creamy in colour and apodous in form.

Life-cycle. The flies are seen as soon as the host crops are available in the field. The slender, white eggs are laid singly in holes made on the upper surface of young leaves, especially near the petiole end of the leaf. On hatching, the maggot forms a short linear leaf mine and further on it tunnels underneath the epidermis of the leaf until it reaches one of the veins which leads it to the midrib and then to the leaf stalk and the stem. Pupation takes place inside the stem. The barrel shaped pupae are black and about 3 mm long. The total life cycle takes 2-3 weeks. As many as seven generations of this pest have been reported during the active season of the pest infestation.

Damage. As a result of severe infestation, the leaves turn yellow, giving the plants a dry appearance. The stems turn brown, become swollen and break down. The spring crops usually suffer less infestation than the late summer crops in which infested plants may constitute over 70 per cent of the plant population. The attacked plants bear less pods which are mostly empty or else their seeds may be very small.

Management. (*i*) Apply 625 ml of monocrotophos 40EC or 750 ml of oxydemeton methyl 25EC in 625 litres of water per ha at 15-day intervals during flowering stage. (*ii*) Soil application of 10 kg of phorate 10G is effective upto 40 days of sowing.

14. Green Nettle Slug Caterpillar, *Thosea aperiens* (Walker) (Lepidoptera: Cochlidiidae)

In India this insect is a pest of sorghum, pigeonpea, red gram, *Eleusine coracana* (finger millet), *Cassia auriculata* (Tanner's cassia), cowpea and tamarind. In Sri Lanka, it is reported to feed on *Dunbaria heynei,* an herbaceous climber.

The greenish larvae which are 30 mm in length, feed on leaves. They have stinging hairs on their body and are thus a nuisance to the field workers, in addition to damaging the plants. Fore wings of the adult moths are dark, cinereous-brown, whereas the hind wings are palish in colour. The wing expanse may vary from 30 to 35 mm.

Life-cycle. The moths are active during October-November and lay eggs on host plants. The caterpillars on emergence feed on leaves till the end of December. When full-grown, they bore into the soil, construct a cocoon and hibernate there in the larval stage. During the following August-September, they pupate and subsequently emerge as moths. The period from the onset of pupation and the emergence as moths takes about 30 days. Normally, one life-cycle is completed in a year.

Damage. The larvae defoliate the plants. The field workers when accidentally touch the stinging hair of the caterpillars may develop painful rashes on their hands and arms.

Management. Chemical control measures are same as in case of gram pod borer.

15. Girdle Beetle, *Obereopsis brevis* (Gahan) (Coleoptera : Cerambycidae)

This is an important pest of soybean and also attacks lablab and cowpea. The adult is a small black beetle.

Life-cycle. The ovipositing female beetle girdles the stem twice and makes 3 punctures just above the lower ring before inserting a single egg through the largest whole into the pith. This results in dropping of the upper part of the stem. A female beetle lays 7-13 eggs and they hatch in 4-5 days. The larva tunnels upwards and downwards within the stem and a single larva can destroy the whole plant. The larval period lasts 34-47 days. Over wintering takes place as the full- grown larva within the feeding tunnel in a gall-like chamber near the base of the plant in the girdled portion of the stem which has fallen out or under plant debris. The pupal period is 8-11 days.

Damage. The female of this insect feeds on the xylem of the stem. The larvae further damage the stem and make tunnels inside and fill these with excreta. The leaves and the growing points dy up. The broken stems can be seen in the field.

Management. Same as in case of bean fly.

16. Leafhopper, *Empoasca kerri* Pruthi (Hemiptera : Cicadellidae)

Leaf hoppers belonging to genus *Empoasca* comprise over 100 species, but *E. kerri* is the most common species feeding on various pulse crops like pigeonpea, cowpea, mungbean, ricebean, etc.

Adults are small green insects, 2.5 mm long and fly when disturbed. The nymphs and adults have a similar shape and colour, but the nymphs do not have wings and run sideways when disturbed.

Life-cycle. Eggs are laid singly within the leaf veins on the upper surface of the leaf which hatch in 3-10 days. Nymphs feed on the lower surface of the leaves. Nymphal period lasts for 7-20 days. Adults live up to 3 months. Life-cycle is completed in 15-45 days and up to 10-12 generations are found in a year. Intermittent rainfall, cloudy weather and high temperature favour the multiplication of this pest and hence the damage caused during *kharif* season is severe than *rabi* or summer/spring crop.

Damage. Both nymphs and adults suck the sap from the plants which results in characteristic yellow discolouration of leaf edges and tips followed by cupping of the leaves. In case of heavy infestation, the leaves roll down at the edges, dry and fall down. Seedlings that have sustained considerable feeding by jassids may be stunted and have red-brown leaflets followed by defoliation.

Management. Same as in case of tur pod bug.

17. Groundnut Aphid, *Aphis craccivora* Koch (Hemiptera : Aphididae)

Several species of aphids are known to attack pulse crops, among which *A. craccivora* is the most prevalent species on pigeonpea, chickpea, lentil, peas and beans. Aphids colonize the young shoots, flowers and pods. Young leaves of seedlings become twisted under heavy infestation. Seedlings may wilt, particularly under moisture-stressed conditions. In case of severe infestation, leaves and shoots get deformed and stunted, and sticky honeydew may be deposited over the leaf surface. Chickpea stunt disease caused by pea leaf-roll virus is transmitted by *A. craccivora.* Stunt disease limits plant growth, rendering leaflets small and reddish brown. The details of the pest are given in chapter 15.

18. Cotton Whitefly, *Bemisia tabaci* (Gennadius) (Hemiptera : Aleyrodidae)

The pest has been discussed in chapter 16.

Management. Spray 100 g of thiamethoxan 25WG or 1.5 litres of triazophos 40EC in 200-250 litres of water/ha.

Other Pests of Pulses

The other insects, which attack pulses include leaf miner, *Chrotomyia horticola* (Goureau) (Diptera : Agromyzidae) : scale insects, *Ceroplastodes cajani* Maskell (Hemiptera : Coccidae); dusky cotton bug, *Oxycarenus laetus* Kirby (Hemiptera : Lygaeidae); the pea thrips, *Caliothrips indicus* (Bagnall) and bean thrips, *Megalurothrips distalis* (Karny) (Thysanoptera: Thripidae); termites, *Odontotermes obesus* (Rambur) (Isoptera : Termitidae); the polyphagous caterpillars, *Spilarctia obliqua* Walker, *Amsacta moorei* (Butler) and *A. albistriga* (Walker) (Lepidoptera: Arctiidae); the Tussock caterpillars, *Euproctis fraterna* Moore and *E. scintillus* Walker (Lepidoptera: Lymantriidae); the foliage caterpillars, *Agrotis* spp., *Spodoptera exigua* (Hubner), *S. litura* (Fabricius) **(Plate 14.1F)**, *Thysanopulsia orichalcea* (Fabricius) and *Anomis flava* (Fabricius) (Lepidoptera : Noctuidae); *til* hawk-moth, *Acherontia styx* (Westwood) (Lepidoptera: Sphingidae); leaf binder, *Cydia critica* Meyrick (Lepidoptera: Tortricidae); pod wasp, *Tanaostigmodes cajaninae* La Salle (Hymenoptera : Tanaostigmatidae) and bean mite, *Polyphagotarsonemus latus* (Banks)(Acarina: Tarsonemidae).

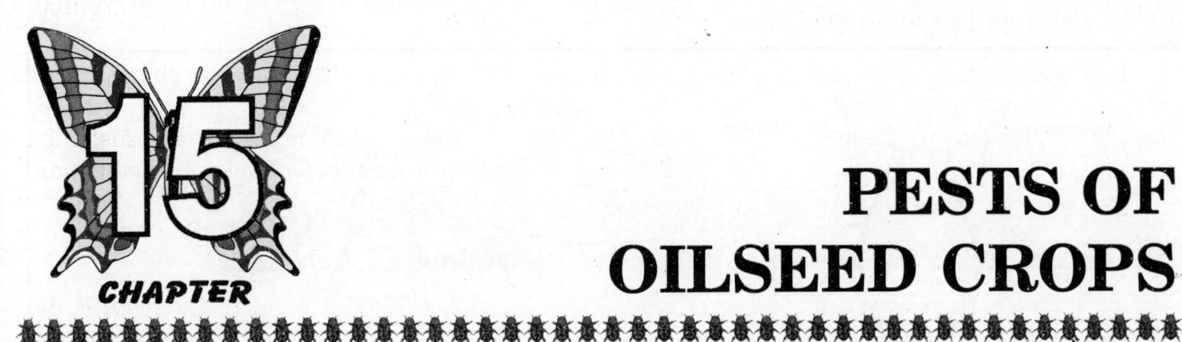

PESTS OF OILSEED CROPS

INTRODUCTION

Among the principal commercial crops grown in India, the oilseeds occupy a prominent place. They play an important role in the economy of the country because of their value in import substitution. The area under oilseeds has slightly declined from 27.52 million ha in 2004-05 to 27.22 million ha in 2010-11. The area further decreased to 26.31 million ha in 2011-12 and slightly increased to 26.53 million ha in 2012-13. The total production of oilseeds during 2010-11 touched a record level of 32.48 million tonnes but decreased to 29.8 million tonnes in 2011-12 and increased to 31.01 million tonnes in 2012-13. The production of oilseeds is estimated to be 32.41 million tonnes in 2013-14 and the target has been increased to 33 million tonnes for 2014-15. At present, India accounts for 8.4 per cent of the world's total output of oilseeds and it has world's largest area under oilseeds. However, its average yield of oilseeds is only 1000 kg/ha as against world average of 1,275 kg/ha and a yield of 2,500 kg/ha in USA.

The losses caused by insect pests of oilseeds are alarming and can be reduced to a great extent by adopting suitable plant protection meaures. These crops are damaged by a number of pests, of which mustard aphid, mustard sawfly and the painted bug are more serious. The aphid is the most serious pest on *Brassica* oilseeds throughout India. On the groundnut crop, the white grub has recently assumed serious proportions in Rajasthan, Gujarat, Maharashtra, parts of Karnataka and Uttar Pradesh. The leafminer and the red hairy caterpillar are serious in central and southern India. The groundnut aphid is a menace throughout the groundnut growing areas. Its incidence during different years varies with rainfall. Intermittent rains have a *depressing* effect on the aphid population.

BRASSICA CROPS (*Brassica* spp.; Family: Cruciferae)

1. Mustard Aphid, *Lipaphis erysimi* (Kaltenbach) (Hemiptera: Aphididae) (Fig. 15.1, Plate 15.1A)

The mustard aphid is worldwide and is a serious pest of cruciferous oilseeds like *toria, sarson, raya, taramira* and *Brassica* vegetables like cabbage, cauliflower, knol-khol, etc. The damage is caused by nymphs and adults, which are louse like, pale-greenish insects. They are seen feeding in large numbers, often covering the entire surface of flower-buds, shoots, pods, etc.

Life-cycle. This insect is most abundant from December to March when it infests various cruciferous oilseeds and vegetables. During summer, it is believed to migrate to the hills and there is some evidence that aphids also survive on abandoned stray plants of cabbage and on cruciferous weeds in the plains. The pest breeds parthenogenetically and the females give birth to 26-133 nymphs. They grow very fast and are full-fed in 7-10 days. About 45 generations are completed in a year. Cloudy and cold weather (20°C or below) is very favourable for the multiplication of this pest. The winged forms **(Fig. 15.1)** are produced in autumn and spring, and they spread from field to field and from locality to locality.

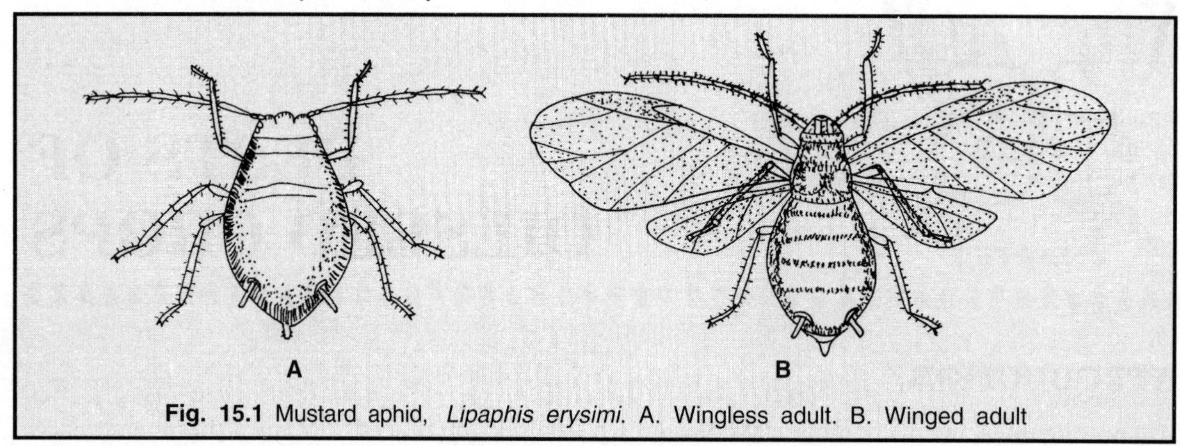

Fig. 15.1 Mustard aphid, *Lipaphis erysimi.* A. Wingless adult. B. Winged adult

Some of the parasitoids of the pest include *Ischiodon scutellaris* (Fabricius) and *Syrphus serarius* (Wiedmann) (Syrphidae). *Brinckochrysa scelestes* (Banks) (Chrysopidae), *Coccinella septempunctata* Linnaeus and *Menochilus sexmaculatus* (Fabricius) (Coccinellidae) predate on the mustard aphid. *Diaeretiella rapae* M'Intosh (Braconidae) and *Lipolexis gracilis* Forester (Aphididae) parasitize the aphid. *Entomophthora coronata* and *Cephalosporium aphidicola* infect *L. erysimi*.

Damage. Both the nymphs and adults suck cell-sap from leaves, stems, inflorescence or the developing pods. Due to the very high population of the pest, the vitality of plants is greatly reduced. The leaves acquire a curly appearance, the flowers fail to form pods and the developing pods do not produce healthy seeds. The yield of an infested crop is reduced to one-fourth or one-fifth.

Management. (*i*) Sow the crop early wherever possible, preferably upto third week of October. (*ii*) Apply recommended dose of fertilizers. (*iii*) Apply any one of the following insecticides when the population of the pest reaches 50-60 aphids per 10 cm terminal portion of the central shoot or when an average of 0.5-1.0 cm terminal portion of central shoot is covered by aphids or when plants infested by aphids reach 40-50 per cent: (*a*) *Foliar sprays.* One litre of oxydemeon methyl 25EC, dimethoate 30EC, quinalphos 25EC, malathion 50EC; 1.5 litres of chlorpyriphos 20EC; 100 *g* of thiamethoxam 25WG in 200-315 litres of water per ha, depending on the stage of the crop. (*b*) *Granular insecticides.* 10 kg of phorate 10G, 33 kg of carbofuran 3G per ha followed by a light irrigation.

2. Painted Bug, *Bagrada hilaris* (Burmeister) (Hemiptera: Pentatomidae) (Fig. 15.2, Plate 15.1B)

The painted bug is a serious pest of cruciferous crops and is widely distributed in Myanmar, Sri Lanka, India, Iraq, Arabia and East Africa. Besides cruciferous crops, it has also been observed feeding on rice, sugarcane, indigo and coffee.

Damage is caused by nymphs as well as by adults. The full-grown nymphs are about 4 mm long and 2.66 mm broad **(Fig. 15.2)**. They are black with a number of brown markings. The adult bugs are 3.71 mm long and 3.33 mm broad. They are sub-ovate, black and have a number of orange or brownish spots.

Life-cycle. The painted bug is active from March to December and during this period all the stages can be seen. It passes the winter months of January and February in the adult stage under heaps of dried oilseed plants lying in the fields. These bugs lay oval, pale-yellow eggs singly or in groups of 3-8 on leaves, stalks, pods and sometimes on the soil. Eggs may be laid during day or night. A female bug may lay 37-102 eggs in its life-span of 3-4 weeks. The eggs hatch in 3-5 days during summer and 20 days during December. The nymphs develop fully in five stages and transform themselves into adults in 16-22 days during the summer and 25-34 days during the winter. The entire life-cycle is completed in 19-54 days and it passes through 9 generations in a year.

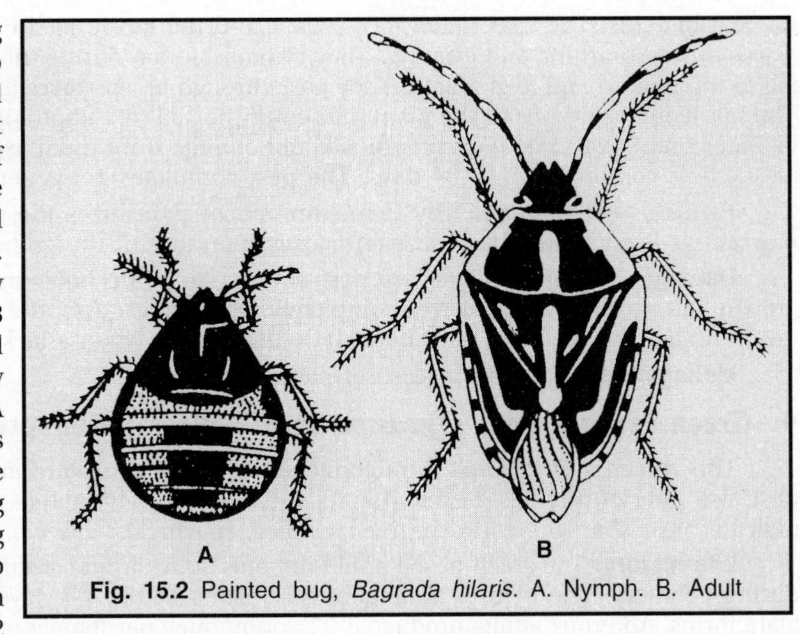

Fig. 15.2 Painted bug, *Bagrada hilaris*. A. Nymph. B. Adult

The eggs are parasitized by *Gryon* sp. (Scelionidae), while *Alophora* sp. (Tachinidae) parasitize the adults.

Damage. Both nymphs and adults suck cell sap from the leaves and developing pods, which gradually wilt and dry up. The nymphs and adult bugs also excrete a sort of resinous material which spoils the pods.

Management. (*i*) Give first irrigation 3-4 weeks after sowing as it reduces the bug population significantly. (*ii*) Spray one litre of malathion 50EC or 625 ml of quinalphos 25EC in 150-200 litres of water per ha once in October and again in March-April.

3. Mustard Sawfly, *Athalia lugens* (Klug) (Hymenoptera : Tenthredinidae) (Fig. 15.3)

The mustard sawfly is widely distributed in Indonesia, Formosa, Myanmar and the Indian Sub-continent. It feeds on various cruciferous plants like mustard, *toria* (*Brassica campestris*), rapeseed, cabbage, cauliflower, knol-khol, turnip, radish, etc.

Damage is done by the larvae which are dark green and have 8 pairs of abdominal prolegs. There are five black stripes on the back, and the body has a wrinkled appearance. A full-grown larva measures 16-18 mm in length (**Fig. 15.3**). The adults are small orange yellow insects with black markings on the body and have smoky wings with black veins.

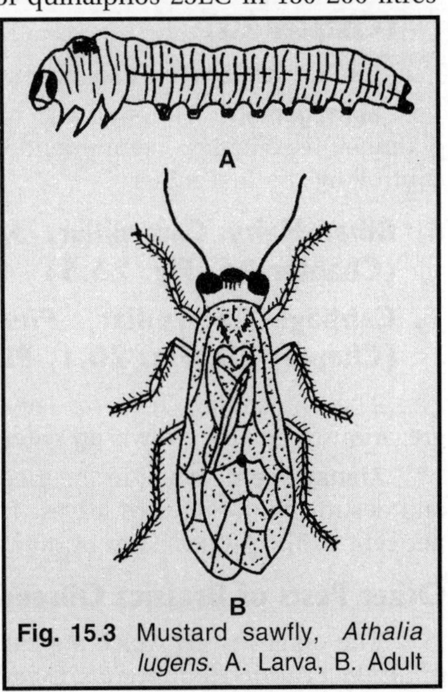

Fig. 15.3 Mustard sawfly, *Athalia lugens*. A. Larva, B. Adult

Life-cycle. The mustard sawfly breeds from October to March and the larvae rest in their pupal cocoons in the ground during summer. The adults emerge from these cocoons early in October. They live for 2-8 days and lay 30-35 eggs singly, in slits made with saw like ovipositors along the underside of the leaf margins. The eggs hatch in 4-8 days and the larvae feed exposed in groups of 3-6 on the leaves during morning and evening. They remain hidden during the day time and, when disturbed, fall to the ground and feign death. They pass through seven stages and are full-grown in 16-35 days. The full-fed larvae descend the plant and enter the soil to a depth of 25-30 mm. There, they pupate in water proof oval cocoons made of silk and emerge from them as adults in 11-31 days. Thus, the life-cycle is completed in 31-34 days. The pest completes 2-3 generations from October to March.

Perilissus cingulator Morby (Ichneumonidae) parasitizes the grubs. The bacterium, *Serratia marcescens* Bizio (Enterobacteriaceae) causes mortality of the grubs.

Damage. The grubs alone are destructive. They bite holes into leaves preferring the young growth and skeletonize the leaves completely. Sometimes, even the epidermis of the shoot is eaten up. Although the seedlings succumb; the older plants, when attacked, do not bear seed.

Management. Same as in case of painted bug.

4. Green Peach Aphid, *Myzus persicae* (Sulzer) (Hemiptera : Aphididae)

This insect is distributed throughout India. Besides mustard and other cruciferous plants, the pest also attacks peaches, beans, potato, tobacco, etc. The aphids are minute (2.0-2.5 mm long), delicate, pear-shaped, yellowish-green winged or wingless insects.

Life-history. The green peach aphid remains active from December to March with peak activity during February. The nymph undergoes 4-5 instars taking 4-7 days for apterous and 5-8 days for alate forms. Apterous adults produce 5-92 young ones per female while the alate forms produce 8-49 nymphs. Longevity of adult is 15-27 days for alate and 10-25 days for apterous forms.

Damage. Both nymphs and adults damage plants by actively sucking their sap. After the appearance of inflorescence, the aphid congregates on terminal buds and feeds there. As a result, there is flower shedding, poor-pod formation and shrivelling of grains. The insect also transmits virus diseases. The honeydew attracts sooty mould.

Management. (*i*) Sow the crop in first week of October. (*ii*) Spray 500 ml of dimethoate 30EC or 625 ml of oxydemeton methyl 25EC in 250 litres of water/ha when aphids start congregating on top flower buds. Only one spray is needed.

5. Pea Leafminer, *Chromatomyia horticola* (Goureau) (Diptera: Agromyzidae (Chapter 20)

The larvae feed by making mines into the leaves and cause heavy damage.

Management. The systemic insecticides (oxydemeton methyl 25EC or dimethoate 30EC or granular insecticides) recommended for the control of mustard aphid should also be used for controlling the leaf-miner.

6. Bihar Hairy Caterpillar, *Spilarctia obliqua* (Walker) (Lepidoptera: Arctiidae) (Chapter 26, Fig. 26.5)

7. Cabbage Caterpillar, *Pieris brassicae* (Linnaeus) (Lepidoptera : Pieridae) (Chapter 20, Fig. 20.1, Plate 15.1C)

The caterpillars feed on leaves, young shoots and green pods. When young, they feed gregariously but the grown-up caterpillars migrate from one field to another.

Management. When in the gregarious stage, the caterpillars can be easily controlled by picking and destroying the infested leaves. The grown-up caterpillars should be controlled with malathion 5 per cent @ 37.5 kg per ha or by spraying 500 ml of dichlorvos 85SL in 200-300 litres of water per ha.

Other Pests of Brassica Oilseeds

The other insects which appear as minor pests of *Brassica* crops include the jassid, *Empoasca binotata* Pruthi (Hemiptera : Cicadellidae); the cabbage butterfly, *Pieris brassicae* (Linnaeus)

(Lepidoptera : Pieridae) (Plate 6); the diamondback moth, *Plutella xylostella* (Linnaeus) (Lepidoptera: Yponomeutidae); *Crocidolomia binotalis* Zeller and *Hellula undalis* (Fabricius) (Lepidoptera: Pyralidae) (See chapter 14); the noctuid caterpillars, *Agrotis ipsilon* (Hufnagel), *Mythimna loreyi* (Duponchel) and *Helicoverpa armigera* (Hubner) (Plate 7) (Lepidoptera : Noctuidae); the flea beetles, *Phyllotreta crucifereae* (Goeze) and *Phaedon brassicae* Baly (Coleoptera : Chrysomelidae) and the leaf-miner, *Chromatomyia horticola* (Goureau) (Diptera : Agromyzidae).

GROUNDNUT (*Arachis hypogaea* L.; Family : Leguminosae)

1. Groundunt Aphid, *Aphis craccivora* Koch (Hemiptera : Aphididae) (Fig 15.4)

This is one of the most serious pests of groundnut. It also attacks peas, beans, pulses, safflower and some weeds. Its distribution is throughout India. It has also been recorded in Africa, Argentina and Chile. This species is also present in non-groundnut areas of the USA, Europe and Australia.

Colonies of aphids suck the sap from the underside of the leaves, top shoots and stem. In severe cases of infestation, the crop gets withered and blighted. The aphid is also a vector of a virus disease known as the rossette of groundnut. The winged adults (**Fig 12.4**) have black wings and they reach the freshly germinated groundnut plants after overwintering on collateral host plants.

Life-cycle. The offspring of the winged form may be wingless. Even without fertilization the females may produce 8-20 young ones in a life span of 10-12 days. The young nymphs are brownish and they pass through four moults to become adults in 5-8 days. The apterous females start producing brood within 24 hours of attaining that stage. Breeding occurs almost throughout the year and both alatae and apterae are present.

The coccinellid beetle, *Menochilus sexmaculatus* (Fabricius) (Coleoptera : Coccinellidae) and *Ischiodon javana* (Wiedemann) (Diptera : Syrphidae) are the main predators of this aphid pest.

Damage. The nymphs and adults suck the sap, usually from the underside of leaves. Infestation in the early stages causes stunting of the plants as well as reducing their vigour. When the attack occurs at the time of flowering and pod formation, the yield is reduced considerably. Infestation on the groundnut crop usually occurs 4-6 weeks after sowing.

Management. As soon as the pest appears on growing points, spray 625 ml of malathion 50 EC or 425 ml of dimethoate 30EC or 425 ml of oxydemeton methyl 25EC in 200 litres of water per ha.

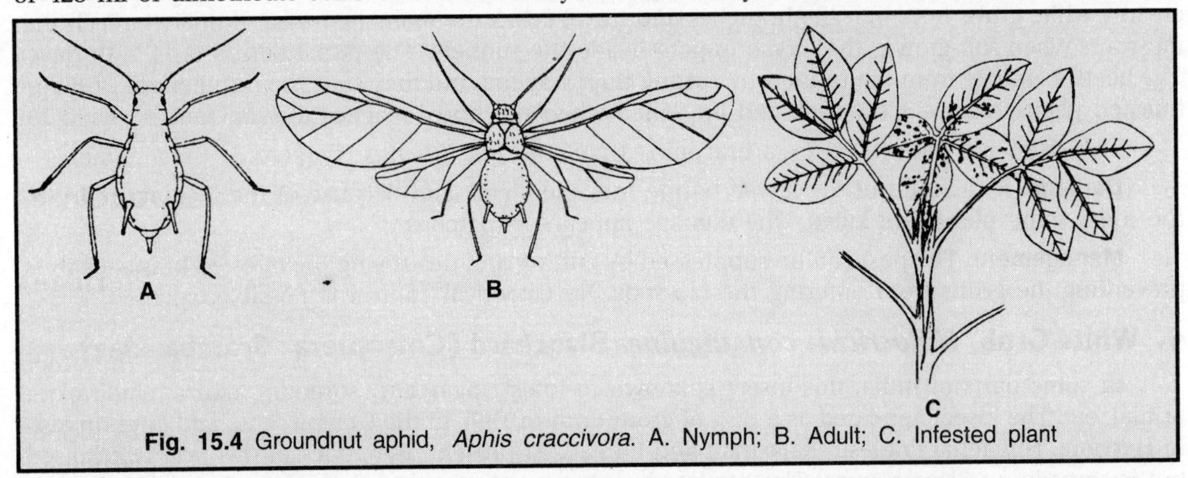

Fig. 15.4 Groundnut aphid, *Aphis craccivora*. A. Nymph; B. Adult; C. Infested plant

2. Groundnut Leafminer, *Aproaerema modicella* (Deventer) (Lepidoptera : Gelechiidae)

This pest is distributed all over India, Pakistan, Sri Lanka, Myanmar and South Africa. In India, it is most serious in Tamil Nadu. Although it has been reported feeding on other plants, it is considered a serious pest of groundnut only.

Damage is caused by tiny caterpillars which mine into the leaves, skeletonize them and web them together. The larva is small and green, and has a conspicuous dark head. The adult is a small bronze moth with a wing expanse of about 1 cm.

Life-cycle. During the day the moths remain concealed under clods of soil or in the crevices. They become active at night when they are also strongly attracted to light. The female moth lays up to several hundred eggs singly on leaves and on shoots. The eggs hatch after about 3 days. The newly hatched larva measures about 1.5 mm in length. After wandering about for some time, it mines into the leaves and, later on, bites its way out. It then webs together a number of leaflets and feeds inside the chamber, thus formed. The larvae are full-grown in 9-17 weeks, when they measure 6-8 mm in length and turn dirty green. They prepare silken cocoons to pupate in. The pupal period lasts about 4 days. There are several generations during a year.

The eggs of this pest are parasitized by *Trichogramma* sp. (Trichogrammatidae). The braconids, *Bracon hebetor* Say and *Apanteles javensis* Rohwer (Braconidae) and eulophids, *Asympiesiella indica* Girault, *Stenomesioideus ashmeadi* Subha Rao & Sharma and *Euplectrus* sp. (Eulophidae), parasitize its larvae.

Damage. As a result of the mining, skeletonizing and webbing together of the leaves, the crop suffers serious losses. In badly infested fields the plants present a look as if the crop has been scorched. The insect also attacks soybean and *arhar* (*Cajanus cajan* L.)

Management. Spray 2.0 kg of carbaryl 50WP or 350 ml of oxydemeton methyl 25EC in 250 litres of water per ha.

3. Groundnut Stem Borer, *Sphenoptera perotetti* Guenee (Coleoptera : Buprestidae)

This pest infests the groundnut crop in Andhra Pradesh, Bihar, Delhi, Gujarat, Kerala, Tamil Nadu, Madhya Pradesh, Maharashtra and Karnataka. It has not been recorded anywhere outside India. Besides groundnut, it attacks sesamum, gram and other pulses.

The grub of this beetle bores into the stem and root. The adult is a small jewel-like beetle, 10-12 mm in length with a striking metallic shine over a dark brown background. The full-grown grub is whitish in appearance.

Life-cycle. The beetles lay scale like eggs and glue them on to the branches. The slender, creamy white grubs bore into the branches and travel down the main stem and ultimately reach the tap-root. When full-grown, the larvae pupate inside the tunnels. The pupal period lasts 7-10 days. The beetles emerge from the tunnels by cutting their way out and they start the next generation. The infested plants receive a set-back and are killed when the borer reaches the tap-root.

The grubs are parasitized by a braconid, *Glyptomorpha smenus* (Cameron) (Braconidae)

Damage. As a result of the larvae boring into the stems and because of the feeding injury to the roots some plants are killed. The damage appears in patches.

Management. The pest can be suppressed by cutting and destroying the infested branches thus preventing the grubs from entering the tap-root. No Chemical control is possible.

4. White Grub, *Holotrichia consanguinea* Blanchard (Coleoptera : Scarabaeidae)

In some parts of India, this insect is known to infest sugarcane, sorghum, maize, chilli, okra, brinjal, etc. The insect appeared as a pest of groundnut in 1957 in the Gujarat State and later on also in Haryana, Himachal Pradesh, Rajasthan and Punjab. At present, the attack of this pest is localised but is spreading to larger areas. The damage becomes evident only when the entire plant dries up due to the grubs feeding on fibrous roots. The grubs are mostly found in the upper 5-10 cm layer of soil. When full-grown, they are about 35 mm long and are white, having a brown head and prominent thoracic legs. The adult beetles are dull brown and measure about 18 mm in length and 7 mm in width.

Life-cycle. The presence of a large number of newly hatched larvae in June shows that this insect becomes active with the onset of the monsoon. The adult beetles lay eggs singly up to a depth of 10 cm. The eggs hatch in 7-10 days. The newly hatched grubs measure about 12 mm in length and their development is completed in 8-10 weeks. After the monsoon, the full-grown larvae migrate to a considerable depth in the soil for pupation. The pupa is semicircular and creamy white and the pupal stage lasts about a fortnight. The beetles remain in the soil at a depth of 10-20 cm and come out for feeding at night. Adults formed in November remain in soil till next June. Their population is maximum in the rainy season and there seems to be only one generation in a year.

The grub is parasitized by *Scolia aureipennis* (Scoliidae). A fungus, *Metarrhizium anisopliae* (Metchnikoff) (Moniliaceae) parasitizes the adults and the common Indian toad, *Bufo melanostictus*, and the wall lizard, *Gecko gecko* feed on the beetles.

Damage. The grubs eat away the nodules, the fine rootlets and may also girdle the main root, ultimately killing the plants. At night, the beetles feed on foliage and may completely defoliate even trees like *neem* (*Azadirachta indica*) and banyan (*Ficus bengalensis*)

Management. (*i*) Plough the fields twice during May-June. It would help in exposing the beetles resting in the soil. (*ii*) Wherever possible, sow the crop early, i.e. between June 10 and 20. (*iii*) Treat the seed before sowing with 12.5 ml of chlorpyriphos 20EC per kg of kernels. (*iv*) Kill the beetles by spraying 500 g of carbaryl 50WP in 250 litres of water per ha on the preferred host plants like *ber*, guava, *rukmanjani*, grapevines, almond, etc. The spray should be carried out in the afternoon and repeated after every rainfall till the middle of July. (*v*) Apply 10 kg of phorate 10G or 30 kg of carbofuran 3G per ha in the soil at or before sowing.

Other Pests of Groundnut

The groundnut crop is also attacked by a number of minor insect pests including the grasshoppers, *Atractomorpha crenulata* (Fabricius), *Cyrtacanthacris ranacea* S., *Chrotogonus trachypterus* (Blanchard), *Acrida turrita* Linnaeaus and *Oxya velox* (Fabricius) (Orthoptera : Acrididae). In South India, the groundnut earwig. *Euborellia stali* Dohrn (Dermapetra : Anisolabiidae) feeds on the kernels of pods in soil causing up to 20 per cent loss to the yield; *Microtermes* sp. (Isoptera : Termitiidae) damages the pods; *Elasmolomus sordidus* (Fabricius) (Hemiptera: Lygaeidae); the red hairy caterpillar, *Amsacta moorei* (Butler) and the Bihar hairy caterpillar, *Spilarctia obliqua* Walker (Lepidoptera : Acrtiidae) defoliate the plants. Several species of thrips also infest the groundnut leaves and they cause yellowing of the plants.

CASTOR (*Ricinus communis* L; Family : Euphorbiaceae)

1. Castor Whitefly, *Trialeurodes ricini* Misra (Hemiptera : Aleyrodidae)

The castor whitefly has been reported from India and Pakistan. It is primarily a pest of castor but the nymphs also feed on the leaves and stems of *Breynia rhamnoides* and *Achras zapota*. The nymphs are transluscent, light yellow in colour and are covered with thick waxen filaments. The adults are pale yellow with white wings covered with waxy powder.

Life-cycle. This pest is active from February to November. The eggs are laid in small clusters on the undersurface of tender leaves. On emergence, the nymphs attach themselves to leaves and become full-grown after four moultings. The life-cycle is completd in 19-21 days during July, August and September.

The lady birds, *Cryptognatha flavescens* Motsch., *Brumoides suturalis* (Fabricius) and *Sticholotis* sp. (Coccinellidae); a drosophilid fly, *Acletoxenus indica* (Drosophilidae) and the lygaeid bug, *Geocoris bicolor* Fabricius (Lygaeidae) are predacious on the nymphs of this pest. *Encarsia lahorensis* Howard and *Aphelinus fuscipennis* (Aphelinidae) parasitize the nymphs.

Damage. As a result of sucking of sap by the nymphs the leaves show yellow patches initially. Later on, there is a gradual drying of the leaves and ultimately the plants die.

Management. Spray one litre of malathion 50 EC or 625 ml of dimethoate 30 EC or oxydemeton methyl 25 EC or formothion 25 EC or 200 ml of phosphamidon 100 EC in 625 litres of water per ha and repeat at 2-3 week intervals.

2. Castor Semilooper, *Achaea janata* (Linnaeus) (Lepidoptera : Noctuidae) (Fig. 15.5)

This is a serious pest of castor in all parts of India and Pakistan and has also been reported from Sri Lanka and Thailand. While occurring in large numbers the semiloopers strip the castor plant bare of all foliage. Unlike most of other noctuid moths which do not feed on fruits, the adults of this species are fruit sucking moths and cause serious damage to citrus.

The adult of *A. janata* is a pale reddish brown moth (Fig. 15.5) with a wing expanse of 6-7 cm. The wings are decorated with broad zig-zag markings, a large pale area and dark brown patches. The full-grown larva is dark and is marked with prominent blue-black, yellow and reddish stripes and has a pair of reddish processes and also a dorsal hump near the head end of the body. There is a characteristic white mark on the head. The colour patterns of the larvae of the third and fourth instars are so variable that they have been mistaken for four different varieties of caterpillars. They are highly conspicuous on the green plants.

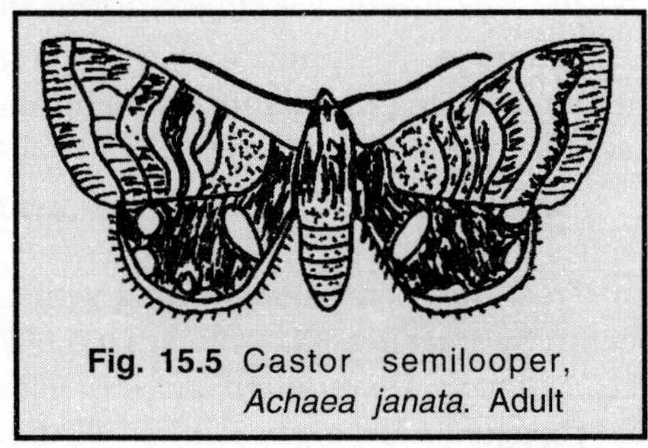

Fig. 15.5 Castor semilooper, *Achaea janata.* Adult

Life-cycle. The moths are active at dusk and lay scattered eggs on tender leaves. A female can lay up to 450 eggs during its life span. The egg, being about 1 mm in length, is fairly large and also has on its surface a few ridges and furrows which radiate from the circular depression at the apex. Yet it is difficult to locate it on the foliage as it is bluish-green. The larva emerges by cutting a hole in the egg-shell in 2-5 days and devours it immediately. Then, it starts feeding on the foliage. The first stage larva measures 3.5 mm. It feeds and moults 4-5 times and becomes full-grown in 15-20 days. The grown-up larva prepares a loose cocoon of coarse silk and some soil particles, and pupates under the fallen leaves on the soil, usually at the edge of the field. In some cases, pupation also takes place within the folded leaves on the plant itself. The pupal stage lasts 10-15 days and the moths, on emergence, feed on the soft fruits of citrus, mango, etc. There are 5-6 generations in a year.

Trichogramma minutum Riley (Trichogrammatidae) parasitizes the eggs. Various larval parasitoids recorded on this pest are *Apanteles sudanus* Wilk., *A. ruidus* Wil., *Microgaster maculipennis* Szep., M. eusirus Lyle, *M. similislyle* and *Rhogas percurrens* Lyle (Braconidae); *Euplectrus leucostomus* Rohwer and *Tetrastichus ophiusae* (Eulophidae).

Damage. The caterpillars feed voraciously on castor leaves, starting from the edges inwards and leaving behind only the midribs and the stalks. Damage is maximum in August-September and with the excessive loss of foliage, the seed yield is reduced considerably. Although the semilooper feeds on a variety of plants, it seems to prefer castor.

Management. Apply 625 ml of methyl parathion 50EC in 625 litres of water per ha.

3. Castor Capsule Borer, *Conogethes punctiferalis* Guenee (Lepidoptera: Pyralidae) (Fig. 15.6)

This borer is distributed throughout India wherever castor is grown. The damage is caused by the caterpillars, which bore into the main stem of a young plant and ultimately into the capsules. The full-grown caterpillar measures 25-30 mm in length, is reddish brown, with black blotches all over the body and a pale stripe on the lateral side. The moths are orange yellow, with black markings on both the wings (Fig. 12.6).

Life-cycle. The moths lay eggs on leaves and other soft parts of the plant. The eggs hatch in about a week. The larvae pass through 4-5 instars and are full-fed in 2-3 weeks. Pupation takes place inside the seed or sometimes in the frass that collects after feeding. The pupal stage lasts about one week. The life-cycle is completed in 4-5 weeks and 3 generations are completed in a year. The pest is active on castor from September to March.

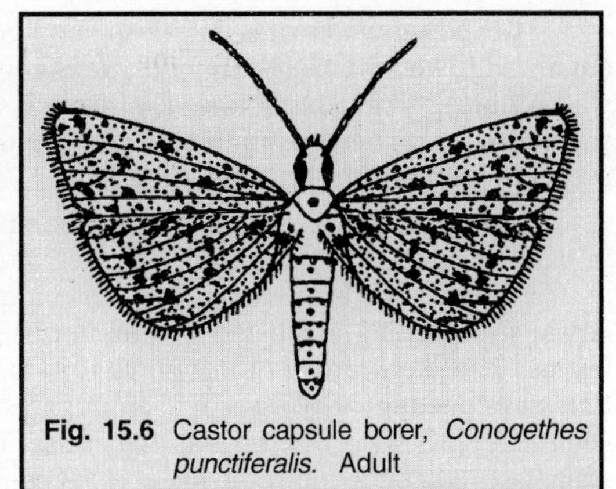

Fig. 15.6 Castor capsule borer, *Conogethes punctiferalis.* Adult

The larvae are parasitized by *Diadegma ricini* Row & Kurian, *D. trochanterata* (Morley) and *Theronia* sp. (Ichneumonidae); and *Apanteles* sp. and *Bracon hebetor* Say (Braconidae).

Damage. The attack by this borer is recognized from a distance by the webbed capsule heads. The yield is reduced considerably since the capsules and the seeds within are damaged.

Management. (*i*) It is advisable that the infested shoots and capsules may be collected and destroyed. (*ii*) Spray 2.5 kg of carbaryl 50WP in 625 litres of water per ha and repeat at 15-day intervals.

4. Castor Hairy Caterpillar, *Euproctis lunata* Walker (Lepidoptera : Lymantriidae)

The castor hairy caterpillar, which is widely distributed in India along with an allied species, *Euproctis fraterna* Moore , is a serious pest of this oilseed. It has also been observed feeding on linseed, groundnut, *guara* (*Cyamposis tetragonoloba*) and grapevine.

Damage is caused by the larvae which measure 35-40 mm when full-grown and are dark grey, with a wide white dorsal stripe, and have long hair all over body. The moths are pale yellow and are seen sitting on castor and other plants during the day time.

Life-cycle. This pest is active throughout the year but its speed of development is considerably reduced during the winter. Moths emerging in February lay a large number of eggs in clusters on the underside of leaves. The eggs are covered with the female anal tuft of brown hair. They hatch in 5-7 days and the young larvae feed gregariously for the first few days. Later on, they disperse and feed individually. They pass through six stages and are full-fed in 2-3 weeks. The full-grown caterpillars make loose, silken cocoons in the plant debris lying on the ground and pupate inside. The pupal stage lasts about one week in the summer. During the winter months, the egg, larval and pupal stages may last 10, 85 and 20 days, respectively. The pest passes through several generations in a year.

The parasites, *Telenomus euproctiscidis* (Mani) (Scelionidae) and *Trichogramma minutum* Riley (Trichogrammatidae) are associated with the eggs. *Apanteles colemani* Viereck and *A. euproctisiphagus* Muzaffar (Braconidae) parasitize the larvae of this pest.

Damage. Caterpillars feed on the leaves of various host plants and in case of severe infestation, they may cause complete defoliation. The attacked plants remain stunted and produce very little seed.

Management. Spray 625 ml of methyl parathion 50EC in 625 litres of water per ha.

5. Castor Slug, *Parasa lepida* (Cramer) (Lepidoptera: Limacodidae) (Fig.15.7)

Although commonly called slug, it is an insect larva that causes damage to the leaves of host plants. It is a serious pest of sporadic occurrence on castor, mango, pomegranate, citrus, coconut, palm, rose, wood apple, country almond, etc., in India, Malaysia and Sri Lanka.

The full-grown larva is flat, fleshy and greenish in colour with white lines on the body. It is covered with spines having red or black tips. The moth is short and stout with forewings predominantly green in the middle and brownish at the end (**Fig. 15.7**).

Fig. 15.7 Castor slug, *Parasa lepida.* Adult

Life-cycle. The female moths lay flat shining eggs in batches of 15- 35 on the under surface of the leaves of host plants. The eggs hatch in 7 days and the young larvae start feeding in clusters on the under surface of leaves. The larva passes through five instars and is fully developed in 40-42 days. It then pupates in a hard shell-like greyish cocoon on the tree trunk. The pupal stage lasts 20-28 days and the life cycle is completed in about 10 weeks. The pest remains active throughout the year.

The larvae are heavily parasitized by *Clinocentrus* sp. (Braconidae), *Eurytoma parasae* G., and *E. monemae* Rusch (Eurytomidae) and *Apanteles* sp. (Braconidae).

Damage. The young caterpillars feed gregariously by scraping the undersurface of the leaves. The loss of sap from plant tissue reduces the vitality of plants and the leaves dry up. As the larvae grow, they get scattered and feed on the entire leaf and cause defoliation.

Management. Same as in case of castor hairy caterpillar.

Other Pests of Castor

The other pests of castor are the grasshoppers, *Chrotogonus* spp. and *Cyrtacanthacris ranacea* S. (Orthoptera : Acrididae); the jassids, *Amrasca devastans* Distant , *Empoasca flavescens* (Fabricius), *E. notata* Mel., *E. kerri* Pruthi and *E. parathea* Pruthi (Hemiptera : Cicadellidae); the coccids, *Aspidiotus destructor* Signoret and *A. orientalis* Newst. (Hemiptera : Diaspididae) and *Parasaissetia nigra* (Nietner) (Hemiptera : Coccidae); *Nezara viridula* (Linnaeus) and *Scutellera nobilis* Fabricius (Hemiptera : Pentatomidae); the thrips, *Scirtothrips dorsalis* Hood (Thysanoptera: Thripidae); *Clania crameri* West (Lepidoptera : Psychidae); *Phycita clientella* Z. (Lepidoptera: Pyralidae); the tobacco caterpillar, *Spodoptera litura* Fabricius (Lepidoptera : Noctuidae); *Amsacta moorei* Butler and *Spilarctia obliqua* Walker (Lepidoptera : Arctiidae), the gallfly, *Asphondylia ricini* Mani (Diptera : Cecidomyidae); the red spider mite, *Tetranychus telarius* Linnaeus and *Oligonychus coffeae* (Nietner) (Acari : Tetranychidae).

SESAME (*Sesamum indicum* L.; Family : Pedialiaceae)

1. Til Leaf and Pod Caterpillar, *Antigastra catalaunalis* (Duponchel) (Lepidoptera : Pyralidae) (Fig. 15.8)

The sesame leaf and pod caterpillar is a serious and regular pest of *til* (*Sesamum orientale* and *S. indicum*) and is also distributed throughout india. This species has also been reported from Europe, Africa, Cyprus, Malta, Indonesia and South-east Africa.

The caterpillars damage apical shoots and young pods. They are pale yellow, when young, but gradually become green and develop black dots all over the body. The full-grown larva measures 14-17 mm. The moth is a small insect with a wing span of about 2 cm having dark brown markings on the wing-tips (**Fig. 15.8**).

Life-cycle. In northern India, the moths appear in August. Females lay up to 140 eggs singly on the tender portions of plants at night. The eggs are shiny, pale-green and they hatch in 2-7 days, depending upon the season. On emerging, the young larva, which measures about 2 mm in length, feeds for a little while on the leaf epidermis or within the leaf tissue. Soon after, it binds together the tender leaves of the growing shoot with the help of silken threads and continues to feed in the webbed mass. The size of this rolled mass increases gradually as the caterpillar grows older. It becomes full-grown in about 10 days in summer, but the period may be prolonged to 33 days in winter. The grown-up larvae creep to the ground and pupate in silken cocoons in soil. Sometimes, pupation also takes place in the plant itself. Pupal development is completed in 4-20 days, depending upon the season. In summer, a generation is completed in about 23 days but in the winter it takes about 67 days. The moths start laying eggs 4-5 days after emergence. They are active in the cool hours of the day and live for about 3 weeks. All stages of the insect are seen in autumn and sometimes up to December. During January and February, the insect hibernates as a larva inside pods, stubble or the harvested stalks. There are nearly 14 generations in a year.

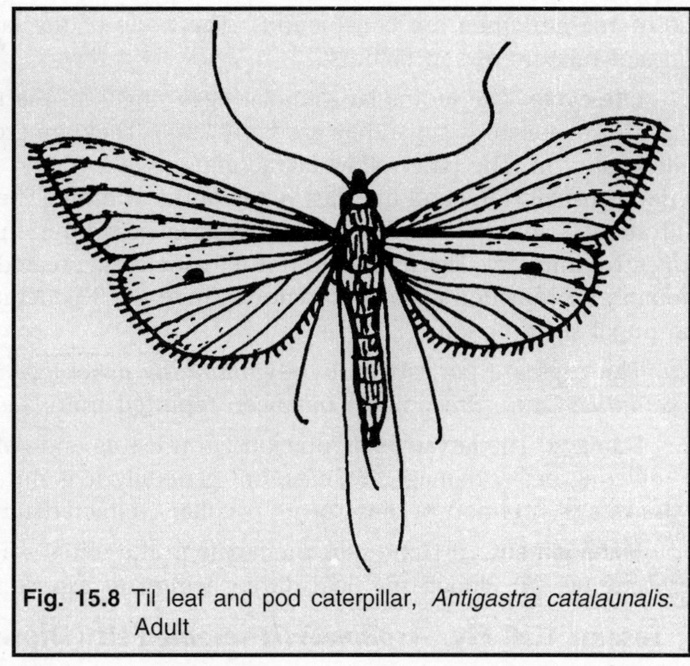

Fig. 15.8 Til leaf and pod caterpillar, *Antigastra catalaunalis*. Adult

The braconids, *Bracon hebetor* Say, *B. kitcheneri* D and G (Braconidae) and Ichneumonids, *Cremastus flavoortictalis Cam.* and *Temelucha biguttula* Matsumura (Ichneumonidae) are known to parasitize its larvae.

Damage. Young caterpillars feed on leaves. They also bore into the shoots, flowers, buds and pods. An early attack kills the whole plant, but infestation of the shoots at a later stage hampers further growth and flowering. One larva can destroy 2-3 young plants. The maximum infestation occurs in May and again in September-October. The pest is also capable of sustaining itself on some wild species of *Sesamum*.

Management. (*i*) Collect and destroy the webbed leaves and infested pods. (*ii*) Spray the crop twice (first at pest appearance and then at flowering stage) with 250 ml of fenvalerate 20EC or 375 ml of deltamethrin 2.8EC or thrice with 500 ml of cypermethrin 10EC at pest appearance, flowering and pod formation in 250 litres of water per ha.

2. Til Hawk Moth, *Acherontia styx* (Westwood) (Lepidoptera : Sphingidae)

This pest of sesamum is common in India and is also distributed in Indonesia, Sri Lanka, the Philippines and Myanmar. Besides sesamum, it has been reported on potato, brinjal, lablab and ornamental plants such as balsam. The moth is also harmful because it sucks honey from honey combs in apiaries, although at times it is stung to death by the bees.

This insect is variously known as hawk moth, sphinx-moth or death's head moth based on its structural and behavioural characteristics. The adult is a large reddish brown, robust thick-set moth with a wing span of about 10 cm. The forewings are decorated with a mixture of dark-brown and grey patterns with dark or black wavy markings and a prominent yellow spot on each wing. There is a prominent Death's head mark on the thorax. The moths are swift fliers and often make hawk like darts to a source of light at dusk.

The full-grown caterpillar, which measures about 5 cm in length and 1 cm width, often retracts some of its anterior body segments and looks like a sphinx. The horn-like projections on the hind end of the abdomen are conspicuous. The body of the larva is plump and is decorated with a pleasant mixture of soft colours.

Life-cycle. The moths lay globular eggs singly on the underside of leaves of food plants. The eggs are conspicuous since they are fairly large. They are greenish white when freshly laid but turn yellow later on. The pale-yellow larvae emerge in 2-5 days and start feeding on leaves. The larval period is usually long and may last two months or more. The pupal stage lasts 2-3 weeks in summer and about 7 months in winter. The changes in coloration of the larvae and adults aid them in protective mimicry. The impression of ferociousness created by the larva as well as by the moth is probably a protection against predators. There are 3 generations in a year. The winter is passed in the pupal stage in soil.

The eggs are parasitized by *Agiommatus acherontiae* Ferr (Pteromalidae) while *Apanteles acherontiae* Cam. (Braconidae) has been reported from the larvae.

Damage. The larvae feed voraciously on leaves and defoliate the plants. The insect is capable of inflicting heavy damage at times, but generally it is not a very serious pest in India. However, it draws our attention because of its peculiar characteristics.

Management. (*i*) Hand-pick the larvae in the initial stages of attack and destroy by keeping in kerosene oil. (*ii*) Plough the field during winter to expose the hibernating larvae.

3. Sesame Gall Fly, *Asphondylia sesame* Felt (Diptera: Cecidomyiidae)

The sesame gall-fly, also known as the sesamum gall-midge, occurs all over India and sometimes assumes serious proportions, particularly in southern India. It has been recorded in Tamil Nadu, Andhra Pradesh, Kerala, Madhya Pradesh, Gujarat and Maharashtra. It has also been reported as a serious pest in Rajasthan.

Life-cycle. The adult is mosquito-like and small. The female lays eggs singly in buds, flowers and developing capsules. The eggs hatch in 2-4 days. The young maggots feed on floral parts and cause malfomation of the buds which fail to develop into seeds. The larvae complete their development in 14-21 days and pupate inside the galls. The moths emerge from the galls in 7-12 days. The life-cycle is completed in 23-37 days. There are several generations in a year.

Eurytoma dentipectus Gohan and *E. nesiotes* (Eurytomidae) parasitize its maggots.

Damage. As a result of feeding by maggots, the buds develop into galls and produce no fruits and seeds. When infestation is severe, the crop may be a total failure. However, some varieties are comparatively resistant to its attack.

Management. (*i*) Clip the galls and pick and burn the shed buds. (*ii*) Spray 2.5 kg of carbaryl 50WP in 625 litres of water per ha.

Other Pests of Sesame

Besides the three insect pests described, the sesame crop is also damaged by various minor pests like the cricket, *Brachytrypes portentosus* Licht (Orthoptera : Gryllidae); the lygaeid bugs, *Elasmolomus sordidus* (Fabricius) (Hemiptera: Lygaeidae) suck milk out of the seeds; *Nysius inconspicuous* Distant (Hemiptera : Lygaeidae) feed on the tender shoots and capsules; the pentatomid bugs, *Eusarcocoris ventralis* (Westwood), *Nezara viridula* (Linnaeus), *Aspongopus janus* Fabricius and *Dolycoris indicus* Stal (Hemiptera : Pentatomidae); the caterpillars, *Spilarctia obliqua* Walker, *Amsacta moorei* (Butler) and *Pericallia ricini* (Fabricius) (Lepidoptera : Arctiidae); *Spodoptera exigua* (Hubner) and *Pulsia signata* (Fabricius) (Lepidoptera : Noctuidae); *Gitonides perspicax* Knab (Diptera : Drosophilidae) and the weevils, *Tanymecus chloroleucas* Wied and *T. indicus* Marshall (Coleoptera: Curculionidae).

LINSEED (*Linum usitatissimum* L.; Family : Linaceae)

1. Linseed Bud Fly, *Dasineura lini* Barnes (Diptera: Cecidomyiidae) (Plate 15.1D)

The adult of this bud fly is a small orange fly. Its tiny maggots feed and produce galls. They destroy flower-buds and thus prevent pod formation. This insect appears as a serious pest of linseed in some parts of India, including Andhra Pradesh, Madhya Pradesh, Bihar, Uttar Pradesh, Delhi and Punjab.

Life-cycle. The female lays 29-103 smooth, transparent eggs in the folds of 8-17 flowers or in tender green buds, either singly or in clusters of 3-5. The eggs hatch in 2-5 days. Just after emergence, the larvae are transparent, with a yellow patch on the abdomen. These larvae feed inside flower buds and eat the contents. They pass through four instars in 4-10 days and when full-grown become deep pink and measure about 2 mm in length. The full-grown maggots drop to the ground, prepare a cocoon and pupate in the soil. The pupal period lasts 4-9 days. A generation is completed in 10-24 days. There are four overlapping generations during the season.

The maggots are parasitized by *Systasis dasyneurae* Mani (Miscogasteridae). Other larval parasites include *Elasmus* sp. (Elasmidae), *Eurytoma* sp. (Eurytomidae), *Torymus* sp. (Torymidae) and *Tetrastichus* sp. (Eulophidae).

Damage. Damage is the result of feeding by maggots on buds and flowers. Consequently, no pod formation takes place **(Plate 15.1D).**

Management. (*i*) The adult flies can be killed by using light traps. The flies are also attracted in day-time to molasses or *gur* added to water. (*ii*) As the incidence of this pest is more on the late-sown crop as compared with the normal-sown crop, the practice of normal-sown crops should be adopted if possible. (*iii*) Dust 5 per cent carbaryl @ 15-20 kg/ha or spray carbaryl 50WP @ 1.125 kg/ha in 200-250 litres of water/ha.

2. Beet Armyworm, *Spodoptera exigua* (Hubner) (Lepidoptera : Noctuidae) (Chapter 16)

The beet armyworm may cause damage by feeding on leaves. Spray the crop with 1.125 kg of carbaryl 50WP or 1.0 litre of malathion 50EC in 200-250 litres of water per ha.

Other Pests of Linseed

Other pests of linseed are the jassid, *Empoasca kerri* var. *motti* Pruthi (Hemiptera : Cicadellidae); the whitefly, *Bemisia tabaci* (Gennadius) (Hemiptera : Aleyrodidae); *Creontiades pallidifer* Walker (Hemiptera : Miridae); *Piezodorus hybneri* (Gmel.) (Hemiptera : Pentatomidae); *Caliothrips indicus* (Bagnall) (Thysanaptera : Thripidae); the hairy caterpillar, *Spilarctia obliqua* Walker (Lepidoptera: Arctiidae); *Grammodes stalida* Fabricius (Lepidoptera : Noctuidae); *Euproctis scintillans* Walker (Lepidoptera : Lymantriidae) ; *Spodoptera litura* (Fabricius), *Thysanoplusia orichalcea* (Fabricius) and *Helicoverpa armigera* (Hubner) (Lepidoptera : Noctuidae).

SAFFLOWER (*Carthamus tinctorium* L.; Family : Compositae)

1. Safflower Caterpillar, *Perigaea capensis* (Guenee) (Lepidoptera : Noctuidae)

It is a serious pest of safflower throughout India. The adult is a dark-brown medium-sized moth with white wavy markings on the forewings. The full grown caterpillar is about 25 mm long, smooth, greenish with purple markings and humped on the anal segment.

Life-cycle. A female lays about 60 eggs singly or in small clusters on leaves and stems. The eggs hatch in 4-5 days. The larva grows feeding on the leaves and becomes full-grown in 2-3 weeks. It pupates in the soil for 10-15 days.

Damage. The larvae feed on the leaves and defoliate the plants which lose their vigour and become stunted.

Management. Spray the crop with 2.5 kg of carbaryl 50WP in 250 litres of water/ha.

2. Safflower Bud Fly, *Acanthiophilus helianthi* (Rossi) (Diptera : Tephritidae)

This pest has been reported from Delhi, Uttar Pradesh, Haryana and Madhya Pradesh in India. The maggots that feed on flower buds are destructive and when full-grown they are 5 mm long. The adult fly is ash coloured with light brown legs.

Life-cycle. The adults are active from March to May. The females lay eggs in clusters of 6-24 within the flower buds or the flowers. The eggs hatch in about one day in April and young maggots start feeding on the florets and the thalamus. Within one week they grow to the full and attain a size of 5 × 1.5 mm. They pupate inside the buds. The pupal stage lasts 7 days. The adults emerge out of the bud through the holes made by the larvae before they pupate. Three generations are completed during a crop season.

The maggots are parasitized by *Ormyrus* sp, (Ormyridae), *Eurytoma* sp. (Eurytomidae) and *Pachyneuron muscarum* (Linnaeus) (Braconidae). *Chrysopa virgestes* Banks (Chrysopidae) is a predator of the maggots.

Damage. The injury is caused by the maggots which feed upon the floral parts including the thalamus. The infested buds begin to rot and an offensive smelling fluid oozes at the apices giving a soaked appearance to the buds. The pest causes reduction in the yield of safflower seed.

Management. The early removal and destruction of infested buds is helpful in checking the spread of the pest.

3. Safflower Aphid, *Uroleucon compositae* (Theobald) (Hemiptera : Aphididae)

This pest causes considerable damage to safflower in Karnataka, Uttar Pradesh, Madhya Pradesh, Punjab and Haryana. The aphids are small shining black, soft bodied insects. The nymphs are smaller in size and are reddish brown in colour.

Life-cycle. The aphid is active from December to April. A female produces 6-56 young ones with an average of 21. It completes its life cycle in 11-16 days. The adult aphid has a life span of 17 days.

The aphid is parasitised by *Aphidencyrtus aphidivorus* (Mayr) and preyed upon by *Brumoides suturalis* (Fabricius).

Damage. The aphids suck the sap from leaves, twigs, flowers and capsules. In infested plants, the height, number of leaves and shoots reduce significantly. The plants become weak, remain stunted and sometimes dry up. Seed production is seriously affected. The aphids secrete honeydew which attracts a black sooty mould.

Management. Spray 250 ml of dimethoate 30EC or monocrotophos 36SL or 625 ml of chlorpyriphos 20EC in 250 litres of water/ha and repeat the spray after 15 days, if necessary.

Other Pests of Safflower

Some other insects which appear as minor pests of safflower include, *Empoasca punjablensis* Pruthi (Hemiptera: Cicadellidae); the green peach aphid, Myzus perrical (Sulzer) ; (Hemiptera: Aphididae); the lace wing, *Monanthia glubulifera* W. (Hemiptera : Tingidae); *Dolycoris indicus* Stal (Hemiptera : Pentatomidae); the safflower caterpillar *Perigaea capensis* Guenee; *Spodoptera exigua* (Hubner); *Helicoverpa armigera* (Hubner) and *Eublemma rivula* Moore (Lepidoptera : Noctuidae); the leaf *miner*,Chromatomyia *horticola* (Goureau) (Diptera : Agromyzidae) and surface weevil, *Tanymecus indicus* Fst. (Coleoptera : Curculionidae).

SUNFLOWER (*Helianthus annuus* L.; Family : Compositae)

1. Head Borer, *Helicoverpa armigera* (Hubner) (Lepidoptera : Noctuidae) (Plate 15.1E)

The head or capitulum borer causes considerable damage to developing grains in the head capsule. The young larvae first attack the tender parts like bracts and petals, and later on shift to reproductive parts of the flower heads. Bigger larvae mostly feed on seeds by making tunnels in the body of the flower heads and often remain concealed. They may also shift to the backside of the heads and even leaves, and feeding may continue upto maturity. Star bud stage of the crop is most vulnerable and suffers maximum yield loss. The detailed account of life history of the pest has been given in Chapter 14.

Management. Spray 2.5 kg of carbaryl 50WP or 2.0 kg of acephate 75SP or 2.5 litres of chlorpyriphos 20EC or 1.25 litres of monocrotophos 36SL in 250 litres of water per ha at the initiation of starbud stage, and repeat after two weeks if necessary.

2. Tobacco Caterpillar, *Spodoptera litura* (Fabricius) (Lepidoptera : Noctuidae) (Chapter 20)

3. Cabbage Semilooper, *Thysanoplusia orichalcea* (Fabricius) (Lepidoptera : Noctuidae) (Chapter 20, Plate 15.1F)

Management. Spray or 1.250 litres of endosulfan 35EC or 500 ml of dichlorvos 85SL in 250 litres of water per ha. Repeat after two weeks, if necessary.

4. Bihar Hairy Caterpillar, *Spilarctia obliqua* (Walker) (Lepidoptera : Arctiidae) (Chapter 26)

5. Cutworms, *Agrotis* spp. (Lepidoptera : Noctuidae) (Chapter 26)

The cutworms may be serious during March-April in fields where sunflower follows potato. Caterpillars cut the seedlings at the ground level.

Management. (*i*) Sow the crop in ridges to avoid cutworm damage in the germinating seedlings. (*ii*) Where flat sowing is practiced, apply 5 litres of chlorpyriphos 20EC per ha before sowing. The insecticide should be mixed in 25 kg fine soil and broadcasted uniformly in the field after last ploughing but before planking.

Other Pests of Sunflower

The other insect pests of sunflower include the leafhopper, *Amrasca bigutulla* (Ishida) (Hemiptera: Cicadellidae); whitefly, *Bemisia tabaci* (Gennadius) (Hemiptera : Aleyrodidae); cotton aphid, *Aphis gossypii* Glover (Hemiptera : Aphididae); green potato bug, *Nezara viridula* (Linnaeus) and yellow plant bug, *Dolycoris indicus* Stal (Hemiptera : Pentatomidae); thrips , *Thrips tabaci* Lindeman (Thysanoptera : Thripidae); pea leaf-miner, *Chromatomyia horticola* (Goureau) (Diptera: Agromyzidae); red hairy caterpillar , *Amsacta moorei* (Butler) (Lepidoptera: Arctiidae); castor hairy caterpillar, *Euproctis subnotata* (Walker) (Lepidoptera : Lymantriidae); lucerne caterpillar, *Spodoptera exigua* (Hubner) and greasy cutworms, *Agrotis ipsilon* (Hufnagel) and *Ochropleura flammatra* (Denis & Schiffermuller) (Lepidoptera : Noctuidae). Besides these insect pests, some birds like parrot, *Psittacula krameri* (Scopoli); house sparrow, *Passer domesticus* (Linn.); dove, *Streptopelia risoria* (Linn.) and crow, *Corvus splendens* Vieillot have also been found feeding on sown seeds, green head and seeds from the maturing heads.

PESTS OF FIBRE CROPS

INTRODUCTION

Cotton, jute and sunnhemp are the major fibre crops of India. Two types of cottons, viz. *desi* cotton (*Gossypium arboreum* L.) and the American cotton (*Gossypium hirsutum* L.) are commonly grown in the Indian Sub-continent. The *desi* cotton varieties are short-linted and are grown in the submontane and high-rainfall areas. Their leaves are comparatively narrow, more hairy, and are less infested by jassid and whitefly. In areas, where *desi* cotton is grown, the boll-worm populations are also lower and, therefore, the plants escape the attack. The American cottons are generally grown in Pakistan, in the southern districts of the Punjab and Haryana, in the irrigated areas of Rajasthan, Gujarat, Maharashtra, parts of Madhya Pradesh, Andhra Pradesh and Tamil Nadu, particularly where there is some source of irrigation. Their leaves are broad and less hairy. They are heavily attacked by jassid and whitefly and in certain years, by cotton thrips, aphids and leaf-rollers. It is generally believed that heavily manured fields of cotton are more severely attacked by jassid, whereas whitefly is more serious where climate is comparatively dry. In these areas, the bollworms, in general, cause heavy damage and of the 40 per cent of the flowers and fruits which drop off, 25 per cent are shed, because of the attack of bollworms.

Jute (*Corcharus capsulairs* L. and *C. alitarius* L.) is the most important commercial bast fibre crop grown in India. It is cultivated in the traditional jute growing tracts of eastern India consisting mainly in the states of West Bengal, Bihar, Orissa, Assam and Central U.P. The hot and humid climate necessary for jute cultivation favours the development of vairous pests from seedling to harvest. Sunnhemp is grown in almost all states of India either as a fibre crop, manure or fodder crop. The states of Bihar, Madhya Pradesh, Maharashtra, Rajasthan, Orissa and Uttar Pradesh grow this crop mainly for fibre. India contributes 23 per cent of production and 27 per cent of world's area under cultivation.

Although India occupies the largest area in the world under cotton, it ranks second in production. The area under cotton has increased from 8.79 million ha in 2004-05 to 12.18 million ha in 2011-12, but declined to 11.38 million ha in 2012-13 and increased to 11.55 million ha in 2013-14. The production of cotton which was 17.90 million bales (170 kg each) in 2004-05 has more than doubled and touched 35.20 million bales in 2011-12. The production decreased to 34.00 million bales 2012-13 and reached a record level of 36.50 million bales in 2013-14. The target for 2014-15 has been lowered to 35 million bales. The area under jute and mesta which was 0.92 million ha in

2004-05, slightly declined to 0.80 million ha in 2010-11. The production of jute and mesta has increased from 10.27 million bales (180 kg each) in 2004-05 to 11.82 million bales in 2009-10. The production of jute and mesta was 11.40 million bales on 0.90 million ha area in 2011-12. The area decreased to 0.87 million ha with a production of 11.30 million bales in 2012-13. The target for 2014-15 has been fixed at 11.2 million bales.

COTTON (*Gossypium* spp.; Family : Malvaceae)

More than 1326 species of insects have been reported attacking cotton in the world. However, in India, only 162 species have been recorded, among which only 15 species may be called as major pests due to their occurrence in serious proportions almost every year. The introduction of Bt cotton has caused changes in insect pest complex in cotton ecosystem. The incidence of bollworms viz. American bollworm, pink bollworm and spotted bollworms is low on Bt cotton. However, Bt cotton does not provide effective control of tobacco caterpillar and sucking pests, viz. jassid, aphid, whitefly and mealy bug.

1. Cotton Jassid, *Amrasca biguttula biguttula* (Ishida) (Hemiptera: Cicadellidae) (Fig.16.1, Plate 16.1A)

The cotton jassid is widely distributed in India and is the most destructive pest of American cotton in the north-western regions. Besides cotton, is also feeds on okra, potato, brinjal and some wild plants, like hollyhock, *kangi buti* (*Abutilon indicus*), etc. Damage to the crop is caused by the adults as well as by the nymphs, both of which are very agile and move briskly, forward and sideways. Adults are about 3 mm long and greenish yellow during the summer, acquiring a reddish tinge in the winter **(Fig. 16.1)**. The winged adults jump or fly away at the slightest disturbances and are also attracted to light at night.

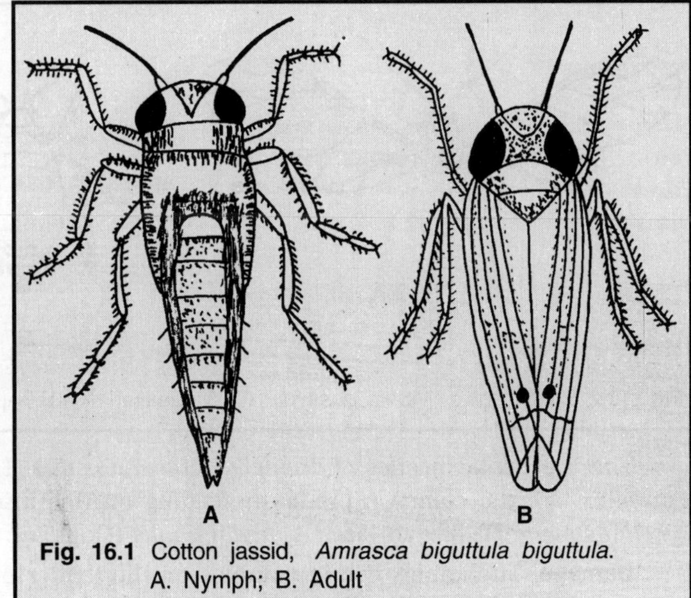

Fig. 16.1 Cotton jassid, *Amrasca biguttula biguttula*. A. Nymph; B. Adult

Life-cycle. The pest breeds practically throughout the year, but during the winter months, only adults are found on plants such as potato, brinjal, tomato, etc. In spring, they migrate to okra and start breeding. The females lay about 15 yellowish eggs on the underside of the leaves, embedding them into the leaf veins. The eggs hatch in 4-11 days and give rise to nymphs which are wedge-shaped and are very active. They suck cell-sap from the underside of the leaves and pass through six stages of growth in 7-21 days. On transformation into winged adults, they live for 5-7 weeks, feeding constantly on the plant juice. The pest completes seven generations in a year.

Chrysoperla sp. (Chrysopidae) and spiders like *Distina albida* L. feed on nymphs and adults. No other parasitoids are known for this insect.

Damage. Injury to plants is due to the loss of sap and probably also due to the injection of toxins. The attacked leaves turn pale and then rust-red **(Plate 16.1A)**. With change in appearance, the leaves also turn downwards, dry up and fall to the ground. Owing to the loss of plant vitality, the cotton bolls also drop off, causing up to 35 per cent reduction in yield.

Management. (*i*) At the time of sowing, smear the cotton seed with imidacloprid 70WS @ 5g/kg seed or thiomethoxam 70WS @ 3 g/kg seed. (*ii*) Spray against jassid should be done at economic threshold level of 1-2 nymphs per leaf or when second grade injury symptoms (yellowing and curling at margins of leaves) appear in 50 per cent of the plants. Any one of the following insecticides can be used in 250 litres of water per ha : 100 ml of imidacloprid 200 SL, 100 ml of imidacloprid 555, 100 ml of imidacloprid 17.8SL, 50 g of acetamiprid 20SP, 100g thiomethoxam 25WG, 200 g flonicamid 50WG.

2. Cotton Aphid, *Aphis gossypii* Glover (Hemiptera: Aphididae) (Fig. 16.2)

The insect is distributed throughout India and is considered as a potential pest of cotton. It also infests many other crops like okra, brinjal, guava, gingelly, etc. The adults are small, soft and greenish-brown insects found in colonies on the tender parts of the plants. The adults exist both in winged and wingless forms **(Fig. 16.2)**.

Life-cycle. The alate as well as apterous females multiply parthenogenetically and viviparously. The female may give birth to 8-22 nymphs in a day. The nymphs moult four times to become adults completing the life-cycle in 7-10 days. The aphids also lay eggs which overwinter.

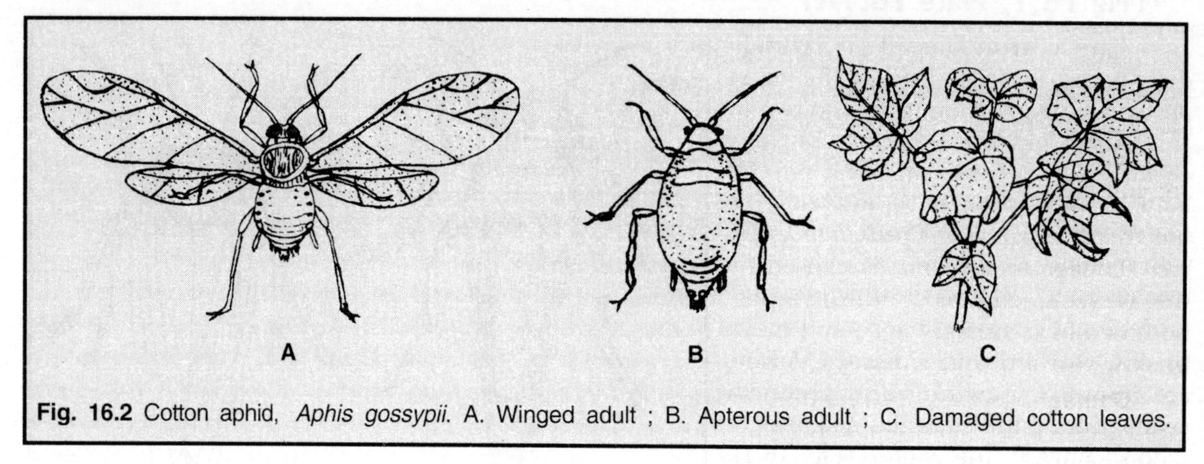

Fig. 16.2 Cotton aphid, *Aphis gossypii*. A. Winged adult ; B. Apterous adult ; C. Damaged cotton leaves.

The coccinellid beetles, *Coccinella septempunctata* Linnaeus and *Menochilus sexmaculatus* (Fabricius) are the common predators feeding on this insect. The aphids are also parasitized by *Aphelinus mali* (Haldeman) and *A. abdominalis* (Dolman).

Damage. The aphids live in colonies on the tender portions of the plants and suck the sap. In case of severe infestation, the plants become weak, leaves curl up and wither **(Fig. 16.2C)**. There is stunted growth, gradual drying and death of the plants. The younger plants are more susceptible than the older ones. The honey-dew secreted by the aphid encourages sooty mould growth on the leaves. The dry conditions favour rapid increase in pest population.

Management. Spray 750 ml of oxydemeton methyl 25EC or 625 ml of dimethoate 30EC in 250 litres of water/ha, when honeydew appears on 50 per cent plants.

3. Cotton Whitefly, *Bemisia tabaci* (Gennadius) (Hemiptera : Aleyrodidae) (Fig 16.3, Plate 16.1B)

This pest is distributed throughout the northern and western regions of the Indian Sub-continent and is a very serious pest of American cotton, particularly in the dry areas. Apart from cotton, this insect also feeds on various other plants such as cabbage, cauliflower, *sarson, toria* (*Brassica campestris*), melon, potato, brinjal, okra and some weeds.

The louse-like nymphs, which suck the sap, are sluggish creatures, clustered together on the under surface of the leaves and their pale-yellow bodies make them stand out against the green background. In the winged stage, they are 1.0-1.5 mm long and their yellowish bodies are slightly dusted with a white waxy powder. They have two pairs of pure white wings and have prominent long hind wings.

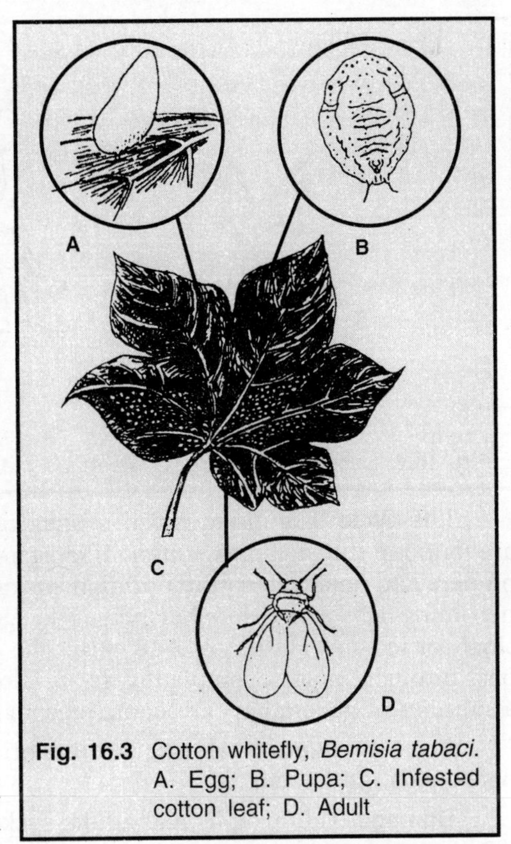

Fig. 16.3 Cotton whitefly, *Bemisia tabaci*. A. Egg; B. Pupa; C. Infested cotton leaf; D. Adult

Life-cycle. The insect breeds throughout the year and all the developmental stages are noticed, although adults predominate in the cold season. The females lay eggs singly on the underside of the leaves, averaging 119 eggs per female. The eggs are stalked, sub-elliptical and light yellow at first, turning brown later on. They hatch in 3-5 days in April-September, 5-17 days in October-November and 33 days in December-January. The nymphs, on emergence, look elliptical and soon fix their mouthparts into the plant tissues. They feed on cell-sap and grow into three stages to form the pupae within 9-14 days in April- September, and in 17-81 days in October-March. In 2-8 days, the pupae change into white flies. The life-cycle is completed in 14-122 days and 11 generations are completed in a year.

Eretmocerus massii Silv. (Aphelinidae) is associated with nymphs and pupae, and incidence of parasitization reaches up to 33 per cent while *Encarsia* sp. is responsible for 5-10 per cent parasitization of nymphs and pupae. *Chrysoperla* sp. (Chrysopidae) and *Brumus* sp. (Coccinellidae) prey upon nymphs and adults.

Damage. Damage by this pest is caused in two ways: (*a*) the vitality of the plant is lowered through the loss of cell sap, and (*b*) normal photosynthesis is interfered with due to the growth of a sooty mould on the honeydew excreted by the insect. From a distance, the attacked crop gives a sickly, black appearance. Consequently, the growth of the plants is adversely affected and when the attack appears late in the season, the yield is lowered considerably. *B.tabaci* is known to transmit a number of virus diseases including the leaf curl disease of tobacco, the vein clearing disease of okra and the leaf-curl of sesame.

Management. Chemical control measures are same as in case of cotton aphid. In addition, spray 500 g of diafenthiuron 50WP or 500 ml of spiromesifen 240SC or 1.5 litres of triazophos 40EC or 2.0 litres of ethion 50EC during boll formation phase. The economic threshold level is 20 nymphs or 8-10 adults/leaf or leaf stickiness due to deposition of honeydew by insects in 50 per cent of plants.

4. Dusky Cotton Bug, *Oxycarenus laetus* Kirby (Hemiptera : Lygaeidae) (Fig. 16.4)

This is a minor pest of cotton in India and its chief importance lies in the fact that the adults and nymphs get crushed at the time of ginning, thus staining the lint and lowering the market value of cotton. Besides cotton, it also feeds on okra, hollyhock and other malvaceous weeds.

The adults are 4-5 mm in length, dark brown and have dirty white transparent wings **(Fig. 16.4)**. The young nymphs have a rotund abdomen and, as they grow older, they resemble the adults, except for being smaller and having prominent wing pads instead of wings.

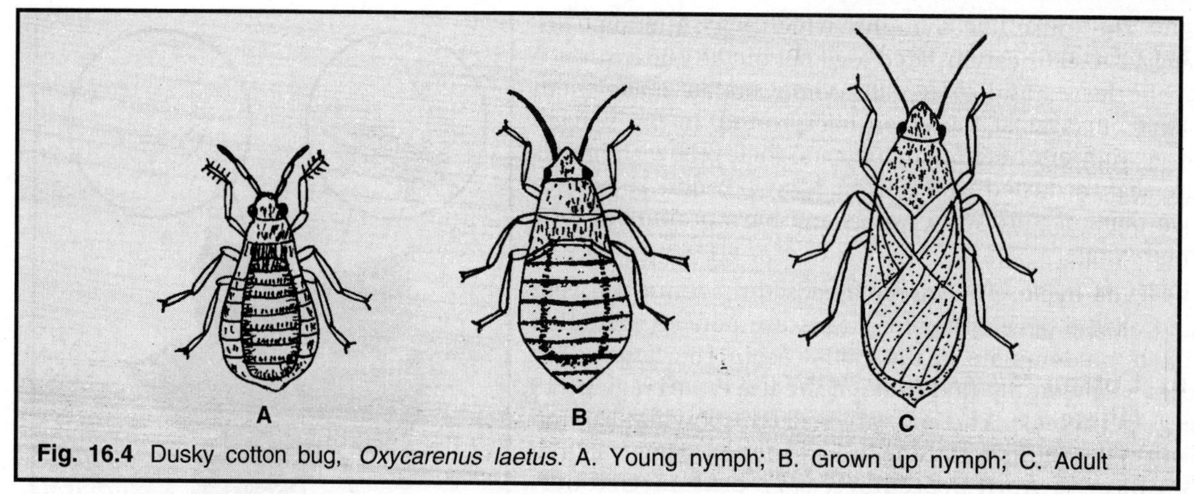

Fig. 16.4 Dusky cotton bug, *Oxycarenus laetus*. A. Young nymph; B. Grown up nymph; C. Adult

Life-cycle. The insect is active practically throughout the year, but during winter, only adults are found in the unginned cotton. The cigar-shaped eggs are laid in the spring on *Hibiscus* and then on okra and finally on cotton during the monsoon. Initially, they are whitish turning pale and finally becoming light pink before hatching. The eggs are usually laid in the lint of half-opened bolls, either singly or in small clusters of 3-18 each. The egg stage lasts 5-10 days and the nymphs, on emerging, pass through 7 stages, completing the development in 31-40 days. The life cycle lasts 36-50 days and a number of generations are completed in a year.

Triphleps tantilus Motsch (Anthocoridae) feeds on the nymphs of this bug and no other parasitoids are known.

Damage. The nymphs and adults suck the sap from immature seeds, whereupon these seeds may not ripen, may lose colour and may remain light in weight. The adults found in the cotton are crushed in the ginning factories, thus staining the lint and lowering its market value.

Management. Same as in case of cotton aphid.

5. Red Cotton Bug, *Dysdercus koenigii* (Fabricius) (Hemiptera : Pyrrhocoridae) (Fig. 16.5)

This insect is widely distributed in India and is a minor pest of cotton in the Punjab and Uttar Pradesh. Apart from cotton, it also feeds on okra, maize, pearl millet, etc. The bugs are elongated slender insects, crimson red with white bands across the abdomen (**Fig. 16.5**). The membraneous portion of their fore wings, antennae and scutellum is black. Both adults and nymphs feed on the cell-sap of cotton, hollyhock, wheat, maize, pearl-millet, clovers, etc.

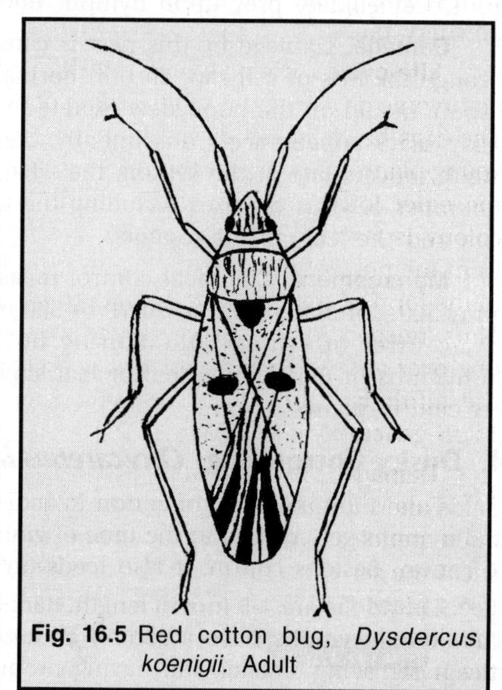

Fig. 16.5 Red cotton bug, *Dysdercus koenigii.* Adult

Life-cycle. This insect is active throughout the year and passes winter in the adult stage. In spring, the bug becomes active and lays, on an average, 100-130 eggs in moist soil or in crevices in the ground. The eggs are spherical, bright yellow and are laid in clusters or in loose irregular masses of 70-80 eggs each. They hatch in 7-8 days and the young nymphs have flabby abdomens, but as they grow older, they become more slender and develop black markings on the body. There are 5 nymphal stages and the development is completed in 49-89 days. In summer, the life of an adult is very variable, but in winter, it may live up to three months.

The predacious bugs like *Antilochus cocqueberti* Fabricius (Pyrrhocoridae) and *Harpactor costalis* (Stal) (Reduviidae) feed on nymphs and adults.

Damage. The insects suck cell-sap from leaves and green bolls of cotton. Heavily attacked bolls open badly and the lint is of poorer quality. The seed produced may also have low germination and less oil. Moreover, the bugs stain the lint with their excreta or body juices as they are crushed in the ginning factories. The staining of lint by the growth of certain bacteria inside the bolls is also believed to be initiated by these bugs.

Management. Same as in case of cotton aphid.

6. Cotton Mealybug, *Phenacoccus solenopsis* Tinsley (Hemiptera : Pseudococcidae) (Plate 16.1C)

Cotton mealybug has recently emerged as a serious pest of cotton in several Asian countries including India and Pakistan. Besides cotton, it also feeds on tomato, mustard, sorghum and *bakain* (*Melia azedarach*) trees, and several weed plants like congress grass(*Parthenium hysterophorus*), *kanghi buti* (*Abutilon theophrasti*), *peeli buti* (*Abutilon indicum*), *puthkanda* (*Achyranthes aspera*), *gutputna* (*Xanthium strumarium*), *bhakhra* (*Tribulus terrestris*), *itsit* (*Trianthema portulacastrum*) and *tandla* (*Digera arvensis*).

This insect severely infected the cotton crop in the cotton growing provinces (Sindh and Punjab) of Pakistan in 2005 and 2006. In India mealybug appeared in serious proportions in 2006 in cotton growing areas of Punjab, Haryana, Rajasthan, Maharashtra and Gujarat, and caused widespread damage to cotton. It struck in epidemic form in 2007 and at many places, farmers had to plough down their badly infested crop.

The adult is pink and elongated oval (about 3 mm in length), flattened and covered with a white mealy powder. The female adults are wingless, however, male adults are having one pair of wings. The nymphs are pale yellow with reddish eyes, which are later on covered with white powdery mass.

Life-cycle. A female lays 150-600 eggs usually in a cotton like ovisac beneath her body. The embryonic development of the young ones takes place within the body of the female (ovoviviparity). The eggs hatch in 6-9 days. The newly emerged nymphs (crawlers) crawl out and start feeding on young plants. They prefer the growing shoot of the plant which is soft and juicy. Later on, they attack the older lower portion of the cotton plant. Its population increases tremendously and soon a colony is formed. Mature females are relatively sessile with a large, outpocketing cottony pouch from the posterior end, stuffed with its offspring. In case of severe infestation, the mealybug occurs in the form of clusters with white cottony crawler sacs protruding from the anal region of their body. Thus, unspecified protective layers are formed due to the overlapping and clinging of one mealy-bug over another. The female mealybugs pass through four instars in 10-15 days. The longevity of the adult is 7-14 days. The total life-cycle is completed in 25-30 days and there may by 14-15 generations in a year.

Damage. The mealy-bug first attacks the alternate hosts growing in outer region of the cotton fields and then shifts to the cotton crop. Both nymphs and adults suck the sap from leaves, flower buds, petioles, twigs, internodes and even from the stem of the cotton plants. The insect heavily sucks the sap from the plant and renders it weak, feeble and dehydrated. In severe cases, defoliation takes place by the development of sooty mould on honeydew produced by crawlers. The sooty mould reduces the photosynthetic ability of the plants. In addition, mealybug also injects toxins into the host tissues.

The flower buds, flowers, immature bolls and even the leaves fall down, and the growth of the plant is retarded. The infested plants bear fewer bolls of smaller size and there is incomplete opening of bolls. The entire plant remains stunted and shoot tips develop a bushy appearance. The cotton plant exhibits a die-back of growing shoots and twigs, ultimately reducing the yield. This damage of the pest is initiated from the borders and spreads to the main cotton crop. White cottony heads appear on the infested plants where intensive infestation of all the stages can be observed. Its infestation starts in patches and in case of severe infestation, the infested plants give the appearance of a field sprayed with a defoliant or herbicide.

Management. Management of mealy-bug is difficult due to its wide host range, wax coating on the body, dense colonies, nature of hiding in cracks and crevices in the bark, crawling off the host plants and easy spread to other areas. However, the following IPM strategy helps to keep the pest under check : (*i*) To check the spread of mealybug, remove regularly the weeds growing adjacent to road sides, pathways, water channels and waste lands. (*ii*) Destroy the infested host plants alongwith the pest very carefully, preferably by burning/deep ploughing, but never throw in open water channels. (*iii*) Avoid growing cotton near the fields of tomato, sorghum and bakain trees, as they serve as alternate hosts of mealybug. (*iv*) Grubs of green lace wing, *Chrysoperla carnea* (Stephens) and adults of *Cryptolaemus montrouzieri* Mulsant could be used as biological control agents as they feed on the nymphs of the mealy-bug. (*v*) In case of severe infestation, spray any one of the following insecticides in 300-400 litres of water per ha : (*a*) *Carbamates.* 2.5 kg of carbaryl 50WP or 625 g of thiodicarb 75WP. (*b*) *Organophosphates.* 1.25 litres of profenophos 50EC or 2.0 litres of quinalphos 25EC or 2.0 kg of acephate 75SP or 5.0 litres of chlorpyriphos 20EC. (*c*) *Insect growth regulator.* 1.25 litres of buprofezin 25EC. (*d*) The first spray should preferably be done with carbamates and if required, repeat spray after 5-7 days with organophosphates. (*e*) Since damage of mealybug is normally localised to border rows, treat only these rows with the above pesticides.

7. Pink Bollworm, *Pectinophora gossypiella* (Saunders) (Lepidoptera : Gelechiidae) (Fig.16.6, Plate 16.1D)

The pink bollworm is one of the most destructive pests of cotton in the world and is found in America, Africa, Australia and Asia. It is highly destructive in the Punjab, Haryana and Pakistan.

The damage is caused by the caterpillars only. They are pink and are found inside flower buds, panicles and the bolls of cotton or the fruits of okra and other allied plants. In the adult stage, the insect is a deep brown moth, measuring 8-9 mm across the spread wings (**Fig. 16.6**). There are blackish spots on the fore wings, and the margins of the hind wings are deeply fringed.

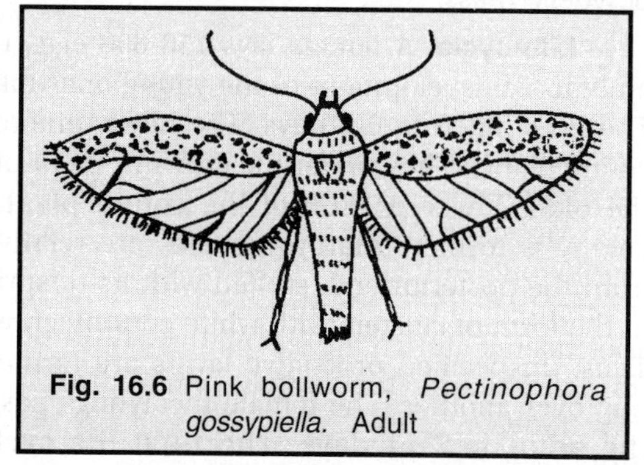

Fig. 16.6 Pink bollworm, *Pectinophora gossypiella.* Adult

Life-cycle. The yearly life-cycle begins with the emergence of moths in the summer. The emergence takes place at two distinct times: in May-June, and then in July-August. The females lay whitish, flat eggs singly on the underside of the young leaves, new shoots, flower buds and the young green bolls. The eggs hatch in one week and the caterpillars, on emergence, are white and turn pink as they grow older. Soon after emergence, the larvae enter the flower-buds, the flowers or the bolls. The holes of entry close down, but the larvae continue feeding inside the seed kernels. They become full-grown (8-10 mm) in about two weeks and come out of the holes for pupation on the ground, among fallen leaves, debris, etc. Within one week, the moths emerge to start the life-cycle all over again. By October-November, 4-6 generations are completed.

Full-grown larvae of the last generation do not, however, pupate. Just a few of them reach the ground, but the great majority keep feeding inside the bolls. They cut window holes in the two adjoining seeds and join them together, forming what are known as the 'double seeds'. Such damaged bolls are generally left unpicked in the field. Later, they fall to ground and form a major source of infestation for the next year. Some are picked along with healthy cotton and reach the ginning factories, from where just a few return to the fields along with the seed. The hibernating larvae lie curled in double seeds for many months and after passing the winter, they emerge as moths. The last life-cycle is very long covering 5-10 months, although during the active season, the life cycle is short, taking only 3-4 weeks. The pest passes through several broods during its active period.

Trichogramma achaeae Nagaraja & Nagarkatti (Trichogrammatidae) parasitizes the eggs of this pest throughout India and the incidence is 6-27 per cent in North India. It appears late in the season and is common in area where pesticides are used sparingly. Releases of *Trichogramma* spp. at 1,50,000 parasitized eggs ha^{-1} at weekly intervals have proved promising for bollworm control. *Apanteles angaleti* Muesebeck (Braconidae) parasitizes 1-17 per cent of the larvae of the host. It is widely distributed in India and the parasitoid is associated with the host throughout the year.

The other hymenopterous parasitoids associated with larvae of the pink bollworm are *Bracon greeni* Ashmead and *Chelonus pectinophorae* Cushman (Braconidae), *Elasmus johnstoni* Ferriere (Elasmidae), *Goniozus* sp. (Bethylidae), *Rogas aligharensi* Quadri and *Bracon lefroyi* D. (Braconidae). An eulophid parasitizes the pupae of this pest. The anthocorid bug, *Triphles tantilus* Motsch. (Anthocoridae) also feeds on eggs and first instar larvae.

Damage. Damage is caused in various ways. When flower is infested, a typical rosette-shaped bloom harbouring the larvae can be seen (**Plate 16.1D**). There is excessive shedding of the fruiting bodies. Of the total shedding, 52.4-88.8 per cent is caused by all the bollworms collectively, one half may be due to the attack of pink bollworm. The attacked bolls fall off permaturely and those which do mature do not contain good lint. The damaged seed-cotton gives a lower ginning percentage, lower oil extraction and inferior spinning quality. It is considered that by controlling the pink and spotted bollworms, the cotton yield can be increased upto 50 per cent.

Management. (*i*) The destruction of off-season cotton sprouts, alternative host plants or the burning of plant debris from cotton fields, minimizes the incidence of this pest. (*ii*) Deep ploughing with a furrow-turning plough by the end of February is also helpful in reducing the carry over of this pest to the next season. (*iii*) Release of *Trichogramma chilonis* Ishii @ 1,50,000/ha starting from 70th day after sowing at weekly interval with a total of 8-10 releases is effective against bollworm. (*iv*) In case the bollworm damage exceeds 5 per cent, the crop should be sprayed immediately and thereafter at 10-day interval with any of the following insecticides in 315-375 litres of water per ha : (*a*) *Organophosphates.* 2.0 litres quinalphos 25EC, 1.25 litres monocrotophos 36SL, 5.0 litres chlorpyriphos 20EC, 1.5 litres triazophos 40EC, 2.0 litres ethion 50EC, 2.0 kg acephate 75SP, 1.25 litres of profenophos 50EC. (*b*) *Carbamates.* 2.5 kg carbaryl 50WP, 625 g thiodicarb 75WP. (*c*) *Synthetic pyrethroids.* 500 ml cypermethrin 10EC, 200 ml cypermethrin 25EC, 400 ml deltamethrin 2.8EC, 250 ml fenvalerate 20EC, 250 ml alphamethrin 10EC, 750 ml β-cyfluthrin 0.25SC, 750 ml fenpropathrin 10EC. (*d*) *Miscellaneous.* 150 ml spinosad 48SC, 100 ml flubendiamide 48SC, 500 ml indoxacarb 15SC, 625 g thiodicarb 75WP, 375 ml novaluron 10EC. (*v*) At least 5-6 sprays are required for effective control of bollworms. Same insecticide should not be sprayed repeatedly to avoid the development of pesticide resistance and appearance of secondary pests. Also, avoid using insecticides of the same group in more than three sprays.

8. Spotted Bollworms, *Earias insulana* (Boisduval) and *E. vittella* (Fabricius) (Lepidoptera : Noctuidae) (Fig.16.7)

These two species of bollworms are widely distributed in North Africa, India, Pakistan and other countries, and are serious pests of cotton. In Punjab, they cause heavy damage to American cotton and are also found on okra (*Abelmoschus esculentus* L.), *sonchal* (*Malva parviflora*), *gulkhaira*

(*Althaea officinalis*), hollyhock (*Althaea rosea*) and some other malvaceous plants. In the larval stage, they bore into the growing shoots, the flower buds, flowers and fruits of cotton and okra, either killing the plants or causing heavy shedding of the fruiting bodies. In the attacked bolls, the lint is spoiled by larval feeding. Okra fruits become distorted and are rendered unfit for human consumption. The full-grown dull-green caterpillars are 20 mm long having tiny stout bristles and a series of longitudinal black spots on the body. The moths are yellow green and measure about 25 mm across the wings. *E. vittella* moths are of the same size and have a narrow light longitudinal green band in the middle of the fore wing **(Fig. 16.7)**.

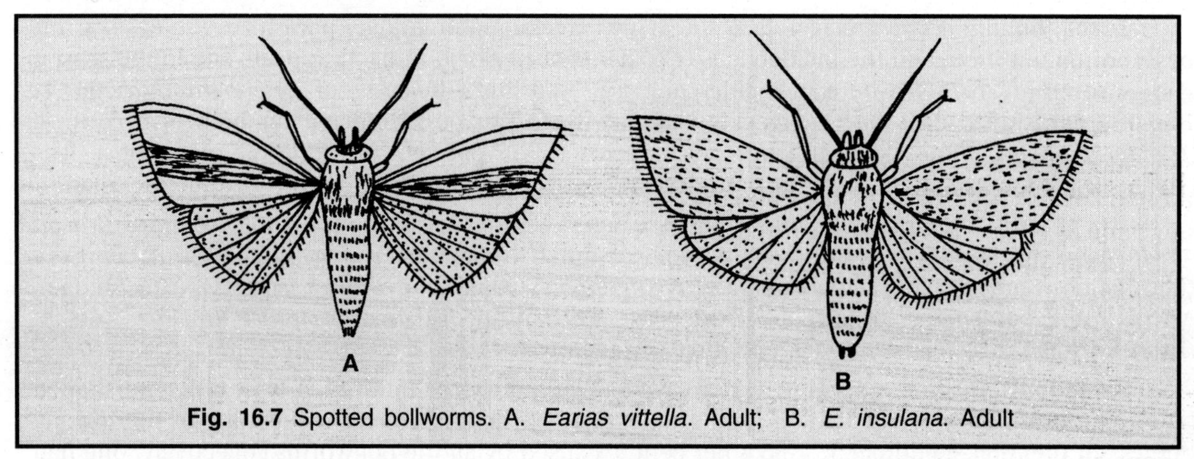

Fig. 16.7 Spotted bollworms. A. *Earias vittella*. Adult; B. *E. insulana*. Adult

Life-cycle. The pests breed practically throughout the year but during the winter only the pupae are found hiding in plant debris. The moths that appear in April live for 8-22 days and lay 200-400 eggs at night, singly on flower buds, brackets and tender leaves of okra or cotton plants. The hairy parts of the plants are preferred for oviposition. In warm weather, the eggs hatch in 3-4 days and the caterpillars pass through 6 stages, becoming full-grown in 10-16 days. They pupate either on the plants or on the ground among fallen leaves and the moths emerge in 4-9 days. The life-cycle is completed in 17-29 days during the summer. In winter, the eggs hatch in about one week and pupal stage is greatly prolonged, taking 6-12 weeks. Several overlapping generations are completed in a year. The roots of cotton plants sprouting in early spring and the fruits of neglected okra left in the field, are the two important sources of early infestation and multiplication of this pest.

Trichogramma chilonis Ishii (Trichogrammatidae), an egg parasite is widely distributed in India and Pakistan. In North India, the incidence varies from 3 to 15 per cent. The release of parasitoids in cotton fields is very useful in reducing the incidence of pest.

Trichogramma brasiliensis (Ashmead) (Trichogramma-tidae), the exotic parasitoid, has been tried in the Punjab and other cotton growing areas. The releases of this parasitoid reduced the incidence by 50 per cent in experimental fields. The unilative releases in July-August and inoculative releases afterwards are recommended.

Another egg parasitoid, *Trichogrammatoidea* sp. near *guamensis* (Trichogrammatidae) is also associated with this pest. *Brachymeria nephantidis* Gahan (Chalcididae) has also been recorded from pupae of *Earias* sp. and parasitism ranges from 13 to 57 per cent in different months. The parasitoids associated with larvae of *P. gossypiella* are also associated with this pest.

Damage. When cotton plants are young, the larvae bore into the terminal portions of the shoots, which wither away and dry up. Later on, they cause 30-40 per cent shedding of the fruiting bodies. The infested bolls open prematurely and produce poor lint, resulting in lower market value.

Management. (*i*) The pest can be suppressed with clean cultivation and the destruction of alternative food plants, particularly when cotton or okra is not growing in that locality. (*ii*) Chemical control measures are same as in case of pink bollworm. The economic threshold level is 10 per cent incidence in shoots or reproductive parts.

9. American Bollworm, *Helicoverpa armigera* (Hubner) (Lepidoptera : Noctuidae) (Plate 16.1E)

The American bollworm has become a serious pest of cotton in northern parts of India. It was first reported on cotton in 1977 in Punjab. In 1983, it appeared in serious form in the northern cotton growing belt and reduced yield by 40-50 per cent. Again in 1990, it caused severe losses to cotton in Haryana, Punjab and Rajasthan. The incidence of this pest has decreased after the introduction of Bt cotton The insecticides effective against pink bollworm are also effective against this pest. The economic threshold is 10 per cent incidence in reproductive parts. The pest has been described in detail in Chapter 14.

Management. (*i*) Avoid the cultivation of okra, moong, arhar, dhaincha and castor in and around cotton crop. (*ii*) Avoid growing American cotton in orchards. (*iii*) Chemical control measures are same as in case of pink bollworm. Prefer to use acephate/chlorpyriphos against grown up larvae (more than 1.25 cm long) of American bollworm. Spinosad and indoxocarb should be used only in emergency at higher incidence of pest when it becomes difficult to control with other insecticides.

10. Cotton Leaf Roller, *Sylepta derogata* (Fabricius) (Lepidoptera : Pyralidae) (Fig. 16.8)

The cotton leaf-roller is widely distributed in the Orient and Africa. In India, it occurs in all the cotton-growing tracts and is an important sporadic pest. Besides cotton, it has also been recorded on various other malvaceous plants like okra, *gulkhaira* (*Althaea officinalis*), *kanghi buti* (*Abutilon indicum*) and some forest trees.

Damage is done by caterpillars, which measure about 25-30 mm in length when full- grown **(Fig. 16.8)**. They are greenish grey or pink and are usually found feeding inside the rolled leaves. Moths are yellowish-white, with black and brown spots on the head and the thorax. They measure about 28-40 mm across the spread wings and have a series of dark brown wavy lines on the wings.

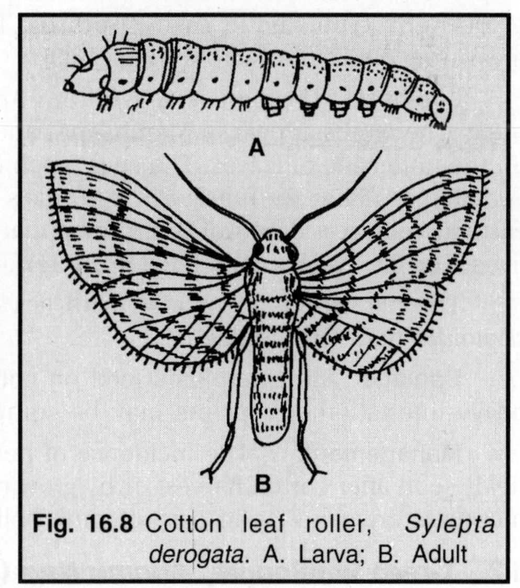

Fig. 16.8 Cotton leaf roller, *Sylepta derogata*. A. Larva; B. Adult

Life-cycle. The leaf-roller is active from March to October and passes the winter as a full-grown caterpillar among plant debris or in the soil. The hibernating larvae pupate by the end of February and the moths emerge during March. The moths are active at night, then they mate and lay 200-300 eggs singly on the underside of the leaves. The eggs hatch in 2-6 days and the young caterpillars feed first on the lower surface of the leaves. The older larvae roll leaves from the edges inwards up to the midrib and feed on leaf tissues from the inside. The larvae grow through seven stages and are full-fed in 15-35 days. They pupate either on the plant, inside the rolled leaves or among plant debris in the soil. They emerge as moths in 6-12 days and live for about a week. The life-cycle is completed in 23-53 days and the pest passes through 5 or 6 generations.

From March to June, the infestation develops mostly in thick plantations of forest trees but during the monsoon, it spreads to nearby fields of American cotton. The low maximum temperature, high relative humidity and cloudy weather are considered to be favourable for the rapid multiplication and spread of this pest.

Trichogramma spp. (Trichogrammatidae) attack the eggs while *Brachymeria tachardiae* Cam. (Chalcididae), *Elasmus indicus* Rohn. (Elasmidae), *Goryhus nursei* Cam. (Ichneumonidae) and *Trichospilus pupivora* Ferrieri (Eulophidae) are associated with the larvae of this pest. *T. pupivora* was also introduced in Pakistan from India and releases were made but it failed to establish there.

Damage. The larvae feed on cotton leaves and in years of serious outbreaks, the cotton plants may be completely defoliated. American cotton is preferred over *desi* cotton by this pest.

Management. Same as in case of pink bollworm.

11. Cotton Semilooper, *Tarache notabilis* (Walker) (Lepidoptera : Noctuidae) (Fig. 16.9)

This insect is found throughout the plains of Pakistan and India as a minor pest of cotton, but in certain years, it assumes a serious form. Besides cotton, it also feeds on the leaves of *sonchal* (*Malva parviflora*), brinjal, etc.

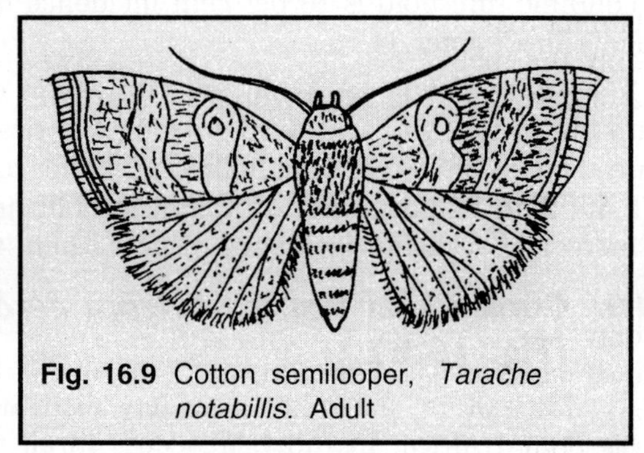

Fig. 16.9 Cotton semilooper, *Tarache notabillis.* Adult

The full-grown caterpillars are about 40 mm long and dark green, having six pairs of black and bright-yellow spots on the back. As they move, they bend the body, forming half loops and hence the name semi-looper. The adults are stoutly built, white moths, with prominent grey and brown spots on the wings, measuring 30-32 mm across (**Fig. 16.9**).

Life-cycle. The pest is active from April to October and passes the winter as an hibernating pupa. The moths appear in March-April and are active at night. They lay green, rounded eggs singly, on the underside of leaves. The eggs hatch in 2-4 days and the caterpillars immediately start feeding voraciously. They are full-fed in 9-16 days and then descend to the ground, where they construct earthen cells at a depth of 3 cm for pupation. They emerge as moths in 5-14 days and are short-lived. The life-cycle is completed in 16-34 days. By October-November, they complete 4-5 generations.

The parasitoids associated with larvae of this pest are *Actia monticola* Mall. and *Exorista seviloides* Bar. (Tachinidae).

Damage. The caterpillars feed on cotton leaves and skeletonize them altogether. In years of heavy infestation, the plants may be completely denuded of leaves.

Management. (*i*) The incidence of pest on thc next crop can bc minimized by ploughing the fields soon after cotton harvest or by growing clovers in rotation with this crop. (*ii*) Chemical control measures are same as in case of pink bollworm.

12. Green Semilooper, *Anomis flava* (Fabricius) (Lepidoptera : Noctuidae) (Fig. 16.10)

The green semi-looper which feeds on leaves is widely distributed throughout the cotton growing areas of Africa, Asia and Australia. In India, it is a sporadic pest of cotton and sometimes causes a serious damage to the crop particularly in Gujarat and Rajasthan. Besides cotton, the larvae also feed on *Hibiscus esculentus* L., *Hibiscus cannabinus* L., and *Phaseolus radiatus* L. among the cultivated plants and on *Hibiscus rosasinensis* L., *Sida cardifolia* L., *Althaea rosea* CAV., *Bombyx malabaricum* DC. and *Malachra capitata* L. among the forest trees.

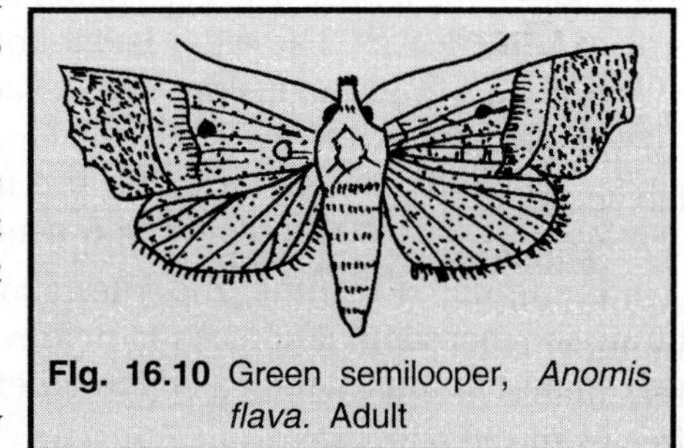

Fig. 16.10 Green semilooper, *Anomis flava.* Adult

The full-grown larva is 25-30 mm long and is pale-yellowish green with five white lines arranged longitudinally on the dorsal surface. The adult is an attractive, small moth with reddish-brown fore wings traversed by two darker zig-zag bands. The hind wings are pale brown **(Fig. 16.10)**.

Life-cycle. The emergence of moths coincides with the warming up of the season. The female lays up to 600 eggs on leaves during an oviposition period of 8-12 days. The eggs hatch in 4-5 days. The larvae which move around with a characteristic semi-loop crawl and feed on leaves. They complete development by passing through 5 instars within a period of 18-20 days. They make a loose cocoon before pupation, either within the leaf folds or among debris in the soil. The pupal period lasts 8-9 days. The moths have a life-span of 12-13 days and the life-cycle is completed in 4-6 weeks. More than one generation is completed in a crop season.

Damage. The young larvae congregate in small groups and move actively and feed on leaf lamina making small punctures. The grown-up larvae feed voraciously on the entire leaves, leaving only the main veins. The caterpillars also eat the tender shoots, buds and bolls. Sometimes, the attack is heavy and the plants may be completely defoliated.

Management. Same as in case of pink bollworm.

13. Bud Moth, *Phycita infusella* (Meyrick) (Lepidoptera : Pyralidae)

This insect is found throughout India as a minor pest of cotton and is more serious on *desi* than on the American varieties. It is sporadic in occurrence and becomes serious in certain localities only. Apart from cotton, it also feeds on okra and other allied plants. In the caterpillar stage, it makes characteristic webbings around the terminal young growth of leaves, where it feeds. The full-grown greenish caterpillars are 13-16 mm long, having a black front and faint brown stripes on the rest of the body. The adult is a greyish-yellow moth, speckled with fuscous, and measures 20 mm across the wing expanse. The front wings are faint yellow at the base with red wavy lines and the hind wings are white, with brown marginal lines.

Life-cycle. The pest is active from April to October and passes winter as a hibernating caterpillar in debris on the ground. The moths appear in April and lay small rounded eggs on tender leaves of cotton or other host plants. The eggs hatch in about 7 days and the young larvae start feeding on bud or young leaves. As the larvae grow, they web together the terminal leaves with silken threads and feed inside. They are full-fed in 21 days and make silken cocoons in the midst of twisted leaves. The pupal stage lasts 11 days and the moths, on emergence, live for a few days only. The life-cycle is completed in 5-6 weeks and the pest completes 4-5 generations in a year.

Bracon brevicornis Wesmael and *B. kitcheneri* Will. (Braconidae) are associated with larvae of this pest.

Damage. Due to the webbing of leaves and feeding by the caterpillars, further growth of the plants is inhibited. The attacked leaves wither away, dry up and drop off. The flowering and fruiting are very poor and the reduction in yield is considerable.

Management. Spray 2.0 kg of carbaryl 50WP in 315-375 litres of water per ha.

14. Tobacco Caterpillar, *Spodoptera litura* (Fabricius) (Lepidoptera : Noctuidae) (Chapter 20, Plate 16.1F)

Tobacco caterpillar is a polyphagous pest and the larvae cause serious damage to cotton from August to November. The young larvae feed gregariously and skeletonize the foliage. The grown up larvae disperse and feed singly. Besides leaves, they also damage the buds, flowers and green bolls.

Management. (*i*) Avoid growing castor, *moong, dhaincha (Sesbania aculeata)* and okra in and around cotton as these are the most preferred hosts of tobacco caterpillar and help the pest to multiply before shifting to cotton. (*ii*) Plough the *dhaincha* crop immediately when it has attained reasonable height and bury the larvae alongwith the straw deep into the soil. (*iii*) Keep the fields free from the weed *itsit (Trianthema portulacastrum),* as it acts as an alternate host of the insect.

(*iv*) Collect and destroy the egg masses and young larvae feeding gregariously on the cotton crop. (*v*) Spray 2·0 kg of acephate 75SP or 5·0 litres of chlorpyriphos 20EC or 2·0 litres of quinalphos 25EC or 625 g of thiodicarb 75WP or 375 ml of novaluron 10EC or 750 ml of pyridalyl 10EC or 150 ml of chlorantraniliprole 18.5SL in 250 litres of water per ha.

15. Cotton Grey Weevil, *Myllocerus undecimpustulatus* Faust (Coleoptera: Curculionidae) (Fig. 16.11)

This insect is found throughout India and is a minor pest of cotton, particularly of the *desi* cotton varieties. It also feeds on a number of other cultivated plants such as *bajra* (*Pennisetum typhoides*), maize, sorghum, guava, *arhar* (*Cajanus cajan*), groundnut, etc. The plants are attacked by the weevils which are prominent above the ground as well as the grubs which feed on the underground parts. The weevils are grey and are 3-6 mm long. The grubs are white, legless, cylindrical and are about 8 mm in length **(Fig. 16.11)**.

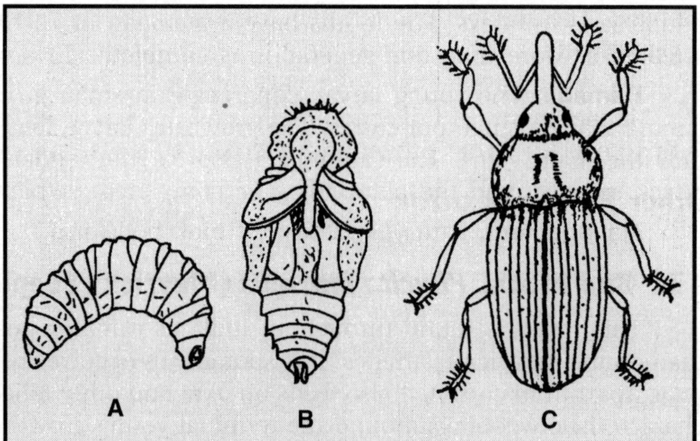

Fig. 16.11 Cotton grey weevil, *Myllocerus undecimpustulatus*. A. Larva; B. Pupa; C. Adult

Life-cycle. The pest is active from April to November and passes winter in the adult stage hidden in debris. The weevils appear in April-May and lay ovoid, light yellow eggs in the soil. A female lays, on an average, 360 eggs over a period of 24 days. The eggs hatch in 3-5 days in May-September and the young grubs feed on the roots of cotton and other plants. The grubs complete their development in 1-2 months. They pupate in the soil inside earthen cells, forming creamy-white pupae, which change into adults in about one week. The adults live for 8-11 days in the summer and 4-5 months in the winter. During the active period, the life-cycle is completed in 6-8 weeks and the pest probably breeds 3-4 times in a year.

Dinocampus mylloceri Walker (Braconidae) parasitizes the larvae of this insect.

Damage. Both adults and grubs cause damage. The grubs feed underground on the roots of cotton seedlings and destroy them. One grub can destroy 9 seedlings in 40 days. The adults which feed on leaves, buds, flowers and young bolls cut prominent round holes.

Management. (*i*) The pest can be suppressed by disturbing the soil up to a depth of 7.5 cm and destroying the eggs, grubs and pupae. (*ii*) Spray 2.5 kg of carbaryl 50 WP in 375 litres of water per ha.

16. Cotton Stem Weevil, *Pempherulus affinis* (Faust) (Coleoptera : Curculionidae)

The grubs of cotton stem weevil feed on many species of cotton in India, Myanmar, Thailand and the Philippines. In India, it occurs in Tamil Nadu, Andhra Pradesh, Karnataka, Kerala, Bihar, Orissa, Rajasthan, Uttar Pradesh, Gujarat and Assam. The most favoured host of this insect is *Triumfetta rhomboidea* Jacq.

The grubs feed on the soft tissues of cotton stems. Greyish-black weevils emerge from the stem killing the plants. The adult is a dirty brown or greyish-black weevil, about 3 cm in length. The grub is slightly curved, creamy white, with a distinct head.

Life-cycle. The weevils emerge from the cotton sticks wherein they spend the summer. After mating, the female makes a cavity in the hypocotyle region of the plant and lays an oval, globular, smooth, milky white egg. During an oviposition period of 60-80 days, a female may lay up to 121 eggs visiting from plant to plant. The eggs hatch in 6-10 days and the grubs feed inside the soft stem tissue. They continue to feed inside till the development is completed in 35-57 days. The pupation takes place inside the stem and this stage lasts from 25 to 30 days. Three generations are completed in a year from October to April and the summer is passed as an adult in the cotton sticks.

Damage. The pest causes serious damage to Cambodia cotton in South India. As the grubs tunnel within the stem, that portion swells, and such symptoms are generally seen at the base of the plant. The younger plants when attacked succumb, while the older plants may survive but suffer in vigour. Under strong winds, the affected plants may break at the swellings. The pest causes plant mortality upto 25 per cent, especially during the early stages of growth.

Management. (*i*) Collection and destruction of affected plants is recommended for the control of this pest. (*ii*) Avoid growing of alternate or collateral host plants during the off season.

Other Pests of Cotton

The other pests of cotton include the grasshopper, *Cyrtacanthacris ranacea* Stal (Orthoptera: Acrididae); tailed mealybug, *Ferrisia virgata* (Cockerell) (Hemiptera : Pseudococcidae); black scale, *Parasaissetia nigra* (Nietner) and white scale, *Pulvinaria maxima* Green (Hemiptera : Coccidae); yellow scale, *Cerococcus hibisci* Green (Hemiptera: Asterolecaniidae); thrips, *Thrips tabaci* Lindeman and *Scirtothrips dorsalis* Hood (Thysanoptera : Thripidae); red hairy caterpillar, *Amsacta albistriga* (Walker) and woolly bear, *Pericallia ricini* (Fabricius) (Lepidoptera: Arctiidae); tussock caterpillar, *Euproctis fraterna* Moore (Lepidoptera : Lymantriidae); beet armyworm, *Spodoptera exigua* (Hubner) (Lepidoptera : Noctuidae) ; cotton stem borer, *Sphenoptera gossypii* Banks (Coleoptera : Buprestidae); blister beetle, *Mylabris pustulata* (Thunberg) (Coleoptera : Meloidae); flower weevil , *Amorphoidea arcuata* M.A. and surface weevil, *Attactogaster finitimus* (Coleoptera : Curculionidae).

JUTE (*Corchorus* spp.; Family : Tiliaceae)

As many as 32 pests have been reported damaging jute in India, but only five pests are of major significance.

1. Jute Mealybug, *Maconellicoccus hirsutus* (Green) (Hemiptera : Pseudococcidae)

Besides jute, this sap-sucking mealy-bug attacks the roselle fibre crop (*Hibiscus sabdariffa* L.), an important fibre allied to jute, and is responsible for an appreciable decrease in the yield of fibre.

Life-cycle. The mated females lay pink cylindrical eggs which are rounded at the ends. The eggs are laid on plants inside the ovisacs and measure 0.3 mm in length and 0.2 mm in breadth. The incubation period varies from 7 to 14 days in different seasons. On completion of the incubation period, the emergence of the nymphs starts from the ovisacs in batches, corresponding with the sequence of egg-laying. The tiny nymphs crawl out on the host and select a suitable spot to settle down. They are light pinkish and secrete both a white mealy powder and honey-dew. They develop distinctive sex characters after undergoing a few early moultings. The full-grown larva secretes fine white mealy fibres with which it forms a cocoon and then pupates in it. The females remain wingless, and on maturity, they develop ovisacs in which eggs are laid. The bugs suck sap from the stem and leaves. The female is a rotund, sack-like, light pink creature and measures about 3 mm in length. The males are slender and have a pair of delicate wings.

Scymnus pallidicollis (Coccinellidae) is the most efficient predator and feeds vigorously on the eggs, nymphs and adult females.

Damage. The nymphs and females feed on the apical parts of a plant which becomes stunted and shows bushy-top symptoms. The petiole becomes shortened, the lamina crumples and the internodal length is reduced, resulting in fibre deterioration and yield reduction.

Management. Spray 1.25 litres of dimethoate 30EC in 625 litres of water per ha.

2. Jute Semilooper, *Anomis sabulifera* (Guenee) (Lepidoptera : Noctuidae) (Fig. 16.12)

This is a specific pest of jute and is the most destructive in the jute tracts of India and Bangladesh. It is also found in Myanmar and Sri Lanka and in parts of Africa. Only the caterpillars cause damage by feeding on the foliage and being green, they camouflage but are easily noticed when they crawl by producing a loop in the middle.

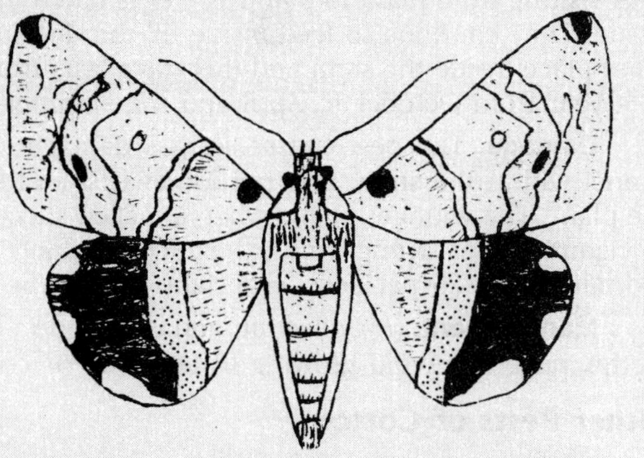

Fig. 16.12 Jute semilooper, *Anomis sublifera.* Adult

Life-cycle. The pest passes winter in soil in the pupal stage and the moths appear in May-June, when the crop begins to grow in the field. They lay eggs singly on the underside of young leaves; a female may lay more than 150 eggs which look like water droplets. When the weather is warm and moist, the egg stage lasts about two days. The tiny caterpillars, on emerging, start feeding on the apical leaves and buds. They attain a full length of about 4 cm after 5 moults, in about 17 days. The pupation may take place on the plant or in the soil. In summer, the pupae emerge in about a week, but those, which diapause, spend the entire winter in that stage. The life-cycle is completed in about one month and several generations are completed in a year.

Litomastix gopimohani Mani (Encyrtidae), *Tricholyga sorbillans Wiedemann* and *Sisyropa formosa* Linnaeus (Tachinidae) are associated with the larvae of this pest.

Damage. The attack is severe on half-grown plants which are one metre high. The second generation is the most damaging and sometimes up to 90 per cent of the leaves may be eaten up. Generally, the top 7-9 leaves are damaged and plant growth is adversely affected, resulting in a considerable reduction in the yield of fibre.

Management. (*i*) The pest can be suppressed by ploughing the infested fields after harvest and thus killing the pupae. (*ii*) The caterpillars can be dislodged into kerosenized water by drawing a rope across the young crop. (*iii*) Spray 500 ml of fenitrothion 50EC in 500 litres of water per ha. Repeat the treatments three times at 15- day interval from mid June or at first appearance of the pest.

3. Beet Armyworm, *Spodoptera exigua* (Hubner) (Lepidoptera : Noctuidae) (Fig. 16.13)

This polyphagous insect is a serious pest of young jute plants and at one time, was a major pest of indigo (*Indigofera arrecta*) and hence also called indigo caterpillar. Also, it is a minor pest of chillies, onion, binjal, sweet pepper, gram, linseed, lentil, cabbage, maize, cotton, safflower (*Carthamus tinctorius* L.), sunflower, etc. Geographically, it is found in Europe, South Africa, America and the Orient. In India, USA, Canada and other countries, it has been recorded as a minor pest of lucerne (*Medicago sativa* L.) and hence, is also known as the lucerne caterpillar.

The caterpillars feed on leaves, making webs, within which they feed gregariously for 2-3 days. The moths have dark spotted fore wings and white hind wings. They are active at night but remain hidden under various shrubs in the day time. The colour of larvae depends on the crop on which they feed.

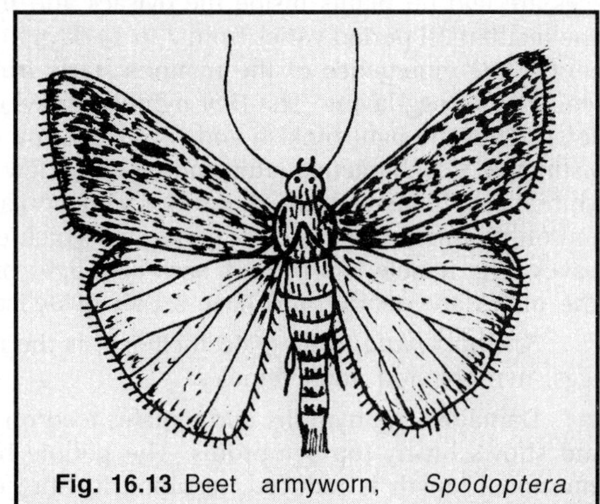

Fig. 16.13 Beet armyworn, *Spodoptera exigua.* Adult

Life-cycle. The pest passes winter as a pupa, which is enclosed in webbings, forming a rough pupal cocoon. Some of them emerge as moths in February, but a number of them continue in that state till the beginning of the monsoon. A female lays up to 200 eggs in clusters. The eggs are spherical and resemble poppy seeds in shape and size, having lines radiating from the centre. The egg-clusters are covered with buff hairs. The eggs hatch in 1-3 days and the young caterpillars start feeding in groups. They feed voraciously on the epidermis and also make webs. Within 2-3 days, the gregarious larvae spread out and hide under various shelters, from where they come out in the morning and evening to feed on the leaves.

The larval stage is completed in 15-20 days and the full-grown larvae seek shelter in debris, on the ground. They spin silken webbings and prepare rough cocoons with bits of leaves and other material. The pupal stage lasts 5-7 days and the life cycle is completed in about 30 days. From November onwards, the pest hibernates in the pupal stage. During the active season, a number of generations are completed.

Euplectus sp. (Pselaphidae) is associated with this pest.

Damage. The damaged crop, on which larvae have fed, gives a webbed appearance. The older caterpillars, which feed in the morning and evening, have a voracious appetite and strip off quite large patches on the foliage. However, the young jute plants, less than two months of age, suffer the most. The early sown *capsularis* varieties suffer greater losses than those sown later.

Management. (*i*) The pest can be suppressed by collecting and destroying the egg masses and the caterpillars which are feeding within the webs. (*ii*) Spray 625 ml of fenitrothion 50EC in 625 litres of water per ha.

4. Jute Stem Girdler, *Nupserha bicolor postbrunnea* Dutt (Coleoptera : Lamiidae)

Originally, this pest was confined to a number of host plants, among which *Sesbania aegyptiaca* (*dhaincha*) was the most important. However, in the past two decades, it has become a major pest of only one species of jute (*Corchorus olitorius* L.) and also of mesta, an allied fibre crop. The main damage is caused by the adult beetle while preparing sites for egg-laying on the stem. The geographic distribution is throughout the jute belt in India and Bangladesh.

Life-cycle. The female beetle makes two rings by cutting a strip, the space in between the rings being 1.0-1.4 cm. In this area, a slit is made, which reaches as far as the pith, where the beetle deposits one egg. Since a female lays, on an average, 35 eggs, many jute plants are damaged within 2-3 weeks. The eggs are yellowish and are about 1.5 mm long and 0.5 mm broad. They hatch in 3-4 days and the emerging larvae start feeding and travelling downwards along the central hollow of the stem, feeding on the pith.

The larvae become full-grown in 30-50 days and measure about 1.4 cm. During the active season, the larvae pupate in a chamber made in the hollow of the stem. With the advent of winter, the larvae cut out small portions of the stem in which they encase themselves and diapause. These encasements are detached from the main stem and when the jute plants are placed under water, they float, carrying with them, the larvae. Thus, the diapausing larvae escape drowning. The females prefer stems of 2.5-5.0 mm in diameter, which are available at different heights of the jute stem, depending on the age of the crop. The larval stage continues up to the next spring and pupation takes place only after the rains have started. The pupation and the emergence of beetles seem to synchronize with the availability of jute plants. There is only one generation in a year.

The common and important parasitoids associated with the larvae of this pest are *Neocatolaccus nupserhae* L. and *Norbanus acuminatus* (Chalcididae)

Damage. The main damage occurs because of oviposition, resulting in the breakage of fibre length at several places. Thus, both the quality of fibre and the yield suffer. The damage is estimated at 6-30 per cent, being more in younger plants than in the older ones. Not much damage is caused by the feeding of larvae or adults.

Management. (*i*) In areas where the girdler is a severe pest, the growing of resistant species of jute (*capsularis*) is useful. (*ii*) Mix 25 kg of phorate 10G per ha in the top soil followed by light irrigation.

5. Jute Stem Weevil, *Apion corchori* Marshall (Coleoptera : Apionidae)

The jute stem weevil can cause appreciable damage to the early-sown jute or the crop grown for seed. It has been recovered in all the jute-growing parts of India and Bangladesh. The small weevil is only 1.8 mm in length and 0.8 mm in breadth, brown or dull black and has small whitish setae on its body. It bores holes for oviposition with the rostrum in the jute stem and the larvae feeding inside the stem also injure the fibre.

Life-cycle. In the active season, the weevils make holes in the stem and lay eggs which are about 0.4 mm in length and 0.3 mm in breadth. A single female may lay as many as 675 eggs during its life-span of about 120 days. The eggs are generally laid at the base of the petiole, in which the larvae bore and damage the top leaves. In older plants, the weevils prefer to oviposit in the basal region and the lower part of the stem is damaged. The larvae emerge after 3-5 days and start feeding on the surrounding tissues. They are full-fed in 8-11 days when they measure 2.8 mm in length and about 1 mm in breadth. Then, they make a rough chamber in the stem and pupate.

The pupa measures 2.1 mm in length and about 1 mm in breadth. The pupal stage is completed in 4-6 days and the weevil emerges from the pupal case through an exit which is either made by the larva before pupation or by the weevil itself. Although the adults may live for about 200 days, the normal life-cycle is completed in 15-24 days. A number of overlapping generations are completed during the jute season. The winter is passed in the adult stage and the weevils seek shelter in bushes, shrubs and hedges, and start laying eggs on the new crop next year.

Damage. The main damage to the quality of fibre is caused by weevils making oviposition holes. A female may make a number of holes before laying an egg and damages numerous stems in her life time. The fields suffering the most are those with nitrogen fertilizer and those which are sown early. The weevil has a number of alternate host plants but the *capsularis* varieties seem to be relished the most.

Management. (*i*) The pest may be suppressed by the removal and destruction of infested plants at the time of thinning the crop and by collecting and destroying the stubble after harvest. (*ii*) Spray 2.5 kg of carbaryl 50WP in 625 litres of water per ha.

6. Yeloow Tea Mite, *Polyphagotarsonemus latus* (Banks) (Acari : Tarsonemidae) (Chapter 23)

Other Pests of Jute

Several other pests associated with jute are the mealy- bugs, *Ferrisia virgata* (Cockerell) and *Pseudococcus filamen-tosus* var. *corymbatus* Green (Hemiptera : Pseudococcidae); scale, *Parasaissetia nigra* (Nietner) (Hemiptera : Coccidae); leaf thrips, *Ayyaria chaetophora* Karny (Thysanoptera : Thripidae); hairy caterpillar, *Spilosoma obliqua* Walker (Lepidoptera : Arctiidae); cutworm, *Spodoptera litura* (Fabricius) and armyworm, *Spodoptera exigua* (Hubner) (Lepidoptera: Noctuidae); leafminer, *Trachys dasi* Thery (Coleoptera : Buprestidae) and beetles, *Pachnephorus bretinghami* Baly and *Nodostoma bengalensis* Du. (Coleoptera : Eumolpidae).

SUNNHEMP (*Crotalaria juncea* L.; Family : Leguminosae)

Sunnhemp is attacked by a dozen insect pests but only a few are important.

1. Sunnhemp Capsid, *Ragmus importunitas* Distant (Hemiptera : Miridae)

This is a small, active, green bug which sucks plant sap and when present in large swarms, causes appreciable damage. The pest is widely distributed in southern India.

Life-cycle. The bugs lay white cylindrical eggs in the plant tissue, generally under the surface layer of leaves. The numphs hatch out from the eggs in 7-8 days. The young nymphs, on emergence, begin to feed on the plants. The nymphs become adults in 10-12 days. The total life-cycle is completed in 35-62 days. All stages of the insect may be found simultaneously.

Damage. The nymphs and adults suck the sap from tender leaves and shoots, and cause yellowing of leaves. In case of severe attack, there is death of the plants.

Management. (*i*) The population of the pest can be suppressed by collecting the bugs with nets or sticky boards. (*ii*) Spray 2.5 kg of carbaryl 50WP or 625 ml of malathion 50WP in 625 litres of water per ha.

2. Sunnhemp Hairy Caterpillar, *Utetheisa pulchella* (Linnaeus) (Lepidoptera: Arctiidae)

This insect is the most important pest of sunnhemp in Tamil Nadu. The caterpillars feed on leaves and bore into the capsules. The adult moth is pale, whitish with red black spots on the upper wings and black marginal blotches on the lower wings. The full-grown caterpillar is about 3.8 mm in length and has red, dark and white markings on its body and a brownish head.

Life-cycle. The moths lay small whitish eggs on the tender leaves and shoots. On emergence from the eggs, the larvae feed on leaves. As the crop matures and pods appear, the caterpillars feed by thrusting the head in and leaving the rest of the body exposed. Pupation takes place either in the leaf folds or in the soil. The life-cycle is completed in about 5 weeks and a number of generations are completed in a year.

Damage. The caterpillars feed on leaves and also cause severe damage by feeding on the contents of developing pods. They defoliate the crop and cause a decrease in seed production. When the population is high and the attack severe, the plants become pale and appear unhealthy.

Management. (*i*) The pest can be suppressed by hand picking and killing the caterpillars as well as by collecting the moths with nets during the day time. (*ii*) Spray 2.5 kg of carbaryl 50WP in 625 litres of water per ha.

3. Flea Beetle, *Longitarsus belegaumensis* Fabricius (Coleoptera : Chrysomelidae)

The yellowish brown beetle has enlarged hind femur.

Life-cycle. The eggs are laid in the soil. The grubs feed by mning into tender roots of sunnhemp plants and when full-grown, pupate in the earthen cells in the soil. The life-cycle is completed in 23-28 days .

Damage. Both adults and larvae cause damage. The adult beetles bite holes in the leaves and cause severe damage. The larvae feed on the roots and reduce the vitality of the plants.

Management. Apply carbofuran 3G @ 12.5 kg/ha in soil followed by spray of carbarcyl 50WP @ 2.5 kg/ha.

Other Pests of Sunnhemp

The minor pests of sunnhemp include cotton white fly, *Bemisia tabaci* (Gennadius) (Hemiptera: Aleyrodidae); armoured scale, *Pinnaspis tempororia* Ferris (Hemiptera : Diaspididae); stem borer, *Cydia pseudonectis* Meyrick (Lepidoptera : Tortricidae); pea pod borers, *Etiella zinckenella* (Treitschke) and *Hedylepta indicata* (Fabricius) (Lepidoptera : Pyralidae); pea blue butterfly, *Lampides boeticus* (Linnaeus) (Lepidoptera : Lycaenidae); tiger moths, *Argina* spp. and *Spilosoma obliqua* Walker (Lepidoptera : Arctiidae); cutworm, *Spodoptera litura* (Fabricius) and semi-lopper, *Chrysodeixis eriosoma* (Doublay) (Lepidoptera: Noctuidae); tussock moth, *Dasychira mendosa* Hubner. (Lepidoptera: Lymantriidae) and pod bruchid, *Bruchus pisorum* (Linnaeus) (Coleoptera : Bruchidae).

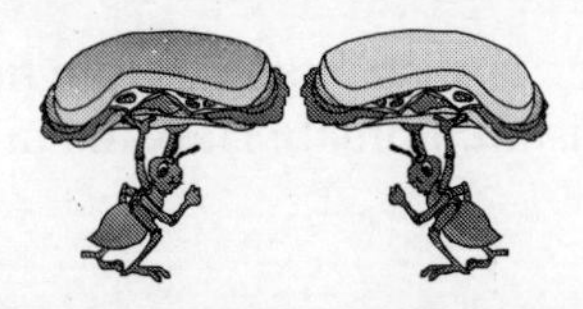

INTRODUCTION

Sugarcane is grown throughout the sub-tropical and tropical parts of South and South-East Aisa. India is world's largest producer of sugar and sugarcane. In India, the area under sugarcane has increased from 3.66 million ha in 2004-05 to 5.04 million ha in 2011-12 and 5.06 million ha in 2012-13. Sugarcane production reached a record level of 355.52 million tonnes during 2006-07, thereafter its production declined in subsequent years but has started witnessing an increasing trend in recent years. The total production of sugarcane during 2010-11 was 342.38 million tonnes which decreased to 341.04 million tonnes during 2011-12 and 338.96 million tonnes during 2012-13. The production is estimated at 348.38 million tonnes in 2013-14 and the target has been lowered to 345 million tonnes for 2014-15.

As many as 288 species of insect pests have been reported to cause damage to the sugarcane crop at one stage or the other of crop growth. However, only a dozen are recognized as major pests. The root borer, *Emmalocera depressella* Swinhoe, is more abundant in eastern India, whereas the early shoot borer, *Chilo infuscatellus* Snellen and the top borer, *Scirpophaga nivella* (Fabricius), are more or less uniformly distributed throughout India and more serious in the early stages of the crop. In the north-western parts of the Indian Sub-continent, the Gurdaspur borer, *Acigona steniellus* (Hampson), is far more destructive than the other species. However, in mid seventies its population started declining and by the early eighties it became an insignificant pest. In the Punjab, western Uttar Pradesh, Haryana and northern Rajasthan, the stalk borer, *Chilo auricilius* Dudgeon, has gained great importance in recent years. Again, the mealybug, *Saccharicoccus sacchari* (Cockerell), is more prevalent in tropical India than in the northern regions. The pyrilla, *Pyrilla perpusilla* (Walker), and sugarcane whitefly, *Aleurolobus barodensis* (Maskell), appear as extremely destructive pests in certain years in one region or another.

SUCKING PESTS

1. Sugarcane Pyrilla, *Pyrilla perpusilla* (Walker) (Hemiptera: Lophopidae) (Fig. 17.1, Plate 17.1A)

Pyrilla or the sugarcane leafhopper is distributed throughout India and appears periodically as a destructive pest of sugarcane in Pakistan, Punjab, Haryana and Rajasthan. Apart from sugarcane, it feeds on wheat, barley (*Hordeum vulgare* Linn.), oats (*Avena sativa* Linn.), maize, sorghum, *baru* (*Sorghum halepense*), Guinea grass, *swank* (*Echinolchloa* spp.) and Sudan grass (*Sorghum vulgare sudanese*).

Both the nymphs and adults suck the cell sap usually from the underside of the leaves. When full-grown, a nymph is pale yellow, 10-15 mm long, and has two white prominent feather like filaments at the tail end of its body. This leafhopper is very agile and jumps around in large numbers,making a faint noise when a person walks through a heavily infested field. The adult, equally active, is about 20 mm long and has a straw-coloured body with dark patches or spots on the wings. At the front end it has a snout like prolongation and prominent red eyes **(Fig. 17.1)**.

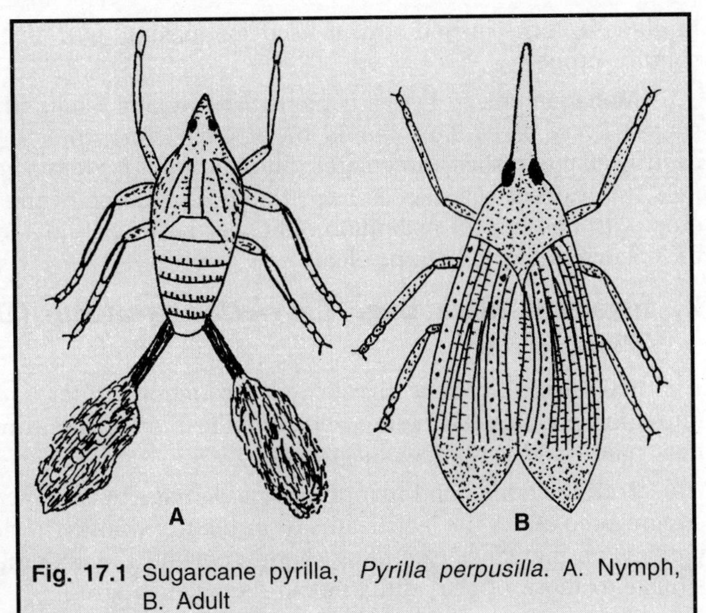

Fig. 17.1 Sugarcane pyrilla, *Pyrilla perpusilla*. A. Nymph, B. Adult

Life-cycle. The insect breeds throughout the year, migrating from one crop to another for fresh food. The adults lay 300-536 eggs in clusters on the underside of leaves during the summer and within the leafsheaths during winter. The clusters are covered with a white, fluffy material from the anal tufts of the females. As this fluff is removed, oval, pale-white eggs are seen in 3-5 longitudinal rows of 35-50 each. The eggs hatch in 8-10 days in summer and in about 3-4 weeks during November or December. The numphs which hatch out from the eggs are pale brown in colour and 1.3 mm long. At this stage they do not possess snout, wings and tufts. Within a week they are characterised by the possession of two long tufts of waxy secretion at the end of the abdomen (Fig. 17.1). They start sucking the sap of canes and grow to maturity through 5 stages within 8 weeks in summer and in 5-6 months in winter. The adults live for 27-52 days in the summer and 18-20 weeks in the winter. In all, 3-4 generations are completed in a year.

The common parasitoids of this pest are *Ooencyrtus papilionis* Ashmead (Encyrtidae), *Tetrastichus pyrillae* J.C. Crawford (Eulophidae), *Lestodrynus pyrillae* Kieffer, *Drynius pyrillae* Kieffer and *Epiricania melanoleuca* Fletcher (Epipyropidae). The eggs and nymphs are attacked by lady bird beetles namely *Brumus suturalis* (Fabricius), *Menochilus sexmaculatus* Fabricius and *Coccinella septempunctata* Linnaeus (Coccinellidae) as well as by chrysopods.

The conservation of egg parasitoids and introduction of *E. melanoleuca* in areas where it is absent is very useful and significant control of this pest can be achieved by the management of natural enemies. The white muscardine fungus, *Metarhizium anisopliae* (Metchnikoff) Sorokin, has also been reported from this pest.

Damage. Succulent varieties of sugarcane with broad leaves are preferred by this pest but when it occurs in abundance, no variety is spared. Owing to the loss of cell-sap the leaves turn pale-yellow and shrivel up later. Even the canes dry up and die when the attack is very severe. The insects excrete a thick transparent liquid, known as honeydew which falls on the leaves and makes a good medium for the growth of a black mould. Therefore, at times, the leaves acquire a sickly black appearance and the attacked crop can be spotted from a distance. The black coating interferes with photosynthesis and very little food is manufactured by the plants. The existing sucrose in the canes is also used up and about 35 per cent reduction in sugar yield is not uncommon. The cane juice becomes high in glucose, turns insipid and, if used for making *gur*, gives rise to a soggy mass which does not solidify properly.

Management. (*i*) Pyrilla is parasitised by many natural parasites during, egg, nymph and adult stages. Every precaution should be observed for protecting the natural enemies. (*ii*) Collect and destroy all the trashes immediately after harvest. (*iii*) Remove sprouts and stubble from the sugarcane field. (*iv*) Collect egg masses from the lower surface of the leaves and destroy them. (*v*) Spray the crop with one litre of malathion 50EC per ha in 250 litres of water at an economic threshold level of 3-5 nymphs or adults per leaf:.

2. Sugarcane Black Bug, *Cavelerius excavatus* (Distant) (Hemiptera : Lygaeidae) (Plate 17.1B)

This pest is widely distributed throughout India and Pakistan, and it is believed that the ratooning practice has helped its multiplication in the Punjab. It has also been found feeding on rice, maize and a number of grasses.

Both the adults and nymphs cause damage by sucking cell sap from the plants. They are seen congregated under the leafsheaths or in the top whorl. The adults are 6-7 mm long, black with white patches on the wings extending slightly beyond the abdomen. The shape and colour of nymphs is similar to those of the adults but are smaller in size.

Life-cycle. The pest actively breeds throughout the year but during winter the adults and eggs are more noticeable. In summer, each bug lays 55-478 eggs in clusters of 14-67 on the inner side of the leafsheaths. The eggs are creamy white and hatch in 9-17 days. The nymphs grow through five stages and complete their development in 4-6 weeks. The adults are long-lived; males die earlier than the females. The insects complete approximately three generations in a year. In winter, the eggs are laid in the soil at a depth of 5-7 cm and they hatch during the next spring. Thus, in March-April the new nymphs and the old adults die and only the nymhs remain and they mature by June. From then on, the generations overlap and all stages are found in the field.

Nardo cumaeus Nixon and *N. phaeax* Nixon parasitize the eggs of this bug.

Damage. On young plants, the nymphs and adults suck cell-sap from the central whorl. On the grown up plants they prefer to feed within the leaf-sheaths, and varieties having broad and loosely attached sheaths are preferred by this pest. The attacked leaves become paler and also show holes after feeding.

Management. Spray 875 ml chlorpyriphos 20EC in 1000 litres of water per ha. Direct the spray material into the leaf-whorl.

3. Sugarcane Whitefly, *Aleurolobus barodensis* (Maskell) (Hemiptera : Aleyrodidae) (Fig. 17.2)

The white fly of sugarcane is found throughout the Indian Sub-continent and is a notorious pest in Gurdaspur, Jalandhar and Yamunanagar districts of Punjab and Haryana. In the absence of sugarcane, it can also survive on *sarkanda* (*Saccharum munja*). Damage is caused by the nymphs which suck the cell-sap from the leaves. The grown up nymphs are about 3 mm long, oval in outline but flattened and scale like in form, thus remain sticking to the same spot on a leaf. They are black and have a silvery grey waxy coating on the body. The adults are small, delicate, pale-yellow insects about 3 mm long, and their wings have a white mealy appearance, mottled with black dots. They flutter about briskly, but they are not easily noticed in the field.

Life-cycle. The pest breeds practically throughout the year, except during winter when there are mostly nymphs and pupae present. Winged adults appear in the spring and soon after emergence, they copulate in an end to end position for 30-40 seconds. The female then lays 60-65 creamy white conical eggs which are glued to the surface of the leaves. The eggs are found in groups of 15-20, arranged in a single file. Within a couple of hours, the eggs turn black and hatch in 8-10 days. On emergence, the young nymphs are pale yellow and they move away from the egg-shell to find a suitable place for feeding by the insertion of their piercing mouthparts. Their movements are very restricted and they complete their development in 25-30 days after passing through 4 instars. The pupal stage lasts 10-11 days and the adults, on emergence, live only for 24-48 hours. Among adults, the females are in preponderance and they out number the males by 2 : 1. The insect completes 9 generations in a year and in March-April, they migrate from the old to the new sugarcane plants. Their life is prolonged in autumn when the nymphal and pupal stages are completed in 3-4 months.

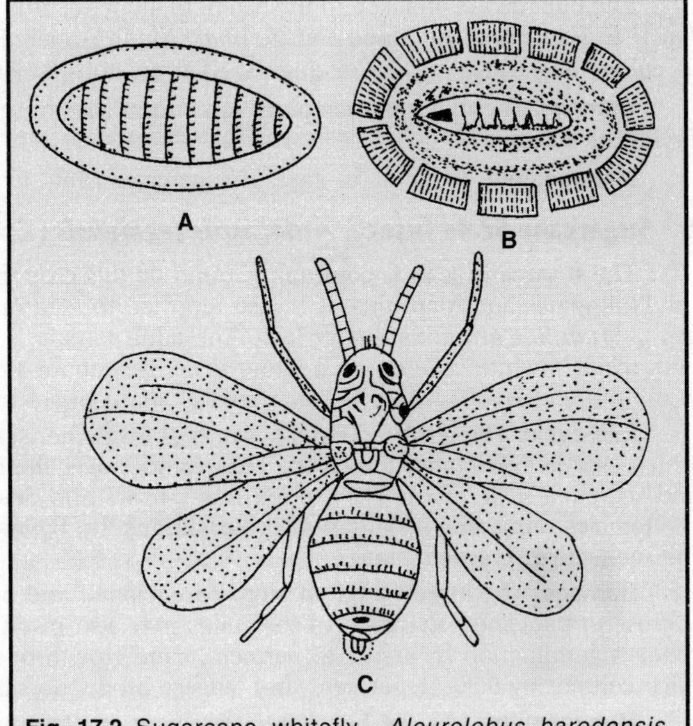

Fig. 17.2 Sugarcane whitefly, *Aleurolobus barodensis*. A. Nymph; B. Pseudo-pupa; C. Adult

Several species of parasitoids have been collected from white flies and the commonly known from nymphs of this pest are *Azotus delhiensis* Lall, *Encarsia issaci* Mani, *E. muliyali* Mani and *Eretmocerus delhiensis* Mani (Aphelinidae).

Damage. Only the nymphs cause damage by sucking the cell-sap. Yellow streaks appear on the attacked leaves and the crop acquires a palish green appearance. The general vitality of the plants is reduced, and the quality and quantity of *gur* production is poor because of subnormal crystalization of sugar. Sugar recovery is reduced by about 15-25 per cent. A black mould develops on the honeydew excreted by the pest and it interferes with proper functioning of the leaves and renders them unfit as fodder. A comparatively poor crop with a thin stand is attacked more readily than a well-manured and heavy crop.

Management. (*i*) Avoid ratooning especially in low lying and water logged areas. (*i*) Detrash lower leaves containing the puparia. (*ii*) Spray 2.50-3.75 litres of malathion 50EC or 100 ml of imidacloprid 200SL or 1.5 litres of triazophos 40EC in 1000 litres of water per ha.

4. Sugarcane Spottedfly, *Neomaskellia bergii* (Signoret) (Hemiptera : Aleyrodidae)

The sugarcane aleyrodid is found in India, Sri Lanka, Java, the Philippines, Reno Manila, Mauritius and Fiji. The pest is injurious in the nymphal stage only and apart from sugarcane it also attacks sorghum. The nymphs are pale brown to dark in colour and are covered with white-wax. The adult flies are pale brown with black markings on the fore wings. The hind wings are short, thin and more or less transparent.

Life-cycle. This pest is most active on sugarcane during September-October. The female lays 120-150 eggs in circular or semi-circular masses on the underside of the leaves during an oviposition period of 2-3 days. The freshly laid white egg is eliptical in shape and it turns black later on. The eggs hatch in 2-3 days. There are four moultings during larval period of 5-7 days. The pupal stage lasts 7-10 days and the life-cycle is completed in 15-18 days in September and 18-25 days from October onwards.

The puparia are attacked to the extent of 45 per cent by parasities, *Eretmocerus delhiensis* Mani, *Encarsia neomaskelliae* Prasad and *E. isaaci* Mani (Aphelinidae). The parasitized puparia turn dark in colour and can easily be distinguished from non-parasitized ones.

Damage. The nymphs suck plant sap and in case of severe attack, cause stunting of canes and drying of leaves.

Management. Same as in case of sugarcane white fly.

5. Sugarcane Scale Insect, *Aulacaspis tegalensis* (Zehntner) (Hemiptera : Diaspididae)

The sugarcane scale is commonly found on this crop in Malaysia, Thailand, Indonesia, Taiwan, the Philippines and Mauritius. It is also reported to feed on sorghum in Tanzania and on the wild grass, *Erianthus arundinaceus* in Java. The adult female is sedentary having no legs or wings and lives inside a white covering or a white scale. It is about 1.8 mm long and 0.9 mm wide. The male is minute and is free living. It has a special sharp organ to penetrate the female's scale covering.

Life-cycle. The fertilized female lays eggs inside her scale and these hatch into mobile nymphs which escape from under the cover. They wander over the host plant and eventually fix themselves and secrete a scale cover. They moult many times and develop differentially according to the sex; the females remaining sedentary. Dry conditions are believed to favour survival and dispersion of the ineffective "crawler" stage.

Damage. The insects feed on stem parenchyma and by doing so prevent the accumulation of sucrose in the cane. Drying up of the canes may also result from heavy infestation. The infestation reduces germination by about 20 per cent, cane growth by 5.5 per cent, cane yield by 43 per cent, juice content by 0.30-41 per cent and jaggery production by 10 per cent.

Management. (*i*) Give hot water treatment at planting time. (*ii*) Dip setts for 15 minutes in 0.1 per cent malathion 50 EC before planting. (*iii*) Spray 1.25 litres of malathion 50 EC or 2.0 litres of dimethoate 30 EC in 1250 litres of water per ha.

6. Sugarcane Mealybug, *Saccharicoccus sacchari* (Cockerell) (Hemiptera : Pseudococcidae)

This species and a number of others are minor pests of sugarcane. They also feed on reeds and some of the grasses. Mealybugs are seen clustered at the basal nodes of canes and are exposed when a leaf sheath is removed. Nymphs and the wingless female adults cause damage by sucking the cell-sap. They are inert pink insects, having a rotund, sack-like, segmented body covered completely with a white mealy powder. The size, even among full-grown females, varies a great deal, but a large female measures about 5 mm in length and 2.5 mm in width. The males, whose only function is to fertilize the females, are sluggish having only one pair of wings and are short-lived.

Life-cycle. The pest breeds practically throughout the year. The females are highly fecund and lay a large number of eggs at short intervals. The eggs are yellowish, smooth, cylindrical and rounded at both ends and measure 0.35 mm in length and 0.16 mm in width. Within a few hours the eggs become soft and elongated, and the crawlers emerge. The tiny young ones are transparent, pink and very active. They wander about for some time and spread all over the field or may even be blown away by the wind to the adjoining fields of sugarcane. Finding a suitable host plant, they force themselves underneath the leafsheaths near the basal nodes. As the canes grow taller, the older bugs remain at the lower end and the crawlers reach the higher nodes. The nymphs feed voraciously and pass through 6 stages before they are full-grown in 2-3 weeks. The life of a full-grown winged female is approximately 3-5 days. The entire life-cycle is completed in about a month during the summer. The pest completes several generations in a year.

The coccinellids reported from this mealy bug are *Pharoscymnus grimeti* Mulsant, *Scymnus coccivora* Ramakrishna Ayyar, *S. andrewsi* Sie. and *S. nubilis* Mulsant A coccinellid, *Hyperaspis trilineala* Mulsant was introduced in Tamil Nadu and Karnataka in 1970.

Damage. Mealy-bugs are first noticed in appreciable numbers when canes are four months old and from then on, they remain on the plants till harvest. Canes having tight fitting sheaths are more or less free from the attack, whereas a drought affected crop is more severely damaged. The bugs drain away large quantites of sap from the canes and befoul them by their mealy secretions and honeydew. A sooty mould develops on these secretions giving a blackish appearance to the canes. It is also suspected that the mottling disease of sugarcane which is serious in certain parts of India, is transmitted by these bugs. In severe cases of infestation, sucrose content decreases by 24 per cent and brix by 16 per cent.

Management. Spray 1.25 litres of malathion 50EC in 1250 litres of water per ha.

7. Sugarcane Woolly Aphid, *Ceratovacuna lanigera* Zehntner (Hemiptera : Aphididae) (Fig. 17.3)

The sugarcane wooly aphid was first reported on sugarcane in 1897 from Java, Indonesia and later on reported as a serious pest of sugarcane in Asia. It is known to occur in India, Pakistan, Korea, Japan, China, Taiwan and the Philippines. In India, it was reported on sugarcane for the first time in West Bengal in 1958 and later it was reported from Assam, Sikkim, Tripura and Uttar Pradesh. The pest appeared in serious proportions in Maharashtra in 2002 and has spread to other states including Karnataka, Andhra Pradesh, Tamil Nadu, Kerala, Bihar and Uttarakhand.

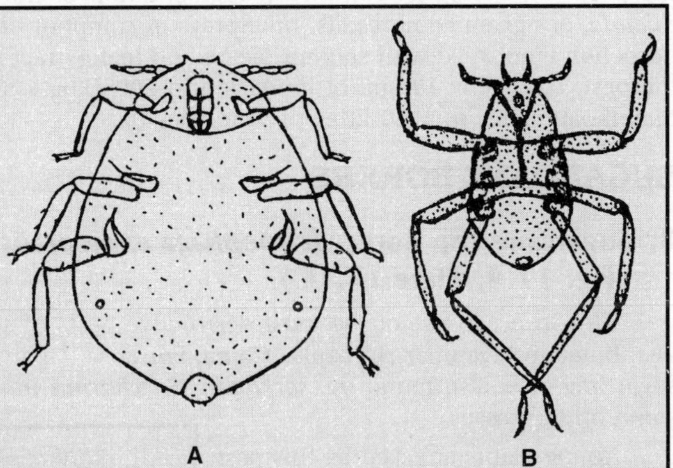

Apterae are small to medium-sized, pale green to brown, covered with white woolly wax, forming dense colonies on the under surface of leaves, often attended by ants. Nymphs have little wax coating on body. Wax filaments increase gradually with age, third instar with clear intersegmental lines and siphunculi, full-grown aphid completely covered with white wooly covering. Alatae have brown-black head and thorax, and dusky transverse bands on abdominal dorsum. Soldier aphids are often seen in the colony **(Fig. 17.3)**.

A B

Fig. 17.3 Sugarcane woolly aphid, *Ceratovacuna lanigera*. A. Apterae adult ; B. Soldier aphid

Life-cycle. The aphid undergoes an anholocyclic life-cycle and no sexuals have been noticed in nature in Taiwan, The nymphal stage occupied 10 days and the alate about 14 days in Taiwan. The optimum temperature for aphid development ranged from 20 to 23°C and the aphids became inactive at temperatures below 15°C and above 28°C. The nymphal development period varies under different temperature and photoperiod regimes. In Japan, apterous nymphal stages occupied 22-32 days. The average longevity of apterous and alate adults was 36 and 8.3 days, respectively. Average fecundity of apterous adult was 60 aphid nymphs, while it was 10 aphid nymphs in alate. In Maharashtra, aphid population peaks from early August to September and again from November to February. The season from March to July appears to be the lean period. In Uttar Pradesh, the pest occurred during October to March and the population was low in June.

Several species of ladybird beetles like *Synochyta grandis* (Thunberg), *Pseudoscymnus kurohime* (Miyatake), *Anisalemnia dilatata* (Fabricius). [*Megalocaria dilatata* (Fabricius)], *Coelophora biplagiata* (Swartz) and *C. saucia* Mulsant are quite effective predators of sugarcane woolly aphid. In addition, *Chrysoperla carnea* (Stephens), *Dipha aphidivora* Meyrick and *Micromus igorotus* (Banks) are potential predators of this pest and are promising candidates for augmentation. Two major parasitoids, namely *Encarsia flavoscutellum* Zehntner and *Diaerelus oregmae* Gahan, have been recorded from Indonesia (Java) and the Philippines, respectively, the former also occurring in India.

The common entomofungal pathogens, *Beauveria bassina, Metarhizium anisophiae* and *Verticillium lecanii* have been found to have only limited to moderate efficacy against the aphid.

Damage. Initial infestation is seen on the under surface of the leaves along the midrib and then over the entire under surface, covering it with flocculent, waxy secretion. Copius honeydew secretion often covers the entire upper surface of the leaves, leading to growth of sooty mould. Due to continuous sap sucking, the crop becomes stunted and continuous infestation leads to reduction in the length, circumference, weight and sugar content of the stalk, loss in tonnage and sugar recovery. The aphid infestation adversely affects the yield and cane quality.

Management. (*i*) Maintain proper sanitation and avoid transport of infested leaves from one area to another. (*ii*) Stripping of infested leaves and burning them help to avoid further spread of the pest. (*iii*) Biological control by augmentation of the coccinellid predators, *S. grandis* and *A. dilatata*, has given good results. (*iv*) Sprays of camphor oil (1%), yam and bean oil, tea seed oil, tea decoction emulsified with sodium oleate and leaf extract of tobacco have provided good control of the pest. (*v*) Spray 100 ml of imidacloprid 200SL or 1.25 litres of malathion 50EC or 2.0 litres of dimethoate 30EC in 1250 litres of water per ha.

SUGARCANE BORERS

8. Sugarcane Top Borer, *Scirpophaga excerptalis* (Walker) (Lepidoptera : Pyralidae) (Fig. 17.4, Plate 17.1C)

This insect is one of the most destructive pests of sugarcane in India and Pakistan. It is also distributed in Myanmar, Sri Lanka, China, Formosa, Japan, the Philippines and Thailand. Apart from sugarcane, it is also found on *sarkanda* (*Saccharum munja*), *kahi* (*Saccharum spontaneum*) and some other grasses.

Young plants attacked by this pest show characteristic reddish streaks on the mid-ribs. They also show a number of shot holes in the leaves which ultimately cause dead-hearts. After cane formation the attacked plants show peculiar bunchy tops.

Damage is causd by caterpillars which are generally found in the top portion of a cane. When full-grown, a caterpillar is 25-30 mm long, creamy white and rather sluggish. The moths are pure white (**Plate 17.1C**). Males are considerably smaller than females. A female measures 25-40 mm across the wings spread and carries a brownish or reddish tuft of silken hairs at the tip of its abdomen.

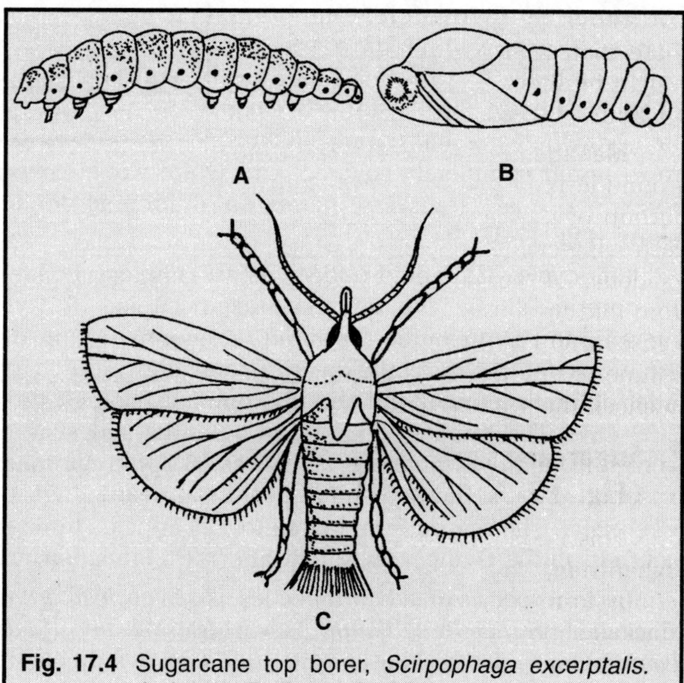

Fig. 17.4 Sugarcane top borer, *Scirpophaga excerptalis*. A. Larva; B. Pupa; C. Adult

Life-cycle. This pest is active from March to November and passes the winter as a full-grown larva in cane tops.

These larvae pupate in the second half of February and the moths emerge during March. The female moths lay eggs on the lower surface of the leaves at the rate of about 150, in clusters of 30-60 eggs. These clusters are covered with a brown tuft of hairs and are quite prominent. Eggs hatch in 5-7 days and the young larvae bore into the mid-rib of a leaf, mining their way to the base. From there, they enter the spindle, feeding on the growing-point and the soft portion of the cane. They grow to maturity in five stages within 4-5 weeks. When full-fed, a larva constructs a chamber with an emergence hole in the rind just above a node. It pupates inside the chamber and emerges as a moth within 7-9 days. The moths are commonly seen sitting on cane tops early in the morning. They become active at night when they mate and the females lay eggs. The moths are short-lived; hardly live more than 4-5 days. Four to five generations are completed during a year. Full-grown caterpillars of the last generation, however, do not pupate. They hibernate in the cane tops throughout the winter in northern India.

The pest is parasitized in the egg stage by *Telenomus beneficiens* (Zehntner) (Scelionidae) in all the sugarcane growing areas of India. The parasitization is 3-5 per cent in North India and may reach as high as 90 per cent in Tamil Nadu.

Isotima javensis Rohw.(Ichneumonidae) is a parasitoid found in North India and Myanmar, and it attacks the larvae and pupae of this pest. The degree of parasitization varies from 5-20 per cent in sub-tropical areas to 70 per cent in tropical areas. During 1960s it was introduced from Uttar Pradesh to Tamil Nadu where it has established and now controls the pest effectively.

The other parasitoids reported are *Trichogramma chilonis* Ishii (Trichogrammatidae) and *Telenomus dignoides* Nixon (Scelionidae) on eggs; *Goryphus* sp., *Xanthopimpla nursei* Cameron and *X. predator* (Fabricius) (Ichneumonidae), *Rhaconotus roslinensis* Lal, *R. scirpophagae* Wilkinson, *R. signipennis* Wilkinson, *Stenobracon deesae* Cameron and *S. nicevillei* (Bingham) (Braconidae), *Goniozus indicus* Ashmead (Bethylidae) and a tachinid parasitoid, *Sturmiopsis inferens* Townsend (Tachinidae) on larvae.

Damage. The first two broods of this pest attack young plants before the formation of canes. These plants are killed and are a total loss. In subsequent broods, the pest attacks the terminal portions of the canes, causing bunchy tops. Damage by the third and fourth broods may result in more than 25 per cent reduction in weight and a decrease in the quality of the juice.The loss in weight in different varieties may vary from 21 to 37 per cent and loss in sugar recovery from 0.2 to 4.1 units.

Management. (*i*) Collect and destroy moths and egg clusters. (*ii*) Cut the attacked shoots at the ground level from April to June. (*iii*) Since about one-third of the larvae remain in the underground portion of the stem, the sharp edge of a sickle should be inserted to kill them. (*iv*) Cutting and destroying cane tops which harbour the over-wintering larvae before mid-February, also reduce the incidence of this pest. (*v*) Release *Trichogramma chilonis* Ishii @ 50,000 per ha at 10 day interval from mid April to end June. Normally 8 releases are required. (*vi*) Apply 25 kg of chlorantriniliprole 0.4GR or 30 kg of carbofuran 3G or phorate 10G per ha at the base of the shoots in the last week of June or first week of July only if the top borer damage exceeds 5 per cent level. Earth up slightly to check the granules from flowing with irrigation water and irrigate the crop immediately.

9. Sugarcane Early Shoot Borer, *Chilo infuscatellus* Snellen (Lepidoptera : Pyralidae) (Fig. 17.5, Plate 17.1D)

This is one of the very serious pests of sugarcane in India and Pakistan. It is also found in Afghanistan, Myanmar, Indonesia, Formosa and the Philippines. In Punjab, infestation occurs in serious form from April to June, i.e. before the rainy season. The central whorl of leaves in the attacked shoots dries up, forming a "dead-heart". In addition to sugarcane, it feeds on maize, *bajra* (*Pennisetum typhoides*), *sarkanda* (*Saccharum munja*), *kahi* (*S. spontaneum*), *baru* (*Sorghum halepense*) and some other grasses.

Damage is caused by the caterpillars which measure 20-25 mm in length at maturity. They are dirty white and have five light- violet longitudinal stripes on the body. The moths have straw-coloured fore wings and whitish hind wings with apical light-buff areas **(Fig. 17.5)**. They measure 25-40 mm in wing expanse and are attracted to light at night.

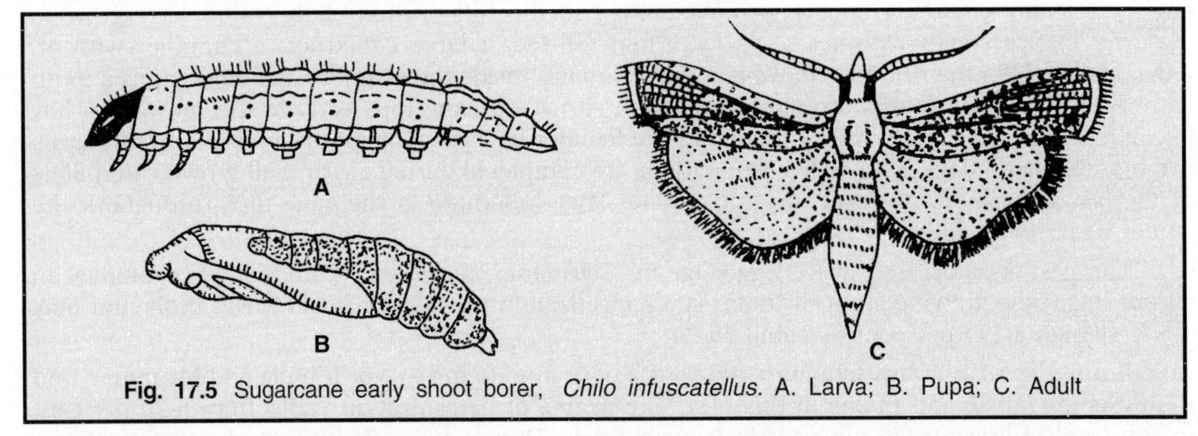

Fig. 17.5 Sugarcane early shoot borer, *Chilo infuscatellus*. A. Larva; B. Pupa; C. Adult

Life-cycle. The pest is active from March to November and passes winter as a full-grown larva in the stubble. The larvae pupate some time in February and emerge as moths during March. The moths are active at night and the females lay creamy white scale like eggs in clusters of 11-36 on the lower surface of leaves. On an average, a female moth lays 300-400 eggs which hatch in 4-5 days. The larvae reach the plant base, bore into the shoot and feed there. They grow through five stages and complete their development in 3-4 weeks. Then each larva constructs a chamber for pupation in the cane and makes an exit from which it emerges as moth after 6-7 days. The moths live only for 2-4 days. The life-cycle is completed in 5-6 weeks and the pest breeds 4 or 5 times a year.

Trichogramma chilonis Ishii (Trichogrammatidae) has given up to 50 per cent parasitization of eggs in Tamil Nadu. The parasitoid, *Telenomus alecto* Crawford (Scelionidae) was introduced from Columbia and releases were made in the Punjab and at Plassey (West Bengal). It failed to establish in the Punjab and parasitization increased from 2-6 per cent in 1967 to 53 per cent in 1968 in Plassey area.

The other egg parasitoids associated with this borer are, *Trichogramma exiguum* Pinto & Platner, *T. intermedium* How. and *Trichogrammatoidea nana* (Zehntner) (Trichogrammatidae), *Telenomus beneficiens* (Zehntner) and *T. dignoides* Nixon (Scelionidae).

The parasitoids associated with larvae are *Campyloneurus mutator* Fabricius, *Apanteles flavipes* (Cameron), *Stenobracon* sp. (Braconidae), *Goniozus indicus* Ashmead and *G. rephoterycis* Kurian (Bethylidae). The tachinid, *Sturmiopis inferens* Townsend and *S. semiberbis* Bezzi (Tachinidae) also parasitize the larvae of this pest.

The only parasitoid associated with pupae is *Tetrastichus ayyari* Rohw. (Eulophidae)

Damage. The plants, which are attacked by this pest, produce dead hearts (Plate 17.1D) from April to June, and completely dry up. A loss of 10-20 per cent of young shoots is not uncommon during this period and in years of serious infestation, it may be as high as 70 per cent. After the formation of canes, the attack does not produce dead-hearts and the damage is confined to a few internodes only. Even then, there is considerable reduction in cane yield and sugar content. At harvest, losses of 22-33 per cent in yield, 12 per cent in sugar recovery, 2 per cent in commercial cane sugar (CCS) and 27 per cent in jaggery have been estimated.

Management. (*i*) To control the early shoot borer, plant the crop early, i.e. before the middle of March. (*ii*) Release *Trichogramma chilonis* Ishii @ 50,000 per ha from mid-April to end-June at 10 days interval. Normally 8 releases are required. (*iii*) Apply 25 kg granules of fipronil 0.3G per ha, before the cane setts are covered with earth by planking. (*iv*) Apply 25 kg granules of cartap hydrochloride 4G or 25 kg of fipronil 0.3G mixed in 50 kg sand or 110 ml of imidacloprid 17.8SL or 5 litres of chlorpyriphos 20EC in 1000 litres of water per ha with sprinkling can along the rows at post-germination stage (about 45 days after planting). Earth up slightly and follow up with light irrigation. The economic threshold level is 15 per cent incidence.

10. Stalk Norer, *Chilo auricilius* Dudgeon (Lepidoptera : Pyralidae)

The stalk-borer, commonly known as the Tarai borer, is perhaps the most destructive pest of sugarcane in northern India. In addition, it feeds on paddy, wheat, oats and *baru* (*Sorghum halepense*). It is widespread in Bihar, the Tarai areas of Uttar Pradesh, and Haryana and Punjab. A large-scale movement of sugarcane within and between the States has helped to spread this pest.

It causes damage in the larval stage by feeding inside the stem. The full-grown caterpillar is 25-30 mm long, with a light bluish pink body and has a dark brown head and five longitudinal violet dorsal stripes. The moth is staw-coloured and the female has a wing expanse of 22-30 mm and the male has a wing-span of 16-25 mm.The fore wings have golden spots and the hind wings are silvery-white.

Life-cycle. The pest is found practically throughout the year but breeds from March to October. The winter months are passed as full-grown larvae in canes or stubble. They pupate sometimes during January and emerge as moths at the beginning of February. Eggs are laid in leafsheaths or on the underside of leaves of the late water-shoots and early ratoon sprouts. A female may lay 200-300 scale-like overlapping eggs in clusters of 60-70. The young larvae, measuring about 1 mm, emerge from these eggs in about a week. They feed on leafsheaths or on mid ribs for the next week or so and then bore into stalks by making circular holes in the rind. Dead-hearts are produced in young plants but after cane formation, the attack is not easily discernible from a distance. The larvae develop through 5 stages and are full-grown in 3-6 weeks. Then the mature larva constructs a chamber with an exit and begins pupation. Moths emerge in about one week and the life-cycle is completed in 5-9 weeks. In all, 5 or 6 over-lapping generations are completed in a year. The winter life-cycle is rather prolonged and is completed in 17-19 weeks, out of which 14-16 weeks are spent in the larval stage and 2-3 in the pupal stage.

Sturmiopsis inferens Townsend (Tachinidae) is found in Haryana and its presence was also reported from Uttar Pradesh. The incidence of parasitoids varies from 21 to 40 per cent. *Apanteles flavipes* (Cameron) (Braconidae) and *Centeterus alternecoloratus* Chushman are associated with larvae and pupae of this pest, respectively.

Damage. In spring, when the pest first appears on the ratoon crop, the late "water-shoots" play an important role in its multiplication. By the time the canes are formed in August-September, 75 per cent of them may be infested, the heavily manured fields and soft varieties suffering more. The lodged crop and the waterlogged fields are also more severely infested. The caterpillars have the habit of boring into one internode after another and moving from plant to plant, thus infesting up to 90 per cent of the canes in certain fields. According to one estimate, this pest causes, on an average, 16 per cent reduction in cane yield and a loss of 2.16 units in sugar recovery.

Management. (*i*) Grow resistant varieties in the area where the stalk borer is a serious pest. (*ii*) Do not use the cane-seed from the infested field. (*iii*) Spread of the pest to uninfested areas may be prevented by restricting the movement of infested canes. (*iv*) The pest population can be suppressed by burning the trash in the fields after harvest and by removing and destroying the water-shoots over large areas during February-March. This practice also induces tillering and is useful. (*v*) Do not ratoon a heavily infested crop; plough up the affected fields, collect the stumps and destroy them. (*vi*) At harvest, do not leave the water-shoots in the field. (*vii*) Release *Trichogramma chilonis* Ishii @ 50,000 per ha from July to October at 10 days interval. Normally 10-12 releases are required.

11. Sugarcane Root Borer, *Emmalocera depressella* Swinhoe (Lepidoptera : Pyralidae) (Fig. 17.6, Plate 17.1E)

This is a very serious pest of sugarcane in eastern India but in Pakistan, Haryana and Punjab it is of minor importance. Apart from sugarcane, it feeds on *sarkanda* (*Saccharum munja*), *baru* (*Sorghum halepense*), Napier grass, etc. Damage is caused by the caterpillars which feed on the underground portions of plants, resulting in drying up of the central whorl of leaves. A full-grown caterpillar measures about 30 mm in length, creamy white with yellowish-brown head and a rather wrinkled body. The moths are pale yellow-brown and have white hindwings. The female has a wing expanse of 30-35 mm.

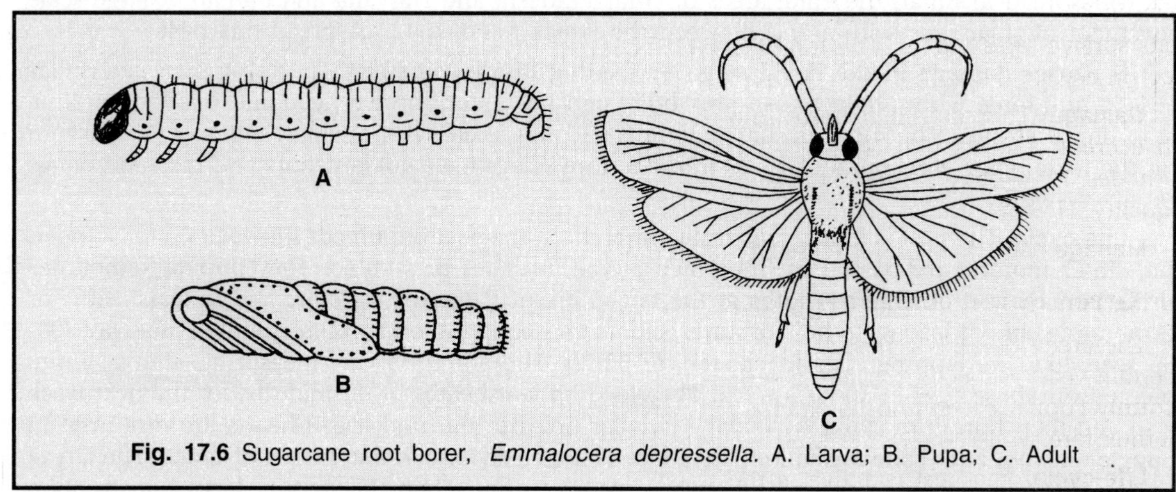

Fig. 17.6 Sugarcane root borer, *Emmalocera depressella.* A. Larva; B. Pupa; C. Adult

Life-cycle. The root borer is active from April to October and passes winter as full-grown larva within the stubble. It pupates sometimes in March and emerges as a moth within 2-3 weeks. The moth lives for 5-7 days. A female lays, on an average, 277-355 scale-like, creamy-white eggs singly on the leaves, the stem or on the ground. The eggs hatch in 5-8 days and the young larvae bore into the stem below the soil surface. As they feed, they cut right across the stem, reaching the adjoining tillers. Central leaves of the attacked plants dry up and form dead-hearts before the cane-forming stage. These dead-hearts are not easily pulled out. The larvae complete their development through five stages in about four weeks. When full-grown, they pupate inside the canes after making emergence holes just above the soil surface. The pupal stages lasts 9-14 days and moths emerge from these holes. The life-cycle is completed in 6-7 weeks. During a year, four generations are completed and the caterpillars of the fifth generation hibernate in winter.

In nature, this pest is attacked by *Trichogramma chilonis* Ishii (Trichogrammatidae), *Apanteles flavipes* (Cameron) and *Stenobracon* sp. (Braconidae).

Damage. This pest is primarily destructive to young plants and the attack is particularly severe from April to June. Plants attacked after the formation of canes are not killed, although their weight and sugar content are reduced. At harvest, a decrease in yield upto 10 per cent and reduction of sucrose in juice by about 0.3 unit have been reported.

Management. (*i*) The pest can be suppressed by ploughing up and burning the stubble in the fields not kept for ratooning. (*ii*) The canes should be harvested below the soil surface in order to kill the caterpillars. (*iii*) Chemical control is not feasible. (*iv*) Staple 40 Tricho-cards (5.0 cm × 2.5 cm hard paper piece glued with 7 days old eggs of laboratory host, *Corcyra cephalonica* parasitized by *Trichogramma chilonis*) to the under-sides of sugarcane leaves from July to October at 10 days intervals. Each card should have approximately 500 parasitized eggs and be spread uniformaly at 100 spots per ha. Normally, 10-12 releases are required.

12. Internode Borer, *Chilo sacchariphagus indicus* (Kapur) (Lepidoptera : Pyralidae) (Plate 17.1F)

The insect is found throughout India and usually occurs on sugarcane late in its growing phase during June to December. It is a serious pest in Andhra Pradesh, Karnataka, Kerala, Tamil Nadu and Uttar Pradesh. The moth is pale brown with white hind legs and the larva has a white body with dark spots and a brown head.

Life-cycle. The females lay white scale-like eggs in masses of 2-60 near the midrib of the leaves, on leaf sheaths and on stems. A maximum of 400 eggs are laid by a female. The incubation period is 5-6 days. The larva bores at the nodal region and enters the stem. The larva becomes full grown in 37-53 days and pupates in the leaf sheath. The pupal period lasts for 8-10 days and the adults survive for 3-4 days. The total life-cycle occupies 50-70 days and there are six broods of the insect in a year.

Damage. The caterpillar bores into the canes near the nodes, the entry holes being plugged with excreta. A larva may attack 1-3 internodes and mostly the attack is seen in the top five internodes. Its feeding causes the tissues turn red. The loss to cane is more to its tonnage than to its quality. However, juice quality is affected if more than 10 per cent of the cane is affected.

Management. Same as in case of stalk borer.

13. Green Borer, *Raphimetopus ablutellus* Zeller (Lepidoptera : Pyralidae)

The green borer is a regular pest of sugarcane in Bihar, Uttar Pradesh, Haryana and Punjab. In certain years, it becomes quite a serious pest. The insect is named after the caterpillar which is uniformly copper green and has a characteristic greyish prothoracic shield. The adult moths have ocherous fore wings and white hind wings.

Life-cycle. The pest is active in the sugarcane fields from February to June only and the rest of the year, it lives as a hibernating larva in the stubble. The moths appear from the pupae towards the end of February or early March. After mating, the females lay, on an average, 37 oval-shaped eggs, in small clusters, scattered in cracks and crevices in the soil or among loose soil particles. The dull white eggs are seen just below the surface of the ground, near the host plants.

The eggs hatch in 7-8 days in April. On emergence, the larvae bore into the soft cane shoots and feed inside. They undergo five moultings in 20-25 days during April and May. The total life-cycle is completed in 35-40 days. They pupate in the cane, leaving an exit hole and emerge as moths in about 7 days. Thus, the total life-cycle is completed in 35-40 days. There are three generations in a year, the larvae of the last generation undergo diapause for many months.

The larvae are parasitized by *Stenobracon deesae* Cameron (Braconidae) which results in some natural control. The timely removal and burning of stubble further help to suppress this pest.

Damage. The damage by this borer is generally found in association with other shoot borers, e.g. *Chilo infuscatellus, Sesamia* sp. and *Emmalocera depressella*, during the summer months. The larvae attack the growing point of the plants, causing dead-hearts. In Uttar Pradesh, it has been found to kill 24-76 per cent of the mother shoots from April to June.

Management. Rake 35 kg of carbaryl 10 per cent dust per ha into the soil at the time of planting of sugarcane.

14. Gurdaspur Borer, *Acigona steniellus* (Hampson) (Lepidoptera : Pyralidae) (Fig. 17.7)

The Gurdaspur borer was considered the most destructive pest of sugarcane in Pakistan and India. It had also been recorded in Vietnam. It was noticed for the first time in District Sialkot in 1923 and was subsequently noticed as a pest at Gurdaspur (Punjab) in 1925. Later, it spread to newer sugarcane areas in Punjab, Haryana, Western Uttar Pradesh and the northern districts of Rajasthan. In the fifties and sixties it was considered to be the most destructive pest of sugarcane in northern India. Its importance as a pest started declining in the late sixties. The exact causes of its natural population decline are not known, although a multiple of factors are considered to have contributed to it, namely (*i*) removal and destruction of infested canes through campaigns in the sugar factory areas during July-September, (*ii*) introduction of new varieties, viz. COJ 46 and COJ 64, replacing the susceptible variety CO 312, and (*iii*) activity of natural enemies like larval parasite, *Stenobracon nicevillei* (Bingham) (Braconidae) and the egg parasite, *Trichogramma* spp. (Trichogrammatidae).

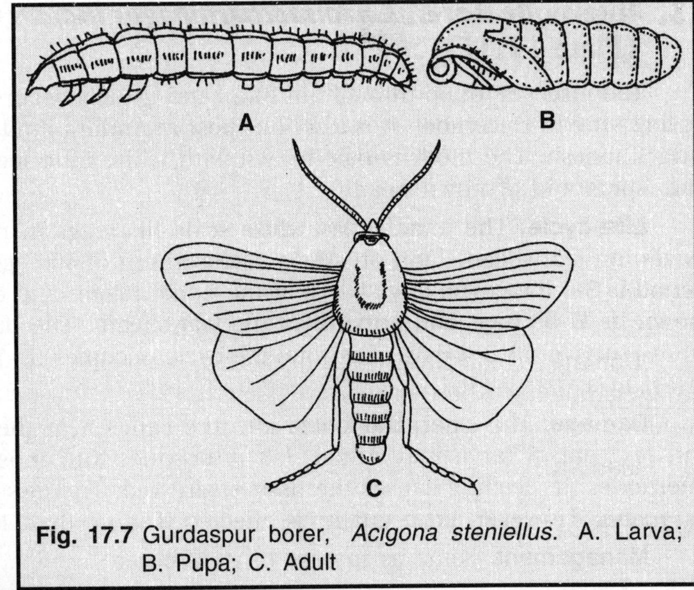

Fig. 17.7 Gurdaspur borer, *Acigona steniellus*. A. Larva; B. Pupa; C. Adult

A full-grown caterpillar is 30-35 mm long, creamy white, with orange brown head. There are four prominent longitudinal violet stripes on the body. The moth is dull brown, 25-45 mm across the spread wings and has a number of dark spots along the outer margins of the fore wings. The hind wings are white.

Life-cycle. The pest is active from July to October and passes the winter and the early part of summer as a full-grown larva in sugarcane stubble. The larvae pupate sometimes in June and start emerging as moths in the end of June or in the first week of July. The moths are nocturnal. A female lays 90-300 flattened scale-like eggs in clusters of 3-22. The eggs are laid on the upper surface of the leaves along the mid-rib and they hatch in 4-9 days. In 4 or 5 hours, the young larvae enter the top portion of a cane through a single hole just above a node. There, they feed gregariously by making spiral galleries which run upwards. After about 7-10 days, when the cane top has dried up, the larvae (in the third stage) come out and enter the adjoining canes single or in twos. They may again come out of these canes and attack more plants. The larvae grow through five stages and are full-fed in 19-27 days. When mature, they make exit holes and pupate inside. The pupal stage lasts 6-12 days and, on emergence, the moths live for 4-5 days. The pest pupates in stubble and the life-cycle is completed in 5-6 weeks. It breeds two or three times in a year. In the beginning of September, the full-grown caterpillars start migration from the upper part by coming out and re-entering the cane near the base. They enter the stubble and hibernate there till June of the next year.

Damage. In the initial stages, the larvae feed gregariously in the top porton of the canes by making spiral galleries thus killing the plants. At about the time the larvae leave these canes and move on to new ones, the dried cane tops can be spotted in a field. Later on, large patches of dried canes appear. The pest destroys 20-25 per cent of the crop. In case of severe infestation, the loss may be as high as 70-75 per cent and sugar recovery from the affected crop is also greatly reduced. A loss of 17 per cent in total solids, 29 per cent in sucrose and an increase of 84 per cent in glucose has been reported due to borer infestation.

Management. (*i*) Rogue out the canes showing withered tops in the afternoon every week from June to September. The tops should be cut off well below the point of attack. (*ii*) Do not ratoon a heavily affected crop. (*iii*) Plough up the fields not meant for ratooning and destroy the stubble before June. (*iv*) Chemical control is not feasible.

15. Plassey Borer, *Chilo tumidicostalis* Hampson (Lepidoptera : Pyralidae)

Plassey borer is a sporadic pest of sugarcane in many areas of India, Bangladesh, Nepal and Thailand. The adults are slender body moths, measuring 16.28±4.32 mm from the head to the tip of forewing. The general color of forewings is brown to pale brown with some darker marking. Hind wings are white in female and dirty white to light brown in male. The larva is creamy white with big dark spots on the body and a dark brown head.

Life-cycle. The adult is nocturnal in habit and mating occurs at dusk. The longevity of adult is about 5 to 7 days. Eggs are laid in batches on both sides of leaf blades. Individual egg is oval-shaped, flat and overlaps each other. Larvae prefer to feed on time before pupation. The larval period is 26.4 days. The pupal period is 7.5 days. The total life-cycle is completed in 43 days.

Damage. The central whorl of the leaves dries up in the damaged plant. Newly hatched larvae are gregarious and bore in the top three to five internodes. After third or fourth stage, they migrate to new canes or bore in the lower internodes of the same cane.

Management. Chemical control measures are same as in case of early shoot borer.

MISCELLANEOUS PESTS

16. Termites, *Odontotermes obesus* (Rambur) and *Microtermes obesi* Holmgren (Isoptera : Termitidae) (Chapter 26)

The termites enter planted setts from the cut ends and make them partly or wholly hollowed inside, filled with mud galleries. The buds are destroyed leading to poor germination. They also occasionally attack roots, hollow the interior and ascend upwards into the stalk that are filled with mud galleries. The affected plants show drying of leaves and death of plants.

Management. (*i*) Use well rotton manure only. (*ii*) Remove the stubble and debris of the previous crop from the field. (*iii*) Application of well rotton neem cake manure @ 60 cartloads/ha reduces the population of termites. (*iv*) Chemical control measures are same as in case of early shoot borer.

17. White Grub, *Holotrichia consanguinea* Blanchard (Coleoptera : Scarabaeidae)

The white grub is prevalent in Bihar, Uttar Pradesh, Punjab, Karnataka, Tamil Nadu, Andhra Pradesh and Kerala

Life-cycle. The adults remain hidden up to one meter deep in soil during winter and emerge out with rains in May-June. Up to 6 eggs are laid by a female in soil of cane fields to a depth of 15 cm. Egg period lasts 7-10 days. The larva feeds for some time on grass roots and then moves to the cane roots and feeds. The grub becomes fully fed in 8 to 10 weeks and pupates deep in soil at depths of 30-150 cm. The pupal stage lasts up to one month. The adults are active at night. There is only one generation in a year.

Damage. As a result of grubs feeding on roots and underground stems, the first symptom is a yellowing (chlorosis) of the leaves. This is usually followed by stunted growth, lodging, plant uprooting and death. In heavily infested areas the crop starts drying up and large-scale withering of the crop is observed. The attack is severe in light sandy soils.

Management. Chemical control measures are same as in case of groundnut in chapter 15.

18. Sugarcane Mite, *Oligonychus indicus* (Hirst) (Acari : Tetranychidae)

About a dozen species of mites infest the sugarcane crop in the world. Of these only four occur in India and Pakistan, viz. *Oligonychus indicus* and *Schizotetraychus andropognii* (Acari : Tetranychidae), *Tarsonemus spinipes* (Acari : Tarsonemidae) and an unidentified species of eriophid mite. The red leaf mite, *O. indicus*, is the most abundant and is a minor pest of sugarcane in Punjab and Haryana and other adjoining States. Besides sugarcane, it infests *baru* (*Sorghum halepense*), *kahi* (*Saccharum spontaneum*) *jowar* (*Sorghum vulgare*) and *bajra* or pearlmillet (*Pennisetum typhoides*)

Both the nymphs and adults cause damage by sucking cell-sap from the undersurface of the leaves. The mites are microscopic and their damage is characterized by red streaks and webbings on the undersurface of leaves.

Life-cycle. This mite remains active on *baru* throughout the year and migrates to sugarcane in April becoming serious in June. A female mite spins a web on the underside of a leaf in which it lays 35-69 dirty-white spherical eggs during its average life-span of 12-33 days. The eggs hatch in 3-4 days and the resulting nymphs develop within the webs by feeding on the leaf. In summer, the male nymphs grow to maturity through three stages completing their development in 2-3 weeks. The females are full-fed in 3-5 weeks. The nymphal development is considerably slowed down in the winter, the duration being 9-23 days in males and 16-24 days in the females. The life cycle of the mite during summer is completed in 3-6 weeks and the pest breeds three times in a year.

Damage. The mite feeds by sucking plant sap with its stylets. The males feed rarely and the damage is done mainly by the females and nymphs. As a result of their feeding, the leaves turn red and gradually dry up. Sugarcane varieties with soft leaves are attacked more readily and damage is noticed to the greater extent during the pre-monsoon period.

Management. (*i*) The pest can be suppressed by the destruction of *baru* (*Sorghum helepens*) grass from the bunds around sugarcane fields. (*ii*) Spray the crop with 1.0 litre of malathion 50EC or 875 ml 35EC in 250 litres of water per ha.

Other Pests of Sugarcane

A number of polyphagous insects are also known to infest sugarcane. These include the grasshoppers, *Hieroglyphus banian* (Fabricius), *Gastrimargus transversus* Thunberg and Phadka grasshopper, *Hieroglyphus nigrorepletus* Bolivar (Orthroptera : Acrididae); black bugs, *Cavelerius excavatus* (Distant) and *Dimorphopterus gibbus* (Fabricius) (Hemiptera : Lygaeidae); leafhopper, *Pruthiana sexnotata* Izzard (Hemiptera : Cicadellidae); bugs, *Tropidocephala signata* Distant, *T. serendiba* Melich and *T.marginepunctata* Melich (Hemiptera : Delphacidae); black leafhopper, *Proutista moesta* (Westwood) (Hemiptera : Derbidae); white fly, *Neomaskellia andropogonis* Corbett (Hemiptera: Aleyrodidae); aphids, *Melanapis indosachhari sacchari* (Zehntner) (Hemiptera : Aphididae); root aphid, *Tetraneura javensis* Goot (Hemiptera : Aphididae); scale insect, *Aclerda japonica* Newstead (Hemiptera : Aclerdidae); thrips, *Sorghothrips jonnaphilus* (Ramkrishna), *Anaphothrips sudanensis* Trybom and *Chloethrips saccharicidus* (Ramakrishna & Margabandhu) (Thysanoptera : Thripidae); bag-worm, *Acanthopsyche* spp. (Lepidoptera : Psychidae); slug caterpillar, *Parasa bicolor* (Walker) (Lepidoptera : Limacodidae); spotted stalk borer, *Chilo partellus* (Swinhoe); leaf rollers, *Cnaphalocrocis medinalis* (Guenee) and *Marasmia trapezalis* (Guenee) (Lepidoptera : Pyralidae); Bihar hairy caterpillar, *Spilarctia obliqua* Walker (Lepidoptera : Arctiidae); armyworm, *Mythimna separata* (Walker); pink stem borer, *Sesamia inferens* (Walker) and cutworm, *Agrotis ipsilon* (Hufnagel) (Lepidoptera : Noctuidae); skipper butter fly, *Telicota augias* Linnaeus (Lepidoptera : Hesperidae); tussock caterpillar, *Psalis pennatula* (Fabricius) (Lepidoptera : Lymantriidae); leaf scraper, *Asmangulia cuspidata* Maulik (Coleoptera : Hispidae); grey beetle, *Tanymecus sciurus* (Olivier) and grey weevil, *Myllocerus* sp. (Coleoptera : Curculionidae).

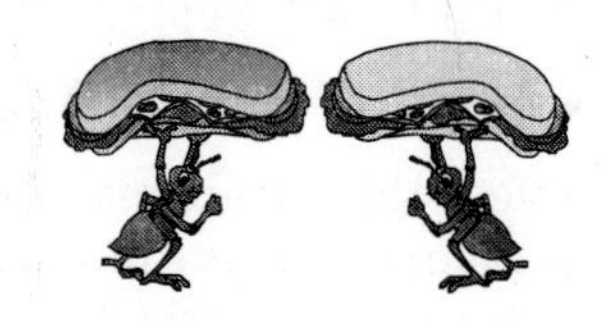

PESTS OF TROPICAL AND SUB-TROPICAL FRUITS

INTRODUCTION

The diverse agro-climatic conditions in India provide tremendous scope for cultivation of horticultural crops (fruits, vegetables, flowers, etc.) of tropical, sub-tropical and temperate regions. Production of horticultural crops has increased considerably as compared to the situation two decades ago. The area under horticultural crops has increased from 16.59 million ha in 2001-02 to 23.69 million ha in 2012-13, with a corresponding increase in production from 145.78 million tonnes to 268.85 million tonnes. During 2012-13, the area under fruit crops was 6.98 million ha with a production of 81.28 million tonnes. While India is the second largest producer of fruits in the world, it is the largest producer of fruits like mango, banana, papaya, sapota, pomegranate and aonla.

Most of these fruits are Asiatic in origin and their improved varieties have been introduced in India and many of the pest problems have found route through these introductions. Some of the local insect populations become opportunists to invade new introductions from one orchard to other and from one locality to another. These match their life-cycle pattern with the tree phenology, becoming active during active growth of the tree and entering into the state of diapause or hibernation during its dormancy so as to tide over ecological adversaries. Orchard ecosystem provides relatively a stable and somewhat permanent ecosystem for insect pests as well as their natural enemies. More than 1000 insect and mite species are known to infest the fruit trees in India.

CITRUS (*Citrus* spp.; Family : Rutaceae)

A number of citrus fruits are grown in the sub-tropics; practically throughout the Indian Sub-continent and also in the rest of South and South-East Asia. In India, citrus is commercially grown in over 1.04 milllion ha with production of 10.09 million tonnes (2012-13). The sweetlime, *Citrus limetta*; *C. limettioides*, are more common in the north-western parts. These fruits are rich source of vitamin C. There are a large number of insect pests of citrus which are widely distributed. Their attack and that of the nematodes are some of the factors contributing to the problem of citrus decline observed in various parts of India and South Asia. By controlling these pests, the citrus decline can be arrested to a considerable extent, at least in those places where pests are a serious problem. Plant protection measures are beneficial, when adopted right from the beginning of the planting of the trees; waiting for 5-6 years, till the decline actually sets in, is not advisable.

1. Citrus Psylla, *Diaphorina citri* Kuwayama (Hemiptera : Aphalaridae) (Fig. 18.1, Plate 18.1A)

The citrus psylla is distributed throughout the Orient and has been reported from India, China, Formosa, Japan, Myanmar, Sri Lanka, the East Indies and New Guinea. In India, it has been recorded on all species of citrus and a number of other plants of the family Rutaceae. It is the most destructive of all the citrus pests.

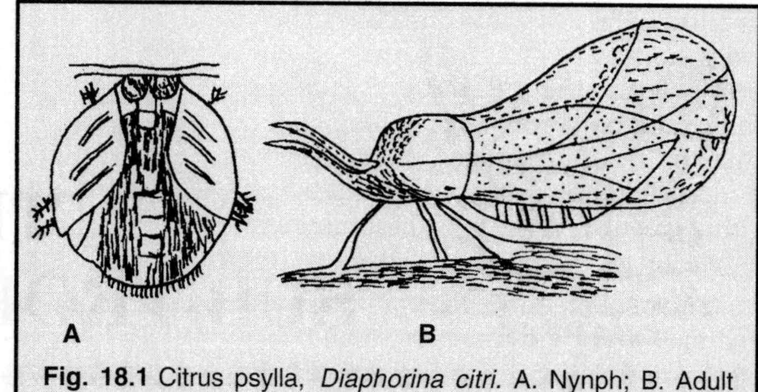

Fig. 18.1 Citrus psylla, *Diaphorina citri.* A. Nynph; B. Adult

Damage is caused by nymphs and adults. The adult is small, 3 mm in length (**Fig. 18.1**), active in habits, and rests on the leaf surface with closed wings, the tail end of the body being turned upwards. The insect is brown with its head lighter brown and pointed. The wings are membraneous, semi-transparent, with a brown band in the apical half of the fore wings. The hind wings are shorter and thinner than the fore wings. The nymphs are flat, louse-like and orange yellow creatures, and are seen congregated in large numbers on young leaves and buds.

Life-cycle. The pest is active throughout the year but its life-cycle is greatly prolonged in the winter. Only adults are to be found when it is very cold. They resume breeding in February-March and lay, on an average, 500 almond-shaped, orange and stalked eggs on tender leaves and shoots of citrus trees. The lower end of the egg stalk is embedded in plant tissue. The eggs are laid either singly or in groups of two or three and are arranged in straight line, there being as many as 50 eggs in one place. The eggs hatch in 10-20 days in winter and 4-6 days in summer. On emerging, the light-yellow nymphs have a tendency to stick close to the egg-shell. They are found congregated on young half-open leaves, but in the later stages of development they may also migrate to the older leaves. There are five nymphal stages and the development is completed in 10-11 days from April to September, 15-20 days in the spring and autumn, and 34-36 days in December-January.

Natural mortality among nymphs varies in different seasons and during the greater part of the year, 20-30 per cent of nymphs die a natural death. However, in December and January, the mortality may be as high as 58 per cent. The most favourable conditions for development are found in March, when there is only about 4 per cent mortality. When the nymphs are full-grown they migrate to the lower surface of leaves, where they change into adults.

The adults copulate 4-8 days after emergence and the females start laying eggs immediately afterwards. There is a great variation in the longevity of adults in various seasons. The females live longer than males, and the duration may be as long as 190 days in winter and only 12-26 days in the summer. During the hot period of May-June, the adults are predominant although other stages may also be found. In July, they start reproducing once again, there being a second lower peak of population in August. Towards autumn, there is again a decline in the population. In summer, the life-cycle is completed in 14-17 days. There are 8-9 overlapping generations in a year.

Tetrastichus radiatus Waterston (Eulophidae) is an important parasitoid of nymphs and is distributed in all the citrus grooves. A number of lady bird beetles, viz. *Coccinella septempunctata* Linnaeus, *C. transversalis* Fabricius, *Menochilus sexmaculatus* (Fabricius), *Chilocorus nigrita* (Fabricius) and *Brumoides suturalis* (Fabricius) (Coccinellidae), and *Chrysoperla carnea* (Stephens) (Chrysopidae) larvae also feed on nymphs.

Damage. This is the most destructive and, consequently, the most important of all the insect pests of citrus. Although there is a visible difference in the rise and fall of its population in various seasons, yet the ill-effects of its damage are so long-lasting that the trees may look sickly even when the population is not high. Thus, sooty and sickly plants seen in the winter are the victims of insects which caused damage during the previous summer.

Only the nymphs are harmful to the plants. With the help of their sharp, piercing mouthparts, they suck the cell-sap in millions. The vitality of the plants deteriorates, and the young leaves and twigs stop growing further. The leaf-buds, flower buds and leaves may wilt and die. Whatever little fruit is formed in the spring, falls off prematurely. Moreover, the nymphs secrete drops of a sweet thick fluid on which a black fungus develops, adversely affecting photosynthesis. It is also thought that the insect produces a toxic substance in the plants as a result of which the fruits remain undersized and poor in juice and insipid in taste. This insect is also responsible for spreading the citrus greening disease. If the pest is not checked in time, the entire orchard may be lost, and after a year or two of continued damage, the plants may be killed.

Management. (*i*) Pest population may be reduced to a great extent by conserving the natural enemies which are quite abundant in citrus ecosystem. (*ii*) Spray 400 g of thiamethoxam 25WG or 500 ml of imidacloprid 17.8SL or 3.125 litres of dimethoate 30EC or 1.5 litres of monocrotophos 36SL or 2.5 litres of oxydemeton methyl 25EC or 250 ml of imidacloprid 200SL in 1250 litres of water per ha during March with the appearance of the pest and again in the first week of September.

2. Citrus Whitefly, *Dialeurodes citri* (Ashmead) (Hemiptera : Aleyrodidae) (Fig. 18.2, Plate 18.1B)

This insect is often found in association with other allied species of whiteflies, namely *Dialeurodes citrifolii* (Morgan), *Aleurocanthus husani* (Corbitt), *A. spiniferus* (Quaintance),etc. The pest is very widely distributed in America, India, Pakistan, Vietnam and China. Although it is a pest of citrus, the insect prefers to feed on certain deciduous plants such as persimmon and *dharek* (*Melia azedarach*). It is often seen on these trees in very large numbers but perishes in the autumn when the leaves are shed. It is said that probably the adults migrate to deciduous plants from evergreens, and that perhaps there is no reverse migration.

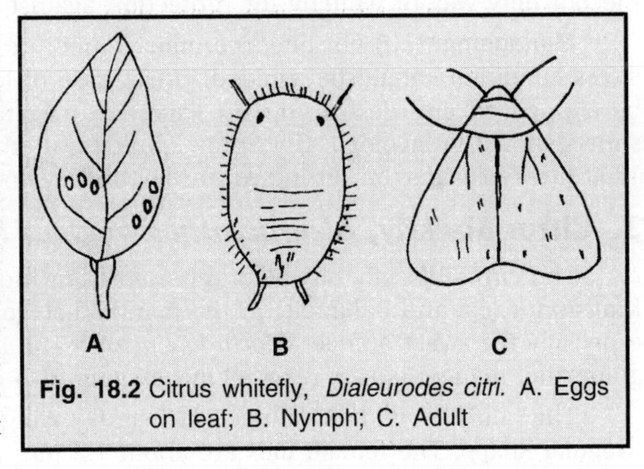

Fig. 18.2 Citrus whitefly, *Dialeurodes citri*. A. Eggs on leaf; B. Nymph; C. Adult

Damage is caused by adults as well as by nymphs. The adult is a minute insect, measuring 1.02-1.52 mm, the males being smaller than the females **(Fig. 18.2)**. The antennae of this insect are six-jointed, the basal joint being thick and stout, and the last segment ending in a sharp point. The eyes are transparent, red and kidney-shaped, with the lower half covered over with bristles. The head is somewhat pointed. The wings are more than twice the length of the body and extend beyond the tip of the abdomen. Both the wings and body are completely covered with a white waxy powder. The nymph is pale yellow, with purple eyes and its body is marginally fringed with bristles. It is distinguished from the pupa which is broadly oval, pale yellow, with an orange or yellow band in the middle of the body.

Life-cycle. The pest is active practically all the year round, but the pupae are generally found from October to February and in June-July. In February, the adults emerge from the pupae. They lay eggs singly on the underside of soft young leaves. The eggs are oval, pale yellow and rest on small stalks. A single female may lay 200 or more eggs during its life which may last from 7 to 10 days. When there is severe infestation, there may be as many as 2,000 eggs on a single leaf. The greatest number of eggs is found in March-April and again in August-September. It may take 10-20 days for the eggs to hatch.

The young larva, on emergence, crawls about for a few hours and then inserts its proboscis into the succulent portion of a twig. The larvae remain there in a stationary state and transform themselves into pupae at the same spot. A nymph is full-fed in 25-71 days and then changes into a pupa. The pupal stage is of the longest duration during the hottest part of summer and the coldest part of the winter, when it lasts 114-159 days.

The adults, on emergence, avoid bright sunlight and settle on the underside of leaves, preferring the north side of a tree. They are quite common in spring and early autumn. From March to August, all stages of the pest are found. In Punjab, there are probably two generations in a year. However, in Japan, where the climate in much milder, three generations a year have been reported.

Encarsia lahorensis (Howard) and *E. citufila* Silv. (Aphelinidae) are the major parasitoids on this insect. A number of predators are recorded feeding on the eggs as well as nymphs of this insect, viz. *Brumoides suturalis* (Fabricius), *Cryptognatha flavescens* Motsch, *Verania cardoni* Weire; a predatory thrips, *Aleurothrips fasciatipennis* and a lacewing, *Chrysopa* sp.

Damage. The pest causes damage in the larval and adult stages. It sucks the cell-sap from leaves which curl over and fall off. The honeydew excreted by the nymphs is a very good medium for the growth of a sooty mould, which interferes with photosynthesis. Thus, the trees infested with this pest deteriorate further. It has been noticed in California that a heavy infestation of whitefly is apt to be followed by an increase in the red scale of citrus, because the young scales collect under the powdery wax of whitefly for protection against bright light.

Management. (*i*) For effective management of flies, close planting, water logging or any other stress condition should be avoided. (*ii*) In case of localised infestation, the affected shoots should be clipped off and destroyed. (*iii*) Excessive irrigation and application of nitrogen and pesticidal sprays should be avoided. (*iv*) Spray 2.5 litres of ethion 50EC or 3.125 litres of triazophos 40EC in 1250 litres of water/ha during April-May and again during September-October.

3. Citrus Blackfly, *Aleurocanthus woglumi* Ashby (Hemiptera : Aleyrodidae)

The citrus blackfly has been reported from India, Sikkim, Sri Lanka, the Philippines, Jamaica, Kingston, Cuba and Bahamas. In the nymphal stage particularly, it is a serious pest of citrus fruits, especially the sweet orange, *Morinda tinctoria* and malta, *Murraya koeniglii*. In addition, avocado, grapevine, mango, guava, pear, plum, etc. are also attacked.

The adult fly is dark orange with smoky wings and fore wings having four whitish areas of irregular shape. The female flies are about 1.2 mm long and the males are 0.8 mm in length. The nymphs are scale-like, shiny black and spiny, and are bearded by a white-fringe of wax.

Life-cycle. The adults emerge in March-April and the females lay yellowish brown oval shaped eggs which are arranged in a spiral on broad leaves. There may be 15-22 eggs in a cluster. The eggs hatch in 7-14 days and the nymphs on emergence start feeding on cell-sap. They pass through four nymphal instars and the nymphal stage is completed in 38-60 days. They pupate on the leaf surface and this stage lasts 100-131 days. The pupa is oval, black in colour and its dorsum is arched with long black spines, and the margins have rounded black teeth. There are two distinct broods in a year. The first brood adults emerge in March-April and those of the second brood emerge in July-October.

The hymenopterous parasites recorded on this pest are *Encarsia divergens* Silv., *E. merceti* Silv. and *Eretmocerus serius* Silo. (Aphelinidae).

Damage. Both adults and the nymphs suck plant sap, reducing the vitality of trees. It results in the curling of leaves and also the premature fall of flower buds and the developing fruits.

Management. Same as in case of citrus white-fly.

4. Citrus Red Scale, *Aonidiella aurantii* (Maskell) (Hemiptera : Diaspididae)

Citrus plants and fruits in various countries of the world are known to be attacked by three species of armoured scales, viz. *Aonidiella aurantii, A. citrina* (Coquillett) and *Chrysomphalus dictyospermi* (Morgan). Of these, the red scale (*A. aurantii*) is the most destructive and it also feeds on *Acacia, Eucalyptus*, fig, grape, rose, willow, *shisham* (*Dalbergia sissoo*) and many other plants.

The scales infest all the above ground parts of the tree. The leaves, branches and fruits may be covered with them. These scales are 2 mm in diameter and have a distinct central exuviae.

Life-cycle. The pest is active throughout the year, but its attack is maximum during autumn. The female scales, instead of laying eggs, give birth to young nymphs. The first stage nymphs or crawlers have well developed legs and antennae, and move about for an hour or so before settling down and cover themselves with a white waxy secretion. The female scales moult two times at 10-20 days intervals, lose their legs and antennae, and incorporate their cast skins into waxy coverings, which become circular, depressed scales. They reach sexual maturity in 10-15 weeks, do not acquire wings and may live for several months. The male scales, on the other hand, are elongated and develop into winged adults in 1-2 months. They fly around and fertilize the sedentary females. This pest passes through a number of broods in a year.

Aphelinus sp. (Aphelinidae) parasitizes a high percentage of scales. The coccinellids, *Chilocorus nigrita* (Fabricius) and *Scymnus quadrillum* are recorded as predators on this scale throughout India.

Damage. Injury to the infested plants is two-fold; first the scale insects feed on plant juice and devitalize the plants; second, they inject a toxic substance into the plant sap. Yellow spots appear at the point of feeding which may be on the leaves, twigs or fruits. When there is a severe infestation, all the leaves may turn pale. It is not practicable to detach all the scales from infested fruits before marketing and, therefore, their value is lowered.

Management. Same as in case of citrus psylla.

5. Cottony Cushion Scale, *Icerya purchasi* Maskell (Hemiptera : Margarodidae)

The cottony cushion scale is a serious pest of citrus plants in certain tropical and sub-tropical regions of the world. It is a native of Australia, where it lives mostly on *Acacia* spp. In India, it is present in southern States, mostly on wild plants such as wattle, casurina and gorse. It has also been recorded damaging apple, almond, walnut, peach, apricot, fig, grapevine, guava, pomegranate, etc.

The adult female is a flat oval (4.5 × 3.5 mm), brown to reddish-brown, soft bodied scale. The most conspicuous part of the insect is the large, white, fluted egg-sac which is secreted by the female. The full-grown larva is broadly oval, 3.0 × 1.5 mm, reddish-brown to brick-red in colour.

Life-cycle. This insect pest is active throughout the year but during the dry-hot months, its multiplication is the maximum. The males are rare and the scale generally reproduces parthenogenetically. The female lays up to 700 eggs in the ovisac held behind the body. The reddish coloured young nymphs emerge from the eggs within 24 hours during the hot season, but may take several weeks when the weather is cold. The newly hatched nymphs move some distance before fixing themselves on leaves and twigs for feeding. They become adult females after moulting three times. The life-cycle is completed in 46-240 days, depending upon the different environmental conditions.

In south India, this pest was brought under check by the introduction and periodic releases of a large number of predaceous beetles, *Rodolia cardinalis* (Mulsant) (Coccinellidae) and by banning the movement of wattle from Nilgiris and Kodai Kanal.

Damage. As a result of severe feeding by scales, the leaves and twigs firstly turn pale and then fall prematurely. Sometimes, the heavily infested young shoots and small nursery plants are also killed.

Management. Same as in case of citrus psylla.

6. Citrus Mealybug, *Pseudococcus filamentosus* Cockerell (Hemiptera : Pseudococcidae)

In appearance, habits and habitat the fluffy mealy bug, *P. filamentosus* resembles other allied species, namely *P. citriculus* Green found in China, and USA; *P. calceolariae* (Maskell) in California and *Planococcus citri* (Risso) found in Malaysia and Indonesia. *P. filamentosus* is more common in southern India than in the north. The mealy-bugs are known to feed on a number of plants, often not closely related to citrus. In the gardens, they are seen on *Cactus* spp., ferns, begonia, gardenia, poinsettia and other flowers.

Damage is caused by nymphs and females. The adult female is wingless, with a flattened body and short, waxy filaments along the margins. It has piercing and sucking mouthparts. The male is winged, midge-like, with long antennae and no mouthparts; consequently, it does not feed. The nymphs are amber with a whitish waxy coating and filaments.

Life-cycle. The females lay eggs in clusters on citrus plants, which are found in protective cotton-like masses. There may be 300 eggs or more in one mass. The eggs hatch in 10-20 days and the nymphs crawl out and start feeding by inserting their mouthparts in the lower surface of the leaves. A waxy white covering is soon formed on their bodies. A female nymph is full-grown in 6-8 weeks. The male nymphs spin cotton-like cocoons, two or three weeks after hatching, and pupate before transforming themselves into winged adults. Mealy bugs are usually found on the underside of the leaves in clusters. Since they multiply rapidly, all stages of development may be present at the same time.

Damage. The insects feed on cell-sap and the plants become pale, wilted and the affected parts eventually die. The insects also excrete honeydew on which a black mould grows, which interferes with photosynthesis. Black ants are attracted to the honeydew and they become a nuisance. In severe cases of infestation, the citrus flowers do not set fruit.

Management. (*i*) For effective management of these coccids, orchard sanitation is extremely important. Weeds act as additional hosts and these must be removed. (*ii*) The infested shoots should be pruned and destroyed. (*iii*) The ant colonies should be destroyed by ploughing the soil around trees and by application of quinalphos or carbaryl dust. (*iv*) The coccinellid beetle, *Cryptolaemus montrouzieri* Mulsant, should be released @ 10 beetles per plant, to control mealy-bugs. Inoculative releases of an exotic parasite, *Leptomastix dactylopii* Howard, are also recommended. (*v*) Spray 4.5 litres of chlorpyriphos 20EC in 1250 litres of water/ha. Repeat the spray after 15 days.

7. Citrus Caterpillar, *Papilio demoleus* Linnaeus (Lepidoptera : Papilionidae) (Fig. 18.3, Plate 18.1C)

The citrus caterpillar or the lemon butterfly, is found in Africa, greater part of Asia as far as Formosa and Japan. It has been reported all over India and causes very severe damage. It can feed and breed on all varieties of cultivated or wild citrus and various other species of the family Rutaceae.

Only the caterpillars cause damage by eating the leaves. The full-gown caterpillar is yellowish green, has a horn-like structure on the dorsal side of the last body segment and, is 40 mm long and 6.5 mm wide. The adult is a large beautiful butterfly, 28 mm in length and 94 mm in wing expanse (**Fig. 18.3**). Its head and thorax are black, there being a creamy-yellow coloration on the underside of the abdomen. Its wings are dull-black, ornamented with yellow markings. The general coloration on the underside of the wings is slightly paler and the markings are also larger. The antennae are black and have club like structures at their ends.

Life-cycle. In the plains, this pest is found throughout the year, whereas in the mountains where winter is very severe, it hibernates in the pupal stage. The butterflies appear in March and lay eggs on tender shoots and fresh leaves, mostly on the undersurface. The eggs are placed singly or in groups of 2-5. Sometimes, the eggs are also noticed on thorns of citrus or on other plants. The eggs are glued firmly on to the surface of the leaf and are pale or greenish yellow, when freshly laid, but later turn brown, becoming dark grey just before hatching. They hatch in 3-4 days during summer and in 5-8 days during winter. The young larva emerges by cutting a round hole through the egg-shell, which forms the first food of the larva. This scavenging habit persists throghout life and the larvae eat their own exuviae after each moulting.

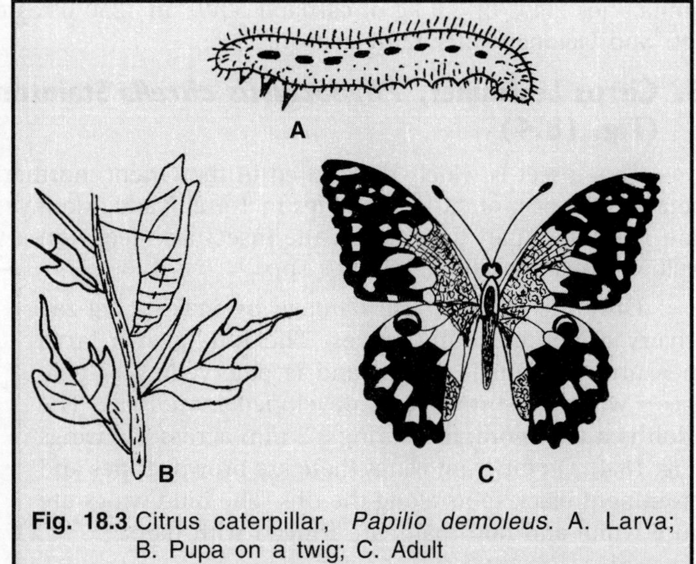

Fig. 18.3 Citrus caterpillar, *Papilio demoleus.* A. Larva; B. Pupa on a twig; C. Adult

The larval life lasts 8-16 days in the summer and about 4 weeks during November-December. The larvae show perference for young and shiny leaves of citrus. After making a full meal, they remain motionless while exposed, usually near the mid-rib. The black or brown and white markings make the larvae look like bird droppings. In addition to its protective coloration, the caterpillar is also equipped with two reddish sacs posterior to the head and they are extended as a sign of danger, discharging a repelling fluid. When full-grown, the larvae have a tendency to crawl away from the host plant, and it is rare that they pupate on the plant, on which they had been feeding. The mature larva spins a supporting girdle around its body and pupates on a twig, a dry stick or any other raised structure. The pupal stage lasts about 8 days in the summer, and 9-11 days in the spring and the autumn. The butterflies usually appear in the morning and begin to fly within one to three hours of emergence. A female usually mates once and lays, on an average, 75-120 eggs within 2-5 days. A smaller number of eggs are laid in the summer. A male lives for three or four days whereas a female lives for about a week.

The pest passes through three or four generations in a year. The butterflies that appear in March give rise to larvae which are seen in abundance towards the end of April. During the intense heat of June, the population gets a setback, but there is fresh egg-laying in July and the numbers increase in August. The pest is most active in September and after the butterflies have laid eggs, there is again a preponderance of caterpillars in October. These larvae pupate in November and may enter hibernation in cold regions. Since there are over-lapping generations, all stages of the pest are found throughout the year, except in severe winter when only pupae may be found. In the laboratory, ten generations have been reared in a year.

The egg parasitoids associated with this pest are *Trichogramma evanescens* Westwood (Trichogrammatidae), *Pteromalus luzonensis* (Pteromalidae) and *Telenomus* sp. (Scelionidae), while larvae are parasitized by *Erycia nymphalidaephaga* Bar., *Charops* sp. (Ichneumonidae) and *Brachymeria* sp. (Chalcididae).

Damage. The young larvae feed only on fresh leaves and terminal shoots. Habitually, they feed from the margin inwards to the midrib. In later stages, they feed even on mature leaves and sometimes the entire plant may be defoliated. The pest is particularly devastating in nurseries and its damage to foliage seems to synchronize with fresh growth of citrus plants in April and August-September. Heavily attacked plants bear no fruits.

Management. (*i*) Hand picking of various stages of the pest and their destruction especially in nurseries and new orchards helps to suppress the population of the pest. (*ii*) Spraying of entomogenous fungus, *Bacillus thuringiensis* Berliner, nematode DD-136 strain or neem seed extract (3%) also gives quite high motality of caterpillars. (*iii*) In severe infesation, spray 1.5 litres of quinalphos 25EC or 2.0 kg of carbaryl 50WP in 1250 litres of water per ha during April (after fruit set) and October (after rainy season).

8. Citrus Leafminer, *Phyllocnistis citrella* Stainton (Lepidoptera : Phyllocnistidae) (Fig. 18.4)

This insect is widely distributed in the Orient, northern Australia and India, and is known to be a serious pest of citrus nurseries in Tamil Nadu, Madhya Pradesh, Assam, Uttar Pradesh, Punjab and Pakistan. Apart from citrus, the insect also feeds on a variety of other plants such as pomelo, willow, cinnamon and *Loranthus* spp.

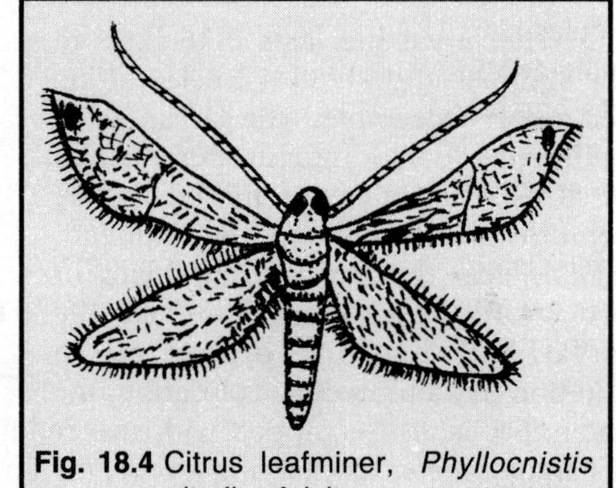

Fig. 18.4 Citrus leafminer, *Phyllocnistis citrella.* Adult

Only the larvae cause damage by making zig-zag silvery mines in young leaves. The full- grown larva measures 5.1 mm in length and is pale yellow or pale green with light-brown well developed mandibles. The adult is a tiny moth, measuring 4.2 mm across the wings (**Fig. 18.4**). On the front wings there are brown stripes and prominent black spots along the tips. The hind wings are pure white and both pairs are fringed with hairs.

Life-cycle. The pest is active throughout the year and breeds on young growth. The duration of various stages depends upon the prevalent temperature. Late in spring or summer, development may be five or six times as fast as in autumn or early spring. The moths lay minute, flattened, transparent eggs on young leaves or tender shoots, usually on the lower surface, particularly near the midrib. The eggs which are laid singly, generally two or three per leaf, hatch in 2-10 days, giving rise to legless larvae. Soon after emerging, the larvae mine into the leaf tissue and form galleries within which they remain confined for the rest of their immature life. Larvae are full-fed in 5-30 days, and when mature, they settle down in enlargements of the galleries near the leaf margin. By the time they spin cocoons for pupation, the leaves get twisted or folded over. In body form, the pupa is not much different from the larva, but turns slightly brownish. It is partly exposed through the gallery wall and has a spine on its head with the help of which it pierces through the wall as it emerges as a moth. The pupal stage lasts 5-25 days and the moths are commonly seen resting on the trunks of the trees near the ground. The life-cycle is completed in 12-55 days and several overlapping generations are produced in a year.

The larvae are parasitized by *Cirrospiloideus phyllocnistoides* (Narayan), *Scotolinx quadristriata* Rao & Rama and *Eurytoma* sp. (Eurytomidae). The parasitization exceeds 30 per cent during September- October.

Damage. Damage by this mining pest is serious on young leaves. The injured epidermis takes the shape of twisted silvery galleries. On older leaves, brownish patches are formed which serve as focii of infection for citrus canker. The attacked leaves remain on the plants for a considerably long time and the damage gradually spreads to fresh leaves. Heavily attacked plants can be spotted from a distance and young nurseries are most severely affected; the young plants of orange and grape-fruit may not even survive. In larger trees, the photosynthesis is adversely affected, vitality is reduced and there is an appreciable reduction in yield.

Management. (*i*) Spray of 2 per cent neem seed extract has been found quite effective and safe. (*ii*) Spray 400 g of thiamethoxam 25WG or 500 ml of imidacloprid 17.8SL or 1.25 litres of fenvalerate 20EC or 2.5 litres of cypermethrin 10EC or 3.125 litres of triazophos 40EC or 4.5 litres of chlorpyriphos 20EC or 500 ml of imidacloprid 200SL in 1250 litres of water per ha during April-May and August-September. Synthetic pyrethroids should be avoided on full-grown trees.

9. Bark Caterpillar, *Indarbela quadrinotata* (Walker) (Lepidoptera : Metarbelidae)

The bark-eating caterpillars of the moth, *I. quadrinotata* feed on citrus, mango, guava, *jamun,* loquat, mulberry, pomegranate, *ber,* drumstick, litchi, amla, rose and a number of forest and ornamental trees. The pest is widely distributed in Myanmar, Bangladesh, Sri Lanka and India.

The freshly hatched larvae are dirty brown while the full-grown caterpillars (50-60 mm) have pale brown bodies with dark brown heads. The adults are pale brown moths with rufous head and thorax. The fore wings are pale rufous with numerous dark rufous bands. Their hind wings are fuscous.

Life-cycle. With the start of the summer season the moths emerge and become active. The females start laying eggs in clusters of 15-25 eggs each, under the loose bark of the trees. The egg laying continues throughout the summer. As many as 2000 eggs may be laid by a single female. The eggs hatch in 8-10 days and the freshly hatched larvae nibble at the bark and after 2-3 days bore inside. The larvae have the habit of making webs along the feeding galleries and above the holes where they bore deeper into the wood. The galleries and the webs above them have a zig-zag shape and contain wooden frass and excreta. The larvae take as many as 9-11 months to complete development. When full-grown, they make a hole into the wood and pupate inside. The pupal stage lasts 3-4 weeks. The moths emerge in summer and they are short lived. Only one generation is completed in a year.

The parasitoid, *Zenillia haterusiae* (Tachinidae) has been noticed attacking the larvae of this pest in Sri Lanka.

Damage. Thick, ribbonlike, silken webs are seen running on the bark of the main stem especially near the forks. The larvae also make holes and as many as 16 holes may be seen on a tree, one caterpillar or pupa occupying each hole. A severe infestation may result in the death of the attacked stem but not of the main trunk. There may be interference with the translocation of cell sap and thus arrestation of growth of the tree is noticed with the resultant reduction in its fruiting capacity.

Management. (*i*) Clean cultivation is essential to prevent infestation of these borers. (*ii*) As soon as infestation is noticed, kill the caterpillars mechanically by inserting an iron spike into the holes made by these caterpillars. (*iii*) During February-March, insert into the borer holes insecticide-soaked cotton plugs (with the help of a metallic spike) and plaster on the outside with mud. The insecticides for 100 litres of water are 40 g carbaryl 50WP or 2 ml dichlorvos 100EC or 5 ml methyl parathion 50EC or 10 ml monocrotophos 40EC. These chemicals should be applied after removing webbings. (*iv*) Treat all alternate host plants in the vicinity of the orchard.

10. Fruit Sucking Moths, *Ophideres* spp. (Lepidoptera : Noctuidae) (Plate 18.1D)

The fruit moths are minor pests of citrus, mango, grapes and apple, and are distributed throughout India. *Ophideres conjuncta* Cramer, *O. Fullonica* Linnaeus *O.materna* Cramer and *O. ancilla* Cramer are the commonest species found in Tamil Nadu, Madhya Pradesh, western Uttar Pradesh and Punjab. They are reported to be in abundance near the forests or other natural vegetation.

The presence of moths in a locality is observed from the characteristic pin-hole damage in citrus and other fruits. These moths are large and stoutly built and their prominent palpi are turned upwards (**Fig. 18.5**). The piercing mouthparts are very well developed and are provided with sharp spines which help in puncturing the fruits. The general body of *O. conjucta* is faint orange brown. Its fore wings are dark grey and the hind wings are orange red, having two black curved patches. The forewings in *O. materna* are pale greenish-grey with palish-white markings and the hind wings are orange brown, having marginal dark bands mixed with white spots.

The larvae are typical semiloopers and have a stout appearance. Their velvety dark brown background along with other patterns make them cryptic. They have distinct eye-spots on the head, yellow or red lateral spots and a dorsal hump on the last segment of the body. A full-grown larva is 50-60 mm in length, and when disturbed, it assumes a characteristic posture by curving round the head and raising the hind part of its body.

Life-cycle. The moths are nocturnal and are not seen during the day. They lay eggs on a number of wild plants and weeds, namely *Tinospora cardifolia, T. smilacina* Benth, *Cocculus hirsutus* L., *Cirsampelos pareira* L., *Convolvulus aruensis, Trichisia pattens* Oliv. and *Pericampylus glancus* Blatter, which are often found growing near citrus orchards. The eggs are round, translucent, measuring about 1 mm in diameter. They hatch in about two weeks and within 24 hours of emergence, the young larvae start feeding on the foliage of host plants. A larva passes through five instars in four weeks. When full-grown, it makes a pupal case by webbing together pieces of leaves and soil particles. The pupa is thick-set and is dark reddish brown. This stage lasts about two weeks.

The moths, on emergence, fly to nearby orchards for feeding on fruit-juice. The exact duration of the life of the moth is not known. Most probably, the moths emerge in spring, when they start breeding. By July, the moths of the second brood are found in large numbers. They damage citrus plants up to October. The pupae of the third brood probably hibernate during winter.

Damage. Unlike most moths and butterflies, the fruit-piercing moths cause damage in the adult stage. With the help of its strong, piercing mouthparts, moth punctures the fruit for sucking juice. Bacterial and fungal infections take place at the site of attack, with the result that the brownish mouth of a puncture becomes pale and eventually the whole fruit turns yellow (Plate 18.1D). It drops off the tree and apparently looks like a premature fruit. If the damaged fruit is squeezed, the juice spurts from the hole. In severe cases of infestation, almost all the fruits are lost.

Management. (*i*) Systematic destruction of alternate host plants in the vicinity of the orchard is suggested to control this pest. (*ii*) Dispose off fallen fruits which attract the moths. (*iii*) Bagging of fruits is effective but very laborious and expensive. (*iv*) Creating smoke in the orchards after sunset may keep the pest at bay, but this method is also cumbersome and not feasible on large scale. (*v*) Spray trees with 2.5 kg of carbaryl 50WP in 500 litres of water per ha at the time of maturity of fruits. (*vi*) Kill moths with a bait containing *gur* 1 kg + vinegar 60 g + lead arsenate 60 g + water 10 litres. Wide-mouthed bottles (1 bottle per 10 trees) containing bait solution should be tied to the plants when the fruits are in unripe condition.

11. Citrus Blossom Midge, *Dasineura citri* Grover (Diptera : Cecidomyiidae)

This midge infests the blossom of many varieties of citrus in Punjab, Assam, Sikkim, Bengal, Andhra Pradesh, Maharashtra and Karnataka. The maggots cause damage to the developing flower buds which fall off. In the adult stage the minute flies are prominent owing to their orange colour.

Life-cycle. During the flowering season in February-March, the orange flies insert their stalked eggs in between the petals of flower buds. The eggs hatch in 32-40 hours and the larvae start feeding on unopened petals. They grow through four larval instars and this developmental stage is completed in 10-12 days. Then they descend to the soil, spin a silken cocoon and pupate there. The pupal stage lasts 4-6 days. During the flowering period of various citrus plants, two or three generations may be completed. The rest of the year is passed in the soil as a pupa.

Damage. The attack of this pest is usually heavy during February-March and the infested blossom looks abnormal in shape. The attacked buds and flowers when shaken by wind drop off easily. Naturally the fruit bearing capacity of the trees is reduced very much.

Management. Spray 1.70 litres of dimethoate 30EC in 1250 litres of water per ha during August-September, when the pest may appear in severe infestation.

12. Citrus Leaf Folder, *Psorosticha zizyphi* (Stainton) (Lepidoptera : Oecophoridae)

This pest has a country-wide distribution and is a common pest of citrus plants in central and northern parts.

Life-cycle. The eggs are laid singly or in groups along midribs of leaves. Up to 404 eggs are laid by a female moth. The egg, larval and pupal periods last 3-5, 9-11 and 5-10 days, respectively. The larva pupates in the leaf-folds. The adult lives for 5-16 days and the total life-cycle occupies 20-31 days.

Damage. The pest is active in the nursery and young plantations from May to October. The larvae web together and fold leaves and start feeding from top to downwards. They feed from within on the epidermis first and on the whole leaves later. The plants become stunted.

Management. Spray 1.5 litres of monocrotophos 36SL or 3.125 litres of chlorpyriphos 20EC or 2.5 litres of quinalphos 25EC in 1250 litres of water/ha.

13. Citrus Aphid, *Toxoptera* spp. (Hemiptera : Aphididae)

The brown citrus aphid, *Toxoptera citricidus* (Kirkaldy) and the black citrus aphid, *T. aurantii* (Boyer de Fonscolombe) are important pests of citrus, the former being a vector of citrus tristeza virus. *T. citricidus* is widely distributed in South America, Central America, South and Southeast Asia. In India, *Toxoptera* spp. is widely present on Coorg mandarin and also on Nagpur mandarin. The host plants of these aphids include various species of the Rutaceae and Rosaceae families. The adult wingless forms (apterae) of *T. citricidus* are shiny black and nymphs are dark reddish brown. The adult winged forms (alatae) can be recognized by the conspicuous black antennal segments I, II and III. The nymphs of brown aphid give out yellow haemolymph and black aphid, red haemolymph on squashing.

Life-cycle. The aphids are usually present on citrus throughout the year. However, they are most abundant in the spring and autumn. They are active from first weak of February to first week of May, with their critical period of infestation from first week of March to first week of April.

The citrus aphid produces about 5 young ones daily for a period of 1-3 weeks parthenogenetically which attain maturity in about 6 days. Winged colonizing adults called 'stem mothers' give birth directly to nymphs with no egg stage and no mating required. When citrus trees flush, the alate adults fly into the orchards, infesting the new shoots and producing nymphs by means of viviparity. The nymphs take 6-42 days to complete 4 molts before becoming adults. Each female can produce 5-68 nymphs. The adults live for 5-25 days. Alate forms appear as a result of crowding. It completes 9 overlapping generations in a year.

Damage. The nymphs and adults suck the cell sap from young leaves and tender twigs. This impairs the vitality of the trees. The affected leaves in severe cases curl up and get deformed. The growth of young shoots is adversely affected resulting into stunted growth. Blossoms and newly set fruits are also attacked. The honeydew excreted by the aphids also provides a good substrate for the growth of sooty mould, which affects the photosynthetic activity of the plants.

Management. (*i*) Release of the coccinellid predator, *Menochilus sexmaculatus* (Fabricius) @ 50 per tree helps to suppress the aphid population. (*ii*) Chemical control measures are same as in case of citrus psylla.

14. Citrus Mite, *Oligonychus citri* McGregor (Acari : Tetranychidae) (Fig. 18.6)

This pest of citrus is of international importance and has been recorded in USA (especially in Florida and California), Indonesia, the Philippines, Israel, Iraq, Iran, UAR, Sri Lanka and India. It was first recorded in southern India where it was found along with *Tetranychus sexmaculata* Riley, *T. urticae* Koch and *T. hindustanicus* Hirst. In the Punjab, it is found infesting sweet-orange, lemon, grapefruit and sour-lime. No alternative hosts of this mite are known.

The citrus red mite has been recorded as a pest of citrus at Abohar in Punjab where there are large citrus orchards. All the feeding stages cause damage by sucking cell-sap from the leaves. The adult is small, plump and orange with thick deep brown patches on the dorsal side of its body and measures 0.33 mm in length (**Fig. 18.5**). Its body is covered with prominent bristles, each borne on a whitish tubercle. The antennae of the females are three segmented and are bright carmine pink. The newly hatched larva is light yellowish brown and has only three pairs of legs. The protonymph is orange brown and the duetonymph is orange brown with a greenish tinge.

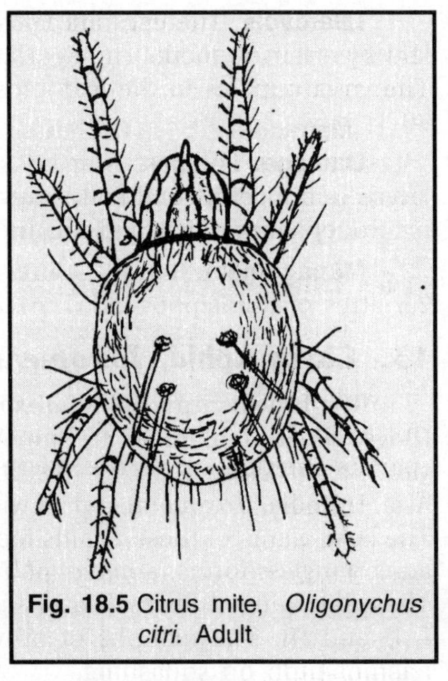

Fig. 18.5 Citrus mite, *Oligonychus citri.* Adult

Life-cycle. This pest is active throughout the year but exhibits slower breeding in the winter. The mites are most active during May-June when they lay about 50 eggs each, arranged singly along large veins on the underside of the leaves. The eggs are minute, round and orange, and are embedded in the leaf tissues with ornamentations of threads which extend from the top outwards. They hatch in about one week. Those eggs laid by unmated females, develop into males, whereas those from the mated females, develop into a mixture of females and males. The first instar larva has three pairs of legs and after feeding on the cell-sap for three or four days, it moults into a protonymph which has four pairs of legs. In 3-4 days, it develops further to form the duetonymph which also has four pairs of legs. It is full-fed in 4-5 days and after moulting, transforms itself into an adult mite. The female mites live for about ten or more days. The life-cycle in summer is completed in 17-20 days and the pest passes through several overlapping generations in a year.

Damage. Mites are considered to be highly destructive pests of citrus plantations. Injury to leaves, tender fruits and green bark is caused by their constant feeding on chlorophyll, resulting in a speckled appearance of the leaves. Heavy infestation may result in complete defoliation, especially of the young nursery plants. The affected fruits become yellow and remain undersized.

Management. Spray 2.5 litres of ethion 50EC or 1.875 litres of fenzaquin 10EC in 1250 litres of water/ha as soon as the mite population appears on the underside of the leaves. Repeat the spray if needed.

15. Citrus Nematode, *Tylenchulus semipenetrans* (Cobb) (Tylenchida : Tylenchulidae) (Fig. 1.5)

This nematode is associated with the problem of citrus decline. The plants become unprofitable because their fruit set is lowered and the incidence of fruit-drop is increased. This citrus nematode was detected as a pest as early as 1913 in the USA and is widely distributed in practically all the citrus-growing regions of the world. In India, it has been recorded in Punjab and Delhi. The moving of nursery stock is the commonest means of carrying this nematode from one country to another.

Life-cycle. The female nematodes are found on thick, stunted rootlets to which a layer of soil particles might be adhering. These particles are held in place by a gelatinous mucus secreted by the females. This mucus normally forms a protective covering around them. The females retain the eggs inside till segmentation in the embryo is demarcated. The eggs hatch in about two weeks and the larvae start feeding on the rootlets. The larvae destined to be males or females have two distinct periods of development. Those which are short and rather broad, develop to form mature males without any feeding; the others are slender and feed on plant roots for 6-8 weeks before developing into females (Fig. 1.3). The most interesting point about this pest is that, in the absence of mating with males, the females are capable of reproducing. These females produce eggs which further give rise to normal males and females. In citrus-growing areas, several generations are completed in a year. It is estimated that if there are 1,500 nematodes per 500 g of soil, the pest can cause appreciable damage to the citrus plants and produce symptoms of tree decline.

Damage. The first indication of injury by this pest is a reduction in terminal growth, followed by a general reduction in vigour, yellowing and drying of leaves and twigs. The trees just survive and produce a reduced crop of inferior fruits.

Management. (*i*) The introduction of this pest into new areas can be checked by using clean nursery stock obtained from those places where the soil has been throroughly fumigated. (*ii*) Use Nemagon @ 25 litres per ha. Pulverise soil in the basin area around a plant and mix the chemical thoroughly followed by flood irrigation.

16. Lance Nematode, *Hoplolaimus indicus* Sher (Nematoda : Hoplolaiminae)

This plant parasitic nematode was first recorded in India during 1963 and its presence and prevalence in citrus orchards and tomato fields was confirmed in 1966. Further studies have proved its widespread distribution on a variety of crops in Punjab. These nematodes feed on root tissues for a short time, like an ectoparasite, and then they retreat into the surrounding soil medium. No part of the development is completed inside the plant tissues.

Life-cycle. The oblong eggs which measure 70-75 microns in length and 30 microns in width are deposited in the two-celled stage. Further development of the embryo continues inside the egg and the first stage larva is visible in about 13 days after the deposition of eggs. The larva emerges from the egg when it is in the second stage and starts feeding immediately after emergence. To attain full growth, the larva passes through five stages. The life-cycle is completed within 42-99 days (temperature 20-35°C.). The development of various stages is quickest at 30°C, which is also the optimum temperature for the survival of various stages and for egg-laying. The optimum pH range for the development of various stages is 6-8 and the maximum increase in population takes place at pH 7. The optimum soil moisture is about 16 per cent. Light soils are more suitable for reproduction. Sandy-loam soil provides the optimum conditions.

Damage. The population of this nematode is highest in April in citrus orchards. Old citrus plants (10 years of age or more) carry more population. The maximum population occurs upto a depth of 15 cm from the surface. The nematode has a high population in tomato fields, often reaching pest proportions. The incidence in citrus orchards has not yet attained the pest level in most places. Thus, there is great need to suppress this nematode before it imposes itself as a serious pest on citrus plants. Of various other plants, the nematode seems to prefer tomato, brinjal, maize and sugarcane. Tobacco, mustard, peas, sugar-beet, water-melon and guava are unsuitable hosts.

Management. Same as in case of citrus nematode

Other Pests of Citrus

The minor pests on citrus in India are the aphids, *Aphis gossypii* Glover and *Myzus persicae* Sulzer (Hemiptera: Aphididae); mealybug, *Nipaecoccus viridis* (Newstead) (Hemiptera : Pseudococcidae) **(Plate 18.1E)**, *Coccus hesperidum* Linnaeus and *Saissetia coffeae* (Walker) (Hemiptera: Coccidae); the bark eating caterpillar, *Indarbela tetraonis* (Moore), *I. quadrinotata* (Walker) and *I. humeralis* Thunberg (Lepidoptera : Metarbelidae); *Papilio memnon* Linnaeus and *P. polytes* Linnaeus (Lepidoptera: Papilionidae); *Spodoptera litura* (Fabricius) (Lepidoptera : Noctuidae); *Bactrocera cucurbitae* (Coquillett) (Diptera : Tephritidae); flower feeding beetle, *Oxycetonia versicolor* (Fabricius) (Coleoptera : Scarabaeidae) **(18.1F)**; *Selenopsis germinata* (Fabricius) and *Oecophylla smaragdina* (Fabricius) (Hymenoptera : Formicidae).

GRAPEVINE (*Vitus vinifera* L.; Family : Vitaceae)

Grape can be grown in sub-tropical and temperate regions. It is an important fruit in parts of southern Asia and in the north-western parts of the Indian Sub-continent. Its cultivation is spreading throughout the Indo-Gangetic plains as well. In India, the area under this fruit plant is 0.12 million ha with a production of 2.48 million tonnes. The fruits are mostly used for table purpose and some quantity is used for wine making also. Indian varieties of grapevine have low sugar content ranging from 13 to 22 per cent and hence are not so good for making raisins.

A large number of insect pests attack the grapevine and over 85 species have been reported in India. However, only about half a dozen pests require attention for their management.

1. Grapevine Leafhopper, *Erythroneura* spp. (Hemiptera : Cicadellidae)

Wherever grapes are grown, various species of leafhoppers are almost invariably found sucking sap from the lower surface of leaves.

Life-cycle. The adults pass the winter in protected places, usually under plant remnants on the ground. During spring, they become active and feed to some extent on any green plant before the grape foliage appears. Eggs are laid in leaf tissues and they hatch in about 14 days. The pale wingless nymphs feed on the lower surface of the leaves and moult fives times before changing into adults. The developmental period requires 3-5 weeks, depending on the temperature. There are 2-3 generations in the season.

Damage. Both adults and nymphs suck cell-sap causing the foliage to become blotched with tiny white spots. Under heavy infestation, the leaves turn yellow or brown and fall from the vines. Since their feeding seriously interferes with normal photosynthesis of the plant, both the quantity and quality of fruits are greatly affected.

Management. Spray 625 ml of fenitrothion 50EC or 3.75 kg of carbaryl 50WP in 1250 litres of water per ha, after rainy season when the jassid damage increases.

2. Grapevine Thrips, *Rhipiphorothrips cruentatus* Hood (Thysanoptera : Heliothripidae)

This is the most destructive pest of grapevine in India and is also seen feeding on rose, *jamun* (*Syzygium cuminii*) and *ak* (*Calotropis procera*).

Both the adults and nymphs cause damage by sucking cell-sap from the leaves. The adults are minute, being 1.4 mm long, blackish brown, with yellowish wings. The nymphs are yellowish brown, just visible to the unaided eye as minute fast-moving streaks on the underside of leaves or in the centre of various flowers, particularly the rose.

Life-cycle. The pest breeds during most of the year, except in winter, when it is found as a pupa in the soil at a depth of 8-18 cm under the host plants. The adults appear in March and lay eggs on the undersurface of leaves by making small slits in the plant tissues, placing one egg in one slit. A female, on an average, lays 50 eggs which are dirty white, bean-shaped and can be spotted as little bright specks when affected leaf is held against a source of strong light. The eggs hatch in 3-8 days and young nymphs appear as reddish active creatures which become yellowish brown, as they grow older. They feed on the underside of leaves by rasping the surface and sucking the oozing cell-sap. They are full-fed in 9-20 days and during the season of active breeding, they pupate on leaves. The pupae possess power of locomotion and crawl away when disturbed. In 2-5 days, they change into adults which also feed like the nymphs. The females can reproduce with or without fertilization; the fertilized eggs hatch into females and unfertilized eggs hatch into males. The insect hibernates as a pupa in the soil from December to March and, during the active period, many generations are completed.

The parasitoid associated with this pest is *Thripocentus maculatus* Wat.

Damage. Plants suffer because of constant feeding by large number of insects. The attacked leaves take a whitish hue, acquire a withered appearance, and then turn brown. The leaves ultimately curl up and drop off the plant. Such vines either do not bear fruit or the fruit drops off prematurely. Even the somewhat mature fruits are of poor quality.

Management. (*i*) Remove grasses from orchard and prune infested leaves. (*ii*) Rake the soil periodically. (*iii*) Spray 500 ml of malathion 50EC in 500 litres of water per 100 vines, once before flowering and again after the fruit set.

3. Grapevine Leafroller, *Sylepta lunalis* Guenee (Lepidoptera : Pyralidae)

The leaf-roller is a serious pest of grapevine, especially during August-October. Its occurrence has been reported earlier from southern India and it has also assumed pest proportions in many Punjab vineyards. Besides feeding on the grapevine, this insect has also been found on eleven different species of the family Vitaceae.

Only the green caterpillars cause damage. The adults are dirty brown, with white spots on the fore and hind wings. The wing span of the male moth is 22-24 mm, whereas that of the female is 25-28 mm. The abdomen is light brown, the terminal segment being dark brown. The full-grown caterpillars are cylindrical and measure 26 mm. The body is covered with hairs and the head is brownish-black.

Life-cycle. The pest is active from August to October. The moths lay creamy white oval (1.4 mm across) eggs on the lower surface of the leaves. The number of eggs laid by a female varies from 98 to 120, the pre-oviposition period being 1-3 days and the oviposition period 2-3 days. The eggs hatch in 2-3 days. The first instar larva is tiny, dirty green and measures 3-5 mm in length. After 2 days, it changes to the second instar, which also lasts 2-3 days. The body length of the 2nd, 3rd and 4th instar larvae is 6, 11 and 18 mm, respectively. There are five larval instars and pupation takes place inside the leaf-rolls. Later on, the pupa falls to the ground on fallen leaves and debris. The pupal stage lasts 6-7 days.

Damage. The young caterpillars (1-3 instars) feed on the lower epidermis of the leaves and skeletonize them. The grown-up larvae (4th and 5th instars) roll up the leaf margins towards the mid-rib with one caterpillar in each roll.

Management. (*i*) In the initial stage of attack, remove the rolled up leaves and destroy them with larvae/pupae within. (*ii*) Spray 1.25 litres of malathion 50EC in 1250 litres of water per ha as soon as the attack starts.

4. Grapevine Beetle, *Sinoxylon anale* Lesne (Coleoptera : Bostrychidae)

In India, the grapevine beetle is commonly known as 'ghun' and is one of the most important pests of grapevine. Its distribution covers France, Italy, erstwhile USSR, Japan and China. It is primarily a pest of grapevine but also feeds on *sal* (*Shorea robusta*), teak (*Ouglinia dalbergioides, Boswellia serrata*), *shisham* (*Dalbergia sissoo*), wooden packing cases, etc.

The adult is sturdy, walks slowly and flies rarely. Typically, it is dark brown and measures 4.25 mm in length and 1.8 mm in breadth. The diagnostic three-bladed antennae and a pair of spines on the posterior elytral extremity are present. The full-grown larvae of the beetle are thick yellowish-white curved grubs, often found feeding along with the adults, when the infested vines are split open.

Life-cycle. This pest is active on dormant vines. After winter sets in, the adult beetles bore into the living woody stems and branches. Eggs are laid in galleries constructed by adult beetles inside the attacked portions. After hatching, the grubs continue feeding inside. The activity of the pest is accompanied with a peculiar crackling noise, followed by the ejection of a dusty material from the exits of the feeding-galleries. The multiplication of the pest continues in the dead vines throughout the year.

Damage. Damage is done only to the dormant grapevine. The adult beetle constructs a circular hole, extending to the centre of the stem and then makes longitudinal galleries and forms a number of exits. The damage is always found about 30 cm above the ground level. Both adults and grubs cause damage by feeding inside the vine-stem. All plant parts above the point of attack dry up completely. The pest prefers the late sprouting varieties, such as Anab-E-Shahi and Selection 7 and its incidence varies from 26.6 to 63.5 per cent

Management. (*i*) Clean cultivation including removal of loose bark coupled with careful pruning and destruction of infested parts will be very helpful to prevent infestation by the beetle. (*ii*) Spray the vines with 1.5 litres of monocrotophos 40EC or 2.5 kg of carbaryl 50WP in 1250 litres of water per ha.

5. Grapevine Girdler, *Sthenias grisator* Fabricius (Coleoptera : Cerambycidae) (Fig. 18.6)

The grapevine-girdler is becoming a serious pest throughout the grape growing areas in India. Beside grapes, the stem-girdler feeds on rose-bushes, mulberry, various garden shrubs, creepers, crotons, etc. The other host plants of this pest include mango, almond, jack-fruit, *Bougainvillea*, yellow oleander and Indian ash-tree. Girdling by the adults is followed by drying up of the wood in which the grubs can then tunnel easily.

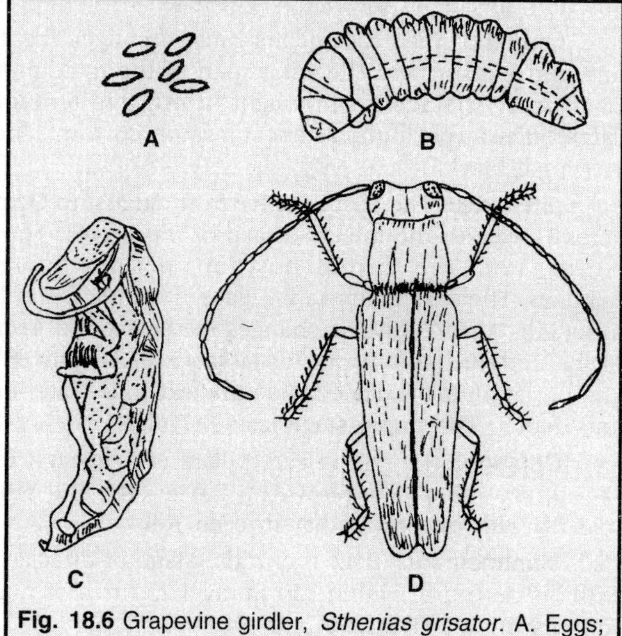

Fig. 18.6 Grapevine girdler, *Sthenias grisator*. A. Eggs; B. Larva; C. Pupa; D. Adult

Life-cycle. In spring, the adults become active at night, mate and deposit eggs in clusters of 2-4 underneath the bark of girdled branches (**Fig. 18.6**). A slight bulge on the bark adjacent to the transverse cut indicates the location of eggs underneath. The eggs are oval, 4 mm long and 1 mm wide in the middle. Each egg is enveloped in a white parchment-like covering and hatches in 8 days. On hatching, the tiny grubs (2-4 mm in length) immediately tunnel into the wood. Their head is dark brown and the mouth has a pair of prominent mandibles, each with two teeth. The most conspicuous part of the grub is the globular thorax, having a few chitinous spines on top. These probably aid in tunnelling into the wood. The full-grown grub is 10-12 mm long. The life-cycle is completed in more than a year. The adults appear again late in summer, but they hibernate during winter.

Damage. During the day, adults hide on the lower side of leaves, under the forkings of branches. The girdling of green branches is an essential event before egg-laying. This results in considerable damage to the vines, as they dry up above the point of girdling. The bark and wood are cut right up to the centre and, at times, even the branches are cut into two bits. Girdling is done at any place from 15 cm to 3 metres above the ground. Branches varying from 1.25 to 2.50 cm in thickness are preferred.

Management. (*i*) Cutting and burning of attacked branches below girdling point, and hand collection and destruction of beetles may help in mitigating the beetle. (*ii*) Chemical control measures are same as for grapevine beetle.

6. Flea Beetle, *Scelodonta strigicollis* Moltschulsky (Coleoptera : Chrysomelidae)

This is one of the most destructive pests of grapevine in India and is found throughout the country. The adult is a shining beetle with metallic bronze colour and 6 dark spots on the elytra and is about 4.5 mm long.

Life-cycle. The female oviposits in the bark or in the soil and eggs may be laid singly or in groups of 20-40. A female lays 220-569 eggs during its life of 8-12 months. Eggs hatch in 4-8 days and the grubs feed on the roots and the larval period lasts 35-45 days. The full grown grubs pupate 6-8 cm deep in the soil and emerge as adults in 7-10 days. The life-cycle is completed in about two months.

Damage. Both the adults and the grules cause damage, the former being voracious feeder are very destructive. The adults eat up and bore into the buds, nibble the leaves making a number of holes on them, scratch the tendrils and eat the epidermis of the branches. The affected buds or sprouts soon dry up. The grubs feeds on the cortex of the roots and cause considerable damage.

Management. (*i*) Remove the loose bark after pruning. (*ii*) Adult beetles may be collected and killed. (*iii*) Spray 2.5 kg of carbaryl 50WP or 3.5 litres of chlorpyriphos 20EC in 1250 litres of water/ha, after pulling out the loose bark.

Other Pests of Grapevine

Some other insects recorded as minor pests on grapes are jassid, *Arboridia viniferata* Sohi & Sandhu (Hemiptera: Cicadellidae); mealybugs, *Nipaecoccus viridis* (Newstead) and *Maconellicoccus hirsutus* (Green) (Hemiptera : Pseudococcidae); lac insect, *Kerria communis* (Mahdihassan) (Hemiptera: Lacciferidae); the horn worm, *Theretra alecto* Linnaeus (Lepidoptera : Sphingidae); the grapevine sphinx, *Hippotion celerio* (Linnaeus) (Lepidoptera: Sphingidae); defoliating beetles, *Anomala dimidiata* (Hope), and *Adoretus* spp. (Coleoptera: Scarabaeidae); wasps, *Polistes hebraeus* (Fabricius) and *Vespa orientalis* Linnaeus (Hymenoptera : Vespidae) and mite, *Oligonychus punicae* (Hirst) (Acari : Tetranychidae).

MANGO (*Mangifera indica* L.; Family : Anacardiaceae)

Mango is considered to be the king of all fruits in South Asia. India is the largest producer and exporter of mangoes in the world. India produces some 18.00 million tonnes of mangoes annually accounting for more than 40 per cent of the world output. The area under mango is 2.50 million ha. The seedling varieties are very tall, whereas the grafted ones are short and commercially more acceptable. In addition to being sweet and succulent, the fruit is a rich source of vitamin A and also vitamin C. There are a number of insect pests of this fruit and over 175 species of insects have been reported damaging mango tree but the most abundant and destructive at the flowering stage are the mango hoppers. It is almost a necessity to control these pest species otherwise there is a heavy fruit drop and the trees may remain without any fruit.

1. Mango Hoppers, *Idioscopus clypealis* (Lethiery); *Amritodus atkinsoni* (Lethiery) (Fig.18.7) (Hemiptera : Cicadellidae)

These are the most destructive pests of all the varieties of mango. Three species of mango-hoppers recorded as pests are *Idioscopus clypealis* (Lethiery), *Amritodus atkinsoni* (Lethiery) (**Fig. 18.7**) and *I. niveosparus* (Lethiery). They are widely distributed in India, Malaysia, Indonesia and Formosa. No alternative host plants of these insects are known.

Injury is caused by nymphs and adults, when they suck cell-sap from the inflorescence and tender shoots. The nymphs of *I. clypealis* are dull yellow or dust yellow, whereas those of *A. atkinsoni* are pale yellow, elongated and more active. Adults of the larger mango-hopper, *I. clypealis,* measure 6.3 mm length and are greyish. There are three dark brown spots on the head, a median band and two black spots on the pronotum. The black triangular marking on scutellum and a central longitudinal dark streak dilated anteriorly and posteriorly, are the characteristics of this species. *A. atkinsoni* adult is about 5.1 mm long. It differs from the larger species by the absence of a central longitudinal dark streak on the scutellum.

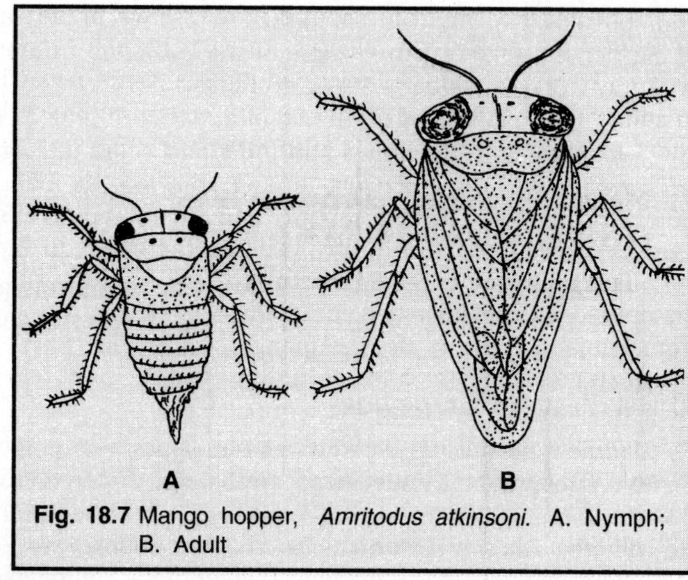

Fig. 18.7 Mango hopper, *Amritodus atkinsoni*. A. Nymph; B. Adult

Life-cycle. This pest is active practically throughout the year but during the hot months of May-June and the cold months of October-January, only the adults are found sitting in thousands on the bark of trunks, branches, etc. The adults surviving winter emerge in February from underneath the bark of trees and other places of shelter. They cluster on the floral buds and start sucking the cell-sap. When the infloresence appears, they start laying eggs in them in the second or third week of February and continue to do so for some weeks. The eggs are deposited singly and since they are embedded in plant tissues, it is extremely difficult to observe them under natural conditions. A female deposits, on an average, 200 eggs, a moderate temperature being more conducive to egg laying.

Within 4-7 days, the eggs hatch and the newly emerged nymphs are first seen at the end of February or in early March. They commence feeding on the inflorescence, quickly suck the cell-sap and excrete honeydew, which serves as a medium for the development of sooty mould, *Chaetothyrium mangiferae.* This gives a dull blackish look to the mango trees. The smoky-black appearance persists for many months until the fruits are mature and the crop is almost over.

By the time the majority of nymphs are mature, the fruit has set. The nymphs then migrate to the stems and the young leaves, and become full-grown in three stages, in 8-13 days. The full-fed nymphs then moult and give rise to winged hoppers. The life-cycle, from the time eggs are laid to the time the adults appear, takes 15-19 days.

The adults are mostly seen congregated on the lower portions of branches and trunks. During the hottest period of summer, they congregate in the shade of large mango groves. There is no feeding or egg-laying in May-June, but the insects remain very agile and hop readily at the slightest disturbance.

A second cycle of brood-rearing starts with the monsoon. Eggs and nymphs of this generation are found during July-August in the submontane districts of the Punjab. Adults of this generation emerge in September and they hibernate during winter.

The natural enemies associated with mango hopper, *I. clypealis* are *Isyndus heros* Fabricius, *Epipyrops fuliginosa* Tams (Epipyropidae) and *Pipunculus annulifemur* Bruvelli

Damage. Mango hoppers are the most destructive pests of fruit trees. Injury to the inflorescence and young shoots is caused by egg-laying and feeding. The voracious feeding nymphs are particularly harmful. They cause the inflorescence to wither and turn brown. Even if the flowers are fertilized, the subsequent develpment and fruit-setting may cease. In thick and protected gardens where the atmosphere is humid, a sooty mould develops on patches of honeydew exuded by the nymphs. As the wind blows, young fruits and dried inflorescences break off at the axil and fall to the ground. The growth of young trees is much retarded and the older trees do not bear much fruit. Damage to the mango crop may be as high as 60 per cent.

Management. (*i*) Do not go for high density planting as it provides favourable habitat for hopper multiplication. (*ii*) Do not encourage plants to put intermittant flushes by regular irrigations and split doses of nitrogenous fertilizers. (*iii*) Avoid waterlogged or damp conditions. (*iv*) In case of old dense orchards, prune some of the branches during winter to have better light interception. (*v*) Spray 2.5 kg of carbaryl 50WP or 2 litres of malathion 50EC in 1250 litres of water per ha, once in end of February and again in end of March. Spraying with malathion LVC @ 1.4 litres per ha with aerial or ground equipment is also effective.

2. Mango Mealybug, *Drosicha mangiferae* (Green) (Hemiptera : Margarodidae) (Fig.18.8, Plate 18.2A)

The mango mealy-bug is widely distributed in the Indo-Gangetic plains from Punjab to Assam. Besides mango, it also attacks 62 other plants, including such trees as the jack-fruit (*Artocarpus heterophyllus* Lam.), the banyan (*Ficus bengalensis*), guava (*Psidium guajava* L.), papaya (*Carica papaya* L.), *Citrus* spp. and *jamun* (*Syzygium* spp.)

Damage is caused by nymphs and wingless females which are oval, flattened and have body covered with a white mealy powder. The males have one pair of black wings and are crimson red (**Fig. 18.8**).

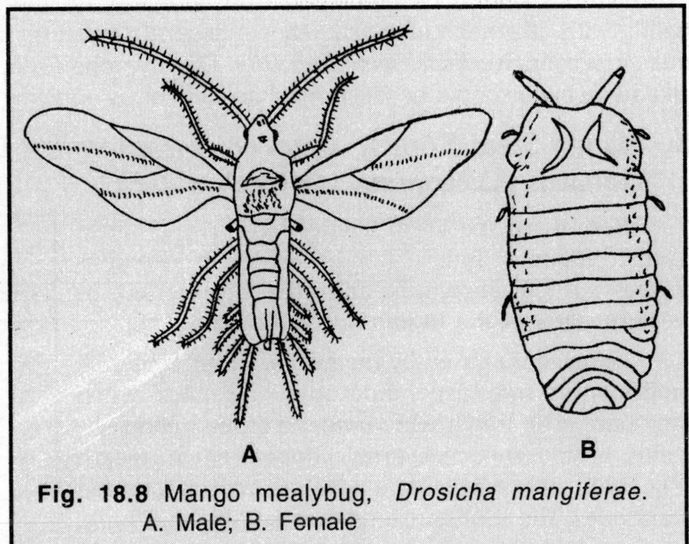

Fig. 18.8 Mango mealybug, *Drosicha mangiferae*. A. Male; B. Female

Life-cycle. This pest is active from December to May and spends rest of the year in the egg stage. The eggs are generally deposited in April-May in soil up to 15 cm within silken purses. The dead body of the female is often found sticking to them. In the crevices or in loose soil, the egg purses may be found as deep as 60 cm. The eggs, 1 mm long and 0.7 mm broad, are oval, shining pink when newly laid and become paler later on. They hatch at the end of December or in January and, in a given locality, continue to hatch for about one month. Thus, the nymphs appear before the fresh growth of flowers on mango-trees. It has been recorded that 70-80 per cent of nymphs ascend the trees immediately. The remaining 20-30 per cent wander about, feeding on weeds and general undergrowth in the orchards. They also ascend the trees and while ascending, they are found clustering near the lichen patches on the bark. Eventually, they congregate on the panicles where they feed on the cell-sap and pass through three stages. The duration of the first stage runs from the middle of December to early February; of the second stage from February to the middle of March and that of the third stage from March to April. At the time of moulting the younger nymphs wander away from the original feeding places in search of suitable shelters such as cracks and crevices in trunks. After moulting, they again seek suitable feeding-sites. The third-stage nymphs, however, stick to their original feeding places and those destined to be females, continue feeding.

Mating takes place soon after emergence of the males at a time when the females are not fully developed. The males fly about in large numbers, apparently in search of their mates. They have a very strong sex instinct and during the life-span of about one week they mate frequently. The females mature after 15-35 days and lay eggs for 22-47 days during April-May.

The nymphs are parasitized by *Phygadeuon* sp. (Ichneumonidae), *Getonides perspicax* Knal, and larvae of *Brinckochrysa scelestes* (Banks) (Chrysopidae), and grubs of *Rodolia fumida* (Mulsant) (Coccinellidae) are predaceous on this mealy-bug.

Damage. Among insect pests of mango, the mealy-bug occupies an important place, second only to the mango-hoppers, with respect to the amount of damage caused. Only the nymphs are destructive and they suck plant juice, causing tender shoots and flowers to dry up. The young fruits also become juiceless and drop off. The pest is responsible for causing considerable loss to the mango growers and when there is a serious attack, the trees retain no fruit at all.

Management. (*i*) Remove weeds from orchards which act as additional hosts for mealy bug. (*ii*) Ploughing of orchards during summer exposes eggs to natural enemies and extreme sun heat. (*iii*) Nymphs should be prevented from crawling up the trees by applying 15-20 cm wide sticky bands with alkathene or plastic sheets around the trunk about one meter above the ground level during second week of December. (*iv*) The nymphs found congregating below the lower edge of alkathene band should be killed mechanically or by applying 50 g of methyl parathion 2 per cent dust.

3. Mango Stem Borers, *Batocera rufomaculata* DeGeer (Fig. 18.9) and *B. rubus* Linnaeus (Coleoptera : Cerambycidae)

Both these species of beetles have a wide range of distribution in India. They have been recorded as serious pests of mango, fig and other trees in north-western parts of the Indian Sub-continent.

Damage is caused by the grubs, killing a branch or the entire tree, depending upon the place which they bore. The full-grown larva is a stout, yellowish-white, fleshy grub, measuring about 6 cm in length (**Fig. 18.9**). Its head is dark with strongly developed mandibles. The adults are longicorn beetles, well built, large and pale greyish, measuring about 5 cm in length and 2 cm in breadth. The first named species is slighly larger than the second. The beetle is provided with long legs and antennae and a dirty white band, extending from the head to tip of the body on each side. A number of dirty yellowish spots are present on the elytra. The head is distinct, with large prominent eyes, and the pronotum is ornamented with two crescent orange-yellow spots.

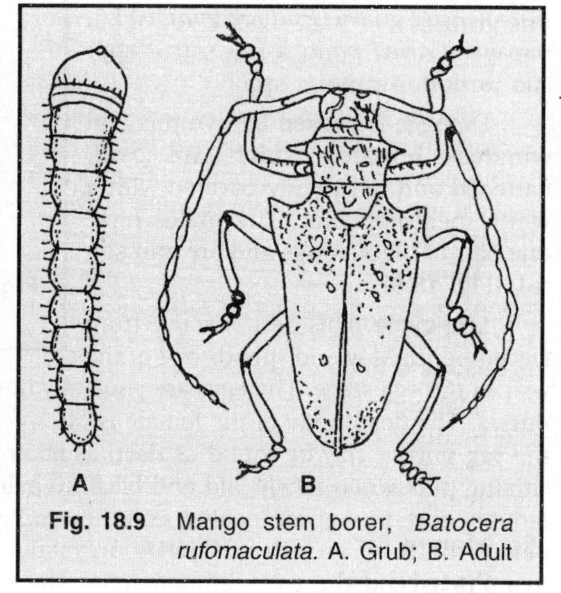

Fig. 18.9 Mango stem borer, *Batocera rufomaculata*. A. Grub; B. Adult

Life-cycle. The life-cycle is prolonged and the adults generally appear during the monsoon. They deposit eggs under the loose bark in a wounded or diseased portion of the trunk or a branch. The grubs are equipped with strong biting mouthparts and they penetrate into the stem or even the roots, feeding on woody tissues. Winter is passed in the grub stage in that very burrow. They again start feeding as soon as the weather warms up in spring and during the feeding process, they bore through the wood by cutting large galleries. The full-grown larvae then hollow out a cell for pupation. The larval stage probably lasts more than a year and the pupal stage lasts about one month. The life-cycle may be completed in 1-2 years.

Damage. Although the borer is not very common, yet whenever it appears in the main trunk or a branch, it invariably kills the host. Though the external symptoms of attack are not always visible, the site can be located from the sap or frass that comes out of the hole. The mango stem-borer is also found in newly fallen trees.

Management. (*i*) Cut and destroy the infested branches with grubs and pupae within. (*ii*) Remove frass near the holes on main stem and inject 4 ml of methyl parathion 50EC mixed in one litre of water into the hole and plug it with mud. In case these holes open, these may be treated again.

4. Mango Stone Weevil, *Sternochetus mangiferae* (Fabricius) (Coleoptera: Curculionidae) (Fig. 18.10)

This is a short stoutly built, ovoid, dark brown weevil which is found inside the stone of mango fruit or in its pulp. It is widely distributed throughout the tropics. The export of mango fruits from India to the USA has been banned to prevent the entry of this weevil.

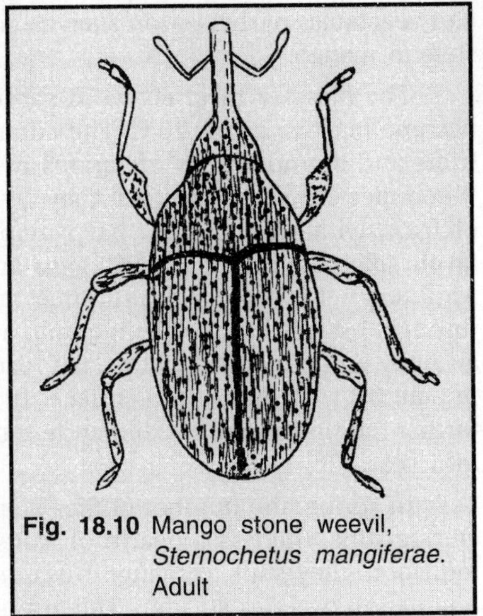

Fig. 18.10 Mango stone weevil, *Sternochetus mangiferae*. Adult

Life-cycle. The weevils are inactive from July-August onwards when they remain concealed in the soil or underneath the bark of mango trees. They become active as soon as the formation of mango fruits takes place. The weevils lay eggs in the skin or ripening fruit. The wound caused by the ovipositor heals soon after and the fruit does not exhibit any outward sign of infestation. On emergence from the egg, the grub moves further inwards, eating its way through the unripe tissue until it bores into the embryo of the mango-stone. When full-grown, the larva forms a cell inside the stone in which it pupates. The weevil **(Fig. 18.10)** cuts its way through the stone and the pulp comes out. The generation is completed in 40-50 days, but the emerging adults become inactive and resume breeding only in the next season. Thus, there seems to be a single generation of this pest in one year.

Damage. The insect attacks mango varieties with a relatively soft flesh. However, it is not very serious in any part of the country. The injury caused by the larvae feeding in pulp sometimes heals over but a certain number of fruits always get spoiled when the weevils make an exit through ripe or near-ripe mangoes.

Management. (*i*) The pest can be suppressed by destroying all fallen fruits, weeviled mangoes and by disposing off refuse, stone, debris, etc. (*ii*) The weevil, being an internal feeder throughout its development, is not amenable to control with any of the insecticides. (*iii*) Raking of soil below the tree in October/November and March can contribute partially to weevil management.

5. Mango Fruit Fly, *Bactrocera dorsalis* (Hendel) (Diptera : Tephritidae) (Fig. 18.11, Plate18.2B)

The mango fruitfly or the Oriental fruit-fly, is the most serious of all fruit-flies and is widely distributed in India and South-East Asia. It has also been recorded in Malaysia, Indonesia, Formosa, the Philippines, Australia and the Hawaii Islands, In many countries, it has displaced the Mediterranean fruitfly, *Ceratitis capitata* (Wiedemann). Apart from mango, the pest also feeds on guava, peach, apricot, cherry, pear, *chiku* (*Achras zapota*), *ber* (*Zizyphus* spp.), citrus and other plants, totalling more than 250 hosts.

Damage is caused by grubs only and they feed on pulp, making the fruit unfit for human consumption. The legless maggots, when full-grown, measure 8-9 mm long and 1.5 mm across the posterior end, and are yellow and opaque. The adult is stout, a little larger than the ordinary housefly and measures 14 mm across the wings and 7 mm in body length **(Fig. 18.11)**. It is brown and has almost transparent wings with yellow legs and dark rust-red and black patterns on the thorax.

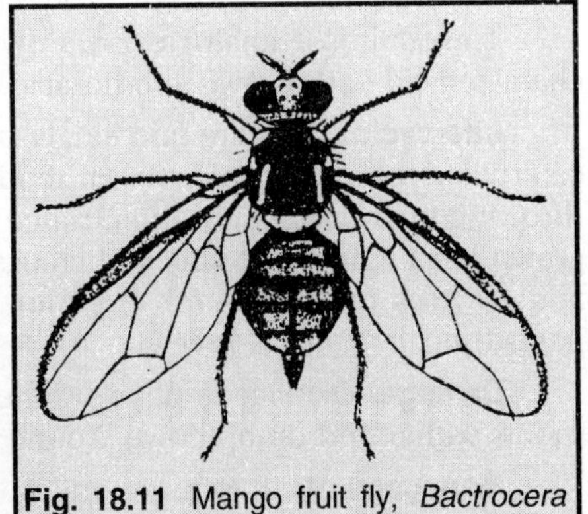

Fig. 18.11 Mango fruit fly, *Bactrocera dorsalis*. Adult

Life-cycle. This pest is active during the summer months and passes the winter (November to March) as a hibernating pupa in the soil. The adult flies emerge in April and reach ripening fruits and vegetables of the season such as guava, loquat, apricot, plum, brinjal, chillies, etc. Later on, they shift to mango.

The flies are most active in gardens when the temperature ranges from 25 to 30°C and they become inactive below 20°C. The adults live for about 4 months and feed on the exudations of ripe fruits and the honeydew of various insects. Mating takes place at dusk and lasts an hour or more. When flies are 10-15 days old, they lay 2-15 eggs at a time in clusters, 1-4 mm deep in the soft skin of fruit, with the help of sharp ovipositors. A female lays, on an average, 50 eggs but under favourable conditions, 150-200 eggs are laid in one month. The eggs hatch in 2-3 days in March-April and 1-1.5 days in the summer and 10 days during winter. As the maggots develop they pass through 3 stages in the ripening pulp and are full-grown in 6-29 days. They leave the fruit and move away by jumping in little hops. On reaching a suitable place, they bury themselves into the soil and pupate 8-13 cm below the surface. In 6-44 days, they emerge as flies and reach the ripe fruit for further multiplication. The life-cycle is completed in 2-13 weeks and many generations are completed in a year.

In spring, the number of flies is quite small, but by successive breeding they multiply steadily. During June, there is a dearth of fruits and the flies resort to brinjal as the only food. During the monsoon, they shift to mango trees and increase in number enormously. During the winter, their population declines steadily. The autumn is another period of food shortage and the flies breed on ripe citrus, guava, etc.

The parasitoids associated with this pest are *Opius compensatus* Silvestri, *O. persulcatus* Silvestri, *Biosleres arisanus* (Sonan), *O. incisus* Silvestri and *O. manii* (Braconidae); *Spalangia philippinensis* Mill., *S. afra, S. stomyoxysine* Gir. and *S. grotiuse* Gir. (Pteromalidae); *Dirhinus giffardi* Silvestri (Chalcididae); *Pachycrepoideus dubiers* Ashmead and *Trybliographa daci* Weld (Eucoilidae).

Damage. Maggots are very destructive and cause heavy losses to all kinds of fruits. The infested fruits become unmarketable and at times almost all of them contain maggots **(Plate 18.2B)**.

Management. (*i*) Avoid infestation of fruit flies by early harvesting of mature fruits. (*ii*) To prevent the carry over of the pest, collect and destroy all fallen infested fruits twice in a week. (*iii*) Plough round the trees during winter to expose and kill the pupae. (*iv*) Monitor the fruit-fly population in orchards by using methyl eugenol traps. (*v*) Spray 1.25 litres of malathion 50EC + 12.5 kg *gur* or sugar in 1250 litres of water per ha and repeat sprays at 7-10 days interval if infestation continues. (*vi*) After harvest, dip the fruits in 5 per cent sodium chloride solution for 60 minutes to kill the eggs, if any and also to decontaminate them of insecticide residue if at all present.

6. Mango Shoot Borer, *Chlumetia transversa* Walker (Lepidoptera : Noctuidae)

The adult is a small greyish brown moth. Young caterpillars are yellowish orange in colour with characteristic dark brown prothoracic shield. Full grown caterpillars are dark pink, with dirty spots.

Life-cycle. Eggs are laid singly on tender leaves and they hatch in 2-3 days. Freshly hatched caterpillars bore into mid-ribs of tender shoots near the growing point tunnelling downwards and throwing their excreta out of entrance hole. Larva has five instars completed in 11-13 days. The full-grown larva enters into slits and cracks in the bark of tree, dried malformed inflorescences or cracks and crevices in the soil for pupation for 12-15 days. The life-cycle occupies 30-42 days. Larva is parasitised by Bracon greeni.

Damage. Damage is done by the larvae by boring into the growing shoots. Leaves of affected shoots wither and droop down. Young grafted seedlings are severely affected and may even be killed.

Management. (*i*) Remove and burn the dried shoots. (*ii*) Spray the new growth with 2 litres of malathion 50 EC in 1250 litres of water/ha.

7. Mango Gall Psyllid, *Apsylla cistella* (Buckton) (Hemiptera : Psyllidae) (Fig. 18.12)

This insect appears occasionally as a serious pest in several parts of North India. The adult psyllid is 3-4 mm long with black thorax and head, and light brown abdomen.

Life-cycle. Eggs are laid partly embedded within the midribs on the lower side of the tender new leaves. Up to 150 eggs are laid by a female during February to April. Nymphs appear during August and September. They feed on the vegetative and reproductive buds causing the formation of cone shaped galls on them. Gall formation is noted in September and October. The nymphs become adults within the galls, the nymphal period lasting for about five months. The adults appear in end of March and one generation is completed in a year.

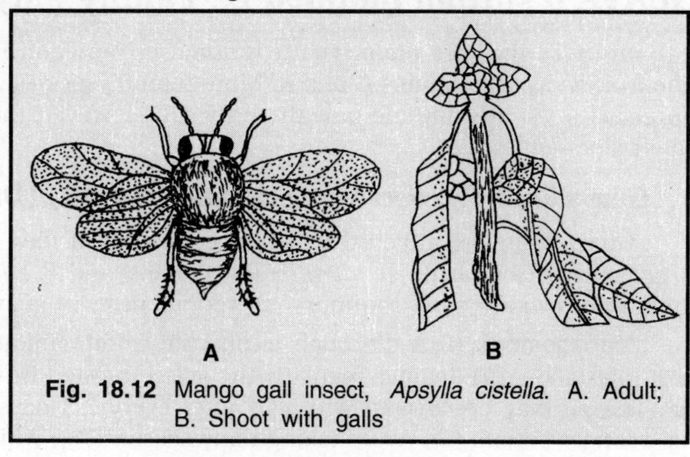

Fig. 18.12 Mango gall insect, *Apsylla cistella.* A. Adult; B. Shoot with galls

Damage. The damage is caused by the nymphs which feed on vegetative and reproductive buds. The attack causes malformation of affected parts and reduces fruit set.

Management. (*i*) Collect and destroy the galls. (*ii*) Spray 1.70 litres of dimethoate 30EC or 1.85 litres of monocrotophos 36SL in 1250 litres of water/ha.

8. Mango Bud Mite, *Aceria mangiferae* Sayed (Acari: Eriophyidae)

The bud mite is a pest of mangoes not only in India but also in Pakistan and USA. In India, the mite is serious particularly in Punjab, Haryana and Uttar Pradesh.

Life-cycle. The detailed life-history of this eriophyid bud mite has not been studied so far. Some observations on the population dynamics of the mango-bud mite have revealed that there is a seasonal variation in its population. In the Punjab, the highest population is found during February after which there is a gradual decrease until it becomes very low in May. There is again an increase in population in June and another peak in July. In August-September, there is a further decrease in population which remains low during October-December. A rapid increase in January results in the peak population being reached in February, which coincides with the maximum new malformation, hence the suspicion that the two are related.

Damage. The bud mite sucks the sap from inside the buds and causes necrosis of tender tissues. When the population is high, the entire bud may be killed. This mite infests all varieties of mango and none has shown resistance to it.

Management. (*i*) Remove and destroy all the panicles bearing infested inflorescences. (*ii*) Spray one litre of dimethoate 30EC in 1250 litres of water per ha, preferably during summer.

Other Pests of Mango

Some other insects recorded as minor pests of mango are the soft scales of mango, *Chloropulvinaria psidi* (Maskell), *Coccus formicarii* and *Rastrococcus iceryoides* Green (Plate 22) (Hemiptera : Coccidae); the hard scales, *Aspidiotus destructor* Signoret and *Chionaspis vitis* Green (Hemiptera : Diaspididae); leaf webber, *Orthaga euadrusalis* Walker (Lepidoptera : Pyralidae); the bark-eating caterpillar, *Indarbela quadrinotata* (Walker) (Lepidoptera : Metarbe-lidae); the mango leaf- mining weevil, *Rhynchaenus mangiferae* Marshall (Coleoptera : Curculionidae); the mango leaf twisting weevil, *Apoderus tranquebaricus Fabricius* (Coleoptera : Anthribidae); the mango-leaf galls,

Amaraemyia spp. (Diptera : Cynipidae); the red tree ant, *Oecophylla smaragdina* (Fabricius) (Hymenoptera : Formicidae).

GUAVA (*Psidium guajava* L.; Family : Myrtaceae)

In India, the area under guava is about 0.24 million ha with production of 3.20 million tons. The fruit is rich in vitamins C and A. More than 80 species of insects and mites have been recorded on guava trees affecting the growth and yield. However, the major pests are fruit flies, bark eating caterpillars and scales.

1. Guava Fruit Fly, *Bactrocera dorsalis* (Hendel) (Diptera : Tephritidae) (Plate 18.2C)

Guava fruits are attacked by five species of fruit flies, viz., *B. dorsalis, B diversus* (Coquillett), *B. cucurbitae* (Coquillett), *B. nigrotibialis* (Perkins) and *B. zonata* (Saunders). However, among these, the former two are most common and serious pests of guava. The details are given under 'Mango'.

Management. (*i*) A thorough clean cultivation/sanitation of orchard is essential in reducing pest infestation. (*ii*) Remove regularly the fallen infested fruits and bury them in at least 60 cm deep pit. The pit may be covered with clay after every 2-3 days and should not be allowed uncovered for a longer period. (*iii*) Avoid taking rainy season crop which may get heavily infested by the pest. (*iv*) Plough and stir the soil well before June-July to expose and kill pupae. (*v*) Harvest the fruits (**Plate 18.2D**) when slightly hard and green in colour. (*vi*) Fix methyl eugenol based PAU fruit fly traps @ 40/ha in the first week of July and recharge the same if required. (*vii*) In orchards with history of severe fruit fly infestation, spray 3.125 litres of fenvalerate 20EC in 1250 litres of water per ha at weekly intervals on ripening fruits commencing from July onwards till the rainy season crop is over. Fruits should be harvested at least on 3rd day after spray.

2. Bark Caterpillar, *Indarbela tetraonis* (Moore), *I. quadrinotata* (Walker) (Lepidoptera: Metarbelidae)

The details are given under 'Citrus'.

3. Guava Mealy Scale, *Chloropulvinaria psidii* (Maskell) (Hemiptera : Coccidae)

The guava mealy scale is widely distributed throughout the tropical regions of the world including India, Bangladesh and Sri Lanka. In India, it is a major pest of guava in Punjab, Uttar Pradesh, Maharashtra, Andhra Pradesh, Karnataka and Tamil Nadu. Apart from guava, the scale feeds on coffee, tea, citrus, mango, *gular,* jack-fruit, *jamun,* litchi, loquat, sapota and many other shrubs and trees.

Life-cycle. The adult scales are shield-shaped, oval, yellowish green and measure 3 mm in body length. Eggs are laid beneath the body of mature female in a conspicuous egg-sac and later the female dies. The first instar nymphs or crawlers are the active dispersive phase responsible for starting new infestation.

Two parasitoids, viz., *Coccophagus cowperi* Girault and *C. bogoriensis* (Koningsberger) (Aphelinidae) and two species of predators, viz., *Scymnus coccivora* Ramakrishna Ayyar and *Crytolaemus montrouzieri* Mulsant (Coccinellidae) are known to feed on *C. psidii.* Among these, *C. montrouzieri* is the most common and effective predator.

Damage. The scale insects are found in large numbers sticking to leaves on ventral side, tender twigs and shoots. They suck sap from ventral side of leaves, petioles, tender shoots and occasionally from fruits. They cause leaf distortion and growth disturbance. The females feed voraciously and also exude copious quantity of honeydew. The honeydew excreted by the scales encourages the development of sooty mould on foliage which interferes with photosynthetic activity of plants and spoils the market value of fruits. Severe infestation could kill the branches.

Management. (*i*) Prune and destroy the infested shoots at initial stage of infestation. (*ii*) In case of severe attack, spray 1.25 litres of monocrotophos 40EC in 1250 litres of water per ha. (*iii*) After two weeks, release 20 *C. montrouzieri* beetles per tree preferably close to the site of infestation.

4. Castor Capsule Borer, *Conogethes punctiferalis* Guenee (Lepidoptera : Pyralidae) (Plate 18.2E)

The damage is done by the caterpillars which bore into the fruits and render these unfit for consumption. The details of this pest are given under 'Castor'.

5. Helmet Scale, *Saissetia coffeae* (Walker) (Hemiptera : Coccidae)

The details of this pest are given under 'Coffee'.

Other Pests of Guava

Several other insects recorded as minor pests on guava include tea mosquito bugs, *Helopeltis antonii* Signoret, *H. febriculosa* Bergroth (Hemiptera : Miridae) ; citrus blackfly, *Aleurocanthus woglumi* Ashby (Hemiptera : Aleyrodidae); cotton aphid, *Aphis gossypii* Glover (Hemiptera : Aphididae) ; mealy bugs, *Ferrisia virgata* (Cockerell), *Planococcus citri* (Risso) and *P. lilacinus* (Cockerell) (Hemiptera : Pseudococcidae); olive soft scale, *Saissetia oleae* (Olivier) (Hemiptera: Coccidae); red banded thrips, *Selenothrips rubrocincuts* (Giard) (Thysanoptera : Thripidae); guava shoot borer, *Microcolona technographa* Meyrick (Lepidoptera : Cosmopterygidae) ; guava stem borer, *Microcolona leucosticta* Meyrick (Lepidoptera: Gelechiidae); pomegranate butterfly, *Deudorix isocrates* (Fabricius) (Lepidoptea : Lycaenidae); castor semilooper, *Achaea janata* (Linnaeus) (Lepidoptera : Noctuidae); cockchafer beetles, *Holotrichia consanguinea* Blanchard, *H. insularis* Brenske (Coleoptera : Scarabaeidae); grey weevils, *Myllocerus blandus* Faust, *M. undecimpustulatus maculosus* Desbrocher and cherry stem borer, *Aeolesthes holoserica* Fabricius (Coleoptera : Curculionidae).

BER (*Ziziphus mauritiana* Lamk.; Family: Rhamnaceae)

The ber, which is one of the most common fruit trees of Indian Sub-continent, is grown on about 90,000 hectares with annual production of 7,50,000 tonnes. It is often called poor man's fruit. Ber fruits are nutritious and are rich in vitamins A, B and C. In India, as many as 80 insect species feeding on *ber* tree have been reported, out of which fruit fly and *ber* beetle are important. The former causes serious damage to fruits and the latter is a foliage feeder and shows preference for *ber* trees. Both are basically polyphagous insects and have also been recorded feeding on a number of fruit trees, but *ber* seems to be the preferred host.

1. Ber Fruit Fly, *Carpomyia vesuviana* Costa (Diptera : Tephritidae) (Fig. 18.13)

This pest is widely distributed in India, Pakistan and southern Italy. It is most destructive to *ber* fruits of the species *Ziziphus mauritiana* Lam. and *Z. jujuba* Mill. in India and *Z. sativa* China in Italy. Also, *C. vesuviana*, *Bactrocera dorsalis* (Hendel) and *B. correcta* (Bezzi) have been recorded as minor pests of *ber* fruits.

Damage is caused only by the larvae which are creamy white and slightly smaller than those of other fruit-flies. The adults, smaller than the housefly, are brownish yellow, with brown longitudinal stripes on the thorax, being surrounded on the sides and the back with black spots (**Fig. 18.14**). There are greyish brown spots on the wings and bristly hair on the tip of the abdomen.

Life-cycle. The pest is active during winter and hibernates in the soil from April to August in the pupal stage. The flies emerge from the pupae during August to mid-November, synchronizing with the blossoming and fruit setting of the ber trees. Flies that emerge early in the season are small and generally die without reproducing. The flies are very fast fliers but the females can easily be caught when ovipositing on very young fruits. At the age of one month, the flies make cavities in the skin of fruit and lay one or two spindle-shaped creamy-white eggs, 1 mm below the skin, leaving behind a resinous material. There is no further growth of the fruit in the vicinity of this puncture and, hence, the fruits become deformed. The eggs hatch in 2-3 days and the maggots feed on the flesh of the fruit, making galleries towards the centre. Such fruits invariably rot near the stones, and as many as 18 maggots have been recorded from one attacked fruit. The larvae are full-grown in 7-10 days and they come out of the fruit by cutting one or two holes in the skin. They move away, making jumps of 15-26 cm and reach a suitable place to pupate generally 6-15 cm below the soil surface. The pupal stage lasts 14-30 days and the shortest life-cycle, from egg to the emergence of the adult, is completed in 24 days. There are 2-3 broods in a year.

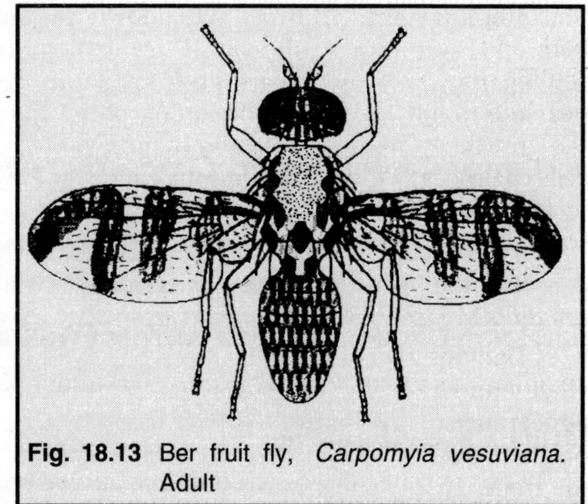

Fig. 18.13 Ber fruit fly, *Carpomyia vesuviana.* Adult

The pest becomes active in the autumn and builds up population in the winter, reaching a peak in February-March. At that time, all the late-maturing ber fruits are found riddled with maggots.

Damage. Fleshy varieties of ber are more seriously damaged than the less fleshy ones. The attacked fruits are rotten near the stones and emit a strong smell. Late maturing fruits are destroyed almost entirely.

Management. (*i*) Collect and destroy the fallen infested fruits at alternate days. (*ii*) Rake the soil around the trees during summer to expose the pupae to heat and natural enemies. (*iii*) Do not allow wild bushes, trees, isolated growing or abandoned plants of other *Ziziphus* species to grow in the vicinity of ber orchards, as their fruits serve as good source for multiplication and help in carry over of fruit flies to cultivated ber fruits. (*iv*) To escape egg laying on fruits, do not allow the fruits to ripe on the trees, harvest at green and firm stage. (*v*) Spray 1.25 litres of dimethoate 30EC in 750 litres of water per ha during February- March, and repeat sprays at 10 days interval. Stop spraying at least 15 days before fruit picking.

2. Lac Insect, *Kerria lacca* (Kerr) (Hemiptera : Kerriidae)

The lac insect is found in India, Pakistan, Sri Lanka, Myanmar, Malaysia, China and Thailand. The insect is used for commercial production of lac in Bihar, Madhya Pradesh, West Bengal, Maharashtra, Uttar Pradesh, Orissa and Assam. However, it is considered as a pest when it feeds on plants of economical significance. The pestilent aspect of this insect has largely been overshadowed by its usefulness. The major host plants of lac insect are ber, *kusum (Schleichera oleosa), palas (Butea monosperma)* and *khair* (*Acacia catechu*). It also attacks grapevine, pomegranate, *amla* or Indian gooseberry (*Emblica officinalis*), litchi, sugar apple, tamarind and several forest and avenue trees.

The newly hatched nymph (crawler) is 0·60 × 0·25 mm in size and pinkish in colour. The female is apterous, pinkish in colour and about 1·5 mm in length. The ventral surface of the body is flat while dorsal surface is convex. The males are pinkish-red in colour and are of two types-winged and wingless. Winged male possesses only one pair of translucent membranous forewings.

Life-cycle. A female lays about 200-500 eggs underneath its body in lac encrustation. The eggs generally hatch within a few hours as the embryos are fully developed in them. Thus the lac insect has ovoviviparous mode of reproduction. The first instar nymphs (crawlers) emerge in large numbers and come out of lac encrustations. The emergence of nymphs, called swarming, may continue for 5 weeks or more. The nymphs crawl about on the branches for some time and on reaching soft succulent twigs, settle down gregariously and thrust their needle-like proboscis into the bark and start sucking the sap. Once settled, the nymphs do not move. After one or two days of settling, they start secreting resin from glands distributed under the cuticle all over the body, except near the mouth parts, two breathing pores and anus. The resinous covering increases in size with the growth of the insect. As the nymphs settle very close together, the lac secretion from adjacent cells coalesces to form a continuous encrustation. Males produce smaller quantity of lac in comparison to females which produce the commercial lac. The nymphs molt thrice before reaching the adult stage. The peak periods of activity of the lac insect are January-February and June-July.

The two most important predators of the lac insect are the large white lac moth, *Eublemma amabilis* Moore (Noctuidae) and the smaller black lac moth, *Holcocera pulverea* Meyr (Gelechiidae).

Damage. De-sapping by large population of lac insect over a long period of time reduces the vitality and vigour of the trees resulting in substantial loss in fruit yield and quality. The cottony appearance of certain healthy encrustations is due to long wax filaments, while the sooty appearance on the leaves is due to black mould growth on the honeydew which imparts ugly appearance to the trees. The severely infested branches dry up. Neglected ber orchards having close plantations, intermixed with other host trees and devoid of regular summer pruning are more prone to lac infestation. Old branches harbouring lac encrustations of the previous generation serve as source of infestation for the next brood.

Management. (*i*) After fruit harvest or when trees are in dormancy, prune all severely infested branches and burn without delay. (*ii*) Avoid overcrowding and growing other lac insect host plants in the vicinity of fruit trees. (*iii*) Spray 625 ml of dimethoate 30EC in 625 litres of water per ha in April and again in September.

3. Ber Beetles, *Adoretus pallens* Arrow and *A. nitidus* Arrow (Coleoptera : Scarabaeidae)

These two defoliating beetles are widely distributed in northern India and Pakistan, and are polyphagous, but they prefer the *ber* tree (*Zizyphus jujuba*) and the grapevine. Only the adult beetles are destructive and can be recognized from their bright yellow colour and yellowish-brown shiny wings. Tarsi on the tips of their legs are deep red and their size varies from 9 to 13 mm. The beetles are attracted to light and appear in large numbers late in spring or early in summer and again during the monsoon.

Life-cycle. This pest is active during summer and passes the winter in larval stage. The adults appear in April-May and lay white, smooth, elongate eggs, singly in the soil near the host plants. Egg laying continues at irregular intervals from May to August and the eggs hatch in 6-9 days in that season. The whitish grubs feed on soil humus, roots of grasses and other vegetable matter found under or near the *ber* trees. When full-grown, the grubs measure 15 mm in length and are creamy white. They make an earthen cell in the autumn and hibernate through the whole of winter. Pupation takes place sometimes during next April and the adults emerge after 11-12 days. There is only one generation in a year.

Damage. Damage is characterized by round holes cut in the leaves by beetles during the night. The *ber* trees are sometimes so heavily attacked that the entire foliage may disappear and such trees do not bear any fruit. The attack starts early in the spring and continues up to August.

Management. (*i*) Light traps are quite effective in trapping the adult beetles. (*ii*) Raking around the trees in useful in exposing the hibernating grubs and killing them. (*iii*) Spray 2.5 kg of carbaryl 50WP in 750 litres of water per ha in the evening as soon as the damage starts. If the damage continues, repeat the spraying after one week.

Other Pests of Ber

Some minor pests of ber tree include the mealybug, *Drosichiella tamarindus* Green (Hemiptera: Coccidae); *Indarbela tetraonis* Moore (Lepidoptera : Metarbelidae); *ber* fruit borer, *Meridarchis scyrodes*

Meyrick (Lepidoptera : Carposinidae); leaf eating caterpillar, *Euproctis* spp. (Lepidoptera : Lymantriidae); grey and black weevils, *Xanthochelus superciliosus* Gyll, *Myllocerus transmarinus* (Herbst) (Coleoptera : Curculionidae); tortoise beetle, *Oocassida pudibanda* (Boheman) Coleoptera: Chrysomelidae) **(Plate 18.2F)**; spittle bug, *Machaerota planitae* Distant (Hemiptera : Cercopidae); leaf webber, *Tonica zizyphi* Stanton (Lepidoptera : Oecophoridae); leaf butterfly, *Tarucus theophrastus* (Fabricius) (Lepidoptera : Lycaenidae) ; hairy caterpillar, *Thiacidas postica* Walker (Lepidoptera: Noctuidae); tussock moth, *Dasychira mendosa* Hubner (Lepidoptera : Lymantriidae); fruit fly, *Bactrocera dorsalis* (Hendel) (Diptera : Tephtritidae) the mite, *Eutetranychus banksi* McGregor (Acari: Tetranychidae) and gall mite, *Eriophyes cernuus* Massee (Acari : Eriophyidae) **(Plate 18.2G)**.

POMEGRANATE (*Punica granatum* L.; Family : Punicacae)

Pomegranate grows as an evergreen tree in the plains and as deciduous during winter in hilly areas at upto 1500 metres height. The annual production of pomegranate in India is 0.74 million tons from an area of 0.11 million ha.

1. Pomegranate Butterfly, *Deudorix isocrates* (Fabricius) (Lepidoptera : Lycaenidae) (Fig. 18.14, Plate 15.2H)

The caterpillars of the *Anar* butterfly cause such a heavy damage to the fruits that this pest alone is responsible for the failure of pomegranate crop in certain areas. This insect is widely distributed all over India and the adjoining countries. It is a polyphagous pest having a very wide range of host plants including *Aonla*, apple, *ber*, citrus, guava, litchi, loquat, mulberry, peach, pear, plum, pomegranate, sapota and tamarind. The caterpillars bore inside the developing fruits and feed on pulp and seeds just below the rind.

The full-grown caterpillars are 17-20 mm long, dark brown in colour and have short hair and whitish patches all over the body. The adult butterflies are glossy-bluish-violet (males) to brownish violet (females) in colour with an orange patch on the fore wings **(Fig. 18.14)**. The wing expanse is 40-50 mm.

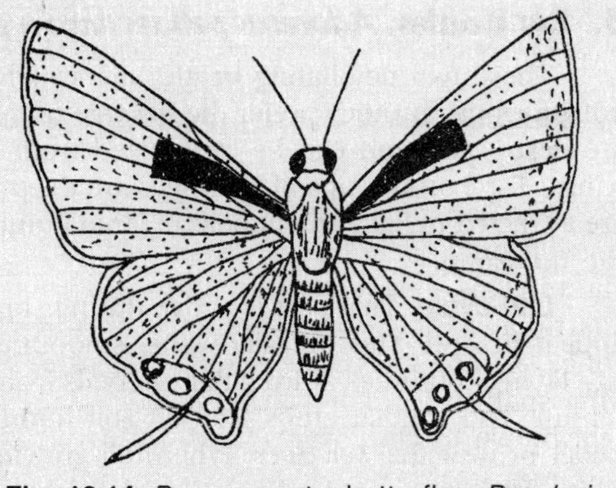

Fig. 18.14 Pomegranate butterfly, *Deudorix isocrates*. Adult

Life-cycle. The pest breeds throughout the year on one fruit or the other. The female butterfly lays shiny white, oval shaped eggs singly on the calyx of flowers and on small fruits. The eggs hatch in 7-10 days and the young larvae bore into the developing fruits. They feed there for 18-47 days till they are full-grown. Then they pupate inside the fruit but occasionally may pupate outside even, attaching themselves to the stalk of the fruit. The pupal stage lasts 7-34 days. There are four overlapping generations in a year.

The larva is parasitized by *Brachymeria euploeae* Westwood (Chalcididae).

Damage. The caterpillars damage the fruit by feeding inside and riddling through the ripening seeds of pomegranate. As many as eight caterpillars may be found in a single fruit. The infested fruits are also attackd by bacteria and fungi which cause the fruits to rot. The affected fruits ultimately fall off and give an offensive smell. This pest may cause from 40 to 90 per cent damage to the fruits.

Management. (*i*) Bagging of fruits before maturity will help in checking damage. (*ii*) Collection and destruction of fallen infested fruits prevents build up of the pest. (*iii*) Remove flowering weeds especially of compositae family. (*iv*) Spray young fruits with 2.5 kg of carbaryl 50 WP in 1250 litres of water per ha during May-June at 15 days interval. Two to three sprays are enough.

2. Bark Caterpillar, *Indarbela quadrinotata* (Walker) (Lepidoptera : Metarbelidae)

The details are given under 'Citrus'.

Other Pests of Pomegranate

Some of the other pests attacking pomegranate include the whitefly, *Siphoninus phillyreae* (Haliday) (Hemiptera : Aleyrodidae); the aphid, *Aphis punicae* Passerini (Hemiptera : Aphididae); striped mealy bug, *Ferrisia virgata* (Cockerell) and mealy bug, *Pseudococcus lilacinus* Cockerell (Hemiptera : Pseudococcidae); leaf thrips, *Retithrips syriacus* (Mayet) and *Rhipiphorothrips cruentatus* Hood (Thysanoptera : Thripidae); bag worm, *Clania cramerii* Westwood (Lepidoptera: Psychidae) ; bark borer, *Indarbela tetraonis* (Moore) (Lepidoptera : Metarbelidae); leaf eating caterpillar, *Parasa lepida* (Cramer) (Lepidoptera : Limacodidae); castor capsule borer, *Conogethes punctiferalis* (Guenee) (Lepidoptera : Pyralidae); castor semi-looper, *Achaea janata* (Linnaeus) (Lepidoptera : Noctuidae); tussock caterpilars, *Euproctis fraterna* (Moore) and *E. scintillans* (Walker) (Lepidoptera : Lymantriidae) and fruit fly, *Bactrocera zonata* (Saunders) (Diptera : Tephritidae).

BANANA (*Musa paradisiaca* L.; Family : Musaceae)

India is the largest producer of banana in the world. Banana is widely grown in South Asia, the south Pacific Islands and tropical countries on both sides of the Atlantic Ocean. In India, banana is grown on about 0.78 million ha and the output is 26.51 million tonnes (2012-13).

There are more than 182 insect pests of banana in India. The banana weevil, *Cosmopolites sordidus* (Germar) is the most destructive pest of banana in India and South-east Asia. Most of the other insects recorded feeding on this plant are minor pests of local importance and are not specific to banana. The aphid, *Pentalonia nigronervosa* Coquerel is, however, important, not as pest but as a vector of a very serious disease called bunchy top of banana.

1. Banana Scale Moth, *Nacoleia octasema* (Meyrick) (Lepidoptera : Pyralidae) (Fig. 18.15)

The caterpillars of the banana scale moth feed on the inflorescence of banana. They can also develop on Manila hemp, maize and some wild plants. This pest occurs in Indonesia, eastern Australia, New Guinea, Solomon Islands, Fiji, Tonga, Samoa and Queensland. The moth has a wing span of 22 mm and its colour varies from light brown to dark brown (Fig. 18.15).

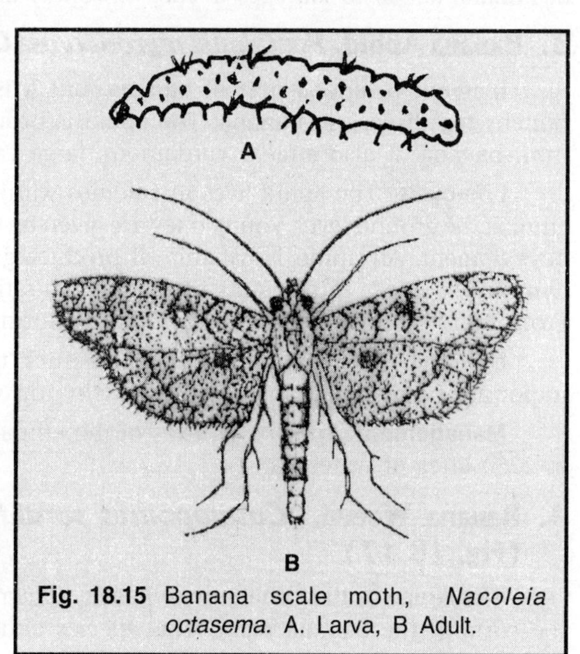

Fig. 18.15 Banana scale moth, *Nacoleia octasema*. A. Larva, B Adult.

Life-cycle. The activity of the moths coincides with the growth of banana plants. The female moths lay pale greenish-white eggs on or near the flag leaf of banana plant just before the bunch emerges. The eggs hatch in about 4 days and the small transparent yellow caterpillars crawl under the closed bracts of the young banana bunch and begin feeding there. After completing this development the larvae may pupate right on the plant or more often in the debris near the base.

The caterpillars are attacked by *Sisyropa panci* (Tachinidae), *Macrocentrus* sp. (Braconidae) and *Argyrophylae* sp. (Tachinidae) which are the larval parasites of this pest.

Damage. The gregarious caterpillars feed on the young female flowers and leave scale-like scars on the young fruits. They feed so voraciously that the inflorescence is soiled with frass and refuse of the larvae, and gives an ugly appearance.

Management. Spray 2.5 kg of carbaryl 50WP in 1250 litres of water per ha directing the spray towards the bracts of the inflorescence.

2. Banana Stem Borer, *Odoiporus longicollis* (Olivier) (Coleoptera : Curculionidae) (Fig. 18.16)

It is a serious pest of banana in North-east India. The adult is a robust reddish brown weevil about 1.5-2.0 cm long. The grub is apodous, yellowish with reddish head.

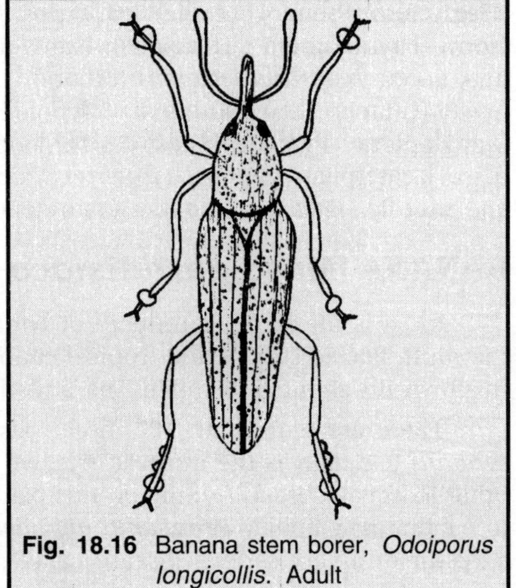

Fig. 18.16 Banana stem borer, *Odoiporus longicollis*. Adult

Life-cycle. The eggs are laid in small burrow in the rhizome or within leaf sheaths just above the ground. Eggs hatch in 3-5 days in summer and 5-8 days in winter. The larva bores into pseudostem making tunnels within and cutting holes on its outer surface. The larval duration is 26 days in summer and 68 days in winter with five larval instars. It pupates in tunnel towards the periphery. The pupal period lasts 20-24 days in summer and 37-44 days in winter. The adult lives for a period of upto two years.

Damage. Both the grubs and adults cause the damage. Grubs bore into the rhizome and make tunnels within it. Adults also tunnel within the stem, feeding on its internal tissues as a result of which the internal shoot is killed. Infested plants show premature withering, leaves become scarce, fruits become undersized and their suckers are killed outright.

Management. (*i*) Uproot and burn the infested plants. (*ii*) Suckers should be dipped in a solution of lindane (0.2%) to kill eggs inside. (*iii*) Spray 2.5 kg of carbaryl 50WP in 1250 litres of water/ha.

3. Banana Aphid, *Pentalonia nigronervosa* Coquerel (Hemiptera : Aphididae) (Fig. 24.1)

It is not a serious pest of banana, but it is notorious as the vector of the virus causing the bunchy top disease in banana. The aphid is brownish in colour and has black-veined wings. Apart from banana, it also attacks cardamom, large cardamom and Colocasia antiquorum.

Life-cycle. The aphid lives in colonies within the leaf-axils, or tender leaves and at base of the culm at the ground level. Young ones are given birth by alate and apterous females by parthenogenetic development. An aphid is capable of producing 30-50 nymphs in its life time of 27-37 days. The nymph undergoes four instars of 2-3 days duration each. Total life-cycle is completed in 8-9 days. From 30 to 40 generations are completed during a year under South Indian conditions.

Damage. Both nymphs and adults suck the sap from the tender leaves. The pest is very important because it transmits the bunchy top virus.

Management. Spray 1.70 litres of dimethoate 30EC or 1.25 litres of oxydemeton methyl 25EC in 1250 litres of water/ha.

4. Banana Weevil, *Cosmopolites sordidus* (Germar) (Coleoptera: Curculionidae) (Fig. 18.17)

The larvae of the banana weevil, also known as the banana beetle and banana borer, bore into the corm of the banana plant. This species although native to South-east Asia, has attained wide distribution on cultivated bananas of the world. Its present distribution, in addition to India and South-east Asia, covers parts of Australia, the Hawaii Islands, tropical and South Africa and tropical America.

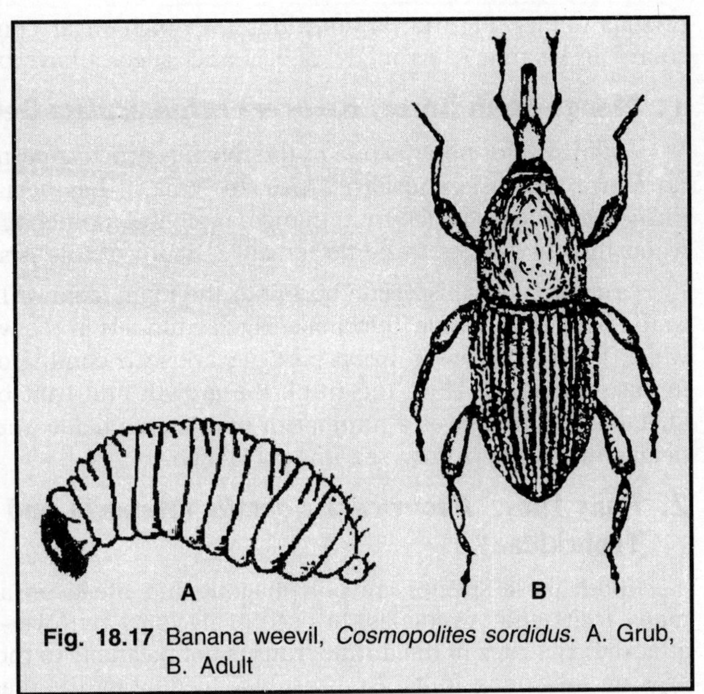

Fig. 18.17 Banana weevil, *Cosmopolites sordidus*. A. Grub, B. Adult

Life-cycle. The female weevil (**Fig. 18.17**) bites a small hole in the corm, lays a single egg and continues this activity on other plants throughout the year. A female can lay 10-50 eggs during its life of a few months. The larvae emerge from the eggs in about a week and bore into the corm, where they feed, making a tunnel. When full-grown in 2-6 weeks, the larvae pupate in the same tunnel. The pupal stage lasts about a week. The adults, on emergence, remain in the soil for sometime, feeding on the underground parts of the plants. Later on, they visit the growing point for oviposition.

Damage. This insect attacks only members of the genus *Musa*. The damage done by the weevil is through the destruction of the corm tissue. The seedlings are killed as a result of the borer attack when the larvae approach the growing-point. The tunnelled corms sometimes break out, since the weevils do not attack the roots, they do not cause toppling over.

Management. (*i*) Out of the various methods of control, cultural operations are the most important. They consist in destroying the sheltering and feeding places of the adult weevils. The pseudostems, from which the bunches have been cut, are chopped up and scattered in the plantation so that they rot quickly. Also, the pesudostems should be cut close to the ground and the infested suckers should not be used for planting. (*ii*) Suckers should be dipped in 0.1 per cent quinalphos emulsion before planting. In case of attack, spray 315 ml of phosphamidon 100 or 625 ml of dimethoate 30EC in 625 litres of water per ha, around the base of the plants or clumps.

Other Pests of Banana

Some other insects which appear as minor pests of banana include the banana mealy-bug, *Pseudococcus* spp. (Hemiptera : Pseudococcidae); thrips, *Heliothrips kadaliphila* Ramakrishna Ayyar & Margabandhu and *Scirtothrips signipennis* and *Chaetanaphothrips signipennis* (Bagnall) (Thysanoptera : Thripidae); the leaf-eating caterpillar, *Parasa lepida* (Cramer) (Lepidoptera : Limacodidae) and *Spodoptera litura* (Fabricius) and *Tiracola plagiata* (Walker) (Lepidoptera : Noctuidae); *Erionota thrax* (Linnaeus) (Lepidoptera : Hesperidae) and banana scale moth. *Nacoleia octasema* (Meyrick) (Lepidoptera : Pyralidae).

FIG (*Ficus carica* L.; Family : Moraceae)

Fig. is a moderately important fruit crop with an annual estimated global production of one million tonnes of fruit, of which about 30 per cent is produced in Turkey. The commercial cultivation of fig is not well developed in India and is hardly grown in about 1500 hectares with a production of 3000 tonnes. Fig fruit is nutritious with high sugar content, low in acid content and is a good source of vitamin A. As many as 50 insect species have been recorded feeding on the fig tree.

1. Mango Stem Borer, *Batocera rufomaculata* DeGeer (Coleoptera : Cerambycidae)

Mango stem-borer is one of the most destructive pests of fig and is widely distributed in India, Sri Lanka, Malaysia and East Africa. In India, it has been recorded on more than 30 host plants including fig, guava, jack-fruit, mango, mulberry, pomegranate, walnut, etc. Since fig has a very soft wood this tree seems to be the greatest victim of this pest.

The grubs of this beetle bore into the main stem or branches without causing outward injury to the bark. Once inside, they make zig-zag tunnels in the wood. The affected stems show holes from which frass and chewed wood particles are seen coming out. Ultimately, the branches dry up and in case the attack is on the trunk the growth and fruit bearing capacity of the tree are severely affected. The tree gives a number of new shoots below the attacked site but it ultimately dies. For details on its life-history see under 'Mango.'

2. Fruit Flies, *Bactrocera dorsalis* (Hendel) and *B. zonata* (Saunders) (Diptera : Tephritidae)

Both these species are polyphagous and breed on a large number of host plants including many fruits and vegetables. In case of fig, only the over-ripe fruits that remain on the tree, are attacked. The latex of the unripe fruits is not palatable to the grubs and the fruitflies also avoid laying eggs on the young fruits. As a result of feeding by the maggots, the pulp of fruits becomes rotten and the fruits fall down. For further details, see under 'Mango'.

3. Fig Midge, *Anjeerodiplosis peshawarensis* Mani (Diptera : Cecidomyiidae)

It is a serious pest of fig in North India; it specifically attacks the country fig (*Ficus carica* Roxb.), which is its only known host. The adult midge is a small light brown fly. The full-grown larva measures 3 mm long and 1 mm broad, is creamy white and flattened dorsoventrally, being narrow anteriorly. The head is small and bears two jointed antennae.

Life-cycle. The midges are one of the most abundant insects and are seen under the host trees in the mornings and at dusk. They breed throughout the year except in the summer from April to July. The eggs are laid in the fruits when these are one week old and are the size of a pea. The eggs are minute, hyaline, unsculptured, oval and pedicellate. They are laid by the females in clusters, the average number of eggs per cluster being 16. The eggs hatch in 3 days in September and March, and 5 days in December and January. The larval stage is completed inside the fig fruits, each of which may contain 200-300 larvae. The larvae pass through four instars and the total larval duration varies from 3 to 4 weeks. From the middle of April until July the fourth instar larvae remain inside the fruit. The full-grown larvae bore out of the fruit and drop to the ground and pupate in the soil without forming a cocoon. The pupal period is completed in 10-26 days. The total duration of the life-cycle varies from 5 to 7 weeks. In all, seven generations are completed in a year.

Damage. The attack by this pest is confined exclusively to the fruits. The infested fruits harbouring larvae become hard, hollow, deformed and irregular in outline. An infested fig may grow as large as a normal one, but it remains yellowish green and fails to ripen. In case of severe infestation, the shrivelled and deformed dry fruits are to be seen scattered in large numbers on the ground under the tree.

Management. (*i*) All infested flowers and fruits should be removed and destroyed. (*ii*) Plough the soil around trees frequently so as to expose and kill the pupae. (*iii*) Spray 2.0 litres of carbaryl 50WP or 1.70 litres of dimethoate 30EC in 1250 litres of water per ha at flower bud formation and again when the fruits are 'pea' size.

4. Fig Mites

Many species of mites have been recorded feeding on fig foliage. These include *Aceria ficus* (Cotte), *Rhyncaphytopus ficifoliae* Keifer and *Eriophyes ficivorus* Channa Basavanna (Eriophyidae); *Melichāres fici* Narand Ghai (Ascidae); *Cheletogenus ornatus* (Canestrini & Farzago) (Cheyletidae); *Eutetranychus hirsti* Pritchard & Baker and *Tetranychus neocaledonicus* Andre (Tetranychidae). All these are eight-legged spiders, very small creatures, which are seen moving among fine webs spun on the ventral surface of the leaves from where they suck cell sap. As a result of their attack, the affected leaves show blotches, turn downwards and ultimately they dry up and fall off. The severely infested trees may be seen almost denuded of the foliage.

Management. For management, see under 'Citrus.'

Other Pests of Fig

The other pests which attack fig are grasshopper, *Poekilocerus pictus* (Fabricius) (Orthoptera: Acrididae); *Cosmoscarta niteara* D. (Hemiptera : Cercopidae); *Pauropsylla depressa* Crawford and *Dinopsylla grandise* Crawford (Hemiptera : Psyllidae); the coccids, *Drosicha stebbingi* (Green) and *Drosichiella tamarindus* (Green) (Hemiptera : Margarodidae); *Aspidiotus cydoniae* C. and *A. lataniae* Signoret (Hemiptera : Diaspididae); *Saissetia oleae* (Olivier) (Hemiptera : Coccidae); *Pseudococcus lilacinus* Cockerell (Hemiptera : Pseudococcidae); *Gigantothrips elegans* Zimmermann and *Thrips tabaci* Lindeman (Thysanoptera : Thripidae); *Phycodes radiata* Ochs. (Lepidoptera : Glyphipterygidae); the leaf defoliator, *Ocinara varians* (Walker) (Lepidoptera : Bombycidae); *Spodoptera litura* (Fabricius), *Helicoverpa armigera* (Hubner) and *H. assulta* (Guenee) (Lepidoptera : Noctuidae); *Hypsa ficus* Fabricius (Lepidoptera: Hypsidae); *Adoretus duvauceli* Blanchard, *A. horticola* Arrow and *A. versutus* Harold (Coleoptera : Scarabaeidae); *Olenecamptus bilobus* Fabricius (Coleoptera : Cerambycidae) and *Aclees cribratus* Gyllenhal (Coleoptera : Curculionidae) which bores into the stems during June and July causing at times serious damage in Assam area.

JACKFRUIT (*Artocarpus heterophyllus* Lamarck; Family : Moraceae)

The estimated area under jackfruit cultivation in India is about 236,000 hectares with production of 1,176,000 tonnes. The fruit contains some minerals and also vitamins A and C. Over 35 species of insects attack jack-fruit trees and inflict some losses.

1. White-tailed Mealybug, *Ferrisia virgata* (Cockerell) (Hemiptera : Pseudococcidae)

Full account of this insect is given in Chapter 18.

Another mealybug, *Nipaecoccus viridis* (Newstead) also attacks jackfruit. It is a polyphagous insect recorded all over India on ber, citrus, fig, grapevine, guava, mango, tamarind, mulberry, etc.

The adult females are dark castaneous and 4.0-4.5 mm long. They are covered with a sticky cretaceous white ovisac. The nymphs are deep chocolate in colour, having their dorsum covered thinly with a whitish mealy material.

This pest is very active during dry season and the peak period of activity is around November. The female lays 200-400 eggs which are cylindrically rounded, flat at both ends and chestnut in colour. The eggs are enclosed in the ovisac. The eggs hatch in 7-10 days. Male and female nymphs take 15 and 20 days to complete development, respectively.

Management. Spray 750 ml of monocrotophos 40 WSC or 625 ml of methyl parathion 50EC in 625 litres of water per ha.

2. Pink Waxy Scale, *Ceroplastes rubens* Maskell (Hemiptera : Coccidae)

The pink waxy scale has been reported so far from East Africa, India, Sri Lanka, parts of China, Japan, Malaysia, the Philippines, Solomon Isles, East Australia, Pacific Islands, New Zealand and Hawaii.

It feeds on *Citrus* spp., jack-fruit, fig, mango, pear and other fruits. The adult females are 3-4 mm long and convex in shape. They are covered with a pink waxy shell often with vertical stripes. Life history is not known in detail. There is only one generation in a year.

The colonies of these scales may be seen covering the shoots and the fruit stalks. They suck cell sap and excrete copious amount of honeydew, which keeps dripping on the leaves and fruits, attracting ants. Sooty mould develops on honey-dew.

Management. Same as in case of white-tailed mealy-bug.

3. Jackfruit Leaf Webber, *Perina nuda* Fabricius (Lepidoptera : Lymantriidae)

The caterpillars of this insect attack jack-fruit and also fig and mango. It is a sporadic pest and causes heavy damage in certain years. The male moth has half ocherous and half transparent fore wings with a wing expanse of 33-36 mm, and the female is dull ocherous white in colour having 38-42 mm wing expanse. The full-grown larvae are 22-25 mm long, having short erect tufts of dusky grey to brownish hairs on their body.

Life-cycle. This pest becomes active during rainy season. The female moth lays 50-400 eggs in clusters or rows on the the leaves of host plants. The eggs are light pink when freshly laid, turning red and brick red later on. The eggs are 0.7 mm long, cylindrically round, tapering towards the end that is attached to the leaf. The larvae undergo six moultings in 16-20 days before entering the pupal stage. The pupae are hairy and brownish green dorsally, and pale yellow ventrally and are 16-18 mm long. The moths emerge from pupae in 5-9 days. The adults live for 3-11 days and the life-cycle is completed in 27-39 days.

Damage. The caterpillars fold or web the leaves together and feed on the green matter, causing extensive defoliation.

Management. (*i*) Hand picking and mechanical destruction of caterpillars in the initial stage of attack prevent further build up of population of leaf-webber. (*ii*) Spray 750 ml of monocrotophos 40WSC or 1.50 litres of chlorpyriphos 20EC or 1.25 litres of quinalphos 25EC in 625 litres of water per ha. Spray should be directed on the upper surface of foliage.

4. Bark Borer, *Batocera rufomaculata* DeGeer (Coleoptera : Cerambycidae)

This pest has been discussed under 'Mango.'

Other Pests of Jackfruit

There are a large number of insects that have been recorded as minor pests on jack-fruit, namely, the spittle bugs, *Cosmoscaria relata* Distant, *Clovia lineaticollis* Marshall and *Ptyleus* sp. (Hemiptera : Cercopidae), which feed on young shoots and leaves; the scale insects, *Icerya aegyptiaca* (Douglas) and *Drosicha stebbingi* (Green) (Hemiptera : Margarodidae); *Hemiberlesia lataniae* (Signoret) (Hemiptera Diaspididae); *Chloropulvinaria psidii* (Maskell) and *Coccus actuesismum* (Green) (Hemiptera : Coccidae), and all of them suck plant sap; the jack-fruit aphid, *Greenidea artocarpi* (Westwood) and the black citrus aphid, *Toxoptera aurantii* Boyer de Fonscolombe (Hemiptera : Aphididae); the thrips, *Pseudodendrothrips divivasna* (Ramakrishna & Marga-bandhu) (Thysanoptera: Thripidae), which produce whitish patches on leaves; the shoot borer, *Palpita caesalis* (Walker) (Lepidoptera : Pyralidae), which bores into tender shoots, flowering buds and developing fruits, and makes them shed; the bark borers, *Indarbela tetraonis* (Moore) (Lepidoptera : Metarbelidae); *Sthenias grisator* Fabricius (Coleoptera : Cerambycidae) and *Platypus indicus* Str. (Coleoptera : Platypotidae) which bore into branches and trunks; the castor capsule borer, *Conogethes punctiferalis* (Guenee) (Lepidoptera : Pyralidae) which bores into the buds and young fruits; the bud weevil, *Ochyromera artocarpi* Marshall (Coleoptera : Curculionidae), whose grubs bore into the tender buds and the developing fruits, and cause rotting and premature dropping.

LOQUAT (*Eriobotrya japonica* Lindley; Family : Pomaceae)

Total world production of loquat is about 500,000 tonnes and China with more than 80 per cent of the world production is the main producing country. In India, loquat occupies an area of 3,000 hectares. The fruit is used for desert purpose and for making jam and jelly. Bark eating caterpillar and fruit fly are the pests of major importance.

1. Bark Caterpillar, *Indarbela quadrinotata* (Walker) (Lepidoptera: Pyralidae)

It is a notorious pest found all over India on a variety of fruit trees including loquat. The damage is conspicuous by the presence of chewed particles and excreta hanging outside the webbed galleries on the bark of tree trunks. Full account of this pest has been given under 'Citrus'.

2. Fruit Fly, *Bactrocera dorsalis* (Hendel) (Diptera: Tephritidae)

It is a polyphagous pest and attacks a number of fruits including citrus, guava, mango, loquat, apricot, fig, peach and plum. The loquat fruits attacked by the maggots do not fall down but remain on the tree. Brownish syrupy juice oozes out of the punctures. Full account of this pest has been given under 'Mango.'

Other Pests of Loquat

The loquat tree is also attacked by other pests like the aphid, *Aphis malvae* (Koch) (Hemiptera: Aphididae); scale insects, *Coccus viridis* (Green), *Chloropulvinaria psidii* (Maskell) and *Saissetia coffeae* Walker (Hemiptera : Coccidae); the flower thrips, *Heliothrips* sp. (Thysanoptera : Thripidae); the *anar* butterfly, *Deudorix isocrates* Fabricius (Lepidoptera: Lycaenidae); the chafer beetles, *Adoretus duvauceli* Blanchard, *A. lasiopygus* Burmeister, A. *horticola* Arrow and A. *versutus* Harold (Coleoptera: Scarabaeidae) and grey weevils, *Myllocerus lactivirens* Maskell and *M. discolor* Bohemann (Coleoptera: Curculionidae).

PINEAPPLE (*Ananas comosus* L. Merr.; Family : Bromeliaceae)

Pineapple is grown on an area of about 0.10 million ha in India with production of 1.57 million tonnes. The fruits are high in sugar content and rich in vitamins A and C. Pinapple does not suffer much on account of pest damage but insects as vectors of virus disesases take a heavy toll.

1. Pineapple Thrips, *Thrips tabaci* Lindeman (Thysanoptera : Thripidae)

This pest has been discussed in detail in Chapter 20.

2. Slug Caterpillar, *Parasa lepida* (Cramer) (Lepidoptera : Limacodidae)

This pest has been discussed in Chapter 15.

The other pests recorded on pineapple are the coccids, *Pseudococcus bromiliae* Bouche and *P. brevipes* Cockerell (Hemiptera : Pseudococcidae); the root-knot nematode, *Meloidogyne* spp. (Tylenhida: Heteroderidae) and the pine-apple mite, *Stigmaeus floridanus* (Acari : Stigmaeidae).

DATE PALM (*Phoenix dactylifera* L.; Family : Palmaceae)

Date palm is grown on an area of about 13000 hectares in India. There are about 1.9 million date palm trees with a production of about 85,000 tonnes of fresh fruits. The fruit has a high food value with sugar content of about 54 per cent and protein 7 per cent. In India, only 13 insect species have been recorded causing damage to date palm.

1. Date Palm Scale, *Aspidiotus destructor* Signoret (Hemiptera : Diaspididae)

This pest has been discussed in Chapter 23.

2. Black Palm Beetle, *Oryctes rhinoceros* (Linnaeus) (Coleoptera : Scarabaeidae)

Full account of this pest has been given in Chapter 23.

3. Red Palm Weevil, *Rhynchophorus ferrugineus* (Olivier) (Coleoptera : Curculionidae)

The pest has been discussed in Chapter 23.

In addition to the above mentioned pests, the date palm is also attacked by other minor insect pests like termites, *Odontotermes obesus* (Rambur) (Isoptera : Termitidae) and the black headed caterpillar, *Opisina arenosella* Walker (Lepidoptera : Xyloryctidae)

JAMUN (*Syzygium cuminii* Skeel; Family: Myrtaceae)

Jamun (Indian blackberry or black plum) tree is found either growing wild or cultivated in backyards, road sides, borders of fields and around orchards. Organised orcharding of jamun is still lacking in India. The fruit contains calcium, iron and proteins in small quantities. About 36 insect species have been recorded either feeding or breeding on *jamun* tree.

1. Bark Caterpillar, *Indarbela tetraonis* (Moore) (Lepidoptera : Metarbelidae)

It is a polyphagous pest of major importance on *jamun,* guava and citrus, specially in north-western India. It is also a minor pest of litchi, mango, loquat, etc. The adult is a stout pale brown moth with grey wavy markings on the wings. The brownish larva is about 3.8 mm long.

Life-cycle. The female moth lays eggs in clusters under the loose bark, during summer. The eggs hatch in 8-10 days and the larvae feed for some time on the bark and then bore inside the trunk and stem, and feed within galleries, from July to March next. Pupation takes place from mid-March to April and lasts 3-4 weeks. The adult moths survive for hardly three days. Only one generation is completed in a year.

Damage. As a result of larval feeding inside the branches and the trunk, the translocation of cell-sap is disrupted and growth as well as fruiting capacity of the tree is reduced.

Management. Management practices are given under 'Citrus'.

2. Jamun Leafminer, *Acrocercops phaeospora* Meyrick (Lepidoptera : Gracilliariidae)

Four species of leaf-miner, viz. *Acrocercops phaeospora* Meyrick, *A. laxias* Meyrick, *A. syngramma* Meyrick and *A. telestis* Meyrick have been reported on jamun in India. Of these, *A phaeospora* is specific to jamun and is comparatively more common and regular in occurrence. *A. laxias* and *A. telestis,* on the other hand, are sporadic, while *A. syngramma* is widely distributed in Indo-Pakistan plains. These are polyphagous in nature and attack jamun, mango, litchi, sweet potato, etc.

Life-cycle. The moths lay eggs singly along the mid rib on the underside of the tender leaves. The newly hatched, delicate, pale yellow caterpillar mines a narrow thread like silvery gallery on the leaf along the mid rib. The mine is slowly transformed into a tubular blister-like swelling which extends from 3 to 6 cm in length on the dorsal surface of the leaf. The caterpillar feeds within the lacerated gallery. When full-fed, it forms a circular flat cocoon in depressed portion or cavity in the mine, usually near the raised conspicuous leaf vein and pupates there. The moth, on emergence, cuts a circular hole in the swollen fragile mine to get an easy exit.

The insect remains active from April to October and passes through several overlapping generations. It overwinters from November to February as pupa inside the tubular mine and the tiny moths emerge during March.

Damage. The damage is caused by the caterpillars which mine into the leaves and form tubular blister like swellings. There may be a single blister covering both halves of the leaf blade or one blister on each half of the blade. After emergence of the moths, the swollen leaf tissues start drying and eventually rupture and drop off, leaving conspicuously big longitudinal holes in the infested leaves. Heavily mined leaves later appear as if scortched due to heat. Sometimes, as high as 66 per cent leaves are found riddled by the larvae. Besides causing decrease in photosynthetic activity, the yield of the plants is adversely affected. Most of the damage takes place during summer and rainy season, coinciding with the onset of new vegetative flush and also fruiting stage of the crop.

Management. (*i*) Clip off and burn the mined leaves during winter to reduce the incidence of leaf-miner on new growth in March-April.

(*ii*) Spray 3·125 litres of dimethoate 30EC or 1·5 litres of monocrotophos 36SL or 2·5 litres of oxydemeton methyl 25EC in 1250 litres of water per ha.

3. Barkeating Caterpillar, *Indarbela quadrinotata* (Walker) (Lepidoptera : Metarbelidae) (Plate 18.21)

This pest has been discussed under 'Citrus'.

Other Pests of Jamun

The minor pests recorded on *jamun* trees are the mulberry bug, *Halys dentatus* Fabricius (Hemiptera : Pentatomidae); the whiteflies, *Dialeurodes eugeniae* (Maskell), *D. citrii* (Ashmead) and *Singhiella bicolor* (Singh) (Hemiptera : Aleyrodidae); the scale insects, *Aonidiella aurantii* (Maskell), *A. orientalis* (Newstead) and *Aspidiotus destructor* Signoret (Hemiptera : Diaspididae) and *Coccus discrepans* Green (Hemiptera : Coccidae); mango mealy bug, *Drosicha mangiferae* (Green) (Hemiptera : Margarodidae); the thrips, *Rhipiphorthrips cruentatus* Hood, *Mallothrips indicus* Ramakrishna and *Thrips florum* Schmutz (Thysanoptera : Thripidae); the *jamun* leaf rollers, *Polychrosis cellifera* Meyrick and *Olethreutes aprobela* Meyrick (Lepidoptera : Tortricidae); leaf eating caterpillar, *Carea subtilis* Walker (Lepidoptera : Noctuidae); fruit flies, *Dacus diversus* Coquillett and *D. correctus* (Bezzi) (Diptera: Tephritidae); leaf beetle, *Holotrichia insularis* Brenske (Coleoptera : Scarabaeidae); fruit weevil, *Curculio Calbum* Scopoli (Coleoptera : Curculionidae) and *Oligonychus mangiferae* Rahman & Sapra (Acari: Tetranychidae).

TAMARIND (*Tamarindus indica* L.; Family : Leguminosae)

Tamarind is grown in tropical countries extending from Africa to South-east Asia including UAR, India, Pakistan, Sri Lanka, Myanmar and Bangladesh. In India, there are extensive tamarind orchards producing 250,000 tonnes of tamarind. Tamarind is attacked by about 40 insect species in India.

1. Coconut Scale, *Aspidiotus destructor* Signoret (Hemiptera : Diaspididae)

Cocount scale is found in India throughout the plains and low hills, on a number of host plants including coconut, tamarind, avocado, banana, *ber*, citrus, datepalm, fig, grapevine, guava, *jamun*, mango, papaya, peach, pear, sapota, etc.

For details on its life-cycle and the damage caused by it see Chapter 23.

Two more scale species, namely *Aonidiella orientalis* (Newstead) and *Aspidiotus tamarindi* (Green) also infest the tamarind tree. Both the young and the adults of these insects suck cell sap. The affected parts are devitalized and there is premature shedding of buds and flowers, affecting ultimately the fruit setting capacity of the tree.

2. Citrus Mealybug, *Nipaecoccus viridis* (Newstead) (Hemiptera : Pseudococcidae)

Full account of this mealy bug is given under 'Jack-fruit'

Another mealy-bug, *Planococcus lilacinus* (Cockerell) (Hemiptera : Pseudococcidae) has been found mainly on tamarind and pomegranate. Both the adults and the nymphs are found in large numbers on the ventral surface on leaflets and the base of leaf petioles, on tender shoots and even on fruits. The pest causes damage by sucking the cell sap and devitalizing the plant. In case of severe attack the leaflets become chlorotic and fall off. There may also be immature fruit fall.

3. Tamarind Fruit Borer, *Phycita orthoclina* Meyrick (Lepidoptera : Phycitidae)

The tamarind fruit borer has been reported from Marquesas island and the main Sub-continent of India. The larval stage causes considerable damage to tamarind fruits and dry stored fruits. Besides tamarind, it also bores into dry fruits and seeds of *Bombax ceiba* Linn.

The moths are small, delicate insects, having elongate fore wings. The hind wings are broad and bear hairs on the dorsal side. The full-grown larva is cylindrical, pink and measures 14 mm long.

Life-cycle. The tamarind fruit borer is most active during December-April. The female moth lays flat, oval shaped and white eggs, singly on the pulpy portion inside the rough shelled pods, through cracks and crevices found on their surface. On an average, the female lays 190 eggs in three days. The larvae emerge from eggs after 4-5 days and enter into the fruit pulp and feed there by making a silken web. The larval stage is completed in 27-40 days. The full-grown larva makes a silken cocoon inside the infested pod and pupates there. The moths emerge in about 6-8 days.

Damage. The larvae feed on the pulp and their castings, excrements and webbings, render the fruit unfit for culinary purposes, hence these cannot be marketed.

Management. Spray 425 ml of monocrotophos 40EC or 315 ml of dichlorvos 100EC or 1.25 kg of carbaryl 50WP in 625 litres of water per ha.

4. Pomegranate Butterfly, *Deudorix isocrates* (Fabricius) (Lepidoptera : Lycaenidae)

Full account of this pest has been given under 'Pomegranate'

5. Castor Capsule Borer, *Conogethes punctiferalis* Guenee (Lepidoptera: Pyralidae)

For details on its life-cycle, see Chapter 15.

6. White Grub, *Holotrichia insularis* Brenske (Coleoptera : Scarabaeidae)

It is one of the most abundant and injurious species of beetles found feeding on a large number of host plants including *ber*, tamarind, *falsa*, guava, *jamun*, *karaunda*, mango, pomegranate, etc. Within India, the pest is more common in Gujarat, Rajasthan, Haryana and Punjab. The adults are brownish black convex beetles. The full-grown grubs are white, fleshy, curved, 38-44 mm long and 6-9 mm wide and are found in the soil.

Life-cycle. The beetles emerge in June, after the break of monsoons. They lay eggs in the soil at a depth of 30-150 mm. The eggs are shiny white in colour and oval in shape. The young grubs emerge after 8-12 days of incubation and they start feeding on the roots of host plants. The larval stage is completed in 55-80 days. The full-grown grubs make earthern cells in the soil, taking 8-12 days to complete the job. Pupation takes place in the cells and the insect hibernates from November to June, first as pupa and then after transformation as an adult. There is only one generation in a year.

Damage. The nymphs feed on rootlets resulting in gradual withering and drying up of seedlings and young plants. The beetles feed on leaves of a number of trees and defoliate them.

Management. Soil treatment with lindane 2 per cent dust @ 50 kg per ha is effective.

Other Pests of Tamarind

The tamarind trees and fruits are also attacked by other pests like the aphid, *Toxoptera aurantii* (Boyer de Fonscolombe) (Hemiptera : Aphididae); the whitefly, *Acaudaleyrodes rachispora* (Singh) (Hemiptera : Aleyrodidae); *Drosicha mangiferae* (Green) and *Perissopneumon tamarinda* (Green) (Hemiptera: Margarodidae); the scale insect, *Saissetia oleae* (Olivier) (Hemiptera : Coccidae); the lac insect, *Kerria lacca* (Kerr) (Hemiptera : Kerriidae); the thrips, *Ramaswamiehiella submudula* Karny and *Scirtothrips dorsalis* Hood (Thysanoptera: Thripidae) and also *Haplothrips ceylonicus* Schmutz. (Thysanoptera : Phlaeothripidae); *Pseudohypatopa pulverea* (Meyrick) (Lepidoptera : Blastobasidae); the leaf eating caterpillars, *Thosea aperiens* (Walker) (Lepidoptera : Cochlidiidae); *Cryptophlebia illepidia* (Butler) and *Cydia palamedes* Meyrick (Lepidoptera : Tortricidae); the rice moth, *Corcyra cephalonica* (Stainton) (Lepidoptera : Pyralidae); *Thalassodes quadraria* Guenee (Lepidoptera : Geometridae); the beetle, *Schizonycha ruficollis* Fabricius (Coleoptera : Melolonthidae); the lesser grain borer, *Rhyzopertha dominica* (Fabricius) (Coleoptera : Bostrichidae); the cigarette beetle, *Lasioderma serricorne* (Fabricius) (Coleptera : Anobiidae); *Caryoborus gonagra* (Fabricius) (Coleoptera: Bruchidae); the rice weevil, *Sitophilus oryzae* (Linnaeus) and tamarind seed borer, S. *linearis* (Herbst) (Coleoptera : Curculionidae).

LITCHI (*Litchi chinensis* Sonnerat; Family : Sapindaceae)

In India, more than 82,700 hectares of area is under litchi cultivation. India's annual production of litchi is about 5.80 lakh tonnes, of which 2.56 lakh tonnes is produced in Bihar alone. Litchi fruit contains 15 per cent soluble sugar and 1-5 per cent protein. Litchi fruits are rich in vitamin E and also contain a fair amount of calcium, iron and phosphorus. In nature, about 40 insect species attack litchi tree. However ,only litchi bug and leaf curl mite cause serious damage.

1. Litchi Bug, *Chrysocoris stolii* Wolff (Hemiptera: Pentatomidae)

The litchi bug is present in India, Myanmar, Formosa and South China. In India, it feeds on litchi, *Logan, Lantana*, banana and a number of berries, particularly in States of Uttar Pradesh, Bihar and West Bengal. The one-cm long adult bug, on emergence, is red but gradually acquires greenish colour. There are six distinct black spots arranged in rows on the thorax and seven on the scutellum, which covers the abdomen completely. The bright green nymphs measure about 9.5 mm in length.

Life-cycle. The bug is available in abundance during March-April and again from August to October. The female bug lays 22 spherical and creamy white eggs either in a single batch or in two or three batches. The young nymphs emerge in 5-7 days. The nymphs moult five times and the duration of successive nymphal instars is 2.5-3.5, 3.5-6.0, 3-6, 5-6 and 5-7 days. Thus, the bug completes its life-cycle in 25-37.5 days.

Damage. Both adults and nymphs suck sap from stems and young fruits, as a result of which the vitality of the plants is reduced and the attacked fruits wither way.

Management. Spray 1.25 litres of monocrotophos 40EC or 2.5 litres of dimethoate 30EC in 1250 litres of water per ha.

2. Leafcurl Mite, *Aceria litchi* Keifer (Acari : Eriophyidae)

The leaf curl mite is the most destructive and specific pest of litchi and has been reported from almost all the litchi growing countries of the world. In India, it has been noticed in West Bengal, North Bihar and Uttar Pradesh. The nymphs and adults are similar in appearance, the nymphs being smaller in size and having lesser number of lateral setae. Both are minute, whitish in colour, and veriform four-legged mites. The adults are 0.15-0.20 mm long, while the young nymphs are microscopic.

Life-cycle. The overwintering adults start multiplying by the end of March and the peak activity is noticed around July. The extremely small (0.04 mm in dia) eggs are laid singly on the ventral side of leaves. These are round in shape and whitish in colour. The eggs hatch in 2-3 days and the nymphs feed on soft leaves. The nymphal stage is completed in 8-12 days. The adults on emergence mate and lay eggs within their short life of 2-3 days.

Damage. Both nymphs and adults puncture and lacerate the tissues of leaf with their stout rostrum and suck the cell sap. Chocolate brown velvety growth on the ventral surface of leaves indicates the presence of this pest. As a result of severe feeding, the leaves curl up apically forming hollow cylinders. The attacked leaves ultimately wither and fall off. The attack generally begins from the lower portion of the tree and gradually extends upwards. The young plants and seedlings in nursery are particularly liable to the attack and severe damage.

Management. Same as in case of litchi bug.

Other Pests of Litchi

The minor pests of litchi include the whiteflies, *Aleurocanthus husaini* Corbett and *Dialeurodes elongata* Dozier (Hemiptera : Aleyrodidae); the aphid, *Toxoptera aurantii* (Boyer de Fonscolombe) (Hemiptera : Aphididae); the scales, *Kerria albizziae* (Green) (Hemiptera : Kerriidae); leaf thrips, *Dolichothrips indicus* Hood and *Megaleurothrips usitatus* (Bagnall) (Thysanoptera : Thripidae); the bark borer, *Indarbela tetranonis* (Moore) (Lepidoptera : Metarbelidae); the leaf roller, *Olethreutes aprobola* Meyr, *O. illepida* Meyr, *O. leucaspis* Meyrick (Lepidoptera : Tortricidae); *Deudorix isocrates* (Fabricius) (Lepidoptera : Lycaenidae); the tussock caterpillar, *Lymantria mathura* Moore (Lepidoptera: Lymantriidae) and leaf weevil, *Ptochus* sp. (Coleoptera : Curculionidae).

PAPAYA (*Carica papaya* L.; Family : Caricaceae)

India is producing about 5.38 million tonnes of papaya from an area of 0.13 million ha under this fruit. Karnataka accounts for more than 50 per cent of India's total output of papaya. The fruit is rich in sugar and contains vitamins A and B and also digestive enzymes. Papaya trees generally do not suffer much loss from insect pests. However, *Ak* grasshopper and mites are considered to be the major pests.

1. Ak Grasshopper, *Poekilocerus pictus* (Fabricius) (Orthoptera : Acrididae) (Fig. 1.8)

The *Ak* grasshopper is distributed in India, Pakistan, Baluchistan and Africa. In India, it is widespread throughout the plains and the desert areas. Although the milkweed *Ak* (*Calotropis* spp.) is its main food, the nymphs and adults are also seen feeding on papaya, citrus, fig, banana, cotton, oleander castor, cowpea and a number of vegetables. The stout adult is blue-green with yellow markings (**Fig. 1.8**). It has red wings and measures 45-60 mm in length. The nymph is yellowish with orange and black stripes all over the body.

Life-cycle. On papaya this pest is most active during July-August and passes winter in the egg stage. The whitish young nymphs start emerging from the soil from the end of March to early April. They feed voraciously and become adults in 30-42 days. When they reach the adult stage, the male rides on the females and mating takes place for 5-7 hours. After a preoviposition period of 25-30 days, the females penetrate their abdomen deep into the soil and lay eggs at a depth of 18-20 cm. About 145-170 orange coloured elongate eggs are laid in a spiral manner to form a compact mass which is covered with a frothy secretion which hardens laters on. The eggs laid in summer overwinter for nearly four months.

Damage. Both adults and nymphs feed voraciously on leaves and skeletonize them. In case of severe infestation even the bark of the trees is not spared.

Management. Grasshopper can be managed by dusting carbaryl 10 per cent @ 25 kg/ha or malathion 5 per cent @ 50 kg/ha.

2. Red Spider Mite, *Tetranychus urticae* Koch (Acari : Tetranychidae)

It is one of the most serious pests found on ventral leaf surface and occasionally on fruits as well. Both nymphs and adults remain protected under webs and suck the cell sap therefrom causing the leaves to turn yellow and fruits to become rough and brownish. For details about life-cycle, see Chapter 20.

Other Pests of Papaya

The minor pests of papaya include the whitefly, *Bemisia tabaci* (Gennadius) (Hemiptera : Aleyrodidae); scale insects and mealy bugs, *Aspidiotus destructor* Signoret (Hemiptera : Diaspididae), *Drosicha mangiferae* Green (Hemiptera : Margarodidae); the aphids, *Myzus persciae* (Sulzer), *Aphis gossypii* Glover and *A. malvae* (Koch) (Hemiptera : Aphididae) act as vectors of papaya *mosaic* virus; fruit flies, *Dacus diversus* (Coquillett), *Bactrocera cucurbitae* (Coquillett), *Toxotrypana curvicauda* Gerstaeker and the Mediteranean fruit fly, *Ceratitis capitata* (Wiedemann) (Diptera : Tephritidae). The nematodes like *Heterodera schachtii* (Schmidt) and *Meloidogyne incognita* (Kofoid and White) (Tylenchida : Heteroderidae) feed on roots of papaya trees. As a result, the leaves turn yellow and curl, and ultimately the tree dies.

SAPOTA (*Manilkara achras* (Mill.) Forberg; Family : Sapotaceae)

Sapota or *chickoo* is grown on about 0.16 million ha in India with production of 1.49 million tonnes. The sapota fruits are delicious and the milky latex is the source of chicle used to make chewing gum. As many as 25 insect species have been reported to attack sapota trees, out of which sapota leaf-webber or *chickoo* moth is the most destructive.

Sapota Leafwebber, *Nephopteryx eugraphella* Ragonot (Lepidoptera : Pyralidae)

The leaf-webber is a major pest of sapota and occurs widely in India. The moth is grey in colour. The destructive larva is 25 mm long, slender in body shape, and is pinksh with a few longitudinal lines on the dorsal surface.

Life-cycle. The pest is found throughout the year but the activity increases with the appearance of new shoots and buds. The maximum activity is seen during June-July and the minimum during winter. With the onset of spring season, the female moths start laying pale-yellow, oval shaped eggs singly or in batches of 2 or 3, on leaves and buds of young shoots. A female may lay as many as 374 eggs in 7 days. The eggs hatch in 2-11 days. The larvae feed for 13-60 days and complete development. They undergo pupation in the leaf-webs and this stage is completed in 8-29 days. The life-cycle is completed in 26-92 days depending upon the varying environmental conditions. There are 7-9 generations of this pest in a year.

Damage. The larvae clump the leaves together and feed on green matter of leaves, often on buds and flowers and sometimes on tender fruits as well. The larvae bore into the buds, which wither, and then they move on to the next buds thus, damaging many of them. The infestation of this pest can be easily spotted by the presence of webbed shoots, the appearance of dark brown patches on leaves and clusters of dead leaves.

Management. (*i*) Removal and destruction of infested leaves and affected fruits by chickoo moth will reduce the infestation. (*ii*) Spray 750 ml of phosphamidon 85EC or 2.5 kg of carbaryl 50WP in 1250 litres of water per ha.

Other Pests of Sapota

This fruit plant is also attacked by a number of minor pests including *Idioscopus* spp. (Hemiptera: Cicadellidae); the coccids, *Pseudococcus lilacinus* Cockerell, *P. iceryoides* Green and *Monophlebus* sp., the soft scales, *Coccus langulum* Dougl., *Saissetia oleae* (Olivier) and *Chloropulvi-naria psidii* Maskell (Hemiptera : Coccidae); the hard scales, *Aspidiotus transparens* Green and *A. cinerea* Green (Hemiptera : Diaspididae); the blossom thrips, *Franklinieella dampfi* Priesner (Thysanoptera : Thripidae); *Anarsia* sp. (Lepidoptera : Gelechiidae), which bores into the flower buds and causes considerable drop; the larvae of *Deudorix isocrates* Fabricius (Lepidoptera : Lycaenidae) bore into fruits; the fruit-flies, *Bactrocera dorsalis* (Hendel) and *Dacus correctus* Bezzi (Diptera : Tephritidae) and the weevil, *Myllocerus maculosus* Desbrochers des Lages (Coleoptera: Curculionidae).

PHALSA (*Grewia asiatica* L.; Family : Tiliceae)

Phalsa is grown on about 400 hectares in India mainly in Uttar Pradesh, Rajasthan, Madhya Pradesh, Gujarat, Punjab and Haryana. The fruit contains 55-65 per cent juice, 11.7 per cent sucrose, 2.8 per cent citric acid and traces of vitamin C. The tree is reported to be attacked by about 20 insect species in India.

1. Plum Hairy Caterpillar, *Euproctis fraterna* (Moore) (Lepidoptera : Lymantriidae)

The plum hairy caterpillar is distributed in India, Myanmar, Bangladesh, Pakistan and Sri Lanka. The insect is known to feed on a number of trees including phalsa, apple, apricot, *ber*, castor, citrus, grapevine, mango, mulberry, peach, pear, pomegranate and strawberry. Full-grown caterpillars are 35-40 mm long. They have red-head, darkish-brown body with white hairs on the head and a tuft of long hairs at anal end. The adults are yellow moths with pale transverse lines on the fore wings. The wing expanse is 24-28 mm.

Life-cycle. The eggs are flat, circular and yellow, and are laid in masses covered with yellow hairs on the under surface of the leaves. A single female may lay as many as 150-300 eggs. They hatch in 4-10 days and caterpillars feed gregariously. The larvae are full-grown in 13-29 days. Pupation takes place in cocoons of hairs on the leaves or on the branches. Pupal period lasts 9-25 days. The total life-cycle is completed in 45-57 days. There are three generations in a year. During winter, the insect hibernates in the larval stage.

Damage. The caterpillars feed gregariously on leaf lamina skeletonizing it completely. Subsequently, the caterpillars segregate and gnaw the leaves. In case of severe infestation, the entire plant may be denuded.

Management. (*i*) Collect and destroy the egg masses and the gregariously feeding young caterpillars. (*ii*) Spray 2.5 kg of carbaryl 50 WP in 1250 litres of water per ha.

2. Mango Mealybug, *Drosicha mangiferae* (Green) (Hemiptera : Margarodidae)

The details of this pest are given under 'Mango'.

3. Bark Caterpillar, *Indarbela tetraonis* (Moore) (Lepidoptera : Metarbelidae)

The details of this pest are given under 'Jamun'.

Other Pests of Phalsa

Several other pests attacking *phalsa* include the pentatomid bug, *Scutellera nobilia* (Fabricius) (Hemiptera: Pentatomidae); cotton whitefly, *Bemisia tabaci* (Gennadius) (Hemiptera : Aleyrodidae); *phalsa* bug, *Gargara mixta* Buckton and membracid bug, *Leptocentrus taurus* (Fabricius) (Hemiptera: Membracidae); *ber* mealy bug, *Perissopneumon tamarindus* (Green) (Hemiptera : Margarodidae); *phalsa* caterpillar, *Giaura sceptica* Swinhoe (Lepidoptera : Noctuidae); leaf eating beetle, *Anomala bengalensis* Blanchard; *phalsa* beetles, *Oxycetonia versicolor* (Fabricius) and *O. albopunctata* (Fabricius); white grubs, *Holotrichia consanguinea* Blanchard and *H. insularis* Brenske (Coleoptera : Scarabaeidae) and almond beetle, *Mimestra cyanura* Hope (Coleoptera : Chrysomelidae).

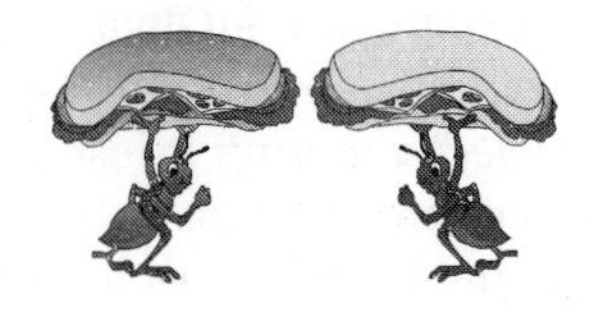

PESTS OF TEMPERATE FRUITS

INTRODUCTION

The temperate fruits, namely apple, pear, cherry, plum, peach, apricot, etc. are grown in the Himalayan region and other mountainous areas of the Indian Sub-continent. Good quality cherries and apples are generally grown above 1,500 metres. The other fruits are grown at lower altitudes and even in the plains. In the plains, only those varieties are successful which have low chilling requirements for breaking dormancy needed for successful fruiting. Temperate fruits are grown in all hilly areas of north western Himalayam region and hilly aras of Rajasthan, Madhya Pradesh and Tamil Nadu.

Temperature fruits are grown over 15 per cent of the total area under fruit crops, contributing to 3.9 per cent to the fruit production. Out of it, 42.2 per cent area is under apple only, which contributes to 75.1 per cent to the production in these regions. The global area under apple is 4.86 million ha with a production of 75.63 million tonnes. China is the largest apple producing country with annual production of 29.85 million tonnes from an area of 2 million ha. India ranks seventh in the world with production of 1.91 million tonnes on an area of 0.32 million ha. The production of pear is 0.29 million tonnes on 0.04 million ha. The production of peach and plum is 98,000 and 74,000 tonnes from an area of 19,000 and 105,000 ha, respectively. India produces 233,000 and 9300 tonnes of walnut and apple on an area of 123,000 and 21,000 ha, respectively.

Since these fruits are grown universally, some of the serious cosmopolitan pests, namely the San Jose scale and the woolly aphis, have also been introduced accidentally. Another dreaded world-wide pest of apple, the codling moth, *Cydia pomonella* (Linnaeus), was recorded in Ladakh in 1964, but its further spread has not been reported. On the stone fruits, the peach leafcurl aphis is the most destructive pest and it is rarely that any variety of peach and almond escapes its attack. Walnuts as stray trees are grown at higher altitudes only and it is not uncommon to see more than half of the fruits dropping off a tree, as a result of the ravages caused by the grubs of the walnut weevil which bore into the fruit and cause damage. In recent years, dwarf walnut varieties, pecon nuts and almond varieties with low chilling requirements have been developed in the world, which should have a good scope in the lower Himalayas and even in the Indo-Gangetic plains. Likewise, *Pistachio* may also become a successful crop in this region.

SAP SUCKERS

These insects suck the cell-sap and adversely affect the growth of the trees.

1. San Jose Scale, *Quadraspidiotus perniciosus* (Comstock) (Hemiptera : Diaspididae) (Fig. 19.1, Plate 19.1 A,B)

This insect is world-wide and is the most serious pest in temperate regions on nearly 700 different species of fruits, shrubs and ornamental plants. It usually prefers plants belonging to the Rosaceae such as apple, plum (*Prunus domestica* L., *P. salicina* L.), pear, peach (*Prunus persica* L.) and other closely related species such as the currant, willow, hawthorn and rose. The damage is caused by nymphs and female scales which suck the sap from twigs, branches and fruits. The scale which forms a covering on the body of the insect is black or brown. Underneath, a lemon-yellow insect is visible when the covering is lifted (**Fig. 19.1**). The infested fruits also have a scaly appearance and each spot is surrounded by a scarlet or red area.

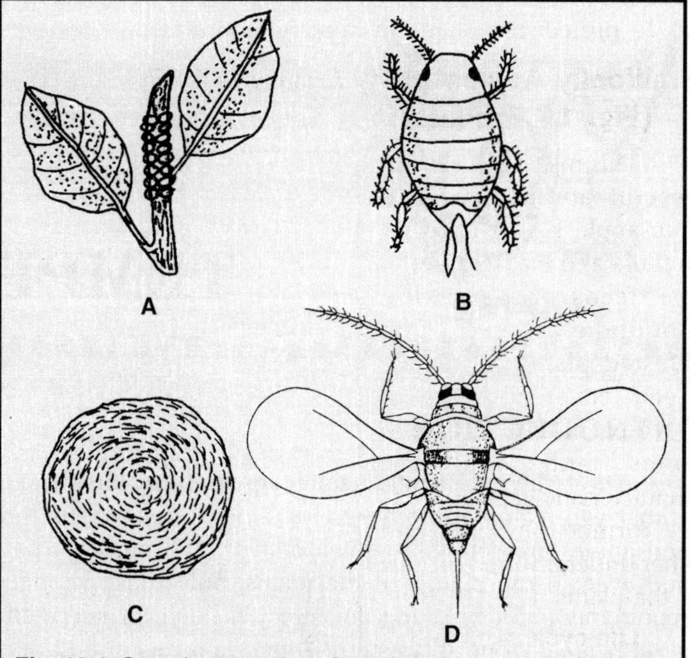

Fig. 19.1 San Jose scale, *Quadraspidiotus perniciosus*. A. Affected twig; B. Crawler; C. Wingless female; D. Winged male

Life-cycle. The pest is active from March to December and passes the winter in the nymphal stage. The overwintering nymphs resume activity in the spring and are full-grown by about April-May. They start to reproduce by mid-May. The San Jose scale gives birth to young ones, which hatch from the eggs developed within the body of the female. Each female may give birth to 200-400 nymphs. The newly-born nymphs crawl out of the parental scale and lead a free life for 12-24 hours. On finding a suitable place, they insert their mouth-parts into the plant tissue and begin to feed by sucking the cell-sap. They become full-grown in 30-40 days and the females again start giving birth to young ones within the next 10-14 days. The gravid mothers live for about 50-53 days. The male nymph has an elliptical or oval scale and develops into a winged adult in 25-31 days. The life of the male adult is short and hardly exceeds 24-32 hours, during which it fertilizes the non-winged females. Four overlapping generations are completed in a year and the fiffh-generation nymphs overwinter.

Augmentative and inoculative releases of two exotic parasitoids, *Encarsia perniciosi* (Tower) and *Aphytis* sp. *proclia* group have given promising results for the suppression of San Jose scale on apples. The Russian strain of *E. perniciosi* proved effective in Himachal Pradesh while Chinese and American strains proved better in Uttar Pradesh.

Damage. All parts of the plant above the ground are attacked and the injury is due to loss of the cell-sap. At first, the growth of the infested plants is checked, but as the scales increase in number, the infested plants may die.

Management. (*i*) For effective management of San Jose scale, orchard sanitation should be given priority. Infested pruned material should be collected immediately and burnt. (*ii*) The parasitoid,

Encarsia perniciosi (Tower) may be released to check the overwintering population on wild host plants growing around. (*iii*) Spray diesel oil emulsion + Bordeaux mixture (diesel oil 68 litres + copper sulphate 15 kg + unslacked lime 3.75 kg) to be emulsified and diluted 5-6 times before spraying or spray 7.5 litres of ESSO tree spray oil emulsion in 250 litres of water per ha during the winter season when the trees are in dormant stage and completely defoliated. (*iv*) Severe scale infestation under neglected orchard management requires additional summer sprays with 2.0 litres of chlorphriphos 20EC or 625 ml of methyl demeton 25EC in 1250 litres of water per ha. (*v*) To protect the plants in nursery, apply carbofuran granules @ 0.75-1.0 g a.i. per plant.

2. Woolly Apple Aphid, *Eriosoma lanigerum* (Hausmann) (Hemiptera : Aphididae) (Fig. 19.2, Plate 19.1 C)

This insect is a serious pest of apple, pear and crab-apple (*Pyrus baccata*) in India and Pakistan and it also feeds on hawthorn, mountain ash, etc. The infested plants have pale green leaves and whitish cottony patches on the stems and branches. Characteristic galls or knots are formed on roots and other underground portions of the plants (**Fig. 19.2**).

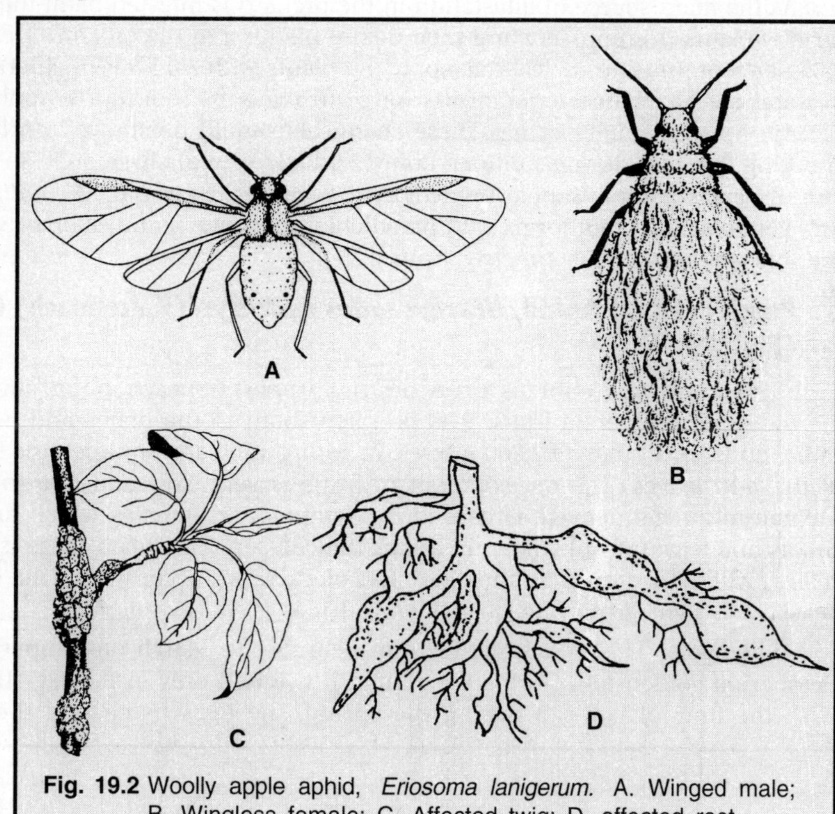

Fig. 19.2 Woolly apple aphid, *Eriosoma lanigerum.* A. Winged male; B. Wingless female; C. Affected twig; D. affected root

Life-cycle. The pest is most active during March-September and multiplies at a reduced pace during October-December. The development from December to February is extremely slow. The aphids reproduce parthenogenetically and the progeny, thus produced, consists of females only. Each female may produce up to 116 young ones in her lifetime at the rate of 1-4 nymphs per day in March- April, 1-5 in May-July, 1-6 in August, and only 1-2 per week in winter. There are four nymphal instars and the duration of each varies according to the season. The total duration of the nymphal period is 35-42 days in February, 29.5 days in August-November, and 10.5-19.5 days in April-July. The winged forms appear in July-September when fresh colonies on new plants or branches are initiated. There may be 13 generations in a year. In December, there is a partial migration from aerial parts to the roots of infested plants and the reverse migration from the roots to the aerial parts takes place in May.

In Himachal Pradesh, the woolly aphid has been controlled by using natural methods by an introduced parasite, *Aphelinus mali* (Haldeman) (Aphelinidae) which attacks the 4th and 5th stage nymphs. The parasite has to be maintained on cut twigs fixed in moist soil inside wire-gauze cages.

Damage. The aphids suck cell-sap from the bark of the twigs and from the roots underground. Swellings or knots appear on the roots which hinder the normal plant functions. Owing to the loss of cell-sap, the twigs also shrivel and the young nursery plants, which are affected the worst, may die quickly.

Management. The management of woolly aphid in apple orchards is a difficult task because of the waxy covering on the body and reinfestation due to local migration from roots. However, good success can be achieved if following strategies are followed : (*i*) Use of resistant root stock seems to be the only answer for subterranean aphid population. Use resistant root stock like Golden Delicius, Northern Spy and Morton Stocks 778, 779, 789 and 793. (*ii*) Biological control of woolly aphid has great potential and it can be achieved by the release of an exotic parasitoid, *A. mali*. (*iii*) As the main source of infestation in the orchard is infested plant material from nurseries, select healthy plants from nursery and then before planting in the orchard, treat them with chlorpyriphos 0.05 per cent. (*iv*) During leaf-fall, spray the plants with 2.0 litres of chlorpyriphos 20EC in 1250 litres of water per ha against aerial forms. For controlling the root forms, apply oxydemeton methyl 25EC in 1250 litres of water per ha. These chemicals should be sprayed during winter months. (*v*) For checking this pest during summer, spray 2.0 litres of malathion 50EC in 1250 litres of water per ha. The insecticidal spray should be avoided where the parasitoid, *A. mali,* is present. (*vi*) Carry out fumigation against root forms with paradichlorobenzene granules in a 15 cm deep trench dug round the infested tree, about 2 metres from it.

3. Peach Leafcurl Aphid, *Brachycaudus helichrysi* (Kaltenbach) (Hemiptera : Aphididae) (Fig. 19.3)

This world wide aphid is a very destructive pest of peach, plum, almond (*Amygdalus communis* L.) and other temperate fruits. It is prevalent both in the plains and in the mountainous areas of India, up to an altitude of 2,200 metres. In spring and early summer, the infested plants show curling of the new leaves. The development of fruits is very slow and due to lack of nutrition, they fall prematurely. This damage is caused by nymphs and females which are confined to the growing shoots and leaves, from where they suck the cell-sap. This plant louse is very small and is generally yellow, with dark stripes on the head. Its colour varies according to the host plant. For example, on peach, it is light green and on golden-rod it is greenish yellow.

Life-cycle. This pest is active from February to March on temperate fruits and from June to October on golden-rod. During the winter, it is found only in the egg-stage at the base of the buds. With the flow of cell-sap during the spring, the eggs hatch and the nymphs move on to the

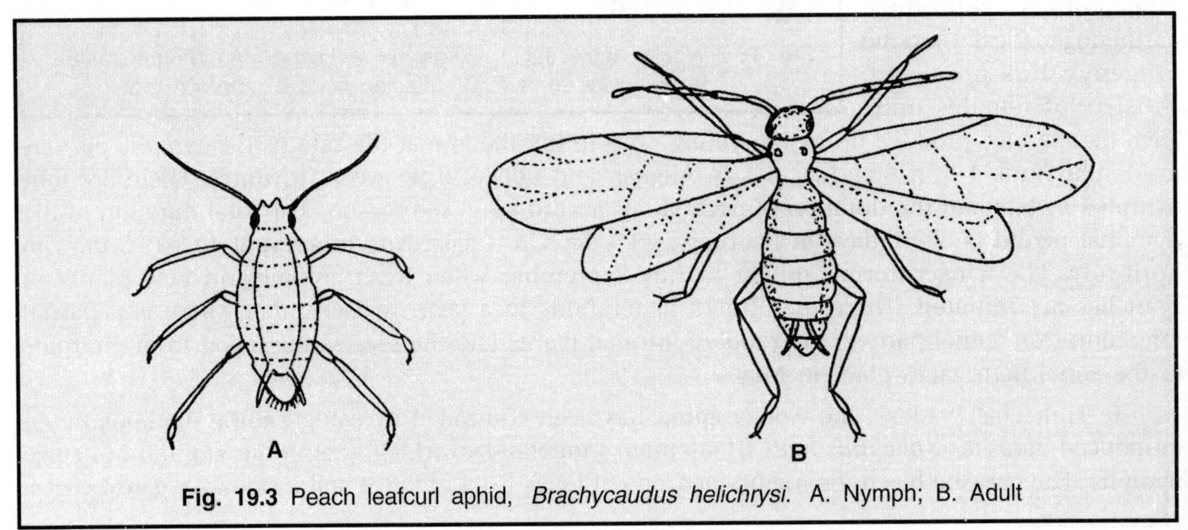

Fig. 19.3 Peach leafcurl aphid, *Brachycaudus helichrysi.* A. Nymph; B. Adult

primordial leaves where they start sucking the sap. In about four weeks time, the nymphs are full-fed and change into non-winged adults. There are no males. The females produce eggs without fertilization and these hatch inside the body of the mother. Thus, the females give birth to young ones instead of laying eggs. Each viviparous female produces about 50 young aphids in her short lifetime of about 13 days. Three or four generations are completed on the fruit plants. With the warming up of the season, winged males and females are also produced. They migrate to other alternative host plants, such as golden-rod, and again start reproducing asexually, as described above. Four or five generations are completed on golden-rod from June to October. Early in November, the winged females are produced again. They migrate back to peach, plum and other fruit-trees. The old foliage has been shed by that time, therefore, the females lay eggs at the base of the buds. Egg-laying is completed by the middle of December, when the females die.

In the plains, another species, *Myzus persicae* (Sulzer) **(Plates 19.2D)**, has been recorded as a serious pest of peach trees.

Damage. The nymphs and females suck the sap from young leaves and cause them to curl. The severely attacked plants have only small crumpled or twisted leaves and bear very little fruit. The curling of the peach leaves is also caused by the fungus, *Taphrina deformans* Berk & Tul, in which a curled leaf becomes thickened near the midrib and is brittle and lustrous.

Management. (*i*) For effective management of aphid, removal of weeds which act as secondary hosts is essential. (*ii*) Spray 2.0 litres of malathion 50EC or dimethoate 30EC in 1250 litres of water per ha. One pre-bloom and a post-bloom spray at 10-day interval should be given.

FRUGIVOROUS PESTS

These pests feed on the fruits or seeds and are often difficult to control.

4. Codling Moth, *Cydia pomonella* (Linnaeus) (Lepidoptera : Tortricidae) (Fig. 19.4)

The larvae of the codling moth cause the heaviest damage and are probably the most notorious of all the apple pests. It is widely distributed throughout Europe, North America, Australia, Baluchistan and Ladakh. In addition to apple, the pear fruits sometimes suffer rather severely from the attack of the larvae. The fruits of quince, walnut and of many wild species of *Pyrus* may also be damaged.

The full-grown larvae are 16-22 mm long and are pinkish or creamy-white in colour with a brown head. They have eight pairs of legs. The adult moth is small, about 12-14 mm in wing span and is 6-8 mm long. The fore wings are dark greyish and are marked with wavy lines and a copper coloured metallic eye like circle towards the outer margin **(Fig. 19.4)**. The hind wings are pale grey.

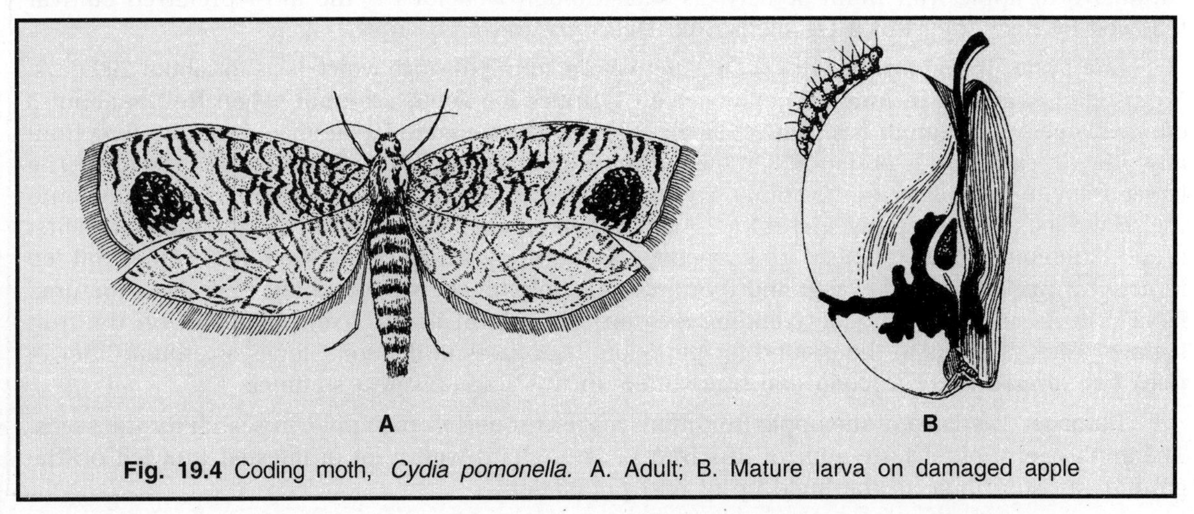

Fig. 19.4 Coding moth, *Cydia pomonella.* A. Adult; B. Mature larva on damaged apple

Life-cycle. The adult moths appear from the end of May to June. They fly at dusk and spend the day at rest on the trunk, branches or leaves of the apple trees. Being cryptic in colour they are difficult to detect.

The females, after pairing, lay white coloured eggs singly on young fruits, leaves and the twigs. The eggs are small, 1 mm in diameter, oval and very flat, appearing, in fact, more like small shining scales than the eggs of an insect. A single female may lay about 100 eggs in her life time. After a period of 4-12 days the young larvae emerge from the eggs and enter the fruit usually through the calyx. Having penetrated into the apple fruit, the larva attacks the core and the flesh around, burrowing further to the centre and the tips, which may also be eaten. After a period of three weeks to one month, the larva is full-fed. At this stage the larva burrows its way out of the apple fruit and falls to the ground. It seeks shelter in cracks or crevices in the bark of the apple trees and having chosen a proper position, it spins a silken cocoon in which it transforms into a yellowish-brown pupa. The pupal period is completed in 8-14 days.

The second brood moths give rise to larvae that attack the apples in the same way as the first brood, though they usually enter the fruit through the side instead of through the calyx. Like the first brood they spin cocoons in which to pass the winter, pupating the following spring and producing moths in due course. In cooler regions, there is only one generation, but in warmer areas there are two generations in a year.

The Indian house-sparrow, *Passer domesticus indicus*, devours a great number of the hibernating larvae. The larvae are also parasitized by *Paralitomastix varicornis* Nees.

Damage. Damage is caused by the larvae which burrow into the fruit and feed on the pulp. The infested fruits lose their shape and fall off prematurely. These fruits cannot be marketed for human consumption.

Management. The control measures are same as in case of tent-caterpillar. Strict quarantine measures should be adopted to check the spread of this pest to other apple growing areas of the country.

5. Apple Fruit Moth, *Argyresthia conjugella* Zeller (Lepidoptera : Yponomeutidae)

It is a pest of common occurrence in Asian, European and American countries falling in the isotherm range of 0 to ± 10°C. In India, though it is present in Himachal Pradesh for over five decades in dry temperate region of Kinnaur within altitude of 2445-2900 m, its identity was authenticated in 1985. The possibility of its occurrence in high altitude of Kashmir can not be ruled out. Short statured trees of vigorous growth and bushy crown with profusely developed leaves are conducive to apple fruit moth activity, as such Golden Delicious is the most preferred cultivar followed by Red Gold, Royal Delicious, Red Delicious and crab apple.

Life cycle. It is a univoltine pest and overwinters in pupal stage which lasts for about 290 days. Moths start emerging in June and emergence continues for about a month, when fruit is about 3 cm in diameter. The moth lays eggs in small clusters. The neonate larvae may enter the fruit from any side but calyx end is preferred. On the entrance hole a crystallized white deposit is formed. The larvae mine in the gallery in the young developing fruit to reach the core and in core it enters into the seed. First, it feeds on perisperm for 2-4 days, then enters cotyledonous part and undergoes first moult. Subsequent 3 instars feed on cotyledons and occupy whole of the seed. It becomes full-fed in about 3 weeks inside the seed and then makes a hole on the testa and after 2-3 days, matured larva tunnels an outgoing gallery ending into an exit hole of about 1 mm diameter on the fruit surface. Then, it drops to the ground for pupation. It pupates under soil, stones or ground litter in thick two layered silken cocoon and overwinters in this stage till next summer.

Damage. The larva of the apple fruit moth makes tunnels in the apple, in search for the seeds. The apples get a bitter taste and rot in advance. An orchard with a lot of infested fruit left on the ground can contribute to the build up of pest population.

Management. Two sprays with 4.5 litres of chlorpyriphos 20EC or 2.5 litres of quinalphos 25EC or 2.0 litres of malathion 50EC in 1250 litres of water per ha are effective in suppression of the infestation. The first spray should be applied on first visible symptoms and the second fortnight later.

6. Peach Fruit Fly, *Bactrocera zonata* (Saunders) (Diptera : Tephritidae) (Fig. 19.5)

The peach fruit-fly is widely distributed in the Indian Sub-continent and feeds on a number of fruits, viz., peach, fig (*Ficus carica*), guava (*Psidium guajava* L.), *ber* (*Zizyphus* spp.), citrus, apple, cucurbits, tomato, brinjal and pear. This fruit-fly causes heavy damage to peach fruits. The damage is caused by the larvae only and the attack is characterized by the dark punctures, oozing of a fluid and rotting or dropping of fruits. The maggots found inside the fruits are dirty white, headless and legless. They are elongated wriggling creatures about 1 cm in length. The fly is a small insect about the size of a housefly and is reddish brown with yellowish cross bands on the abdomen **(Fig. 19.5)**. The wings are transparent and have a small brown spot on the tip of each wing.

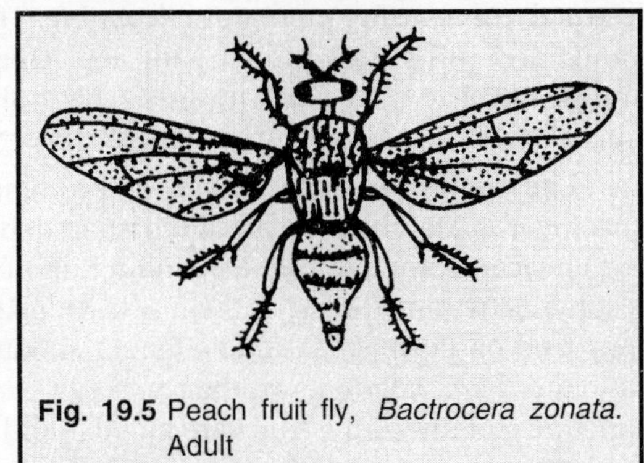

Fig. 19.5 Peach fruit fly, *Bactrocera zonata*. Adult

Life-cycle. This pest is active from March to November or December and passes winter in the pupal stage. The adult flies appear in March and breed in a number of fruits like *ber*, loquat, peach, mango, cucurbits, citrus and guava. When the fruits mature in various seasons, the flies are active and are noticed in the gardens. They mate a number of times and at 20 days of age, they lay white, cylindrical eggs just under the skin of a fruit, in groups of 2-9. A single female may lay, on an average, 137 eggs in her life-time. The place of oviposition is marked by a resinous secretion. The eggs hatch in 2-4 days in May-June and the maggots feed on the fruit pulp by making galleries. These larvae are full-fed in 4-16 days and, at the same time, the fruits also fall from the tree. The larvae crawl out of the rotting fruits and by jumping movements, reach a suitable place for pupation in the soil at a depth of 25.4-76.2 mm. The flies emerge from these pupae in about one week and the life-cycle is completed in 13-27 days. Many generations are completed in a year.

During spring and summer, the population builds up gradually and reaches a peak in the monsoon. Of all the fruit-flies, this species used to be the most abundant in various fruits, but now it is being replaced by the mango fruit fly (chapter 16). The latter is also displacing the Mediterranean fruit-fly in the Hawaiin islands and other places.

Damage. The peaches, apricots or other fruits attacked by this pest are malformed, misshaped, undersized and are rotten within. The damage caused is very heavy and the fruits become unmarketable.

Management. (*i*) Plant early maturing varieties such as Prabhat, Florida Prince, Earli Grande, Partap, Shan-e-Punjab and Flordasun. (*ii*) Hoe the orchard in May-June to expose the pupae which are present mostly at 4-6 cm depth. (*iii*) Harvest the ripening fruits and do not let the ripe fruit remain on the tree. (*iv*) Remove regularly the fallen fruits from around and bury the infested fruits at least 60 cm deep. (*v*) Fix methyl eugenol based PAU fruit fly traps **(Plate 18.2D)** @ 40 traps/ha in the first week of May in peach and first week of June in pear. (*vi*) Apply bait spray of yeast hydrolysate (250 g), crude sugar (2.5 kg) and malathion 50EC (250 ml) in 250 litres of water per ha. Spraying should be started two weeks before harvesting. (*vii*) In orchards, only with history of severe infestation, spray 3.125 litres of fenvalerate 20EC in 1250 litres of water per ha, in the end of June and repeat the spray at weekly intervals if required. Fruits should be harvested at least on third day after spray.

7. Walnut Weevil, *Alcidodes porrectirostris* Marshall (Coleoptera : Curculionidae)

This is the most destructive pest of English walnuts (*Juglans regia* L.) in the Himalyan region at altitudes of 1,000-2,450 metres and has been reported as a serious pest in Kumaon, Kulu and Kashmir. No other host of this insect is known.

The adults feed on buds and flowers but the grubs feed inside the fruits and are extremely destructive in causing premature dropping. The adult weevil is about 10 mm long, pitch black when young, and turns dark brown with age. There is a prominent snout in front which is somewhat straight and is directed downwards. The grub is legless, with a pale-brown head and measures 15 mm in length. Fruits with spots of dark brown or black resinous excretion harbour the grubs inside.

Life-cycle. This pest is active throughout the warm weather and passes the winter (from November to March) as an adult hiding in debris, under stones, under the bark of trees, and in cracks and crevices in the ground. Depending upon the altitude and the weather in a locality, the weevils become active in April and their activity generally coincides with the appearance of flower-buds. They feed on flowers, leaf-buds, tender shoots or young fruits. As the fruits are formed, the weevils puncture them, laying more than one egg. As many as 15 eggs have been noticed in one fruit. The punctures on the young fruits are circular and on older fruits, they are crescent or irregular. The eggs hatch in one week and the creamy white tiny grubs bore deeper, feeding on the kernels. The grub is full-grown in 13-22 days and pupates inside the fruit. The pupa is creamy white and is transformed into an adult in 9-17 days. The adults emerge from the fruits by biting roundish holes. They start the second generation and when weevils of this generation appear in September, the walnut fruits are almost mature. The adults feed on tender leaves and twigs up to the autumn and then hibernate. There are two generations in a year.

Damage. In the Kulu valley, the pest is most destructive at altitudes of 1,000-2,450 metres. The grubs feed exclusively on walnut kernel and reduce it to a useless black mass, but the weevils feed on green twigs, petioles, flowers and young fruits also. The grubs are far more destructive than the adults and as a result of their attack, more than 50-70 per cent of the fruits drop off and even those which remain on the trees, may be unmarketable. It is a common sight to see the ground under walnut trees completely covered with infested and fallen fruits.

Management. (*i*) Collect and destroy the fallen fruits in May-June to check the population build up of the pest. (*ii*) Spray 2.0 kg of carbaryl 50WP in 1250 litres of water per ha at fortnightly intervals when the young fruits are being formed.

XYLOPHAGOUS PESTS

These pests bore and eat into the woody portion of the trees.

8. Apple Stem Borer, *Apriona cinerea* Cheverlot (Coleoptera : Cerambycidae)

The grubs of this pest are destructive stem borers of apple, peach, fig and other fruit trees. The insect has been recorded from Pakistan, Afghanistan, Kashmir, Himachal Pradesh and Uttar Pradesh. The adult beetles are 35-50 mm long, grey in colour and have antennae larger than the body.

Life-cycle. The beetles appear in July-August after the start of the monsoon and mate soon after. The female when ready to lay eggs excavates an oval patch on a shoot and lays eggs inside the cavity. Thus, she moves from tree to tree and lays a number of eggs. The grubs emerge in 7-8 days and start feeding by boring inside the stem. As autumn approaches the feeding activity of the grubs becomes very slow and by October stops altogether. It remains quiscent during the winter and resumes feeding in March. As it feeds it makes a tunnel towards the thicker side of the stem and thus by September-October it reaches close to the tree trunk. As winter sets in the grubs remain quiscent. As the summer season comes it starts activity and when full-fed it pupates inside a tunnel made in the woody tissue. The pupal stage lasts more than a month and the adults emerge in July-August to start procreation. The life-cycle is completed in about two years.

Damage. As a result of feeding by the grubs, the affected trees may not die for many years, but their vitality and productivity is greatly impaired. The adult beetles feed on bark and have an unusual habit of cutting more than they acually consume.

Management. (*i*) Prune and burn all attacked shoots and branches during winter. (*ii*) Location of live holes and injection of carbon disulphide or chloroform or petrol and sealing them with mud will kill the pest. (*iii*) The borers can be effectively controlled by inserting cotton wick soaked in insecticide solution of about 10 ml of dimethoate 30EC or oxydemeton methyl 25EC or dichlorvos 100EC inside the hole and plugging it by moist soil or mud. (*iv*) Placing of 0.2 g aluminium phosphide tablets inside the hole and then plugging with moist soil will also ensure control of the pest.

9. Apple Root Borer, *Dorysthenes hugelii* Redtenbacher (Coleoptera : Cerambycidae) (Plate 19.1E)

The grubs of this beetle are very damaging root borers of apple in the Kumaon region of Himalayas. Occasionally, it also attacks the roots of apricot, cherry, peach, pear, walnut and a number of forest trees. The full-grown grub is creamy-white with black head and mandibles, and may be 75-100 mm long. The adult beetles are chestnut in colour with head and thorax darker than the elytra.

Life-cycle. Adult beetles emerge from the soil with the advent of monsoon, around the end of June and some of them continue to emerge upto the middle of July. Immediately after emergence, they start mating and the males die soon after. The females also live for 10-12 days only, during which time they lay eggs to progenerate. A female lays, on an average, 300 ovoid shaped, yellow-white eggs, 8-12 mm below the soil surface. The eggs hach in 30-40 days.

The grubs, on emergence from the eggs, go down into the soil, 100-250 mm deep and feed on organic matter and also on the roots of a number of plants and trees. The legless, slow moving creatures feed and move gradually, under the soil surface among the root system of host plants. It takes up to 3.5 years for them to mature and become full-grown, storing huge quantities of food reserves. If the grubs are confined without food they can survive for 24-90 days. When full-grown they pupate inside the earthen cocoons and then emerge as beetles after the first shower of monsoons.

Damage. As a result of the grubs feeding on roots, the small as well as the major roots are severed from the base. The trees, if young, die immediately whereas the older ones become weak and fall down eventually owing to the action of strong winds.

Management. (*i*) Avoid dry sandy soils for planting apple orchards. (*ii*) Remove and destroy the grubs from the affected trees. (*iii*) Inter-culturing in the soil under the trees helps in killing the grubs. (*iv*) Once the infestation has occurred, it is imperative to treat the tree basins with phorate granules @ 100 g a.i. per tree or apply 2% lindane dust @ 200-300g/tree soon after monsoon.

10. Peach Stem Borer, *Sphenoptera lafertei* Thompson (Coleoptera : Buprestidae)

This beetle is widely distribued in Afghanistan, Pakistan and India. Its grubs feed by boring into the stem of peach, almond, apricot, cherry, loquat, pear and plum trees. The beetles which feed on the foliage of host plants are blackish-bronze and are 10-13 mm long. The grubs are smoky dark or black, club-shaped and attain 18-24 mm body length.

Life-cycle. The beetles start appearing by the middle of March. After mating they lay small, spherical, white eggs singly, scattered all over the tree trunk and the main branches. The eggs hatch in 20 days. The grubs on emergence feed on the bark and as they grow they bore inside. The larval stage is completed in two months in summer, but those which over-winter take 6 months. The larva makes a small chamber in the woody tissue about 10 mm deep from the surface. The pre-pupal and pupal periods last 1-2 and 8-12 days, respectively in summer. The winter is passed in the grub stage. Three generations of this pest are completed in a year.

Damage. The grubs feed under the bark as well as bore deep into the wood. Gum globules ooze out of the entrance holes. Leaves of the attacked plants turn pale and their growth is arrested. The attacked branches dry up and do not bear fruits. With continued damage the tree dies ultimately.

Management. (*i*) The dead or heavily gummed branches should be cut and destroyed. (*ii*) Drench spray of 2.5 litres of chlorpyriphos 20EC in 1250 litres of water per ha during June after harvest of the crop and in October.

11. Cherry Stem Borer, *Aeolesthes holosericea* Fabricius (Coleoptera : Cerambycidae)

This polyphagous defoliating beetle is widely distributed in India, Sri Lanka, Bangladesh, Myanmar, Malaysia and Thailand. The grubs of the beetle are the most destructive pests of cherry and they also attack apple, guava, apricot, crab apple, mulberry, peach, pear, plum, walnut and other trees.

The adults are dark brown, 38-45 mm long, having short mottled yellowish pubescence on the elytra. Antennae of the male are 1.5 times their body length, while those of the female are about the same length as the body. The flabby grubs are yellowish in colour and are clothed with fine bristles. They feed inside woody portion of the host trees and when full-grown, they measure 70-80 mm long.

Life-cycle. The beetles emerge out of the tunnels in the host trees from May to October and are active throughout the summer. After mating, the females start laying eggs on the dry woody portions of the host trees in the cuts or in cracks and crevices of the bark. A female, on an average, lays about 100 eggs. The eggs are white, elliptical in shape and are about 2.5 mm in length. They hatch in 7-12 days and the young grubs on emergence feed first on the bark and, as they grow in size, they bore deeper inside the woody portions of a tree branch. In summer, they feed at a fast rate, throwing frass from the exit holes. As a result of their feeding and cutting into the wood, sap flow in that portion of the branch or the trunk becomes restricted. It gradually causes death of that portion.

In the winter time, the rate of feeding of the grubs is very slow and it is seen quite often that the full-fed grubs may just rest in the tunnels without feeding.

The larval period is completed in 27-32 months. The pre-pupal period varies from 3 to 150 days and pupation takes place either in October-November or in March-April. The pupal stage lasts from 40 to 100 days. The beetles that emerge from puparia formed in October, remain within the tunnels throughout the winter and the spring, while those beetles that emerge from the puparia formed in April, rest for only 6 weeks. Thus, the life-cycle is completed in three years.

Damage. The newly hatched grubs first feed on bark and make zig-zag galleries. They bore inside and feed on sap wood. As a result of their feeding and the consequential damage to the woody tissue, the vitality of the trees is reduced. Parts of the attacked tree start drying till it becomes unproductive. Within a few years, the tree is dead. The pest can be located from the frass that comes out of the holes in the branches or in the main trunk. There may be more than one such hole, indicating multiple damage.

Management. (*i*) Collect and destroy the grubs and beetles. (*ii*) Since the pest feeds inside the woody portions no insecticide spray can be effective. The only sure way to get rid of this pest is to locate the feeding holes, clear the passage and inject contact poisons or even diluted phenyle. Insert in the holes cotton wicks soaked in dichlorvos (0.1%) or dimethoate (0.03 %) or methyl demeton, phosphamidon or thiometon (all at 0.025%) and seal with mud.

12. Long-horned Walnut Beetle, *Batocera horsfieldi* Hope (Coleoptera : Cerambycidae)

The grubs of this beetle are a serious pest of walnut in Darjeeling, Kumaon hills, Kulu valley and Simla hills. The grubs are 90-150 mm long and pale yellow in colour. The beetles are 45-65 mm long, black in colour with fine ashy or yellow-grey pubescence. Their elytra have numerous shining black tubercles at the base and several rounded white marks extending up to the apex.

Life-cycle. The beetles emerge in June and July, and live for about 4 months during which period they mate and find suitable sites and lay eggs. The eggs are brown and oval, and are laid singly in the bark of the walnut tree. A female may lay 55-60 eggs in her life-time. The egg stage lasts 8-15 days. The gubs at first feed on the bark and then bore inside the wood, completing their development in 20-25 months. There they remain in the pre-pupal stage for 50-182 days and make puparia. Once it pupates the insect remains in that stage for 40-90 days. The life cycle is completed in 23-32 months which is spread over 2-3 winters.

It has been noticed that if the grub is already full-grown before the on-set of winter, it remains quiescent from October to March. The younger grubs which are less than one year old, however, continue to feed slowly during the winter also.

Damage. The young grubs feed on the inner side of the bark making zig-zag tunnels. Later on, the grubs bore down to the surface of sap wood and go even further into the centre of the wood. The attacked timber trees become useless except as fire wood, which hardly fetches the cost of cutting and marketing the wood.

Management. (*i*) Hand picking and mechanical destruction of grubs and adults checks the population of the beetle. (*ii*) Plug the live holes with cotton soaked in kerosene, petroleum or EDCT mixture and then plaster them from outside with mud.

FOLIAGE FEEDERS

These pests feed on foliage of fruit trees and cause considerable damage due to defoliation.

13. Tent Caterpillar, *Malacosoma indicum* Walker (Lepidoptera : Lasiocampidae) (Plate 19.1F)

The insect is an important pest of apple (*Malus pumila*) in north-western India, being more serious in the Simla Hills. It also damages the pear (*Pyrus communis* L.), apricot (*Prunus armeniaca* L.), walnut (*Juglans regia* L.), etc.

The caterpillars feed gregariously on foilage, leaving behind only the midrib and other harder veins. When full-grown, the larva is 40-45 mm long, with its black head and abdomen. The male moth is light reddish and the female moth is light brown, with a wing expanse of 29-32 mm and 35-37 mm in male and female, respectively.

Life-cycle. The pest is active from mid-March to May and passes the remaining 9 months of the year in the egg stage. The eggs hatch by about the middle of March when buds appear on the plants. The larvae live gregariously and, soon after emerging, each spins a silken nest at a convenient and sheltered place on the tree. As the caterpillars grow, the nest is also enlarged until it is 0.3 -0.5 metre across. During the day, the caterpillars rest in their nests and at night they feed on the leaves. The larval stage lasts 39-68 days and when full-fed, they spin oval, white and compact cocoons, each about 25 mm in length. The pupal stage lasting 8-22 days is passed inside these cocoons in some protected place. The moths begin to emerge some time in the 3rd week of May and continue to do so till the beginning of June. The pre-oviposition period lasts 1-3 days and the females lay eggs in broad bands around the branches. Each band may consist of 200-400 eggs. The moths are short-lived and a female in captivity may survive for 3-5 days. The life cycle is completed in one year.

Damage. In case of severe infestation, the entire plant may be defoliated and subsequently the caterpillars may feed even on the soft bark of twigs. When there is serious infestation, 40-50 per cent of the apple plants in an orchard may be defoliated producing a poor harvest.

Management. (*i*) The caterpillars can be killed by mopping up the 'tents' with a pole and some rags dipped in kerosene tied on its end. The best results are obtained when the operation is carried out from 12 noon to 3 p.m. on clear sunny days. Kerosenized water in an open vessel should be placed below the tree so that the larvae that fall may also be killed readily. (*ii*) Destroy all egg bands at the time of pruning in December-January. (*iii*) Spray 2.5 kg of carbaryl 50WP in 1250 litres of water per ha.

14. Apple Leaf Folder, *Archips termias* Meyrick (Lepidoptera: Tortricidae)

Apple leaf roller is a black headed grape-green caterpillar of a small buff coloured torticid moth, which is a polyphagous multivoltine pest with apple as its preferred host.

Life-cycle. The pest overwinters as larva in fallen leaves or cracks and crevices of trees for about 7 to 8 months. The overwintering larvae resume feeding when foliage and blossom appear in April and adult emergence occurs in May. They lay dorsoventrally flattened eggs in clusters of 35-180 on dorsal surface of the leaf and depressed fruit surface. Peak moth activity is observed in May, which is followed by high damage to fruit in June. Late maturing varieties suffer more from its attack. At least 2 generations are completed (with egg, larval and pupal stage of 1-2 weeks, 3-5 weeks and 1-3 weeks, respectively) up to September at about 2000-m altitude and third generation larvae overwinter.

Damage. It not only feeds on foliage after folding young leaves by silken threads, but also damages the fruit on the tree as well as in the storage by scrapping the skin and causes up^to 41 per cent fruit loss during storage.

Management. (*i*) *Bacillus thuringiensis* (Bt) is effective against the larval stage. Bt is only effective on small (<1.25 cm long) caterpillars and usually requires more than one application. (*ii*) Spray of 2.5 kg of carbaryl 50WP or 2.0 litres of malathion 50EC in 1250 litres of water per ha, three weeks prior to the harvest gives good control of the pest. Apply sprays only when there is evidence of a damaging population, such as large number of larvae early in the spring or large number of egg masses.

15. Indian Gypsy Moth, *Lymantria obfuscata* Walker (Lepidoptera : Lymantriidae)

The caterpillars of the Indian gypsy moth defoliate apricot, apple and walnut trees in the Western Himalayas. It is primarily a pest of forest trees and attacks willows (*Salix babylonica* Linn.) and *S. sataz* (Peema) and poplar (*Populus* spp). The caterpillars are 40-50 mm long and are clothed in tufts of hair. The female moths are dark grey and they have atrophied wings. The males are comparatively more active in moving around and in mating.

Life-cycle. After being fertilized by a winged male, the female settles on the bark of a host tree for oviposition. It lays, during June-July, round, shining and light greyish brown eggs, in batches of 200-400. The eggs are laid under the loose bark and are covered over with yellowish-brown hairs. The eggs over-winter as such and hatch in March-April when the season warms up a little. The young larvae feed gregariously and as they grow through five instars they become more mobile but remain together in large groups. The larval period is completed in 66-100 days. The pupal formation takes place in soil among debris and this stage lasts 9-21 days from May to mid-July. The male moths live for 4-10 days and the females for 11-31 days. One generation is completed in a year.

Damage. The caterpillars are gregarious and they eat voraciously at night time. Their habit to defoliate the host trees completely results in the failure of fruit formation.

Management. Same as in case of tent-caterpillar.

16. Almond Weevil, *Myllocerus lactivirens* Marshal (Coleoptera : Curculionidae)

The almond weevil is polyphagous and besides almond, it has been reported to feed on apple, apricot, *ber*, citrus, falsa, loquat, mango, peach, plum and pomegranate. Almond and pear are its preferred hosts. The weevils are small 3-4 mm long and pale metallic green in colour. The full-grown grubs are creamy white, 4 mm long, are stout in body which is without legs. They have short erect setae which help them in locomotion.

Life-cycle. The adults appear in May and lay eggs from the end of July till the beginning of September in the soil in batches of 40-50 each. The eggs are broadly oval in shape, creamy-yellow, smooth, transparent and shiny. They hatch in 4-5 days and the grubs burrow deep in the soil up to 200-300 mm. They feed on roots of the host plants. When full-grown, they come up on soil surface to pupate in the upper 25 mm of the soil. The grub stage is completed in 10 months and the pupal period lasts 4-5 days. The pest passes winter in the pupal stage in the soil.

Damage. The weevils congregate on ventral surface on leaves, nibble irregular holes and gradually eat away the entire leaf laminae, leaving only the mid-ribs. The tender leaves are eaten first and then the older leaves are also skeletonized. The pest causes the maximum damage to foliage during the rainy season.

Management. Soil application of lindane 2 per cent dust @ 30 kg per ha will effectively control grubs.

17. Defoliating Beetles

In north-western parts of the Indian Sub-continent, there are many species of nocturnal beetles that defoliate a large number of plants in the summer. Apart from the forest nurseries and ornamental plants which they damage, the pome fruit-trees also fall prey to the ravages of beetles. The damaged leaves present a characteristic sieve-like appearance on account of the shot- holes. Several species have been recorded in the western Himalayan and adjoining regions (Table 19.1).

Table 19.1 Common species of defoliating beetles in the western Himalayas

S.No.	Common name	Scientific name
1.	Apple beetle	*Brahmina coriacea* (Hope) (Coleoptera : Scarabaeidae)
2.	Apple blackish green beetle	*Anomala rufiventris* Redtenbacker (Coleoptera : Scarabaeidae)
3.	Cherry gold lustre beetle	*Anomala flavipes* Arrow (Coleoptera : Scarabaeidae)
4.	Plum beetle	*Anomala lineatepennis* Blanchard (Coleoptera : Scarabaeidae)
5.	Grape green beetle	*Anomala dimidiata* (Hope) (Coleoptera : Scarabaeidae)
6.	Almond chrysomelid	*Mimastra cyanura* (Hope) (Coleoptera : Scarabaeidae)
7.	Apple weevil	*Myllocerus discolor* Boheman (Coleoptera : Curculionidae)
8.	Broom apple beetle	*Mylabris mecilenta* Linnaeus (Coleoptera : Meloidae)

Life-cycle. Not much is known about the life-histories of these insects. Presumably, they feed in the grub stage on humus, roots of grasses, weeds and other forest plants. The life-cycles are completed in one or sometimes more than one year. The adult beetles are known to defoliate fruit trees, causing serious damage to the young flowers. The damage is most severe on the marginal trees of an orchard. The newly planted trees close to forest areas are also more severely damaged.

Management. Spray 2.5 kg of carbaryl 50WP in 1250 litres of water per ha as soon as the damage is noticed. The spray should be done in the evening and repeated after 7 days if the damage continues or if the insecticidal deposit is washed away by rain.

Other Pests of Temperate Fruits

Some other insects which appear as minor pests of temperate fruits include the aphids; green aphid, *Aphis pomi* DeGeer; melon aphid, *Aphis gossypii* Glover; peach mealy aphid, *Hyalopterus pruni* (Geoffroy); peach black aphid, *Pterochlorus persicae* Cholodkovisky; the cherry aphid, *Myzus cerasi* (Fabricius) and peach mealy aphid, *Hyalopterus pruni* (Geoffryoy) : and walnut aphids, *Chromaphis juglandicola* (Kaltenbach) and *Callaphis juglandis* (Goeze) (Hemiptera: Aphididae) ; the gypsy moth, *Lymantria dispar* (Linnaeus) ; hairy caterpillars, *Euproctis* spp. (Lepidoptera : Lymantriidae) ; the plum caseworm, *Cremastopsyche pendula* Joann (Lepidoptera: Psychidae) ; apple leaf-miner, *Gracillaria zachrysa* (Meyrick) (Lepidoptera : Gracillariidae); fruit sucking moths, *Caliptera ophideroids* Guenee and *Othreis fullonia* (Clerck) (Lepidoptera: Noctuidae); peach twig borer, *Anarsia lineatella* Zeller (Lepidoptera: Gelechiidae); the flat-headed apple tree-borer, *Chrysobothris femorata* (Olivier) (Coleoptera : Buprestidae); the apple root-borer, *Lophosternus hugelii* Redtenback (Coleoptera : Cerambycidae) and *Lachnosterna* sp. (Coleoptera : Scarabaeidae) which attack apple roots; the apricot chalcid, *Eurytoma samsonovi* Vasiljev (Hymenoptera : Eurytomidae).

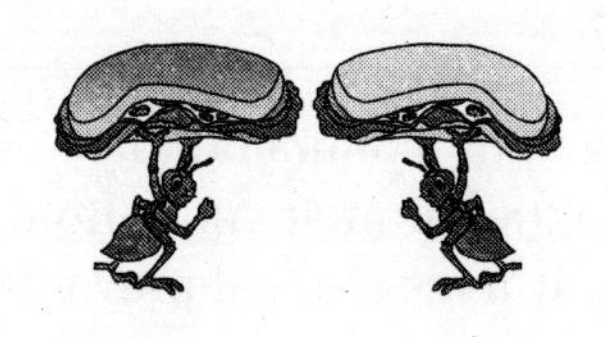

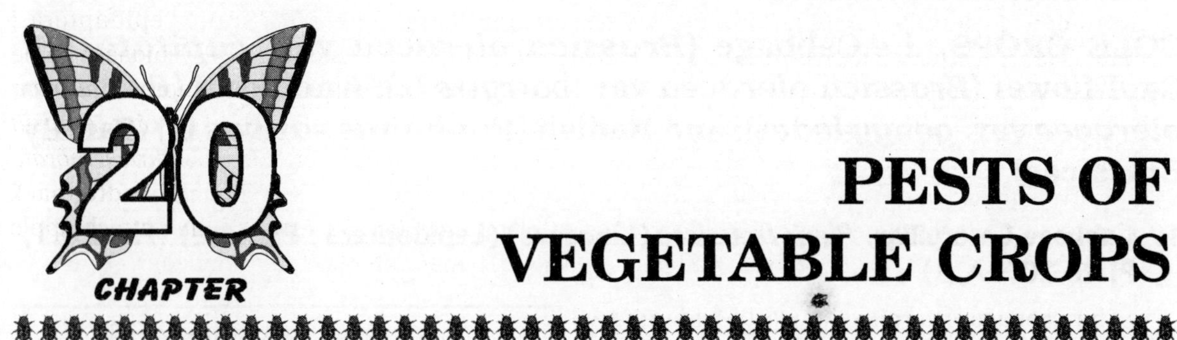

PESTS OF VEGETABLE CROPS

INTRODUCTION

The importance of vegetables as protective food and as supplier of adequate quantities of vitamins, proteins, carbohydrates and minerals is well known. The per capita annual consumption of vegetables in India is only 230 kg as against 300 g recommended dictary allowance. The area under vegetables in India increased from 6 million ha in 2011-02 to 9.20 million ha during 2012-13. The total production of vegetables in India during 2012-13 was 162.19 million tonnes. India is the second largest producer of vegetables after China and is a leader in the production of peas and okra. India occupies the second position in the production of brinjal, cabbage, cauliflower, onion, tomato and potato. India is growing about 14 per cent of the world's vegetables and would have to produce 225 million tonnes to meet the requirements of the expanding population.

Among the winter vegetables, the *Brassica* crops are the most important, the acreage under cauliflower (*Brassica oleracea* var. *botrytis* L.) and cabbage (*Brassica oleracea* var. *capitata* L.) being much larger than that under others. Cabbage is grown over an area of 0.37 million ha with a production of 8.53 million tonnes. The area under cauliflower is 0.40 million ha and production is 7.89 million tonnes. Onions are grown over an area of 1.05 million ha with a production of 16.81 million tonnes. The area under peas is 0.42 million ha with a production of 4.01 million tonnes. These crops are attacked by a number of pests of palearctic origin. Therefore, they are active during the winter only. It has been noticed that in recent years the diamondback moth is gaining importance by becoming a serious pest of the early sown cauliflower in September. Other vegetables belonging to the Crucifereae, grown in the winter are turnip (*Brassica rapa* L.) and radish (*Raphanus sativus* L.). These vegetables suffer heavy losses late in the winter when the aphids breed very fast.

Among the spring vegetables, the most important are potato, brinjal (*Solanum melongena*) and tomato (*Lycopersicon esculentum* Mill.), all of which belong to family Solanaceae. Okra is also important summer vegetable belonging to the family Malvaceae. Potato is grown over an area of 1.99 million ha with a production of 45.34 million tonnes. The area under brinjal is 0.72 million ha and production is 13.44 million tonnes. Tomato is grown over an area of 0.88 million ha with a production of 18.23 million tonnes. The area under okra is 0.53 million ha and production is 6.35 million tonnes. They are attacked by leafhoppers, leaf-eating beetles and lepidopteran tissue borers, the last mentioned group being the most difficult to control.

The cucurbits share about 5.6 per cent of the total vegetable production in India. The cucurbits, as a group, are attacked the most in their early stages by the red pumpkin beetle, the attack of which can be so heavy that the crop might have to be resown. Cucurbits and egg-plant (brinjal) are also heavily attacked by a number of mites, and their population particularly builds up in those fields which are sprayed with chlorinated hydrocarbons.

I. WINTER VEGETABLES

COLE CROPS, *i.e.*Cabbage (*Brassica oleracea* var. *capitata* L.); Cauliflower (*Brassica oleracea* var. *botrytis* L.); Knol-khol (*Brassica oleracea* var. *gongy-lodes*); and Radish (*Raphanus sativus* L.) (Family Brassicae)

1. Cabbage Caterpillar, *Pieris brassicae* (Linnaeus) (Lepidoptera : Pieridae) (Fig. 20.1, Plate 20.1A,B)

The cabbage caterpillar is world-wide and is found wherever cruciferous vegetables are grown. It is a serious pest of cabbage, cauliflower, knol-khol and it may also attack turnip, radish, *sarson, toria* (*Brassica campestris*) and other cruciferous plants.

Damage is caused by caterpillars only. When full-grown, they measure 40-50 mm in length (**Fig. 20.1**). The young larvae are pale yellow, and become greenish-yellow later on. The head is black and the dorsum is marked with black spots. The body is decorated with short hair. The butterflies are pale white and have a smoky shade on the dorsal side of the body. The wings are pale white, with a black patch on the apical angle of each fore wing and a black spot on the costal margin of each hind wing. The females measure 6.5 cm across the spread wings and have two conspicuous black circular dots on the dorsal side of each fore wing. Males are smaller than the females and have black spots on the underside of each fore wing

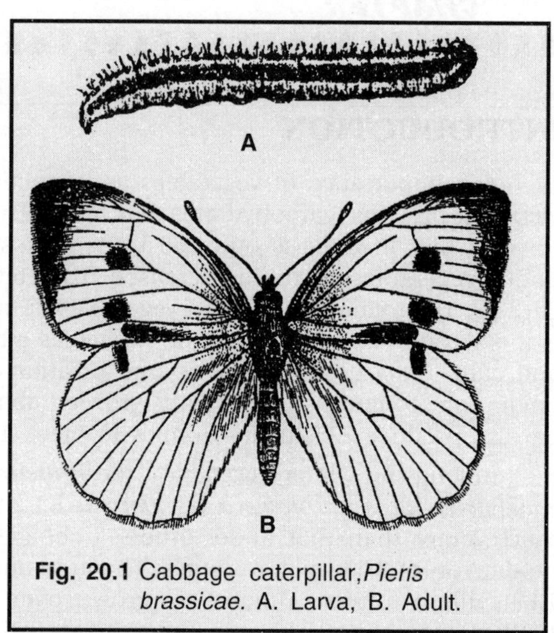

Fig. 20.1 Cabbage caterpillar, *Pieris brassicae*. A. Larva, B. Adult.

Life-cycle. In the Indo-Gangetic plains, this pest appears on cruciferous vegetables at the beginning of October and remains active up to the end of April. From May to September, the pest is not found in the plains but breeding takes place in the mountains. The butterflies are very active in the field and lay, on an average, 164 yellowish conical eggs in clusters of 50-90 on the upper or the lower side of a leaf. The eggs hatch in 11-17 days in November-February and 3-7 days in March-May. The caterpillars feed gregariously during the early instars and disperse as they approach maturity. They pass through five stages and are full-fed in 15-22 days during March-April and 30-40 days during November-February. The larvae pupate at some distance from the food plants, often in barns or on trees. The pupal stage lasts 7.7-14.4 days in March-April and 20-28 days in November–February. The butterflies live for 2.5-12.5 days and the pest breeds four times during October-April.

The larvae of this insect are parasitized by *Apanteles glomeratus* (Linnaeus) (Braconidae) in the natural populations.

Damage. The caterpillars alone cause damage. The first instar caterpillars just scrape the leaf surface, whereas the subsequent instars eat up leaves from the margins inwards, leaving intact the main veins. Often, entire plants are eaten up.

Management. (*i*) Handpicking and mechanical destruction of caterpillars during early stage of attack can reduce infestation. (*ii*) Spray one litre of malathion 50EC in 250 litres of water per ha. Repeat spraying at 10-day intervals if necessary. Do not spray the crop at least one week before the harvest.

2. Diamondback Moth, *Plutella xylostella* (Linnaeus) (Lepidoptera : Yponomeutidae) (Fig. 20.2, Plate 20.1C,D)

This world-wide moth is a serious pest of cauliflower and cabbage, but also feeds on many other cruciferous, solanaceous and liliaceous plants, all over India.

Damage is caused by the caterpillars which, in the earlier stages, feed in mines on the lower side of cabbage leaves and, in the later stages, feed exposed on the leaves. When full-grown, the larvae measure about 8 mm in length and are pale yellowish green with fine black hair scattered all over the body. The moths measure about 8-12 mm in length and are brown or grey, with conspicuous white spots on the fore wings, which appear like diamond patterns when the wings lie flat over the body (**Fig. 20.2**).

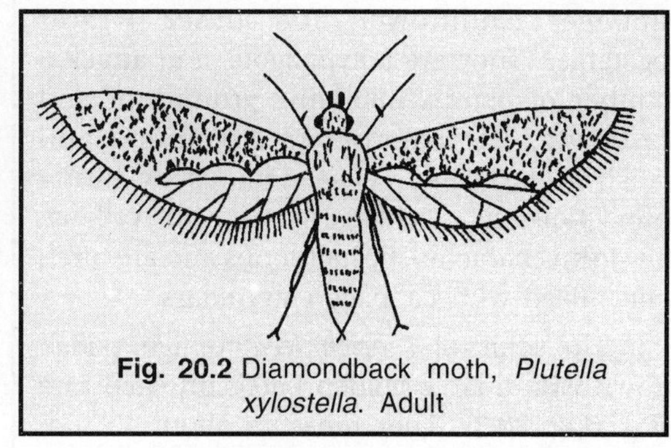
Fig. 20.2 Diamondback moth, *Plutella xylostella*. Adult

Life-cycle. This insect is active throughout the year. Yellowish eggs of the size of pin-heads are laid singly or in batches of 2-40 on the underside of leaves. A female may lay 18-356 eggs in her life-time. The eggs hatch in 9.1, 3.7 and 2.5 days at 13.9, 17.9 and 23.4° C, respectively. The newly hatched caterpillars bore into the tissue from the underside of leaves and feed in these tunnels. At first, their presence is detected only from the blackish excreta that appears at the mouth of each tunnel, but in the second instar, the mines become more prominent. In the third instar, the caterpillars usually feed outside the tunnels. The larvae of the fourth instar feed from the underside of leaves, leaving intact a parchment-like transparent cuticular layer on the dorsal surface. The larvae are very sensitive to touch, wind or other physical disturbances and readily feign death. They become full-grown in 16.6, 14.3 and 8.6 days at 17, 20 and 25°C, respectively. Before pupating, the larva constructs a barrel-shaped silken cocoon which is open at both ends and is attached to the leaf surface. The pupal stage lasts 4-5 days at 17-25°C and the moths may live for as long as 20 days. The life-cycle is completed in 15-18 days during September-October and there are several generations in a year.

Larval stage is parasitized by *Voria ruratis* Fab., *Itoplectis* sp. (Ichneumonidae), *Apanteles sicarius* (Braconidae) and *Tetrasticus sokolowskii* (Eulophidae). *Brachymeria excarinata* Gahan (Chalcididae) is associated with pupa. A larval-pupal parasitoid, *Diadrumus collaris* (Gravenhorst) (Ichneumonidae) is also found to be associated with this pest. *Apanteles plutellae* Kurdyumov effectively checks the population of diamondback moth on cabbage in Gujarat and Karnataka under favourable environmental conditions. In the hill regions of Tamil Nadu, *Diadegma semiclausum* Hellen exercises 70 per cent parasitization of diamondback moth in winter months.

Damage. Caterpillars damage the leaves of cauliflower, cabbage and rape-seed (*Brassica napus*), particularly in the heart of the first two. Central leaves of cabbage or cauliflower may be riddled (Plate 20.1D) and the vegetables rendered unfit for human consumption. The pest is most serious when it appears on the early crop in August-September.

Management. (*i*) Remove and destroy all the remnants, stubble, debris, etc. after the harvest of the crop and plough the fields. (*ii*) Tomato, when intercropped with cabbage, inhibits or reduces egg laying by diamond-back moth. (*iii*) Indian mustard, which attracts 80-90 per cent diamondback moths for colonisation, can be used as a trap crop. (*iv*) Spray 750 ml of Dipel 8L or 750 g of Halt WP (*Bacillus thuringiensis* var. *kurstaki*) in 200-250 litres of water in the evenings at one week interval. (*v*) Spray 625 ml of spinosad 2.5SC or 175 g emamectin benzoate 0.58G or 325 ml indoxacarb 15.8EC or 500 g cartap hydrochloride 50SP or quinalphos 35EC or 250 ml of fenvalerate 20EC in 250 litres of water per ha. Repeat sprays after 10 days, if necessary.

3. Cabbage Semilooper, *Thysanoplusia orichalcea* (Fabricius) (Fig. 20.3) and *Autographa nigrisigna* (Walker) (Lepidoptera : Noctuidae)

These two species are widely distributed in north-western India and are minor pests of cabbage, cauliflower and other winter vegetables. They are polyphagous and attack a number of plants, including groundnut and sunflower. The caterpillars are plump and palish green. They cause damage by biting round holes into cabbage leaves. On walking, they form characteristic half-loops and are often seen mixed with cabbage caterpillars.

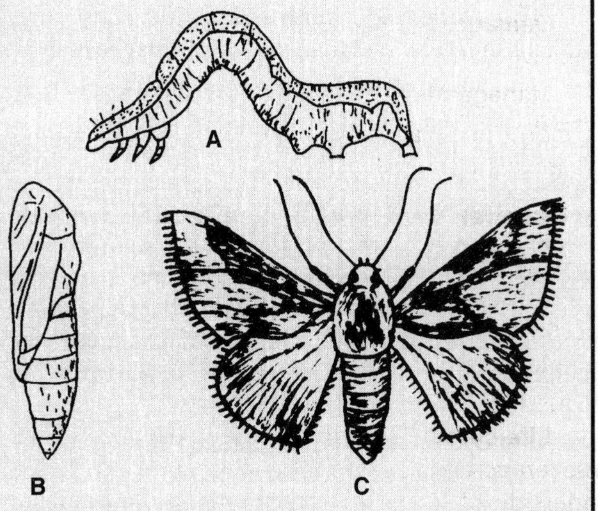

Fig. 20.3 Cabbage semilooper, *Thysanoplusia orichalcea*. A. Larva; B. Pupa; C. Adult

The adults of *T. orichalcea* are light palish brown with a large golden patch on each fore wing **(Fig. 20.3)**. They measure about 42 mm across the spread wings. The adults of *A. nigrisigna* are darker and have dark-brown and dirty-white patches on the fore wings.

Life-cycle. These insects are active during the winter and it is not known how they survive the heat of summer. During the active period, they lay eggs on leaves of host plants and the caterpillars feed individually, biting holes of varying size according to the stage of their development. When full-grown, they pupate in the debris lying on the ground. The moths are very active at dusk on flowers in gardens and public parks, where they are seen in hundreds during the spring season.

Management. Same as in case of cabbage caterpillar.

4. Tobacco Caterpillar, *Spodoptera litura* (Fabricius) (Lepidoptera : Noctuidae) (Fig 20.4)

The tobacco caterpillar is found throughout the tropical and sub-tropical parts of the world. It is widespread in India and besides tobacco (*Nicotiana tabacum* L.), feeds on castor (*Ricinus communis* L.), groundnut (*Arachis hypogaea* L.), tomato, sunflower, cabbage and various other cruciferous crops.

The damage is done only by the caterpillars, which measure 35-40 mm in length at maturity **(Fig. 20.4)**. They are velvety black with yellowish-green dorsal stripes and lateral white bands. The moths are about 22 mm long and measure 40 mm across the spread wings. The fore wings have beautiful golden and greyish brown patterns.

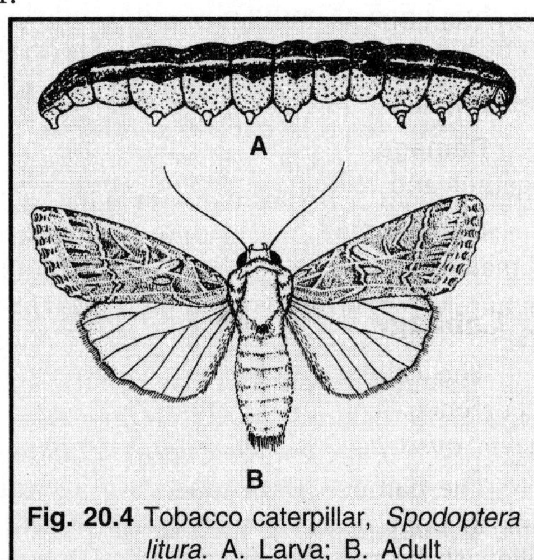

Fig. 20.4 Tobacco caterpillar, *Spodoptera litura*. A. Larva; B. Adult

Life-cycle. This pest breeds throughout the year, although its development is considerably retarded during winter. The moths are active at night when they mate and the female lays about 300 eggs in clusters. These clusters are covered over by brown hair and they hatch in about 3-5 days. The larvae feed gregariously for the first few days and then disperse to feed individually. They pass through 6 stages and are full-fed in 15-30 days. The full-grown larvae enter the soil where they pupate. The pupal stage lasts 7-15 days and the moths, on emergence, live for 7-10 days. The life cycle is completed in 32-60 days and the pest completes eight generations in a year.

The natural enemies, *Compoletis* sp., *Eriborus* sp., *Rogas* sp. (Braconidae) and *Strobliomyia orbata* W. (Anthomyiidae) are associated with larvae of this pest.

Damage. The larvae feed on leaves and fresh growth. They are mostly active at night and cause extensive damage, particularly in tobacco nurseries.

Management. (*i*) Remove the egg masses and clusters of larvae and destroy them.

(*ii*) Chemical sprays recommended against diamondback moth will also work against this pest, but in this case spot application may be enough if the attack is not widespread in the field.

5. Crucifer Leaf Webber, *Crocidolomia binotalis* Zeller (Lepidoptera : Pyralidae) (Fig. 20.5)

It is a serious pest of cabbage, radish, mustard and other cruciferous plants in India, Myanmar and Sri Lanka. The larvae web the leaves together and feed on them from the lower surface, often completely skeletonising them. The larva is green with a red head and it has longitudinal red stripes on the body. It is 2 cm in length.

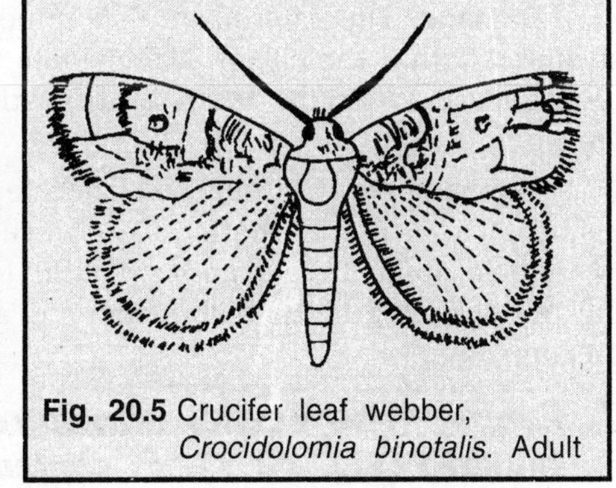

Fig. 20.5 Crucifer leaf webber, *Crocidolomia binotalis*. Adult

Life-cycle. The activity of the pest coincides with the cruciferous crops. The moth (**Fig. 20.5**) lays eggs on the underside of leaves in masses of 40-100 each. The eggs hatch in 5-15 days.In the early stages the larvae feed gregariously on the leaf parenchyma. As they grow, they spread out and start webbing the leaves and feeding on them. The larval stage is completed in 24-27 days in the summer and about 50 days in winter. When full-grown, the larva descends to the ground and pupates in the soil after making an earthen cocoon. The adult moth emerges in 14-40 days and the life-cycle is completed in 43-82 days. More than one generation may be completed in the season.

The larvae are parasitized by *Microbracon mellus* Ram and *Apanteles crocidolmiae* Ahmed (Braconidae).

Damage. The caterpillars cause considerable damage to the crops by webbing the leaves together and feeding on them. They also feed on flower buds and bore into the pods.

Management. (*i*) Remove and destroy the webbed leaves with larvae within. (*ii*) Spray 940 ml of malathion 50 EC or 320 ml of dichlorvos 100 EC in 250-300 litres of water per ha.

6. Cabbage Borer, *Hellula undalis* (Fabricius) (Lepidoptera : Pyralidae) (Fig. 20.6)

The cabbage borer is one of the serious pests of cruciferous crops, having a world wide occurrence. The larvae attack cabbage, cauliflower, radish, knol-khol, beet-root and the weed, *Gynadropsis pentaphylla* (Capparidaceae).

The damage is casued by the caterpillar which is 12-25 mm long and creamy yellow with a pinkish tinge and has seven purplish brown longitudinal stripes. The adult moth is slender, pale yellowish-brown, having grey wavy lines on the fore wings. Its hind wings are pale dusky.

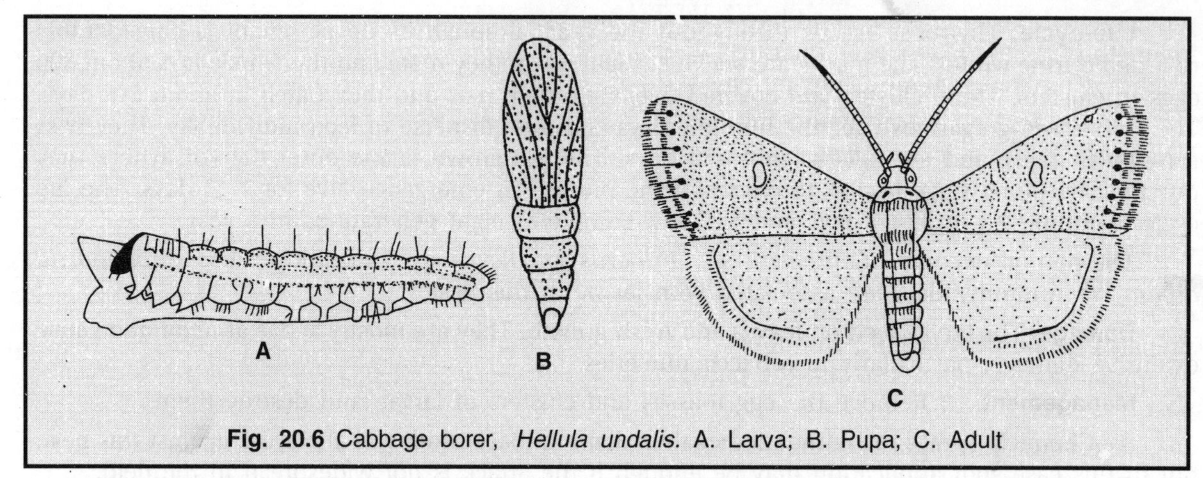

Fig. 20.6 Cabbage borer, *Hellula undalis.* A. Larva; B. Pupa; C. Adult

Life-cycle. This pest breeds throughout the year but comes to the notice in autumn when the cruciferous vegetables are sown. The female moths lay eggs singly but more often in clusters, on the under surface of the leaves or some other parts of the plant. The colour of the egg is pinkish and their shape is oval. The eggs hatch in 2-3 days. The caterpillars feed in the heart of the cabbage and become full-grown in 7-12 days, after undergoing four moultings. The full-grown caterpillar spins a cocoon among the leaves touching the ground or even inside larval burrows. The pupal period is about 6 days and the life cycle is completed in 15-25 days.

Damage. The caterpillars first mine into the leaves. Later on, they feed on the leaf surface, sheltered within the silken passages. As they grow bigger they bore into the heads of cauliflower and cabbage. When the attack is heavy, the plants are riddled with worms and outwardly the heads look deformed.

Management. (*i*) Collection and mechanical destruction of caterpillars in early stage of attack helps to check the infestation. (*ii*) Since the attack is mostly in the nursery and on young plants in the field, spray 375 g of carbaryl 50WP in 150 litres of water per ha. Repeat spray at 10-day interval, pick all flowering heads before spraying and observe 7 days waiting period after spray for the next picking.

7. Cabbage Flea Beetles, *Phyllotreta cruciferae* (Goeze) (Fig. 20.7), *P. chotanica* Duvivier, *P. birmanica* Harold, *P. oncera* Maulik and *P. downesi* Baly (Coleoptera : Chrysomelidae)

The cabbage flea beetles attack almost all the cruciferous plants in Europe, erstwhile USSR, North and South America, Australia, Japan and India. The common field crops like mustard, *raya, taramira, toria* and vegetables like radish, turnip, cabbage, cauliflower and knol-khol are severely damaged by adult beetles in West Bengal and Uttar Pradesh. Some winter flowering plants, namely dahlia, sweet sultan, antirrhinum and sweet peas also provide food for the beetles.

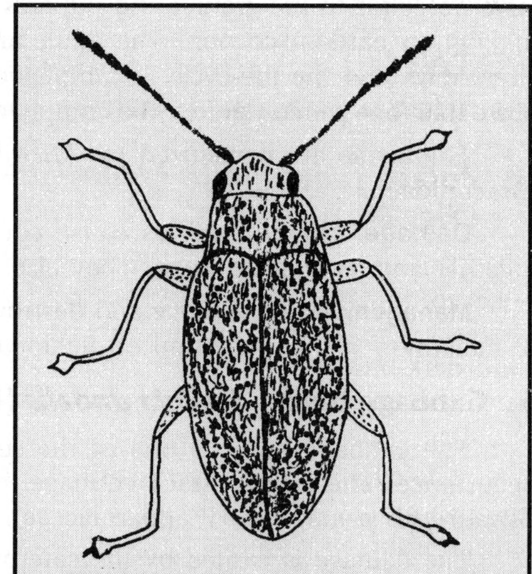

Fig. 20.7 Cabbage flea beetle, *Phyllotreta cruciferae.* Adult

The dorsum of the adult beetle, *P. cruciferae,* is metallic blue in colour, with a greenish hue (**Fig. 20.7**). The body is elongate narrow in front but broad distally. The beetle is round at the anal end. The head is finely punctate and the antennae extend beyond the middle of elytra. The female beetle measures 2.0 mm in length while the male beetle measures 1.8 mm. The larva is dirty white in colour with pale white head and measures about 5 mm in length.

Life-cycle. This pest is active throughout the year except during the winter months, when it hibernates in the soil or among plant debris. The over-wintered adult beetles emerge in the last week of February or in the beginning of March and settle on the growing cruciferous plants. The female beetle lays 50-80 creamy white eggs singly in the soil around the host plants, during the oviposition period of 25-30 days. The incubation period ranges between 5-10 days. The larva is very active and feeds on the tender roots of the host plant. It moults thrice during a total larval period of 9-15 days. The larva before entering into the pre-pupal stage wriggles out of the mined roots and prepares an earthern cell, 0.5 mm long, in the vicinity of the infested plants. There it pupates. The pre-pupal period is 2-4 days and the pupal stage lasts 8-14 days. There are 7-8 generations of this pest in a year.

The adults are parasitized by *Microctonus indicus* (Braconidae)

Damage. The adults mostly feed on the leaves by making innumerable round holes in the host plants. The stem, the flowers and even pods may also be attacked. The old, eaten away leaves dry up, while the young leaves are rendered unfit for consumption. A special kind of decaying odour is emitted by the cabbage plants attacked by this pest.

Management. Spray 2.5 kg of carbaryl 50WP in 250 litres of water per ha.

Other Pests of Cabbage, Cauliflower, Knol-khol and Radish

The minor pests of cruciferous vegetables include the painted bug, *Bagrada hilaris* (Burmeister) (Hemiptera: Pentatomidae); aphids, *Brevicoryne brassicae* (Linnaeus) **(Plate 20.1E)**; *Lipaphis erysimi* (Kaltenbach) and *Myzus persicae* (Sulzer) (Hemiptera : Aphididae); thrips, *Thrips tabaci* Lindeman and *Caliothrips indicus* (Bagnall) (Thysanoptera: Thripidae); Bihar hairy caterpillar, *Spilosoma obliqua* Walker (Lepidoptera: Arctiidae); cutworms, *Agrotis ipsilon* (Hufnagel), *Agrotis segetum* (Denis & Schiffermuller), *A. c-nigrum* (Linnaeus) and cabbage green looper, *Trichoplusia ni* (Hubner) (Lepidoptera : Noctuidae); pea leaf-miner, *Chromatomyia horticola* (Goureau) (Diptera : Agromyzidae); mustard saw fly, *Athalia lugens* (Klug) (Hymenoptera : Tenthredinidae); flea beetle, *Chaetocnema basalis* Bally (Coleoptera : Chrysomelidae) and mite, *Tetranychus neocaledonicus* Andre (Acari : Tetranychidae).

POTATO (*Solanum tuberosum* L.) and TOMATO (*Lycopersicon esculentum* Mill.) (Family : Solanaceae)

1. Potato Tuber Moth, *Phthorimaea operculella* (Zeller) (Lepidoptera : Gelechiidae) (Fig. 20.8)

This pest occurs especially in hot and dry climates. It is destructive to potato and also attacks tobacco, tomato, brinjal and solanaceous weeds. It is particularly serious on potato in Himachal Pradesh. The larvae cause the damage and are recognized as pinkish-white or geenish caterpillars, with dark-brown heads. They are about 20 mm in length. The adult is very small narrow-winged nocturnal moth, about 13 mm across the wings when spread **(Fig. 20.8)**. It is greyish brown with mottling of dark brown. Another species of tuber moth attacking potato is *P. heliopa* (Lower)

Life-cycle. If food is available and the climatic conditions are favourable, this pest may breed throughout the year. In cold weather, the life-cycle is much prolonged. Early in the spring, the moths escape from store-houses and start breeding in fields. A female, on an average, lays 150-200 eggs singly on the underside of leaves or on exposed tubers. The larvae first produce blotch mines on leaves but subsequently, they work their way into the stems. The larval stage in summer lasts 2-3 weeks and the mature larva pupates in a greyish silken dirt-covered cocoon, which is about 13 mm in length. The moths emerge in 7-10 days. They complete their life-cycle in about one month and there are usually 5-6 generations in a year.

Damage. In warm dry climates extensive damage may be done to the crop; but the potato tubers kept in cold stores escape damage. Later generations in the field infest the tubers also. At the time of digging, the moths may lay eggs on tubers. The larvae, on hatching, may work their way just under the skin and, later, may make tunnels through the flesh, causing damage to the tubers.

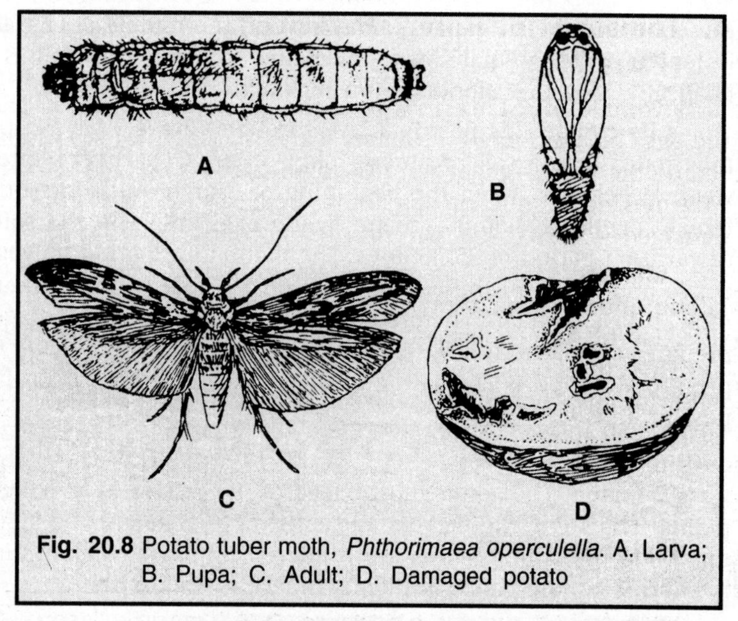

Fig. 20.8 Potato tuber moth, *Phthorimaea operculella*. A. Larva; B. Pupa; C. Adult; D. Damaged potato

Management. (*i*) All the infested tubers should be removed and destroyed. (*ii*) Cold storage of tubers also helps in suppressing the pest. (*iii*) The pest can be checked by spraying 2.5 kg of carbaryl 50 WP in 625 litres of water per ha or by dusting seed potato with 150 g of malathion 2 per cent per 100 kg. Repeat spraying 2-3 times at fortnightly intervals and do not dust the edible potatoes. (*iv*) If potato tubers in the stores get infested, fumigation with carbon disulphide @ 2-3 litres per 100 m^3 should be done.

2. Jassid, *Amrasca biguttula biguttula* Ishida (Hemiptera : Cicadellidae) (Chapter 16)

3. Aphid, *Aphis gossypii* Glover (Chapter 16) and *Myzus persicae* (Sulzer) (Hemiptera: Aphididae) (Chapter 15)

The leaves curl, turn pale bronze and dry up as a result of jassid and aphid attack. The crop is stunted and has blighted appearance. *M. persicae* appears late in the season. Besides sucking the cell sap, it transmits viruses and lowers the quality of the seed crop.

Management. (*i*) Spray the crop with 750 ml of dimethoate 30EC or oxydemeton methyl 25EC or 200-250 ml of phosphamidon 85SL in 250 litres of water/ha on the appearance of pest. Give another spray after 10 days. (*ii*) In case of the crop meant for seed, apply 12.5 kg of phorate 10G to the soil at the time of first earthing up.

4. Tobacco Caterpillar, *Spodoptera litura* (Fabricius) (Lepidoptera : Noctuidae)

Spray 875 ml of endosulfan 35EC in 200-250 litres of water/ha at 10 days intervals beginning soon after germination. The details of the pest have been discussed under 'Cole crops'.

5. Cutworms, *Agrotis ipsilon* (Hufnagel) and *Ochropleura flammatra* (Denis & Schiffermuller) (Lepidoptera : Noctuidae) (Chapter 26)

The cutworms cause considerable damage from February to March by cutting the young plants at the ground level and later on by making holes into the tubers.

Management. Drench the soil around the plant and the ridges with chlorpyriphos 20EC. @ 2.5 litres/ha in 1000 litres of water at the appearance of the pest in January to February.

6. Tomato Fruit Borer, *Helicoverpa armigera* (Hubner) (Lepidoptera: Noctuidae) (Plate 20.1 F)

Tomato fruit borer is one of the most destructive pests of tomato. The adults lay majority of the eggs on the upper and lower leaf surfaces of the first four leaves in the top canopy. The larvae scrape the tomato foliage until early or late second instar stage. Thereafter, the larva bores into the fruit making it unfit for marketing. In severe cases of infestation, more than 80 per cent fruits get damaged (Plate 20.1E). The detailed account of the pest has been given in Chapter 14.

Management. (*i*) Deep ploughing after harvesting the crop to expose the pupae for natural killing affords good protection. (*ii*) Hand picking of larvae in small area is also recommended. (*iii*) Use of African marigold (*Tagitus erecta*) as a trap crop is useful for control of fruit borer. (*iv*) Give three sprays at 2 week intervals starting from the initiation of flowers with any of the following insecticides using 250 litres of water/ha : 2.0 litres of malathion 50EC, 2.0 kg of carbaryl 50WP, 250 ml of fenvalerate 20EC, 500 ml of cypermethrin 10EC, 400 ml of deltamethrin 2.8EC, 1.5 litres of profenophos 50EC, 75 ml of flubedamide 480 SL.

7. Golden Cyst Nematode, *Globodera rostochiensis* Wollenweber (Tylenchida : Heteroderidae)

This nematode is wide-spread in Europe, the Americas, Algeria, Israel and Japan. It has recently been recorded in India in the Nilgiri Hills and near Simla. It is a notorious pest of the potato crop but is also found infesting tomato, brinjal, chillies and some solanaceous weeds. The infested plants, when pulled out, show clusters of white and golden spherical bodies attached to the roots and tubers, about the size of a sand grain. These are the cysts of mature female bodies and generally appear when the potato crop is 2-3 months old. If these cysts are crushed and examined under a microscope, 200-600 eggs will be released. The other eelworms attacking potatoes include *Meloidogyne incognita* and *M. javanica*.

Life-cycle. When host plants are not available, the eggs remain dormant in the cyst. In the presence of the crop, the plant roots secrete some sort of exudate which is believed to stimulate the hatching of eggs within cysts. The young larvae remain within the cysts and moult once before they are set free. They wriggle through the soil and reach the roots of the host plant and bore into them near the tips, the head being directed towards the conducting tissue. There, they feed and undergo three more moults. When full-fed, the males measure about 1 mm in length. At that stage, they leave the roots and move around in search of females.

The female larvae, which become enlarged and swollen after each moulting, remain sedentary with the neck and head embedded in root tissue and the main portion of the body protruding. The males approach the females for mating and after the latter have been fertilized, they swell up further becoming almost spherical. It is at this stage that the pest is visible to the unaided eye as white or golden specks attached to the roots. At first, the colour is white but gradually turns golden and finally dark brown. The body wall then becomes thickned and forms a cyst. When the plants are harvested, these cysts get dislodged, drop into the soil and remain dormant till the next potato season. The life-cycle is completed in about 5-7 weeks. Only one generation is normally completed in a year.

Damage. Larvae of the golden cyst nematode feed on the juice sucked from the roots. The attacked plants remain stunted and give a dull and unhealthy appearance. The lower leaves turn yellow, wither and drop off. With young leaves at the top remaining intact, the plants give a 'tufted head' appearance. The root system is poorly developed and the size of tubers is reduced considerably. Badly infested plants may give little or no harvest. The repeated cultivation of potatoes leads to a rapid build up of the nematode population.

Management. (*i*) The preventive measures useful in checking its spread include avoiding the cultivation of potato, brinjal, tomato and chillis in infested fields; adjusting the crop rotation so that potato and related crops in the same field are sown after 4 years and using uninfested potatoes as seed. (*ii*) The nematode can be controlled by using 450 litres of DD mixture or 225 litres of EDB or 25 kg of Nemagon per ha.

Other Pests of Potato and Tomato

Many other insect pests known to attack potato and tomato are the leafbug, *Nesidiocoris tenuis* (Reuter) (Hemiptera : Miridae); white flies, *Trialeurodes vaporariorum* (Westwood) and *Bemisia tabaci* (Gennadius) (Hemiptera : Aleyrodidae); aphids, *Aulacorthum solani* (Kaltenbach), *Aphis gossypii* Glover, *Aphis nasturtii* (Kaltenbach), *Macrosiphum euphorbiae* (Thomas), *Myzus ornatus* Laing, *Brevicoryne brassicae* (Linnaeus) (Plate 20.1F) and *Myzus pesicae* (Sulzer) (Hemiptera : Aphididae); mealybug, *Ferrisia virgata* (Cockerell) (Hemiptera : Pseudococcidae); lunate fly, *Eumerus* sp. (Diptera : Syrphidae); beetle, *Holotrichia conferta* Sharp (Coleoptera : Scarabaeidae); spotted leaf beetle, *Henosepilachna vigintioctopunctata* (Fabricius) (Coleoptera : Coccinellidae); flea beetle, *Chalaenosoma metallicum* F. (Coleoptera : Alticidae) and leaf weevil, *Myllocerus subfasciatus* Guerin–Meneville (Coleoptera : Curculionidae).

ONION (*Allium cepa* L.; Family : Alliaceae)

1. Onion Thrips, *Thrips tabaci* Lindeman (Thysanoptera : Thripidae) (Fig. 20.9)

Onion thrips is world-wide and is found throughout India as a major pest of onion and garlic (*Allium fistulosam* L.). It also feeds on many other plants, including cotton (*Gossypium* spp.), cabbage, cauliflower, potato, tobacco, tomato, cucumber (*Cucumis sativus* L.), etc.

Damage is done by adults as well as by nymphs. The adults are slender, yellowish brown and measure about 1 mm in length (**Fig. 20.9**). The males are wingless whereas the females have long, narrow strap-like wings, which are furnished with long hair along the hind margins. In shape and colour, the nymphs resemble the adults but are wingless and slightly smaller. The insects are just visible to the unaided eye and are seen moving briskly on the flowers and leaves of onion and garlic plants.

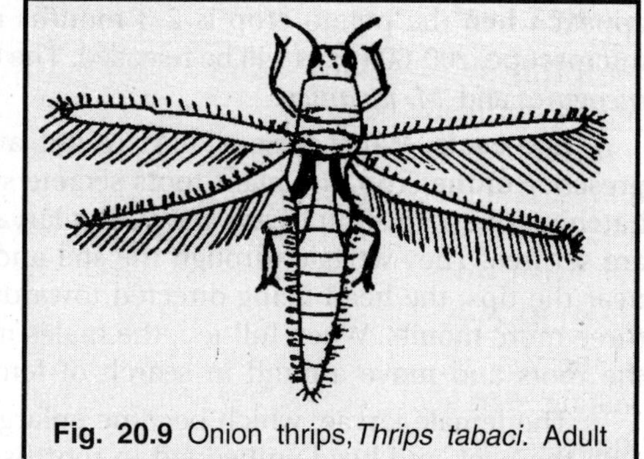

Fig. 20.9 Onion thrips, *Thrips tabaci*. Adult

Life-cycle. This pest is active throughout the year and breeds on onion and garlic from November to May when it migrates to cotton and other summer host plants and breeds there till September. In October, it is found on cabbage and cauliflower. The adult female lives for 2-4 weeks and lays 50-60 kidney-shaped eggs singly in slits which are made in leaf tissue with its sharp ovipositors. The eggs hatch in 4-9 days and the nymphs start feeding on plant juices by lacerating the leaf tissues. On onion and garlic, they are usually congregated at the base of a leaf or in the flower. On cotton, cabbage and cauliflower, they generally feed on the lower surface of leaves. The nymphs pass through four stages and are full-fed in 4-6 days, after which they descend to the ground and pupate at a depth of about 25 mm. The pre-pupal and pupal stages last 1-2 and 2-4 days, respectively. Several generations are completed in a year.

Damage. In onion and garlic, the leaves of attacked plants become curled, wrinkled and gradually dry up. The plants do not form bulbs nor do the flowers set seed. In cotton, the leaves become wrinkled and fall off and the plants bear very few bolls. In Hawaii, *T. tabaci* is known to act as a vector of the streak-virus disease of peas and yellow-spot of pineapple.

Management. Spray 625 ml of malathion 50 EC in 200 litres of water per ha as soon as the pest appears. A waiting period of 7 days should be observed before harvest.

2. Onion Fly, *Delia antiqua* (Meigen) (Diptera: Anthomyiidae)

The onion fly is widely distributed in France, Germany, Canada, USA, Japan, erstwhile USSR and England. This pest also attacks onions in northern India. The flies are slender, greyish, large-winged, rather bristly and about 6 mm in body length. The maggots are small, white and about 8 mm in length.

Life-cycle. The female lays elongate, white eggs near the base of the plant, in cracks in the soil. The eggs hatch in 2-7 days. The maggots crawl up to the plant, and enter the inside of the leaf sheath and reach the bulb. They feed there and become full-grown in 2-3 weeks. The maggots then crawl out of the bulb and pupate in the soil. After 2-3 weeks, the adults emerge and start the new generation. In the third generation, the pest often attacks the onions shortly before the harvest. It initiates the process of rotting of the onions in storage.

Damage. The maggots bore into the bulbs, causing the plants to become flabby and yellowish. They mine through the small bulbs completely, leaving only the outer sheath and thus causing a thin stand of the crop. At times, it destroys 8-9 per cent of the plants. The larger bulbs are attacked by several maggots at a time, making cavities. The attack is not completely destructive to the bulbs but it causes subsequent rotting in storage. It has been indicated that this insect causes the initial damage which leads to the development of soft rot of onion, caused by *Bacillus carotovorus*

Management. Apply 10 kg of carbaryl 4G or phorate 10G to the soil followed by light irrigation.

PEAS (*Pisum sativum* L.; Family : Leguminosae)

1. Pea Leafminer, *Chromatomyia horticola* (Goureau) (Diptera : Agromyzidae)

This pest is widely distributed in northern India and feeds on a large number of cruciferous plants, antirrhinum, nasturtinum, pea, linseed (*Linum usitatissimum* L.) and potato (*Solanum tuberosum* L.). Only the larvae are destructive and they make prominent whitish tunnels in the leaves. If the attacked leaves are held against bright light, the minute slender larvae can be seen feeding within the tunnels. The adults are two-winged flies having greyish black mesonotum and yellowish frons.

Life-cycle. The pest is active from December to April or May and is believed to pass the rest of the year in soil, in the pupal stage. The adults emerge at the beginning of December and after mating, start laying eggs singly, in leaf tissues. The eggs hatch in 2-3 days and the larvae feed between the lower and upper epidermis by making zig-zag tunnels. They are full-grown in about 5 days and pupate within the galleries. The adults emerge from the pupae in 6 days and the life-cycle is completed in 13-14 days. The pest passes through several broods from December to April-May.

Damage. The large number of tunnels made by the larvae interfere with photosynthesis and proper growth of the plants, making them look unattractive.

Management. Spray one litre of dimethoate 30EC in 250 litres of water per ha and repeat spray at 15-day interval. A waiting period of 20 days should be observed for picking of pods.

2. Pea Stem Fly, *Ophiomyia phaseoli* (Tryon) (Diptera : Agromyzidae)

The pest causes damage in the larval stage. It is sporadic and is widely distributed in India, Sri Lanka, the Philippines, and China. It attacks peas, *Phaseolus mungo* L., *Phaseolus aconitifolius* Jacq., soybean, *Glycine max* Mer., cowpeas, *Vigna catjans* Walp. and *Lablab niger* L. The larvae are leaf-and stem-miners. The adult flies are metallic black.

Life-cycle. The flies are active in summer and mate 2-6 days after emergence. The female lays 14-64 elongate, oval and white eggs into the leaf tissue with the help of its elongated ovipositor. The eggs hatch in 2-4 days and the maggots on emergence feed on leaf tissue at first but later on move to the terminal stems. They pass through three instars and the larval development is completed in 6-7 days in March-April and 9-12 days in November and December. The larva pupates within its gallery and the pupal period lasts 5-9 days in March and April, and 18-19 days in November and December. The female flies live for 8-22 days and the males for 11 days. The pest completes 8-9 generations from July to April and shifts from one host plant to the other in various seasons. It passes winter as larva or as pupa.

Damage. The maggots bore into the stem thereby causing withering and ultimate drying of the affected shoots, thus reducing the bearing capacity of the host plants. The adults also cause damage by puncturing the leaves, and the injured parts turn yellow. The damage is more severe on seedlings than on the grown up plants.

Management. (*i*) Avoid sowing of the crop earlier than mid-October to check the attack of the pest. (*ii*) Remove and destroy all the affected branches during the initial stages of attack. (*iii*) Sow the crop in the second fortnight of October to escape the damage of the pest. (*iv*) Apply 7.5 kg of phorate 10G or 25 kg of carbofuran 3G per ha in furrows at the time of sowing (*iv*) On the crop, spray three times 750 ml of oxydemeton methyl 25EC in 250 litres of water per ha. The first application should be just after germination and the other two at an interval of 2 weeks each.

3. Pea Aphid, *Acyrthosiphon pisum* (Harris) (Hemiptera : Aphididae)

Pea aphid is cosmopolitan in distribution in both Palaearctic and Nearctic regions and has been recorded from practically all the areas where the peas are grown. Adult aphids are soft-bodied, long legged, pear-shaped, green yellow or pink in colour with long conspicuous cornicles.

Life-cycle. Both alate as well as apterous forms are present and these are generally females; males are rare. Winged and wingless males have been reported from Europe and USA but not from India. Reproduction is parthenogenetic and viviparous. It takes about a week to complete one generation and there are several overlapping generations in a year.

Damage. Both nymphs and adults suck the sap from young shoots, ventral surface of tender leaves, inflorescence and even on stems. There is curling of leaves, which become irregularly distorted, while the shoots become stunted and malformed. The leaves turn pale and dry. Honeydew secreted by the aphids encourages growth of sooty mould and this superficial black coating on leaves and stems hinders the photosynthetic activity of the plants, which become weak, thus affecting adversely the pod formation. Aphids are carriers of pea mosaic.

Management. Spray 1·0 litre of dimethoate 30EC in 250 litres of water per ha when the attack starts and repeat after 15 days if necessary.

4. Pea Pod Borer, *Etiella zinckenella* (Treitschke) (Lepidoptera : Pyralidae) (Chapter 14, Fig. 14.2)

The larvae damage the crop by feeding on flowers and pods.

Management. Spray 750 ml of endosulfan 35EC or 2.25 kg of carbaryl 50WP in 250 litres of water per ha when the attack starts. Repeat after 15 days if necessary.

5. American Bollworm, *Helicoverpa armigera* (Hubner) (Lepidoptera : Noctuidae) (Chapter 14, Fig. 14.1)

The damage is caused by the caterpillar by feeding on flowers and pods.

Management. Spray 5 litres of chlorpyriphos 20EC or 2.0 kg of acephate 75SP in 250 litres of water per ha.

II. SUMMER VEGETABLES

BRINJAL (*Solanum melongena* L.; Family : Solanaceae)

1. Brinjal Lacewing Bug, *Urentius sentis* Distant (Hemiptera : Tingidae) (Fig. 20.10)

The lace-wing bug is distributed in the north-western parts of the Indian Sub-continent and is common in the plains. Except for brinjal, it has not been recorded feeding on any other plant. Both the nymphs and adults are destructive. The full-grown nymphs are about 2 mm long and 1.35 mm broad. They are pale ochraceous and are stoutly built, with very prominent spines. The adult bugs measure about 3 mm in length and are straw coloured on the dorsal side and black on the ventral side (**Fig. 20.10**). On the pronotum and wings, there is a network of markings and veins.

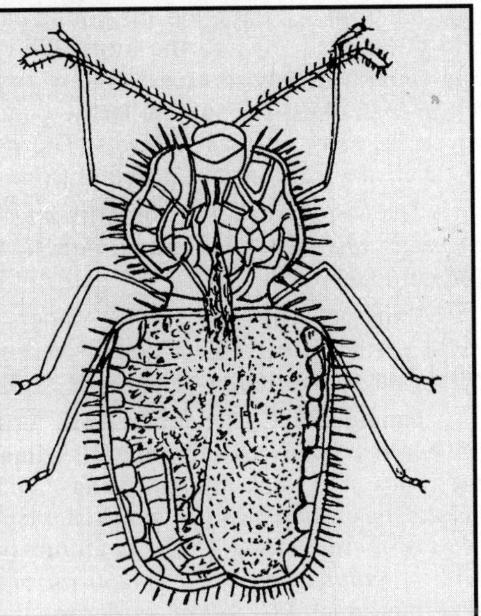

Fig. 20.10 Brinjal lacewing bug, *Urentius sentis*. Adult

Life-cycle. The pest is active from April to October and hibernates as an adult from November-March in cracks and crevices in the soil under the brinjal plants or other protected places. The bugs live for 30-40 days and lay 35-44 shining white nipple-shaped eggs singly in the tissues on the underside of the leaves. The eggs hatch in 3-12 days and the young nymphs feed gregariously on the lower surface of the leaves, but the fully-developed nymphs are found feeding and moving about individually on the lower surface as well as on the upper surface. They grow through five stages and transform themselves into adults in 10-23 days. The insect passes through 8 overlapping generations in a year.

Damage. The damage caused by this insect is very characteristic. The adults and the nymphs suck the sap from leaves and cause yellowish spots which, together with the black scale-like excreta deposited by them, impart a characteristic mottled appearance to the infested leaves. The pest is most abundant in August-September. When the attack is severe, about 50 per cent of the crop may be destroyed.

Management. Apply one litre of dimethoate 30EC in 325 litres of water per ha.

2. Brinjal Fruit and Shoot Borer, *Leucinodes orbonalis* Guenee (Lepidoptera : Pyralidae) (Fig. 20.11, Plate 20.2A, B)

This pest of brinjal fruits and shoots is widely distributed in Malaysia, Myanmar, Sri Lanka, India, Pakistan, Germany and East Africa. Besides brinjal, it has also been recorded feeding on many other solanaceous plants and occasionally on the green pods of peas also. The damage is done by the caterpillars which are creamy white when young, but light pink when full-grown. They measure about 18-23 mm in length. The moth is white but has pale brown or black spots on the dorsum of the thorax and abdomen (**Fig. 20.11**). Its wings are white with a pinkish or bluish tinge and are ringed with small hair along the apical and

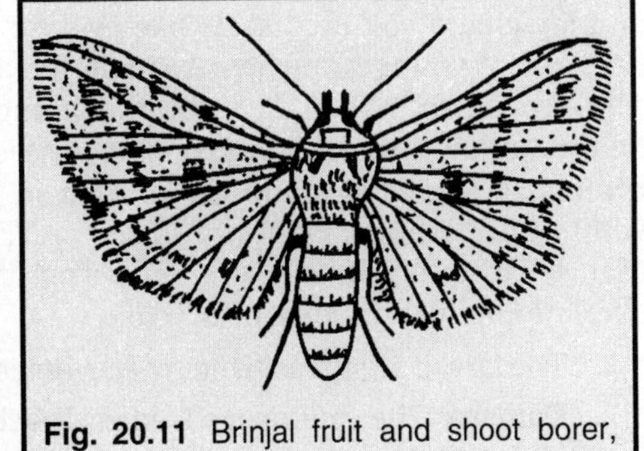

Fig. 20.11 Brinjal fruit and shoot borer, *Leucinodes orbonalis*. Adult

anal margins. The fore wings are ornamented with a number of black, pale and light brown spots. The moth measures about 20-22 mm across the spread wings.

Life-cycle. The caterpillars hibernate in winter and pupate early in spring. The moths appear in March-April and during their life span of 2-5 days, lay 80-120 creamy white eggs, singly or in batches of 2-4 on the underside of leaves, on green stems, flower buds or the calyces of fruits. The eggs hatch in 3-6 days and the young caterpillars bore into tender shoots near the growing points into flower buds or into the fruits. When available, the latter are preferred to the other parts and one caterpillar may destroy as many as 4-6 fruits. The larvae grow through 5 stages and are full-fed in 9-28 days. The mature larvae come out of their feeding tunnels and pupate in tough silken cocoons among the fallen leaves. The pupal stage lasts 6-17 days and the life-cycle is completed in 20-43 days during the active season. There are five overlapping generations in a year.

The parasitoids associated with larvae of this pest are *Pristomerus testaceus* Morl. and *Cremastus flavoorbitalis* Cam. (Ichneumonidae), *Bracon* sp., *Shirakia schoenobii* Vier and *Iphiaulax* sp. (Braconidae)

Damage. When the terminal shoots are attacked, the growing points are killed. Damage to the fruits, particularly in the autumn, is very severe and it is not uncommon to see the whole of the crop destroyed by the borers (Plate 20.2 A, B).

Management. (*i*) Remove and destroy all the affected shoots and fruits with borers inside. (*ii*) Avoid continuous cropping of brinjal crop. (*iii*) Do not ratoon the brinjal crop. (*iv*) Spray 3-4 times at 2-week interval using 250-300 litres of water per ha with any one of the following insecticides as soon as the attack starts : (*a*) *Organophosphates.* Two litres of quinalphos 25EC, 1.4 litres of monocrotophos 36SL, 1.25 litres of triazophos 40EC. (*b*) *Carbamate.* Two kg of carbaryl 50 WP. (*c*) *Synthetic pyrethroids.* 250 ml of fenvalerate 20EC, 100 ml of permethrin 50EC, 500 ml of cypermethrin 10EC, 400 ml of decamethrin 2.8EC. Insecticides of same group should not be used repeatedly in order to avoid development of pesticide resistance and appearance of secondary pests. A waiting period of 7 days should be observed after the spray.

3. Brinjal Stem Borer, *Euzophera perticella* Ragnot (Lepidoptera : Pyralidae)

The stem-borer is a minor pest of brinjal and is widely distributed in India. It may also attack chillies (*Capsicum annum* L.), potato and tomato plants.

Damage is caused by the caterpillars which measure about 20-22 mm in length, when full-grown. They are creamy white and have a few bristly hairs. Their bodies taper posteriorly. The moths measure about 32 mm across the spread wings and have pale-yellow abdomens. The head and thorax are greyish, the fore wings are pale straw-yellow and the hind wings are whitish.

Life-cycle. This pest is active from March to October and passes the winter as a hibernating caterpillar in the stems of old plants. The overwintered larvae pupate at the beginning of March and emerge as moths in the second half of the month. The moths are active at night when they mate and lay cream-coloured scale like eggs singly or in batches on the underside of young leaves or in the axils of young branches. A single female may lay 104-363 eggs in its life span of about a week. The eggs hatch in 3-10 days and the young larvae feed for a few minutes on exposed parts of plants before boring into the stem where they feed on the pith by making longitudinal tunnels. They pass through 4 or 5 stages and are full-fed in 26-58 days. When full-grown, they make silken cocoons within the feeding galleries or in the cracks and crevices in the soil. After pupation, they transform themselves into adults in 6-8 days. The life-cycle is completed in 35-76 days and the pest has 5-6 overlapping generations in a year.

The larvae are parasitized by *Pristomerus testaceus* Morl. and *P. euzopherae* Vier (Ichneumonidae)

Damage. The caterpillars feed exclusively in the main stem and have never been observed to bore into the fruits. As a result of their attack in the field, stray plants are seen withering and drying up.

Management. (*i*) When the attack of this borer is serious, the ratooning of brinjal plants should be discontinued. The withered plants should be uprooted and burnt. (*ii*) Four sprays of 500 ml of malathion 50EC or 315 ml of dichlorvos 100 EC in 325 litres of water per ha should be given at 15-day intervals.

4. Brinjal Hadda Beetles, *Henosepilachna dodecastigma* (Wiedemann) and *H. vigintioctopunctata* (Fabricius) (Fig. 20.12, Plate 20.2C) (Coleoptera: Coccinellidae)

Two species of *hadda* beetles, viz., *Henosepilachna dodecastigma* and *H. vigintioctopunctata*, attack different solanaceous vegetables like brinjal, tomato and potato. Another species, *Epilachna demurili,* attacks cucurbitaceous vegetables exclusively.

Damage is caused by the beetles as well as by the grubs. Beetles of all the three species are about 8-9 mm in length and 5-6 mm in width (**Fig. 20.12**). *H. vigintioc-topunctata* beetles are deep red and usually have 7-14 black spots on each elytron whose tip is somewhat pointed. Beetles of *H. dodecastigma* are deep copper-coloured and have six black spots on each elytron whose tip is more rounded. *E. demurili* beetles have a dull appearance and are light copper-coloured. Each of their elytron bears six black spots surrounded by yellowish rings.

Grubs of all the three species are about 6 mm long, yellowish in colour and have six rows of long branched spines.

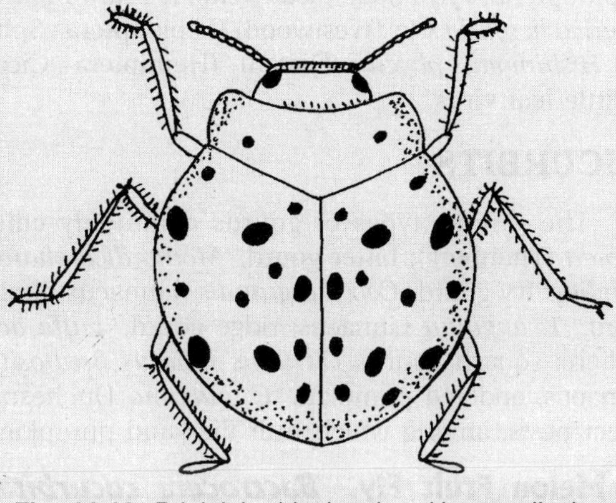

Fig. 20.12 Brinjal hadda beetle, *Henosepilachna vigintioctopunctata.* Adult

Life-cycle. The life cycle and mode of damage of the three species of *hadda* beetles are very similar. Considering their abundance, *H. vigintioctopunctata* is the most important. It passes the winter as a hibernating adult among heaps of dry plants or in cracks and crevices in the soil. It resumes activity during March–April and lays yellow cigar-shaped eggs, mostly on the underside of leaves, in batches of 5-40 each. A single female can lay up to 400 eggs in her lifetime. The eggs hatch in 5, 3.3 and 2.9 days at 25, 30 and 35°C, respectively. The grubs feed on the lower epidermis of leaves and are full-grown in 17.8, 8.7 and 7.1 days at 25, 30 and 35°C, respectively. The pupae are darker and are found fixed on the leaves, stems and, most commonly, at the base of the plants. The pupal stage lasts 13.4, 6.7 and 5.1 days at 25, 30 and 35°C, respectively. The pest passes through several broods from March to October and its population is at a maximum at the end of April or in early May. During the hot and dry months, the number declines greatly but the population again builds up in August.

Eggs are parasitized by *Tetrastichus ovulorum* Ferr. and *Chrysonotomyia appannai* L. (Eulophidae) while *Pleurotropis epilachnae* Rohw., *P. foveolatus* C., *Chrysocharis johnsoni* S. Rao, *Solindenia vermai* Bhet and *Tetrastichus* sp. (Eulophidae) parasitize the larvae, and *Pediobius foveolatus* (J.C.Crawford) (Eulophidae) parasitize the pupae.

Damage. Both the adults and grubs cause damage by feeding on the upper surface of leaves. They eat up regular areas of the leaf tissue, leaving parallel bands of uneaten tissue in between. The leaves, thus, present a lace-like appearance. They turn brown, dry up and fall off and completely skeletonize the plants.

Management. (*i*) Collect and destroy the infested leaves along with insects in the initial stages. (*ii*) Varieties like Arkashirsh, Hissar, Sel. 1-4 and Shankar Vijai have shown resistance to this insect.

(*iii*) Spray 625 ml of malathion 50EC or 2.5 kg of carbaryl 50WP in 325 litres of water per ha, at 10 day intervals as soon as the pest appears.

Other Pests of Brinjal

The minor pests of brinjal include the aphid, *Aphis gossypii* Glover (Hemiptera : Aphididae); the brinjal leaf-roller, *Eublemma olivacea* Walker; leaf roller, *Antoba olivacea* (Walker). and hairy caterpillar, *Selepta celtis* M. (Lepidoptera : Noctuidae); the green potato bug, *Nezara viridula* Linnaeus (Hemiptera : Pentatomidae); the pumpkin stink bug, *Aspongopus janus* Fabricius (Hemiptera: Pentatomidae); the brinjal mealy-bug, *Phenacoccus insolitus Green* (Hemiptera : Pseudococcidae); bug, *Anoplocnemis phasiana* (Fabricius) (Hemiptera : Coreidae); leaf folder, *Phycita clientella* Zeller (Lepidoptera : Phycitidae); leaf webber, *Psara bipunctalis* (Fabricius)(Lepidoptera : Pyraustidae); leaf roller, *Acherontia styx* (Westwood) (Lepidoptera : Sphingidae); the jassids, *Amrasca devastans* (Distant) and *Hishimonus phycitis* (Distant) (Hemiptera : Cicadellidae). The last mentioned species is a vector of little-leaf virus.

CUCURBITS

The various types of gourds commonly cultivated in India include ash gourd, *Benincasa hispida* (Thunberg); bitter gourd, *Momordica charantia* Linnaeus; bottle gourd, *Lagenaria siceraria* Standley; Ivy gourd, *Coccina grandis* (Linnaeus); pointed gourd, *Trichosanthes dioida* Roxberg; snake gourd, *T. anguina* Linnaeus; ridge gourd, *Luffa acutangula* Roxberg; sponge gourd, *L. aegyptiaca* Milliere; squash gourd, *Citrullus lanatus fistulosus* Duthie and Fuller; pumpkin, *Cucurbita pepo* Linnaeus, and red pumpkin, *C. maxima* Duchesne. Cucurbits are attached by several species of insect pests, among which fruit flies and pumpkin beetles are important.

1. Melon Fruit Fly, *Bactrocera cucurbitae* (Coquillett) (Diptera : Tephritidae) (Fig. 20.13, Plate 20.2D)

This is the commonest and most destructive fruit-fly of musk melon and other cucurbits throughout India. It is also found in Pakistan, Myanmar, Malaysia, China, Formosa, Japan, East Africa, Australia and the Hawaiian Islands. In addition to melons it has been found feeding on tomato, chillies, guava, citrus, pear, fig, cauliflower, etc. In north-western India, it is very common on late-sown melons that ripen after the monsoon rains begin. Two other allied species common in India are *Dacus ciliatis* Loew and *Bactrocera dorsalis* (Hendel).

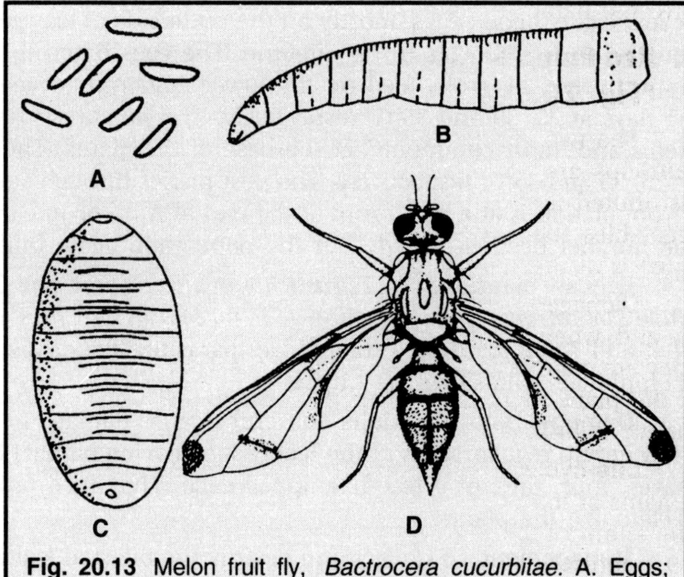

Fig. 20.13 Melon fruit fly, *Bactrocera cucurbitae.* A. Eggs; B. Larva; C. Pupa; D. Adult

Only the maggots cause damage by feeding on near-ripe fruits, riddling them and polluting the pulp. The maggots are legless and appear as headless, dirty-white wriggling creatures, thicker at one end and tapering to a point at the other. A full-grown maggot is 9-10 mm long and 2 mm broad in the middle. The adult flies are reddish brown with lemon-yellow markings on the thorax and have fuscous areas on the outer margins of their wings **(Fig. 20.13)**

Life-cycle. This pest is active throughout the year, but the life cycle is prolonged during winter. The adult flies emerge from pupae in the morning hours and mate at dusk. It takes a few days for the eggs to mature inside the body of a female which starts laying them within 14 days. During winter, the pre-oviposition period is prolonged. They oviposit in comparatively soft fruits avoiding those with hard rind. The selection of a suitable site and the actual laying of eggs take about 6-8 minutes. A cavity is made by the sharp ovipositor and about a dozen white cylindrical eggs are laid, mostly in the evening hours. After laying the eggs, the female releases a gummy secretion which cements the tissues surrounding the puncture and makes the entrance water proof. The secretion solidifies to form a shiny brown resinous material. The female, on an average, lays 58-95 eggs in 14-54 days.

The eggs hatch in 1-9 days and the maggots bore into the pulp, forming galleries. The attacked fruits decay because of secondary bacterial infection. The larvae are full-grown in 3 days during summer and 3 weeks during winter. The mature larvae come out of the rotten fruits and move away in jumps of 12-20 cm. These are made possible by folding and unfolding the two ends of the elongated body. After reaching a suitable place, they bury themselves about 5 mm deep in the soil and pupate. The pupae are barrel-shaped, light brown and they transform themselves into winged adults in 6-9 days in the rainy season and 3-4 weeks in the winter. There are several generations in a year.

Pupae are parasitized by *Opius fletcheri* Silvestri, *O. compensatus* Silvestri and *O. insisus* Silv. (Braconidae), *Spalangia philippinensis* Full. and *Pachycepoideus debrius* Ashm. (Pteromalidae), *Dirhinus giffardi* Silvestri and *D. lzonensis* Rohw. (Chalcididae).

Damage. The maggots pollute and destroy fruits by feeding on the pulp. The damage caused by this fruit-fly is most serious in melons and after the first shower of the monsoon, the infestation often reaches 100 per cent. Other cucurbitaceous fruits may also be infested up to 50 per cent.

Management. (*i*) The regular removal and destruction of the infested fruits helps in the suppression of this pest. (*ii*) Frequent raking of the soil under the vine or ploughing the infested field after the crop is harvested can help in killing the pupae. (*iii*) Apply the bait spray containing 50 ml of malathion 50 EC + 0.5 kg of *gur*/sugar in 50 litres of water per ha. When the attack is serious, it should be repeated at weekly intervals. (*iv*) Spraying the bait on the lower surface of the leaves of maize plants grown at distance of 8-10 cm as trap crop has been found to be effective as the flies have the habit of resting on such tall plants.

2. Red Pumpkin Beetle, *Raphidopalpa foveicollis* (Lucas) (Coleoptera : Chrysomelidae) (Fig. 20.14)

The two species, red pumpkin beetle, *R. foveicollis* and blue pumpkin beetle, *R. atripennis* Fabricius are common in north-western India, the former being more important. It is widely distributed in Asia, Australia, southern Europe and Africa. It is a serious pest of cucurbitaceous vegetables such as ash gourd (*Benincasa hispida*), pumpkin (*Cucurbita pepo* L.), *tinda* (*Citrullus vulgaris* var. *fistulosus*), *ghia tori* (*Luffa aegyptica*), cucumber and melon.

Damage is caused by grubs as well as by beetles (**Fig. 20.14**). The grubs lead a subterranean life and, when full-grown, they measure about 12 mm in length and 3.5 mm across the mesothorax. They are creamy white, with a slightly darker oval shield at the back. The beetles feed on those parts of the plant which are above the ground. They are oblong and 5-8 mm long. Their dorsal body surface is brilliant orange red and ventral surface is black, being clothed in short white hair.

Life-cycle. The beetles are found concealed in groups under dry weeds, bushes and plant remains or in the crevices of soil. They resume activity as soon as the season warms up and in their life span of 60-85 days, they lay about 300 oval yellow eggs singly or in batches of 8-9 in moist soil, near the base of the plants. The eggs hatch in 6-15 days and the grubs remain below the soil surface feeding on roots, underground stems of creepers and on fruits lying in contact with

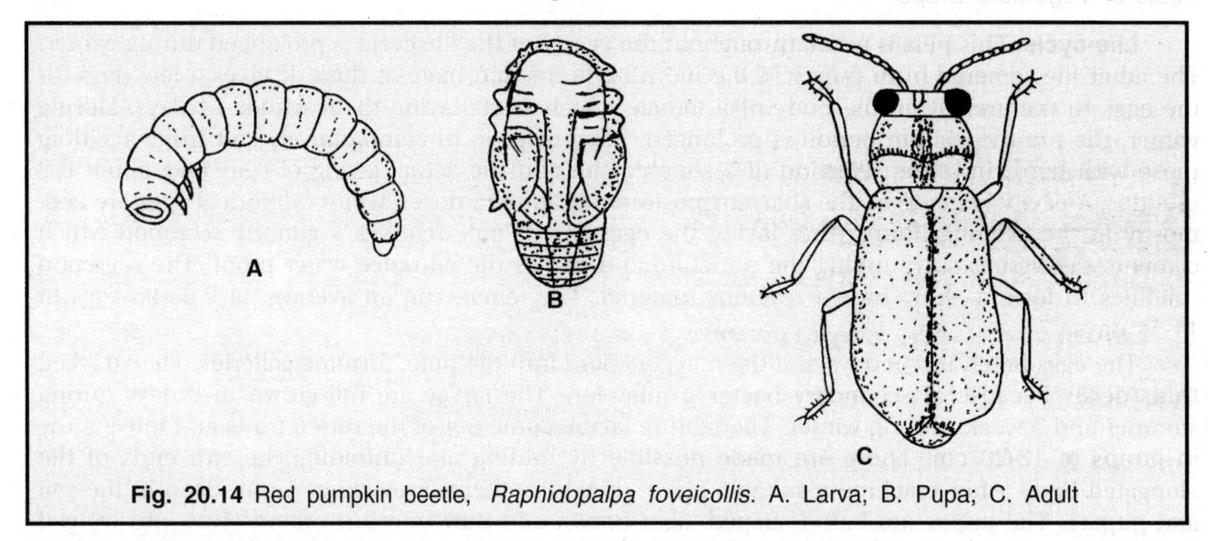

Fig. 20.14 Red pumpkin beetle, *Raphidopalpa foveicollis*. A. Larva; B. Pupa; C. Adult

the soil. They are full-grown in 13- 25 days and pupate in thick-walled earthen chambers in the soil, at a depth of about 20-25 cm.

The pupal stage lasts 7-17 days and the beetles, on emergence, begin to feed and breed. The life-cycle is completed in 26-37 days and the pest breeds five times from March to October.

Damage. The beetles are very destructive to cucurbitaceous vegetables, particularly during March-April when the creepers are very young. The grubs damage the plants by boring into the roots, underground stems and sometime into the fruits touching the soil. The beetles injure the cotyledons, flowers and foliage by biting holes into them. The early sown cucurbits are so severely damaged that they have to be resown.

Management. (*i*) After harvesting the infested fields, plough deep to kill the grubs in the soil. (*ii*) Sow the crop in November to avoid damage by this pest. (*iii*) Apply 7.0 kg of carbofuran 3G per ha 3-4 cm deep in the soil near the base of the plants just after germination and irrigate or spray 375 g of carbaryl 50WP in 250 litres of water per ha.

OKRA (*Abelmoschus esculentus* L; Family : Malvaceae)

1. Cotton Jassid, *Amrasca biguttula biguttula* (Ishida) (Hemiptera : Cicadellidae) (Chapter 16) (Fig. 16.1, Plate 20.2E)

The plants attacked by jassid show yellowing and curling along the margins, turn pale to bronze colour and finally premature defoliation occurs.

Management. (*i*) Spray once or twice at fortnightly interval with 1.4 litres of malathion 50EC in 300 litres of water/ha. (*ii*) As soon as flowering starts, give three sprays at fortnightly interval with 1.25 kg of carbaryl 50WP or 250 ml of fenvalerate 20EC or 200 ml of cypermethrin 25EC in 250-300 litres of water/ha. (*iii*) For seed crop, apply 20 kg of phorate 10G at sowing in furrows or spray twice at fornightly interval starting 15 days after sowing with 625 ml of dimethoate 30EC or 50 g of acetamiprid 20SP in 250-300 litres of water/ha.

2. Spotted Bollworms, *Earias insulana* (Boisduval) and *E. vittella* (Fabricius) (Lepidoptera: Noctuidae) (Chapter 16) (Fig. 16.4, Plate 20.2F)

The shoots infested with borer droop downwards and dry up. The infested fruits have a varying number of holes.

Management. (*i*) Plant resistant varieties, viz. Vaishali Madhu and Sel-6-1. (*ii*) Remove regularly the attacked fruits and bury deep in the soil. (*iii*) Uproot hollyhock and the ratooned cotton, which are host plants for bollworms. (*iv*) Spray the crop when 20-30 per cent shoots show borer damage with 1.0 litre of monocrotophos 36SL or 250 ml of fenvalerate 20EC or 400 ml of deltamethrin 2.8EC or 200 ml of cypermethrin 25EC in 250-300 litres of water/ha.

3. Cotton Whitefly, *Bemisia tabaci* (Gennadius) (Hemiptera : Aleyrodidae) (Chapter 16)

4. Dusky Cotton Bug, *Oxycarenus laetus* Kirby (Hemiptera : Lygaeidae) (Chapter 16)

5. Cotton Leaf Roller, *Sylepta derogata* Fabricius (Lepidoptera : Pyralidae) (Chapter 16)

WATER NUTS (*Trapa bispinosa* Roxb.; Family : Trapaceae)

Singhara Beetle, *Galerucella birmanica* (Jacoby) (Coleoptera : Chrysomelidae) (Fig. 20.15)

The singhara beetle is widely distributed in Pakistan, Sri Lanka, Myanmar and India, and is a serious pest of water-nuts. So far it has not been recorded feeding on any other food plant.

Damage is caused by the grubs as well as by the beetles (**Fig. 20.15**). The full-grown grubs are about 6 mm in length, the upper surface of the body being black and the lower surface being yellow. They are seen feeding on the upper epidermis of floating leaves. The beetles are about 6 mm long and 3 mm broad and are yellowish brown to dark brown, with black eyes and a large hump in the middle of the body.

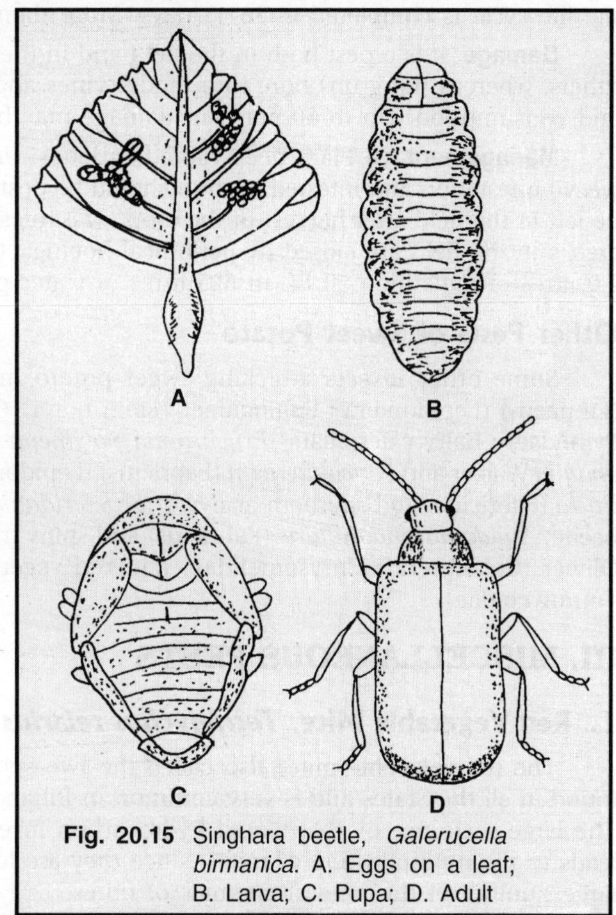

Fig. 20.15 Singhara beetle, *Galerucella birmanica*. A. Eggs on a leaf; B. Larva; C. Pupa; D. Adult

Life-cycle. This pest is active throughout the year, although the speed of development is considerably retarded during winter. The beetles lay reddish-brown rounded eggs on the upper surface of leaves, in clusters of 6-10. The eggs are securely glued to the leaves. A beetle may lay, on an average, 254 eggs which hatch in 7-8 days and the grubs start feeding on the upper epidermis of the leaves. They grow through 3 stages and become mature in 7- 22 days. Pupation takes place on the leaf surface and, like the eggs, the pupae are also firmly glued to the leaves. The pupal stage lasts 3-7 days and the adult beetles live for about a month; although some of them live up to three months. The life-cycle is completed in 17-37 days and there may be as many as 12-17 overlapping generations in a year.

Damage. Both the grubs and adult beetles feed on leaves. Hence, there is an untimely drying up of the plants. The integument of the *singhara* fruits at the surface of the water may also be damaged and this results in rotting of the fruits.

Management. The pest can be controlled by dusting the crop with carbaryl 5 per cent or malathion 2 per cent @ 25 kg per ha.

SWEET POTATO (*Ipomea batatas* Poir; Family : Solanaceae)

Sweet Potato Weevil, *Cylas formicarius* (Fabricius) (Coleoptera : Apionidae)

This weevil is a serious pest of sweet potato in India and it also infests the allied species *I. litoralis* Choisy, *I. learii* (L.), *I. purpurea* (L.), *I. prescaprae* (L.),. *I. trifida* (L.), *I. trichocarpa* (L.), *I. paundurata* (L.), *I. palmata* (L.), *I. sepiaria, Calonyction aculeatum* (L.) and *Jacquemontia tamnifolia* Gray. Both the grubs and the weevils bore into the tubers and make them unfit for consumption.

The adult weevils are small, 5.0-6.5 mm in length, bluish black in colour, with reddish brown prothorax and a long snout. The apodous grub is whitish with a brown head and is 8.22 mm long.

Life-cycle. The pest is particularly active during the rainy season, coinciding with the growth of sweet potatoes. The female weevil lays 97- 216 eggs in cavities made on the vines or the tubers. The oviposition period of the adults may range from 51 to 102 days. The eggs hatch in 3-7 days. The grubs feed both on vines and the tubers, and the larval development is completed in 21-26 days. The pupation also takes place either in the vine or in the tuber. The pupal stage lasts 7-11 days and the life-cycle is completed in 36-43 days. More than one generations are completed in a year.

Damage. It is a pest both in the field and in the storage. The weevils feed on leaves, vines and tubers, whereas the grubs bore into tender vines and tubers, making the latter unfit for marketing and consumption. Up to 60 per cent damage may be caused by the pest.

Management. (*i*) Plant deep-rooted varieties. (*ii*) Vines used for planting should be free from weevil infestation and infested portion should be destroyed. (*iii*) No infested or rotting tuber should be left in the field after harvest of the crop. (*iv*) Infestation may be minimised to some extent if the cracks in the field are closed by periodical hoeing. (*v*) Spray 2.5 kg of carbaryl 50 WP followed by 2.0 litres of malathion 50 EC in 625 litres of water per ha.

Other Pests of Sweet Potato

Some other insects attacking sweet potato include sphinx hawk moth, *Agrius convolvuli* (Linnaeus) (Lepidoptera : Sphingidae) ; stem borer, *Omphisa anastomosalis* (Guenee) (Lepidotpera: Pyralidae) ; hairy caterpillars, *Eruchromia polymena* Linnaeus (Lepidoptera: Amatidae) ; *Spilosoma obliqua* Walker and *Pericallia ricini* (Fabricius) (Lepidoptera : Arctiidae); tobacco caterpillar, *Spodoptera litura* (Fabricius) and southern armyworm, *S. eridania* (Cramer) (Lepidoptera : Noctuidae) ; tortoise beetle, *Aspidomorpha miliaris* (Fabricius) and spiny triangular brown beetle,*Oncocephala tuberculata* Olivier (Coleoptera : Chrysomelidae) and red vegetable mite, *Tetranychus urticae* Koch (Acari : Tetranychidae).

III. MISCELLANEOUS PESTS

1. Red Vegetable Mite, *Tetranychus telarius* Linnaeus (Acari : Tetranychidae)

The red vegetable mite, also called the two-spotted spider mite, is world-wide. In India, it is found in all the States and is very common in Bihar, Mysore, Rajasthan, Uttar Pradesh and Punjab. The large scale use of chlorinated hydrocarbon insecticides for the control of various other pests leads to the multiplication of mites, since they are less toxic to mites and especially since they kill large number of the natural enemies of mites.

The mite is a polyphagous pest and is known to feed on 183 species of plants including cucurbits, brinjal and okra on which it is occasionally very serious. The damage is caused both by the nymphs and adults. A large number of webs are formed on the leaves giving an unhealthy appearance. A fully developed nymph is microscopic and measures about 0.33 mm in length. It is light brown and has two eye-spots, four pairs of legs and is quite active. The adult male measures about 0.52 mm in length and 0.30 mm in breadth. The body of the female is oval, pyriform and variable in colour. It may be ferruginous red, greenish amber or rusty green. Two large pigmented spots are present on the body.

Life-cycle. The mite is active from March to October and passes the winter as a gravid female. As the season warms up in March, it spins webs on the undersurface of leaves of various host plants and lays 60-80 eggs. The eggs are spherical and hatch in 2-6 days. The emerging larvae are light brown and have three pairs of legs. They feed underneath the webs and, within 3-4 days, change into nymphs which have four pairs of legs. The nymphs grow to maturity in two stages within 4-9 days and the adults live for 9-11 days. The life-cycle during the active period is completed in 9-19 days. In Punjab, this mite is believed to complete 32 generations in a year.

Damage. All the active stages usually feed on the underside of the leaves by sucking cell-sap. Gradually, the infested leaves dry up and the webbing interferes with plant growth. There is a poor setting of the fruits and the yield is considerably reduced.

Management. Spray 625 ml of dimethoate 30EC or oxydemeton methyl 25EC in 250 litres of water per ha and repeat spray at 10-day interval.

2. Root-knot Nematode, *Meloidogyne incognita* Chitwood (Tylenchida : Heteroderidae)

This root-knot nematode is world-wide and is distributed throughout India. It is found with other allied species namely, *M. javanica, M. arenaria, M. hapla* and *M. brevicauuda*. It has been recorded on at least 45 host plants, including a number of economic crops and ornamental plants. It is a serious pest of vegetables such as brinjal, and tomato.

The root-knot nematode feeds on the roots of its host plants and forms galls on them. The attacked plants remain stunted and show symptoms of yellowing and shortening of the leaves. The appearance of flowers is also delayed.

Life-cycle. The gravid females of this nematode are always found in tissue inside the galls and they die there after laying eggs, which are oval and are found in yellowish clusters. A single female may lay as many as 768 eggs. The optimum temperature for hatching of the eggs ranges from 20 to 30°C. There is no hatching during winter. In February-March, the larvae are liberated into the soil after the root-knots disintegrate. They live in the soil temporarily and ultimately reach the roots of host plants. They feed by making repeated thrusts into the cortex into which they enter eventually. The larval stage is completed in 3 weeks and the females reach maturity in 33 days. It has been determined experimentally that pH of 7.1-7.25 is very congenial to the development and multiplication of this nematode. During the active period, many overlapping generations are completed on one crop or another.

Damage. At the site of feeding, the internal tissues of the roots become modified to form galls. Sometimes, the galls also appear on other parts of the plant, which are above the ground. The typical symptoms of the disease caused by this nematode include the discoloration and shortening of the leaves, stunting of plants and, occasionally, their death. The diseased plants in a field can be spotted quite easily. The brinjal crop suffers more than other vegetables.

Management. (*i*) The intercropping of okra with sesame reduces the attack of this nematode on the former crop. (*ii*) Dip the roots of nursery plants in 0.03 per cent dimethoate for six hours before transplating.

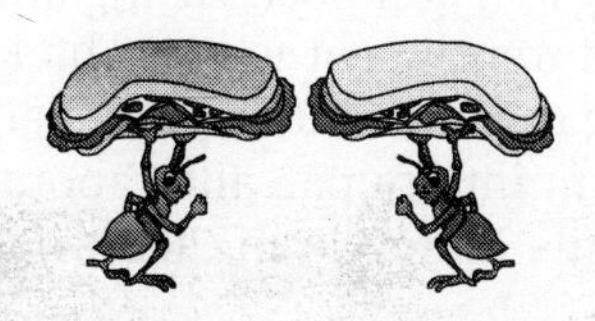

PESTS OF MUSHROOMS

INTRODUCTION

Mushroom cultivation is a new phenomenon across the world. The annual global production of all types of mushroom is estimated to be over 20 million tonnes and production of button mushroom has reached 3.5 million tonnes. India has registered twenty-fold increase in production of mushrooms in the last four decades. At present, the annual production of mushroom in India is about 1.2 lakh tonnes of which 85 per cent is the button mushroom. India contributes about 3 per cent of the total world button mushroom production. In the wake of increasing population, increase in awareness about health benefits of mushrooms and changing food habits, the demand for various mushrooms is likely to increase sharply.

In India, mainly three species of edible mushroom, viz. white button mushroom (*Agaricus* spp.), paddy straw mushroom (*Volvariella* spp.) and oyster mushroom (*Pleurotus* spp.) are commercially cultivated. About 20 insect and mite pests are reported to attack mushroom all over the world. In India, the dipteran flies, viz. sciarid fly, phorid fly and cecid fly have been found causing considerable damage.

INSECT PESTS

1. Sciarid fly, *Bradysia tritici* (Coquillet) (Diptera: Sciaridae) (Fig. 21.1)

Identification. The adults are small, delicate, greyish black, two winged insects measuring 2.5-3.0 mm in length with conspicuous bead-like antennae. Females have a pointed abdomen that is frequently swollen with eggs, while males have prominent claspers at the end of their abdomen. The larvae (maggots) are dirty white with black head and are 6.5 mm long.

Life-history. The females lay eggs on compost, casing and mushroom in clusters of chain. The eggs are small, 1.5 mm long, oval and translucent white. The larvae hatch from the eggs after about 4-6 days. The larvae feed on decaying organic matter, mycelial attachments below casing and at the spawning surface. The larvae transform into pupae after about 20 days and pupation generally lasts about a week. The insect completes its life-cycle in 26-28 days. There are five generations during the cropping period.

Damage. The larvae feed on mushroom, enter from the stem end and move towards the cap portion, The most serious injury is, however, caused to the developing pin-heads, which turn

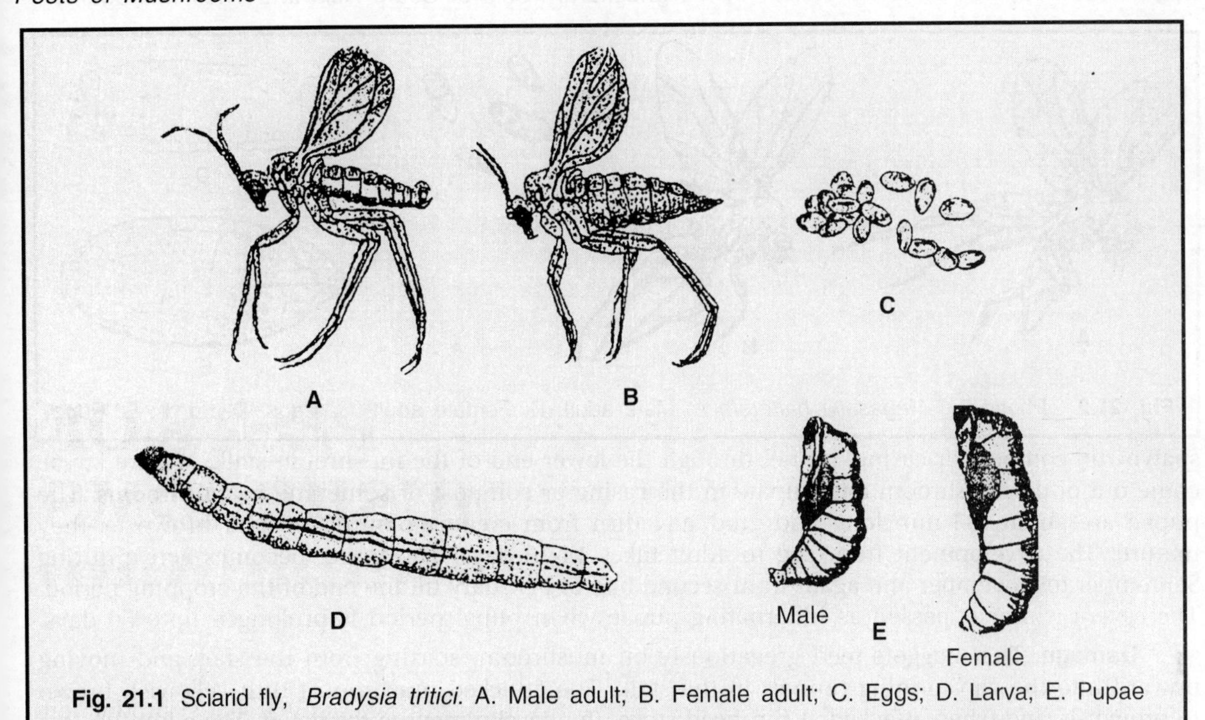

Fig. 21.1 Sciarid fly, *Bradysia tritici.* A. Male adult; B. Female adult; C. Eggs; D. Larva; E. Pupae

yellowish brown and never develop into normal mushrooms. Very high numbers of larvae feeding in the compost during spawn run can also inhibit fruit body production through destruction of the compost and the mycelium. The incidence of this pest starts in the first week of February and reaches its peak in the second week of March, with an overall average infestation of 40 per cent.

Management. (*a*) ***Preventive measures.*** (*i*) Compost should not be prepared directly on soil surface. (*ii*) Sterilization of compost should be done at 60°C to kill insects which come through compost. (*iii*) Screen room exteriors or interiors with nylon net of 35-40 mesh size. (*iv*) The trap consisting of 15 watt bulb and polythene sheet (50 × 75 cm) coated with white grease and then hanged over the wall below bulb during second fortnight of January in mushroom houses is most efficient in reducing the fly population. (*v*) Mix the compost thoroughly at last turning after spraying 15-20 ml of chlorphriphos 20EC or 100 ml of Nimbecidine 0.003% in 20 litres of water. This quantity is used for the compost prepared from 100 kg of wheat/paddy straw. If the flies are present in the mushroom house before casing, mix thoroughly 10 ml of chlorphriphos 20EC in 4-5 litres of water in 100 kg of ready to use casing material before its use. (*b*) ***Remedial measures***. Spray premises with dichlorvos 100EC at the rate of 6 ml per 10 litres of water on window panes, walls and ceiling of mushroom house. After spraying, close the mushroom house for 2 hours. Observe an interval of 48 hours between spraying and picking of mushrooms. Direct spraying on beds should be avoided.

2. Phorid fly, *Megaselia halterata* (Loew) (Diptera: Phoridae) (Fig. 21.2)

Identification. The adult flies are small, hump-backed and light to brown in colour. The adults are two winged flies, measuring 2-3 mm in length and the bristle like antennae are incospicuous. They appear stockier than sciarid flies, and are very active, running and hopping with quick and jerky movements. The larvae are dirty white maggots, 3.5 mm long with narrow head and visible blackish mouth hooks.

Life-history. The female flies lay about 50 eggs, which are very small, elongated, cylindrical and whitish. They prefer spawned compost and gills of opened mushrooms for egg laying. The larvae after hatching directly bore into the mushroom cap, whereas those hatching from the eggs laid on

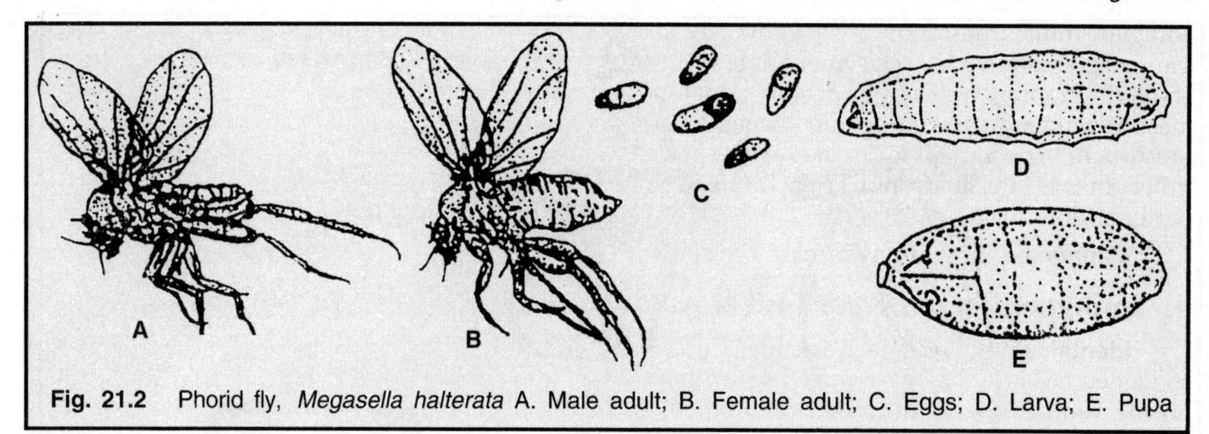

Fig. 21.2 Phorid fly, *Megasella halterata* A. Male adult; B. Female adult; C. Eggs; D. Larva; E. Pupa

spawn-run compost or casing, tunnel through the lower end of the mushroom stalk. Mature larvae come out of the mushroom and pupate in the casing or compost or sometime in mushrooms. The pupae are about 2-3 mm long and gradually turn from cream coloured to dark brown as they mature. The development from egg to adult takes 14-16 days. The insect becomes active during September to November and again from second half of February till the end of the cropping period. The severe winter is passed as hibernating pupae when pupal period is prolonged upto 50 days.

Damage. The maggots feed gregariously on mushroom, starting from the stalk and moving upward into the cap forming tunnels in the stalk. The attacked mushrooms turn yellowish brown to brownish, and when attacked at pin-head stage, the development is restricted. The adult flies are capable of transmitting fungal and bacterial diseases. Since they are active fliers, phorid flies can be a significant irritant to picking crews.

Management. Same as in case of sciarid fly.

3. Cecid fly, *Heteropeza pygmaea* Winnertz (Diptera : Cecidomyiidae) (Fig. 21.3)

Identification. These are tiny flies and orange black in colour. The cecid fly larvae are white legless maggots, about 2 mm long and bluntly pointed at both ends. The head and tail are not easily distinguished except by the direction of travel.

Life-history. The adults of cecid flies are rarely seen because under most conditions, larvae become 'mother larvae' giving birth directly to 10-30 daughter larvae. This species usually does not become a pupa and subsequent adult that must mate before laying eggs. Reproduction is accomplished without mating and gives rise to daughter larvae directly (paedogenesis). When conditions are optimal, this method of reproduction can result in very rapid multiplication of this pest, leading to very large number of larvae, tens of thousands per square metre.

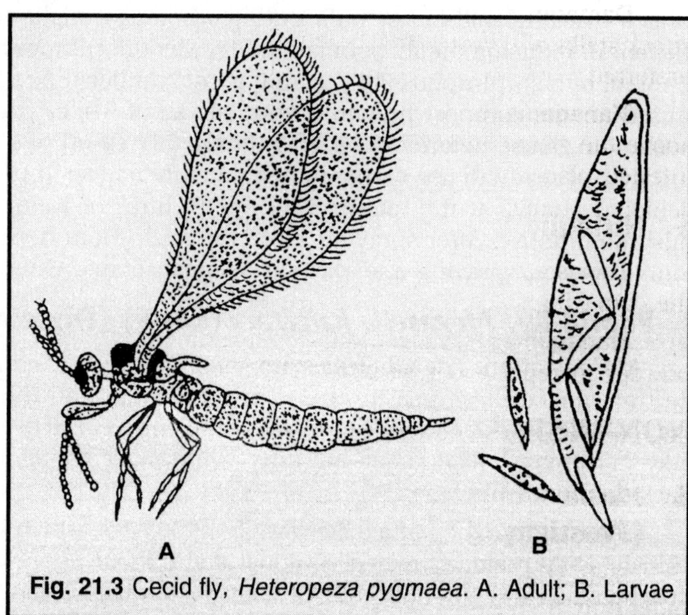

Fig. 21.3 Cecid fly, *Heteropeza pygmaea*. A. Adult; B. Larvae

Damage. Larvae are the damaging stage which feed on mycelium and cause damage to stalk and gills. Brown coloured stripes can be seen on the stipe and gills. Initial infestation of cecid fly may take place by transport of infested peat or substrate. As the maggots are sticky, they might spread through trays, tools, clothes, shoes, etc. of the workers. Small infestations may not be readily apparent at first. If large populations develop, the larvae may mass together on the floor and disperse in large groups. Larvae can also be found on mature mushroom caps packed for market. This species has the potential to significantly reduce yield, when it becomes established on a farm. Damage to mushrooms has been recorded to be as high as 50 per cent due to this pest.

Management. Same as in case of sciarid fly.

4. Springtail, *Seira iricolor* Tosii & Ashraf (Colembolla : Entomobryidae)

Identification. Springtails are tiny, 1 mm long, silver coloured, stout, having long tails. They are so named because they have forked structure or furculum with which they jump. The furculum arises on the ventral side of the fourth abdominal segment. During rest period, the furculum is folded forward under the abdomen. The adults are of ground colour with light violet band alongsides of the body. Intensely dark and rounded scales are present all over the body. The mouth parts are elongate and stylet like and are concealed within the head. Antennae are short and with a few segments.

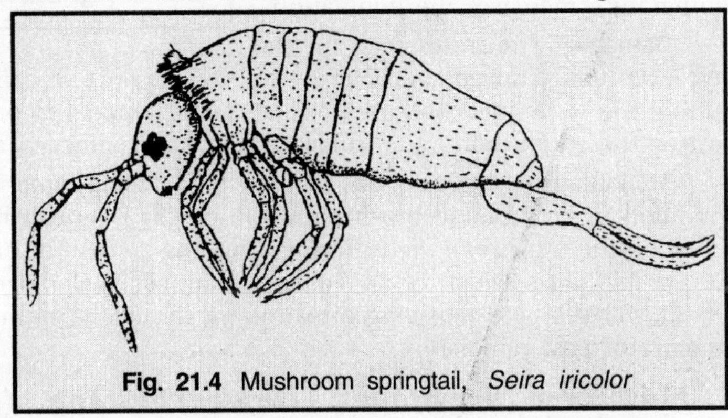

Fig. 21.4 Mushroom springtail, *Seira iricolor*

Life-history. Eggs are laid singly or in small groups on paddy straw pieces. A female lays 9-16 eggs which hatch in 5-20 days. The pre-reproductive period is 10-15 days and adult longevity is between 22-30 days. *S. iricolor* is active throughout the year on button, oyster and paddy straw mushrooms. Maximum activity occurs during July-August on tropical mushroom (*Volvariella volvacea*). In the absence of mushrooms, *S. iricolor* survives on moist organic matter near mushroom houses. They are active in the dark and remain hidden either under the casing soil compost or fruit bodies and move by springing several centimetres, when disturbed.

Damage. Both nymphs and adults feed on mycelium and destroy planted spawn. They also attack stalks and caps. They are found in straw, compost and casing layer. Oyster mushroom is most preferred.

Management. (*i*) Follow proper sanitation and pasteurization. (*ii*) Crop should be raised above the ground level. (*iii*) Never use spent compost. (*iv*) Spray malathion 50EC at 0.05% on the infested place. (*v*) Use diazinon 20EC @ 15 ml in 100 kg wheat straw compost.

5. Nitidulid Beetle, *Cyllodes indicus* Grouvelle (Coleoptera : Nitidulidae)

This is a serious pest of oyster mushroom. Female lays eggs singly between the gills. Grubs are the damaging stage which feed between gills and then on stipes of mushroom making them hollow.

Management. (*i*) Over matured mushroom should not be left unharvested. (*ii*) Apply bleaching powder to repel these beetles.

NON-INSECT PESTS

6. Mushroom Mites, *Pigmophorus* spp. (Prostigmata : Pyemotidae), *Tarsonemus* spp. (Prostigmata : Tarsonemidae), *Tyroglyphus* spp. (Stigmata : Acaridae) (Fig. 21.5)

Sixty species of mites have been found associated with mushrooms world wide. However, in India only 8 species have been identified to be economically important. The mushroom mites are

0.3 to 0.5 mm long, *i.e.* pin head size and varying in colours. The saprophagous mite, *Tyrophagus putrescentiae* (Schrank), is the most demaging species of mite occurring in Himachal Pradesh, Delhi, West Bengal and other states.

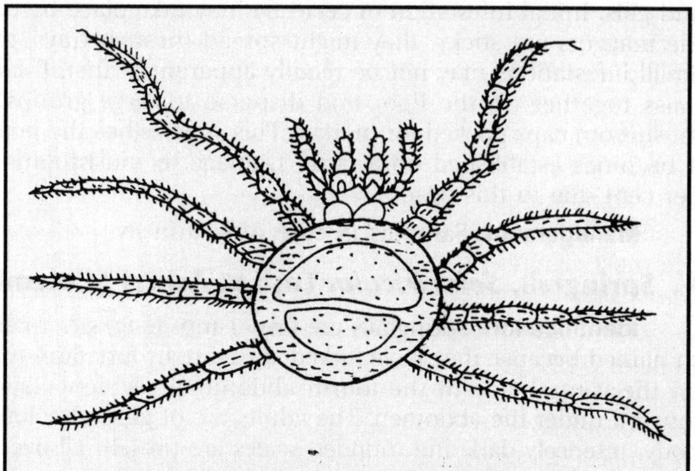

Life-cycle. The mites are present in compost and migrate to casing material. They enter into mushroom bed via flies, on which the migratory stages of mites are clung by means of suckers. Female lays about 100-200 eggs. Fertile eggs develop into females and unfertile ones into males. Life-cycle lasts for 10 days at 20-30°C. Prolonged summer temperature increases the population.

Fig. 21.5 Mushroom mite, *Tyrophagus* spp.

Damage. The mites feed on the mycelium and damage sporophores by causing shrunk caps and brown rusted spots on buttons. Buttons are sometimes hollowed out or cavities appear in on stalks and caps. They may completely destroy the young buttons of mushrooms, when attacked at early stage of development.

Management. (*i*) Pasteurize comport and casing material properly. (*ii*) Maintain strict hygienic condition in and around mushroom house. (*iii*) Disinfect mushroom houses by kelthane 50EC at 0.01% on the mushroom beds during cropping in case of mite infestation. (*iv*) Sterilize compost at 60°C to kill pests which come through compost. (*v*) Burn sulphur in empty mushroom houses. (*vi*) All infested and decaying mushrooms should be removed and destroyed to prevent further increase of pest population.

7. Mushroom Nematodes, *Ditylenchus* spp. (Tylenchida : Anguinidae) and *Aphelenchoides* spp. (Aphelenehida: Aphelenchoididae) (Fig. 21.6)

The nematodes encountered in mushroom compost/beds have been categorized as myceliophagous, saprophagous predaceous, plant parasites and animal parasites on the basis of their nature of parasitism. Among these, only myceliophagous and saprophagous forms hold significance as they effect the crop yield through direct feeding or indirectly. Predeaceous forms usually occur in low counts and are beneficial as they feed upon the harmful myceliophagous and saprophagous nematodes. Plant and animal parasitic nematodes rarely prevail in small numbers, merely as contaminants.

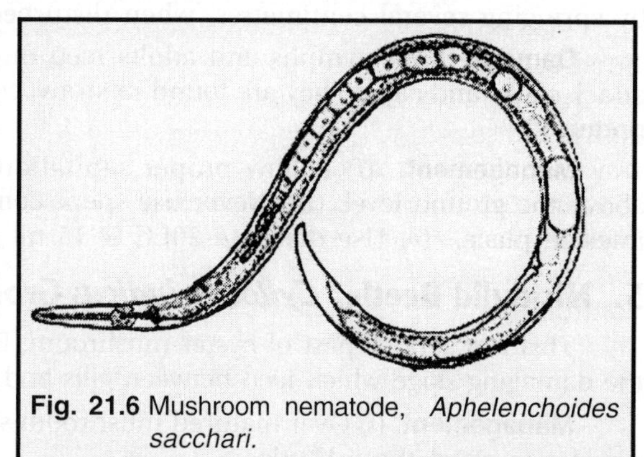

Fig. 21.6 Mushroom nematode, *Aphelenchoides sacchari.*

Myceliophagous nematodes have often been observed to infest the mushroom beds and reduce productivity of compost and its market value. Nematodes have often been observed to infest the mushroom beds and reduce productivity of compost and its market value. Nematode infestation in mushroom is unique in a way that the mushroom nematodes have not only adapted themselves fully with ecological requirements of the crop and multiply very fast to occur in high number in mushroom beds, but these are only examples of ectoparasitic forms, capable of inflicting 100 per cent crop losses.

These are small microscopic thread like organisms which can be found in almost any type of environment, all types of soils and areas, where organic matter is found and are capable of rapid multiplication. The myceliophagous nematodes are the ones which suck the mycelial sap and leave it devitalized. Till today, six genera, two belonging to order Tylenchida (*Ditylenchulus* spp. and *Pseudhalenchus* spp.) and four belonging to order Aphelenchida (*Aphelenchoides* spp., *Aphelenchus avenae, Paraphelenchus* spp. and *Scinura* spp.) have been found to be associated with mushroom cultivation. Among tylenchids, the predominant is *Ditylenchus myceliophagous* Goodey, and among aphelenchids, *Aphelenchoides* with 20 species reported from mushroom farms located globally are the most significant species. Out of these, 10 species have been reported from India, including *A. compositicola* Franklin, *A. sacchari* Hooper and *A. agarici* Seth & Sharma.

Life-cycle. Aphelenchids, showing a mycelophagous behaviour, generally have a short life-cycle of 7-12 days, with temperature playing the major role. It has been reported that *A. composticola* took 18 days to complete its life-cycle at 13°C, it was completed in 10 and 8 days at 18 and 23°C, respectively. The mean sex ratio was found to be 0.69 and was always less than one, the number of females being always more than males. The developmental period of *A. sacchari* from egg to egg has been reported to be 12 days. The longevity of females and males was 35-80 and 35-100 days, respectively. *A. agaricus* is known to complete one life-cycle in eight days at 28°C. It is bisexual species with an average female to male sex ratio of 1.5 : 1. The average life span of females and males, with sufficient food availability, was 89 and 100 days, respectively. However, the females in water survived between 19-32 days.

Damage. Myceliophagous nematodes (which feed on mycelium) are known to be the most damaging. They have needle like stylet on their mouth part, with which they penetrate and puncture the mycelial cell. These nematodes suck the sap and leave the cell devitalized. After feeding upon one cell, the nematode shifts to another with the help of moist film present in the compost. The cell sap oozed out due to leakage from the damaged hyphae may also assist the nematode movement from one cell to another.

The myceliophagous nematode affected beds show various types of symptoms like mycelium disappearance, browning and stunting of water soaked pinheads and the fruit bodies appear in patches. The compost emits a particular odour and yield declines drastically.

Management. (*i*) Maintain strict hygienic conditions in the growing rooms by providing doors fitted with fly proof wire mesh and complete sanitation around the mushroom farm. (*ii*) The peat heat temperature during pasteurization must be maintained at 58-59°C for four hours. (*iii*) The casing soil should be sterilized properly either with steam heat (65°C) or with formatin. (*iv*) Wooden trays should be sterilized properly either with steam, formalin or dipped in hot water (80°C) for 2-3 minutes. (*v*) Growing of nematode-resistant strains in rotation with the crop of common white button mushroom helps to reduce the inoculum level of the nematodes in the mushroom house. (*vi*) The nematode trapping fungi like *Arthrobotrys* spp. can be used as biological control agents for controlling mushroom nematodes. The fungus traps the nearby nematodes and kills them.

PESTS OF ORNAMENTAL PLANTS

INTRODUCTION

The importance of ornamental plants from aesthetic, environmental and economic points of view has been realized since quite recently. India has made a noticeable advancement in the production of flowers, particularly the cut flowers, which have a high potential for export. During 2012-13, floriculture covered an area of 0.23 million ha with a production of 1.73 million tons of loose flowers and 76.73 million tons of cut flowers. The leading cut flower producing states are West Bengal (33.1%), followed by Karnataka (12.3%), Maharashta (10.3%), Andhra Pradesh (9.07%) and Orissa (7.9%). The leading loose flower producing states are Tamil Nadu (18.1%), followed by Andhra Pradesh (12.98%), Karnataka (12.0%), Madhya Pradesh (11.16%) and Gujarat (8.63%).

Ornamental plants are attacked by insects, mites, nematodes, millipedes, molluscs, earthworms and rodents. The insects eat away or bore into the leaves, flowers, buds, fruits and roots. They may suck the sap, making the plant look pale and unhealthy. The growth and the beauty of the ornamental plants receive a set-back. Various species of thrips, aphids, leafhoppers, scale insects, mealy-bugs, leaf-miners, caterpillars, cutworms and chaffer beetles attack the common ornamental plants including rose, chrysanthemum, hibiscus, hollyhock, sunflower, iris, jasmine, etc. As many as 33 species of mites have been recorded feeding on ornamental plants in India. The nematodes cause the swelling of stems, deformation of the leaves, suppression of the blossoms and production of galls in roots. In the case of severe attack, the plants remain stunted or even die. The millipedes, molluscs and earthworms are harmful, when their population becomes very high. The rodents, including field rats, do not spare the ornamental plants and they may spoil the plants as well as the lawns while making their burrows. Most of these miscellaneous pests are discussed under various crops and others are mentioned in this chapter.

INSECT PESTS

1. Rose Aphid, *Macrosiphum rosaeiformis* Das (Hemiptera : Aphididae) (Fig. 22.1)

The rose aphid, has been found infesting roses from November to April in northern India. The population of the aphids often increases to pest level. This insect is generally neglected by the garden owners owing to its minute size and the absence of the visible signs of its attack at the early stages. The vigour of the plant is reduced and the quality of flowers deteriorates. This aphid has been recorded in Punjab, Delhi, Mysore, Andhra Pradesh and the Nilgiri Hills. In addition to this species, the cotton aphid, *Aphis gossypii* Glover, also infests rose plants from September to December in northern India.

Life-cycle. The non-winged rose aphid has an elongated body measuring 2.5-2.6 mm, large red eyes, black cornicles and a yellowish green tip of the abdomen. The form of species occurring in southern India has a purple abdomen, yellow head and dark legs. Nymphal development is completed in 11-14 days of the non-winged forms and 14-19 days of the winged forms, the growth being the quickest in March. In northern India, the insect appears in the middle of November. The population increases progressively and is the highest in March, declining early in April as the season warms up. The winged forms are present throughout the period from November to April, but there is an increase in their number from December onwards, reaching the peak in March, when about 90 per cent of the adults become winged. The aphid multiplies most rapidly in late spring but cannot withstand the summer heat. Consequently, its population declines.

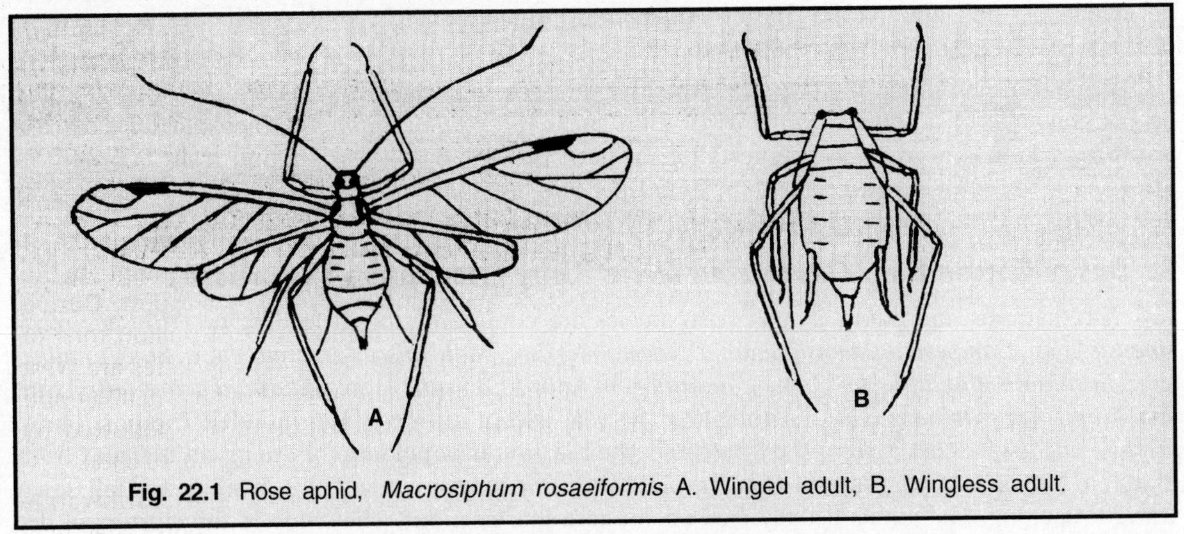

Fig. 22.1 Rose aphid, *Macrosiphum rosaeiformis* A. Winged adult, B. Wingless adult.

Damage. The aphid sucks cell sap from the tender leaves, buds and twigs. It is particularly injurious to tender buds, resulting in the disfigurement and withering of flowers. Each aphid makes several punctures, producing wounds, which leave their mark as the flower opens. A black fungus may also develop on the honey-dew excreted by these insects, giving an ugly appearance to the plant. Some varieties are most susceptible than others, e.g. Damask is susceptible whereas Hawaii is comparatively resistant.

Management. Spray 500 ml of malathion 50EC or methyl demeton 25EC or 200 ml of phosphamidon 85WSC in 500 litres of water per ha. The treatment may be repeated at the reappearance of the pest.

2. Red Scale, *Lindigaspis rossi* (Maskel) (Hemiptera: Aphididae)

The tiny insects look like reddish brown waxy scales. Reddish brown encrustations are seen on the lower portion of the old stems and at times on the younger shoots also. The tiny specks in scurvy like patches on the affected stems appear like spots of pox. The affected plant parts become disfigured, dry and wither away. In case of severe infestation, the entire plant dies.

Management. (*i*) The affected part should be rubbed with a piece of rough cloth soaked in methylated spirit. (*ii*) Spray 500 ml of dimethoate 30EC or endosulfan 35EC in 500 litres of water per ha.

3. Groundnut Aphid, *Aphis craccivora* Koch (Hemiptera: Aphididae)

This black coloured aphid species attacks *Bougainvillea* spp. from December to February. Both adults and nymphs infest flowers, tender leaves and the terminal portions of the plant and suck cell sap. Its attack results in the premature fall of the flowers and the downward cupping of the leaves. The details of its life-cycle are given in Chapter 15.

4. Cotton Aphid, *Aphis gossypii* Glover (Hemiptera: Aphididae)

Both adults and nymphs suck plant sap of many ornamental plants like *Hibiscus rosa-sinensis* Linn., *Cassia glauca* Linn., *Tecoma capensis* Lindl. and *Rosa* spp. from September to April in northern India. The maximum population is observed in *H. rosa-sinensis* during March-April. On *Rosa* spp., it is also observed in September-October and on *C. glauca* in March-April. As a result of its attack, a sooty mould develops which results in a blackened look of the plants. The adults are small, soft and greenish-brown insects. They are parthenogenetic and viviparous giving birth to 8-20 nymphs per female per day. The nymphs moult 4 times to become adults in 7-10 days. The detailed account of the pest has been given in Chapter 16.

The other common aphid species which also attack ornamental plants are *Aphis nerii* Boyer de Fonscolombe on *Asclepias curassavica* Linn. and *Aphis fabae solanella* Theobald on *Castrum nocturnum* Linn. The maximum infestation of both the aphid species is found from February to April. As a result of excessive feeding by adults and nymphs on the plant sap, the leaves curl downwards and the plants give a very shabby look.

5. Dusky Cotton Bug, *Oxycarenus laetus* Kirby (Hemiptera : Lygaeidae)

The nymphs and adults of this pest species are commonly found feeding on *Hibiscus rosa-sinensis* Linn., *Dambeya natalensis* Send., *D. spectabilis* Cav., *Malvaviscus arboreus* Fabr., *Bougainvillea* spp., *Jasminum grandiflorum* Linn., *J. multiflorum* Andr., *J. humile* Linn., *Bauhinia acuminata* Linn. and *Plumeria acuminata* (Linn.) throughout the year except during winter months. The host plant, *H. rosa-sinensis* is most preferred by this bug. The maximum population of the insect is found from March to May. The flower buds of *Hibiscus* plants become pale as a result of its feeding and fall down without opening. The adults usually feed on the terminal portions. They hide in the clusters of dry leaves and flowers during December-January. The life-cycle of this pest has been mentioned in Chapter 16.

6. Milk Weed Bug, *Lygaeus civilies* Wolff (Hemiptera : Lygaeidae)

It is a red bug with black spots on corium and clavus, and three white spots on the black membrane of the hemelytra. The adults of this species are found feeding on plant sap of *Hibiscus rosa- sinensis* Linn., *Asclepias curassavica* Linn., *Carissa carandas* (Linn.), *Dombeya spectabilis* Cav. and *D. natalensis* Sind. througout the year. The maximum population is observed in May-June on its most preferred host plant, *A. curassavica*.

Management. Spray 500 ml of malathion 50EC in 500 litres of water per ha.

7. Hollyhock Tingid Bug, *Urentius euonymus* Fabricius (Hemiptera : Tingidae) (Fig. 22.2)

The tingid bug infests the garden hollyhock, *Althaea rosea* (Linn.), *Abutilon indicum* Sweet, *Sida cardifolia* Linn. and *Chrozophora rottleri* A. Juss. The bugs are easily recognized by their densely reticulate body and wings, and measure about 5-6 mm in length. The nymphs are spiny in appearance.

Life-cycle. The hollylock tingid bug appears on hollylock plants from March to June. The adult female lays eggs on the upper surface of leaves. The incubation period of the eggs is 8-10 days.

There are five nymphal instars and development is completed in 15-27 days. The full developmental cycle is completed on a single leaf. The pest overwinters in egg stage.

Damage. Both adults and nymphs suck plant sap from the under-surface of leaves. The infested leaves become pale-yellow and turn brown. Ultimately they shrivel and dry up.

Management. Spray 500 ml of dimethoate 30EC in 500 litres of water per ha.

A B

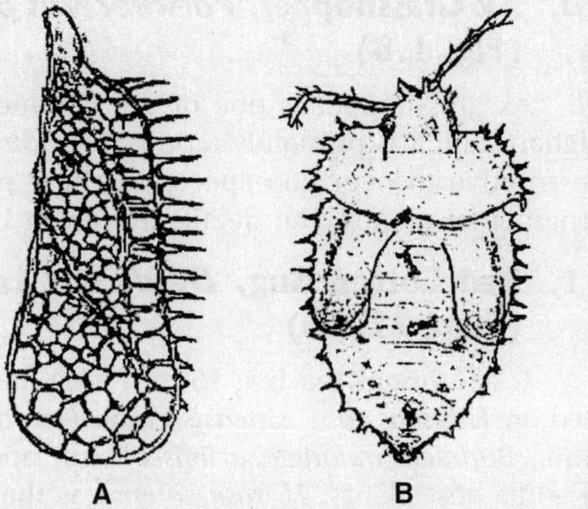

Fig. 22.2 Hollyhock tingid bug, *Urentius euonymus.* A. Nymph; B. Adult

8. Sunflower Lacewing Bug, *Cadmilos retiarius* Distant (Hemiptera : Tingidae)

The sunflower lacewing bug is a small insect measuring about 4 mm, with transparent, shiny, reticulated wings and black body. Both adults and nymphs cause damage to many garden plants like sunflower, *Gaillardia*, daisy, *Chrysanthemum*, marigold, vernonia, launea and the weed, *Argemone mexicana* Linn.

Life-cycle. The pest appears during July and remains active upto September. The adult female bugs lay eggs mainly on the upper surface of leaves of the host plants. The eggs are inserted slantingly into plant tissue leaving the opercula exposed, which appear like white or brown dots. The eggs hatch in 5-7 days and the young nymphs moult five times during nymphal period of 2-3 weeks. Then they become adults.

The eggs are parasitized by *Trichogramma* sp. (Trichogrammatidae). A mite, *Leptus* sp. (Erythraeidae) is parasitic both on nymphs and adults.

Damage. The nymphs and adults suck plant sap and the infested leaves turn yellowish-brown and finally dry up.

Management. Spray 500 ml of malathion 50EC in 500 litres of water per ha.

9. Ak Butterfly, *Danais chrysippus* Linnaeus (Lepidoptera : Nymphalidae) (Fig. 22.3, Plate 22.1A, B)

Ak butterfly (also called plain tiger or African monarch) is widerspread in Asia and Africa. It is a medium sized butterfly with a wingspan of about 7-8 cm. The apical half of the forewing is black with a white band. The hindwing has three black spots around the centre. The body of caterpillars is covered with bands of black and white intersegmented with thick, yellow dorsolateral spots. There are three pairs of long and black tentacle-like appendages on third, sixth and twelfth segments **(Plate 22.1B)**. The larval host plants are from several families, most importantly Asciepiadoideae (Apocynaceae). The caterpillars of the *Ak* butterfly feed on *Asclepias curassavica* Linn. from September to December. Their population becomes high during the months of October and November. The larvae feed on the leaves and flowers of this plant, and in many cases, the entire leaves are eaten up.

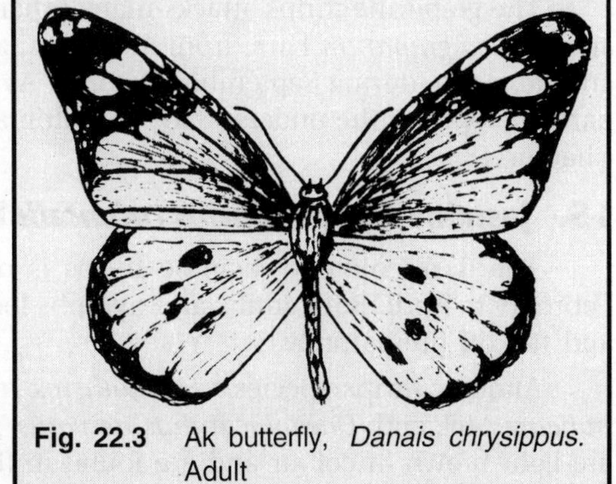

Fig. 22.3 Ak butterfly, *Danais chrysippus.* Adult

Management. Spray one litre of methyl parathion 50EC in 625 litres of water per ha.

10. Ak Grasshopper, *Poekilocerus pictus* (Fabricius) (Orthoptera : Pyrgomorphidae) (Fig. 1.8)

Ak grasshopper is one of the common grasshopper species occurring in India, Pakistan and Afghanistan. it is primainly a pest of *Calotropis* spp., viz. *C. gigantea* and *C. procera*. It is also highly destructive to over 200 species of host plants including field crops, vegetables, fruit trees and ornamental plants. The details are given in chapter 18.

11. Red Cotton Bug, *Dysdercus koenigii* (Fabricius) (Hemiptera : Pyrrhocoridae) (Plate 22.1C)

It is a bright red bug 15 mm long, having black spots on the wings. Both adults and nymphs feed on *Hibiscus rosa- sinensis* Linn., *Dombeya spectabilis* Cav., *D. natalensis* Sind., *Carissa carandas* Linn., *Buddleia madagascariensis* Lamk. and *Jasminum sambac* Linn., from April to December. Out of all its host plants, *H. rosa-sinensis* is the most preferred one. As a result of severe sucking of sap from the leaves and flower-buds, vitality of the plants and their flowering capacity is reduced. The details about the life-cycle have been discussed in Chapter 16.

12. Citrus Psylla, *Diaphorina citri* Kuwayana (Hemiptera : Aphalaridae)

The citrus psylla is a serious pest of *Murraya paniculata* Jack. from April to July. It also feeds on *Plumeria acuminata* Linn., *Dombeya spectabilis* Can. and *Duranta plumieri* Jacq. The nymphs and adults suck sap of the fresh leaves and tender parts of the plants, and thus vitality of infested plants is reduced, giving them a sickly appearance. For details on the life-cycle, see Chapter 18.

13. Cotton Whitefly, *Bemisia tabaci* (Gennadius) (Hemiptera : Aleyrodidae)

The cotton whitefly is found feeding on *Euphorbia pulcherima* Wild., *Hibiscus rosa-sinensis* Linn., *Lantana camara* Linn., *Dombeya spectabilis* Cav. and *Poinciana pulcherima* (Linn.) from May to October. The maximum population is found in August-September on all these host plants. The attack of the whitefly results in downward cupping of the leaves. A sooty mould develops on the leaves which gives blackish look to the plants. With excessive sap sucking, the leaves fall down. Full account of the life-history of this pest has been given in Chapter 16.

14. Grapevine Thrips, *Rhipiphorothrips cruentatus* Hood (Thysanoptera : Thripidae)

The grapevine thrips attacks many ornamental plants like *Rosa* spp., *Lagerstroemia indica* Linn. and *Punica granatum* Linn. from August to January. The infestation becomes heavy on *P. granatum* and *Rosa* spp. during September-October. As a result of severe feeding, the leaves turn rusty brown, particularly from the undersurface and ultimately they fall down. For details about the life-cycle, see Chapter 18.

15. Jasmine Thrips, *Thrips orientalis* Bagnall (Thysanoptera : Thripidae)

The dark coloured jasmine thrips is mainly found on *Jasminum multiflorum* Andrs. from February to April. Both adults and nymphs feed on flowers. The attacked flowers give a decayed look and fall off prematurely.

Another thrips species, *Taeniothrips traegardhi* Tryban, also attacks the flowers of *Nerium indicum* Mill. and *Daedolacanthus nervosus* T. Sandwss. in the months of March-April. The insects are light brown in colour and are found in flowers, giving them a decayed look caused by feeding.

Management. Same as in case of grapevine thrips.

16. Bihar Hairy Caterpillar, *Spilarctia obliqua* (Walker) (Lepidoptera : Arctiidae)

The caterpillars of this polyphagous insect attack the ornamental plant, *Barleria cristata* Linn. during October-November and December-January. They feed on leaves and floral parts of the plant.

In many cases, the whole leaf is eaten away by the caterpillars. Full account of the life-history of this pest is given in Chapter 26.

17. Castor Hairy Caterpillar, *Euproctis lunata* Walker (Lepidoptera : Lymantriidae)

The castor hairy caterpillar, in addition to castor, is also found on *Lagerstroemia indica* Linn., *Punica granatum* Linn., *Hibiscus rosa-sinensis* Linn. and *Rosa* spp. during September-October. The young larvae eat the leaf margins of the host plants. The full-grown caterpillars, however, feed on the entire leaf lamina and cause maximum damage during these months. The life-cycle of this pest has been given in Chapter 15.

18. Jasmine Leaf Webworm, *Nausinoe geometralis* (Guenee) (Lepidoptera: Pyraustidae) (Fig. 22.4)

The larvae of this moth attack the terminal portion of the jasmine plants. This pest is widely distributed in West Africa, India, Pakistan, Sri Lanka, Myanmar, Java, Formosa, China and Australia.

The moths are brown with hyaline patches on the wings. Their abdomen is purplish brown, which is interspersed with lateral patches of lighter shades on each segment. The young caterpillars are light yellow in colour but as they grow they become darker.

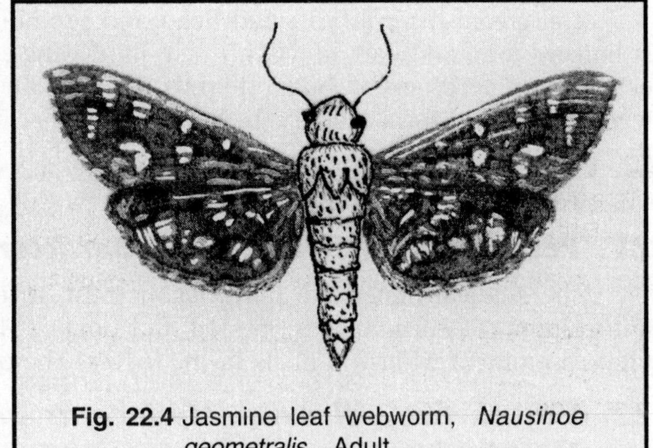

Fig. 22.4 Jasmine leaf webworm, *Nausinoe geometralis.* Adult

Life-cycle. This pest is most active during the rainy season. The moths are seen resting on the lower surface of leaves on the periphery of the plants, thus avoiding the inner shaded portions. The female lays 15-20 eggs singly on the leaf lamina. The eggs are greenish yellow in colour which hatch in 3-4 days. The larvae become full grown in 12-15 days after passing through five instars. The larvae spin extensive webbings in the shaded portion of the plant around the leaves on which they feed and pupate within that area. The pupal stage lasts 6-7 days and the life-cycle is completed in 22-24 days during July-August.

Damage. During the rainy season, the caterpillars attack leaves of the plant mostly in the lower bushy and shaded portions. During the dry and sunny weather the terminal shoots are attacked. The leaves are webbed in an open and loose manner. The silk threads are seen as a cobweb on the surface of leaves. The larvae skeletonize the leaves by eating away the parenchyma. As a result of the gregarious attack of the caterpillars, the vitality of the plant is reduced which tells upon its further growth and the production of buds, which are so important for getting a good commercial crop of jasmine flowers.

Management. Spray 500 ml of dimethoate 30EC in 500 litres of water per ha.

19. Jasmine Budworm, *Hendecasis duplifascialis* Hampson (Lepidoptera : Pyraustidae) (Plate 22.1D, E)

The larva is yellowish green in colour with a distinct black head and prothoracic shield. The moth is pale white.

Life-cycle. The freshly laid eggs are round, creamy and glued to the flower buds. The eggs hatch in 2.6-3.5 days and the larva passes through five instars. The total larval duration varies from 11.5 to 17.0 days depending on the jasmine species. The pupal stage lasts 6.5-9.0 days and the life cycle is completed in 21-29 days.

Damage. The larva bores into immature bud and feeds on the internal contents. The larva feeds voraciously on the corolla leaving only the corolla tube in mature buds. In the case of younger buds, the larva remains outside the buds and feeds on the inner floral whorl through a small hole in the corolla tube. The infested flower turns violet and eventually dries out. A single larva may damage upto 6 flower buds. During heavy infestation, the adjacent buds along the inflorescence are webbed together by silken thread.

Management. Spray 500 ml of dimethoate 30EC or 200 ml of cypermethrin 25EC in 500 litres of water per ha.

20. Jasmine Gallery Worm, *Elasmopalpus jasminophagus* Hampson (Lepidoptera: Pyralidae)

The greenish larva has a red head and prothorax with brown streaks on its body. The moth is narrow, long and dark grey with pale hind wings, grows to a length of 2.5 cm. The larva in the early instars feeds inside buds. It webs together the terminal leaves, shoots and flower head and feeds on them. Faecal matter can be seen attached to the silken web. Pupation takes place inside the web itself. All varieties of jasmine are attacked.

Management. Same as in case of jasmine bud worm.

21. Pea Leafminer, *Chromatomyia horticola* (Goureau) (Diptera : Agromyzidae)

The pea-leaf-miner is a polyphagous pest. It feeds on a large number of cruciferous plants, antirrhinum, nasturtium, pea, linseed and potato. The larvae of this pest are very destructive and make prominent whitish tunnels in the leaves. The details of the life-cycle are given in Chapter 20.

22. Cotton Grey Weevil, *Myllocerus undecimpustulatus* Faust (Coleoptra: Curculionidae)

The cotton grey weevil is commonly found from July to September on *Hibiscus rosa-sinensis* Linn., *Lagerstroemia indica* Linn., *L. speciosa* Ders., *Bauhinia acuminata* Linn., *Dombeya spectabilis* Cur., *D. natalensis* Sond., *Thuja occidentalis* Linn. and *Malvaviscus arboreus* Fabr. The adults feed on the leaves of these host plants in an irregular fashion, starting from the leaf margins. For details on the life-cycle, see Chapter 16.

In addition, the other weevil species belonging to family Curculionidae, namely *Tanymecus* sp. also feeds on *Rosa* spp. during July-September; *Alcidodes* sp. feeds on *Lagerstroemia indica* in April-May and *Amblyrrhinus* sp. feeds on various ornamental shrubs like *Lagerstroemia indica* L., *L. speciosa* Pers., *Lawsonia inermis* Linn., *Rosa* spp., *Bauhinia acuminata* Linn. and *Poinciana pulcherrima* Linn. from March to October. The adults of all these weevils feed on the leaves and flowers of the host plants.

23. Cotton Mealybug, *Phenacoccus solenopsis* Tinsley (Hemiptera: Pseudococcidae) (Plate 22.1G)

Cotton mealybug has recently emerged as a serious pest of a number of agricultural crops including cotton, tomato, mustard, sorghum and ornamental plants. The details are given in chapter 16.

24. Lily Moth, *Polytela gloriosae* Fabricius (Lepidoptera : Noctuidae) (Fig. 22.5, Plate 22.1H)

The lily moth is a sporadic and specific pest of lilies in India and Sri Lanka. The larvae have chocolate brown head and they defoliate the lily plants. The full-grown larva measures 39-42 mm and possesses black, white and red mosaic patterns on the body. The moth has mosaic patterns of red, yellow and black on fore wings, with a row of black and yellow dots on the apical margins. The hind wings are black.

Life-cycle. The adult moths emerge from the hibernating pupae after the first heavy shower during July. The female moth lays round, yellowish eggs on the apical portion of the undersurface of leaves in clusters of 13-42 eggs. The larvae emerge from eggs after 3-6 days and they feed on leaves for 16-20 days. When full fed, they pupate in the soil in an earthen cocoon and the adults emerge within 15-20 days. The insect has two generations in a year and the pupae of second generation hibernate.

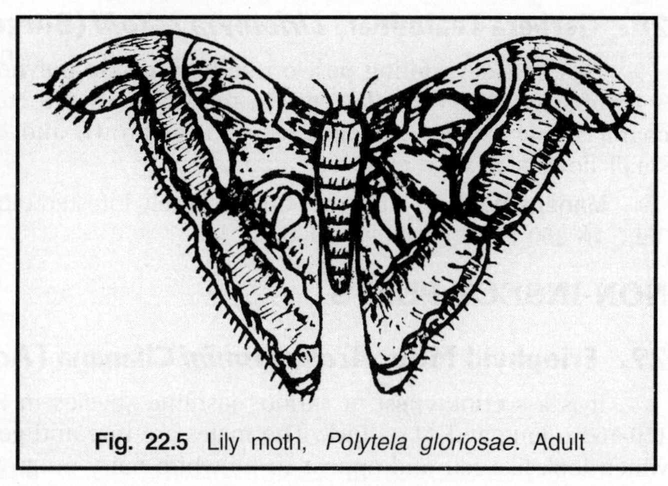

Fig. 22.5 Lily moth, *Polytela gloriosae.* Adult

Damage. The larvae feed on green matter of leaves which may result in complete defoliation of the lily plants.

The other noctuid capterpillars namely *Brithys crini* (F.) and *Spodoptera litura* (Fabricius) occasionally defoliate the lily plants.

Management. Spray 500 ml of malathion 50EC in 500 litres of water per ha.

25. White Grub, *Holotrichia consanguinea* Blanchard (Coleoptera : Scarabaeidae)

The adults of the white grub feed on *Acalypha tricolor* Seem., *Bauhinia acuminata* Linn., *Jatropha intergrima* Linn., *Lagerstroemia indica* Linn., *L. speciosa* Pers., *Punica granatum* Linn., *Rosa* spp. and *Thuja occidentalis* Linn. during the period June to August and they completely defoliate the host-plants. Full account of the life-cycle has been given in Chapter 15.

Another dark-brown chaffer beetle, *Schizonycha* sp. also commonly feeds voraciously on the leaves and flowers of *Lagerstroemia indica* L., *L. speciosa* Pers., *Rosa* spp. and *Punica granatum* Linn. from May to July. This beetle is nocturnal in habit and starts its activity soon after sunset.

26. Blister Beetle, *Mylabris pustulata* (Thunberg) (Coleoptera: Meloidae) (Plate 22.11)

This prominent large beetle has six alternating, bright orange and black bands, against the general dark background of the body. It is 3 cm in length. The adult beetles attack the flowers of *Hibiscus rosa-sinensis* and *Ruellia indica* and other plants from July to September, and devour them completely. In August, the population becomes high and they appear to be more prominent than the flowers.

Management. (*i*) Handpick and destroy the beetles. (*ii*) Spray 1.25 kg of carbaryl 50WP in 625 litres of water per ha.

27. Citrus Red Scale, *Aonidiella aurantii* (Maskell) (Hemiptera : Diaspididae)

Scales mainly infest the matured shoots of rose and spread to young shoots in case of severe infestation. Infested shoots bear reddish-brown encrustations. Both young and adult stages suck the sap. The damaged shoots lose vigour, produce small and few flowers, and dry in case of severe attack. The detailed account of the pest has been given in Chapter 18.

Management. (*i*) Select scale-free planting material. (*ii*) Cut and burn heavily infested shoots. (*iii*) Apply carbofuran 3G @ 20 kg/ha at 5 cm depth to soil after digging. (*iv*) Apply Pongamia oil (1%) to pruned shoots soon after pruning.

28. Gerbera Leafminer, *Liriomyza trifolii* (Burgess) (Diptera : Agromyzidae)

This is a devastating pest on gerbera inside polyhouse. Infestation starts when plants are at 3-4 leaf stage. Adult flies lay eggs in the leaves by puncturing. Larvae feed inside leaves by making characteristic mines. Damaged leaves turn brown and dry while plants get stunted and produce small flowers in case of severe attack.

Management. (*i*) Pluck and burn severely infested older leaves. (*ii*) Spray one litre of dimethoate 30EC in 250 litres of water per ha.

NON-INSECT PESTS

29. Eriophyid Mite, *Aceria jasmini* Chanana (Acarina : Eriophyidae)

It is a s erious pest of various jasmine species in South India. The female is cylindrical and 150-160 μ long and 44 μ thick. The males are rare and generally 95 μ long. These mites make webs which look like felt and appear to be white hairy outgrowth on the leaf surface, tender stems and flower buds. As a result of this activity, growth of plants is reduced and the production of flowers is suppressed.

Management. Spray 625 ml of dimethoate 30EC or malathion 50EC in 375 litres of water per ha.

30. Red Spider Mite, *Tetranychus urticae* Koch (Acari : Tetranychidae)

Red spider mite is a serious pest of carnation inside polyhouse. Heavy infestation of the mite is observed during summer (March-June). Colonies of mite are mainly seen on leaves and young buds covered with dirty coloured webs. Mite spreads to the entire plant in case of heavy infestation. Attacked leaves look dirty, discoloured and dry. Damaged plants lose vigour, become stunted and dry in severe cases.

Management. (*i*) Cut and burn the severely infested and dried shoots. (*ii*) Wash plants with force to dislodge webs and reduce mite population. (*iii*) Watering of polyhouse and providing proper ventilation during summer are essential to reduce mite population. (*iv*) Spray 500 ml of dimethoate 30EC or oxydemeton methyl 25EC in 250 litres of water/ha.

31. Snails and Slugs

Snails and slugs are soft-bodied animals belonging to class Gastropoda of the phylum Mollusca. Their body is asymmetrical, spirally coiled and enclosed in a shell. They have a large flat foot which is used for creeping and they do not have separate sexes.

The common snails found in Himachal Pradesh, Uttar Pradesh, Andhra Pradesh, Bihar, Maharashtra and Orissa are *Helix* spp. Another phytophagous species, *Achatina fulica* Ferussae, the giant African snail, has been reported as a serious pest of fruits, vegetables and ornamental plants in coastal areas of Orissa, West Bengal, Assam, Tamil Nadu and Kerala.

The common garden slug, *Laevicantis alte* Ferussae, in Punjab and Himachal Pradesh, has been observed feeding on a number of ornamental plants, including balsam, portulaca, pot- marigold, verbena, dahlia, cosmos, narcissus and lily. Another slug, *Limax* sp., occurs all over India.

Life-history. The common snail breeds in the spring and the summer. It makes a hole of 1.24 cm in diameter and 3 cm in depth in damp soil, in which it lays eggs in a loose mass of about 60. The eggs hatch within two weeks and the young snails start feeding upon tender plants. The shell increases in size with age and the snail is full-grown in about two years. Snails are seen at all hours, except during mid day when it is hot and dry. In winter, they stay in colonies and are found among rockeries, loose boards of fences, at the bottom of hedges, in rubbish heaps, etc.

The slugs feed at night in damp places and destroy young shoots of various plants. Under dry conditions, their population is reduced considerably.

Damage. Snails and slugs appear as sporadic pests in those places where damp conditions prevail. They may also appear in large number on roads and runways, creating problems during the taking- off or the landing of the aircraft. The slugs are known to feed on celery, lettuce, cabbage and a number of ornamental plants. When their population in high, they may do serious damage.

Management. (*i*) Low population can be collected and destroyed. (*ii*) Dust 15 per cent metaldehyde dust or spray 20 per cent metaldehyde liquid or sprinkle 5 per cent metaldhyde pellets around infested fields.

32. Root-lesion Nematodes, *Pratylenchus* spp. (Tylenchoidea : Tylenchidae)

Of the various ornamental plants, roses are the most affected by parasitic nematodes. More than a dozen species have been recorded from the roots of rose plants in India. Out of these, the more important are three species of the root-lesion nematodes, belonging mainly to the genus *Pratylenchus* and some to the genera *Tylenchorynchus, Hoplolaimus, Xiphinema* and *Trichodorus.* These root-lesion nematodes are vagrant parasites of plant roots and inhabit the aboveground portions only in rare cases. All these groups of nematodes have world-wide distribution.

Life-cycle. Both adults and larvae go in and out of the roots. The penetration usually occurs in the mature region of the rootlets and not from the root-tips. A female usually lays one egg per day. The egg stage lasts 16-20 days. The development and reproduction are rather slow in *P. pratensis* taking 54 days to complete the life-cycle. In other species, like *P. zeae,* the life-cycle is completed in 35-40 days. During periods of drought, these nematodes lie quiescent, but they resume growth as soon as free moisture is available. The population of the root-lesion nematodes is high in October, whereas the infestation of *Tylenchorynchus* and *Hoplolaimus* occurs more frequently in the rainy season.

Damage. Lesion nematodes feed on the parenchyma of the root and cause lesions, especially when a large number of them feed together. The root injury results in decreased growth of the aboveground portions. The plants bear small or no flowers at all.

Management. (*i*) The roots of several varieties of French and African marigold and of sesame are known to release nematicidal compounds in rootlets, which are left in the soil and are effective against nematodes. The cultivation of marigold in rotation or as an intercrop is a useful practice. (*ii*) Mix phorate 10G @ 10 kg/ha or carbofuran 3G @ 30 kg/ha in soil at the time of planting. (*iii*) Apply Nemagon @ 25 litres per ha, sprinkle thoroughly chemical on soil followed by flood irrigation.

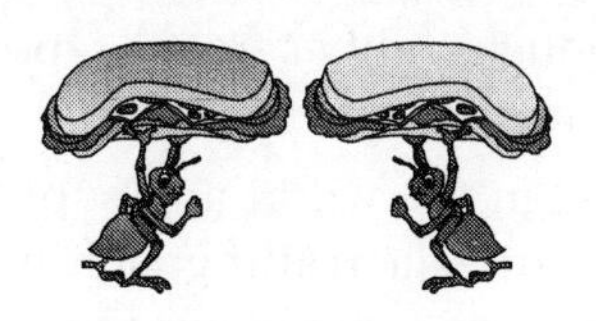

PESTS OF PLANTATION CROPS

INTRODUCTION

The important plantation crops grown commercially in Asia are the various palm trees, namely coconut, palmyra, areca and other trees and shrubs like rubber, coffee, tea and cashewnut. The total production of plantation crops in India during 2012-13 has been 15.61 million tonnes from an area of 3.64 million ha. Coconut accounts for the major share of the production of plantation crops, followed by cashewnut and arecanut. At present, 15.61 million tonnes of coconut are produced in India from an area of 2.14 million ha. The area under arecanut is 0.40 million ha with annual production of 0.48 million tonnes. The coconut and other palm-trees are attacked by some specific pests and also by a number of polyphagous insects.

The cashewnut tree is grown in about 0.99 million ha area with a production of 0.75 million tonnes of raw nuts. India ranks first in production of cashew and conributes about 30 per cent to world cashew production. The cashew tree is attacked by many pests, but about a dozen of them cause serious damage when they get favourable conditions. Arecanut is grown in 0.45 million ha with a production of 0.61 million tons. In India, rubber plantations occupy 0.71 million ha with a production of 0.90 million tonnes. Although, rubber tree is reported to be attacked by many insects, yet none of them has attained pest status.

India is the largest producer of tea in the world and supplies about 50 per cent of the world's requirement, with Sri Lanka and Indonesia trailing behind. In India, the area under tea is 0.60 million ha with an annual production of 1.20 million tonnes. About 75 per cent of the total area under tea plantations lies in the North-east India with a total production of about 70 per cent. Owing to the availability of abundant succulent shoots on this perennial plant, which may live for more than one hundred years, a number of pests have established themselves. As many as 147 insect pests have been recorded on this crop. Coffee is also one of the important plantation crops grown in India on 0.4 million ha. The country produces 0.32 million tonnes of coffee. Coffee is attacked by about one dozen insect pests, only a few of which are serious, some of them being specific to one or other variety. The coffee stem borer, *Xylotrechus quadripes* Chevrolat, is the most important pest of the *arabica* coffee, whereas the coffee shot hole borer, *Xylosandrus compactus* Eichhoff, prefers the *robusta* coffee and is a major pest of this crop.

ARECANUT (*Areca catechu* L.; Family : Aracaceae)

Arecanut Mirid Bug, *Carvalhoia arecae* Miller & China (Hemiptera : Miridae) (Fig. 23.1)

This bug is one of the important pests of *Areca catechu, A. lutescens* and *Loxococcus* sp. in South India. The adult is a red and black bug measuring 6 mm long and 2.5 mm wide (**Fig. 23.1**). The nymphs are greenish with reddish brown patches. The adults and the nymphs suck sap from central spindles and the young fronds.

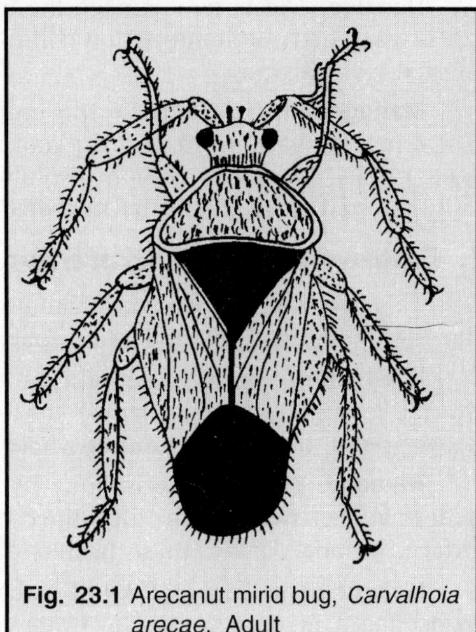

Fig. 23.1 Arecanut mirid bug, *Carvalhoia arecae.* Adult

Life-cycle. The pest is active during February. The female bug lays eggs singly into the tissues of the tender unopened leaves of the palm. The eggs are oval, 13.6 × 0.34 mm with two-bristle like structures arising from the operculum. The eggs hatch in nine days and the young nymphs suck cell sap from tender parts. The nymphs become adults in 2-3 weeks after undergoing 5 moultings.

Damage. The adults and the nymphs suck sap from the tender leaves of arecanut palm. The bugs remain clustered together within the top-most of the leaf axils. In case of severe infestation, the leaves get shredded and stand erect. Continued attack over a number of years results in stunted growth of the palm and the yield is also reduced.

Management. Spray 250 ml of monocrotophos 40EC or 375 ml of quinalphos 25EC in 250 litres of water per ha at monthly intervals especially on the spindles and young fronds.

Other Pests of Arecanut

The other pests of Areca-palm include the grasshopper, *Aularches miliaris* Linnaeus (Orthoptera: Acrididae); the termite, *Odontotermes obesus* (Rambur) (Isoptera : Termitidae); the aphids, *Cerataphis lataniae* (Boisduval) and *C. variabilis* H.R.L. (Hemiptera : Aphididae); *Coccus hesperidum* Linnaeus (Hemiptera : Coccidae); *Dysmicoccus brevipes* (Cockerell) (Hemiptera: Pseudococcidae); the leaf thrips, *Rhipiphorothrips cruentatus* H. and the flower thrips, *Thrips hawaiiensis* (Morgan) (Thysanoptera : Thripidae); *Elymnias caudata* Butl. (Lepidoptera: Nymphalidae); the white-grub, *Leucopholis lepidophora* Bl. (Coleoptera : Melolonthidae); the nut beetle, *Araecerus asciculatus* (DeGeer) (Coleoptera: Anthribidae); *Rhynchophorus ferrugineus* (Olivier) and *Diocalandra stigmaticollis* Gyll. (Coleoptera : Curculionidae); the stored arecanut beetle, *Coccotrypes carpophagus* Horn. (Coleoptera : Scolytidae) and mites, *Typhlodromus ovalis* Evans (Acari : Phytoseiidae), *Oligonychus indicus* (Hirst) and *O. biharensis* Hirst (Acari : Tetranychidae), and *Raoiella indica* Hirst (Acari : Tenuipalpidae).

CASHEWNUT (*Anacardium occidentale* L.; Family: Anacardiaceae)

1. Cashew Tree Borer, *Plocaederus ferrugineus* (Linnaeus) (Coleoptera : Cerambycidae)

This pest is the most important pest of the cashew-tree in South India. The adult is a medium-sized dark-brown beetle. In the grub stage, it feeds inside the tree trunk or branches, making tunnels.

Life-cycle. The beetle lays eggs under the loose bark on the trunk. The newly emerged grubs bore into the bark and feed on soft tissues, making tunnels in all directions. The grown-up grubs may also feed on wood. The openings of tunnels are seen plugged with a reddish mass of chewed fibre and excreta. When full-grown, the larva measures 7.5 cm and tunnels its way to the root region, where it forms a calcareous shell for pupation. The life-cycle is completed in more than a month and there are several overlapping generations in a year.

Damage. The borers damage the cambial tissues and hence the flow of sap is arrested. The tree is weakened and if infestation continues it may die. Plantations over 15 years old are often seen infested with this pest.

Management. (*i*) Remove the grubs by peeling the bark mechanically and destroy them. (*ii*) Drench the basal trunk and the root region with 2.0 litres of chlorpyriphos 20EC in 250 litres of water per ha. (*iii*) Inject carbon disulphide into the tunnels and plaster them with mud. (*iv*) The badly infested trees should be uprooted and destroyed by burning.

2. Cashew Leafminer, *Acrocerops syngramma* Meyrick (Lepidoptera : Gracillariidae)

This pest is distributed throughout the cashew growing areas. The adult is a silvery-grey moth. The freshly hatched caterpillars are pale white, turning to reddish-brown when fully grown.

Life-history. The eggs are laid on very tender leaves. The larvae hatch in 2-3 days and mine into the leaves. Up to eight larvae are found to attack a single leaf. The larva becomes full-grown in two weeks. It drops to ground where it pupates for 7-9 days.

Damage. The damage is done by the caterpillars by mining through the leaves. The thin epidermal peel swells up in the mined areas and appear as whitish blistered patches on the leaf-surface. In older leaves, these blisters dry and drop off leaving big holes.

Management. Spray 2.0 kg of carbaryl 50WP or 1.25 litres of malathion 50EC or 1.5 litres of endosulfan 35EC in 625 litres of water/ha.

Other Pests of Cashewnut

The other pests of cashewnut include cashew aphid, *Toxoptera odinae* (van der Goot) (Hemiptera : Aphididae); **(Plate 23.1A)** mealybug, *Planococcus lilacinus* (Cockerell) (Hemiptera: Pseudococcidae); the tea mosquito bug, *Helopeltis antonii* Signoret (Hemiptera : Miridae); the wax scale, *Ceroplastes floridensis* Comstock and *Coccus hesperidum* Linnaeus (Hemiptera : Coccidae); *Catacanthus* sp. (Hemiptera : Pentatomidae); the castor thrips, *Retithrips syriacus* M., grapevine thrips, *Rhipiphorothrips cruentatus* Hood and the cashew thrips, *Selenothrips rubrocinctus* (Giard) (Thysanoptera : Thripidae); the bark borer, *Indarbela tetraonis* M. (Lepidoptera: Metarbelidae); the castor slug caterpillar, *Parasa lepida* (Cramer) (Lepidoptera: Limacodidae); cashew leaf and blossom webber, *Lamida moncusalis* Walker **(Plate 23.1B)**; the Indian meal moth, *Plodia interpunctella* (Hubner) and the fig moth, *Ephestia cautella* (Walker) (Lepidoptera : Pyralidae); the leaf caterpillar, *Cricula trifenestrata* H. (Lepidoptera : Saturniidae); the weevil, *Myllocerus viridanus* Fabricius (Celeoptera : Curculionidae) and the shoot weevil, *Apion ampulum* Fst. (Coleoptera : Apionidae).

COCONUT (*Cocos nucifera* L.; Family : Palmae)

1. Coconut Scale, *Aspidiotus destructor* Signoret (Hemiptera : Diaspididae)

The coconut scale is one of the most dangerous pests of coconut palm in most of the coconut growing regions. It occurs from Iran to Japan, and USA, and southwards down to South Africa and Australia. In addition, this pest also feeds on other palms, bananas, avocado, cocoa, citrus, ginger, guava, *Artocarpus*, *Pandanus*, papaya, rubber, sugarcane, yam and many wild plants. The scale of the female is circular, flat, transparent, whitish to grey white and about 1.8 mm in diameter. The scale of the male is oval and much smaller than that of the female.

Life-cycle. The female deposits about 20-25 yellow, tiny eggs under her scale. Incubation takes 7-8 days. On hatching, the crawler takes up a position on the leaf and starts feeding. The male nymph moults three times and the female twice. The larval development takes 24 days. The total life-cycle is completed in 31-35 days and there are about 8-10 generations per year.

The coccinellid predators including *Chilocorus* sp., *Azya trinitalis* Mshl., *Cryptognatha nodiceps* Mshl., *Rhyzobius lophanthae* (Blaisdell) and *Pentilia castanea* Muls. (Coccinellidae) play a significant role as natural limiting factor for the coconut scale. The nymphal parasites belonging to genera *Comperiella* (Encyrtidae) and *Encarsia* sp. (Aphelinidae) have a more local significance.

Damage. The scale affected leaves first show a yellow discoloration around areas of the sucking activity, followed by brown necroses. In extreme cases, the leaves dry up, entire fronds drop off, the crown dies and the whole crop is lost.

Management. Spray 500 ml of malathion 50EC in 250 litres of water per ha.

2. Black-headed Caterpillar, *Opisina arenosella* Walker (Lepidoptera : Xyloryctidae) (Fig 23.2)

Next to the rhinoceros beetle, this caterpillar is the most important pest of coconut palm in Kerala as well as along the western coast of India, Sri Lanka and Myanmar. The moth is ash grey (Fig. 23.2). It is medium-sized, measuring 10-15 mm, with a wing expanse of 20-25 mm. The caterpillars feed hidden inside silk galleries on the underside of leaves.

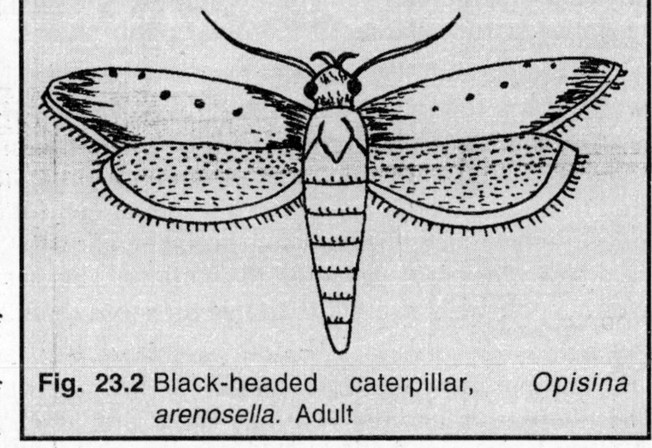

Fig. 23.2 Black-headed caterpillar, *Opisina arenosella.* Adult

Life-cycle. A female moth lays 125 scale-like eggs in small batches on the underside of the tips of the old leaves. Freshly laid eggs are creamy white and turn pink on completion of embryonic development. The incubation period lasts about 3-5 days in summer and 10 days in winter. A young larva is about 1.5 mm in length and increases to 15 mm before pupation. It feeds on green matter and simultaneously constructs a silken gallery in which it feeds and voids frass. When full-grown in about 40 days, the larva transforms itself into a brownish pupa inside the gallery. Within about 12 days, a whitish moth emerges and starts the life-cycle all over again.

The caterpillar is parasitized by *Apanteles taragammae* Vier, *Bracon brevicornis* Wesmael (Braconidae) and *Elasmus nephantids* Gahan (Elasmidae). Inundative releases of native parasitoids, *Goniozus nephantidis* (Muesebeck) and *B. brevicornis* at 3000 and 4500 ha^{-1}, respectively have given encouraging results for management of black-headed caterpillar. The pupa is parasitized by *Trichospilus pupivora* Ferrieri (Eulophidae), *Stomatoceras sulecatiscutellum* Gir and *Brachymeria nephantidis* Gahan (Chalcididae) and *Xanthopimpla punctata* Fabricius (Ichneumonidae). A bacterium, viz. *Serratia marcescens* Bizio (Entero-bacteriaceae) has also been reported from this pest.

Damage. As a result of the numerous galleries made by the feeding caterpillars (**Plate 23.1C**), the foliage dries up. Infested trees can be recognized from the dried up patches in the fronds. In certain years, the population of the pest is quite high and damage to foliage becomes very prominent, resulting in a considerable reduction in yield.

Management. (*i*) Cut and destroy the first infested fonds by burning. (*ii*) Apply 2.5 litres of chlorphyriphos 20EC in 625 litres of water per ha.

3. Rhinoceros Beetle, *Oryctes rhinoceros* (Linnaeus) (Coleoptera : Scarabaeidae) (Fig. 23.3, Plate 23.1D)

The rhinoceros beetle is one of the most important pests of the coconut and other palms. It is found throughout South-east Asia, the Philippines and southern China. It was introduced into the South Pacific Islands during the first decade of this century and reached Mauritius in the sixties.

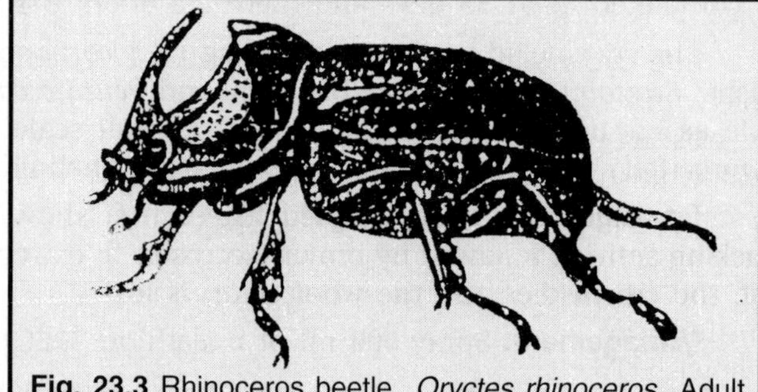

Fig. 23.3 Rhinoceros beetle, *Oryctes rhinoceros.* Adult

The stoutly built beetle has a pointed horn on its head, is elongate and cylindrical and measures about 4-5 cm (**Fig. 23.3**). It has well-developed wings and can fly long distances. It is harmful only in the adult stage when it feeds on the crown of the coconut tree. The larvae, however, feed on decaying organic matter in the ground.

Life-cycle. The oval, white, seed-like eggs are laid 5-15 cm below the soil surface in decaying organic matter. The early stages of the beetle are, therefore, not found on trees, but are generally passed in manure pits and decomposing vegetable matter, chiefly in the dead palm trunks.

A female may lay 100-150 eggs which hatch in 8-18 days and the grubs start feeding on the decaying matter found in the vicinity. The larvae pass through three instars to complete their development in 99-182 days (mean 130 days). The optimum temperature for development of the larvae is 32-40°C. Pupation takes place in chamber at a depth of about 30 cm and the beetles emerge after 10-25 days. They remain in the pupal cell for about 11-20 days before coming out of the soil and, on emergence, they are soft-bodied creatures. Soon, they fly to the nearest palm-tree and start the attack. They lay eggs after 20- 60 days. The beetles are active at night and may be attracted to a source of light. The adults can live for more than 200 days. Generally, one generation is completed in a year.

The natural enemies of the grubs are *Sarcophaga fuscicauda* Botlcher (Sarcophagidae) and *Pheropsophus hilaris* var. *sobrinus* Daj (Carabidae). The exotic predatory bug, *Platymeris laevicollis* (Distant) has shown considerable promise against this pest. The frogs, toads, birds, rats and squirrels also prey upon grubs of this pest. The grubs are also infected by *Metarrhizium anisopliae* ((Metchnikoff) and *Beauveria bassiana* (Balsamo) Vuillemin (Monillaceae).

Damage. The beetles throw out a fibrous mass while feeding in the burrows made in the young fronds. The injury is seen as a series of holes on the fronds when they open out. As a result, the growing point is soon cut off and the tree dies (**Plate 23.1E**). The damage caused by this beetle is more serious on young trees. Besides the coconut and date-palm, the beetle also attacks palmyra and some other plants.

Management. (*i*) Maintenance of sanitation in coconut gardens by proper disposal of decaying organic debris is an important step. (*ii*) Hook out the beetles from the affected crown by means of hooks during July-August (A hooked metal rod about 0.6 metre long and 0.8 mm thick with a hook at one end and a handle at the other end will serve the purpose). (*iii*)Treat the breeding places with 2.0 kg of carbaryl 50WP in 250 litres of water per ha. (*iv*) Drench the manure heaps upto a depth of 60 cm at quarterly intervals. (*v*) Baculovirus inoculated beetles @ 10-15 per ha can be released to bring down the pest population.

4. Red Palm Weevil, *Rhynchophorus ferrugineus* (Olivier) (Coleoptera : Curculionidae) (Fig. 23.4, Plate 23.1F)

The pest is distributed in India, Pakistan, Bangladesh, Sri Lanka, Malaysia, the Philippines and New Guinea. It is one of the most destructive pests of coconut palm in Maharashtra, Assam, Kerala, Tamil Nadu, Mysore and Orissa. It is also found on date palm.

This weevil is reddish-brown, cylindrical, with a long curved snout (**Fig. 23.4**). The male has tuft of hairs along the dorsal surface of the snout, whereas the female is without it. The weevil is incapable of causing direct damage but in the early stage, it is harmful.

Life-cycle. The mother weevil scoops out a small hole with its snout in the soft tissues of the trees or in the existing wounds, in the crown or trunk and lays an oval, whitish egg. A weevil may lay up to 200-500 eggs in its life-span of 3-4 months. The eggs hatch in 2-5 days and the soft whitish grubs on hatching, feed on the soft tissues and tunnel into the tree trunk. When full-fed, the grubs measure about 65 mm and are yellowish. The larval period ranges from 2 to 4 months. The grub changes into a pupa after spinning a cocoon. The weevil emerges from this cocoon on the completion of the pupal stage in about 14 days. The adults remain within the case for 11-18 days out of life-span of 50-113 days, the male surviving for a longer period than the female. The adult is capable of flight and is diurnal in habits.

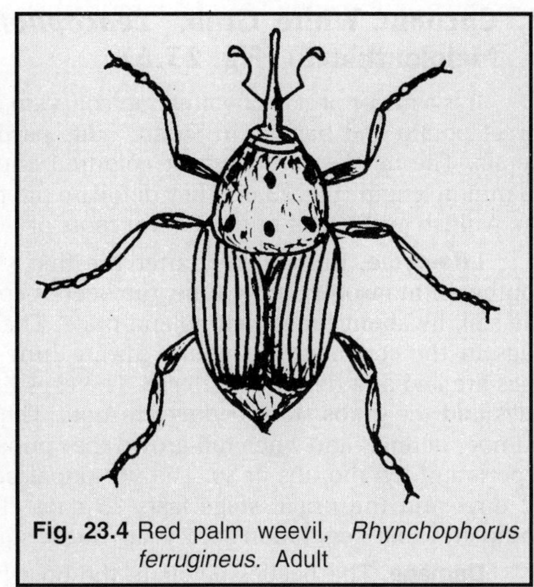

Fig. 23.4 Red palm weevil, *Rhynchophorus ferrugineus.* Adult

Damage. The larvae feed on the soft tissue of trees and often cause very severe damage, especially when a large number of them bore into the soft growing parts. The borer attacks all palms, including the coconut, the date-palm and the sago-palm (*Metroxylon* spp.). The weevil is attracted to the trees by the smell of palm juice, which flows as a result of the wounds caused by man or other agents. The dead palms also attract these insects. The pest multiplies quickly in young coconut plantations.

Management. (*i*) Destroy the infected, dying and dead palms. (*ii*) Cut petioles 120 cm away from the trunks to prevent entry of weevil. (*iii*) Inject emulsion of dimethoate 30EC @40 ml or chlorptriphos 20EC @ 2.0 litres or carbaryl 50WP @ 250 g in 25 litres of water into the live holes and plaster them with mud after application of chemicals. (*iv*) Apply 2.0 litres of chlorpyriphos 20EC in 250 litres of water to the damaged trunks or spray on crowns after every three months.

5. Coconut Weevil, *Diocalandra frumenti* (Fabricius) (Coleoptera : Curculionidae) (Fig. 23.5)

The grubs of this weevil attack coconut palm, date-palm, the oil and nipa-palms and sorghum. The pest is recorded from Sri Lanka, Myanmar, South India, Malaysia, Thailand, Indonesia and the Philippines.

The adults are small weevils (6-8 mm in length) shiny blackish with four large reddish spots in the elytra.

Life-cycle. Eggs are laid in crevices at the base of the adventitious roots at the foot of the trunk, flowers, petiole or at the base of the peduncle. The eggs hatch in 4-9 days. Larval development takes 8-10 weeks and the pupal period lasts 10-12 days. The life-cycle is completed in 10-12 weeks.

Damage. The grubs attack all parts of the coconut palm particularly the roots, the leaves, and the fruit stalks. As a result of this attack there is premature fruit-fall. The loss of yield is appreciable.

Management. Spray 2.0 kg of carbaryl 50WP in 500 litres of water per ha.

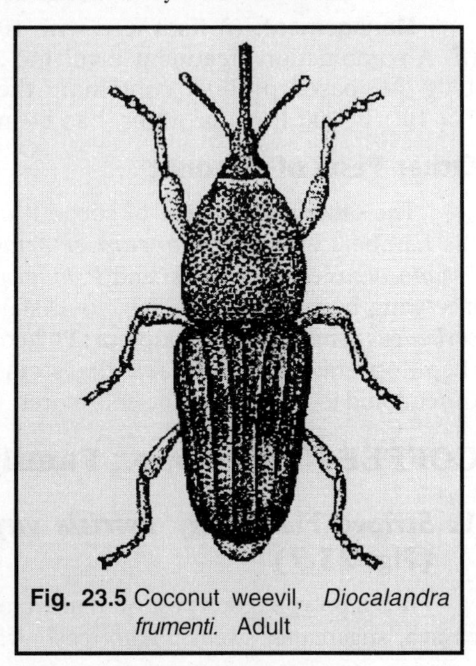

Fig. 23.5 Coconut weevil, *Diocalandra frumenti.* Adult

6. Coconut White Grub, *Leucopholis coneophora* Burmeister (Coleoptera : Melolonthidae) (Fig. 23.6)

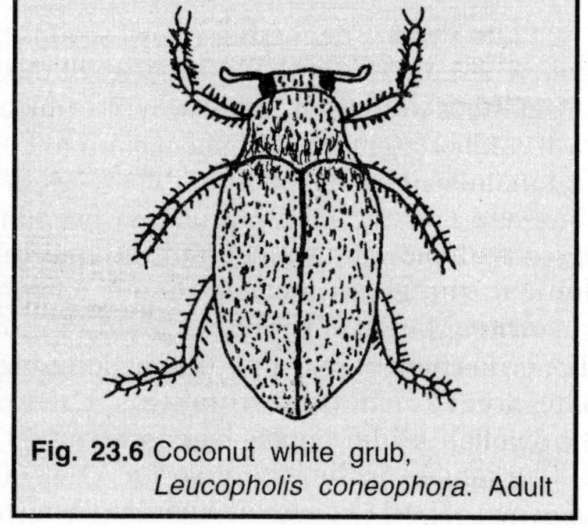

Fig. 23.6 Coconut white grub, *Leucopholis coneophora*. Adult

It is a major pest of coconut, tapicoa, yam, colocasia, sweet potato and banana in South India, particularly in Kerala. The beetles are chestnut coloured and measure 16 mm in length (Fig. 23.6). They defoliate the plants and the whitish grubs feed in the soil on roots of host plants.

Life-cycle. Immediately after the first showers of southwest monsoon during June, the beetles come out of the soil, fly about for sometime and mate. Then they lay eggs in the soil near the palms during June-July. The eggs are laid at a depth of 7-15 cm. The eggs hatch in 20 days and the grubs start feeding on roots. They moult a number of times and when full-grown they pupate, within a period of ten months or so. The pre-pupal period is 9-12 days and the pupal stage lasts 25 days. The insect completes one generation in a year. Birds, squirrels, dogs, cats and bats are the natural enemies.

Damage. The beetles defoliate the host plants. As a result of continued feeding of the grubs on roots, the vitality of the palms is reduced and their colour becomes yellowish. In the end trees become .very weak and show the symptom of pemature falling of the nuts.

Management. Dust the soil with chlorpyriphos 2 per cent @ 30 kg per ha.

7. Coconut Eriophyid Mite, *Aceria guerreronis* Keifer (Acarina : Eriophyidae)

This mite has been reported as a serious pest of coconut in Kerala and Tamil Nadu.

Life-cycle. A female lays about 18 eggs and the mite completes a generation in 9-12 days.

Damage. The mites are seen in colonies in the perianth of developing nuts and live in the white tender portion covered by the inner bract of perianth and suck the sap from the tender meristematic mesocarp tissues. Maturing nuts up to nine month old harbour the mites. The feeding results in warts and numerous longitudinal fissures on the husk of developing nuts.

Management. (*i*) Root feed with 10 ml of monocroto-phos 36WSC in 10 ml water per palm. (*ii*) A combination treatment involving 500 ml of 0·03 per cent azadirachtin oil formulation and 100g talc based product containing the fungus, *Hirsutella thompsonii* (having spore count of 1×10^7) in 200 litres of water, has been found promising.

Other Pests of Coconut

The other minor pests of coconut palm in India include the aphids, *Cerataphis variabilis* Hill Ris Lambers and *Hysteroneura setariae* (Thomas) (Hemiptera : Aphididae); the mealy bugs, *Pseudococcus coccotis* Mask. and *P. longispinus* (Targioni-Tozzetti) (Hemiptera : Pseudococcidae); the lace-wing bug, *Staphantis typicus* Distant (Hemiptera : Tingidae); the flower thrips, *Haplothrips ceylenicus* Schmutz (Thysanoptera : Phlaeothripidae); the castor slug caterpillar, *Parasa lepida* (Cramer) (Lepidoptera: Limacodidae); the stem weevil, *Discalandra stigmaticollis* Gyll (Coleoptera : Curculionidae) and the shot-hole borer, *Xyleborus parvulus* Eichhoff (Coleoptera : Scolytidae).

COFFEE (*Coffea* spp.; Family : Rubiaceae)

1. Striped Mealybug, *Ferrisia virgata* (Cockerell) (Hemiptera : Pseudococcidae) (Fig. 23.7)

It is a polyphagous pest on many crops like coffee, cocoa, citrus, cotton, jute, groundnut, beans, casava, sugarcane, sweet potato, cashew, guava, tomato, etc. This bug is found in India, Pakistan,

Bangladesh, Myanmar, Sri Lanka, Malaysia, the Philippines, Java and New Guinea. The adult female is a distinctive mealy bug with a pair of conspicuous longitudinal submedian dark stripes and long glossy wax threads and a pronounced tail (**Fig. 23.7**).

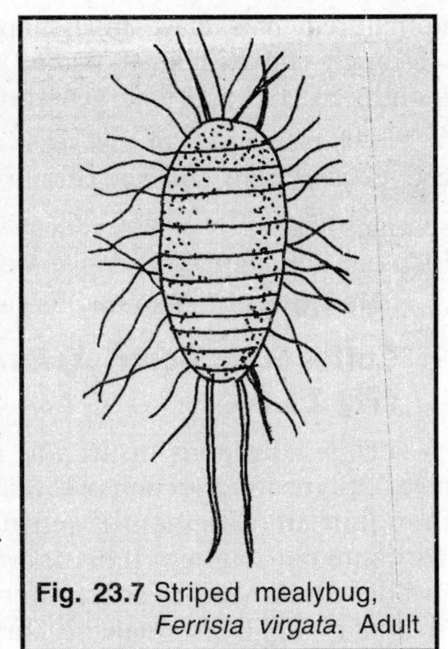

Fig. 23.7 Striped mealybug, *Ferrisia virgata*. Adult

Life-cycle. The sessile female lays 300-400 eggs, which hatch in a few hours and the young nymphs move away quite rapidly. The nymphs feed by sucking cell sap. They become full-grown in six weeks and the life cycle is completed in about 40 days. This pest multiplies in large numbers and becomes abundant during drought conditions.

Damage. The mealy bugs suck plant sap from the young shoots, berries and leaves, resulting in the withering and yellowing of plants. It is also reported to act as a vector of virus which causes the swollen shoot disease of cocoa.

Management. Spray 500 ml of quinalphos 25EC in 250 litres of water per ha.

2. Coffee Green Bug, *Coccus viridis* (Green) (Hemiptera: Coccidae)

The soft green scale is widespread in the tropics. It is found in India, Sri Lanka, Bangladesh, Myanmar, Thailand and Malaysia. It is a major pest of coffee in the States of Kerala, Andhra Pradesh, Karnataka and Tamil Nadu. In addition to coffee, this green coloured scale also feeds on citrus, guava, loquat, mango and other plants. The crawlers are flat, ovate, slightly convex, pale-green to yellowish-green in colour. When mature these are 2-3 mm long. They feed by sucking sap from a leaf or a green shoot.

Life-cycle. The scales are ovoviviparous and the males are very rare. The female lays 300-500 eggs in her life time of 2-5 months. The hatching of eggs occurs within a few hours but the young ones remain under the scale for a few days and then crawl out. They get fixed at some succulent spot for feeding. There are three nymphal instars and then they become adults. A generation is completed in 1-2 months and there are many generations in a year.

The ants, *Oecophylla smaragdina* (Fabricius), *Cremastogaster* sp. and *Camponotus* sp. (Formicidae) feed on this insect. The entomophagus fungi, *Empusa lecanii* Zimm. and *Cephalosporium lecanii* Zimm. infect the scale and thus play an effective role in the natural control of the pest.

Damage. The scales are present on the under surface of leaves, crowding along the midribs and veins. Tender branches and developing fruits are also attacked. As a result of heavy feeding, the leaves and fruits become discoloured and malformed and drop off. The plants become very weak and unproductive.

Management. Spray 750 ml of malathion 50EC or 500 ml of dimethoate 30EC or 375 ml of methyl parathion 50EC or 200 ml of phosphamidon 85WSC in 500 litres of water per ha.

3. Helmet Scale, *Saissetia coffeae* (Walker) (Hemiptera: Coccidae)

The helmet scale is a cosmopolitan pest. It is widely distributed in the tropics and in some sub-tropical areas. It attacks both *arabica* and *robusta* coffee. It has a wide range of alternative hosts including tea, citrus, guava, mango and many other plants, both wild and cultivated. The scale is green in colour when young but becomes dark brown later on. Mature scales are about 2 mm long.

Life-cycle. The mature female may lay upto 600 eggs beneath her helmet. The newly hatched crawlers are flat and oval in shape. They get fixed at suitable spots and may feed on leaves, berries, or the green shoots. The nymphs undergo three moultings and then they become adults. The immature females have an H-shaped yellow mark on their body. The dark brown adult females have a strongly convex helmet-shaped carapace and are immobile. The males are unknown. It takes six months to complete one generation.

The parasitoids of this scale are *Aneristus ceroplastae* Howard, *Coccophagus flavescens* Howard and *Coccophagus cowperi* Girault (Aphelinidae).

Damage. The scales cluster on the shoots, leaves and green berries. Because of their feeding the chlorotic leaves and fruits begin to drop.

Management. Same as in case of coffee green bug.

4. Coffee Stem Borer, *Xylotrechus quadripes* Chevrolat (Coleoptera : Cerambycidae) (Fig 23.8)

This indigenous insect was not a recorded pest until the introduction of *arabica* coffee in the first half of nineteenth century. It is now distributed in southern India as well as in Assam. Besides India, the pest has been reported in Myanmar, Sri Lanka, Thailand, Indonesia and the Philippines.

The white larvae of this beetle bore into the coffee stem, killing the young plant and giving the older ones an unhealthy appearance. The adult is a blackish-brown beetle, about 1.25 cm long with prominent antennae. There is a characteristic pattern of yellowish bands on the elytra (**Fig. 23.8**).

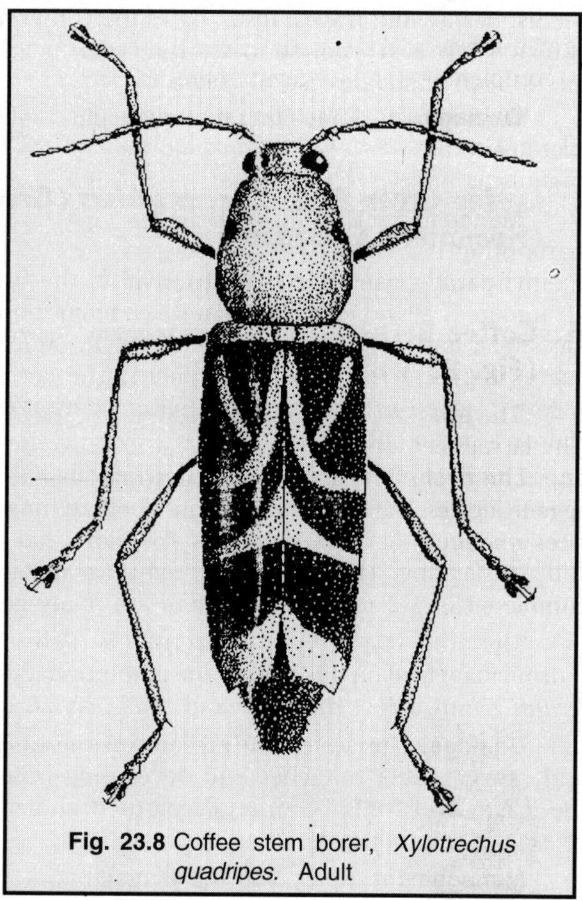

Fig. 23.8 Coffee stem borer, *Xylotrechus quadripes.* Adult

Life-cycle. A female beetle lays about 100 eggs in 3-4 weeks in the cracks and crevices of trunk bark or primary branches. They hatch in about ten days. After feeding on soft bark for sometime, the caterpillars start boring into the woody tissue. The borer continues to feed and tunnel in all directions for about ten months. When full-grown, it cuts an exit hole and pupates near it, inside the tunnel. The pupal stage is completed in about one month after which the beetle emerges. Although only one generation is completed during the year, there are two distinct periods during which the adults emerge, one occurring in April-May and the other during September-December. The emergence in autumn is greater and results in heavy egg-laying.

The grub is parasitized by *Metapelma* sp. (Encyrtidae).

Damage. The branches tunnelled by larvae wilt and break easily and the young plants often succumb to this injury. When infestation is severe, the plantation may have to be abandoned or replanted.

Management. (*i*) The pest can be suppressed by removing and destroying the affected shoots along with borers. (*ii*) The incidence can also be reduced by catching and killing the beetles, especially during their egg-laying period. (*iii*)Dislodging the eggs and young borers with tough brushes of coconut husk is also useful. (*iv*) Carry out swabbing of the main stem and thick primary branches with chlorpyriphos 20 EC emulsion @ 1.25 litres mixed in 125 litres of water per ha, once during April-May and twice during September-January.

5. Coffee Shothole Borer, *Xylosandrus compactus* (Eichhoff) (Coleoptera: Scolytidae)

At times, the borer can be a serious pest of coffee in southern India. The adult is a cylindrical, dark-brown beetle. Both the adult and the larva produce a large number of pin holes in the bark, generally on the underside of tertiary branches. This borer prefers *robusta* coffee, which it damages seriously. Both the larvae and adults tunnel through the bark and cause wilting of the branch.

Life-cycle. The female beetle bores into the bark of tertiary branches and lays eggs in the tunnels so formed. Many tunnels are made by a female which lays up to 50 eggs. These hatch in about one week. The white apodous larvae feed on ambrosia, a fungal growth developed on the beetle excreta inside the tunnel. The larval development is completed in about 20 days and the full-grown larvae make cocoons near the exits. The pupal period lasts about 10 days and the life-cycle is completed in 35-40 days. There are several generations during a year.

Damage. The tunnelling by larvae and adults results in wilting, defoliation and dieback of the plant. The area around a shot-hole becomes discoloured and secondary infection due to micro-organisms hastens the death of the affected shoots.

Management. (*i*) The pest can be suppressed by removing and destroying the affected shoots harbouring the borers, early in the season. (*ii*) Apply 2.5 litres of malathion 50EC or 1.5 litres of methyl parathion 50EC in 250 litres of water per ha.

6. Coffee Berry Borer, *Hypothenemus hampei* (Ferrari) (Coleoptera : Scolytidae) (Fig. 23.9)

The grubs of this insect are a serious pest of *robusta* as well as the low-altitude *arabica* coffee. The larvae feed on many plants of *Rubiaceae* and Leguminoseae, including *Phaseolus* and *Hibiscus* spp. This pest has been recorded from tropical Africa, Sri Lanka, South-east Asia and Indonesia. The white, legless, brown-headed grubs feed by tunnelling in the tissues of beans. The adult female beetle is about 2.5 mm long and the male about 1.6 mm **(Fig. 23.9)**.

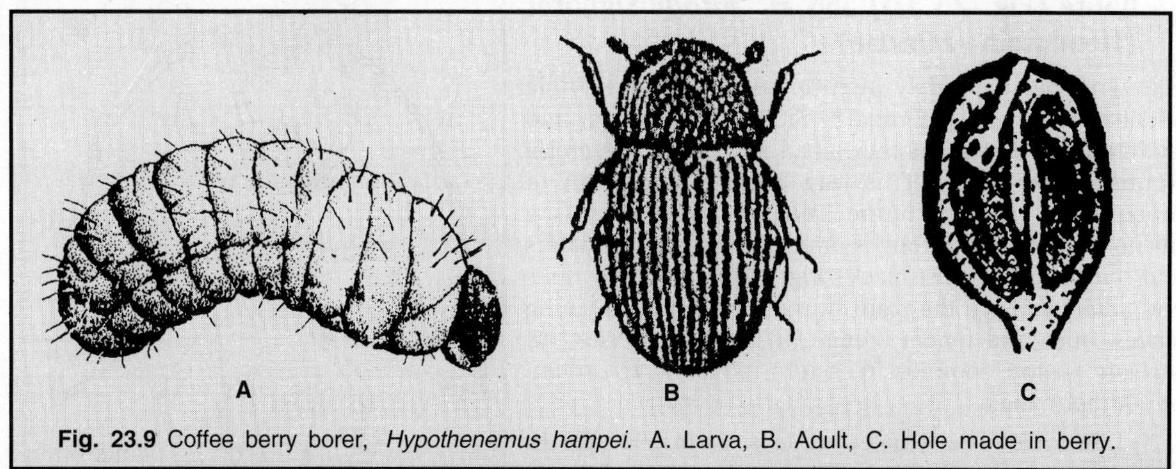

Fig. 23.9 Coffee berry borer, *Hypothenemus hampei.* A. Larva, B. Adult, C. Hole made in berry.

Life-cycle. The activity of the beetles coincides with the appearance of shoots which bear the berries. The females lay 30-60 eggs in batches of 8-12 on the maturing coffee-beans during an oviposition period of 3-7 weeks. The eggs hatch in 8-9 days. The white, legless, brown- headed grubs feed by tunnelling into the beans. The larval period is completed in 2-3 weeks, after undergoing 2-3 moultings. The pupal stage lasts 7-8 days and is spent in the larval galleries. The females mate with the flightless males inside the galleries. They fly from tree to tree, depositing the eggs.

The Uganda wasp, *Prorops nasuta* Waterson (Bethylidae) is a predator of the grubs that bore inside the shoots and the beans.

Damage. A number of beetles at a time make one or more round holes near the apex of green or the ripe berries. The damaged beans contain upto 20 larvae of different sizes. The damage caused by this pest makes the coffee beans unfit for marketing.

Management. Spray 1.25 litres of quinalphos 25 EC in 625 litres of water per ha.

Other Pests of Coffee

Other insects which appear as minor pests on the *arabica* coffee are the green leaf-hopper, *Nephotettix nigropictus* (Stal) (Hemiptera : Cicadellidae); the coffee whitefly, *Dialeurodes vulgaris* Singh (Hemiptera : Aleyrodidae); the aphid, *Toxoptera aurantii* (Boyer de Fonscolombe) (Hemiptera: Aphididae); the mealy bug, *Planococcus citri* (Risso) and *P. lilacinus* (Cockerell) (Hemiptera : Pseudococcidae); the guava scale, *Chloropulvinaria psidii* (Maskell) (Hemiptera : Coccidae); the coffee shoot and berry bug, *Antestiopsis cruciata* (Ghesquiere & Carayon) and *Nezara viridula* (Linnaeus) (Hemiptera : Pentatomidae); thrips, *Haplothrips ceylonicus* Sch. (Thysanoptera : Phlaeothripidae); leaf thrips, *Scirtothrips bispinosus* Bagnall *Heliothrips haemorrhoidalis* (Bouche) and *Thrips florum* Schmutz (Thysanoptera : Thripidae); the red borer, *Zeuzera coffeae* Nietner (Lepidoptera: Cossidae) **(Plate 23.1G, H)** ; the flush worm, *Homona coffearia* (Nietner) (Lepidoptera: Tortricidae); *Agrotis segetum* (Denis & Schiffermuller) and *Spodoptera litura* (Fabricius) (Lepidoptera: Noctuidae); coffee leaf-miner, *Melanagromyza coffeae* (Koningsberger). (Diptera: Agromyzidae); the root grub, *Holotrichia coniferta* S. and *Serica pruinosa* Saylor (Coleoptera: Scarabaeidae); mites, *Oligonychus coffeae* (Nietner) (Acari : Tetranychidae). The *robusta* is comparatively less susceptible to insect pests. Besides the shot-hole borer, it is occasionally attacked by the green bug, the mealy bug and the mites.

TEA (*Camellia* spp.; Family : Theaceae)

1. Tea Mosquito Bug, *Helopeltis theivora* Waterhouse (Fig. 23.10) and *H. antonii* Signoret (Hemiptera : Miridae)

This pest is widely distributed in southern India and has also been recorded in Sri Lanka, Vietnam and Indonesia. In Africa, another allied species, *H. bergrothi* Reuter is prevalent. This bug bears no relation to mosquitoes and the name 'tea mosquito bug' is a misnomer. The female bug is orange across the shoulders and the male is almost black **(Fig. 23.10)**. Both nymphs and adults damage the plant by sucking sap from young leaves, buds and tender stems. Of the two species, *H. theivora* is more common in eastern India and *H. antonii* in southern India.

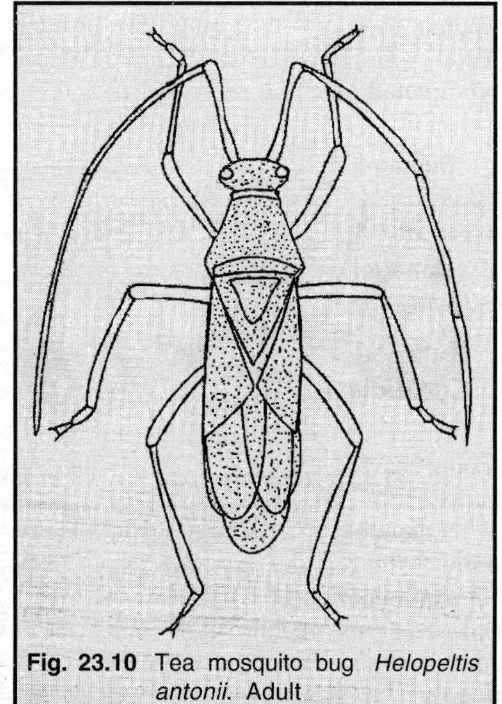

Fig. 23.10 Tea mosquito bug *Helopeltis antonii.* Adult

Life-cycle. The female bug lays up to 500 eggs in the buds and axils of the leaves or often in the broken end of the plucked shoot. The eggs are elongate, sausage-shaped, having two long unequal protruding filaments. They hatch in 5-27 days, depending upon the prevailing temperature. A freshly emerged nymph is wingless and on account of its long appendages looks like a spider. In summer, it completes its development in about two weeks after passing through five moults. In winter, the nymphal stage is prolonged to eight weeks or more. The adult is slender, 6-8 mm long, agile and a good flier. It feeds at night on tea and several alternative host plants. During flight, the adults may be carried long distances by the wind and get dispersed over wide areas. Throughout the day, they are seen concealed on the underside of the leaves. The nymphs are parasitized by a nematode, *Agamermis paradecaudate* Steiner.

Damage. The nature and extent of damage caused by various species of the genus *Helopeltis* vary a great deal. The attack of *H. theivora* in India and Sri Lanka is mostly on young leaves. The bug punctures the leaf frequently to suck the juice. While feeding, the saliva is injected and the tissues around the site of feeding are necrosed, becoming brownish or black. By the coalescence of a large number of such patches, the entire leaf may become black and shrivelled, and may fall off eventually. A single bug is capable of destroying two shoots per day. In severe attack, causing defoliation, the shoots are killed and the plants appear like brooms. The bugs prefer a moist, warm atmosphere and the tea bushes that provide this microclimate are preferred by the pest. *H. antonii* has a number of alternative host plants and damages the guava fruit, cashewnut and many medicinal plants in the Karnataka State.

Management. (*i*) Collect nymphs and adults with hand nets early in the morning or in the evening and destroy them. (*ii*) Spray the bushes with 500 ml of malathion 50EC in 500 litres of water per ha.

2. Bunch Caterpillar, *Andraca bipunctata* Walker (Lepidoptera : Bombycidae)

The pest is widespread in India, Indonesia, Formosa and Vietnam. The larvae are gregarious and from the third instart onwards, they congregate on the branches of food plants during day-time. The brown moths have a wing span of 40-50 mm and the fore wings have wavy cross lines with two white spots near the outer margins.

Life-cycle. The female moth lays up to 500 yellowish eggs in linear clusters on the underside of leaves. The larvae emerge in about ten days and are light yellow. At first they eat up their egg-shell and later on, they start feeding on the leaf tissues. The larvae pass through five instars and complete their development in 20-30 days. The full-grown larvae descend to the ground and spin cocoons among dry leaves. The cocoons are noticed on the ground in batches. The pupal period is completed in 15-30 days. In the north-eastern region, the pest passes through four generations in a year.

Damage. The caterpillars feed on leaves and are potentially a dangerous pest. The population of this insect fluctuates tremendously and is kept under check by a fly parasite and a bacterial disease which are spread easily owing to gregarious nature of the caterpillars.

Management. (*i*) The pest can be suppressed by collecting the caterpillars manually and destroying them. (*ii*) Spray 500 ml of malathion 50EC in 500 litres of water per ha.

3. Humped Slug Caterpillar, *Spatulicraspeda castaneiceps* Hampson (Lepidoptera : Cochlidiidae)

This caterpillar is quite common on tea in Sri Lanka and India. It also feeds on castor and coconut leaves. The male moth is dark brown with bipectinate antennae and has a wing expanse of 15-17 mm. The female moth has chestnut brown fore wings and smoky black hind wings with a wing expanse of 20-22 mm. The full-grown caterpillars are small, dark brown or pale brown with a reddish band.

Life-cycle. The pest is active from June to September. The female moth lays 190-200 pale yellow and oval eggs on the under surface of leaves. The incubation period of eggs is 5-6 days. The young larvae start feeding on tender leaves. The larval stage is completed in 20-22 days. The larva pupates in a broadly oval globular cocoon attached to the plant. The pupation is completed in

13-14 days. The duration of life-cycle from egg to adult stage ranges from 38 to 42 days. The larvae are parasitized by *Apanteles* sp. (Braconidae), *Eucepsis* sp. (Chalcididae) and *Chrysis* sp. (Chrysididae)

Damage. The caterpillars mainly feed on the young leaves of tea which are the primary commercial product. Hence, the pest causes a significant damage to the crop.

Management. Spray 625 g of malathion 50WP in 625 litres of water per ha.

4. Red Borer, *Zeuzera coffeae* Nietner (Lepidop-tera : Cossidae) (Fig. 23.11, Plate 23.1G, H)

The pest is distributed in all the tea growing states. The moth is white with many small black spots on forewings and marginal dots on hindwings, and measures 28-40 mm across stretched wings. (**Fig. 23.11**). The full grown larva is purplish brown with brown head and measures 38 mm.

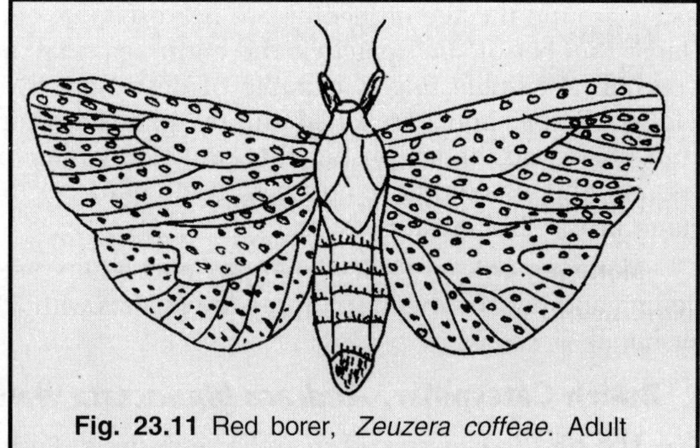

Fig. 23.11 Red borer, *Zeuzera coffeae.* Adult

Life-cycle. The eggs are deposited in rows on the surface of leaves, cracks and crevices of the bark or even in the soil. The eggs hatch in about 10 days and the larva bores into the bark. The caterpillar tunnels down entering the main stem, reaching up to the tap root in young plants. The larva becomes full grown in 4-5 months and pupates within its tunnel for a period of 3-4 weeks. The emergence of moths takes place during April-May and August-September.

Damage. Leaves of the attacked branches wither and the branches themselves die eventually. The attacked plants show pinkish excreta emerging out of the holes on the stem. Young plants when attacked are killed outright.

Management. (*i*) Prune and remove the affected branches. (*ii*) Chemical control measures are same as in case of red palm weevil.

5. Red Crevice Tea Mite, *Brevipalpus phoenicis* (Geijskes) (Acari : Tenuipalpidae) (Fig. 23.12)

This mite is a sporadic pest of tea in India, Sri Lanka and Malaysia. It also attacks citrus, coffee, rubber, *Phoenix* spp., *Grevillea,* some medicinal plants and *Parthenium hyterophorus.* The adult is rather flat, elongate and oval. It is scarlet red in colour with black marks dorsally (**Fig. 23.12**). It measures 0.3 mm in length.

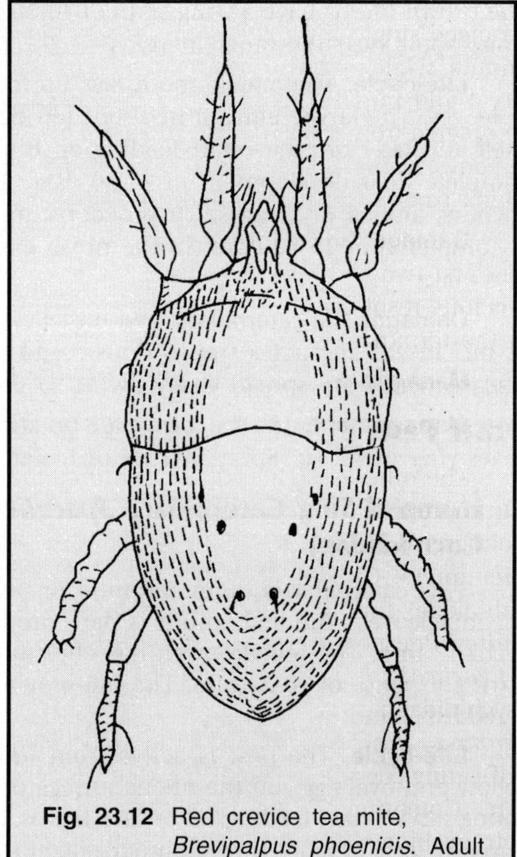

Fig. 23.12 Red crevice tea mite, *Brevipalpus phoenicis.* Adult

Life-cycle. A female has been observed to lay up to 47 bright red, oval shaped eggs during an oviposition period of 40 days. The eggs are stuck firmly to the underside of leaves or in the crevices in comparatively young bark. They hatch in 6-13 days. The nymphs are flat-bodied, oval in outline and are scarlet in colour. The life-cycle is completed in 21-28 days. The predacius mite, *Typhlodromus pyri* Scheuten (Acari : Phytoseiidae) feed on different stages of this acarine pest.

Damage. All active stages feed on the underside of leaves, especially along the midrib and the base. The loss of cell sap causes yellowing of the leaves. The bark and the leaf petioles of the affected shoots split, turn brown and dry up.

Management. Spray 1.25 litres of methyl demeton 25EC or 1.5 litres of dimethoate 30EC in 625 litres of water per ha.

6. Yellow Tea Mite, *Polyphagotarsonemus latus* (Banks) (Acari : Tarsonemidae) (Fig. 23.13)

The yellow tea mite (broad mite) is widely distributed in tea and cotton growing areas of the world. In Asia, it is found in Bangladesh, India, Malaysia, the Philippines and Sri Lanka. In addition to tea and cotton, this sporadic pest also feeds on coffee, jute, tomato, potato, chillies, sesame, castor, beans, peppers, avocado, citrus, mango and rubber. The adult is yellow with white stripes on the dorsal side and is about 1.5 mm long **(Fig. 23.13)**

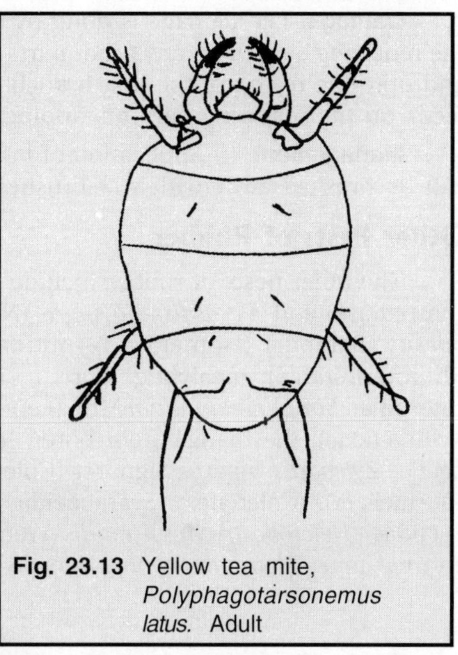

Fig. 23.13 Yellow tea mite, *Polyphagotarsonemus latus.* Adult

Life-cycle. The male mites emerge earlier than the females. The latter lay oval but flattened eggs on the underside of the young flush of leaves. The eggs are arranged singly and they hatch in 2-3 days. The larvae are minute, white and pear-shaped which feed on cell sap. The larval and pupal stages last 2-3 days and the life-cycle is completed from 4 to 5 days. The female mite lives for about ten days, laying 2-4 eggs per day. The mite transmits a virus which causes leaf curl or 'Murda' disease of chillies.

Damage. The mite is a serious pest in tea nurseries. The first two or three leaves and the buds are infested by this mite resulting in the browning of leaves. The shoot growth is also slowed down and production of the terminal tea-leaves is reduced very much.

Management. Same as in case of red crevice tea mite.

Other Pests of Tea

The other pests of tea include the dark brown cricket, *Brachytrypes portentosus*) Lichtenstein (Orthoptera: Gryllotalpidae); the termites, *Odontotermes assamensis* Holmgren and *O. parvidens* Holmgren & Holmgren. Holmgren (Isoptera : Termitidae); the tea jassid, *Empoasca flavescens* Fabricius (Hemiptera: Cicadellidae); tea aphid, *Toxoptera aurantii* (Boyer de Fronscolombe) (Hemiptera : Aphididae); thrips, *Scirtothrips dorsalis* Hood (Thysanoptera: Thripidae) and black thrips, *Haplothrips andresi* Priesner and *H. tenuipennis* Bag. (Thysanoptera : Phlaeothripidae); the tea tortrix, *Homona coflearia* (Nietner) (Lepidoptera : Tortricidae); various bag worms, *Clania* spp. (Lepidoptera: Psychididae); castor semi- looper, *Achaea janata* (Linnaeus) (Lepidoptera: Noctuidae); tea leaf miner, *Agromyza theae* Meij (Diptera : Agromyzidae); the white grub, *Holotrichia impressa* (Brumeister) and leaf-eating cockchafer, *Serica assamensis* Brenske (Coleoptera: Scarabaeidae); the root borer, *Batocera* spp. (Coleoptera : Cerambycidae); the red spider mite, *Oligonychus coffeae* (Nietner) (Acari : Tetranychidae) and a number of species of root-knot nematodes.

RUBBER (*Havea brasiliensis* Muell. Arg; Family : Euphorbiaceae)

Rubber Bark Caterpillar, *Aestherastis circulata* Meyrick (Lepidoptera : Hyponomentidae)

The pest is distributed in rubber growing areas like Tamil Nadu and Karnataka. The adult is small, white and black spotted moth. The larva is bright red flat caterpillar, measuring about 2.5 cm when full grown.

Life-cycle. The eggs are laid on the bark of the rubber tree. The eggs hatch in 2-4 days. The larva feeds under a web on the bark. It has a duration of about three weeks and it pupates under a piece of bark in a web for 10 days. There are 2-3 generations in a year.

Damage. The damage is done by the caterpillar by feeding on the bark of the tree, either on the renewing bark or on any other part of the tree but generally at a height of 90 cm from the ground and upto the region of the first branch. Usually the caterpillar feeds on the dead bark, but when it feeds on the renewing bark or tapping surface, it becomes troublesome.

Management. (*i*) Application of tar on the bark surface prevents the attack. (*ii*) The caterpillars can be brushed away with stiff brushes and destroyed.

Other Pests of Rubber

The other pests of rubber include scale insects, *Abgrallaspis cyanophylli* Signoret (Hemiptera: Diaspididae) and *Parasaissetia nigra* (Nietner) (Hemiptera : Coccidae). the termites, *Odontotermes obesus* (Rambur) (Isoptera : Termitidae) and *Coptotermes curvignathus* Holmgren (Isoptera : Rhinotermitidae); mealybug, *Ferrisia virgata* (Cockerell) (Hemiptera : Pseudococcidae); mulch caterpillar, *Simplicia simulata* Mre. (Lepidoptera : Noctuidae); *Ptochoryctis rosaria* Meyr. (Lepidoptera: Xyloryctidae); the mango stem borer, *Batocera rubus* Linnaeus (Coleptera : Cerambycidae); bark beetle, *Zyleborus biporus* Signoret (Coleptera : Scolytidae); cockchafer beetle, *Holotrichia bidentata* (Burmeister). (Coleoptera: Scarabaeidae); boring beetles, *Minthea rugicollis* (Walker) (Coleoptera : Lyctidae), *Heterobostrychus aequalis* (Waterhouse) and *Sinoxylon anale* Lesne (Coleptera : Bostrychidae) and red mite, *Hemitarsonemus dorsalis* Ewing (Acari: Tarsonemidae).

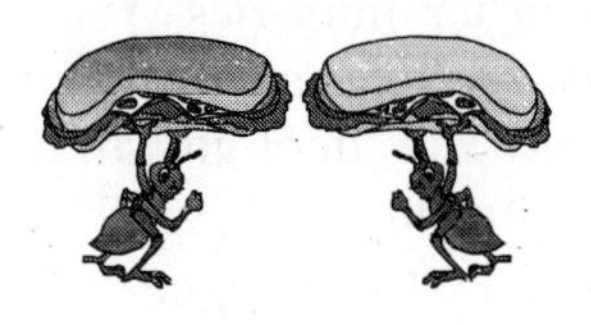

PESTS OF SPICES

INTRODUCTION

India is the largest producer, consumer and exporter of spices and spice products in the world. Over 100 plant species are known to yield spices and spice products, among which about 50 are grown in India. The spice production in India is currently (2012-13) at 5.74 million tonees from an area of about 3.08 million ha.

Pepper is considered to be the 'king of spices,' and cardamom the 'queen of spices'. In India, pepper covers 1,25,000 ha with an annual production of 53,000 tonnes. Kerala, Karnataka and Tamil Nadu produce more than 90 per cent pepper. Cardamom (small) is also grown in these three states of India over 92,000 hectares of land, with a total production of 18,000 tonnes in a year. In India, large cardamom is mainly cultivated in Sikkim and West Bengal over an area of 30.,039 ha producing 5401 tonnes of cardamoms. Ginger is grown in almost all the states and the total acreage of this crop is about 1.36 lakh ha and the annual production is 6.83 lakh tonnes of dry ginger. Turmeric is grown in many states and covers an area of 1,94,000 hectares with an annual production of 9,71,400 tonnes. Chilli occupies an area of 7.94 lakh hectars in India with an average annual production of 13.04 lakh tonnes. India produces half the production of chilli in the world.

Since spices come from different kinds of plants, belonging to a number of families, the pests that have adapted themselves to feeding on these aromatic plants are also of different types. Apart from the field pests of the spice plants, a number of insect pests of stored products also damage spices in storage.

CARDAMOM (*Ellettaria cardamomum* L.; Family : Zingiberaceae)

About 56 species of insects and mites have been reported to attack cardamom in India.

1. Banana Aphid, *Pentalonia nigronervosa* Coquerel (Hemiptera : Aphididae) (Fig. 24.1)

The banana aphid is more or less pantropical in distribution and is of common occurrence in southern parts of India, Sri Lanka and Australia. In addition to banana, it also feeds on small and large cardamom, *Colocasia* sp., *Alocasia* sp. and *Caladium* sp.

The wingless aphid is dark brown pyriform measuring 1.34 mm in length and with six segmented antennae which are longer than the body. Abdomen is dark brown, shining and slightly bulged. The winged form is dark brown, elongated and pyriform. They are longer than the wingless forms but with less body width.

Life-cycle. The reproduction takes place parthenogenetically. The longevity of adult varies from 8 to 26 days with an average of 14 days. A single female lays 8-28 offsprings with an average of 14. A single female may produce as many as four offsprings in 24 hours. The development is completed through three and four moults taking 12.6 and 15 days, respectively from November till January. There are 21-24 generations in a year.

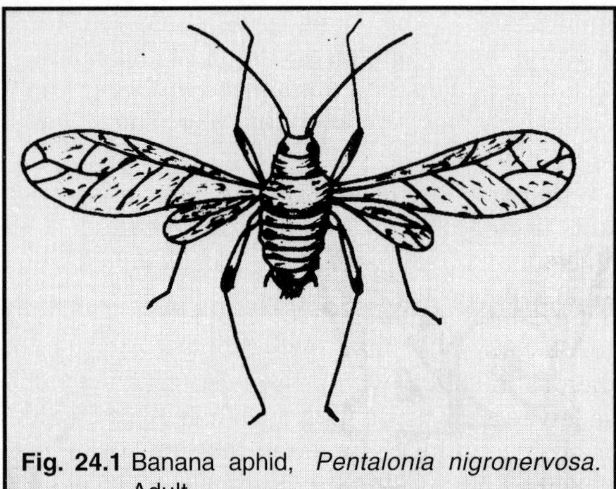

Fig. 24.1 Banana aphid, *Pentalonia nigronervosa*. Adult

Damage. The aphids feed on the leaf sheath and pseudostem. The insect causes little direct damage but is of considerable significance being vector of cardamom mosaic (*Kattle* disease), *Amomum* mosaic and *Foorky* disease of large cardamom. The aphids are disseminated in cardamom plantation mostly by crawling from plant by the contact of foliage at the top and along the soil.

Management. Spray 300 ml of phosphamidon 85WSC or 875 ml of dimethoate 30EC in 250 litres of water per ha at an interval of two weeks.

2. Cardamom Thrips, *Sciothrips cardamomi* (Ramkrishna Ayyar) (Thysanoptera : Thripidae, Plate 24.1A)

The cardamom thrips is the most destructive pest of cardamom in South India. The adult is greyish-brown and measures 1.25-1.50 mm in length.

Life-cycle. The pest is active throughout the year except during the monsoon season. The female lays 5-71 eggs at random on all the feeding areas of the plant. The young nymphs emerge from the eggs in 9-12 days. The first two nymphal instars are active and grow by feeding on the plant sap. Life-cycle is completed in 25-30 days.

Damage. The thrips cause damage by sucking cell sap. It feeds on tender blossoms and the bunch pods of cardamom. The attack on the flower stalk results in shedding of flowers. The panicle stalks also become stunted and do not bear flowers.

Management. Spray 875 ml of dimethoate 30EC or 1.0 litre of quinalphos 25EC in 250 litres of water per ha.

3. Castor Capsule Borer, *Conogethes punctiferalis* Guenee (Leidoptera : Pyralidae)

The castor capsule borer is a serious pest of nursery plants and young green pods of cardamom also. In the nursery plants, it bores into the stem and causes the death of the central shoot. It also eats away the tender seeds of the young berries. The full account of this pest is given in Chapter 15.

4. Cardamom Hairy Caterpillars

(*i*) *Lenodera vittata* Walker (Lepidoptera : Lesiocampidae)

The hairy caterpillars are commonly found feeding on cardamom in South India. The moth is stout and fairly big and densely covered with scales. The larvae are clothed with a dense felt of capitate hairs and measure 106-110 mm in length.

Life-cycle. The moths emerge in June and lay cream coloured dome shaped eggs in rows on both the upper and the lower surface of leaves. A female lays 100-130 eggs during an oviposition period 6-9 days. The young larvae emerge from the eggs in 10-13 days. The larvae start feeding on the leaves and other tender parts of the plant and moult six times during the larval period of 112-118 days. Pupation takes place in the soil in an earthen cell, in which it stays for 5-7 months. There is only one generation in a year. The larvae are parasitized by a tachinid fly, *Carcelia kockiana* (Tachinidae).

Damage. The caterpillars are voracious feeders and cause extensive damage to the cardamom plants from August to December by feeding on leaves. Only the pseudostems and midribs remain un-eaten.

(*ii*) *Eupterote cardamomi* Ranga Ayyar (Lepidoptera: Bombycidae)

E. cardamomi is a sporadic pest of cardmom in South India. The adults are large moths, ocherous in colour, with post medial lines on the wings. They measure 70-80 mm in wing expanse. The larvae are hairy, dark grey in colour with pale brown head, bearing conical tufts of hairs on the dorsal side of the body. When full-grown the caterpillar measures 90 mm in length.

Life-cycle. The moths emerge with the commencement of the South-West monsoon rains in June and July. The female moths lay 400-500, yellowish and dome-shaped eggs in flat masses on the undersurface of leaves. Each egg mass contains about 50-160 eggs. The hatching of eggs occurs in 15-17 days. The larva passes through ten instars in 140-151 days. It pupates in the soil in a silken cocoon at a depth of 5-8 cm for 7- 8 months. The moth lives for about 20 days. There is only one generation in year.

Larvae are parasitized by *Sturmia sericariae* (Tachindae) and *Aphanisles eupterote* D. (Ichneumonidae).

Damage. The larvae feed on leaves of the shade trees up to the 6th or 7th instar and then they drop drown on the cardamom plants growing underneath, with the help of silken threads. They start feeding on the leaves voraciously and defoliate the cardamom plants causing heavy reduction in the yield.

(*iii*) *Eupterote canarica* Moore, *E. testacea* Walker. and *E. fabia* Cramer (Lepidoptera : Bombycidae)

These hairy caterpillars also appear sporadically and cause damage to cardamom plants.

The life-cycle and habits are more or less similar to the above mentioned species.

Management. Spray 500 ml of malathion 50EC or 500 g of carbaryl 50WP in 250 litres of water per ha.

5. Rhizome Weevil, *Prodioctes haematicus* Chevrolat (Coleoptera : Curculionidae)

The rhizome weevil is found widely on cardamom plants in various States of South India. The adult is a brown weevil measuring 12 mm in length.

Life-cycle. The weevils emerge in large numbers in April, soon after an early shower of the monsoons. The eggs are laid in cavities made on rhizomes. The young grubs come out of the eggs in 8-10 days and bore into the rhizome, making tunnels. The larvae feed inside the rhizomes and become full-fed in three weeks. They pupate within the feeding tunnels for another 3 weeks. On emergence, the adult weevils live for 7-8 months. There is only one generation in a year.

Damage. The severe tunnelling and feeding by grubs inside the rhizomes results in the death of entire clumps of the cardamom plants.

Management. (*i*) Destroy affected plants/seedings.

(*ii*) If the grub population is more in the soil, drench the base of the clamp with 1.25 litres of malathion 50EC or 1.25 kg of carbaryl 50WP in 625 litres of water per ha.

Other Pests of Cardamom

The cardamom plants are also attacked by a number of pests like the root grub, *Basilepta fulvicorne* Jacoby (Coleoptera : Chrysomelidae) **(Plate 24.1B)** wingless grasshopper, *Orthacris* sp. (Orthoptera : Acrididae); the leaf-hopper, *Tettigoniella ferruginea* (Fabricius) (Hemiptera: Cicadellidae); the spittle bug, *Aphrophora nuwarans* Distant (Hemiptera : Aphrophoridae); the banana lacewing bug, *Stephanitis typica* (Distant) (Hemiptera : Tingidae); *Riptortus pedestris* (Fabricius). (Hemiptera: Coreidae); the cardamom thrips, *Leewania maculans* Pr. and Sesh. (Thysanoptera: Thripidae); the bag worm, *Acanthopsyche bipar* Walker (Lepidoptera : Psychidae); the root borer, *Hilarographa caminodes* Meyr (*Lepidoptera :* Plutellidae); the cutworm, *Areilasisa plagiata* M. (Lepidoptera : Noctuidae); *Attacus atlas* Linnaeus (Lepidoptera : Saturnidae); *Homona* sp. (Lepidoptera : Tortricidae); the looper caterpillar, *Anisodes denticulatus* Hampson (Lepidoptera: Geometridae); *Lampides elpis* Godart (Lepidoptera: Lycaenidae); *Euproctis lutifacia* Hampson (Lepidoptera: Lymantriidae) and the root gall midge, *Hallomyia cardamomi* Nayar (Diptera: Cecidomyiidae).

LARGE CARDAMOM (*Amomum subulatum* Roxburgh; Family: Zingiberaceae)

Although there is no major pest attacking large cardamom but aphids and thrips are considered to be quite important.

1. Banana Aphid, *Pentalonia nigronervosa* Coquerel (Hemiptera : Aphididae)

For the details on its life-cycle, see under 'Cardamom'.

2. Grapevine Thrips, *Rhipiphorothrips cruentatus* Hood (Thysanoptera : Heliothripidae)

This thrips infests the leaves, which turn brown and wither gradually. For full account of the life-cycle of this pest see Chapter 18.

In addition, the other minor pests of large cardamom include the aphid, *Rhophalosiphum maidis* Fitch (Hemiptera : Aphididae), which transmits the virus causing the "Chirke" disease in West Bengal ; the leaf caterpillar, *Clelea plumbiola* Hmp. (Lepidoptera: Zygaenidae), which feeds on the green matter of leaves.

CHILLIES (*Capsicum annuum* L.; Family: Solanaceae)

More than 20 insect species have been recorded attacking both leaves and fruits of chilli.

1. Cotton Whitefly, *Bemisia tabaci* (Gennadius) (Hemiptera : Aleyrodidae)

For full details on the life-cycle of this pest, see Chapter 16.

2. Chillies Thrips, *Scirtothrips dorsalis* Hood (Thysanoptera : Thripidae) (Fig. 24.2)

The chillies thrips is a polyhagous pest and is widely distributed in India. It feeds on a number of plants including chillies, tomato, castor, sunflower, cotton, mango, citrus and *Acacia arabica* L. The adults are slender, yellowish brown in colour, having apically pointed wings, and they measure about 1 mm in length. The females possess long, narrow wings with the fore margin

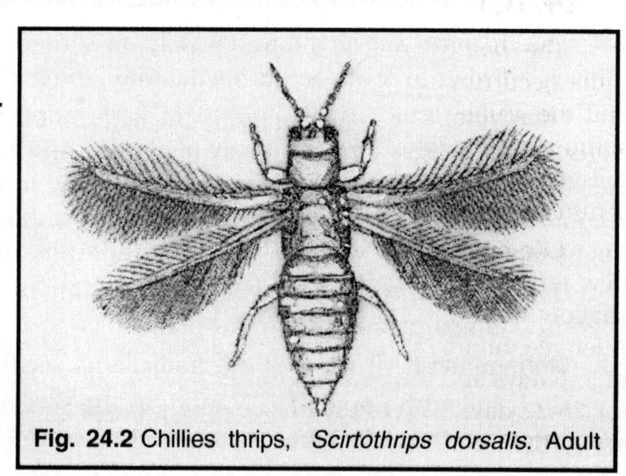

Fig. 24.2 Chillies thrips, *Scirtothrips dorsalis.* Adult

fringed with long hairs. The nymphs resemble the adults in shape and colour but are wingless and smaller in size.

Life-cycle. This pest is active throughout the year except during the rainy season. The female thrips lays 45-50 eggs inside the tissues of the leaves and shoots. The eggs hatch in 5 days. The larvae feed for 7-8 days and pupate in 2-4 days. The adult thrips lives for about 31 days. There are several over-lapping generations of this pest in a year.

Damage. Damage is caused by the adults as well as by the nymphs. They suck the cell sap from tender regions and cause the leaves to shrivel. In case of severe infestation, there is malformation of leaves, buds and fruits, which may damage half the crop. The attacked plants are stunted and may finally dry up. The insect is also responsible for transmitting the virus causing leaf curl disease of chillies.

Management. Spray 1.0 litre of malathion 50EC or 1.0 kg of carbaryl 50WP in 250-300 litres of water per ha.

3. White Grubs, *Holotrichia consanguinea* Blanchard and *H. insularis* Brenske (Coleoptera: Scarabaeidae)

The chillies crop is seriously damaged by these insects in Rajasthan. Damage is caused by the grubs that feed on the roots. A full account of the life-cycle is given in Chapter 15.

Other Pests of Chillies

The chillies crop is also attacked by a number of minor pests like *Tricentrus bicolor* Distant (Hemiptera : Membracidae); the aphid, *Aphis gossypii* Glover (Hemiptera: Aphididae) (Plate 28); the thrips, *Coliothrips indicus* (Bagnall) and *Frankliniella sulphurea* (Schmutz) (Thysanoptera: Thripidae); the stem borer, *Euzophera perticella* Ragonot (Lepidoptera : Pyralidae); fruit borers, *Spodoptera exigua* (Hubner), *S. litura* (Fabricius) and *Helicoverpa armigera* (Hubner) (Lepidoptera: Noctuidae); *Anomala bengalensis* (Blanchard) (Coleoptera : Scarabaeidae); the root grub, *Arthrodeis* sp. (Coleoptera: Tenebrionidae) and the mite, *Polyphagotarsonemus latus* (Banks) (Acari : Trasonemidae).

BLACK PEPPER (*Piper nigrum* L.; Family: Piperaceae)

About 20 insect species have been recorded damaging pepper plantations.

1. Pollu Beetle, *Longitarsus nigripennis* (Motschulsky) (Coleoptera : Chrysomelidae) (Fig. 24.3, Plate 24.1C)

The *pollu* beetle is a specific pest of black pepper in India occurring regularly in the plantations. Both the adults and the grubs cause damage to berries. The adult is a small shining, yellow and blue flea-beetle with stout hind legs. The full-grown grub is yellowish with a black head and it measures 5 mm in length.

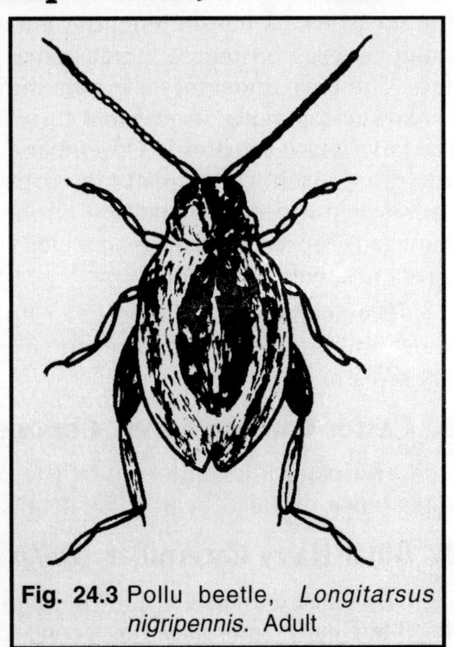

Fig. 24.3 Pollu beetle, *Longitarsus nigripennis*. Adult

Life-cycle. The adult beetles appear in July when the new tender berries appear on the plants. The females make shallow holes on the berries and lay 1-2 eggs in each hole. A female, on an average, lays about 100 eggs. The eggs hatch in 5 -8 days and the young grubs bore into the berry and feed for 20-32 days. Then they drop to the ground and pupate in an earthen cell in the soil at 5.0-7.6 cm depth. The adults emerge in 6-7 days after pupation. The total life-cycle is completed in 39-50 days. The insect completes four overlapping generations in a year.

Damage. The grubs cause damage by boring into the berries and eating the contents completely in about 10 days. Each grub destroys at least 3-4 berries during the larval period. The attacked berries appear dark in colour, are hollow inside and crumble when pressed (**Plate 24.1C**). The grubs may also eat into the spike and cause the entire distal region to dry up. The adults feed voraciously on tender leaves and make holes in them.

Management. (*i*) Tilling the soil at the base of vines at regular intervals can reduce the population considerably. (*ii*) Spray 1.5 litres of dimethoate 30EC or 2.0 litres of quinalphos 25EC in 500 litres of water per ha in late July and again in early October.

Other Pests of Black Pepper

The pepper plant is also attacked by the whitefly, *Aleurocanthus piperis* Maskell (Hemiptera: Aleyrodidae); *Ferrisia virgata* (Cockerell) and *Planococcus citri* (Risso) (Hemiptera : Pseudococcidae) (**Plate 24.1D**); the scales, *Lepidosaphes piperis* (Green), *Marsipococcus marsupiale* Green, *Pinnaspis aspidistrae* S. and *Aspidiotus destructor* Signoret (Hemiptera: Diaspididae); the thrips. *Gynaikothrips karnyi* Bagnall and *Andrethrips flavipes* Karny (Thysanoptera : Phlaeothripidae); the vine shoot borer, *Cydia hemidoxa* Meyrick (Lepidoptera : Tortricidae); *Cricula trifenestrata* Heffer. (Lepidoptera: Saturnidae); *Cecidomyia malabarensis* Felt (Diptera : Cecidomyiidae); *Neculla pollinaria* Baly and *Pagria costatipennis* Jacoby (Coleoptera: Alticidae).

TURMERIC (*Curcuma longa* L.; Family : Zingiberaceae)

1. Skipper Butterfly, *Udaspes folus* Cramer (Lepidoptera: Hesperiidae) (Fig. 24.4, Plate 24.1E)

This is a common pest of turmeric. The adult is a brownish-black butterfly with 8 white spots on forewings and one large patch on hindwing. The full-grown larva is dark-green and measures 36 mm in length. A female lays about 50 eggs on leaves which hatch in 3-4 days. The larva undergoes 5 instars during 12-21 days and pupates in leaf-fold for 6-7 days. The larva which pupates in December emerges only in March. The insect is present in abundance during August to October. The damage is caused by the larvae which fold the leaves and feed on them.

Fig. 24.4 Skipper butterfly, *Udaspes folus*. Adult

Management. (*i*) The pest can be kept under check by collecting the butterflies with the help of net and destroying them. (*ii*) In case of severe infestation, spray 1.5 litres of quinalphos 25EC in 500 litres of water per ha.

2. Castor Capsule Borer, *Conogethes punctiferalis* Guenee (Lepidoptera: Pyralidae)

The caterpillar enters into the aerial stem killing the central shoot which results in the appearance of 'dead heart'. For details on the life-cycle of this insect see Chapter 15.

3. Bihar Hairy Caterpillar, *Spilarctia obliqua* (Walker) (Lepidoptera : Arctiidae)

This pest damages the turmeric plants extensively in Bihar and Bengal States. The life-cycle of the Bihar hairy caterpillar has been described in Chapter 26.

Other Pests of Turmeric

The other insects which attack turmeric include the coccids, *Aspidiotus hartii* Cockerell (Hemiptera: Diaspididae) which infests the rhizome under storage; *Aspidiotus cucumae* Gr.(Hemiptera : Diaspididae) which infests the plants; the banana lacewing-bug, *Stephanitis typica* (Distant) (Hemiptera : Tingidae); the leaf thrips, *Anaphothrips sudanensis* Trybom, *Asprothrips indicus* (Bagnall) and *Panchaetothrips indicus* (Bagnall) (Thysanoptera : Thripidae); the skipper butterfly, *Udaspes folus* Cr. (Lepidoptera : Hesperidae) and the beetle, *Lema pracusta* (F.) (Coleoptera : Chrysomelidae).

GINGER (*Zingiber officinale* Roscoe; Family : Zingiberaceae)

Ginger is attacked by insect pests both in field and in storage.

1. Castor Capsule Borer, *Conogethes punctiferalis* Guenee (Lepidoptera: Pyralidae)

The caterpillar bores into the aerial stem and kills it. For detailed account see Chapter 12.

2. Scale, *Aspidiotus hartii* Cockerell (Hemiptera : Diaspididae)

This is a small circular hard scale which infests the rhizome in large numbers. A female lays about 180 eggs under the scale which hatch within a day. The nymph takes about 30 days for its development. The male forms a pupa before emergence as the winged adult. The insect multiplies in large numbers on stored ginger rhizomes which shrink and dry up. Rhizomes in the field are also attacked by the scale and the infested plants look pale and dried up.

Management. Same as in case of coffee green bug (Chapter 19).

Other Pests of Ginger

The ginger crop is also attacked by other minor pests like; *Acrocercops irradians* Meyr. (Lepidoptera : Gracillariidae); the turmeric skipper, *Udaspes folus* Cramer (Lepidoptera: Hesperidae); the maggots of *Calobata* sp. (Diptera : Micropezidae) which bore into the rhizomes and roots; the maggots of *Chalcidomyia atricornis* Mall. and *Formosina flavipes* Mall. (Diptera : Chloropidae), *Celyphus* sp. (Diptera : Celyphidae) and rhizome fly, *Mimegralla coeruleifrons* (Dipteta: Micropezidae) **(Plate 24.1F)** which feed on the rhizomes and cause their rotting; the weevil, *Hedychrous rufofasciatus* M. (Coleoptera : Curculionidae) which feeds on the leaves.

CORIANDER (*Coriandrum sativum* L; Family : Umbelliferae)

The important pest of coriander is only whitefly.

Cotton Whitefly, *Bemisia tabaci* (Gennadius) (Hemiptera : Aleyrodidae)

The nymphs suck sap of the plants and adversely affect their growth. The life-cycle of this pest has been described in Chapter 16.

Other pests which are found on coriander plants are the aphid, *Hyadophis coriandri* (Das) (Hemiptera : Aphididae); the pentatomid bug, *Agonoscelis nubila* F. (Hemiptera : Pentatomidae) and indigo caterpillar, *Spodoptera exigua* (Hubner) (Lepidoptera : Noctuidae).

CINNAMON (*Cinnamomum zeylanicum* Blume; Family : Lauraceae)

The cinnamon crop is atacked by a number of insect pests, but only cinnamon butterfly is considered to be the most destructive.

Cinnamon Butterfly, *Chilasia clytia* Linnaeus (Lepidoptera : Papilionidae) (Fig. 24.5)

Cinnamon butterfly is widely distributed in the cinnamon tracts of Sri Lanka and South India. The insect has been reported to feed on a number of wild species of cinnamon and other forest plants, viz. *Alseodaphne semicarpifolia* Nees, *Litsaea sebifera, L. tomentosa, Machilis gamblei* and *Phoebe lanceolata* Nees.

Freshly hatched larva is jet black in colour with white patches which later undergoes various changes in colour pattern. The upper side of adult moth is rich velvety brown, while on underside of the body, the colour varies from soft pale brown to rich dark velvety brown.

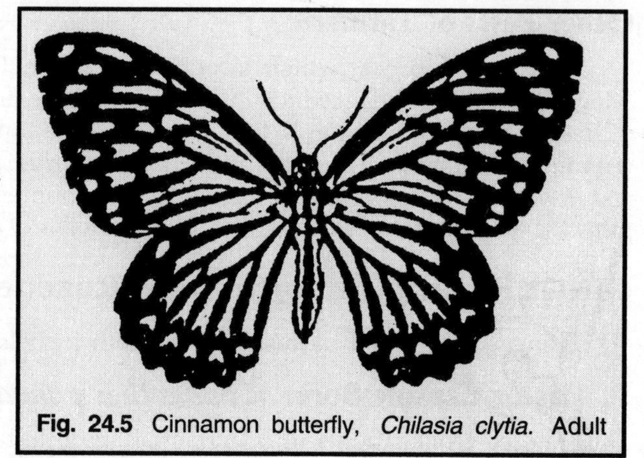

Fig. 24.5 Cinnamon butterfly, *Chilasia clytia.* Adult

Life-cycle. The female butterfly lays eggs singly on the upper and lower surface of young leaves, petioles and even tender shoots. Eggs are small round and pale yellow in colour. The larvae hatch out in 3-5 days. The larva moults five times to complete its development in 12-18 days. The pupation takes place in rough silken padding on the stem prepared by the larva. The pupal period is completed in 11-13 days. The adults live for 3-5 days and the total life-cycle is completed in 24-36 days.

Damage. After hatching, the first instar larva starts feeding on the lamina of the freshly emerged leaves. The later instars feed voraciously on the tender leaves leaving only the mid ribs with portion of veins. In case of severe infestation, the growth of the plant is adversely affected.

Management. Same as in case of skipper butterfly.

Other Pests of Cinnamon

Other pests which attack cinnamon are the leaf psyllid, *Pauropsylla depressa* Crawford (Hemiptera: Psyllidae); the leaf-miner, *Phyllocnistis chrysophthalina* Stainton (Lepidopetra: Phyllocnistidae) and the tussock caterpillar, *Dasychira mendosa* (Hubner) (Lepidoptera : Lymantriidae).

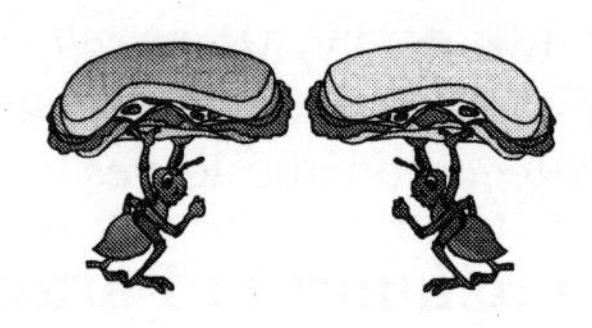

PESTS OF FOREST TREES

INTRODUCTION

Forests provide a number of ecological and economic services. They support energy flow and chemical cycling, reduce soil erosion, absorb and release water, purify water and air, influence local and regional climate, store atmospheric carbon and provide numerous wild life habitats. They also provide fuelwood, lumber, pulp to make paper, mining, livestock grazing, recreation and jobs. Forests are not only the most important natural resource for mankind but also the best protector of our climate and our earth's ecology.

According to the India State of Forest Report 2013, the total forest and tree cover of the country is 78.92 million ha, which is 24.01 per cent of the geographical area. This cover comprises of very dense forest (2.54%), moderately dense forest (9.70%) and open forest (8.99%). Madhya Pradesh has the largest forest cover of 77,522 km^2 in terms of the area in the country followed by Arunachal Pradesh with forest cover of 67,321 km^2. Further, Mizoram has the highest percentage of forest cover with respect to total geographical area with 90.38 per cent followed by Lakshadweep with 84.56 per cent.

Next to man, the greatest destroyer of forests are insects which damage almost all parts of the forest trees. The insect pests damage about 100,000 ha of forest in India. The annual loss caused by insects to seeds, transplants, standing trees, wood and finished products has been estimated to be about 10 per cent of the total revenue of the forests.

DEFOLIATORS

The foliage of trees provides food to a large varietaties of insects such as the leaf-miners, leaf rollers, leaf feeders, sap suckers, gall producers, etc. The overall effects of all these insects on the health of the trees are the same, namely, a serious attack may kill the trees, a moderate one debilate them and adversely affect the quality of wood they produce.

1. Teak Defoliator, *Hyblaea puera* (Cramer) (Lepidoptera : Hyblaeidae) (Fig. 25.1, Plate 25.1A, B)

The teak defoliator is commonly found in South India, Western Ghats, Maharashtra and Uttarakhand. This pest has also been reported from Australia, Myanmar, Sri Lanka, Java, Southern USA and Africa. This is oligophagous pest, but teak (*Tectona grandis*) is its princial host. Its alternative hosts are scattered in the families Verbenaceae, Bignonaceae, Araliaceae, Jugalandaceae and Oleaceae. The moths are with greyish brown fore wings, and hind wings are with black and orange-yellow markings (**Fig. 25.1**). It has wing span of 3-4 cm. Larvae are defoliators. Pupae are brownish and obtect type.

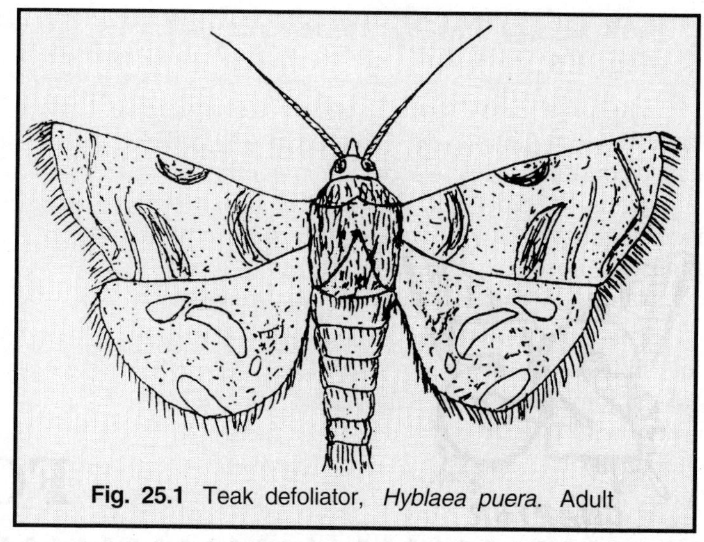

Fig. 25.1 Teak defoliator, *Hyblaea puera.* Adult

Life-cycle. Mated female lays her eggs singly near the veins on the under surface of the tender leaves. A single female can lay about 400 eggs. The eggs hatch within two days and newly emerged tiny larvae feed on the tender leaves of the teak plant. The larva moults for 5 times and prefers only young and tender leaves of teak for feeding. Larval stage lasts for 10-12 days or may extend up to 20 days in cold weather. The full-grown larva descends down on the ground with the help of silken threads and pupates in soil or in litter. Depending upon climatic conditions, pupal stage lasts for 4-8 days. The whole life-cycle is completed within 19-36 days. The life-cycle period is extended in cold weather. Many generations are possible in a year.

Various parasitoids, viz. *Brachymeraia lasus* (**Walker**). *Eriborus gardneri* (Cushman), *Plalexorista* sp. and *Symplensis* sp. cause about 70 per cent mortality in the larval stage of the pest.

Damage. Only the caterpillars are destructive. They feed on tender leaves of teak and affect the growth of the plant. The pest causes about 44 per cent damage.

Management. (*i*) Collect and destroy the egg masses, larvae and pupae. (*ii*) Clean cultivation and digging the field for exposing pupae to natural mortality factors, is helpful in reducing the incidence of the pest. (*iii*) *Bacillus thuringiensis* and nuclear polyhedrovirus play an important role in suppression of the pest. (*iv*) Pheromone traps may be utilized to attract and kill the pest.

2. Teak Skeletonizer, *Eutectona machaeralis* (Walker) (Lepidoptera : Pyralidae) (Plate 25.1C)

The teak skeletonizer is prevalent in Western Ghats, Maharashtra, Uttarkhand, Andhra Pradesh, Tamil Nadu and Karnataka, Outside India, it has been recorded from Australia and Indonesia. The pest attacks teak (*Tectona grandis*) and some species of *Callicarpa.* The moths are bright yellow with fulvous or pink transverse marking of zigzag pattern or serrate lines on the white fore wings. The hind wings are pale with ocherous or reddish marginal line or band. The pattern and colour of wings vary with the temperature and humidity.

Life-cycle. The mated female lays her eggs singly on tender leaves of the teak plant. A single female can lay about 250 eggs on the underside of the leaves. The eggs hatch within few days and the newly emerged larvae feed on tender leaves of teak. The larvae feed on the entire green matter of the leaf leaving veinlet network intact, thereby riddling the leaf. All instars can feed on young and older leaves of the plant. There are five larval instars. The last instar larvae pupate along with fallen leaves. The pest hibernates from January to March. The life-cycle is completed within one month.

Trichogramma minutum Riley, *T. brasilensis* (Ashmead) and *T. evanscens* Westwood attack eggs of the pest, whereas *Cedria paradosa* parasitizes the larvae.

Damage. The caterpillars feed voraciously on the tender and old leaves, and skeletonize the plant affecting the growth.

Management. Same as in case of teak defoliator.

3. Pink Gypsy Moth, *Lymantria mathura* Moore (Lepidoptera : Lymantriidae) (Fig. 25.2)

The pink gypsy moth is widely distributed in Western Ghats, Assam, Himalaya region, Uttar Pradesh, Uttarakhand and Sikkim. Outside India, it has bean recorded from Japan, USA and several Asiatic countries. This pest occurs in cool, temperate to warm climates and in fixed forests, temperate coniferous, tropical and subtropical broad leaf dry forests and moist broad leaf forests. It is a polyphagus pest and damages a large number of plants belonging to 45 genera of 24 familiers. Important host plants include sal, asna, Australian red cedar, zelkowa, cherry, willow, peach, apple, ash, walnut, oak, pine, chestnut, wax tree, etc.

The male moth shows black spots on vertex. The ground colour of fore wing is paler and dorsal side is black with zigzag markings (**Fig. 25.2**). Hind wings are orange with black spot at the end of cell and conjoined series of sub-marginal spots showing a curved band. Females are characterised by having head and thorax white, frons fuscus and two black spots on vertex. Abdomen is crimson red with small black spots on vertex. Fore wings are white with crimson and black spots and with typical zigzag marks. Hind wings are crimson with fuscus spots at the end of the cell. A sub-marginal maculate band is present on hind wing with some spots on centre of margin. The wing expanse of male is 40-51 mm and that of female is 96-112 mm. The larva is brownish with full of hairs on the body.

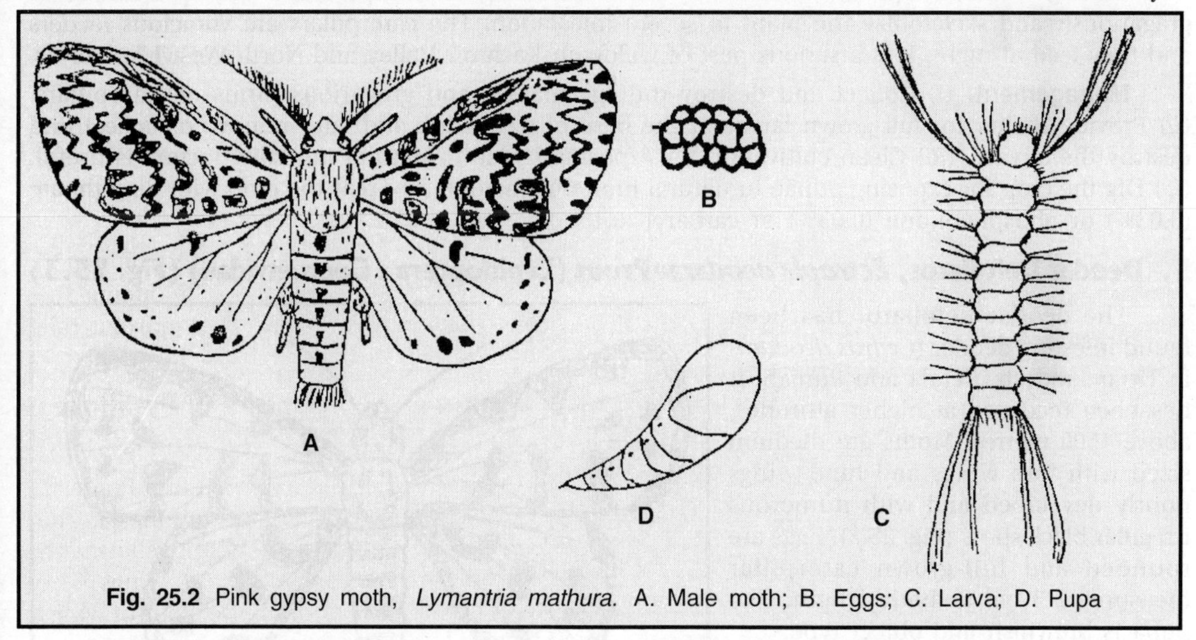

Fig. 25.2 Pink gypsy moth, *Lymantria mathura.* A. Male moth; B. Eggs; C. Larva; D. Pupa

Life-cycle. The females lay eggs on leaves of the host plants. Hundreds of eggs are laid by a single female in batches or in clusters. Eggs hatch in March/April. The larvae are gregarious feeders. The larva passes through 5-6 instars, and the total larval period is about 85 days. The full-grown larva pupates in soil and the pupal period lasts for 10-15 days. Only one generation is completed in a year.

Damage. The caterpillars feed gregariously on flowers and leaves, and skeletonize the plant completely in severe infestation. Young shoots are completely destroyed and many times total loss is not uncommon.

Management. (*i*) Collect and destroy the eggs masses, and gregarious and solitary larvae. (*ii*) Dig soil for exposing pupae to natural mortality factors. (*iii*) Use pheromone traps to catch and kill the moths. (*iv*) Use light traps; the moths are attracted to black coloured traps. (*v*) Spray the crop with azadirachtin (0.03%) or DDVP (0.03%)

4. Indian Gypsy Moth, *Lymantria obfuscata* Walker (Lepidoptera : Lymantriidae)

The Indian gypsy moth is distributed in Himachal Pradesh, Uttarakhand, North West Himalaya, and Jammu and Kashmir. The insects feed on a number of hosts including willow, poplar, apricot, apple and walnut. The male moth is greyish brown with wing expanse of 32 mm, Fore wings are with post-medial double lines more regular and hind wings are with dark lunule at the end of cell. The larvae are pale brown with short dorsal tufts of hair and long lateral tufts. The larva also shows a dark brown dorsal band with lines down the centre and on each side. Full-grown larva measures about 40-50 mm in body length.

Life-cycle. The female lays eggs in batches on the bark in June and July. Eggs are rounded, shining and light greyish brown. Each batch consists of 200-400 eggs and covered by yellowish brown hair. The winter is passed in egg stage and the eggs hatch in March or early April. Newly emerged larvae feed gregariously on leaves. The larva moults five times and is full-grown within 6-14 weeks depending on climatic conditions. The full-grown larvae gather on the bark of the tree on under surface of branches and under the debris and stones, and pupate in the soil within the debris. The pupal stage lasts for 9-21 days. Only one generation is completed in a year Males survive for 4-10 days and the females for 11-31 days.

Damage. The caterpillars are the only destructive stage of the pest and they feed on leaves gragariously and skeletonize the plant in severa infestation. The caterpillars are voracious feeders and they feed at night. It is a serious pest of willow in Kashmir Valley and North West Hkmalayas.

Management. (*i*) Collect and destroy the egg masses and gregarious forms of caterpillars. (*ii*) Provide shelter for full-grown larvae at the base of tree trunk and later remove or collect and destroy the larvae. (*iii*) Clean cultivation for exposing larvae to natural mortality factors is useful. (*iv*) Dig the crop for exposing pupae to natural mortality factors. (*v*) Treat the crop with dimethoate (0.03%) or phosphamidon (0.03%) or carbaryl (0.15%).

5. Deodar Defoliator, *Ectropis deodarae* Prout (Lepidoptera : Geometridae) (Fig. 25.3)

The deodar defoliator has been found infesting deodar (*Cedrus deodara*) in Uttar Pradesh, Kerala and Punjab. It has been recorded at higher altitudes, above 1500 metres. Moths are medium sized with fore wings and hind wings poorly developed and with numerous irregular black spots (**Fig. 25.3**). Eggs are rounded and full-grown caterpillar measures 2-3 cm in body length. The pupa is brownish and obtect type.

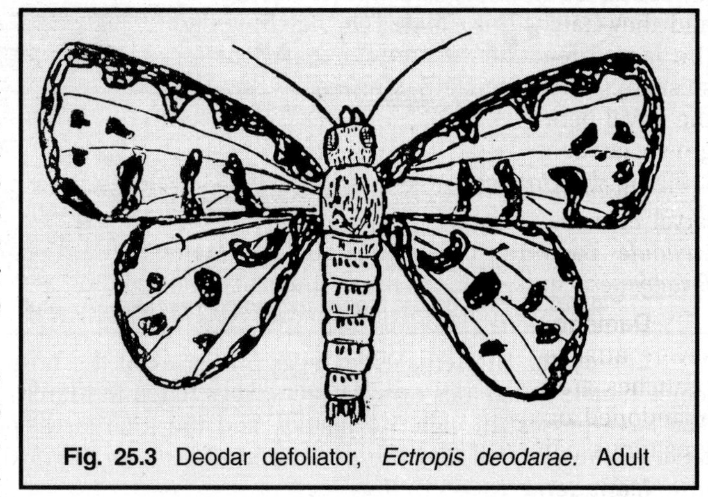

Fig. 25.3 Deodar defoliator, *Ectropis deodarae.* Adult

Life-cycle. Drapausing pupae emerge as moths in spring and moths mate immediately after emergence. Mated females then start climbing on the deodar tree for egg laying on the needles of the tree. Eggs are laid in mass and they hatch within 2-5 days in spring. Newly emerged larvae start feeding on the needles of the deodar and moult 5-6 times. Full-grown larvae descend to the ground for pupation. The larvae pupate on the ground within the fallen needles or in soil under a layer of humus. The pest hibernates in pupal stage.

Damage. The caterpillars are the only destructive stage of the pest. The caterpillars are most injurious to deodar tree since they feed on the needles. In case of severe infestation, they skeletonize the whole plant.

Management. (*i*) Collect and destroy the egg masses and the caterpillars. (*ii*) Dig field under the shade of the tree for exposing the pupae for natural mortality factors. (*iii*) Clean cultivation for exposing pupae to biotic and abiotic factors helps in suppression of the pest. (*iv*) Use shrews in forest against mature larvae and pupae of this pest. (*v*) Encourage birds against larvae and pupae of the pest. (*vi*) Spray the crops with azadirachtin (0.03%) or dust the crop with carbaryl 5 per cent @ 25 kg per ha.

6. Shisham Defoliator, *Plecoptera reflexa* Guenee (Lepidoptera : Noctuidae) (Fig. 25.4)

The shisham defoliator occurs on shisham in Punjab, Maharashtra, West Bengal, throughout North India and Andamans. The moth is grey brown with bright fulvous head and collar, and has a wing span of 30-35 mm **(Fig. 25.4)**. Fore wing contains a large reniform spot with black centre and with a rufos spot on the costa. The marginal series of minute dark specks is the characteristic of the fore wing. The hind wing is fuscus brown with outer area slightly darker. Males show fuscus area on costal margin of hind wing. The caterpillar is a green semilooper turning pinkish and measuring about 25 mm long when grown.

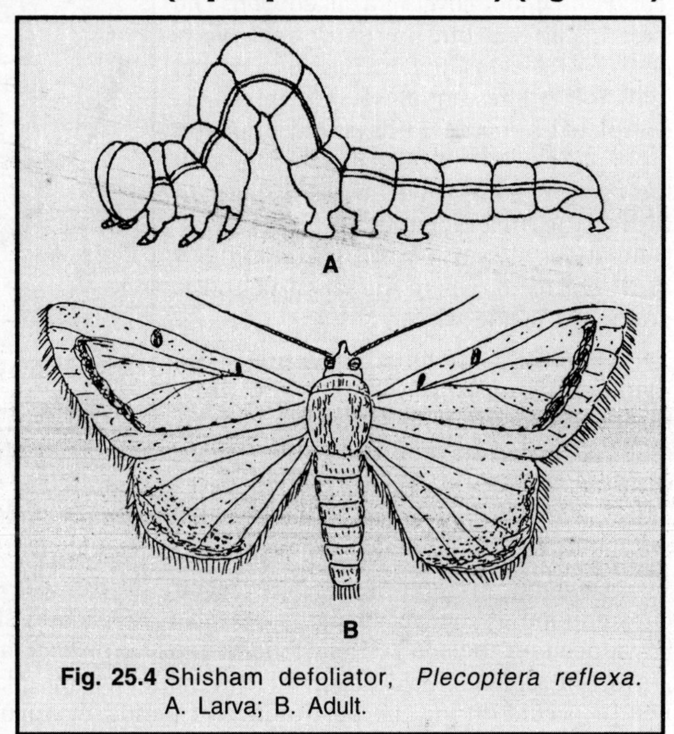

Fig. 25.4 Shisham defoliator, *Plecoptera reflexa.* A. Larva; B. Adult.

Life-cycle. The female moth lays eggs on tender leaves and a single female can lay about 400 eggs. Eggs are laid in clusters and they hatch in 4-6 days. On hatching, the larvae feed on the tender leaves of shisham. There are 5-6 larval instars and the larval period lasts for 14-21 days. Full-grown larvae descend to the ground and pupate within the fallen leaves.

Shisham defoliator is attacked by several parasitoids, viz. *Euplectrus parvulus* Ferriere (early larval instars), *Disophrys sissoo* Wilkinson and *Microgaster plecopterae* Wilkinson (larvae), *Exorista civiloides* Baranov, *E. picta* Baranov and *Podomyia setosa* Doleschall (both larvae and pupae), and *Brachymeria nursei* (Cameron)(pupae). The levels of parasitism range from 2-24 per cent.

Damage. Plantations about three years old and above are attacked by this pest. In repeated severe attackes, tree remains leafless for the greater part of the growing season and epicormic branches are produced. In case of severe defoliation, poor quality plantations have often been abandoned or converted. Defoliation is severe in roadside and canal-bank plantations but not in gregarious natural stands.

Management. (*i*) Collect and destroy clusters of egg masses and gregarious and solitary caterpillars from the leaves. (*ii*) Dig the soil near and under the shade and base of the tree for exposing pupae to natural mortality factors. (*iii*) Use natural enemies like shrews and predatory birds. (*iv*) Spray carbaryl (0.15%) or azadirachtin (0.03%).

7. Kadam Defoliator, *Arthroschista hilaris* (Walker) (Lepidoptera : Pyralidae) (Fig. 25.5)

The kadam defoliator has been recorded in India, Malaysia and the Philippines on kadam (*Neolamarckia cadamba*). The bluish green moth has a wing span of about 34 mm **(Fig. 25.5)**. The mature larva is pale green with a dark brown head capsule, and about 25 mm long with inconspicuous hair.

Life-cycle. The female moth lays 60-70 eggs, singly or in groups of two or three, on leaves. There are five larval instars. The first and second instar larvae feed on soft leaf tissue under cover of a silken web. The later instars eat out the entire leaf blade between the veins, under cover of a partial leaf fold. The larval development is completed in about 15 days and pupation takes place inside the silken web. The total life-cycle is completed in 21-26 days. In India, the insect can complete 11-12 generations a year in West Bengal and 8-9 in Uttarakhand, where the larval period is prolonged in the winter.

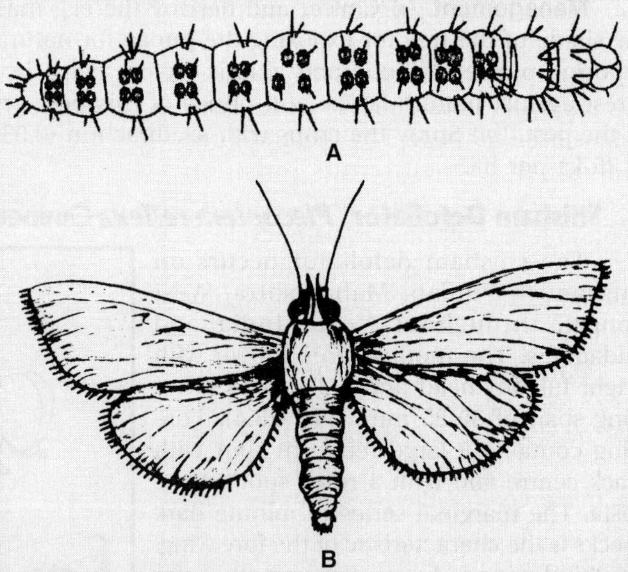

Fig. 25.5 Kadam defoliator, *Arthroschista hilaris*. A. Larva; B. Adult.

Natural enemies include six hymenopteran larval parasitoids, three hymenopteran purpal parasitoids and a few reduvid, carabid and ant predators. *Apanteles balteatae* Lal was reported to parasitize up to 60 per cent of larvae during peak incidence of the pest in West Bengal and *A. stantoni* Ashmead up to 50 per cent of larvae in Sabah, Malaysia. Other larval parasitoids include *Cedria paradoxa* Wilkinson, *Macrocentrus philippinensis* Ashmead and *Sympiesis* sp.

Damage. Feeding of the early instars on the leaf surface causes browning of leaves, while consumption of the leaf blade by older larvae leads to shedding of leaves. In defoliated trees, the larvae feed on the soft terminal shoot, causing dieback and formation of epicormic branches. The growth of saplings is adversely affected, although the plants seldom die. Peak infestation in West Bengal occurs during the post-monsoon period in August-September, during which moderate to heavy defoliation occurs. At Saba in Malaysia, population peaks have been recorded twice a year, in April-June and November-January.

Management. (*i*) Clean cultivation to expose the perpae to natural mortality factors helps to check this pest. (*ii*) Encourage the natural enemies to suppress the pest. (*iii*) Spray carbaryl (0.15%) or azadirachtin (0.03%).

8. Gamhar Defoliator, *Craspedonta leayana* (Latreille) (Coleoptera : Chrysomelidae) (Fig. 25.6)

The gamhar defoliator has been recorded as a pest of *Gmelena arborea* in India, Bangladesh, Myanmar and Thailand. In India, it is prevalent in the morthern region but also occurs in central and southern regions. The beetle is 12-16 mm long and has a brilliant metallic colouration, with coarsely wrinkled, bluish green to violet elytra and pale yellow to reddish brown pronotum and legs (**Fig. 25.6**). The larva has a typical appearance, with lateral spines. The excrement, instead of being ejected, is extruded in fine, black filaments, longer than the body, and formed into bunches attached to the anal end. When disturbed, the larva flicks the anal filaments up and down, and assumes a defensive posture.

Life-cycle. Eggs are laid in clusters of about 10-100 (average 68), on the under-surface of leaf or on tender stem and are covered by a stsicky, frothy secretion which solidifies to form a domed, brownish ootheca. The oviposition period may range up to 45 days, with an average fecundity of

874. The larvae are gregarious and there are five larval instars. The larval period is completed in about 18 days under optimal conditions. The full-grown larva fastens itself to the leaf by the first three abdominal segments and pupation occurs on the leaf itself. Under favourable conditions, the life-cycle is completed in 35-50 days, but third generation adults enter into hibernation in winter.

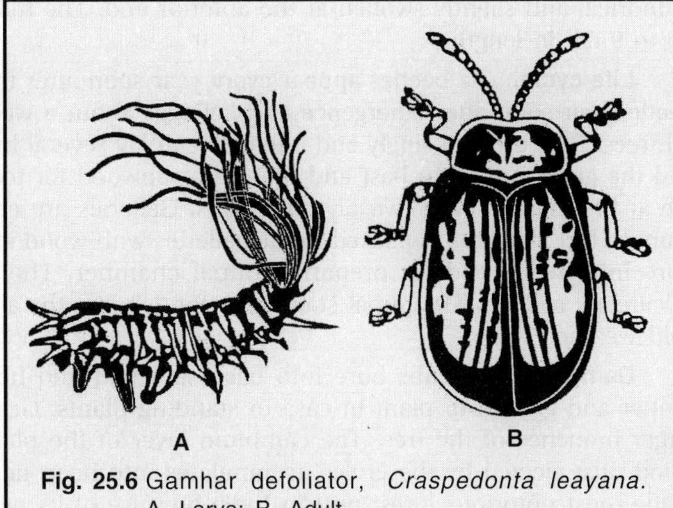

Fig. 25.6 Gamhar defoliator, *Craspedonta leayana*. A. Larva; B. Adult.

Damage. Both the adult and the larvae feed on leaves. The early instars feed by scraping the surface of the leaf but later instars and the adult feed by making large, irregular holes on the leaf. Even young shoots are eaten up when the population density is high. Heavy attack causes total defoliation and drying up of the leader shoots in young trees, leading to severe growth retardation. With two or mor consecutive complete defoliations, the tree is likely to be killed.

Management. (*i*) Collect and destroy the egg masses and gregarious larvae. (*ii*) The population of adult beetles can be checked by trapping them in artificial hibernation shelters. (*iii*) Hand-picking of beetles returning to the plantation after overwintering helps in pest suppression. (*iv*) Commercial preparations of *Bacillus thuringiensis* subsp. *kurstoki* and *Beauveria bassiana* have been shown to be effective against the larvae. (*v*) Spray the crop with DDVP (0.05%) or quinalphos (0.04%).

BORERS

The larvae of many beetles and moths cause considerable damage to forest trees by boring into the trunks and feeding on their living tissues. They bore holes and tunnels of various sizes, thus reducing the vitality of the trees and often causing their decay and death. In any case, the quality of the timber is affected and the value of the plantation goes down.

9. Sal Borer, *Hoplocerambyx spinicornis* Newman (Coleoptera : Cerambycidae) (Fig. 25.7, Plate 25.1D)

The sal borer is known to occur on sal (*Shorea robusta*) in Western Ghats, Maharashtra, Assam, West Bengal and Uttar Pradesh. Outside India, this pest has been reported from Afghanistan, Nepal, Indonesia, Singapore Borneo and the Philippines. The adult beetle is dark brown and variable in size measuring 20-60 mm in length and 5-16 mm in breadth. In the male, entennae are much longer than the body and in the female, antennae are shorter than the body (**Fig. 25.7**). Eggs are elongate,

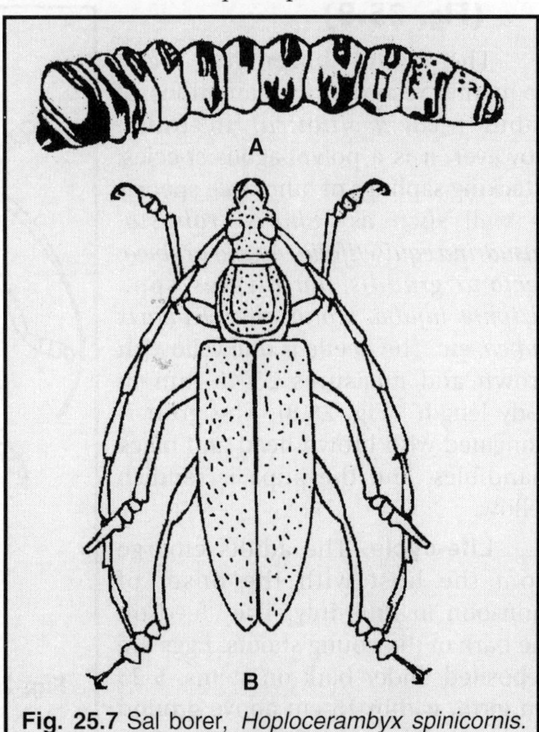

Fig. 25.7 Sal borer, *Hoplocerambyx spinicornis*. A. Larva; B. Adult female.

cylindrical and slightly swollen at the anterior end. The full-grown larva is yellowish and measures up to 9 cm in length.

Life-cycle. The beetles appear every year soon after the monsoon rainfall in June or July. The beetles pair soon after emergence and lay eggs about a week later, on cuts or holes in the bark of sal trees. Eggs are laid singly and a female can lay several hundred eggs. Eggs hatch within 5-8 days and the grubs bore into bast and then into sapwood for feeding. They prepare the galleries, which are at first narrow with two or more arms. Galleries are entirely packed with wood excreta of the grub. In fact, the grub is packed in the galleries with wood excreta ejected by the grub. Grubs further bore into heartwood for preparing pupal chamber. The pupal chamber is covered with white calcareous cocoon. The pupal stage lasts for 2-3 months and the pest hibernates in pupal state in cold weather.

Damage. The grubs bore into bast, sapwood and heartwood and thus affect the quality of timber and growth of plant in case of standing plants. Grubs can bore into roots, main stem and larger branches of the tree. The cambium layer of the plant is usually completely destroyed. The wood dust ejected by the grules accumulates into heap near the base of the tree **(Plate 25.1E)**. It is the most notorious forest pest in India because of its periodic outbreaks, during which millions of sal trees are killed **(Plate 25.1F)**.

Management. (*i*) Remove felled trees from the forest and bark every tree within a week of felling. (*ii*) Collect and destroy the beetles when they appear for infestation on the trees. (*iii*) Encourage natural enemies like birds, parasitoids and predators. Woodpeckers feed on the grubs by tunnelling the wood. (*iv*) Cotton balls soaked in chloroform or petroleum or kerosene be placed in boring holes and sealed with mud for killing the pest inside the tunnel. (*v*) Treat felled trees with carbaryl 5 per cent dust.

10. Babul Borer, *Celosterna scabrator* (Fabricius) (Coleoptera : Cerambycidae) (Fig. 25.8)

The babul borer has been frequently recorded in plantations of babul (*Acacia nilotica*) in India. However, it is a polyphagous species, attacking saplings of other tree species as well such as *Shorea robusta, Casuarina equisetifolia, Acacia arabica, Tectona grandis, Eucalyptus* spp., *Ziziphus jujuba, Morus alba, Tamarix indica*, etc. The beetle is dull yellowish brown and measures 25-40 mm in body length **(Fig. 25.8)**. The grub is elongated with brown head and black mandibles, and the pupa is reddish yellow.

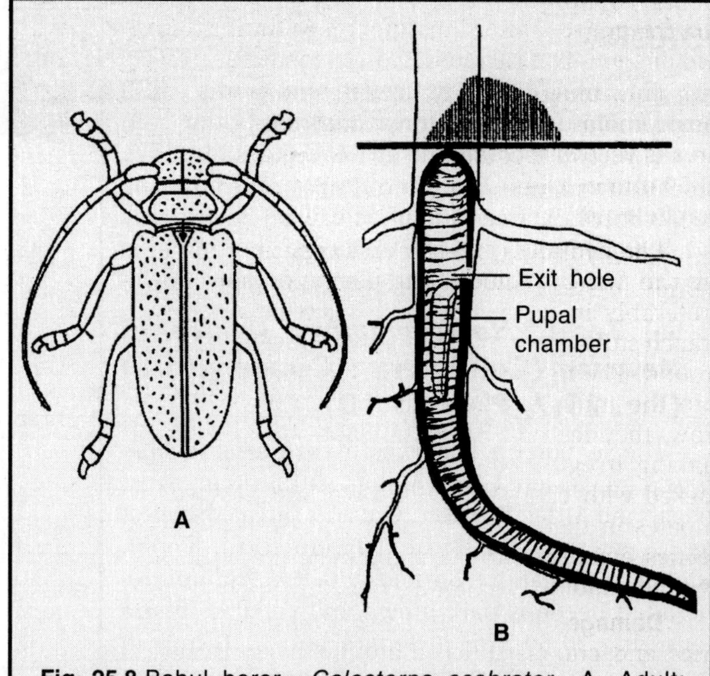

Fig. 25.8 Babul borer, *Celosterna scabrator.* A. Adult; B. Diagrammatic view of an infested *Acacia nilotica* sapling, showing the larval tunnel.

Life-cycle. The adults emerge from the host with the onset of monsoon in June-July. They feed on the bark of the young shoots. Eggs are deposited under bark on stems, 5-23 cm girth, within 15 cm above ground level, usually one egg per stem. The newly hatched larva bores into the stem and as it grows older, it tunnels downwords hollowing out the main root (Fig. 25.8B). The larva ejects the frass through a hole in the stem just above the ground level and the frass accumulates at the base of the stem. The larval period is about 6-8 months and the full-grown grub pupates in the pupal chamber. The grubs block the pupal chamber with fibrous material and the pupal period lasts four weeks. The beetles emerge out through the fibrous material of the exit tunnel. Only one generation is completed in a year.

Damage. The grubs cause damage by tunnelling down the centre of stem and later the thicker portion of the stem, base of the stem and subsequently the roots. In this process, they cause death of the tree.

In addition to the larvae tunnelling into the stem and root, adult beetles feed on the bark of stems and branches of saplings, in irregular patches, often girdling the shoots and causing them to dry up.

Management. (*i*) Collect and destroy the beetles when they freshly emerge from pupal chambers. (*ii*) Collect and destroy the infested plant parts along with the pest stages. (*iii*) Fill wounds or bored portions with cotton boll soaked with petroleum oil or chloroform or kerosene and plug the same with mud to kill the pest inside the tunnel.

11. Semul Borer, *Xystrocera globosa* Olivier (Coleoptera : Cerambycidae) (Fig. 25.9)

The semul borer is a serious pest of semul (*Bombax malabaricum*) in Assam, Uttarakhand, Tamil Nadu, Maharashtra and Western Ghats. It has also been recorded from Egypt, Myanmar, Sri Lanka, Indonesia and the Philippines. Other hosts of this pest include *Xylia dolabriformis*, *Albezzia bebbek*, *Populus cuphratica* and *Salix* spp. The medium-sized beetle, 30-35 mm in body length, is reddish brown in colour with dark blue-green lateral stripes on the prothorax and elytra (**Fig. 25.9**). The larva is yellowish green and grows up to 40-50 mm in length. Pupa is whitish and exarate type.

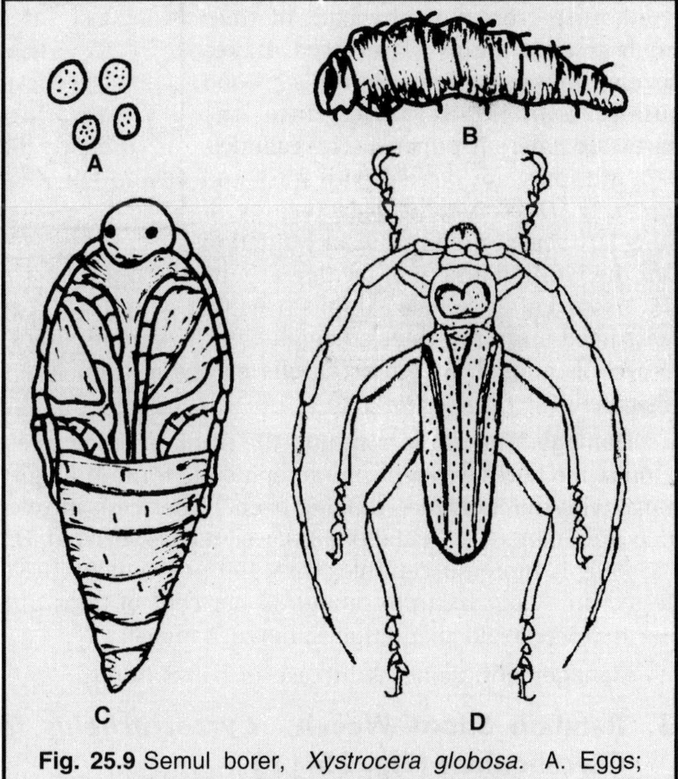

Fig. 25.9 Semul borer, *Xystrocera globosa*. A. Eggs; B. Larva; C. Pupa; D. Adult (male)

Life-cycle. The adult beetles appear on the wing in April and lay eggs, preferably in crevices on the stem or branch stubs, generally 3-4 metres above ground. The newly emerged larvae bore into the inner bark and as the larvae grow, they feed on the outer sapwood, making irregular downward galleries, packed with frass. The full-grown larva pupates in the pupal chamber and the beetles emerge from the trees in April. Two generations are completed in a year.

Damage. The larvae damage bast and sapwood by making galleries, with the result the bark dries and cracks, and the heavily infested trees dry up. Weakened stems are sensitive to wind, particularly during the rainy season. Even when the trees are not killed, borer attack reduces the growth rate and timber quality.

Management. Same as in case of sabul borer.

12. Shisham Borer, *Aristobia horridula* Hope (Lepidoptera: Cerambycidae) (Fig. 25.10)

The shisham borer is an emerging serious pest of *Dalbergia cochinchinensis* in Thailand and it also attacks *D. sissoo* in India (West Bengal and Uttar Pradesh). It has also been recorded on *Pterocarpus indicus* and *P. macrocarpus*. The adult beetle, 27-32 mm long, is brownish, with bluish hair on the elytra. A characteristic of the species is the presence of a dense tuft of hair on the distal portion of the first and second antennal segments, those on the second segment being longer. (**Fig. 25.10**). The prothorax has a pair of lateral spines. The full-grown larva is 55-60 mm long and is creamy white.

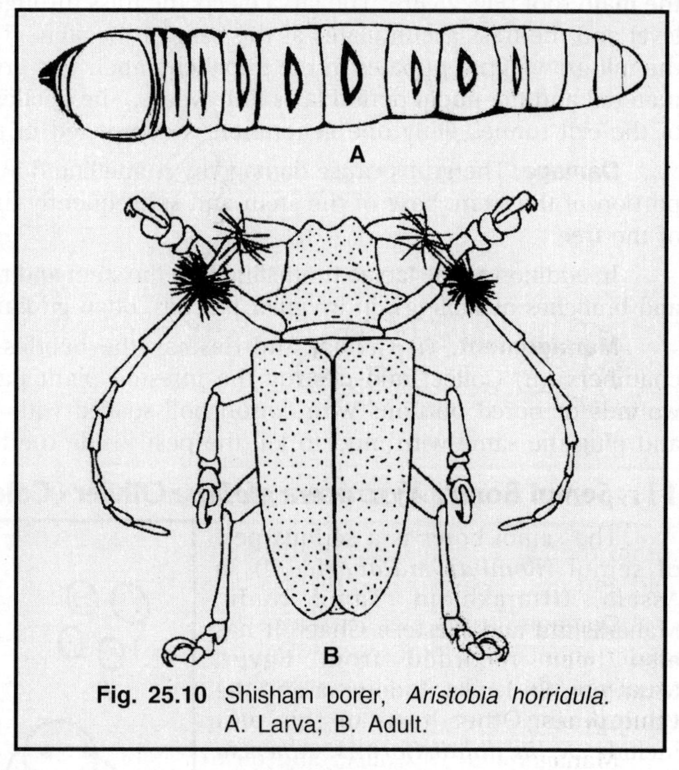

Fig. 25.10 Shisham borer, *Aristobia horridula*. A. Larva; B. Adult.

Life-cycle. The female beetle makes a transverse groove on the bark of trees and lays eggs singly. The larva makes irregular, upward galleries in the sapwood initially and finally bores into the heartwood where it pupates. The galleries, 50-75 cm long, are packed with frass and excreta. In *D. cochinchinensis,* feeding of young larva causes reddish resin exudation from the bark. In addition, feeding around the inner bark causes swelling of bark around the stem. In *D. sissoo,* entrance hole exhibits 'weeping symptom' (oozing of black fluid). In *D. sissoo,* the attack is restricted up to four metre height, with maximum attack taking place at about breast height. The life-cycle is annual, with most adults emerging from July to September in India and April to June in Thailand.

Damage. The grubs bore into the trunk of living trees and borer damage degrades the timber. In India, the borer was reported in epidemic form in all girth classes of *D. Sissoo* plantation in West Bengal with the incidence ranging from 10 per cent in one-year-old plantation to 80-90 per cent in older plantations. In Thailand, 33 per cent of 8-year-old *D. cochinchinensis* plantation was infested. *P. indicus* is more susceptible, with 100 per cent of 10-year-old roadside plantation being found infested. In *P. macrocarpus,* about 33 per cent of trees in 8-year-old plantation and 83 per cent of trees in 16-year-old plantation suffered damage.

Management. Same as in case of babul borer.

13. Bamboo Shoot Weevil, *Cyrtotrachelus longipes* Fabricius (Coleoptera : Curculionidae (Fig. 25.11)

The bamboo shoot weevil is prevalent in bamboo (*Bamboo* spp.) in Chittagong hill tracts of India. The adult weevil is 20-40 mm long with long rostrum and legs (**Fig. 25.11**). Females are much larger than the males. Grubs are legless, curved and whitish, and pupa is whitish in colour.

Life-cycle. The adult weevils become active with the onset of monsoon in May-June. They mate soon and the female then finds out young sprouting bamboos for egg laying. They suck the sap of tender shoots and lay eggs on the culm. The larva on hatching bores into the culm, making a long tunnel, passing internally through several internodes and perforating each. The larva pupates

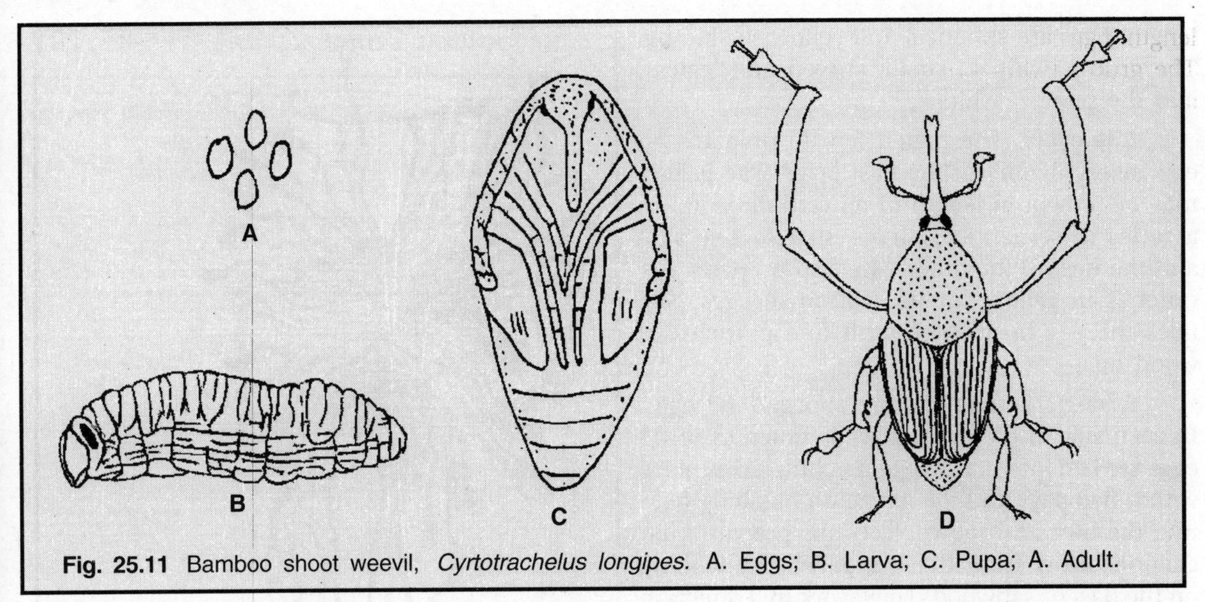

Fig. 25.11 Bamboo shoot weevil, *Cyrtotrachelus longipes.* A. Eggs; B. Larva; C. Pupa; A. Adult.

within the fallen buried end of the shoot at depth of 8-10 cm. The pupa stays in the chamber for cold and hot seasons, and emerges as an adult at the initiation of rains. The life-cycle is annual.

Damage. The female damages the shoots by ovipositing upon them and sucking the sap of tender shoots. The grubs bore into the shoot and affect growth and vigour of the tree. Feeding usually results in death of the culm or sometimes development of multiple shoots of little commercial value. A single larva can destroy a developing culm.

Management. (*i*) Collect and destroy adult weevils and other stages. (*ii*) Treat the crop with carbaryl 5 per cent dust.

14. Deodar Beetle, *Scolytus major* Stebbing (Coleoptera : Scolytidae) (Fig. 25.12)

The deodar beetle (also called scolytid beetle) is widespread in Northwest Himalaya in India, as a pest of deodar (*Cedrus deodara*). The black coloured beetle measures 4.0-4.5 mm in body

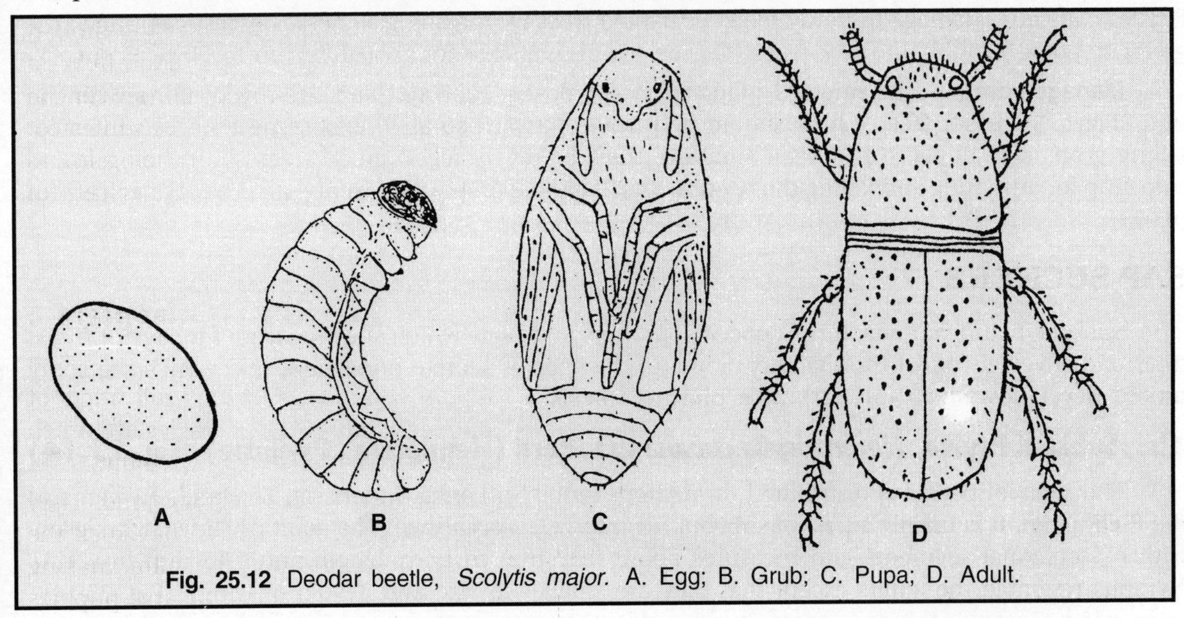

Fig. 25.12 Deodar beetle, *Scolytis major.* A. Egg; B. Grub; C. Pupa; D. Adult.

length. Eggs are spherical and yellowish shining. The grub is whitish, small, curved and legless, and the pupa is whitish.

Life-cycle. The mated female prepares an egg gallery mainly in the bast layer. The gallery may be 5-7 cm in length in upward direction in a series of serpentine curves, first to one side and the next to the other. She also prepares egg notches in gallery for deposition of eggs. She fixes the egg in a notch with fine particles of wood dust.

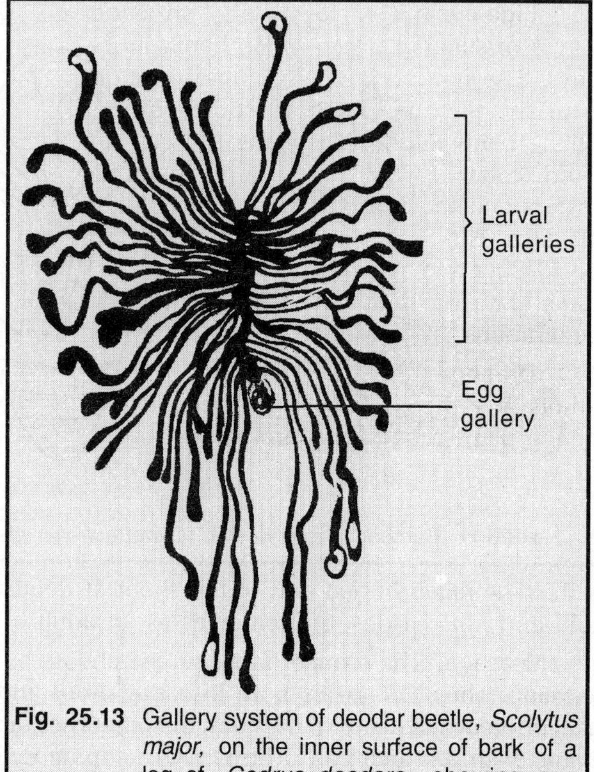

Fig. 25.13 Gallery system of deodar beetle, *Scolytus major,* on the inner surface of bark of a log of *Cedrus deodara,* showing egg gallery and larval galleries.

A single female can lay about 70-85 eggs. In each side of the egg galery, as much as 35-43 eggs are laid in notches. Egg hatching takes place within two days and the larvae on hatching bore into the bast and sapwood. Grubs prepare their galleries away from the egg gallery by feeding on the bast or sapwood. They work in a direction more or less right angle and in an upward or downward direction, and hence they have a special type of damage pattern (Fig. 25.13). The larval period lasts for four weeks. Grubs bore down sapwood in poles, large trees and saplings. The full-grown gurb prepares a pupal chamber in sapwood and pupates in it. The pupal stage lasts for about two weeks. The pest completes its life cycle within 6-7 weeks and there are four generations in a year.

Damage. This is a serious pest of deodar and damages the tree by making galleries in the bast and sapwood. The beetle can attack trees of all sizes from smallest sapling to the largest tree. Upward flow of sap is affected due to tunnelling the plant, which adversely affects the vitality of the plant. Beetles can cause damage to newly felled unbarked trees by making galleries. The pest can attack the green standing healthy and sick trees. The pest is responsible for cosualities up to 50 per cent.

Management. (*i*) The infested plant parts should be collected and destroyed along with the pest stages. (*ii*) Inner face of bark should be turned outward so as to expose pest stages to sun for killing grubs and pupae. (*iii*) Green standing trees or newly felled green trees should be used as trap crop for attracting and killing the beetles. (*iv*) Treat the crop with carbaryl (0.15%) or azadirachtin (0.03%).

SAP SUCKERS

Suckers include aphids, thrips, coccids, psyllids and bugs which suck the sap of foliage of forest trees and adversely affect their vitality. Some of the species secrete honeydew, over which the sooty mould developes which interfers with photosynthesis.

15. Subabul Psyllid, *Heteropsylla cubana* Crawford (Hemiptera : Psyllidae) (Fig. 25.14)

The subabul psyllid is distributed throughout India, Sri Lanka, Indonesia, Thailand, Taiwan and the Philippines. It is mainly a pest of subabul (*Leucaena leucocephala*). The adult psyllid is light yellow with 4-5 antennal segments and measures about 0.32 mm in body length and 0.15 mm in width. Nymphs resemble the adults except that they are smaller in size and devoid of wings (**Fig. 25.14**).

Lige-cycle. Females lay eggs on tender leaves of subabul twigs, which hatch in 2-3 days. The newly emerged nymphs wander here and there in search of tender portions of the plant and start sucking the cell sap from tender leaves. There are five nymphal instars and the nymphal period lasts for 8-10 days. The life-cycle is completed within 10-13 days and the adults can survive for 7 days. During a year, as many as 8-10 generations are completed on subabul.

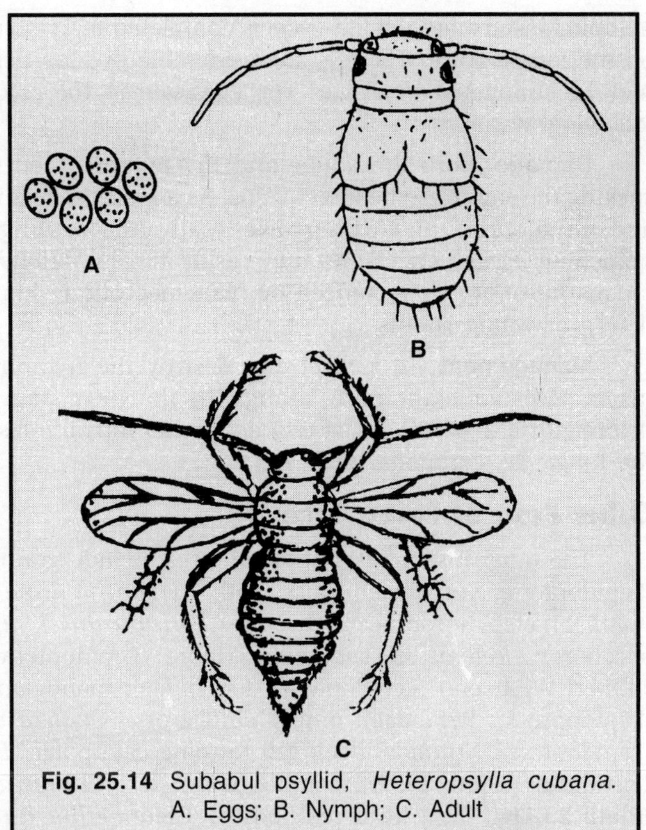

Fig. 25.14 Subabul psyllid, *Heteropsylla cubana.* A. Eggs; B. Nymph; C. Adult

Damage. Both the nymphs and the adults suck the cell sap from tender portion of the plant especially leaves, buds, stems, flowering and fruiting bodies. The psyllids inject toxins into the plant body when they suck the cell sap. This results in curling and yellowing of leaves, and dropping of flowering bodies. The nymphs and adults secrete honeydew on the leaves and other plant parts, which attracts sooty moulds to develop. The sooty moulds on leaves affect the photosynthesis of the crop, and affect the growth and quality of the plant.

Management. (*i*) Collect and destroy the infested plant parts along with the pest stages. (*ii*) The fungus, *Beauveria bassiana* causes white muscardine disease in the pest and controls about 64-100 per cent of pest population. (*iii*) A ladybird beetle, *Menochilus sexmaculatus* (Fabricius) feeds on nymphs and adults of the psyllid, and can be manipulated to control the pest. (*iv*) Grow pest resistant varieties such as K6, K8, K500 and K636. (*v*) Spray the crop with malathion (0.03%) or DDVP (0.03%) or phosphamidon (0.03%) or carbaryl (0.15%) or monocrotophos (0.05%) or quinalphos (0.05%).

16. Gamhar Lace Bug, *Tingis beesoni* Drake (Hemiptera : Tingidae) (Fig. 25.15)

The gamhar lace bug is an occassionally serious pest of young gamhar (*Gmelina arborea*) seedlings in India, Myanmar and Thailand. In India, it is prevalent in Uttarakhand, Madhya Pradesh, Uttar Pradesh, Maharashtra and Kerala. The small, dark, lace bugs (4.5 × 1.7 mm) aggregate in large numbers on the stems and branches of saplings and feed gregariously.

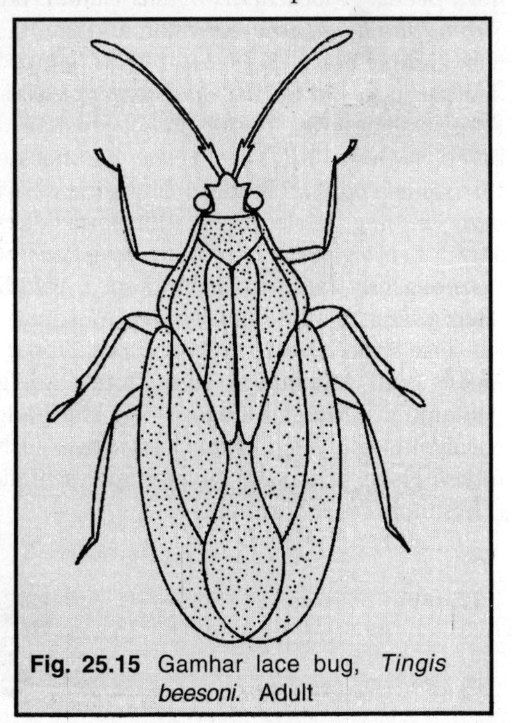

Fig. 25.15 Gamhar lace bug, *Tingis beesoni.* Adult

Life-cycle. Eggs are inserted in a vertical row into the tender shoot tissue. The nymphs congregate at the base of the leaf lamina and the axilla. There are five nymphal instars and the life-cycle is completed in 11-40 days. between April and October, depending on the temperature. There is a considerable overlap between generations, up to seven generations may be completed in a year. The eggs laid in the cold weather overwinter, hatching only in the following March.

Damage. Both the adults and the nymphs feed gregariously at the base of the leaf blade, sucking the sap from the larger veins. As a result of feeding by the insect, the leaves become spotted and discoloured, and wither. Eventually, the saplings show dieback of shoots and epicormic branching. Heavy infestation may result in total defoliation. The insect attack may be followed by fungus infestion, characterized by black necrotic lesions at leaf bases, followed by defoliation and drying of young shoots.

Management. (*i*) Collect and destroy the nymphs and adults. (*ii*) Remove and destroy the insect infested plant parts alongwith the pest stages. (*iii*) Spray monocrotophos (0.02%) or deltamethrin (0.005%). In case of additional fungal infestion, spray a mixture of the insecticide and the fungicide, carbendazim (0.1%).

Other Pests of Forest Trees

The other insect pests of forest trees include gram pod borer, *Helicoverpa armigera* (Hubner) (Lepidoptera : Noctuidae); gypsy moth, *Lymantria dispar* (Linnaeus); oak trabala caterpillar, *Trabala vishnou* (Lefebvre); spruce budworm, *Choristoneura fumiferana* (Clemens) (Lepidoptera : Tortricidae); teak borer, *Alceterogystia cadambae* (Moore) (Lepidoptera : Cossidae); chir pine defoliator, *Cryptothelia crameri* Westwood (Lepidoptera : Psychidae); mahogamy shoot borer, *Hypsiphyla robusta* Moore (Lepidoptera : Pyralidae); poplar defoliators, *Clostera cupreata* Butler and *C. fulgurita* (Walker) (Lepidoptera : Notodontidae); gall forming caterpillar, *Betousa stylophora* (Swinhoe) (Lepidoptera : Thyrididae) **(Plate 25.1G)**; bamboo leaf roller, *Crysiptya coclesalis* Walker (Lepidoptera : Pyralidae) **(Plate 25.1H)**; long horn teak borer, *Gelonaethia huta* Fairm ; deodar long horn bast eater, *Trinophyllum cribratum* Bates; oak timber longicorn borer, *Lophosternus hugelii* Redtenb; sal long horn beetle, *Plocaederus obesus* Gahan; poplar stem borer, *Apriona cinerea* Chevrolat ; teak borer, *Stromatum longicorne* Newman and quetta borer, *Aeolesthes saita* Solsky (Coleptera: Cerambycidae); teak pinhole borer, *Xyleborus noxius* Sampson; mahogamy beetle, *Scolytoplatypus brahma* Blandford; deodar buprestid beetle, *Sphenoptera aterrima* Kerremams and poplar buprestid, *Capnodis miliaris* Klug (Coleoptera : Buprestidae); shisham bostrichid, *Sinoxylon anale* Lesne and bamboo shoot borer, *Dinodeus pilifrons* Lesne; gamhar defoliator; *Calopepla leayana* (Latreille); poplar defoliator, *Chrysomela populi* Linnaeus; blue pine weevils, *Rhyncholus himalayensis* Stebbing and *Cyptorhynchus raja* Stebbing (Coleoptera : Chrysomelidae); oak aphid, *Cervaphis quercus* Takahashi and bamboo aphid, *Ceratoglyphia bambusae bengalensis* Ghosh (Hemiptera : Aphididae) ; sugarcane woolly aphid, *Ceratovacuna lanigera* (Hemiptera : Aphididae) ; pentatomid bug, *Urostylus puntigera* Westwood (Hemiptera : Pentatomidae) ; tea mosquito bug, *Helopeltis antonii* Signoret (Hemiptera : Miridae) ; San Jose scale, *Quadraspidiotus persiciosus* (Comstock) (Hemiptera : Coccidae) : pine woolly aphid, *Pineus pini* (Macquart) (Hemiptera : Adelgidae) ; Cottony cushion scale, *Icerya purchasi* Maskell (Hemiptera : Margarodidae) ; twig gall midge, *Aspondylia tectonae* Mani (Diptera : Cecidomyiidae); eucalyptus gall wasp, *Leptocybe invasa* Fisher & LaSalle **(Plate 25.1I)**; blue gum chalcid, *Leptocybe invasa* Fisher & LaSalle (Hymenoptera : Eulophidae) ; thrips, *Heliothrips haemorrhoidalis* (Bouche)) (Thysanoptera : Thripidae).

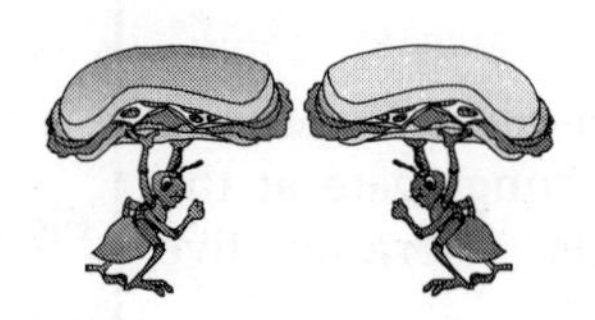

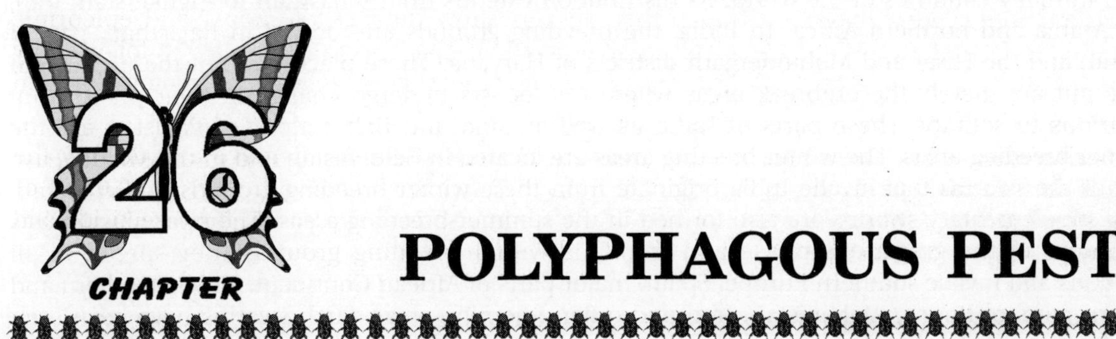

POLYPHAGOUS PESTS

INTRODUCTION

Polyphagous pests are those which feed on plants belonging to diverse taxonomic groups. Generally, these pests multiply in large number, are voracious feeders and cause considerable damage to agricultural crops. The most important polyphagous pests are termites, grasshoppers, hairy caterpillars, cutwwoms, etc. Other polyphagous pests include aphids, whiteflies, white grubs, mealybugs, tobacco caterpillar, gram caterpillar, etc. These polyphagous pests damage a number of crops and cause substantial losses. The magnitude of damage and loss caused by some of the polyphagous pests such as locusts is gigantic and beyond imagination as they have the capacity to cause starvation due to their voracious feeding ability. On an average, a small locust swarm eats as much food in one day as about 10 elephants, 25 camels or 2500 people.

LOCUSTS AND GRASSHOPPERS

1. Locusts

The larger grasshoppers which form swarms are called locusts and there are three species of the family Acrididae found in India.

(*i*) Desert locust, *Schistocerca gregaria* (Forskal)

(*ii*) Bombay locust, *Patanga succincta* Linnaeus

(*iii*) Migratory locust, *Locusta migratoria* (Linnaeus)

Of these, the desert locust is the most important. There are very few records of swarms involving other locusts, particularly of the migratory locusts. The first and second species are important in Maharashtra, Gujarat and Rajasthan but the desert locust is of all India importance. In fact, it is an international pest and efforts are being made constantly to control it through the International Locust Control Organization of the Food and Agriculture Organization.

(i) Desert Locust, *Schistocerca gregaria* (Forskal) (Orthoptera : Acrididae)

The desert locust is found in two phases, *i.e.* the solitary phase and the gregarious phase. Individuals having characteristics in between the two are often placed under the transient phase. The characteristics of the first two phases are different from each other, particularly in the colour of their nymphs. The nymphs of the gregarious phase are yellow or pink, with distinct black markings, whereas the colour of the solitary hoppers varies according to the colour of the surrounding vegetation. The gregarious adults are pink on emergence, gradually turning grey and ultimately yellow, when sexually mature. The adults of the solitary phase remain greenish grey throughout their life.

Distribution. The desert locust is an inhabitant of the dry grasslands of desert areas and is found in many countries of the world. Its distribution extends from Rajasthan to Afghanistan, Iran, Iraq, Arabia and northern Africa. In India, the breeding grounds are located in Rajasthan, part of Gujarat, and the Hisar and Mohindergarh districts of Haryana. These places are not the permanent home but are merely the outbreak areas where the locusts undergo change in their phase from gregarious to solitary. These parts of India as well as Sind and Bahawalpur of Pakistan are the summer breeding areas. The winter breedng areas are located in Baluchistan and in the Middle East. Most of the swarms that invade India originate from these winter breeding grounds but in certain years, supplementary swarms are also formed in the summer breeding areas. The swarming locusts are always in the gregarious phase and from the winter breeding grounds they spread in all directions and invade southern Europe, Spain, major parts of African Continent, Israel, southern and western parts of Russia, Afghanistan, Iran and other adjoining areas. Such swarms have been seen 1930 km at sea and, in one instance, a swarm was recorded 6440 km away from the place of the origin. It is on record that some swarms flew straight from the Arabian coast to Gujarat. The swarms fly quite high and they are known to have crossed mountains as high as 4,600 metres.

In India, the locusts invade the north-western region, penetrating the hilly tracts of the north. Sometimes, they reach as far east as Assam and as far south as Tamil Nadu. However, their breeding-grounds do not extend beyond the western districts of Uttar Pradesh and northern Madhya Pradesh. In 2006, the locusts from China invaded ther Zansker valley of Ladakh and destroyed crops over an area of 240 km^2.

Locust plagues and upsurges. The attack of the desert locust used to occur earlier in phases of plague cycles (a period of more than two consecutive years of widespread breeding, swarm production and thereby damaging of crops is called a plague period) followed by a period of 1-8 years of very little locust activity called as the recession period, again to be followed by another spell of plague. India witnessed several lucust plagues, and locust upsurges and incursions during the last two centuries (**Table 26.1**).

Table 26.1 Locust plague cycles and upsurges in India

Years	Years	Years	Years
Plagues		Upsurges	
1812-1821	1900-1907	1964 (4)	1986 (3)
1843-1844	1912-1920	1968 (167)	1989 (15)
1863-1867	1926-1931	1970 (2)	1993 (172)
1869-1873	1940-1946	1973 (6)	1997 (4)
1876-1881	1949-1955	1974 (6)	1997 (SSLLB)
1889-1891	1959-1962	1975 (19)	1998 (SSLLB)
		1976 (2)	2002 (SSLLB)
		1978 (20)	2005 (SSLLB)
		1983 (26)	2010 (SSLLB)

Figures in parentheses indicate the number of swarms. SSLLB - Small scale localized locust breeding.
Source : www.ppqs.gov.in

Breeding season and migration of swarms. The breeding of locust depends upon the rainfall and the subsequent vegetation. The eggs are laid in the sandy soil and adequate moisture is required before they can hatch. In India, there are two breeding seasons during the year, viz. (*i*) the summer breeding-season, and (*ii*) the monsoon breeding-season. In countries where the rainfall is in winter and early spring, as for example, south-eastern Arabia, the Red Sea coast, southern Iran, Baluchistan, etc., and wherever vegetation is available, the locust breeds late in winter or in the spring. There may be one or two broods during that season. In Baluchistan, for example, breeding starts in January and then again in April. Thus, the swarms that originate in Africa or Arabia breed in Baluchistan and the surrounding countries during the spring and they migrate to Pakistan and India in the summer.

In countries such as India and the Sudan where the rainfall occurs in summer, the locusts breed during that season. However, even in these countries, at places where there is some rainfall in winter alongwith plentiful supply of wild grasses, the locust can breed in the spring. Punjab is such an area. Ordinarily, in western India there is vegetation after the monsoon and locust swarms breed up to the end of September, October or even November, provided the temperature is fairly high, as it was in 1949. That year the locusts completed two broods during the summer.

Some of the swarms produced in the monsoon season may fly westwards to Baluchistan, erstwhile Russia and Eastern Arabia. A great many of the swarms formed in Rajasthan and Sind fly north, east and south, thus invading all parts of India. They damage the *kharif* crops. Some swarms may over winter in north western India, and when the temperature conditions are suitable early next spring, they may damage the *rabi* crops. Thus, the locust cycle continues from season to season, sometimes for a number of years at a stretch. Eventually, in a given year, if the rainfall is low and vegetation poor or the conditions are otherwise unsuitable, the locust population dwindles and is transformed into the solitary phase, scattered throughout the area of its invasion. Even in this solitary phase, the locust continues to breed for sometime and damage some of the crops.

Origin of a new locust cycle. The following sequence of events generally mark the beginning of the locust cycle:

- Extensive breeding on coastal areas of Arabian countries as a result of heavy winter and spring rainfall, and the formation of gregarious swarms.
- Migration of locusts from the Arabian coast into the interior (Baluchistan, Afghanistan) in spring.
- Migration of the these swarms, in summer, to Sind and Rajasthan.
- Extensive breeding of these swarms in Rajasthan in July-August and September.
- Migration of the subsidiary gregarious swarms from Rajasthan to Punjab in late summer.

Life-cycle. The locust has three stages in its life cycle, viz. the egg stage, the hopper stage and the adult stage. When sexually mature, the adults are yellowish, sluggish, reluctant to fly and cluster on the ground. While mating, the male clings to the back of the female who takes him around. Copulation lasts 8-24 hours and egg-laying starts soon after mating, sometimes even when the male is clinging to the back of the female. Egg-laying continues for many weeks. A single female may lay up to 11 egg-pods, each pod containing up to 120 eggs. A female normally lays 500 eggs in about 5 pods. Before egg-laying, the female, with the help of her ovipositor, bores a hole into the loose sandy soil, 5-10 cm deep (**Fig. 26.1**). It takes 1-4 hours to dig a hole. Having laid a pod, she secretes a frothy material over the eggs, which hardens on drying and makes the pod water-proof. While laying eggs, the females may be sitting very close to one another and as many as 5,000 eggs may be laid in one square metre. The ground used for laying eggs is easily recognized by numerous holes, which are of the diameter of an ordinary lead-pencil.

The egg, resembling grain of rice, is lightly curved and 7-9 mm long. The duration of the egg stage depends upon the soil conditions, temperature and moisture. The eggs laid in February and

March hatch in 3-4 weeks and those laid in May-September hatch in 12-15 days. In very dry soils, the eggs may remain unhatched for a long time until there is a shower of rain.

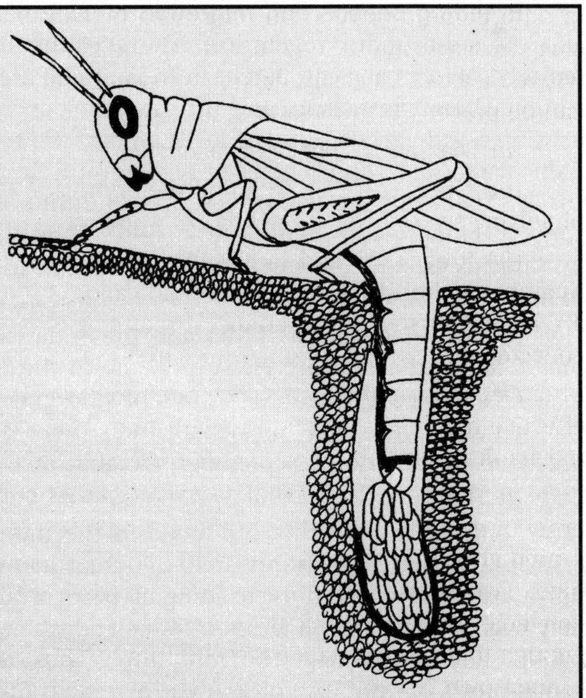

The nymphs, at the time of emerging, break the egg-shell and creep out of the holes. These freshly hatched hoppers are light yellow but soon turn black (in the gregarious brood). They exhibit the gregarious instinct and march in swarms, feeding on all kinds of vegetation as they move. The duration of the nymphal (hopper) stage lasts 6-8 weeks in spring and 3-4 weeks in summer. The young adults are bright pink and when sexually mature, they turn bright yellow. The pink locust adults are very active and are likely to cause much damage to crops, whereas the yellow swarms, though not so destructive, are equally dangerous as they lay eggs which give rise to destructive nymphs.

Fig. 26.1 Desert locust, *Schistocerca gregaria*. Female depositing eggs in the soil

The locust swarms and the hopper bands generally rest during the night on bushes, crops, trees, etc. In the morning, they hop around at first and fly to form a swarm only when the temperature rises to an appropriate degree. The flight temperature varies from day-to-day and is determined by the maximum temperature on the previous day. During summer, if the day temperature is too high, the locusts rest again.

There are two broods in a year and their life-cycle is completed in various months, viz. summer brood (February-April : Pairing and egg laying ; March-May : Hoppers ; May-June : Adults) and monsoon brood (July-September : Pairing and egg laying ; July-October : Hoppers ; September-October : Adults). Adults of the monsoon brood either overwinter in India when the locust cycle is manifested or they migrate to other north-western countries.

Damage. The locust is harmful in both the adult and the hopper stages. These gregarious and voracious feeders eat almost any vegetation, except a few plant species such as *ak* (*Calotropis procera*), *dharek* (*Melia azedarch*), *neem* (*Azadirachta indica*), *dhatura* (*Datura stramonium*), etc. When in swarms, they can consume all the green vegetation and cause a famine. In spite of some expensive control measures, the damage to crops caused by locusts during the 1926-31 cycle was estimated at 100 million rupees. In addition to the damage to crops, orchards, forests, etc., the locust can be nuisance in houses, as these creatures climb over the walls, invade kitchens, store-rooms and even enter into the beds. They fall into wells by the million and thus make water unfit for drinking. If an army of hoppers or adults marches on to the railway lines, all traffic is suspended because the crushed hoppers cause slippery rails.

Management. Management operation can be carried out against all stages of the locust, the most practicable and effective measures are against the nymphs. Operations are directed towards the destruction of all the locust population rather than towards protecting a certain area or a given crop field.

(*i*) The adults can be beaten to death with thorny sticks, brooms or can be swept together and buried underground in heaps. These things are easy to do when the females are laying eggs. When resting on trees or bushes in the waste-lands, they should be scorched to death with fire torches or with flame-throwers. Burning is particularly effective at night or early morning when adults are sluggish because of the cold. During the cold weather, the adults rest at night on top of trees from where they can be easily shaken off, swept together and burnt or buried. During the day, the swarms can be prevented from settling on crops by waving white pieces of cloth or by the beating of drums.

(*ii*) Diazinon 2 per cent, if dusted on crops, trees and the ground, is very effective.

(*iii*) Diazinon (2%), in an oil medium, is also effective when sprayed with an aeroplane on top of a flying swarm. Diazinon 25EC in water suspension, can also be sprayed on locusts, on the ground.

(*iv*) If eggs are laid in a well-defined area, a trench may be dug around it, so that the young nymphs on emerging drop into it and can be buried alive, filling the ditch with soil. If these trenches are heavily dusted with lindane, it may not be necessary to bury the nymphs. The eggs can also be collected by sifting the soil with sieves but this method is not very practicable.

Destruction of nymphs. The hopper stage is the most vulnerable and control measures are most effective before the second moult. The nymphs may be destroyed either with chemicals or by using mechanical methods :

(*i*) The principal mechanical method of control lies in digging trenches in front of the moving army of nymphs and driving them into these trenches, with brooms or with twigs of trees and then, buried alive. The nymphs can also be guided to the trenches along metal or canvas barriers, 45 cm high. Two barriers, one on each side of the army of the marching hoppers are so fixed that they converge on a narrow gap that leads to a trench in which they drop. The trench should be deep enough to accommodate a large number of hoppers most of which would then die under the weight of their own fellow-creatures. Later, the trench can be filled with earth. In the early stages, a trench, 30-45 cm wide and 60 cm deep, is sufficient but when the hoppers are older, the width should be 75 cm and the depth more than 60 cm

(*ii*) At night when the hoppers rest on bushes, they can be burnt with flame-throwers.

(*iii*) Poison baits such as the poisoned bran mash or sawdust, if scattered in the early morning or in the evening, are effective. During day, the bait dries quickly so the hoppers do no eat it.

(*iv*) Diazinon dust and diazinon spray (0.2%) are very effective.

(*v*) Diazinon as a spray can also be applied shortly before the emergence of the hoppers, so that as soon as they come out, they come into contact with the insecticide and die.

(*vi*) Spraying the crop with neem seed kernel powder suspension (1%) has been found to be very promising.

(*vii*) A number of birds attack locusts and of these, the common *myna* and the *tiliar* (starling) are the most important. During the locust cycle, if practicable, these birds should be protected.

Anti-Locust Organization

The Anti-Locust Organi-sation in India consists of (*a*) The Central Anti-Locust Organization, and (*b*) the State Anti-Locust Organizations.

(*a*) *Central Anti-Locust Organization.* This organization is called the Locust Warning Organization and was established in 1939. It is handled by the Plant Protection Adviser to the Government of India, who is also the Director of Locust Control, having his headquarters at Faridabad (Haryana). The Directorate is requred to :

- Record the weekly density of locusts per unit area in the breeding areas located in India and to carry out control operations there.
- Interpret the records and pass on the information on locust movements to the various State Locust Control and Warning Officers and to the revenue authorities of the concerned districts before the locust reaches there so that arrangements for control can be made in advance.

- Keep watch on the coming swarms of extra-Indian origin, their direction and size.
- Give technical and material assistance to the various States.
- Coordinate the anti-locust work in India by issuing a fortnightly bulletin on the locust situation both in India and in foreign countries.

In India, the work of this organization extends both to the scheduled desert areas where locusts breed and to the cultivated areas where they do the damage. In the scheduled desert areas, the Central Anti-Locust Organization operates over an area of 2,05,785 square kilometers in the States of Rajasthan, Haryana, Maharashtra and Gujarat. This area is divided into 23 locust outposts which are grouped into 5 circles with headquarters at Bikaner, Jodhpur, Barmer, Jaisalmer and Palanpur (Gujarat). The responsibility of the control work is entrusted to the Locust Entomologist incharge of the circle. During the 10th Five Year Plan, it has been proposed to restructure the organisation with 10 circles each at Bikaner, Jaisalmer, Barmer, Palanpur, Churu, Jalore, Nagaur, Sikar, Phalodi and Bhuj besides one field headquarter at Jodhpur and a Central Headquarter at Faridabad by merging existing 5 circles and 23 locust outposts for

(*i*) monitoring of locust activity,

(*ii*) issuance of fortnightly locust situation bulletins,

(*iii*) organizing Indo-Pak border meetings on locust situation,

(*iv*) conducting trainings to the farmers, state functionaries and locust staff on locust control, and

(*v*) conducting research on locusts and grasshoppers.

(*b*) *State Anti-Locust Organizations.* In the scheduled area of the desert as well as in the adjoining States where locusts cause damage, there are State Anti-Locust Organizations to take suitable measures against the swarms. The State organization is headed by the Locust Control and Warning Officer who is well connected with the local revenue authorities, from the Deputy Commissioner to the Tehsildar and Patwari. When the latter official receives information from district headquarters warning him of a locust invasion, he alterts the villagers to be ready with machines and insecticides, kerosene, flame-throwers, spades etc. When the swarm actually arrives, the farmers try to kill it as well as they can and then the Patwari intimates the Locust Control Officer and his own Deputy Commissioner of the extent of success achieved. In case the swarm settles down to lay eggs, the Locust Control Officer or his staff visits the place to devise suitable control measures.

During the locust invasion of 1949-53 and subsequently in 1960-64, the organization proved to be very efficient and effective. They have been admired as one of the best in the world.

(ii) Bombay Locust, *Patanga succincta* Linnaeus (Orthoptera : Acrididae)

The Bombay locust is present in India, Sri Lanka and Malaysia. In India, its activities are generally confined to the area extending from Gujarat in the north to Tamil Nadu in the south, though in certain years its swarms have reached as far as Bengal. The adult is smaller in size than the desert locust and it has an acute ventral tubercle. There are conspicuous stripes laterally, on pronotum and elytra. The hind wings become rose coloured at base after the insects have spent one month as fledgings. The elytra have oblique dark markings.

Life-cycle. In the desert areas of India, the adults of this locust species are found during autumn, winter or spring but none during summer. They are generally met with amongst a tall sedge (*Cyperus tuberosus*), locally known as *chiya*. Studies have shown that the adults which fledge during September-October, undergo diapause and do not mature till June-July next year. Thus, there is only one brood in a year. During July and August, the minimum incubation and nymphal periods recorded were 34 and 56 days, respectively. The hoppers pass through 7-9 instars before becoming adults.

Damage. Both adults and hoppers normally feed on wild grasses but when the locust plague occurs they inflict heavy damage to crops. Unlike the desert and the migratory locusts, hoppers of this species do not congregate to form bands but remain scattered among crops or grasses.

(iii) Migratory Locust, *Locusta migratoria* (Linnaeus) (Orthoptera : Acrididae)

The geographical distribution of the migratory locust is Europe, Africa, Pakistan, Eastern and Southern Asia and Australia. In India, it develops swarms in certain years; swarms were observed in 1876 in Madras (Tamil Nadu) and in June 1954 in Mysore (Karnataka).

The main distinguishing characteristics of migratory locust from Bombay locust are that the ventral tubercle and eye-stripes are absent in the former. The hind wings are hyaline in the young forms, but become yellow subsequently.

Life-cycle. It breeds during spring in Baluchistan (Pakistan) and the resultant adults migrate into the desert areas of India as individuals. They breed there during summer and again in September-October. Under the laboratory conditions, the eggs of this locust species hatch in 12 days during July-August. The hoppers moult 5-6 times during the nymphal period of 13 days. Under favourable environmental conditions, overlapping generations of this locust are met with. Unlike the Bombay locust, the adults of this species do not undergo diapause.

Damage. During swarming period, both adults and hoppers feed on all kinds of vegetation, destroying all crops. It may cause famine in the areas of its abundance.

2. Sporadic Grasshopper, *Oxya nitidula* Walker (Orthoptera : Acrididae)

Sporadic grasshopper is widely distributed in the Orient and is a pest of sporadic occurrence. In Punjab, it is serious in the low-lying areas of Amritsar, Gurdaspur and Hoshiarpur which are flooded occasionally. Damage is done by nymphs and adults which feed on various *kharif* crops, such as maize, sorghum, rice, sugarcane, etc. The nymphs are shiny green and have a broad, black band extending from the base of the eyes to the wing pads. The legs are greenish-yellow and the full-grown nymphs measure about 15-25 mm in length. The adults are similar in body colour but are larger (about 25-30 mm) and have green wings.

Life cycle. The pest is active from March to December and passes winter in egg stage in the soil. The over-wintered eggs hatch in March and the emerging nymphs feed on *berseem* (*Trifolium alexandrinum*), *khabbal* grass (*Cynodon dactylon*) and other available crops. They moult 6 or 7 times by June and are transformed into winged adults in 60-64 days. The adults also feed on various *kharif* crops and live for 40-49 days. They lay brown eggs in clusters on the underside of leaves of various food plants, particularly sugarcane. A female may lay 29-143 eggs, which hatch in 16-24 days. The nymphs of the second generation developing during July-August are transformed into winged adults in 46-57 days. These adults live for about 60-71 days and lay eggs in the soil instead of on plants. Some of these eggs hatch in October and give rise to third-generation nymphs which never complete their development and eventually die of cold in December. A majority of eggs, however, do not hatch and remain in the soil till next March. Thus, the pest completes two broods in a year.

Damage. The nymphs and adults feed on *berseem*, maize, sorghum, *bajra* and sugarcane. When there is a serious infestation, the plants may be completely defoliated and the crops, such as *berseem*, may be wiped out.

Management. Dust the crop with fenvalerate 0.4 per cent @ 25 kg per ha.

3. Surface Grasshopper, *Chrotogonus trachypterus* (Blanchard) (Orthoptera : Acrididae) (Fig. 26.2)

This species of surface grasshoppers is widely distributed in the Orient and Africa. In India, it is common in the north whereas *C. oxypterus* occurs in the southern regions. The pests are very common in the field and assume serious proportions in certain years. The former causes considerable damage to the germinating cotton in April-May and the germinating wheat in November. The grasshoppers are polyphagous and, besides cotton and wheat, they feed on a number of other cultivated crops such as *berseem* (*Trifolium alexandrinum*), sugarcane, *kharif* fodders, barley, etc.

Damage is done by the nymphs as well as the adults. The adults are stoutly built, about 20 mm long and 8 mm broad, with a rough brownish appearance (Fig. 21.1). There are two longitudinal rows of black dots on the underside of the white body. The nymphs are similar in body form but are smaller and have wing pads instead of wings.

Life-cycle. *C. trachypterus* remains active throughout the year but its activity is considerably reduced during winter when only eggs and adults are found. A female may lay 36-434 eggs in 2-15 egg pods, at a depth of 3-6 cm in the soil. There are generally 16-46 eggs in a pod. In summer, the eggs hatch in 19-33 days whereas in winter they hatch in about 5 months. The developing nymphs feed on grasses or other soft foods. They moult 5-7 times and are full-fed in 40-70 days. Adults are long-lived, the female for about three months and the male about 1.5 months. There are two complete generations in a year and sometimes the eggs laid in October also hatch giving rise to a third partial brood, which dies in the nymphal stage during the winter.

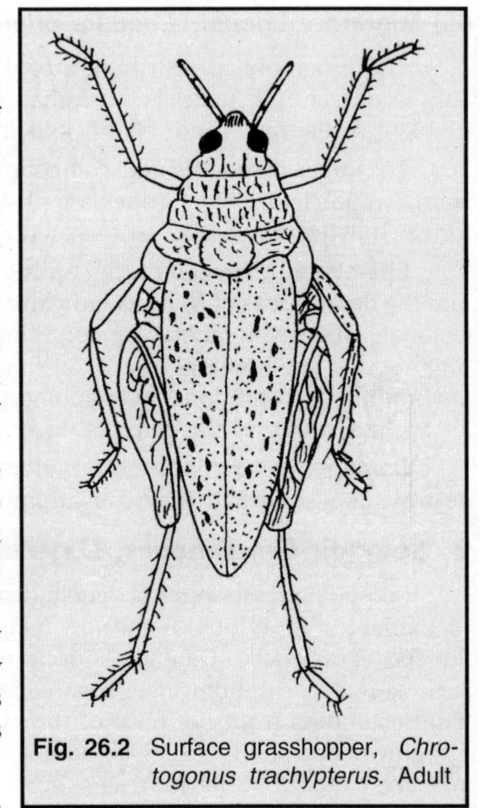

Fig. 26.2 Surface grasshopper, *Chrotogonus trachypterus.* Adult

Damage. Both nymphs and adults feed on leaves by cutting germinating plants of cotton, wheat, etc. particularly in areas adjoining the waste-lands. Often the damaged fields have to be resown. Unlike *Oxya* sp., the attack by this grasshopper is more serious in low-rainfall areas.

Management. Same as in case of sporadic grasshopper.

TERMITES

The termites, commonly called white-ants, are among the common insect pests belonging to the order Isoptera. They are found abundantly and widely in tropical and sub-tropical regions of the world. They live in large communities, mostly in underground nests and are familiar because of their depredations. They make small earthen mounds or earthen passages that are visible above the ground. On opening a portion of an earthen passage, greyish white, wingless insects are seen moving towards or away from the centre of their nest, where the queen of the colony resides. They belong to many families and their nesting behaviour is characteristic of every group. All of them have one thing in common that they are social creatures.

The termites are social insects and their colony organization is based on a caste-system. In a colony, there are numerous workers, lots of soldiers, one queen, a king and a good number of complementary or the colonizing forms of true but immature males and females (**Fig. 26.3**). The various castes and their duties are described as under.

A. Productive Castes

1. Colonizing individuals. These are winged individuals of both sexes and are produced in large numbers during the rainy season. When the temperature and moisture conditions are optimum, they emerge from the parent colony and hover over street lamps early in the evening, usually after a shower of rain. The wings are meant for the nuptial flight only and when they have mated, the wings usually drop off. Most of the winged insects are eaten up by frogs, lizards and snakes. A pair that happens to escape may start a new colony in a crevice in the soil as the queen and the king of a colony. To start with, they themselves attend to the foraging and other duties which, later on, are performed by the workers.

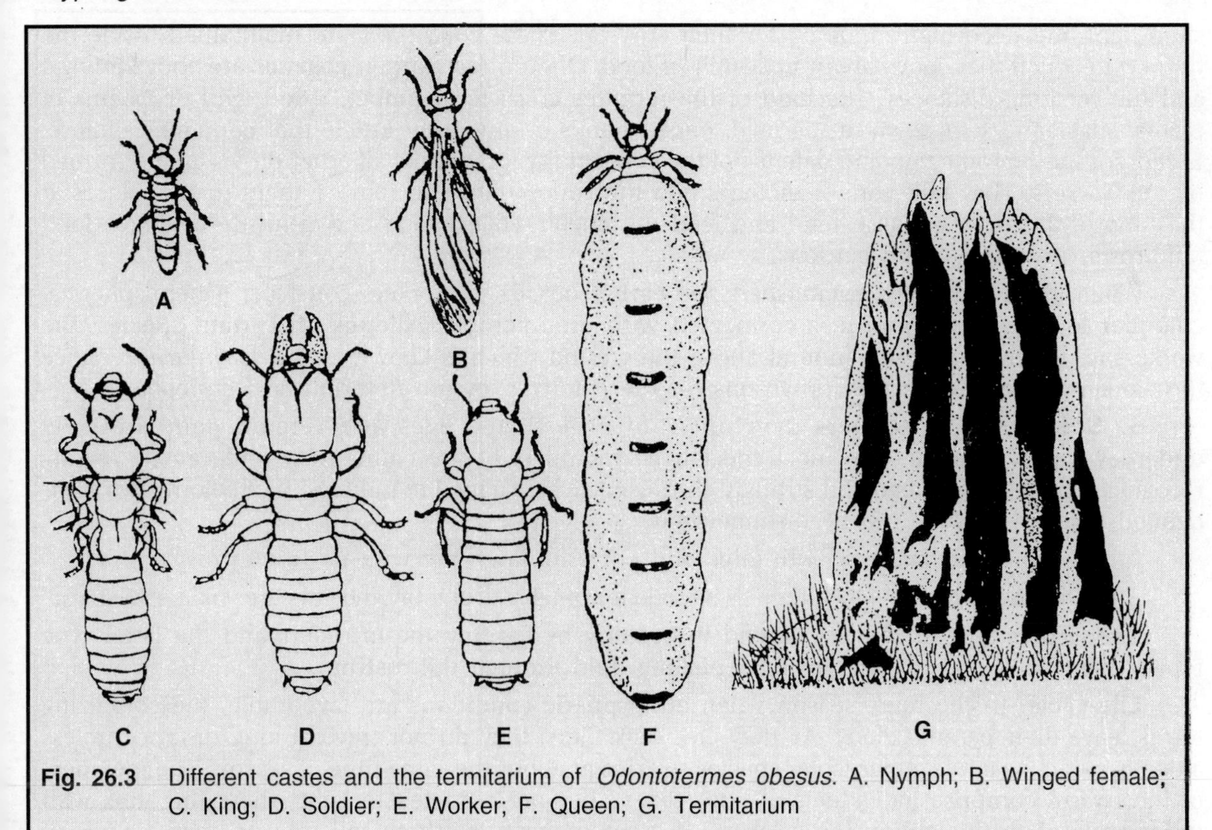

Fig. 26.3 Different castes and the termitarium of *Odontotermes obesus*. A. Nymph; B. Winged female; C. King; D. Soldier; E. Worker; F. Queen; G. Termitarium

2. Queen. This is the only perfectly developed female in the colony. She develops either from colonizing individuals described above or from the wingless forms (complementals) present in an established colony. She attains a much larger size when developed from the former caste. The queen of *Microtermes obesi* Holmgren measures 5.0-7.5 cm in length. She has a creamy-white abdomen marked with transverse dark brown stripes. Her reproductive system occupies almost the entire abdomen. She is a phenomenal "egg-laying machine", laying one egg per second or 70,000-80,000 eggs in 24 hours. There is only one queen in a colony and normally she lives from 5 to 10 years, but there are cases on record in which the queen lived for 20 years. The queen is fed by the workers on the choicest food, and is housed in a special area referred to as the 'royal chamber' which is situated in the centre of the nest, at depth of about 0.5 metre below the ground surface.

3. King. A king develops from an unfertilized egg and becomes fully developed by consuming a superior diet. He is the father of the colony and is a constant companion of the queen, living with her in the 'royal chamber'. He is much smaller than the queen and is slightly bigger than the colonizing individuals. He mates with the queen from time to time and, thus, aids her in laying fertilized eggs from which the colonizing forms and workers develop. The king's life is much shorter than that of the queen and, when he dies, he is replaced by a new one.

4. Complementary castes. They are short-winged or wingless creatures of both sexes and lead a subterranean life. In the event of the untimely death of the king or queen in a colony, the complementary castes replace them. They are induced by the workers to undergo sexual development.

B. Sterile Castes

5. Workers. The workers develop from the fertilized eggs but remain stunted as they are reared on ordinary food. Numerically, they are the most abundant in a colony but are smaller than the soldiers. Their mandibles are well developed and are adapted to the gnawing of wood. The workers shun light and need high humidity for their survival. These conditions are maintained inside the tunnels in which they move about in search of food. Often these earthen galleries are above ground and run for long distances. The food of the termites consists of timber, wood, and dry stems of plants, sugarcane setts sown in the field, paper, books or any other article that contains cellulose. Except for the reproduction and defence of the community, practically all other duties are performed by the workers. They take care of the eggs and the young ones and remove them to safe places at the time of danger. They also tend and feed the queen, collect food and cultivate a fungus food (ambrosia) in underground gardens.

When a new colony is established, the earlier broods of workers construct a hemispherical chamber for the queen and it is connected with innumerable galleries. In certain species, the workers also construct a high mound above the ground which is known as the *termitarium*. Since the workers have to collect all provisions for the colony, they are notoriously destructive.

6. Soldiers. The soldiers develop from unfertilized eggs and remain comparatively underdeveloped. They are the most specialized members of the community and can be readily recognized by the large head and strongly chitinized sickle-shaped mandibles. In a colony, two well-defined types of soldiers can be distinguished.

(*a*) The *mandibulate type* with large and powerful jaws, but with no frontal rostrum.

(*b*) The *nasute type* in which there is a median frontal rostrum, but the jaws are small or vestigeal.

The former type of soldiers defend the colony by fighting the intruders and the latter type repels them by spraying an abnoxious smelling fluid through the rostrum.

Life-cycle. In the rainy season when atmospheric conditions are favourable, the colonizing forms leave their parent colony. As they are weak fliers, they do not travel a long distance unless aided by wind. As a rule, a particular species swarms at about the same time of season. The members of the swarm comprise individuals of both sexes They are attracted to light where they mix with individuals of the neighbouring swarms. A great majority of them fall prey to many types of predators and only a few individuals survive in the end. Sooner or later the survivors fall down and cast their wings and mate before or after shedding them. Both the male and female participate in the early operations of forming a nest by excavating small burrows or galleries or the nuptial chambers. In the beginning only a few eggs are laid and are looked after by them and the newly hatched nymphs are fed by the parents themselves. They develop into workers and then take over all the brooding. During the first season, the reproductive castes are usually not produced. Gradually, the queen grows in size and the number of eggs laid increases.

The eggs hatch after one week during the summer and within 6 weeks the larvae develop to form soldiers or workers, as the case may be. The reproductive castes, when produced, mature in 1-2 years. The queen is capable of laying many millions of eggs during her life, which is very difficult to estimate, but probably 6-9 years is the approximate span.

The nests of many termites grow fungus gardens in the centre, near about the 'royal chamber'. The fungus grows into a comb-like structure and is fed to the royal pair and the larvae.

In addition to the various castes in a termite nest, there live such insects as beetles, flies, thrips, some species of Collembola, Acarina, Diplopoda and Chilopoda. A termite mound may also afford shelter to lizards, snakes, scorpions and certain birds. All these creatures live either as symbionts for mutual benefits or as guests.

The family Termitidae contains over one-half the known species of termites and is economically very important. A level account of the inportant species is given below :

(*i*) *Globitermes sulphureus* (Haviland) occurs from Myanmar to Malaysia and Vietnam. It attacks dead wood. The nest is largely underground, with a small dome-shaped part (cemented earth) projecting out; this visible 'mound' is 80 cm high and 60 cm in diameter at the base. Swarming occurs from January to May.

(*ii*) *Eremotermes* spp. are small, dull white, delicate looking termites which live underground and usually attack roots. *E. dehraduni* Roonwal & Sen-Sarma attacks the "Camphor"-yielding bush, *Ocimum kilimandscharicum*, in nurseries in North India. *E. paradoxalis* Holmgren is a pest of sugarcane in India.

(*iii*) *Microcerotermes* spp. are small, slender termites which live underground and often build globular carton nests which are either fully or partially buried in the ground. *M. minor* Holmgren is a pest of eucalyptus saplings in nurseries in India.

(*iv*) *Macrotermes gilvus* (Hagen) is a large, mound-building species, which occurs throughout South-east Asia. Its large, domeshaped earthen mound has a broad base and it goes up to about 3.5 m in height above the ground. The termite attacks wood-work in buildings in Malaysia and Indonesia and is also a sporadic pest of sugarcane there.

(*v*) *Odontotermes* spp. include several small to medium-sized species. They build large earthen mounds which sometimes rise to 2 or 3 metres in height above the ground. Some of the important species are mentioned here:

1. *O. assanuthi* Holmgren is widespread in India and is a mound-builder in the South. It is an important pest of sugarcane. Swarming in this species occurs in June and July.
2. *O. ceylonicus* (Wasmann) infests buildings in Sri Lanka and it does considerable damage to wood-work. It nests in the ground.
3. *O. feae* (Wasmann) occurs in India, Sri Lanka and South-east Asia. It is a serious pest of wood-work in buildings in India and Myanmar and also attacks eucalyptus seedlings in India. Swarming occurs from June to September.
4. *O. formosanus* (Shiraki) occurs from eastern India to Vietnam and Taiwan. It attacks living plants and is a serious pest of sugarcane and young camphor trees in Taiwan.
5. *O. microdentatus* Roonwal & Sen-Sarma, occurs in Uttar Pradesh, Madhya Pradesh and Bihar. It builds large, domeshaped mounds and is a pest of eucalyptus seedlings.
6. *O. obesus* (Rambur) is one of the most common termites in Pakistan, India, Bangladesh and Myanmar. It is an important pest of several economic crops like sugarcane, wheat, barley, cotton, groundnut, tea, coconut palm, sunn-hemp, vegetables including chillies and several fruit trees including mango, citrus, seedlings of *eucalyptus* and many grasses. It builds tall, sub-cylindrical, earthen mounds which rise above the ground up to about 2.4 m which often possess a number of flat, vertical butteresses all around. Swarming occurs at the beginning of the monsoons.
7. *O. parvidens* Holmgren & Holmgren extends from Pakistan via India to Myanmar. It attacks the bark of several trees and may also kill standing plantations of teak, by girdling the trunk. It is also a pest of tea in north-eastern India. The nest lies underground and the species swarms from February to April.
8. *O. redemanni* (Wasmann) occurs in Bengal, South India and Sri Lanka. It attacks wood-work in buildings and coconut palm. It builds tall, subconical earthen mounds which go up to about 2 metres in height. Swarming occurs during the monsoon in eastern India, and in November and December in Sri Lanka.
9. *O. wallonensis* (Wasmann) is widespread in India, especially in deciduous forests. It builds large, irregularly-shaped earthen mounds with a number of rounded, open mouthed chimneys jutting out from the mound surface. It attacks wood-work in buildings and is also a pest of sugarcane as well as *Casuarina* and shisham (*Dalbergia sissoo*) trees.

(*vi*) *Hypotermes obsuriceps* (Wasmann) is found in eastern India, Sri Lanka and Vietnam. It builds large, broad-based, sub-conical earthen mounds in Sri Lanka, and it attacks timber in houses and also the living coconut palm. In Vietnam it is a pest of living palms and rubber trees.

(*vii*) *Microtermes obesi* Holmgren is widespread in South Asia. It is a serious pest of wheat, barley, oats, millets (bajra), maize, sugarcane, groundnut, tea, jute, mulberry, several vegetables and garden plants. Swarming occurs mostly in July and August.

(*viii*) *M. mycophagus* (Desneux) is an arid zone species and occurs from Pakistan to western India. Swarming occurs from late June to early September.

(*ix*) *M. pakistanicus* Ahmad is found from India to Malaya. It is a serious pest of tea and other plants in Malaysia. It nests underground.

(*x*) *M. insperatus* Kemmer is a serious pest of wood- work in houses and also attacks dead portions of living trees in Java.

(*xi*) *Nasutitermes* spp. are small to medium-sized termites. Soldiers generally have a rounded, nasutiform head which projects forward into an elongated, pointed rostrum. Nests are generally rounded and ball-like which are made of wood-carton and are lodged in branches of trees. A member of the genus *Nasutitermes* attacks wheat in India. In Borneo and Vietnam, *N. matangensis matangensoides* (Holmgren) is highly destructive to timber in houses, while *N. panayensis* (Oshima) occupies a similar position in the Philippines.

(*xii*) *Trinervitermes biformis* (Wasmann) is found in India and Sri Lanka. It is serious pest of several plants, e.g. wheat, sugarcane, cotton, groundnut, and several fruit trees, vegetables and grasses. The nest is generally subterranean, but in South India a small mound is built.

Damage. The termites live on cellulose which they obtain from dead and living vegetable matter. To obtain their food, they destroy wood-work, household articles, fences and wooden poles that come into contact with the soil. They also damage fruit and shade-trees, such crops as wheat, maize (**Plate 26.1A**), sorghum, sugarcane, chillies and peas. They also make complimentary colonies in houses near the roof and damage the books and other material stored in cup-boards.

Management of Termites

1. When a colony is established, it is not so easy to eradicate the pest. The only sure method is to reach the centre of the nest and kill the queen and the complementary forms. Since the termite tunnels run for hundreds of metres, it is often difficult to locate the royal chamber. However, in the mound-forming species of termites, it is much easier to locate the centre which lies in the mound or just below it.
2. To avoid the attack of white-ants in cultivated fields, care should be taken not to use green manure or raw farmyard manure.
3. *Insecticidal control*

   (*a*) *Fruit trees*

   (*i*) In new plantations, the pits should be treated with 0.2 per cent lindane emulsion or crude-oil emulsion before planting the trees. This is done by thoroughly mixing 0.25 kg of crude-oil emulsion and a little arsenic in about 4 baskets of subsoil taken from the pit. The treated soil is returned to the pit.

   (*ii*) To protect the tree trunks, spraying them with 1 per cent lindane is effective.

   (*iii*) To protect the roots, 0.5 per cent chlorpyriphos or lindane or 3 per cent sanitary fluid in the irrigation basin should be applied.

   (*b*) *Field crops*

   (*i*) For protecting chillies in small plots, 3 per cent sanitary fluid should be applied to the soil. When large areas are to be treated, the sanitary fluid is put in a canvas bag at the rate of 25-35 litres per hectare. The bag is suspended in the irrigation channel.

(*ii*) Soaking the sugarcane setts in 0.5 per cent chlorpyriphos suspension or 0.25 per chlorpyiphos emulsion in furrows at the time of planting, saves them from termite attack. Alternatively, apply 25 kg of fibronil 0.3G mixed in 50 kg of sand per ha in the furrows and cover the sugarcane setts by planking.

(*iii*) Wheat seed treatment before sowing, with 400 ml emulsion of chlorpyriphos 20EC or 600 ml of fipronil 5SC in 2.5 litres of water per 100 kg seed remains effective for the crop season. Soil application of chlorpyriphos 4G @20 kg per ha or fipronil 0.3G @ 12.5 kg per ha at the time of field preparation is quite effective.

(*c*) *Buildings*

A galvanised sheet of iron with its outer edge turned downwards when placed just above the damp-proof layer makes the house white-ant-proof. Wooden structures such as door frames should not directly touch the ground and should be raised on a cement layer. An insecticidal barrier between the ground and wood-work in building should be made by treating the soil beneath the building and around foundations with 0.5 per cent chlorpyriphos. The solution should be applied at the rate of 5 litres per m^2.

4. To protect wood-work, paint it with solignum. The cupboards, almirahs, shelves, etc. should be sprayed with chlorpyriphos frequently. The place from where the galleries originate in the house should be either sprinkled over or injected with 0.5 per cent lindane or chlorpyriphos suspension in water. If wooden structures have already been attacked the injection of 0.5 per cent chlorpyriphos emulsions with a hypodermic needle into the wood and in the crevices is the only remedy. For a lasting relief the nests should be located in the vicinity and destroyed by flooding them with the insecticide emulsions.
5. Mounds of termite, if any, in the area should be treated with 0.5 per cent chlorpyriphos after breaking open the earthen structure and making holes with an iron bar. The insecticidal emulsion should be used at the rate of 4 litres per m^3 of the mound.

HAIRY CATERPILLARS

1. Red Hairy Caterpillar, *Amsacta moorei* (Butler) (Lepidoptera : Arctiidae) (Fig. 26.4)

The red hairy caterpillar or *kutra* is widely distributed in the Orient, including India. It is a polyphagous insect and feeds practically on all kinds of vegetation growing during the *kharif* season. Its attack is particularly serious on sunnhemp, maize, jowar (*Sorghum vulgare*), *guara* (*Cyamopsis tetragonoloba*), *mung* (*Phaseolus aureus*), *moth* (*Phaseolus aconitifolius*) and sesamum.

Damage is caused by full-grown caterpillars which measure about 25 mm in length. Their colour varies from reddish- amber to olive green and the body is covered with numerous long hairs arising from the fleshy tubercles. The moths are stoutly built and have white wings with black spots (**Fig. 26.4**). The outer margins of the fore wings, the anterior margin of the thorax and the entire abdomen are scarlet red. There are black bands and dots on the abdomen.

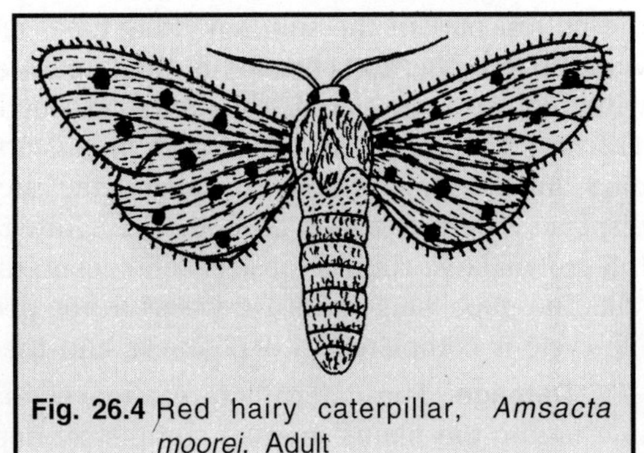

Fig. 26.4 Red hairy caterpillar, *Amsacta moorei*. Adult

Life-cycle. This pest is active from mid-June to the end of August and passes rest of the year in pupal stage in the soil. Moths from these pupae appear usually with the first shower of the monsoon. They are nocturnal in habit and lay light-yellow spherical eggs in clusters of 700-850 each on the under surface of the leaves of host plants. A single female may lay up to 1,500 eggs, which hatch in 2-3 days. The young caterpillars feed gregariously and, as they grow older, they march in bands destroying field after field of various *kharif* crops. The caterpillars grow through six stages and complete their development in 15-23 days. They enter the soil, shed their hair and make earthen cocoons at a depth of about 23 cm. Here they pupate and remain in this stage for many months till they emerge next year from the cocoons. In a given population, probably more than one generation is completed in a year.

Damage. The young *kutra* caterpillars prefer to eat the growing points of plants. The older ones have no such discrimination and they feed voraciously on all vegetation resulting in disaster. Field after field is devastated by the moving army of caterpillars. In the years of severe infestation, there may be a complete failure of the *kharif* crops.

Management. (*i*) The moths are strongly attracted to artificial light. Therefore, light traps of electric or petromax lamps placed just above a broad flat basin full of kerosenized water, should be put on the night following the first shower of the monsoon and continued throughout the period of emergence for about one month. (*ii*) Young larvae are gregarious. They can be destroyed by pulling out the infested plants and burying them under-ground. (*iii*) The grown up caterpillars may be destroyed by crushing them under feet or picking and putting them into kerosenoized water. (*iv*) The pupae may be collected at the time of summer ploughing and destroyed. (*v*) In case of serious attack, spray 1.25 litres of quinalphos 25EC or 250-500 ml of dichlorvos 100EC in 500 litres of water per ha.

2. Bihar Hairy Caterpillar, *Spilarctia obliqua* (Walker) (Lepidoptera : Arctiidae) (Fig. 26.5, Plate 26.1B)

The Bihar hairy caterpillar is a sporadic pest and is widely distributed in the Orient. In India, it is very serious in Bihar, Madhya Pradesh, Uttar Pradesh and the Punjab as a polyphagous pest, particularly of sesamum, *mash* (*Phaseolus mungo*), *mung* (*P. aureus*), linseed, mustard, sunflower and some vegetables.

Damage is caused by full-grown caterpillars which measure 40-45 mm in length and are profusely covered with long greyish hair. The moth measures about 50 mm across the wing spread (Fig 26.5). The head, thorax and under side of the body are dull yellow. The antennae and eyes are black.

Fig. 26.5 Bihar hairy caterpillar, *Spilarctia obliqua*. Adult

Life-cycle. The pest breeds from March to April and again from July to November. It passes the hottest part of the summer (May-June) and winter (December to February) in the pupal stage amidst plant debris. Adults emerge from the over-wintering larvae in March. The moths are nocturnal and they mate during the night. The female lays 412-1241 light-green, spherical eggs in clusters on the underside of leaves. The eggs hatch in 8-13 days and during the first two stages, the tiny caterpillars feed gregariously, but afterwards they disperse widely in search of food. They grow to maturity through 7 stages, within 4-8 weeks. When full-grown, larva spins a loose silken cocoon in which pupation takes place in plant debris or in the soil. The pupal stage lasts 1-2 weeks in the active period and the moths live for about a week. The life-cycle is completed in 6-12 weeks and the pest passes through 3 or 4 broods in a year.

Damage. The caterpillars eat leaves and soft portions of stems and branches. In severe infestation, the plants may be completely denuded of leaves.

Management. (*i*) The young caterpillars can be killed easily by dusting the infested crop with malathion 5 per cent @ 25 kg/ha. (*ii*) When they are full-grown, it is difficult to kill them and very high doses of the pesticides are needed. The chemical control measures are same as in case of red hairy caterpillar.

CUTWORMS

The term cutworm or surface caterpillar is applied to the larvae of several species of noctuid moths which have, in common, the habit of biting through the stems of seedlings at ground level and eating the leaves or the entire seedlings. The various species of cutworms have a wide range of food plants, both cultivated and wild.

The majority of cutworms found in India fall under the genera *Agrotis* and *Euxoa*. Of the recorded species, only four are widespread in the plains and have considerable economic importance. *Agrotis ipsilon* (Hufnagel) is the commonest and *Ochropleura flammatra* (Denis & Schiffermuller) is occasionally abundant in India, the latter being particularly destructive in the north. The moths of these species have the curious habit of hiding in sheltered spots or in thatched roofs of houses during cold weather. *Agrotis segetum* (Denis & Schiffermuller) and *A. biconica* Kollar are also quite common in India, being more so in the South. Their life-history seems to be very similar to that of the *Agrotis* spp.

1. Gram Cutworm, *Ochropleura flammatra* (Denis & Schiffermuller) (Lepidoptera : Noctuidae) (Fig. 26.6)

The gram cutworm is one of the most important pests of chickpea or gram in northern India, particularly in the Punjab and in the sub- Himalayan region. Its distribution extends to Europe, Syria, Persia, western Siberia and Pakistan. It is a polyphagous pest and besides gram, it feeds on the seedlings of many vegetables and other plants such as potato, cucurbits, peas, okra, wheat, *piazi* (*Asphodelus tenuifalius*), tobacco, opium, poppy, etc.

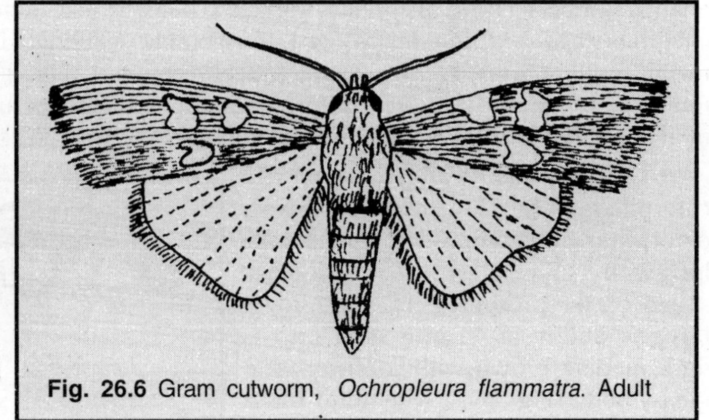

Fig. 26.6 Gram cutworm, *Ochropleura flammatra*. Adult

Only the caterpillars cause damage. The adult is a heavy-bodied, greyish-brown or wheat-coloured insect which measures 5.0-6.2 cm in wing expanse **(Fig. 26.6)**. Its fore wings have characteristic markings and smoky patches, two-thirds of the proximal areas being pale. On each wing, there is a semi-circular spot below the pale area and a greyish-brown kidney-shaped spot towards the apical area. The caterpillars are dark grey or dull green and their skin is smooth and greasy. The full-grown caterpillar is 40-50 mm in length.

Life-cycle. This pest is active from October to April in the plains and it probably passes the summer in aestivation in the adult stage. Since it is active in the mountains during summer, it is sometimes thought that perhaps there is a migration of moths to and back from the mountains. The moths probably appear in the plains sometime in October, to lay yellowish-white eggs on the under surface of leaves, on shoots, stems, or in the soil, within 5-9 days of their short life-span of 7-13 days. A female lays up to 980 eggs in its life-time. The eggs hatch in 4-7 days during summer and 10-14 days during winter. The larvae remain hidden in soil during the day time and feed at night on young shoots or underground tubers. They are full grown in 4-7 weeks and then make earthen cells in the soil for pupation. The pupal stage lasts 12-15 days but during winter it extends up to 5 weeks. The life-cycle is completed in 7-11 weeks and there are generally two generations in a year. Judging from the sudden appearance and disappearance of this pest, it would seem that it is a migrant. This insect is a strong flier and is active only at night.

The adult moths are reported to aestivate in the thatched roofs of houses in summer.

Damage. This cutworm is a sporadic pest of major importance. In some years, 50 per cent of the gram crop may be destroyed. The caterpillars spend the day hiding near and about the plant bases. They remain in the top 5-10 cm of the soil near the plants that might have been cut the night before. At night, they come out and become active, cutting down the young plants of gram, potato, vegetable seedlings, etc. just above or slightly below the surface of the soil. They seem to be very voracious eaters and they fell more plants than they can consume.

Management. (*i*) The pest can be controlled with fenvalerate 0.4 per cent dust @ 25 kg per ha. (*ii*) Alternatively drench the soil around the plant and the ridges with chlorpyriphos 20EC @ 2.5 litres in 1000 litres of water/ha, at the appearance of the pest.

2. Greasy Cutworm, *Agrotis ipsilon* (Hufnagel) (Lepidoptera : Noctuidae) (Fig. 26.7)

This is a pest of world wide occurrence and is found in America, Europe, North Africa, Syria, Japan, China, Indonesia, Australia, New Zealand, Hawaii, Sri Lanka, Myanmar and India. It has been reported from almost all the potato growing regions of northern India, forming a continuous belt from the Punjab in the west to Bengal in the east and Madhya Pradesh in the south. It causes considerable damage to potato. In the Punjab, it is not as common as *O. flammatra*, but along with other cutworms, it causes much damage to the gram crop. Besides gram, this polyphagus pest has been reported feeding on potato, tobacco, peas, wheat, lentil, mustard, linseed, maize, sugarcane, cucurbits, *bhang* (*Cannabis sativa* L.), vegetable seedlings and several weeds. The other cutworm, turnip moth, *Agrotis segetum* (Denis & Schiffermuller) **(Fig. 26.8)**, is a European species, but it is also found in Asia and Africa. The caterpillars attack the roots and lower parts of stem of a wide range of plants, and can be particularly serious pest of root vegetables and cereals.

Damage is caused by the caterpillars only. The slightly yellowish caterpillar, on emergence, is 1.5 mm long with a shiny, black head and a black shield on the prothorax. The full-grown larva is about 42-45 mm long and is dark or dark-brown with a plump and greasy body **(Fig. 26.7)**. The adult moth measures about 25 mm from the head to the tip of the abdomen and looks dark or blackish with some greyish patches on the back and dark streaks on the fore wings.

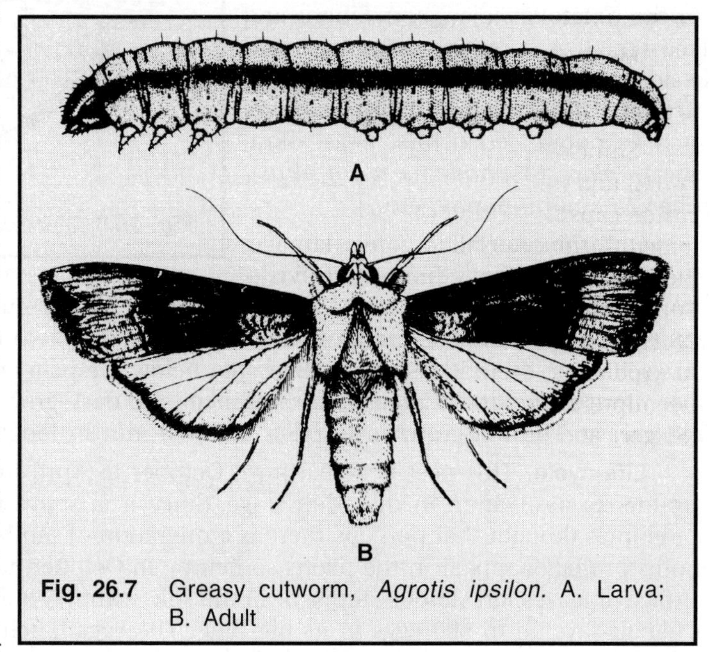

Fig. 26.7 Greasy cutworm, *Agrotis ipsilon*. A. Larva; B. Adult

Life-cycle. The pest is active from October to April and probably migrates to the mountains for further breeding during summer. The moths appear in the plains in October and come out at dusk and fly about until darkness sets in. They oviposit at night and lay creamy-white, dome-shaped eggs in clusters of about 30 each, either on the undersurface of the leaves of food plants or in the soil. The number of eggs laid by a female varies from 199 to 344. Oviposition continues from 5 to 11 days and the duration of the egg stage varies from 2 days in summer to 8-13 days in winter. The newly hatched larvae feed on their egg-shells and move like a semilooper. The larval stage varies from 30 to 34 days in February-April. The advanced-stage larvae may become cannibalistic. The caterpillars are found throughout the winter and become active at night when they cut off and fell the young plants. During the day, they hide in cracks and crevices in the soil. When full-grown, they make earthen chambers in the soil and pupate underground. The pupal stage varies from 10 days in summer to 30 days in winter. The moths usually emerge at night. The life-cycle is completed in 48-77 days and generally three generations are completed in a year. It is a cold-weather pest and is active from October to March in the plains. It suddenly disappears with the onset of summer during April and is not traceable during the off-season, from April to August-September.

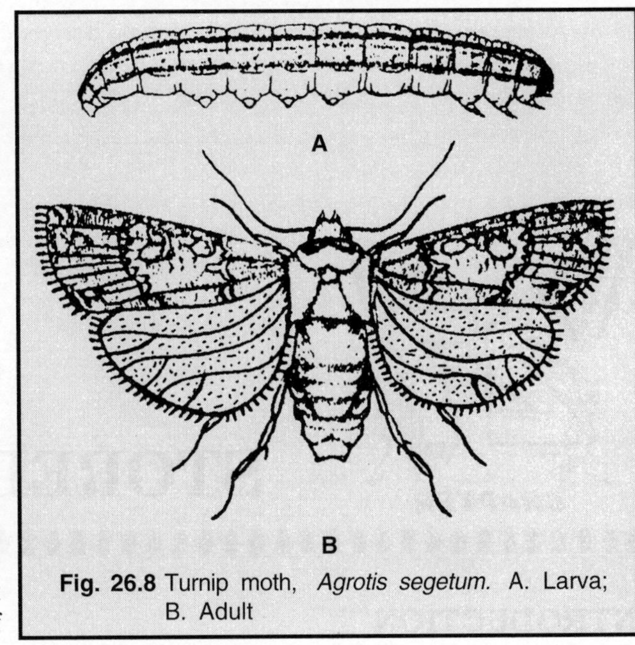

Fig. 26.8 Turnip moth, *Agrotis segetum.* A. Larva; B. Adult

Damage. The young larvae feed on the epidermis of the leaves. As they grow, their habit changes. During the day-time they live in cracks and holes in the ground and come out at night and fell the plants by cutting their stems, either below the surface or above the ground. The cut branches are sometimes seen to have been dragged into the holes where the leaves are eaten at leisure. The larvae may also bore into cabbages. Damage to the *rabi* crops may go as high as 37 per cent and to the potato crop may be as high as 80 per cent.

Management. Same as in case of gram cutworm.

Other Polyphagous Pests

Some other polyphagous pests include the gram caterpillar, *Helicoverpa armigera* (Hubner) **(Plate 26.1C)** and tobacco caterpillar, *Spodoptera litura* (Fabricius) **(Plate 26.1D)** (Lepidoptera : Noctuidae); castor capsule borer, *Conogethes punctiferalis* Guenee (Lepidoptera : Pyralidae); serpentine leaf-miner, *Liriomyza trifolii* (Burgess) Diptera : Agromyzidae); white grub, *Holotrichia consanguinea* Blanchard (Coleoptera : Scarabaeidae) **(Plate 26.1E)** ; fruit fly, *Bactrocera dorsalis* (Hendel) (Diptera: Tephritidae) ; mealybug, *Phenacoccus solenopsis* Tinsley and *Ferrisia virgata* (Cockerell) **(Plate 26.1F)**;(Hemiptera: Pseudococcidae) cotton whitefly, *Bemisia tabaci* (Gernadius) (Hemiptera : Aleyrodidae) ; lac insect, *Kerria lacca* (Kerr) (Hemiptera : Kerriidae) and several species of aphids (Hemiptera : Aphididae), viz. cotton aphid, *Aphis gossypii* Glover ; green peach aphid, *Myzus persicae* (Sulzer) and groundnut aphid, *Aphis craccivora* Koch.

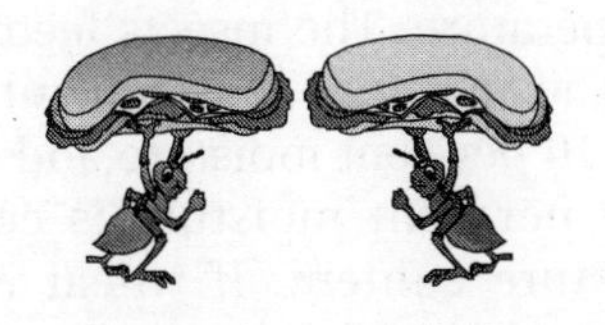

STORED GRAIN PESTS

INTRODUCTION

Since the cereals and pulses are the staple food of people in India, they have to be stored by the producers in their homes and by the traders and the government agencies in godowns for one year or more. The foodgrains are constantly exposed to physical factors such as temperature and moisture, and biological agencies such as insects, mites, rodents, birds, fungi, etc. Almost all the insect pests of stored grains have a remarkably high rate of multiplication and within one season, they may destroy 10-15 per cent of the grain and contaminate the rest with undesirable odours and flavours. Nearly one thousand species of insects have been associated with stored products in various parts of the world. The majority of insect pests belong to the orders Coleoptera and Lepidoptera, which account for about 60 and 8-9 per cent of the total number of species, respectively. Under Indian conditions, there are about a dozen species of insect pests of stored grains which are most important among a large number of species recorded. Out of these, some are capable of damaging all kinds of stored foodgrains while others attack the broken or milled grains only. The former are known as primary insect pests and the latter as secondary pests.

ENVIRONMENT AND STORAGE PESTS

Temperature is one of the most important factors of the environment, as the multiplication of insect pests of stored grains depends on it. The minimum temperature at which these insects are able to develop and multiply is between 15.5 and 18.3°C. Many of the insects can live for long periods at lower temperature but their activity is reduced very much. The optimum temperature for most of the species lies between 28.0 and 32.0°C.

The effect of humidity on the development of insect pests of grains and other dry products is intimately associated with that of temperature. The insects feeding on dry materials need a certain minimum of moisture in their food. The moisture requirement varies from species to species but practically all of them need more than 10 per cent moisture and the optimum is around 14 per cent. Therefore, grain having less than 10 per cent moisture is considered safe for storage. Micro-organisms also develop at high moisture content. If wheat is stored at a moisture content of 14.0-14.8 per cent at 21.1°C, it will be attacked by *Aspergillus restrictus* while at higher moisture content, *A. ruber* will predominate. Attack of *A. candidus* takes place at moisture content of 16-17 per cent and of *A. flavus* at 18 per cent or above, with relative humidity between 85 and 90 per cent. The moist grains in storage sometimes contain toxins such as aflatoxin, ochratoxin, F-2 toxin and rubrotoxin, produced by the fungi, *A. flavus, A. cohraceus, Fusarium* sp. and *Penicillium rubrum*, respectively. The mycotoxins found in foodgrains are possible hazards to human and animal health.

The grain in storage also undergoes chemical changes with a change in the moisture. Increased respiration of stored grain under high temperature increases fatty acids and reduces sugar content. At high moisture levels, however, carbohydrate fermentation may occur with the production of alcohol or acetic acid resulting in a characteristic sour odour. Storage of grain above 16 per cent moisture content at comparatively high temperatures may reduce seed viability to a great extent. Some other known effects of storage of moist grain are discoloration of the kernel, mustiness, problems of milling and separation of starch. Grain with high moisture content not only deteriorates in quality but also occupies more space as its bulk increases.

At the time of harvest, therefore, the grain should be dried until the moisture content is less than 9 per cent, which is ideal for storage; but sometimes, uncertain weather would not allow the proper drying up of grain. Again, the grain kept in damp godowns and improper receptacles may absorb moisture from the ground or from the atmosphere, particularly during the monsoon. In general, warm season and high moisture content of the grains are highly conducive to the proper development and rapid multiplication of insect pests of stored grains and pulses.

LEPIDOPTERN PESTS

1. Angoumois Grain Moth, *Sitotroga cerealella* (Olivier) (Lepidoptera : Gelechiidae) (Fig. 27.1)

The distribution of this pest is world-wide. In the Indian Sub-continent, the pest is more abundant in the mountainous areas or where the climate is rather mild. It is considered an important pest of stored grain such as wheat, maize, sorghum, barley, oats, etc.

Only the larvae cause damage by feeding on the grain kernels. A full-grown larva is about 5 mm long, with a white body and yellow-brown head **(Fig. 27.1)**. The adult is a buff, grey-yellow, brown or straw-coloured moth, measuring about 10-12 mm in wing expanse. The characteristic feature of this insect is the presence of the narrow pointed wings fringed with long hair, most prominent along the posterior margin.

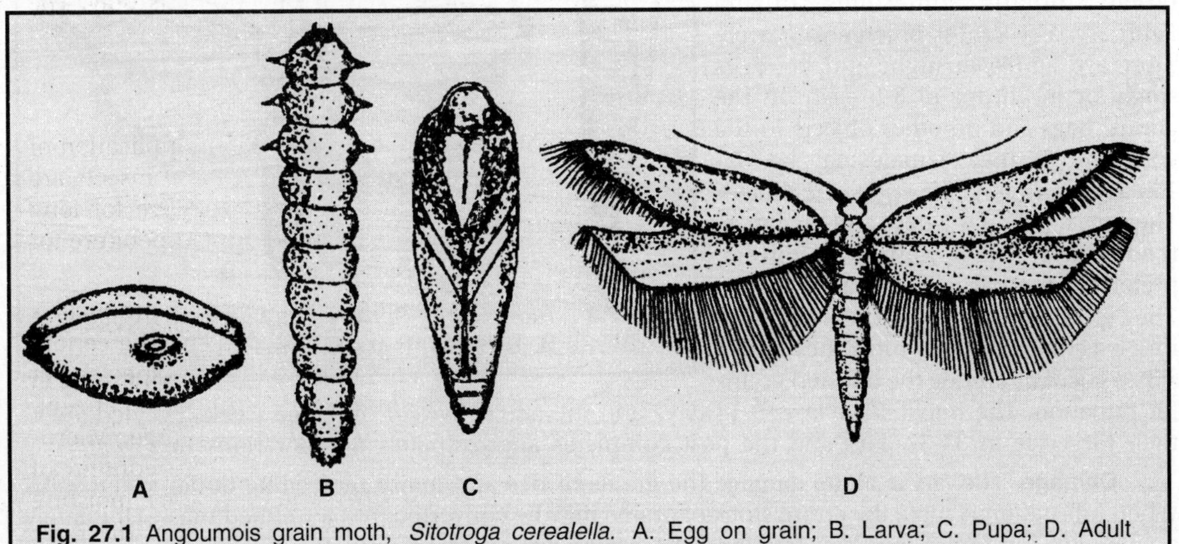

Fig. 27.1 Angoumois grain moth, *Sitotroga cerealella.* A. Egg on grain; B. Larva; C. Pupa; D. Adult

Life-cycle. Breeding takes place from April to October. The insect overwinters as a hibernating larva and as the season warms up, it pupates in early spring. After emergence, moths mate within 24 hours and the females start laying eggs singly or in batches on or near the grain. The eggs are small and white, when freshly laid, turning reddish later on. A single female lays, on an average, 150 eggs, usually within a week after mating. The incubation period is about 4-8 days in summer. Experimentally, the eggs hatch in 8.5, 6.0, 4.1 and 4.6 days at 20, 25, 30 and 35°C, respectively. The newly emerged larva soon bores into the grain and feeds on its contents. The larval stage may last about 3 weeks. Before pupation, the larva constructs a silken cocoon in a cavity made during feeding and then turns into reddish-brown pupa. Later, the adult emerges by pushing aside the seed-coat that covers the exit. At the constant temperatures of 20, 25, 30 and 35°C, the duration of the larval and pupal stages is 23.5, 20.2, 19.4, and 22.2 days, and 12.5, 9.1, 6.5 and 12.2 days, respectively. The female adults have a longer life-span than the males, the duration being 10.3, 7.5, 6.6 and 4.3 days at 20, 25, 30 and 35°C, respectively. During the active season, the life-cycle is completed in about 50.6 days. Several generations are completed in a year.

Damage. The damage is at its maximum during the monsoon. The larva bores into the grain and feeds on its contents. As it grows, it extends the hole which partly gets filled with pellets of excreta. Usually, about 30-50 per cent of the contents are consumed, but sometimes the larva finishes off the entire grain. With infestation the grains give out an unpleasant smell and present an unhealthy appearance, each grain being covered with scales shed from the moths. In a heap of grain, it is the upper layers that are most severely affected.

2. Rice Moth, *Corcyra cephalonica* (Stainton) (Lepidoptera : Pyralidae) (Fig. 27.2)

The rice moth is distributed in Asia, Africa, North America and Europe. In the larval stage, it is an important stored-grain pest in both India and Pakistan. It also infests gram, sorghum, maize, groundnut and cotton-seed.

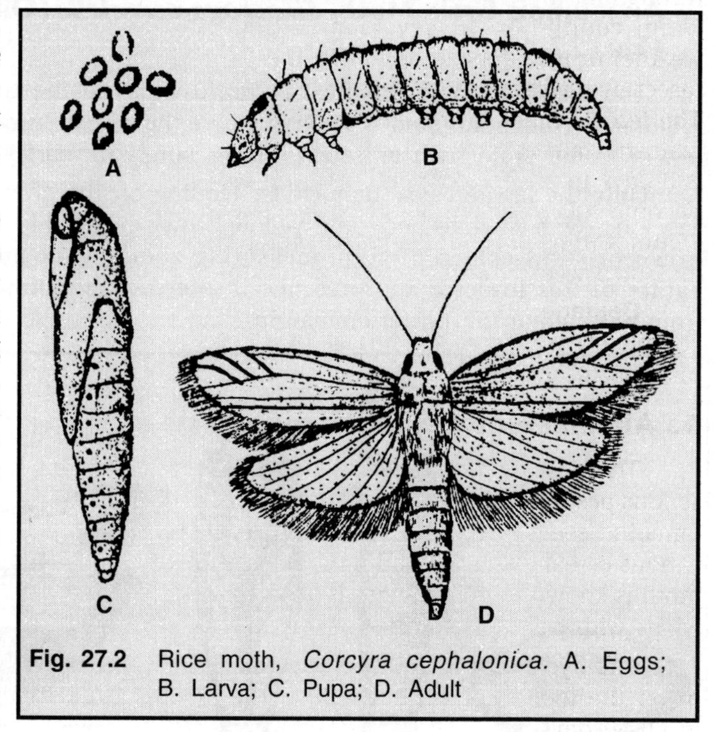

Fig. 27.2 Rice moth, *Corcyra cephalonica*. A. Eggs; B. Larva; C. Pupa; D. Adult

Life-cycle. The rice moth is active from March to November when all stages are noticed (**Fig. 27.2**). It passes winter in the larval stage and the overwintered larvae pupate sometimes during February. The moths emerge in March. They are active at night and lay eggs singly or in groups of 3-5 each on the grains, bags and on other objects in the godowns. A single female may lay 62-150 eggs during its life-span of 2-4 days. The eggs hatch in 4-7 days and the larvae feed under silken web-like shelters, preferring the partially damaged grains. They grow in five stages and are full-fed in 21-41 days, after which they make silken cocoons among the infested grains for pupation. The pupal stage lasts 9-14 days and the adults live for over one week. They complete their life-cycle in 33-52 days and the pest completes approximately 6 generations in a year.

Damage. The larvae alone damage the grains of rice and maize by feeding under silken webs. When infestation is high, the entire stock of grains may be converted into a webbed mass. Ultimately, a characteristic foul odour develops and the grains are rendered unfit for human consumption.

3. Indian Meal Moth, *Plodia interpunctella* (Hubner) (Lepidoptera : Pyralidae) (Fig. 27.3)

The pest is worldwide. It infests grains, meals, breakfast foods, soybean, dried fruits, nuts, seeds, dried roots, herbs, dead insects, etc.

Only the larva causes damage and may be recognized in the early stages as a small whitish caterpillar, often tinged with green or pink, a light-brown head, a prothoracic shield and an oval plate. On reaching maturity, the larva is 8-13 mm in length (**Fig. 27.3**). The adult moth is about 13-20 mm in wing expanse. It has a coppery lustre on the outer two-thirds and terminal whitish grey on the inner portions and on the end of the body. The palps form a characteristic cone-like beak in front of the head.

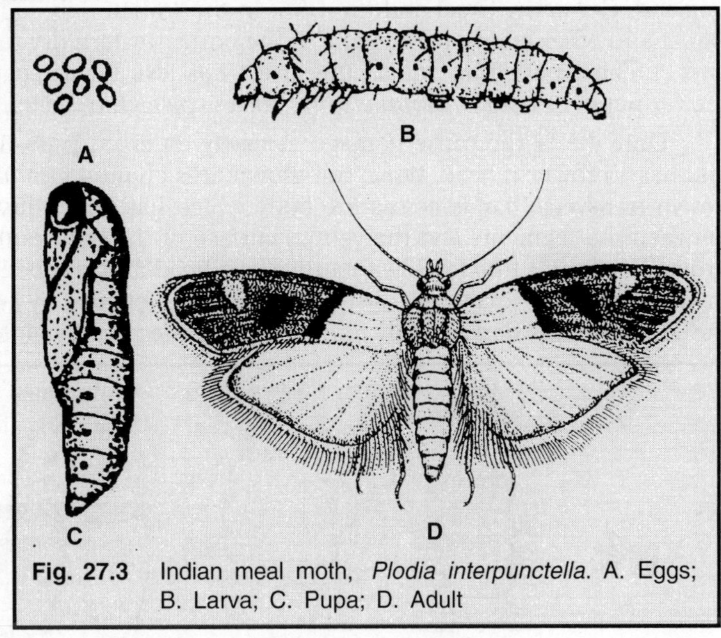

Fig. 27.3 Indian meal moth, *Plodia interpunctella*. A. Eggs; B. Larva; C. Pupa; D. Adult

Life-cycle. Breeding continues throughout the year under favourable warm conditions; however, in cold weather or in unheated buildings, the insect over winters in the larval stage. The female moth lays 30-350 minute whitish ovate eggs, singly or in clusters, on or near the appropriate foodstuffs. The incubation period is 2 days to 2 weeks depending upon weather. The larvae feed upon the grains or other foods and become full-grown in 30-35 days. They crawl up to the surface of the food material and pupate within a thin silken cocoon. The pupal stage lasts 4-35 days. In summer, the life-cycle is completed in 5 or 6 weeks and there are about 4-6 generations in a year.

Damage. Extensive damage may be caused by the active, crawling caterpillars which may completely web over the surface of a heap of grains with silken threads. The adults fly about from one bin to another and spread the infestation.

4. Almond Moth, *Ephestia cautella* (Walker) (Lepidoptera : Pyralidae)

The pest is also known as the dried currant moth and the fig moth. It is worldwide and is a serious pest of dried fruits such as currants, raisons, dried apples, dates, berries, figs, almonds, walnuts, tamarind seeds, etc. It has also been recorded on lac, malted milk, dried mango, pulp, garlic bulbs, various cereal grains and grain products. Although it is primarily a pest of stored products, the almond moth is known to maintain itself on figs and date-palm in some countries. The adult moth has greyish wings with transverse stripes on the outer region and the wing expanse is about 12 mm.

Life-cycle. The female lays whitish eggs indiscriminately in cracks and crevices of the receptacles or on the food stuff. While feeding, the larvae spin tubes in the food material and are full-grown in 40-50 days. The full-grown larva is white with pinkish tinge and measures 1.5 cm. The larvae pupate inside the cocoons and pupal stage lasts about 12 days. The life-cycle is completed in about two months and there are 5-6 generations in a year.

Damage. The caterpillars make tunnels in the food materials. The number of silken tubes is sometimes extremely high and these clog the mill machinery where the infested grains have been sent for milling.

BEETLES AND WEEVILS

5. Khapra Beetle, *Trogoderma granarium* Everts (Coleoptera : Dermestidae) (Fig. 27.4)

Although a native of India, this pest has spread to many countries and has been reported from England, Germany, Israel and the USA. In the Indian Sub-continent, it is a very destructive pest of wheat and other grains, particularly in the north-western dry regions of Pakistan, Rajasthan, Haryana and Punjab. Apart from wheat, the insect has also been recorded on sorghum, rice, barley, gram, maize, poppy (*Papaver samniferum*), pulses, pistachio, walnut and other dried fruits.

Only the larvae cause damage. A newly emerged yellowish-white larva is about 1.5 mm long and has a brownish head. When full-grown, it is about 4 mm in length and is brownish, with yellow-brown transverse bands across the body which has long hairy bristles (**Fig. 27.4**). The integument between the segments and the ventral surface of the body is pale yellow. The adult is a small dark-brown beetle, 2-3 mm long, with a retractile head and clubbed antennae. The entire body is clothed in fine hairs. The males are distinguished from females by being smaller (usually half the size of the females) and darker, with more elongated terminal points of the antennae.

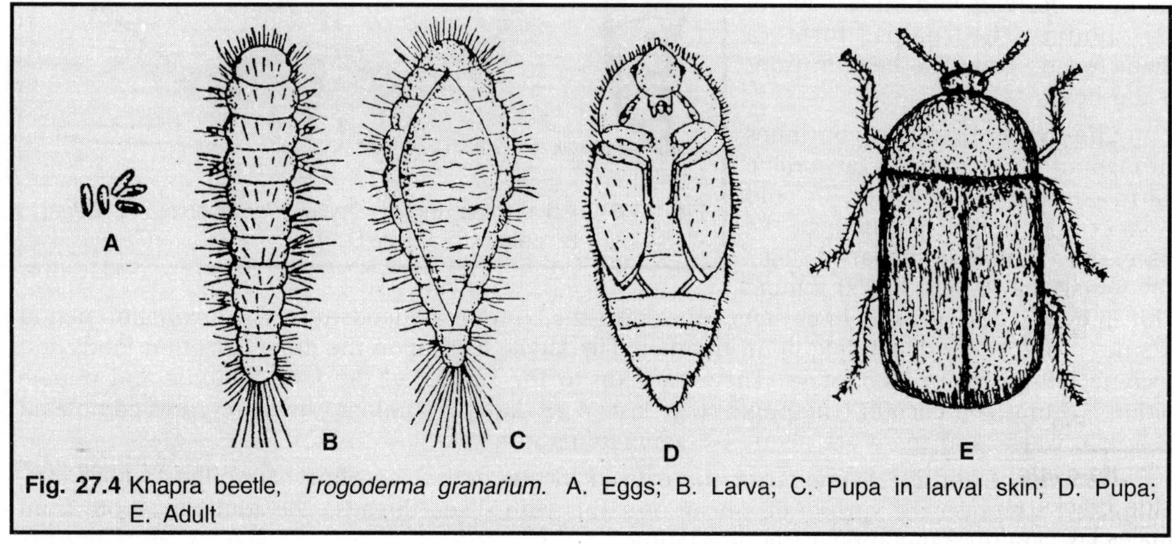

Fig. 27.4 Khapra beetle, *Trogoderma granarium.* A. Eggs; B. Larva; C. Pupa in larval skin; D. Pupa; E. Adult

Life-cycle. The insect breeds from April to October and hibernates in the larval stage from November to March in cracks and crevices of walls and floors or in other sheltered places. Copulation takes place 2-3 days after emergence, a male being capable of fertilizing more than one female. One to three days after copulation, the female begins to lay white translucent eggs on the grains, singly or sometimes in clusters of 2-5. The eggs are rather cylindrical, rounded at one end and narrow at the other. A female may lay 13-35 eggs in 1-7 days at the rate of 1- 26 eggs per day, the largest number being laid on the first day. The incubation period varies from 3-5 days in June to 6-10 days in October. The viability of the eggs varies from 86 per cent in September to 58 per cent in October. The male larva is full-fed in 20-30 days and the female larva in 24-40 days. Pupation takes place in the last larval skin among the grains. This stage lasts 4-6 days. The adults are incapable of flying. There are 4-5 generations in a year.

Damage. The greatest damage is done in summer from July to October. The grubs eat the grain near the embryo or at any other weak point and from there proceed inwards. They usually confine themselves to the upper 50 cm layer of grains in a heap or to the periphery in a sack of grains. If the infestation is severe, the devastation is complete, reducing the grain to a mere frass. Since the larvae are positively thigmotactic, they can be collected by merely placing gunny bags on a heap of grain.

6. Rice Weevil, *Sitophilus oryzae* (Linnaeus) (Coleoptera : Curculionidae) (Fig. 27.5)

The pest is world-wide and is found practically throughout India. It is the commonest and, perhaps, the most destructive pest of stored grain throughout the world. It is distinct from another allied species, the grain weevil, *Sitophilus granarius* (Linnaeus). Both the species are similar in size and appearance and are found together feeding upon rice, wheat, maize and other grains. The rice weevil may, however, be found in the paddy fields as well.

Both the adults and the grubs cause damage. The full-grown larva is 5 mm in length and is plump, fleshy legless creature, having a white body and a yellow-brown head (**Fig. 27.5**). The adult is a small reddish-brown beetle, about 3 mm in length, with a cylindrical body and a long, slender, curved rostrum. Its elytra bears four light reddish or yellowish spots and thorax is fitted with round depressions. Unlike *S granarius*, the metathoracic wings of this insect are well developed.

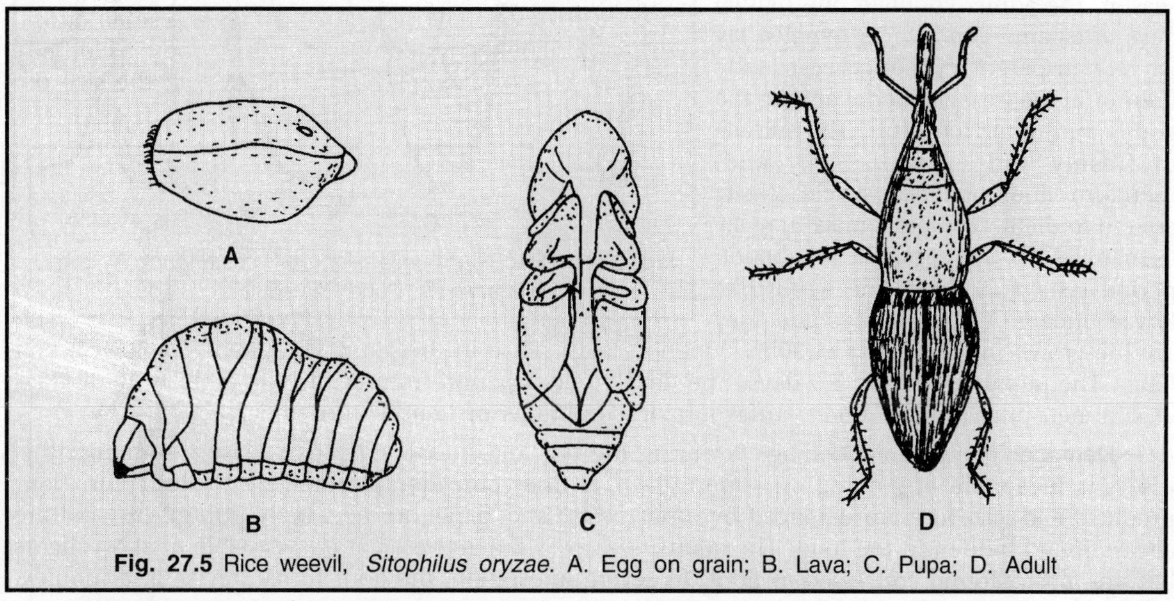

Fig. 27.5 Rice weevil, *Sitophilus oryzae.* A. Egg on grain; B. Lava; C. Pupa; D. Adult

Life-cycle. The rice weevil breeds from April to October and hibernates in winter as an adult inside cracks and crevices or under wheat bags in the godowns. During the active season, the females lay eggs on the grain by making a depression with the help of their mandibles. After an egg has been laid, the hole is sealed with a gelatinous secretion. A single female may lay as many as 400 eggs. The eggs hatch in 6-7 days and the young larvae bore directly into grain, where they feed and grow to maturity. Then, they pupate inside the grain. The pupa, at first, is dirty white, but later on becomes dark brown. The pupal stage lasts 6-14 days. On emergence, the adult weevil cuts its way out of the grain and lives for about 4-5 months. At least 3-4 generations are completed in a year.

Damage. Heavy damage may be caused by this pest to wheat, rice, maize and sorghum grains, particularly in the monsoon. It has also been reported feeding on oats, barley, cotton-seed, linseed and cocoa. The weevils destroy more than what they eat.

7. Red Flour Beetle, *Tribolium castaneum* (Herbst) (Coleoptera : Tenebrionidae) (Fig. 27.6)

This insect is world-wide and is the most common pest of wheat-flour. It also feeds upon dry fruits, pulses and prepared cereal foods, such as cornflakes. Both the larvae and adults cause damage. The young larva is yellowish white and measures 1 mm in length. As it matures, it turns reddish yellow, becomes hairy and measures over 6 mm in length. Its head, appendages and the last abdominal segment are darker. The adult is a small reddish-brown beetle, measuring about

3.5 mm in length and 1.2 mm in width (**Fig. 27.6**). Its antennae are bent and bear a distinct club formed by the three enlarged terminal joints. The last antennal segment is transversely rounded. Another allied species, *T. confusum* Duval, is also often present in the wheat flour.

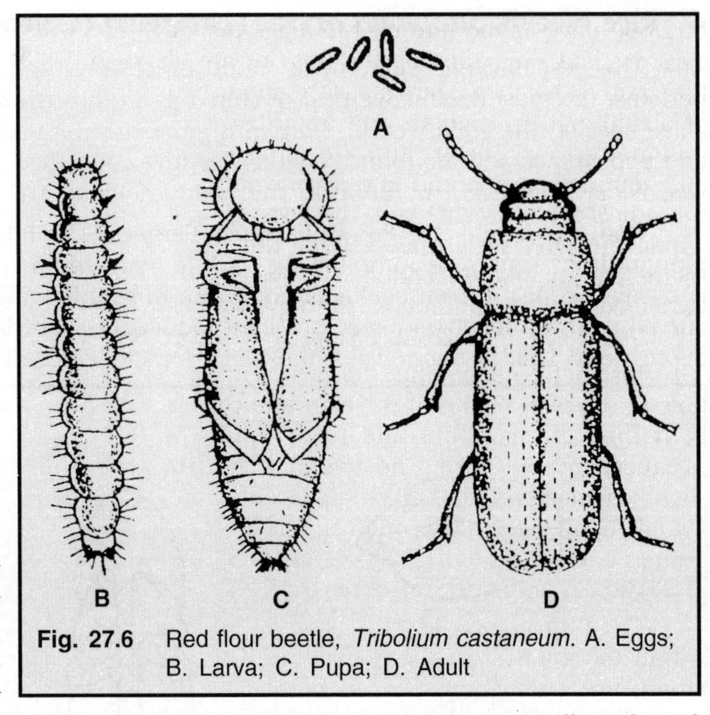

Fig. 27.6 Red flour beetle, *Tribolium castaneum*. A. Eggs; B. Larva; C. Pupa; D. Adult

Life-cycle. The insect breeds from April to October and passes the winter mostly in the adult stage. In the active season, the adults copulate one or two days after emergence. The females lay white, transparent, cylindrical eggs in the flour or in the frassy material among the grains and other foodstuff. The surface of freshly laid eggs is sticky and, therefore, flour or dust particles easily adhere to them. A single female may lay as many as 327-956 eggs. The incubation period lasts 4-10 days. The worm like larvae undergo 6-7 moultings and they are full-grown in 22-25 days at 30°C. Pupation takes place in the flour. The pupa is yellowish and hairy. The pupal stage lasts 5-9 days. The development period from egg to the adult is 26-30 days in summer, but is longer under unfavourable conditions of temperature and food.

Damage. The greatest damage is during the hot and humid monsoon season. Although this insect is incapable of feeding on sound grain, it does considerable damage to flour and flour products and also to grain damaged by other pests. The larvae are negatively phototactic and are always found hidden in the food. The adults, however, are active creatures, capable of short flights but are mostly found concealed in flour. In severe infestation, the flour turns greyish and mouldy, and has a pungent, disagreeable odour making it unfit for human consumption.

8. Lesser Grain Borer, *Rhyzopertha dominica* (Fabricius) (Coleoptera : Bostrichidae) (Fig. 27.7)

Originally, an inhabitant of India, this insect has now spread to the rest of world. It has been reported as a pest from Algeria, Greece, United States, New South Wales (Australia), Japan and China. It is known as a destructive pest of wheat and has also been recorded on other grains and foods such as rice, maize, sorghum, barley, lentils, army biscuits, ship biscuits, stored grains, dried potatoes, corn flour, beans, pumpkin seeds, tamarind seeds and millets.

Both the adults and larvae cause damage. The larva is about 3 mm long, dirty white, with a light-brown head and a constricted elongated body (**Fig. 27.7**). The adult is a small cylindrical beetle measuring about 3 mm in length and less than 1 mm in width. It is shining dark brown or black with a deflexed head, covered over by a crenulated hood-shaped pronotum. The antennae are 10-jointed and terminate in a prominent tripartite club, the first two segments of which are triangular and the apical segment oval. No morphological difference separates the two sexes.

Life-cycle. The pest breeds from March to November and in December, it enters hibernation as an adult or as a larva. In March, it resumes activity and by the first week of April, the females begin to lay eggs. Polyandry and polygamy are common in this insect. During oviposition, there are spurts of vigorous feeding, alternating with heavy egg-laying. A single female can lay 300-400 eggs in 23-60 days at the rate of 4-23 eggs per day. The eggs are laid singly among the frass or are glued to the grain in batches. They measure about two-thirds of a millimetre in length and about one-fifth of a millimetre in diameter, and are almost cylindrical and rounded at one end and somewhat pointed at the other. When freshly laid, the eggs are glistening white, but later on a pink opaque line appears on them. The incubation period is about 5-9 days but under controlled conditions, the eggs hatch in 9.7, 6.4, 4.7 and 5.7 days at 25, 30, 35 and 37°C, respectively.

Just before emerging, the larva cuts a circular hole in the pedicel end of the eggs and comes out of it. It then crawls about and feeds upon the flour produced by the boring beetles or it burrows into slightly damaged grains and passes its entire life inside, feeding upon the starchy contents. The larva moults 4-5 times and is full-fed in 52.7, 31.3, 23.7 and 25.6 days at 25, 30, 35 and 37°C, respectively.

The pupal stage, like the larval stage, is passed within the grain or in the grain dust. The adult, on emergence, remains inside the grain for a few days until it is mature. It then cuts its way through the side of the grain and comes out. The period from the egg to adult stage is completed in 30 days in the summer. Experimentally, the duration of the pupal and adult stages at 25, 30, 35 and 37°C was observed to be 6.7, 4.6, 3.8 and 3.9 days, and 87.0, 79.0, 54.9 and 39.9 days, respectively. The males have a longer life-span than the females. There are 5-6 generations in a year.

A B C D

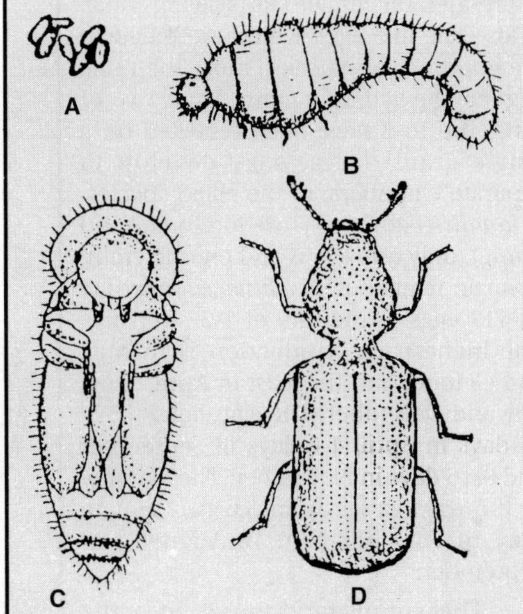

Fig. 27.7 Lesser grain borer, *Rhyzopertha dominica*. A. Eggs; B. Larva; C. Pupa; D. Adult

Damage. The adults and grubs cause serious damage to the grains by feeding inside them and reducing them to mere shells with many irregular holes. The adults are powerful fliers and migrate from one godown to another, causing fresh infestation. When the infestation is severe, the adults produce a considerable amount of frass, spoiling more than what they eat. The flour, so produced, serves as nourishment for the young grubs until they are ready to bore into the grain.

BRUCHIDS

9. Gram Dhora, *Callosobruchus chinensis* (Linnaeus) (Coleoptera : Bruchidae) (Fig. 27.8)

The gram *dhora* or pulse beetle has been reported from the USA, Mauritius, Formosa, Africa, China, the Philippines, Japan, Indonesia, Sri Lanka, Myanmar and India. It is a notorious pest of gram, *mung* (*Phaseolus aureus*), *moth* (*Phaseolus aconitifolius*), peas, cowpeas, lentil and *arhar* (*Cajanus cajan*). It has also been reported on cotton seed, sorghum and maize.

The larva does the damage by feeding inside the grain. In the early stages, it is whitish with a light-brown head and later on it acquires a creamy hue. The mature larva is 6-7 mm long (**Fig. 27.8**). The adult beetle measuring 3-4 mm in length, is oval, chocolate or reddish brown and has long serrated antennae. There is a pair of white elongate prominences in the middle of the hind margin of the thorax, a spine on each of the inner and outer edges of the end of the hind femur, and truncate elytra, not covering the pygidium.

Life-cycle. The pest breeds actively from March to the end of November. It hibernates in winter in the larval stage. At the end of March, the adults appear and copulate immediately after emergence.

A day later, the female starts laying small, oval, scale-like eggs which are glued to the grain. In this species, more than one egg may be laid on a grain. Thus, two or three (up to 8 have been reported on a single grain) larvae may develop in separate chambers. In the allied species, *C. analis* (Fabricius) (predominantly on *mung*), only one egg is laid on one grain. A single female of *C. chinensis* may lay 34-113 eggs at the rate of 1-37 per day. The highest egg production is in May and October and the least in April, June, July and December. The eggs hatch in 7-14 days in April, 4-6 days in September and 8-16 days in November. The viability of the eggs varies from 3.6 per cent in May to 76.9 per cent in August and September.

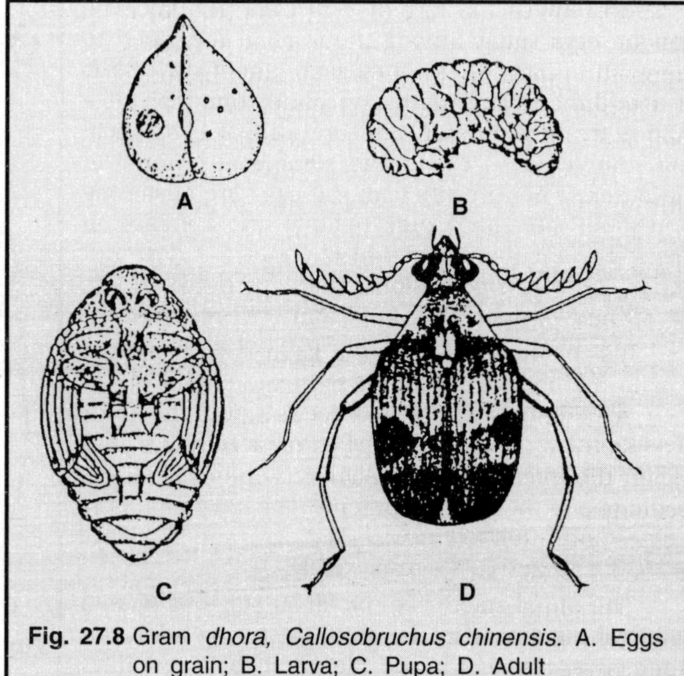

Fig. 27.8 Gram *dhora, Callosobruchus chinensis.* A. Eggs on grain; B. Larva; C. Pupa; D. Adult

The young larva bores into the grain and completes its development inside. The larval stage is completed in 10-12 days in August and September, and 26-38 days in November. The hibernating larvae take 117-168 days to complete their development. The full-grown larva migrates towards the periphery and comes to lie next to the seedcoat where it turns into an oval white pupa. The pupal stage lasts 4-28 days, depending upon the season. The adult escapes by cutting a circular hole in the seedcoat and such grains can be spotted easily. The average life-span of an adult is 5-20 days. A preponderance of males occurs throughout the active season. The insect passes through 7-8 overlapping generations in a year.

Damage. The damage is at its peak from April to September and is considerably reduced in October-November. Damage to the pulses infested with this insect is very high and quite often each and every grain is infested so that the pulses become unfit for human consumption. Infested grain is often converted by the traders into flour which has a characteristic off-flavour and should not be marketed.

10. Mung Dhora, *Callosobruchus analis* (Fabricius) (Coleoptera : Bruchidae)

The distribution of *mung dhora* or pulse beetle is comparatively limited. So far, it has been recorded from Germany, Rhodesia, Myanmar and India. It is a pest of *mung* (*Phaseolus aureus*), *mash* (*Phaseolus mungo*), *moth* (*Phaseolus aconitifolius*), peas, cowpeas and other pulses. Both the larvae and adults cause damage. The larva is recognized by its creamy-white, oval, flabby body. The adult is an oval beetle, somewhat smaller than *C. chinensis.* The female is of chocolate colour with a black trapezoidal area surrounded by white streaks on three sides. This area is located in the centre of each elytra towards the outer margin. The exposed pygidium is black with a central longitudinal white streak. The male is uniformly chocolate with a tinge of straw. The other species of pulse beetle common in India are *C. maculatus* (Fabricius) and *C. phaseoli* Gyll.

Life-cycle. The insect breeds from March to December and then over-winters in the larval stage within the infested grain. In the first week of March, the adults escape by cutting out a circular hole in the seedcoat. Soon after emergence, copulation takes place which lasts 3-15 minutes. The females lay oval, white eggs singly or in batches within 72 hours of mating. The eggs are glued to the grain and the number laid in 2-22 days by a female varies from 11 to 150. The highest number of eggs is laid in August (average 95 eggs) and the lowest in July or December (average being 63). Viability among eggs varies from 14.5 per cent in August to 72.4 per cent in May, which probably depends upon the prevailing humidity. The incubation period is 3-6 days in May-August, 4-8 days in April to October, 8-13 days in November and 18 days in December. Experimentally, the eggs have been hatched in 4.9, 3.8, 4.1 and 5.3 days at 25, 30, 35 and 37°C, respectively. There was no hatching at 40°C. The young larva, on emergence, bores into the grain and feeds there. The development is fast during the summer but during the winter larvae take as long as 95-145 days to complete their development. When full-grown, the larvae pupate inside the grain. Experimentally, the duration of the larval and pupal stages at 25, 30, 35 and 37°C was 15.8, 10.2, 19.9 and 26.1 days, and 10.4, 5.6, 6.1 and 7.9 days, respectively. The life of adults varies from 5 to 10 days, the male having longer life. The insect completes 9-19 overlapping generations and the sex ratio varies at different times of the year.

Damage. This pest causes the maximum damage from April to August, when all stages of development are present. The larvae feed and breed inside the grain, consuming the entire contents. When the infested grain happens to be in a poorly aerated receptacle, a foul smelling fungus also develops.

Other Pests of Stored Grains

The other insect species recorded as minor pests on stored grains and products in India are the saw-toothed beetle, *Oryzaephilus surinamensis* (Linnaeus) (Coleoptera : Silvanidae); the flat grain beetle, *Cryptolestes pusillus* (Schonherr) (Coleoptera : Cucujidae); the cadelle, *Tenebroides mauritanicus* (Linnaeus) (Coleoptera : Trogossitidae); the long-headed flour beetle, *Latheticus oryzae* Waterhouse (Coleoptera : Tenebrionidae); the tobacco beetle, *Lasioderma serricorne* (Fabricius) (Coleptera: Anobiidae) and the black fungus beetle, *Alphitobius laevigatus* (Fabricius) (Coleoptera: Tenebrionidae)

MANAGEMENT OF STORAGE PESTS

The effective management of storage pests may be ensured by drying the grains properly before storage, storing new grains in the clean godowns or receptacles and plugging all cracks, crevices and holes in the godowns thoroughly. If infestation of grain has already taken place or is imminent on account of a peculiar situation in which the grain is stored, then application of chemicals becomes necessary. The chemical controls to be practised under different situations are variable.

1. Surface treatment. The grain kept in jute bags can be protected by spraying insecticides on the surface of the stack. The godowns are disinfested before. storing the grain and new infestation can be prevented by surface treatment. Disinfect old gunny bags by dipping them in 0.0125 per cent fenvalerate 20EC or cypermethrin 25EC for 10 minutes and drying them in shade before filling with grains or use new gunny bags. Disinfect empty godowns or receptacles by spraying 0.05 per cent malathion emulsion on the floor, walls and ceiling.

2. Seed treatment. The grain meant for seed can be protected by mixing dusts with it. At present, mixing of malathion 5 per cent at the rate of 250 g per quintal of seed is recommended. The grains may also be treated with 25 ml of malathion 50 EC or 2 ml of fenvalerate 20EC or 1.5 ml of cypermethrin 25EC or 14 ml of deltamethrin 2.8EC per quintal of seed by diluting in 500 ml of water. Before treatment, the grains should be spread in thin layer on *pucca* floor or polythene sheet. After treatment, the grains should be mixed thoroughly and then put into the containers. Against pulse beetle (*dhora*), cover the pulses stored in bulk with 7 cm layer of sand or sawdust or dung ash.

3. Fumigation. Fumigation is a process of creating lethal concentration of a toxic gas for the time necessary to destroy insect infestation in a closed space. If there is any sign of infestation in the grain, whether stored in receptacles or in godowns, it is best to fumigate it with one of the poisonous gases available for the purpose and then to sieve and dry it before returning to the receptacles. To ensure a complete mortality of the infesting pests, it is advisable to repeat the fumigation after one month. Since the fumigants are highly poisonous to human-beings and animals, fumigation should be carried out only if the godowns are away from the residential buildings and only if it is possible to make them air-tight. If the godowns are in an awkward position and these conditions are not met with, the grain, whether in bags or in bulk, should be taken out in the open and fumigated under a gas-proof tarpaulin or a tent. Metalic drums or wooden boxes can also be used for fumigating small quantities of grain.

The chemicals used are available as liquids, gases or solids. In India, ethylene dichloride and carbon tetrachloride mixture has been recommended for fumigation of foodgrains in storage at farm level, and hydrogen phosphide in the form of aluminium phosphide or methyl bromide for protection in warehouses, godowns and silos.

Suitable fumigant which can be used under ordinary conditions of storage is the mixture of ethylene dichloride and carbon tetrachloride at the rate of 1 litre for 20 quintals of grain or 35 litres per 100 m^3 of space with exposure period of 4 days. This fumigant-mixture is comparatively safe. Methyl bromide is also becoming popular because of its high toxicity to insects and low cost. It is used at the rate of 3.5 kg per 100 m^3 of space with 10-12 hours exposure. The fumigant, hydrogen phosphide (aluminium phosphide), is currently used for large scale fumigation of foodgrains in those storage structures which are away form living premises. It is available in tablet form and can be used at the rate of one tablet (3 g) per metric tonne or 25 tablets per 100 m^3 of space with an exposure period of 7 days.

Procedure for fumigation. It is desirable to do mud plastering of storage structures in advance, leaving only exit door or openings in the wall. For small quantities of grain, if an EDB ampoule has to be used, it is inserted a little below the surface of the grain. The aluminium phosphide tablets when used are inserted into the grain kept in bulk or distributed uniformly among bagged grains. The opening is then immediately sealed with mud plaster and polythene sheet.

The storage structures thus sealed should be left undisturbed for a period of seven days or as long as the grain is not wanted. After the required period of exposure to the fumigant, grain should be aerated well for a few hours until no smell of the fumigant is left. The treated grains should be cleaned and sieved properly before they are consumed.

Precautions. The following precautions should be strictly followed during fumigation:

- The fumigants should be used only in airtight stores or under tarpaulins in the open by specially trained persons, because these compounds are deadly poisonous.
- Inhalation of fumigant vapours should be avoided by wearing gas masks and protective clothing.
- The fumigants should be applied at the recommended dosages. EDB ampoules should not be used for the fumigation of milled products, oilseeds and moist grains, to avoid excessive absorption.
- Splashing of the liquid fumigants over hands and face should be avoided.
- Aluminium phosphide must not be used in living quarters. Its use in a godown next to the living rooms may also be hazardous.

4. Use of improved storage receptacles. The grains can be best protected by using improved insect-proof receptacles of various types **(Fig. 27.9)** :

A. Indoor Bins

(*i*) **Domestic metal bins.** Several types of domestic designs of metal bins have been developed, the capacities of which range from 3 to 27.5 quintals. These are indoor bins and may be kept in a room or verandah under a roof. The bins are made from galvanized iron sheets and are moisture,

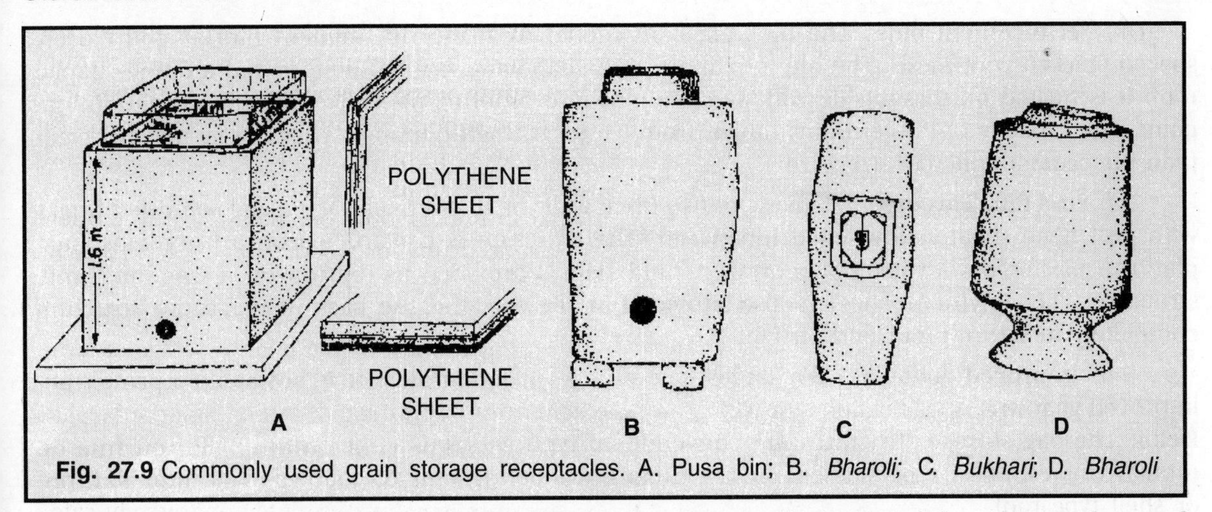

Fig. 27.9 Commonly used grain storage receptacles. A. Pusa bin; B. *Bharoli*; C. *Bukhari*; D. *Bharoli*

rodent and insect-proof. These are found suitable for storage of wheat, paddy, maize, pulses and seed grains.

(*ii*) **Gharelu theka.** This type of storage structure consists of rat-proof metal base, rubberized cloth containers and bamboo posts, having different storage capacities ranging from 2 to 3 metric tonnes.

(*iii*) **Pucca kothi.** The improved structure called 'Pucca kothi' is constructed using burnt bricks plastered with cement mortar. The structure is constructed in two compartments with a capacity of 1 metric tonne.

(*iv*) **Welded wire-mesh bin.** The bin is fabricated using wire-mesh with hessian cloth lining inside so that the air may circulate freely through it. The structure is mounted on prefabricated steel elevated base to prevent the entry of rodents. The storage capacity of bin is 2.8 metric tonnes. The foodgrains like maize and paddy can be stored in this bin even at slightly higher moisture level without deterioration in quality.

(*v*) **Reinforced cement ring bins.** This design consists of pre-fabricated RCC rings placed one over the other with gripping joints at the edges.

(*vi*) **Paddy straw-mud structure.** The improved paddy straw-mud structure of 400 kg capacity is made from paddy straw rope plastered on both sides with specially prepared mud. Externally, the structure is further plastered with water proof mud to prevent the entry of moisture. The structure is mounted on a raised brick masonary platform to prevent the entry of rats. It is commonly used for paddy storage.

B. Outdoor Bins

(*i*) **Flat and hopper bottom-metal bins.** These bins are fabricated using either steel or aluminium metal sheets of different capacities ranging from 2.0 to 10.5 metric tonnes. The bin can be either erected on brick masonary base, brick masonary columns or prefabricated steel elevated base. These bins are suitable for storage of wheat, paddy and maize under different climatic conditions.

(*ii*) **Composite bins.** These are the outdoor flat bottom storage bins fabricated with steel and timber in combination with varying storage capacities ranging from 3 to 14.5 metric tonnes.

(*iii*) **Partly underground and above ground structures.** A prototype structure of 10 m^3 or 7.5 metric tonnes capacity which is partly underground and partly above ground has been constructed. The underground part of the structure is of RCC or brick, while the above ground part of the structure is of galvanized iron sheet. This bin is suitable for construction in areas where water table is low.

(*iv*) **Seed storage bins.** Quite popular in rural areas, these bins are either circular or square in shape with storage capacities ranging from 1 to 15 quintals.

(*v*) **Ferro-cement bins.** The bins are constructed by using rich cement mortar and closely spaced chicken wire-mesh. The bin is cylindrical in shape and has flat bottom and a dome-shaped roof. It is treated on the outside with suitable moisture proofing paint like bituminous aluminium paint. The capacity of the ferrobins ranges from 0.6 to 3.0 metric tonnes. They are lighter in weight than the cement concrete structure.

(*vi*) **Pusa bin.** This design consists of two brick walls of 10 cm thick each using sun dried bricks with polythene sheet sandwitched inbetween. The structure is constructed on a brick masonary platform plastered with cement mortar. A mud slab is provided as the top on a wooden frame structure. The polythene sheet is also provided at the top and the base to make the structure completely moisture-proof and airtight.

(*vii*) **Improved godowns.** For larger bag storage of foodgrains, it is advisable to construct improved godowns, specifications for which are available from the Bureau of Indian Standards, New Delhi. The bag storage structures are conventional type godowns either with angle iron trusses, tubular trusses, raiser type trusses, precast trusses, wooden trusses or godowns with RCC flat roof or shell type roof.

(*viii*) **Bulk storage installations.** In addition to these godowns, large scale storage installations with adequate storage and mechanical grain handling facility are used by grain handling authorities in the country like Food Corporation of India, Marketing Cooperative Federations, Central and State Warehousing Corporations, etc. For bulk food storage, silo elevators have been installed in India at Hapur, Khurja, Jagraon, Gobindgarh, Moga, seaports and many other places. The storage bins or silos have a capacity of 10,000 tonnes each and provide for bulk storage of grain. The elevators are provided with mechanical operations for receiving and issuing out grain with the help of a mechanical system on scientific lines. The super structure consists of twenty cylindrical silos with a capacity of 500 tonnes each. The silo is about 6.7 metres in diameter and 23 metres high. At the head of the silos stands the head house which is 9.1 metres in diameter and 47.3 metres high. The auxillary shipping tanks, grain cleaning and disinfesting equipment and automatic weighing machines are housed in this building.

The grain can be stored up to five years without deterioration or loss. There is practically no access to insects or rodents. They are also provided with aeration and fumigation system whereby the grain can be periodically aerated and fumigated to maintain good quality.

(*ix*) **Vacuum process storage.** The vacuum process storage (VPS) is a modern concept in storage system. Under the VPS system, the PVC containers, manufactured for the purpose, are filled with foodgrains and air is sucked out of them instantly. Thus, vacuumised containers could be operated like stone boxes and stacked like gunny bags in the open field. The foodgrains in such containers never get infected or catch fungus in the absence of air. Besides, the foodgrains thus stored never get discoloured and suffer damage in quality even though the stocks are kept for 3-4 years.

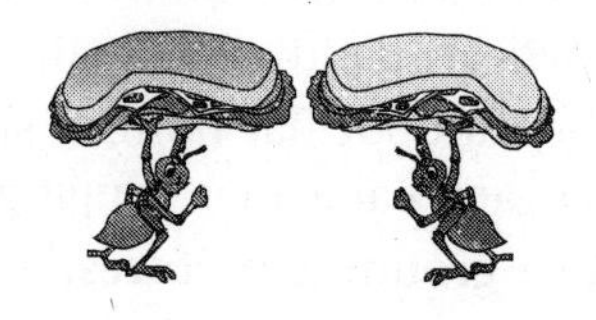

HOUSEHOLD PESTS

INTRODUCTION

Many of the household pests are cosmopolitan in distribution because wherever man went he took these pests alongwith his belongings. Some of these pests, apart from causing annoyance and discomfort to man, also transmit diseases. The housefly transmits a number of bacterial and virus diseases like dysentery, cholera, typhoid, skin diseases, eye infections, etc. by contaminating the food or by sitting on the body. Similarly, cockroaches are also known to pollute the food. Another group of insects namely the rat-flea and mosquitoes act as intermediatory hosts and transmit bacterial and protozoan diseases, respectively. The dreaded bubonic plague transmitted by the rat-flea has now been controlled in the world but in the past it has killed millions of people. The pneumonic plague struck in epidemic form in Surat (Gujarat) in 1994 and caused widespread panic throughout the country. Mosquitoes, apart from being most annoying to man, they also transmit the malarial parasite in many parts of the world. Whereas malaria has been eradicated from Europe, it is still the killer of millions of people in many other tropical and sub-tropical parts of the world.

PESTS ASSOCIATED WITH MAN

1. House Fly, *Musca nebulo* Wiedemann (Diptera: Muscidae) (Fig. 28.1)

The house fly is distributed all the world over and it assumes alarming proportions in hot and humid climates. In India, the common housefly belongs to two main species, *Musca vicina* and *M. nebulo*, the latter being more common. *M. domestica* Linnaeus occurs only in the temperate climate obtained in the Himalyas and other places. Of all the familiar household pests, the house fly is the commonest and is found wherever there is human habitation and wherever unhygenic conditions prevail. It is not only a nuisance, but it assumes prodigious importance in transmitting micro- organisms of certain infectious diseases notably diarrhoea, dysentery, cholera, typhoid and other enteric fevers. It is one of the four F's in the epidemiology of these infectious diseases; faeces, fingers, flies and food. It also serves as an intermediate host for Helminthes, of which 3 species of tape worms are parasitic on poultry and 3 species of nematodes are parasitic on horses, mules and donkeys.

The adult house fly, about 6-7 mm in length, 13-15 mm across the wings, has a dull colour, pale grey wings, with a yellow base, greyish dorsum of the thorax, with 4 equally broad longitudinal lines and a plumose arista on the antenna (**Fig. 28.1**).

Life-cycle. The life-cycle of a fly, under favourable conditions of hot and moist weather, may take only 12-14 days from the egg to the adult stage. However, in colder weather, this cycle is prolonged. Copulation takes place 24 hours after emergence and the females lay 15-150 small, white, elongate eggs in batches in heaps of manure, faeces or any other type of filth. Each female lays about 500 eggs and the incubation period varies from 12 hours to 12 days, depending upon the season. The larvae or maggots, as they are commonly called, crawl to the margins of the breeding material and pupate in brown, barrel-shaped puparia for 4-5 days. The adult flies live for 20-30 days and there are 10-12 generations in one summer.

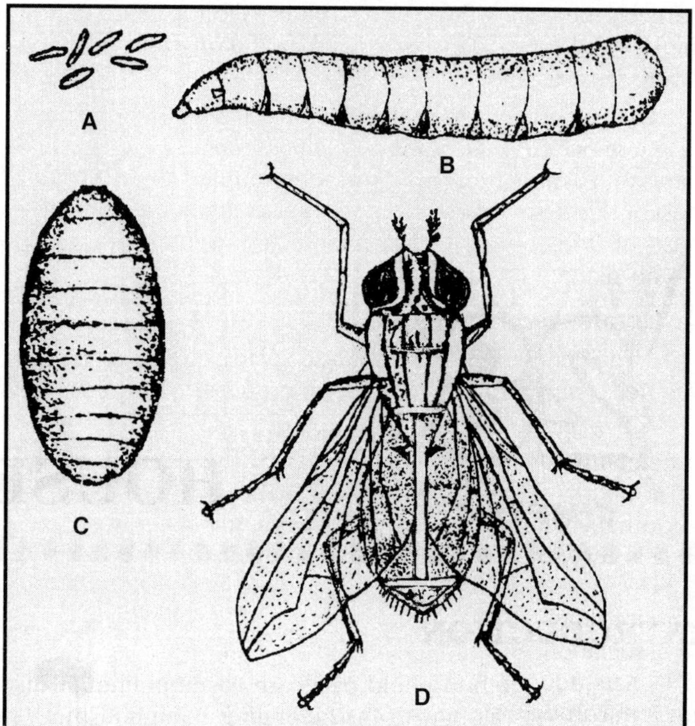

Fig. 28.1 House fly, *Musca nebulo*. A. Eggs; B. Larva; C. Pupa; D. Adult

Damage. The greatest damage done by flies is the contamination of food, resulting in the transmission of important infectious diseases to human beings and in the transmission of parasitic diseases to certain mammals and birds of agricultural importance.

Management. (*i*) The population of house flies can be kept suppressed by the proper disposal of manures, garbage, sewage, food waste, human excrement, dead animals and other organic materials. (*ii*) Inside the houses, spraying the flies with malathion/diazinon 2 per cent or trichlorphon 0.5 per cent is effective in killing them. The deposits of the sprays left on the walls also continue to be effective. (*iii*) Poison baits and sticky paper strips attract and kill a large number of flies. (*iv*) Fly-swatters should always be handy to knock off and kill the flies. (*v*) The surfaces of windows and doors where the flies rest, may be smeared with 3 per cent malathion or 1.5 per cent diazinon emulsion with the help of a paint-brush.

2. Mosquitoes

Mosquitoes belong to the sub-order Nematocera of the order Diptera and all the 2,500 species described in the world so far, belong to the family Culicidae. Mosquitoes inhabit almost all parts of the world except the polar regions. These insects are small and delicate having 3-6 mm their body length, although some species like *Psorophora ciliata* (Fabricius) and *Psorophora howardii* may measure as much as 9 mm in length having a wing span of 13 mm. They inhabit up to 4,200 metres altitudes in Kashmir and 1020 metres below the sea level in gold mines of South India. Their egg, larval and pupal stages are spent in water and, therefore, the presence of water in the environment is essential for their breeding and existence. They commonly breed in marshy lands, near filthy stagnant ponds, cesspools, dump cellars and standing rain or canal waters.

The mosquitoes are among the most unwelcome biting pests because they cause irritation and itching. It is the female that gives bites. Mosquitoes as a group, transmit, in human being malaria (protozoan pathogen), Bancroftian and Malayan filariasis (nematode infection), and the arboviruses namely, the yellow fever virus, the breakbone or dengue haemorrhagic fever (DHF) and the dengue shock syndrome (DSS) viruses, the Chikungunya virus, Japanese encephalitis virus, the West Nile fever virus, etc.

The species of mosquitoes which are of importance in medical entomology are those belonging to the genera *Anopheles, Culex* and *Aedes*. The various species of *Anopheles* transmit the malarial parasite, *Plasmodium* spp. from man to man. *Culex quinquefasciatus* say is the principle intermediate host of *Wuchereria bancrofti* and spreads filaria in urban areas, while *Mansonia* spp. are the intermediate hosts of *Brugia malayi* and spread filaria in rural areas. *Aedes* spp. are responsible for the transmission of the dengue fever, various types of encephalitis and the yellow fever.

A. Genus-*Aedes*

The genus *Aedes* has many species but among them, four species are very important. These are *Aedes aegypti* (Linnaeus), *Aedes albopictus* (Skuse), *Aedes vittatus* (Bigot) and *Aedes variegatus.*

A. aegypti, originally an African species, was introduced in India in the beginning of the nineteenth century. This is also called the tiger mosquito as it has silvery white stripes and bands on a nearly black background. The broad, flat, imbricated scales completely cover the head and abdomen, which are also present on the scutellum. The scales, thus, give a characteristic satiny appearance. It can be easily spotted by the lyre-shaped white marking over the dorsum of the thorax. It is purely domestic mosquito, rarely found far away from towns or villages. Like other bright coloured mosquitoes, it bites chiefly during the day and only occasionally at night. *A. aegypti* breeds in association with *A. stephensi* and *C. quinquefasciatus* in domestic and peridomestic habitats. The pH of its aquatic habitat generally varies from 7.3 to 8.4 with an average of 7.4. Females can live for a maximum period of 225 days with a mean longevity of 70-116 days. Each female deposits from 20 to 75 eggs at a time. The number of eggs in one genotrophic cycle, laid in one, two or more layings averages about 140. The total number of eggs laid by a single female may be 752 in as many as 16 layings within 72 days.

The eggs may be deposited along the sides of the container above the water surface, where they are closely attached. At places, eggs can be seen floating on the water surface separately, instead of being cemented together to form rafts as in *Culex* mosquitoes. The eggs are very minute, black, cigar-shaped and extremely resistant to desiccation owing to the presence of highly chitinised shell. The eggs can survive without water at least for one year. They are generally found in cisterns, rain water barrels, unused tins, earthenware pots and other small artificial water collections.

The egg stage lasts 2 days whereas the larvae complete their development within 4-9 days and the pupal stage lasts 1-5 days. The larvae are bottom feeders and can dive as deep as 1.5 metres in water. The larval head and thorax are broad and they appear strong and stout. The larvae have a characteristic worm like movement. There are lateral chitinous hooks, one pair on each side of thorax. As many as 8-10 scales are arranged in a row on the lateral aspects of 8th abdominal segment. The siphon tube is short and stumpy. Adult *A. aegypti* mosquitoes attack from behind or under tables or desks, its sound is not quite audible. It has the habit of hiding itself behind pictures or furniture.

Yellow fever. It is transmitted by *Aedes* and other mosquitoes and is an acute specific viral fever of short duration, varying in severity. The severe cases are characterised by toxic jaundice, albuminuria and haemorrhages (apistaxis, haematemesis, melaena), etc. It occurs endemically or epidemically in certain geographical areas of Africa and South America. The limited outbreaks occur in forest areas. The yellow fever has not been recorded in Asia, although the potential vectors like *A. aegypti* abound there. The strict control of immigrants, disinfection/disinfestation of aircrafts and ships has so far prevented the entry of this virus in the Asian continent. The virus can be eradicated in man as well as in the reservoir hosts. Vector mosquitoes can be destroyed and the human beings can be vaccinated against yellow fever. The persons going from or coming to infected areas require international certificate of yellow fever vaccination. The effect of vaccination is valid 10 days after the date of vaccination and extends up to 10 years.

Dengue fever. It was Rush, who in 1780, gave the first clear cut account of dengue fever from Philadelphia and called it 'break-bone' fever. Graham in 1905 was the first to prove the mosquito transmission of dengue. *A. aegypti* as the carrier of dengue was proved in 1906 by Bancrofti in Australia. The disease (domestic dengue) is caused by an arbovirus (B group) existing in four forms, viz. Dengue 1, Dengue 2, Dengue 3 and Dengue 4, all of which are transmitted chiefly by *A. aegypti*. Other species of the genus *Aedes* (*A. albopictus, A. squtellaris, A. albimanus* and *A. hebrideus*) and *Armigeres obtarbans* also transmit this viral infection. The domestic mosquito (*A. aegypti*) maintains the disease cycle in man, while *A. albopictus* and others living in the bush or forests help in the maintenance of infection among monkeys (jungle dengue).

The disease is widely distributed in the tropical and sub-tropical areas. It occurs in the Eastern Mediterranean countries, Africa, India and South-east Asian countries, in the Far East and in the Hawaiian and Caribbean Islands. It also occurs in southern United States and Australia. The disease may be benign or may carry a serious prognosis. The former is a classical or normal dengue fever which occurs in infants and young children with mild febrile illness and maculopapular rash. Older children or adults may suffer from classical symptoms of high fever, headache, myalgia and rashes. The dengue haemorrhagic fever (DHF) and its more severe form, the dengue shock syndrome (DSS) are more serious diseases. In epidemic form, Dengue 3 and Dengue 4 may cause high morbidity and even high mortality in children, the aged and the infirm. Calcutta city has already experienced outbreaks of haemorrhagic fever of the type Dengue 3 and Dengue 4 during 1963. In 1996, there was an epidemic of dengue fever in Delhi with more than 7000 DHF cases and about 300 mortalities. The dengue appeared in epidemic form in 2006 with more than 13000 cases and 200 deaths. Recently, in 2009-12, dengue incidence and mortality rose by more than 100 per cent in India. Between 2009 and 2012, cases grew from 15,000 to 37,070 and deaths from 110 to 227.

The dengue fever is an acute febrile illness clinically characterized by haemorrhagic phenomenon and a tendency to develop a shock syndrome which may be fatal. The diagnostic symptoms are acute onset, high continuous fever lasting to 2-7 days, with various haemorrhagic manifestations like petechiae, purpura, achymosis epistaxis, gum-bleeding, haematemesis and/or melena, the enlargement of liver and shock manifested by rapid and weak pulse, narrow pulse pressure or hypotension, etc. The patient is restless with cold clammy skin. Incubation period in man is 4-10 days. The mosquito becomes infected only during the first 3 days of patient's illness. Incubation period in mosquito varies from 8 to 11 days. Once infected, the mosquito remains so for its life and when the mosquito introduces saliva into the man's skin during feeding it transmits infection. There is no transovarial transmission with respect to the mosquitoes. Although in India, it is becoming endemic in some parts, it still exists in epidemic form in certain parts of South-east Asia like Thailand, Myanmar and Malaysia. The preventive measures include the vector control and the screening of all early cases so as to avoid the mosquitoes becoming infective.

Chikungunya. It is a viral (arborvirus-A) disease characterized by high fever and severe pains in the trunks and joints. This disease occurs in Africa and Asia. In India, it was detected in Calcutta during 1963 during an outbreak of haemorrhagic fever. Subsequently, this disease was also reported to occur in Chennai and Vellore in South India. The recent outbreak of this disease occured in 2006 and the affected states were Andhra Pradesh, Delhi, Gujarat, Karnataka, Kerala, Madhya Pradesh, Maharashtra, Pondicherry, Rajasthan and Tamil Nadu. Many mosquitoes act as vectors for the transmission of this disease. *A. aegypti, A. africanus* (Theobald), *C. quinque-fasciatus Manosnia.* spp. are active in Africa, while *A. aegypti, C. quinquefasciatus, C. tritaeniorhynchus, Giles* and *C. gelidus* Theobald are the vector species incriminated in Thailand. In India, *A. aegypti* remains the sole vector of this disease. The incubation period in mosquitoes is 14 days. Besides man, monkeys and apes play some part as reservoirs of this virus.

B. Genus-*Culex*

This is a domestic mosquito, bites human beings and thus causes irritating lesions. Certain species also transmit filaria, Japanese encephalitis, yellow fever and dengue fever. *Culex pipiens* Linnaeus is cosmopolitan in its distribution. It has several varieties including *Culex pipiens fatigans* Wiedemann (now called *C. quinquefasciatus*), *C. pipiens pipiens* Linnaeus and *C. pipiens molestus*. The former two varieties can be differentiated on the basis of their male genitalia.

The most common mosquito from this complex in our country is represented by *C. quinquefasciatus*. The other important species occurring in India include *C. vishnui* Theobald, *C. pseudovishnui* Colless, *C. tritaeniorhynchus, C. bitaenioryhynchus* Giles, *C. sinensis, C. gelidus* Theobald, *C. sitiens* Wiedemann, etc. *C. quinquefasciatus* is found up to 1930 metres altitude above sea level. The rapid and indiscriminate spread of urbanization is creating favourable habitats for the ever-increasing spread of this mosquito in Africa and Asia. Its proboscis is of uniform colour without any pale ring, while the abdomen has brownish yellow basal bands.

C. quinquefasciatus breeds in enormous numbers in the spring, developing in rain barrels, horse troughs, ground pools, ditches, cement pits, open drains, masonary tanks, tin cans, cisterns, cesspits, flooded latrines, catch basins, pits, in which coconut husks are soaked in South India, or in any standing water near houses, sewers with high organic pollution. It may remain alive up to 210 days. Average life span is 2.5-3.0 months. Adults can fly to a distance of 5 kilometres from their breeding places.

Adult females deposit their brownish black eggs in a raftlike mass, appearing like a speck of soot on the surface of water. From 50 to 400 eggs may be found in these masses. The individual egg is ovoid and floats with the narrow and pointed end upwards. A single female may deposit several masses during her lifetime. The egg stage lasts 24-36 hours; larvae take 7-10 days to mature; pupae take about 2 days to become adults. The life-cycle is completed in 10-14 days. The larvae undergo four moults. The female mosquito is a night-biter and the oviposition takes place at night. Their maximum biting reflex is at 31 per cent relative humidity.

There is no proof of hibernation or aestivation in this species. The highest activity has been noticed in the month of March when the overall average density in 6.5 man-hour. The adults take shelter in cowsheds, dark corners of dwelling houses, behind dumped clothes and cupboards, and also under fallen leaves in gardens. They take rest by day on clothes and furnishings in houses, as well as on walls and roofs. Deforestation, establishment of plantations, urbanization, thatched roofs and mud huts are associated with increased mosquito population. Construction of dams, ditches, ponds, lakes, reservoirs, roadways and railways are apt to encourage mosquito breeding.

C. Genus-*Anopheles*

One of the characteristics of the *Anopheles* is the occurrence of the sub-species complex. *A. maculipennis* Meigen complex of Europe has been divided into seven sub-species. There are 5 members of *A. gambiae* Giles complex of Africa. *Anopheles hyrcanus* (Pallas) forms a complex in central and southern Asia, north Mediterranean regions and Libya. Two distinct races are found in *A. stephensi* Liston in India. The various Indian mosquitoes exhibiting races within the species are *A. stephensi* Liston, *A culicifacies* Giles , *A. philippinensis* Ludlow, *A. maculatus* Theobold, *A. aconitus* Donitz, *A. fluviatilis* James, *A. leucosphyrus* Donitz and *A. sinensis* Wiedemann.

Anopheline mosquitoes are found in all parts of India, from the sea level up to an altitude of 3,530 metres whereas *A. gigas* var. *simlensis* has been found at Kedarnath in U.P. In general, the number of species along the sea coast or in the plains is smaller than in the hills and near the foot-hills. There is a gradual falling-off in the number of species from east of India, bordering South-east Asian countries, to the west in areas bordering Pakistan.

Most species can tide over the unfavourable seasons by either finding ecological niches with microclimates which are suitable for their survival or through larval survival in habitats which are not normal to them. Unlike *Aedes* eggs, eggs of *Anopheles* are unknown to survive prolonged desiccation. The larvae of *A. jeyporiensis* bury themselves in wet sand and emerge again after rains. Hibernation in the adult stage may occur in temperate zone species. *A. culicifacies* and *A. aconitus* are able to survive the dry season in Sri Lanka and Myanmar. Hibernating mosquitoes may live up to six months. *Anopheles* adults, on an average, survive for about one month. The life span of female *A. culicifacies* is 8.34 days. The male may live up to 8 days. Males do not fly a considerable distance. They usually remain near the breeding places. *A. sundaicus* (Rodenwaldt) can fly 2.5 km. In Panama, the flight range of *A.albimanus* Wiedemann is 19 km.

A typical anopheline egg is boat-shaped with lateral floats attached to both sides. Certain *Anopheles* species and their races can be indentified by the characteristic floats and egg- colour. Usually, they lay eggs at night, and oviposition may continue for 2 or 3 successive nights in various species. Eggs are not so resistant to drying in comparison to *Aedes* but may survive three weeks on drying muds. Eggs generally hatch within 2-3 days. Very few *Anopheles* species pass the winter in egg stage with the exception of *A. walkeri*. *Anopheles* larvae lie horizontal to the water surface. Four larval stages are usually completed in a week in the tropics. The larval stage may be delayed in some *Anopheles* species. By examining the larva, various species of the mosquito can be identified. The morphological characters useful in the identification of anopheline and culicine mosquitoes **(Fig. 28.2)** are given in **Table 28.1.** The larval habitat may be different for different species. The antennae of *Anopheles* larvae gradually taper from base to apex with a small number of hairs. The dorsal aspect of abdominal segments possesses plamate hairs on the lateral aspects. Thoracic plamate hairs may be present in some species. On the eighth abdominal segment, there is a pair of respiratory openings. Diapause in the larval stage occurs in some species. In the Punjab and North West Frontier region, *Anopheles* mosquitoes probably pass the winter in larval form. The pupal stage usually lasts about 2 days.

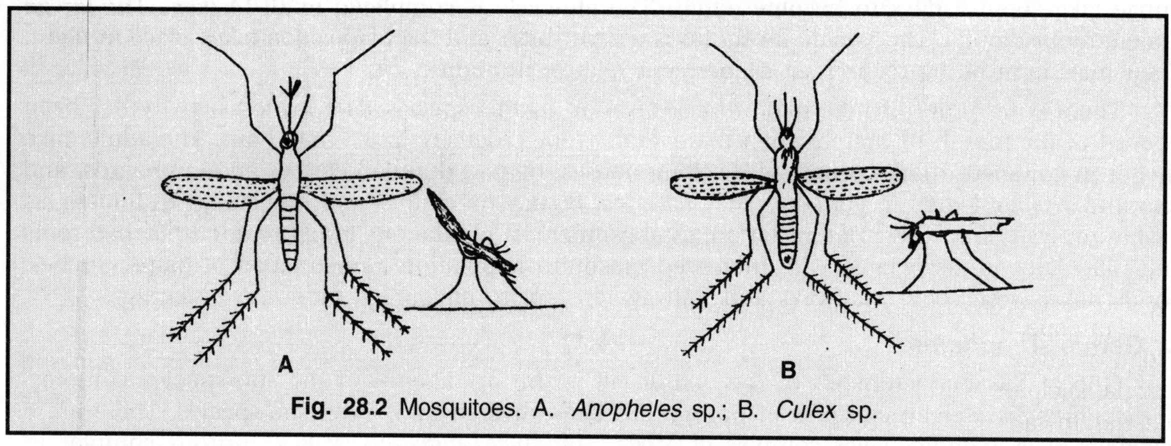

Fig. 28.2 Mosquitoes. A. *Anopheles* sp.; B. *Culex* sp.

Malaria. Human malaria is caused by infection of sporozoa of *Plasmodium* transmitted by the bite of female anopheline mosquitoes. An Italian word, *mala aria* means bad air. In Latin, it is called *paludism-palus,* means a marsh. At present, about 10 million people in 102 countries around the globe including India fall sick due to this disease every year, of which one million ultimately succumb. In Afica alone malaria kills one million children each year. In India, malaria killed 800,000 people in 1953 out of the 75 million who got the disease. Thereafter, fever cases declined to 60,000 with no deaths in 1962. The *National Malaria Control Programme* and the subsequent *National Malaria Eradication Programme* in our country, generated considerable hopes to at least contain the disease from the Indian Sub-continent. But it was during the year 1969 that the malarial resurgence took place. During 1975, 1976 and 1977, about 5.1, 6.4 and 4.7 million people suffered from malaria, respectively. In India, deaths due to malaria were 3 in 1974, 99 in 1975, 59 in 1976 and 55 in the year 1977.

Table 28.1 Differences between the various life stages of *Anopheles* and *Culex*

Stage of life	*Anopheles*	*Culex*
Egg	• Boat-shaped • Laid singly and often form patterns on surface of water. • Have float and frill • Usually found in fresh clean water	• Ovate • Found cemented together in rafts • Prominent micropillar process • Usually found in still water, often collections of dirty water in and near houses
Larva	• Has no siphon tube • Lies horizontally with surface of water • Has palmate hairs (except *A. umbrosus*) • Anterior clypeal hair, 2 pairs	• Has conspicuous siphon tube • Lies at an angle with surface of water • Has no palmate hairs • Anterior clypeal hair, one pair
Pupa	• Breathing trumpet (also known as siphon tube) short and broad (funnel shaped) • Accessory paddle hair lies above the origin of paddle hairs	• Breathing trumpet long and narrow (trumpet shaped) • Accessory paddle hair placed beside the paddle hair or may be absent
Adult	• Body at an angle with surface on which it rests (**Fig. 28.2A**) • Head, thorax and abdomen are in the same straight line • Proboscis in straight line with body • Scutellum bar-shaped • Abdomen often not scaled; scales if present, not closely imbricated • Palpi of female as long as proboscis • Palpi of male clubbed • Wings usually spotted	• Body more or less parallel with surface on which it rests (**Fig. 28.2B**) • Thorax makes an angle with the addomen • Proboscis not in straight line with body • Scutellum tri-lobed • Abdomen always clothed with dense layers of closely imbricated scales • Palpi of female always shorter than proboscis • Palpi of male not clubbed (except *Theobaldia*) • Wings usually unspotted

Four species of the protozoa, *Plasmodium,* cause malaria. These are *Plasmodium vivax* (benign tertian or vivax malaria); *Plasmodium falciparum* (malignant tertian, subtertian, aestivoautumnal malaria, falciparum malaria); *Palsmodium malariae* (quartan malaria); and *Plasmodium ovale* (ovale malaria). In case of *P. vivax* and *P. falciparum* there are races and strains according to which clinical picture, geographical distribution and immunological responses vary. In India, *P. vivax* is responsible for 65-90 per cent, whereas *P. falciparum* can cause 25-30 per cent of total malaria. *P. malariae* represents only 1 per cent of the cause whereas *P ovale* is distributed sporadically.

These protozoa invade the parenchyma cells of the liver and after passing through a development phase, attack and reside inside the red blood corpuscles. The female anophelines while injecting saliva during their act of biting and feeding, also inject sporozoites. Other irregular or accidental means of malarial transmission can be either through blood transfusion or congenital transmission (due to placental defect) or even by the use of syringe from an infected person to another.

The period between the bite of an infective mosquito and the first attack or paroxysm of fever due to *P. vivax, P. falciparum* and *P. ovale* is within 10-14 days. In *P. malariae* it varies between 18 days and 6 weeks. A series of febrile paroxysms follow in three stages; the cold stage (20 minutes to one hour), the hot stage (1-4 hours) and the sweating stage (2-3 hours). The paroxysm recurs every third day in *tertian* fever, and every fourth day in quartan fever. Since the red blood corpuscles are destroyed gradually anaemia develops. There is an enlargement of the spleen.

For completing the life-cycle of malarial parasites, two quite different hosts are involved. Asexual reproduction takes place in man (intrinsic phase or intermediate host) and the sexual reproduction takes place in the insect vector (extrinsic phase or definitive host). *P. vivax, P. ovale* and *P. falciparum* have no animal reservoir. *P. malariae* is present in chimpanzees at low levels, in Africa. A person who harbours the gametocytes in the peripheral blood serves as the reservoir of infection. In *P. vivax* and *P. ovale* relapses may occur for up to 3 years. *P. malariae* infection may persist for 30 years or more. In the absence of reinfection, recrudescence of *falciparum* malaria usually disappears completely in 6-8 months.

Malaria can be regarded as endemic in an area where there is a measurable incidence and there is natural transmission over the years. A periodic or occasional sharp increase in the morbidity or mortality of the human community due to malaria can be regarded as epidemic malaria. Out of 364 known species of *Anopheles,* only 91 are known malaria vectors. In India, Pakistan and Bangladesh, only 10 species are regarded as vectors of malaria. These are *A. stephensi, A. culicifacies, A. philippinensis, A. fluviatilis, A. varuna, A sundaicus, A. minimus, A. leucosphyrus* Donitz., *A. maculatus* and *A. annularis.* The heaviest salivary gland infection with *P. vivax* in *A. stephensi* can occur at 80°F and 50 per cent relative humidity. The hibernating *Anopheles* passing the winter in warm stables and homes may play an important role in the transmission of malaria in temperate regions. During the winter, plasmodia may remain dormant in a mosquito but renew their development in the insect in the spring.

The most important points in malaria control are; reducing to the minimum the breeding nitches or their insecticidal treatment, and surveillance of the human population and the use of prophylactic drug therapy, apart from making the dwellings mosquito proof. Smallpox and plague have been eradicated because chances of death in a patient are very high. Malaria, on the other hand, gives intermittent relief and the humans get cheated. Biologically speaking, malaria can be far greater scourge than any other disease because *Plasmodium* is potentially a perfect parasite and hence has the ability to live for ever, at least as long as man.

The following strategies would go in a long way to provide protection against malaria :

- If possible, the stagnant waters should be drained. Otherwise these should be treated with 0.025 per cent malathion emulsion. Kerosene oil may also be used for killing mosquitoes in water.
- Grasses and weeds around the buildings should be cut or sprayed with 1 per cent malathion after every week in the season when mosquitoes are very active.
- Mosquito bites can be prevented by using the mosquito nets or repellents such as citronella oil and cremes available under various trade names.
- The adults may be killed by using space sprays of certain proprietary products of pyrethrins, synthetic pyrethroids or dichlorvos.
- A standard practice for the control of mosquitoes under National Malaria Eradication Programme had been to spray human dwellings, cattle sheds and barns with DDT @ 1 g per m^2 or technical HCH equivalent to 0.2 g *gamma* isomer per m^2. However, several other residual insecticides such as propoxur, fenitrothion and malathion (2g/m^2) have also been found effective.

3. Sand Flies, *Phlebotomus argentipes* Annandale & Brunetti (Diptera : Psychodidae)

Several species of sand-flies are known in the world. They bite man and act as carriers of such diseases as kala-azar, three-day fever, tropical ulcers, etc. In India, *P. argentipes* is the most widespread and acts as a vector of kala-azar virus. Only the female flies suck blood. Besides man, they bite some other warm-blooded animals.

The flies are dark, very small (2 mm) and look like moths. The legs, body and wings are covered with long coarse hairs. There are no cross veins in the wings. The larvae bear a number of spines in transverse rows on each segment. The terminal segment has two long dorsal bristles.

Life-cycle. The fly remains active throughout the year. The females lay eggs in dark damp places where there is an abundance of organic matter. The larvae emerge in about one week and they feed on decaying nitrogenous matter and grow to maturity in four stages in about one month. They transform themselves into pupae and emerge as adults in about 10 days. The flies live for 8-10 days and the life-cycle is completed in two months. They pass through a number of generations in a year.

Damage. The flies are active at night when they attack silently and inflict a very painful bite while sucking blood. The ankles, dorsum of the feet, wrists, inner elbows and the knee-joints of human beings are preferred for biting. The act of biting is followed by intense itching and wheel-like swelling.

Management. (*i*) Preventive measures to control sand-flies include cleanliness in and around human habitations, so that flies may not find any suitable place to breed. (*ii*) The sand-flies can be killed by surface spraying with lindane 5 per cent to serve as residue spray. In addition, the insecticidal application recommended for mosquito control is also effective. Pyrethrum ointment can be used to repel the sand flies.

PESTS OF HOUSEHOLD MATERIALS

4. Cockroaches, *Periplaneta americana* (Linnaeus) (Fig. 28.3) and *Shelfordella tartara* Sauss (Dictyoptera : Blattidae)

Although these pests are world-wide, they thrive better in the tropical and sub-tropical climates. The housewife considers them the most obnoxious and filthy insects, especially in the kitchen where these nocturnal omnivorous creatures come out of their hiding-places at night and crawl all over the uncovered foods devouring and ruining them with the foul odour and excreta.

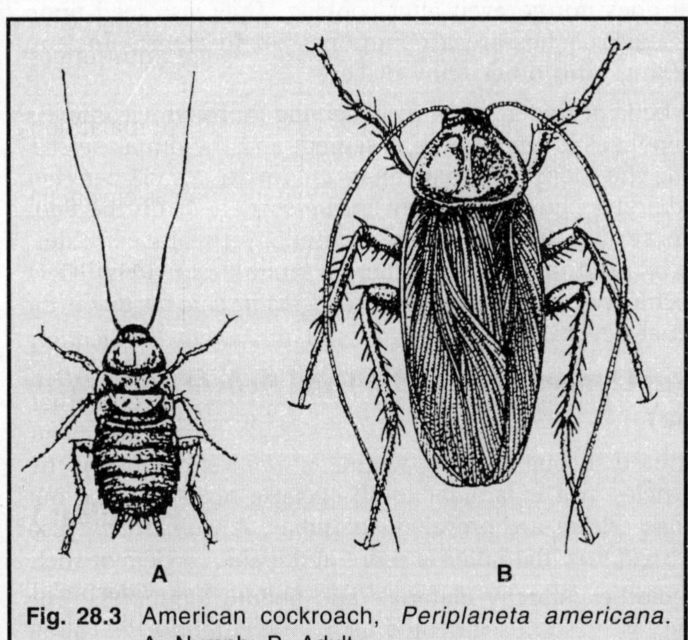

Fig. 28.3 American cockroach, *Periplaneta americana*. A. Nymph; B. Adult

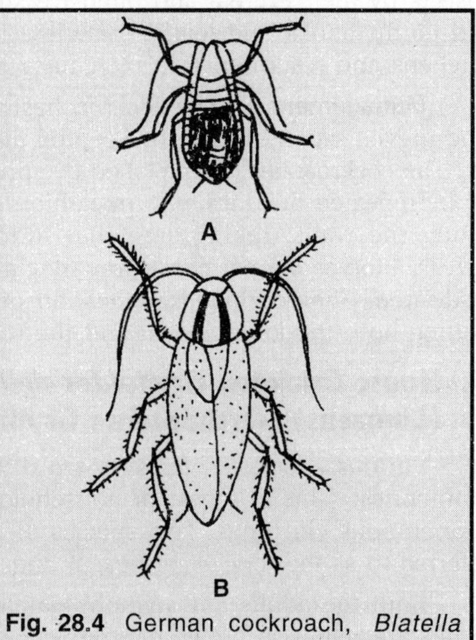

Fig. 28.4 German cockroach, *Blatella germanica*. A. Nymph; B. Adult

The damage is done both by the nymphs and the adults. In northen India, *S. tartara* is the commonest species. It is a shining flattened black insect, about 30 mm in length, with its head bowed downwards, so that it is almost covered by the pronotum from above. There are long slender antennae which are used as tactile organs. The head, the body and the legs are so arranged as to enable it to pass through tiny cracks and holes. *P. americana* is confined to South India. The other important species in India are the German cockroach, *Blattella germanica* (Linnaeus) **(Fig. 28.4)** and *B.orientalis* Linnaeus **(Fig. 28.5)**.

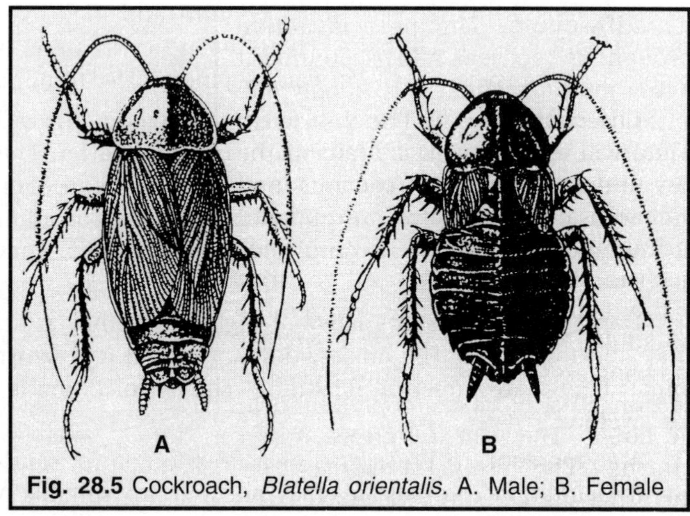

Fig. 28.5 Cockroach, *Blatella orientalis*. A. Male; B. Female

Life-cycle. These pests are most active during the hot and humid season. The females of *S. tartara* lay cigar-shaped eggs in a bean-shaped dark-brown capsule which resembles a bulging purse. The incubation period is about 2 days in the summer and several weeks in the winter. The emerging nymphs resemble the adults except for their being smaller and lighter in colour. They are often brooded over and cared for by the mother. The nymphal stage lasts over a month and finally the adults emerge from the moult. They are long-lived. The life-cycle is completed in 4-5 weeks during the summer. There are several generations in a year.

Eggs of *P. americana* hatch in three weeks during the summer and three months in the winter. The nymphal period is 3-10 months, depending on prevailing temperature. Generally, there is one generation.

Damage. The sugary and starchy substances in the houses or godowns are ruined by these insects by their excreta and offensive smell does not go even after cooking. They also feed upon old damp books and leather articles. The cockroaches are a common sight in trains, unclean kitchens and restaurants, in old musty buildings and other filthy places.

Management. (*i*) The cockroaches can be kept under check by observing thorough cleanliness and preventing infestation by keeping the pipelines leading to the basement and the drains sealed. (*ii*) The cockroaches can be killed by spraying the room with malathion/ chlorpyriphos 0.5 per cent or by repeated dustings with malathion/carbaryl 5 per cent at night in the corners or on the floor along the walls, taking care that food materials are not contaminated by the insecticides. (*iii*) Dichlorvos 0.5 per cent spray may also be used in situations where a rapid knock-down effect is desired. The combined application of dichlorvos and a persistent insecticide can be made for getting both the knock-down and the residual effects.

5. House Crickets, *Grylloides sigillatus* Linnaeus (Fig. 28.6) and *Acheta domesticus* (Linnaeus) (Orthoptera : Gryllidae)

Various species of crickets are distributed throughout the world, A. *domesticus* being the commonest. This is a pest of household articles and is familiar to all because of its musical but monotonous chirpings. Two species of house-cricket are prevalent in India ; *A. domesticus,* also referred to as the 'cricket of hearth' and *G. sigillatus*; the latter is more abundant.

Both the adults and nymphs damage clothes, starchy materials and pollute food-stuffs. The adults are yellowish brown or straw-coloured, with 2 or 3 segmented tarsi, long antennae and stout hind legs which are used for jumping **(Fig. 28.6)**.

Life-cycle. The pest is active throughout the year and is abundant during the monsoon. There is only one generation in a year. The yellowish cylindrical eggs are usually laid in the late summer and autumn, and are found in clusters in the soil or in crevices and dark moist corners inside houses. They hatch in about one week and give rise to tiny active nymphs which resemble the adults except that they are smaller and wingless. They are omnivorous and feed on all sorts of materials found in the house. They are full-grown in 3-4 months. The crickets are solitary nocturnal insects and hide during the day-time under boxes, behind curtains, pictures, books, etc.

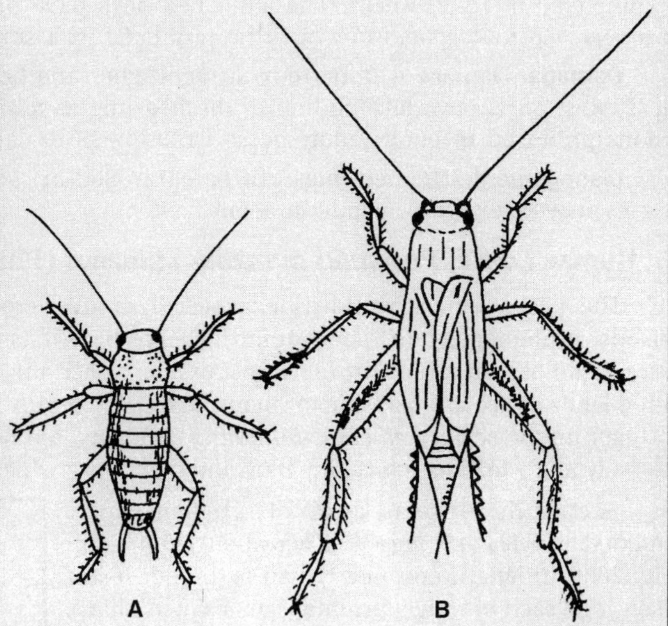

Fig. 28.6 House cricket, *Grylloides sigillatus*. A. Nymph; B. Adult

Damage. At home, these insects are chiefly a nuisance and they disturb by their monotonous chirping produced at night. They will also eat food and clothing. In the field, they occasionally damage crops by chewing roots, underground stems of various plants and the fruits touching the ground.

Management. The crickets can be killed by dusting the corners and floors of the house with malathion/carbaryl 5 per cent at night, taking care that the food-stuffs are placed at a safe height.

6. Bed Bug, *Cimex lectularius* Linnaeus (Fig. 28.7) and *C. hemipterus* (Fabricius) (Hemiptera : Cimicidae)

This pest is world-wide and is found in the slums, poultry sheds and barns. Once the infestation gets a firm hold, it is very difficult to eradicate it. This is the most repulsive nocturnal bug residing chiefly in dirty old homes. It is a parasite that sucks the blood of human beings as well as that of other mammals and poultry.

The bed-bug is a small bristly mahogany-brown, wingless, rather pear-shaped insect, about 13 mm long, having a distinct odour (**Fig 28.6**). Before a blood meal, it is flat and active, but following the meal, it turns oval and becomes sluggish.

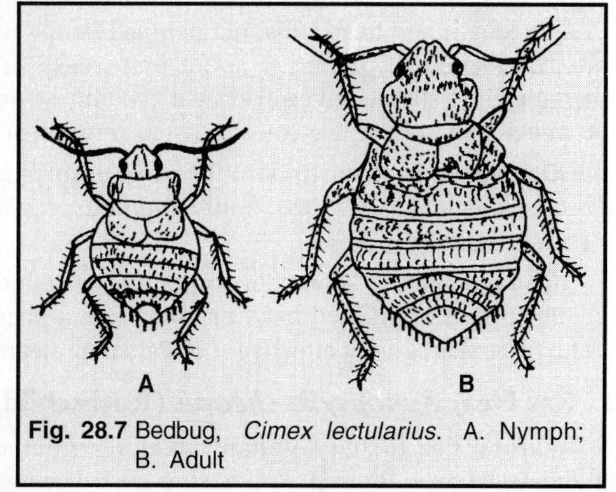

Fig. 28.7 Bedbug, *Cimex lectularius*. A. Nymph; B. Adult

Life-cycle. Depending upon the favourable warm climate and food supply, the bed-bug may breed throughout the year, although egg-laying is suspended in cold weather. The female lays elongate, whitish eggs in cracks of furniture, behind base boards, under loose edges of wallpaper, etc. She may lay 75-500 eggs at the rate of 3-4 per day. The eggs hatch in 6-17 days and the tiny nymphs moult 5 times during 6-8 weeks. Normally, an adult female lives 2-10 months. The life-cycle is completed in 15-50 weeks. The adult bed-bugs have been observed to withstand starvation up to a year. Up to 4 generations may be produced in a single year.

Damage. Damage is done only at night when the bed-bug prowls about, looking for a blood meal. Most of us have felt the bite of the bed-bug as an intensely itching inflammatory wale. The lesions produced in poultry may be so irritating as to cause the birds to abandon their nests.

Management. The bed-bugs can be controlled by treating the furniture and beds in infested houses repeatedly with malathion 1 per cent.

7. Human Louse, *Pediculus humanus* Linnaeus (Phthiraptera : Pediculidae) (Fig. 28.8)

This pest is almost world-wide, especially in over-crowded slum tenements, military barracks, prisons, orphanages, etc. The human lice are blood-sucking insects which cause irritating skin lesions and are important in the transmission of certain rickettsial diseases, notably typhus fever. Three kinds of lice are parasitic on man : *Pediculus capitis* DeGeer, the head louse living on the skin and hair on the scalp; *Pediculus humanus* Linnaeus, the body louse living on garments adjacent to the body and *Phthirus pubis* Linnaeus, the crab louse which infests human hair in the pubic region.

The adult louse is a small, greyish-white, flattened, wingless, six-legged insect, about 1.5-4.0 mm (**Fig. 28.8**) in length and nearly half as broad. It has heavy legs, each of which terminates in a single, sharp, curved claw for grasping. The body louse is typically larger than the head louse and is lighter in colour.

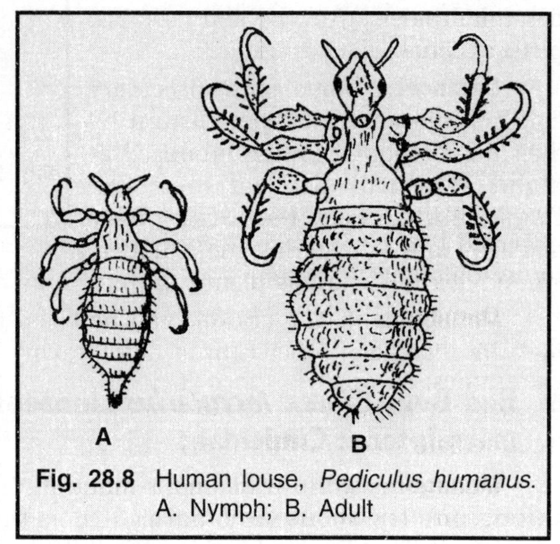

Fig. 28.8 Human louse, *Pediculus humanus*. A. Nymph; B. Adult

Life-cycle. Breeding usually goes on throughout the year in over-crowded, poorly ventilated and comparatively warm dwellings. The female of the head louse lays its eggs (nits) glued to the hairs on the head, whereas the body louse lays her eggs on the seams of clothes. A female lays 8-10 eggs daily until 50-100 are laid by the head louse and 200-300 by the body louse. The eggs are oval, whitish, about 1 mm in length and have a distinct pebbled lid at one end. The incubation period is about one week. The emerging tiny, slender nymphs begin feeding at once and undergo 3 moultings over a period of 1-4 weeks. The adults live for about one month or more.

Damage. The damage done by these insects is of three types; the stigma of having this pest which is associated with dirty slum conditions, the irritating itchy skin lesions and the transmission of certain human diseases.

Management. (*i*) A powder containing malathion 2 per cent is useful as a delousing treatment. For the control of infested head or body, the application of malathion 5 per cent dust two times at 10-day intervals is very effective. (*ii*) Personal cleanliness is essential for obtaining constant relief.

8. Rat Flea, *Xenopsylla cheopis* (Rothschild) (Siphonaptera : Pulicidae) (Fig. 28.9)

The rat-flea or the Oriental flea is distributed throughout the world. Fleas, in general, are cutaneous feeders, shifting from host to host. The rat-flea is notorious for being the direct transmitter of the bubonic plague (caused by the bacterium *Pasteurella pestes*) among human beings. In the past, the bubonic plague was responsible for almost wiping out certain sections of the human population.

The adult rat-flea is a hard-skinned, very spiny insect, with piercing-sucking mouthparts, concealed antennae, no wings and with very large legs for jumping (**Fig. 28.9**).

Life-cycle. Breeding takes place throughout the year but somewhat slowed down in cold weather. The eggs are either deposited in dust, dirt or bedding of the human host or laid while the female is on an animal. The eggs lie loosely on the hairs and usually slip into the ground where they incubate for 2-14 days. They are short, ovoid, relatively large and are laid in small numbers, although the total egg production by a single female may be several hundreds. The young larvae are legless, eyeless and very slender and are whitish. With their chewing mouthparths, they feed on the excreta of adult fleas or on that of mice, rats and other rodents. The larval stage lasts 1-5 weeks or more. The full-grown larva spins a small, oval, silky cocoon from which it emerges as an adult after 1-5 weeks, although it may pass the entire winter in the pupal stage. It has been observed that it may take 2 weeks to 2 years for a generation of fleas to complete its development under different conditions.

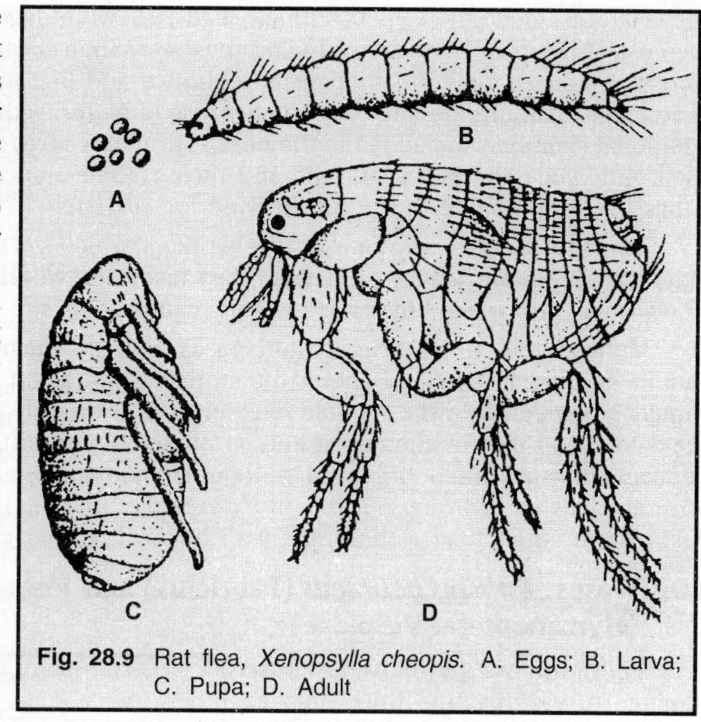

Fig. 28.9 Rat flea, *Xenopsylla cheopis*. A. Eggs; B. Larva; C. Pupa; D. Adult

Damage. The adult fleas, which feed only on blood, cause damage to human beings in 2 ways: first, by their piercing skin bites which cause irritating and itchy skin lesions, especially on the extremities; second, as the principal vectors in the transmission of certain important infectious diseases of man, such as the bubonic plague and murine endemic typhus. The fleas are also intermediate hosts of dog tapeworm and rodent tapeworm.

Management. (*i*) The houses should be kept rat-free by keeping cats or by frequent poison-baiting. (*ii*) The houses should also be kept clean, well swept and ventilated, with occasional spraying of floors with malathion 0.5 per cent.

9. Black Ants, *Monomorium indicum* Forel and *M. destructor* (Jerdon) (Hymenoptera: Formicidae)

Being world-wide, black ants are the bore of every house-wife, especially in the tropical and sub-tropical humid climates. There are numerous species of black ants distributed throughout the world. This is one of the commonest troublesome household pests because of its lust for almost every human food and its habit of forming nests in cracks of walls, floors and the roofs of buildings.

The foremost species which invade houses and godowns in India are the small red house ants, *Dorylus labiatus* Shuckard, *Monomorium indicum* Forel, *M. destructor* (Jerdon) and the large black ant, commonly known as carpenter ant, *Componotus compressus* (Fabricius).

Life-cycle. The ants are fascinating social insects. Each colony consists of 3 castes; the workers, the males and the queen. The workers are modified females. In certain seasons (e.g. the monsoon), the swarming of ants takes place and the winged males and females (young queens) mate in the air, following which the males may or may not die and the queens descend, shedding their wings and start excavating a small nest. When the cavity is complete, the queen seals the opening and isolates herself for sometime until the eggs of her ovaries mature. This may take months and then she lays eggs. Once the eggs hatch, she feeds the white legless grubs on her salivary secretions at the cost of her own body fat and wing muscles. When full-fed, they pupate and the first batch of workers emerges. They break open the entrance and henceforth supply the food. The subsequent generations of adults are large in body because of better feeding, and as the colony increases in size, additional chambers are added to the nest. It probably takes several years for the colony to establish itself. Ants generally live a long life and their colonies, once established, flourish for many years. It has been reported that one queen lived for 15 years.

Damage. Damage is done by the workers which get into almost everything that is edible, although they prefer sweets and fats. They also remove grain and seed from stores. They attack wood-works and make holes in the roofs, causing leakage during the monsoon.

Management. (*i*) All the holes and cracks through which ants may enter a house or make a nest should be closed. The food should be stored in ant-proof containers. The legs of food cabinets should be dipped in dishes containing water. (*ii*) The only sure method is to locate the centre of the colony and destroy the queen and its progeny by pouring large quantities of chlorpyriphos 0.5 per cent emulsion into it. (*iii*) In the outdoor situations, it is easy to find out nests of such ants which form ant-hills by throwing out bits of earth while building nests in the ground. Such nests can be destroyed by drenching with about 10-15 litres of 0.5 per cent chlorpyriphos.

10. Wasps, *Polistes hebraeus* (Fabricius) and *Vespa orientalis* Linnaeus (Fig. 28.10) (Hymenoptera: Vespidae)

The adult wasps in India cause damage to various ripe fuits such as pear, peach, plum, grapes and they may also attack the honeybees or cause nuisance in houses and buildings, by inflicting pain upon human beings through stinging. There are two species common in the north-western plains and two are found exclusively in the western Himalayan region. *P. hebraeus* and *V. orientalis* are the species found in the plains. The former is uniformly yellow and the latter is larger and deep brownish with yellow bands across the abdomen which gives a shiny appearance. Both the species make their nests in protected places in houses on the ceiling or in hollows inside walls or trees. Of the mountain wasps, *Vespa ducalis* Smith is of the same size as *V. orientalis* but is rust-red and hairy. It makes large nests atop the trees and can be easily seen in forests in the winter when there are no leaves on some of the tall trees such as *simbal* (*Bombyx malabaricum*). The other species, *Vespa magnifica* Smith, makes its nests in the ground and is the largest wasp in northern India. It is dark brown and hairy, and has a robust look. It is a very devastating predator of honeybees and is seen hovering over the beehives, snapping them with the greatest ease. This species is more common in the higher altitudes, whereas *V. ducalis* is more abundant at the lower altitudes.

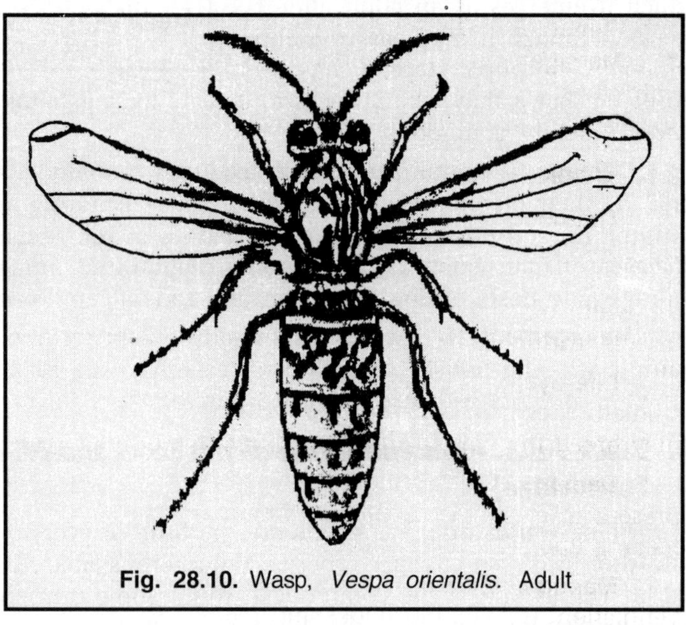

Fig. 28.10. Wasp, *Vespa orientalis.* Adult

Life-cycle. The wasps remain active during the warm summer months and hibernate as fertilized females (queens) during the winter, in cracks and crevices in the ground or in other places of protection. The three species which make exposed nests, desert them during the winter and try to come back to the old nests next season. If, however, the nests have been destroyed in the meantime, they make new ones. The nests of *P. hebraeus* are simple and rounded and are often seen in ceilings of verandahs and that of *V. orientalis* are a series of irregular simple nests, generally found in hidden places in walls or in the hollows of trees. The nests of *V. ducalis* are conical multistoreyed structures, covered with a protective outer wall with a ventral entrance hole. These nests are common on tall trees. *V. magnifica* makes nests in the ground on a raised place which is not likely to be flooded. There is an entrance hole in the ground that leads to the multistoreyed nests inside.

In all these species, the first brood in the spring is raised in the old or new nests by the queen alone which lays eggs in the cells. The larvae on hatching remain in the cells and are fed and reared there. This brood consists entirely of workers which, on emergence from the pupae, take over foraging and brood-rearing. From then onwards, the brood is reared more or less continually. The young larvae are fed on masticated animal tissues obtained by predation on insects or by biting off bits from the dead bodies of animals. Apparently, the sugary foods are also used. The full-grown larvae pupate in the cells which are then capped by the workers. Before emergence, the adults bite their way out of the cells. There are several broods in a year and the last brood- reared in the beginning of the autumn consists of both male and fertile females which mate, leaving behind the fertilized queens for wintering over. During the cold season, all the workers and the males die, and only the queens survive till the next spring.

Damage. The damage caused by these wasps is of various types. All of them feed on ripe pears and other sweet fruits, sometimes causing heavy damage. The mountainous species are very serious predators of honeybees and become a limiting factor in the success of beekeeping in certain areas.

Management. (*i*) The honeybees can be protected against the attack of wasps by killing them with clubs. (*ii*) Their nests should be located and destroyed. (*iii*) They can also be killed in the houses or orchards by hanging sugar baits containing trichlorphon as the poison.

11. Book Louse, *Liposcelis transvaalensis* Enderlein (Psocoptera : Liposcelidae)

The book-louse, which is world-wide, is a minor household pest. It is found in collections of old books and papers where they remain hidden and feed upon paste and paper. Most of us have seen these creatures on opening an old book where like a tiny speck, a pale insect rushes across the width of a page and disappears quickly. These are tiny, soft-bodied, wingless, light-coloured insects, with chewing mouth-parts and relatively long legs and antennae.

Life-cycle. The biology of the order, to which this insect belongs, needs to be thoroughly studied. They usually overwinter in the egg stage, but under warm, damp and dark conditions, they may breed throughout the year. The emerging nymphs are the adults in miniature.

Damage. The destruction caused by this insect is not usually very significant, but if they are present in enormous numbers, they may damage old rare books. Occasionally, they do harm by feeding upon the flour and cereal products in the kitchen and store-houses.

Management. (*i*) The best way to eliminate the pest is to dry out the houses and provide ventilation. (*ii*) The old books and paper files should be brushed regularly. (*iii*) The articles likely to be infested should be kept in tight containers. (*iv*) The paste meant for binding books should contain 5 per cent copper sulphate. (*v*) The almirahs and shelves containing books and paper records should be treated with 5 per cent carbaryl dust. (*vi*) Heavily infested libraries should be made airtight and fumigated with carbon disulphide or phosphine gas.

12. Silverfish, *Lepisma saccharina* Linnaeus (Fig. 28.11) and *Thermobia domestica* Packard (Thysanura : Lepismatidae)

These insects occur all over the world and are minor household pests. They are usually found in hot and moist parts of a house, such as the basement, among old books and magazines.

There are two species and the more familiar is the silverfish or 'fish-moth', *L. saccharina*, which is a carrot-shaped, soft-bodied, glistening, wingless insect, about 13 mm in length and covered with silvery scales. The other species is the fire-brat, *T. domestica*, which resembles the silverfish, except for mottled black and white patches on the upper surface of the body and its love for warm places such as fire-places, chimneys and places near the ovens.

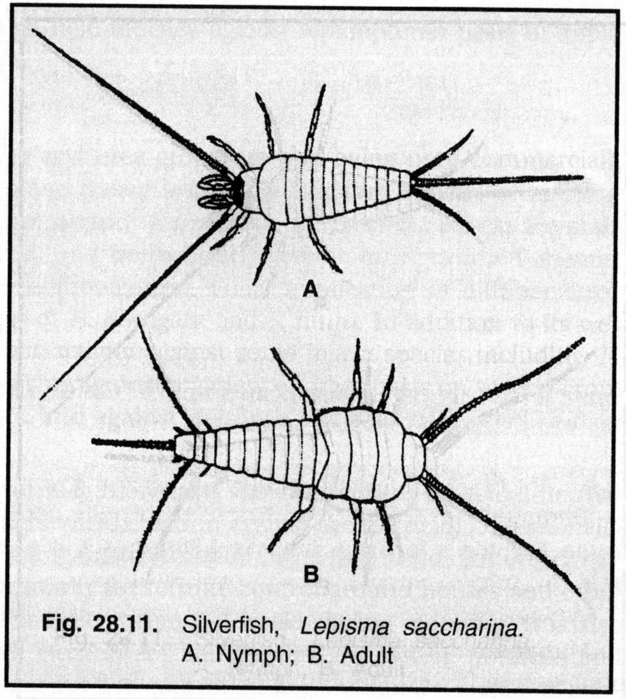

Fig. 28.11. Silverfish, *Lepisma saccharina*. A. Nymph; B. Adult

Life-cycle. These pests are active throughout the year in the warmer climates. All stages may be found throughout the year and the number of generations may vary according to the temperature and humidity of a locality. The female lays a few whitish oval eggs in warm and damp places. The incubation period is 6-10 days in India. The young ones resemble the adults except that they are smaller. The development is slow and an indefinite number of moultings take place. The insects reach maturity in 7-9 months and the generations overlap.

Damage. The damage done by this pest is not much. It is, however, a nuisance. It prefers glue and starchy materials, and the insect with its chewing mouthparts may eat book bindings, paper, wall- paper, fabrics, starched clothes, etc. Occasionally, they cause considerable destruction in libraries and to certain fabrics.

Management. (*i*) The population of the pest may be checked by regular cleansing and ventilation of the places. The cracks, if any, should be sealed. (*ii*) Poisoned glue pasted on small cardboard pieces to be placed among the books, serves an effective bait against these insects. The bait may be prepared by mixing wheat-flour (200 parts), sodium fluoride or white arsenic (16 parts), powdered sugar (10 parts) and powdered common salt 5 parts. (*iii*) Dusting the haunts of the pest with malathion 5 per cent is also useful. Alternatively, naphthalene balls may also be kept in boxes.

13. Greater Wax Moth, *Galleria mellonella* (Linnaeus) (Lepidoptera : Pyralidae) (Fig. 28.12)

The honey-bee comb or beeswax in beehives or in stores is attacked by two species of wax-moths, the greater wax-moth, *G. mellonella* and the lesser wax-moth, *Achroia grisella* (Fabricius). Of the two, the former is more important in the world. In the plains and lower hills of India, it presents a serious threat to the beekeeping industry, but is rare at high altitudes.

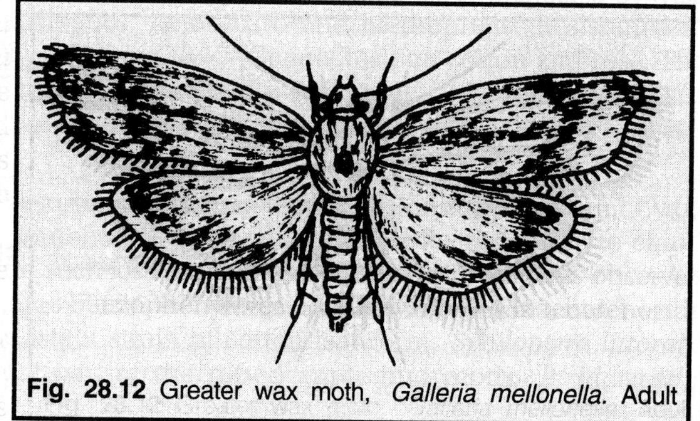

Fig. 28.12 Greater wax moth, *Galleria mellonella*. Adult

Damage is caused only by the caterpillars which feed on old combs, propolis, pollen, larval skins and other proteinaceous matter, but they cannot live on beeswax alone. A full-grown caterpillar is dirty grey with a brownish head. It is about 33 mm long and is seen moving actively in silken tunnels made in the comb and on the bottom board in beehives. The moth is generally brownish grey and measures 25-40 mm across the spread wings.

Life-cycle. The greater wax-moth is active from March to October and passes the winter mostly as a hibernating larva and sometimes as a pupa. The moths emerge during March-April and mate outside the beehive. The females re-enter the hives usually at night when the bees are not active. They lay spherical, smooth and creamy white eggs in clusters which are hidden in cracks and crevices of the hive or are found right in the combs. A single female may lay about a thousand eggs during its life span of two weeks. The eggs hatch in 7-18 days and the young caterpillars feed on gnawed pieces of the comb or other debris. As a protection against the attack of bees, they make silken tunnels in the combs or on the bottom board, wherein they feed unhindred. They pass through 5-7 stages and are full-fed in about 4 weeks in the summer. The full-grown larvae spin thick silken cocoons in which they pupate. The pupal stage lasts about a week and the life-cycle is completed in 6-7 weeks during the active period and it passes through several overlapping generations in a year.

Damage. The caterpillars eat combs and interfere with brood-rearing by making silken galleries through the cells. In case of severe infestation, the whole comb becomes a mass of webbings in which excreta of the caterpillars is enmeshed. The infested bee colonies often abscond. In the offseason, when the combs are stored, the caterpillars damage them also.

Management. Caterpillars of the wax-moths can be killed with fumigants, but this method cannot be adopted in the inhabited beehives. In that case, only preventive and mechanical measures of control can be followed : (*i*) All the cracks and crevices in the beehive should be closed with the moulding-clay. (*ii*) The hive entrances should be reduced to widths which can be easily guarded by bees. (*iii*) All debris on the bottom boards should be removed and cleaned regularly and the caterpillars removed and killed. (*iv*) The beeswax and combs in stock should be kept in airtight boxes. (*v*) If infestation develops in a store, it should be fumigated with aluminium phosphide, EDCT mixture or methyl bromide.

Other Household Pests

Some of the other household pests are the clothes and fur moths, *Tinea pellionella* Linnaeus and *T. pachyspita* (Lepidoptera : Tineidae); the carpet beetles, *Attagenus* spp. and *Anthrenus* spp. (Coleoptera: Dermestidae); the furniture beetles, *Heterobostrychus aequalis* (Waterhouse) and *Lyctus africanus* Lesne (Coleoptera : Lyctidae).

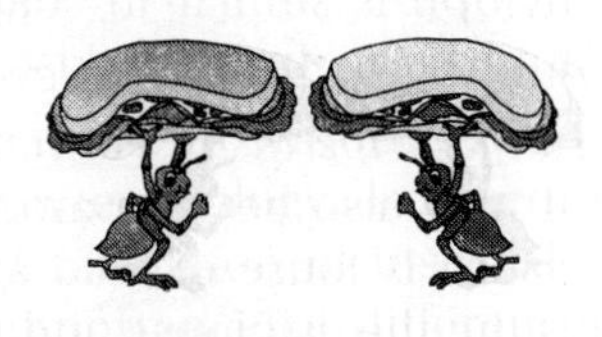

PESTS OF FARM ANIMALS

INTRODUCTION

Farm animals are attacked by a number of ectoparasites from among insects, ticks and mites. The insect pests generally breed on carrion, wounds of diseased animals and sometimes even on healthy animals. They are good fliers and attack the host for sucking their blood. In this process, they cause a great annoyance to cattle, buffaloes, horses and other domestic animals. The ticks and mites are wingless and live, more or less, a sedentary life. In their case, infestation from one animal to another is through contact or when an animal touches the objects on which the parasite might be awaiting its arrival. Some of the parasites are also known to transmit diseases.

LICE

1. Sucking Cattle Louse, *Haematopinus surysternus* Denny (Phthiraptera : Haematopinidae)

It is a universal parasite of cattle. The adult is wingless and slate grey. It is about 3 mm long, has a flat body and pointed legs. The nymph is similar in shape. In addition, many other species of blood-sucking lice infest domestic animals in India, Pakistan and adjoining countries.

Other important species include *H. quadripertusus* Fahrenhoiz, the tail switch louse of cattle, which attacks mainly the tail switch, but may also be seen among the long hairs, around the anus, vulva and eyes; *H. tuberculatus* (Burmeister) is louse of wild and domestic pigs but it also occurs on goats; *Linognathus vituli* (Linnaeus) is the the calf louse and is more often seen around the horns and on the neck; goat is also infested with *Haematopinus* sp. and *Linognathus stenpsis* as well as *L. africanus*.

Life-cycle. Lice are active throughout the year but breed more actively during summer. A female deposits 25-30 peg-shaped yellow eggs (nits) which are glued fast to the hairs at the base. They hatch in 15-18 days. The nymphs feed frequently, moult three times and are mature in two weeks, when they start laying eggs. They move about very little except when laying eggs. The life-cycle is completed in three weeks and several generations are completed in a year.

Damage. The lice cling to the hairs with their claw-like legs and thrust their mouthparts into the skin for sucking blood. These lice suck blood of the host and while doing so they inject saliva which has toxic properties. Louse bite produces irritation and causes itching as a result of which the animal attacked rubs itself vigorously against any available rough surface. Patches of skin become devoid of hairs and sores might develop there. Haematophagus flies which get attracted to these wounds are a further source of annoyance to the animal. The animal becomes restless, loses appetite and does not gain weight. The loss of blood weakens the young host, with the result that the normal production of milk and meat is adversely affected.

Management. (*i*) Malathion or carbaryl 5 per cent dust or 0.5 per cent suspension as a dip are effective. (*ii*) The application of raw linseed oil all over the surface of the body is also effective in killing the sucking lice.

2. Chewing Lice

They mostly infest poultry but a few species are parasitic on sheep, goats and buffalo calves. The more common species of chewing lice that infest fowl in India are : *Menacanthus stramineus* (Nitzsch) (Phthiraptera : Menoponidae) (chicken body louse) occurring mostly on the skin; *Menopon gallinae* (Linnaeus) (shaft louse) present on shaft of feathers; *Cullutogaster hetrographus* (chicken head louse) occurring on the skin and among feathers of head and neck; *Goniocotes gallinae* (DeGeer) (Goniodidae) (fluff louse) occurring on fluffy basal parts of feathers; *G. dissimilis* (brown louse) occurring on wing covers and large feathers of body; *Goniodes gigas* Taschenberg (large hen louse) occurring mostly on feathers; *Lipeurus caponis* (Philopteridae) (wing louse) occurring on large wing feathers and *L. tropicalus* (tropical hen louse) occurring generally on the base of head and neck feathers. The chewing lice that infest sheep, goat and buffalo calves are *Damalinia ovis* (Schrank), *D. caprae* (Gurlt) and *D. bovis* (Linnaeus) (Trichodectidae), respectively. Among the poultry lice, the shaft louse, *Menopon gallinae* (Linnaeus), is a universal parasite of domestic fowl.

Poultry Shaft Louse, *Menopon gallinae* (Linnaeus) (Phthiraptera : Menoponidae) (Fig. 29.1)

It is called the 'shaft louse' of fowl because it is seen running towards the body along the shaft of a feather when the bird parts its wings. Its body is dorso-ventrally flattened and wingless. It is about 1.5 mm in length and has biting mouth parts. The louse is more abundant on poultry than on other species.

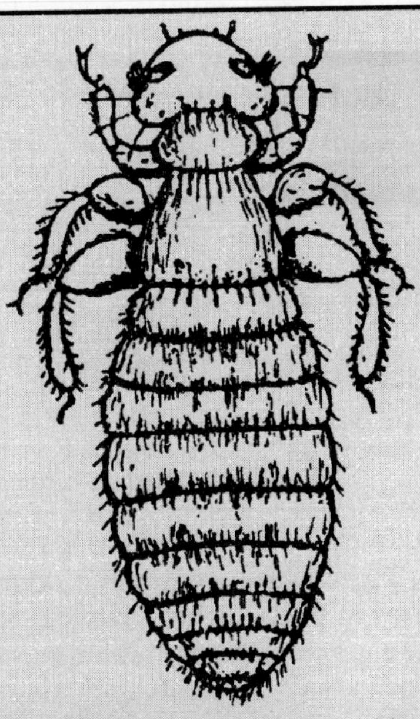

Fig. 29.1 Poultry shaft louse, *Menopon gallinae.* Adult

Life-cycle. M. gallinae moves about quickly in the feathers. The female deposits eggs singly on the down or on the feathers. The eggs hatch in 4-7 days. The nymphs feed on the epidermal scales and moult three times. They are mature in 4 weeks. The life-cycle is completed in 4-5 weeks and a number of generations are completed in a year, the eggs being more abundant from November to February.

Damage. These lice feed on feathers, downs and epidermal scales resulting in itching and falling of feathers. Heavily infested birds lose healthy condition, become weak and egg production is lowered. The young birds suffer from retarded growth and development.

Management. (*i*) The chickens may be sprayed individually or in groups with water-based sprays of 0.5 per cent carbaryl or malathion @ 5 litres per 100 birds. (*ii*) The dusts of 5 per cent malathion or carbaryl may be thoroughly applied preferably to the individual birds 500 g per 100 birds.

FLIES

3. Biting Gnats

They include sand-flies, biting midges and black-flies.

(*a*) **Sand flies.** Sand-flies of the *Phlebotomus* spp. are vicious biters and suck blood mostly of cattle and buffalo but they also bite human beings. Thirty species of this genus have been reported from India but *P. argentipes* Annadale & Brunetti, *P. papatasi* (Scopoli), *P babu* and *P. minutus* (Rondani) (**Fig. 29.2**) (Diptera : Psychodidae), are more important. These are minute, greyish brown flies about 1.5-3.5 mm in length. They are widely distributed in India. Their bodies and wings are thickly covered with hairs but are devoid of scales. The larva is hairy and carries 2 pairs of long bristles at the last abdominal segment.

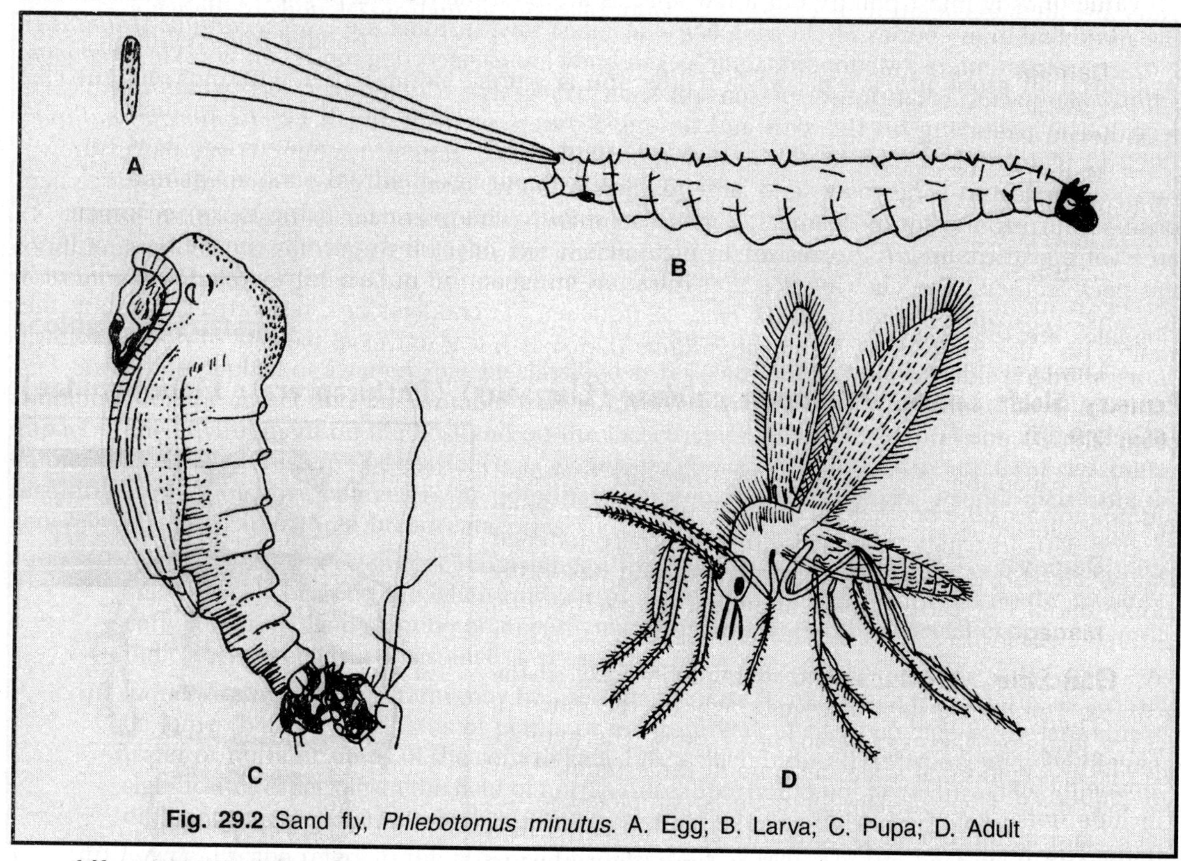

Fig. 29.2 Sand fly, *Phlebotomus minutus.* A. Egg; B. Larva; C. Pupa; D. Adult

Life-cycle. Sand-flies are active at night and remain hiding in the day time. They breed in any damp, moist place, containing organic matter but away from light. The biting activity of the flies increases in those nights when the temperature is high and there is little or no breeze. The eggs laid in organic matter hatch in about 3-7 days. The larval and pupal stages together last about 30 days. The larvae feed on solids and they moult four times. The pupa always carries the last larval skin around two distal segments of the abdomen, which is bent upwards. The adult flies live for 40 days.

Damage. Sand-flies are vicious biters and suck blood mostly of cattle and buffalo. In addition to debilitating the animals by sucking their blood they transmit canine leishmaniosis. All of them also transmit kala-azar, oriental sore and sand-fly fever in human beings.

Management. (*i*) The population of sand-flies may be reduced by keeping the dark corners in and around sheds dry and clean of organic matter. (*ii*) The flies are attracted to light at night and, thus, may be killed on the wall around lamps. (*iii*) Dust the dark corners in and around sheds with malathion 5 per cent dust. (*iv*) Spray the sheds to the point of run-off, using water based sprays @ 5-10 litres per 100 m^3 containing 0.5 per cent diazinon or 1 per cent malathion or 1 per cent ronnel.

(*b*) **Biting midges.** Biting midges of the genus *Culicoides* are presistent and irritating pests of animals and human beings. About 30 species of *Culicoides* have been reported from India which are more common in Assam and Bengal.

These midges are dark brown and 1.0-1.5 mm long. The wings are usually hairy and are marked with spots and circles.

Life-cycle. The body of larva is slender, has a few hairs and bears 4 pairs of appendages at the last segment. *Culicoides* breed in water or semi-aquatic situations, containing manure or urine of cattle or washings from the stables. Minute eggs are deposited in several batches. The egg stage is completed in 3-4 days. The larvae feed for 7-70 days and then enter into pupal stage for 3-7 days.

Damage. The females suck blood and by their bites cause severe irritation. Some species of *Culicoides* are intermediate hosts of the helminth parasite, *Onchocera gibsoni,* which causes worm nodules in cattle.

Management. (*i*) The breeding of biting-midges can be reduced by not allowing the water flowing out of sheds to stagnate. (*ii*) Spray 2 per cent oil emulsion of malathion at the places where these insects are breeding.

(*c*) **Black-flies.** Of the 100 species of black-flies reported so far, about 32 of the genus *Simulium* occur in India. These flies are small, robust and dark in colour and measure 3-6 mm in length. Females are vicious biters and suck blood mainly of cattle and horses.

Life-cycle. These gnats breed mainly in running water. Eggs are laid in masses on submerged objects such as aquatic plants, rocks and sticks. A female gnat lays about 500 eggs which are enveloped in a gelatinous substance. Eggs hatch after 7 days. The larvae feed on algae and animalcules for 4-5 weeks and undergo pupation inside cocoons. The pupal stage lasts 5-14 days. The adults escape through a slit in the cocoon and are carried to the surface of the water on a bubble of air collected during the transformation period.

Damage. Black-flies are vicious biters. Their bites cause severe irritation and may also cause swelling which is due to the toxic properties of their saliva.

Management. Same as in case of sand-flies.

4. Gad Flies, *Tabanus* spp. (Diptera : Tabanidae) (Fig. 29.3)

There are more than 2,000 species of blood sucking gad flies or horse flies in the family Tabanidae, of which about 150 occur in India. The female fly bites and sucks blood from its shoots, which include horse, cattle, camel, elephant and sometimes man. The male feeds on nectar, honeydew, oozing tree-sap, etc. The fly is characterized by a thick-set body, a large head and big eyes (**Fig. 29.3**).

Different species have their feeding preference for certain sites on their hosts. For instance, *Tabanus rubidus* Wiedemann (Diptera : Tabanidae) settles on the lower parts of the back of the hind legs; *T. striatus* Fabricius on the undersurface of the abdomen; *T. macer* Bigot on the sides of the neck; *T. ditaeniatus* Macquart on the udder; *Chrysops dispar* Fabricius on the inner side of the fore legs and *Haematopota javana* Wiedemann on the hump or the neck.

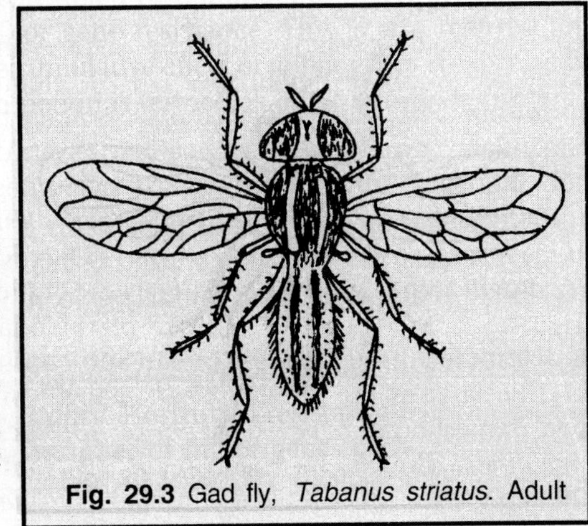

Fig. 29.3 Gad fly, *Tabanus striatus.* Adult

Life-cycle. The tabanid flies are abundant in those areas where temperatures are moderately high, water is available and forests are close by. They are particularly common during the monsoon when they are seen sitting on walls or on window panes. They love the sunshine and are active during the day-time only. It appears they find their hosts by sight and not by smell and they even attack animal dummies. They are also attracted by moving objects such as railway trains. Females generally attack the animals in the sun and a blood meal is essential for the maturation of eggs which are deposited in masses on aquatic and subaquatic vegetation as well as logs of wood near water. Quite often, the egg masses are cemented together with a thin layer of transparent material which is water-proof. The eggs are torpedo-shaped, 1-2 mm long and are white when freshly laid but become darker later on. In one cluster, there may be 300-600 eggs which hatch in 4-7 days.

A larva, on emerging from the egg, is white and tapers at both ends and measures about 3.5 mm. It is aquatic and feeds on small crustaceans or on the maggots of flies. The larvae of *Chrysops* sp. or of *Hamatopota* sp. may also be found in mud, moist soil or manure. They pass through 7-8 instars and may complete their development in 9 days to almost 7 months, depending upon the species, season and food supply. On completing their development, the larvae move to drier soil on the banks of streams to the edge of ponds or to similar aquatic habitats. The pupa is generally yellowish brown, finely wrinkled and has a lateral tuft of hairs on each abdominal segment. The pupal stage lasts from 3 to 21 days and the flies, on emergence, have irridescent eyes. In various species, the life-cycle is completed in 4-5 months and there are usually two broods in a year.

Damage. The female flies are parasitic on warm-blooded animals and suck blood which may ooze from the site of feeding. It is not uncommon to see as man as 50 flies on a single host. While taking blood meal, the female inserts and withdraws the beak several times, shifting to a new place every time. Thus, she completes her meal in 3-11 minutes from 5-6 places and at the end of that period her abdomen is so fully distended that occasionally a drop of host blood comes out of its anus. Generally, there is an interval of 1-2 days between the meals. The flies are responsible for the dissemination of a number of diseases of cattle, viz. surra, anaplasmosis, anthrax and swamp fever and a kind of filariasis in human being.

Management. (*i*) Stagnant water in the area should be drained away to prevent the breeding of the pest. (*ii*) The aplication of sprays containing 0.1 per cet pyrethrins +1 per cent piperonyl butoxide at 1-2 litres per animal, twice or thrice a week, is effective.

5. Bot and Warble Flies, *Oestrus ovis* Linnaeus (Fig. 29.4) and *Hypoderma lineatum* (Villers) (Diptera : Oestridae)

The bot flies in the larval stage act as parasites on the flesh of various domestic animals. They attack cattle, horse, sheep, goat, elephant, rhinoceros, etc., and feed in the alimentary tract, the nasal and pharyngeal cavities and in subcutaneous tumours formed by their bites. In India, these flies belong to the genera *Oestrus, Hypoderma, Cobboldia, Rhinoestrus* (Oestridae), *Gyrostigma* and *Gasterophilus* (Gasterophilidae). The sheep or goat-bot fly, *Oestrus ovis* Linnaeus, is distributed throughout the Indian Sub-continent and its larva is the familiar head maggot of the two domestic animals. The larva of *Rhinoestrus purpureus* (Brauer) has been recorded from the nasal passages of horses in the Punjab. *Hypoderma*

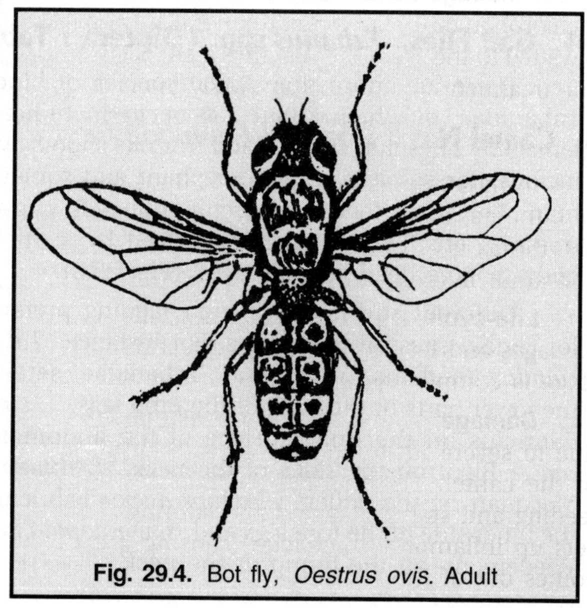

Fig. 29.4. Bot fly, *Oestrus ovis.* Adult

lineatum (Villers) is the well-known warble fly on cattle, which is occasionally found on goat also. It attacks the skin and is responsible for deterioration in quality of the hide. There are instances when larvae of *Hypoderma* and *Gasterophilus* have been collected from the skin of man and are known to cause 'creeping myiasis'.

Life-cycle. The parasitic forms of bot flies which live in the alimentary canal of animals deposit their eggs among the body hair, from where their larvae are swallowed through licking. The larvae of the forms that parasitize the naso-pharynx are deposited in the nostrils. Those found in the skin lay their eggs on the legs of host. On emergence from eggs, the larvae penetrate inside and after passing through various tissues reach the back of the host.

The warble fly is common in the western parts of India and causes tumours in the skin of cattle and buffaloes, thus spoiling the hide. The adult fly is 12-14 mm long and has a wing expanse of 23-25 mm. The body is black, banded with yellowish and orange hairs. Legs are well covered with hairs of black and orange colour. The wing veins are black. The dull-yellowish-white eggs, which have a smooth shining surface and are about three quarters of a millimetre in length, are deposited on the body hair in groups of 5-12. The total number of eggs laid by a female fly varies from 200 to 500.

The eggs hatch in 4-5 days and the young larvae penetrate the skin through hair follicles. Within 2-3 months, they reach the wall of the oesophagous and later on by way of the thoracic cavity they reach the back where characteristic warbles are produced in about 7 months. There are three larval instars. The final instar larva cuts a small hole in the skin covering the warble for breathing air from outside. This results in damage to the hide. The mature light brown larva wriggles out of the warbles through a hole and drops on the ground to pupate in soil. After a pupal period of 6-8 days, the adult fly emerges. It lives only for a few days for mating and reproducing. In the plains of north-west India, the egg-laying season of ox-warblefly extends from March to June. Warbles appear on the back of the cattle from October to January. There is only one generation of this fly in a year.

Damage. The penetration of the larvae into skin causes irritation and later on hypodermal rash is produced, which may be mistaken for mange. Cows infested by warble-fly yield less milk. The flesh around the mature larva gets inflammed and becomes unfit for consumption.

Management. (*i*) Hair close to the hoofs may be cut in order to destroy the eggs during the egg laying season, i.e. monsoon months. (*ii*) The larvae can be easily squeezed out from the cyst and this removal decreases the suffering of the animal. (*iii*) The water-based sprays containing 1 per cent trichlorphon or 0.05 per cent rotenone are applied for controlling the pest. The entire body of the animal is wetted to the skin. The first treatment is given 40-45 days after appearance of first warbles on the back. The treatment is repeated at 45-day intervals as long as the grubs on appearing on the back.

6. Camel Nasal Bot Fly, *Cephalopina titillator* (Clark) (Diptera : Oestridae)

In India, the camel nasal bot-fly is present in Rajasthan. The adult fly looks somewhat like *O. ovis*, the nasal bot-fly of sheep. The upper part of the head is orange and lower part is pale yellow. The thorax is reddish brown. The legs are yellowish brown and the terminal parts of tarsi are dark. The adult fly measures 8-11 mm long.

Life-cycle. The female fly deposits its larvae in the nostrils. The larvae bore into the nasal passages and feed for 3-11 months and then fall down on the ground to pupate in soil. The pupal stage is completed in 18-25 days. The adult fly lives for 38 days,

Damage. When the flies deposit larvae in the nostrils, the animal sneezes and snorts violently due to severe irritation. The mature larvae leave the nasopharynx and crawl down to the nostrils of the camel. The animal becomes restless, shakes its head continuously, becomes dull, stops feeding and starts sneezing and snoring. The presence of larvae in the nasal passage of the camel sets up inflammatory changes causing great discomfort. An offensive discharge tinged with blood comes out of the nostrils.

Management. This malady can be controlled by applying inside the nostrils 0.2 per cent suspension of trichlorphon.

7. Stable Fly, *Stomoxys calcitrans* (Linnaeus) (Diptera : Muscidae) (Fig. 29.5)

This fly and many other species of the subfamily Stomoxydinae are true blood-suckers and are commonly known as the stable-flies or biting flies. They are distinguished from other muscids by their having characteristic elongated apical portion of the proboscis. It is longer than the length of the head, swollen at the base and tapering at the apex. *S. calcitrans* is cosmopolitan and feeds on the blood of cattle, horse and other domestic animals.

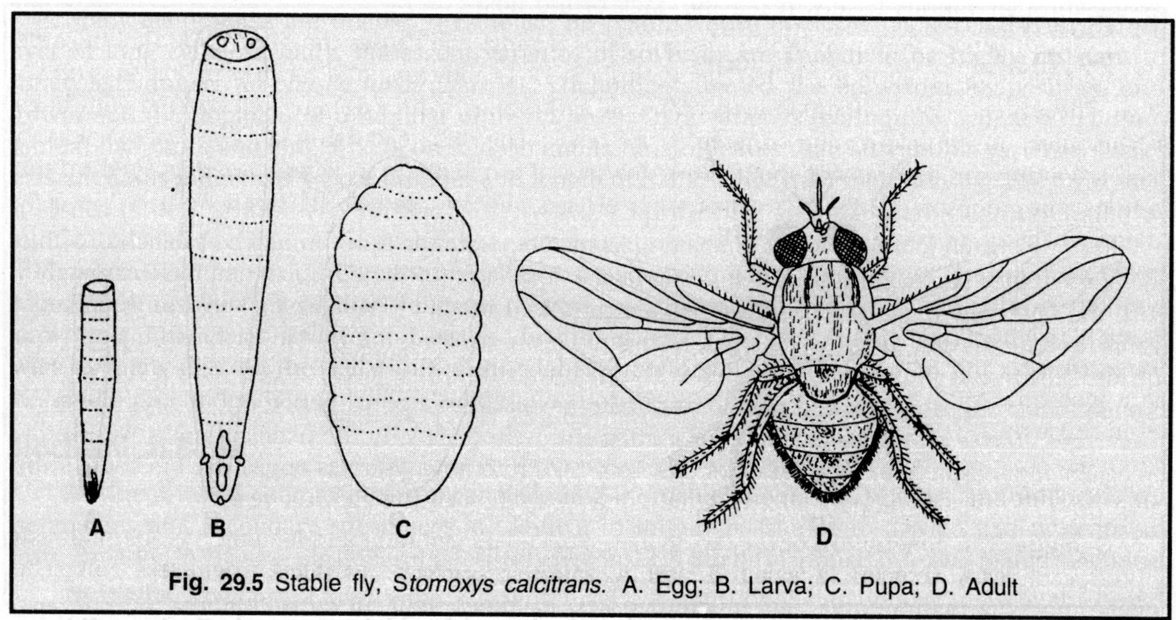

Fig. 29.5 Stable fly, *Stomoxys calcitrans*. A. Egg; B. Larva; C. Pupa; D. Adult

Life-cycle. Both sexes suck blood from the host becoming engorged in 2-5 minutes. They feed two times a day. The fly breeds in litter mixed with dung and urine, in sewage mud or fermenting tidal deposits. A female starts egg-laying one week after emergence and may lay up to 600 eggs. The eggs measure 1.3 mm in length and are white, when freshly laid. They hatch in 36-48 hours. The larval stage is completed in 2-3 weeks and the full-grown larva measures about 10 mm. The puparium is mahogany brown and 4.5 mm in length. The pupal stage lasts about 10 days and the life-cycle is completed in 4-5 weeks.

Damage. The stable-flies are notorious for biting, irritating and causing restlessness in cattle. According to one estimate, these flies cause 9.26 per cent reduction in milk yield during a month's infestation as compared to 3.33 per cent reduction caused by *Musca nebulo*. *Stomoxys niger* Macquart, a related species, is the vector of surra in Mauritius, but it is not known to be an efficient vector of this disease in India. However, it may spread anthrax.

Management. (*i*) The animal can be sprayed with 0.05 per cent pyrethrins +0.5 per cent piperonyl butoxide (1-2 litres per head) once or twice a week. Alternatively, the animal may be treated with malathion 5 per cent dust @ 50 g/ dairy cattle. (*ii*) The pest can be controlled by spraying 1 per cent diazinon on the surfaces in the cattle sheds.

8. Buffalo Fly, *Siphona exigua* (de Meijere) (Diptera : Tachinidae)

The buffalo-fly is yellowish grey and is smaller than the stable-fly. The proboscis is very slender and projects downwards. Though normally a pest of cattle and buffaloes, it may also attack other species of livestock when the latter are moving with cattle and buffaloes. When feeding or resting on the host, the wings of this fly are held at an angle from its body.

Life-cycle. The buffalo-fly lays eggs only on freshly dropped dung of cattle and buffaloes. About 20 eggs are laid by a female. Eggs hatch within a day. The larvae feed for 3-5 days and pupate in the bottom layer of the dropping or in nearby soil. The adult emerges after a pupal period of 3-5 days.

Damage. The fly is active generally during the day and feeds on the flank and neck of the cattle. The bite of this fly is painful. The fly sucks considerable amount of blood of the host. A constant attack of hundreds of flies seriously affects milk production and work efficiency of the animals.

Management. Same as in case of stable-fly.

9. Blow Flies, *Calliphora vicina* Robineau-Desvoidy (Fig 29.6), *Lucilia* spp. (Fig 29.7) and *Chrysomya bezziana* Villeneuve (Diptera : Calliphoridae)

These flies have a bright metallic appearance, being shining blue or green and are known as blue-bottles or green bottles. They are small to medium, some being smaller and other larger than the common house fly **(Fig. 29.6)**. Three of the genera of the family Calliphoridae namely *Chrysomyia, Calliphora* and *Lucilia* include the largest number of flies that feed on carrion and many of them even cause myiasis in man and domestic animals. *C. vicina, Lucilia cuprina* (Wiedemann), *L. sericata* (Meigen) and *C. bezziana* may be considered the type species, being also very common in various parts of India. All these species are widely distributed in various parts of the world, particularly in Eurasia.

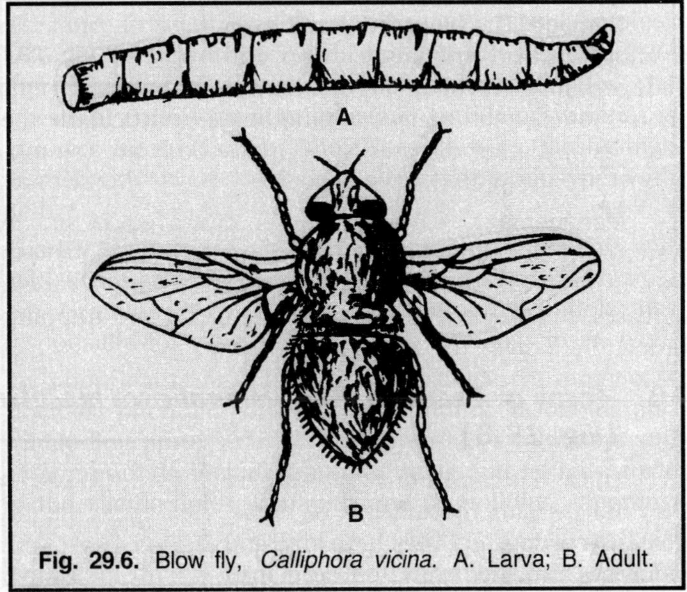

Fig. 29.6. Blow fly, *Calliphora vicina.* A. Larva; B. Adult.

Life-cycle. *C. vicina* is the well known European blow-fly (blue-bottle) and is found in the Himalayas and other places in India where the climate is mild. The fly breeds in cadavers and meat. At 23°C, the eggs hatch in 24 hours and the larval and pupal stages are completed in about 10 days.

L. cuprina and *L. sericata* are allied species and can be crossbred. The former is the most important blow-fly on sheep in Australia and the latter in England. The colour is variable in different parts of the world. For example, *L. sericata* (green-bottle) is bright green in Europe and Africa, bronze-green in India and coppery in New-Zealand.

The adults of these species are commonly seen on sweets and meat in Indian bazars. A female deposits clusters of 400-500 eggs in summer. These hatch within 12 hours. The larval and pupal stages last 3-4 and 4-5 days, respectively. A full-grown larva is dirty white and is 12 mm long and the puparium is light brown. The life-cycle, including the pre-oviposition period, is completed in a fortnight. Sometimes, these flies cause muscular myiasis in animals having wounds. Excretions of *L. sericata* are known to contain a potent bactericide.

C. bezziana is the commonest fly that causes myiasis in India. The site of the disease may be the mouth, ears, eyes, skin or vagina of the farm animals, camels, elephants or even humans. The flies may lay eggs either right on the diseased tissue or near it, and even in the pus-soaked dressing material. As many as 522 eggs have been recorded from one wound. The egg is cylindrical, 1.5 mm long, and hatches in 1.0-1.5 days. The mature larva is about 12 mm long white with an orange tinge, darker at either extremity. The body segments have transverse rows of short, stout, light brown, black-tipped, posteriorly directed spines. The projections on the body give the appearance of a screw, hence the name screw worm. The larva fully matures in 5-6 days and drops to the ground where it pupates in soil. The puparium is deep mahogany. The pupal period is completed in 7-9 days. The life cycle is completed in a fortnight.

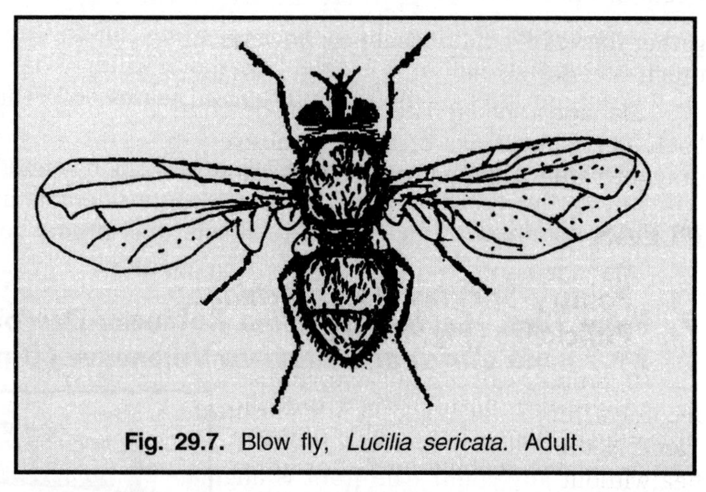

Fig. 29.7. Blow fly, *Lucilia sericata*. Adult.

Damage. The blow-flies apart from feeding on dead animals, carrion and cadavers also infest meat in butcher shops. Some of the species of blue-bottles, particularly the screw worms, cause myiasis in animals. Wounds and other diseased sites on the body are the most liked for oviposition. The sick cattle in a herd are the preferred victims.

Management. (*i*) Larvae may be destroyed *in situ* by spraying the wounds with chloroform-water to induce them to leave the tissue or by removing them with forceps. Thereafter, the wounds should be dressed with some disinfectant and a repellent like pinetar. (*ii*) Proper disposal of carcases, treatment with larvicidal chemicals and trapping of flies with baits may be carried out in order to suppress the population of flies.

10. Horny or Leathery Flies, *Hippobosca maculata* Leach (Diptera : Hippoboscidae) (Fig. 29.8)

These flattened horny or leathery flies are permanent parasites of horse, goat, sheep, dog, etc. They have long and stout legs, provided with well developed tarsal claws adapted for clinging to the hair of their hosts **(Fig. 29.8)**.

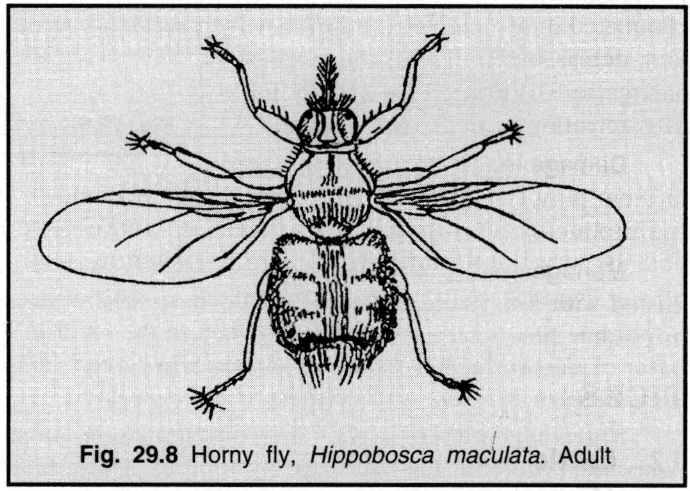

Fig. 29.8 Horny fly, *Hippobosca maculata*. Adult

The common parasites of farm animals are placed in four genera, viz. *Lipoptena, Melophagus, Pseudolynchia* and *Hippobosca*. *Lipoptena caprina* Austen is found on goat, *Melophagus ovinus* (Linnaeus) on sheep and *Pseudolynchia maura* Bigot on pigeons. The more abundant species in north-western India are *Hippobosca maculata* on bovines and equines, *H. capensis* Olfers on dogs and *H. camelina* Leach on camels. The flies are extremely tough and difficult to kill, except by decapitating them.

Life-cycle. Their most characteristic feature is that they retain their young inside the body till they are ready to pupate. The matured larvae are deposited on the ground for pupation. The pupal period lasts 24 days and the adult flies live for 28 days. They feed on the blood of the host throughout their life.

Damage. It appears that the bite of these flies is not painful to the host and it only causes annoyance. Since blood does not ooze from the site of the bite, the muscids and other flies do not gather there. In certain localities, however, the fly population may be so high that the host may lose much blood and die.

Management. (*i*) The pest can be controlled by sprinkling malathion 5 per cent dust on the back, neck and flanks of the animal every 10-14 days, if needed. (*ii*) A single spray treatment of draught cattle with 0.05 per cent lindane is very effective for controlling the leathery flies.

FLEAS

11. Poultry Stickfast Flea, *Echidnophaga gallinacea* (Westwood) (Siphonaptera : Pulicidae) (Fig. 29.9)

It probably occurs all over India, and has been recorded from Uttar Pradesh, Maharashtra, Karnataka, Tamil Nadu and Kerala. The adult is a small delicate flea without any comb. The front is angulate and the thoracic tergites together are shorter than the first abdominal tergite (**Fig. 29.9**).

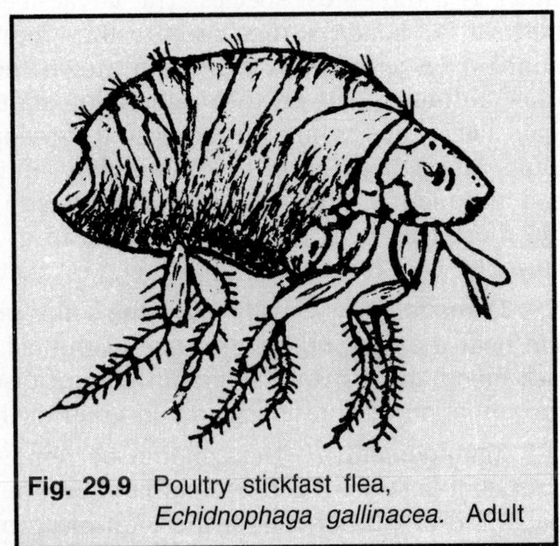

Fig. 29.9 Poultry stickfast flea, *Echidnophaga gallinacea.* Adult

Life-cycle. The adult flea attacks the bare parts of the body of a fowl, such as the comb, wattles, and around the eyes and beak, particularly of the young chicks. These insects bury their mouthparts in the flesh of the bird and remain fixed in that position. The egg-laying by adults begins after 6-10 days of attachment of the host. The oviposition usually takes place at night. The eggs hatch in 6-8 days. The larval period is completed in 14-31 days. When mature the larvae bury into debris and pupate inside cocoons. The adults emerge in 9-19 days. When the adult becomes 5-8 days old, it attaches to the host. The male flea lives for about 2-6 days after mating but the female lives up to 6 weeks on the host.

Damage. When the infestation is heavy, the young birds become dull and droopy, and some of them may even die. In older birds, prolonged infestation makes them anaemic which decreases egg production.

Management. The infested birds may be dipped in 0.05 per cent carbaryl or malathion or dusted with 5 per cent carbaryl or malathion dust. These dusts may also be sprinkled on the litter in poultry houses.

TICKS

12. Cattle Ticks, *Hyalomma anatolicum* Koch (Fig. 29.10) and *Boophilus microplus* (Canestrini) (Acari : Ixodidae)

Ticks are very important external parasites of cattle, sheep, goat, horse and other farm animals. Many species attack cattle. Of these, *B. microplus* and *H. anatolicum* are very common in India. Unlike insects in which the body is divided into three parts (head, thorax and abdomen), ticks have a fused head and thorax.

Damage is done by larvae, nymphs and adults. They suck blood from their host. Larvae are small creatures having three pairs of legs, whereas nymphs and adults have four pairs. The ticks measure about 2.5 × 1.5 mm in size (**Fig. 29.10**). They increase in size enormously after gorging themselves with blood.

Life-cycle. Ticks remain active throughout the year. Eggs are laid in cracks and crevices in buildings or in the soil. A single female may lay more than 2,000 eggs in clusters. They hatch in 2-6 weeks, depending upon the season. The six-legged larvae attach themselves to a suitable host, as and when it might happen to pass by their abode. *B. microplus*, which feeds on cattle, is a single host species and the larvae moult into nymphs and finally into adults on the same infested animal. Mating takes place on the host and the fully engorged adults drop to the ground. *H. anatolicum* which feeds on cattle, goat, horse, donkey and camel, is a three-host species. The larvae feed on the blood of the host for 4-7 days, then drop to the ground and moult into nymphs. These nymphs may then attach themselves to some other animals, feed on their blood for about two weeks and again fall to the ground to moult and develop into adults. The adults attach themselves to still another animal on which they feed and mature. The life-cycle is completed in about 3-5 weeks and there may be a number of broods in a year. *H. isaaci, H. ferozedini* and *H. hussaini* Sharif feed on cattle and buffalo.

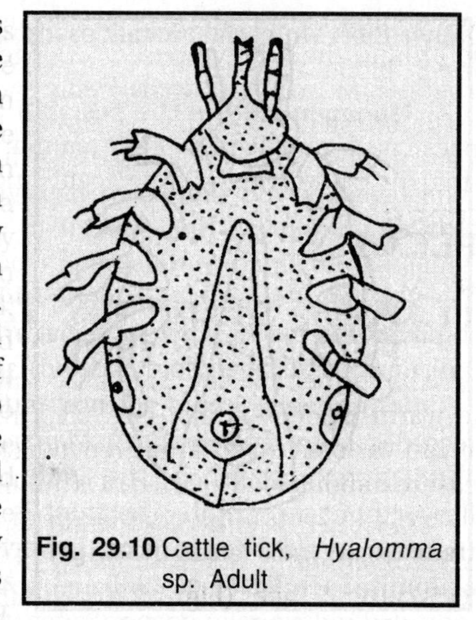

Fig. 29.10 Cattle tick, *Hyalomma* sp. Adult

Damage. Ticks usually infest the softer portions of the host animal. Most of them are found on skin near the ears, udder and base of the tail, with their heads embedded deep into the skin. They suck blood and transmit several important diseases like tick fever and Texas fever. Tick bites lead to the formation of abrasion marks in grain of the hide which fetches a lower price in the market.

Management. (*i*) Ticks should be removed by hand and killed. (*ii*) Dust powders containing 2 per cent lindane, 5 per cent carbaryl or 5 per cent malathion are effective. (*iii*) Spray of 1 per cent malathion or 0.2 per cent suspension of bromocyclen (Alugan) is very effective against *Hyalomma* ticks on cattle, camel and dog.

13. Fowl Tick, *Argas persicus* (Oken) (Acari : Argasidae) (Fig. 29.11)

The fowl tick is distributed throughout the world and it is known to transmit spirochaetosis. This tick also attacks turkey, goose, duck and pigeon. Whereas the adults and nymphs feed at night, the larvae remain continuously on the skin of the host till they are fully developed (**Fig. 29.11**)

Life-cycle. The females lay eggs in batches of 30-100 in cracks and crevices of fowl-houses. Up to 900 eggs may be laid by single female. Depending upon the season, the eggs hatch in 10-90 days. The first stage larva is full-fed in 3-10 days after which it drops off and moults into the first nymphal stage, within one week. In a few days, it refixes itself to the host and becomes replete in a couple of hours and drops off. After two weeks, it moults and emerges as the second-instar nymph which again engorges itself with blood for 2-3 hours, drops to the ground and after about two weeks moults into an adult. The female deposits eggs only after taking a meal of blood from the host. The adults are very tenacious and may live without food for as long as 3 years. Normally, the life cycle is completed in 5-6 weeks.

Damage. Apart from causing spirochaetosis through the tranmission of *Spirochaeta anserina*, the tick causes considerable injury to poultry, resulting in the reduction of egg production.

Management. (*i*) The poultry houses should be cleaned thoroughly. The hiding places of ticks should be eliminated by removing the rubbish and by closing the cracks and crevices. (*ii*) The

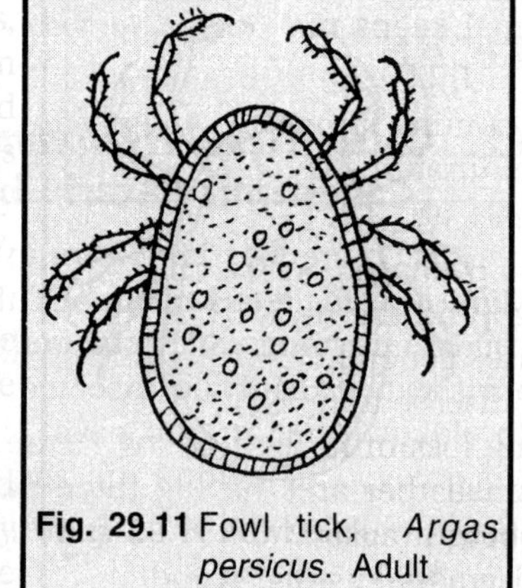

Fig. 29.11 Fowl tick, *Argas persicus*. Adult

fowls should not be allowed to roost in trees. (*iii*) The wild birds should not be allowed to nest in and around the poulty houses. (*vi*) The pest can be controlled by spraying malathion 3 per cent or carbaryl 2 per cent on walls, ceiling and floor. The spray fluid should be forced into the cracks. (*v*) The infested fowls may be dipped in 0.01 per cent solution of dichlorvos or malathion or may be given dust bath containing these insecticides.

MITES

14. Mange Mite, *Sarcoptes scabiei* (Linnaeus) (Acari : Sarcoptidae) (Fig. 29.12); Scabies Mite, *Psoroptes communis* (Furstenberg) (Acari: Psoroptidae)

The former species of mite (*S. scabiei*) lives in subcutaneous galleries made in the skin of man and warm-blooded animals. It has a remarkable range of hosts and in the course of its evolution, various varieties have appeared which are confined to specific hosts. On an average, the male measures about one-quarter of a millimetre and the female about two-fifth of a millimetre.

There are several described varieties of these mites, *S. s.* var. *hominus* is found on man; *S. s.* var. *bubulus* Oudemans infests cattle; *S. s.* var. *caprae* Furstenburg attacks primarily goats but also horses and mules; *S. s.* var. *equi* Gerlach is found on horses, mules and donkeys; *S. s.* var. *dromedarii* Gervais is common on camels and *S. s.* var. *canis* Gerlach appears on dogs. The allied species, *P. communis* has various varieties, namely *bos bubulus, ovis, caprae* and *equi*, which attack buffalo, sheep, goat and horse, respectively.

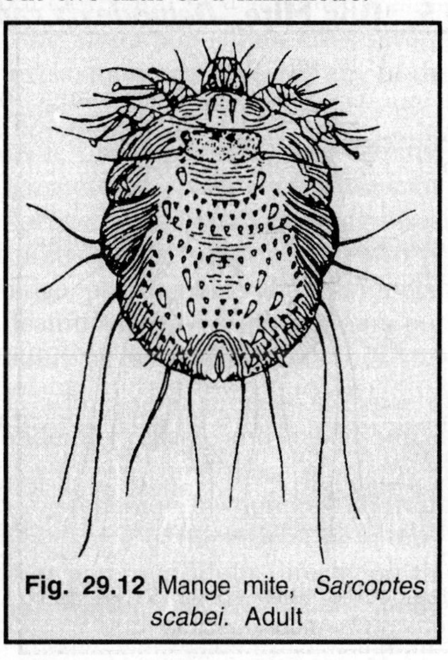

Fig. 29.12 Mange mite, *Sarcoptes scabei.* Adult

Life-cycle. The mites causing *Sarcoptes* mange mate on the host skin and then the mated females burrow inside the skin, making galleries. The eggs are deposited along with excrement. The nymphs, on emergence, feed inside the galleries and come out after changing into adults. The cast-off skins may be seen at various points in the skin. At room temperature, the mites may survive for about a week and, in this stage, they infest new hosts on contact.

S. scabiei on cattle and buffaloes appear at the junction of the horns and skin, and then they work their way into the distal extremity, reducing the horny tissue to powder, exposing the sensitive core. The diseased animals keep striking their horns against the manger. In sheep, the scabies appears on skin, matting the hair at first and later causing the fleece to drop. On the affected area, a greasy crust is formed. The affected animal keeps moving its lips, shaking its head and trying to bite its own skin.

Unlike the mange mite, which makes galleries, the scabies mite remains in the skin superficially. On coming into contact with the skin, the female lays about a hundred eggs which hatch in 2 days. The durations of the different stages namely larva, female nymph, male nymph and the pubescent female are 2.2, 2.3, 5.0 and 2.2 days, respectively. The life cycle is completed in about 11 days.

Damage. *Sarcoptes* mange appears on cattle and buffaloes in the perineal region, the upper thigh and the neck which the animals keep rubbing against any hard surface. In goats, the mange causes general debility and sudden death, sometimes producing a high mortality in herds. In horse, the disease appears on those areas where skin is thin. Thus, encrustation appears in patches, there being loss of hair. At places the oozing of blood is also noticed. In camel, the mange is noticed in cold weather and the animals that suffer remain in continual agitation and keep biting the skin or rubbing it against the trunks of trees or any other hard surface. The disease first starts in the groin and gradually affects the flanks, belly, legs, head, neck and back. In pigs, the mange first appears around the eyes and ears and later extends to other external parts of the body, including the inner side of the thighs. The disease is rare in pet dogs, but is sometimes noticed in stray dogs, which, if touched, can infest human beings also.

The cattle and buffaloes suffering from psoroptic scabies keep striking their horns on the manger, caring little for food and, thus, their milk yield is reduced.In sheep, the scabies spreads over the body and causes loss of valuable fleece and also reduces weight of the animal. The goats infested with this mite give up feeding, keep scratching their ears and eventually start going round and round, showing occasional symptoms of fits.They ultimately die. In horse, the upper part of the neck and the tail show this disease.

Management. (*i*) Repeated application of powdererd sulphur in vegetable oil base controls the disease. (*ii*) Two bodywashes at 7-day intervals with trichlorphon 0.25 per cent are recommended against mite on buffalo calves. Five washes of 0.4 per cent carbaryl or 0.5 per cent malathion emulsion on dogs at 7-day interval are also effective.

15. Dog Mite, *Demodexis canis* Leydig (Acari : Demodicidae) (Fig. 29.13)

This mite measures about 0.25 mm in length and causes a serious disease, follicular mange, of dogs all over the country. Although this species lives predominantly on the domestic dogs, but can occasionally infest humans.

The mite occurs in the hair follicles and occasionally in sebacious glands of various other mammals as the dogs lack sebacious glands.

As the mite is located deep in the dermis, transmission is usually only possible via prolonged direct contact, such as mother-to-pup transmission during suckling. As a result, the most common sites for early appearance of demodicodic lesions are the face, muzzle, forelimbs and periborbital regions. Demodicosis can manifest as lesions of two types : squamous-which causes dry alopelia and thickening of the skin; and pustular-which is the more severe form, causing secondary infection (usually by *Staphylococus*), resulting in the characteristic red, numerous pustules and wrinkling of the skin.

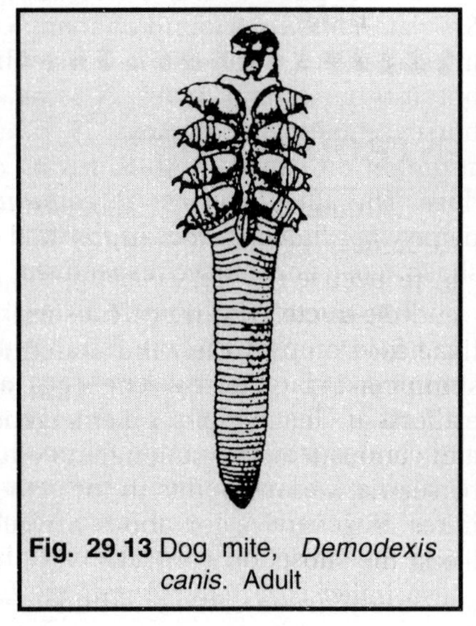

Fig. 29.13 Dog mite, *Demodexis canis*. Adult

Management. There is no satisfactory treatment although the following preparations are said to have given good results : (*i*) Rotenone 1g, alcohol 50 ml, acetone 10 ml and water 40 ml (*ii*) An ointment made of β-naphthol 4g, sublimed sulphur 8 g, balsam of Peru 30 ml and petroleum 30 ml. The mixture is to be applied daily or alternate days on the affected parts.

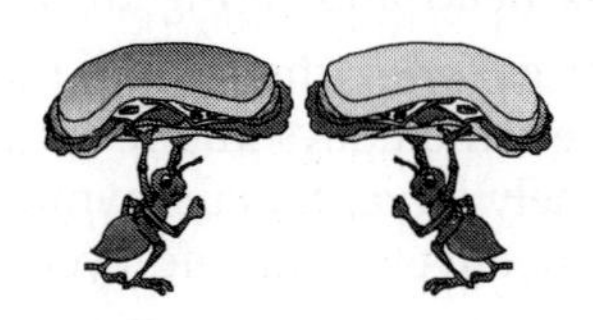

NON-INSECT PESTS

INTRODUCTION

A number of non-insect pests like mites, nematodes, rodents, birds, snails and slugs are known to damage agricultural crops. The mites and nematodes have been discussed under various crops in respective chapters. Among the mammal pests, rats and mice are at present the most abundant and destructive in the field as well as inside the houses and godowns. They damage practically all the field crops but some food crops like wheat, rice and groundnut seem to be their favourites. A number of large mammals which used to destroy crops have now dwindled in number and, in fact, efforts are being made to preserve some of this wild life in sanctuaries and national parks established in various states of India. It is interesting to note that only 25 of about 1200 species of birds found in India (i.e. only 2.1%) have been reported to inflict damage to crops and fruits. It would appear that, in general, the granivorous birds became abundant with the increase in area under grain crops and the subsequent storage of grain in godowns or bulk storage. Several species of snails and slugs also inflict considerable damage to cultivated crops.

BIRDS

1. House Crow, *Corvus splendens* Vieillot (Passeriformes : Corvidae)

The house-crow is distributed throughout the plains of the Indian Sub-continent, Sri Lanka and Myanmar. From the tip of the beak to the end of the tail, it is about 45 cm in length. It is to be distinguished from another allied species, the jungle-crow, *Corvus macrorhynchos* Wagler, which is more abundant in the mountains. The jungle-crow is larger and is glossy jet black with a heavy bill. The house-crow is grey and has black and grey wings with a black area on the throat and the forehead. It is perhaps the commonest bird in India and abounds in the vicinity of towns and villages, living in close association with man.

Habits. The house-crow is a very intelligent and cunning bird, and is usually not afraid of women and children. It is a constant nuisance in the houses and is always on the look-out for stealing titbits. Crows have the habit of community roosting in selected trees in the out-skirts of towns and villages to which they are seen flying in large numbers every evening. A crow's nest is in the form of a platform made of twigs with a cup-like depression in the centre. The twigs are intermixed with iron wires, threads or other fibres and the depression is lined with coir fibre, etc.

It is 3 metres or more above the ground and sometimes there are several nests in a tree. The nesting season is from April to June and 4-5 eggs are laid at a time. The eggs are pale blue green and are streaked with brown. They are incubated and the feeding of the young ones is shared by both sexes. The shrewd or cunning *koel* (*Eudynamys scolopacea*) is often seen hanging around the nests of crows in which she lays her own eggs and throws out of the nest an equal number of crow eggs. The young ones of the *koel* are brought up by the naive crows as their own. Thus, the *koel* acts as a biological parasite on the latter.

The house-crow is an omnivorous bird and may feed on dead rats, carrion, kitchen scraps, locusts, termites, the eggs of other birds and the ripening grains of maize and fruits. At times, damage to the ripening crop of maize can be very serious. The crows are particularly attracted to maize when it is exposed on the cob. They are often seen in flocks in maize and other fields.

Management. (*i*) The crows have a great sense of communal alarm and, if one of them is killed, the whole group creates a tremendous noise; therefore, a dead crow hanging on the top of a pole can effectively be used as a scare-crow. (*ii*) Destroy eggs and nests during June-August. (*iii*) The maize cobs on plants can be protected by wrapping one or two of the nearby leaves around them. (*iv*) A large wire-gauze case, 2 × 1 × 1 metres, having on one side a converging entrance, can be used as a trap for crows if some attractive food is kept inside. As soon as two or three are caught, they should be removed through a side-window, otherwise they will deter the others from getting into the trap. (*v*) A piece of *chapatti* dipped in 0.3 per cent methyl parathion placed on top of a roof is good bait.

2. Weaver Bird, *Ploceus philippinus* (Linnaeus) (Passeriformes : Ploceidae)

The common weaver-bird is distributed throughout India, Pakistan, Myanmar and Sri Lanka. It is the size of the house-sparrow and its colour resembles the female sparrow when it is not in the breeding season. It is dark-streaked, fulvous brown above and plain white fulvous below.

Habits. The male builds a number of intricately and compactly woven nests which are retort-shaped with a long vertical entrance tube. The nests are made of paddy leaves and grasses and are seen in clusters hanging from *ber* (*Zizyphus* spp.) or other medium-sized trees near ponds or water holes. Mud is used for plastering the egg-chamber on the inside. By completing these nests, the male attracts a number of females who then incubate their respective clutches of 2-4 pure-white eggs. Breeding takes place from May to September.

The birds are active during the paddy season because this crop provides food as well as nesting material. In paddy areas, they roost in very large numbers as governed by their seasonal local migration. They can cause quite heavy damage particularly to the early-ripening crop.

Management. The periodic collection and destruction of nests would lead to reduction in their population.

3. House Sparrow, *Passer domesticus* (Linnaeus) (Passeriformes : Ploceidae)

The house-sparrow has always followed human civilization and is practically world-wide, the races from the temperate regions being somewhat larger. In India, it is equally common on the plains and the mountains up to 2,150 metres. The female is ash to greyish brown above and fulvous ash-white below. The male is 15 cm long, darker above, with blackish streaks on the wings and a black patch on the throat and breast. It has prominent white cheeks.

Habits. This is the most familiar bird in cities and villages alike. Its constant chirping, tsi,tsi,tsi or cheer, cheer, cheer is too well known to us all. It breeds practically all year round and makes nests in the holes in walls or ceilings of houses by using straw, rubbish and feathers. The eggs are greenish white, blotched with brown. Three to five eggs are incubated at a time and several successive broods may be raised.

The sparrow is omnivorous and feeds on insects, fruit-buds, flower nectar, kitchen scraps and grains of all sorts. The non-breeding birds roost in large groups in leafy medium-sized trees. They visit the ripening fields, particularly those of wheat in the spring season and cause much damage both by feeding and by causing the grains to shed.

Management. (*i*) Destroy eggs and nests during April-May and September-October. (*ii*) Bajra (*Pennisetum typhoides*) seeds soaked in 0.3 per cent methyl parathion emulsion and dried, are placed in small cups and hung from the rafters or branches of trees. The poison bait is expected to be more effective in that season when no ripe grain crop is available in the fields. (*iii*) Spraying the wheat crop when ears are in the milk stage with Thiurum (TMTD) 0.6 per cent, repels the sparrows and protects the crop. (*iv*) In China, people were encouraged to catch and eat sparrows and thus the effort was successful in eliminating the pest.

4. Common Myna, *Acridotheres tristis* (Linnaeus) (Passeriformes : Saturnidae)

The common myna is distributed throughout the Indian Sub-continent up to 2,750 metres and is also found is Myanmar and Sri Lanka. It is about 22 cm in length, being larger than the *bulbul* (*Pycnonotus cafer*) and smaller than the pigeon. It is a dark-brown bird, with a bright yellow bill, legs and patches around the eyes. As it flies, large white patches become visible on the wings. The birds live in pairs or in groups and are common near human habitations or in the countryside.

Habits. The bird is known for its sharp chattering calls which are quite annoying to the ear. It breeds from April to August, raising two broods in succession. The nests are built in tree-hollows and holes in walls or ceilings. During the nesting season, the birds are seen collecting twigs, roots, rubbish and paper or tin foils from cigarette boxes. Four or five glossy blue eggs are incubated at a time and both sexes participate in raising the young ones.

The birds are omnivorous and feed on insects, earthworms, grasshoppers, fruits and kitchen scraps. They are pests and nuisance around houses and pick up all sorts of grains from the threshing-floors or from the fields. They are seen in ripening maize and wheat fields feeding on grains and their flocks are found alongside those of crows and parakeets.

Management. (*i*) Destroying the nests goes a long way in checking their multiplication. (*ii*) Since it is a clean bird, its shooting as a game should be encouraged in the field. (*iii*) They can be killed by feeding them on *chapatti* soaked in 0.3 per cent methyl parathion emulsion.

5. Green Bee Eeater, *Merops orientalis* Latham (Croaciformes : Meropidae)

This bird, the size of the house-sparrow, is found throughout the Indian Sub-continent, Myanmar and Sri Lanka. It is grass green, tinged with reddish brown on head and neck. It has a slender, long, slightly curved bill and the central pair of tail feathers have blunt extremities. It is distinguished from the larger allied species, the chestnut-headed bee-eater, *Merops leschenaulti* Vieillot, which has a bright chestnut head and anterior back, without the long tail feathers. The blue tailed bee-eater, *Merops philippinus* Linnaeus, is larger with the crown, throat and breast a deep chestnut and a blue tail.

Habits. The common green bee-eater, like the other species, is a resident or a local migrant.It inhabits open country and is found near gardens and woodlands. It is often seen sitting on electric or telephone wires snapping at honeybees and other wild bees, flies and wasps. In the evening, it roosts in large numbers in its favourite leafy trees. It makes a horizontal or oblique tunnel-like nest in the side of earth cuttings, sides of hillocks in the Shiwaliks or in uneven sandy ground. The eggs are pure white, roundish oval and 4-7 of them are placed in the widened egg-chamber. Both sexes participate in excavating nests and feeding the young.

This bird is extremely destructive to domesticated honeybees and is the cause of the failure of beekeeping in certain areas. In the plains their numbers become numerous in April-May and again in August-September.

Management. (*i*) Locating the burrows and destroying the eggs may give some relief. (*ii*) They should not be allowed to sit on trees and wires near apiaries and should be killed or chased away with air-guns.

6. Rose-ringed Parakeet, *Psittacula krameri* (Scopoli) (Psittaciformes : Psittacidae)

The rose-ringed parakeet, commoly known as *tota,* is one of the commonest birds in India, Pakistan and Myanmar, and is distributed from the Himalayan foothills right up to Sri Lanka in the south. In the hills of southern India,it has been recorded up to 1500 metres.It is slightly larger than the common *myna,* is grass-green and possesses a typical short massive and hooked red bill. The male has a black and rose-pink collar, whereas the female is plain.It is distinct from the Alexandrine or the larger Indian parakeet (*rai-tota*), *Psittacula eupatria* (Linnaeus), which is larger, somewhat the size of a pigeon and has a more powerful voice. In addition, the male of *rai-tota* has a conspicuous maroon patch on each shoulder and also has a rose pink and black- collar. The female of this species is also grass-green but is devoid of the collar. There is yet another species, the blossom-headed parakeet (*tuiatota*), *Psittacula cyanocephala* (Linnaeus) which is the size of the common myna and is an inhabitant of the more wooded country and is also found in the Himalayas up to 1,850 metres. The head is bluish at the crown and red below and the male has a small maroon patch on each shoulder. The female does not have the shoulder patches and its head is greyer with a bright yellow collar round the neck. This parakeet is a swift flier and it easily flies in between the trees.

Habit. The rose-ringed and the large Indian parakeets are found living together in lightly wooded country on top of leafy trees, near orchards or cultivated areas and they are common in towns and villages alike. Quite often, they form large flocks and roost near the source of food which is primarily the ripening fruits, maize and other grains.They make much noise and chatter and the sharp screams (*keeak, keeak, keeak*) of the rose-ringed parakeet are quite characteristic. Both the parakeets can be caged and they learn to repeat a few words or perform various circus tricks. Their nests consist of natural or dug-out hollows in trees and they may also utilize the holes in walls or rocks for this purpose. The nests are at moderate heights and there may be 4-6 pure-white rounded eggs in one clutch. Both male and female share the domestic duties of incubation and feeding during the breeding season which is December to April.

The parakeets are among the most wasteful and destructive birds. They gnaw at and cut into bits all sorts of near-ripe fruits such as guava, *ber*, mango. plums, peaches, etc. The ground underneath the fruit-trees is sometimes literally covered with dropped fruits and refuse. The birds are equally destructive to ripening cobs of maize and *jowar* (*Sorghum vulgare*) ears, and are often seen carrying and flying with ears of wheat. In the maize field, they are seen in large flocks and cause tremendous damage to the crop.

Management. (*i*) The control measures include the location of nests and the destruction of eggs or the fumigation of holes with aluminium phosphide at the rate of 0.5 g tablet per hole, so as to kill the young ones. (*ii*) Orchard owners often throw stones with *gulel* (slingshot) or try to frighten them with crackers or make noise with empty kerosene tins, with some success. (*iii*) Killing them with shot-guns is very effective but expensive. (*iv*) The parakeets do not accept poison baits. They can be repelled from maize and sunflower fields using malathion 5 per cent dust on cobs, sunflower heads, especially on outer rows in maize.

7. Blue Rock Pigeon, *Columba livia* Gmelin (Columbiformes : Columbidae)

The blue rock pigeon is about 32 cm long, widely distributed in Eurasia, and in India it has been recorded up to 4,000 metres in the Himalayas. It is the most familiar bird near churches, temples, historical buildings with minarets, power-houses, railway station, grain warehouses, etc. It is slatey grey with a glistening metallic green, purple and magenta sheen on the neck and upper breast. There are two dark bars on the wings and a band across the end of the tail.

Habits. In the wild state, it abounds in mountain cliffs or old and crumbling battlements, outlying cultivations, but in the semi-domesticated state as a commensal of man, it has learnt to live in towns and even in noisy cities. It breeds practically throughout the year and makes nests by collecting a few sticks and lines them with rags and feathers. The nests are located in the holes on cliffs, rafters and ceilings of the houses, particularly those uninhabited. The projections outside ventilators or sunbreakers in modern houses are commonly utilized by them for building nests and raising their broods. There are generally two white elliptical eggs in a clutch and both sexes share domestic duties.

Pigeons generally roost together in large numbers and visit, in huge flocks, the ripening grain fields, threshing-floors, etc., where they feed. They use up quite large quantities of wheat, *bajra* (*Pennisetum typhoides*), sorghum, maize, etc. Individually or in small groups, they are always around to pick up grains from here and there. In large buildings, they are a nuisance as they cause noise and make the floors and walls dirty with their droppings. In powerhouses, they are a special hazard. While sitting around on wires and insulations they cause short circuits and power failures.

Management. (*i*) The meat of the rock pigeon as well as that of the dove is quite tasty and, therefore, they can be killed as game. (*ii*) The strategic places in power-houses should be protected by installing chicken wire-netting around the weak spots. (*iii*) Destroy eggs and nests during March-June and September-October. (*iv*) They can also be killed by feeding on poisoned *bajra* grains (see under 'House-sparrow').

RATS AND MICE

Rodents including rats, mice and squirrels are gnawing animals having chisel like teeth. They have highly developed sense of smell, taste and hearing. They are omnivorous and feed on grains, vegetables, fruits, meat and other products in the houses or in the field. Rodents are responsible for causing enormous losses to crops and stored grains. Besides feeding on these products, they destroy a substantial quantity by spillage and contamination with their droppings, urine, body hair, etc. Rats daily consume food equal to about 10 per cent of their body weight and the figure may go up to 30 per cent in case of mice. They damage about 20 times the amount they actually consume, by their gnawing activity and by polluting the foodgrains. Rodents represent about 35 per cent of the total mammalian fauna but their population outnumbers the total population of living mammals put together. They inflict, on an average, about 5-20 per cent loss in food production alone.

Of the 84 species of order Rodentia in India, 50 belong to the family Muridae.The important ones belong to two sub-families, viz., Murinae and Gerbillinae.Murinae has 18 species of genus *Rattus*, 2 of *Bandicota*, 7 of *Mus*, 2 of *Apodemus* and a few each under eight other genera.The sub-family Gerbillinae has 2 species of *Gerbillus*, one each of *Tatera* and *Meriones*. Family Sciuridae has 10 species of flying squirrels and 13 species of tree squirrels in India.

On the basis of the reports on the damage inflicted by rodents to crops and stored food materials, about 10 rodent species are of major importance.The commonest species found in houses throughout India are *Rattus rattus, Mus musculus* and *Rattus norvegicus*. The species of field rats most widely distributed throughout the country and causing damage to crops are *Bandicota bengalensis, Rattus meltada, Tatera indica* and *Mus* spp. However, in certain regions they may be outnumbered by some other species, for example, *Meriones burrianae* in the Indian desert and *T. indica* in peninsular India. The population structure and intensity of rodents differ with time and agro-climatic conditions of the region.In crop fields of Punjab, *T. indica* dominated in early seventies and was replaced by *R. meltada* in mid and late seventies. With intensification of paddy cultivation during the last two decades, *B.bengalensis* has become the most predominant rodent in summer.According to the estimates made in wheat and groundnut fields, the population of *Tatera indica* is 50-75 individuals per hectare in the peak season.

1. Indian Mole Rat, *Bandicota bengalensis* (Gray) (Rodentia : Murinae)

This is a fierce animal and grunts when angry. A god swimmer, it is covered with thick harsh fur, greyish-brown or black on top and greyish -white on the belly. It is heavily built and has a pig-like face and a short stumpy head. It measures 16-24 cm in length (without the tail) and the average male weighs 326 g and the female 287 g. The tail, which is scaly and has 160-170 rings, is often shorter but is sometimes as long as the body.

It breeds throughout the year and the number of the young per litter vary from 6 to 15. In southern India, the mole-rat breeds during September-October and January-March, coinciding with the maturity of paddy. The non-breeding adult individuals live separately in their own burrows. The mole-rat hoards large amount of food in its burrow. Up to 7.3 kg of wheat ears has been found hoarded in a single burrow. The burrows are underground and shallower than those made by *Tatera indica.* There are 2-12 openings in a burrow and they are generally closed with loose excavated soil during the day- time. The tunnel has 2-5 lanes of which some are blind and have food-chambers at the end. The food materials, such as groundnut, paddy, wheat, etc. are placed in these chambers in an orderly manner and they are sealed with balls of soil. The brood chamber is at the end of a zig-zag alley, having a lining of grass and straw.

The presence of the mole-rat in its burrow is detected from a freshly dug earth piled on the closed opening. The burrows are generally found in the fields of groundnut, wheat, gram, sugarcane, maize, sorghum, cotton, paddy and in fruit orchards. They are also concentrated in uncultivated areas where there may be *sarkanda* (*Saccharum munja*) growth. The mole-rat cuts the entire paddy ear but eats only part of it. Like other species, it migrates from the harvested fields to the newly sown crops.

2. Soft-furred Field Rat, *Rattus meltada* (Gray) (Rodentia : Murinae)

It is smaller than the mole-rat and its fur is soft and dark brownish-grey except the belly which is pale grey. This species is found in wetlands as well as in drylands having many habits similar to those of *B.bengalensis.* These rats make simple burrows in the field as well as in the bunds. It is 10.0-15.6 cm long and weighs about 64 g. The tail is a little shorter in length than the rest of the body. It has two subspecies- *R.m.pallidor* occurs in northern and eastern india, and *R.m. meltada* in southern India.In northern India, this nocturnal and gregarious rat produces young ones in two breeding seasons, viz. March and August. A female produces 1-4 litters during the breeding season and the interval between two litters is 20-44 days. In captivity, the number of young ones per litter varies from 1 to 8. In the spring, they attain sexual maturity in 3-4 months, whereas in the autumn in 7 months. This rat makes burrows which are smaller and narrower than those of *B. bengalensis* and *T.indica.*There are 1-4 openings to a burrow and they are always kept open, but a few bits of grasses are kept in front of the openings.

In southern India, the breeding season coincides with the maturity of paddy in September-October and again in January-April. They may live alone, in pairs or in groups of 1-9.The number of young ones per litter is 2-12 and there may be more than one litter in a season.In the summer when food is not available, they feed on grass or on crabs and snails.The length of the burrow ranges from 88 to 240 cm and the width, 3 to 7 cm. There are 1-4 bed chambers, each of the size 13 × 3 cm.

In southern India, these rats outnumber *B.bengalensis* in July-August but their population declines steadily and they are hardly seen in the spring. They are fairly good swimmers and have the characteristic habit of cutting a paddy ear and swimming back to the bund with the booty.

3. Indian Gerbil, *Tatera indica* (Hardwicke) (Rodentia : Gerbillinae)

It is also known as the antelope-rat and is found in sandy soils. A poor swimmer,it is light brown or reddish from above and white on the belly.It is 11.6-20.3 cm long and weighs 115 g.The tail is 20-30 per cent longer than the body and it is hairy, being of dark shade on the upper and lower surfaces and a pale shade on the sides. The tail ends in a tassel. It is a very sensitive and fast-moving rat and lives in colonies.In captivity, it has two breeding seasons; (*i*) from last week of March to mid May ; (*ii*) from the last week of August to mid October. A female produces 1-3 litters, there being 3-8 young ones in each.In the field, there are 5-10 young per litter.

Unlike the mole-rat and the meltad, the Indian gerbil has its burrows in sandy areas, dryland patches near the crop fields or river beds. The burrow opening is often located under the cover of a thorny bush or some vegetation ,but the scattered freshly excavated soil helps in its detection. The burrows are deeper and shorter in length than those of the mole-rat. Like other gerbils, *T.indica* feeds on seeds in the winter; stems, rhizomes and insects in the summer, and on leaves and flowers of plants in the monsoon.

4. Indian Desert Gerbil, *Meriones hurrianae* (Jerdon) (Rodentia : Gerbillinae)

It inhabits the arid and semi-arid regions of Punjab, Haryana, Rajasthan and spreads up to Pakistan, Afghanistan and Iran. It is a diurnal rat and digs extensive burrows with 10-15 openings. Its head and body is 100 mm, tail is equal to or slightly shorter than head and body is covered with black tassel of hairs at its tip. Merion gerbil breeds all the year round with peaks during February-March and July-September. Litter size varies from 1 to 9 (average 4.4). The gestation period averages 30 days. It causes damage to crops and debarks the trees.

5. Indian Field Mouse, *Mus booduga* (Gray) (Rodentia: Murinae)

It is found in dry as well as in wetlands and is a fairly good swimmer. It is brown from above and whitish or dull grey on the underside. The tail is equal to or shorter than the body and is dark above and pale below. The mouse measures 5-8 cm and weights 16-19 g. In southern India, it breeds in September-October and February-June. It makes small burrows without any branches, having only one bed chamber. The field-mice live in pairs and are common in the same habitat as of the mole- rat. The length of the burrows ranges from 50-116 cm and its diameter is 2-3 cm. The number of young ones per litter is 6-13. It feeds on paddy grains and at times steals from the food chambers of *B.bengalensis.*

Damage by Rodents

The nature and extent of damage vary from crop to crop. Rats prefer cereals to dicot seeds and grains. In the wheat crop, little damage is done to the seedlings and most of the damage is caused at the ripening stage. The estimated loss to cereal yield is 52 kg per hectare in the Punjab. The presence of alternative crops and the lack of shelter reduce the damage to some extent. Damage to the groundnut crop in the seedling stage is high. Later on, only branches are cut but at maturity,the loss is again heavy because the rats dig out pods and eat the kernel, leaving behind empty shells. At places where no alternative food crops are available, the damage can be the maximum during the growing season. The estimated loss of groundnut yield is 47 kg per hectare in the Punjab.

The damage to sugarcane in the seedling stage is negligible and it starts from October onwards and continues up to harvest. The attacked canes are not a complete loss, but are gnawed usually from the Ist internode to the 5th. In the lodged crop, however,all internodes may be damaged, resulting in greater loss. The rats uproot paddy nurseries to eat the seeds and later on they also cut down the plants.In the vegetative phase, the damage starts when the plants form 3-4 nodes and damage continues in the reproductive phase till harvest.In the advanced tillering stage, damage can be heavy but some of the early-attacked tillers recover and produce late earheads which may mature.After maturity, the damaged patches of the crop are easily detected as the plants are still in the vegetative or in the flowering stage. It has been noticed that the rats restrict their activity within 5-10 metre radius around the burrow. Damage by rats to coconut fruits is characteristic in that they make one or two holes near the stalk and feed on the carpel.

The loss to *rabi* crops by the rodents has been worked out at 11.5 per cent in wheat, 5-8 per cent in barley and 9.9 percent in gram. The loss to the *kharif crops* is still greater.The percentage loss varies from 4.6-5.4 in paddy, 4.1-25.8 in groundnut, 14.0 in maize, 2.2 in sugarcane and 5.0 in coconut. Rodents cause more damage at seedling and ripening stages of the crops. Average damage to wheat is 2.9 per cent at seedling and 4.5 per cent at ripening stage, and to pea is 1.1 per cent at seedling and 5.9 per cent at ripening.In winter maize they cause 10.7 per cent damage. The rat damage in wheat crop near sugarcane fields,waste land,canals and roads may reach up to 25 per cent. The extent of damage varies with fluctuation of the rat population in an area, which changes from year to year.

Apart from eating the grains and other crops, rats are known to spread deadly diseases among man and livestock. *Rattus rattus, Bandicota bengalensis* and *Tatera indica* carry the bubonic plague. Moreover,they gnaw electrical wiring, disfigure furniture, burrow inside the houses and cause the sinking of the floors and walls.In the field,their burrows interfere with irrigation, causing reduction in crop yields.

Management of Rodents

The various techniques to manage rodents include cultural, mechanical, biological and chemical control practices.

Cultural control. Deep ploughing upto 45 cm, reduction of size and trimming of field bundhs at the time of land preparation would certainly go a long way in controlling rodents. Weed management and removal of burrows reduce availability of alternative food and shelter. Simultaneous planting prevents rodent migration from harvested to unharvested fields.

Mechanical control. Guarding of rat attacks by means of rodent-proof containers and plastering storage structures help in checking rodent infestation. The rodent problem can be solved to a large extent by obstructing the entry of rats in houses and stores. Trapping is an economical and effective way of reducing rodent population. Trapped rats should be killed by drowning cages in ponds and dead rats should be buried deep in the soil. The trapped rats should not be released in the fields as they usually come back to their original place.

Biological control. A number of predators like snakes, owles, eagles, mangooses, etc. contribute to natural regulation of rodent population. Keeping cats in houses also checks the rat population.

Chemical control. The two most commonly employed chemical control measures include poison baits and fumigation.

Poison baits. The most effective method of controlling rats is the use of poison baits. The poisons used in the baits are of two types: (*a*) acute poisons which are used in a single dose, e.g. strychnine hydrochloride,zinc phosphide, norbormide (Raticate),sodium fluoro-acetate, thallium sulphate, ANTU (Alphanaphthyl thiourea); and (*b*) chronic poisons which act as blood anticoagulants and are used in multiple doses. They include hydroxy coumarins (Warfarin, Fumarin, Tomarin, Recumin) and indandions (Pival, Radione, Valone). These poisons are lethal when consumed for several days,as they cause external and internal haemorrhage.The other anticoagulant rodenticides (Brodicacoum, Bromadiolane) are lethal in a single dose but the rats die after several days of poisoning.

Zinc phosphide (2% bait) is the most commonly used rodenticide in India.The bait is prepared by mixing 1 part of zinc phosphide with 40 parts of whole or cracked grains of wheat, gram, maize, *bajra* or sorghum smeared with vegetable oil. Racumin bait (0.0375% bait) is more effective against the bandicoot rat than other species. The bait is prepared by mixing 50 g of 0.75 per cent racumin powder, 20 ml of groundnut or sunflower oil and 20 g of powdered sugar in 1 kg of cracked wheat or any other cereal. Another bait (0.005% biomadiolone bait) can be prepared by mixing 20 g of 0.25 per cent bromadiolone powder, 20 g of oil and 20 g of powdered sugar in 1 kg of any cereal flour.The mixture is wrapped in paper packets which are distributed in the field at 10-15 m distance.For better results in the field, the poison operations with zinc phosphide should be preceded by careful prebaiting, i.e., false baiting with non-poisoned bait for 1-2 days. Field operations with zinc phosphide result in 70-80 per cent reduction of rodent population. The rest of the population acquires bait shyness after a single exposure to the bait.Poison baiting with zinc phosphide should not be repeated within 2-3 months.The residual population should be controlled by the fumigation of burrows or by baiting with warfarin.

Fumigation of burrows. Another effective method of rat control is killing them inside the burrows by fumigation.Aluminium phosphide tablets (2 tablets of 0.6 g or half tablet of 3 g per burrow) have been found to be very effective and safe. After introducing a tablet into a live burrow, the opening is closed tight with soil. The chemical reacts with soil moisture and deadly phosphine gas is generated.

These methods of killing rats are effective only when carried out on a large scale, covering large contiguous areas, and are repeated time and again. The aim should be to kill more than 90 per cent of the population, otherwise they breed so fast that population reaches the same level within a few months.

FRUIT BATS

Indian Fruit Bat, *Pteropus giganteus* Brunnich (Chiroptera: Pteropodidae)

Bats are the only flying mammals and the Indian fruit-bats are often seen in the Punjab plains hanging on electric wires, after having been electrocuted accidentally. Their roosting abodes lie at various places in the Kangra Valley (Himachal Pradesh) and with the ripening of fruits, they migrate temporarily to higher or lower altitudes.They are confined to the tropical and sub-tropical regions of India and do not live permanently in the temperate zones.These bats are to be distinguished from the dark-brown or smoky smaller insect-eating bats,which are not harmful except for being a nuisance in and around houses. In South India, *P. edwardsii* is the common species, *P.vampyrus* and *P. edulis* are present in Malaysia.

The fruit-bats or flying foxes have a long range flight and are known to visit fruit orchards in Amritsar at night and return to their roosting places in the lower Himalayas before daybreak. These bats have long-snouted fox-like faces and light-brown fur with darker patches.In between the fingers of their fore-arms, there are loose folds of skin extending backwards along the legs. These are the 'wings' used for flight and the claw-like toes are used for hanging upside down. Their body length is 18-25 cm and they measure 1.2 metres across the wing spread.

Habits. Bats are unique mammals as they suckle their young ones like other mammals but fly like birds. There are few barriers that hinder their flight and yet their distribution is limited by climate. Bats prefer to live in tropical, semi-evergreen, moist and dry deciduous zones.They do not penetrate into evergreen rain forests.In the desert or semi-desert areas, they are found only at those places where man has planted fruit-trees. The distribution and movements of these bats are influenced by the flowering and fruiting of trees that provide them with food.According to the availability of ripe fruits,the bats may be abundant at a given place in one season and absent in another.Their breeding also coincides with the fruiting season.

During day-time, the Indian fruit-bats rest in large numbers on tall trees, such as *banyan*, fig, tamarind,etc. They persistently stick to the same location, except when they migrate temporarily, following the warmer weather and the ripening fruits.Their winged arms and feet can be used for holding food and for walking, but during the active season, they eat, nibble and drop the damaged fruits or take them away to their roosting places and thus they cause large number of seeds to accumulate under the trees. During cold weather, they remain inactive and suspend their destructive feeding activities.

Fruit-bats are essentially nectar-or fruit-juice feeders. They munch the fruit pith and throw away most of the morsel while in the upside down hanging position. They are voracious feeders and, in this manner, use or destroy more than their own weight of fruits every day. All this is done at night in the dark and it is possible for them to do by their remarkable power of echo- sounding (precursor of radar invention), unmatched hearing power and the most sensitive sense of touch in their wing membranes.

Damage. Fruit bats are active at night, flying long distances to reach the orchards. They munch, drop and destroy huge quantities of fruit such as pear, plum, peach, guava, mango and apple. In South India, they damage coconuts.The orchards invaded by flying foxes are littered with remnants of fruits picked and destroyed.

Management. (*i*) Many methods,such as keeping the orchards well-lit at night, creating noises, shooting, etc. have been tried by individual orchardists with varying degrees of success. (*ii*) The most accepted and perhaps the most effective method is shooting them at the roosting places. (*iii*) Killing them by the detonation of explosives on the roosting trees has also been tried with good results under the supervision of the staff of the Inspector-General of Explosives, Government of India.

OTHER MAMMALS

1. Monkey, *Macaca mulatta* Zimmerman (Primates: Cercopithecidae)

The common monkey is found in the Himalayas, Assam, Myanmar, northern and central India (as far south as the Tapti and the Godavari). It is an animal with the crown hair radiating backwards and with orange-red fur on its loins and rump. Monkeys live in large troops and destroy fruits, maize cobs and steal away eatables from shops, temples and houses.They prove themselves to be a great nuisance. They pluck and throw away much more than what they actually eat.

Habits. They live in large groups near villages, towns, in groves or on the outskirts of forests and show a preference of the open country. They invade gardens and fields of crops in the morning and evening. In the afternoon, they rest in the shade of trees on the roadside near temples, ponds and shady places. They are often seen bathing in water, with the babies clinging to their mothers. The monkeys can swim and they keep themselves cool by frequent dips. During winter, they acquire a thicker coat as protection against cold.

The social troop life of monkeys has not been studied thoroughly, but among them there seems to be a male leader who apparently dominates others.The majority in a given group are females and young ones. In the western Himalayas, mating commonly takes place in August-September and the young ones are born from March to May. However, sex life is persistent and mating may take place at any time of the year and this instinct seems to be the main driving force for social and community living. Fur-picking is another social habit that binds the members together. This habit is not meant to search and remove lice and fleas as is commonly thought because monkeys are generally free from these external parasites.

Damage. When the maize crop and fruits ripen in the lower Himalayas, large troops of monkey descend from the forest edges to invade towns and villages. The devastation caused by them is often complete and extremely wasteful.

Management. (*i*) Owing to the religious sentiments of the people, shooting, killing and poison baiting are not recommended. (*ii*) Under contract with the Government, authorized persons may be allowed to catch monkeys with various types of nets and snares, and the captives can then be kept in laboratories for experiments in medicine or exported for the same purpose.

2. Indian Porcupine, *Hystrix indica* Kerr (Rodentia: Hystricidae)

The Indian porcupine is found throughout the Indian Sub-continent and parts of western Asia.In the western Himalayas, it is found up to 2,400 metres above sea-level. This rodent is very characteristic in having quills or spines (modified hairs) on its body. The quills may be solid or hollow, depending upon their location and they possess deep-brown or black end rings. In southern

India, the animal may acquire a bright rusty red or orange hue and is often given the misnormer 'red porcupine'.In the central and eastern Himalayas up to an elevation of 1,500 metres, another species, *H.hodgsoni* (Gray) is prevalent.

Habits. The Indian porcupine is found in all types of habitat,dry or humid, open land or forest, rocky hillsides and the undulating plains. Although burrows are not essential, the porcupine generally lives in burrows they dig. There is considerable earth lying around the mouth of a burrow and some pieces of chewed-up bones or horns may also be seen. The bones and horns supply the much needed calcium for the proper growth of the quills.When irritated or alarmed, the porcupine erects its spines, rattles the hollow quills and charges backwards with great speed. The quills are driven deep into the enemy and can be fatal even to some of the large animals, like the tiger. New quills grow under the old ones which are shed. Porcupines breed in the spring and both parents live with 2-4 young ones, which are born with their eyes open.

Damage. This rodent causes damage to vegetables, grains, fruits and crop roots. During day-time, it hides in the vegetation near the fields, in grass, bushes and scrub. At night it comes out and causes damage. Sometimes, it even tunnels through the walls of gardens to reach food.

Management. (*i*) The porcupine may best be killed with a shot-gun. (*ii*) Trapping is obsolete and not humane.The animal has a strong sense of smell, therefore, poison -baiting is not so effective. (*iii*) It can also be killed by fumigating the burrow with phosphine and closing the entrance.

3. Jackal, *Canis aureus* Linnaeus; *Dhole, Cuon alpinus* (Pallas); Wolf, *Canis lupus* Linnaeus (Carnivora: Canidae)

(*a*) The jackal or *giddar* has a wide distribution ranging from south eastern Europe to India, Myanmar and Thailand.It belongs to the dog family and there can be successful crossing between a jackal and a wolf or a dog, although under natural conditions, free interbreeding is not noticed because of their characteristic habits.

Jackals are notorious for their damaging activities in fields of ripe sugarcane, maize, musk-melon, water-melon, etc. A full-grown jackal is about 38-43 cm in height, 60-75 cm in body length excluding the tail and weighs about 8-11 kg. It is the size of a small dog and has a coat which is a mixture of black, white and mud-buff.It is shy during the day-time.

Habits. Just before dusk, the jackals are heard howling with a long-drawn *eerie* voice. Although jackals are found in mountainous areas, they are common in the plains, near towns and villages and near the places of cultivation. During the day,they hide in holes in the ground, unused buildings or in dense grass.They generally go about alone but sometimes form packs of two or three, living as scavangers on carcasses. They are essentially omnivorous creatures and and also feed on maize cobs, musk-melon, water-melon , sugarcane, fallen *ber* fruits,etc. They breed all the year round and lay their young in hollows in the ground or other sheltered places. The average life of a jackal is about 12 years.

(*b*) The Indian wild dog or *dhole* is widely distributed in central eastern Asia, Manchuria, Malaysia and India. Unlike the jackal, the *dhole* is essentially a carnivorous animal. It is bigger than the jackal and is much like a domestic dog in general appearance,the body being rather long like that of a wolf. It is relatively short in the leg and the muzzle, with ears rounded at the tips and the trail quite bushy. It differs from wolf in having six molar teeth in the lower jaw, whereas the wolf has seven. The *dhole* has also 12-14 teats as compared to 10 of the dog. It has a distinctive red coat which may vary in tone with the season. The wild dogs live in forests and sometimes in the open countryside.In India, however, they keep entirely to the forest, there being plenty of food, shade and water. The wild dogs have a social life because of their prolonged association between the parents and young ones.One or more families may comprise the pack. They are very effective in killing quite large animals such as the deer, the buffalo, the bear or even the tiger. Their main breeding season in southern India is November- December and 4-6 young ones are born in January-February. The young ones are sheltered by the mother in a cave or in some other place of protection. Several families may breed together in a colony.

Although this jungle creature mostly preys upon deer, antelopes, etc., sometimes it may also attack domesticated animals like cattle, buffaloes, sheep and goats.

(*c*) The wolf or *bheria* is widley distributed in Europe, North America and in a greater part of Asia. In India, they are common in open regions. In the western Himalayas and Kashmir, they have a nomadic life and come down to the valleys in winter. The hollows in mountainsides or small caves and cavities in rocks provide them shelter in the cold weather.In the desert areas of India, the wolf may dig burrows in sand-dunes for protection against the sun heat. They breed at the end of the rainy season and the young ones are born in December.In the Himalayan region, however, they are born in spring or in early summer. There are 3-9 young ones in a litter and the span of life is 12-15 years.

The wolf hunts deer, rodents, wild sheep, gazelle, black-buck and other animals during the day as well as during the night. It occasionally attacks domestic animals such as sheep, goats and cattle. Where human habitations are close to the forest, the wolf may even attack man and carry off small children.

Management. (*i*) All the three pests can be killed by shooting. (*ii*) For the protection of musk-melon and sugarcane against jackals, meat poisoned with strychnine (300:1) is quite effective. (*iii*) For the protection of a flock of sheep against wolves, shephards in the Himalayan region keep one or more dogs of the *Gaddi* breed. The dogs are very strong, alert and faithful. They give a good fight to the wolves and provide effective protection to the flock of sheep.

4. Long-eared Hedgehog, *Hemiechinus auritus* (Gmelin) (Insectivora : Erinaceidae)

The long-eared hedgehog is deep brown or almost black and is found along with another species, the pale, light-coloured hedgehog, *Paraechinus micropus* (Blyth). The hedgehogs are confined to the dry western and north western areas of the Indian Sub-continent. One race of the latter species is also found in the plains of southern India.

The long-eared hedgehog has a snout which resembles with that of a pig; hence the name hedgehog. It has a stout body, a short tail and stubby legs furnished with claws. The eyes and the ears are well-developed and there is a dense mat of spines on the back and sides of the body. At the back and flanks, the skin is loose. At the time of danger,the animal feigns death and the skin covers the rolled-up body. The pointed bristles give protection to the animal.

Habits. The hedgehog is a nocturnal animal and during the day-time, it lives in sand holes or underneath thorny bushes or grass. The young ones are produced inside holes lined with leaves or bits of grass. They are blind at birth and their spines are soft and flexible. The hedgehogs are active throughout the summer and their activity is reduced greatly in winter.

The animal comes out at dusk for feeding and retires at dawn. Its main food consists of insects, worms, lizards, rats, but it also feeds on fruits and roots of tuber crops. At places, the damage to nurseries, vegetables or tuber crops might be quite serious.

Management. The animals can be killed by fumigating them in their holes with phosphine. After placing one tablet (3 g) of aluminium phosphide in the burrow, the entrance hole should be plugged with soil.

5. Wild Boar, *Sus scrofa* Linnaeus (Artiodactyla : Suidae)

The wild boar is widely distributed in the Indian Sub-continent, Myanmar, Thailand, Malaysia and Sri Lanka. The male has well developed tusks, measures 0.9 metres at the shoulder and may weigh 250 kg.The animal is black rusty brown and has a sparse coat of white hair and a mane of black bristles.The new-born pigs are brown with black stripes.

Habits. The wild boar lives in scant jungles or grass and has high intelligence and great courage.Its sense of smell is acute and eyesight and hearing are moderate.It breeds in all seasons and is very prolific. While pairing, the animals collect in groups of 150 or more,the master boar being in the centre. The gestation period is four months and there are 4-6 young ones in a litter, which are protected by the mother. After breeding the males may live alone or in the company of one or two males or females.

After the rains,the boars are commonly seen in tall crops of sugarcane, maize, etc. They feed early in the morning or late in the evening and are extremely destructive. They are essentially omnivorous and may feed on crops, tubers, roots, insects and carrion. Their population has declined very much in areas where jungles are receding, but at places they are still a great nuisance.

Management. (*i*) The best and most effective method is to shoot them at sight. (*ii*) The old method of ensnaring by leading them into pitfalls with sharp bamboo sticks at the bottom is not a humane method of killing.

6. Sloth Bear, *Melursus ursinus* (Shaw) (Carnivora: Ursidae)

The sloth bear is found in the forest areas of the Indian Sub-continent from the base of the Himalayas to Sri Lanka. This species is distinct from the larger Himalayan black bear (*Solenarctos thibetanus*) which normally lives near the tree-line. During winter, it comes down to 1,500 metres or so.

The sloth bear is 0.6-0. 75 metre tall at the shoulder and 1.35-1.67 metre in length. The male is bigger than the female and weighs 125-145 kg. It is dull black with a brownish tinge. It has an elongated muzzle with long, dull, unkempt hair, short hind legs and is the most unattractive of all bears.There is a white V-shaped breast mark and the muzzle and tips of the feet are dirty white or yellow.

Habits. By habit, the sloth bear prowls at night when it looks for food. During day-time, it retires to the shade of boulders or projecting rocks. When the weather is cloudy or cool, it may remain active even during the day. The bear lives in forests and comes near human habitation, particularly in the autumn season when there may be a dearth of food. Mating takes place in the summer and after a gestation period of 7 months, the female gives birth to cubs in December or January. When they grow sufficiently strong and have a good hold, they are carried by the mother on her back. The mother cares for the young ones for 2-3 years till they attain maturity.In captivity, the sloth bear is known to live up to 40 years.

The sloth bear is omnivorous. It feeds on fruits, berries, roots, insects, termites, sugarcane, maize and in time of dearth, even on carrion. The combs of the domesticated and wild honeybees make their most prized food and they take great pains in climbing up the trees for getting hold of the combs or they may shake the bush or branch of the tree till the combs fall to the ground. The Indian honeybee, which lives in tree hollows, is also attacked by this animal. The domesticated bees kept in modern hives have to be protected as the bear has the habit of turning over the hive in order to reach the comb and honey. Towards autumn, when natural food in the forest is scanty, the bear comes out to the cultivated areas and does considerable damage to sugarcane and the maize crops, destroying more than it actually eats.In that season, it is also likely to encounter human beings whom it may attack.

Management. (*i*) Killing with shot-guns is the only sure method of getting rid of this animal. (*ii*) It is afraid of the human voice and fire or a strong source of light. When moving about late in the evening, people living near jungles walk in groups talking loudly and invariably carry a burning torch or lantern.

7. Blue Bull, *Boselaphus tragocamelus* (Pallas) (Artiodactyla : Bovidae)

The blue-bull or *nilgai* is found in the Indian Sub-continent, from the base of the Himalayas to Karnataka. It is not found in Assam or Bangladesh. The males of this antelope are 1.2-1.4 metres high, have short horns and are horse-like in build with a low hump. There is a tuft of black hair hanging from the throat. Their colour is iron grey with a white ring below each fetlock and two white spots on each cheek. The females are smaller and lighter in colour and are not so ungainly in appearance. The blue-bulls avoid dense forests and roam about in the hills or in the plains covered with grass or patches of scrub where there might be a few trees. Like other antelopes, the bule-bull keeps visiting the same spot to rest. Generally, there is a large accumulation of droppings at such places. They do not drink water for quite some time, even in hot weather, and generally they live in herds of 4-10, comprising bulls, cows and calves. They breed practically in all seasons and the gestation period is 8-9 months.

In many parts of India, people refrain from killing the blue-bulls because of the misconception that it is a close relative of the cow which is considered sacred. The animal breeds unhindered and may become quite a destructive pest of crops. It feeds on freshly fallen flowers of *mohawa* (*Madhuca latifolia*), leaves and fruits of *ber* (*Zizyphus* spp.), grass, and may also freely enter the cultivated fields. It feeds late in the morning or early in the evening and may cause much damage to the wheat crop.

Management. It is rare now that this animal is a pest because man has killed it indiscriminately for sport and food. Thus, like many other species of birds and mammals, its population is dwindling so fast that it may become extinct.The spotted deer or the *chital* has met a similar fate in the Terai region of Uttar Pradesh.

SNAILS AND SLUGS

Snails and slugs are soft-bodied animals belonging to the order Stylommatophora and class Gastropoda of the phylum Mollusca. They are the animals without backbone, having asymmetrical, unsegmented and spirally coiled body. While snails have a well developed shell, slugs have only a rudimentary shell often enclosed in a visceral hump. Snails and slugs are hermaphrodites but there is reciprocal exchange of spermatozoa as they mature before development of eggs. Self-fertilization is prevented. They lack good protection against dehydration, hence they avoid direct sunlight and environments with a low relative humidity. They hide during day time in moist places or under debris and feed mainly at night when the temperature drops and the humidity rises. Snails secrete light yellow slime and slugs secrete colourless slime which becomes silvery after drying.

Snails have been reported to cause serious menace to the cultivated crops in many parts of the world. In India, 1500 species of land snails occur but the number of species of slugs are limited. Among these nine species of snails and 12 species of slugs have been reported as pests of ornamental plants, vegetables, fruits and field crops. The common snail, *Helix* spp., is found in Himachal Pradesh, Uttar Pradesh, Andhra Pradesh, Bihar, Maharashtra and Orissa. Another phytophagous species, the giant African snail, *Achatina fulica* Ferussac, has been reported as a serious pest of fruits, vegetables and ornamental plants in coastal areas of Orissa, West Bengal, Assam, Tamil Nadu and Kerala. The common garden slug, *Laevicaulis alte* Ferussac, has been observed feeding on a number of ornamental plants including balsam, portulaca, pot-marigold, verbena, dahlia, cosmos, narcissus and lily in Punjab and Himachal Pradesh. The black slug, *Filicaulis alte* Ferussac feeds on the seedlings of several economically important plants. Another slug, *Limax* spp. (Fig. 1.4B), occurs all over India.

Life-cycle. The giant African snail, *A. fulica* lays 50-200 eggs once a year on the soil surface or a little below. It lays about 1000 eggs in its life span of 5 years, out of which about 80 per cent are viable and hatch within a week. The shell and body of the snail keep on growing till about 8-12 months. The life span of this snail is 3-5 years.

The garden slug, *L. alte* lays eggs in groups of 6 to 45 in moist soil. Maximum egg laying is observed in the month of September. Eggs are oval, whitish or creamish in colour and transparent, strung along a thread. Incubation period ranges from 9 to 18 days with an average of about 13 days. Adults mature within a period of 240 to 323 days with an average of 271 days and measure 55 mm in breadth and 50 mm in length. They are found in soil under debris and survive during unfavourable conditions like food scarcity and low moisture during summer months. They remain active throughout the year but their severity is noticed in cool and damp situation.

The brown slug, *F. alte* lays dirty creamish white spongy eggs in masses (74-80 eggs/mass) on damp soil in polythene bags containing nursery plants. Newly hatched juveniles resembling adults in colour and appearance tend to remain close to the hatching spot and start feeding immediately. They become mature and start egg laying at the age of about 8-9 months and lay eggs twice a year. The adult is about 8·0–8·5 cm in length, 1·5–2·0 cm in breadth and 7-8 g in weight. Average life span is 390 days with the longest being 567 days.

Damage. Snails and slugs appear as sporadic pests in those places where damp conditions prevail. They may also appear in large number on roads and runways, creating problems during the taking off or the landing of the aircraft. Snails and slugs are polyphagous, feeding on a wide range of host plants. The giant African snail is known to feed on 227 host plants. The slugs have been reported to feed on celery, lettuce, cabbage, tomato and a number of ornamental plants.

Snails and slugs completely devour the small leaves while mature leaves show holes on them or are eaten away around the edge. Thicker leaves are mostly rasped on the lower surface. Mine-like holes and tunnels are bored on tuber, roots and bulbs. Sown seeds of wheat in soil are completely hollowed out starting from the embryo. In case of slug damage to tomato fruit, complete pericarp is eaten away within an overnight period leaving behind the inner core. The damage from feeding by slugs can be quite extensive, resulting in large irregular holes in plant leaves, debarking the stem near ground, cutting stem of seedlings and roots of plants.

Snails and slugs act as carriers of propagules of plant pathogens. Spores of *Alternaria* sp., *Fusarium* sp. and *Phytophthora* sp. have been found in the faeces and the slime. The snail, *A. fulica* is known to transmit black rot disease caused by *Phytophthora palmivora* Butl. on cocoa. The infestation of snails and slugs in the field can be detected by the slimy trails left behind by them as they crawl about. *A. fulica* has been reported to cause economic damage to crops of Cucurbitaceae (46%), Basellaceae (39%) and Dioscoriaceae (35%). *F.alte* is known to cause significant damage to seedlings of marigold (97 %), cabbage (75 %), balsem (69 %), coriander (69 %), portulaca (59 %), zinnia (58 %), sponge gourd (53 %), brinjal (51 %), spinach (48 %) and cauliflower (20 %).

Management. (*i*) Collect snails especially during midnight and both snails and slugs before dawn and after dusk. These should be destroyed in 10 per cent solution of common salt or in boiling water and buried in the fields away from the populated area. (*ii*) Keep area under crops free from weeds, creating a belt of clear land around garden or farm and use barrier strip of dehydrating chemical like common salt, quick lime or copper sulphate. (*iii*) Smooth copper or zinc sheets (0·8 mm thick) can be used as mechanical barriers. The copper sheets must reach a height of 5cm above the soil surface, the zinc 20-25 cm. The upper edge of the barrier is bent through a right angle twice. It should be burried to a depth of about 30cm below the soil surface. (*iv*) Poison bait consisting of 10 per cent carbaryl 50WDP in wheat bran having 6 per cent mango flavour can be used to attract and kill the pests. Offer 2·0 g of this bait on paper pieces to pests at 2 m distance in the evening. Collect dead animals for the next 2-3 mornings and bury them in soil. (*v*) Spray copper sulpate (3%) @ 12kg dissolved in 400 litres of water per ha. (*vi*) Dust 15 per cent metaldehyde @ 50 kg per ha or spray 50 per cent metaldehyde powder @ 10 kg per ha in 500 litres of water per ha or sprinkle 5 per cent metaldehyde pellets around infested fields.

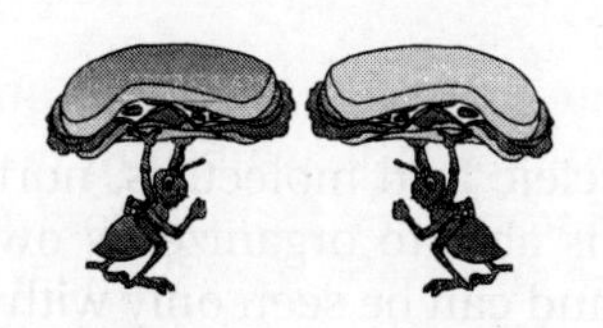

INSECT VECTORS OF PLANT DISEASES

INTRODUCTION

A vector is an organism capable of transmitting pathogens from one host to another. The insects, besides directly damaging the crops, sometimes become responsible for the spread of pathogens in plants. All those insects which acquire the disease causing organisms by feeding on the diseased plants or by contact and transmit them to healthy plants are known as insect vectors of plant diseases. Plant pathogens are transmitted either through contact, by contamination through soil or other biological agencies. Majority of the plant diseases are transmitted by insects and a few by other arthropods like mites and only a small percentage by mechanical means or contamination of the soil.

Insects having both piercing and sucking mouthparts and biting and chewing mouthparts are associated with disease transmission. Most the insect vectors belong to the order Hemiptera (aphids, leafhoppers, whiteflies and mealy bugs), but a few others belong to Thysanoptera (thrips), Coleoptera (beetles), Orthoptera (grasshoppers) and Dermaptera (earwigs). Homopteran insects alone are known to transmit about 90 per cent of the plant diseases. The salient features of homopterans (aphids and leafhoppers), which make them efficient vectors are as follows :

- They make brief but frequent probes with their mouthparts into host plants.
- As the population density reaches a critical level, winged migratory individuals are produced.
- In many species, winged females deposit a few progeny on each of the many plants.
- These insects do not cause wholesome destruction of cells during feeding and viruses require living cells for their subsistence and multiplication.

A number of plant diseases caused by viruses, phytoplasmas, bacteria and fungi are transmitted by insects.

VIRUSES

A virus is a set of one or more nucleic acid molecules, normally encased in a protective coat or coats of protein or lipoprotein that is able to organize its own replication only within suitable host cells. It is ultramicroscopic in size and can be seen only with the aid of an electron microscope. It is spherical or rod-shaped, nucleoproteinaceous (chief constituents being ribonucleic acid

5–35% and proteins 65–95% by weight) in composition and can live and multiply only in living cells. Viruses are responsible for many diseases in man (influenza, measles, mumps, polio, pox, etc.) and plants (mosaic, leaf curl, etc.).

Plant virus diseases have become more prevalent and destructive in recent years. This is mainly because of better recognition of the virus diseases, exchange of plant material from region to region facilitating spread of the virus to new areas, and distribution of many insect vectors in new regions in the world. There are over 850 described plant virus species. About half of the insect vectors are aphids, a third are the leafhoppers. Mealy bugs and whiteflies transmit some viruses, and six are transmitted by thrips. The main aphid vectors are *Myzus persicae* (Sulzer), *Aphis gossypii* Glover and *Aphis craccivora* Koch (Table 31.1). In addition, whitefly, *Bemisia tabaci* (Gennadius) and leafhoppers are also responsible for transmission of plant viruses. Whitefly mostly transmits

Table 31.1 List of important insect vectors of plant viruses

S.No.	Insect vector	Virus(Hosts)
1.	*Acyrthosiphon pisum* (Harris) (Hemiptera : Aphididae)	Alfalfa mosaic (alfalfa), bean mosaic (beans), onion yellow dwarf (onion), pea mosaic (peas), soybean mosaic (beans, peas)
2.	*Aphis craccivora* Koch (Hemiptera : Aphididae)	Alfalfa mosaic (alfalfa), cowpea mosaic (cowpea), onion yellow dwarf (onion), papaya mosaic (papaya)
3.	*Aphis gossypii* Glover (Hemiptera : Aphididae)	Alfalfa mosaic (alfalfa, potato), chilli mosaic (chillies), cucumber mosaic (cucumber), dahlia mosaic (dahlia, zinnia), lettuce mosaic (lettuce, sweet pea), papaya mosaic (papaya), sugarcane mosaic (sugarcane)
4.	*Bemisia tabaci* (Gennadius) (Hemiptera : Aleyrodidae)	Cotton leaf curl (cotton), dolichos yellow mosaic (dolichos), okra yellow vein mosaic (okra), papaya leaf curl (papaya), sesame leaf curl (sesame), tobacco leaf curl (tobacco), tomato leaf curl (tomato)
5.	*Laodelphax striatella* Fallen (Hemiptera : Delphacidae)	Rice streaked dwarf (rice), rice stripe (rice)
6.	*Myzus persicae* (Sulzer) (Hemiptera : Aphididae)	Alfalfa mosaic (alfalfa), beet yellows (spinach, sugarbeet), cabbage black ring spot (cabbage), cauliflower mosaic (cabbage, cauliflower), cucumber mosaic (cucumber), dahlia mosaic (calendula, dahlia, zinnia), lettuce mosaic (garden pea, lettuce, sweat pea), onion yellow dwarf (onion), pea mosaic (broadbean, garden pea, sweet pea), potato virus Y (potato, tobacco, tomato), soybean mosaic (soybean), sugarcane mosaic (maize, sorghum, sugarcane)
7.	*Nephotettix nigropictus* (Stal), *N. virescens* (Distant) (Hemiptera : Cicadellidae)	Tungro (rice), yellow-orange leaf (rice)
8.	*Nilaparvata lugens* (Stal) (Hemiptera : Delphacidae)	Grassy stunt (rice), ragged stunt (rice)
9.	*Pentalonia nigronervosa* Coquerel (Hemiptera : Aphididae)	Banana bunchy top (banana), cardamom mosaic (cardamom)
10.	*Rhopalosiphum maidis* (Fitch) (Hemiptera : Aphididae)	Barley yellow dwarf (barley, oat, rye, wheat), onion yellow dwarf (onion), sugarcane mosaic (maize, sorghum, sugarcane)

Source : Modified after Gillot (2005)

mosaics and leaf curls in pulses, vegetables and other crops like cotton, tobacco and papaya. The leaf- and planthoppers transmit tungro, yellow-orange leaf, grassy stunt and ragged stunt in rice. Tomato spotted wilt is known to be transmitted by thrips. Mandibulate insects like grasshoppers, earwigs and chrysomelid beetles transmit turnip yellow mosaic. Several species of mites are also responsible for transmission of viruses of cereals and fruit crops (Table 31.2).

Table 31.2 List of important mite vectors of virus diseases of plants

S.No.	Mite vector	Virus	Host(s)
1.	*Abacarus hystrix* (Nalepa) (Acari : Eriophyidae)	Agropyron mosaic	Wheat, switch grass
2.	*Aceria ficus* (Corte) (Acari : Eriophyidae)	Fig mosaic	Fig
3.	*Aceria tulipae* (Keifer) (Acari : Eriophyidae)	Wheat streak mosaic	What, oats, barley, maize
4.	*Aculus fockeui* (Nalepa & Trouessart) (Acari : Eriophyidae)	Prunus necrotic ring spot	Plum, peach, cherry
5.	*Eriophyes inaequalis* Wilson & Oldfield (Acari : Eriophyidae)	Cherry leaf mottle	Sweet cherry
6.	*Eriophyes insidiosus* Keifer & Wilson (Acari : Eriophyidae)	Peach mosaic	Peach, nectarine

Source : Gillot (2005)

Types of Viruses

On the basis of the method of transmission and persistence in the vector, viruses may be classified into three categories :

(*i*) **Non-persistent viruses.** These are those viruses which are believed to be transmitted as contaminants of the mouthparts. Such viruses are also called stylet-born viruses and the type of transmission is mechanical. The vector is able to acquire the virus from a disease source and transmit to a healthy plant by feeding for a few seconds. These viruses do not persist longer within the insect vectors which can transmit them soon after feeding on infected plant but the ability to transmit fresh infection soon disappears after the insect feeds on healthy or immune plants. The efficiency of transmission of non-persistent virus is greatly affected by modifying the time of feeding and by starving the vectors before and after feeding. Aphids are the vectors of a great majority of such viruses which are carried only in their stylets.

The following are the main features of the non-persistent viruses :

- Vectors are optimally infective when they have fed for approximately 30 seconds on the infected plant.
- Transmission is improved if vector is starved for a period before an infection feed.
- If the vector is starved after an acquisition, it begins to lose ability to transmit within 2 minutes.
- After acquisition feeding, infectivity is rapidly lost when the vectors feed on healthy plants.

(*ii*) **Semi-persistent viruses.** These viruses are carried in the anterior regions of the gut of a vector, where they may multiply to a certain extent. Vectors do not normally remain infective after a molt, presumably because the viruses are lost when the foregut intima is shed. Several of the leafhopper transmitted viruses fall under this category.

(*iii*) **Persistent viruses.** Persistent viruses are those that persist longer within the infective agent, *i.e.* vector. These viruses, when acquired by a vector, pass through the midgut wall to the salivary glands from where they can infect new hosts. In case of these viruses, the insect has to feed on the source of virus for comparatively longer periods. The insect, after such acquisition of virus, becomes infective only after a certain period, ranging from several hours to 10-20 days, which is called the *incubation period* or *latent period.* Such viruses may multiply within tissues of a vector, which retains the ability to transmit the virus for several days and in some instances the rest of its life. Therefore, the vector need not feed on the virus source again and again to retain its infective capacity.

Thus, the vector insect feeds on the diseased plant (acquisition feed), requires some time after acquisition feed to transmit the virus (latent or incubation period), feeds on healthy plant (inoculation feed) and in the process transmits the virus acquired earlier. Such viruses are also called circulative or circulative-propagative viruses and the type of transmission as non-mechanical. Many of the leafhopper transmitted viruses belong to this category.

Mechanism of Transmission

For inoculation of virus into a plant by sucking insects, the puncture is initiated by a number of forward and backward movements of the inner pair of stylets. During the forward movements, the fluid flows into them, during the backward movements, saliva is ejected. Generally, an insect injects by feeding on any part of the plant, but in some cases the virus is only found in the phloem and has to be injected into the phloem, the movement of which is perhaps controlled by the pH gradient between the mesophyll and the phloem. Some viruses are concentrated in the epidermal cells and others in the mesophyll or xylem.

The mandibulate insects like grasshoppers and beetles regurgitate during feeding. The regurgitated fluid containing the virus is brought into contact with the healthy plant, thus transmitting the virus.

Virus-Vector Relationship

Irrespective of the type of transmission, virus-vector relationship is highly specific. Generally, one type of virus disease is transmitted only by insects belonging to one particular group, *i.e.* mosaics by aphids and leaf curls by whiteflies. In case of leafhoppers, among 110 species known to be vectors, about 100 species transmit only one virus. Similarly, there are viruses which are transmitted by a particular species of an insect and not by others of the same genus. For instance, cabbage ring spot is transmitted only by *M. persicae* and not by *M. ornatus.* A vector can also acquire and transmit more than one virus to the respective hosts. For example, the aphid, *Pentalonia nigronervosa* Coquerel, transmits banana bunchy top and cardamom mosaic. Similarly, the whitefly, *Bemisia tabaci* (Gennadius) transmits okra yellow vein mosaic, dolichos yellow mosaic, tomato leaf curl, papaya leaf curl, etc. Onion yellow dwarf is known to be transmitted by 60 insect vectors.

For transmission of viruses, activity of insect vectors is more important rather than their number. In case of aphids, it is the activity and number of migrant insects that is important in the efficiency of virus transmission rather than the number of apterous individuals which are, of course, important in respect of their direct injury to the crop.

PHYTOPLASMAS

Phytoplasmas (originally called mycoplasma-like organisms) are non-culturable degenerate gram-positive prokaryotes closely related to mycoplasmas and spiroplasmas. They are without a visible cell wall, whose place is taken by a thin elastic cytoplasmic membrane which cannot withstand osmotic pressure. Phytoplasmas are pleomorphic and may be spherical or oval, varying from 80 to 800µ in diameter.

Phytoplasmas are important insect-transmitted pathogenic agents causing more than 700 diseases in plants. The single most successful order of insect phytoplasma vectors is the Hemiptera.

This group collectively possesses several characteristics that make its members efficient vectors of phytoplasmas.

- They are hemimetabolous, thus nymphs and adults feed similarly and are in the same physical location-often both immatures and adults can transmit phytoplasmas.
- They feed specifically and selectively on certain plant tissues, which makes them efficient vectors of pathogens residing in these tissues.
- Their feeding is nondestructive, promoting successful inoculation of the plant vascular system without damaging the conductive tissues and eliciting defensive responses.
- They have a propagative and persistent relationship with phytoplasmas. They have obligate symbiotic prokaryotes that are passed to the offspring by transovarial transmission, the same mechanisms that allow the transovarial transmission of phytoplasmas.

Mechanism of Transmission

Phytoplasmas are phloem-limited, therefore, only phloem-feeding insects can potentially acquire and transmit the pathogen. Most phytoplasma vectors are members of the family Cicadellidae (**Table 31.3**). Phloem-feeding insects acquire phytoplasmas passively during feeding in the phloem of infected plants. The feeding duration necessary to acquire a sufficient titer of phytoplasma (acquisition access period), may range from a few minutes to several hours, the longer the period, the greater the chance of acquisition. The time that elapses from initial acquisition to the ability to transmit the phytoplasmas (latent period or incubation period) is temperature dependent and ranges from a few to 80 days.

Table 31.3 List of important vectors of phytoplasma-transmitted diseases

S.No.	Insect vector	Mycoplasma(s)	Host(s)
1.	*Cestius phycitis* (Distant)(Hemiptera : Cicadellidae)	Brinjal little leaf	Brinjal
2.	*Macrosteles quadrilineatus* Forbes (Hemiptera : Cicadellidae)	Aster yellows	Aster, barley, carrot, celery, cucumber, wheat
3.	*Nephotettix nigropictus* (Stal), *N. virescens* (Distant) (Hemiptera : Cicadellidae)	Yellow dwarf	Rice
4.	*Orosius albicinctus* Distant (Hemiptera : Cicadellidae)	Sesame phyllody	Sesame
5.	*Recilia dorsalis* (Motschulsky) (Hemiptera : Cicadellidae)	Yellow dwarf	Rice
6.	*Scaphytopius acutus* (Say), *S. irroratus* (Orthoptera : Gryllotalpidae)	Aster yellows	Aster, barley, carrot, celery, cucumber, wheat
7.	*Stephanitis typica* (Distant) (Hemiptera : Tingidae)	Coconut root wilt	Coconut

Source : Modified after Gillot (2005)

During the latent period, phytoplasmas move through and replicate in the vector's body. They can pass intracellularly through the epithelial cells of the midgut and replicate within a vesicle or they can pass between two midgut cells and through the basement membrane to enter the hemocoel. Phytoplasmas circulate in the haemolymph, where they may infect other tissues such as the Malpighian tubules, fat bodies and brain or reproductive organs. The replication in these tissues, albeit not essential for transmission, may be indicative of a longer coevolutionary relationship between host and pathogen. To be transmitted to plants, phytoplasmas must penetrate specific cells of the salivary glands and high levels must accumulate in posterior acinar cells of the salivary gland before they can be transmitted.

Vector-Phytoplasma Relationship

The interaction between insects and phytoplasmas is complex and variable. The complex sequence of events required for an insect to acquire and subsequently transmit phytoplasmas to plants suggests a high degree of specificity of phytoplasmas to insects. However, numerous phytoplasmas are transmitted by several different insect species. In addition, a single vector species may transmit two or more phytoplasmas, and an individual vector can be infected with dual or multiple phytoplasma strains. Vector-host plant interactions also play an important role in determining the spread of phytoplasmas. Polyphagous vectors have the potential to inoculate a wider range of plant species, depending on the resistance to infection of each host plant.

It has been found that leafhoppers are not able to acquire equally phytoplasmas from different infected plant species. Chrysanthemum yellows (CY) phytoplasma is successfully transmitted by three leafhoppers, viz. *Euscelidius variegatus, Macrosteles quadripunctulatus* and *Euscelis incisus.* All three species acquire from and transmit to CY-infected chrysanthemum and uninfected chrysanthemum, respectively. However, only *M. quadripunctulatus* and *E. variegatus* acquire CY after feeding on CY-infected periwinkle and subsequently transmit CY to uninfected plants. None of the leafhoppers acquire the phytoplasma from CY-infected celery, a dead-end host. *Dead-end hosts* are plants that can be inoculated and subsequently become infected with phytoplasma, but from which insects can not acquire phytoplasma.

BACTERIA

Bacteria are microscopic single celled organisms increasing by fission, they have a cell membrane, a rigid cell wall and often one or more flagella. Of the total about 1800 known bacterial species, most are saprophytes living on dead plant or animal tissues or organic wastes. There are about 200 species of bacteria which are parasitic on plants and many of them consisting of numerous pathovars. Bacterial diseases fall into three categories : (*i*) Wilting, due to invasion of the vascular system or water-conducting vessels, e.g. cucumber wilt. (*ii*) Necrotic blights, rots and leaf spots, where the parenchyma is killed, e.g. fire blight. (*iii*) Hyperplasia or over growth, e.g. crown gall.

Pathogenic bacteria apparently cannot enter plants directly through unbroken cuticle but get in through insect or other wounds, stomata, hydathodes, lenticels and flower nectaries. The fact that the plant pathogenic bacteria are unable to penetrate plant tissue without a court of entry has led to the realization of significance of the role of insects in transmission. The insect contributes through feeding and oviposition wounds, as a mechanical carrier of the organism on its body and in some cases, by virtue of a mutualistic relationship between the organism and the insect which insures a continuing association among pathogen, insect and the host plant.

A number of plant diseases caused by bacteria are known to be transmitted by insects **(Table 31.4).** Fire blight of apple and pear, caused by *Erwinia amylovora* is carried by aphids, leafhoppers, etc. Potato blackleg, caused by *Erwinia carotowora,* is transmitted by seedcorn maggot, *Hylemyia cilicrura* (Rondani). Bacterial wilt of cucurbits, caused by *Erwinia tracheiphila* is transmitted by cucumber beetle, *Diabrotica duodecimpunctata* (Olivier). Bacterial wilt of maize, caused by *Xanthomonas stewarti* is transmitted by the flea beetle, *Chaetocnema pulicaria* (Meisheimer). Black rot of crucifers, caused by *Xanthomonas campestris* is transmitted by several insects and slugs.

FUNGI

Fungi are organisms having no chlorophyll, reproducing by sexual and asexual spores, not by fission like bacteria and typically possessing a mycelium or mass of interwoven threads (hyphae) containing well marked nuclei. There are about 4300 valid genera of fungi and about 70,000 species living as parasites or saprophytes on other organisms or their residues. More than 8,000 species are known to cause plant diseases.

Table 31.4 List of major vectors of bacterial diseases of crop plants

S.No.	Insect vector	Disease	Host(s)
1.	*Chaetocnema pulicaria* (Meisheimer), *C. denticulata* (Coleoptera : Chrysomelidae)	Bacterial wilt (or Stewart's bacterial wilt)	Maize
2.	*Diabrotica duodecimpunctata* (Olivier), *D. vittata* (Coleoptera : Chrysomelidae)	Cucurbit wilt	Cucumber, muskmelon
3.	*Diaphorina citri* Kuwayama (Hemiptera : Aphalaridae)	Citrus canker	Citrus
4.	*Hylemya cilicrura* (Rondani), *H. trichodactyla* (Rondani) (Diptera : Anthomyiidae)	Potato blackleg	Potato
5.	Wide range of species of bees, wasps, flies, ants and aphids	Fire blight	Apple, pear, quince, etc.

Source : Modified after Gillot (2005)

There are several insects associated with the spread of fungal diseases **(Table 31.5)**. Many flies mechanically transmit the ergot of cereals caused by *Claviceps purpurea*. The ergot disease of *bajra*, caused by *Sphacelia microcephala*, is mechanically carried by insects that visit the flowers attracted by the sugary secretion found on the fungus infected earheads. The cotton wilt, caused by *Fusarium vasinfectum*, is transmitted through the faecel pellets of many grasshoppers like *Melanoplus differentialis* (Thomas), after they have fed upon infected plants. The common sooty mould fungus (*Capnodium* spp.) grows on the honeydew excreted by several homopteran insects like aphids, leafhoppers, mealy bugs, whiteflies, etc.

Table 31.5 List of principal vectors of fungal diseases of crop plants

S.No.	Insect vector(s)	Disease	Host
1.	*Hylurgopinus rufipes* (Eichhoff) (Coleoptera : Scolytidae)	Dutch elm disease	Elm
2.	*Melanoplus differentialis* (Thomas) (Orthoptera : Acrididae)	Cotton wilt	Cotton
3.	*Scolytus multistriatus* (Marsham), *S. scolytus* (Fabricius) (Coleoptera : Scolytidae)	Dutch elm disease	Elm
4.	About 40 species of insects especially flies, beetles and aphids	Ergot of cereals	Cereals
5.	Several species of insect visiting flowers	Ergot of bajra	Bajra

Source : Modified after Gillot (2005)

MANAGEMENT OF VECTORS AND DISEASES

Among the various types of plant diseases transmitted by insects, virus diseases are considered to be the most serious. Hence a multi-pronged strategy needs to be adopted to manage the vectors and virus diseases. Some of the important components of such a strategy should involve selection of healthy seed, cultural practices, biological measures, resistant varieties and use of chemicals.

Healthy Seed

Management of virus diseases starts with obtaining healthy seed, cuttings or plants. Care should be taken to obtain only certified seed, *i.e.* seed obtained from the plants which have been inspected during growing season and found free of certain diseases. Virus-free foundation stock can be built up by heat treatment, *i.e.* growing plants at high temperatures for weeks or even months. The production of virus-free stocks can also be achieved by taking advantage of the fact that some plants grow and elongate faster than the virus can occupy the new tissue. Therefore, the virus can be eliminated by using meristem or tip cultured plants. Virus free stock is tested by indexing (growing a part of the cutting or plant in a pot or greenhouse and recording its condition with respect to disease symptoms), bioassays and/or serological assays.

Cultural Control

Several cultural practices have proved to be helpful in reducing the incidence of vectors and vector-borne diseases. Intercropping with a barrier crop has provided encouraging results to reduce the incidence of several diseases. For example, the incidence of yellow vein mosaic of okra is reduced by intercropping with soybean. Similarly, intercropping of tomato with coriander and lobiabean reduces the incidence of tomato leaf curl virus in tomato. Plant spacing such as close spacing reduces the incidence of French bean crinkle stunt disease. Manipulation in planting dates is another way of reducing the disease incidence. Rogueing also helps the removal of the source of disease causing organism. Removal of weeds and alternate hosts of viruses and vectors helps to reduce the incidence of diseases.

Resistant Varieties

Growing resistant/tolerant varieties is another effective way of managing vectors and vector-transmitted diseases. A number of genotypes have been identified under various All India Coordinated Research Projects, sponsored by the Indian Council of Agricultural Research and various State Agricultural Universities, which have resistance against virus borne diseases in pulses, tomato, cotton, etc.

Biopesticides

The use of biopesticides such as parasitoids and predators, microbials and plant extracts is an eco-friendly approach to manage the vectors and vector born diseases. A fungus, *Paecilomyces farinosus* has been found parasitic on *Bemisia tabaci* (Gennadius). Several neem-based formulations have provided effective control of *B. tabaci* on cotton. Aqueous extracts of leaves of *Clerodendron frageans* and *Aerva anguinolenta* and roots of *Boerhavia diffusa*, sprayed at 4 per cent concentrations at 3-4 days starting from germination, was found to reduce yellow mosaic incidence in blackgram and mungbean.

Chemical Control

The control of insect vectors by application of insecticides appears to be a difficult task as few survivors would be able to transmit the disease. Still insecticidal control of insect vectors is the most practicable method of control of plant viruses. The timely application of insecticides restricts the spread of the disease by reducing the vector population. Several systemic and non-systemic insecticides have been reported to control the insect vectors. The prominent insecticides reported to be effective are sprays of oxydemeton methyl, malathion, dimethoate, carbaryl, dichlorvos, fenitrothion, phosphamidon, monocrotophos, triazophos and ethion in doses ranging from 300 ml to 1·5 litres per ha. The soil application of carbofuran and phorate granules @ 10-12 kg per ha has also proved useful.

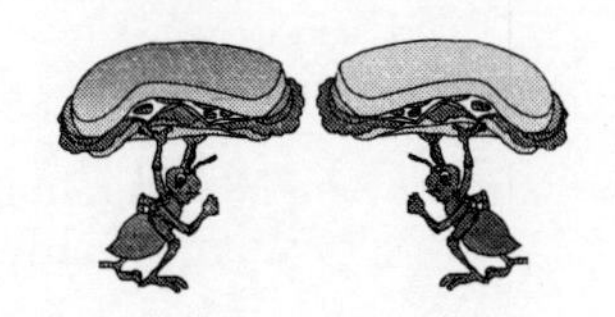

PART-IV

PEST MANAGEMENT IN GLOBAL PERSPECTIVE

- Climate Change and Pest Management
- Pest Management: Global Scenario
- Pest Management: Future Perspectives

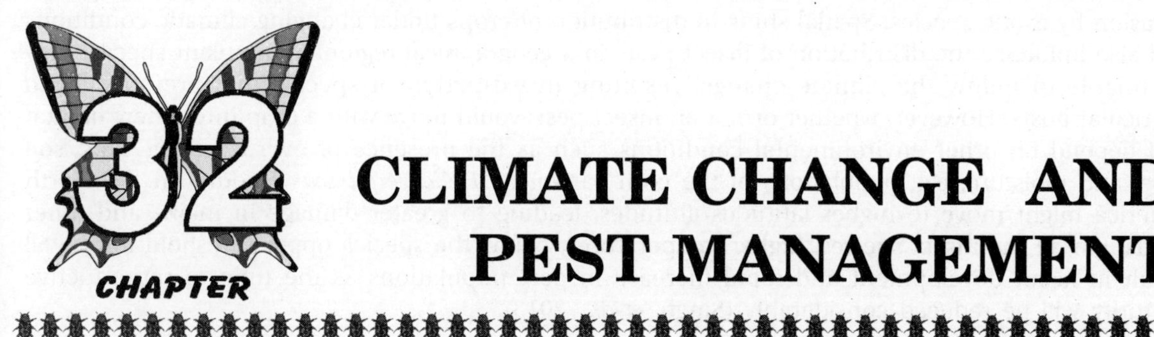

CHAPTER 32 CLIMATE CHANGE AND PEST MANAGEMENT

INTRODUCTION

The earth's climate has always been subject to natural variations with shifts between cold and warm. However, this is the first instance when man appears to accelerate the change enabling warming to take place quickly. The things that normally happen during geological time are happening during the span of human life, thus leaving limited time for species to adapt and adjust, and avoid extinction. Since different species adapt to climate change in different ways, the natural cycles of interdependent species tend to fall out of synchrony. Habitats are changing along with the timing of annual processes like flowering, leaf coverage, migration and birth. Uncoupling of certain associations and establishment of new ones are in the offing, as species respond to changing climatic conditions. Loss of genetic variability in the newly established host ranges tends to influence the ability of insect populations to adapt to changing climatic conditions involving new food plants, competitors and natural enemies.

The key factors in changing earth's climate are the greenhouse gases (mainly CO_2) released by industry, agriculture, automobiles and coal fired electrical generation plants. The concentration of CO_2 in the atmosphere helps to determine the earth's surface temperature, and CO_2 and temperature have risen sharply during the 20th century. The global temperature had risen by only 4°C over the last 18,000 years. However, it has increased by about 0.5-1.0°C since the last century, when weather records started being maintained on the basis of data collected from about 2,000 weather stations located around the world. Similarly, the concentration of CO_2 has risen from about 280 ppm since the early 19th century to about 360 ppm by the end of 20th century. The current estimates of changes in climate indicate an increase in global mean annual temperatures of 1°C by 2025 and 3°C by the end of 21st century. If the present emission trend continues, the level of CO_2 in atmosphere could reach up to 500 ppm by 2050 and 700 ppm in 2100. Such changes in the climate would drastically affect insects in terms of geographical distribution, over wintering, population growth, number of generations per annum, length of growing season, crop-pest synchronization, intraspecific interaction, dispersal, migration and availability of host plants and refugia. The losses due to insect damage are likely to increase as a result of changes in crop diversity and increased incidence of insect pests due to global warming (Dhaliwal *et al.,* 2013).

GEOGRAPHICAL DISTRIBUTION OF INSECTS

Climate change will have major effect on geographical distribution of many insect pests, and low temperatures are often more important than high temperatures in determining geographical distribution of insect pests. Increasing temperatures may result in a greater ability to overwinter in insect species limited by low temperatures at higher latitudes, extending their geographical range and sudden outbreaks of insect pests can wipe out certain crop species, and also encourage the invasion by exotic species. Spatial shifts in distribution of crops under changing climatic conditions will also influence the distribution of insect pests in a geographical region. Some plant species may be unable to follow the climate change, resulting in extinction of species that are specific to particular hosts. However, whether or not an insect pest would move with a crop into a new habitat will depend on other environmental conditions such as the presence of overwintering sites, soil type, and moisture, e.g. populations of the corn earworm, *Helicoverpa zea* (Boddie) in the North America might move to higher latitudes/altitudes, leading to greater damage in maize and other crops. For all the insect species, higher temperatures, below the species upper threshold limit, will result in faster development and rapid increase in pest populations as the time to reproductive maturity will be reduced considerably (Singh *et al.,* 2013).

In addition to the direct effects of temperature changes on development rates, improvement in food quality due to abiotic stress may result in dramatic increases in growth of some insect species, while the growth of certain insect pests may be adversely affected. Pest outbreaks are more likely to occur with stressed plants as a result of weakening of plants' defensive system, and thus, increasing the level of susceptibility to insect pests.

Global warming will lead to early infestation by *H. zea* in North America, and *Helicoverpa armigera* (Hubner) in North India, resulting in increased crop loss. Rising temperatures are likely to result in availability of new niches for insect pests. Temperature has a strong influence on the viability and incubation period of *H. armigera* eggs. Egg incubation period can be predicted based on day degrees required for egg hatching, which decreases with an increase in temperature from 10 to 27 °C, and egg age from 0 to 3 days. An increase of 3°C in mean daily temperature would cause the carrot fly, *Delia radicum* (Linnaeus) to become active a month earlier than at present, and temperature increases of 5 to 10°C would result in completion of four generations each year, necessitating adoption of new pest control strategies. An increase of 2°C will reduce the generation turnover of the bird cherry aphid, *Rhopalosiphum padi* (Linnaeus) by varying levels, depending on the changes in mean temperature. An increase of 1° to 3°C in temperature will cause northward shifts in the potential distribution of the European corn borer, *Ostrinia nubilalis* (Hubner) up to 1,220 km, with an additional generation in nearly all regions where it is currently known to occur (Sharma, 2010).

Overwintering of insect pests will increase as a result of climate change, producing larger spring populations as a base for a build-up in numbers in the following season. These may be vulnerable to parasitoids and predators if the latter also overwinter more readily. Diamondback moth, *Plutella xylostella* (Linnaeus) overwintered in Alberta in 1994, and if overwintering becomes common, the status of this insect as a pest in North America will increase dramatically. Many insects such as *Helicoverpa* spp. are migratory and, therefore, may be well adapted to exploit new opportunities by moving rapidly into new areas as a result of climate change. There may also be increased dispersal of airborne insect species in response to atmospheric disturbances.

Conditions are more favourable for the proliferation of insect pests in warmer climates. Longer growing seasons will enable insects such as grasshoppers to complete a greater number of reproductive cycles during the spring, summer, and autumn. Warmer winter temperatures may also allow larvae to winter-over in areas where they are now limited by cold, thus causing greater infestation during the following crop season. Altered wind patterns may change the spread of both wind-borne pests and of the bacteria and fungi that are the agents of crop diseases. Crop-pest interactions may shift as the timing of developmental stages in both hosts and pests are altered. Some of the insect pests which are likely to assume serious proportions due to the changing climate and cropping patterns include American bollworm, *Helicoverpa armigera* (Hubner); tobacco caterpillar, *Spodoptera litura* (Fabricius); cotton whitefly, *Bemisia tabaci* (Gennadius); brown planthopper, *Nilaparvata lugens* (Stal); gall midge, *Orseolia oryzae* (Wood-Mason); pink stem borer, *Sesamia inferens* (Walker); wheat aphid, *Sitobion miscanthi* (Takahashi); pyrilla, *Pyrilla perpusilla* (Walker); sugarcane woolly aphid, *Ceratovacuna lanigera* Zehntner; serpentine leaf miner, *Liriomyza trifolii* (Burgess); diamondback moth, *Plutella xylostella* (Linnaeus); rhinoceros beetle, *Oryctes rhinoceros* (Linnaeus) and tea mosquito bug, *Helopeltis antonii* Signoret. The possible increases in pest infestations may bring about greater use of chemical pesticides to control them, a situation that will require the further development and application of integrated pest management techniques (Dhaliwal *et al.*, 2010).

EFFECT OF TEMPERATURE ON INSECTS

The rate at which most insect pests develop is dependent on temperature as insects are poikilothermic. Thus, temperature is probably the single most important environmental factor influencing insect behaviour, distribution, development, survival and reproduction and every species has a particular 'threshold temperature' above which development can occur. It has been estimated that with a 2°C temperature increase, insects might experience 1-5 additional life cycles per season. This effect may be most noticeable in insects with short life cycles such as aphids and diamondback moth.

It was predicted that *Chilo suppressalis* (Walker), which is presently confined to the easternmost part of Hokkaido Island, would occur throughout Hokkaido and produce two generations a year after 2°C warming. The outbreaks of *C. suppressalis* are taking place in association with cool and wet summers or in July, because of the high survival rate of the first generation larvae. The expansion of its distribution ranges has usually occurred though increased density mediated by increased reproduction. As long as *C. suppressalis* remains at the current low population density under the conditions of global warming, the northward range expansion of this species would scarcely be expected.

Very low and very high temperatures will be critical in determining the geographical distribution of *H. armigera*. Climate change will possibly shift the northern boundary of *H. armigera* by 500-700 km, while the southern edge of ranges may have little effect as this pest is currently distributed throughout Australia and Pacific islands, but with little chance to invade the Antarctica. The abrupt climate change during cropping season or in the preceding period triggers the outbreak of *H. armigera*. The off season rains during May-June in north India promote the growth of alternate hosts. It also helps in the carryover and population increase during off seasons. In Andhra Pradesh, *H. armigera* outbreaks have been recorded in years experiencing un-seasonal rains/hurricanes during November-December. Studies suggest that temperature increases may extend the geographical range of some insect pests currently limited by temperature. The European corn borer is a major pest of grain maize in many parts of the world. It is multivoltine (multigenerational) and, depending on climatic conditions, can produce up to four generations per year. Using degree-day (thermal) requirements, the potential distribution of the European corn borer in Europe has been mapped under present (1951-80) temperatures. With a 1°C increase in temperature a northward shift in distribution of between 165 and 500 km is indicated for all generations. In addition to favourable climatic conditions, the distribution of any pest is dependent on the availability of a host plant. The potential limit of grain maize cultivation is also likely to shift northwards with increasing temperatures providing suitable conditions for the European corn borer. This example serves to highlight the need to examine crop-pest interactions in any climate impact assessment.

Under a warmer climate at mid-latitudes, there would be an increase in the overwintering range and population density of a number of important agricultural pests, such as the potato leafhopper which is a serious pest of soybeans and other crops in the USA. Assuming planting dates did not change, warmer temperatures would lead to invasions of pest earlier in the growing season (i.e. at more susceptible stages of plant development) and probably lead to greater damage to crops. In the US corn belt, increased damage to soybeans is also expected due to earlier infestation by the corn earworm, which could result in serious economic losses. Temperature increases already have caused changes in species diversity and distribution. For example, the mountain pine beetle, a major forest pest in the United States and Canada, has extended its range northward by approximately 300 kilometers with the temperature increase of approximately 1.9°C. Temperature has shown profound effect on some biological parameters of insects.

EFFECT OF CO_2 ON INSECTS

Insect host plant interactions will change in response to the effects of CO_2 on nutritional quality and secondary metabolites of the host plants. Increased levels of CO_2 will enhance plant growth, but may also increase the damage caused by some phytophagous insects. In the enriched CO_2 atmosphere expected in the next century, many species of herbivorous insects will confront less nutritious host plants that may induce both lengthened larval developmental times and greater mortality.

A gradual continual rise in atmospheric CO_2 will affect pest species directly (CO_2 fertilization effects) and indirectly (via interactions with other environmental variables). Generally, elevated CO_2 levels induce increased consumption rates in insect herbivores. The CO_2 effects on insects are usually indirect in terms of insect damage that results from changes in the host crop. Among the probable effects of elevated atmospheric CO_2 are changes in plant nitrogen balance that will adversely affect host plant quality for many herbivorous insects. Plants grown in elevated CO_2 typically have lower N content per unit of plant tissue. This results in higher C/N ratio. Insects that feed on plants with lower N per unit of plant tissue generally respond by increasing consumption, but may still suffer longer developmental times and higher mortality. In atmospheres experimentally enriched with CO_2, the nutritional quality of leaves declined substantially due to dilution of nitrogen by 10 to 30 per cent.

Increased CO_2 may also cause a slight decrease in nitrogen-based defenses (e.g. alkaloids) and a slight increase in carbon-based defenses (e.g. tannins). Acidification of water bodies by carbonic acid (due to high CO_2) will also affect the floral and faunal diversity. Lower foliar nitrogen content due to CO_2 causes an increase in food consumption by the herbivores up to 40 per cent, while unusually severe drought increases the damage by insect species such as spotted stem borer, *Chilo partellus* (Swinhoe) in sorghum. Cotton is a good example of a plant with phenolics that can reduce insect feeding. Elevated carbon dioxide levels allow many plant species to greatly increase their carbon-based defenses. For example, potatoes and plums are plants containing defensive compounds based on nitrogen.

Larval growth and development of *Spodoptera litura* (Fabricius) on groundnut plants subjected to elevated (550-750 ppm CO_2) and reduced (350 ppm in chamber and open conditions) levels of CO_2 revealed that the larvae consumed significantly higher quantity of foliage under elevated CO_2 than under reduced and ambient conditions. An increase of nearly two days in larval duration was observed with elevated CO_2 level. Rising carbon dioxide will increase the carbon-to-nitrogen balance in plants, which in turn will affect insect feeding, concentrations of defensive chemicals in plants, compensation responses by plants to insect herbivory, and competition between pest species.

PEST MANAGEMENT STRATEGIES

Host plant resistance. microbial pesticides, natural enemies, and synthetic chemicals are some of the potential options for integrated pest management. However, the relative efficacy of many of these pest control measures is likely to change as a result of influence of global warming on extension of geographical range of insect pests, increased over-wintering and rapid population growth, changes in insect – host plant interactions, increased risk of invasion by migrant pests, impact on arthropod diversity and extinction of species, changes in synchrony between insect pests and their crop hosts, introduction of alternative hosts as green bridges, and reduced effectiveness of crop protection technologies.

Host Plant Resistance

Host plant resistance to insects is one of the most environment friendly components of pest management. However, climate change may alter the interactions between the insect pests and their host plants. Global warming may also change the flowering times in temperate regions, leading to ecological consequences such as introduction of new insect pests, and attaining of pest status by non-pest insects. However, many plant species in tropical regions have the capability to withstand the phenological changes as a result of climate change. Global warming may result in breakdown of resistance to certain insect pests. Sorghum varieties exhibiting resistance to sorghum midge, *Stenodiplosis sorghicola* (Coquillett), in India become susceptible to this pest under high humidity and moderate temperatures near the Equator in Kenya. There will be increased impact on insect pests which benefit from reduced host defenses as a result of the stress caused by the lack of adaptation to sub-optimal climatic conditions. Chemical composition of some plant species changes in direct response to biotic and abiotic stresses; as a result, their tissues become less suitable for growth and survival of insect pests. However, problems with new insect pests will occur if climatic changes favour the introduction of insect susceptible cultivars or crops. The introduction of new crops and cultivars to take advantage of the new environmental conditions is one of the adaptive methods suggested as a possible response to climate change (Sharma, 2010).

Insect-host plant interactions will change in response to the effect of CO_2 on nutritional quality and secondary metabolites of the host plants. Plants, grown under low-nutrient conditions, do have higher concentrations of carbon-based allelochemicals than plants grown under high-nutrient conditions. Host plants growing under enriched- CO_2 environments exhibited significantly larger biomass (+38.4%), increased C/N ratio (+26.57%), and decreased nitrogen concentration (-16.4%), as well as increased concentrations of tannins (+29.9%) and other phenolics (Heagle, 2003). In contrast to the C/N balance hypothesis, plants grown in elevated (700 ppm) CO_2 conditions had similar, or lower, concentrations of carbon-based allelochemicals than plants grown in ambient (350 ppm) CO_2 conditions. Larvae fed with foliage grown in elevated CO_2 with low N fertilization consumed significantly more plant material than insects fed with foliage grown in ambient CO_2; but, again, no differences were observed with high N fertilization and the insects fed on low N plants had significantly higher mortality in elevated CO_2.

The production of the nitrogen-based toxin was affected by an interaction between CO_2 and N; elevated CO_2 decreased N allocation but the reduction was largely alleviated by the addition of nitrogen, thus indicating that future expected elevated CO_2 concentrations by climate change, alter plant allocation to defensive compounds and have enough impact on plant-herbivore interactions. Increases of carbon-defensive compounds by elevated CO_2 or low N availability or both, adversely affected growth and survival of *Spodoptera exigua* (Hubner) in Bt cotton. It was observed that feeding guild, in which some species have shown increases in population density in elevated CO_2, are the phloem feeders. It is likely that climate change will not minimize the outbreaks; on the contrary, it might benefit some pests, which might increase the consumption of pesticide. Chewing insects have shown no change or reduction in abundance, though relative abundance may be greatly affected since compensatory feeding is common in these groups.

Densities of leaf miner species on host species were lower in every year in elevated CO_2 than they were in ambient CO_2. The results showed that elevated CO_2 significantly decreased herbivore abundance (-21.6%), increased relative consumption rates (+165%), development time (+3.87%) and total consumption (+9.2%), and significantly decreased relative growth rate (-8.3%), conversion efficiency (-19.9%) and pupal weight (-5.03%). No significant differences were observed among herbivore guilds. To the contrary, thrips population size was not significantly affected by CO_2, but laminar area scraped by thrips feeding was 90 per cent greater at elevated than at ambient CO_2. Because of increased growth, however, undamaged leaf area was approximately 15 per cent greater at elevated than at ambient CO_2. Endophytes, which play an important role in conferring tolerance to both abiotic and biotic stresses in grasses, may also undergo a change in response to disturbance in the soil due to climate change.

Transgenic Crops

Environmental factors such as soil moisture, soil fertility, and temperature have strong influence on the expression of *B. thuringiensis* (Bt) toxin proteins deployed in transgenic plants. Cotton bollworm, *Heliothis virescens* (Fabricius) destroyed Bt-transgenic cottons due to high temperatures in Texas, USA. Similarly, *H. armigera* and *H. punctigera* (Wallengren) destroyed the Bt-transgenic cotton in the second half of the growing season in Australia because of reduced production of Bt toxins. Cry1Ac levels in transgenic plants decrease with the plant age, resulting in greater susceptibility of the crop to insect pests during the later stages of crop growth. Possible causes for the failure of insect control in transgenic crops may be due to inadequate production of the toxin protein, effect of environment on transgene expression, Bt-resistant insect populations, and development of resistance due to inadequate management. It is, therefore, important to understand the effects of climate change on the efficacy of transgenic plants for pest management.

Natural Enemies

Climate change can have diverse effects on natural enemies of pest species. The fitness of natural enemies can be altered in response to changes in herbivore quality and size induced by temperature and CO_2 effects on plants. The susceptibility of herbivores to predation and parasitism could be decreased through the production of additional plant foliage or altered timing of herbivore life cycles in response to plant phenological changes. The effectiveness of natural enemies in controlling pests will decrease if pest distributions shift into regions outside the distribution of their natural enemies, although a new community of enemies might then provide some level of control. The abundance and activity of natural enemies will be altered through adaptive management strategies adopted by farmers to cope with climate change. These strategies may lead to a mismatch between pests and enemies in space and time, decreasing their effectiveness for biocontrol.

Relationships between insect pests and their natural enemies will change as a result of global warming, resulting in both increases and decreases in the status of individual pest species. Changes in temperature will also alter the timing of diurnal activity patterns of different groups of insects, and changes in interspecific interactions could also alter the effectiveness of natural enemies for pest management. Quantifying the effect of climate change on the activity and effectiveness of natural enemies for pest management will be a major concern in future pest management programs. Oriental armyworm, *Mythimna separata* (Walker) populations increase during extended periods of drought (which is detrimental to the natural enemies), followed by heavy rainfall because of the adverse effects of drought on the activity and abundance of the natural enemies of this pest. Aphid abundance increases with an increase in CO_2 and temperature, however, the parasitism rates remain unchanged in elevated CO_2. Temperatures up to 25°C will enhance the control of aphids by coccinellids. Temperature not only affects the rate of insect development, but also has a profound effect on fecundity and sex ratio of parasitoids.

Effects of predators can be encouraged or discouraged by temperature increases. For instance, below 11°C, the pea aphid reproduction rate exceeds the rate at which the lady bird beetle, *Coccinella septempunctata* Linnaeus, can consume it. Above 11°C, the situation is reversed. In contrast, natural enemies of the spruce budworm, *Choristoneura fumiferana* (Clemens), are less effective at higher temperatures. The growth and development of insect herbivores varied with the nutritional quality of their diet (host plant) and the dietary differences showed varied effects on parasitoids. The population size of the insects significantly differed under elevated CO_2 and in turn influencing the insect fecundity. Thus, any dietary differences that prolong developmental time, increase food consumption, and reduce growth by herbivores serve to increase the susceptibility of herbivores to natural enemies.

Increasing CO_2 concentrations could alter the preference of lady beetle to aphid prey and enhance the biological control of aphids by lady beetle in cotton crop. This study provided the first empirical evidence that changes in prey reared on host plant grown at different levels of CO_2 altered the feeding preferences of the predator. Some studies suggest that higher temperatures increase the probability that a host will kill its parasitoid. For instance, parasitism of the caterpillar, *Spodoptera littoralis* (Boisduval), by the parasitoid, *Microplitis rufiventris* Kak, is less efficient at 27°C than at 20°C.The interactions between insect pests and their natural enemies need to be studied carefully to devise appropriate methods for using natural enemies in pest management.

Microbial Pesticides

There will be an increase in variability in insect damage as a result of climate change. Higher temperatures will make dry seasons drier, and conversely, may increase the amount and intensity of rainfall, making wet seasons wetter than at present. Natural plant products, entomopathogenic viruses, fungi, bacteria, and nematodes, and synthetic pesticides are highly sensitive to the environment. Although, microbial biopesticides are among the widely used ecofriendly methods of pest control, their efficiency is highly variable across microorganisms with the change in climatic factors (**Table 32.1**). The available information indicated that high relative humidity is associated with high mortality in host insects infected with fungal and viral pathogens, however, such effects on host infected with bacteria are variable, as environmental influence on host-pathogen interactions operates through biology of the pathogen, immune response of the host. Temperature, relative humidity, elevated atmospheric CO_2, UV radiation, pH, rainfall, soil moisture, etc., are the most

Table 32.1 Effect of environmental factors on the efficacy of microbial pesticides in pest management

Climate component(s)	Fungi	Bacteria	Viruses	Nematodes
Temperature	Optimum, 25-30°C	Optimum, Up to 30°C	>90°C decreases efficacy	Optimum, 25°C; 10-15°C host searching problem; >30°C results in mortality
Relative humidity (RH)	Positively correlated with RH	Effect variable	Positively correlated with RH	Optimum soil moisture, 75%; >75% causes oxygen depletion and mortality; <75% decreases searching ability and survival
Elevated atmospheric CO_2	No information	Increase in efficacy	Increase in efficacy	No information
UV light	Causes degradation and mortality	Causes degradation and mortality	Causes degradation and mortality	Causes degradation and mortality
Soil, water and plant pH	Variable	Variable	Variable	Variable

Source: Abbaszadeh *et al.* (2011)

influential climatic factors determining the efficiency of microbial biopesticides. Therefore, there is greater need to understand their interactions with biopesticides, identify robust and adaptive microbial strains, modify their beneficial traits as per the requirements using molecular techniques and devise appropriate formulations and strategies for their effective use in insect pest management under diverse array of environmental conditions.

Toxins from *B. thuringiensis* have been used as pest management tools for more than 50 years. The effect of these toxins depends on the quantity of Bt toxins ingested by susceptible insects. Food ingestion is affected by CO_2 concentration; plants grown in elevated CO_2 often have increased carbon/nitrogen ratios (C/N), resulting in greater leaf area consumption. Elevated CO_2 would improve the efficacy of foliar applications of *B. thuringiensis*.

Botanical Pesticides

Among the botanical pesticides, azadirachtin, an important limnoid of neem tree (*Azadirachta indica* A. Juss.) has been found to be quite promising. However, the azadirachtin content of neem tree is highly variable across genotypes, regions, habitats, fruit bearing season and climatic conditions. The neem genotypes belonging to coastal, arid and semi- arid ecosystems synthesize higher azadirachtin A content than those from the sub-humid regions. The neem trees growing in Andhra Pradesh, Karnataka, and Tamil Nadu in India synthesize higher azadirachtin content as compared to those growing in Orissa, Delhi, and Rajsthan; whereas those growing in Maharashtra, Madhya Pradesh, Uttar Pradesh and Punjab produce intermediate azadirachtin. The neem trees growing in hot sub-humid, hot arid, and hot semi-arid with cold winter regions synthesize lower azdirachtin content compared to those grown in hot semi-arid with mild winter region, indicating that moderate climatic conditions are favourable for azadirachtin synthesis as compared to extreme climatic condition.

Similarly, there are significant variations in azadirachtin content of neem seeds of Asian and African origin. The neem seeds of Nicaragua and Indonesia produced more azadirachtin followed by those from India, Myanmar and Mauritius. It has been found that the winter stress favours synthesis of azadirachtin B and F in the neem tree seeds. Moreover, azadirachtin A is highly influenced by environmental factors compared to oil in the seeds and a significant and positive correlation was found for azadirachtin A and oil content with relative humidity, and significant and negative association with atmospheric temperature. The neem genotypes bearing fruits in November produce 2-4 times more azadirachtin A than those producing fruits in July, suggesting that the neem tree genotypes bearing fruits in November should be identified and commercially deployed for producing neem based biopesticide formulations.

FUTURE OUTLOOK

Climate change will have serious consequences on diversity and abundance of insect pests and extent of losses. Pest outbreaks might occur more frequently, particularly during extended periods of drought, followed by heavy rainfall. Some of the components of pest management such as host plant resistance, microbial biopesticides and natural enemies will be rendered less effective as a result of increase in temperatures and UV radiation, and decrease in relative humidity. Adverse effects of climate change on the activity and effectiveness of natural enemies will be a major concern in future pest management programmes. The immediate solutions include:

- Careful monitoring of pest damage, including associated pathogen and insect populations. This would be the first step in a strategy to understand and deal with the effect of global climatic changes as they occur. The acquired knowledge would provide tools to enhance resistance/tolerance to biotic stresses, which would lead to improved IPM and sustained crop productivity.

- Development of prediction models for insect pest outbreaks and understand the influence of climatic change on species diversity and cropping patterns, and their influence on the abundance of insect pests and their natural enemies.
- Transgenic crops are likely to play a prominent role in future IPM programmes. Therefore, studies on the effect of climate change on the efficacy of transgenic crops should receive priority.
- Studies on the expression of resistance/virulence genes under different temperature regimes, and the identification and use of resistance genes that are stable under high temperature regimes, should be intensified.
- The possible increased use of insecticides resulting from an increase in pest outbreaks will likely have negative environmental and economic impacts for agriculture. The best economic strategy for farmers to follow is to use integrated pest management practices to closely monitor insect occurrence. Keeping pest and crop management records over time will allow farmers to evaluate the economics and environmental impact of pest control and determine the feasibility of using certain pest management strategies or growing particular crops.

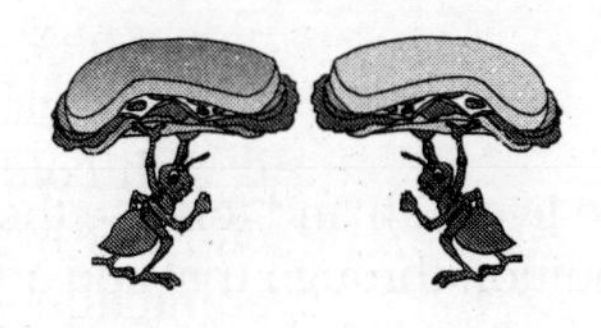

PEST MANAGEMENT : GLOBAL SCENARIO

INTRODUCTION

During ancient times, humans had to live with and tolerate the ravages of insects and other pests, but gradually learned to improve their condition through trial and error experiences. Over the centuries, farmers developed a number of mechanical, cultural, physical and biological control measures to minimize the damage caused by phytophagous insects. Synthetic organic insecticides developed during the mid-twentieth century initially provided spectacular control of these insects and resulted in the abandonment of traditional pest control practices. This was followed by the development of high yielding varieties of important crop plants. The intensive cultivation of these varieties, together with the application of increasing amounts of fertilizers and pesticides, has resulted in a manyfold increase in productivity. However, this technology package has also resulted in aggravation of pest problems in agricultural crops. The importance of achieving sustainable food production through the use of eco-friendly sustainable pest management techniques is being realized more and more in the recent past. Hence the increasing problems encountered with insecticide use resulted in the origin of the integrated pest management (IPM) concept (Dhaliwal and Koul, 2010).

SALIENT EVENTS IN PEST MANAGEMENT

Though it is not generally recognized, evolution of the concept of pest management spans a period of more than a century (**Table 33.1**). Many components of integrated pest management (IPM) were developed in the late 19th and early 20th century. The rapidly developing technologies and changing societal values had their impact on the pest control tactics also. The modern concept of pest management is based on ecological principles and involves the integration and synthesis of different components/control tactics into a pest management system. IPM, in turn, is a component of the agroecosystem management technology for sustainable crop production. The history of agricultural pest control, thus, has three distinct phases, viz. the era of traditional approaches, the era of pesticides and the era of IPM.

Era of Traditional Approaches (Ancient-1938)

Cultural and mechanical practices like crop rotation, field sanitation, deep ploughing, flooding, collection and destruction of damaging insects/insect infested plants, etc. developed by farmers through experience were among the oldest methods developed by humans to minimize the damage

Table 33.1 Landmarks in the history of pest management

Period	Landmark(s)
(1)	(2)
Era of Traditional Approaches	
Ancient	The Chinese used chalk and woodash for the control of insect pests in enclosed spaces and botanical insecticides for seed treatment. They also used ants for biological control of stored grain as well as foliage feeding insects. In India, neem leaves were placed in grain bins to keep away troublesome pests. In Middle and Near East, powder of chrysanthemum flowers was used as an insecticide.
900 AD	The Chinese used arsenic to control garden pests.
1690	The tobacco extract was used as a plant spray in parts of Europe.
1762	*Mynah* (a bird) from India was imported for the control of locusts in Mauritius.
1782	"Underhill" variety of wheat reported resistant to Hessian fly in USA.
1831	"Winter Majetin" variety of apple reported resistant to woolly apple aphid in USA.
1848	Derris (Rotenone) reported to be used in insect control in Asia.
1855	A. Fitch reported the role of lady bird beetles, green lacewings and other predaceous insects in the control of insect pests of crops.
1858	Pyrethrum first used for insect control in the USA.
1887	Balfour published his book *The Agricultural Pests of India and of Eastern and Southern Asia.* It was the first publication dealing with agricultural insect pest control in this region.
1889	Biological control of cottony cushion scale on citrus in the USA by use of Vedalia beetle imported from Australia.
1890	Control of grape phylloxera in Europe by grafting of European grapevine scions to resistant North American rootstocks.
1892	Lead arsenate used for the control of Gypsy moth in the USA.
1898	The coccinellid, *Cryptolaemus montrouzieri* Mulsant from Australia was released against coffee green scale, *Coccus viridis* (Green) in India. It established but failed to control the scale.
1911	Cotton Pest Act was enforced in the erstwhile Madras State, India. Under the Act cotton stalks had to be removed by 1st August every year to minimise the incidence of pink bollworm.
1923	Multiple component suppression techniques involving the use of resistant varieties, sanitation practices and need-based application of insecticides developed for the control of bollweevil in the USA.
1931	The cottony cushion scale attacking wattle of commerce, *Acacia decurrens* Willd. was controlled in India by release of predatory beetle, *Rodolia cardinalis* Mulsant from California.
Era of Pesticides	
1939	• Insecticidal properties of DDT reported by Paul Muller in Switzerland. • *Bacillus thuringiensis* Berliner first used as a microbial insecticide.
1940	Work on the development of jassid-resistant varieties was undertaken in India resulting in the release of varieties like LSS, 4F, 289F, etc., during 1940s.
1941	Insecticidal activity of Hexachlorocyclohexane (HCH) discovered in France.
1946	Parathion, the first organo-phosphatic insecticide developed.
1948	• Use of DDT and HCH on agricultural crops in India. • "Doom" based on *Bacillus popilliae* Dutky and *B. lentimorbus* Dutky registered in USA for the control of Japanese beetle larvae on turf.

Table 33.1 contd...

(1)	(2)
1951	• R.H. Painter published his classic book *Insect Resistance in Crop Plants.* • Introduction of first carbamate insecticide, Isolan
1952	First plant to produce HCH established at Rishra (India).
1959	• Concept of integrated control involving integration of chemical and biological control introduced. • Concept of economic injury level and economic threshold developed by V.M. Stern and coworkers.
1962	Publication of the book *Silent Spring* by Rachel Carson which dramatized the impact of misuse and overuse of pesticides on the environment.
1964	Publication of the book *Biological Control of Insect Pests and Weeds* by Paul DeBach, which established biological control as a separate discipline in Entomology.
Green Revolution Era	
1970	Establishment of US. Environmental Protection Agency (EPA), armed with responsibility for registration of pesticides.
1972	The first major IPM project, commonly called the Huffaker Project (1972-1978), was launched in USA, and covered six crops (alfalfa, citrus, cotton, pines, pome and stone fruits, and soybean).
1973	Development of first photo-stable pyrethroid, permethrin.
1975	• Elcar (*Helicoverpa* NPV) registered for the control of bollworm and tobacco budworm on cotton. • First insect growth regulator (Methoprene) registered for commercial use in USA. • Publication of the book *Introduction to Insect Pest Management* by R.L. Metcalf and W.H. Luckman, which was the first comprehensive treatise on IPM and established the concept on a firm footing.
1980	The interest in botanical pesticides revived and the First International Conference on Neem was held at Rottach-Egern, Germany.
1985	Structure of azadirachtin determined.
Post-Green Revolution Era	
1987	• The concept of 'push-pull' strategy was proposed by B. Pyke and his colleagues, for the management of *Helicoverpa* spp. on cotton in Australia. • Development of first transgenic plant, reported by M. Vaeck and coworkers of Belgian Biotechnology Company, Plant Genetic Systems by transferring *B. thuringiensis* δ-endotoxin gene to tobacco for the control of *Manduca sexta* (Johannsen).
1989	An IPM Task Force was established to garner international support for development and implementation of IPM programmes. A team of consultants appointed by the Task Force reviewed the status of IPM and made recommendations. The Task Force was later reconstituted as the Integrated Pest Management Working Group (IPMWG) in 1990.
1992	• Concept of Environmental Economic Injury Levels proposed by L.P. Pedigo and L.G. Higley. • Dr Edward F. Knipling and Dr Raymond C. Bushland were awarded the World Food Prize for developing sterile insect technique. • United Nations Conference on Environment and Development (Rio de Janeiro, Argentina) assigned a pivotal role to IPM in the agricultural programmes and policies envisaged as part of its Agenda 21
1994	A Task Force consisting of FAO, the World Bank, UNDP and UNEP co-sponsored the establishment of the Global IPM Facility with the Secretariat located at FAO, Rome.

Table 33.1 contd...

(1)	(2)
1995	Dr Hans R. Herren was awarded the World Food Prize for developing and implementing the world's largest biological control project for cassava mealybug which had almost destroyed the entire cassava crop of Africa.
1997	Dr Ray F. Smith and Dr Perry L. Adkisson were awarded the World Food Prize for their pioneering work in development and implementation of integrated pest management (IPM) concept.
Gene Revolution Era	
2002	Bt cotton approved for commercialisation in India by the Genetic Engineering Approval Committee (GEAC) with the release of three hybrids, viz. MECH 12Bt, MECH 162Bt and MECH 184Bt.
2004	*Helicoverpa zea* (Boddie) reported to develop resistance to Bt cotton in South-Eastern USA.
2005	The first decade (1996-2005) of commercialisation of transgenic crops completed. The transgenic crops were grown on an area of 90 million ha.
2006	• The area under transgenic crops crossed 100 million ha. These crops were grown by 10.3 million farmers in 22 countries on an area of 102 million ha. • Bt brinjal was approved for large-scale field trials in India by GEAC. • *Busseola fusca* (Fuller) reported to develop resistance to Bt maize in South Africa.
2007	• *Spodoptera frugiperda* (J.E. Smith) reported to develop resistance to Bt maize in Puerto Rico. • Azadirachtin was synthesised in the laboratory by Dr Steven V. Ley and colleagues, University of Cambridge, Cambridge (UK).
2008	• The increase in resistance allele frequency to Bt cotton in field populations of *Helicoverpa armigera* (Hubner) was reported in China. • GEAC approved the Bt cotton variety developed by using an indigenous variety Bikaneri Narma. This is the first public sector transgenic crop in India, and has been approved for North, Central and South zones of the country. • *Pectinophora gossypiella* (Saunders) developed resistance to Bt cotton in some regions of Gujarat, India.
2009	• Bt brinjal approved for environmental release in India by GEAC.
2010	• Government of India withheld the commercial release of Bt brinjal till more scientific data on biosafety are generated. • Genetic Engineering Approval Committee renamed as Genetic Engineering Appraisal Committee. • SmartStax maize, stacking 8 genes (6 for insect resistance and 2 for herbicide tolerance), released for commercial cultivation in USA and Canada. • After 15 years of commercialization, transgenic crops were grown on 148 million ha by 15.4 million farmers in 29 countries.
2011	India celebrated the first decade of commercialization of Bt cotton, which was planted on an area of 10.6 million ha.
2013	GEAC approved more than 200 varieties of transgenic crops for field trials in March.
2014	• The Government of India allowed confined field trials of more than 200 verieties of transgenic crops including wheat, maize, castor and cotton, among others, in February. • GEAC approved field trials of 15 varieties of transgenic crops, including rice, brinjal, chickpea, mustard and cotton in July. However, the Government of India put the field trials on hold.

***Source*:** Modified from Dhaliwal *et al.* (2013)

caused by insect pests. This was followed by the use of plant products from neem, chrysanthemum, rotenone, tobacco and several other lesser known plants in different parts of the world. The Chinese were probably the pioneers in the use of botanical pesticides as well as biological control methods for the management of insect pests of stored grains and field crops. However, systematized work on many important tactics of pest control including the use of resistant varieties, biological control agents and botanical and inorganic insecticides was done in the USA from the end of the 18th to the end of the 19th century. Remarkable success was achieved in the management of grape phylloxera caused by *Daktulosphaira vitifoliae* (Fitch) by grafting of European grape vine scions to resistant North American rootstocks during the 1880s. At around the same time, cottony cushion scale, *Icerya purchasi* Maskell which was causing havoc to the citrus industry in California, USA was successfully controlled by release of the Vedalia beetle, *Rodolia cardinalis* (Mulsant) imported from Australia.

A number of synthetic inorganic insecticides containing arsenic, mercury, tin and copper were also developed towards the end of the nineteenth and the beginning of the twentieth century. With the development of these insecticides, the focus of research in entomology slowly shifted from ecological and cultural control to chemical control, even before the development of synthetic organic insecticides.

Era of Pesticides (1939-1975)

The synthetic inorganic insecticides were broad spectrum biocides and were highly toxic to all living organisms. These were followed in due course by the synthetic organic insecticides like alkyl thiocyanates, lethane, etc. The era of pesticides, however, began with the discovery of the insecticidal properties of DDT [2, 2-(*p*-chlorophenyl)–1, 1, 1-trichloroethane] by Paul Muller in 1939. The impact of DDT on pest control is perhaps unmatched by any other synthetic substance and Muller was awarded Nobel Prize for this work in 1948. DDT was soon followed by a number of other insecticides like HCH, chlordane, aldrin, dieldrin, heptachlor (organochlorine group) ; parathion, toxaphene, schradan, EPN (organophosphorus group) and allethrin (synthetic pyrethroid) during the 1950s and a large number of other popularly used organophosphates and carbamates in the ensuing decade.

Due to their efficacy, convenience, flexibility and economy, these pesticides played a major role in increasing crop production. The success of high yielding varieties of wheat and rice that ushered in the 'green revolution' was partially due to the protection umbrella of pesticides (Pradhan, 1983). The spectacular success of these pesticides masked their limitations. The intensive and extensive use, misuse and abuse of pesticides during the ensuing decades caused widespread damage to the environment. In addition, insect pest problems in some crops increased following the continuous application of pesticides. This, in turn, further increased the consumption of pesticides resulting in the phenomenon of the pesticide treadmill. The combined impact of all these problems together with the rising cost of pesticides provided the necessary feedback for limiting the use of chemical control strategy and led to the development of the IPM concept.

Era of IPM (1976 onwards)

Although many IPM programmes were initiated in late 1960s and early 1970s in several parts of the world, it was only in late 1970s that IPM gained momentum. The first major IPM project in USA, commonly called the Huffaker Project, spanned 1972-78 and covered six crops, *i.e.* alfalfa, citrus, cotton, pines, pome and stone fruits, and soybean. This was followed by another large scale IPM project called CIPM, the Consortium for Integrated Pest Management (1979-85), which focused on alfalfa, apple, cotton and soybean. The national IPM programmes were launched in late 1980s and early 1990s in several developing countries. The most outstanding success has been the FAO-IPM programme for rice in Southeast Asia. Several other programmes were launched to promote IPM, which included Global IPM Facility, CGIAR SP-IPM, Integrated Pest Management Working Group (IPMWG), IPMnet, USAID IPM Collaborative Research Support Programme (CRSP), IPM Europe, Africa IPM Forum, etc. The details of these programmes are given under 'Pest Management Organizations'.

IPM IN DEVELOPED COUNTRIES

Many developed countries took a lead in developing IPM systems in major agricultural crops and achieved varying levels of success. The pioneering efforts were made in USA for the development and implementation of IPM in major agricultural crops. (Dhaliwal *et al.,* 2013)

IPM in Apple Orchards

One of the earliest and classical examples of IPM in fruit trees is the control of apple pests in Nova Scotia, Canada. Ecological studies were initiated in 1943 to determine the long-term influence of spray chemicals on the fauna of apple orchards. Main emphasis was laid on field projects, and laboratory and insectary studies when possible. Preliminary tests with insecticides on specific pests and beneficial species were carried out in small plots and extended to whole orchards when the initial results justified it. Within two years, it was demonstrated that the use of sulphur as fungicide was the main factor causing population increase of oystershell scale, *Lepidosaphes ulmi* (Linnaeus). As a result of elimination of sulphur from the spray programme, the pest was reduced to minor importance.

Studies also indicated that many spray chemicals, particularly DDT and sulphur were deterimental to some important predators of the European red mite, *Panonychus ulmi* (Koch). The long-term studies have provided a basis that makes it possible to select spray programmes that allow natural control of phytophagous mites and, at the same time, provide adequate chemical control of injurious insects and diseases. Sulphur also proved deterimental to predaceous thrips, *Haplothrips faurei* Hood and *Leptothrips mali* (Fitch), both active predators of the codling moth, *Cydia pomonella* (Linnaeus). The spray schedule was changed from an intensive to a modified programme; in some cases no insecticides were used and in others only one or two applications were made. As a reduction of number of insecticidal sprays, codling moth populations decreased and after about 4 years on this programme, many of the orchards produced apples with less than 10 per cent total insect damage to the fruit. The integration of biological and chemical control programmes solved all the problems concerned with the control of apple orchard pests in Nova Scotia.

Huffaker Project

The first major IPM project in the USA, commonly called the Huffaker Project, spanned 1972-1978 and covered six crops, viz. alfalfa, citrus, cotton, pines, pome and stone fruits, and soybean. The crops were selected on the basis of three criteria, *i.e.*(*i*) current level of insecticide use (very high in cotton and citrus, and very low in soybean), (*ii*) potential for successful biological control (alfalfa, citrus, pome and stone fruits), and (*iii*) representation of a non-agricultural system (pines). An anticipated outcome of the programme was 40-45 per cent reduction in the use of more environmentally polluting insecticides within a 5 year period and 70-80 per cent in 10 years. Advances were made in many aspects implementing improved IPM strategies for all systems. The project improved integrated insect control programmes that lessen the dependence on chemical insecticides especially on cotton and apples.

Consortium for IPM

After the success of the Huffaker Project, a group of scientists under the leadership of P.L. Adkisson secured funding for the second large scale IPM project in the United States, called the Consortium for Integrated Pest Management (CIPM) (1979-1985). The project focused on four major crops, viz. alfalfa, apple, cotton and soybean. It was claimed that the average adoption of IPM for four crops was 66 per cent over 5·76 million ha. The main indicators of adoption were the use of economic thresholds and economic injury levels for spray decisions, use of selective pesticides or application of lower dosages of broad spectrum insecticides. A significant achievement of the programme was the genuine attempt to integrate weed science and plant pathology, and the emphasis on economic assessment of IPM adoption.

National IPM Initiative

The U.S. Department of Agriculture (USDA) launched National IPM Initiative in 1993, the goal of which was to implement IPM practices on 75 per cent of the nation's crop area by 2000. To measure the level of adoption, USDA put forth the PAMS concept, the acronym for prevention, avoidance, monitoring and suppression. To qualify as an IPM practitioner, a farmer was required to utilize at least three of the four PAMS components. It has been estimated that some level of IPM has been implemented on about 70 per cent of the US crop acreage. The highest percentage of cropland under IPM was in case of cotton (86) and vegetables (86), followed by soybean (78), maize (76), barley (71), wheat (65), other crops and pasture (63), fruits and nuts (62), and lucrene hay (40).

Integrated Fruit Production System

The pome fruit growers in Australia are funding a national system for integrated fruit production (IFP). This system incorporates whole-farm planning, site-specific selection of scion/rootstock combinations, IPM, irrigation and nutrition, crop management, quality assurance, food safety and occupational health and safety. IFP takes a broad approach to pest management decision making by encouraging integration and understanding of the interactions occurring in the orchard and their impacts on crop quality. On the basis of a survey conducted in all production areas, it has been found that slightly more growers selected pesticides on the basis of compatibility with predators (86%) than efficacy against the target pest (84%), suggesting that they are prepared to balance pest control with the value of maintaining a predator population. Nearly all (92%) based spraying decisions on the basis of monitoring. Overall, 45 per cent had their own staff trained to monitor pest populations, while 57 per cent used a consultant for their pest management.

The search for apple IPM system in New Zealand gained momentum in the 1980s and 1990s with a focus to minimize environmental and human health impact. Solutions that have been adopted for individual pests include a decision support model for European red mite, biological control, chemical control and pheromone traps to reduce insecticide applications against leafrollers. Recently, the introduction of insect growth regulators for leafroller and codling moth has enabled the development of selective pest management and increased the focus on integrating natural enemies within the apple IPM programme. This development has been an integral part of the apple industry's IFP programme that requires pest monitoring and justification of all pesticide use, backed up by auditing and certification systems. There has been very significant reductions in the use of broad-spectrum insecticides after several years of development and implementation of IFP. For example, from 1997 to 2000, there was a 72 per cent reduction in organophospate use in New Zealand apple orchards, with a 90 per cent reduction in the use of azinophos methyl. By 2000-01 season, this was equivalent to an overall 90 per cent reduction in organophosphate use, wide application of the IFP programme by growers occurring in the 2001 season.

IPM in Greenhouses

The greenhouse crops are grown over an area of 300,000 ha worldwide. Vegetables are produced in two-third of these greenhouses and ornamentals in one-third. Although this represents a tiny fraction (0.02%) of the 1.5 billion ha under crops worldwide, greenhouses offer the opportunity to grow larger quantities of high-quality crops on a very small area. For example, in the Netherlands 30 per cent (by value) of agricultural output is produced in greenhouses that occupy only 0.5 per cent of the total agricultural land. IPM is used on a large scale in all the main vegetable crops grown in greenhouses in Europe. In the Netherlands, for example, more than 90 per cent of all tomatoes, cucumbers and sweet peppers are produced under IPM. Worldwide 5 per cent of the greenhouse area is under IPM and there is a potential for increase to about 20 per cent of the area in the next decade.

A good example of an IPM programme is the one used for tomato in Europe. It involves more than 10 natural enemies and several other methods like host plant resistance, climate control and cultural control. When tomatoes are grown in soil, soil sterilization by steaming is used shortly before planting the main crop to eliminate the soil-borne diseases such as tomato mosaic virus (TMV), *Fusarium* and *Verticillium*, and insect pests such as tomato moth, *Lacanobia oleracea* (Linnaeus) and three species of leaf-miners, *Liriomyza* spp. Therefore, only foliage pests and *Botrytis cinerea* Pers. ex. Nocca. & Balb. require direct control measures.

IPM IN DEVELOPING COUNTRIES

The developing countries are characterized by tropical and sub-tropical climates. The integrated pest management strategy appears to be more feasible in these regions because of the following factors :

- Climate and physical environment in tropical and sub-tropical areas generally allow populations of both pests and natural enemies to proliferate throughout the year. Regulation of numbers by natural biotic mechanisms thus can function optimally and in relative abundance.
- Agroecosystems, which have not yet been subjected to rigorous economic development generally have been only little disturbed by pesticide applications. They offer a relatively sound base for the development of an IPM system and farmers do not have to be re-educated.
- If the agricultural production is destined for home consumption, farmers usually pay little attention to light infestation and they would hardly spray for cosmetic reasons. However, transformation to market-oriented production leads to excessive use of pesticides, particularly on vegetables, fruits and flowers.
- Where new areas are opened up for agricultural production, as may be the case in sparsely inhabited parts of Africa, Asia and South America, there would be an opportunity to introduce IPM from the beginning and to ban unwanted chemical control practices.

Growing concern among the general public for environmental issues, particularly pesticide misuse, has prompted many developing countries to formally and explicitly advocate the use of IPM as an environmentally friendly form of crop protection. For example, the Governments of India and Malaysia adopted IPM as official policy in 1985 by Ministerial Declaration. Similarly, Governments of Indonesia and Philippines promulgated Presidential Decree/Declaration to adopt IPM as official policy. China introduced the Green Certificate programme and banned highly toxic pesticides on vegetable crops. Biological control is a national priority for Cuba ; the new policy is intended to make IPM biointensive, with 80 per cent of pests managed through biological control. Iran formed the High Council of Policy and Planning for Reduction of Agricultural Pesticides. Nepal and Sri Lanka have adopted national IPM policies. Under such conditions, the cases for funding IPM programmes will be heard more sympathetically and the agencies empowered by government to allocate resources will give proper attention to the needs of IPM (Koul *et al.*, 2004).

The philosophy of IPM did not percolate down to the farmers for quite a long time after its presentation and prescription for solving pest problems in modern agriculture. It was also suggested that the illiterate farmers of the developing countries were unable to grasp the concept of IPM and, therefore, could not implement it. However, the pessimists have been proven wrong and the same farmers have now demonstrated that they are quite capable of understanding the intricacies of IPM. The success of *Farmer Field Schools* (FFSs) in the implementation of IPM in many Asian countries proves that farmers are quite responsive to appropriate technologies which give due weightage to their traditional wisdom, local conditions and socioeconomic constraints.

Participatory IPM led by farmers is now practised in over 50,000 communities found mostly in Indonesia, Vietnam, Philippines, Bangladesh, China, Sri Lanka, India, Cambodia, Laos, Republic of Korea, Ghana, Kenya, Cote d'Ivoire, Burkina Faso, Mali, Egypt, Sudan, Honduras, Nicaragua, Senegal and Zimbabwe. This is over 2 per cent of all rural villages in all developing countries. There are over 30,000 competent IPM trainers any of whom can facilitate a Farmers Field School through an entire crop season, and then the resulting farmers' IPM group through the remainder of their year-round production agroecosystem. Farmers practicing IPM have increased yields per hectare from 1 to over 10 per cent, reduced pesticide use by 30 to over 95 per cent (and often eliminating insecticide use) and substantially lowered occupational health risks. Most of the communities practicing IPM grow rice, but IPM is also practiced by farmers of maize, soybean and other field beans, cabbage, tomato, groundnut, coconut, cacao, coffee, peppers, sweet potato, cotton, mango and cucumber (Dhaliwal and Arora, 2012).

The Government of India adopted IPM as the official guiding principle of plant protection strategies in 1985. The IPM programme has been strengthened in India since 1994 and 12931 Farmers' Field Schools have been conducted in rice, cotton, vegetables, pulses and oilseeds, where 54,349 Agriculture/Horticulture Extension Officers and 3,88,863 farmers have been trained up to March 2010 (Plate 16.1). The yield increase in rice and cotton in IPM-areas, varied from 7 to 40 per cent and 23 to 27 per cent respectively, as compaired to non-IPM areas. There was a reduction in pesticide consumption to the tune of 50-100 per cent and 30-50 per cent in rice and cotton, respectively, in IPM-areas. The use of Bt-based and neem-based pesticides increased from 123 metric tonnes during 1994-95 to 1262 metric tonnes during 2009-10. Overall consumption of chemical pesticides in the country decreased from 75033 metric tonnes (technical grade) during 1990-91 to 41822 metric tonnes during 2009-10.

Rice

Major efforts in implementing IPM in irrigated rice have been carried out in Asia by the United Nations' Food and Agriculture Organization (FAO) through the Inter Country Programme for the Development and Application of Integrated Pest Control in Rice in South and South-east Asia. This programme remains one of the best examples of IPM implementation in the tropical region. It involves purposeful, direct efforts to change farmers' practices in contrast to some more indirect routes of IPM technology diffusion in many industrialized, temperate environments. The programme itself has evolved into its present transnational form from a relatively small project supported by Australia in the late 1970s, following the large-scale pest outbreaks in several South-east Asian countries.

The first phase of the FAO programme (1980-86) focused on developing and testing the technical aspects of the IPM concept in its seven participating countries, viz. Bangladesh, India, Indonesia, Malaysia, Philippines, Sri Lanka and Thailand. More recently, the project has been directed towards enhancing farmers' adoption of IPM. The programme is supported by Australia, the Netherlands and the Arab Gulf Fund. One significant accomplishment of the programme has been to cause policy changes within several governments, in the form of official support of IPM as the means for national plant protection in the Philippines, Indonesia, India, Sri Lanka and Malaysia.

Indonesia. A case study of the National IPM Programme in Indonesia as a part of the regional programme during 1989-1991 provides an interesting scenario. Indonesia subsidised up to 85 per cent of the cost of pesticides during the early 1980s. Following research findings showing the relation between brown planthopper outbreaks and high pesticide use, the Indonesian government banned the use of 57 of the 66 broad-spectrum pesticides used on rice. By 1989, pesticide subsidies were no longer offered and IPM was declared as the national pest control strategy. These measures, created a favourable climate for the large scale implementation of IPM (**Fig. 33.1**). Consequently, there was dramatic increase in Indonesian rice production ; production rose from 38 to 45 million tonnes, while Indonesia went from a net importer of pesticides (US$ 47 million) to a net exporter (US$ 15 million). The savings as a result of elimination of subsidies exceeded US$ 1 billion. It has been estimated that FAO training of 200,000 farmers in IPM techniques has resulted into a saving of $160 million in pesticide subsidies. By 2002, rice production reached 51 million tonnes, providing a good example of what progressive policies combined with massive farmer training can accomplish.

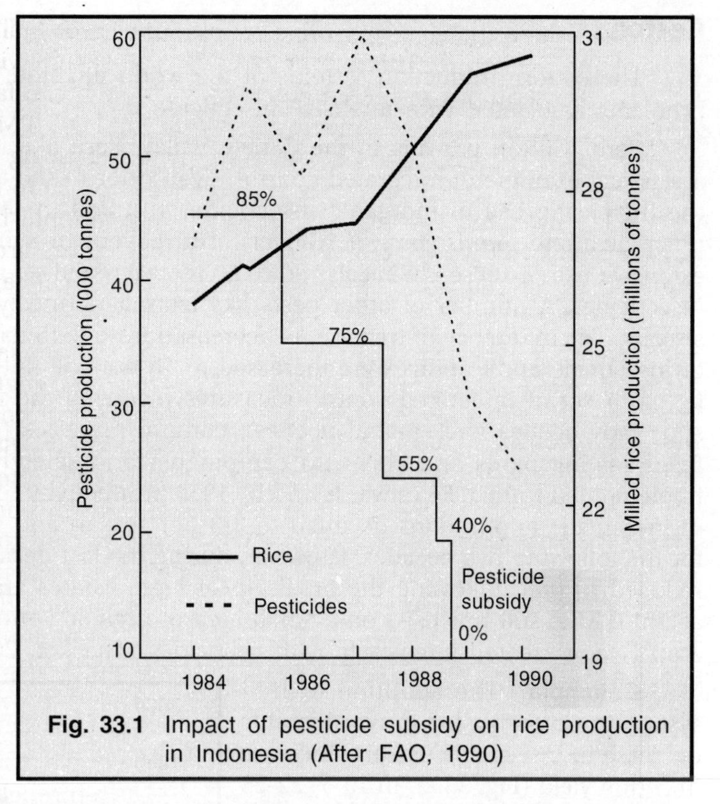

Fig. 33.1 Impact of pesticide subsidy on rice production in Indonesia (After FAO, 1990)

China. China, which had been experimenting with IPM for the control of rice pests since early 1980s, was also invited to join the FAO project in 1989. During 1989-90 alone, nearly 1,60,000 farmers from over 2000 villages received IPM training. Compared with untrained farmers, IPM-trained farmers saved roughly a third of the pesticides in rice cultivation and still obtained 7 per cent higher yield. It was estimated that the investment in IPM training generated a return of more than 400 per cent. Encouraged by these results, the Ministry of Agriculture has set up a National Steering Committee for the comprehensive prevention and control of diseases and insect pests to protect the nation's rice crop and increase profits. The Committee conducts IPM tests, gives demonstrations and makes appraisals.

Philippines. In Philippines, National IPM Programme was launched in 1993 and a total of 40,024 farmers have been trained during 1993-95, among whom 36,024 are rice farmers. Among 1,632 FFSs, 1470 were devoted to rice farmers. The increase in rice yield obtained by IPM farmers varied from 4.7 to 62 per cent. The expenditure on pesticide use (15% of total cost) was almost eliminated in case of IPM farmers.

India. In India, the first IPM programme in rice was started at Cuttack in 1975 covering an area of about 1000 ha in 10 villages. As a result of implementation of a number of cultural practices and ETL based applications of insecticides, the number of applications was reduced from 3-4 to usually single one. Subsequently, an Operational Research Project (ORP) was instituted on integrated control of rice pests at six locations in five provinces. This resulted in a reduction in the number of sprays and increased yield of the crop. On the basis of 5 years of ORP at Kuttanad in Kerala, it has been observed that, on an average, one-third of expenditure on plant protection operations can be saved by practising pest management methods. Prophylactic sprays and spray schedules have been abandoned by the cultivators and plant protection measures were taken only when the pest population exceeded economic threshold.

The validation of IPM in Basmati rice (Pusa Basmati 1) was demonstrated in village Shikohpur in Distt Baghpat (UP) during 1999. This village had the history of indiscriminate use of pesticides on this crop and some farmers had applied 10-12 rounds of highly toxic chemicals. The success of IPM approach during 1999 led entire village to follow and adopt this technology in an area of 120 ha in 2000 and 160 ha during kharif 2001 and 2002. By following the IPM tactics, the pesticide use was drastically cut, other input costs like fertilizers, number of irrigations, etc. reduced and farmers could secure residue free higher produce. The cost : benefit ratios were 1:3.18 (IPM) and 1:2.28 (non-IPM) in 2000, 1:3.16 (IPM) and 1:2.12 (non-IPM) in 2001, and 1:2.21 (IPM) and 1:1.64 (non-IPM) in 2002.

Cotton

The cotton production systems of the world illustrate well the ecological and environmental problems associated with intensive insecticide use.

Peru. Cotton growing in the Canete Valley, Peru is a well documented case that has become a classical example of integrated control. Even prior to the advent of synthetic organic insecticides, the increasing use of inorganic insecticides resulted in serious outbreaks of *Heliothis virescens* (Fabricius) and *Aphis gossypii* (Glover). The advent of synthetic organic insecticides resulted in extensive use of these chemicals. Both *H. virescens* and *A. gossypii* soon became resistant to these insecticides. A number of other pests like leafrollers, mealybugs, leaf perforator, etc. also became serious. The frequency of treatments decreased from 8-15 days to every 3 days. The average number of insecticide applications were increased to 16 but still lint yields touched the lowest levels of 332 kg/ha. A set of integrated control measures involving the use of natural enemies (*Trichogramma* spp., lady beetles and carabid beetles), cultural practices (early maturing varieties, crop rotation, fixed sowing dates and clean up campaigns) and selective use of inorganic insecticides were implemented from 1956 onwards and by 1959 productivity reached a level of 800 kg/ha. The number of insecticide applications declined to 2-3 per season and increase in cotton yields was sustained for the following two decades. However, during the last decade, the area planted to cotton has been reduced to half, following the break-up of large estates and pesticide use has begun to increase again. IPM is still practised on a small area of organic cotton, but a large proportion of the cotton crop is now treated intensively with pesticides (Thacker, 2002).

Colombia. The adoption of IPM system in Colombia led to considerable decrease in insecticide use and increase in cotton yield (**Fig. 33.2**). In 1977, 22-28 insecticide treatments were carried out in cotton, most of them to combat *Heliothis virescens* (Fabricius). This pest had developed resistance to methyl parathion and all other available insecticides. With the introduction of pyrethroids it again became possible to control the pests effectively, with a much smaller quantity of active ingredient per hectare than had been needed before. At the same time, an IPM programme was set up–the pest populations were carefully monitored and measures to control them only implemented if relevant threshold values were exceeded. Massive numbers of beneficial arthropods were then released. Rigid calender based treatments were replaced with crop monitoring, which allowed pest attacks to be detected at an early stage. This in turn, allowed insecticides to be employed in a targeted manner. Insecticide use was optimized in this IPM system, and developed as a complement to biological control measures (Bellotti *et al.*, 1990).

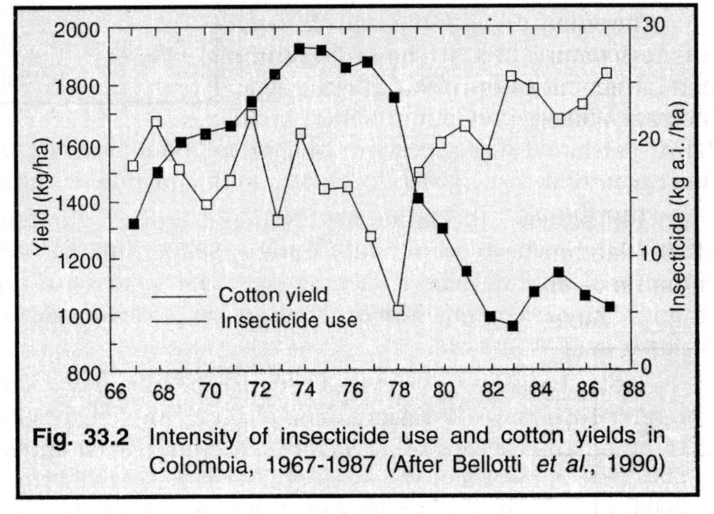

Fig. 33.2 Intensity of insecticide use and cotton yields in Colombia, 1967-1987 (After Bellotti *et al.*, 1990)

India. The Indian Council of Agricultural Research, sponsored a village level IPM project to test and demonstrate the efficacy, practicability and economics of IPM in cotton. The IPM programme called the Operational Research Project (ORP), was initiated in Punjab and Tamil Nadu in 1975 and 1980, respectively. In Punjab, 15 villages during different years were covered up in ORP to test and demonstrate the efficacy, practicability and economics of IPM strategy. The adoption of IPM technology resulted in 73.7 and 12.4 per cent reduction in the number of insecticide sprays for control of sucking pests and bollworms, respectively. The reduction of bollworm incidence in ORP villages was 38.55

per cent higher than non-ORP villages leading to 23.1 per cent higher yield and 31.1 per cent higher net income per unit area. The yield of seed cotton obtained in demonstration fields was 20.45 q/ha as against 10.17 q/ha on other fields. ORP project on IPM has also been successful at Coimbatore in Tamil Nadu in reducing the use of insecticides without affecting the yield of seed cotton, at the same time encouraging the natural enemies and reducing the environmental pollution. The mean quantity of insecticides used in the project villages was 3.82 kg a.i./ha in 6 applications at spray interval of 15-18 days, which was 58.6 per cent less than the amount (9.23 kg a.i./ha) used in non-project villages with 11 sprays at an interval of 7-8 days. The IPM system increased the abundance of native natural enemies by three fold, reduced the cost of insecticide and environmental pollution by 50.3 and 53.4 per cent, respectively.

The Insecticide Resistance Management programme was launched in 2002 in 26 cotton-growing districts in 10 states, which consume 80 per cent of India's insecticide use on cotton. The basic objectives of the IRM strategy are to disseminate available technology related to IPM, to monitor the level of resistance to commonly used insecticides, to assess the impact of IRM strategy on economic status of the farmers and to validate the IRM technology. In Punjab, three districts (Bathinda, Ferozepur and Mansa) have been covered, which account for more than 70 per cent of the total area under cotton cultivation. Three more districts, *viz.* Muktsar, Faridkot and Barnala were adopted in 2006. The district Bathinda is considered to be the hot-spot for pesticide consumption, where 50 per cent of the cost of cultivation is on pesticides. From about 2-3 insecticide applications in cotton in 1970s, farmers have been reported to resort to more than 30 insecticide applications towards the beginning of the twenty-first century.

The impact of IRM based IPM strategy, has been summarised in Table 33.2. In general, there was less damage by the bollworm complex, lower incidence of sucking pests, higher number of natural enemies, less number of insecticidal sprays, lowest cost of production and increase in net income. With the adoption of IRM strategies, farmers were able to get an additional profit of Rs. 10097 per ha. The decrease in pesticide load led to increase in the population of natural enemies during early season which helped in reducing the pest population and ultimately the damage caused to the crop (Dhawan *et al.*, 2011).

A landmark in cotton IPM has been the validation of the cropping system based holistic community approach of IPM at village Ashta (1998-2001) in Nanded district (Maharashtra). The

Table 33.2 Impact of IRM/IPM strategy on cotton in Punjab, India, 2002-2010

Year	Number of		Area under IPM (ha)	Seed cotton yield (kg/ha)	No. of sprays*	Spray cost (Rs/ha)	Additional profit (Rs/ha)
	Villages	Farmers					
2002	32	300	1152	1582	10.6	5460	7670
2003	30	567	2882	1969	9.5	7937	7844
2004	45	900	19429	2441	5.4	4021	13111
2005	121	10044	37032	2452	3.4	1793	5509
2006	213	14804	44848	2435	3.1	2163	5371
2007	230	16237	45262	2260	4.7	2714	9479
2008	825	42852	150875	2433	4.2	1761	12456
2009	225	12120	35182	2586	3.6	1955	15699
2010	53	4282	14048	2205	3.8	2076	13732
Total/ Average	**1774**	**102106**	**350710**	**2263**	**5.4**	**3320**	**10097**

*The average number of sprays in 2001 was 21; Bt cotton was adopted as component of IPM in 2006-10

Source : Dhawan *et al.* (2011)

baseline information indicated less than a quintal average seed cotton yield per ha in the *Helicoverpa armigera* (Hubner) epidemic year of 1997 when the farmers had sprayed more than 12-13 chemical pesticide spray rounds. All the farmers of the village were involved and the IPM approach covered 180 ha cotton area. The off-season practices included management of *H. armigera* on pigeonpea and chickpea through use of neem seed kernel extract (NSKE) and HaNPV, field sanitation and deep ploughing. In the pre-sowing practices, multiplicity of cultivars was avoided by selecting only two sucking pest moderately resistant cultivars and treating the seed with imidacloprid. The sowing of the entire village was completed within a week to avoid vulnerability of crop over a long period. The IPM interventions included use of *Trichogramma chilonis* Ishii, HaNPV and NSKE. Lastly, chemical pesticides were used when needed and these included spray of endosulfan/bavistin for the management of bollworms or grey mildew in certain pockets. The cost of plant protection was reduced by 31.3 per cent, chemical pesticide consumption reduced by 94.3 per cent and seed cotton yield increased by 56.9 per cent. The system has become self-sustainable as the farmers of Ashta have themselves become decision-makers and have started adopting many IPM practices on their own. The average cost : benefit ratio over the four year period was 1:1.473 (IPM) and 1:1.018 (non-IPM) (Singh *et al.*, 2006).

Pakistan. The research efforts had been focused to develop IPM programme in Pakistan since 1971. In 1997, Asian Development Bank supported the CABI Bioscience to run a pilot FFS project on cotton that formed the basis of IPM success. Based on success of the project, the National IPM programme was established in 2000, which led to scale up of farmers-led IPM through integration of international and national efforts. This National IPM programme with the help of international support took further initiatives to execute three projects to have sustainable IPM programme. These included FAO-EU Regional Project "Cotton IPM Programme for Asia" (2000-2004); ADB-FAO Pakistan Project, "Cotton IPM Programmes" (2002-04); and AGFUND-FAO Pakistan Project, "Pesticide Risk Reduction for Women in Pakistan" (2002-03). By the end of 2004, under different projects, a total of 425 IPM facilitators (8 women) were trained in 12 ToF courses, a total of 525 crop season long FFS were conducted in which 12,999 farmers were benefited (Khan *et al.*, 2005). The impact assessment carried out in 2003 showed better cotton yield (30%), reduced cost of pesticides (55%), reduced use of chemical pesticides (43%), reduced use of highly toxic pesticides (54%), increase in number of farmers joining community organization (33%), and reduced poverty of the target farmers group (16%).

Vegetable Crops

The practical application of IPM in vegetable cultivation is generally still limited, although simple packages do exist for some key pests. China is the only country in the region where IPM in major vegetable crops has been carried out through an exclusive technical body since late 1970s. In 1987, more than 1,07,000 ha of non-polluted vegetables were cultivated in 200 cities of 22 provinces, producing more than 6.4 million tonnes of vegetables. The Chinese have used the cultural control method as the foundation, while giving priority to biocontrol agents followed by judicious use of safer insecticides. During the past decade, efforts have been made to develop IPM technology in situations where chemical insecticides have failed to provide effective control in other countries of Asia.

The diamondback moth (DBM) has emerged as the most serious pest on cabbage, cauliflower and Chinese cabbage throughout South and South-East Asia. The research conducted at the Asian Vegetable Research and Development Centre (AVRDC) on the control of DBM without pesticides has resulted in the development of an IPM programme. A collaborative Vegetable Research Network for Southeast Asia (AVNET) was jointly established by Indonesia, Malaysia, the Philippines and Thailand in 1989. Another network, the South Asian Research Network (SAVERNET) has linked Bangladesh, Bhutan, India, Nepal, Pakistan and Sri Lanka since 1992. Both networks are supported by the Asian Development Bank with AVRDC as the executive agency and both have IPM of DBM as one of their sub-networks.

On the basis of decisions arrived at joint meetings of two networks, a master plan is developed for implementation in all the participating countries. It involves releases of a parasitoid, *Diadegma semiclausum* Hellen ; a bacterial insecticide *Bacillus thuringiensis* Berliner or neem kernel extract sprays, pheromone traps and growing mustard as a trap crop. To date, parasitoid rearing facilities have been established in all South-east Asian countries. Introduction and release of *D. semiclausum* has reduced the use of chemical insecticides by up to 86 per cent in the highlands of Malaysia. In Philippines, when cabbages were contaminated by over use of pesticides, IPM training resulted in farmers reducing from over 20 applications per season to three, including one treatment with an insect pathogen. It is estimated that in the Philippines highland area of Cordillera, extension and adoption of the IPM technology by all farmers in the 7000 ha cabbage area would result in a cost reduction of US$ 10.5 million over three cropping seasons. In Thailand, adoption of IPM technology resulted in a more than 145 per cent increase in net profits in IPM fields over non-IPM fields in crucifers. In Vietnam, the number of pesticide applications in cabbage have been reduced from 6 of chemical pesticides to 1 of Bt in IPM areas, in tomato from 14 (3 insecticides, 10 fungicides, 1 of 2, 4-D as growth stimulator) to no application in IPM.

Researchers at the Indian Institute of Horticultural Research (IIHR), Bangalore have developed an IPM combination to control DBM and it is becoming increasingly popular among the farmers. It involves growing of paired rows of bold-seeded Indian mustard as a trap crop at the beginning and after every 25 cabbage rows. Among the paired rows, the first is sown 15 days prior to cabbage planting while the second is sown 25 days after planting. This attracts upto 80-93 per cent of DBM for colonization. Depending upon the severity of infestation some atack of DBM and other pests is recorded on cabbage. To control these, 4 per cent neem seed kernel extract (NSKE) or neem soap (1%) or pongamia soap (1%) is applied at primordial or head initiation stage; 2-3 post-heading sprays may be applied at 10-15 days interval depending on population pressure. NSKE is safe to *Apanteles plutellae* Kurdyumov which is the dominant parasitoid of DBM found in the cabbage ecosystem. The insects colonizing Indian mustard may defoliate the entire plant and move on to cabbage. To control these, 0.1 per cent dichlorvos is applied at 9-15 days interval after first mustard sowing. This technology has been validated at three locations near Bangalore (50ha), Varanasi (15.6ha) and Ranchi (6.3ha) during 2001-2004. The overall cost: benefit ratios were 1:3.13 (IPM) and 1:2.11 (non-IPM).

Tomato is another vegetable crop in which successful IPM technology has been developed for the management of fruit borer, *Helicoverpa armigera* (Hubner) at the IIHR. It involves growing of African tall marigold cv. Golden Age as a trap crop. Marigold nursery is raised 15 days earlier to tomato and 40-day old marigold and 25-day old tomato seedlings are simultaneously planted in the field. One row of marigold is alternated after every 16 tomato rows. Fruit borer moths lay eggs predominantly on the buds and flowers of marigold. It has also been observed that marigold has an added attraction potential for the serpentine leafminer, *Liriomyza trifolii* (Burgess), which infests tomato. The tomato fruits may suffer about 10 per cent infestation in the trap crop fields. This infestation may be controlled by sprays of NPV at 250 larval equivalents (LE)/ha or by use of conventional insecticides on tomato crop. The marigold crop is not sprayed to maintain its attraction of fruit borer moths for oviposition and colonization. It also helps in conserving the natural enemies. The grown up *H. armigera* may be hand collected from the marigold flowers and put in a pail of water containing any insecticide solution. The above technology has been validated at three locations near Bangalore (40 ha), Varanasi (25.4 ha) and Ranchi (8.9 ha) during 2001-2004. The overall cost : benefit ratios were 1:1.4.06 (IPM) and 1:1.32 (non-IPM).

Plantation Crops

IPM considerations are especially important in tropical plantation crops, where, unlike annual crops, biological interactions have a chance to reach equilibrium or near equilibrium conditions. Adoption of IPM is feasible for the effective management of rhinoceros beetle, *Oryctes rhinoceros*

(Linnaeus), one of the major pests of coconut palm. The beetle suffers from natural epizootics produced by a fungus, *Metarhizium anisopliae* (Metchnikoff) Sorokin and a baculovirus of *Oryctes* (OBU). Release of baculovrius infested beetles has been successfully used for the biosuppression of the beetle in several countries including South Pacific Islands, Fiji, Mauritius, Seychelles, Papua New Guinea and Mauritius. In India, the damage by *O. rhinoceros* was substantially checked when the OBV infected beetles were released in Minicoy and Androth islands of Lakshadweep, Chhittilappilly in Trichur and Sipighat of Andaman islands. The release of *O. rhinoceros* beetles infected with Kerala isolate of the virus resulted in over 99 per cent reduction in palm damage within three years of augmentation in Andaman islands.

There have been reports of resurgence of the beetle in some other regions including Kerala and Western Samoa. There is also the possibility of development of resistance to baculovirus in Indonesia, East Java and South Sulawese. To overcome these problems, an IPM has been proposed for the management of the beetle which includes planting legumes as cover crops to conceal the potential breeding sites of the beetle; extracting the adult beetles by means of a metal rod (about 0.5 m long) with a hook at one end during peak period of pest abundance (June-September) from the infested sites; treatment of breeding places of the beetle with the entomopathogen, *M. anisopliae*; promoting the spread of the virus by leaving some dead standing palms; increasing the prevalence of the virus by releasing virus-infected beetles (10-15 per ha) preferably during dusk; and treating the breeding sites with carbaryl (0.01%) where biological control is not possible (Singh *et al.*, 2006).

PEST MANAGEMENT ORGANIZATIONS

There are several organizations involved in promoting the cause of pest management at international and national levels.

A. INTERNATIONAL ORGANIZATIONS

Food and Agriculture Organization of the United Nations

Food and Agriculture Organization (FAO), founded in 1945, leads international efforts to defeat hunger, with a focus on developing rural areas, home to about 70 per cent of the world's poor and hungry people. FAO was one of the first organizations to promote IPM in the 1960*s* and has held several important symposia and workshops. It has organized regional IPM implementation programmes and also supported national IPM programmes. A panel of experts was formed in 1967 to advise FAO/UNEP in promoting and implementing IPM projects. The contributions of two programmes sponsored by FAO have been widely acclaimed.

FAO Intercountry Programme for IPM in Rice. The Intercountry Programme for the Development and Application of Integrated Pest Control in Rice in South and South-east Asia remains one of the best examples of IPM implementation in the tropical region. It involves purposeful, direct efforts to change farmers' practices in contrast to some more indirect routes of IPM technology diffusion in many industrialized temperate environments. The first phase of the programme (1980-86) focused on developing and testing the technical aspects of the IPM concept in its seven participating countries. More recently, the project has been directed towards enhancing farmers' adoption of IPM. One significant accomplishment of the programme has been to cause policy changes within several governments, in the form of official support of IPM as the means for national plant protection in the Philippines, Indonesia, India, Sri Lanka and Malaysia.

Global IPM Facility. The global IPM Facility has been established on June 30, 1995 at FAO, Rome and co-sponsored by FAO, UNDP, UNEP, CABI and the World Bank. This development is in response to the UN Conference on Environment and Development, held at Rio de Janeiro, Brazil, June 3-14, 1992, which assigned a central role for IPM in agriculture as part of 'Agenda 21'. Its focus is to assist interested governments and NGOs in initiating, developing, and expanding IPM programmes that aim to reduce pesticide use and associated negative impact on health and environment, while increasing production and profits through improved crop and pest management. The Facility serves as coordinating, consulting, advising and promoting agency for the advancement of IPM on global basis. The Facility helps governments in several ways to promote IPM :

- Assistance with project identification, needs assessment and preparatory studies.
- Support for pilot projects demonstrating farmer-led IPM.
- A forum for policy consultation and coordination.
- Mobilization of national expertise.
- Workshops in training methods and international exchanges.
- Resource materials including field guides.
- An on-line global IPM database and reference library.

Consultative Group on International Agricultural Research

The Consultative Group on International Agricultural Research (CGIAR), established in 1971, is a strategic partnership of diverse donors that support 15 international centres, working in collaboration with hundreds of government and non-government organizations around the world. The Group's focus is to reduce poverty and hunger, improve human health and nutrition, and enhance ecosystem resilience through high quality international agricultural research, partnership and leadership. The centres which have made significant contribution to promote IPM include Africa Rice Centre (Formerly West Africa Rice Development Association), International Centre for Tropical Agriculture (CIAT), International Maize and Wheat Improvement Centre (CIMMYT), International Centre for Agricultural Research in the Dry Areas (ICARDA), International Centre for Research in Arid and Semi-Arid Tropics (ICRISAT) and the International Rice Research Institute (IRRI). The contributions made by the IRRI in development and implementation of IPM in rice are spectacular. IRRI has developed improved varieties of rice which possess multiple resistance to several insect pests and diseases. For example, IR36 is resistant to brown planthopper, green leafhopper, stem borers, gall midge, blast, bacterial blight and tungro. This variety has been grown on about 11 million ha in South-east Asia and yielded an additional annual income of one billion dollars annually to rice growers and processors. In addition, IRRI has provided technology in biological control, botanical pesticides and application techniques of pesticides. IRRI has also provided training in IPM tactics to students, scientists and farmers from South and South-east Asian countries. The CGIAR has also initiated several IPM programmes.

CGIAR SP-IPM. The system wide programme on IPM (SP-IPM) is an intercentre initiative of the CGIAR. The programme was launched in 1995 in recognition of the need for a radically different approach if IPM research is to be more responsive to sustainable agricultural development. The SP-IPM is a global network of CG and other International Agricultural Research Centres. The SP-IPM target zones are in Africa, Asia and Latin America. The beneficiaries of the programme are farmers, national and international agricultural research organizations, extension programmes and NGOs who benefit from exchange of expertise, information as well as genetic and other resources to increase their capacity to manage pest problems, stabilize productivity and income, and foster a pesticide safe environment.

Integrated Pest Management Task Force. An Integrated Pest Management Tast Force (IPMTF) was set up in mid-1989 by a group of interested donors to assist the Technical Advisory Committee (TAC) of the CGIAR in its consideration of international support for plant protection. Its mission was to report on the status of IPM and make recommendations for improving its implementation and then proceed as necessary to execute those recommendation. The Task Force appointed a team of consultants to carry out a review and report the findings. The Consultants' Report submitted in December 1989 provides a broad overview of the major issues which influence the pace of IPM adoption in the developing world. As the work of the Task Force moved into the area of implementation, it was reconstituted as the Integrated Pest Management Working Group (IPMWG) in May 1990.

Integrated Pest Management Working Group. The Integrated Pest Management Working Group (IPMWG), set up in 1990 to promote the implementation of IPM, is a loose affiliation of organizations who meet twice a year, in May and October. Its current focus is to identify initiatives for implementation through a series of regional workshops involving key developing-country researchers, extension specialists and policy makers. The Group has been active in improvement of information exchange including publication of a quarterly newsletter IPM Working for Development since November 1993. During the last several years, IPMWG has organized a series of regional workshops, where country representatives and their development partners discussed the different constraints in implementation, and identified plans and initiatives for action. As a follow up, Regional IPM Working Groups were set up to build on the momentum created by the workshops, to carry out identified plans and to provide a structure for IPM implementation in each region, in the context of sustainable development. As a result of the efforts of the IPMWG, both as a group and through its individual members, a new financing group called IPM Facility has been established. This facility has grown out of cooperation between WB, FAO, UNDP and UNEP, in an effort to mobilize the multilateral donor system for IPM. At the same time, the research centres of CGIAR have established an Inter-centre Working Group to ensure that IPM becomes central to their research.

Commonwealth Agricultural Bureau International

The Commonwealth Agricultural Bureau Inter-national (CABI) is a non-profit international organization that improves the people's lives by providing information and applying scientific expertise to solve problems in agriculture and the environment. CABI traces its origin back in 1910, when it began as Entomological Research Committee, followed by formation of Imperial Bureau of Entomology (1913) and Imperial Agricultural Bureau (IAB) (1930), formed as Commonwealth Organization. IAB became Commonwealth Agricultural Bureaux in 1947 and truly became an international organization, viz. Commonwealth Agricultural Bureau International (CABI). In 1998, CABI Publishing came into existence; in the same year, International Institute of Entomology, International Institute of Biological Control, International Institute of Parasitology merge to form CABI Bioscience. In 2006, CABI Publishing and CABI Bioscience became united under one single CABI brand. CABI has international centres in the UK, Kenya, Pakistan, Switzerland, Malaysia, and Trinidad and Tobago, and country offices in Costa Rica, India, Vietnam and China.

CABI has become an international service in agricultural information, pest identification and biological control. It has provided a range of services to national programmes that represent critical inputs for the effective planning and implementation of IPM programmes. Its diagnostic services in identification of pathogens and insect pests are invaluable. Its comprehensive information services in agriculture are also unmatched. Recently, CABI has brought out many valuable publications relating to different aspects of IPM. Of particular relevance to IPM is CABI's International Institute of Biological Control (IIBC). For nearly 70 years, IIBC has been providing international service in biological methods of pest and weed control.

IPMnet

The global IPM Information Service was created through the joint efforts of the Consortium for International Crop Protection (CICP) and the National Biological Impact Assessment programme (NBIAP) of the US Department of Agriculture in late 1993. IPMnet offers IPM support almost immediately on line through Internet 24 hours a day, seeks to assemble and disseminate global information that will support IPM research, teaching, technology implementation and policy development. In addition to serving as an information source, IPMnet also provides an effective channel for contacting others working on the same problem, and serves as a source for finding specialized equipment or materials. There is also a forum to use for engaging colleagues around the world in discussion and debate related to IPM development and strengthening issues.

B. NATIONAL ORGANIZATIONS

These are a number of organizations in India, which are engaged in promoting the cause of IPM to reduce the pesticide load in the environment.

Directorate of Plant Protection, Quarantine and Storage

Directorate of Plant Protection, Quarantine and Storage (DPPQS) was established at Faridabad under the Ministry of Agriculture, Government of India in 1946 to bring awareness about plant protection problems and monitor pests and diseases. In this connection, Central Plant Protection Stations were established in 1957-58. The strategy adopted by these stations was to educate the farmers and extension officers of the State Department of Agriculture through regular training programmes and demonstration on the use of plant protection measures. The notable initiatives taken by the Government of India for the promotion of IPM on sustainable basis are : (*i*) Setting up of 31 Central IPM Centres (CIPMC) for promotion of IPM approach in 28 states and 1 union territory, (*ii*) Assistance to state for setting up of State Biocontrol Laboratories and emphasis on production and releases of biocontrol agents. (*iii*) Allocation of 50 per cent state funds on plant protection to promote IPM.

The Government of India has adopted integrated pest management (IPM) as cardinal principle and main plank of plant protection in the overall crop production programme since 1987, with the following objectives :

- Maximise crop production with minimum input costs.
- Minimise environmental pollution in soil, water and air due to pesticides.
- Minimise occupational health hazards due to chemical pesticides.
- Preserve ecosystem and maintain ecological equilibrium.
- No or less use of chemical pesticides for minimum pesticide residues.
- To improve farming systems.

The Directorate is undertaking the following activities to achieve the above objectives.

- Popularizing IPM approach among farming community.
- Conducting regular pest surveillance and monitoring to assess pest/disease situation, for need based and timely application of selective insecticides.
- Conservation and augmentation of beneficial species such as parasitoids, predators and microorganisms.
- Promotion of biopesticides and improved cultural practices including the use of tolerant/resistant varieties and other non-chemical methods of pest control.
- Play a catalytic role in spread of innovative IPM skills to extension workers and farmers equally to resource-poor and resource-rich states.
- Human resource development in IPM by imparting training to master trainers, extension workers and farmers through Farmer Field Schools through Central IPM Centres, Krishi Vigyan Kendras and ICAR institutes.

National Bureau of Agriculturally Important Insects

An All India Coordinated Research Project on Biological Control of Crop Pests and Weeds was initiated in 1977 under the Indian Council of Agricultural Research. When in 1988, the Commonwealth Institute of Biological Control (CIBC), Indian Station, Bangalore was closed, the Project Co-ordinator's cell was merged with that unit taken over by the ICAR. The new headquarters called Biological

Control Centre was shifted to the premises of this erstwhile CIBC, Indian Station. The ultimate recognition of the importance of biological control came during the VIII plan with the upgradation of the centre to the Project Directorate of Biological Control (PDBC) during October 1993 with headquarters at Bangalore (Bengaluru). PDBC has been further upgraded to National Bureau of Agriculturally Important Insects (NBAII) during the XI plan.

NBAII is a nodal institute at national level for research and development on all aspects of work on harnessing resources of insects including biological control of crop pests and weeds, training, information repository, technology dissemination, and national and international cooperation. Under All Indian Coordinated Research Protect on Biological Control of Crop Pest and Weeds (located at NBAII), with 20 centres carries out work on basic and applied aspects of biological control of crop pests and weeds.

The following thrust areas have been proposed under the Bureau's programme :

- Exploration, identification, characterization, conservation and utilization of biodiversity of beneficial insect resources.
- National and international exchange of beneficial insects.
- Development of digitized inventories, information storage and retrieval systems on insect resources.
- Consultancy and human resource development for exploitation of insect sources.
- Utility of effective symbionts and toxins produced in insect cell line cultures and cloned products of biopesticides with increased virulence and persistence.
- Establishment of repository for insect derived beneficial bacteria and testing for crop use.
- Exploitation of acarine microbial associates.
- Insect symbiosis for pest and disease management in crop plants.

National Centre for Integrated Pest Management

The National Centre for Integrated Pest Management (NCIPM) was established by the Indian Council of Agricultural Research in February 1988 to cater to the plant protection needs of the country. The mandate of NCIPM is

- To develop and promote IPM technologies for major crops so as to sustain higher crop yields with minimum ecological implications.
- To develop information base on all aspects of IPM to advise on related national priorities and pest management policies.
- To establish linkages and collaborative programmes with other national and international institutes in the area of IPM.
- To extend technical consultancies in IPM.

The Centre aims for effective cooperation with All India Co-ordinated Crop Improvement Programmes, Research Institutes, State Agricultural Universities, Department of Science and Technology, Department of Biotechnology, National Remote Sensing Agencies, Indian Meteorology Department, National Information Centre Network, National Horticultural Board and Department of Plant Protection, Quarantine and Storage for implementation of its programmes. The headquarter of the Centre has recently been shifted from Faridabad to New Delhi. The Centre has started publishing a NCIPM newsletter with effect from June, 1995.

Department of Biotechnology

The Department of Biotechnology (DBT), Government of India is providing financial assistance to various state agricultural universities and research centres for developing and producing biopesticides and biocontrol agents such as nuclear polyhedrosis virus, granulosis virus, *Trichogramma, Chrysoperla* and *Trichoderma.* Presently, 10 biopesticide production units are able to cover an area of one million ha per annum on 10 crops. Under this programme, biopesticide production units and plant protection clinical centres have also been established at regional research stations. DBT has made concerted efforts towards the development of biopesticide technology in a systematic manner during the last more than 20 years by launching various projects and programmes.

National Biocontrol Network Programme. DBT had established a National 'Biocontrol R & D Network Programme in 1989 to study the control of insect pests, diseases and weeds of economically important crops. Almost 300 R & D projects have been implemented at various ICAR/CSIR national institutes and state agricultural universities throughout the country. This is a continued programme of DBT and its main aim is to develop better formulations as well as cost effective, commercially viable mass production technologies of various biopesticides. Based on the technology developed through this programmes, two biocontrol units were set up at Coimbatore and Madurai in Tamil Nadu.

Integrated Pest Management for Sustainable Agriculture. A major R & D programme on Biological Control of Crop Pests and Diseases under IPM as a Component of INM Programme' was launched during 1998-99. This programme was launched at 14 centres in 12 states. The main aims of this programme were (*i*) to develop effective modules/package of practices, which are cost effective, sustainable and ecofriendly in different ecosystems ; (*ii*) to demonstrate the sustained preservation of ecosystem by conducting frontline demonstrations in identified clusters by adopting IPM and INM techniques ; (*iii*) to develop suitable techniques for packaging, storage and application of bioagents to enhance their field efficacy; and (*iv*) to promote involvement of NGOs towards the use of biopesticides. This programme has been concluded and finally reviewed in 2003.

National Institute of Plant Health Management

The National Institute of Plant Health Management (NIPHM), Hyderabad, is an autonomous institute under the Department of Agriculture and Cooperation, Ministry of Agriculture, Government of India, which came into being in 2008. NIPHM has its origin in the Central Plant Protection Training Institute (CPPTI) which was established in 1966 as a training agency of the Directorate of Plant Protection, Quarantine and Storage. The basic objective of CPPTI was to develop human resources in various plant protection disciplines. The institute was renamed as the National Plant Protection Training Institute (NPPTI) in 1994. Based on the recommendations of an evaluation study by National Productivity Council, Chandigarh, Department of Agriculture and Cooperation, decided to make this institute as an autonomous body in October 2008. In accordance with its mandate, NIPHM offers its services in training, research, consultancy and plant health management education, with the following objectives :

- Human resource development in plant protection technology, plant quarantine and biosecurity with special emphasis on crop-oriented IPM approaches, and protecting the biosecurity borders, both in public and private sectors.
- Human resource development in analysis of pesticide formulations and pesticide residues for monitoring the quality status of pesticides in states/Uts.
- Develop systematic linkages between state, regional, national and international institutions of outstanding accomplisments in the field of plant protection technology.
- Function as a nodal agency/forum for exchange of latest information on plant protection technology.
- Collect and collate information on plant protection technology for dissemination among the state extension functionaries and farmers.
- Identify, appreciate and develop modern management tools, techniques in problem-solving approaches and utilizing the mechanism of personnel management, resource management, input management and finally conflict management at the organization level.

- Develop need-based field programmes for training and retraining of senior and middle level functionaries for executing plant protection programmes, and using training of trainer approaches to ensure maximum reach of the programmes.
- Conduct programme-oriented research in the areas of plant protection, integrated pest management, pesticide management, plant quarantine and pesticide delivery systems and residues, to provide feedback to training programmes.
- Serve as repository of ideas and develop communication and documentation services at national, regional and international levels with regard to the subject of plant protection management.
- Forge linkages with national and international institutions and create networks of knowledge sharing through a programme of institutional collaboration and employment of consultants.
- Function as policy support to the Central Government in various sectors of plant protection, viz. integrated pest management, pesticide management, plant quarantine, biosafety, etc.

FUTURE OUTLOOK

The green revolution, which addressed the food security crisis of the 1960s and 1970s, involved scientific innovation and active promotion of high-yielding varieties that could transform fertilizer inputs into grain production, but it created many ecological problems. The future IPM programmes should be so designed that they minimise the risks associated with the green revolution technology and maximise the benefits of the gene technology. The pest management system should blend the traditional ecological principles and frontier technologies. This implies that the IPM programmes should represent a sustainable approach to manage pests combining biological, chemical, cultural and biotechnological tools to ensure favourable economic, ecological and sociological consequences. For the evergreen revolution to succeed in feeding the world through sustainable cropping systems, a concerted and integrated effort on the part of scientists, farmers and policy makers will be required to create awareness and a political and regulatory environment in which sustainable technologies like IPM are embraced to meet the food and fibre needs of the 21st century.

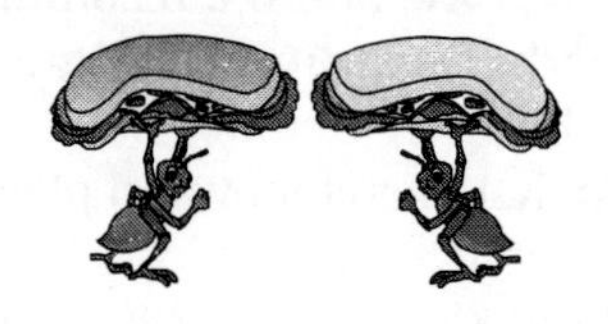

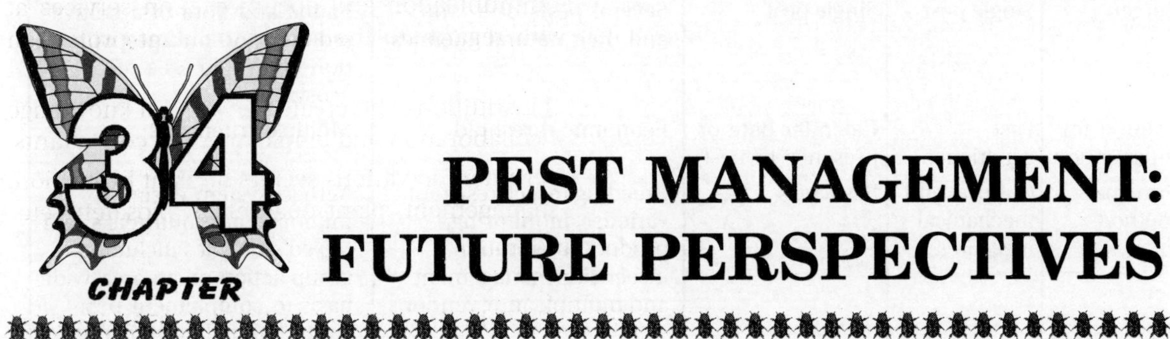

PEST MANAGEMENT: FUTURE PERSPECTIVES

INTRODUCTION

The introduction and widespread use of high-external-input modern agriculture has caused many problems. Firstly, these problems are socio-economic in nature, *i.e.* technologies improved the condition of larger farmers and did not take enough care of small farmers; land tenure patterns became more skewed; position of women often became problematic as their role was not taken into consideration at all. Secondly, problems were encountered in the areas of environment and health; the magnitude of these problems varied from direct poisoning of farmers by use/misuse of agrochemicals to deterioration in quality of air, water and food for human consumption. The third and most recently talked about problem is that the green revolution approach in itself was not sustainable. After initial yield increases, introduction of green revolution packages often led to such disturbances in the agroecological system that subsequent yield reductions were inevitable.

The term 'sustainable agriculture' first appeared in the literature in 1978 but was formally introduced in 1985 when the US Congress "enacted the Food Security Act that initiated a programme in 'low input sustainable agriculture' to help farmers use resources more efficiently, protect the environment and preserve rural communities." Sustainable agriculture may be considered as an agriculture that can evolve indefinitely toward greater human utility, greater efficiency of resource use, and a balance with the environment that is favourable both to humans and to most other species. Sustainable agriculture emphasizes that agricultural systems must be economically viable and they must contribute to desirable environmental qualities over the long term. IPM employs ecologically based management processes developed with an understanding of natural cycles and natural regulators of those species that compete with humans for resources in agricultural production systems. Therefore, successful IPM programmes are those that enhance agricultural enterprises and protect the environment for the indefinite future (Koul and Cuperus, 2007).

There have been considerable variations in pest management approaches from traditional to sustainable agriculture (**Table 34.1**). The modern integrated pest management (IPM) approach aims at systemic adjustment, i.e. it looks for more efficient use of pesticides and product substitution (biorationals and botanicals in place of conventional pesticides) within an agricultural system that essentially remains unchanged. However, sustainable agriculture aims at structural changes, i.e. it seeks to change the way we look at agriculture through better understanding of ecological processes and maximizing synergy between crops.

Table 34.1 Approaches to pest management : Retrospect and prospect

Parameter	Traditional agriculture	Green revolution technology	Current IPM approach	Sustainable agriculture
Goal	Reduce pest damage	Eliminate or reduce pest species	Reduce cost of production	Multiple-social, economic and ecological goals
Target	Single pest	Single pest	Several pests of a crop and their natural enemies	Fauna and flora of a cultivated area and linkages with non-cultivated area of ecosystem
Criteria for intervention	Past experience	Calendar date or presence of pest	Economic threshold	Multiple criteria
Principal method	Cultural and mechanical measures	Pesticides	Breeding insect-resistant varieties, monitoring, product substitution, insecticide management and multiple interventions	Agroecosystem design to minimize pest outbreaks and mixed strategies including group action on an area-wide basis to complement pest control aimed at individual fields
Diversity	High	Low	Low to medium	High
Spatial scale	Single farm	Single farm	Single farm or small region defined by pests	Biogeographic regions
Time scale	Single season	Immediate	Single season	Long term, steady state, dynamic equilibrium
Effect on enviornment	Usually negligible	Highly detrimental	Moderately detrimental	Negligible

Source: Modified from Dhaliwal and Koul (2010)

IPM: SPRINGBOARD TO SUSTAINABLE AGRICULTURE

The proponents of the sustainability concept for crop production have found great affinity with principles and approaches of IPM. Indeed, IPM provided both a conceptual approach and an implementation paradigm for sustainable agriculture. IPM and sustainable agriculture share the goal of developing agricultural systems that are ecologically and economically sound. IPM may be considered a key component of sustainable agriculture. From an IPM perspective, the concept of sustainable agriculture provides a platform for launching IPM to higher levels of integration. Therefore, the future developments and successes in IPM are quite important to the sustainability of agriculture in the twenty-first century. The contributions of IPM are critical to meet the economic, environmental and social mandates in assuring sustainability of agricultural systems (Koul *et al.,* 2004).

Economic Mandate

Economic considerations basic to IPM are consistent with the requirement for profitability in sustainable agricultural systems. IPM is based on the concept that low population densities of pests usually do not threaten the profitable production of agricultural commodities. This aspect of IPM philosophy has resulted from the development of concepts of 'economic injury level' and 'economic threshold'. The sequence of sampling to assess the prevalence of pests followed by decision making using criteria of economic thresholds supports the basic tenet of both IPM and sustainable agriculture, i.e. production systems must be economically viable over the long term. In this context, the economic injury level and economic threshold are the parameters that not only serve as criteria for decision making, but are also important for defining the contribution of effective pest management to the sustainability of production systems. As the frequency of pest occurrence at population densities exceeding the cost of available controls and the profitability of pest control measures are summarized over time, the value of pest management in production systems can be estimated.

Environmental Mandate

The idea that profitability in agricultural production and protection of the environment are jointly attainable goals, is central to the philosophy of IPM and sustainable agriculture. It is critical to sustainability that the use of off-farm inputs such as agrochemicals which may be potential pollutants, and farming practices such as tillage operations, which may contribute to soil erosion, be employed in a manner that does not result in the degradation of soil and water. The judicious use of chemical pesticides in agricultural production systems has been central to the development of IPM since its inception. Through IPM, the judicious use of chemical pesticides is emphasized to (*i*) preserve natural control agents, such as entomophagous insects and beneficial microorganisms, (*ii*) decrease the potential for mortality of an array of non-target organisms such as wildlife species, and (*iii*) limit the accumulation of toxic residues in the environment.

Social Mandate

Health and well-being are highly valued in societies all over the world, resulting in demands for a safe, wholesome food supply that is produced without harm to the environment or hazards to those who work in agriculture. Clearly, the demands of people around the world for safe, abundant food supply can be met only by the development of sustainable agricultural systems that can effectively combine pest controls with profitability and the maintenance of a safe environment and human food supply. In addition to providing a safe and effective means for reducing damage to commodities by all types of pests, IPM is also the primary means by which hazards of both chemical and microbial contamination of food commodities may be greatly decreased. The perception of an unlimited supply of cheap food that exists in many countries could rapidly be provided an illusion if effective means of pest management are not maintained through judicious use of existing control technologies and consistent investment in research to develop new avenues for pest control for the future. Through deliberate and persistent educational efforts, the public must come to appreciate the development and implementation of IPM, even when programmes involve the use of products of biotechnology such as genetically modified crops, are a sound investment of resources in support of sustainable agricultural production of the future.

PERSPECTIVES IN IPM

Pest management on high-value crops relies heavily on the use of pesticides, often to the exclusion of other methods. With an increasing restraint on insecticide use due to development of resistance in insect populations and environmental contamination, integration of several management techniques has become essential to reduce reliance on pesticides and prolong the utility of valuable molecules. In order to overcome the toxic and chronic effects of pesticides, as well as pest resurgence, intensive research efforts are needed to develop a balanced programme for IPM. This would require exploring and refining of various components of IPM, and devising strategies for their suitable integration (Dhaliwal 2008; Dhaliwal *et al.*, 2013).

Parasitoids and Predators

Entomophages, which feed on insects, play an important role in regulating pest populations. However, their potential has largely remained underestimated and under-exploited. It has been estimated that only 15 per cent of entomophages have so far been identified. Parasitoids belonging to the insect order Hymenoptera have been involved in about 66 per cent of all successful biocontrol programmes. Most of the common predators occur in insect orders Coleoptera, Hemiptera and Neuroptera. In addition, there are several species of mites and spiders feeding on a wide range of insects and mites. More than 125 species of natural enemies are commercially available at global level for augmentative biological control programmes, including 37 commonly used species such as the moth egg parasitoid, *Trichogramma* spp.; whitefly parasitoid, *Encarsia formosa* Gahan, and the spider mite predator, *Phytoseiulus persimilis* Athias-Henriot. The potential of classical biocontrol has been demonstrated in Africa where the cassava mealybug, *Phenacoccus manihoti* Matile-Ferrero, has been virtually eliminated in 30 countries, by the exotic parasitoid, *Epidinocarsis lopezi* (De Santis) (Dhaliwal and Koul, 2007; Sivastava and Dhaliwal, 2010).

Parasitoids and predators can provide long term regulation of pest species provided proper management practices are followed to make the environment conducive to furthering their abundance and efficiency in target agro-ecosystems. Biological control can potentially become a self-perpetuating strategy, providing economic control with the least environmental hazards. However, much work needs to be done to optimize the utilization of parasitoids and predators in integrated pest management (Jindal *et al.,* 2013).

- There is an urgent need to establish a network of large scale multiplication units so that the natural enemies are available to the farmers at reasonable prices. A new industry of mass propagation of natural enemies is born as costs of mass rearing are reduced, making this process commercially competitive. As the technology of mass propagation of natural enemies develops, more arthropod pest species will become amenable to biological control.
- Heat-and cold-tolerant strains have to be selected/developed in the case of a number of natural enemies. The environmental implications of releases of these organisms, especially in cases of introductions and genetically engineered organisms should be, investigated.
- Since the performance of most biocontrol agents is known to be affected by physical or chemical attributes of the target crops, it is necessary to identify species or strains adapted to major crop systems.
- Major improvements in biological control of insect pests can be made through habitat management. Increasing genetic diversity could provide useful means of augmenting natural enemy populations. However, the response varies across crops and cropping systems. Therefore, appropriate cropping systems should be identified for specific predators and parasitoids to increase their efficacy.
- The allelochemicals produced by the host plant mediated through the pests, affect the biology and efficiency of natural enemies. Hence, a thorough understanding of tritrophic interactions among crop plants, insect pests and their natural enemies is essential to derive strategies for biological control of insect pests.
- There is a need to develop entomophage parks, i.e. undisturbed habitats of natural vegetation near agricultural areas to protect and enhance specific natural enemies and provide them with resources such as nectar, pollen, physical refuge, alternative prey, alternative hosts and mating sites. This will improve natural enemy fitness and effectiveness.

Microbial Pesticides

So far, over 3000 microorganisms have been reported to cause diseases in insects. However, scientists familiar with specific pathogen groups agree that a very large number of insect pathogens remain undiscovered or unidentified. More than 100 bacteria have been identified as arthropod pathogens, among which *Bacillus thuringiensis* Berliner (Bt) has received the maximum attention as microbical control agent. Viruses have been isolated from more than 1000 species of insects from 13 different insect orders. World over, about 525 insect species in 52 families and 8 orders are known to be infected by nuclear polyhedrosis virus (NPV) and of this large number of species, 455 belong to Lepidoptera. Till now, over 800 species of entomopathogenic fungi have been identified. More than 1000 species of protozoa pathogenic to insects have been described and many more remain to be discovered. The two major groups of entomopathogenic nematodes are *Steinernema* (55 species) and *Heterorhabditis* (12 species) (Koul *et.al.* 2013).

In spite of their great potential, there are several constraints in the use of microbial pesticides, which require a focus in the future. Depending upon the country, registration of a microbial pesticide is a lengthy and expensive procedure. Compared to chemical pesticides, microbial pesticides are expensive, slow acting and field efficacy is inconsistent. Microbial pesticides require special procedures for formulation, packaging, storage and application, and their efficacy is also influenced by environmental conditions. Narrow host range, necessity to ingest Bt toxins by the target insects, ability of insect larvae to avoid lethal doses of Bt by penetrating into the plant tissue, inactivation by sunlight, and effect of plant surface chemicals on its toxicity limit its widespread use in crop protection. Similarly, narrow range, slow rate of insect mortality, difficulties in mass production, stability under sunlight and farmers' attitude have limited the use of NPVs as commercial pesticides.

Owing to the early successes and continuing growth of biopesticide market, expectations for the performance of microbial biopesticides are quite high. However, there are many challenges that will need to be overcome so that the self-perpetuating nature of most of the microbial pathogens may prove to be an asset in sustainable agriculture.

- In order to increase the utility of microbial pesticides in IPM programmes, systematic surveys are required in different agroecological regions to identify naturally occurring pathogens. Detailed studies are necessary on the properties, mode of action and pathogenicity of such organisms.
- Ecological studies on the dynamics of diseases in insect populations are necessary because the environmental factors play a significant role in disease outbreaks and ultimate control of the pests.
- There is a need to develop and standardize mass production technologies of microbial biopesticides in order to solve potential problems associated with contamination, formulation potency, alternation of pesticidal activity and shelf-life.
- Suitable formulations should be developed to increase their residual activity and improve shelf-life. Commercially dry formulations are preferred over liquid formulations. Lyophilization and encapsulation should be explored to produce stable formulations with persistent toxicity. The use of formulations that include stilbene-derived optical brighteners, increase efficiency of NPV formulations.
- The relatively slow speed with which microbial pathogens kill their host has hampered their effectiveness as well as acceptance by potential users. Genetic improvement with conventional and biotechnological tools would lead to the production of strains with improved pathogenesis and virulence.

Botanical Pesticides

The use of plants and their crude extracts for the protection of crops and stored products from insect pests has been a part of traditional agriculture for generations. Neem leaves and kernel powder have been traditionally used by farmers against pests of household, agricultural and medical importance. More than 6000 plant species from at least 235 plant families have been screened for pest control properties. A large number of plant products derived from neem, custard apple, tobacco, pyrethrum, etc. have been used as safer pesticides for pest management. Phytochemicals from Meliaceae family have shown remarkable feeding deterrency, repellency, toxicity, sterilant and growth disruptive activities. Azadirachtin, the major bioactive principle of *Azadirachta indica* A. Juss. and azadirachtin based formulations show wide array of pest control properties and are now globally available. Efforts are needed to identify more molecules of plant origin so that they can be used successfully in pest management in the future (Koul and Dhaliwal, 2001; Dhaliwal and Koul, 2007).

Several problems have been encountered while commercializing the botanical pesticides, which are related to quantity of raw material, thermal and photostability, as well as quality control and product standardization. Like synthetic pesticides, the repeated and excessive use of botanical pesticides may also lead to pest resistance. The possibility of insects developing resistance looms large if single botanical pesticide like azadirachtin is allowed to be used too frequently. The phytotoxicity observed with the botanicals is also a matter of concern. Neem oil based formulations are often phytotoxic to tomato, brinjal and ornamental plants at oil levels above 1 per cent. Although plant products are considered to be relatively safe to humans, however, this cannot be assumed for all plant species. Some of the most toxic substances known to man, *e.g.* aconitine and ricin are produced by plants. Some of the plant species such as *Taxus* spp., *Aconitum* spp. and *Ricinus communis* Linnaeus have notoriously high toxicity to man. Some of the commonly used plant materials in Africa such as *Tephrosia vogelii* Hook f. have well-known environmental impacts, particularly against fish (Koul *et al.*, 2012).

To ensure there is a future for pesticidal plants, there are many issues that need to be addressed by the scientific community, policy makers and institutions involved in knowledge dissemination.

- Scientists need to provide better information that explains how pesticidal plants work, which arthropod species are affected, how the bioactive chemicals may vary according to season, locality or variety and how best plants should be harvested and processed to conserve and deliver bioactivity.
- Furthermore, scientists need to engage with policy makers to tackle issues such as conservation of wild habitats and survey of unexplored plant biodiversity for pesticidal plants.
- There is a need of special set of guidelines for registration of plant products, which should be less stringent than other chemical pesticides.
- Quality control in botanical pesticides is a major problem. The active ingredient levels are often affected by the agroecological factors in different regions of plant growth. There is an urgent need to define quality standards for botanical pesticides in order to obtain consistent results.
- The threshold of active ingredients such as azadirachtin in neem, pyrethrins in chrysanthemum, their biomass etc. may be raised through natural selection, exotic introductions, tissue culture and other biotechnological manipulations.
- The photo- and heat-liability of botanical pesticides is another area which requires serious attention. Because of these undesirable traits, repeated outdoor applications of products are necessitated. Appropriate methodologies need to be developed to improve both residual and shelf life. Suitable stabilizers, UV-screens and antioxidants need to be identified for incorporation in the formulations.

Semiochemicals

Semiochemicals or behaviour modifying chemicals have gained prominence since quite recently in pest management programmes. There was a rapid growth in the identification of insect pheromones during 1970*s*, and by the end of 1980s, pheromones and pheromone mimics were known for about 1000 species of insects. Today, more than 1500 moth sex pheromones and hundreds of other pheromones have been identified, including sex and aggregation pheromones from beetles and other groups. The first field trials that involved assaying pheromones for pest control were carried out in mid 1980*s*. Since then, hundreds of other pheromones have been identified and there are more than 50 (in over 300 different formulations) that can be used in pest management programmes. Over 30 target species have been controlled successfully by mating disruption. It has been estimated that at least 20 million pheromone lures are produced for monitoring or mass trapping every year.

The world mean sale of semiochemical products is about US$ 70-80 million or 1 per cent of the agrochemical market. The fact that producers rely on the deployment of air permeation and attract- and kill- techniques for pest control over 1 million ha would justify continued efforts to develop and implement management programmes based on the use of these semiochemicals (Witzgall *et al.*, 2010; Dhaliwal *et al.*, 2013).

Development of semiochemical-based techniques for pest management needs stimulus from the scientific community, industry and the policy makers (Jindal *et al.*, 2013).

- Basic studies on understanding of the mechanisms underlying communication systems in insects, coupled with a good working knowledge of biology, behaviour and mating systems to target insects, should be undertaken. The effect of various meteorological and physiochemical factors on chemical language of insects and plants should be understood to achieve success with semiochemicals.
- Many of the semiochemicals are photodegradable and, therefore, rapidly loose their efficacy following their applications on the crop. Therefore, suitable protected and controlled release formulations should be developed, which retain their effectiveness for a considerable length of time.
- Insects are believed to be less prone to development of resistance to semiochemicals because of their novel mode of action. However, several cases of semiochemical resistance have been documented and therefore, timely and effective measures should be undertaken so that we may not loose the useful attributes of these safe compounds.
- The best successes with semiochemicals have been achieved where large, contiguous areas have been treated with these compounds. Therefore, an area-wide approach will have to be followed to control the target pests in a defined area. For this, cooperation of the farmers is essential and there is a need for more efficient technology transfer for those who will benefit by application of control methods based on semiochemicals.
- Commercialization of semiochemical-based products is strongly affected by the size of the potential market, cost of registration and the product's price competitiveness. The lack of commercial interest by major agrochemical firms has clearly hampered the development of semiochemical-based products. Therefore, the industry should be given proper incentives and the commercial successes achieved with semiochemicals should energize the level of commitment by industry in developing new products.

Chemical Control

Pesticides have played a pivotal role in bringing about green revolution in many countries. The potential of high yielding varieties was realized under the pesticide umbrella. Pesticides are the most powerful tool in pest management. Pesticides are highly effective, economical, rapid in action, adaptable to most situations and flexible enough to meet the changing agronomic situations. Pesticides are the most reliable means of reducing crop damage when the pest populations exceed economic threshold levels (ETLs). When used properly based on ETLs, pesticides provide a dependable tool to protect the crops from the ravages of insect pests. Despite their effectiveness, much pesticide use has been unsound, leading to problems of development of resistance, pest resurgence, pesticide residues in the food commodities, non-target effects in the environment, and direct hazards to human beings. More than 577 species of insects and mites have developed resistance to different groups of pesticides. Resurgence of insect pests not only leads to increased use of pesticides, but also increases the cost of cultivation, greater exposure of the operators to toxic chemicals, and failure of the crop in the event of poor control of target pests. Many scientific studies have proved biomagnification of pesticide residues in human tissues, and products of animal origin. Over 100,000 cases of accidental exposure to pesticides are reported every year, of which a large number prove fatal. Hence, there is a need to look for new molecules, which are effective against insect pests but cause minimum environmental hazards (Dhaliwal and Singh, 2000).

Many industrialized countries have enforced stringent pesticide regulations and developed alternative pest management approaches as a result of which pesticide use in these countries has shown a declining trend. Consequently, the magnitude of contamination of food materials has also slowed down. However, many developing countries continue to use persistent pesticides in agriculture and public health programmes, and the contamination of different components of the environment continues to be excessive and pervasive. In addition, pesticide subsidies coupled with improper pesticide application and use has further accentuated the problems. Therefore, there is an urgent need to rationalise the use of pesticides in the context of IPM.

- Development of resistance to pesticides has often resulted in widespread failure of chemical control. Pesticide resistance management strategies have aimed either at preventing the development of resistance or to contain it. All rely on a strict temporal restriction in the use of certain pesticides and their alteration with other pesticide groups to minimize selection for resistance. Because of economic advantages and safety to non-target organisms, all efforts should be directed towards developing management strategies aimed at prolonging the life of useful molecules.
- Resurgence of insect pests in several species on various crops has posed a serious problem. This phenomenon not only leads to increased use of pesticides, but also increases the cost of cultivation, greater exposure of the operators to toxic chemicals, and failure of the crop in the event of poor control of the target pests. Therefore, mechanisms underlying resurgence must be taken to avoid/or delay insect resurgence.
- There is an urgent need for improvements in pesticide application methods, timing and placement. The refinement of spray technology will result in improved efficacy with reduced pesticide residues in raw agricultural commodities. Some of the application equipment does not give the desired performance for specific crop-pest, climatic, and topographic conditions. There is a need to devise suitable application equipment to meet the farmers' needs in rain-fed agriculture. Moreover, dry areas need different types of pesticide formulations, which require a minimum amount of water. Hence, research efforts should be concentrated on developing the right type of plant protection equipment vis-a-vis pesticide formulations.
- Efforts should continue to search and identify newer compounds that can be successfully used in pest management programmes. The pesticide industry must emphasize the development of new products with greater selectivity for natural enemies and minimal environmental hazards.
- In view of the environmental hazards of the pesticides, there is an urgent need to rationalize the use of pesticides in pest management. This would require vigilant efforts on the part of policy planners, government implementing agencies, scientists, farmers and the consumers, to reduce the pesticide load in the environment. Pesticides would remain an indispensable part of modern agriculture and must be used in combination with other approaches in integrated pest management.

Host Plant Resistance

Inspite of the significance of host plant resistance as an important component of IPM, breeding for plant resistance to insects has not been as rapidly accepted as breeding of disease-resistant cultivars. This is partially due to the relative ease with which insect control is achieved with the use of insecticides and slow progress in developing insect resistant cultivars because of the difficulties involved in ensuring adequate insect infestation for resistance screening. High levels of plant resistance are available against a few insect species only. In fact, very high levels of resistance are not a pre-requisite for use in IPM. Varieties with low to moderate levels of resistance or those that can avoid pest damage can be deployed for pest management in combination with other components of IPM. Deployment of pest-resistant cultivars should be aimed at conservation of natural enemies and minimizing the number of pesticide applications. Resistant cultivars can be used as a principal method of pest control, an adjunct to other management tactics, and a check against the release of susceptible cultivars. Resistant crop varieties developed in recent years represent some of the greatest achievements of modern agriculture in increasing and stabilizing world food and fibre supplies (Dhaliwal and Singh, 2005; Smith and Clement, 2012).

Considerable progress has been made in identification and utilization of plant resistance to insects. The current global economic value of plant resistance is several hundred million dollars per year. The ecological value of plant resistance has greatly decreased world pesticide usage, contributing to healthier environment for humans, livestock and wildlife. Agricultural producers have benefited from crops with arthropod resistance through decreased production costs. Consumer benefits derived from insect-resistant crops include safer and more economically produced food. Plant resistance to insects should form the backbone of pest management programmes in integrated pest management.

- Multilocational testing of the identified sources and breeding material need to be strengthened to identify stable and diverse sources of resistance or establish the presence of new insect biotypes. Resistance to insects should be given as much emphasis as yield to identify new varieties and hybrids.
- New and improved insect infestation techniques and devices that safely and efficiently place test insects onto the crop plants will also be essential to future progress. The development and refinement of standardized rating scales to determine insect damage to more crops will greatly facilitate the development of insect-resistant cultivars to several additional crop plant species.
- More information is needed on mechanisms of resistance, genetic regulation of resistance traits, and biochemical pathways and their physiological effects. Our knowledge of how plants recognize insect-feeding attacks and the elicitors they produce in response to insect feeding is increasing rapidly The evolving model of the differences in plant defense-response elicitors must be researched, challenged and modified to better understand induced resistance function and how plant metabolism can possibly be modified to use induced crop-plant resistance in insect pest management programmes.
- The complexicity of tritrophic interactions plays a vital role in host plant resistance. Elucidation of these interactions can help further understanding, and provide greater potential for manipulation of these systems to specific crop species and varieties. The possibility of using compounds from plants to reduce herbivore damage and increase the effectiveness of biological control agents is quite attractive. Ideally, plant resistance should strive to reduce substances attractive to herbivores, while increasing the substances attractive to natural enemies.
- Occurrence of new biotypes of the target pest may limit the use of certain insect resistant varieties in time and space. In such situations, we should go for polygenic resistance or continuously search for new genes and transfer them into high yielding varieties.

Transgenic Crops

The introduction of transgenic technology has added a new dimension to pest management. The global area under transgenic crops has increased from 1.7 million ha in 1996 to 175.2 million ha in 2013. Of the 27 countries growing transgenic crops, 19 were developing and the remaining 8 were developed countries. A total of 18 million farmers grew transgenic crops in 2013, about 90 per cent were small resource–poor farmers from developing countries. India celebrated a decade of successful cultivation of Bt cotton in 2011, when Bt cotton occupied 88 per cent of the total 12.1 million ha of cotton crop. The increase from 50,000 ha of Bt cotton in 2002 to 10.6 million ha in 2011 represents an unprecedented 212-fold increase in 10 years. The area under Bt cotton further increased to 11 million ha in 2013. India enhanced farm income from Bt cotton by US$ 12.6 billion in the period 2002 to 2011 and US$ 3.2 billion in 2011 alone. Thus, Bt cotton has transformed cotton production in India by increasing yield substantially, decreasing insecticide applications by about 50 per cent, and through welfare benefits, contributed to the alleviation of poverty of 7.3 million small resource-poor farmers and their families in 2013 alone.

In addition to higher yield, the benefits to farmers of transgenic crops include the lower input costs in terms of pesticide use, and ease of crop management. The reduction in pesticide usage would lead to reduced exposure of farm labour to pesticides, reduction in harmful effects of pesticides on non-target organisms, and reduced amounts of pesticide residues in food and food products. The additional benefits to farmers would be to control insect pests which have become resistant to commonly used pesticides, and reduction in crop protection costs. These factors are likely to have substantial impact on the livelihood of farmers in both developed and developing countries. In many developing countries, small-scale farmers suffer pest-related yield losses because of technical and economic constraints. Pest-resistant genetically modified crops can contribute to increased yields and agricultural growth in such situations. Available impact studies of insect-resistant and herbicide-tolerant crops show that these technologies are beneficial to farmers and consumers, producing large aggregate welfare gains as well as positive effects for the environment and human health. The advantages of future applications could be even much bigger. Transgenic crops can contribute significantly to global food security and poverty alleviation (Sharma, 2012).

Despite numerous future promises, there are number of ecological and economic issues that need to be addressed when considering the development and deployment of transgenic crops for pest management. There is a multitude of concerns about the real or conjectural effects of transgenic plants on non-target organisms, including human beings, and evolution of resistant strains of insects. As a result, caution has given rise to doubt because of lack of adequate information. One of the risks of growing transgenic plants for pest management is the potential spread of the transgene beyond the target area. There is a feeling that the genes introduced from outside the range of sexual compatibility might present new risks to the environment and humans, and will lead to development of resistance to herbicides in weeds, and to antibiotics. The biosafety issues related to the development of transgenic plants include risks for animal and human health, such as allergies, toxicity, and food quality and safety. While some of these concerns may be real, others seem to be conjectural and highly exaggerated.

Future research on development and deployment of transgenic crops should focus on the following issues:

- Effects of transgenic plants on the activity and abundance of non-target herbivore arthropods, natural enemies, and fauna and flora in the rhizosphere and aquatic systems should be thoroughly investigated. Development of transgenic crops with wide spectrum of activity against insect pests feeding on a crop, but harmless to natural enemies and other non-target organisms should be given top priority.
- There is a need for having a detailed understanding of resistance mechanisms, insect biology, and plant molecular biology to tailor gene expression in transgenic plants for efficient pest management. Future researches should focus on pyramiding of novel genes with different modes of action with conventional host plant resistance, and multiple resistance to insect pests and diseases.
- The potential of RNAi, a technique to study the function of particular gene by silencing that gene in an organism, has been established in insects. The research efforts must be intensified to identify the potential insect genes which are important for biological functions of the target insects. The identified potential genes should be used for development of transgenic plants against that particular insect.

- One of the risks of growing transgenic crops is the potential spread of the transgene beyond the target area. There is a feeling that genes introduced from outside the range of sexual compatibility might present new risks to the environment. Therefore, studies should be undertaken to determine the extent and implications of gene transfer. Appropriate measures should be devised to contain gene flow where its likely consequences may be deleterious to the environment.
- The need for identification and detection of transgenic crops and food products derived from them has increased with the rapid expansion in cultivation of transgenic crops over the past decade. Labeling and traceability of transgenic material is important to address the concerns of the consumer. Establishment of reliable and economical methods for detection, identification and quantification of genetically modified food continues to be a great challenge at the international level.

Integrated Pest Management

Integrated pest management (IPM) programmes were initially evolved as a result of the pest problems caused by repeated and excessive use of pesticides and increasing cases of pest resistance to these chemicals. It is only during the past two decades that economic and social aspects of IPM have also received increasing attention. If the environmental and social costs of pesticide use are taken into account, IPM appears to be more attractive alternative with lower economic costs. Production, storage, transport, distribution and application of pesticides involves greater health hazards than the safer inputs used in IPM. The IPM programmes do not endanger non-target organisms nor do they pollute soil, water and air. IPM builds upon indigenous farming knowledge, treating traditional cultivation practices as components of location-specific IPM practices. The incorporation of IPM into traditional practices helps the farmers to modernize while maintaining their cultural roots. The inputs used in IPM are usually based on local resources and outside dependence is minimized. This helps in maintaining social and political stability. It is now being increasingly realized that modern agriculture cannot sustain the present productivity levels with the exclusive use of pesticides. Increasing pest problems and disruption in agroecosystems can only be corrected by use of holistic pest management programmes (Koul *et al.*, 2004).

Pest management practices may not be sustainable for a variety of reasons: (*i*) The control tactic may no longer be effective over time due to selection against pests that are susceptible to the tactic. (*ii*) The control tactic leads to distruption in the ecosystem that may result in further outbreak of the target pest or outbreaks of new pests. (*iii*) The cost of the practice may be too expensive to maintain indefinitely. (*iv*) The practice may degrade the quality of human health, environment or agronomic resources over time. (*v*) New pest problems may arise due to introduction of pests or natural enemies that attack existing biological control agents and thereby increase pest populations. (*vi*) As the types and the abundance of pests change due to crop intensification, the previous management tactics may not adequately control pest population. Therefore, pest management decisions will have to be taken, keeping in view the dynamics of pest population, sustainability of the management tactics, compatibility of the tactics and stability of the agroecosystem. As control measures are generally disruptive to the ecosystems, preventing the pest problem from arising in the first place is preferable to control and promote sustainability. If pesticides are part of the IPM system, a pesticide resistance management strategy is essential, so that the target pest's susceptibility to pesticide does not decline over time. Other management tactics like pest-resistant cultivars, biological control agents and cultural practices are not necessarily sustainable over time, which may require periodical monitoring. Farmers' own management practices need to be incorporated in IPM systems to make them more acceptable and sustainable (Dhaliwal and Koul, 2010).

Many of the IPM strategies can be implemented effectively only on an area-wide basis. This is possible through increased farmers' awareness and enactment of suitable legislative measures.

IPM also needs to be integrated with other components of crop production and rural development. Ultimately, IPM is to be used at the farmers' level and, therefore, it needs to be converted from a scientist-oriented to a farmer-oriented concept. The recent advances in information and communication technology have provided us a unique opportunity to achieve these objectives. Computer-based interaction systems installed at the village level can help the farmers in pest identification, forecasting of pest populations, range of options available for pest management with advantages and limitations of each of these options. This will help the farmers in identifying the best option based on their requirements and resources (Koul *et al.*, 2008; Dhaliwal and Arora, 2015).

IPM programmes gained momentum during 1980s and since then many major food and fibre crops were covered under IPM technology in many countries. However, many crops of extreme importance to subsistence and resource-poor farmers around the world have not received due attention. These crops, often referred to as 'orphan crops' because of relative lack of research and development applied to them, include root and tuber crops such as cassava, sweet potato and yam; millets such as pearl millet, finger millet and foxtail millet; and several legumes and tree crops. Moreover, the package of practices in many developing countries still lay emphasis on pesticide based pest management programmes. Therefore, the future IPM programmes need to be ecologically based in order to achieve sustainable crop protection.

- IPM programmes have been developed and validated for almost all the major crops in different parts of the world. However, their widespread acceptance by the farmers in many developing countries is far from satisfactory. Therefore, farmers must be involved in devising and refining IPM schedules so that they are convinced of the benefits of the IPM technology. Viewing farmers as an equal partner in technology development and testing will foster ownership of IPM technologies and increase adoption.
- Different tactics of IPM may not always be complementary to each other. There have been situations where host plant resistance and chemical control, host plant resistance and biological control, chemical control and biological control, and transgenic crops and biological control have been incompatible. Therefore, the interactions among various tactics of IPM should be thoroughly investigated before applying them to IPM programmes.
- Generally IPM programmes have been devised taking into consideration the major target pest. Efforts should be made to follow a holistic approach by taking into consideration the entire insect pest and disease complex of the agroecosystem.
- A field-to-field approach is followed by individual farmer to manage pests on their farms. There are always chances of movement of insects from the adjoining untreated fields to colonise the treated crop after a few days of the control operation. Therefore, area-wide pest management approach should be followed where the farmers practice the IPM schedule in contiguous blocks.
- Pesticides have been dominating the scene of pest management even after the concept of IPM became popular and widely accepted. There is a need to shift the IPM paradigm from focusing on pest management strategies relying on pest management to a system approach relying primarily on biological knowledge of pests and their ecological interactions with the crops. Digital technology and high-speed telecommunications can enable access to recent information in the Internet. The use of GPS will compliment web networks by providing researchers and extension workers with tools that will enable them to define regions where production constraints are most acute, develop targeted technologies for those regions and monitor their use. Modeling and computer programmes can aid in understating the dynamics of pest populations and devising sustainable pest management strategies.

FUTURE OUTLOOK

The global population reached 7 billion in 2011, increasing from 5 billion in 1987 to 6 billion in 12 years in 1999. The world population stood at 7.24 billion in mid-2014. Although the overall growth rate has declined from 2.1 per cent per year in late 1960s to 1.2 per cent at present, the population is still growing, particularly in Asia and sub-saharan Africa. It is entirely possible that that 8th billion would be added in 12 years as well. This would place us squarely in the middle of history's most rapid population expansion. Therefore, strenuous efforts will have to be made to increase world food supplies to ensure environment and food security. Ecostrategies are likely to play a prominent role to achieve the above objectives. In this context, integrated pest management, which relies on suppression of pest problems while causing minimum disruption to the agroecosystem, is one of the viable alternatives.

The current methodology for assessing insect damage to undertake control measures is cumbersome, and the farmers are not able to properly understand and practice the methods. Simple techniques to assess insect damage and population density would be useful for timely application of appropriate control measures. There is a need to develop economical high-resolution environmental and biological monitoring systems to enhance our capabilities to predict pest incidence, estimate damage, and identify valid economic thresholds. Economic threshold levels (ETLs) are available for a limited number of insect species. ETLs developed without taking into consideration the potential of naturally occurring biological control agents and levels of resistance in the cultivars to the target pests are of limited value. The ETLs have to be developed for specific crop-pest-climatic situations. The ETLs developed in one region are not applicable in other areas where the crop-pest and socioeconomic conditions are different. Simple methods of assessing ETLs could help avoid unnecessary pesticide applications.

Nanotechnology is a promising field of research which has opened up a wide array of opportunities and is expected to give major impulse to technical innovations in future. These include enhancement of agricultural productivity involving nanoporous zeolites for slow release and efficient dosage of water and fertilizer, nanocapsules for herbicide delivery, and vector and pest management, and nanosensors for pest detection. Nanoparticles help to produce new pesticides and insect repellents. Nanoencapsulation (a process through which a chemical is slowly but efficiently released to a particular plant), with nanoparticles in the form of pesticides allows proper absorption of the chemical into plants unlike in the case of larger particles. This process can also deliver DNA and other desired chemicals into plant tissues for protection of host plants against insect pests. It is known that aluminosilicate filled monotube can stick to the plant surfaces while nanoingredients of nanotube have the ability to stick to the surface hair of insect pests, and ultimately enter the body and influence certain physiological functions. Nanoencapsulation is currently the most promising technology for protection of crop plants against insect pests. Research on nanoparticles and insect control should be directed towards production of faster and ecofriendly pesticides to deliver into the target host tissue through nanoencapsulation. This will control pests efficiently, prolong the protection time and lead to sustainable crop protection. Thus, nanotechnology is likely to revolutionize agriculture in general and pest management in particular in the near future.

Foods derived from genetically modified plants are now appearing in the market and many more are likely to emerge in the future. It is important to ensure the safety of food derived from transgenic crops based on the principle of nutritional equivalence. Strenous efforts should be made to make this technology available to farmers who cannot afford the high cost of seeds and chemical pesticides in developing countries. Transgenic crops would play a significant role in integrated pest management in future reducing the number of pesticide applications and pesticide residues in food. Concerted efforts are required involving international and advanced research institutes, and the national research organizations to harmonize the regulatory requirements to assess the biosafety of the food derived from genetically engineered crops and their effects on non-target organisms for sustainable crop production and food security.

The goals of the future IPM programmes are to improve the economic benefits related to the adoption of IPM practices and to reduce potential human health risks and unreasonable environmental effects from pests and from the use of pest management practices.

- A major determining factor in the adoption of IPM programmes is whether the economic benefit outweighs the cost to implement an IPM practice. Conducting a "cost benefit" analysis of the proposed IPM strategies is not based solely on the monetary costs; it is based on four main parameters, i.e. monetary, environmental/ecological health and function, aesthetic benefits, and human health. While there may be many benefits from adoption of IPM practices, if new IPM programmes do not appear to be as economically beneficial as practices already in place, they are not likely to be adopted. Evaluation of the short and long term risks and benefits is needed. Therefore, improving the overall economic benefit to humans and the broader natural systems, resulting from the adoption of IPM practices, is a critical component of the future IPM programme.
- IPM programmes need to be designed with the goal of reducing potential human health risks by reducing exposure of both the general public and workers to pests as well as high-risk pest management practices, whether mechanical, chemical or biological in nature. IPM protects human health through its contribution to food security by reducing potential health risks and enhancing worker safety. In the past, success in achieving the goal of reduced risk from pest management practices was generally measured by tracking changes in the annual amount of pesticides used. However, when used alone it is a poor indicator of human health risk, and more advanced systems of measurement are required.
- IPM programme should be designed to protect agricultural, urban and natural resource environments from the encroachment of native and non-native pest species while minimizing unreasonable adverse effects on soil, water, air and beneficial biological organisms. For example, in agriculture, IPM practices should promote a healthy within crop environment, and conserve organisms that are beneficial to agricultural systems, including natural enemies and pollinators. By reducing off-target impacts, IPM should help to maximize the positive contributions that agricultural land use can make to watershed health and function. Greater IPM efforts are required to maintain functional and aesthetic standards in natural and recreational environments such as lakes, streams, parks and sport facilities.

Classical integrated management programmes for apple pests in Canada and cotton pests in Peru provided some of the early models for successful implementation of IPM in the field. The FAO subsequently provided the coordination to spread the IPM concept in developing countries. The success of an IPM programme in rice in Southeast Asia was based on linking outbreaks of the brown planthopper with application of broad-spectrum insecticides, and the realization of the fact that the brown planthopper populations were kept under check by the natural enemies in the absence of insecticide applications. Much of the impact of this programme was brought out through field demonstrations, training programmes, and farmers' field schools. Subsequently, many more developing countries launched their own national IPM programmes. The success of some of these programmes has led to the establishment of the Global IPM Facility, under the auspices of FAO, UNDP and the World Bank, which will serve as a coordinating and promoting entity for IPM worldwide. Currently many IPM programmes have been developed in which different control tactics are combined to suppress pest numbers below a threshold. These vary from judicious use of insecticides based on ETLs and regular scouting to ascertain pest population levels to sophisticated systems using computerized crop and population models to assess the need, optimum timing, and selection of insecticides for sprays. The increase in our knowledge about insect-plant-environment interactions and advances in modern technologies like biotechnology and nanotechnology, would give further impetus to IPM in the future.

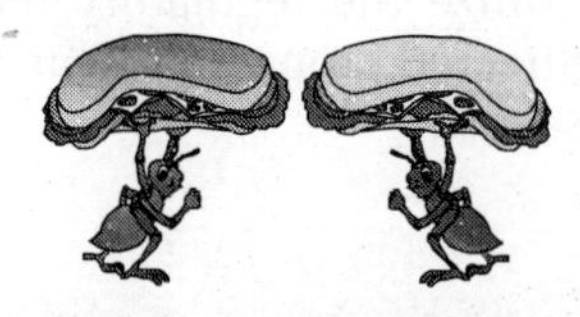

SELECTED BIBLIOGRAPHY

Abbaszadeh, G.; Dhillon, M.K.; Srivastava, C.; Gautam, R.D. (2011) Effect of climate factors on the bioefficacy of biopesticides in insect pest management. *Biopestic, Int* 7 : 1–14.

Abdulaa Koya, K.M.; Devasahayam, S.; Prem Kumar, T. (1991) Insect pests of ginger (*Zingiber officinale* Rosc.) and turmeric (*Curcuma longa* Linn.) in India. *J. Plant. Crops* **19** (1) : 1-13.

Abraham, C.C. (1994) Pests of coconut and arecanut. In : K.L. Chadha and R. Rethinam (eds). *Advances in Horticulture.*Vol. 10. *Plantation and Spice Crops* : Part 2. Malhotra Publishing House, New Delhi, pp. 709-726.

Abrol, D.P. (ed.) (2014) *Integrated Pest Management : Current Concepts and Ecological Perspective.* Elsevier Inc., San Diego, USA.

Abrol, D.P., Shankar, U. (2012) *Integrated Pest Management: Principles and Practice.* CAB International, Wallingford, U.K.

Agarwal, R.A.; Gupta, G.P.; Garg, D.O. (1984) *Cotton Pest Management in India.* Researchco Publications, New Delhi.

Aggarwal, R.; Singh, J.; Shukla, K.K. (2004) Biology of pink stemborer, *Sesamia inferens* (Walker) on rice crop. *Indian J. Ecol.* 31 (1) : 66-67.

Ahmad, B.; Shankar, U.; Monobrullah, M.; Kaul, V.; Singh, S. (2007); Bionomics of cabbage butterfly, *Pieris brassicae* (Linn.) on cabbage. *Ann. Pl. Prot. Sci.* 15 (1) : 47-52.

Altieri, M.A. (ed.) (1993) *Crop Protection Strategies for Subsistence Farmers.* Intermediate Technology Publications, London.

Altieri, M.A.; Nicholls, C.I. (2004) *Biodiversity and Pest Management in Agroecosystems.*Food Products Press, Binghamton, New York.

Aronson, A.T. (1994) *Bacillus thuringienesis* and its use as a biological insecticide. *Pl. Breed. Rev.* 12: 19-45.

Atwal, A.S. (1962a) Insect pests of citrus in the Punjab, I. Biology and control of citrus psylla *Diaphorina citri* Kuwayama (Hemi : Psyllidae). *Punjab Hort. J.* 2(2): 104- 108.

Atwal, A.S. (1962b) Insect pests of citrus in the Punjab, II. Biology and control of citrus whitefly, *Dialeurodes citri* Ashmead (Hemi : Aleyrodidae). *Punjab Hort. J.* 2(3) : 149-152.

Atwal, A.S. (1962c) Insect pests of citrus in the Punjab, III. Biology and control of citrus mealy bug, *Pseudococcus filamentosus* Cki. (Hemi : Coccidae). *Punjab Hort. J.* 2(4): 230-232.

Atwal, A.S. (1963a) Insect pests of mango and their control. *Punjab Hort. J.* 3(2-4) : 235-258.

Atwal, A.S. (1963b) Insect pests of citrus in the Punjab, IV. Biology and control of fruit piercing moth. *Ophideres fullonica* L. (Lepidoptera). *Punjab Hort. J.* 3(1) : 43- 45.

Atwal, A.S. (1964a) Insect pests of citrus in the Punjab, V. Biology and control of citrus caterpillar. *Papilio demoleus* L. (Lep: Papilionidae). *Punjab Hort. J.* 4(1): 40-44.

Atwal, A.S. (1964b) Insect pests of citrus in the Punjab, IV. Biology and control of citrus leaf miner. *Phyllocnistis citrella* Stainton (Lep : Phyllocnistidae). *Punjab Hort. J.* 4(2): 100-103.

Atwal, A.S. (1964c) Insect pests of citrus in the Punjab, VII. Biology and control of citrus mite. *Punjab Hort. J.* 4 (3 & 4) : 142-144.

Atwal, A.S. (1967) Diapause among insect pests of crops. *Trop. Ecol.*8 (1 & 2) : 1-16.

Atwal, A.S. (1974) Ecology of pest complex in stored grains. In : *Postharvest Prevention of Waste and Loss of Food Grains.* Asian Productivity Organization, Tokyo, pp. 131-142.

Atwal, A.S. (1986) Future of pesticides in plant protection. *Proc. Indian Natn. Sci. Acad.* 52 : 77-90.

Atwal, A.S. (2013) *Mellifera Beekeeing and Pollination.* Kalyani Publishers, New Delhi.

Atwal, A.S.; Dhaliwal, G.S. (1971) The rate of increase of the population of *Trogoderma granarium* Everts on different varieties of wheat. *Indian J. Entomol.* 33(2): 172-175.

Atwal, A.S.; Bains, S.S. (1974) *Applied Animal Ecology.* Kalyani Publishers, Ludhiana.

Atwal, A.S.; Dhingra, S. (1971) Biological studies on the rose aphid, *Macrosiphum rosaeiformis* Das. *Indian J. Entomol.* 33 (3-4) : 136-142.

Atwal, A.S.; Mangar, A. (1971) Biology, host range and abundance of *Meloidogyne incognita* (Kofoid and White 1919) Chitwood 1949, *J. Zool. Soc. India.* 23: 113-118.

Atwal, A.S.; Singh, B. (1990) *Pest Population and Assessment of Crop Losses.* Indian Council of Agricultural Research, New Delhi.

Atwal, A.S.; Singh, K. (1969a) Carry over of the pink bollworms (*Pectinophora gossipiella* Saund.) from one crop to the other and its cultural control. *Indian J. Entomol.* 31(2) : 116-120.

Atwal, A.S.; Singh, K. (1969b) Comparative efficacy of *Bacillus thuringiensis* Berliner and some contact insecticides against *Pieris brassicae* L. and *Plutella maculipenis* (Curtis) on cauliflower. *Indian J. Entomol.* 31 : 361-363.

Atwal, A.S.; Singh, R. (1972) Development of *Euproctis lunata* Walker (Lepidoptera : Lymantridae) in relation to different levels of temperature and relative humidity. *J. Res. Punjab Agric. Univ.* 9(1) : 50-54.

Atwal, A.S.; Singh, S. (1971) The influence of different levels of temperature and humidity on the speed of development, survival and fecundity of *Macropes excavatus* Distant (Hemiptera : Lygaeidae). *Indian J. Entomol.* 33(2) : 166-171.

Atwal, A.S.; Sethi, S.L. (1963) Predation by *Coccinella septumpunctata* L. on the cabbage aphid, *Lipaphis erysimi* (Kalt.) in India, *J. Anim. Ecol.* 32 : 481-488.

Atwal, A.S.; Sohi, B.S. (1969) Effectiveness of *Bacillus thuringiensis* Berliner against *Bissetia steniellus* Hampson (Lep. : Pyralidae). *J. Res. Punjab Agric. Univ.* 6(1) Suppl : 174-176.

Atwal, A.S.; Verma, N.D. (1972) Development of *Leucinodes orbonalis* Guen. (Lepidoptera : Pyraustidae) in relation to different levels of temperature and humidity, *Indian J. Agric. Sci.* 42 : 849-854.

Atwal, A.S.; Bhatti, D.S.; Sandhu, G.S. (1969) Some observations on the ecology and control of mango mealy bug, *Drosicha mangiferae* Green. *J. Res. Punjab Agric. Univ.* 6(1): 107-114.

Atwal, A.S.; Chahal, B.S.; Grewal, G.S. (1966) Persistence of *Bacillus thuringiensis* Berliner in sugarcane field against *Bissetia steniellus* Hampson (Lep. : Pyralididae). *J. Res. Punjab Agric. Univ.* 3(4): 421-424.

Atwal, A.S.; Chaudhary, J.P.; Ramzan, M. (1969a) Studies on the biology and control of the pea leaf minor, *Phytomyza atricornis* Meigen. *J. Res. Punjab Agric. Univ.* 6(1) Suppl. : 163-169.

Atwal, A.S.; Chaudhary, J.P.; Ramzan, M. (1969b) Some preliminary studies on the bionomics and rate of parasitization of *Diaeretiella rapae* Curtis (Braconidae : Hymenoptera) – a parasite of aphids. *J. Res. Punjab Agric. Univ.* 6(1) Suppl. : 177-182.

Atwal, A.S.; Chaudhary, J.P.; Ramzan, M. (1969b) Studies on the biology and control of the pea leaf miner, *Phytomyza atricornis* Meigen (Agromyzidae : Diptera). *J. Res. Punjab Agric. Univ.* 6(1) Suppl. : 214-219.

Atwal, A.S.; Chaudhary, J.P.; Ramzan, M. (1970) Studies on the development and field population of citrus psylla, *Diaphorina citri* Kuwayama (Psyllidae : Homoptera). *J. Res. Punjab Agric. Univ.* 7(1) : 333-338.

Atwal, A.S.; Chaudhary, J.P.; Ramzan, M. (1971). Mortality factors in the natural population of cabbage aphid, *Lipaphis erysimi* (Kalt.) (Aphididae : Homoptera) in relation to parasites, predators and weather conditions. *J. Agric. Sci* 41(5) : 507-510.

Atwal, A.S.; Chaudhary, J.P.; Sohi, B.S. (1967) Studies on the biology and control of *Sogatella furcifera* Horv. (Delphacidae : Homoptera). *J. Res. Punjab Agric. Univ.* 4 : 547-555.

Atwal, A.S.; Chaudhary, J.P.; Sohi, B.S. (1969) Effect of temperature and humidity on development and population of cotton jassid, *Empoasca devastans* Distant. *J. Res. Punjab Agric. Univ.* 6(1) Suppl.: 255-261.

Atwal, A.S. Sidhu, A.S.; Gupta, J.C. (1968) Studies on the growth of population of *Trogoderma granarium* Everts and *Callosobruchus analis* (Fabricius). *Indian J. Entomol.* 30(3): 185-191.

Avasthy, P.N. (1967) The problem of white grubs of sugarcane in India. *Proc. Intomol. Sec. Sug. Cane Technol.* 12: 1321-1333.

Awasthi, V.B. (2011) *Agricultural Insect Pests and Their Control.* Scientific Publishers (India), Jodhpur.

Awasthy, V.B. (2012) *Introduction to General and Applied Entomology.* Scientific Publishers (India), Jodhpur.

Backle, A.P.; Smith R.H. (eds) (1994) *Rodent Pests and Their Control.* CAB International, Wallingford, UK.

Barbosa, P.; Schultz, J.C. (eds) (1987) *Insect Outbreaks,* Academic Press, San Diego.

Barbosa, P., Letoureau, D.; Agrawal, A. (eds) (2012) *Insect Outbreaks Revisited.* Blackwell Publishing Ltd., Oxford, U.K.

Barker, G.M. (ed.) (2002) *Molluscs as Crop Pests.* CAB International, Wallingford, Oxon, UK.

Barnett, S.A.; Prakash, I. (1975) *Rodents of Economic Importance in India,* Arnold-Heinemann, New Delhi.

Beckage, N.E.; Thompson, S.N.; Frederici, B.A. (eds) (1993a) *Parasites and Pathogens of Insects,* Vol. I : *Parasites.* Academic Press, London.

Beckage, N.E.; Thompson, S.N.; Frederici, B.A. (eds) (1993b) *Parasites and Pathogens of Insects.*Vol. 2 *Pathogens.* Academic Press, London.

Bedding, R.A. (2006) Entomopathogenetic nematodes : From discovery to application. *Biopestic. Int.* 2 (2): 87-119

Bellotti, A.C.; Smith, L.; Lapointe, S.L. (1999) Recent advances in cassava pest management. *Annu. Rev. Entomol.* 44 : 343-370.

Bernays, E.A.; Chapman, R.F. (1994) *Host-Plant Selection by Phytophagous Insects.* Chapman & Hall, London.

Bhat, O.K.; Bhagat, K.C.; Koul, V.K.; Bhan, R. (2005) Studies on the biology of spotted bollworms *Earias vittella* (Fab.) on okra. *Ann. Pl. Prot. Sci.* 13 (1) : 65-68

Bindra, O.S. (1965) Biology and bionomics of *Clavigralla gibbosa* Spinola, the pod bug of pigeonpea. *Indian J. Agric. Sci.* 35: 322-334.

Bindra, O.S.; Singh H. (1977) *Pesticide-Application Equipment.* Oxford & IBH Publishing Co. Pvt. Ltd., New Delhi.

Boethel, D.J.; Eikenbary, R.D. (eds) (1986) *Interactions of Plant Resistance and Parasitoids and Predators of Insects.* Ellis Horwood Limited, Chichester, England.

Bonning, B.C., Boughton, A.J., Jin, H., Harrison, R.L. (2003) Genetic enhancement of baculovirus insecticides. In : R.. Upadhyay (ed.) *Advances in Microbial Control of Insect Pests.* Kluwer Academic/Plenum Publishers. New York, pp. 109-125.

Boulter, D. (1989) Genetic engineering of plants for insect resistance. *Outlook Argic.* 18 : 2-6.

Boulter, D. (1993) Insect pest control by copying nature using genetically engineered crops. *Phytochemistry* 34 : 1453-1466.

Brar, D.S.; Aneja, A.K.; Singh, J.; Mahal, M.S. (2005) Biology of whitefly, *Bemisia tabaci* (Gennadius) on American cotton, *Gossypium hirsutum* Linnaeus. *J. Insect Sci.* 18 (1): 48-59.

Burges, H.D. (ed.) (1981) *Microbial Control of Pests and Plant Diseases 1970-1980.* Academic Press, London.

Burgess, N.R.H. (1990) *Public Health Pests.*Chapman & Hall, London.

Burn, A.J.; Coaker, T.H.; Jepson, P.C. (eds) (1987) *Integrated Pest Management.* Academic Press, London.

Busvine, J.R. (1980) *Insects and Hygiene.*Chapman & Hall, London.

Busvine, J.R. (1993) *Disease Transmission by Insects.*Springer. Verlag, Berlin, Germany.

Carson, R. (1962) *Silent Spring.* Riverside Press, Boston.

Carter, W. (1973) *Insects in Relation to Plant Disease.* John Wiley & Sons, New York.

Chadwick, D.J. (ed.) (1993) *Crop Protection and Sustainable Agriculture.* John Wiley & Sons, New York.

Chakravarthy, A.K.; Shashank, P.R.; Doddabasappa, B.; Thyagaraj, N.E. (2012) Studies on shoot and fruit borer, *Conogethes* spp. (Lepidoptera : Crambidae) in the Orient : Biosystematics, bioecology and management. *J. Insect Sci.* 25(2) : 107-117.

Chihillar, B.S.; Gulati, R., Bhatanagar, P (2007) *Agricultural Acarology.* Daya Publishing House, New Delhi.

Ciancio, A.; Mukerji, K.G. (eds.) (2007) *General Concepts in Integrated Pest and Disease Management.* Kluwer Academic/Plenum Publishers, New York.

Clark, J.M.; Yamaguchi, 1. (eds) (2002) *Agrochemical Resistance : Extent, Mechanisms and Detection.* ACS Symp. Series 808, American Chemical Society, Washington, DC.

Cook, S.M.; Khan, Z.R.; Picket, J.A. (2007) The use of push-pull strategies in integrated pest management. *Annu. Rev. Entomol.* **52** : 375-400

Corey, S.A.; Dall, D.J.; Milne, W.M. (eds) (1993) *Pest Control and Sustainable Agriculture.* CSIRO, Melbourne, Australia.

Cotton, R.T. (1963) *Insect Pests of Stored Grain and Grain Products.* Burges Publishing Co., Minneapolis.

Cramer, H. (1967) Plant protection and world crop production. *Pflariz. Nach.* 20 : 1-24.

David, B.V. (ed.) (1992) *Pest Management and Pesticides : Indian Scenario.* Namrutha Publications, Madras.

David, B.V. (2002) *Elements of Economic Entomology.* Popular Book Depot, Chennai.

DeBach, P.; Rosen, D. (1991) *Biological Control by Natural Enemies.* Cambridge University Press, Cambridge.

De Barrow, P.J.; Liu, S-S.; Boykin, L.M.; Dinsdale, A.B. (2011) *Bemisia tabaci* : A statement of species status. *Annu. Rev. Entomol.* **56** : 1-19.

Deka, D.; Byjesh, K.;Kumar, U.; Choudhary, R. (2009). Climate change impacts on crop pests. In: S. Panigrahy, S.S. Ray and J.S. Parihar (eds.) *Impact of Climate change on Agriculture.* International Society of Photogrammetry and Remote Sensing, Ahmedabad, India, pp. 147-149.

Denholm, I.; Pickett, J.A.; Devonshire, A.L. (eds) (1999) *Insecticide Resistance : From Mechanisms to Resistance Management.* CAB International, Wallingford, UK.

Dent, D. (1991). *Insect Pest Management.* CAB International, Wallingford, UK.

Dent, D. (ed.) (1995) *Integrated Pest Management.* Chapman & Hall, London.

Dent, D. (2000) *Insect Pest Management.* CAB International, Wallingbord, U.K.

Dhaliwal, G.S. (2008) Pest management in global context : From green revolution to gene revolution. *Key Note Address, 2nd Congress on Insect Science.*Indian Society for the Advancement of Insect Science, Ludhiana.

Dhaliwal, G.S. (2015) *An Outline of Entomology.* Kalyani Publishers, New Delhi.

Dhaliwal, G.S.; Arora, R. (eds) (1994) *Trends in Agricultural Insect Pest Management.* Commonwealth Publishers, New Delhi.

Dhaliwal, G.S.; Arora, R. (2015) *Integrated Pest Management: Concepts and Approaches.* Kalyani Publishers, New Delhi.

Dhaliwal, G.S.; Dilawari, V.K. (eds) (1993) *Advances in Host Plant Resistance to Insects.* Kalyani Publishers, New Delhi.

Dhaliwal, G.S.; Heinrichs, E.A. (eds) (1998) *Critical Issues in Insect Pest Management.* Commonwealth Publishers, New Delhi.

Dhaliwal, G.S.; Koul, O. (2007) *Biopesticides and Pest Management. Conventional and Biotechnological Approaches.* Kalyani Publishers, New Delhi.

Dhaliwal, G.S.,; Koul,O. (2010) *Quest for Pest Management: From Green Revolution to Gene Revolution.* Kalyani Publishers, New Delhi.

Dhaliwal, G.S.; Singh, B. (eds) (2000) *Pesticides and Environmental.* Commonwealth Publishers, New Delhi.

Dhaliwal, G.S.; Singh, R. (eds) (2005) *Host Plant Resistance to Insects : Concepts and Applications.* Panima Publishing Corporation, New Delhi.

Dhaliwal, G.S.; Arora, R.; Dhawan, A.K. (2004) Crop losses due to insect pests in Indian agriculture: An update. *Indian J. Ecol.* 31(1) : 1-7.

Dhaliwal, G.S.; Jindal, V., Dhawan, A.K. (2010) Insect pest problems and crop losses : Changing trends. *Indian J. Ecol.* 37(1) : 1-7.

Dhaliwal, G.S.; Singh, R.; Chillar, B.S. (2015) *Essentials of Agricultural Entomology.* Kalyani Publishers, New Delhi.

Dhaliwal, G.S.; Singh, R.; Jindal, V. (2013) *A Textbook of Integrated Pest Management.* Kalyani Publishers, New Delhi.

Dhawan, A.K.; Kumar, V.; Singh, K.; Saini, S. (2011) *Pest Management Strategies in Cotton.* Society for Sustainable Crop Production, Ludhiana.

Dhawan, A.K.; Singh, B.; Bhullar, M.B.; Arora, R. (eds) (2013) *Integrated Pest Management.* Scientific Publishers (India), Jodhpur.

Dhindsa, M.S.; Saini, H.K. (1994) Agricultural ornithology : An Indian perspective. *J. Biosci.* 19(4) : 391-402.

FAO (1967) *Report of the First Session of the FAO Panel of Experts on Integrated Pest Control.* Food and Agriculture Organization, Rome, Italy.

FAO (1988) *Integrated Pest Management in Rice in Indonesia.*United Nations Food and Agriculture Organization, Jakarta, Indonesia.

FAO (1990) *Mid-term Review of FAO Intercountry Program for the Development and Application of Integrated Pest Control in Rice in South and South-east Asia.* Mission Report, Food and Agriculture Organization, Rome.

FAO (2007) *Overview of Forest Pests–India.* Working Paper FBS/18E, Food and Agriculture Organization, Rome.

Fenemore, P.G.; Prakash, A. (1995) *Applied Entomology*. Wiley Eastern Ltd., New Delhi.

Fitzwater, W.D.; Prakash, I. (1989) *Handbook of Vertebrate Pest Control.*Indian Council of Agricultural Research, New Delhi.

Gillot, C. (2005) *Entomology*. Spinger, Dordrecht, The Netherlands.

Gour, T.B. and Sridevi, D. (2012) *Chemistry, Toxicity and Mode of Action of Insecticides*. Kalyani Publishers, New Delhi.

Green, M.B.; Hedin, P.A. (eds.) (1986) *Natural Resistance of Plants to Pests : Roles of Allelochemicals.* ACS Symp. Ser. 296, Am. Chem. Soc., Washington, DC.

Green, M.B.; LeBaron, H.M.; Moberg, W.K. (eds.) (1990) *Managing Resistance to Agrochemicals : From Fundamental Research to Practical Strategies.* ACS Symp. Ser. 421, Am. Chem. Soc., Washington, DC.

Gurr, G.M.; Wratten, S.D.; Snyder, W.E.; Read, M.Y. (eds) (2012) *Biodiversity and Insect Pests: Key Issues for Sustainable Management.* John Wiley & Sons, Ltd., New York.

Haines, C.P. (ed.) (1991) *Insects and Arachnids of Tropical Stored Products : Their Biology and Identification.* Natural Resources Institute, Kent, United Kingdom.

Hall, D.W. (1980) *Handling and Storage of Food Grains in Tropical and Subtropical Areas.* Food and Agriculture Organization. Rome.

Hardie, J.; Mink, A.K. (eds) (1999) *Pheromones of Non-Lepidopteran Insects Associated with Agricultural Plants.* CAB International, Wallingford, UK.

Hedin, P.A. (ed.) (1983) *Plant Resistance to Insects.* ACS Symp. Ser. 208, Am. Chem. Sec., Washington, DC.

Hedin, P.A. (ed.) (1994) *Bioregulators for Crop Protection and Pest Control.* Am. Chem. Soc., Washington, DC.

Hedin, P.A.; Menn, J.J.; Hollingworth, R.M. (eds) (1988) *Biotechnology for Crop Protection.* Am. Chem. Soc., Washington, DC.

Hedin, P.A.; Menn. J.J.; Hollingworth, R.M. (eds) (1994) *Natural and Engineered Pest Management Agents.* ACS Symp. Ser. 551, Am. Chem. Soc., Washington, DC.

Heinrichs, E.A. (ed.) (1994a) *Biology and Management of Rice Insects.* Wiley Eastern Limited, New Delhi.

Heinrichs, E.A. (1994b) Development of multiple pest resistant crop cultivars. *J. Agric. Enotomol.* 11 (3) : 225-257.

Heinrichs, E.A.; Miller, T.A. (eds) (1991) *Rice Insects : Management Strategies.* Springer-Verlag, New York.

Highley, E.; Wright, E.J.; Banks, H.J.; Champ. B.R. (eds) (1994) *Stored-Product Protection.* CAB International, Wallingford, U.K.

Higley, L.G.; Wintersteen, W.K. (1992) A novel approach to environmental risk assessment of pesticides as a basis for incorporating environmental costs into economic injury level. *Am. Entomol.* 38(1) : 34-39.

Hilder, V.A.; Gatehouse, A.M. R. (1990) Transforming plants as means of crop protection against insects. *Outlook Agric.* 19(3) : 179-183.

Hill, D.S. (1994) *Agricultural Entomology.* The Timber Press, Portland, USA.

Hill, D.S. (1997) *The Economic Importance of Insects.* Chapman & Hill, London.

Hill, D.S. (2002) *Pests of Stored Foodstuffs and Their Control.* Kluwer Academic Publishers, Dordrecht, The Netherlands.

Hill, D.S. (2008) *Pests of Crops of Warmer Climates and Their Control.* Springer Science, London, UK.

Hodgson, C.J.; Abbas, G.; Arif, M.J.; Saeed, S.; Karar, H. (2008). *Phenacoccus solenopsis* Tinsley (Sternorrhyncha: Coccoidea: Pseudococcidae), an invasive mealybug damaging cotton in Pakistan and India, with a discussion on seasonal morphological variation. *Zootaxa* **1913:** 1-35.

Hokkanel, H.M.T. (1991) Trap cropping in pest management. *Annu. Rev. Entomol.* 36 : 119-138.

Howse, P.; Stevens, I; Jones, O. (1994) *Insect Pheromones and Their Use in Pest Control.* Chapman & Hall, London.

Hoy, M.A.; Herzog, D.C. (eds) (1985) *Biological Control in Agricultural IPM Systems.* Academic Press, London.

Jagtap, A.B.(2000) Snails and slugs problem of crops and their integrated management. In : R.K. Upadhyay, K.G. Mukerji and O.P. Dubey (eds) *IPM System in Agriculture.* Vol. 7. *Key Animal Pests.* Aditya Books Pvt. Ltd., New Delhi, pp. 309-322.

Jain, P.C. and Bhargava, M.C. (eds.) (2007) *Entomology Novel Approaches.* New India Publishing Agency, New Delhi.

James, C. (2013) *Global Status of Commercialized Biotech/GM Crops : 2013.* ISAAA Briefs No. 46. International Service for the Acquisition of Agri-biotech Applications, Ithaca, New York.

Jayaraj, S. (ed.) (1987) *Resurgence of Sucking Pests.*Tamil Nadu Agricultural University, Coimbatore, India.

Jindal, V.; Dhaliwal, G.S.; Koul, O (2013) Pest management in 21st century : Roadmap for future. *Biopestic. Int.* **9**(1) : 1-22.

Jutsum A.R.; Gordon, R.F.S. (eds) (1989). *Insect Pheromones in Plant Protection.* John Wiley & Sons, Chichester.

Kadir, A.A.S.A.; Barlow, H.S. (eds) (1992) *Pest Management and the Environment in 2000.* CAB International, Wallingford, JK.

Kadam, U.K.; Chandele, A.G. (2013) Biology of fruit fly, *Bactrocera dorsalis* (Hendel) on guava. *J. Insect Sci.* **26**(2) : 233-235.

Kalmakoff, J. and Ward, V.K. (2004) *Baculoviruses.* University of Otago, Dunedin, New Zealand.

Kettle, D.S. (1995) *Medical and Veterinary Entomology.* CAB International, Wallingford, UK.

Khare, B.P. (1994) *Stored Grain Pests and Their Management.* Kalyani Publishers, Ludhiana.

Khater, H.F. (2012) Prospects of botanical biopesticides in insect pest management. *J. Appl. Pharmacol. Sci.* **2** (5) : 244-259.

Khush, G.S.; Brar, D.S. (1991) Genetics of resistance to insects in crop plants. *Adv. Agron.* 45 : 223-274.

Khush, G.S.; Brar, D.S. (1999) Transgenic plants for pest management. Progress and future outlook. In : G.S. Dhaliwal and Ramesh Arora (eds). *Environmental Stress in Crop Plants.* Commonwealth Publishers, New Delhi, pp. 281-307.

Khush, G.S. and Virk, P.S. (2005) *IRRI Varieties and Their Impact.* International Rice Research Institute, Los Banos, Philippines.

Knipling, E.F. (1979) *The Basic Principles of Insect Population Suppression and Management.* USDA Agricultural Handbook No. 512, Washington, DC.

Kogan, M. (1982) Plant resistance in pest management. In: R.L. Metcalf and W. Luckmann (eds). *Introduction to Insect Pest Management.* John Wiley & Sons, New York, pp. 93-134.

Kogan, M. (ed.) (1986) *Ecological Theory and Integrated Pest Management Practice.* John Wiley & Sons, New York.

Kogan, M. (1998) Integrated pest management : Historical perspectives and contemporary developments. *Annu. Rev. Entomol.* 43 : 243-270.

Kogan, M.; Ortman, E.F. (1978) Antixenosis. A new term proposed to define Painter's 'non-preference' modality of resistance *Bull. Entomol. Soc. Am.* 24 : 175-176.

Koppenhofer, A.M. and Kaya, H.K. (2002) Entomopatho-genic nematodes and insect pest management. In: O. Koul and G.S. Dhaliwal (eds) *Microbial Biopes-ticides.* Taylor & Francis, London, pp. 277-305.

Koul, O. (2012) Plant biodiversity as a resource for natural products for insect pest management. In : G.M. Gurr; S.D. Wratten and W.E. Snyder (eds) *Biodiversity and Insect Pests; Key Issues for Sustainable Management.* John Wiley & Sons, New York, pp. 85-105.

Koul, O.; Cuperus, G.W. (eds) (2007) *Ecologically Based Integrated Pest Management.* CAB International, Wallingford, UK.

Koul, O.; Dhaliwal, G.S. (eds) (2001) *Phytochemical Biopesticides.* Harwood Academic Publishers, Amsterdam, The Netherlands.

Koul, O.; Dhaliwal, G.S. (eds) (2002) *Microbial Biopesticides.* Taylor & Francis, London, UK.

Koul, O.; Dhaliwal, G.S. (eds) (2003) *Predators and Parasitoids.* Taylor & Francis, London, UK.

Koul, O.; Dhaliwal, G.S. (eds) (2004) *Transgenic Crop Protection : Concepts and Strategies.* Science Publishers, Inc., Enfield, USA.

Koul, O.; Cuperus, G.; Elliot, W. (eds) (2008) *Areawide Pest Management: Theory and Implementation.* CAB International, Wallingford, UK.

Koul, O.; Dhaliwal, G.S.; Cuperous, G.W. (eds) (2004) *Integrated Pest Management : Potential, Constraints and Challenges.* CAB International, Wallingford, UK.

Koul, O.; Dhaliwal, G.S.; Kaul, V.K. (eds) (2009) *Sustainable Crop Protection: Biopesticide Strategies.* Kalyani Publishers, New Delhi.

Koul, O.; Dhaliwal, G.S.; Khokhar, S.; Singh, R. (eds) (2012) *Biopesticides in Environment and Food Security.* Scientific Publishers (India), Jodhpur.

Koul, O.; Dhaliwal, G.S.; Khokhar, S.; Singh R. (eds) (2014) *Biopesticides in Sustainable Agriculture: Progress and Potential.* Scientific Publishers (India), Jodhpur.

Krishnaiah, N.V.; Jhansi Lakshmi, V. (2012) Rice brown planthopper migration in India and its relevance to Punjab, *J. Insect Sci.* **25** (3) : 231-236.

Kumar, S.; Kular, J.S. (2013) Biology of *Phenacoccus solenopsis* Tinsley on Bt cotton. *Indian J. Ecol.* **40**(1) : 132-136.

Mani, M.S. (1995) *Insects.* National Book Trust India, New Delhi.

Maredia, K.M., Dakouo, D.; Mota-Sanchez, D. (eds) (2003) *Integrated Pest Management in the Global Arena.* CAB International, Wallingford, UK.

Maxwell, F.G.; Jennings, P.R. (eds) (1980) *Breeding Plants Resistant to Insects.* John Wiley & Sons, Chichester.

Maxwell, F.G.; Jenkins, J.N.; Parrott, W.L. (1972) Resistance of plants to insects. *Adv. Agron.* 24 : 187-265.

Metcalf, R.L. ; Luckmann, R.L. (eds) (1994) *Introduction to Insect Pest Management.* John Wiley & Sons, New York.

Metcalf, R.L.; Metcalf, E.R. (1992) *Plant Kairomones in Insect Ecology and Control.* Chapman & Hall, New York.

Metcalf, R.L. ; Metcalf, R.A. (1993) *Destructive and Useful Insects.* McGraw Hill, New York.

Miller, J.R.; Cowles, R.S. (1990) Stimulo : deterrent diversion: A concept and its possible application to onion maggot control. *J. Chem. Ecol.* **16** : 3197-3212.

Monro, H.A.V. (1969) *Manual of Fumigation for Insect Control.* Food and Agriculture Organization, Rome.

Moscardi, F. (1999) Assessment of the application of baculoviruses for control of Lepidoptera. *Annu. Rev. Entomol.* 44 : 257-289.

Munro, J.W. (1966) *Pests of Stored Products.* Hutchinson, London, UK.

Nair, K.S.S. (2007) *Tropical Forest Insect Pests: Ecology, Impact and Management.* Cambridge University Press, Cambridge, UK.

Nair, M.R.G.K. (1995) *Insects and Mites of Crops in India.* Indian Council of Agricultural Research, New Delhi.

NAS (1969) *Principles of Plant and Animal Pest Control.* Vol. 3. *Insect Pest Management and Control.* National Academy of Sciences, Washington, DC, USA.

Nayar, K.K. ; Ananthakrishnan, T.N. ; David, B.V. (2000) *General and Applied Entomology.* Tata McGraw Hill Publication Co. Ltd., New Delhi.

Nester, E.W.; Thomsashow, L.S.; Metz, M.; Gordon, M. (2002) *100 Years of Bacillus thuringiensis : A Critical Scientific Assessment.* American Academy of Microbiology, Washington, DC.

Nordlund, D.A.; Jones, P.L.; Lewis, W.J. (eds) (1981) *Semiochemicals : Their Role in Pest Control.* John Wiley & Sons, New York.

Norris, R.F.; CasWell-Chen, E.P.; Kogan, M. (2002). *Concepts in Integrated Pest Management.* Prentice-Hall of India Pvt. Ltd., New Delhi.

Norton, G.W.; Heinrichs, E.A.; Luther, G.L.; Irwin, M.E. (eds) (2005) *Globalizing Integrated Pest Management : A Participatory Research Process.* Blackwell Publishing Ltd. Oxford, UK.

NRC (1996) *Ecologically Based Pest Management : New Solutions for a New Century.* National Academy Press, Washington, DC.

NRI (1992) *Integrated Pest Management in Developing Countries. Experience and Prospects.* Natural Resources Institute, Chatham, Maritime, UK.

Oerke, E-C. (2006) Crop losses to pests. *J. Agric. Sci.* **144:** 31-43.

Oerke, E-C. ; Dehne, H-W. ; Schonbeck, F. ; Weber, A. (1994) *Crop Production and Crop Protection.* Elsevier Science B.V., Amsterdam. The Netherlands.

Onstand, D.W. (ed.) (2008) *Insect Resistance Management : Biology, Economics and Prediction.* Elsevier, London.

Painter, R.H. (1951) *Insect Resistance in Crop Plants.* The Macmillan Company, New York.

Panda, N. (1979) *Principles of Host-Plant Resistance to Insects.* Hindustan Publishing Co., New Delhi.

Panda, N. ; Khush, G.S. (1995) *Host Plant Resistance to Insects.* CAB International, Wallingford, UK.

Pant, N.C. (ed.) (1964) *Entomology in India.* Entomological Society of India, New Delhi.

Panwar, V.P.S. (2006) *Agricultural Insect Pests of Crops and Their Control.* Kalyani Publishers, New Delhi.

Parsley, G.J. (ed.) (1995) *Biotechnology and Integrated Pest Management.* CAB International, Wallingford, U.K.

Pathak, M.D. ; Dhaliwal, G.S. (1981) *Trends and Strategies for Rice Insect Problems in Tropical Asia.* IRRI Research Paper Series No. 64, International Rice Research Institute, Los Banos, Philippines.

Pathak, M.D. ; Khan, Z.R. (1994) *Insect Pests of Rice.* International Rice Research Institute, Manila, Philippines.

Pedigo, L.P. (2002) *Entomology and Pest Management.* Prentice-Hall of India Pvt Ltd., New Delhi.

Pedigo, L.P.; Rice, M.E. (2009) *Entomology and Pest Managment.* Prentice Hall, New Jersey.

Pedigo, L.P. ; Higley, L.G. (1992) The economic injury level concept and environmental quality. *Am. Entomol.* 38(1) : 12-21.

Pedigo, L.P. ; Hutchins, S.H. ; Higley, L.G. (1986) Economic injury levels in theory and practices. *Annu. Rev. Entomol.* 31 : 341-368.

Pena, J. (ed.) (2013) *Invasive Pests of Agricultural Crops.* CAB Inernational, Wallingford, U.K.

Peshin, R.; Dhawan, A.K. (eds) (2009) *Integrated Pest Management: Innovation-Development.* Spinger, Dordrecht, The Netherlands.

Petroski, R.J.; Tellez, M.R.; Behle, R.W. (eds) (2005) *Semiochemicals in Pest and Weed Control.* ACS Symp. Ser. 906, American Chemical Society, Washington, DC.

Pimentel, D. (ed.) (1975) *Insects, Science and Society.* Academic Press, New York.

Pimentel, D. (ed.) (1993) *CRC Handbook of Pest Management in Agriculture.* Vols. 1, 2, 3. CBS Publishers and Distributors, New Delhi.

Pradhan, S. (1967) Strategy of integrated pest control. *Indian J. Entomol.* 29(1) : 105-122.

Pradhan, S. (1983) *Agriculture Entomology and Pest Control.* Indian Council of Agricultural Research, New Delhi.

Pradhan, S.; Jotwani, M.G.; Rai, B.K. (1962) The neem seed deterrent to insects. *Indian Fmg.* 12 : 7-11.

Prakash, I. ; Mathur R.P. (1987) *Management of Rodent Pests.* Indian Council of Agricultural Research, New Delhi.

Price, P.W. (1986) Ecological aspects of host plant resistance and biological control. In: D.J. Boethel; R.D. Eikenbary (eds) *Interactions of Plant Resistance and Parasitoids and Predators of Insects.* Ellis.Harwood Limited, Chichester, UK, pp. 11-30.

Pruthi, H.S. (1969) *Textbook on Agricultural Entomology.* Indian Council of Agricultural Research, New Delhi.

Pyke, B.; Rice, M.; Sabine, B.; Zalucki, M.P. (1987). The push-pull strategy-behavioural control of *Heliothis. Aust. Cotton Grow.* May-July : 7-9.

Rabb, R.I.; Guthrie, F.E. (eds) (1970) *Concepts of Pest Management.* North Carolina State University, Raleigh.

Radcliffe, E.B.; Hutchinson, W.D.; Cancelado, R.E. (eds) (2009) *Integrated Pest Management : Concepts, Tactics, Strategies and Case Studies.* Cambridge University Press, New York.

Ramesh, P. (1994) *Pests of Floriculture Crops and Their Control.* Kalyani Publishers, New Delhi.

Reddy, P.P. (2014) *Biointensive Integrated Pest Management in Horticultural Ecosystems.* Scientific Publishers (India), Jodhpur.

Reddy, P.R. (2010) *Insect, Mite and Vertebrate Pests and Their Management in Horticultural Crops.*Scientific Publishers (India), Jodhpur.

Reddy, V.D.; Rao, P.N.; Rao, K.V. (eds) (2010) *Pests and Pathogens: Management Strategies.* B.S. Publications, Hyderabad.

Richards, O.W.; Davies, R.G. (1977) *Imm's General Textbook of Entomology,* Chapman & Hall, London.

Ridgway, R.L.; Silverstein, R.M. ; Inscoe, M.N. (eds) (1990) *Behaviour Modifying Chemicals for Insect Management.* Marcel Dekker, Inc., New York.

Russell, G.E. (1978) *Plant Breeding for Pest and Disease Resistance.* Butterworth, New York.

Sadasivam, S.; Thayumanavan, B. (eds) (2003) *Molecular Host Plant Resistance to Pests.* Marcel Dekker, New York.

Saha, L.R.; Dhaliwal, G.S. (2006) *Handbook of Plant Protection.* Kalyani Publishers, New Delhi.

Saini, R.K.; Mrig, R.K.; Sharma, S.S. (eds) (2011) *Advances in Diagnosis of Arthropod Pests', Damage and Assessement of Losses.*CCS Haryana Agricultural University, Hisar.

Sathe, T.V. (2009) *A Textbook of Forest Entomology.* Daya Publishing House, New Delhi.

Saxena, R.C.; Barrion, A.A. (1987) Biotypes of insect pests of agricultural crops. *Insect Sci. Applic.* 8: 453-458.

Schaller, A. (ed.) (2008) *Induced Plant Resistance to Herbivory.* Springer, Dordrecht, The Netherlands.

Schmutterer, H. (1990) Properties and potential of natural pesticides from the neem tree *Azadirachta indica. Annu. Rev. Entomol.* 35 : 271-297.

Schoonhoven; L.M., van Loon, J.L.A.; Dicke, M. (2006) *Insect-Plant Biology.* Oxford University Press, New Delhi.

Sharma, H.C. (2009) *Biotechnological Approaches for Pest Management and Ecological Sustainability.* CRC Press, Boca Raton, Florida, USA.

Shelton, A.M.; Badenes-Perez, F.R. (2006) Concepts and applications of trap cropping in pest management. *Annu. Rev. Entomol.* 51 : 285-308.

Singh, A; Sharma, O.P.; Garg, D.K. (eds) (2005) *Integrated Pest Management : Principles and Applications.* Vol. 1. *Principles.* CBS Publishers & Distributors, New Delhi.

Singh, A.; Sharma, O.P.; Garg, D.K.(eds) (2006) *Integrated Pest Management : Principles and Applications.* Vol.2. *Applications.* CBS Publishers & Distributors, New Delhi.

Singh, A.; Trivedi, T.P.; Sardana, H.R. ; Sharma, O.P.; Sabir, N. (eds) (2003) *Recent Advances in Integrated Pest Management.* National Centre for Integrated Pest Management, New Delhi.

Singh, D.P.; Singh, A (2005) *Disease and Insect Resistance in Plants.* Science Publishers, Enfield, New Hampshire.

Singh, H. (1984) *Household and Kitchen Garden Pests : Principles and Practices.* Kalyani Publishers, New Delhi.

Singh, R.; Dhawan, A.K.(2006) Development, survival and reproduction of *Helicoverpa armigera* (Hubner) in relation to temperature, relative humidity and photoperiod. *Indian J. Ecol.* 33 (1) : 48-52.

Singh, R.; Jindal, V., Dhaliwal, G.S. (2013) Climate change : Impact on insect pests and their management. *Indian J.Ecol.* **40** (2) : 178-186.

Singh, S.; Sharma, D.R. (2014) PAU fruit fly trap : An eco-friendly way to control fruit flies. *Prog. Fmg.* **50** (5) : 14, 26.

Singh, S.R. (ed.) (1990) *Insect Pests of Legumes.* Longman and sons Ltd. New York.

Singh, Y. (ed.) (1995) *Termite Management in Buildings.* Oxford & IBH Publishing Co. Pvt. Ltd., New Delhi.

Smith, C.M. (2005) *Plant Resistance to Arthiopods : Molecular and Conventional Approaches.* Spunger, Dordrecht, The Netherlands.

Smit, C.M.; Clement, S.J. (2012) Molecular bases of plant resistance to arthropods. *Annu. Rev. Entomol,* **57:** 309-328.

Smith, R.F. (1975) History and complexity of integrated pest management. In : E.H. Smith and D. Pimentel (eds). *Pest Control Strategies.* Academic Press, New York, pp. 41-53.

Sridhara, S. (2006) *Vertebrate Pests in Agriculture: The Indian Scenario.* Scientific Publishers (India), Jodhpur.

Srivastava, K.P.; Dhaliwal, G.S. (2010) *A Textbook of Applied Entomology.* Vol. I *Concepts in Pest Management.* Kalyani Publishers, New Delhi.

Srivastava, K.P.; Dhaliwal, G.S. (2011) *A Textbook of Applied Entomology.* Vol. II, *Insects of Economic Importance.* Kalyani Publishers, New Delhi.

Stern, V.M.; Smith, R.F.; van den Bosch, R.; Hagen, K.S. (1959) The integrated control concept. *Hilgardia* **27**: 81-101.

Swinton, S.M.; Norton, G.W. (2009) Economic impact of IPM. In : E.B. Radcliffe, W.D. Hutchinson and R.E. Cancelado (eds.) *Integrated Pest Management : Concepts, Tactics, Strategies and Case Studies.* Cambridge University Press, Cambridge, pp. 14-24.

Thacker, J.R.M. (2002) *An Introduction to Arthropod Pest Control.* Cambridge University Press, Cambridge.

Tong-Xian, L.; Le, K. (eds)(2012) *Recent Advances in Entomological Research : From Molecular Biology to Pest Management.* Springer, Dordrecht, The Netherlands.

Upadhyay, R.K. ; Mukerji, K.G. ; Rajak R.L. (eds) (1996) *IPM System in Agriculture.* Vol. I. *Principles and Perspectives.* Aditya Books Pvt. Ltd., New Delhi.

Verma, L.R.; Verma, A.K.; Gautam, D.C. (eds) (2005) *Pest Mangement in Horticulture Crops : Principles and Practices.* Asiatech Publishers Inc., New Delhi.

Vincent, C.; Goettel, M.S.; Lazarovits, G.(eds) (2007) *Biological Control: A Global Perspective.* CAB International, Wallingford, UK.

Vreysen, M.J.B.; Robinson, A.S.; Hendrichs, J (eds) (2007) *Area-Wide Control of Insect Pests: From Research to Field Implementation.* Springer, Dordrecht, The Netherlands.

Walter, D.E. ; Proctor, H.C. (1999) *Mites : Ecology, Evolution and Behaviour.* CAB International, Wallingford, UK.

Whalon, M.E.; Mota-Sanchez, D.; Hollingworth, R.M. (eds) (2008) *Global Pesticide Resistance in Arthropods.* CAB International, Wallingford, UK.

Witzgall, P.; Kirsh, P.; Cock, A. (2010) Sex pheromones and their impact in pest management. *J. Chem. Ecol.* **36**: 80-100.

Ziska, L.W.; Bluementhal, D.M.; Runion, G.B.; Hunt, E.R. Jr. and Diaz-Soltero, H. (2011) Invasive species and climate change. An agronomic perspective. *Climate Change* 105: 13-42.

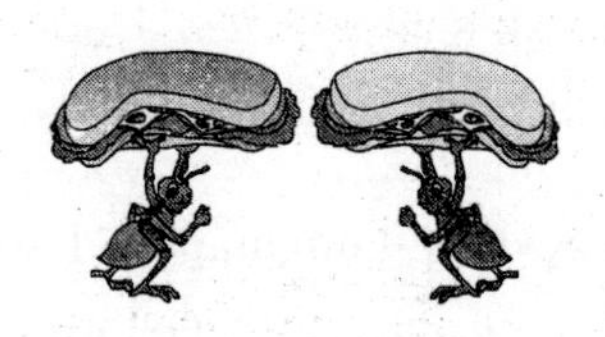

SUBJECT INDEX

A

Abiotic factors, 42
air currents, 45
carbon dioxide, 45, 614
edaphic factors, 45
light, 44
moisture, 43
oxygen, 45
temperature, 43, 237, 613
water currents, 45
Absolute estimates, 48
Acanthiophilus helianthi (Rossi), 360
Acaricides, 167
Acceptable daily intake, 186
Aceria guerreronis Keifer, 494
Aceria jasmini Chanana, 486
Aceria litchi Keifer, 433
Aceria mangiferae Sayed, 417
Achaea janata (Linnaeus), 101, 354
Acherontia styx (Westwood), 357
Acheta domesticus (Linnaeus), 564
Acigona steniellus (Hampson), 76, 391, 392
Acrocercops phaeospora Meyrick, 430
Acrocerops syngramma Meyrick, 490
Acyrthosiphon pisum (Harris), 226, 230, 239, 462, 601
Adisura atkinsoni Moore, 340
Adoretuc nitidus Arrow, 421
Adoretus pallens Arrow, 421
Aedes aegypti (Linnaeus), 109, 137, 557
Aeolesthes holosericea Fabricius, 446
Aerosols, 172
Aestherastis circulata Meyrick, 502
African snail, 100
Aggregation pheromones, 194
Agricultural aircrafts, 177
Agricultural pests, 250
Agrobacterium–mediated gene transfer, 268
Agrotis ipsilon (Hufnagel), 275, 299, 458, 540
Agrotis segetum (Denis & Schiffermuller), 115, 540, 541
Air pollution, 238
Ak butterfly, 481
Ak grasshopper, 434, 482
Alarm pheromones, 194, 278
Alcidodes porrectirostris Marshall, 444
Aldrin, 156, 157
Aleurocanthus woglumi Ashby, 398
Aleurolobus barodensis (Maskell), 382, 383
Alkylating agents, 249
Allelochemicals, 199, 209, 226, 228, 236
Allomones, 199
Allopatric resistance, 223
Almond moth, 545
Almond weevil, 448
Aluminium phosphide, 167, 190
American bollworm, 212, 298, 371, 462, 613
American cockraoch, 212, 563
Amrasca biguttula biguttula (Ishida), 44, 238, 239, 288, 298, 363, 458, 468
Amritodus atkinsoni (Lethiery), 299, 411
Amsacta moorei (Butler), 87, 93 96, 136, 330, 537
α-Amylase inhibitors, 277
Andraca bipunctata Walker, 499
Angoumois grain moth, 100, 543
Anguina tritici (Steinback), 337
Animal diversity, 4
Animal world, 3
Anjeerodiplosis peshawarensis Mani, 426
Anomis flava (Fabricius), 372
Anomis sabulifera (Guenee), 376
Anopheles maculipennis Meigen, 559
Anthonomus grandis Boheman, 41, 86, 212, 215, 228, 238, 243, 244, 259
Antibiosis, 224, 231
Anticarsia gemmatalis (Hubner), 114
Antifeedants, 209, 210
Antigastra catalaunalis (Duponchel), 356
Anti-locust organization, 529
Antimetabolites, 249
Antixenosis, 224, 231
Aonidiella aurantii (Maskell), 94, 203, 217, 399, 485
Apanteles glomeratus (Linnaeus), 452
Apanteles plutellae Kurdyumov, 139, 453
Aphelinus mali (Haldeman), 124, 439
Aphis craccivora Koch, 90, 298, 346, 351, 480, 601
Aphis gossypii Glover, 194, 226, 298, 364, 458, 480, 601
Apion corchori Marshall, 378
Apple fruit moth, 442
Apple leaf folder, 448
Apple root borer, 445
Apple stem borer, 444
Application of pesticides, 154
Apriona cinerea Cheverlot, 444
Apsylla cistella (Buckton), 417
Arecanut mirid bug, 489
Archips termias Meyrick, 448
Argas persicus (Oken), 582
Argyrestha conjugella Zeller, 442
Aristobia horridula Hope, 520

Armyworm, 334, 335
Arthropoda, 8
Arthropods, *102*
Arthroschista hilaris (Walker), 515, 516
Asian maize borer, 329, 330
Asparagine, 226
Asphondylia sesame Felt, 358
Aspidiotus destructor Signoret, 429, 431, 490
Aspidiotus hartii Cockerell, 509
Athalia lugens (Klug), 95, 349
Atherigona soccata Rondani, 89, 226, 239, 240, 298, 330, 332
Attainment of pest status, 40
Augmentation, 122
Augmentative biocontrol, 123
Aulacaspis tegalensis (Zehntner), 298, 384
Avermectins, 208
Azadirachta indica A. Juss., 129, 618
Azadirachtin, 131, 132, 133, 141, 146, 618, 645

B

Babul borer, 518
Bacillus lentimorbus (Dutky), 106
Bacillus popilliae (Dutky), 106
Bacillus thuringriensis Berliner, 107, 108, 271, 616, 618, 621, 644
Bacteria, 106, 605
Bactrocera cucurbitae (Coquillett), 130, 216, 258, 259, 466
Bactrocera dorsalis (Hendel), 215, 216, 258, 259, 415, 418, 426, 429
Bactrocera zonata (Saunders), 215, 426, 443
Baculoviruses, 113, 115
Bagrada hilaris (Burmeister), 348, 349
Bamboo shoot weevil, 520, 521
Banana aphid, 424, 503, 504, 506
Banana scale moth, 423
Banana stem borer, 424
Banana weevil, 424
Bandicota bengalensis (Gray), 590
Bark borer, 428
Bark caterpillar, 403, 418, 423, 429, 430, 431, 436
Bases of resitance, 225
Batocera horsfieldi Hope, 446
Batocera rufomaculata DeGeer, 414, 426, 428
Bean fly, 344
Beauveria bassiana (Balsamo) Vuillemin, 110, 111
Bed bug, 565
Beet armyworm, 359, 376
Bemisia tabaci (Gennadius), 94, 135, 182, 183, 218, 230, 298, 346, 364, 365, 369, 482, 506, 509, 601, 603, 613
Ber beetles, 421
Ber fruit fly, 215, 419, 420
BHC, 156, 157
Bihar hairy caterpillar, 84, 350, 361, 482, 508, 538
Binomial sampling, 65
Biochemical bases, 226
Biological control, 100, 118, 242
Biological control agents, 102
Biological control techniques, 118
Biophysical bases, 225
Biorational approaches, 192
Biotechnological approaches, 264
Biotechnological methods, 264, 265
Biotic factors, 46
food, 47
natural enemies, 46
Biotypes, 230
Birds, 586
Black ants, 567
Black-headed caterpillar, 491
Black palm beetle, 430
Blatella germanica (Linnaeus), 138, 182, 212, 563, 564
Blatella orientalis Linnaeus, 564
Blister beetle, 342, 485
Blow flies, 211, 579
Boll weevil, 41, 86
Bombay locust, 530
Book louse, 569
Boophilus microplus (Canestrini), 182, 581
Bot fly, 576
Botanical pest control, 128
Botanical pesticides, 618, 645
Brachycaudus helichrysi (Kaltenbach), 94, 440
Bradysia tritici (Coquillet), 472, 473
Brain hormone, 202, 203
Brevicoryne brassicae (Linnaeus), 236
Brevipalpus phoenicis (Geijskes), 500
Brinjal fruit and shoot borer, 212, 214, 232, 299, 463
Brinjal hadda beetles, 95, 465
Brinjal lacewing bug, 463
Brinjal stem borer, 464
Brown planthopper, 52, 71, 85, 123, 135, 183, 184, 217, 226, 229, 230, 232, 233, 239, 240, 242, 244, 277, 298, 314
Bt brinjal, 623
Bt cotton, 284, 623
Bt endotoxins, 271
Bt toxins, 289
Bud moth, 373
Buffalo fly, 578
Bunch caterpillar, 499
Busseola fusca (Fuller), 202, 272, 273

C

Cabbage borer, 455, 456
Cabbage caterpillar, 350, 452
Cabbage flea beetles, 456
Cabbage semilooper, 361, 454
Cadmilos retiarius Distant, 481
Calliphora vicina Robineau-Desvoidy, 579
Callosobruchus analis (Fabricius), 550
Callosobruchus chinensis (Linnaeus), 135, 275, 549, 556
Calocoris angustatus Lethiery, 331
Camel nasal fly, 577
Capture-recapture technique, 48
Carbamates, 141, 164
Carbaryl, 164
Carbofuran, 164
Carbon dioxide, 45, 614

Cardamom hairy caterpillars, 504
Cardamom thrips, 504
Carpomyia vesuviana Costa, 215, 419, 420
Cartap hydrochloride, 166
Carvalhoia arecae Miller & China, 489
Cashew aphid, 490
Cashew leaf and blossom webber, 490
Cashew leafminer, 490
Cashew tree borer, 489
Cassava mealybug, 121
Castor capsule borer, 354, 419, 432, 504, 508, 509
Castor hairy caterpillar, 355
Castor semilooper, 354
Castor slug, 355, 356
Castor whitefly, 353
Cattle ticks, 581
Cavelerius excavatus (Distant), 298, 382
Cecid fly, 474
Celosterna scabrator (Fabricius), 518
Central Insecticides Board, 78
Central Insecticides Laboratory, 79
Cephalopina titillator (Clark), 577
Cephus cinctus Norton, 225, 239, 240
Ceratitis capitata (Wiedemann), 216, 258, 259, 262
Ceratovacuna lanigera Zehntner, 385, 613
Cereal crops, 313
Ceroplastes rubens Maskell, 427
Changing status of pests, 52
Chemical control, 144, 240, 647
Chemosterilants, 248
Cherry stem borer, 446
Chilasia clytia Linnaeus, 509, 510
Chillies thrips, 506
Chilo auricilius Dudgeon, 85, 389
Chilo infuscatellus Snellen, 124, 298, 387, 388
Chilo partellus (Swinhoe), 88, 89, 94, 136, 202, 272. 273, 298, 327, 333
Chilo sacchariphagus indicus (Kapur), 100, 124, 391
Chilo suppressalis (Walker), 213, 214, 228, 272, 273, 275, 276, 319, 613
Chinaberry, 133
Chitin synthesis inhibitors, 204, 218
Chlorinated hydrocarbons, 155
Chloropulvinaria psidi (Maskell), 418
Chlorpyriphos, 159
Chlumetia transversa Walker, 416
Chordata, 4
Chromosome translocation, 256
Chrotogonus trachypterus (Blanchard), 531, 532
Chrotomyia horticola (Goureau), 350, 461, 484
Chrysanthemum, 133, 163
Chrysocoris stolii Wolff, 433
Chrysomya bezziana Villeneuve, 579
Cimex hemipterus (Fabricius), 565
Cimex lectularius Linnaeus, 565
Cinnamon butterfly, 509, 510
Citrus aphid, 405
Citrus blackfly, 398
Citrus blossom midge, 404
Citrus caterpillar, 299, 400, 401
Citrus leaf folder, 404
Citrus leafminer, 94, 299, 402
Citrus mealybug, 400, 431
Citrus mite, 405
Citrus nematode, 406
Citrus psylla, 94, 396, 482
Citrus red scale, 94, 399, 485
Citrus whitefly, 299, 397
Classical biological control, 118, 119
Classification of insecticides, 147
Clavigralla gibbosa Spinola, 343
Clean culture, 86
Clean seed, 85
Climate change, 53, 611
Cnaphalocrocis medinalis (Guenee), 108, 217, 272, 273, 298, 321
Coccinella septempunctata Linnaeus, 103, 104, 348, 364, 396, 616
Coccus viridis (Green), 111, 495
Cochliomyia hominivorax (Coquerel), 248, 251, 253, 259
Cockroaches, 563
Coconut eriophyid mite, 494
Coconut scale, 431, 490
Coconut weevil, 491
Coconut white grub, 494
Codlemone, 213
Codling moth, 67, 212, 213, 217, 218, 258, 259, 260, 299, 441, 625
Cofana spectra (Distant), 316
Coffee berry borer, 42, 497
Coffee green bug, 111, 495
Coffee shothole borer, 497
Coffee stem borer, 496
Colorato potato beetle, 41, 70, 71, 182, 274
Commercialization of transgenic crops, 282
Computer programming, 69
Concept of injury levels, 295
Conogethes punctiferalis Guenee, 354, 355, 419, 432, 504, 508, 509
Conservation, 121
Consortium for IPM, 625
Contact proisons, 148
Constraints in IPM implementation, 307
Corcyra cephalonica (Stainton), 130, 205, 544
Corn rootworms, 210
Corvus splendens Vieillot, 585
Cosmopolites sordidus (Germar), 424, 425
Cost : benefit ratios, 300
Cotton, 124, 271, 363, 375, 630
Cotton aphid, 364, 480
Cotton boll weevil, 212, 215, 243
Cotton grey weevil, 93, 374, 484
Cotton jassid, 44, 232, 239, 298, 363, 468
Cotton leaf roller, 371, 469
Cotton mealybug, 367, 485
Cotton semilooper, 372
Cotton stem weevil, 374
Cotton whitefly, 298, 232, 346, 364, 365, 469, 482, 506, 509, 613
Cottony cushion scale, 42, 101, 203, 217, 399

Craspedonta leayana (Latreille), 516, 517
Crocidolomia binotalis Zeller, 455
Crop losses, 39, 56
Crop residues, 94
Crop rotation, 88
Crop spacing, 87
Crucifer leaf webber, 455
Cryptolaemus montrouzieri Mulsant, 368, 400
Cucurbitacins, 228
Culex quinquefasciatus Say, 40, 109, 251, 255, 259, 559
Cultural control, 75, 82, 243
Custard apple, 136
Cutworms, 299, 361, 458, 539
Cydia pomonella (Linnaeus), 67, 115, 205, 212, 213, 217, 218, 258, 259, 260, 275, 299, 441, 625
Cylas formicarius (Fabricius), 470
Cypermethrin, 165, 183
Cyrtorhinus lividipennis Reuter, 138, 141, 184, 242, 301, 315
Cyrtotrachelus longipes Fabricius, 520, 521
Cysteine protease inhibitors, 276
Cytoplasmic incompatibility, 255

D

Daktulosphaira vitifoliae (Fitch), 41, 220, 239
Danais chrysippus Linnaeus, 481
Dasineura citri Grover, 404
Dasineura lini Barnes, 359
Database management, 69
Date palm scale, 429
DDT, 144, 145, 153, 155, 156, 179, 186, 188, 189, 621
Decision making, 302, 303
Decision support, 197
Decision support systems, 70
Defoliating beetles, 449
Delia antiqua (Meigen), 258, 259, 261, 461
Deltamethrin, 141, 165
Demodexis canis Leydig, 584
Dengue fever, 558
Deodar beetle, 521, 522
Deodar defoliator, 514
Desert locust, 525, 528
Decision support systems, 70
Deudorix isocrates (Fabricius), 96, 422, 432
Development inhibitors, 202
Development of IPM programme, 304
Dialeurodes citri (Ashmead), 203, 217, 299, 397
Diamondback moth, 141, 182, 299, 453, 613
Diaphorina citri Kuwayama, 94, 396, 482, 606
Diazinon, 159, 160
Dichlorvos, 159, 161
Dicladispa armigera (Olivier), 136, 298, 324
Diflubenzuron, 166, 204, 205, 218
DIMBOA, 227, 241
Diocalandra frumenti (Fabricius), 493
Direct gene transfer methods, 269
Direct uptake of DNA, 269
Diversity patterns, 3
Dog mite, 584
Dorysthenes hugelii Redtenbacher, 445
Double sampling, 65
Drosicha mangiferae (Green), 87, 96, 413, 436
Durable plant resistance, 234
Dusky cotton bug, 95, 365, 366, 469, 480
Dusters, 171
Dysdercus koenigii (Fabricius), 205, 366, 482

E

Earias insulana (Boisduval), 44, 369, 370, 468
Earias vittella (Fabricius), 130, 212, 369, 370, 468
Ecdysone, 202
Echidnophaga gallinacea (Westwood), 584
Echinocnemus oryzae (Marshall), 324
Ecological impact, 285
Ecological resistance, 222
Ecological sustainability, 309
Ecologically based pest management, 293
Economic analysis, 248
Economic constraints, 308
Economic impact, 236, 285
Economic importance of insects, 35
Economic injury level, 295, 296
Economic mandate, 642
Economic threshold levels, 296, 298
Economic sustainability, 309
Ectropis deodarae Prout, 514
Elasmopalpus jasminophagus Hamson, 484
Electroporation, 265, 269
Emmalocera depressella Swinhoe, 390
Empoasca kerri Pruthi, 345
Encarsia perniciosi (Tower), 438
Endosulfan, 156, 158
Entomology in India, 37
Entomopathogenic nematodes, 117
Environmental economic injury levels, 297, 299
Environmental impact, 140
Environmental impact of pesticides, 181
Environmental mandate, 643
Enzymes, 278
Ephestia cautella (Walker), 545
Epidinocarsis lopezi (De Santis), 121, 644
Epiricania melanoleuca (Fletcher), 100
Era of IPM, 624
Era of pesticides, 621, 624
Era of traditional approaches, 620, 621
Eriophyid mite, 486
Eriosoma lanigerum (Hausmann), 42, 94, 101, 220, 230, 439
Essential oils, 137
Estimation of losses, 55
Etiella zinckenella (Treitschke), 340, 341, 462
Eugenol, 137, 215, 216

Euproctis fraterna (Moore), 436
Euproctis lunata Walker, 355
Eupterote canarica Moore, 505
Eupterote cardamomi Ranga Ayyar, 505
Eupterote fabia Cramer, 505
Eupterote testacea Walker, 505
Eutectona machaeralis (Walker), 512, 513
European corn borer, 67, 86, 89, 137, 206, 227, 236, 239, 240, 273, 328, 329, 612
Euzophera perticella Ragnot, 464
Evolutionary concept, 222
Exelastis atomosa (Walsingham), 339
Expert systems, 71
Extensive studies, 47
Extrinsic resistance, 224

F

Factors affecting plant resistance, 237
Factors affecting survey, 62
Farmers Field Schools, 627, 628
Farmers' participation, 305
Fenvalerate, 165, 183
Ferrisia virgata (Cockerell), 427, 494, 495
Fertilizers, 86
Fibre crops, 362
Field bean pod borer, 340
Fig midge, 426
Fig mites, 427
Flea beetle, 299, 379, 410
Food and Agriculture Organization, 634
Food safety, 289
Formulations, 150
Fowl tick, 582
Framework of IPM programme, 308
Frugivorous pests, 441
Fruit flies, 215, 260, 426, 429
Fruit sucking moths, 403
Fumigants, 149, 166
Fungi, 110, 111, 505
Future outlook, 285, 618, 640
Future perspectives, 641

G

Gad fly, 575
Galerucella birmanica (Jacoby), 469
Galleria mellonella (Linnaeus), 183, 570, 571
Gamhar defoliator, 516, 517
Gamhar lace bug, 523
Gene pyramiding, 234, 281
Gene revolution era, 623
General equilibrium position, 300
Genetic control, 246
Genetic resistance, 223
Genetics of resistance, 228, 229
Geographical distribution of insects, 612
Gerbera leafminer, 486
German cockroach, 212, 563, 564
Ghujia weevil, 335
Girdle beetle, 345
Global IPM Facility, 634
Global warming, 612, 615
Globodera rostochiensis Wollenweber, 459
Glucosinolates, 227
Golden cyst nematode, 459
Gossyplure, 211
Gossypol, 227, 228
Government support, 306
Gram cutworm, 539
Gram dhora, 549, 550
Gram pod borer, 50, 336, 338, 339
Grape phylloxera, 41, 220, 239
Grapevine beetle, 409
Grapevine girdler, 410
Grapevine leafroller, 409
Grapevine thrips, 408, 482, 506
Greasy cutworm, 540
Greater wax moth, 570, 571
Green borer, 391
Greenbug, 230, 239, 242, 243
Green leafhopper, 88, 136, 229, 230, 232, 233, 239, 316
Green nettle slug caterpiller, 345
Green peach aphid, 217, 236, 350
Green potato bug, 343, 344
Green revolution era, 622
Green semilooper, 372
Groundnut aphid, 346, 351, 480
Groundnut leafminer, 351
Groundnut stem borer, 352
Grylloides sigillatus Linnaeus, 564, 565
Guava fruit fly, 215, 418
Guava mealy scale, 418
Gurdaspur borer, 108, 391, 392
Gypsy moth, 71, 212, 213

H

Haematopinus surysternus Denny, 572
Hairy caterpillars, 330, 537
HCH, 145, 147, 156, 157, 179, 187, 189, 621
Head borer, 361
Helicoverpa armigera (Hubner), 50, 53, 89, 94, 108, 115, 135, 182, 205, 211, 212, 218, 239, 243, 272, 274, 275, 278, 282, 298, 299, 302, 336, 338, 339, 369, 371, 459, 462, 612, 613, 616, 633
Heliothis virescens (Fabricius), 226, 243, 255, 262, 271, 272, 275, 276, 277, 287, 616, 630
Hellula undalis (Fabricius), 455, 456
Helmet scale, 419
Helopeltis antonii Signoret, 131, 498, 613
Helopeltis theivora Waterhouse, 498
Hendecasis duplifascialis Hampson, 484
Henosepilachna dodecastigma (Wiedemann), 465
Henosepilachna vigintioctopunctata (Fabricius), 136, 465
Hessian fly, 86, 94, 220, 229, 230, 236, 239, 240
Heterodera avenae Wollen Weber, 336
Heteropeza pygmaea Winnertz, 474
Heteropsylla cubana Crawford, 125, 522, 523
Hieroglyphus banian (Fabricius), 325
Hieroglyphus nigrorepletus Bolivar, 325

High expression level, 279
Hippobosca maculata Leach, 580
Hollyhock tingid bug, 480
Holotrichia consanguinea Blachard, 86, 352, 393, 485
Holotrichia insularis Brenske, 432
Hoplocerambyx spinicornis Newman, 517
Horizontal resistance, 223
Hormonal control, 26
Horny fly, 580
Horticultural crops, 124
Host evasion, 222
Host-marking pheromones, 195
Host plant resistance, 220, 239, 615, 648
House crickets, 564, 565
House crow, 585
House fly, 555, 556
House sparrow, 586
Household pests, 555
Huffaker Project, 625
Human louse, 566
Humped slug caterpillar, 499
Hyalomma anatolicum Koch, 581
Hyblaea puera (Cramer), 511, 512
Hybrid sterility, 254
Hyperparasitism, 99
Hypoderma lineatum (Villers), 576
Hypothenemus hampei (Ferrari), 42, 497

I

Icerya purchasi Maskell, 42, 76, 101, 203, 399
Idioscopus clypealis (Lethiery), 411
Imidacloprid, 166
Implementation phase, 305
Improved awareness, 307
Indarbela quadrinotata (Walker), 403, 418, 423, 429, 431
Indarbela tetraonis (Moore), 418, 430, 436
Indian field mouse, 591
Indian fruit bat, 593
Indian gypsy moth, 448, 514
Indian meal moth, 109, 212, 545
Indian mole rat, 590
Induced resistance, 222, 235
Inert dusts, 148
Informational constraints, 308
Insect classification, 27
Insect growth and development, 24
Insect structure, 13
Insect vectors, 600
Insect-resistant varieties, 232
Insecta, 11
Insecticide resistance management, 631
Inoculative releases, 123
Insecticide resistance, 181
Insecticides, 155
Insecticides Act, 1968, 77
Insects, 11, 35, 103
Institutional constraints, 307
Institutional infrastructure, 307
Integrated Fruit Production System, 626
Integrated pest management, 58, 292, 641, 651
    cotton, 630
    future outlook, 653
    perspectives, 643
    plantation crops, 633
    potential, 309
    rice, 628
    sustainable agriculture, 642
    vegetable crops, 632
Integrated Pest Management Task Force, 635
Integrated Pest Management Working Group, 636
Integration of tactics, 141, 240, 261, 282, 301
Intensity of resistance, 221
Intensive studies, 47
Intercropping, 89, 90
International organizations, 634
Internode borer, 124, 391
Intrinsic resistance, 224
Inundative releases, 123, 262
Invasive pests, 41
Ionising radiations, 250
IPM in apple orchards, 625
IPM in developed countries, 625
    Australia, 626
    Canada, 625
    Europe, 627
    Netherlands, 626
    New Zealand, 626
    USA, 625
IPM in developing countries, 627
    China, 629
    Colombia, 630
    India, 629, 630
    Indonesia, 628, 629
    Pakistan, 632
    Peru, 630
    Philippines, 629
IPM in greenhouses, 626
IPMnet, 636
Irrigation, 85

J

Jackfruit leaf webber, 428
Jacobiella facialis (Jacobi), 239
Jamun leafminer, 430
Japanese beetle, 41, 106
Japanese encephalitis virus, 557
Jasmine budworm, 484
Jasmine gallery worm, 484
Jasmine leaf webworm, 483
Jasmine thrips, 482
Jasmolins, 134
Jute mealybug, 375
Jute semilooper, 376
Jute stem girdler, 377
Jute stem weevil, 378
Juvenile hormone, 202, 203
Juvenile hormone analogues, 203, 216

K

Kadam defoliator, 515, 516
Kairomones, 192, 193, 200, 226
Karanjin, 135
Keiferia lycopersicella (Walsingham), 214, 272
Kerria albizziae (Green), 434
Kerria communis (Mahdihassan), 411

Kerria lacca (Kerr), 420
Khapra beetle, 546
Knapsack sprayer, 174, 175
Knipling's SIRM Model, 251, 252
Knipling's technique, 250
Knockdown sampling, 50

L

Labelling insects, 33, 34
Lac insect, 420
Lampides boeticus (Linnaeus), 341
Lance nematrode, 407
Leafcurl mite, 433
Lectins, 277
Legislative control, 75
Legislative measures, 306
Lenodera vittata Walker, 504
Lentil pod borer, 340, 341
Lepisma saccharina Linnaeus, 569
Leptinotarsa decemlineata (Say), 41, 70, 71, 130, 182, 228, 272, 274, 275
Leptocorisa acuta (Thunberg), 326
Lesser grain borer, 548, 549
Lethal factors, 257
Leucinodes orbonalis Guenee, 212, 214, 274, 299, 463
Leucopholis coneophora Burmeister, 494
Lily moth, 484, 485
Lindigaspis rossi (Maskel), 479
Linseed bud fly, 359
Lipaphis erysimi (Kaltenback), 130, 194, 200, 225, 275, 299, 347, 348
Liposcelis transvaalensis Enderlein, 569
Liriomyza trifolii (Burgess), 42, 486, 613, 633
Litchi bug, 433
Locusts, 525
Long-horned walnut beetle, 446
Long-term forecasting, 66
Longitarsus belegaumensis Fabricius, 379
Longitarsus nigripennis (Motschulsky), 507
Lucilia cuprina (Wiedeman), 579
Lucilia sericata (Meigen), 579, 580
Lure and infect, 198
Lure and kill, 198
Lycosa pseudoannulata (Bosenberg & Strand), 138, 141, 242, 301, 315
Lygaeus civilies Wolff., 480
Lymantria dispar (Linnaeus), 71, 115, 212, 213
Lymantria mathura Moore, *513*
Lymantria obfuscata Walker, 448, 514

M

Maconellicoccus hirsutus (Green), 299, 375
Macrosiphum rosaeiformis Das, 478, 479
Maize, 273, 327, 330
Maize borer, 94, 232, 327
Major gene resistance, 223, 234
Major groups of pesticides, 155
Malacosoma indicum Walker, 447
Malaria, 560
Malathion, 146, 159
Mange mite, 583
Mango bud mite, 417
Mango fruit fly, 415
Mango gall psyllid, 417
Mango hoppers, 299, 411, 412
Mango mealybug, 87, 96, 413, 436
Mango shoot borer, 416
Mango stem borer, 414, 426
Mango stone weevil, 415
Marker genes, 289
Maruca testulalis (Geyer), 340
Mass trapping, 197
Mating disruption, 199
Mayetiola destructor (Say), 86, 94, 220, 230, 239, 240
Mechanical control, 75, 95
Mechanical traps, 96
Mechanisms of resistance, 224
Mediterranean fruit fly, 216, 258, 259
Megaselia halterata (Loew), 473, 474
Melanagromyza obtusa (Malloch), 299, 341
Melia azedarach Linnaeus, 133
Meloidogyne incognita Chitwood, 471
Melon fruit fly, *466*
Menochilus sexmaculatus (Fabricius), 103, 104, 348, 351, 364, 396
Menopon gallinae (Linnaeus), 573
Metaldehyde, 150, 171, 599
Metamorphosis, 26
Metarhizium anisopliae (Metchnikoff) Sorokin, 110, 381, 634
Methods of forecasting, 66
Methoprene, 203
Methyl bromide, 167, 189
Microbial control, 105
Microbial pesticides, 617, 644
Microorganisms, 104
Microinjection, 265
Microprojectile bombardment, 269
Micropropagation, 266
Microtermes obesi Holmgren, 336, 393
Milk weed bug, 480
Millets, 313
Minor gene resistance, 223
Mites, 103
Modelling, 68
Molluscicides, 171
Mollusca, 5
Molya nematode, 336
Monitoring, 196
Monogenic resistance, 223
Monomorium destructor (Jerdon), 567
Monomorium indicum Forel, 567
Mosaics, 282
Mosquitoes, 556
Moulting hormone, 202, 206
Moulting hormone analogues, 217
Mounting insects, 33, 34
Multiline varieties, 235
Multiple parasitism, 99
Multiple pest resistant varieties, 233
Multitrophic interactions, 224
Mung dhora, 550
Mus booduga (Gray), 591

Musca nebulo Wiedemann, 555, 556
Mushroom mites, 475
Mushroom nematodes, 476
Mustard aphid, 347, 348
Mustard sawfly, 95, 349
Mylabris pustulata (Thunberg), 342, 485
Myllocerus lactivirens Marshal, 448
Myllocerus undecimpustulatus Faust, 93, 374, 485
Mythimna separata (Walker), 116, 133, 142, 322, 334, 335, 616
Myzus persicae (Sulzer), 90, 141, 200, 205, 217, 236, 275, 350, 601, 603

N

Nacoleia octasema (Meyrick), 423
Nanotechnology, 653
National IPM Initiative, 626
National organizations, 637
Natural enemies, 61, 101, 138, 616
Nausinoe geometralis (Guenee), 483
Neem, 129
Neem-based pesticides, 132
Nemathelmenthes, 6
Nematicides, 168
Nematodes, 116
Neomaskellia bergii (Signoret), 383
Nephopteryx eugraphella Ragonot, 435
Nephotettix nigropictus (Stal), 316, 601, 604
Nephotettix virescens (Distant), 230, 239, 275, 301, 316, 604
Nezara viridula (Linnaeus), 343, 344
Nicotiana tahacum Linnaeus, 134
Nicotine, 131, 134
Nilaparvata lugens (Stal), 66, 71, 133, 136, 226, 230, 239, 240, 242, 275, 277, 314, 601
Nipaecoccus viridis (Newstead), 431
Nomenclature, 32
Non-insect pests, 475, 486, 585
Nomenclature, 32
Nonpreference, 224
Non-target organisms, 184, 286
Novaluron, 166, 205
Novel genes, 278
Nuclear polyhedrosis viruses, 114
Nutrients, 226, 237
Nupserha bicolor postbrunnea Dutt, 377
Nymphula depunctalis Guenee, 322

O

Obereopsis brevis (Gahan), 345
Ochropleura flammatra (Denis & Schiffermuller), 458, 539
Odontotermes obsesus (Rambur), 336, 393, 533
Odoiporus longicollis (Olivier), 424
Oestrus ovis Linnaeus, 576
Oilseeds, 348
Oligogenic resistance, 223
Oligonychus citri McGegor, 405, 406
Oligonychus indicus (Hirst), 393
Onion fly, 258, 259, 261, 461
Onion thrips, 460
Ophideres fullonica Linnaeus, 403
Ophideres materna Cramer, 403
Ophiomyia phaseoli (Tryon), 344, 461
Opisina arenosella Walker, 125, 491
Organophosphates, 141, 158, 189
Organotin compounds, 249
Origin of IPM, 292
Orseolia oryzae (Wood-Mason), 230, 239, 298, 323, 613
Oryctes rhinoceros (Linnaeus), 430, 491, 492, 613, 633
Ostrinia nubilalis (Hubner), 67, 86, 89, 137, 206, 227, 239, 240, 272, 273, 275, 280, 286, 328, 329, 612
Oxadiazines, 209
Oxya nitidula Walker, 531
Oxycarenus laetus Kirby, 95, 365, 366, 469, 480

P

Painted bug, 348, 349
Papilio demoleus Linnaeus, 136, 299, 400, 401
Parapheromones, 193, 216
Parasa lepida (Cramer), 355, 356, 429
Parasitoids, 99, 104, 125, 287, 643
Passer domesticus (Linnaeus), 586
Patanga succincta Linnaeus, 530
Pathogens, 125
Pea aphid, 230, 462
Pea blue butterfly, 341
Pea leafminer, 232, 350, 461, 484
Pea pod borer, 232, 462
Pea stem fly, 461
Peach fruit fly, 215, 443
Peach leafcurl aphid, 94, 440
Peach stem borer, 445
Pectinophora gossypiella (Saunders), 97, 108, 206, 211, 212, 213, 243, 249, 259, 260, 272, 298, 368
Pediculus humanus Linnaeus, 566
Pempherulus affinis (Faust), 374
Pentalonia nigronervosa Coquerel, 424, 503, 504, 506, 601, 603
Peregrinus maidis (Ashmead), 331
Perigaea capensis (Guenee), 359
Perina nuda Fabricius, 428
Periplaneta americana (Linnaeus), 212, 563
Perspectives in IPM, 643
Pest forecasting, 59, 65, 66
Pest management, 611, 620, 641
Pest management organizations, 634
Pest management strategies, 614
Pest management system, 294
Pest populations, 39, 42, 47
Pest resistance, 141
Pest resurgence, 183
Pest surveillance, 59, 60
Pesticidal plants, 129
Pesticide application equipment, 171
Pesticide consumption, 179, 180
Pesticide poisoning, 188
Pesticide residues, 185, 186
Pesticides Management Bill, 2008, 82
Pests of crops and animals, 37
    cereals and millets, 313
    farm animals, 572
    fibre crops, 362
    forest trees, 511
    fruits, 395, 437

household, 555
mushrooms, 472
oilseed crops, 347
ornamental plants, 478
plantation crops, 488
pulse crops, 338
spices, 503
stored grains, 542
sugarcane, 380
vegetable crops, 451
Phenacoccus manihoti Matile-Ferrero, 121, 644
Phenacoccus solenopsis Tinsley, 288, 367, 484
Phenylpyrazoles, 208
Pheromones, 193, 194, 211, 213
Phlebotomus argentipes Annandale & Brunneti, 562
Phlebotomus minutus (Rondani), 574
Phorate, 159, 160, 170
Phorid fly, 473, 474
Photoperiod, 237
Phthorimaea operculella (Zeller), 42, 85, 88, 212, 272, 274, 457, 458
Phycita infusella (Meyrick), 373
Phycita orthoclina Meyrick, 432
Phyllocnistis citrella Stainton, 94, 299, 402
Phyllotreta cruciferae (Goeze), 456
Physical control, 95
Physical factors, 96
Phytochemicals, 130, 141
Phytoplasmas, 603
Pieris brassicae (Linnaeus), 108, 206, 227, 228, 350, 452
Pineapple thrips, 429
Pink bollworm, 97, 211, 212, 213, 249, 259, 260, 298, 368
Pink gypsy moth, 513
Pink stem borer, 232, 320, 321, 333, 336, 613
Pink waxy scale, 427
Plantation crops, 125
Plant-derived genes, 274
Plantation crops, 633
Platygaster oryzae Cameron, 323
Plecoptera reflexa Guenee, 515
Plocaederus ferrugineus (Linnaeus), 489
Plodia interpunctella (Hubner), 109, 206, 212, 545
Plum hairy caterpillar, 436
Plumbagin, 205
Plume moth, 239
Plutella xylostella (Linnaeus), 47, 108, 133, 138, 141, 152, 205, 272, 274, 286, 299, 453, 613
Pneumatic sprayers, 173
Poekilocerus pictus (Fabricius), 13, 21, 22, 23, 434, 482
Polistes hebraeus (Fabricius), 13, 568
Political constraints, 308
Pollinators, 36, 287
Pollu beetle, 507
Polygenic resistance, 223
Polynactins, 208
Polyphagotarsonemus latus (Banks), 378, 501
Polyphagous pests, 526
Polytela gloriosae Fabricius, 484,485
Pomegranate butterfly, 96, 422, 432
Pongram, 135
Popillia japonica Newman, 41, 106, 217
Population dynamics, 247
Population indices, 51
Post-green revolution era, 622
Potato, 274
Potato tuber moth, 42, 85, 88, 212, 274, 457, 458
Potential and constraints, 309
biological control, 126
biorational approaches, 219
biotechnolgocial approaches, 290
botanical pest control, 142
chemical control, 190
cultural control, 97
genetic control, 262
host plant resistance, 244
Poultry shaft louse, 573
Poultry stickfast flea, 581
Predators, 99, 103, 125, 286, 643
Preserving insects, 33, 35.
Prevention of Food Adulteration Act, 1954, 81
Primary pest resurgence, 183
Problem definition phase, 304
Prodioctes haematicus Chevrolat, 505
Propesticides, 207
Protease inhibitors, 235, 274
Prothoracicotropic hormone, 202
Protozoa, 112
Pseudococcus filamentosus Cockerell, 400
Psittacula krameri (Scopoli), 588
Psoroptes communis (Furstenberg), 583
Psorosticha zizyphi (Stainton), 404
Pteropus giganteus Brunnich, 593
Public health pests, 250
Pulses, 338, 346
Push-pull strategy, 201, 622
Pyrethrins, 124, 141
Pyrethrum, 124, 153
Pyridine insecticides, 209
Pyrilla perpusilla Walker, 52, 67, 111, 121, 298, 380, 381, 613
Pyrrole insecticides, 208

Q

Quadraspidiotus perniciosus (Comstock), 41, 42, 76, 94, 101, 217, 299, 438
Quadrat method, 48
Qamyl, 169
Qualitative survey, 60
Quantitative seasonal studies, 65
Quantitative survey, 60
Quarantine, 41, 75, 76
Quassia amara Linnaeus, 137
Quassia, 137
Quassin, 137
Quassinoids, 137
Quneensland fruit fly, 260
Quercetin, 210
Quercitrin, 210
Quinalphos, 159, 163, 183

R

Random sampling, 64
Raphidopalpa foveicollis (Lucas), 95, 467, 468
Raphimetopus ablutellus Zeller, 391
Rat flea, 566, 567
Recilia dorsalis (Motschulsky), 317, 604
Recognition pheromones, 195
Recombinant DNA technology, 267
Recruitment pheromones, 195
Red borer, 500
Red cotton bug, 366, 482
Red crevice mite, 500
Red flour beetle, 547, 548
Red gram pod fly, 341
Red hairy caterpillar; 87, 93, 96, 537
Red palm weevil, 212, 215, 250, 430, 492, 493
Red pumpkin beetle, 95, 467, 468
Red scale, 479
Red spider mite, 434, 486
Red vegetable mite, 470
Refuge strategy, 279
Registration of insecticides, 79
Relative estimates, 49
Remote sensing, 50
Repellents, 209
Research phase, 305
Resistance to biorationals, 218
Resistance to Bt, 109
Resistance to transgenics, 279
Retinue pheromones, 195
Rhinoceros beetle, 491, 492, 613, 633
Rhipiphorothrips cruentatus Hood, 408, 482, 506
Rhizome weevil, 505
Rhynchophorus ferrugineus (Olivier), 212, 215, 250, 430, 492, 493
Rhyzopertha dominica (Fabricius), 135, 548, 549
Rice, 123, 273, 314, 628
Rice bug, 298, 326
Rice caseworm, 322
Rice earcutting caterpillar, 322
Rice gall midge, 88, 230, 232, 233, 239, 298, 323
Rice grasshoppers, 325
Rice hispa, 298, 324
Rice leaf folder, 123, 298, 321
Rice mealybug, 326
Rice moth, 544
Rice root weevil, 298, 324
Rice weevil, 251, 547
Rickettsiae, 112
RNA interference, 265, 270
Rodenticides, 170
Rodents, 589
Rodolia cardinalis (Mulsant), 101, 399
Root-knot nematode, 471
Root-lesion nematode, 487
Rose aphid, 478, 479
Rose-ringed parakeet, 588
Rotation of varieties, 234
Rotations, 281
Rotenone, 131, 135
Rubber bark caterpillar, 502
Ryania, 136
Ryanodine, 136

S

Sabadilla, 136
Saccharicoccus sacchari (Cockerell), 384
Safflower aphid, 360
Safflower bud fly, 360
Safflower caterpillar, 359
Saissetia coffeae (Walker), 419, 495
Sal borer, 517, 518
Sampling habitat, 64
Sampling pattern, 64
San Jose scale, 41, 42, 94, 101, 299, 438
Sand flies, 562, 574
Sapota leaf webber, 435
Sarcoptes scabei (Linnaeus), 583
Scabies mite, 583
Schistocerca gregaria (Forskal), 133, 525, 528
Schizaphis graminum (Rondani), 230, 239, 242, 244, 298
Sciarid fly, 472, 473
Sciothrips cardamomi (Ramakrishna Ayyar), 504
Scirpophaga excerptalis (Fabricius), 298, 386
Scirpophaga incertulas (Walker), 94, 212, 214, 238, 273, 298, 318
Scirtothrips dorsalis Hood, 298, 506
Sclerotization disruptors, 207
Scolytus major Stebbing, 521, 522
Screwworm fly, 248, 251, 253, 258, 259
Secondary insect pests, 287
Secondary pest resurgence, 183
Seed rate, 85
Seira iricolor Tosii & Ashraf, 475
Semiochemicals, 192, 201, 212, 646
Semul borer, 519
Sequential release, 234
Sequential sampling, 64
Serine protease inhibitors, 270
Sesame gall fly, 358
Sesamia inferens (Walker), 276, 320, 321, 333, 335, 613
Sex pheromones, 194
sex-ratio distortion, 257
Shelfordella tartara Sauss, 563
Shisham borer, 520
Shisham defoliator, 515
Short-term forecasting, 66
Silverfish, 569, 570
Simulation techniques, 72
Singhara beetle, 183, 469
Sinoxylon anale Lesne, 409
Siphona exigua (de Meijere), 578
Sitobion miscanthi (Takahashi), 334, 613
Sitophilus oryzae (Linnaeus), 136, 547
Sitotroga cerealella (Olivier), 100, 135, 275, 543
Skipper butterfly, 508
Slug caterpillar, 429
Slugs, 486, 598
SmartStax maize, 273
Snails, 486, 598
Social mandate, 643
Sociological constraints, 308
Sogatella furcifera (Horvath), 66, 238, 315
Soil biota, 288

Soil moisture, 237
Soil pH, 238
Somaclonal variation, 265, 266
Somatic hybridization, 266
Sorghum, 331, 333
Sorghum earhead bug, 331
Sorghum midge, 89, 226, 229, 232, 238, 239, 333
Sorghum shoot bug, 331
Sorghum shoot fly, 89, 226, 229, 232, 239, 241, 298, 330, 332
Soybean caterpillar, *114*, *238*
Spatulicraspeda castaneiceps Hampson, 499
Species replacement, 258
Sphenoptera laferteri Thomson, 445
Sphenoptera perotetti Guenee, 352
Spiders, *102*
Spilarctia obliqua (Walker), 84, 88, 116, 138, 299, 330, 350, 361, 482, 508, 538
Spinosad, 166
Spinosyns, 208
Spodoptera exigua (Hubner), 359, 376
Spodopteta litura (Fabricius), 53, 86, 108, 130, 135, 183, 205, 212, 275, 298, 299, 361, 373, 454, 613, 614
Sporadic grasshopper, 531
Spotted bollworms, 44, 212, 298, 369, 370, 468
Spotted pod borer, 340
Sprayers, 172
Springtail, 475
Stable fly, 578
Stalk borer, 85, 232, 389
Stenodiplosis sorghicola (Coquillett), 89, 238, 239, 333
Sterile insect technique, 248, 251
Sterile male technique, 251
Sternochetus mangiferae (Fabricius), 415
Sthenias grisator Fabricius, 410
Stomach poisons, 147
Stomoxys calcitrans (Linnaeus), 578
Storage structures, 553
Stored grain pests, 542
Stratified sampling, 64
Striped mealybug, 494, 495
Subabul psyllid, 522, 523
Successes in genetic control, 258
Sucking cattle louse, 572
Sugarcane, 124
Sugarcane black bug, 298, 382
Sugarcane early shoot borer, 124, 298, 387, 388
Sugarcane mealybug, 384
Sugarcane mite, 393
Sugarcane pyrilla, 67, 298, 380, 381, 613
Sugarcane root borer, 390
Sugarcane scale insect, 232, 298, 384
Sugarcane spottedfly, 383
Sugarcane top borere, 232, 298, 386
Sugarcane whitefly, 382, 383
Sugarcane woolly aphid, 385, 613
Sunflower lacewing bug, 481
Sunnhemp capsid, 378
Sunnhemp hairy caterpillar, 379
Superparasitism, 99
Surface grasshopper, 531, 532
Sustainable agriculture, 641, 642
Sweet potato weevil, 470
Sylepta derogata (Fabricius), 371, 469
Sylepta lunalis Guenee, 409
Sympatric resistance, 223
Synergism, 125
Synomones, 200
Synthetic pyrethroids, 165, 184
Systematic sampling, 64
Systemic poisons, 148
Systems analysis, 68

T

Tabanus striatus Fabricius, 575
Tamarind fruit borer, 432
Tanacetum cinerariaefolium (Treviranus), 133
Tanymecus indicus Faust, 335
Tarache notabilis (Walker), 372
Tea mosquito bug, 498, 613
Teak defoliator, 511, 512
Teak skeletonizer, 512, 513
Temperature, 43, 237, 613
Temporal expression of toxins, 280
Tent caterpillar, 447
Termitarium, 533
Termites, 336, 393, 532
Tetranychus telarius Linnaeus, *470*
Tetranychus urticae Koch, 90, 182, 194, 200, 434, 486
Thosea aperiens (Walker), 345
Thrips orientalis Bangall, 482
Thrips tabaci Lindeman, 429, 460
Thysanoplusia orichalcea (Fabricius), 361, 454
Till hawk moth, 357
Till leaf and pod caterpillar, 356
Tillage, 84
Time of harvesting, 94
Time of sowing, 94
Tingis beesoni Drake, 523
Tissue culture, 265
Tissue expression of toxins, 280
Tobacco, 134
Tobacco caterpillar, 361, 373, 454
Tolerance, 225
Tomato fruit borer, 232, 459
Tomato pinworm, 214
Toxoptera aurantii (Boyer do Fonscolombe), 405
Toxoptera citricidus (Kirkaldy), 405
Toxoptera odinae (van der Goot), 490
Trail pheromones, 195
Transgenic crop protection, 270
Transgenic crops, 283, 616, 649
Trap cropping, 91, 92
Trap plant strategy, 282
Trialeurodes ricini Misra, 353
Tribolium castaneum (Herbst), 547, 548
Trichogramma chilonis Ishii, 123, 124, 243, 328, 369, 370, 387, 388, 590, 632
Trichogramma japonicum Ashmead, 123, 318, 322
Trogoderma granarium Everts, 546
Tsetse flies, 211, 259, 261
Tungro, 233, 601
Tur pod bug, 343

Turnip moth, 540, 541
Tylenchulus semipenetrans (Cobb), 406
Types of losses, 54
Types of resistance, 221

U

Udaspes folus Cramer, 508
Ultra low volume sprayers, 172, 176
Unavoidable losses, 54
Underhill, 220
Unidirectional incompatibility, 255
Urentius enonymus Fabricius, 480
Urentius sentis Distant, 463
Uroleucon compositae (Theobald), 360

V

Variable intensity sampling, 65
Varietal mixtures, 235
Vector management, 606
Vector-mediated gene transfer, 267
Vedalia beetle, 100
Vegetables, 274, 632
Vertebrates, 102
Vertical resistance, 223
Vespa orientalis Linnaeus, 568
Virus-phytoplasma relationship, 605
Virus-vector relationship, 603
Viruses, 112, 601

W

Walnut weevil, 95, 444
Warble fly, 576
Warfarin, 170
Wasps, 568
Weeds, 93
Wheat, 334, 337
Wheat aphid, 334, 613
Wheat stem saw fly, 225, 239, 240
Wheat gall nematode, 337
Wheat streak mosaic, 602
White grub, 352, 393, 432, 485, 507
White rice leafhopper, 316
Whitebacked planthopper, 229, 238, 315
White-tailed mealybug, 427
Wild relatives of crops, 288
Winter Majetin, 220
Woolly apple aphid, 42, 94, 101, 220, 230, 439
World Food Prize, 121, 234, 254, 293, 623

X

Xanthochelus superciliosus Gyll, 422
Xanthopimpla nursei Cameron, 387
Xanthopimpla predator (Fabricius), 387
Xanthopimpla punctata Fabricius, 491
Xenopsylla cheopis (Rothschild), 566, 567
Xyleborus noxius Sampson, 524
Xyleborus parvulus Eichhoff, 494
Xylophagous pests, 444
Xylosandrus compactus (Eichhoff), 497
Xylotrechus quadripes Chevrolat, 212, 496
Xystrocera globosa Olivier, 519

Y

Yellow dwarf, 604
Yellow fever, 557
Yellow orange leaf virus, 601
Yellow stem borer, 52, 212, 238, 298, 318
Yellow tea mite, 378, 501
Yellow vein mosaic, 601, 603
Yield loss, 57, 62
Yield loss assessment, 54

Z

Zabrotes subfasciatus (Boheman), 275
Zeuzera coffeae Nietner, 500
Zigzag leafhopper, 229, 317
Zinc phosphide, 170, 189, 592
Zingiber officinale Roscoe, 138, 509
Zonabris phalarata Pallas, 333
Zones of abundance, 68
Zyleborus biporus Signoret, 502

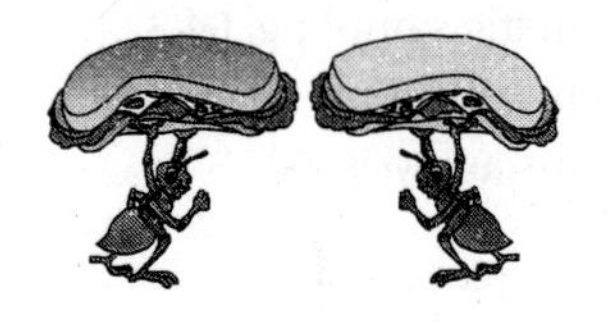

A. Praying mantis, *Creobrote gemmatus* adult

B. Green lacewing, *Chrysoperla carnea* adult

C. Pirate bug, *Orius sp.* feeding on larva

D. Ladybird beetle, *Menochilus sexmaculatus* adult

E. Ladybird beetle, *Coccinella septempunctata* adult

F. Syrphid fly, *Episyrphus viridaureus* adult

Plate 5.1 Biological control agents : Predators

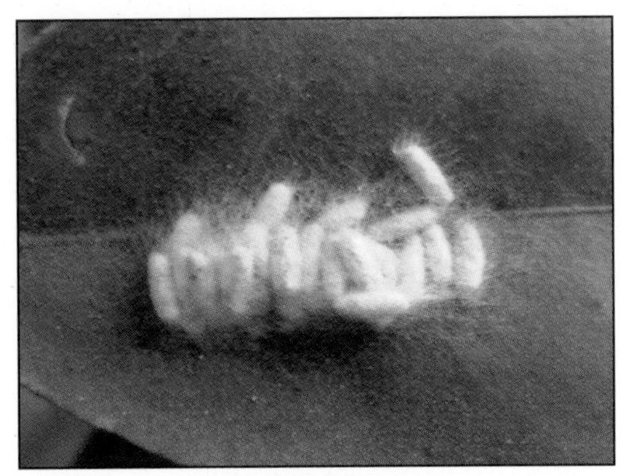

A. *Apanteles* sp. coccon

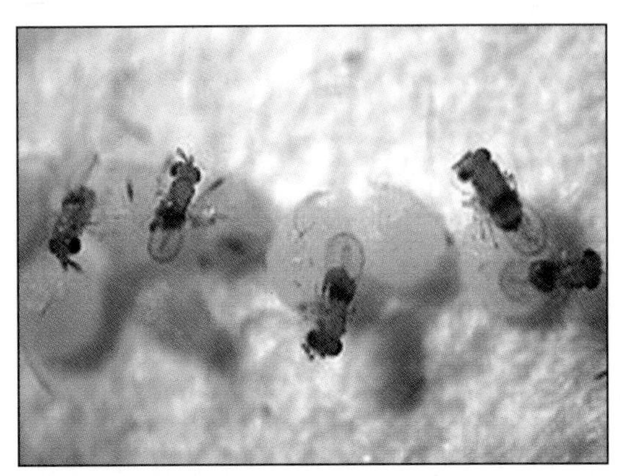

B. *Trichogramma* sp. adults

C. *Aenasius bambawalei* feeding on mealybug

D. *Spodoptera litura* larva infected with entomopathogenic fungus

E. *Spodoptera litura* larva infected with NPV

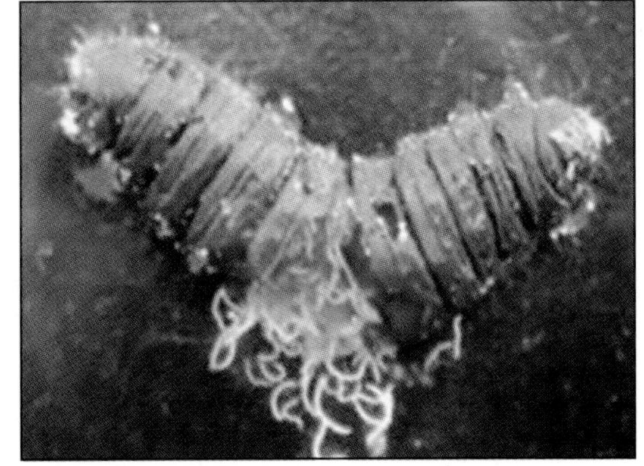

F. Entomopathogenic nematode multiplication inside *Galleria* cadaver

Plate 5.2 Biological control agents : Parasitoids and microbial pathogens

A. Brown planthopper

B. Hopperburn in rice

C. Whitebacked planthopper

D. Green leafhopper

E. Yellow stem borer

F. White stem borer

G. White ears due to stem borer

H. Rice leaf folder adult

I. Rice leaf folder larva and damaged leaves

Plate 13.1 Pests of rice

A. Leaf injury in maize due to stem borer

B. Dead heart in maize due to stem borer

C. Aphid attack in maize

D. Dead heart in sorghum due to shoot fly

E. Leaf injury in sorghum due to stem borer

F. Dead heart in sorghum due to stem borer

G. Termite attack in sorghum

H. Aphid damage in wheat

I. Armyworm larvae on wheat

Plate 13.2 Pests of maize, sorghum and wheat

A. Pod borer in chickpea

B. Spotted pod borer damage in pigeonpea

C. Pea blue butterfly

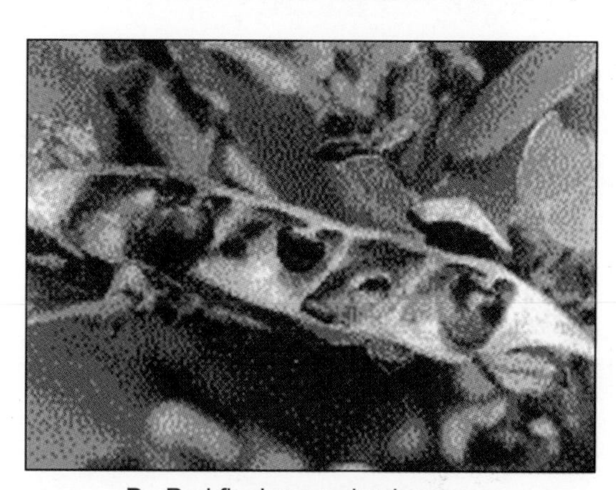

D. Pod fly damage in pigeonpea

E. Blister beetles on pigeonpea

F. Tobacco caterpillar on soybean

Plate 14.1 Pests of pulse crops

A. Aphid damge on mustard

B. Painted bug

C. Cabbage caterpillar on mustard

D. Damage by linseed bud fly

E. Head borer feeding on sunflower

F. Cabbage semilooper feeding on sunflower

Plate 15.1 Pests of oilseed crops

A. Damage caused by cotton jassid

B. Cotton whitefly

C. Mealybug on cotton plant

D. Pink bollworm larva in cotton flower

E. American bollworm feeding in cotton boll

F. Tobacco caterpillar on cotton

Plate 16.1 Pests of cotton

A. Sugarcane pyrilla

B. Black bug feeding on sugarcane

C. Sugarcane top borer

D. Damage due to early shoot borer

E. Sugarcane root borer

F. Internode borer

Plate 17.1 Pests of sugarcane

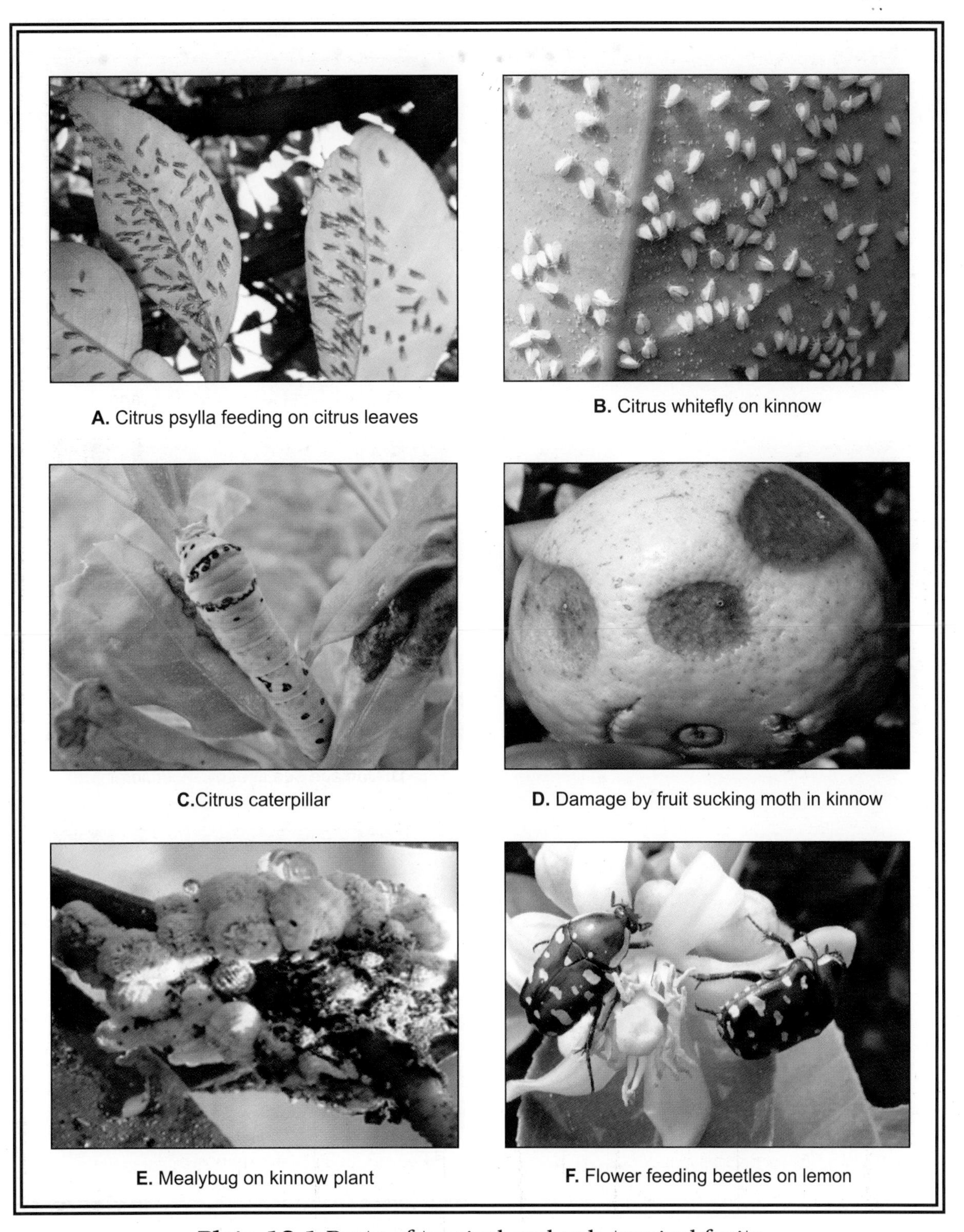

A. Citrus psylla feeding on citrus leaves

B. Citrus whitefly on kinnow

C.Citrus caterpillar

D. Damage by fruit sucking moth in kinnow

E. Mealybug on kinnow plant

F. Flower feeding beetles on lemon

Plate 18.1 Pests of tropical and sub-tropical fruits

A. Mango mealybug

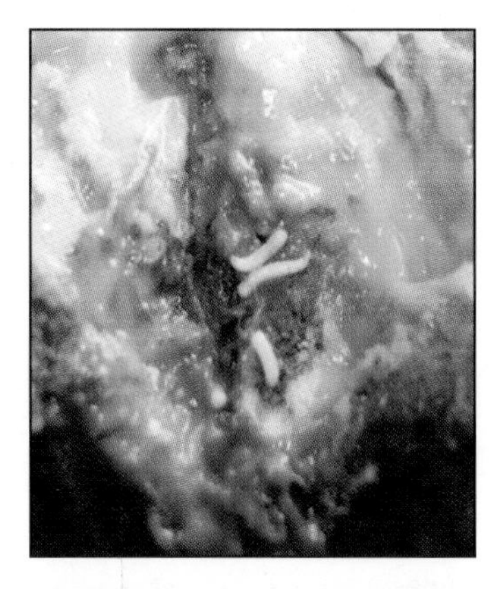

B. Fruit fly larvae inside mango fruit

C. Fruit fly damage in guava

D. PAU fruit fly trap

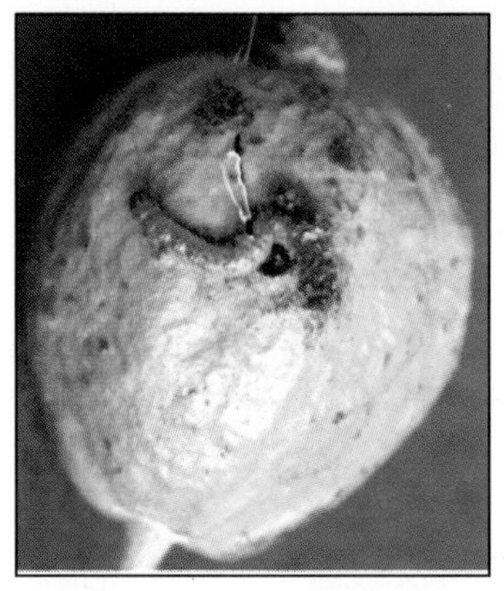

E. Damage by castor capsule borer on guava fruit

F. Tortoise beetle on ber nursery plant

G. Gall caused by ber gall mite

H. Damage by larva of pomegranate fruit fly

I. Bark eating caterpillar on jamun

Plate 18.2 Pests of tropical and sub-tropical fruits

A. San Jose scale attack on apple
B. Different stages of San Jose scale
C. Woolly apple aphid
D. Green peach aphid damaged leaves
E. Apple root borer
F. Tent caterpillar

Plate 19.1 Pests of temperate fruits

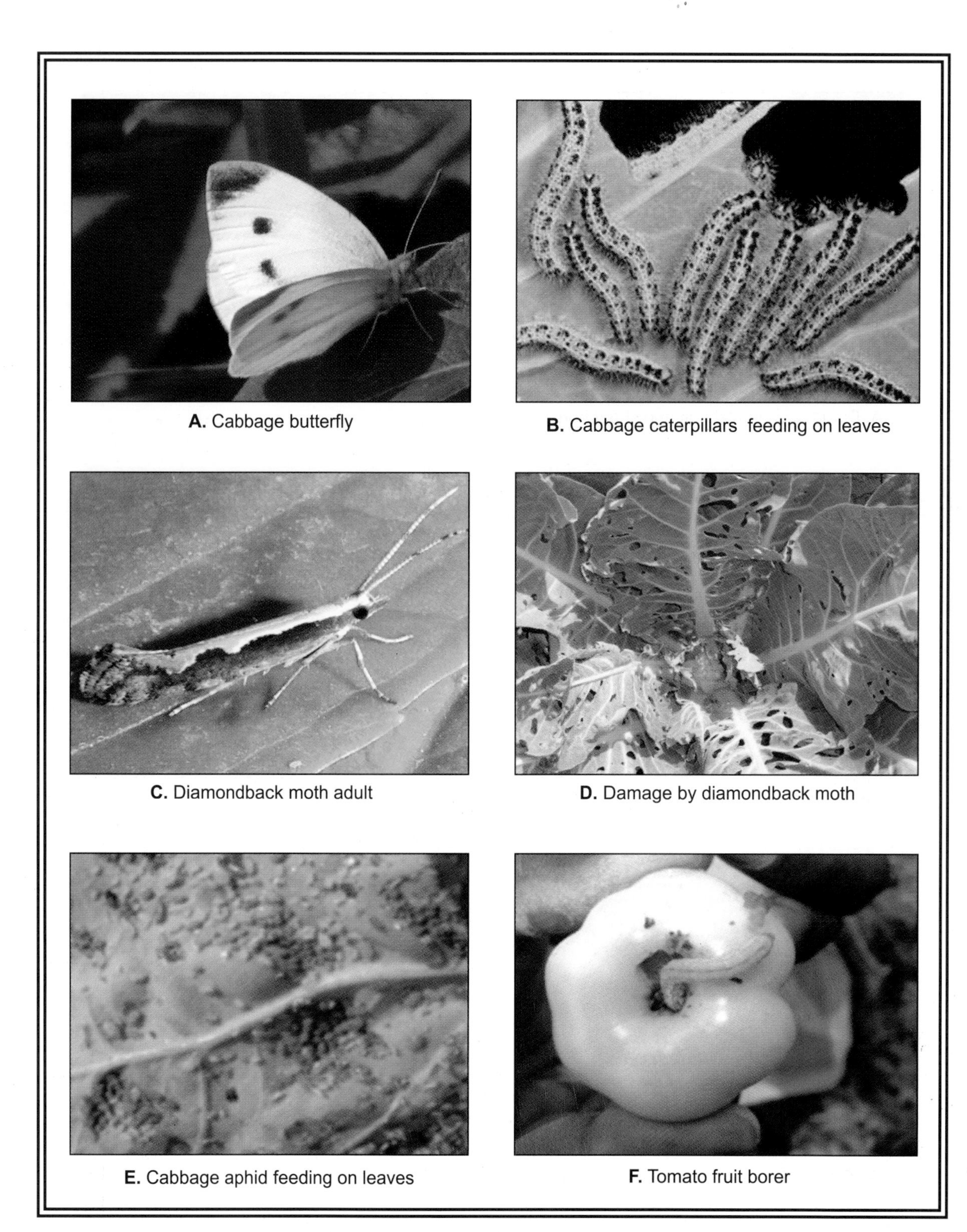

A. Cabbage butterfly

B. Cabbage caterpillars feeding on leaves

C. Diamondback moth adult

D. Damage by diamondback moth

E. Cabbage aphid feeding on leaves

F. Tomato fruit borer

Plate 20.1 Pests of winter vegetables

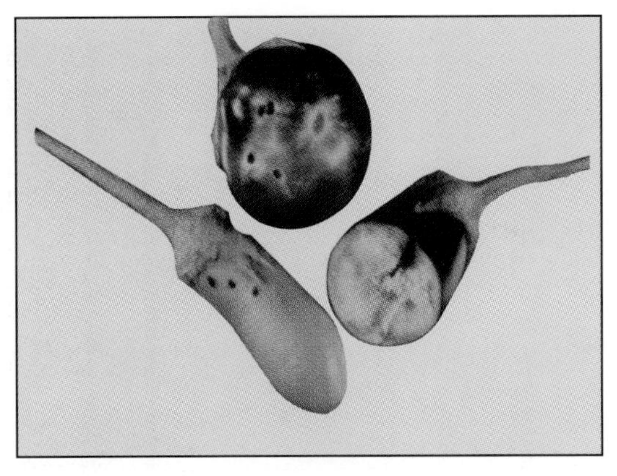

A. Damage by brinjal fruit and shoot borer

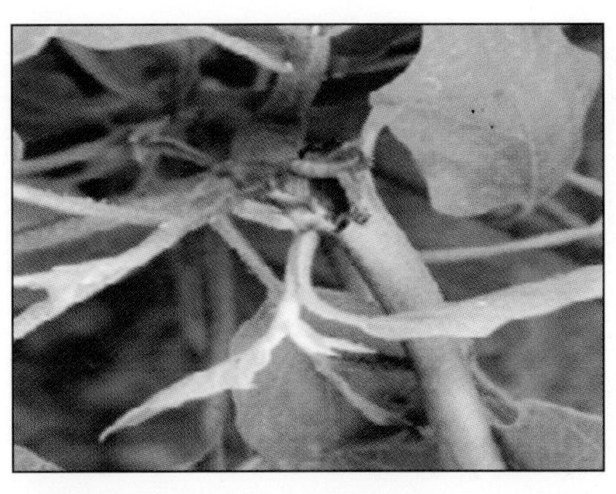

B. Damage by brinjal fruit and shoot borer on plants

C. Different life stages of brinjal hadda

D. Bitter gourd damaged by fruit fly

E. Okra leaves damaged by leafhopper

F. Okra fruits damaged by spotted bollworms

Plate 20.2 Pests of summer vegetables

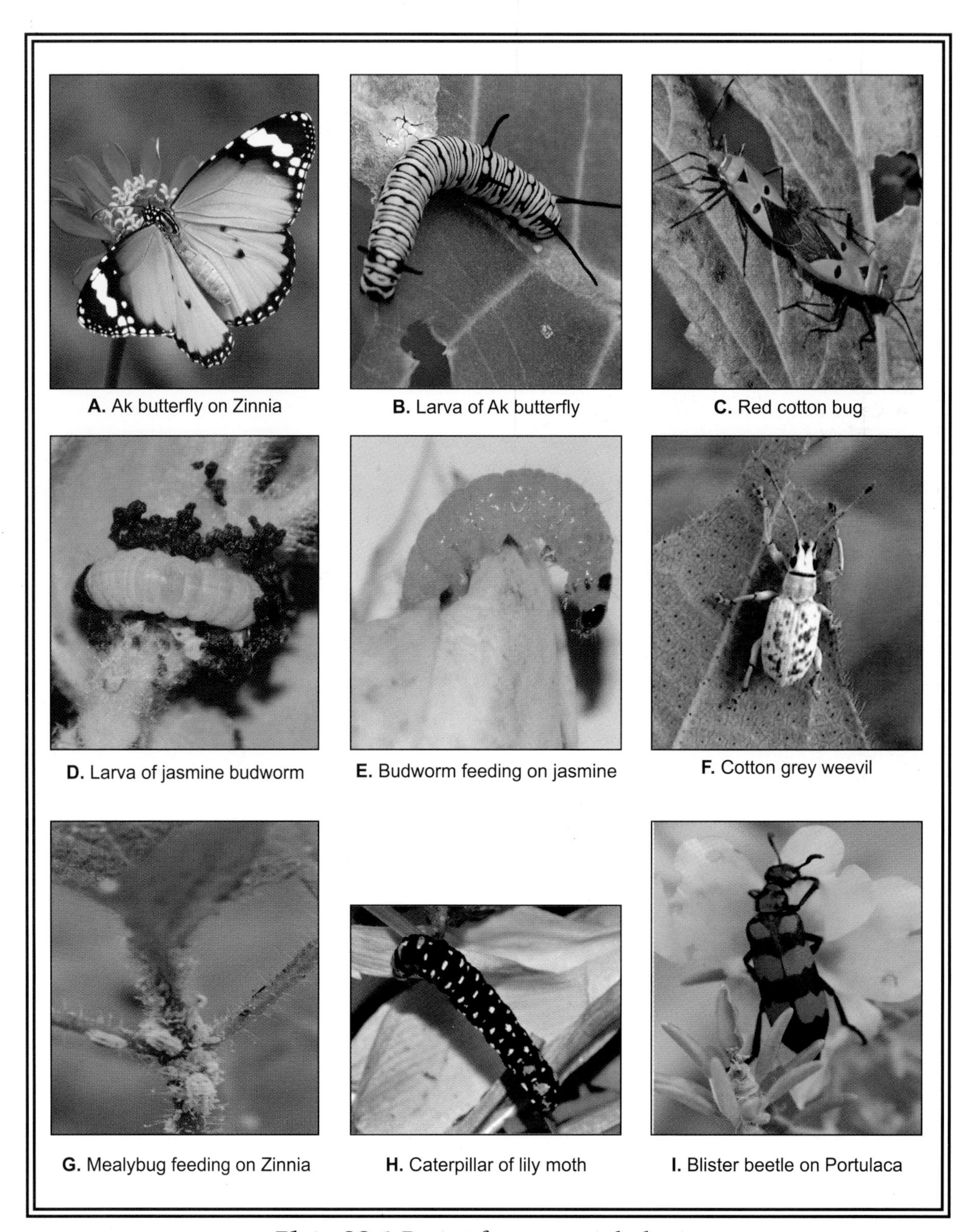

Plate 22.1 Pests of ornamental plants

A. Cashew aphid feeding on cashew leaf

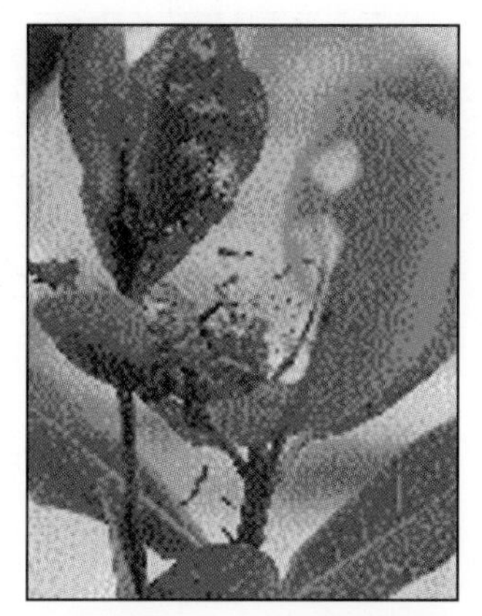

B. Cashew leaf and blossom webber damage

C. Black-headed caterpillar damage on coconut

D.Rhinoceros beetle adult

E. Rhinoceros beetle damage in coconut

F. Red palm weevil damage on coconut

G. Red borer larva on coffee

H. Red borer damage on coffee stem

Plate 23.1 Pests of plantation crops

A. Cardamom thrips damage

B. Cardamom root grub

C. Pollu beetle infested berries on black pepper spike

D. Mealybug on black pepper roots

E. Damage by skipper butterfly on turmeric

F. Damage by rhizome fly on ginger rhizome

Plate 24.1 Pests of spices

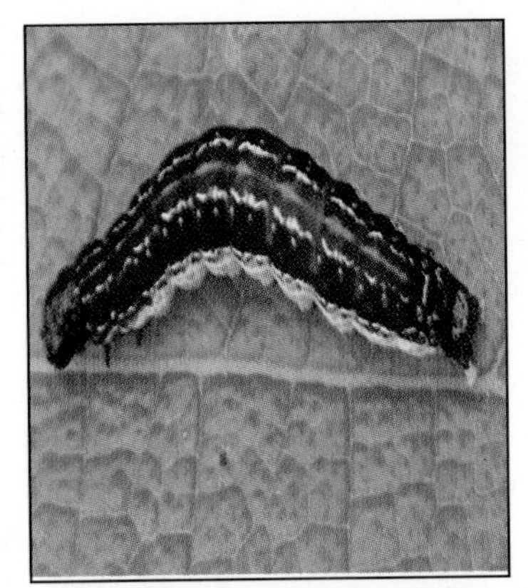

A. Larva of teak defoliator

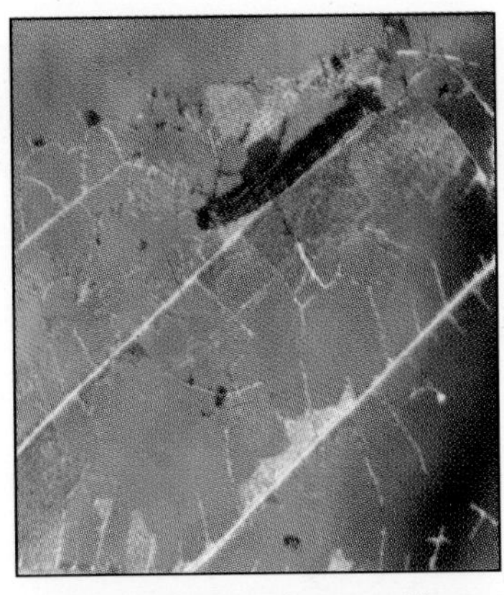

B. Teak defoliator damaged leaves

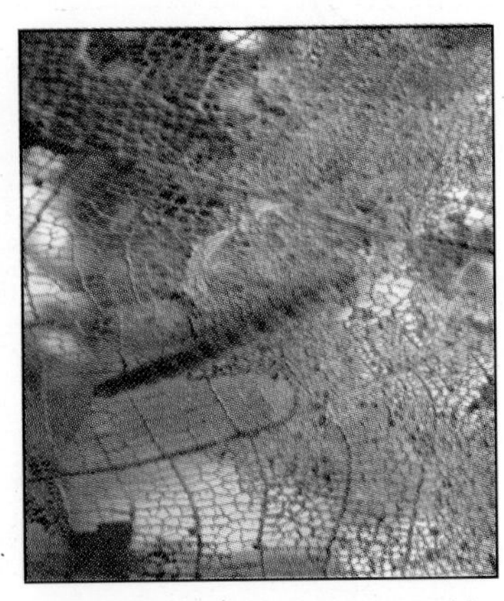

C. Teak skeletonizer damaged leaf

D. Mature grubs of sal borer

E. Wood dust around sal tree due to sal borer

F. Dead sal tree due to sal borer

G. Galls formed in aonla by gal forming catepillar

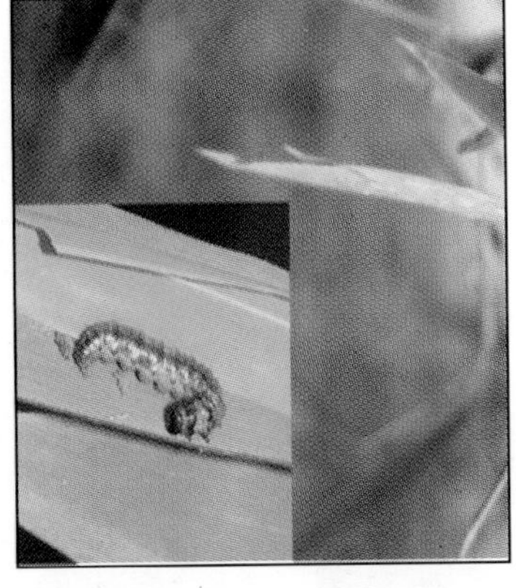

H. Bamboo leaves damaged by leaf roller

I. Leaf and stem galls on eucalyptus due to gall wasp

Plate 25.1 Pests of forest trees

A. Termite attack in maize

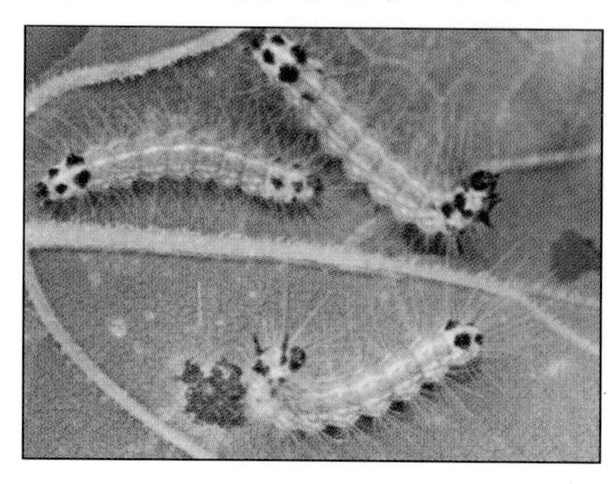

B. Bihar hairy caterpillar on sunflower

C. Pod borer in chickpea

D. *Spodoptera litura* larvae in tomato

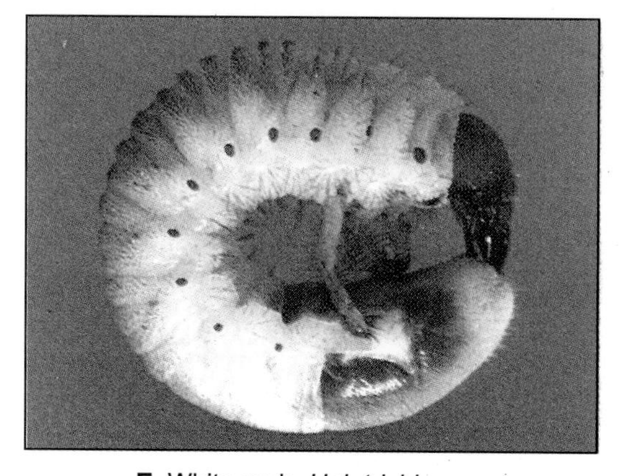

E. White grub, *Holotrichia* spp.

F. Mealybug, *Ferrisia virgata* on guava.

Plate 26.1 Polyphagous pests

PHOTO GALLERY

An Overview of Agricultural Pests and Biocontrol Agents